T0222850

Lecture Notes in Artificial Intelligence 10900

Subseries of Lecture Notes in Computer Science

More information about this series at http://www.springer.com/series/1244

Didier Galmiche · Stephan Schulz
Roberto Sebastiani (Eds.)

Automated Reasoning

9th International Joint Conference, IJCAR 2018
Held as Part of the Federated Logic Conference, FloC 2018
Oxford, UK, July 14–17, 2018
Proceedings

 Springer

Editors
Didier Galmiche
Université de Lorraine
Vandoeuvre-lès-Nancy
France

Roberto Sebastiani
University of Trento
Trento
Italy

Stephan Schulz
Baden-Wuerttemberg Cooperative State
  University
Stuttgart
Germany

ISSN 0302-9743   ISSN 1611-3349   (electronic)
Lecture Notes in Artificial Intelligence
ISBN 978-3-319-94204-9   ISBN 978-3-319-94205-6   (eBook)
https://doi.org/10.1007/978-3-319-94205-6

Library of Congress Control Number: 2018947356

LNCS Sublibrary: SL7 – Artificial Intelligence

Printed on acid-free paper

This Springer imprint is published by the registered company Springer International Publishing AG
part of Springer Nature
The registered company address is: Gewerbestrasse 11, 6330 Cham, Switzerland

Preface

This volume contains the papers presented at the 9th International Joint Conference on Automated Reasoning, IJCAR 2018, held during July 14–17, 2018 in Oxford, UK, as part of the Federated Logic Conference, FLoC 2018.

There were 125 abstracts submitted to IJCAR, resulting in 108 complete submissions. Each submission was assigned to three Program Committee members and received at least three reviews. The committee accepted 46 papers in total, 38 full papers and eight system descriptions. In addition, the program included two invited talks by Erika Abraham and Martin Giese, and accommodated a number of FLoC central events.

IJCAR is the premier international joint conference on all aspects of automated reasoning, including foundations, implementations, and applications, comprising several leading conferences and workshops. It was first held in Sienna, Italy, in 2001, uniting CADE, the Conference on Automated Deduction, TABLEAUX, the International Conference on Automated Reasoning with Analytic Tableaux and Related Methods, and FTP, the Workshop on First-Order Theorem Proving. Since 2004, IJCAR has been held every second year, alternating with separate meetings of its constituent conferences. In 2018, IJCAR united CADE, TABLEAUX, and FroCoS, the International Symposium on Frontiers of Combining Systems, and, for the fourth time, was part of the Federated Logic Conference. IJCAR also hosted the CADE ATP System Competition and 11 workshops.

IJCAR acknowledges the generous sponsorship of EurAI, the European Association for Artificial Intelligence (https://www.eurai.org/), for supporting in part our invited speakers.

We would like to thank the organizers of IJCAR, FLoC, and associated events, but in particular the members of the IJCAR Program Committee (PC) and the additional external reviewers. They provided high-quality reviews.

The PC chairs also would like to acknowledge EasyChair. The system was extremely supportive for most major tasks, including the reviewing and selection of papers, the organization of the program, and creating this proceedings volume.

May 2018

Didier Galmiche
Stephan Schulz
Roberto Sebastiani

Organization

Program Committee

Carlos Areces	FaMAF - Universidad Nacional de Córdoba, Argentina
Alessandro Artale	Free University of Bolzano-Bozen, Italy
Arnon Avron	Tel Aviv University, Israel
Franz Baader	TU Dresden, Germany
Clark Barrett	Stanford University, USA
Peter Baumgartner	CSIRO, Australia
Christoph Benzmüller	Freie Universität Berlin, Germany
Armin Biere	Johannes Kepler University Linz, Austria
Nikolaj Bjorner	Microsoft
Jasmin Christian Blanchette	Vrije Universiteit Amsterdam, The Netherlands
Maria Paola Bonacina	Università degli Studi di Verona, Italy
Torben Braüner	Roskilde University, Denmark
Agata Ciabattoni	Vienna University of Technology, Austria
Leonardo de Moura	Microsoft
Hans De Nivelle	Nazarbayev University, Kazakhstan
Stéphane Demri	LSV, CNRS, ENS Paris-Saclay, France
Clare Dixon	University of Liverpool, UK
François Fages	Inria, Université Paris-Saclay, France
Pascal Fontaine	Loria, Inria, University of Lorraine, France
Didier Galmiche (Co-chair)	LORIA - Université de Lorraine, France
Vijay Ganesh	Waterloo
Silvio Ghilardi	Università degli Studi di Milano, Italy
Jürgen Giesl	RWTH Aachen University, Germany
Laura Giordano	DISIT, Università del Piemonte Orientale, Italy
Valentin Goranko	Stockholm University, Sweden
Rajeev Gore	The Australian National University, Australia
Alberto Griggio	Fondazione Bruno Kessler, Italy
John Harrison	Intel
Moa Johansson	Chalmers University of Technology, Sweden
Cezary Kaliszyk	University of Innsbruck, Austria
Deepak Kapur	University of New Mexico, USA
Konstantin Korovin	The University of Manchester, UK
Laura Kovacs	Vienna University of Technology, Austria
George Metcalfe	University of Bern, Switzerland
Dale Miller	Inria and LIX/Ecole Polytechnique, France
Cláudia Nalon	University of Brasília, Brazil
Albert Oliveras	Universitat Politècnica de Catalunya, Spain
Nicola Olivetti	LSIS, Aix-Marseille University, France

Jens Otten	University of Oslo, Norway
Lawrence Paulson	University of Cambridge, UK
Nicolas Peltier	CNRS, LIG, France
Frank Pfenning	Carnegie Mellon University, USA
Silvio Ranise	FBK-Irst, Italy
Christophe Ringeissen	Inria, France
Philipp Ruemmer	Uppsala University, Sweden
Katsuhiko Sano	Hokkaido University, Japan
Uli Sattler	The University of Manchester, UK
Renate A. Schmidt	The University of Manchester, UK
Stephan Schulz (Co-chair)	DHBW Stuttgart, Germany
Roberto Sebastiani (Co-chair)	University of Trento, Italy
Viorica Sofronie-Stokkermans	University of Koblenz-Landau, Germany
Thomas Sturm	CNRS, France
Geoff Sutcliffe	University of Miami, USA
Cesare Tinelli	The University of Iowa, USA
Alwen Tiu	The Australian National University, Australia
Ashish Tiwari	SRI International, USA
Josef Urban	Czech Technical University in Prague, Czech Republic
Luca Viganò	King's College London, UK
Uwe Waldmann	Max Planck Institute for Informatics, Germany
Christoph Weidenbach	Max Planck Institute for Informatics, Germany

Additional Reviewers

Awad, Ahmed	Del-Pinto, Warren
Backeman, Peter	Delzanno, Giorgio
Baelde, David	Dershowitz, Nachum
Barbosa, Haniel	Docherty, Simon
Berzish, Murphy	Echenim, Mnacho
Bobot, François	Ecke, Andreas
Bouraoui, Zied	Fernandez Gil, Oliver
Bourke, Timothy	Fervari, Raul
Boy de La Tour, Thierry	Fiorentini, Camillo
Brotherston, James	Fiorino, Guido
Brown, Chad	Fitting, Melvin
Brown, Christopher	Flaminio, Tommaso
Cerrito, Serenella	Fleury, Mathias
Chaudhuri, Kaustuv	Foguel, Tuval
Chen, Jieying	Frohn, Florian
Cohen, Liron	Fränzle, Martin
Das, Anupam	Fuhs, Carsten
Dawson, Jeremy	Galmiche, Didier

Genco, Francesco Antonio
Gianola, Alessandro
Gil-Férez, José
Gleiss, Bernhard
Govind, Hari
Hemery, Mathieu
Hoffmann, Guillaume
Hoffmann, Johannes
Hosni, Hykel
Hou, Zhe
Hughes, Dominic
Hupel, Lars
Hustadt, Ullrich
Iosif, Radu
Irfan, Ahmed
Jamet-Jakubiec, Line
Kamburjan, Eduard
Kerhet, Volha
Kimura, Daisuke
Konev, Boris
Kop, Cynthia
Kovtunova, Alisa
Kucik, Andrzej
Leitsch, Alexander
Liang, Jimmy
Lierler, Yuliya
Lisitsa, Alexei
Manighetti, Matteo
Marshall, Andrew
Meel, Kuldeep S.
Mora, Federico
Morawska, Barbara
Nagele, Julian
Nishida, Naoki

Paskevich, Andrei
Pease, Adam
Peñaloza, Rafael
Pimentel, Elaine
Popescu, Andrei
Pozzato, Gian Luca
Racharak, Teeradaj
Ramanayake, Revantha
Reger, Giles
Reynolds, Andrew
Ritirc, Daniela
Rivkin, Andrey
Robillard, Simon
Rodríguez Carbonell, Enric
Schlichtkrull, Anders
Schreck, Pascal
Sebastiani, Roberto
Soliman, Sylvain
Steen, Alexander
Suda, Martin
Thorstensen, Evgenij
Tourret, Sophie
Van Oostrom, Vincent
Vigneron, Laurent
Wang, Ren-June
Wies, Thomas
Winkler, Sarah
Xia, Bican
Yamada, Akihisa
Zeljić, Aleksandar
Zhao, Yizheng
Zohar, Yoni
Zulkoski, Ed

Abstract of Invited Talks

Industrial Data Access

What are the Reasoning Problems?
And is Reasoning the Problem?

Martin Giese

University of Oslo
martingi@ifi.uio.no

Optique (http://optique-project.eu) [2] was an EU FP7 project that ran from November 2012 to October 2016. The main objective was to test the idea of "Ontology Based Data Access" (OBDA) on real industrial applications. Concretely: to support the work of geologists and geophysicists in the oil & gas company Statoil, and the work of turbine engineers at Siemens AG. This line of work now continues in the nationally funded 'Centre for Research-based Innovation' SIR-IUS (http://sirius-labs.no) at the University of Oslo, with participation from the Universities of Oxford and Trondheim, as well a large number of participating companies.

The software produced by the project features elaborate user interfaces, and no $\forall$ or $\exists$ can be seen on the surface. Still, most of the functionality is controlled by an ontology, which is nothing more than a set of axioms in a particular description logic. As a consequence, a variety of reasoning tasks takes place under the hood, all the way from query optimisation [1], via entity alignment [4] and up to the user interface control code [3]. This talk presents a selection of these problems, both solved and as-yet unsolved.

Though logic and reasoning are close to the hearts of many of the researchers involved, the success of the project was also dependent on other factors: inter-disciplinary communication, usability considerations, and many pragmatic compromises, to name some. And sometimes, these would again lead to 'nice' research. The talk also covers some of these extra-logical aspects of the project.

References

1. Calvanese, D., Cogrel, B., Komla-Ebri, S., Kontchakov, R., Lanti, D., Rezk, M., Rodriguez-Muro, M., Xiao, G.: Ontop: answering SPARQL queries over relational databases. Semant. Web, 8(3), 471–487 (2017)
2. Giese, M., Soylu, A., Vega-Gorgojo, G., Waaler, A., Haase, P., Jiménez-Ruiz, E., Lanti, D., Rezk, M., Xiao, G., Ozçep, Ö., Rosati, R.: Optique: zooming in on big data. IEEE Comput. 48(3), 60–67 (2015)
3. Soylu, A., Kharlamov, E., Zheleznyakov, D., Jimenez Ruiz, E., Giese, M., Skjaeveland, M.G., Hovland, D., Schlatte, R., Brandt, S., Lie, H., Horrocks, I.: OptiqueVQS: a visual query system over ontologies for industry. Semant. Web (2017, in press)
4. Xiao, G., Hovland, D., Bilidas, D., Rezk, M., Giese, M., Calvanese, D.: Efficient ontology-based data integration with canonical IRIs. In: Navigli, R., Vidal, M.-E. (eds.) Proceedings of 15th International Extended Semantic Web Conference (ESWC) (2018. to appear)

Symbolic Computation Techniques in SMT Solving: Mathematical Beauty Meets Efficient Heuristics

Erika Ábrahám

RWTH Aachen University, Germany

Checking the satisfiability of quantifier-free real-arithmetic formulas is a practically highly relevant but computationally hard problem. Some beautiful mathematical decision procedures implemented in computer algebra systems are capable of solving such problems, however, they were developed for more general tasks like quantifier elimination, therefore their applicability to satisfiability checking is often restricted.

In computer science, recent advances in satisfiability-modulo-theories (SMT) solving led to elegant embeddings of such decision procedures in SMT solvers in a way that combines the strengths of symbolic computation methods and heuristic-driven search techniques. In this talk we discuss such embeddings and show that they might be quite challenging but can lead to powerful synergies and open new lines of research.

Contents

An Assumption-Based Approach for Solving the Minimal
S5-Satisfiability Problem 1
    Jean-Marie Lagniez, Daniel Le Berre, Tiago de Lima,
    and Valentin Montmirail

FAME: An Automated Tool for Semantic Forgetting
in Expressive Description Logics 19
    Yizheng Zhao and Renate A. Schmidt

Superposition for Lambda-Free Higher-Order Logic 28
    Alexander Bentkamp, Jasmin Christian Blanchette,
    Simon Cruanes, and Uwe Waldmann

Automated Reasoning About Key Sets. 47
    Miika Hannula and Sebastian Link

A Tableaux Calculus for Reducing Proof Size 64
    Michael Peter Lettmann and Nicolas Peltier

FORT 2.0 81
    Franziska Rapp and Aart Middeldorp

Formalizing Bachmair and Ganzinger's Ordered Resolution Prover 89
    Anders Schlichtkrull, Jasmin Christian Blanchette, Dmitriy Traytel,
    and Uwe Waldmann

The Higher-Order Prover Leo-III. 108
    Alexander Steen and Christoph Benzmüller

Well-Founded Unions 117
    Jeremy Dawson, Nachum Dershowitz, and Rajeev Goré

Implicit Hitting Set Algorithms for Maximum Satisfiability
Modulo Theories. 134
    Katalin Fazekas, Fahiem Bacchus, and Armin Biere

Cubicle-$\mathcal{W}$: Parameterized Model Checking on Weak Memory. 152
    Sylvain Conchon, David Declerck, and Fatiha Zaïdi

QRAT$^+$: Generalizing QRAT by a More Powerful QBF
Redundancy Property. 161
    Florian Lonsing and Uwe Egly

A Why3 Framework for Reflection Proofs and Its Application
to GMP's Algorithms   178
   Guillaume Melquiond and Raphaël Rieu-Helft

Probably Half True: Probabilistic Satisfiability over Łukasiewicz
Infinitely-Valued Logic   194
   Marcelo Finger and Sandro Preto

Uniform Substitution for Differential Game Logic   211
   André Platzer

A Logical Framework with Commutative
and Non-commutative Subexponentials   228
   Max Kanovich, Stepan Kuznetsov, Vivek Nigam,
   and Andre Scedrov

Exploring Approximations for Floating-Point Arithmetic Using UppSAT   246
   Aleksandar Zeljić, Peter Backeman, Christoph M. Wintersteiger,
   and Philipp Rümmer

Complexity of Combinations of Qualitative Constraint
Satisfaction Problems.   263
   Manuel Bodirsky and Johannes Greiner

A Generic Framework for Implicate Generation Modulo Theories   279
   Mnacho Echenim, Nicolas Peltier, and Yanis Sellami

A Coinductive Approach to Proving Reachability Properties
in Logically Constrained Term Rewriting Systems   295
   Ştefan Ciobâcă and Dorel Lucanu

A New Probabilistic Algorithm for Approximate Model Counting.   312
   Cunjing Ge, Feifei Ma, Tian Liu, Jian Zhang, and Xutong Ma

A Reduction from Unbounded Linear Mixed Arithmetic
Problems into Bounded Problems   329
   Martin Bromberger

Cops and CoCoWeb: Infrastructure for Confluence Tools.   346
   Nao Hirokawa, Julian Nagele, and Aart Middeldorp

Investigating the Existence of Large Sets of Idempotent Quasigroups
via Satisfiability Testing.   354
   Pei Huang, Feifei Ma, Cunjing Ge, Jian Zhang, and Hantao Zhang

Superposition with Datatypes and Codatatypes   370
   Jasmin Christian Blanchette, Nicolas Peltier, and Simon Robillard

Efficient Encodings of First-Order Horn Formulas in Equational Logic   388
Koen Claessen and Nicholas Smallbone

A FOOLish Encoding of the Next State Relations of Imperative Programs. . .   405
Evgenii Kotelnikov, Laura Kovács, and Andrei Voronkov

Constructive Decision via Redundancy-Free Proof-Search   422
Dominique Larchey-Wendling

Deciding the First-Order Theory of an Algebra of Feature Trees
with Updates   439
Nicolas Jeannerod and Ralf Treinen

A Separation Logic with Data: Small Models and Automation   455
Jens Katelaan, Dejan Jovanović, and Georg Weissenbacher

MædMax: A Maximal Ordered Completion Tool   472
Sarah Winkler and Georg Moser

From Syntactic Proofs to Combinatorial Proofs.   481
Matteo Acclavio and Lutz Straßburger

A Resolution-Based Calculus for Preferential Logics   498
Cláudia Nalon and Dirk Pattinson

Extended Resolution Simulates DRAT.   516
Benjamin Kiesl, Adrián Rebola-Pardo, and Marijn J. H. Heule

Verifying Asymptotic Time Complexity of Imperative Programs
in Isabelle   532
Bohua Zhan and Maximilian P. L. Haslbeck

Efficient Interpolation for the Theory of Arrays   549
Jochen Hoenicke and Tanja Schindler

ATPBOOST: Learning Premise Selection in Binary Setting
with ATP Feedback.   566
Bartosz Piotrowski and Josef Urban

Theories as Types.   575
Dennis Müller, Florian Rabe, and Michael Kohlhase

Datatypes with Shared Selectors   591
*Andrew Reynolds, Arjun Viswanathan, Haniel Barbosa,
Cesare Tinelli, and Clark Barrett*

Enumerating Justifications Using Resolution.   609
Yevgeny Kazakov and Peter Skočovský

A SAT-Based Approach to Learn Explainable Decision Sets   627
Alexey Ignatiev, Filipe Pereira, Nina Narodytska,
and Joao Marques-Silva

Proof-Producing Synthesis of CakeML with I/O and Local State
from Monadic HOL Functions   646
Son Ho, Oskar Abrahamsson, Ramana Kumar, Magnus O. Myreen,
Yong Kiam Tan, and Michael Norrish

An Abstraction-Refinement Framework for Reasoning
with Large Theories   663
Julio Cesar Lopez Hernandez and Konstantin Korovin

Efficient Model Construction for Horn Logic with VLog:
System Description   680
Jacopo Urbani, Markus Krötzsch, Ceriel Jacobs, Irina Dragoste,
and David Carral

Focussing, MALL and the Polynomial Hierarchy.   689
Anupam Das

Checking Array Bounds by Abstract Interpretation
and Symbolic Expressions   706
Étienne Payet and Fausto Spoto

Author Index   723

An Assumption-Based Approach for Solving the Minimal S5-Satisfiability Problem

Jean-Marie Lagniez, Daniel Le Berre, Tiago de Lima,
and Valentin Montmirail$^{(\boxtimes)}$

CRIL, Artois University and CNRS, 62300 Lens, France
{lagniez,leberre,delima,montmirail}@cril.fr

Abstract. Recent work on the practical aspects on the modal logic S5 satisfiability problem showed that using a SAT-based approach outperforms other existing approaches. In this work, we go one step further and study the related minimal S5 satisfiability problem (MinS5-SAT), the problem of finding an S5 model, a Kripke structure, with the smallest number of worlds. Finding a small S5 model is crucial as soon as the model should be presented to a user, displayed on a screen for instance. SAT-based approaches tend to produce S5-models with a large number of worlds, thus the need to minimize them. That optimization problem can obviously be solved as a pseudo-Boolean optimization problem. We show in this paper that it is also equivalent to the extraction of a maximal satisfiable set (MSS). It can thus be solved using a standard pseudo-Boolean or MaxSAT solver, or a MSS-extractor. We show that a new incremental, SAT-based approach can be proposed by taking into account the equivalence relation between the possible worlds on S5 models. That specialized approach presented the best performance on our experiments conducted on a wide range of benchmarks from the modal logic community and a wide range of pseudo-Boolean and MaxSAT solvers. Our results demonstrate once again that domain knowledge is key to build efficient SAT-based tools.

Keywords: Modal logic · S5 · Incremental SAT · Minimisation

1 Introduction

Over the last twenty years, modal logics have been used in various areas of artificial intelligence like formal verification [1], database theory [2] and distributed computing [3] for example. More recently, the modal logic S5 was used for knowledge compilation [4] and in contingent planning [5]. Different solvers for different modal logics have been designed to decide the satisfiability of modal formulas since the 90's [6,7]. Some of them have been designed quite recently [8–11]. Despite the variety of techniques employed, none of them formally guarantees that, when a model is found, it is the smallest model possible (in number of

© Springer International Publishing AG, part of Springer Nature 2018
D. Galmiche et al. (Eds.): IJCAR 2018, LNAI 10900, pp. 1–18, 2018.
https://doi.org/10.1007/978-3-319-94205-6_1

worlds) for the input formula, in fact many of them do not even output a model but simply answer yes/no.

Providing a model, a certificate of satisfiability, is important to check the answer given by the solver. This is true both for the author of the solver or a user of that solver. This is mandatory nowadays in many solver competitions, among them the SAT competition [12]. It has also been shown that those models can help improving NP-oracle based procedures [13]: a procedure requiring a polynomial number of calls to a yes/no oracle can be transformed into a procedure requiring only a logarithmic number of calls when the oracle can provide a model. Another example of the importance for an oracle to provide a model may be found in the model rotation technique [14], a method for the detection of clauses that are included in all MUSes (Minimal Unsatisfiable Sets) of a given formula via the analysis of models returned by a SAT oracle. Even if the theory does not guarantee a reduction of the number of oracle calls, in practice it provides a huge performance gain (up to a factor of 5) [15]. Finding the smallest model may be even more important in some contexts. The provided model usually has a meaning for the user, like in Hardware Verification [1] where the model is in fact an explanation of the bug found in the design of the hardware. The smaller the model, the more precise could be the location of the bug. It could also be the case that the model should be inspected by the user or displayed on a screen. Thus, the smaller, the better. There is a huge literature on minimizing models for SAT [16–18]. We are interested here in minimizing models for S5-SAT.

Our goal in this paper is to propose techniques to compute the smallest S5-model, in number of worlds, for a given input formula. We call this problem MinS5-SAT. We focus exclusively on the modal logic S5, for which the satisfiability problem is NP-complete [19] as for the classical propositional logic (CPL) [20]. We propose and compare different techniques. (1) The first obvious technique is based on a translation into CPL. The parameter given to the translation is the number n of possible worlds that the solution model is assumed to have. Linear or dichotomous search can be used to minimize the S5-model. (2) The second technique adds to the previous encoding selector-variables in order to activate or deactivate worlds. Finding a minimal S5-model in this case amounts to an equivalent MaxSAT problem [21], or, more surprisingly, to a MSS problem. Thus we can rely on off-the-shelves MaxSAT solvers or MSS-extractors to solve the original problem. (3) The last technique goes one step further from the two previous approaches. Thanks to a specific property of modal logic S5, we interpret the set of selectors causing the inconsistency of the formula to reach the theoretical upper bound faster. We compare these different techniques and show empirically which one better suits the benchmarks used. All benchmarks we could find for mono-agent S5 are randomly generated or created automatically following a pattern ("crafted"). (Note that reasoning about knowledge problems, such as those in [22], are all multi-agents.) However, we know from the SAT community [12] that the performance of a solver can be significantly different when the problem is randomly generated, when it is "crafted" or when it models a "real" problem which has some kind of "structure". Thus, we generated new S5 benchmarks translated from planning problems with incomplete information about the initial state (with sensing and full observability) [23] to complete the picture. We choose planning problems to obtain structured benchmarks requiring relatively large Kripke models.

In the reminder of this article, we first present the modal logic S5 and define the MinS5-SAT problem. Then, we provide a first approach to solve MinS5-SAT using a SAT-oracle and selector variables. We provide a translation from MinS5-SAT problems to equivalent MSS-extraction problems. We present a specific property of modal logic S5 that speed up our initial SAT-based approach. Finally, we present the experimental results and conclude.

2   Preliminaries

2.1   Modal Logic S5

A central problem associated with any logic is the satisfiability problem, that is to decide whether a given formula has a model. The first complexity results for satisfiability in modal logic were achieved by Ladner [19]. He showed that the satisfiability problem in modal logic K (K-SAT) is in PSPACE and that the satisfiability problem in modal logic S5 (S5-SAT) is NP-complete. In this paper, we are interested in S5. In what follows, let $\mathbb{P}$ be a countably infinite non-empty set of propositional variables. The language $\mathcal{L}$ of modal logic is the set of formulas ϕ defined by the following grammar in BNF, where p ranges over $\mathbb{P}$:

$$\phi ::= \top \mid p \mid \neg\phi \mid \phi \wedge \phi \mid \phi \vee \phi \mid \Box\phi \mid \Diamond\phi$$

The operators $\rightarrow$ and $\leftrightarrow$, defined by the usual abbreviations, are also used. A formula of the form $\Box\phi$ (*box phi*) means 'ϕ is necessarily true'. A formula of the form $\Diamond\phi$ (*diamond phi*) means 'ϕ is possibly true'.

Example 1. Let $\mathbb{P} = \{a, b\}$. $\phi = ((\Box\neg a \vee \Diamond b) \wedge \Diamond a \wedge \Box b)$ is a modal logic formula.

Formulas in $\mathcal{L}$ are interpreted using S5-structures [24], which are defined as follows:

Definition 1 (S5-Structure). *A S5-structure is a triplet $M = \langle W, R, \mathcal{I} \rangle$, where:*
W is a non-empty set of possible worlds;
R is a binary relation on W which is an equivalence relation ($\forall w.\forall v. (w, v) \in R$);
$\mathcal{I}$ is a function associating, to each $p \in \mathbb{P}$, the set of worlds from W where p is true.

Note that because R is an equivalence relation, we will omit it in the rest of this paper.

Definition 2 (Pointed S5-Structure). *A pointed S5-structure is a pair $\langle M, \omega \rangle$, where M is a S5-Structure and ω, called the actual world, is a possible world in W.*

In the remainder of this article, 'structure' means 'pointed S5-structure'. We define the size of a structure $\langle M, \omega \rangle$, noted $|M|$, as its number of worlds. Below, the satisfaction relation between such structures and formulas in $\mathcal{L}$ is defined.

Definition 3 (Satisfaction Relation). *Let* $M = \langle W, R, \mathcal{I} \rangle$, *an S5-structure. The satisfaction relation* $\vDash$ *between formulas and structures is recursively defined as follows:*

$$\langle M, \omega \rangle \vDash \top \quad \langle M, \omega \rangle \vDash p \text{ iff } \omega \in \mathcal{I}(p)$$
$$\langle M, \omega \rangle \vDash \neg \phi \text{ iff } \langle M, \omega \rangle \nvDash \phi$$
$$\langle M, \omega \rangle \vDash \phi \wedge \psi \text{ iff } \langle M, \omega \rangle \vDash \phi \text{ and } \langle M, \omega \rangle \vDash \psi$$
$$\langle M, \omega \rangle \vDash \phi \vee \psi \text{ iff } \langle M, \omega \rangle \vDash \phi \text{ or } \langle M, \omega \rangle \vDash \psi$$
$$\langle M, \omega \rangle \vDash \Box \phi \text{ iff for all } v \in W \text{ we have } \langle M, v \rangle \vDash \phi$$
$$\langle M, \omega \rangle \vDash \Diamond \phi \text{ iff there exists } v \in W \text{ such that } \langle M, v \rangle \vDash \phi$$

Definition 4 (Satisfiability). *A formula* ϕ *is satisfiable if and only if there exists a structure* $\langle M, \omega \rangle$ *that satisfies* ϕ. *Such a structure is called a 'model of* ϕ'.

Example 2. Here is a structure $\langle M, \omega_0 \rangle$ satisfying the formula ϕ from Example 1: $W = \{w_0, w_1, w_2\}$, $\mathcal{I} = \{\langle a, \{w_0\} \rangle, \langle b, \{w_0, w_1, w_2\} \rangle\}$. The size of $\langle M, \omega_0 \rangle$ equals 3.

As S5-SAT is NP-complete [19], we proposed a reduction from this problem to SAT in [10]. The reduction function takes as parameter the number of worlds n and is defined as follows:

Definition 5 (Translation Function tr). *Let* $\phi \in \mathcal{L}$.

$$\text{tr}(\phi, n) = \text{tr}'(\phi, 1, n)$$

$$\text{tr}'(p, i, n) = p_i \qquad\qquad\qquad \text{tr}'(\neg \psi, i, n) = \neg \text{tr}'(\psi, i, n)$$

$$\text{tr}'(\psi \wedge \chi, i, n) = \text{tr}'(\psi, i, n) \wedge \text{tr}'(\chi, i, n) \quad \text{tr}'(\psi \vee \chi, i, n) = \text{tr}'(\psi, i, n) \vee \text{tr}'(\chi, i, n)$$

$$\text{tr}'(\Box \psi, i, n) = \bigwedge_{j=1}^{n} (\text{tr}'(\psi, j, n)) \qquad\qquad \text{tr}'(\Diamond \psi, i, n) = \bigvee_{j=1}^{n} (\text{tr}'(\psi, j, n))$$

The function 'tr' is satisfiability preserving if n is large enough. Moreover, some additional simplifications are performed to avoid outputting a very large formula (e.g., the Tseitin algorithm). See [10] for more details.

Example 3. Let ϕ the formula in Example 1. Its translation $\text{tr}(\phi, 2)$ is $((\neg a_1 \vee b_1 \vee b_2) \wedge (\neg a_2 \vee b_1 \vee b_2) \wedge (a_1 \vee a_2) \wedge (b_1 \wedge b_2))$.

2.2   Unsatisfiable Cores

Recent SAT solvers are incremental, i.e., they are able to check the satisfiability of a formula "under assumptions" [25] and are able to output a core (a "reason" for the unsatisfiablity of the formula). The use of unsatisfiable cores is key to many applications, such as MaxSAT [21], MCS (Minimal Correction Set) [26], MUS (Minimal Unsatisfiable Set) [27]. The unsatisfiable core is defined as follows:

Definition 6 (Unsatisfiable Core Under Assumptions). *Let Σ be a satisfiable CPL formula in CNF built using Boolean variables from $\mathbb{P}$. Let A be a consistent set of literals built using Boolean variables from $\mathbb{P}$ such that $(\Sigma \wedge \bigwedge_{a \in A} a)$ is unsatisfiable. $C \subseteq A$ is an unsatisfiable core (UNSAT core) of Σ under assumptions A if and only if $(\Sigma \wedge \bigwedge_{c \in C} c)$ is unsatisfiable.*

Definition 7 (SAT Solver Under Assumptions). *Let Σ be a CPL formula in CNF. A SAT solver for Σ, given assumptions A, is a procedure which provides a pair $\langle r, s \rangle$ with $r \in \{\text{SAT}, \text{UNSAT}\}$ such that if $r = \text{SAT}$ then s is a model of Σ, else if $r = \text{UNSAT}$ then s is an UNSAT core of Σ under assumptions A.*

2.3   MSS and co-MSS

The problem of computing a Maximal Satisfiable Set of clauses (MSS problem) consists of extracting a maximal set of clauses from a formula in CNF that are consistent together [28]. The minimal correction subset (MCS or co-MSS) is the complement of its MSS.

Definition 8. *Let Σ be a given unsatisfiable formula in CNF. $S \subseteq \Sigma$ is a Maximal Satisfiable Subset (MSS) of Σ if and only if S is satisfiable and $\forall c \in \Sigma \setminus S$, $S \cup \{c\}$ is unsatisfiable.*

Definition 9. *Let an unsatisfiable formula Σ in CNF be given. $C \subseteq \Sigma$ is a Minimal Correction Subset (MCS or co-MSS) of Σ if and only if $\Sigma \setminus C$ is satisfiable and $\forall c \in C$, $\Sigma \setminus (C \setminus \{c\})$ is unsatisfiable.*

3   The MinS5-SAT Problem

As pointed in [10], the necessary number of worlds to S5-satisfy a formula is bound by $\mathrm{dd}(\phi) + 1$, where $\mathrm{dd}(\phi)$ is given in Definition 10 below.

Definition 10 (Diamond-Degree). *The diamond degree of $\phi \in \mathcal{L}$, noted $\mathrm{dd}(\phi)$, is defined recursively, as follows:*

$$\mathrm{dd}(\phi) = \mathrm{dd}'(\mathrm{nnf}(\phi))$$
$$\mathrm{dd}'(\top) = \mathrm{dd}'(\neg\top) = \mathrm{dd}'(p) = \mathrm{dd}'(\neg p) = 0$$
$$\mathrm{dd}'(\phi \wedge \psi) = \mathrm{dd}'(\phi) + \mathrm{dd}'(\psi)$$
$$\mathrm{dd}'(\phi \vee \psi) = \max(\mathrm{dd}'(\phi), \mathrm{dd}'(\psi))$$
$$\mathrm{dd}'(\Box\phi) = \mathrm{dd}'(\phi) \qquad \mathrm{dd}'(\Diamond\phi) = 1 + \mathrm{dd}'(\phi)$$

We denote by $nnf(\phi)$ the formula ϕ in negation normal form (the negation applies only to propositional variables). Thus, we have that $tr(\phi, dd(\phi) + 1)$ is equisatisfiable to ϕ. Even if, in practice, we obtain very good results using this value as upper-bound, it seems far from the optimal value in all cases. For instance, in the model of Example 1, we can see redundancies: two worlds contain the same valuation. In contexts where the size of the returned model is critical, it makes sense to try to minimize it.

Consequently, in this article, we are interested in finding a model for a formula in S5 with the smallest number of worlds in its S5-structure. We call this problem the minimal S5 satisfiability problem and we define it as follows:

Definition 11 (Minimal S5 Satisfiability). *A formula ϕ is min-S5-satisfied by a structure $\langle M, \omega \rangle$ (noted $\langle M, \omega \rangle \models_{\min} \phi$) if and only if $\langle M, \omega \rangle \models \phi$ and ϕ has no model $\langle M', \omega' \rangle$ such that $|M'| < |M|$.*

Definition 12 (Minimal S5 Satisfiability Problem). *Let a formula ϕ in $\mathcal{L}$ be given. The minimal S5 satisfiability problem (MinS5-SAT) is the problem of finding a structure $\langle M, \omega \rangle$ such that $\langle M, \omega \rangle \models_{\min} \phi$.*

Let us remark that obtaining the minimal model for ϕ is not as simple as merging the worlds with the same valuations into only one world in any model of ϕ. The minimality cannot be guaranteed that way. Let us go back to the Example 2 to illustrate this. There, we have $dd(\phi) + 1 = 3$. If we remove the redundancy, we obtain a model of size 2. However, ϕ is also satisfied by the following structure containing only one world: $W = \{w_0\}$, $\mathcal{I} = \{\langle a, \{w_0\}\rangle, \langle b, \{w_0\}\rangle\}$, which is a minimal model.

A very simple way to tackle this problem is to use the solver S52SAT [10] with a linear search strategy. Roughly, the procedure starts by trying structures of size $b = 1$. If no model is found, it iterates the process, each time increasing the value of b by 1. It iterates until a model of ϕ is found or the upper bound $dd(\phi) + 1$ is reached. This strategy is called 1toN. It is of course also possible to do it in reverse order: the procedure starts with $b = dd(\phi) + 1$ and decreases the value of b by 1 (this is called Nto1). Yet another possibility is to use a binary search (called Dico).

However, these approaches are very naive. If we take, for instance, 1toN when the solution is a model of size m, it will perform m translations from S5 to SAT and then m calls to a SAT solver. The problem here is that such strategy does not take advantage of the previous UNSAT answer of the SAT solver to solve the new formula.

Modern SAT solvers are able to take advantage of previous calls when they are used incrementally [29]. The usual way to do that is to add selectors (assumptions) to the input formula and to get as output, on the suitable cases, some kind of "reason" for its unsatisfiability, in terms of these selectors. We propose here a way to add such selectors in the translation from S5 to SAT.

4   Preliminary Step: An Assumption-Based Translation

The translation 'tr' proposed in [10] is based on a simple, yet effective, idea: let ϕ be the input formula, every sub-formula of the form $\Box\psi$ is translated to $\bigwedge_{i=1}^{n} \mathrm{tr}(\psi, i, n)$, whereas sub-formulas of the form $\Diamond\psi$ are translated to $\bigvee_{i=1}^{n} \mathrm{tr}(\psi, i, n)$. The number n is the number of possible worlds of the model being constructed. If we set $n = \mathrm{dd}(\phi) + 1$, we are guaranteed to have an equi-satisfiable formula on the output.

In order to take advantage of the ability of modern SAT solvers to return a reason for unsatisfiability, we can add selector variables s_i to enable or disable worlds w_i. We update the translation of $\Box\psi$ to $\bigwedge_{i=1}^{n}(\neg s_i \vee \mathrm{tr}(\psi, i, n))$ and the one of $\Diamond\psi$ to $\bigvee_{i=1}^{n}(s_i \wedge \mathrm{tr}(\psi, i, n))$. Worth noticing that due to the simplifications authorized in S5, the modalities cannot be embedded modalities. The modal depth equals 1. While the resulting CNF is be bigger than the original one, the size of the S5-model will now be decided by the number of satisfied selector variables. The complete translation function is given below:

Definition 13 (Translation with Selectors). *Let* $\phi \in \mathcal{L}$.

$$tr_s(\phi, n) = tr'_s(\phi, 1, n)$$

$$tr'_s(p, i, n) = p_i \qquad\qquad tr'_s(\neg\psi, i, n) = \neg tr'_s(\psi, i, n)$$

$$tr'_s(\psi \wedge \delta, i, n) = tr'_s(\psi, i, n) \wedge tr'_s(\delta, i, n) \quad tr'_s(\psi \vee \delta, i, n) = tr'_s(\psi, i, n) \vee tr'_s(\delta, i, n)$$

$$tr'_s(\Box\psi, i, n) = \bigwedge_{j=1}^{n}(\neg s_j \vee (tr'_s(\psi, j, n))) \quad tr'_s(\Diamond\psi, i, n) = \bigvee_{j=1}^{n}(s_j \wedge (tr'_s(\psi, j, n)))$$

An S5-model has to have at least one possible world (the current world). W.l.o.g., we consider that the current world is the world number 1, thus s_1 is always set to true. In the remainder of this article, we denote by $\mathrm{tr_s}(\phi)$ the formula $(\mathrm{tr_s}(\phi, \mathrm{dd}(\phi) + 1) \wedge s_1)$ and the set of all selectors of $\mathrm{tr_s}(\phi)$ is denoted by $\mathcal{S}(\phi)$ (i.e., $\mathcal{S}(\phi) = \{s_i \mid 1 \leq i \leq \mathrm{dd}(\phi) + 1\}$).

Example 4 (Example of '$\mathrm{tr_s}$'). Let us go back to Example 1 and reuse the formula $\phi = ((\Box\neg a \vee \Diamond b) \wedge \Diamond a \wedge \Box b)$. Its translation $\mathrm{tr_s}(\phi, 3)$ is:

$$(\neg s_1 \vee (\neg a_1 \vee (s_1 \wedge b_1) \vee (s_2 \wedge b_2) \vee (s_3 \wedge b_3))) \wedge$$
$$(\neg s_2 \vee (\neg a_2 \vee (s_1 \wedge b_1) \vee (s_2 \wedge b_2) \vee (s_3 \wedge b_3))) \wedge$$
$$(\neg s_3 \vee (\neg a_3 \vee (s_1 \wedge b_1) \vee (s_2 \wedge b_2) \vee (s_3 \wedge b_3))) \wedge$$
$$((s_1 \wedge a_1) \vee (s_2 \wedge a_2) \vee (s_3 \wedge a_3)) \wedge ((\neg s_1 \vee b_1) \wedge (\neg s_2 \vee b_2) \wedge (\neg s_3 \vee b_3)) \wedge s_1$$

Intuitively, every formula with subscript i is a formula that is true at the possible world i. If the selector s_i is false, then the world i is not present in the model (and we do not care about the valuation of the propositions there in). Below, a formula that is equivalent to $\mathrm{tr_s}(\phi, 3)$ but with s_1 and s_2 activated and s_3 deactivated.

$(\neg\top \lor (\neg a_1 \lor (\top \land b_1) \lor (\top \land b_2) \lor (\bot \land b_3))) \land$
$(\neg\top \lor (\neg a_2 \lor (\top \land b_1) \lor (\top \land b_2) \lor (\bot \land b_3))) \land$
$(\neg\bot \lor (\neg a_3 \lor (\top \land b_1) \lor (\top \land b_2) \lor (\bot \land b_3))) \land$
$((\top \land a_1) \lor (\top \land a_2) \lor (\bot \land a_3)) \land ((\neg\top \lor b_1) \land (\neg\top \lor b_2) \land (\neg\bot \lor b_3)) \land \top$

This formula is equivalent to $(\neg a_1 \lor b_1 \lor b_2) \land (\neg a_2 \lor b_1 \lor b_2) \land (a_1 \lor a_2) \land (b_1 \land b_2)$, which is the same as the one presented in Example 3. It corresponds to the problem of deciding if ϕ is satisfiable in a model with 2 worlds.

As we can see, the problem of solving the minimal S5 satisfiability problem is now equivalent to the problem of satisfying $\mathrm{tr}_s(\phi, n)$ and minimize the number of s_i, for $i > 1$, assigned to true (or, equivalently, maximize the number of s_i assigned to false). Obviously, it can be seen as a pseudo-Boolean optimization problem [30], where the optimization function to be minimized is the number of selectors assigned to true. This problem is also often solved nowadays as an instance of the Partial MaxSAT problem [21], which consists in satisfying all the hard clauses (clauses that MUST be satisfied) and the maximum number of soft clauses (the clauses that are not mandatory). In our case, the hard clauses are those generated by the translation function, and the soft-clauses are the unit clauses $\{\neg s_i \mid s_i \in \mathcal{S}(\phi) \text{ and } i > 1\}$ built from the selector variables.

We can thus use state-of-the-art Partial MaxSAT solvers. However, it is not the only way, as we show in the following section. By considering the structure of S5-models, extracting a MSS can also be used to decide the problem.

W.l.o.g., in the following sections, we represent a set of unit soft clauses as the set of selectors composing it (eg.: $\{s_2, s_3, s_4\}$ rather than $\{\neg s_2, \neg s_3, \neg s_4\}$).

5   First Insight: Cardinality Optimality Equals Subset Optimality

In the Sect. 2.3, we defined the MSS problem. It is also possible to define a partial version of the MSS problem, where the objective is to compute a MSS such that some given subset of the clauses (the hard clauses) must be satisfied. This problem is related to the Partial MaxSAT problem. In fact, a solution to a Partial MaxSAT problem is one of the biggest MSS that satisfies the set of hard-clauses. In general, a partial MSS is not a solution to a Partial MaxSAT problem but, in the specific case of MinS5-SAT, a partial MSS is also a solution to its corresponding Partial MaxSAT problem, which means that in that specific context, subset optimality (MSS) is equivalent to cardinality optimality (MaxSAT).

Proposition 1. *Let $\Sigma = tr_s(\phi, \mathrm{dd}(\phi) + 1)$ a CNF, and let χ be the formula $\bigwedge_{i=2}^{\mathrm{dd}(\phi)+1} \neg s_i$. An MSS of $(\Sigma \land \chi)$, where Σ is the set of hard clauses, is also a solution to the Partial MaxSAT problem $(\Sigma \land \chi)$.*

The proof of Proposition 1 uses the following lemma.

Lemma 1. *Let $\Sigma = tr_s(\phi, n)$, and let χ be the formula $\bigwedge_{s_i \in \mathcal{S}'} \neg s_i$, where $\mathcal{S}' \subseteq \mathcal{S}(\phi)$. If $(\Sigma \wedge \chi)$ is satisfiable then so is the formula $(\Sigma \wedge \chi')$, where χ' is obtained from χ by replacing the occurrences of one selector $s \in \mathcal{S}'$ by another selector $s' \in \mathcal{S}(\phi) \setminus \mathcal{S}'$.*

Proof (sketch). The proof is done by an induction on the length of the formula ϕ. In the induction base, $\Sigma = p$, for some $p \in \mathbb{P}$. We have $\Sigma = p_1$, $\chi = \neg s_1$ and $\mathcal{S}(\phi) = \{s_1\}$, which means that the claim is true (because $\mathcal{S}(\phi) \setminus \mathcal{S}' = \emptyset$). We have several cases on the induction step. Since their proofs are all similar, we show only one them here. Let $\phi = \Box \phi'$. We have $\Sigma = \bigwedge_{i=1}^{n} (\neg s_i \vee tr_s(\phi', i, n))$, $\chi = \bigwedge_{s_i \in \mathcal{S}'} \neg s_i$, and $\mathcal{S}(\phi) = \{s_1, \ldots, s_n\}$. Now, let χ' be obtained from χ where s_i is replaced by $s_j \in \mathcal{S}(\phi) \setminus \mathcal{S}'$. If $(\Sigma \wedge \chi)$ is satisfied by a model M then we construct a new model M', which equals M except that the truth assignment of all propositional variables with subscript i are the same as those with subscript j. We immediately have that if $M \models \chi$ then $M' \models \chi'$. We also have that if $M \models \neg s_i$ then $M' \models \neg s_j$. Finally, for each $1 \leq i \leq n$, if $M \models tr_s(\phi', i, n)$ then $M' \models tr_s(\phi', j, n))$, by the induction hypothesis (since the length of ϕ' is strictly smaller than that of ϕ). Therefore, $M' \models \Sigma \wedge \chi'$.

Proof (of Proposition 1). Towards a contradiction, assume that there exists a MSS $\delta_1 = (\Sigma \wedge \chi_1)$, where $\chi_1 = \bigwedge_{s \in S_1} \neg s$, which is not the biggest one. Thus, there exists another MSS $\delta_2 = (\Sigma \wedge \chi_2)$, where $\chi_2 = \bigwedge_{s \in S_2} \neg s$ and such that $|S_1| < |S_2|$. Now, let $S_3 = S_2 \setminus \{s\} \cup \{s'\}$, where $s \in S_2$ and $s' \in S_1$. By Lemma 1, the formula $\delta_3 = (\Sigma \wedge \chi_3)$, where $\chi_3 = \bigwedge_{s \in S_3} \neg s$ is satisfiable, because it is δ_2 with one of the selectors of S_2 in χ_2 replaced by another selector. It is easy to see that one can keep replacing selectors in this set until we have the set S_k, such that $S_1 \subseteq S_k$. The formula $\delta_k = (\Sigma \wedge \chi_k)$, where $\chi_k = \bigwedge_{s \in S_k} \neg s$, is satisfiable, by applying Lemma 1 $|S_1|$ times. Then δ_k a MSS that includes δ_1, which contradicts the assumption. This means that every MSS of the initial formula is one of the biggest ones. Therefore, any MSS of $(\Sigma \wedge \chi)$ is also a solution to the partial MaxSAT problem $(\Sigma \wedge \chi)$. $\square$

As a direct consequence of Proposition 1, we can always find a MSS such that the indexes of the selectors inside it are contiguous. This means that we can consider an optimisation that reduces the search space (breaks the symmetries), by adding the following:

$$\left(\bigwedge_{i=1}^{n-1} \neg s_i \rightarrow \neg s_{i+1} \right) \tag{1}$$

By giving as input $tr_s(\phi, n)$ plus $\mathcal{S}(\phi)$, we can solve the MinS5-SAT problem with a MaxSAT solver, or a Pseudo Boolean (PB) solver. If we also add Eq. 1 to the input, we can then use a MSS-extractor. However, we demonstrate in the following section that we can push the envelope by considering a dedicated approach using an incremental SAT solver with unsatisfiable cores.

6 Second Insight: Only Core Size Matters

Consider the following example: let ϕ be the input formula and let $dd(\phi) + 1 = 10$. We translate ϕ using selectors and start looking for a model for it. Assume that, after some computation, we conclude that 4 worlds cannot be deactivated altogether, i.e., if the selectors s_i, s_j, s_k and s_l are set to false, we have an inconsistency. We can infer from that information that we will need at least 7 worlds in the S5-model for ϕ. This comes from the fact that the '4 worlds which cannot be deactivated altogether' can be, in fact, any group of 4 worlds. Indeed, in the sequel, we demonstrate that if we have a group of m selectors forming an unsatisfiable core and the upper-bound equals n, then we need at least $(n - m + 1)$ worlds in the S5-model of the input formula.

Proposition 2. *Let $\phi \in \mathcal{L}$ such that $dd(\phi) + 1 = n$. If C is an UNSAT core of ϕ under assumptions $\mathcal{S}(\phi)$ then $tr_s(\phi, n')$ is unsatisfiable for all $n' \in \{1, \ldots, (n - |C|)\}$.*

Lemma 2. *If C is an UNSAT core of ϕ with assumptions $\mathcal{S}(\phi)$ then any set of literals $C' = \{\neg s \mid s \in \mathcal{S}(\phi)\}$ such that $|C'| = |C|$ is an UNSAT core of ϕ.*

Proof. Assume that C is an UNSAT core of ϕ with assumptions $\mathcal{S}(\phi)$. We have that $(\phi \wedge \bigwedge_{l \in C} l)$ is unsatisfiable. Now, towards a contradiction, also assume that there exists a set $C' = \{\neg s \mid s \in \mathcal{S}(\phi)\}$ such that $|C'| = |C|$ and $(\phi \wedge \bigwedge_{s \in C} \neg s)$ is satisfiable. By Lemma 1, we can obtain a new set D from C' by replacing the selectors in C' by those in C such that $(\phi \wedge \bigwedge_{s \in D} \neg s)$ is satisfiable. Because $D = C$, we have a contradiction. Therefore, any set of literals C' obtained as such is an UNSAT core of ϕ. □

Proof (of Proposition 2). The formula has n worlds. The SAT solver returns a core C of size m. So one of the selector has to be true. But due to Lemma 2, we have to put at least one selector to true to all the possible unsatisfiable cores of size m. Said otherwise, we must have $(n - m + 1)$ selectors to be true together, or the formula will be necessarily unsatisfiable. This also means that $\forall b' \in [1 \ldots (n - m)]\ tr_s(\phi, b')$ is unsatisfiable. □

Using this property, it is possible to construct an iterative algorithm which is based on incremental SAT. The SAT solver will be able to return an unsatisfiable core, and by interpreting it in the specific case of S5 as explained in Proposition 2, we can refine the bound used in the translation. The procedure starts by trying structures of size $b = 1$. If no model is found, it iterates the process, each time increasing the value of b by $(dd(\phi) + 1 - |s| + 1)$ (where $|s|$ is the size of the core). It iterates until a model of ϕ is found or the upper bound $dd(\phi) + 1$ is reached. Note that $|s|$ strictly decreases at each step, because we strictly increase the number of satisfied selectors. The approach $Dicho_c$ is similar.

7   Experimental Results

We compared several different approaches to the MinS5-SAT problem: S52SAT [10] with five different strategies: 1toN$_c$, 1toN, Nto1, Dicho$_c$, Dicho. CNF plus MaxSAT solver: maxHS-b [31], mscg2015b [32], and MSUnCore [33]. Pseudo-Boolean (PB) translation plus PB solver: NaPS [34], SAT4J-PB [35], SCIP [36]. CNF plus symmetry breaking plus MCS extraction with the LBX solver [37].

To see the impact of our minimisation, we use the state-of-the-art modal logic S5 solver S52SAT with glucose (4.0) as embedded SAT solver [29] (with its caching activated). We selected MaxSAT solver which have shown good performances in the MaxSAT competition 2016 [38]. We also considered LMHS-2016 [39] but, unfortunately, we did not manage to compile it due its multiple links to other software and our configuration environment.

Despite our through research, we could not find benchmarks for modal logic S5. Due to this fact, we chose to use the following benchmarks for modal logics K, KT and S4: TANCS-2000 modalised random QBF (MQBF) formulae [40] complemented by additional MQBF formulae provided by [41]; LWB K, KT and S4 formulae [42], with 56 formulae chosen from each of the 18 parametrized classes, generated from the script given by the authors of [9]; and Randomly generated $3CNF_{KSP}$ formulae [43] of modal depths 1 and 2. The benchmarks are classified as SAT or UNSAT in [9,42]. However, we kept only the benchmarks satisfied in their original logic to see the impact of a potential use of a S5-solving as a preprocessor for other modal logics. We have no interest with the UNSAT benchmarks because the unsatisfiability in K,KT and S4 implies the unsatisfiability in S5. We also proposed new benchmarks based on planning with uncertainties in the initial states, to check the performance of the different approaches on structured benchmarks. In such planning problems, some fluent f may be initially true, initially false, or neither. I the latter case, two different initial situations are possible. As a result, instead of a single initial state s_0, we may have several different initial states, which are consistent with available knowledge about the system (see [23] for more details and applications). By construction, all instances considered here have a plan to minimize. Modal logic S5 formulas are generated with a CEGAR approach [44]. We increase the value of the bounded-horizon until we reach the smallest value for which there exists a plan as explained in [45].

We performed experimental evaluations on a variety of planning benchmarks. It includes the traditional conformant benchmarks, namely: Bomb-in-the-toilet, Ring, Cube, Omelet and Safe (see [46] for more details) modeled here as planning with uncertainties in the initial state. We also performed evaluations on classical benchmarks: Blocksworld, Logistics, and Grid, in which the authors of [47] introduced uncertainty about the initial state. All the benchmarks are available for download[1].

To select the "minimalizable" benchmarks, we set a time-out of 1500 s. We managed to solve 28 benchmarks out of the 119 available. We tried other solvers:

---

[1] https://fai.cs.uni-saarland.de/hoffmann/ff/cff-tests.tgz.

Spartacus [48] solved 15 instances and SPASS [49] solved 5, with both being a subset of the 28 solved by S52SAT. Our generator has negligible execution times and is available for download[2]. Each of these benchmarks has a plan of size N (where N can be different for each benchmark) which has been verified. We then generated modal logic benchmarks from these instances by fixing the horizon at $N, N + 1, \ldots, N + 9$ having thus 280 benchmarks, all S5-satisfiable, to test our minimisation techniques.

The benchmarks and the different solvers (especially S52SAT, which is the one translating the formulas into propositional logic) are available[3]. The experiments ran on a cluster of Xeon 4 cores, 3.3 GHz, running CentOS 6.4. The memory limit is set to 32 GB and the runtime limit is set to 900 s per solver per benchmark. In the following tables, we provide the number of benchmarks for which a minimal S5 model is found. In bold face, the best result for each row/-column. The VBS (Virtual Best Solver) represents the union of the benchmarks solved by all the approaches.

7.1   State-of-the-Art Modal Logics Benchmarks

Logic WorkBench (LWB) Benchmarks. All the results are reported in the Table 1. The difference in the results between the approaches using S52SAT and the MaxSAT solvers came from the fact that MaxSAT solvers cannot take into account inherent properties of modal logic S5. They have embedded cardinality constraints used to count the number of satisfied/falsified clauses to return the smallest model. By comparing the results of 1toN and 1toN_c in number of benchmarks solved, one could think that selectors do not make much difference. But the runtime provides a different picture, as in the scatter plot depicted in Fig. 1. The x-axis corresponds to the time used by 1toN_c while the y-axis corresponds to the time used by 1toN to solve these problems. As expected, 1toN_c performs less iterations and thus calls the SAT solver fewer times. We remark that the solver took less than 10 s for the majority of the instances. It turns out that it makes sense to consider this approach as a pre-processing for a more generic minimal modal logic SAT solver (eg., for logics K, KT and S4). Indeed if we find a minimal model in S5, we obtain in the same way an upper-bound on the size of the minimal model in K, KT and S4.

$3CNF_{KSP}$ Benchmarks. The randomly generated $3CNF_{KSP}$ formulae [43] of depths 1 and 2 consist of 1000 formulae, where 457 are satisfiable in modal logic K and 89 are satisfiable in S5. All the results are reported in the Table 2. As for LWB, 1toN and Dicho are better than Nto1 because the minimal models found are relatively small. It is interesting to notice that the modal depth of the formulas influences the result. This is surprising due to the fact that, in S5, all formulae can be reduced to modal depth 1. In fact, many instances with modal depth 2, that are SAT in modal logic K, are UNSAT in modal logic S5.

---

[2] http://www.cril.fr/~montmirail/planning-to-s5/.
[3] http://www.cril.fr/~montmirail/s52SAT.

Table 1. #Instances solved in LWB

Method	K	KT	S4	Total
# benchs	(185)	(279)	(160)	(624)
1toN	**185**	**279**	**160**	**624**
Nto1	17	34	2	53
Dicho	119	175	78	372
1toN$_c$	**185**	**279**	**160**	**624**
Dicho$_c$	135	201	100	436
maxHS	17	25	72	114
MSCG	74	65	103	242
MSUn Core	19	30	80	129
NaPS	126	64	71	261
SAT4J	18	27	58	103
SCIP	104	158	112	374
LBX	118	173	92	383
VBS	185	279	160	624

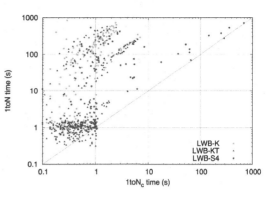

Fig. 1. Scatter-plot of 1toN vs 1toN$_c$

Randomly Modalized QBF (MQBF) Benchmarks. Originally, this benchmark set contains 1016 formulas, among them 617 are SAT while 399 are UNSAT in K. All the results are reported in the Table 3. Dicho, Dicho$_c$, 1toN and 1toN$_c$ approaches are better than the other ones. Moreover, it is interesting to see that the whole **qbf** family is in fact $S5$-satisfiable, even though they are normally used to evaluate modal logic K solvers. It is worth noticing that the performance of a MaxSAT or a PB approach are globally worse than the MSS-extraction approach. However, if we add the symmetry breaking from Eq. 1 then the performances become equivalent.

Structured Benchmarks: Planning with Uncertainties. As in the random and crafted benchmarks before, we can see in Fig. 2c that the use of selectors allows us to solve more benchmarks. But, surprisingly, here the best approach is to use a dichotomic search instead of a linear search from 1 to N. This is mainly due to the size of the smallest model, which is rarely a small number, as it was the case in LWB for example. Moreover, each call to the SAT solver is more time-consuming because the instances are harder to solve in practice. This again reminds us that the benchmarks considered can influence the result obtained.

Minimization Overhead. We can see on Fig. 2a that it requires only an acceptable over-head computation to get the smallest model possible instead of the first one returned by the solver (from less than 10 s to less than 40 s). On the other hand, as we can see on Fig. 2b the minimization can reduce drastically the size of the returned model. Moreover, we can also see the structural difference between randomly generated instances, that can finally be solved with only few worlds, and 'real-world' applications instances that need a larger number of worlds to be solved. There is also a gain against other solvers able to output a model (Fig. 2d), such as Spartacus [48]. Note that we could only compare Spartacus models on planning problems because Spartacus is dedicated to modal logics K, KT and

Table 2. #instances solved $3CNF_{KSP}$

Benchs	1toN	Nto1	Dicho	1toN$_c$	Dicho$_c$	maxHS	MSCG	MSUnCore	NaPS	SAT4J	SCIP	LBX	VBS
md = 1 (62)	55	0	26	**62**	40	40	30	38	42	35	42	47	62
md = 2 (27)	17	0	9	**27**	17	12	12	12	17	12	20	17	27
Total (89)	72	0	35	**89**	57	52	42	50	59	47	62	64	89

Table 3. #instances solved in $MQBF$

Benchs	1toN	Nto1	Dicho	1toN$_c$	Dicho$_c$	maxHS	MSCG	MSUnCore	NaPS	SAT4J	SCIP	LBX	VBS
qbf (56)	**56**	55	**56**	**56**	**56**	**56**	**56**	**56**	55	48	**56**	**56**	56
qbfS (171)	**171**	0	**171**	**171**	**171**	0	156	0	144	140	155	167	171
Total (227)	**227**	55	**227**	**227**	**227**	56	212	56	199	188	211	223	227

(a) S52SAT vs S52SAT-1toN$_c$ (time)　　　(b) S52SAT vs S52SAT-1toN$_c$ (size)

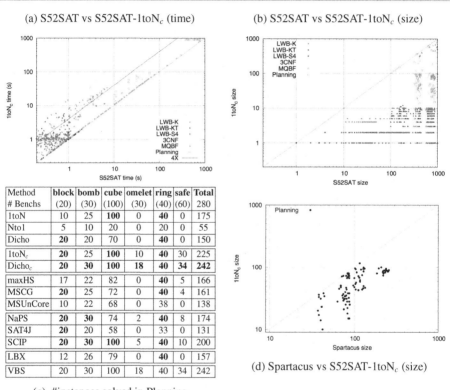

Method	block	bomb	cube	omelet	ring	safe	Total
# Benchs	(20)	(30)	(100)	(30)	(40)	(60)	280
1toN	10	25	**100**	0	**40**	0	175
Nto1	5	10	20	0	20	0	55
Dicho	**20**	20	70	0	**40**	0	150
1toN$_c$	**20**	25	**100**	10	**40**	30	225
Dicho$_c$	**20**	**30**	**100**	**18**	**40**	**34**	**242**
maxHS	17	22	82	0	**40**	5	166
MSCG	**20**	25	72	0	**40**	4	161
MSUnCore	10	22	68	0	38	0	138
NaPS	**20**	**30**	74	2	**40**	8	174
SAT4J	**20**	20	58	0	33	0	131
SCIP	**20**	**30**	**100**	5	**40**	10	200
LBX	12	26	79	0	**40**	0	157
VBS	20	30	100	18	40	34	242

(c) #instances solved in Planning

(d) Spartacus vs S52SAT-1toN$_c$ (size)

Fig. 2. Results on planning and analysis of the overhead

S4, not S5. However, on those specific benchmarks, since the modal depth is one and all K-models on those benchmarks are S5 models, we can compare their size. Note also that Spartacus outputs an open-saturated tableau (which indicates the existence of a model) and not a full model which should be even larger (see [50] for more details).

8   Conclusion

We defined in this article a new optimisation problem that we call the minimal S5 satisfiability problem (MinS5-SAT). It is the problem of finding the smallest S5-model w.r.t. the number of possible worlds. We demonstrated that this problem can be reduced to the problem of extracting a Maximal Satisfiability Set of clauses (MSS) and thus, can be solved with a MSS-extractor or one of the state-of-the-art PB or MaxSAT solvers. We also showed that, thanks to an inherent property of modal logic S5, this problem can also be solved using unsatisfiable cores in an incremental SAT procedure. The latter approach is the one that obtained the best performance in our experiments.

We applied these different techniques to various benchmarks: randomly generated formulas and also formulas expressing planning with uncertainties problems. Experimental results showed that the best technique for one set of benchmarks is not necessarily the best technique for the other, reminding us the importance of the choice of benchmarks in experimental evaluations. The technique used obtained huge gains in the size of the output models, when compared to the other approaches that do not try minimisation. In addition, the overhead imposed by the minimisation is acceptable. Therefore, we believe that finding minimal models for modal logic formulas is an interesting task. We can also mention that smaller models are more user-friendly, they permit to speedup the model checking phase and, in addition, some real-applications may prefer "smaller solutions" (smaller plans, for instance).

One possible future work is the application of these techniques to other NP-complete modal logics such as KD45 which is the belief counterpart of S5. Moreover, one could also try to solve the Minimal Satisfiability Problem without the use of a SAT solver, e.g. To compute the auto-bisimilar of a model retured by a Tableau proved such as Spartacus [48]. Also, one could try to solve the more general Minimal Modal Logic Satisfiability Problem (MinML-SAT), for which the standard satisfiability problem is typically PSPACE-hard. For instance, it would be interesting to try to filter the Minimal modal logic K satisfiability problem with the Minimal S5 satisfiability problem. It is known that satisfiability in S5 entails the satisfiability in K. If one finds a minimal S5-model of size n for a formula ϕ then the minimal K-model for ϕ has at most n possible worlds. This insight may help improving a naive search for the minimal K-model, because the only known bound b for modal logic K is an exponential function on the length the input formula.

Acknowledgments. We thank the anonymous reviewers for their insightful comments. Part of this work was supported by the French Ministry for Higher Education and Research, the Nord-Pas de Calais Regional Council through the "Contrat de Plan État Région (CPER) DATA" and an EC FEDER grant.

References

1. Fairtlough, M., Mendler, M.: An intuitionistic modal logic with applications to the formal verification of hardware. In: Pacholski, L., Tiuryn, J. (eds.) CSL 1994. LNCS, vol. 933, pp. 354–368. Springer, Heidelberg (1995). https://doi.org/10.1007/BFb0022268

2. Fitting, M.: Modality and databases. In: Dyckhoff, R. (ed.) TABLEAUX 2000. LNCS (LNAI), vol. 1847, pp. 19–39. Springer, Heidelberg (2000). https://doi.org/10.1007/10722086_2

3. Murphy VII, T., Crary, K., Harper, R.: Distributed control flow with classical modal logic. In: Ong, L. (ed.) CSL 2005. LNCS, vol. 3634, pp. 51–69. Springer, Heidelberg (2005). https://doi.org/10.1007/11538363_6

4. Bienvenu, M., Fargier, H., Marquis, P.: Knowledge compilation in the modal logic S5. In: Proceedings of AAAI 2010 (2010)

5. Niveau, A., Zanuttini, B.: Efficient representations for the modal logic S5. In: Proceedings of IJCAI 2016, pp. 1223–1229 (2016)

6. Hustadt, U., Schmidt, R.A., Weidenbach, C.: MSPASS: subsumption testing with SPASS. In: Proceedings of DL 1999. CEUR Workshop Proceedings, vol. 22. CEUR (1999)

7. Tacchella, A.: *SAT system description. In: Proceedings of DL 1999, vol. 22. CEUR (1999)

8. Sebastiani, R., Vescovi, M.: Automated reasoning in modal and description logics via SAT encoding: the case study of K(m)/ALC-satisfiability. J. Artif. Intell. Res. **35**, 343–389 (2009)

9. Nalon, C., Hustadt, U., Dixon, C.: $K_S P$: a resolution-based prover for multimodal K. In: Proceedings of IJCAR 2016, pp. 406–415 (2016)

10. Caridroit, T., Lagniez, J.M., Le Berre, D., de Lima, T., Montmirail, V.: A SAT-based approach for solving the modal logic S5-satisfiability problem. In: Proceedings of AAAI 2017 (2017)

11. Lagniez, J.M., Le Berre, D., de Lima, T., Montmirail, V.: A recursive shortcut for CEGAR: application to the modal logic K satisfiability problem. In: Proceedings of IJCAI 2017 (2017)

12. Simon, L., Le Berre, D., Hirsch, E.A.: The SAT2002 competition. Ann. Math. Artif. Intell. **43**(1), 307–342 (2005)

13. Marques-Silva, J., Janota, M.: On the query complexity of selecting few minimal sets. Electron. Colloq. Comput. Complex. (ECCC) **21**, 31 (2014)

14. Marques-Silva, J., Lynce, I.: On Improving MUS extraction algorithms. In: Sakallah, K.A., Simon, L. (eds.) SAT 2011. LNCS, vol. 6695, pp. 159–173. Springer, Heidelberg (2011). https://doi.org/10.1007/978-3-642-21581-0_14

15. Belov, A., Marques-Silva, J.: Accelerating MUS extraction with recursive model rotation. In: Proceedings of FMCAD 2011, pp. 37–40 (2011)

16. Iser, M., Sinz, C., Taghdiri, M.: Minimizing models for tseitin-encoded SAT instances. In: Järvisalo, M., Van Gelder, A. (eds.) SAT 2013. LNCS, vol. 7962, pp. 224–232. Springer, Heidelberg (2013). https://doi.org/10.1007/978-3-642-39071-5_17

17. Soh, T., Inoue, K.: Identifying necessary reactions in metabolic pathways by minimal model generation. In: Proceedings of ECAI 2010, vol. 215, pp. 277–282. IOS Press (2010)

18. Koshimura, M., Nabeshima, H., Fujita, H., Hasegawa, R.: Minimal model generation with respect to an atom set. In: Proceedings of FTP 2009, pp. 49–59 (2009)

19. Ladner, R.E.: The computational complexity of provability in systems of modal propositional logic. SIAM J. Comput. **6**(3), 467–480 (1977)
20. Cook, S.A.: Characterizations of pushdown machines in terms of time-bounded computers. J. ACM **18**(1), 4–18 (1971)
21. Li, C.M., Manyà, F.: MaxSAT, hard and soft constraints. In: Handbook of Satisfiability, pp. 613–631. IOS Press (2009)
22. Fagin, R., Halpern, J.Y., Moses, Y., Vardi, M.Y.: Reasoning About Knowledge. MIT Press, Cambridge (1995)
23. Eiter, T., Faber, W., Leone, N., Pfeifer, G., Polleres, A.: Planning under incomplete knowledge. In: Lloyd, J., Dahl, V., Furbach, U., Kerber, M., Lau, K.-K., Palamidessi, C., Pereira, L.M., Sagiv, Y., Stuckey, P.J. (eds.) CL 2000. LNCS (LNAI), vol. 1861, pp. 807–821. Springer, Heidelberg (2000). https://doi.org/10.1007/3-540-44957-4_54
24. Kripke, S.A.: Semantical analysis of modal logic I. Normal propositional calculi. Z. M. L. G. M. **9**(56), 67–96 (1963)
25. Eén, N., Sörensson, N.: An extensible SAT-solver. In: Giunchiglia, E., Tacchella, A. (eds.) SAT 2003. LNCS, vol. 2919, pp. 502–518. Springer, Heidelberg (2004). https://doi.org/10.1007/978-3-540-24605-3_37
26. Grégoire, É., Lagniez, J., Mazure, B.: An experimentally efficient method for (MSS, CoMSS) partitioning. In: Proceedings of AAAI 2014, pp. 2666–2673 (2014)
27. Belov, A., Lynce, I., Marques-Silva, J.: Towards efficient MUS extraction. AI Commun. **25**, 97–116 (2012)
28. O'Sullivan, B., Papadopoulos, A., Faltings, B., Pu, P.: Representative explanations for over-constrained problems. In: Proceedings of AAAI 2007, pp. 323–328 (2007)
29. Audemard, G., Lagniez, J.-M., Simon, L.: Improving glucose for incremental SAT solving with assumptions: application to MUS extraction. In: Järvisalo, M., Van Gelder, A. (eds.) SAT 2013. LNCS, vol. 7962, pp. 309–317. Springer, Heidelberg (2013). https://doi.org/10.1007/978-3-642-39071-5_23
30. Roussel, O., Manquinho, V.M.: Pseudo-Boolean and cardinality constraints. In: Handbook of Satisfiability, pp. 695–733. IOS Press (2009)
31. Davies, J., Bacchus, F.: Exploiting the power of MIP solvers in MAXSAT. In: Järvisalo, M., Van Gelder, A. (eds.) SAT 2013. LNCS, vol. 7962, pp. 166–181. Springer, Heidelberg (2013). https://doi.org/10.1007/978-3-642-39071-5_13
32. dos Reis Morgado, A.J., Ignatiev, A.S., Silva, J.M.: MSCG: robust core-guided MaxSAT solving. J. Satisfiability Boolean Model. Comput. **9**, 129–134 (2014)
33. Heras, F., Morgado, A., Marques-Silva, J.: Core-guided binary search algorithms for maximum satisfiability. In: Proceedings of AAAI 2011 (2011)
34. Sakai, M., Nabeshima, H.: Construction of an ROBDD for a PB-constraint in band form and related techniques for PB-solvers. IEICE Trans. **98–D**(6), 1121–1127 (2015)
35. Le Berre, D., Parrain, A.: The SAT4J library, release 2.2. JSAT **7**(2–3), 59–64 (2010)
36. Maher, S.J., Fischer, T., Gally, T., Gamrath, G., Gleixner, A., Gottwald, R.L., Hendel, G., Koch, T., Lübbecke, M.E., Miltenberger, M., Müller, B., Pfetsch, M.E., Puchert, C., Rehfeldt, D., Schenker, S., Schwarz, R., Serrano, F., Shinano, Y., Weninger, D., Witt, J.T., Witzig, J.: The SCIP optimization suite 4.0. Technical report 17–12, ZIB, Takustr. 7, 14195 Berlin (2017)
37. Mencía, C., Previti, A., Marques-Silva, J.: Literal-based MCS extraction. In: Proceedings of IJCAI 2015, pp. 1973–1979 (2015)
38. Argelich, J., Min Li, C., Manyà, F., Planes, J.: Max-SAT 2016: eleventh Max-SAT evaluation (2016). http://www.maxsat.udl.cat/16/

39. Saikko, P., Berg, J., Järvisalo, M.: LMHS: a SAT-IP hybrid MaxSAT solver. In: Creignou, N., Le Berre, D. (eds.) SAT 2016. LNCS, vol. 9710, pp. 539–546. Springer, Cham (2016). https://doi.org/10.1007/978-3-319-40970-2_34

40. Massacci, F., Donini, F.M.: Design and results of TANCS-2000 non-classical (modal) systems comparison. In: Dyckhoff, R. (ed.) TABLEAUX 2000. LNCS (LNAI), vol. 1847, pp. 52–56. Springer, Heidelberg (2000). https://doi.org/10.1007/10722086_4

41. Kaminski, M., Tebbi, T.: InKreSAT: modal reasoning via incremental reduction to SAT. In: Bonacina, M.P. (ed.) CADE 2013. LNCS (LNAI), vol. 7898, pp. 436–442. Springer, Heidelberg (2013). https://doi.org/10.1007/978-3-642-38574-2_31

42. Balsiger, P., Heuerding, A., Schwendimann, S.: A benchmark method for the propositional modal logics K, KT, S4. J. Autom. Reason. **24**(3), 297–317 (2000)

43. Patel-Schneider, P.F., Sebastiani, R.: A new general method to generate random modal formulae for testing decision procedures. J. Artif. Intell. Res. **18**, 351–389 (2003)

44. Clarke, E.M., Grumberg, O., Jha, S., Lu, Y., Veith, H.: Counterexample-guided abstraction refinement for symbolic model checking. J. ACM **50**(5), 752–794 (2003)

45. Rintanen, J.: Planning and SAT. In: Handbook of Satisfiability, pp. 483–504. IOS Press (2009)

46. Petrick, R.P.A., Bacchus, F.: A knowledge-based approach to planning with incomplete information and sensing. In: Proceedings of AIPS 2002, pp. 212–222 (2002)

47. Hoffmann, J., Brafman, R.I.: Conformant planning via heuristic forward search: a new approach. Artif. Intell. **170**(6–7), 507–541 (2006)

48. Götzmann, D., Kaminski, M., Smolka, G.: Spartacus: a tableau prover for hybrid logic. Electron. Notes Theor. Comput. Sci. **262**, 127–139 (2010)

49. Weidenbach, C., Dimova, D., Fietzke, A., Kumar, R., Suda, M., Wischnewski, P.: SPASS version 3.5. In: Proceedings of CADE 2009, pp. 140–145 (2009)

50. Lagniez, J.M., Le Berre, D., de Lima, T., Montmirail, V.: On checking Kripke models for modal logic K. In: Proceedings of PAAR@IJCAR 2016, pp.69–81 (2016)

FAME: An Automated Tool for Semantic Forgetting in Expressive Description Logics

Yizheng Zhao[(⊠)] and Renate A. Schmidt

School of Computer Science, The University of Manchester, Manchester, UK
yizheng.zhao@manchester.ac.uk

Abstract. In this paper, we describe a high-performance reasoning tool, called FAME, for semantic forgetting in expressive description logics. Forgetting is a non-standard reasoning service that seeks to create restricted views of ontologies by eliminating concept and role names from ontologies in such a way that all logical consequences up to the remaining signature are preserved. FAME is a Java-based implementation of an Ackermann-based method for forgetting concept and role names from ontologies expressible in the description logic $\mathcal{ALCOIH}$. $\mathcal{ALCOIH}$ is the extension of the basic description logic $\mathcal{ALC}$ with nominals, inverse roles and role inclusions. FAME can be used as a standalone tool or a Java library for forgetting or related tasks. Results of an evaluation of FAME on a corpus of 396 biomedical ontologies have shown that: (i) in more than 90% of the test cases FAME was successful (i.e., eliminated all specified concept and role names) and (ii) the elimination was done within one second in more than 70% of the successful cases.

1 Introduction

Ontologies, exploiting description logics as the representational underpinning, provide a logic-based data model for knowledge representation thereby supporting effective reasoning of domain knowledge for a range of real-world applications, most evidently for applications in life sciences, text mining and the semantic web. However, with their growing utilisation, not only has the number of available ontologies increased considerably, but they are often large in size and are becoming more complex to manage. Capturing domain knowledge in the form of ontologies is moreover labour-intensive work. There is therefore a strong demand for techniques and automated tools for creating restricted views of ontologies. *Forgetting* is a non-standard reasoning service that seeks to create restricted views of ontologies by eliminating concept and role names from ontologies in such a way that complete information is preserved up to the remaining signature. Forgetting allows users to focus on specific parts of ontologies that can be easily reused, or to zoom in on ontologies for in-depth analysis of certain subparts. It is also useful for information hiding, ontology summarisation, explanation generation (abduction), ontology debugging and repair, as well as computing the logical difference between ontology versions [3–5,8,10].

© Springer International Publishing AG, part of Springer Nature 2018
D. Galmiche et al. (Eds.): IJCAR 2018, LNAI 10900, pp. 19–27, 2018.
https://doi.org/10.1007/978-3-319-94205-6_2

Forgetting can be defined in two closely related ways; it can be defined syntactically as the dual of *uniform interpolation* [5] and it can be defined model-theoretically as *semantic forgetting* [10,13]. The two notions differ in the sense that uniform interpolation preserves all *logical consequences* up to certain names, whereas semantic forgetting preserves *equivalence* up to certain names. Hence, semantic solutions are in general stronger than the uniform interpolants; they often require the target language to be extended to express them.

FAME is the first automated tool for semantic forgetting in description logics. It is a Java-based implementation of a semantic forgetting method developed in our recent work [11,12]. Being based on non-trivial generalisations of a monotonicity property, namely, Ackermann's Lemma [1], the method can eliminate concept and role names from ontologies expressible in the description logic $\mathcal{ALCOIH}$, i.e., the basic $\mathcal{ALC}$ extended with nominals, inverse roles and role inclusions. The universal role $\triangledown$ and role conjunction $\sqcap$ are included in the target language, making the language more expressive to represent the forgetting solutions. For example, the semantic solution of forgetting the role name r from the ontology $\{A_1 \sqsubseteq \exists r.B_1, A_2 \sqsubseteq \forall r.\neg B_1\}$ is $\{A_1 \sqsubseteq \exists\triangledown.B_1, A_1 \sqcap A_2 \sqsubseteq \bot\}$, whereas the uniform interpolant is $\{A_1 \sqcap A_2 \sqsubseteq \bot\}$, which is weaker.

The current version of FAME includes several significant improvements, as well as a number of minor ones, over the prototypes used in [11,12]. It has been evaluated on a corpus of biomedical ontologies, including SNOMED CT and NCIT, with the results showing that: (i) in more than 90% of the test cases FAME was successful (i.e., eliminated all specified concept and role names) and (ii) the elimination was done within one second in more than 70% of the successful cases.

In this paper, we describe the top-level design of FAME, the main algorithm used by FAME, and details of an evaluation on a corpus of biomedical ontologies.

2   Semantic Forgetting for $\mathcal{ALCOIH}$

Let N_C, N_R and N_I be countably infinite and pairwise disjoint sets of *concept names*, *role names* and *individual names*, respectively. *Roles* in $\mathcal{ALCOIH}(\triangledown, \sqcap)$ can be a role name $r \in N_R$, the inverse r^- of a role name r, the universal role $\triangledown$, or a conjunction of a finite number of role names. *Concepts* in $\mathcal{ALCOIH}(\triangledown, \sqcap)$ can be of the following forms: $\top \mid \bot \mid a \mid A \mid \neg C \mid C \sqcap D \mid C \sqcup D \mid \exists R.C \mid \forall R.C$, where $a \in N_I$, $A \in N_C$, C and D are any concepts and R is any role. The forgetting method used by FAME works with TBox and ABox axioms in clausal normal form. A *TBox literal* is a concept of the form a, $\neg a$, A, $\neg A$, $\exists R.C$, or $\forall R.C$. A *TBox clause* is a disjunction of a finite number of TBox literals. An *RBox atom* is a role name, an inverted role name, or the universal role. An *RBox clause* is a disjunction of an RBox atom and a negated RBox atom. TBox and RBox clauses are obtained from (TBox and RBox) axioms using the standard clausal normal form transformation, where in the case of role axioms role negation is introduced. Let $\mathcal{S} \in N_C \cup N_R$ be a designated concept or role name. An occurrence of $\mathcal{S}$ is assumed to be *positive* (*negative*) in a clause if it is under an *even* (*odd*) number of negations. The semantics of $\mathcal{ALCOIH}(\triangledown, \sqcap)$ is as expected. For more details of the logics considered in this paper, we refer the reader to [12]

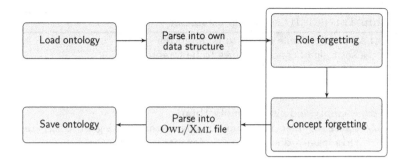

Fig. 1. The top-level design of FAME

By $\mathsf{sig}_\mathsf{C}(X)$ and $\mathsf{sig}_\mathsf{R}(X)$ we denote respectively the sets of the concept names and role names occurring in X, where X ranges over concepts, clauses, sets of clauses and ontologies. By $\mathsf{sig}(X)$ we denote the union of $\mathsf{sig}_\mathsf{C}(X)$ and $\mathsf{sig}_\mathsf{R}(X)$.

Definition 1 (Semantic Forgetting for $\mathcal{ALCOIH}$). *Let $\mathcal{O}$ be an $\mathcal{ALCOIH}$-ontology and let $\mathcal{F}$ be a subset of $\mathsf{sig}(\mathcal{O})$. An ontology $\mathcal{O}'$ is a* semantic solution *of forgetting $\mathcal{F}$ from $\mathcal{O}$ iff the following conditions hold: (i) $\mathsf{sig}(\mathcal{O}') \subseteq \mathsf{sig}(\mathcal{O}) \backslash \mathcal{F}$ and (ii) for any interpretation $\mathcal{I}$: $\mathcal{I} \models \mathcal{O}'$ iff $\mathcal{I}' \models \mathcal{O}$, for some interpretation $\mathcal{I}'$ $\mathcal{F}$-equivalent to $\mathcal{I}$, i.e., $\mathcal{I}$ and $\mathcal{I}'$ coincide but differ possibly in the interpretations of the names in $\mathcal{F}$.*

In this paper, the notation $\mathcal{F}$ is used to denote the *forgetting signature*, i.e., the set of concept and role names to be forgotten. $\mathcal{F}_\mathsf{C}$ and $\mathcal{F}_\mathsf{R}$ are used to denote respectively the concept names and role names in $\mathcal{F}$.

3   Implementation

The top-level design of FAME is shown in Fig. 1. FAME uses the OWL API Version 3.5.6[1] for the tasks of loading, parsing and saving ontologies. The ontology to be loaded must be specified as an OWL/XML file, or as a URL pointing to an OWL/XML file, but internally FAME uses own data structure for efficiency.

FAME defaults to eliminating role names first because during the role forgetting process concept definer names may be introduced (to facilitate the normalisation of the input ontology). These definer names, regarded as regular concept names, can thus be eliminated as part of subsequent concept forgetting. Given an $\mathcal{ALCOIH}$-ontology $\mathcal{O}$ and a forgetting signature $\mathcal{F} = \{r_1, \ldots, r_m, A_1, \ldots, A_n\}$, where $r_i \in \mathsf{sig}_\mathsf{R}(\mathcal{O})$ ($1 \leq i \leq m$) and $A_j \in \mathsf{sig}_\mathsf{C}(\mathcal{O})$ ($1 \leq j \leq n$), the forgetting process in FAME includes four main phases: (i) the conversion of $\mathcal{O}$ into a set of clauses (clausification), (ii) the role forgetting phase, (iii) the concept forgetting phase, and (iv) the conversion of the resulting clause set into an ontology $\mathcal{O}'$ (declausification). The role (concept) forgetting phase is an iteration of several

---

[1] http://owlcs.github.io/owlapi/.

Algorithm 1. FORGET(r_sig, c_sig, clause_set)

    Input   : a set r_sig of role names to be forgotten
                a set c_sig of concept names to be forgotten
                a set clause_set of clauses
    Output: a set clause_set of clauses (after forgetting)

 1 **do**
 2   **if** r_sig is empty and c_sig is empty **then**
          `// clause_set does not contain any names in r_sig or c_sig; in`
              `this case, clause_set is a forgetting solution`
 3   **return** clause_set
 4   **end**
 5   **initialising** final int sig_size_before to (r_sig.size() + c_sig.size())
 6   **initialising** Set⟨Name: ⟩ pure_sig to null
         `// get from r_sig and c_sig all names that are pure in clause_set`
 7   pure_sig := getPureNames(r_sig, c_sig, clause_set)
         `// apply Purify to clause_set to eliminate names in pure_sig`
 8   clause_set := purify(pure_sig, clause_set)
         `// simplify all axioms in clause_set`
 9   clause_set.getSimplified()
10   **initialising** Set⟨Clause⟩ sub_clause_set to null
11   **foreach** RoleName role in r_sig **do**
         `// get from clause_set all axioms that contain role`
12   sub_clause_set := getSubset(role, clause_set)
         `// remove from clause_set all axioms in sub_clause_set`
13   clause_set.removeAll(sub_clause_set);
         `// attempt to transform sub_clause_set into r-reduced form`
14   sub_clause_set.getRReducedForm(role, sub_clause_set);
         `// check whether sub_clause_set is in r-reduced form`
15   **if** isRReducedForm(role, role_clause_set) **then**
            `// apply Ackermann`^{R} `to sub_clause_set to eliminate role`
16   sub_clause_set := ackermann(role, sub_clause_set)
            `// simplify all axioms in sub_clause_set`
17   sub_clause_set.getSimplified()
            `// add the resulting set sub_clause_set back to clause_set`
18   clause_set.addAll(sub_clause_set)
            `// remove role from r_sig`
19   r_sig.remove(role)
            `// add all introduced definer names to c_sig`
20   c_sig.addAll(sub_clause_set.getDefiners())
21   **else**
            `// add the unchanged set sub_clause_set back to clause_set`
22   clause_set.addAll(sub_clause_set)
23   **end**
24   Similar for loop over the concept names in the present c_sig
25   **initialising** final int sig_size_after to (r_sig.size() + c_sig.size())
26 **while** sig_size_before != sig_size_after
    `// clause_set still contains names in r_sig or c_sig`
27 **return** clause_set

---

rounds in which the role (concept) names in $\mathcal{F}$ are eliminated. The elimination is based on the calculi Ack^R and Ack^C, described in detail in [12].

The calculus Ack^R includes four types of rules: (i) two Purify^R rules, (ii) one Ackermann^R rule, (iii) two rewrite^R rules, and (iv) definer introduction rules. The Purify^R rules eliminate a role name when the name occurs only positively or only negatively in the current clause set (i.e., in this case, the name is said to be *pure* in the clause set). The Ackermann^R rule eliminates a role name when the name occurs both positively and negatively in the current clause set in r-*reduced form*, where r is the current role name to be forgotten. The r-reduced form is a specialised normal form suitable for application of the Ackermann^R rule. The rewrite^R rules and definer introduction transform a clause set (not in r-reduced form) into r-reduced form.

The calculus Ack^C includes three types of rules: (i) two Purify^C rules, (ii) one Ackermann^C rule, and (iii) two rewrite^C rules. The purify^C rules eliminate a concept name when the name is pure in the current clause set. The Ackermann^C rule eliminates a concept name when the name occurs both positively and negatively in the current clause set in A-*reduced form*, a specialised normal form suitable for application of the Ackermann^C rule, where A is the current concept name to be forgotten. The rewrite^C rules transform a clause set (not in A-reduced form) into A-reduced form. Note that using the rules in Ack^R (Ack^C), a role (concept) name cannot always be eliminated. This is because there is a gap in the scope of the rewrite rules: transforming a clause set into r-reduced form or A-reduced form is not always possible.

The main algorithm used by FAME is shown in Algorithm 1. FAME performs *purification* prior to other steps (lines 6–9). This is because the Purify rules do not require the clause set to be normalised or in reduced form, and they can be applied at any time (purification is relatively cheap). Moreover, applying the Purify rules to a clause set often results in numerous syntactic redundancies, tautologies and contradictions inside the clauses which are immediately simplified or eliminated, leading to a much reduced set with fewer clauses and fewer names. The getSubset($\mathcal{S}$, $\mathcal{O}$) method extracts from the clause set $\mathcal{O}$ all axioms that contain the name $\mathcal{S}$. $\mathcal{S}$ can thus be eliminated from this subset, rather than from the entire set $\mathcal{O}$. Subsequent simplifications are performed on the resulting subset (i.e., the elimination and the simplification are performed *locally*). This significantly reduces the search space and has improved the efficiency of FAME compared to the early prototypes used in [11,12]. It is found that a name that could not be eliminated by FAME might become eliminable after the elimination of another name [12]. We therefore impose a do-while loop on the iterations of the elimination rounds. The breaking condition checks if there were names eliminated during the previous elimination rounds. If so, FAME repeats the iterations again, attempting to eliminate the remaining names. The loop terminates when the forgetting signature becomes empty or no names were eliminated during the previous elimination rounds.

Table 1. Types of axioms that can be handled by FAME

	Type of axiom	Representation
TBox	SubClassOf(C1 C2)	SubClassOf(C1 C2)
	EquivalentClasses(C1 C2)	SubClassOf(C1 C2), SubClassOf(C2 C1)
	DisjointClasses(C1 C2)	SubClassOf(C1 ObjectComplementOf(C2))
	DisjointUnion(C C1...Cn)	EquivalentClasses(C ObjectUnionOf(C1...Cn))
		DisjointClasses(C1...Cn)
	SubObjectPropertyOf(R1 R2)	SubObjectPropertyOf(R1 R2)
	EquivalentObjectProperties(R1 R2)	SubObjectPropertyOf(R1 R2)
		SubObjectPropertyOf(R2 R1)
	ObjectPropertyDomain(R C)	SubClassOf(ObjectSomeValuesFrom(R owl:Thing), C)
	ObjectPropertyRange(R C)	SubClassOf(owl:Thing ObjectAllValuesFrom(R C))
ABox	ClassAssertion(C a)	SubClassOf(a C)
	ObjectPropertyAssertion(R a1 a2)	SubClassOf(a1 ObjectSomeValuesFrom(R a2))

Table 2. Statistics of ontologies used for evaluation of FAME

	Maximum	Minimum	Mean	Median	90th percentile
$\#(\mathcal{O})$	1833761	100	4651	1096	12570
$\#\mathrm{sig}_C(\mathcal{O})$	847760	36	2110	502	5598
$\#\mathrm{sig}_R(\mathcal{O})$	1390	0	54	12	144
$\#\mathrm{sig}_I(\mathcal{O})$	87879	0	216	0	206

What FAME outputs at the end of the forgetting process is an ontology $\mathcal{O}'$ (i.e., a set of TBox and ABox axioms). If $\mathcal{O}'$ does not contain any names in $\mathcal{F}$, then FAME was *successful* and $\mathcal{O}'$ is a *solution* of forgetting $\mathcal{F}$ from $\mathcal{O}$.

4   Evaluation

We evaluated the current version of FAME on a corpus of real-world ontologies taken from the NCBO BioPortal repository,[2] a resource that currently includes more than 600 ontologies originally developed for clinical research. The repository covers a range of topics in biomedicine such as genomics, organology, and anatomy. Differing in size, structure, and expressivity, the BioPortal ontologies offer a rich, diverse and realistic test data set for the evaluation of FAME. The corpus used for our evaluation was based on a snapshot of the repository taken in March 2017 [9], containing 396 OWL API compatible ontologies.

The expressivity of the ontologies in the snapshot ranges from $\mathcal{EL}$ and $\mathcal{ALC}$ to $\mathcal{SHOIN}$ and $\mathcal{SROIQ}$. Since FAME can handle ontologies as expressive as

---

[2] https://bioportal.bioontology.org/.

Table 3. Results of forgetting 10%, 40% and 70% of concept names

Settings	Results				
#$\mathcal{F}_C$ (avg)	Time (sec)	Timeouts	Success rate	Nominal	Clause growth
211 (10%)	0.307	1.8%	94.9%	7.6%	-10.3%
844 (40%)	0.895	3.4%	93.4%	17.4%	-41.2%
1477 (70%)	1.364	6.6%	90.2%	24.7%	-72.4%

Table 4. Results of forgetting 10%, 40% and 70% of role names

Settings	Results				
#$\mathcal{F}_R$ (avg)	Time (sec)	Timeouts	Success rate	Definer	Clause growth
5 (10%)	0.309	0.0%	100.0%	0.0%	0.9%
22 (40%)	0.977	2.5%	97.5%	0.0%	3.5%
38 (70%)	1.891	6.6%	93.4%	0.0%	6.7%

$\mathcal{ALCOIH}$, we adjusted these ontologies to the language of $\mathcal{ALCOIH}$. This involved easy reformulations as summarised in Table 1, which also lists the types of axioms that FAME can handle. Concepts not expressible in $\mathcal{ALCOIH}$ were replaced by $\top$. Table 2 shows statistical information about the adjusted ontologies, where $\#(\mathcal{O})$ denotes the number of axioms in the test ontologies, and $\#\text{sig}_C(\mathcal{O})$, $\#\text{sig}_R(\mathcal{O})$ and $\#\text{sig}_I(\mathcal{O})$ denote respectively the numbers of the concept names, role names and individual names in the test ontologies.

To reflect real-world application scenarios, we evaluated the performance of FAME for forgetting different numbers of concept names and role names from each test ontology. In particular, we considered the cases of forgetting 10%, 40% and 70% of concept names and role names in their signatures. The names to be forgotten were randomly chosen. The experiments were run on a desktop computer with an Intel® Core™ i7-4790 processor, four cores running at up to 3.60 GHz and 8 GB of DDR3-1600 MHz RAM. We ran the experiments 100 times on each ontology and averaged the results in order to verify the accuracy of our findings. A timeout of 1000 seconds was imposed on each run.

The results obtained for forgetting different numbers of concept names from the ontologies are shown in Table 3. The column headed 'Success Rate' shows that FAME was successful in more than 90% of the test cases (i.e., eliminated all concept names in $\mathcal{F}$ within the timeout). In the cases of forgetting 10% and 40% of concept names the elimination was done within one second and in the cases of forgetting 70% the elimination was done within two seconds (on average); see the Time column. Because of the nature of one rewrite rule in the Ack^C calculus [12], fresh nominals might be introduced during the forgetting process. The column headed Nominal shows that forgetting solutions containing fresh nominals only occurred in a small number of cases ($\leq 25\%$). Compared to the input ontologies, there was a decrease in the number of clauses in the forgetting solutions; see the

Clause Growth column. It can be observed that the forgetting solutions consisted of fewer clauses when more concept names were forgotten.

The results obtained from forgetting different numbers of role names from the ontologies are shown in Table 4. The column headed 'Success Rate' shows that FAME was successful in more than 93% of the test cases (i.e., eliminated all role names in $\mathcal{F}$ within the timeout). In the cases of forgetting 10% and 40% of role names the elimination was done within one second and in the cases of forgetting 70% of role names the elimination was done within two seconds (on average). The column headed Definer shows that all introduced definer names were eliminated from the results in all test cases. Compared to concept forgetting, (i) an increase in the number of clauses in the forgetting solutions was observed; see the Clause Growth column, and (ii) when more role names were forgotten, the forgetting solutions consisted of more clauses.

The most closely related tools to FAME are LETHE [6, 7] and the tool developed by [8]. Both use resolution-based methods to compute uniform interpolants for $\mathcal{ALC}$ TBoxes, and in the case of LETHE several extensions of $\mathcal{ALC}$ TBoxes. A preliminary comparison of a previous version of FAME and LETHE has shown that FAME is considerably faster than LETHE [2]. The current version of FAME can be downloaded via http://www.cs.man.ac.uk/~schmidt/fame/.

Acknowledgements. We would like to thank the EPSRC IAA 204 (AR4MO) and Babylon Health for funding.

References

1. Ackermann, W.: Untersuchungen über das eliminationsproblem der mathematischen logik. Math. Ann. **110**(1), 390–413 (1935)
2. Alassaf, R., Schmidt, R.A.: A preliminary comparison of the forgetting solutions computed using SCAN, LETHE and FAME. In: Proceedings of SOQE 2017. CEUR Workshop Proceedings, vol. 2013, pp. 21–26 (2017)
3. Del-Pinto, W., Schmidt, R.A.: Forgetting-based abduction in $\mathcal{ALC}$. In Proceedings of the SOQE 2017. CEUR Workshop Proceedings, vol. 2013, pages 27–35 (2017)
4. Grau, B.C., Motik, B.: Reasoning over ontologies with hidden content: the import-by-query approach. J. Artif. Intell. Res. **45**, 197–255 (2012)
5. Konev, B., Walther, D., Wolter, F.: Forgetting and uniform interpolation in large-scale description logic terminologies. In: Proceedings of the IJCAI 2009, pp. 830–835. IJCAI/AAAI Press (2009)
6. Koopmann, P.: Practical uniform interpolation for expressive description logics. Ph.D thesis. The University of Manchester, UK (2015)
7. Koopmann, P., Schmidt, R.A.: LETHE: saturation-based reasoning for non-standard reasoning tasks. In: Proceedings of the DL 2015. CEUR Workshop Proceedings, vol. 1387, pp. 23–30 (2015)
8. Ludwig, M., Konev, B.: Practical uniform interpolation and forgetting for $\mathcal{ALC}$ TBoxes with applications to logical difference. In: Proceedings of the KR 2014, pp. 318–327. AAAI Press (2014)
9. Matentzoglu, N., Parsia, B.: BioPortal Snapshot 30.03.2017, March 2017

10. Wang, K., Wang, Z., Topor, R.W., Pan, J.Z., Antoniou, G.: Eliminating concepts and roles from ontologies in expressive descriptive logics. Comput. Intell. **30**(2), 205–232 (2014)
11. Zhao, Y., Schmidt, R.A.: Concept forgetting in $\mathcal{ALCOI}$-ontologies using an Ackermann approach. In: Arenas, M., Corcho, O., Simperl, E., Strohmaier, M., Srinivas, K., Groth, P., Dumontier, M., Heflin, J., Thirunarayan, K., d'Aquin, M., Staab, S. (eds.) ISWC 2015. LNCS, vol. 9366, pp. 587–602. Springer, Cham (2015). https://doi.org/10.1007/978-3-319-25007-6_34
12. Zhao, Y., Schmidt, R.A.: Forgetting concept and role symbols in $\mathcal{ALCOIH}\mu^+(\nabla, \sqcap)$-Ontologies. In: Proceedings of the IJCAI 2016, pp. 1345–1352. IJCAI/AAAI Press (2016)
13. Zhao, Y., Schmidt, R.A.: Role forgetting for $\mathcal{ALCOQH}(\nabla)$-Ontologies using an Ackermann-based approach. In: Proceedings of the IJCAI 2017, pp. 1354–1361. IJCAI/AAAI Press (2017)

Superposition for Lambda-Free
Higher-Order Logic

Alexander Bentkamp[1]([✉]), Jasmin Christian Blanchette[1,2,3], Simon Cruanes[3,4],
and Uwe Waldmann[2]

[1] Vrije Universiteit Amsterdam, Amsterdam, The Netherlands
`a.bentkamp@vu.nl`
[2] Max-Planck-Institut für Informatik, Saarland Informatics Campus,
Saarbrücken, Germany
[3] Université de Lorraine, CNRS, Inria, LORIA, Nancy, France
[4] Aesthetic Integration, Austin, TX, USA

Abstract. We introduce refutationally complete superposition calculi for intentional and extensional λ-free higher-order logic, two formalisms that allow partial application and applied variables. The calculi are parameterized by a term order that need not be fully monotonic, making it possible to employ the λ-free higher-order lexicographic path and Knuth–Bendix orders. We implemented the calculi in the Zipperposition prover and evaluated them on TPTP benchmarks. They appear promising as a stepping stone towards complete, efficient automatic theorem provers for full higher-order logic.

1 Introduction

Superposition is a highly successful calculus for reasoning about first-order logic with equality. We are interested in *graceful* generalizations to higher-order logic: calculi that, as much as possible, coincide with standard superposition on first-order problems and that scale up to arbitrary higher-order problems.

As a stepping stone towards full higher-order logic, in this paper we restrict our attention to a λ-free fragment of higher-order logic that supports partial application and application of variables (Sect. 2). This formalism is expressive enough to permit the axiomatization of higher-order combinators such as pow_τ : $nat \to (\tau \to \tau) \to \tau \to \tau$:

$$\mathsf{pow}\ 0\ h \approx \mathsf{id} \qquad\qquad \mathsf{pow}\ (\mathsf{S}\ n)\ h\ x \approx h\ (\mathsf{pow}\ n\ h\ x)$$

Conventionally, functions are applied without parentheses and commas, and variables are italicized. Notice the variable number of arguments to pow and the application of h. The expressiveness of full higher-order logic can be recovered by introducing SK-style combinators to represent λ-abstractions and proxies for the logical symbols [24,32].

A widespread technique to support partial application and application of variables in first-order logic is to make all symbols nullary and to represent

© Springer International Publishing AG, part of Springer Nature 2018
D. Galmiche et al. (Eds.): IJCAR 2018, LNAI 10900, pp. 28–46, 2018.
https://doi.org/10.1007/978-3-319-94205-6_3

application of functions of type $\tau \to \upsilon$ by a family of binary symbols $\mathsf{app}_{\tau,\upsilon}$. Following this scheme, the higher-order term $\mathsf{f}\,(h\,\mathsf{f})$ is translated to $\mathsf{app}(\mathsf{f},\mathsf{app}(h,\mathsf{f}))$, which can be processed by first-order methods. We call this the *applicative encoding*. The existence of such a reduction explains why λ-free higher-order terms are also called "applicative first-order terms." Unlike for full higher-order logic, most general unifiers are unique for our λ-free fragment, just as they are for applicatively encoded first-order terms.

Although the applicative encoding is complete [24] and is employed fruitfully in tools such as Sledgehammer [9, 27], it suffers from a number of weaknesses, all related to its gracelessness. Transforming all the function symbols into constants considerably restricts what can be achieved with term orders; for example, argument tuples cannot be compared using different methods for different symbols. In a prover, the encoding also clutters the data structures, slows down the algorithms, and neutralizes the heuristics that look at the terms' root symbols. But our chief objection is the sheer clumsiness of encodings and their poor integration with interpreted symbols. And they quickly accumulate; for example, using the traditional encoding of polymorphism relying on a distinguished binary function symbol t [8, Sect. 3.3] in conjunction with the applicative encoding, the term $\mathsf{S}\,x$ becomes $\mathsf{t}(\mathsf{nat}, \mathsf{app}(\mathsf{t}(\mathsf{fun}(\mathsf{nat}, \mathsf{nat}), \mathsf{S}), \mathsf{t}(\mathsf{nat}, x)))$.

Hybrid schemes have been proposed to strengthen the applicative encoding: If a given symbol always occurs with at least k arguments, these can be passed directly [27]. However, this relies on a closed-world assumption: that all terms that will ever be compared arise in the input problem. This noncompositionality conflicts with the need for complete higher-order calculi to synthesize arbitrary terms during proof search [6]. As a result, hybrid encodings are not an ideal basis for higher-order automated reasoning. Instead, we propose to generalize the superposition calculus to *intensional* and *extensional* λ-free higher-order logic. In the extensional version of the logic, the property $(\forall x.\ h\,x \approx k\,x) \longrightarrow h \approx k$ holds for all functions h, k of the same type. For each logic, we present two calculi (Sect. 3). The intentional calculi perfectly coincide with standard superposition on first-order clauses; the extensional calculi depend on an extra axiom.

Superposition is parameterized by a term order, which prunes the search space. If we assume that the term order is a simplification order enjoying totality on ground terms, the standard calculus rules and completeness proof can be lifted verbatim. The only necessary changes concern the basic definitions of terms and substitutions. However, there is one monotonicity property that is hard to obtain unconditionally: *compatibility with arguments*. It states that $s' \succ s$ implies $s'\,t \succ s\,t$ for all terms s, s', t such that $s\,t$ and $s'\,t$ are well typed. We recently introduced graceful generalizations of the lexicographic path order (LPO) [11] and the Knuth–Bendix order (KBO) [3] with argument coefficients, but they both lack this property. For example, given a KBO with $\mathsf{g} \succ \mathsf{f}$, it may well be that $\mathsf{g}\,\mathsf{a} \prec \mathsf{f}\,\mathsf{a}$ if f has a large enough multiplier on its argument.

Our calculi are designed to be refutationally complete for such nonmonotonic orders (Sect. 4). To achieve this, they include an inference rule for argument congruence, which derives $C \vee s\,x \approx t\,x$ from $C \vee s \approx t$. The redundancy criterion is defined in such a way that the larger, derived clause is not subsumed by the premise. In the completeness proof, the most difficult case is the one that normally excludes superposition at or below variables using the induction hypothesis. With nonmonotonicity, this approach no longer works, and we propose two alternatives: Perform some superposition inferences onto higher-order variables, or "purify" the clauses to circumvent the issue. We refer to the corresponding calculi as *nonpurifying* and *purifying*. Detailed proofs are included in a technical report [5], together with more explanations and examples.

The calculi are implemented in the Zipperposition prover [17] (Sect. 5). We evaluate them on TPTP benchmarks [39, 40] and compare them with the applicative encoding (Sect. 6). We find that there is a substantial cost associated with the applicative encoding and that the nonmonotonicity is not particularly expensive.

2   Logic

Refutational completeness of calculi for higher-order logic (also called simple type theory) is usually stated with respect to Henkin semantics [6, 22], in which the universes used to interpret functions need only contain the functions that can be expressed as terms. Since the terms of λ-free higher-order logic exclude λ-abstractions, in "λ-free Henkin semantics" the universes interpreting functions can be even smaller. Unlike other higher-order logics, there are no comprehension principles, and we disallow nesting of Boolean formulas inside terms, as a convenient intermediate step on our way towards full higher-order logic.

Problematically, in a logic with applied variables but without Hilbert choice, skolemization is unsound, unless we make sure that Skolem symbols are suitably applied [28]. We achieve this using a *hybrid logic* that supports both mandatory (uncurried) and optional (curried) arguments. Thus, if symbol sk takes two mandatory and one optional arguments, $\mathsf{sk}(x, y)$ and $\mathsf{sk}(x, y)\, z$ are valid terms. Nevertheless, as in our earlier work [3, 11], we use the adjective "graceful" in the strong sense that we can exploit optional arguments, identifying the first-order term $\mathsf{f}(x, y)$ with the curried higher-order term $\mathsf{f}\, x\, y$.

A type τ, υ of λ-free higher-order logic is either an element ι of a fixed set of atomic types or a function type $\tau \to \upsilon$ of functions from type τ to type υ. In our hybrid logic, a type declaration for a symbol is an expression of the form $\bar{\tau}_n \Rightarrow \tau$ (or simply τ if $n = 0$). We write $\bar{a}_n$ or $\bar{a}$ to abbreviate the tuple $(a_1, \ldots, a_n)$ or product $a_1 \times \cdots \times a_n$, for $n \geq 0$.

We fix a set $\mathcal{V}$ of typed variables, denoted by $x : \tau$ or x. A signature consists of a nonempty set Σ of symbols with type declarations, written as $\mathsf{f} : \bar{\tau} \Rightarrow \tau$ or f. We reserve the letters s, t, u, v for terms and x, y, z for variables and write $: \tau$ to indicate their type. The set of λ-free higher-order terms $\mathcal{T}_\Sigma^X$ over X is defined inductively. Every variable in $X \subseteq \mathcal{V}$ is a term. If $\mathsf{f} : \bar{\tau}_n \Rightarrow \tau$ and $u_i : \tau_i$ for all $i \in \{1, \ldots, n\}$, then $\mathsf{f}(\bar{u}_n) : \tau$ is a term. If $t : \tau \to \upsilon$ and $u : \tau$, then $t\, u : \upsilon$ is a term, called an *application*. Non-application terms ζ are called *heads*. Terms can be decomposed in a unique way as a head ζ applied to zero or more arguments:

$\zeta\, s_1 \ldots s_n$ or $\zeta\, \bar{s}_n$ (abusing notation). Substitution and unification are generalized in the obvious way, without the complexities associated with λ-abstractions; for example, the most general unifier of x b z and f a y c is $\{x \mapsto f\ a, y \mapsto b, z \mapsto c\}$, and that of h (f a) and f (h a) is $\{h \mapsto f\}$.

Formulas φ, ψ are of the form $\bot$, $\top$, $\neg\,\varphi$, $\varphi \vee \psi$, $\varphi \wedge \psi$, $\varphi \longrightarrow \psi$, $t \approx_\tau s$, $\forall x.\ \varphi$, or $\exists x.\ \varphi$, where t, s are terms and x is a variable. We let $s \not\approx t$ abbreviate $\neg\ s \approx t$. We normally view equations $s \approx t$ as unordered pairs and clauses as multisets of such (dis)equations.

Loosely following Fitting [20], an *interpretation* $\mathcal{J} = (\mathcal{U}, \mathcal{E}, \mathcal{J})$ consists of a type-indexed family of nonempty sets $\mathcal{U}_\tau$, called *universes*; a family of functions $\mathcal{E}_{\tau,\upsilon} : \mathcal{U}_{\tau \to \upsilon} \to (\mathcal{U}_\tau \to \mathcal{U}_\upsilon)$, one for each pair of types τ, υ; and a function $\mathcal{J}$ that maps each symbol with type declaration $\bar{\tau}_n \Rightarrow \tau$ to an element of $\bar{\mathcal{U}}_{\tau_n} \to \mathcal{U}_\tau$. An interpretation is *extensional* if $\mathcal{E}_{\tau,\upsilon}$ is injective for all τ, υ. Both intensional and extensional logics are widely used. The semantics is *standard* if $\mathcal{E}_{\tau,\upsilon}$ is bijective. A *valuation* ξ is a function that maps variables $x : \tau$ to elements of $\mathcal{U}_\tau$.

For an interpretation $(\mathcal{U}, \mathcal{E}, \mathcal{J})$ and a valuation ξ, the denotation of a term is defined as follows: $[\![x]\!]_\mathcal{J}^\xi = \xi(x)$; $[\![f(\bar{t})]\!]_\mathcal{J}^\xi = \mathcal{J}(f)([\![\bar{t}]\!]_\mathcal{J}^\xi)$; $[\![s\ t]\!]_\mathcal{J}^\xi = \mathcal{E}([\![s]\!]_\mathcal{J}^\xi)([\![t]\!]_\mathcal{J}^\xi)$. The truth value $[\![\varphi]\!]_\mathcal{J}^\xi \in \{0, 1\}$ of a formula φ is defined as in first-order logic. The interpretation $\mathcal{J}$ is a model of φ, written $\mathcal{J} \models \varphi$, if $[\![\varphi]\!]_\mathcal{J}^\xi = 1$ for all valuations ξ.

3   The Inference Systems

We introduce four versions of the superposition calculus, varying along two axes: intentional versus extensional, and nonpurifying versus purifying. To avoid repetitions, our presentation unifies them into a single framework.

3.1   The Inference Rules

The calculi are parameterized by a partial order $\succ$ on terms that is well founded, total on ground terms, and stable under substitutions and that has the subterm property. It must also be *compatible with function contexts*, meaning that $t' \succ t$ implies both $f(\bar{s}, t', \bar{u})\ \bar{v} \succ f(\bar{s}, t, \bar{u})\ \bar{v}$ and $s\ t'\ \bar{u} \succ s\ t\ \bar{u}$. On the other hand, it need not be *compatible with optional arguments*: $s' \succ s$ need not imply $s'\ t \succ s\ t$. Function contexts are built around *argument subterms*, defined as the reflexive transitive closure of the relation inductively specified by $f(\bar{s})\ \bar{t} \rhd s_i$ and $\zeta\bar{t} \rhd t_i$ for all i. We write $s\langle u \rangle$ to indicate that the subterm u of $s[u]$ is an argument subterm. For example, f and f a are subterms of f a b, but not argument subterms. The literal and clause orders are defined as multiset extensions in the usual way.

Literal selection is supported. The selection function maps each clause C to a subclause of C consisting of negative literals. A literal L is (*strictly*) *eligible* in C if it is selected in C or there are no selected literals in C and L is (strictly) maximal in C.

We start with the **extensional nonpurifying** calculus, which consists of five rules:

$$\frac{\overbrace{D' \vee t \approx t'}^{D} \quad \overbrace{C' \vee [\neg]\, s\langle u \rangle \approx s'}^{C}}{(D' \vee C' \vee [\neg]\, s\langle t' \rangle \approx s')\sigma} \ \text{Sup} \qquad \frac{C' \vee s' \approx t' \vee s \approx t}{(C' \vee t \not\approx t' \vee s \approx t')\sigma} \ \text{EqFact}$$

$$\frac{C' \vee s \not\approx s'}{C'\sigma} \ \text{EqRes} \qquad \frac{C' \vee s \approx s'}{C' \vee s\,\bar{x} \approx s'\,\bar{x}} \ \text{ArgCong} \qquad \frac{C' \vee s\,\bar{x} \approx s'\,\bar{x}}{C' \vee s \approx s'} \ \text{PosExt}$$

In the first three rules, σ denotes the most general unifier of the two grayed terms. For Sup, we assume that D's and C's variables have been standardized apart. For Sup, EqFact, and EqRes, the following standard order conditions apply on the premises after the application of σ: The last literal in each premise is eligible and even strictly eligible for positive literals of Sup. For the last literal of each premise of Sup and the last two literals of the premise of EqFact, the left-hand sides are not smaller than or equal to ($\npreceq$) the respective right-hand sides. For Sup, $C\sigma \npreceq D\sigma$.

Definition 1. A term of the form $x\,\bar{s}_n$, for $n \geq 0$, *jells* with a literal $t \approx t' \in D$ if $t = \tilde{t}\,\bar{y}_n$ and $t' = \tilde{t}'\,\bar{y}_n$ for some $\tilde{t}, \tilde{t}'$ and distinct variables $\bar{y}_n$ that do not occur elsewhere in D.

We add the following *variable condition* as a side condition to Sup, to further prune the search space, using the naming convention from Definition 1 for $\tilde{t}'$:

If u has a variable head x and jells with the literal $t \approx t' \in D$, there must exist a ground substitution θ with $t\sigma\theta \succ t'\sigma\theta$ and $C\sigma\theta \prec C''\sigma\theta$, where $C'' = C[x \mapsto \tilde{t}']$.

This condition generalizes the standard condition that $u \notin \mathcal{V}$. The two coincide if C is first-order. In some cases involving nonmonotonicity, the variable condition effectively mandates Sup inferences at variable positions, but never below.

The last two rules are nonstandard. For ArgCong, $s \approx s'$ must be strictly eligible in the premise, and $\bar{x}$ is a tuple of fresh variables. For PosExt, $s\,\bar{x} \approx s'\,\bar{x}$ must be strictly eligible in the premise, and $\bar{x}$ is a tuple of distinct variables that occur nowhere else in the premise. Furthermore, for every function type $\tau \to \upsilon$ occurring in the input problem, we introduce a Skolem symbol $\mathrm{diff}_{\tau,\upsilon}$: $(\tau \to \upsilon)^2 \Rightarrow \tau$ characterized by the following *extensionality axiom*: $h\,(\mathrm{diff}(h, k)) \not\approx k\,(\mathrm{diff}(h, k)) \vee h \approx k$.

The second calculus is the **intensional nonpurifying** variant. We obtain it by removing the PosExt rule and the extensionality axiom and by replacing the variable condition with "if $u \in \mathcal{V}$, there exists a ground substitution θ with $t\sigma\theta \succ t'\sigma\theta$ and $C\sigma\theta \prec C[u \mapsto t']\sigma\theta$." For monotone term orders, this condition amounts to $u \notin \mathcal{V}$.

By contrast, the purifying calculi never perform superposition at variables. Instead, they rely on purification [14,35] (also called abstraction) to circumvent nonmonotonicity. The idea is to rename apart problematic occurrences of a variable x in a clause to $x_1, \ldots, x_n$ and to add *purification literals* $x_1 \not\approx x, \ldots, x_n \not\approx x$ to connect the new variables. We must then purify the initial clauses and all derived clauses.

In the **extensional purifying** calculus, the purification $pure(C)$ of clause C is defined as the result of the following iterative procedure. Consider the literals of C excluding those of the form $y \not\approx z$. If these literals contain both $x\,\bar{u}$ and $x\,\bar{v}$ as distinct argument subterms, replace all argument subterms $x\,\bar{v}$ with $x_i\,\bar{v}$, where x_i is fresh, and add the purification literal $x_i \not\approx x$. This calculus variant contains the POSEXT rule and the extensionality axiom. The conclusion E of each rule is changed to $pure(E)$, except for POSEXT, which preserves purity. Moreover, the variable condition is replaced by "either u has a non-variable head or u does not jell with the literal $t \approx t' \in D$."

In the **intensional purifying** calculus, we define $pure(C)$ iteratively as follows. Consider the literals of C excluding those of the form $y \not\approx z$. If these literals contain a variable x both applied and unapplied, replace all unapplied occurrences of x in C by a fresh variable x_i and add the purification literal $x_i \not\approx x$. We remove the POSEXT rule and the extensionality axiom. The variable condition is replaced by "$u \notin \mathcal{V}$." The conclusion C of ARGCONG is changed to $pure(C)$; the other rules preserve purity.

Finally, we impose some additional restrictions on literal selection. In the nonpurifying variants, a literal may not be selected if $x\,\bar{u}$ is a maximal term of the clause and the literal contains an argument subterm $x\,\bar{v}$ with $\bar{v} \neq \bar{u}$. In the extensional purifying calculus, a literal may not be selected if it contains a variable that is applied to different arguments in the clause. In the intensional purifying calculus, a literal may not be selected if the literal contains an unapplied variable that also appears applied in the clause.

3.2   Rationale for the Inference Rules

A key restriction of all four calculi is that they superpose only onto argument subterms, mirroring the requirement that the term order enjoy compatibility with function contexts. The ARGCONG rule then makes it possible to simulate superposition onto non-argument subterms. However, in conjunction with the SUP rule, ARGCONG can exhibit an unpleasant behavior, which we call *argument congruence explosion*:

$$\text{SUP} \cfrac{\text{ARGCONG} \cfrac{\mathsf{g} \approx \mathsf{f}}{\mathsf{g}\,x \approx \mathsf{f}\,x} \quad h\,\mathsf{a} \not\approx \mathsf{b}}{\mathsf{f}\,\mathsf{a} \not\approx \mathsf{b}} \qquad \text{SUP} \cfrac{\text{ARGCONG} \cfrac{\mathsf{g} \approx \mathsf{f}}{\mathsf{g}\,x\,y\,z \approx \mathsf{f}\,x\,y\,z} \quad h\,\mathsf{a} \not\approx \mathsf{b}}{\mathsf{f}\,x\,y\,\mathsf{a} \not\approx \mathsf{b}}$$

In both cases, the higher-order variable h is effectively the target of a SUP inference. Such derivations essentially amount to superposition at variable positions (as shown on the left) or even superposition below variable positions (as shown on the right), both of which can be extremely prolific. In standard superposition, the explosion is averted by the condition on the SUP rule that $u \notin \mathcal{V}$. In the extensional purifying calculus, the variable condition tests that either u has a non-variable head or u does not jell with the literal $t \approx t' \in D$, which prevents derivations such as the above. In the corresponding nonpurifying variant, some such derivations may need to be performed when the term order exhibits nonmonotonicity for the terms of interest.

In the intensional calculi, the explosion can arise even for monotonic orders, and it must be tamed by heuristics. The reason is connected to the absence of the POSEXT rule (which would be unsound). The variable condition in the extensional calculi is designed to prevent derivations such as those shown above, but since it only considers the shape of the clauses, it might also block SUP inferences whose side premises do not originate from ARGCONG. Consider a left-to-right LPO [11] instance with precedence $h \succ g \succ f \succ b \succ a$, and consider the following unsatisfiable clause set:

$$g\,(x\,b)\ x \approx a \qquad\qquad g\,(f\,b)\ h \not\approx a \qquad\qquad h\ x \approx f\ x$$

The only possible inference from these clauses is POSEXT, showing its necessity. It is unclear whether POSEXT is necessary for the extensional purifying variant as well, but our completeness proof suggests that it is. Our proof also suggests that to achieve refutational completeness, due to nonmonotonicity, we need either to purify the clauses or to allow some superposition at variable positions, as mandated by the respective variable conditions. However, we have yet to find an example that demonstrates the necessity of these measures.

A significant advantage of our calculi over the use of standard superposition on applicatively encoded problems is the flexibility they offer in orienting equations. The following example gives two definitions of addition on Peano numbers:

$$\mathsf{add_L}\ 0\ y \approx y \qquad\qquad\qquad \mathsf{add_R}\ x\ 0 \approx x$$
$$\mathsf{add_L}\ (\mathsf{S}\ x)\ y \approx \mathsf{add_L}\ x\ (\mathsf{S}\ y) \qquad\qquad \mathsf{add_R}\ x\ (\mathsf{S}\ y) \approx \mathsf{add_R}\ (\mathsf{S}\ x)\ y$$

Let $\mathsf{add_L}\ (\mathsf{S}^{100}\ 0)\ n \not\approx \mathsf{add_R}\ n\ (\mathsf{S}^{100}\ 0)$ be the negated conjecture. With LPO, we can use a left-to-right comparison for $\mathsf{add_L}$'s arguments and a right-to-left comparison for $\mathsf{add_R}$'s arguments to orient all four equations from left to right. Then the negated conjecture can be simplified to $\mathsf{S}^{100}\ n \not\approx \mathsf{S}^{100}\ n$ by rewriting (demodulation), and $\bot$ can be derived with a single inference. If we use the applicative encoding instead, there is no instance of LPO or KBO that can orient both recursive equations from left to right. For at least one of the two sides of the negated conjecture, the rewriting is replaced by 100 SUP inferences, which is much less efficient, especially in the presence of additional axioms.

3.3   Redundancy Criterion

For our calculi, a redundant (or composite) clause cannot simply be defined as a clause whose ground instances are entailed by smaller ($\prec$) ground instances of existing clauses, because this would make all ARGCONG inferences redundant. Our solution is to base the redundancy criterion on a weaker ground logic in which argument congruence does not hold. This logic also plays a central role in our completeness proof, to reason about the nonmonotonicity emerging from the lack of compatibility with optional arguments.

The weaker logic is defined via an encoding $\lfloor\ \rfloor$ of ground hybrid λ-free higher-order terms into uncurried terms, with $\lceil\ \rceil$ as its inverse. Accordingly, we refer to clausal λ-free higher-order logic as the *ceiling logic* and to its weaker relative as the *floor logic*. Essentially, the encoding indexes each symbol occurrence with its argument count. Thus, $\lfloor f \rfloor = f_0$ and $\lfloor f\, a \rfloor = f_1(a_0)$. This is enough to disable argument congruence; for example, $\{f \approx g,\ f\, a \not\approx g\, a\}$ is unsatisfiable, whereas its encoding $\{f_0 \approx g_0,\ f_1(a_0) \not\approx g_1(a_0)\}$ is satisfiable. For clauses built from fully applied ground terms, the two logics are isomorphic, as we would expect from a graceful generalization.

Given a signature Σ in the ceiling logic, we define a signature $\Sigma^\downarrow$ in the floor logic as follows. For each higher-order type τ, we introduce an atomic type $\lfloor\tau\rfloor$ in the floor logic. For each symbol $f : \bar{\tau}_k \Rightarrow \tau_{k+1} \to \cdots \to \tau_n \to \upsilon$ in Σ, where υ is atomic, we introduce symbols $f_m : \lfloor\bar{\tau}_m\rfloor \Rightarrow \lfloor\tau_{m+1} \to \cdots \to \tau_n \to \upsilon\rfloor$ for $m \in \{k, \ldots, n\}$. The translation of ground terms is given by $\lfloor f(\bar{u}_k)\, u_{k+1} \ldots u_m \rfloor = f_m(\lfloor\bar{u}_m\rfloor)$. We extend this mapping to literals and clauses by applying it to each side of a literal and to each literal of a clause. Using $\lceil\ \rceil$, the clause order $\succ$ can be transferred to the floor logic by defining $t \succ s$ as equivalent to $\lceil t\rceil \succ \lceil s\rceil$. The property that $\succ$ on clauses is the multiset extension of $\succ$ on literals, which in turn is the multiset extension of $\succ$ on terms, is maintained because $\lceil\ \rceil$ maps the multiset representations elementwise.

Crucially, argument subterms in the ceiling logic correspond to argument subterms in the floor logic, whereas non-argument subterms in the ceiling logic are not subterms at all in the floor logic. Well-foundedness, totality on ground terms, compatibility with *all* contexts, and the subterm property hold for $\succ$ in the floor logic.

In standard superposition, redundancy relies on the entailment relation $\models$ on ground clauses. We define redundancy of ceiling clauses in the same way, but using the floor logic's entailment relation: A ground ceiling clause C is *redundant* with respect to a set of ceiling ground clauses N if $\lfloor C\rfloor$ is entailed by clauses from $\lfloor N\rfloor$ that are smaller than $\lfloor C\rfloor$. This notion of redundancy gracefully generalizes the first-order notion without making all ARGCONG inferences redundant.

For SUP, EQFACT, and EQRES, we can use the more precise notion of redundancy of inferences instead of redundancy of clauses, a ground inference being redundant if the conclusion follows from existing clauses that are smaller than the largest premise. For ARGCONG and POSEXT, we must use redundancy of clauses.

3.4   Skolemization

A problem expressed in λ-free higher-order logic must be transformed into clausal normal form before the calculi can be applied. This process works as in the first-order case, except for skolemization. The issue is that skolemization, when performed naively, is unsound for λ-free higher-order logic with a Henkin semantics. For example, given $f : \tau \to \upsilon$, the formula $(\forall y. \exists x.\, f\, x \approx y) \wedge (\forall z.\, f\,(z\,a) \not\approx a)$ has a model with $\mathcal{U}_\tau = \mathcal{U}_\upsilon$ that interprets f as the identity function and ensures that none of the functions in the image of $\mathcal{E}_{\upsilon,\tau}$ map $\mathcal{J}(a)$ to $\mathcal{J}(a)$. Yet, naive skolemization would yield the clause set $\{f\,(\mathsf{sk}\,y) \approx y, f\,(z\,a) \not\approx a\}$, whose unsatisfiability can be shown by taking $y := a$ and $z := \mathsf{sk}$. The crux of the issue is that sk denotes a new function that can be used to instantiate z.

Inspired by Miller [28, Sect. 6], we adapt skolemization as follows. An existentially quantified variable $x : \tau$ in a context with universally quantified variables $\bar{x}_n$ of types $\bar{\tau}_n$ is replaced by a fresh symbol $\mathsf{sk} : \bar{\tau}_n \Rightarrow \tau$ applied to the tuple $\bar{x}_n$. For the example above, we obtain $\{f\,(\mathsf{sk}(y)) \approx y, f\,(z\,a) \not\approx a\}$. Syntactically, z cannot be instantiated by sk, which is not even a term. Semantically, the clause set is satisfiable because we can have $\mathcal{J}(\mathsf{sk})(\mathcal{J}(a)) = \mathcal{J}(a)$ even if the image of $\mathcal{E}_{\tau,\tau}$ contains no such function.

4 Refutational Completeness

The proof of refutational completeness of the four calculi introduced in Sect. 3.1 follows the same general idea as for standard superposition [2, 42]. Given a clause set $N \not\ni \bot$ saturated up to redundancy, we construct a term rewriting system R based on the set of ground instances $\mathcal{G}_\Sigma(N)$. From R, we define an interpretation. We show, by induction on the clause order, that this interpretation is a model of $\mathcal{G}_\Sigma(N)$ and hence of N.

To circumvent the term order's potential nonmonotonicity, our SUP inference rule only considers the argument subterms u of a maximal term $s\langle u \rangle$. This is reflected in our proof by the reliance of the floor logic from Sect. 3.3. In that logic, the equation $\mathsf{g}_0 \approx \mathsf{f}_0$ cannot be used directly to rewrite the clause $\mathsf{g}_1(\mathsf{a}_0) \not\approx \mathsf{f}_1(\mathsf{a}_0)$; instead, we first need to apply ARGCONG to derive $\mathsf{g}_1(x) \approx \mathsf{f}_1(x)$ and then use that equation. The floor logic is a device that enables us to reuse the traditional model construction almost verbatim, including its reliance on a first-order term rewriting system.

Following the traditional proof, we obtain a model of $\lfloor \mathcal{G}_\Sigma(N) \rfloor$. Since N is saturated up to redundancy with respect to ARGCONG, the model $\lfloor \mathcal{G}_\Sigma(N) \rfloor$ can easily be turned into a model of $\mathcal{G}_\Sigma(N)$ by conflating the interpretations of the members $\mathsf{f}_k, \ldots, \mathsf{f}_n$ of a same symbol family. For this section, we fix a set $N \not\ni \bot$ of λ-free higher-order clauses that is saturated up to redundancy. For the purifying calculi, we additionally require that all clauses in N are purified. To avoid empty Herbrand universes, we assume that the signature Σ contains, for each type τ, a symbol of type τ.

4.1 Candidate Interpretation

The construction of the candidate interpretation is as in the first-order proof, except that it is based on $\lfloor \mathcal{G}_\Sigma(N) \rfloor$. We first define sets of rewrite rules E_C and

R_C for all $C \in \lfloor \mathcal{G}_\Sigma(N) \rfloor$ by induction. Assume that E_D has already been defined for all $D \in \lfloor \mathcal{G}_\Sigma(N) \rfloor$ with $D \prec C$. Then $R_C = \bigcup_{D \prec C} E_D$. Let $E_C = \{s \to t\}$ if the following conditions are met: (a) $C = C' \vee s \approx t$; (b) $s \approx t$ is strictly maximal in C; (c) $s \succ t$; (d) C is false in R_C; (e) C' is false in $R_C \cup \{s \to t\}$; and (f) s is irreducible with respect to R_C. Otherwise, $E_C = \emptyset$. Finally, $R_\infty = \bigcup_D E_D$. A rewrite system R defines an interpretation $\mathcal{T}_\Sigma^\emptyset / R$ such that for every *ground* equation $s \approx t$, we have $\mathcal{T}_\Sigma^\emptyset / R \models s \approx t$ if and only if $s \leftrightarrow_R^* t$. Moreover, $\mathcal{T}_\Sigma^\emptyset / R$ is term-generated. To lighten notation, we will write R to refer to both the term rewriting system R and the interpretation $\mathcal{T}_\Sigma^\emptyset / R$.

4.2   Lifting Lemmas

Following Waldmann's version of the first-order proof [42], we proceed by lifting inferences from the ground to the nonground level. We also need to lift ARGCONG. A complication that arises when lifting purifying inferences is that the nonground conclusions may contain purification literals (corresponding to applied variables) not present in the ground conclusions. Given an inference I of the form $\bar{C} \vdash pure(E)$, we refer to the ground instances of $\bar{C} \vdash E$ as ground instances of I *up to purification*.

Lemma 2 (Lifting of non-Sup inferences). *Let $C\theta \in \mathcal{G}_\Sigma(N)$, where θ is a substitution and the selected literals in $C \in N$ correspond to those in $C\theta$. Then every* EQRES *or* EQFACT *inference from $C\theta$ and every ground instance of an* ARGCONG *inference from $C\theta$ is a ground instance of an inference from C up to purification.*

The conditions of the lifting lemma for SUP differ slightly from the first-order version. For standard superposition, the lemma applies if the superposed term is not at or under a variable. This condition is replaced by the following criterion.

Definition 3. We call a ground SUP inference from $D\theta$ and $C\theta$ *liftable* if the superposed subterm in $C\theta$ is not under a variable in C and the corresponding variable condition holds for D and C.

Lemma 4 (Lifting of Sup inferences). *Let $D\theta, C\theta \in \mathcal{G}_\Sigma(N)$ where the selected literals in $D \in N$ and $C \in N$ correspond to those in $D\theta$ and $C\theta$, respectively. Then every liftable* SUP *inference between $D\theta$ and $C\theta$ is a ground instance of a* SUP *inference from D and C up to purification.*

4.3   Main Result

The candidate interpretation R_∞ is a model of $\lfloor \mathcal{G}_\Sigma(N) \rfloor$. Like in the first-order proof, this is shown by induction on the clause order. For the induction step, we fix some clause $\lfloor C\theta \rfloor \in \lfloor \mathcal{G}_\Sigma(N) \rfloor$ and assume that all smaller clauses are true in $R_{C\theta}$. We distinguish several cases, most of which amount to showing that $C\theta$ can be used in a certain inference. Then we deduce that $\lfloor C\theta \rfloor$ is true in $R_{C\theta}$ to complete the induction step.

The next two lemmas are slightly adapted from the first-order proof. The justification for Lemma 5, about liftable inferences, is essentially as in the first-order case. The proof of Lemma 6, about nonliftable inferences, is more problematic. The standard argument involves defining a substitution θ' such that $C\theta'$ and $C\theta$ are equivalent and $C\theta' \prec C\theta$. But due to nonmonotonicity, we might have $C\theta' \succ C\theta$, blocking the application of the induction hypothesis. This is where the variable conditions, purification, and the POSEXT rule come into play.

Lemma 5. *Let $D\theta, C\theta \in \mathcal{G}_\Sigma(N)$, where the selected literals in $D \in N$ and in $C \in N$ correspond to those in $D\theta$ and $C\theta$, respectively. We consider a liftable* SUP *inference from $D\theta$ and $C\theta$ or an* EQRES *or* EQFACT *inference from $C\theta$. Let E be the conclusion. Assume that $C\theta$ and $D\theta$ are nonredundant with respect to $\mathcal{G}_\Sigma(N)$. Then $\lfloor E \rfloor$ is entailed by clauses from $\lfloor \mathcal{G}_\Sigma(N) \rfloor$ that are smaller than $\lfloor C\theta \rfloor$.*

Lemma 6. *Let $D\theta, C\theta \in \mathcal{G}_\Sigma(N)$, where the selected literals in $D \in N$ and in $C \in N$ correspond to those in $D\theta$ and $C\theta$, respectively. We consider a nonliftable* SUP *inference from $D\theta$ and $C\theta$. Assume that $C\theta$ and $D\theta$ are nonredundant with respect to $\mathcal{G}_\Sigma(N)$. Let $D'\theta$ be the clause $D\theta$ without the literal involved in the inference. Then $\lfloor C\theta \rfloor$ is entailed by $\neg \lfloor D'\theta \rfloor$ and the clauses in $\lfloor \mathcal{G}_\Sigma(N) \rfloor$ that are smaller than $\lfloor C\theta \rfloor$.*

Using these two lemmas, the induction argument works as in the first-order case.

Lemma 7 (Model construction). *Let $\lfloor C\theta \rfloor \in \lfloor \mathcal{G}_\Sigma(N) \rfloor$. We have*

(i) $E_{\lfloor C\theta \rfloor} = \emptyset$ if and only if $R_{\lfloor C\theta \rfloor} \models \lfloor C\theta \rfloor$;
(ii) if $C\theta$ is redundant with respect to $\mathcal{G}_\Sigma(N)$, then $R_{\lfloor C\theta \rfloor} \models \lfloor C\theta \rfloor$;
(iii) $\lfloor C\theta \rfloor$ is true in R_∞ and in R_D for every $D \in \lfloor \mathcal{G}_\Sigma(N) \rfloor$ with $D \succ \lfloor C\theta \rfloor$; and
(iv) if $C\theta$ has selected literals, then $R_{\lfloor C\theta \rfloor} \models \lfloor C\theta \rfloor$.

Given a model R_∞ of $\lfloor \mathcal{G}_\Sigma(N) \rfloor$, we construct a model $R_\infty^\uparrow$ of $\mathcal{G}_\Sigma(N)$. The key properties are that R_∞ is term-generated and that the interpretations of the members $f_k, \ldots, f_n$ of a same symbol family behave in the same way.

Lemma 8 (Argument congruence). *For all ground terms $f_m(\bar{s})$ and $g_n(\bar{t})$, if $[\![f_m(\bar{s})]\!]_{R_\infty}^\xi = [\![g_n(\bar{t})]\!]_{R_\infty}^\xi$, then $[\![f_{m+1}(\bar{s}, u)]\!]_{R_\infty}^\xi = [\![g_{n+1}(\bar{t}, u)]\!]_{R_\infty}^\xi$ for all u.*

The proof relies on the saturation of N up to redundancy with respect to ARGCONG.

Definition 9. Define an interpretation $R_\infty^\uparrow = (\mathcal{U}^\uparrow, \mathcal{E}^\uparrow, \mathcal{J}^\uparrow)$ in the ceiling logic as follows. Let $(\mathcal{U}, \mathcal{E}, \mathcal{J}) = R_\infty$. Let $\mathcal{U}_\tau^\uparrow = \mathcal{U}_{\lfloor \tau \rfloor}$ and $\mathcal{J}^\uparrow(f) = \mathcal{J}(f_k)$, where k is the number of mandatory arguments of f. Since R_∞ is term-generated, for every $a \in \mathcal{U}_{\lfloor \tau \to \upsilon \rfloor}$, there exists a ground term $s : \tau \to \upsilon$ such that $[\![\lfloor s \rfloor]\!]_{R_\infty}^\xi = a$. Without loss of generality, we write $s = f(\bar{s}_k) \, s_{k+1} \ldots s_m$. Then we have $a = [\![f_m(\lfloor \bar{s}_m \rfloor)]\!]_{R_\infty}^\xi$ and define $\mathcal{E}^\uparrow$ by

$$\mathcal{E}_{\tau,\upsilon}^\uparrow(a)(b) = \mathcal{J}(f_{m+1})([\![\lfloor \bar{s}_m \rfloor]\!]_{R_\infty}^\xi, b) \quad \text{for all } b \in \mathcal{U}_\tau$$

It follows that $\mathcal{E}^{\uparrow}_{\tau,\upsilon}(a)\big(\llbracket u\rrbracket^{\xi}_{R_\infty}\big) = \llbracket \mathsf{f}_{m+1}(\lfloor \bar{s}_m\rfloor,u)\rrbracket^{\xi}_{R_\infty}$ for any term u. This interpretation is well defined if the definition of $\mathcal{E}^{\uparrow}$ does not depend on the choice of the ground term s. To show this, we assume that there exists another ground term $t = \mathsf{g}(\bar{t}_l)\ t_{l+1} \ldots t_n$ such that $\llbracket \lfloor t\rfloor \rrbracket^{\xi}_{R_\infty} = a$. By Lemma 8, it follows from $\llbracket \lfloor s\rfloor \rrbracket^{\xi}_{R_\infty} = \llbracket \lfloor t\rfloor \rrbracket^{\xi}_{R_\infty}$ that

$$\llbracket \mathsf{f}_{m+1}(\lfloor \bar{s}_m\rfloor,u)\rrbracket^{\xi}_{R_\infty} = \llbracket \mathsf{g}_{n+1}(\lfloor \bar{t}_n\rfloor,u)\rrbracket^{\xi}_{R_\infty}$$

indicating that the definition of $\mathcal{E}^{\uparrow}$ is independent of the choice of s.

Since R_∞ is a term-generated model of $\lfloor \mathcal{G}_\Sigma(N)\rfloor$, we can show that $R^{\uparrow}_\infty$ is also term-generated. And using the same argument as in the first-order proof, we can lift this result to nonground clauses. For the extensional variants, we also need to show that $R^{\uparrow}_\infty$ is an extensional interpretation.

Lemma 10 (Model transfer to ceiling logic). $R^{\uparrow}_\infty$ *is a term-generated model of* $\mathcal{G}_\Sigma(N)$.

Lemma 11 (Model transfer to nonground clauses). $R^{\uparrow}_\infty$ *is a model of N.*

Lemma 12 (Completeness of the extensionality axioms). *If N contains the extensionality axioms, $R^{\uparrow}_\infty$ is extensional.*

We summarize the results of this section in the following theorem.

Theorem 13 (Refutational completeness). *Let N be a clause set that is saturated by any of the four calculi, up to redundancy. For the purifying calculi, we additionally assume that all clauses in N are purified. Then N has a model if and only if $\bot \notin N$. Such a model is extensional if N contains the extensionality axioms.*

5   Implementation

Zipperposition [16,17] is an open source superposition-based theorem prover written in OCaml.[1] It was initially designed for polymorphic first-order logic with equality, as embodied by TPTP TFF [10]. We will refer to this implementation as Zipperposition's first-order mode. Recently, we extended the prover with a pragmatic higher-order mode with support for λ-abstractions and extensionality, without any completeness guarantees. Using this mode, Zipperposition entered the 2017 edition of the CADE ATP System Competition [38]. We have now also implemented a complete λ-free higher-order mode based on the four calculi described in this paper, extended with polymorphism.

The pragmatic higher-order mode provided a convenient basis to implement our calculi. It includes higher-order term and type representations and orders. Its ad hoc calculus extensions are similar to our calculi. Notably, they include an ARGCONG rule and a POSEXT-like rule, and SUP inferences are performed

---

[1] https://github.com/c-cube/zipperposition

only at argument subterms. In the term indexes, which are imperfect (overapproximating), terms whose heads are applied variables and λ-abstractions are treated as fresh variables. This could be further optimized to reduce the number of unification candidates.

To implement the λ-free mode, we restricted the unification algorithm to non-λ-terms, and we added support for mandatory arguments to make skolemization sound, by associating the number of mandatory arguments to each symbol and incorporating this number in the unification algorithm. To satisfy the requirements on selection, we avoid selecting literals that contain higher-order variables. Finally, we disabled rewriting of non-argument subterms to comply with our redundancy notion.

For the purifying calculi, we implemented purification as a simplification rule. This ensures that it is applied aggressively on all clauses, whether initial clauses from the problem or clauses produced during saturation, before any inferences are performed.

For the nonpurifying calculi, we added the possibility to perform SUP inferences at variable positions. This means that variables must be indexed as well. In addition, we modified the variable condition. However, it is in general impossible to decide whether there exists a ground substitution θ with $t\sigma\theta \succ t'\sigma\theta$ and $C\sigma\theta \prec C''\sigma\theta$. We overapproximate the condition as follows: (1) check whether x appears with different arguments in the clause C; (2) use an order-specific algorithm (for LPO and KBO) to determine whether there might exist a ground substitution θ and terms $\bar{u}$ such that $t\sigma\theta \succ t'\sigma\theta$ and $t\sigma\theta\,\bar{u} \prec t'\sigma\theta\,\bar{u}$; and (3) check whether $C\sigma \not\succeq C''\sigma$. If these three conditions apply, we conclude that there might exist a ground substitution θ witnessing nonmonotonicity.

For the extensional calculi, we added a single extensionality axiom based on a polymorphic symbol $\mathsf{diff} : \forall\alpha\,\beta.\,(\alpha \to \beta)^2 \Rightarrow \alpha$. To curb the explosion associated with extensionality, this axiom and all clauses derived from it are penalized by the clause selection heuristic. We also added a negative extensionality rule that resembles Vampire's [21].

Using Zipperposition, we can quantify the disadvantage of the applicative encoding on the problem given at the end of Sect. 3.2. Well-chosen LPO and KBO instances allow Zipperposition to derive $\bot$ in 4 iterations of the prover's main loop and 0.04 s. KBO or LPO with default settings needs 203 iterations and 0.5 s, whereas KBO or LPO on the applicatively encoded problem needs 203 iterations and almost 2 s.

6   Evaluation

We evaluated Zipperposition's implementation of our four calculi on TPTP benchmarks. We compare them with Zipperposition's first-order mode on the applicative encoding with and without the extensionality axiom. Our experimental data is available online.[2] Since the present work is only a stepping stone

---

[2] http://matryoshka.gforge.inria.fr/pubs/lfhosup_data/

towards a prover for full higher-order logic, it is too early to compare this prototype to state-of-the-art higher-order provers that support a stronger logic.

We instantiated all variants with LPO [11] (which is nonmonotonic) and KBO [3] without argument coefficients (which is monotonic). This gives us a rough indication of the cost of nonmonotonicity. However, when using a monotonic order, it may be more efficient (and also refutationally complete) to superpose at non-argument subterms directly instead of relying on the ARGCONG rule.

We collected 671 first-order problems in TFF format and 1114 higher-order problems in THF, both groups containing monomorphic and polymorphic problems. We excluded all problems containing λ-expressions, the quantifier constants !! (∀) and ?? (∃), arithmetic types, or the $distinct predicate, as well as problems that mix Booleans and terms. Figures 1 and 2 summarize, for various configurations, the number of solved satisfiable and unsatisfiable problems within 300 s. The average time and number of main loop iterations are computed over the problems that all configurations for the respective logic and term order found to be unsatisfiable within the timeout. The evaluation was carried out on StarExec [37] using Intel Xeon E5-2609 0 CPUs clocked at 2.40 GHz.

Our approach targets large, mildly higher-order problems—a practically relevant class of problems that is underrepresented in the TPTP library. The experimental results confirm that our calculi handle first-order problems gracefully. Even the extensional calculi, which include (graceless) extensionality axioms, are almost as effective as the first-order mode. This indicates that our calculi will perform well on mildly higher-order problems, too, where the proving effort is dominated by first-order reasoning. In contrast, the applicative encoding is comparatively inefficient on problems that are already first-order. For LPO, the success rate drops by 16%–18%; for both orders, the average time to show unsatisfiability roughly quadruples.

Many of the higher-order problems in the TPTP library are satisfiable for our λ-free logic, even though they may be unsatisfiable for full higher-order logic and labeled as such in the TPTP. This is a reason why we postpone a comparison with state-of-the-art higher-order provers until we have developed a prover for full higher-order logic. On higher-order problems, the nonpurifying calculi outperform their purifying relatives. The comparison of the applicative encoding and the nonpurifying calculi, however, is not entirely conclusive. In the light of the results of this evaluation, in future work, we would like to collect benchmarks for large, mildly higher-order problems and to investigate whether we can weaken the selection restrictions of our calculi.

The nonpurifying calculi perform slightly better with KBO than with LPO. This confirms our expectations, given that KBO is generally considered the more robust default option for superposition and that the nonmonotonic LPO triggers SUP inferences at variable positions—which is the price to pay for nonmonotonicity.

		# sat		# unsat		∅ time (s)		∅ iterations	
		LPO	KBO	LPO	KBO	LPO	KBO	LPO	KBO
TFF	first-order mode	0	0	**181**	**220**	**4.0**	4.4	**1497**	**1473**
	applicative encoding	0	0	150	203	19.0	16.0	1698	1916
	nonpurifying calculus	0	0	**181**	219	4.2	4.6	**1497**	**1473**
	purifying calculus	0	0	**181**	218	4.3	4.8	**1497**	**1473**
THF	applicative encoding	**444**	**438**	**676**	671	0.8	**0.2**	**72**	81
	nonpurifying calculus	353	360	675	**676**	**0.6**	0.3	83	**63**
	purifying calculus	338	343	664	666	0.8	1.0	116	231

Fig. 1. Evaluation of the intensional calculi

		# sat		# unsat		∅ time (s)		∅ iterations	
		LPO	KBO	LPO	KBO	LPO	KBO	LPO	KBO
TFF	first-order mode	0	0	**181**	**220**	**2.8**	4.3	1219	1420
	applicative encoding	0	0	151	201	19.0	17.6	1837	1792
	nonpurifying calculus	0	0	179	215	6.2	6.8	1610	1524
	purifying calculus	0	0	180	215	5.0	7.4	1291	1464
THF	applicative encoding	**426**	**421**	**677**	671	0.7	0.8	**78**	89
	nonpurifying calculus	310	327	669	**675**	**0.6**	**0.4**	83	**66**
	purifying calculus	227	261	647	650	1.0	1.0	114	108

Fig. 2. Evaluation of the extensional calculi

7 Discussion and Related Work

Our calculi join a long list of extensions and refinements of superposition. Among the most closely related is Peltier's [30] Isabelle formalization of the refutational completeness of a superposition calculus that operates on λ-free higher-order terms and that is parameterized by a monotonic term order. Extensions with polymorphism and induction, developed by Cruanes [16,17] and Wand [43], contribute to increasing the power of automatic provers. Detection of inconsistencies in axioms, as suggested by Schulz et al. [34], is important for large axiomatizations. Also of interest is Bofill and Rubio's [13] integration of nonmonotonic orders in ordered paramodulation, a precursor of superposition. Their work is a veritable tour de force, but it is also highly complicated and restricted to ordered paramodulation. Lack of compatibility with arguments being a mild form of nonmonotonicity, it seemed preferable to start with superposition, enrich it with an ARGCONG rule, and tune the side conditions until we obtained a complete calculus.

Most complications can be avoided by using a monotonic order such as KBO without argument coefficients, but we suspect that the coefficients will play an important role to support λ-abstractions. For example, the term $\lambda x.\ x + x$ could be treated as a constant with a coefficient of 2 on its argument and a heavy weight to ensure $(\lambda x.\ x + x)\ y \succ y + y$. LPO can also be used to good effect. This technique could allow provers to perform aggressive β-reduction in the vast majority of cases, without compromising completeness.

Many researchers have proposed or used encodings of higher-order logic constructs into first-order logic, including Robinson [32], Kerber [24], Dowek et al. [19], Meng and Paulson [27], and Czajka [18]. Encodings of types, such as those by Bobot and Paskevich [12] and Blanchette et al. [8], are also crucial to obtain a sound encoding of higher-order logic. These ideas are implemented in proof assistant tools such as HOLyHammer and Sledgehammer [9].

Another line of research has focused on the development of automated proof procedures for higher-order logic. Robinson's [31] and Huet's [23] pioneering work stands out. Andrews [1] and Benzmüller and Miller [6] provide excellent surveys. The competitive higher-order automatic theorem provers include LEO-II [7] (based on unordered paramodulation), Satallax [15] (based on a tableau calculus and a SAT solver), AgsyHOL [26] (based on a focused sequent calculus and a generic narrowing engine), and Leo-III [36] (based on a pragmatic extension of superposition with no completeness guarantees). The Isabelle proof assistant [29] and its Sledgehammer subsystem also participate in the higher-order division of the CADE ATP System Competition [38].

Zipperposition is a convenient vehicle for experimenting and prototyping because it is easier to understand and modify than highly-optimized C or C++ provers. Our middle-term goal is to design higher-order superposition calculi, implement them in state-of-the-art provers such as E [33], SPASS [44], and Vampire [25], and integrate these in proof assistants to provide a high level of automation. With its stratified architecture, Otter-λ [4] is perhaps the closest to what we are aiming at, but it is limited to second-order logic and offers no completeness guarantees. In preliminary work supervised by Blanchette and Schulz, Vukmirović [41] has generalized E's data structures and algorithms to λ-free higher-order logic, assuming a monotonic KBO [3].

8   Conclusion

We presented four superposition calculi for intensional and extensional λ-free higher-order logic and proved them refutationally complete. The calculi nicely generalize standard superposition and are compatible with our λ-free higher-order LPO and KBO. Our experiments partly confirm what one would naturally expect: that native support for partial application and applied variables outperforms the applicative encoding.

The new calculi reduce the gap between proof assistants based on higher-order logic and superposition provers. We can use them to reason about arbitrary higher-order problems by axiomatizing suitable combinators. But perhaps more importantly, they appear promising as a stepping stone towards complete, highly efficient automatic theorem provers for full higher-order logic.

Acknowledgment. We are grateful to the maintainers of StarExec for letting us use their service. We want to thank Christoph Benzmüller, Sander Dahmen, Johannes

Hölzl, Anders Schlichtkrull, Stephan Schulz, Alexander Steen, Geoff Sutcliffe, Andrei Voronkov, Petar Vukmirović, Daniel Wand, Christoph Weidenbach, and the participants in the 2017 Dagstuhl Seminar on Deduction beyond First-Order Logic for stimulating discussions. We also want to thank Mark Summerfield, Sophie Tourret, and the anonymous reviewers for suggesting several textual improvements to this paper and to the technical report. Bentkamp and Blanchette's research has received funding from the European Research Council (ERC) under the European Union's Horizon 2020 research and innovation program (grant agreement No. 713999, Matryoshka).

References

1. Andrews, P.B.: Classical type theory. In: Robinson, J.A., Voronkov, A. (eds.) Handbook of Automated Reasoning, vol. II, pp. 965–1007. Elsevier and MIT Press (2001)
2. Bachmair, L., Ganzinger, H.: Rewrite-based equational theorem proving with selection and simplification. J. Log. Comput. **4**(3), 217–247 (1994)
3. Becker, H., Blanchette, J.C., Waldmann, U., Wand, D.: A transfinite Knuth–Bendix order for lambda-free higher-order terms. In: de Moura, L. (ed.) CADE 2017. LNCS (LNAI), vol. 10395, pp. 432–453. Springer, Cham (2017). https://doi.org/10.1007/978-3-319-63046-5_27
4. Beeson, M.: Lambda logic. In: Basin, D., Rusinowitch, M. (eds.) IJCAR 2004. LNCS (LNAI), vol. 3097, pp. 460–474. Springer, Heidelberg (2004). https://doi.org/10.1007/978-3-540-25984-8_34
5. Bentkamp, A., Blanchette, J.C., Cruanes, S., Waldmann, U.: Superposition for lambda-free higher-order logic (technical report). Technical report (2018). http://matryoshka.gforge.inria.fr/pubs/lfhosup_report.pdf
6. Benzmüller, C., Miller, D.: Automation of higher-order logic. In: Siekmann, J.H. (ed.) Computational Logic, Handbook of the History of Logic, vol. 9, pp. 215–254. Elsevier, Amsterdam (2014)
7. Benzmüller, C., Paulson, L.C., Theiss, F., Fietzke, A.: LEO-II—a cooperative automatic theorem prover for classical higher-order logic (system description). In: Armando, A., Baumgartner, P., Dowek, G. (eds.) IJCAR 2008. LNCS (LNAI), vol. 5195, pp. 162–170. Springer, Heidelberg (2008). https://doi.org/10.1007/978-3-540-71070-7_14
8. Blanchette, J.C., Böhme, S., Popescu, A., Smallbone, N.: Encoding monomorphic and polymorphic types. Log. Methods Comput. Sci. **12**(4:13), 1–52 (2016)
9. Blanchette, J.C., Kaliszyk, C., Paulson, L.C., Urban, J.: Hammering towards QED. J. Formaliz. Reason. **9**(1), 101–148 (2016)
10. Blanchette, J.C., Paskevich, A.: TFF1: the TPTP typed first-order form with rank-1 polymorphism. In: Bonacina, M.P. (ed.) CADE 2013. LNCS (LNAI), vol. 7898, pp. 414–420. Springer, Heidelberg (2013). https://doi.org/10.1007/978-3-642-38574-2_29
11. Blanchette, J.C., Waldmann, U., Wand, D.: A lambda-free higher-order recursive path order. In: Esparza, J., Murawski, A.S. (eds.) FoSSaCS 2017. LNCS, vol. 10203, pp. 461–479. Springer, Heidelberg (2017). https://doi.org/10.1007/978-3-662-54458-7_27
12. Bobot, F., Paskevich, A.: Expressing polymorphic types in a many-sorted language. In: Tinelli, C., Sofronie-Stokkermans, V. (eds.) FroCoS 2011. LNCS (LNAI), vol. 6989, pp. 87–102. Springer, Heidelberg (2011). https://doi.org/10.1007/978-3-642-24364-6_7

13. Bofill, M., Rubio, A.: Paramodulation with non-monotonic orderings and simplification. J. Autom. Reason. **50**(1), 51–98 (2013)
14. Brand, D.: Proving theorems with the modification method. SIAM J. Comput. **4**, 412–430 (1975)
15. Brown, C.E.: Satallax: an automatic higher-order prover. In: Gramlich, B., Miller, D., Sattler, U. (eds.) IJCAR 2012. LNCS (LNAI), vol. 7364, pp. 111–117. Springer, Heidelberg (2012). https://doi.org/10.1007/978-3-642-31365-3_11
16. Cruanes, S.: Extending superposition with integer arithmetic, structural induction, and beyond. Ph.D. thesis, École polytechnique (2015)
17. Cruanes, S.: Superposition with structural induction. In: Dixon, C., Finger, M. (eds.) FroCoS 2017. LNCS (LNAI), vol. 10483, pp. 172–188. Springer, Cham (2017). https://doi.org/10.1007/978-3-319-66167-4_10
18. Czajka, Ł.: Improving automation in interactive theorem provers by efficient encoding of lambda-abstractions. In: Avigad, J., Chlipala, A. (eds.) CPP 2016, pp. 49–57. ACM (2016)
19. Dowek, G., Hardin, T., Kirchner, C.: Higher-order unification via explicit substitutions (extended abstract). In: LICS 1995, pp. 366–374. IEEE (1995)
20. Fitting, M.: Types, Tableaus, and Gödel's God. Kluwer, Dordrecht (2002)
21. Gupta, A., Kovács, L., Kragl, B., Voronkov, A.: Extensional crisis and proving identity. In: Cassez, F., Raskin, J.-F. (eds.) ATVA 2014. LNCS, vol. 8837, pp. 185–200. Springer, Cham (2014). https://doi.org/10.1007/978-3-319-11936-6_14
22. Henkin, L.: Completeness in the theory of types. J. Symb. Log. **15**(2), 81–91 (1950)
23. Huet, G.P.: A mechanization of type theory. In: Nilsson, N.J. (ed.) IJCAI 1973, pp. 139–146. William Kaufmann, Burlington (1973)
24. Kerber, M.: How to prove higher order theorems in first order logic. In: Mylopoulos, J., Reiter, R. (eds.) IJCAI 1991, pp. 137–142. Morgan Kaufmann, Burlington (1991)
25. Kovács, L., Voronkov, A.: First-order theorem proving and VAMPIRE. In: Sharygina, N., Veith, H. (eds.) CAV 2013. LNCS, vol. 8044, pp. 1–35. Springer, Heidelberg (2013). https://doi.org/10.1007/978-3-642-39799-8_1
26. Lindblad, F.: A focused sequent calculus for higher-order logic. In: Demri, S., Kapur, D., Weidenbach, C. (eds.) IJCAR 2014. LNCS (LNAI), vol. 8562, pp. 61–75. Springer, Cham (2014). https://doi.org/10.1007/978-3-319-08587-6_5
27. Meng, J., Paulson, L.C.: Translating higher-order clauses to first-order clauses. J. Autom. Reason. **40**(1), 35–60 (2008)
28. Miller, D.A.: A compact representation of proofs. Stud. Log. **46**(4), 347–370 (1987)
29. Nipkow, T., Paulson, L.C., Wenzel, M.: Isabelle/HOL: A Proof Assistant for Higher-Order Logic. LNCS, vol. 2283. Springer, Heidelberg (2002). https://doi.org/10.1007/3-540-45949-9
30. Peltier, N.: A variant of the superposition calculus. Archive of Formal Proofs (2016). https://www.isa-afp.org/entries/SuperCalc.shtml
31. Robinson, J.: Mechanizing higher order logic. In: Meltzer, B., Michie, D. (eds.) Machine Intelligence, vol. 4, pp. 151–170. Edinburgh University Press, Edinburgh (1969)
32. Robinson, J.: A note on mechanizing higher order logic. In: Meltzer, B., Michie, D. (eds.) Machine Intelligence, vol. 5, pp. 121–135. Edinburgh University Press, Edinburgh (1970)
33. Schulz, S.: System description: E 1.8. In: McMillan, K., Middeldorp, A., Voronkov, A. (eds.) LPAR 2013. LNCS, vol. 8312, pp. 735–743. Springer, Heidelberg (2013). https://doi.org/10.1007/978-3-642-45221-5_49

34. Schulz, S., Sutcliffe, G., Urban, J., Pease, A.: Detecting inconsistencies in large first-order knowledge bases. In: de Moura, L. (ed.) CADE 2017. LNCS (LNAI), vol. 10395, pp. 310–325. Springer, Cham (2017). https://doi.org/10.1007/978-3-319-63046-5_19

35. Snyder, W., Lynch, C.: Goal directed strategies for paramodulation. In: Book, R.V. (ed.) RTA 1991. LNCS, vol. 488, pp. 150–161. Springer, Heidelberg (1991). https://doi.org/10.1007/3-540-53904-2_93

36. Steen, A., Benzmüller, C.: The higher-order prover Leo-III. In: Galmiche, D., Schulz, S., Sebastiani, R. (eds.) IJCAR 2018. LNAI, vol. 10900, pp. 108–116. Springer, Cham (2018)

37. Stump, A., Sutcliffe, G., Tinelli, C.: StarExec: a cross-community infrastructure for logic solving. In: Demri, S., Kapur, D., Weidenbach, C. (eds.) IJCAR 2014. LNCS (LNAI), vol. 8562, pp. 367–373. Springer, Cham (2014). https://doi.org/10.1007/978-3-319-08587-6_28

38. Sutcliffe, G.: The CADE-26 automated theorem proving system competition–CASC-26. AI Commun. **30**(6), 419–432 (2017)

39. Sutcliffe, G., Benzmüller, C., Brown, C.E., Theiss, F.: Progress in the development of automated theorem proving for higher-order logic. In: Schmidt, R.A. (ed.) CADE 2009. LNCS (LNAI), vol. 5663, pp. 116–130. Springer, Heidelberg (2009). https://doi.org/10.1007/978-3-642-02959-2_8

40. Sutcliffe, G., Schulz, S., Claessen, K., Baumgartner, P.: The TPTP typed first-order form with arithmetic. In: Bjørner, N., Voronkov, A. (eds.) LPAR 2012. LNCS, vol. 7180, pp. 406–419. Springer, Heidelberg (2012). https://doi.org/10.1007/978-3-642-28717-6_32

41. Vukmirović, P.: Implementation of lambda-free higher-order superposition. M.Sc. thesis, Vrije Universiteit Amsterdam (2018)

42. Waldmann, U.: Automated reasoning II. Lecture notes, Max-Planck-Institut für Informatik (2016). http://resources.mpi-inf.mpg.de/departments/rg1/teaching/autrea2-ss16/script-current.pdf

43. Wand, D.: Superposition: types and polymorphism. Ph.D. thesis, Universität des Saarlandes (2017)

44. Weidenbach, C., Dimova, D., Fietzke, A., Kumar, R., Suda, M., Wischnewski, P.: SPASS version 3.5. In: Schmidt, R.A. (ed.) CADE 2009. LNCS (LNAI), vol. 5663, pp. 140–145. Springer, Heidelberg (2009). https://doi.org/10.1007/978-3-642-02959-2_10

Automated Reasoning About Key Sets

Miika Hannula and Sebastian Link[✉]

Department of Computer Science, University of Auckland,
Auckland, New Zealand
{m.hannula,s.link}@auckland.ac.nz

Abstract. Codd's rule of entity integrity stipulates that every table
in a database has a primary key. Hence, the attributes that form the
primary key carry no missing information and have unique value combinations. In practice, data records cannot always meet such requirements.
Previous work has proposed the notion of a key set, which can identify
more data records uniquely when information is missing. Apart from the
proposal, key sets have not been investigated much further. We outline
important database applications, and investigate computational limits
and techniques to reason automatically about key sets. We establish a
binary axiomatization for the implication problem of key sets, and prove
its coNP-completeness. We show that perfect models do not always exist
for key sets. Finally, we show that the implication problem for unary
key sets by arbitrary key sets has better computational properties. The
fragment enjoys a unary axiomatization, is decidable in time quadratic
in the input, and perfect models can always be generated.

1 Introduction

Keys provide efficient access to data in database systems. They are required
to understand the structure and semantics of data. For a given collection of
entities, a key refers to a set of column names whose values uniquely identify
an entity in the collection. For example, a key for a relational table is a set of
columns such that no two different rows have matching values in each of the key
columns. Keys are fundamental for most data models, including semantic models, object models, XML, RDF, and graphs. They advance many classical areas
of data management such as data modeling, database design, and query optimization. Knowledge about keys empowers us to (1) uniquely reference entities
across data repositories, (2) reduce data redundancy at schema design time to
process updates efficiently at run time, (3) improve selectivity estimates in query
processing, (4) feed new access paths to query optimizers that can speed up the
evaluation of queries, (5) access data more efficiently via physical optimization
such as data partitioning or the creation of indexes and views, and (6) gain new
insight into application data. Modern applications create even more demand for

Research is supported by Marsden funding from the Royal Society of New Zealand.

D. Galmiche et al. (Eds.): IJCAR 2018, LNAI 10900, pp. 47–63, 2018.
https://doi.org/10.1007/978-3-319-94205-6_4

keys. Here, keys facilitate data integration, help detect duplicates and anomalies, guide the repair of data, and return consistent answers to queries over dirty data. The discovery of keys from data sets is a core task of data profiling.

Due to the demand in real-life applications, data models have been extended to accommodate missing information. The industry standard for data management, SQL, allows occurrences of a null marker to model any kind of missing value. Occurrences of the null marker mean that no information is available about an actual value of that row on that attribute, not even whether the value exists and is unknown nor whether the value does not exist. Codd's principle of entity integrity suggests that every entity should be uniquely identifiable. In SQL, this has led to the notion of a primary key. A primary key is a collection of attributes which stipulates uniqueness and completeness. That is, no row of a relation must have an occurrence of the null marker on any columns of the primary key and the combination of values on the columns of the primary key must be unique. The requirement to have a primary key over every table in the database is often inconvenient in practice. Indeed, it can happen easily that a given relation does not exhibit any primary key. This is illustrated by the following example.

Example 1. Consider the following snapshot of data from an accident ward at a hospital [15]. Here, we collect information about the *name* and *address* of a patient, who was treated for an *injury* in some *room* at some *time*.

room	name	address	injury	time
1	Miller	⊥	cardiac infarct	Sunday, 19
⊥	⊥	⊥	skull fracture	Monday, 19
2	Maier	Dresden	leg fracture	Sunday, 16
1	Miller	Pirna	leg fracture	Sunday, 16

Evidently, the snapshot does not satisfy any primary key since each column features some null marker occurrence, or a duplication of some value.

In response, several researchers proposed the notion of a key set. As the term suggests, a key set is a set of attribute subsets. Naturally, we call the elements of a key set a key. A relation satisfies a given key set if for every pair of distinct rows in the relation there is some key in the key set on which both rows have no null marker occurrences and non-matching values on some attribute of the key. The formal definition of a key set will be given in Definition 1 in Sect. 3. The flexibility of a key set over a primary key can easily be recognized, as a primary key would be equivalent to a singleton key set, with the only element being the primary key. Indeed, with a key set different pairs of rows in a relation may be distinguishable by different keys of the key set, while all pairs of rows in a relation can only be distinguishable by the same primary key. We illustrate the notion of a key set on our running example.

Example 2. The relation in Example 1 satisfies no primary key. Nevertheless, the relation satisfies several key sets. For example, the key set $\{\{room\}, \{time\}\}$ is satisfied, but not the key set $\{\{room, time\}\}$. The relation also satisfies the key sets $\mathcal{X}_1 = \{\{room, time\}, \{injury, time\}\}$ and $\mathcal{X}_2 = \{\{name, time\}, \{injury, time\}\}$, as well as the key set $\mathcal{X} = \{\{room, name, time\}, \{injury, time\}\}$.

It is important to point out a desirable feature that primary keys and key sets share. Both are independent of the interpretation of null marker occurrences. That is, any given primary key and any given key set is either satisfied or not, independently of what information any of the null marker occurrences represent. Primary keys and key sets are only dependent on actual values that occur in the relevant columns. This is achieved by stipulating the completeness criterion. The importance of this independence is particularly appealing in modern applications where data is integrated from various sources, and different interpretations may be associated with different occurrences of null markers.

Given the flexibility of key sets over primary keys, and given their independence of null marker interpretations, it seems natural to further investigate the notion of a key set. Somewhat surprisingly, however, neither the research community nor any system implementations have analyzed key sets since their original proposal in 1989. The main goal of this article is to take first steps into the investigation of computational problems associated with key sets. In database practice, one of the most fundamental problems is the implication problem. The problem is to decide whether for a given set $\Sigma \cup \{\varphi\}$ of key sets, every relation that satisfies all key sets in Σ also satisfies φ. Reasoning about the implication of any form of database constraints is important because efficient solutions to the problem enable us to facilitate the processing of database queries and updates.

Example 3. Recall the key sets $\mathcal{X}_1$, $\mathcal{X}_2$, and $\mathcal{X}$ from Example 2. An instance of the implication problem is whether $\Sigma = \{\mathcal{X}_1, \mathcal{X}_2\}$ implies the key set $\varphi = \mathcal{X}$, and another instance is whether Σ implies $\varphi' = \{\{room\}, \{name\}, \{address\}, \{time\}\}$.

Contributions. Our contributions can be summarized as follows.

- We compare the notion of a key set with other notions of keys. In particular, primary keys are key sets with just one element, and certain keys are unary key sets, for which every key is a singleton.
- We illustrate how automated reasoning tools for key sets can facilitate efficient updates and queries in database systems.
- We establish a binary axiomatization for the implication problem of key sets. Here, binary refers to the maximum number of premises that any inference rule in our axiomatization can have. This is interesting as all previous notions of keys enjoy unary axiomatizations, in particular primary keys. What that means semantically is that every given key set that is implied by a set of key sets is actually implied by at most two of the key sets.
- We establish that the implication problem for key sets is *coNP*-complete. Again, this complexity is quite surprising in comparison with the linear time decidability of other notions of keys.
- An interesting notion in database theory is that of Armstrong databases. A given class of constraints, such as keys, key sets, or other data dependencies [13], is said to enjoy Armstrong databases whenever for every given set of constraints in this class there is a single database with the property that for every constraint in the class, the database satisfies this constraint if and

only if the constraint is implied by the given set of constraints. This is a powerful property as multiple instances over the implication problem reduce to validating satisfaction over the same Armstrong database. Consequently, the generation of Armstrong databases would create 'perfect models' of a given constraint set, which has applications in the acquisition of requirements in database practice. We show that key sets do not enjoy Armstrong relations, as opposed to other classes of keys known from the literature.

– We then identify an expressive fragment of key sets for which the associated implication problem can be characterized by a unary axiomatization and a quadratic-time algorithm. The fragment also enjoys Armstrong relations and we show how to generate them with conservative use of time and space.

Organization. We discuss related work in Sect. 2. Basic notions and notation are fixed in Sect. 3. Section 4 discusses applications of key sets in the processing of queries and updates. An axiomatization for key sets is established in Sect. 5. The *coNP*-completeness of the implication problem is settled in Sect. 6. The general existence of Armstrong relations is dis-proven in Sect. 7. A computationally friendly fragment of key sets is identified in Sect. 8. We conclude and briefly discuss future work in Sect. 9.

2   Related Work

We provide a concise discussion on the relationship of key sets with other notions of keys over relations with missing information.

Codd is the inventor of the relational model of data [4]. He proposed the rule of entity integrity, which stipulates that every entity in every table should be uniquely identifiable. In SQL that led to the introduction of primary keys, which stipulate uniqueness and completeness on the attributes that form the primary key. The primary key is a distinguished candidate key. We call an attribute set a *candidate key* for a given relation if and only if every pair of distinct tuples in the relation has no null marker occurrences on any of the attributes of the candidate key and there is some attribute of the candidate key on which the two tuples have non-matching values. The notions of primary and candidate keys have been introduced very early in the history of database research [12]. Candidate keys are singleton key sets, that is, key sets with just one element (namely the candidate key). Hence, instead of having to be complete and unique on the same combination of columns in a candidate key, key sets offer different alternatives of being complete and unique for different pairs of tuples in a relation. Candidate keys were studied in [7]. In that work, the associated implication problem was characterized axiomatically and algorithmically, the automatic generation of Armstrong relations was established, and extremal problems associated with families of candidate keys were investigated. As Example 1 shows, there are relations on which no candidate key holds, but which satisfy key sets.

Lucchesi and Osborn studied computational problems associated with candidate keys [12]. However, their focus was an algorithm that finds all candidate keys implied by a given set of functional dependencies. They also proved that deciding whether a given relation satisfies some key of cardinality not greater than some given positive integer is NP-complete. Recently, this problem was shown to be W[2]-complete in the size of the key [2]. The discovery which key sets hold on a given relation is beyond the scope of this paper and left as an open problem for future work.

Key sets were introduced by Thalheim [14] as a generalization of Codd's rule for entity integrity. He studied combinatorial problems associated with unary key sets, such as the maximum cardinality that non-redundant families of unary key sets can have, and which families attain them [13,15]. Key sets were further discussed by Levene and Loizou [11] where they also generalized Codd's rule for referential integrity. Somewhat surprisingly, the study of the implication problem for key sets has not been addressed by previous work. This is also true for other automated tasks which require reasoning about key sets.

More recently, the notions of possible and certain keys were proposed [8]. These notions are defined for relations in which null marker occurrences are interpreted as 'no information', and possible worlds of an incomplete relation are obtained by independently replacing null marker occurrences by actual domain values (or the N/A marker indicating that the value does not exist). A key is said to be *possible* for an incomplete relation if and only if there is some possible world of the incomplete relation on which the key holds. A key is said to be *certain* for an incomplete relation if and only if the key holds on every possible world of the incomplete relation. For example, the relation in Example 1 satisfies the possible key $p\langle room, name, address \rangle$, since the key $\{room, name, address\}$ holds on the possible world:

room	name	address	injury	time
1	Miller	Dresden	cardiac infarct	Sunday, 19
2	Maier	Pirna	skull fracture	Monday, 19
2	Maier	Dresden	leg fracture	Sunday, 16
1	Miller	Pirna	leg fracture	Sunday, 16

of the relation. In contrast, the key $\{room, name\}$ is not possible for the relation because the first and last tuple will have matching values on room and name in every possible world of the relation. The key $\{address\}$ is possible, but not certain, and the key $\{room, time\}$ is certain for the given relation. Now, it is not difficult to see that an incomplete relation satisfies the certain key $c\langle A_1, \ldots, A_n \rangle$ if and only if the relation satisfies the key set $\{\{A_1\}, \ldots, \{A_n\}\}$. In this sense, certain keys correspond to key sets which have only singleton keys as elements. The papers [8] investigate computational problems for possible and certain keys with NOT NULL constraints. In the current paper we investigate a different class of key constraints, namely key sets. In particular, the computationally-friendly fragment of key sets we identify in Sect. 8 subsumes the class of certain keys as the special case of unary key sets.

Recently, *contextual keys* were introduced as a means to separate completeness from uniqueness requirements [16]. A contextual key is an expression (C, X)

where $X \subseteq C$. These are different from key sets since $X \subseteq C$ is a key for only those tuples that are complete on C. In particular, the special case where $C = X$ only requires uniqueness on X for those tuples that are complete on X. This captures the UNIQUE constraint of SQL. We leave it as future work to investigate contextual key sets.

3   Preliminary Definitions

In this section, we give some basic definitions and fix notation.

A *relation schema* is a finite non-empty set of attributes, usually denoted by R. A *relation r* over R consists of tuples t that map each $A \in R$ to $\mathrm{Dom}(A) \cup \{\bot\}$ where $\mathrm{Dom}(A)$ is the domain associated with attribute A and $\bot$ is the unique null marker. Given a subset X of R, we say that a tuple t is X-*total* if $t(A) \neq \bot$ for all $A \in X$. Informally, a relation schema represents the column names of database tables, while each tuple represents a row of the table, so a relation forms a database instance. Moreover, $\mathrm{Dom}(A)$ represents the possible values that can occur in column A of a table, and $\bot$ represents missing information. That is, if $t(A) = \bot$, then there is no information about the value $t(A)$ of tuple t on attribute A.

In our running example, we have the relation schema

$$\mathrm{WARD} = \{room, name, address, injury, time\}.$$

Each of these attributes comes with a domain, which we do not specify any further here. Each row of the table in Example 1 represents a tuple. The second row, for example, is $\{injury, time\}$-total, but not total on any proper superset of $\{injury, time\}$. The four tuples together constitute a relation over WARD.

The following definition introduces the central object of our studies. It was first defined by Thalheim in [14].

Definition 1. *A key set is a finite, non-empty collection $\mathcal{X}$ of subsets of a given relation schema R. We say that a relation r over R satisfies the key set $\mathcal{X}$ if and only if for all distinct $t, t' \in r$ there is some $X \in \mathcal{X}$ such that t and t' are X-total and $t(X) \neq t'(X)$. Each element of a key set is called a* key. *If all keys of a key set are singletons, we speak of a* unary key set.

In the sequel we write $\mathcal{X}, \mathcal{Y}, \mathcal{Z}, \ldots$ for key sets and $X, Y, Z, \ldots$ for attribute sets, and $A, B, C, \ldots$ for attributes. We sometimes write A instead of $\{A\}$ to denote the singleton set consisting of only A. If $\boldsymbol{X}$ is a sequence, then we may sometimes write simply $\boldsymbol{X}$ for the set that consists of all members of $\boldsymbol{X}$.

As already mentioned in Example 2, the relation in Example 1 satisfies the key sets $\mathcal{X}_1$, $\mathcal{X}_2$, and $\mathcal{X}$. It also satisfies the unary key set $\{\{room\}, \{time\}\}$, but not the singleton key set $\{\{room, time\}\}$.

A fundamental problem in automated reasoning about any class of constraints is the *implication problem*. For key sets, the problem is to decide whether for an arbitrary relation schema R, and an arbitrary set $\Sigma \cup \{\varphi\}$ of key sets over

R, Σ implies φ. Indeed, Σ *implies* φ if and only if every relation over R that satisfies all key sets in Σ also satisfies the key set φ. The following section illustrates how solutions to the implication problem of key sets can facilitate the efficient processing of queries and updates.

4   Applications for Automated Reasoning

The most important applications of data processing are updates and queries. We briefly describe in this section how automated reasoning about key sets can facilitate each of these application areas.

4.1   Efficient Updates

When databases are updated it must be ensured that the resulting database satisfies all the constraints that model the business rules of the underlying application domain. Violations of the constraints indicate sources of inconsistency, and an alert of such inconsistencies should at least be issued to the database administrator. This is to ensure that appropriate actions can be taken, for example, to disallow the update. This quality assurance process incurs an overhead in terms of the time it takes to validate the constraints. As such, users of the database expect that such overheads are minimized. In particular, the time on validating constraints increases with the volume of the database. As a principal, the set of constraints that are specified on the database and therefore subject to validation upon updates, should be non-redundant. That is, no constraints should be specified that are already implied by other specified constraints. The simple reason is that the validation of any implied constraints is a waste of time because the validity of the other constraints already ensures that any implied constraint is valid as well. This is a strong real-life motivation for developing tools that can decide implication. In our running example, the set $\Sigma = \{\mathcal{X}_1, \mathcal{X}_2, \mathcal{X}\}$ of key sets is redundant because the subset $\Sigma' = \{\mathcal{X}_1, \mathcal{X}_2\}$ implies the key set $\mathcal{X}$. Automated solutions to the implication problem can thus automatize the minimization of overheads in validating constraints under database updates.

4.2   Efficient Queries

We are interested in the names of patients that can be identified uniquely based on information about their name and the room and time at the accident ward, or based on information about their injury and the time at the accident ward. In SQL, this may be expressed as follows.

```
       SELECT name
         FROM WARD
        WHERE room IS NOT NULL AND name IS NOT NULL AND
              time IS NOT NULL
     GROUP BY room, name, time
       HAVING count(room, name, time) ≤ 1
        UNION
       SELECT name
         FROM WARD
        WHERE injury IS NOT NULL AND time IS NOT NULL
     GROUP BY injury, time
       HAVING count(injury, time) ≤ 1 ;
```

Knowing that the underlying relation over WARD satisfies the two key sets $\mathcal{X}_1$ and $\mathcal{X}_2$ and that the key set $\mathcal{X} = \{\{room, name, time\}, \{injury, time\}\}$ is implied by $\mathcal{X}_1$ and $\mathcal{X}_2$, one can deduce that every tuple of WARD must be in at least one of the sub-query results of the UNION query. That is, the query above can be simplified to

```
       SELECT DISTINCT name
            FROM WARD ;
```

Note that the DISTINCT word is necessary since the UNION operator eliminates duplicates. When evaluated on the example from the introduction, each query will return the result $\{(name: \text{Miller}), (name: \bot), (name: \text{Maier})\}$.

Motivated by the applications of key sets for data processing and the lack of knowledge on automated reasoning tasks associated with key sets, the following sections will investigate the implication problem for key sets.

Table 1. An axiomatization $\mathfrak{A}$ for key sets

$\dfrac{\mathcal{X}}{\mathcal{X} \cup \mathcal{Y}}$	$\dfrac{\mathcal{X} \cup \{XY\}}{\mathcal{X} \cup \{X, Y\}}$	$\dfrac{\mathcal{X}_1 \quad \mathcal{X}_2}{\{Z_{(X_1, X_2)} \mid (X_1, X_2) \in \mathcal{X}_1 \times \mathcal{X}_2\}}$
		$Z_{(X_1, X_2)} \subseteq X_1 \cup X_2$, and
		$X_1 \subseteq Z_{(X_1, X_2)}$ or $X_2 \subseteq Z_{(X_1, X_2)}$
Upward closure	**Refinement**	**Composition**

5 Axiomatizing Key Sets

In this section we establish axiomatizations for arbitrary key sets as well as unary ones. This will enable us to effectively enumerate all implied key sets, that is, to determine the semantic closure $\Sigma^* = \{\sigma \mid \Sigma \models \sigma\}$ of any given set Σ of key sets. A finite axiomatization facilitates human understanding of the interaction of the given constraints, and ensures all opportunities for the use of these constraints in applications can be exploited.

In using an axiomatization we determine the semantic closure by applying *inference rules* of the form $\dfrac{\text{premise}}{\text{conclusion}}$. For a set $\mathfrak{R}$ of inference rules let $\Sigma \vdash_{\mathfrak{R}} \varphi$ denote the *inference* of φ from Σ by $\mathfrak{R}$. That is, there is some sequence $\sigma_1, \ldots, \sigma_n$ such that $\sigma_n = \varphi$ and every σ_i is an element of Σ or is the conclusion that results from an application of an inference rule in $\mathfrak{R}$ to some premises in $\{\sigma_1, \ldots, \sigma_{i-1}\}$. Let $\Sigma^+_{\mathfrak{R}} = \{\varphi \mid \Sigma \vdash_{\mathfrak{R}} \varphi\}$ be the *syntactic closure* of Σ under inferences by $\mathfrak{R}$. $\mathfrak{R}$ is *sound* (*complete*) if for every set Σ over every R we have $\Sigma^+_{\mathfrak{R}} \subseteq \Sigma^*$ ($\Sigma^* \subseteq \Sigma^+_{\mathfrak{R}}$). The (finite) set $\mathfrak{R}$ is a (finite) *axiomatization* if $\mathfrak{R}$ is both sound and complete.

Table 1 shows a finite axiomatization $\mathfrak{A}$ for key sets. A non-trivial rule is **Composition** which is illustrated by our running example.

Example 4. Recall Example 1 from the introduction, in particular $\Sigma = \{\mathcal{X}_1, \mathcal{X}_2\}$ and $\varphi = \mathcal{X}$. It turns out that φ is indeed implied by Σ, since φ can be inferred from Σ by an application of the Composition rule, and the rule is sound for the implication of key sets. Indeed, $\mathcal{X}_1 \times \mathcal{X}_2$ consists of:

$$(\{room, time\}, \{name, time\}),$$
$$(\{room, time\}, \{injury, time\}),$$
$$(\{injury, time\}, \{name, time\}), \text{ and}$$
$$(\{injury, time\}, \{injury, time\}) \,.$$

and for each element $X = (X_1, X_2)$ we need to pick one attribute set Z_X that is contained in the union $X_1 \cup X_2$ and contains either X_1 or X_2. For the first element we pick $\{room, time, name\}$, and for the remaining three elements we pick $\{injury, time\}$. That results in the key set $\mathcal{X}$.

We now proceed with the completeness proof for the axiom system $\mathfrak{A}$ of Table 1. The proof proceeds in three stages. First in Lemma 1, we show a characterization of the implication problem. This is applied in Lemma 2 to show that $\mathfrak{A}$ extended with n-ary Composition for all $n \in \mathbb{N}$ is complete (see Table 2). At last, we show in Lemma 3 that n-ary Composition can be simulated with the binary Composition of $\mathfrak{A}$.

Table 2. The n-ary Composition rule

$$\dfrac{\mathcal{X}_1 \quad \ldots \quad \mathcal{X}_n}{\{Z_{\mathbf{X}} \mid \mathbf{X} \in \mathcal{X}_1 \times \ldots \times \mathcal{X}_n\}}$$
$$Z_{\mathbf{X}} \subseteq \bigcup \mathbf{X} \text{ and } \bigvee_i X_i \subseteq Z_{\mathbf{X}}$$

Lemma 1. $\{\mathcal{X}_1, \ldots, \mathcal{X}_n\} \models \mathcal{Y}$ *iff for all* $(X_1, \ldots, X_n) \in \mathcal{X}_1 \times \ldots \times \mathcal{X}_n$ *there is* $\mathcal{Z} \subseteq \mathcal{Y}$ *such that* $\bigcup \mathcal{Z} \subseteq \bigcup_i X_i$, *and* $X_i \subseteq \bigcup \mathcal{Z}$ *for some* i.

Proof. Assume first that one finds such an $\mathcal{Z}$. We show that any relation r that satisfies each $\mathcal{X}_i$ satisfies also $\mathcal{Y}$. Let t, t' be two tuples from r. Then for some $(X_1, \ldots, X_n) \in \mathcal{X}_1 \times \ldots \times \mathcal{X}_n$, t and t' are both $\bigcup_i X_i$-total and disagreeing on each X_i. Assume that i is such that $X_i \subseteq \bigcup \mathcal{Z}$, and let $A \in X_i$ be such that $t(A) \neq t'(A)$. Then selecting some $Z \in \mathcal{Z}$ such that it also contains A, we have that t and t' are Z-total and deviate on Z. Thus Z is witness for $r \models \mathcal{Y}$.

For the other direction we assume that no such $\mathcal{Z}$ exists. Then there is $(X_1, \ldots, X_n) \in \mathcal{X}_1 \times \ldots \times \mathcal{X}_n$ such that for $\mathcal{Z} := \{Z \in \mathcal{Y} \mid Z \subseteq \bigcup_i X_i\}$, $X_i \not\subseteq \bigcup \mathcal{Z}$ for all i. Then, selecting an attribute A_i from $X_i \setminus \bigcup \mathcal{Z}$ for all i, we may construct a relation r satisfying $\{\mathcal{X}_1, \ldots, \mathcal{X}_n, \neg\mathcal{Y}\}$. This relation r consists of two tuples t, t' where t is a constant function mapping all of R to 0, and t' maps $\bigcup_i A_i$ to 1, $\bigcup_i X_i \setminus \bigcup_i A_i$ to 0, and all the remaining attributes to $\perp$. Now, obviously r satisfies all $\mathcal{X}_i$. Furthermore, for $Y \in \mathcal{Y} \setminus \mathcal{Z}$, t' is not Y-total, and for $Y \in \mathcal{Y} \cap \mathcal{Z}$ both t and t' are Y-total but with constant values 0. Therefore, r is a witness of $\{\mathcal{X}_1, \ldots, \mathcal{X}_n\} \not\models \mathcal{Y}$ which concludes the proof. $\qquad \square$

Notice that the latter condition of Lemma 1 can be equivalently stated as $X_i \subseteq \bigcup\{Y \in \mathcal{Y} \mid Y \subseteq \bigcup_i X_i\}$ for some i.

Lemma 2. *The axiomatization $\mathfrak{A}$ extended with n-ary Composition is complete for key sets.*

Proof. Assume $\{\mathcal{X}_1, \ldots, \mathcal{X}_n\} \models \mathcal{Y}$. Then we obtain by Lemma 1 for all $\boldsymbol{X} = (X_1, \ldots, X_n) \in \mathcal{X}_1 \times \ldots \times \mathcal{X}_n$ a subset $\mathcal{Z}_{\boldsymbol{X}} \subseteq \mathcal{Y}$ such that $\bigcup \mathcal{Z}_{\boldsymbol{X}} \subseteq \bigcup \boldsymbol{X}$, and $X_i \subseteq \bigcup \mathcal{Z}_{\boldsymbol{X}}$ for some i. Then by Composition we may derive $\{\bigcup \mathcal{Z}_{\boldsymbol{X}} \mid \boldsymbol{X} \in \mathcal{X}_1 \times \ldots \times \mathcal{X}_n\}$. With repeated applications of Refinement we then derive $\bigcup\{\mathcal{Z}_{\boldsymbol{X}} \mid \boldsymbol{X} \in \mathcal{X}_1 \times \ldots \times \mathcal{X}_n\}$. Since this set is a subset of $\mathcal{Y}$, we finally obtain $\mathcal{Y}$ with a single application of Upward closure. $\qquad \square$

Lemma 3. *n-ary Composition is derivable in $\mathfrak{A}$.*

Proof. Assume that $\mathcal{K} = \{Z_{\boldsymbol{X}} \mid \boldsymbol{X} \in \mathcal{X}_1 \times \ldots \times \mathcal{X}_n\}$ is obtained from $\mathcal{X}_1, \ldots, \mathcal{X}_n$ by an application of n-ary Composition. We will perform consecutive applications of (binary) Composition until we have obtained $\mathcal{K}$. Composition is applied incrementally so that the first application of this rule combines $\mathcal{X}_1$ and $\mathcal{X}_2$ to obtain a new key set $\mathcal{X}$, the second combines $\mathcal{X}$ and $\mathcal{X}_3$ to obtain the next key set $\mathcal{X}'$, the third $\mathcal{X}'$ and $\mathcal{X}_4$ to obtain $\mathcal{X}''$, and so forth. Once $\mathcal{X}_n$ is reached the cycle is started again from $\mathcal{X}_1$.

At each step of the aforementioned procedure we have deduced a key set $\mathcal{X}$ such that each $X \in \mathcal{X}$ either is a union $\bigcup \mathcal{Y}_1 \cup \ldots \cup \bigcup \mathcal{Y}_n$ for $\mathcal{Y}_i \subseteq \mathcal{X}_i$, or belongs to the required key set $\mathcal{K}$. In the previous case, provided that each $\mathcal{Y}_i$ is the maximal subset of $\mathcal{X}_i$ such that $\bigcup \mathcal{Y}_i \subseteq X$, we refer to $\mathcal{Y}_1 \cup \ldots \cup \mathcal{Y}_n$ as the *maximal decomposition* of X and $|\mathcal{Y}_1 \cup \ldots \cup \mathcal{Y}_n|$ as the *decomposition size* of X. Furthermore, given a set $Z_{\boldsymbol{X}} \in \mathcal{K}$ where $\boldsymbol{X} \in \mathcal{X}_1 \times \ldots \times \mathcal{X}_n$ we say that a set $X_i \in \boldsymbol{X}$ is *full* in $Z_{\boldsymbol{X}}$ if $X_i \subseteq Z_{\boldsymbol{X}}$. By the prerequisite of the n-ary Composition some member of $\boldsymbol{X}$ is always guaranteed to be full in $Z_{\boldsymbol{X}}$.

Initialization. Consider an instance of n-ary Composition. We initialize the procedure by applying Composition $n-1$ many times so that we obtain the key set $\{\bigcup X \mid X \in \mathcal{X}_1 \times \ldots \times \mathcal{X}_n\}$. This is done by letting $\mathcal{U}_1 := \mathcal{X}_1$ and taking the key set $\mathcal{U}_{i+1} = \{X_1 \cup X_2 \mid (X_1, X_2) \in \mathcal{U}_i \times \mathcal{X}_{i+1}\}$ for $i = 1, \ldots, n-1$.

Inductive Step. After the initial step we have reached a key set $\mathcal{V}_1 := \mathcal{U}_n$ such that all $X \in \mathcal{V}_1 \setminus \mathcal{K}$ have decomposition size at least 1. Assume now that we have reached a key set $\mathcal{V}_m$ such that all $X \in \mathcal{V}_m \setminus \mathcal{K}$ have decomposition size at least m. As the induction step we show how to obtain a key set $\mathcal{V}_{m+1}$ such that every member of $\mathcal{V}_{m+1} \setminus \mathcal{K}$ has decomposition size at least $m+1$. This is done by taking a single round of applications of Composition to $\mathcal{V}_m$ and $\mathcal{X}_1, \ldots, \mathcal{X}_n$. That is, $\mathcal{V}_m$ and $\mathcal{X}_1$ are first combined using Composition, then the outcome is combined with $\mathcal{X}_2$, and its outcome with $\mathcal{X}_3$, and so forth until we have applied this procedure to $\mathcal{X}_n$. All these applications keep the members of $V_m \cap \mathcal{K}$ fixed. For instance, at the first step $Z_{(X,Y)}$ for $X \in \mathcal{V}_m \cap \mathcal{K}$ and any $Y \in \mathcal{X}_1$ is defined as X. We show how this deduction handles an arbitrary $X \in \mathcal{V}_m \setminus \mathcal{K}$.

By induction assumption each $X \in \mathcal{V}_m \setminus \mathcal{K}$ has decomposition size at least m. Let $\bigcup \mathcal{Y}_1 \cup \ldots \cup \bigcup \mathcal{Y}_n$ be the maximal decomposition of X. Now, assume towards a contradiction that for each i there is $Y_i \in \mathcal{Y}_i$ such that Y_i is not full in any $Z_{\boldsymbol{Y}} \in \mathcal{K}$ where $\boldsymbol{Y} \in \mathcal{Y}_1 \times \ldots \times \mathcal{Y}_n$ and Y_i is the ith member of $\boldsymbol{Y}$. Then, however, the diagonal $\boldsymbol{Y}' = (Y_1, \ldots, Y_n)$ must have a member that is full in $Z_{\boldsymbol{Y}'}$. This is a contradiction and hence there is i such that all $Y_i \in \mathcal{Y}_i$ are full in some $Z_{\boldsymbol{Y}} \in \mathcal{K}$ where $\boldsymbol{Y} \in \mathcal{Y}_1 \times \ldots \times \mathcal{Y}_n$ and Y_i is the ith member of $\boldsymbol{Y}$. With regards to X, Composition is then applied as follows. For the first $i-1$ applications X is kept fixed. For the ith application that considers $\mathcal{X}_i$, each pair of X and $Y \in \mathcal{Y}_i$ is transformed to that $Z_{\boldsymbol{Y}} \in \mathcal{K}$ in which Y is full. Furthermore, each pair of X and $Y \in \mathcal{X}_i \setminus \mathcal{Y}_i$ is transformed to XY. Take note that the decomposition size of XY is at least $n+1$. At last, the remaining applications of Composition keep the obtained sets fixed. Since this procedure is applied to all $X \in \mathcal{V}_m \setminus \mathcal{K}$, we obtain that $\mathcal{V}_{m+1} \setminus \mathcal{K}$ has only sets with decomposition size at least $m+1$. This concludes the induction step.

Now, $\mathcal{V}_{M+1}$ where $M = |\mathcal{X}_1 \cup \ldots \cup \mathcal{X}_n|$ is a subset of $\mathcal{K}$. Hence, we conclude that $\mathcal{V}_{M+1}$ yields $\mathcal{K}$ with one application of Upward closure. $\qquad \square$

Note that a simulation of one application of n-ary of Composition to $\{\mathcal{X}_1, \ldots, \mathcal{X}_n\}$ takes at most $(n+1) \cdot |\bigcup_{i=1}^{n} \mathcal{X}_i|$ applications of binary Composition plus one application of Upward Closure.

The previous three lemmata now generate the following axiomatic characterization of key set implication. We omit the soundness proof which is straightforward to check.

Theorem 1. *The axiomatization $\mathfrak{A}$ is sound and complete for key sets.*

Another Important Application. A direct application of an axiomatization is the efficient representation of collections of key sets. Similar to the computation of non-redundant covers during update operations, removing any redundant constraints makes the result easier to understand by humans. This is, for example, important for the discovery problem of key sets in which one attempts to efficiently represent all those key sets that a given relation satisfies. Even more directly, one can understand any sound inference rule as an opportunity to apply pruning techniques as part of a discovery algorithm. A complete axiomatization ensures all opportunities for the pruning of a search space can be exploited.

6   Complexity of Key Set Implication

In this section we settle the exact computational complexity of the implication problem for key sets. While the implication problem for most notions of keys over incomplete relations is decidable in linear time, the implication problem for key sets is likely to be intractable. This should also be seen as evidence for the expressivity of key sets.

Theorem 2. *The implication problem for key sets is coNP-complete.*

Proof. Consider first the membership in $\mathsf{co-NP}$. By Lemma 1, for determining whether $\{\mathcal{X}_1, \ldots, \mathcal{X}_n\} \not\models \mathcal{Y}$, it suffices to choose $X_1, \ldots, X_n$ respectively from $\mathcal{X}_1, \ldots, \mathcal{X}_n$, and then deterministically check that $X_i \not\subseteq \bigcup \mathcal{Z}$ for all i, where $\mathcal{Z}$ is selected deterministically as $\mathcal{Z} := \{Z \in \mathcal{Y} \mid Z \subseteq \bigcup_i X_i\}$.

For the hardness, we reduce from the complement of 3-SAT. Let $C_1, \ldots, C_n$ be a collection of clauses, each consisting of three literals, i.e., propositions of the form p or negated propositions of the form $\neg p$. Let P be the set of all proposition symbols that appear in some C_i, and let $\overline{P}$ consist of their negations. Letting $P \cup \overline{P}$ be our relation schema, we show that $\bigwedge_i \bigvee C_i$ has a solution iff $\{\{p, \neg p\} \mid p \in P\} \not\models \{C_1, \ldots, C_n\}$. Notice that the antecedent is a set of singleton key sets, each of size two.

Assume first that there is a solution. Let $S \subseteq \mathcal{P}(P)$ encode the complement of that solution, i.e., S is such that each C_i contains some $p \notin S$ or some $\neg p$ for $p \in S$. Let $\overline{S} = \{\neg p \mid p \notin S\}$, and define singleton sets $X_p = \{p, \neg p\} \cap (S \cup \overline{S})$, encoding those literals that are set false by the solution. Then $C_i \not\subseteq \bigcup_p X_p$ for all i, implying be Lemma 1 that $\{\{p, \neg p\} \mid p \in P\} \not\models \{C_1, \ldots, C_n\}$.

Assume then that $\{\{p, \neg p\} \mid p \in P\} \not\models \{C_1, \ldots, C_n\}$. By Lemma 1 we find $X_p \in \{p, \neg p\}$ such that for no $\mathcal{Z} \subseteq \{C_1, \ldots, C_n\}$ we have that $\bigcup \mathcal{Z} \subseteq \bigcup_p X_p$ and $\bigvee_p X_p \subseteq \bigcup \mathcal{Z}$. Now, $C_i \subseteq \bigcup_p X_p$ implies $X_p \subseteq C_i$ for three distinct p, and therefore we must have $C_i \not\subseteq \bigcup_p X_p$ for all i. It is now easy to see that the sets X_p give rise to a solution to the satisfiability problem.   □

7   Armstrong Relations

In this section we ask the basic question whether key sets enjoy Armstrong relations. These are special models which are perfect for a given collection of key sets. More formally, a given relation r is said to be *Armstrong* for a given set Σ of key sets if and only if for all key sets φ it is true that r satisfies φ if and only if Σ implies φ. Indeed, an Armstrong relation is a perfect model for Σ since it satisfies all keys sets implied by Σ and does not satisfy any key set that is not implied by Σ. Armstrong relations have important applications in data profiling [1] and the requirements acquisition phase of database design [10].

Unfortunately, arbitrary sets of key sets do not enjoy Armstrong relations as the following result manifests.

Theorem 3. *There are sets of key sets for which no Armstrong relations exist.*

Proof. An example is $\Sigma = \{\{\{A\},\{B\}\},\{\{C\},\{D\}\}\}$ with attributes $A, B,$ C, D. Then $\sigma_1 = \{\{A,C\},\{A,D\},\{B,C\}\}$ and $\sigma_2 = \{\{A,D\},\{B,C\},\{B,D\}\}$ are two non-consequences of Σ, respectively exemplified by the two 2-tuple relations on the left of Fig. 1, where "d" refers to any distinct total value.

These are the only possible types of tuple pairs that satisfy $\Sigma \cup \{\neg\sigma_1\}$ and $\Sigma \cup \{\neg\sigma_2\}$, respectively. Therefore, we observe that any relation r satisfying Σ and refuting both σ_1 and σ_2 has a homomorphism from a relation of the form on the right of Fig. 1 to a subset of r with the condition that this homomorphism preserves nulls and maps domain values to domain values. However, then neither $\{\{A\},\{B\}\}$ nor $\{\{C\},\{D\}\}$ is a key set anymore. □

$A\ B\ C\ D$
$d\ d\perp d$
$\perp d\ d\ d$

$A\ B\ C\ D$
$d\ d\ d\perp$
$d\perp d\ d$

$A\ B\ C\ D$
$d\ d\perp d$
$\perp d\ d\ d$
$d\ d\ d\perp$
$d\perp d\ d$

Fig. 1. Relations used in the proof of Theorem 3

8 Implication for Unary by Arbitrary Key Sets

In this section we identify a fragment of key sets for which automated reasoning is efficient. This is strongly motivated by the results of the previous sections in which the coNP-completeness of the implication problem, and the lack of general Armstrong relations has been established. Indeed, the fragment is the implication of unary key sets by arbitrary key sets. We show that this fragment is captured axiomatically by the Refinement and Upward Closure rules, can be decided in time quadratic in the input, and Armstrong relations always exist and can be computed with conservative use of time and space.

8.1 An Algorithmic Characterization

Our first result establishes that unary key sets must be implied by a single key set from the given collection of key sets.

Theorem 4. *Let $\Sigma = \{\mathcal{X}_1,\ldots,\mathcal{X}_n\}$ be a collection of arbitrary key sets, and let $\varphi = \{\{A_1\},\ldots,\{A_k\}\}$ be a unary key set over relation schema R. Then Σ implies φ if and only if there is some $i \in \{1,\ldots,n\}$ such that $\bigcup \mathcal{X}_i \subseteq \{A_1,\ldots,A_k\}$.*

Proof. If $\bigcup \mathcal{X}_i \subseteq X$ for some $i \in \{1, \ldots, n\}$, Refinement and Upward Closure infer φ from Σ. Due to the rules' soundness, φ is implied by Σ.

Vice versa, assume that $\bigcup \mathcal{X}_i \not\subseteq X$ holds for all $i = 1, \ldots, n$. Let r be defined as $r = \{t, t'\}$ where t and t' are two total tuples that agree on $X = \{A_1, \ldots, A_k\}$ and disagree elsewhere. It follows that r violates φ. Since $\bigcup \mathcal{X}_i \not\subseteq X$ for all $i = 1, \ldots, n$, t_1 and t_2 must differ on some attribute in $\bigcup \mathcal{X}_i$ for $i = 1, \ldots, n$. This means, r satisfies all key sets in Σ. Consequently, Σ does not imply φ. □

A direct consequence of Theorem 4 is the quadratic time complexity of the implication problem for unary by arbitrary key sets. For a collection Σ of key sets let $|\Sigma|$ denote the total number of attribute occurrences in elements of Σ.

Corollary 1. *The implication problem of unary key sets by arbitrary key sets is decidable in time $\mathcal{O}(|\Sigma| \times |\varphi|)$ in the input $\Sigma \cup \{\varphi\}$.*

8.2 A Finite Axiomatization

Our next result establishes a finite axiomatization for the implication of unary by arbitrary key sets that consists of the Refinement and Upward Closure rules. As this fragment is decidable in time quadratic in the input, and the general case is *coNP*-complete, the Composition rule is the source of likely intractability.

Corollary 2. *The implication problem of unary key sets by arbitrary key sets has a sound and complete axiomatization in Refinement and Upward Closure.*

Proof. Let $\Sigma = \{\mathcal{X}_1, \ldots, \mathcal{X}_n\}$ be a set of key sets, and let $\varphi = \{\{A_1\}, \ldots, \{A_k\}\}$ be a unary key set over relation schema R. If φ can be inferred from Σ by a sequence of applications of the Refinement and Upward Closure rules, the soundness of these rules ensures that φ is also implied by Σ.

For completeness we assume that φ cannot be inferred from Σ by means of applications using the Refinement and Upward Closure rules. Hence, $\bigcup \mathcal{X}_i \not\subseteq X$ holds for all $i = 1, \ldots, n$. Theorem 4 shows that Σ does not imply φ. □

8.3 Existence and Computation of Armstrong Relations

Armstrong models relative to unary consequences are also easy to obtain. It merely suffices to take a disjoint union of all of the two tuple relations mentioned in the proof of Theorem 4.

Corollary 3. *The implication problem of unary key sets by arbitrary key sets has Armstrong relations.* □

While the existence of perfect models is easy to come by the disjoint union construction, an actual generation of Armstrong relations by this construction is not efficient. Smaller Armstrong relations can be constructed as follows. Theorem 4 shows that the implication problem of unary key sets $\mathcal{X}$ by a collection $\Sigma = \{\mathcal{X}_1, \ldots, \mathcal{X}_n\}$ of arbitrary key sets only depends on the attributes contained in each given key set of Σ, and not on how they are grouped as sets in a key set. We thus identify, without loss of generality, $\mathcal{X}$ with $\bigcup \mathcal{X}$ and each $\mathcal{X}_i$ with $\bigcup \mathcal{X}_i$.

The idea is then to compute so-called anti-keys, which are the maximal subsets of the underlying relation schema which are key sets not implied by Σ. Given the anti-keys, an Armstrong relation for Σ can be generated by starting with a single complete tuple, and introducing for each anti-key a new tuple that has matching total values on the attributes of the anti-key and unique values on attributes outside the anti-key. This construction ensures that all non-implied (unary) key sets are violated and all given key sets are satisfied. The computation of the anti-keys from Σ can be done by taking the complements of the minimum transversals of the hypergraph formed by the elements of Σ. A transversal for a given set of attribute subsets $\mathcal{X}_i$ is an attribute subset $\mathcal{T}$ such that $\mathcal{T} \cap \mathcal{X}_i \neq \emptyset$ holds for all i. While many efficient algorithms exist for the computation of all hypergraph transversals, it is still an open problem whether there is an algorithm that is polynomial in the output [5]. We can show that this construction always generates an Armstrong relation whose number of tuples is at most quadratic in that of an Armstrong relation that requires a minimum number of tuples.

Corollary 4. *Armstrong relations that are at most quadratic in that of a minimum Armstrong relation can be generated for unary by arbitrary key sets.*

Proof (Sketch). One can show first that a given relation is Armstrong for a given set of key sets if and only if for every anti-key the relation has two tuples which have matching values on exactly those attributes that form the anti-key and for no union over the elements of a key set there is a pair of tuples with matching values on all attributes in the union. Subsequently, one can show that the number of tuples in a minimum-sized Armstrong relation is bounded from below by one half of the square root of 1 plus 8 times the number of anti-keys, and bounded upwards by the increment of the number of anti-keys. Consequently, our construction generates an Armstrong relation that is at most quadratic in a minimum-sized Armstrong relation.   □

Our construction can also be viewed as a construction of Armstrong relations for certain keys by key sets. Note that [8] constructed Armstrong relations for sets of possible and certain keys under NOT NULL constraints, whenever they exist. Our construction here does not require null markers.

Example 5. Consider the set $\Sigma = \{\mathcal{X}_1, \mathcal{X}_2\}$ with $\mathcal{X}_1$ and $\mathcal{X}_2$ from Example 2 over the relation schema WARD. Then $\bigcup \mathcal{X}_1 = \{room, time, injury\}$ and $\bigcup \mathcal{X}_2 = \{name, time, injury\}$. The minimum transversals would be $\mathcal{T}_1 = \{time\}$, $\mathcal{T}_2 = \{injury\}$, and $\mathcal{T}_3 = \{room, name\}$, and their complements on WARD are the anti-keys $\mathcal{A}_1 = \{room, name, address, injury\}$, $\mathcal{A}_2 = \{room, name, address, time\}$, and $\mathcal{A}_3 = \{address, injury, time\}$. The following relation is Armstrong for Σ.

room	name	address	injury	time
1	Miller	24 Queen St	leg fracture	Sunday, 16
1	Miller	24 Queen St	leg fracture	Monday, 19
1	Miller	24 Queen St	arm fracture	Monday, 19
2	Maier	24 Queen St	arm fracture	Monday, 19

The relation satisfies $\mathcal{X}_1$ and $\mathcal{X}_2$, but the relation violates the unary key set $\varphi' = \{\{room\}, \{name\}, \{address\}, \{time\}\}$, so φ' is not implied by Σ.

9   Conclusion and Future Work

We took first steps in investigating limits and opportunities for automated reasoning about key sets in databases. Key sets provide a more general and flexible implementation of entity integrity than Codd's notion of a primary key. We showed that the implication problem for general key sets enjoys a binary axiomatization, is *coNP*-complete, and lacks Armstrong relations. The implication problem of unary key sets by arbitrary key sets enjoys a unary axiomatization, is decidable in quadratic input time, and Armstrong relations can always be generated using hypergraph transversals such that the number of tuples is guaranteed to be at most quadratic in the minimum number of tuples required.

Interesting questions arise in theory and practice. Our *coNP*-completeness result calls for fixed-parameter solutions. A characterization for the existence of Armstrong relations in the general case would be interesting, and their efficient construction whenever possible. The validation of key sets in databases is an important practical issue, for which effective index structures need to be found. The problem of computing all key sets that hold in a given relation is important for data profiling [1]. Automated reasoning about foreign key sets is interesting as they generalize referential integrity [11]. Similar to how functional and inclusion dependencies and independence atoms interact [3,9], automated reasoning for functional, multivalued, and inclusion dependency sets is interesting [6].

References

1. Abedjan, Z., Golab, L., Naumann, F.: Profiling relational data: a survey. VLDB J. **24**(4), 557–581 (2015)
2. Bläsius, T., Friedrich, T., Schirneck, M.: The parameterized complexity of dependency detection in relational databases. In: IPEC 2016, 24–26 August 2016, pp. 6:1–6:13 (2016)
3. Casanova, M.A., Fagin, R., Papadimitriou, C.H.: Inclusion dependencies and their interaction with functional dependencies. J. Comput. Syst. Sci. **28**(1), 29–59 (1984)
4. Codd, E.F.: A relational model of data for large shared data banks. Commun. ACM **13**(6), 377–387 (1970)
5. Eiter, T., Gottlob, G., Makino, K.: New results on monotone dualization and generating hypergraph transversals. SIAM J. Comput. **32**(2), 514–537 (2003)
6. Hannula, M., Kontinen, J., Link, S.: On the finite and general implication problems of independence atoms and keys. J. Comput. Syst. Sci. **82**(5), 856–877 (2016)
7. Hartmann, S., Leck, U., Link, S.: On Codd families of keys over incomplete relations. Comput. J. **54**(7), 1166–1180 (2011)
8. Köhler, H., Leck, U., Link, S., Zhou, X.: Possible and certain keys for SQL. VLDB J. **25**(4), 571–596 (2016)
9. Köhler, H., Link, S.: Inclusion dependencies and their interaction with functional dependencies in SQL. J. Comput. Syst. Sci. **85**, 104–131 (2017)

10. Langeveldt, W., Link, S.: Empirical evidence for the usefulness of Armstrong relations in the acquisition of meaningful functional dependencies. Inf. Syst. **35**(3), 352–374 (2010)
11. Levene, M., Loizou, G.: A generalisation of entity and referential integrity in relational databases. ITA **35**(2), 113–127 (2001)
12. Lucchesi, C.L., Osborn, S.L.: Candidate keys for relations. J. Comput. Syst. Sci. **17**(2), 270–279 (1978)
13. Thalheim, B.: Dependencies in Relational Databases. Springer, Heidelberg (1991). https://doi.org/10.1007/978-3-663-12018-6
14. Thalheim, B.: On semantic issues connected with keys in relational databases permitting null values. Elektronische Informationsverarbeitung und Kybernetik **25**(1/2), 11–20 (1989)
15. Thalheim, B.: The number of keys in relational and nested relational databases. Discret. Appl. Math. **40**(2), 265–282 (1992)
16. Wei, Z., Link, S., Liu, J.: Contextual keys. In: Mayr, H.C., Guizzardi, G., Ma, H., Pastor, O. (eds.) ER 2017. LNCS, vol. 10650, pp. 266–279. Springer, Cham (2017). https://doi.org/10.1007/978-3-319-69904-2_22

A Tableaux Calculus for Reducing Proof Size

Michael Peter Lettmann[1](✉) and Nicolas Peltier[2]

[1] Institute of Logic and Computation, Technische Universität Wien, Vienna, Austria
michael.lettmann@tuwien.ac.at
[2] Univ. Grenoble Alpes, CNRS, Grenoble INP, LIG, 38000 Grenoble, France
Nicolas.Peltier@univ-grenoble-alpes.fr

Abstract. A tableau calculus is proposed, based on a compressed representation of clauses, where literals sharing a similar shape may be merged. The inferences applied on these literals are fused when possible, which can reduce the size of the proof. It is shown that the obtained proof procedure is sound, refutationally complete and can reduce the size of the tableau by an exponential factor. The approach is compatible with all usual refinements of tableaux.

1 Introduction

Tableau methods (see for instance [2] or [6]) always played a crucial role in the development of new techniques for automated theorem proving. They are easy to comprehend and implement, well-adapted to interactive theorem proving, and, therefore, normally form the basis of the first proof procedure for any newly defined logic [4]. Nonetheless, they cannot compete with resolution-based calculi both in terms of efficiency and deductive power (i.e. proof length, see for instance [3]). This is partly due to the ability of resolution-based methods to generate lemmas and to simulate atomic cuts[1] in a feasible way. There have been attempts to integrate some restricted forms of cut into tableau methods, improving both efficiency and proof size (see for instance [6,11]). But, for more general forms of cuts, it is difficult to decide whether an application of the cut rule is useful or not, thus the rule is not really applicable during proof search. Instead, cuts may be introduced after the proof is generated, to make it more compact by introducing lemmas and fusing recurring patterns [8,9].

In this paper, rather than trying to integrate cuts into the tableau calculus, we devise a new tableau procedure in which a proof compression, that is similar to the compressive power of a Π_2-cut, is achieved by employing a shared representation of literals. Formal definitions will be given later, but we now provide a simple example to illustrate our ideas. Consider the schema of clause sets:

M.P. Lettmann—Funded by FWF project W1255-N23.

[1] We recall that the cut rule consists in expanding a tableau by adding two branches with $\neg\phi$ and ϕ respectively, where ϕ is any formula (intuitively ϕ can be viewed as a lemma). A cut is atomic if ϕ is atomic.

D. Galmiche et al. (Eds.): IJCAR 2018, LNAI 10900, pp. 64–80, 2018.
https://doi.org/10.1007/978-3-319-94205-6_5

$\{\bigvee_{i=1}^{n} p_0(a_i), \forall y.\neg p_n(y)\} \cup \{\forall x.\neg p_{i-1}(x) \lor p_i(x) \mid i \in [1,n]\}$. A closed tableau can be constructed by adding n copies of the clauses $\neg p_n(y^j)$ and $\neg p_{i-1}(x_i^j) \lor p_i(x_i^j)$ (for $i, j \in [1,n]$) and unifying all variables x_i^j and y^j with a_j. One gets a tableau of size $O(n^2)$. To make the proof more compact, we may merge the inferences applied for each a_j, since each of these constants are handled in the same way. This can be done by first applying the cut rule on the formula $\exists x.p_0(x)$. The branch corresponding the $\neg\exists x.p_0(x)$ can be closed by using the first clause. In the branch corresponding to $\exists x.p_0(x)$ a constant c is generated by skolemization and the branch can be closed by unifying x_i and y with c. This yields a tableau of size $O(n)$. Since it is hard to guess in advance whether such an application of the cut rule will be useful or not, we investigate another solution allowing the same proof compression. We represent the disjunction $\bigvee_{i=1}^{n} p_0(a_i)$ by a single literal $p_0(\alpha)$, together with a set of substitutions $\{[\alpha \backslash a_i] \mid i \in [1,n]\}$. Intuitively, this literal states that $p_0(\alpha)$ holds for some term α, and the given set of substitutions specifies the possible values of α. In the following, we call such variables α *abstraction variables*. The clauses are kept as compact as possible by grouping all literals with the same heads and in some cases inferences may be performed uniformly regardless of the value of α. In our example we get a tableau of size $O(n)$ by unifying x_i^j and y^j with α, this tableau may be viewed as a compact representation of an ordinary tableau, obtained by making n copies of the tree, with $\alpha = a_1, \ldots, a_n$. If we find out that an inference is applicable only for some specific value(s) of α (e.g., if one wants to close a branch by unifying $p_0(\alpha)$ with a clause $\neg p_0(a_1)$), then one may "separate" the literal by isolating some substitution (or sets of substitutions) before proceeding with the inference.

In this paper, we formalize these ideas into a tableau calculus called *M-tableau*. Basic inference rules are devised to construct M-tableaux and a strategy is provided to apply these rules efficiently, keeping the tableau as compact as possible. We prove that the procedure is sound and refutationally complete and that it may reduce the size of the proofs by an exponential factor. Our approach may be combined with all the usual refinements of the tableau procedure.[2]

2   Notations

We briefly review usual definitions (we refer to, e.g., [13] for details). Terms, atoms and clauses are built as usual over a (finite) set of function symbols Σ (including constants, i.e. nullary function symbols), an (infinite and countable) set of variables $\mathcal{V}$ and a (finite) set of predicate symbols Ω. The set of variables occurring in an expression (term, atom or clause) e is denoted by $\mathcal{V}(e)$. For readability, a term $f(t)$ is sometimes written ft. Ordinary (clausal) tableaux are trees labelled by literals and built by applying Expansion and Closure rules, the Expansion rule expands a leaf by n children labelled by literals $l_1, \ldots, l_n$, where a copy of $l_1 \lor \cdots \lor l_n$ occurs in the clause set at hand, and the Closure rule closes

---

[2] Due to the limited number of pages, we omit proofs that are not necessary for the understanding of the method. A full version can be found in [10].

a branch by unifying the atoms of two complementary literals. A substitution is a function (with finite domain) mapping variables to terms. A substitution mapping x_i to t_i (for $i \in [1, n]$) is written $[(x_1, \ldots, x_n) \backslash (t_1, \ldots, t_n)]$. The identity substitution (for $n = 0$) is denoted by id. The image of an expression e by a substitution σ is defined inductively as usual and written $e\sigma$.

3   A Shared Representation of Literals

We introduce the notion of an M-literal, that is a compact representation of a disjunction of ordinary literals with the same shape. The interest of this representation is that it will allow us to perform similar inferences in parallel on all these literals. We assume that $\mathcal{V}$ is partitioned into two (infinite) sets $\mathcal{V}_o$ and $\mathcal{V}_a$. The variables in $\mathcal{V}_o$ are ordinary variables. They may be either universally quantified variables in clauses, or rigid variables in tableaux. The variables in $\mathcal{V}_a$ are called *abstraction variables*. These are not variables in the standard sense, but can been seen rather as placeholders for a term that may take different values in different literals or branches. These variables will permit to share inferences applied on different literals. The set of ordinary variables (resp. abstraction variables) that occur in a term t is denoted by $\mathcal{V}_o(t)$ (resp. $\mathcal{V}_a(t)$). A *renaming* is an injective substitution σ such that $x \in \mathcal{V}_o \Rightarrow x\sigma \in \mathcal{V}_o$ and $\alpha \in \mathcal{V}_a \Rightarrow \alpha\sigma \in \mathcal{V}_a$.

Definition 1 (Syntax of M-Clauses). *An* M-literal *is either* **true** *or a triple* $\langle L, \bar{t}, \mathcal{S} \rangle$, *where:*

- *L is either a predicate symbol P or the negation of a predicate symbol $\neg P$,*
- *$\bar{t}$ is an n-tuple of terms, where n is the arity of P,*
- *and $\mathcal{S}$ is a set of substitutions σ with the same domain $D \subseteq \mathcal{V}_a(\bar{t})$ (by convention D is empty if $\mathcal{S} = \emptyset$) and such that $\mathcal{V}(\bar{t}\sigma) \cap D = \emptyset$.*

An M-clause *is a set of M-literals, often written as a disjunction.*

With a slight abuse of words, we will call the set D in the above definition the *domain of $\mathcal{S}$* (denoted by $\mathrm{dom}(\mathcal{S})$). The semantics of M-clauses is defined by associating each M-literal with an ordinary clause (or **true**):

Definition 2 (Semantics of M-clauses). *For every M-literal l, we denote by* formula(l) *the formula defined as follows (with the convention that empty disjunctions are equivalent to* **false***):*

$$\mathrm{formula}(\langle L, \bar{t}, \mathcal{S} \rangle) \overset{def}{=} \bigvee_{\theta \in \mathcal{S}} L(\bar{t}\theta)$$
$$\mathrm{formula}(\mathbf{true}) \overset{def}{=} \mathbf{true}$$

For every M-clause C, we denote by formula(C) *the clause $\bigvee_{l \in C}$ formula(l). For every set of M-clauses $\mathcal{C}$, we denote by* formula($\mathcal{C}$) *the formula (in conjunctive normal form) $\bigwedge_{C \in \mathcal{C}}$ formula(C).*

We write $E \simeq E'$ iff formula(E) $=$ formula(E') *(up to the usual properties of $\vee$ and $\wedge$: associativity, commutativity and idempotence).*

Example 1. Let P be a unary predicate, Q be a binary predicate, c be a constant, f be a unary function, x be an ordinary variable, and α, β, γ be abstraction variables. The triples $l_1 = \langle P, \alpha, \{[\alpha \backslash f(c)]\} \rangle$ and

$$l_2 = \langle Q, (\beta, f(\gamma)), \{[(\beta, \gamma) \backslash (f(c), c)], [(\beta, \gamma) \backslash (c, f(c))]\} \rangle$$

are M-literals, and

$$\text{formula}(l_1) = P(f(c))$$
$$\text{formula}(l_2) = Q(f(c), f(c)) \vee Q(c, f(f(c)))$$

The common shape $Q(\cdot, f(\cdot))$ is shared between the two literals in the second clause.

Remark 1. Observe that if $\mathcal{S} = \emptyset$ then $\text{formula}(\langle L, \bar{t}, \mathcal{S} \rangle) = \textbf{false}$, i.e. $\langle L, \bar{t}, \mathcal{S} \rangle$ denotes an empty clause. Moreover, any ordinary literal may be encoded as an M-literal where the set of substitutions is a singleton, e.g., the formula corresponding to $\langle P, (a, x), \{[\alpha \backslash a]\} \rangle$ is $P(a, x)$. Also, an M-literal $\langle L, \bar{t}, \{\sigma\} \rangle$ is always equivalent to $\langle L, \bar{t}\sigma, \{id\} \rangle$.

The application of a substitution σ to an M-literal is defined as follows:

$$(\textbf{true})\sigma \stackrel{\text{def}}{=} \textbf{true}$$
$$\langle L, \bar{t}, \mathcal{S} \rangle \sigma \stackrel{\text{def}}{=} \langle L, \bar{t}\sigma', \{\theta\sigma \mid \theta \in \mathcal{S}\} \rangle$$

where σ' denotes the restriction of σ to the variables not occurring in $\text{dom}(\mathcal{S})$.

Example 2. Let $l = \langle P, (\alpha, x), \{[\alpha \backslash x], [\alpha \backslash y]\} \rangle$ and $\sigma = [x \backslash a]$. Then:

$$l\sigma = \langle P, (\alpha, a), \{[\alpha \backslash a], [\alpha \backslash y]\} \rangle$$

Let $l' = \langle Q, (\alpha), \{[\alpha \backslash a], [\alpha \backslash b]\} \rangle$ and $\theta = [\alpha \backslash a]$. Then $l'\theta = l'$.

Proposition 1. *Let $l = \langle L, \bar{t}, \mathcal{S} \rangle$ be an M-literal. If $\text{dom}(\mathcal{S}) = \emptyset$, then one of the following conditions hold:*

- $\mathcal{S} = \emptyset$ *and* $\text{formula}(l) = \textbf{false}$*;*
- $\mathcal{S} = \{id\}$ *and* $\text{formula}(l) = L(\bar{t})$*.*

Proof. The identity is the only substitution with empty domain.

A given ordinary clause may be represented by many different M-clauses, for instance $P(a) \vee P(b)$ may be represented as $\langle P, (a), \{id\} \rangle \vee \langle P, (b), \{id\} \rangle$ or $\langle P, (\alpha), \{[\alpha \backslash a], [\alpha \backslash b]\} \rangle$, or even $\langle P, (\alpha), \{[\alpha \backslash a], [\alpha \backslash b]\} \rangle \vee \langle Q, (\beta), \emptyset \rangle$. In practice it is preferable to start with a representation in which useless literals are deleted and in which the remaining literals are grouped when possible. This motivates the following:

Definition 3. *An M-clause C is in* bundled normal form *(short: BNF) if it satisfies the following conditions.*

– For every M-literal $\langle L, \bar{t}, \mathcal{S} \rangle \in C$, $\mathcal{S} \neq \emptyset$.
– If **true** $\in C$ then $C = \{\mathbf{true}\}$.
– For all distinct literals $\langle L_1, \bar{t}_1, \mathcal{S}_1 \rangle, \langle L_2, \bar{t}_2, \mathcal{S}_2 \rangle \in C$, L_1 is distinct from L_2.

An M-clause set $\mathcal{C}$ is in BNF if all M-clauses of $\mathcal{C}$ are in BNF.

Example 3. Let

$$l_1 \overset{\text{def}}{=} \langle P, \alpha, \{[\alpha \backslash f(c)]\} \rangle,$$

$$l_2 \overset{\text{def}}{=} \langle P, \beta, \{[\beta \backslash f(c)]\} \rangle,$$

$$l_3 \overset{\text{def}}{=} \langle Q, (\beta, f\gamma), \{[(\beta, \gamma) \backslash (f(c), c)], [(\beta, \gamma) \backslash (c, f(c))]\} \rangle,$$

and

$$l_4 \overset{\text{def}}{=} \langle Q, (\beta, \gamma), \{[(\beta, \gamma) \backslash (f(c), c)], [(\beta, \gamma) \backslash (c, f(c))]\} \rangle$$

be M-literals. The M-clause $\{l_3, l_4\}$ is not in BNF while the M-clauses $\{l_1, l_4\}$ and $\{l_2, l_4\}$ are in BNF.

Definition 4. *An M-clause C is* well-formed *if for all distinct literals $l = \langle L_1, \bar{t}_1, \mathcal{S}_1 \rangle$ and $m = \langle L_2, \bar{t}_2, \mathcal{S}_2 \rangle$ in C, $\mathrm{dom}(\mathcal{S}_1) \cap \mathrm{dom}(\mathcal{S}_2) = \emptyset$.*

Example 4. Consider the two M-clauses $C_1 := \{l_1, l_4\}$ and $C_2 := \{l_2, l_4\}$ of Example 3. C_1 is well-formed, C_2 is not well-formed. By renaming, C_2 can be transformed into C_1.

It is clear that every M-clause can be transformed into an equivalent well-formed M-clause by renaming. In the following, we shall implicitly assume that all the considered M-clauses are well-formed.

Lemma 1. *Let* F *be a formula in conjunctive normal form. Then there is an M-clause set $\mathcal{C}$ in BNF such that formula$(\mathcal{C}) \simeq$ F.*

Example 5. Consider the clause $\{l_3, l_4\}$ of Example 3. It can be written in BNF as $\langle Q, (\beta, \gamma), \mathcal{S} \rangle$, where $\mathcal{S}$ denotes the following set of substitutions:

$$\{[(\beta, \gamma) \backslash (f(c), f(c))], [(\beta, \gamma) \backslash (c, f(f(c)))], [(\beta, \gamma) \backslash (f(c), c)], [(\beta, \gamma) \backslash (c, f(c))]\}$$

4   A Tableaux Calculus for M-Clauses

In this section, we devise a tableaux calculus for refuting sets of M-clauses. This calculus is defined by a set of inference rules, that, given an existing tableau $\mathcal{T}$, allow one to:

1. Expand a branch with new children, by introducing a new copy of an M-clause of the set at hand.

2. Instantiate some of the (rigid) variables occurring in the tableau.
3. Separate shared literals inside an M-clause, so that different inferences can be applied on each of the corresponding branches. The rule can be applied on nodes that are not leaves.

Steps 1 and 2 are standard, but Step 3 is original.

Definition 5 (Pre-Tableau). *A* pre-tableau *is a tree $\mathcal{T}$ where vertices are labelled by M-literals or by* **false**. *We call the direct successors of a node its* children. *The* root *is the (unique) node that is not a child of any node in $\mathcal{T}$ and a* leaf *is a node with no child. A* path *P is a sequence of nodes $(\nu_1, \ldots, \nu_n)$ such that ν_{i+1} is a child of ν_i for $i \in [1, n-1]$. Furthermore, we call ν_1 the* initial node *of P and ν_n the* last node *of P. A* branch *is a path such that the initial node is the root and the last node is a leaf. With a slight abuse of words we say that a branch* contains *an M-literal l if it contains a node labelled by l.*

The descendants *of a node ν are inductively defined as ν and the descendants of the children of ν. The* subtree of root *ν in $\mathcal{T}$ is the subtree consisting of all the descendants of ν, as they appear in $\mathcal{T}$.*

If ν is a non-leaf node with exactly $n > 0$ children $\nu_1, \ldots, \nu_n$ labelled by M-literals $l_1, \ldots, l_n$ respectively, then the formula associated with *ν is defined as: $\bigvee_{i=1}^{n} \text{formula}(l_i)$.*

We say that an M-literal $\langle L, \bar{t}, \mathcal{S} \rangle$ (resp. a node ν labelled by $\langle L, \bar{t}, \mathcal{S} \rangle$) introduces *an abstraction variable α if $\alpha \in \text{dom}(\mathcal{S})$ (as shown in Proposition 2, the abstraction variables are introduced by exactly one M-literal or node in an M-tableau).*

Definition 6. *Let $\mathcal{T}$ be a pre-tableau and σ be a substitution. Then $\mathcal{T}\sigma$ denotes the result of applying σ to all M-literals labelling the nodes of $\mathcal{T}$.*

Definition 7 (Tableau for a Set of M-Clauses). *An M-tableau $\mathcal{T}$ for a set of M-clauses $\mathcal{C}$ is a pre-tableau built inductively by applying the rules Expansion, Instantiation and Separation to an initial tableau containing only one node, labelled by* **true** *(also called the* initial M-literal*).*

In the following, the word "tableau" always refers to an M-tableau, unless specified otherwise (we use the expression "ordinary tableau" for standard ones).

The rules are defined as follows (in each case, $\mathcal{T}$ denotes a previously constructed tableau for a set of M-clauses $\mathcal{C}$).

Expansion Rule. Let λ be a leaf of $\mathcal{T}$, and C be an element of $\mathcal{C}$ not containing **true**. Let C' be a copy of C where all variables that occur also in $\mathcal{T}$ are renamed such that C' share no variable[3] with $\mathcal{T}$. The pre-tableau $\mathcal{T}'$ constructed by adding a new child labelled by l to λ for each $l \in C'$ is a tableau for $\mathcal{C}$.

---

[3] Note that both ordinary and abstraction variables are renamed.

Instantiation Rule. Let t be a term and $x \in \mathcal{V}_o$ such that for all nodes ν, ν', if ν is labelled by an M-literal containing x and ν' introduces an abstraction variable $\alpha \in \mathcal{V}_a(t)$, then ν is a proper descendant of ν'. Then $\mathcal{T}[x\backslash t]$ is a tableau for $\mathcal{C}$.

Remark 2. Observe that if t contains no abstraction variables then the condition always holds, since no node ν' satisfying the above property exists. In practice, the Instantiation rule should of course not be applied with an arbitrary variable and term. Unification will be used instead to find the most general instantiations closing a branch. A formal definition will be given later (see Definition 11).

Example 6. Let

$$\mathcal{T}_1 \overset{\mathrm{def}}{=} \quad \begin{array}{c} \text{true} \\ | \\ \langle \neg P, x, \{id\} \rangle \\ | \\ \langle P, \alpha, \{[\alpha\backslash a], [\alpha\backslash b]\} \rangle \end{array} \qquad\qquad \mathcal{T}_2 \overset{\mathrm{def}}{=} \quad \begin{array}{c} \text{true} \\ | \\ \langle P, \alpha, \{[\alpha\backslash a], [\alpha\backslash b]\} \rangle \\ | \\ \langle \neg P, x, \{id\} \rangle \end{array}$$

be two tableaux for some set of M-clauses $\mathcal{C}$. The pre-tableau $\mathcal{T}_1[x\backslash\alpha]$ is not a tableau, because x is substituted by a term containing an abstraction variable α, and x occurs above the literal introducing α. On the other hand, $\mathcal{T}_2[x\backslash\alpha]$ is a tableau.

Separation Rule. The rule is illustrated in Fig. 1 towards Fig. 3. Let ν be a non-leaf node of $\mathcal{T}$. Let μ be a child of ν, labelled by $l = \langle L, \bar{t}, \mathcal{S} \rangle$. Let $\bar{r} = \bar{t}\theta$ be an instance of $\bar{t}$, with $\mathrm{dom}(\theta) = \mathrm{dom}(\mathcal{S})$ and $\mathcal{V}_a(\bar{t}\theta) \cap \mathrm{dom}(\mathcal{S}) = \emptyset$. Let $\mathcal{S}_1$ be the set of substitutions $\sigma \in \mathcal{S}$ such that there exists a substitution σ' with $\bar{t}\sigma = \bar{r}\sigma'$ and every variable in $\mathrm{dom}(\sigma')$ is an abstraction variable not occurring in $\mathcal{T}$, and let $\mathcal{S}_2 \overset{\mathrm{def}}{=} \mathcal{S} \setminus \mathcal{S}_1$. Assume that $\mathcal{S}_1 \neq \emptyset$. We define the new literal $l' \overset{\mathrm{def}}{=} \langle L, \bar{r}, \{\sigma' \mid \sigma \in \mathcal{S}_1\} \rangle$. The Separation rule is defined as follows:

1. We apply the substitution θ to $\mathcal{T}^4$.
2. We replace the label l of μ by l'.
3. We add a new child to the node ν, labelled by a literal $\langle L, \bar{t}, \mathcal{S}_2 \rangle$.

Observe that if $\mathcal{S}_2 = \emptyset$ then $\mathrm{formula}(\langle L, \bar{t}, \mathcal{S}_2 \rangle) = \textbf{false}$ hence the third step may be omitted, since the added branch is unsatisfiable anyway. The rule does not apply if $\mathcal{S}_1$ is empty.

Example 7. Let

$$l_1 := \langle P, \alpha, \{[\alpha\backslash fc]\} \rangle,$$
$$l_2 := \langle \neg P, \alpha', \{[\alpha'\backslash fc]\} \rangle,$$
$$l_3 := \langle \neg Q, (\beta', \gamma'), \{[(\beta', \gamma')\backslash(fx, y)]\} \rangle, \text{ and}$$
$$l_4 := \langle Q, (\beta, \gamma), \{[(\beta, \gamma)\backslash(fc, c)], [(\beta, \gamma)\backslash(z, fc)], [(\beta, \gamma)\backslash(fz, fc)]\} \rangle$$

be M-literals and $\mathcal{C} = \{\{l_1, l_4\}, \{l_2\}, \{l_3\}\}$ be an M-clause set in BNF. Applying three times the Expansion rule, we can derive the tableau.

---

[4] Actually, due to the above conditions, the variables in $\mathrm{dom}(\theta)$ only occur in the subtree of root μ, hence θ only affects this subtree.

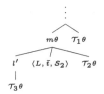

where m is the label of node ν and $\mathcal{T}_1, \mathcal{T}_2$, and $\mathcal{T}_3$ are possibly empty subtrees.

where m is the label of node ν and $\mathcal{T}_1, \mathcal{T}_2$, and $\mathcal{T}_3$ are possibly empty subtrees.

where m is the label of node ν and $\mathcal{T}_1, \mathcal{T}_2$, and $\mathcal{T}_3$ are possibly empty subtrees.

Fig. 1. The initial subtree in the Separation rule.

Fig. 2. The subtree after an application of the Separation rule.

Fig. 3. The subtree without redundant substitutions after an application of the Separation rule.

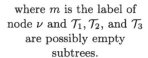

Now, we apply two times the Separation rule. First we choose m of Fig. 1 to be **true** and l to be $\langle \neg P, \alpha', \{[\alpha' \backslash fc]\}\rangle$, where the substitution θ is $[\alpha' \backslash fc]$ (hence we get $\mathcal{S}_1 = \{id\}$). Afterwards, we choose analogously $\langle \neg P, fc, \{id\}\rangle$ (which is the result of the first application) and $\langle P, \alpha, \{[\alpha \backslash fc]\}\rangle$, with the substitution $[\alpha \backslash fc]$. Both times, the tuple $\bar{r}$ of the Separation rule is fc (with $\mathcal{S}_2 = \emptyset$ in both cases). This leads to the tableau:

$$\textbf{true}$$
$$|$$
$$\langle \neg P, fc, \{id\}\rangle$$
$$\langle Q, (\beta, \gamma), \{[(\beta, \gamma)\backslash(fc, c)], [(\beta, \gamma)\backslash(z, fc)], [(\beta, \gamma)\backslash(fz, fc)]\}\rangle \quad \langle P, fc, \{id\}\rangle$$
$$|$$
$$\langle \neg Q, (\beta', \gamma'), \{[(\beta', \gamma')\backslash(fx, y)]\}\rangle$$

Afterwards, we again apply the Separation rule, to modify the node labelled with l_4 where $\bar{r} = (f\delta, \gamma^*)$ and $\mathcal{S}_2 = \{[(\beta, \gamma)\backslash(z, fc)]\}$.

$$\textbf{true}$$
$$|$$
$$\langle \neg P, fc, \{id\}\rangle$$
$$\langle Q, (f\delta, \gamma^*), \{[(\delta, \gamma^*)\backslash(c, c)], [(\delta, \gamma^*)\backslash(z, fc)]\}\rangle \quad \langle Q, (\beta, \gamma), \{[(\beta, \gamma)\backslash(z, fc)]\}\rangle \quad \langle P, fc, \{id\}\rangle$$
$$|$$
$$\langle \neg Q, (\beta', \gamma'), \{[(\beta', \gamma')\backslash(fx, y)]\}\rangle$$

After some further applications of the Separation and Expansion rules, we are able to construct the following tableau by applying the Instantiation rule with the substitutions $[z\backslash fx']$, $[y\backslash fc]$, $[x\backslash \delta]$, $[y\backslash \gamma^*]$.

$$\text{true}$$
$$|$$
$$\langle \neg P, fc, \{id\}\rangle$$

$\langle Q, (f\delta, \gamma^*), \{[(\delta, \gamma^*)\backslash(c, c)], [(\delta, \gamma^*)\backslash(fx', fc)]\}\rangle \qquad \langle Q, (fx', fc), \{id\}\rangle \qquad \langle P, fc, \{id\}\rangle$
$\qquad\qquad | \qquad\qquad\qquad\qquad\qquad\qquad\qquad\qquad\qquad |$
$\qquad \langle \neg Q, (f\delta, \gamma^*), \{id\}\rangle \qquad\qquad\qquad \langle \neg Q, (fx', fc), \{id\}\rangle$

5 Soundness

Definition 8. *Let $\mathcal{T}$ be a pre-tableau or a tableau. A branch B of $\mathcal{T}$ is closed if it contains* **false** *or two nodes labelled by literals $\langle L_1, \bar{t}_1, \mathcal{S}_1\rangle, \langle L_2, \bar{t}_2, \mathcal{S}_2\rangle$ such that $\bar{t}_1 = \bar{t}_2$, $L_1 = P$, $L_2 = \neg P$ for some predicate symbol P. The (pre)-tableau $\mathcal{T}$ is closed iff all branches of $\mathcal{T}$ are closed.*

Example 8. The final tableau of Example 7 contains three branches, i.e.

$$\{\langle \neg P, fc, \emptyset\rangle, \langle Q, (f\delta, \gamma^*), \{[(\delta, \gamma^*)\backslash(c, c)], [(\beta, \gamma)\backslash(c, ffx')]\}\rangle, \langle \neg Q, (f\delta, \gamma^*), \emptyset\rangle\},$$
$$\{\langle \neg P, fc, \emptyset\rangle, \langle Q, (fx', fc), \emptyset\rangle, \langle \neg Q, (fx', fc), \emptyset\rangle\}, \text{ and}$$
$$\{\langle \neg P, fc, \emptyset\rangle, \langle P, fc, \emptyset\rangle\}.$$

All of them are closed and so the tableau is closed. Observe that the inferences closing the branches corresponding to the literals $Q(f(c), c)$ and $Q(f(z), f(c))$ in $\{l_1, l_4\}$ are shared in the constructed tableau (both branches are closed by introduced suitable instances of l_3), whereas the literal $Q(z, f(c))$ is handled separately (by instantiating z by $f(x')$ and using yet another instance of l_3).

Proposition 2. *Let $\mathcal{T}$ be a tableau. If ν_1 and ν_2 are distinct nodes in $\mathcal{T}$, labelled by the M-literals $\langle L_1, \bar{t}_1, \mathcal{S}_1\rangle$ and $\langle L_2, \bar{t}_2, \mathcal{S}_2\rangle$ respectively, then $\mathcal{S}_1$ and $\mathcal{S}_2$ have disjoint domains.*

Proposition 3. *Let $\mathcal{T}$ be a tableau for a set of M-clauses $\mathcal{C}$. For every non-leaf node ν in $\mathcal{T}$, the formula associated with ν (as defined in Definition 5) is an instance of a formula formula(C), where C is a renaming of an M-clause in $\mathcal{C}$.*

Proof. It suffices to show that all the construction rules preserve the desired property.

- **Expansion.** The property immediately holds for the nodes on which the rule is applied, by definition of the rule. The other nodes are not affected.
- **Instantiation.** By definition, the formula associated with a node ν in the final tableau is an instance of the formula associated with ν in the initial one. Thus the property holds.

- **Separation.** The nodes occurring outside of the subtree of root μ are not affected. By definition, the formula associated with the descendants of μ in the new tableau are instances of formulas associated with nodes of the initial tableau. Thus it only remains to consider the node ν. The formula associated with ν in the final tableau is obtained from that of the initial one by removing the formula corresponding to an M-literal $l = \langle L, \bar{t}, \mathcal{S} \rangle$ and replacing it by formula(l') $\vee$ formula$(\langle L, \bar{t}, \mathcal{S}_2 \rangle)$. Since $l' = \langle L, \bar{r}, \{\sigma' \mid \sigma \in \mathcal{S}_1\}\rangle$, we have formula$(l') = \bigvee_{\sigma \in \mathcal{S}_1} L(\bar{r}\sigma') = \bigvee_{\sigma \in \mathcal{S}_1} L(\bar{t}\sigma)$ (since $\bar{t}\sigma = \bar{r}\sigma'$ by definition of the Separation rule). Thus formula(l') $\vee$ formula$(\langle L, \bar{t}, \mathcal{S}_2 \rangle)$ = formula$(\langle L, \bar{t}, \mathcal{S} \rangle)$ and the proof is completed.

Proposition 4. *Let $\mathcal{T}$ be a tableau for $\mathcal{C}$. Let $\alpha \in \mathcal{V}_a$ be a variable introduced in a node ν and assume that α occurs in an M-literal labelling a node μ. Then μ is a descendant of ν.*

Lemma 2. *Let $\mathcal{T}$ be a closed tableau for $\mathcal{C}$ and let $\mathcal{T}'$ be the tableau after applying once the Separation rule to a node ν of $\mathcal{T}$ with a child μ labelled with l. Then there is at most one branch B in $\mathcal{T}'$ that is not closed. Moreover, this branch necessarily contains the node labelled by $\langle L, \bar{t}, \mathcal{S}_2 \rangle$ (see Fig. 3). In particular, if $\mathcal{S}_2$ is empty then $\langle L, \bar{t}, \mathcal{S}_2 \rangle$ is **false** and $\mathcal{T}'$ is closed.*

Theorem 1 (Soundness). *If a set of M-clauses $\mathcal{C}$ admits a closed tableau $\mathcal{T}$ then $\mathcal{C}$ is unsatisfiable.*

Proof. We prove soundness by transforming $\mathcal{T}$ into a closed tableau that contains only M-literals where the substitution set is $\{id\}$ or $\emptyset$. Due to Proposition 3, the resulting tableau then corresponds to an ordinary tableau. The soundness of ordinary tableau then implies the statement.

We transform the tableau $\mathcal{T}$ by an iterative procedure. We always take an arbitrary topmost node ν labelled with an M-literal $l = \langle L, \bar{t}, \mathcal{S} \rangle$ where $\mathcal{S} \neq \{id\}$ and $\mathcal{S} \neq \emptyset$. Then we consider a substitution $\theta \in \mathcal{S}$ and we apply the Separation rule with the tuple $\bar{t}\theta$. We have $\bar{t}\theta = \bar{t}\theta id$ and if $\sigma \in \mathcal{S} \setminus \{\theta\}$, then $\bar{t}\sigma \neq \bar{t}\theta$, hence there is no substitution σ' with $\bar{t}\sigma\sigma' = \bar{t}\theta$, such that dom$(\sigma')$ only contains fresh variables. Consequently, the rule splits $\mathcal{S}$ into a singleton $\mathcal{S}_1 = \{\theta\}$ and $\mathcal{S}_2 = \mathcal{S}_1 \setminus \{\theta\}$. The literal l gets replaced by $l' = \langle L, \bar{t}\theta, \{id\}\rangle$ and we add the node μ labelled with $\langle L, \bar{t}, \mathcal{S}_2 \rangle$. By Lemma 2, there is at most one non-closed branch, i.e. the branch ending with the node μ is the only open branch. We consider a copy $\mathcal{T}_\nu$ of the subtree of root ν in $\mathcal{T}$, renaming all variables introduced in $\mathcal{T}_\nu$ by fresh variables. We replace the root node of $\mathcal{T}_\nu$ by $\langle L, \bar{t}, \mathcal{S}_2 \rangle$ and replace the subtree of root μ in the tableau by $\mathcal{T}_\nu$. It is easy to check that the obtained tableau is a closed tableau for $\mathcal{C}$. Furthermore, the length of the branches does not increase, the number of non-empty substitutions occurring in the M-literals does not increase, and it decreases strictly in l' and $\langle L, \bar{t}, \mathcal{S}_2 \rangle$. This implies that the multiset of multisets $\{|\{\sigma \in \mathcal{S}'_1 \mid \sigma \neq id\}|, \ldots, |\{\sigma \in \mathcal{S}'_n \mid \sigma \neq id\}|\}$ of natural numbers, where $\{\langle L_i, \bar{t}_i, \mathcal{S}'_i \rangle \mid i \in [1, n]\}$ is a branch in $\mathcal{T}$ is strictly decreasing according to the multiset extension of the usual ordering. Since this ordering is well-founded, the process eventually terminates, and after a finite number of applications of this procedure we get a tableau only containing nodes labelled with M-literals whose substitution set is equal to $\{id\}$ or $\emptyset$.

Remark 3. As a by-product of the proof, we get that the size of the minimal ordinary tableau for a clause set formula($\mathcal{C}$) is bounded exponentially by the size of any closed tableau for $\mathcal{C}$. Indeed, we constructed an ordinary tableau from an M-tableau$\mathcal{T}$ in which every branch $(l_1, \ldots, l_n)$ in $\mathcal{T}$ is replaced by (at most) k^n branches, where k is the maximal number of substitutions in l_i. In Sect. 7, we shall prove that this bound is precise, i.e. that our tableau calculus allows exponential reduction of proof size w.r.t. ordinary (cut-free) tableaux.

6   Completeness

Proving completeness of M-tableauis actually a trivial task, since one could always apply the Separation rule in a systematic way on all M-literals to transform them into ordinary literals (as it is done in the proof of Theorem 1), and then get the desired result by completeness of ordinary tableau. However, this strategy would not be of practical use. Instead, we shall devise a strategy that keeps the M-tableau as compact as possible and at the same time allows one to "simulate" any application of the ordinary expansion rules. In this strategy, the Separation rule is applied on demand, i.e. only when it is necessary to close a branch. No hypothesis is assumed on the application of the ordinary expansion rules, therefore the proposed strategy is "orthogonal" to the usual refinements of ordinary tableaux, for instance connection tableaux[5] [12] or hyper-tableaux[6] [1]. Thus our approach can be combined with any refutationally complete tableau procedure [6].

The main idea denoted by *simulate a strategy* is to do the same steps as in ordinary tableau, while keeping M-clauses as compressed as possible. If ordinary tableau expands the tableau by a clause, we expand the tableau with the corresponding M-clause, and if a branch is closed in the ordinary tableau, then the corresponding branch is closed in the M-tableau. This last step is not trivial: Given two ordinary literals $P(\bar{t})$ and $\neg P(\bar{r})$ the ordinary tableau might compute the most general unifier (mgu) of $\bar{t}$ and $\bar{r}$. But in the presented formalism, the two literals might not appear as such, i.e. there are no literals $m = \langle P, \bar{t}, \{id\}\rangle$ and $k = \langle \neg P, \bar{r}, \{id\}\rangle$. In general, there are only M-literals $m' = \langle P, \bar{t}', \mathcal{S}_1\rangle$ and $k' = \langle \neg P, \bar{r}', \mathcal{S}_2\rangle$ such that $\bar{t}'\theta = \bar{t}$ and $\bar{r}'\vartheta = \bar{r}$, where θ and ϑ denote the compositions of the substitutions occurring in the M-literals in the considered branch. Note also that, although $\bar{t}'$ and $\bar{r}'$ are unifiable, the Instantiation rule cannot always be applied to unify them and close the branch. Indeed, the domain of

---

[5] Connection tableaux can be seen as ordinary tableaux in which any application of the Expansion rule must be followed by the closure of a branch, using one of the newly added literals and the previous literal in the branch.

[6] Hyper-tableaux may be viewed in our framework as ordinary tableaux in which the Expansion rule must be followed by the closure of all the newly added branches containing negative literals.

the mgu may contain abstraction variables, whereas the Instantiation rule only handles universal variables. For showing completeness, it would suffice to apply the Separation rule on each ancestor of k' and m' involved in the definition of θ or ϑ, to create a branch where the literals m and k appear explicitly. Thereby, we would lose a lot of the formalisms benefit. Instead, we shall introduce a strategy that uses the Separation rule only if this is necessary for making the unification of $\bar{t}'$ and $\bar{r}'$ feasible (by mean of the Instantiation rule). Such applications of the Separation rule may be seen as preliminary steps for the Instantiation rule. This follows the maxim to stay as general as possible because a more general proof might be more compact.

In the formalisation of the Instantiation rule we ensured soundness by allowing abstraction variables only to occur in descendants of the literal that introduced the variable. This has a drawback to our strategy: The unification process that we try to simulate can ask for an application of the Instantiation rule which would cause a violation of this condition for abstraction variables if we follow the procedure in the former paragraph. We thus have to add further applications of the Separation rule to ensure that this condition is fulfilled.

Definition 9. *Let $B = (\nu_0, \nu_1, \ldots, \nu_n)$ be a path in a tableau $\mathcal{T}$ where ν_0 is the initial node of $\mathcal{T}$ and each node ν_i (with $i > 0$) is labelled by $\langle L_i, \bar{t}_i, \mathcal{S}_i \rangle$.*

An abstraction substitution for B is a substitution $\eta_n \ldots \eta_1$ with $\eta_i \in \mathcal{S}_i$, for $i = 1, \ldots, n$.

A conflict in a branch B is a pair $(\bar{t}_i, \bar{t}_j)$ with $i, j \in [1, n]$, L_i and L_j are dual and $\bar{t}_i$ and $\bar{t}_j$ are unifiable. A conflict is η-realizable if η is an abstraction substitution for B such that $\bar{t}_i \eta$ and $\bar{t}_j \eta$ are unifiable.

In practice, we do not have to check that a conflict is realizable (this would be costly since we have to consider exponentially many substitutions).

If $(\bar{t}, \bar{t}')$ is a conflict then $\bar{t}$ and $\bar{t}'$ are necessarily unifiable, with some mgu θ. As mentioned before, this does not mean that a branch with conflict can be closed. Moreover, according to the restriction on the Instantiation rule, a variable x cannot be instantiated by a term containing an abstraction variable α, if x occurs in some ancestor of the literal introducing α in the tableau. This motivates the following:

Definition 10. *A variable $\alpha \in \mathcal{V}_a$ is blocking for a conflict $(\bar{t}, \bar{t}')$, where $\theta = mgu(\bar{t}, \bar{t}')$ if $\alpha \in \text{dom}(\theta)$ or α occurs in a term $x\theta$, where $x \in \mathcal{V}_o$ and x occurs in a literal labelling an ancestor of the node introducing α.*

Finally, we introduce a specific application of the Separation rule which allows one either to "isolate" some literals in order to ensure that they have a specific "shape" (as specified by a substitution), or to eliminate abstraction variables completely if needed.

Definition 11. *If σ is a substitution, we denote by $\text{dom}_a(\sigma)$ the set of variables $\alpha \in \mathcal{V}_a$ such that $\alpha\sigma \notin \mathcal{V}_a$.*

A tableau is compact if it is constructed by a sequence of applications of the tableau rule in which:

– *The Instantiation rule is applied only if the tableau contains a branch with a conflict $(\bar{t}, \bar{t}')$, with no blocking variable. Each variable $x \in \mathrm{dom}(\sigma)$ is replaced by a term $t\sigma$, where $\sigma = mgu(\bar{t}, \bar{t}')$ (since there is no blocking variable it is easy to check that the conditions on the Instantiation rule are satisfied). Afterwards, it is clear that the branch is closed.*
– *The Separation rule is applied on a node labelled by $\langle L, \bar{t}, \mathcal{S} \rangle$, using a substitution θ only if there exists a conflict $(\bar{t}, \bar{t}')$ with $\sigma = mgu(\bar{t}, \bar{t}')$ such that one of the following conditions holds:*
  1. *$\mathrm{dom}(\mathcal{S})$ contains a blocking variable in $\mathrm{dom}_a(\sigma)$ and θ is defined as follows: $\mathrm{dom}(\theta) = \mathrm{dom}(\mathcal{S}) \cap \mathrm{dom}_a(\sigma)$, $x\theta = x\sigma$ if $x\sigma$ is a variable, otherwise $x\theta$ is obtained from $x\sigma$ by replacing all variables by pairwise distinct fresh abstraction variables.*
  2. *Or $\mathrm{dom}(\mathcal{S})$ contains a blocking variable not occurring in $\mathrm{dom}_a(\sigma)$, and $\theta \in \mathcal{S}$.*

The applications of the Separation rule in Definition 11 are targeted at making the closure of the branch possible by getting rid of blocking variables, while keeping the tableau as compact as possible (thus useless separations are avoided).

Example 9. For instance, assume that we want to close a branch containing two literals $l = \langle L, (\alpha), \{[\alpha \backslash f(a)], [\alpha \backslash f(b)], [\alpha \backslash a] \} \rangle$ and $l' = \langle \neg L, f(x), \{id\} \rangle$. To this aim, we need to ensure that α is unifiable with $f(x)$. This is done by applying the separation rule (Case 1 of Definition 11) with the substitution $[\alpha \backslash f(\beta)]$, so that α has the desired shape. This yields: $\langle L, (f(\beta)), \{[\beta \backslash a], [\beta \backslash b] \} \rangle$ Afterwards, if x does not occur before l in the branch then the branch is closed by unifying x with β. If x occurs before l, then this is not feasible since this would contradict the condition on the Instantiation rule, and we have to apply the Separation rule again (Case 2) to eliminate β, yielding (for instance) $\langle L, (f(a)), \{id\} \rangle$. The direct application of the Separation rule with, e.g., $\theta = [\alpha \backslash f(a)]$ is forbidden in the strategy.

Theorem 2 (Completeness). *Let* C *be a clause set and* $\mathcal{C}$ *be an M-clause set with* formula($\mathcal{C}$) $\simeq$ C. *If* C *is unsatisfiable, then there is a closed compact tableau for* $\mathcal{C}$.

Proof. For an M-clause, $C_i \in \mathcal{C}$, we denote by C_i the ordinary clause in C with formula(C_i) $\simeq$ formula(C_i). W.l.o.g. we can assume that clauses do not appear twice, neither in $\mathcal{C}$ nor in C. Now, we can perform a proof search based on ordinary tableau starting with C (using any complete strategy) yielding a closed tableau, built by applying the usual Expansion and Closure rules. In the following, T will denote an already constructed ordinary tableau for C and $\mathcal{T}$ will represent the corresponding M-tableaufor $\mathcal{C}$. More precisely, the tableau $\mathcal{T}$ is constructed in such a way that there exists an injective mapping h from the nodes in $\mathcal{T}$ to those in T, a function $\nu \mapsto \eta_\nu$ mapping each node ν in $\mathcal{T}$ labelled by some literal $\langle L, \bar{t}, \mathcal{S} \rangle$ to an abstraction substitution η and a substitution ϑ (with $\mathrm{dom}(\vartheta) \subseteq \mathcal{V}_o$) such that the following property holds (denoted by $(\star)$):

1. The root of $\mathcal{T}$ is mapped to the root of T.

2. If ν is a child of ν' then $h(\nu)$ is a child of $h(\nu')$.
3. For any path $(\nu_0, \ldots, \nu_n)$, if ν_n is labelled by an M-literal $\langle L, \bar{t}, \mathcal{S} \rangle$, then $h(\nu_n)$ is labelled by a literal $L(\bar{t}) \eta_{\nu_n} \ldots \eta_{\nu_1} \vartheta$.
4. If a branch of the form $(h(\nu_0), \ldots, h(\nu_n))$ is closed in T, then $(\nu_0, \ldots, \nu_n)$ is closed in $\mathcal{T}$.

Note that the mapping is not surjective in general (T may be bigger than $\mathcal{T}$). By $(\star)$, if T is closed then $\mathcal{T}$ is also closed, which gives us the desired result. It is easy to check that applying the Separation rule on a node ν_i in a branch $(\nu_0, \ldots, \nu_n)$ in $\mathcal{T}$ preserves $(\star)$, provided it is applied using a substitution θ (as defined in the Separation rule) that is more general than $\eta_{\nu_n} \circ \cdots \circ \eta_{\nu_1}$.

The tableau $\mathcal{T}$ is constructed inductively as follows. For the base case, we may take $\mathcal{T} = \text{T} = \textbf{true}$ and $(\star)$ trivially holds.

Expansion: The Expansion rule of ordinary tableaux allows one to expand the tableau by an arbitrary clause C_i of C. We define the corresponding tableau for $\mathcal{C}$ as the tableau expanded by C_i. The mappings h and η_ν may be extended in a straightforward way so that $(\star)$ is preserved (the unique new node ν in $\mathcal{T}$ may be mapped to an arbitrary chosen new node in T).

Closure: Assume that a branch in T is closed by applying some substitution σ, using two literals $L(\bar{t})$ and $\neg L(\bar{r})$, where $\sigma = \text{mgu}(\bar{t}, \bar{r})$. If one of these literals do not occur in the image of a branch of $\mathcal{T}$ then it is clear that the operation preserves $(\star)$ (except that the substitution ϑ is replaced by $\vartheta\sigma$), hence no further transformation is required on $\mathcal{T}$. Otherwise, by $(\star)$, there is a branch $B = (\nu_0, \ldots, \nu_n)$ in $\mathcal{T}$ where for every $i \in [1, n]$, ν_i is labelled by $l_i = \langle L_i, \bar{t}_i', \mathcal{S}_i \rangle$, two numbers $j, k \in \mathbb{N}$ such that $L_j = L$, $L_k = \neg L$, $\bar{t}_j' \eta \vartheta = \bar{t}$ and $\bar{t}_k' \eta \vartheta = \bar{r}$, with $\eta = \eta_{\nu_n} \ldots \eta_{\nu_1}$.

By definition, $(\bar{t}_j', \bar{t}_k')$ is an η-realizable conflict. Let σ' be the mgu of $\bar{t}_j'$ and $\bar{t}_k'$. If there is no blocking variable for $(\bar{t}_j', \bar{t}_k')$, then the Instantiation rule applies, replacing every variable $x \in \text{dom}(\sigma')$ by $x\sigma'$, and the branch may be closed. By definition $\eta\vartheta\sigma$ is a unifier of $\bar{t}$ and $\bar{r}$, hence $\eta\vartheta\sigma = \sigma'\theta'$, for some substitution θ'. By definition, the co-domain of η contains no abstraction variables, thus $\theta' = \eta\vartheta'$, with $\text{dom}(\vartheta') \subseteq \mathcal{V}_o$, and $\vartheta\sigma = \sigma'\vartheta'$. The application of the rule preserves $(\star)$, where ϑ is replaced by the substitution ϑ'. Indeed, consider a node ν in $\mathcal{T}$, initially labelled by a literal $\langle L, \bar{t}, \mathcal{S} \rangle$, where $h(\nu)$ is labelled by $L(\bar{t}) \eta_{\nu_n} \ldots \eta_{\nu_1} \vartheta$. After the rule application, ν is labelled by $\langle L, \bar{t}\sigma', \mathcal{S}\sigma' \rangle$, $h(\nu)$ is labelled by $L(\bar{t}) \eta_{\nu_n} \ldots \eta_{\nu_1} \vartheta\sigma$, and the substitutions η_{ν_i} are replaced by $\eta_{\nu_i}\sigma'$. Since $\text{dom}(\sigma') \cap \mathcal{V}_a = \emptyset$, it is clear that $\sigma' \eta_{\nu_n} \ldots \eta_{\nu_1} \sigma' \vartheta' = \eta_{\nu_n} \ldots \eta_{\nu_1} \sigma' \vartheta' = \eta_{\nu_n} \ldots \eta_{\nu_1} \vartheta\sigma$.

Otherwise, the set of blocking variables is not empty, and since all abstraction variables occurring in the tableau must be introduced in some node, there exists $l \in [1, n]$ such that $\text{dom}(\mathcal{S}_l)$ contains a blocking variable. According to Definition 11, the Separation rule may be applied on ν_l (it is easy to check that all the application conditions of the rule are satisfied). In Case 1 (of Definition 11), θ is more general than σ' by definition, and since $\eta_{\nu_n} \circ \cdots \circ \eta_{\nu_1} \vartheta\sigma$ is a unifier of $\bar{t}_j'$ and $\bar{t}_k'$, the mgu σ' must be more general than $\eta_{\nu_n} \circ \cdots \circ \eta_{\nu_1}$. In Case 2, we can take $\theta = \eta_{\nu_l}$, which is more general than $\eta_{\nu_n} \circ \cdots \circ \eta_{\nu_1}$ (since the substitutions η_ν have disjoint domains). Thus the property $(\star)$ is preserved.

This operation is repeated until the set of blocking variables is empty, which allows us to apply the Instantiation rule as explained before. The process necessarily terminates since each application of the Separation rule either increases the size of the tableau (either by adding new nodes, or by instantiating a variable by a non variable term), or does not increase the size of the tableau but strictly reduces the number of abstraction variables. Furthermore, by $(\star)$, the size of $\mathcal{T}$ is smaller than that of $\mathbf{T}$.

7   An Exponentially Compressed Tableau

In this section, we will show that the presented method is able to compress tableaux by an exponential factor. This corresponds to an introduction of a single Π_2-cut[7] (see [9]). As a simplified measurement of the size of a tableau we consider the number of nodes. Let us consider the schema of M-clause sets $\mathcal{C}^n \overset{\text{def}}{=} \{\{l_1^n\}, \{l_2, l_3\}, \{l_4^n\}\}$ with

$$l_1^n = \langle P, \bar{\alpha}, [\bar{\alpha}\backslash(x, f_1 x)], \ldots, [\bar{\alpha}\backslash(x, f_n x)]\rangle$$
$$l_2 = \langle \neg P, \bar{\beta}, [\bar{\beta}\backslash(x, y)]\rangle$$
$$l_3 = \langle P, \bar{\gamma}, [\bar{\gamma}\backslash(x, fy)]\rangle$$
$$l_4^n = \langle \neg P, \bar{\delta}, [\bar{\delta}\backslash(f x_1, f x_2)], \ldots, [\bar{\gamma}\backslash(f x_{n-1}, f x_n)]\rangle$$

where $\bar{\alpha} = (\alpha_1, \alpha_2), \bar{\beta} = (\beta_1, \beta_2), \bar{\gamma} = (\gamma_1, \gamma_2)$, and $\bar{\delta} = (\delta_1, \delta_2)$ for $n \in \mathbb{N}$. Then we can construct a closed tableau for $\mathcal{C}^n$ whose size is linear w.r.t. n:

The clause set schema $\mathcal{C}^n$ is a simplified variant of the example in [9, Sects. 3 and 9] and one can easily verify that an ordinary tableau method is of exponential size of n. Just consider the term instantiations which are necessary for an ordinary tableau:

---

[7] In a Π_2-cut, the cut formula is of the form $\forall x \exists y A$ where A is a quantifier-free formula.

$x \leftarrow \{fx_1, ff_{i_1}fx_1, \ldots, ff_{i_{n-2}}f \ldots f_{i_1}fx_1 | i_1, \ldots, i_{n-2} \in [1,n]\}$ in l_1^n,

$(x,y) \leftarrow \{(t, f_it) | i \in [1,n] \wedge t$ is a substitution for x in $l_1^n\}$ in l_2 and l_3, and

$(x_1, \ldots, x_n) \leftarrow \{(x_1, f_{i_1}fx_1, \ldots, f_{i_{n-1}}f \ldots f_{i_1}fx_1) | i_1, \ldots, i_{n-1} \in [1,n]\}$ in l_4^n.

Obviously, x_n in l_4^n is substituted with n^{n-1} terms. These correspond to the instantiations defined in [9, Theorem 13].

8 Future Work

From a practical point of view, algorithms and data-structures have to be devised to apply the above rules efficiently, especially to identify conflicts and blocking substitutions in an incremental way. In the wake of this, experimental evaluations based on an implementation are reasonable. While the procedure is described for first-order logic, we believe that the same ideas could be profitably applied to other logics, and even to other calculi, including saturation-based procedures. It would also be interesting to combine this approach with other techniques for reducing proof size, for instance variable splitting [7], or with techniques for the incremental construction of closures [5].

A current restriction of the calculus is that no abstraction variables may occur above their introduction (see the condition on the Instantiation rule in Sect. 4). This restriction is essential for soundness: without it, one could for instance construct a closed tableau for the (satisfiable) set of M-clauses $\{\{l\}, \{l'\}\}$, with $l = \langle L, (x, \alpha), \{[\alpha \backslash a, \alpha \backslash b]\} \rangle$ and $l' = \langle L, (\beta, y), \{[\beta \backslash a, \beta \backslash b]\} \rangle$, by replacing x by α and y by β. We think that this condition can be relaxed by defining an order over the abstraction variables. This would yield a more flexible calculus, thus further reducing proof size. It would be interesting to know whether the exponential bound of Remark 3 still holds for the relaxed calculus. An ambitious long-term goal is to devise extensions of M-tableaux with the same deductive power of cuts, i.e. enabling a non-elementary reduction of proof size.

References

1. Baumgartner, P.: Hyper tableau—the next generation. In: de Swart, H. (ed.) TABLEAUX 1998. LNCS (LNAI), vol. 1397, pp. 60–76. Springer, Heidelberg (1998). https://doi.org/10.1007/3-540-69778-0_14

2. de Rijke, M.: In: D'Agostino, M., Gabbay, D.M., Hähnle, R., Posegga, J. (eds.) Handbook of Tableau Methods (2001). J. Logic, Lang. Inform. **10**(4), 518–523

3. Eder, E.: Relative Complexities of First-Order Logic Calculi. Springer, Heidelberg (1990). https://doi.org/10.1007/978-3-322-84222-0

4. Fitting, M.: First-Order Logic and Automated Theorem Proving. Texts and Monographs in Computer Science. Springer, New York (1990). https://doi.org/10.1007/978-1-4684-0357-2

5. Giese, M.: Incremental closure of free variable tableaux. In: Goré, R., Leitsch, A., Nipkow, T. (eds.) IJCAR 2001. LNCS, vol. 2083, pp. 545–560. Springer, Heidelberg (2001). https://doi.org/10.1007/3-540-45744-5_46

6. Hähnle, R.: Tableaux and related methods. In Robinson, J.A., Voronkov, A. (eds.) Handbook of Automated Reasoning, vol. I, chap. 3, pp. 100–178. Elsevier Science (2001)
7. Hansen, C.M., Antonsen, R., Giese, M., Waaler, A.: Incremental variable splitting. J. Symb. Comput. **47**(9), 1046–1065 (2012)
8. Hetzl, S., Leitsch, A., Reis, G., Weller, D.: Algorithmic introduction of quantified cuts. Theoret. Comput. Sci. **549**, 1–16 (2014)
9. Leitsch, A., Lettmann, M.P.: The problem of Π_2-cut-introduction. Theoret. Comput. Sci. **706**, 83–116 (2018)
10. Lettmann, M.P., Peltier, N.: A tableaux calculus for reducing proof size. In: CoRR, abs/1801.04163 (2018)
11. Letz, R., Mayr, K., Goller, C.: Controlled integration of the cut rule into connection tableau calculi. J. Autom. Reason. **13**(3), 297–337 (1994)
12. Letz, R., Stenz, G.: Model elimination and connection tableau procedures. In: Handbook of Automated Reasoning, pp. 2015–2112. Elsevier Science Publishers B. V., Amsterdam (2001)
13. Robinson, J.A., Voronkov, A. (eds.) Handbook of Automated Reasoning, vol. 2. Elsevier and MIT Press (2001)

FORT 2.0

Franziska Rapp and Aart Middeldorp[(✉)] [iD]

Department of Computer Science, University of Innsbruck, Innsbruck, Austria
{franziska.rapp,aart.middeldorp}@uibk.ac.at

Abstract. FORT is a tool that implements the first-order theory of rewriting for the decidable class of left-linear right-ground rewrite systems. It can be used to decide properties of a given rewrite system and to synthesize rewrite systems that satisfy arbitrary properties expressible in the first-order theory of rewriting. In this paper we report on the extensions that were incorporated in the latest release (2.0) of FORT. These include witness generation for existentially quantified variables in formulas, support for combinations of rewrite systems, as well as an extension to deal with non-ground terms for properties related to confluence.

1 Introduction

In a recent paper [6] we introduced FORT, a decision and synthesis tool for the first-order theory of rewriting induced by finite left-linear right-ground rewrite systems. In this theory one can express well-known properties like termination, normalization, and confluence, but also properties like strong confluence ($\forall s \, \forall t \, \forall u \, (s \to t \wedge s \to u \implies \exists v \, (t \to^= v \wedge u \to^* v))$) and the normal form property ($\forall s \, \forall t \, \forall u \, (s \to t \wedge s \to^! u \implies t \to^! u)$). The decision procedure implemented in FORTis based on tree automata techniques (Dauchet and Tison [3]).

In this paper we present several extensions designed to make the tool more useful. First of all, we added support for combinations of rewrite systems. This is required to express properties like *commutation* ($\forall s \, \forall t \, \forall u \, (s \to_0^* t \wedge s \to_1^* u \implies \exists v \, (t \to_1^* v \wedge u \to_0^* v))$) and *equivalence* ($\forall s \, \forall t \, (s \leftrightarrow_0^* t \iff s \leftrightarrow_1^* t)$) that refer to two or more rewrite systems. Tree automata operate on *ground* terms. Consequently, variables in formulas range over ground terms and hence the properties that FORT is able to decide are restricted to ground terms. Whereas for termination and normalization this makes no difference, for other properties it does, even for the restricted class of *left-linear right-ground rewrite systems* as will be shown below. This brings us to the second extension: How can one use FORT to decide properties on open terms? We show that for properties related to confluence it suffices to add one or two fresh constants. We furthermore provide sufficient conditions which obviate the need for additional constants. The third extension is concerned with increasing the understanding of the yes/no answer provided by FORT in decision mode. For logical formulas with free variables we are not only interested whether they are satisfied by a particular rewrite system,

This research is supported by FWF (Austrian Science Fund) project P30301.
F. Rapp—Currently at Allgemeines Rechenzentrum Innsbruck.

but also which terms act as witnesses. Witness generation is also of interest for existentially quantified variables appearing in formulas.

We assume familiarity with term rewriting. A command-line version of FORT is available.[1] We refer to [6] and the description on the website for the syntax of the commands and formulas that can be passed to FORT.

2 Witness Generation

The usual output of FORT consists of tree automata (or their size) corresponding to subformulas of the given formula, which is hard to read. To help the user in understanding why a property holds or does not hold, we have now implemented witness generation, which provides evidence by generating an n-tuple of ground terms for free variables or (implicitly) existentially quantified ones. For instance, if a given or synthesized TRS is not ground-confluent ($\neg \forall s \forall t \forall u : (s \rightarrow^* t \wedge s \rightarrow^* u \implies \exists v (t \rightarrow^* v \wedge u \rightarrow^* v))$), it is interesting to provide witnessing terms for the variables s, t, and u. Given the TRS consisting of the rules $\mathsf{a} \rightarrow \mathsf{f}(\mathsf{a},\mathsf{b})$ and $\mathsf{f}(\mathsf{a},\mathsf{b}) \rightarrow \mathsf{f}(\mathsf{b},\mathsf{a})$, FORT produces the following terms as witnesses: $s = \mathsf{f}(\mathsf{a},\mathsf{b})$, $t = \mathsf{f}(\mathsf{b},\mathsf{a})$, and $u = \mathsf{f}(\mathsf{f}(\mathsf{a},\mathsf{b}),\mathsf{b})$.

To cope with n-ary relations on terms, FORT uses bottom-up tree automata that operate on encodings of n-tuples of ground terms, subsequently called RR_n automata. Given a signature $\mathcal{F}$ we let $\mathcal{F}^{(n)} = (\mathcal{F} \cup \{\bot\})^n$ with $\bot \notin \mathcal{F}$ a fresh constant. The arity of a symbol $f_1 \cdots f_n \in \mathcal{F}^{(n)}$ is the maximum of the arities of $f_1, \ldots, f_n$. Given terms $t_1, \ldots, t_n \in \mathcal{T}(\mathcal{F})$, the encoding $\langle t_1, \ldots, t_n \rangle \in \mathcal{T}(\mathcal{F}^{(n)})$ is best illustrated on a concrete example. For the ground terms $s = \mathsf{f}(\mathsf{g}(\mathsf{a}),\mathsf{f}(\mathsf{b},\mathsf{b}))$, $t = \mathsf{g}(\mathsf{g}(\mathsf{a}))$, and $u = \mathsf{f}(\mathsf{b},\mathsf{g}(\mathsf{a}))$ we have $\langle s,t,u \rangle = \mathsf{fgf}(\mathsf{ggb}(\mathsf{aa}\bot),\mathsf{f}\bot\mathsf{g}(\mathsf{b}\bot\mathsf{a},\mathsf{b}\bot\bot))$. So for each position occurring in one of the terms, function symbols of all terms are put together, where the fresh symbol $\bot$ is used for missing positions. We refer to [6] for a formal definition.

The recursive algorithm depicted in Fig. 1 generates (encoded) witnesses that reach a given state α of an RR_n automaton. As a side condition, it does not make use of the given set Q_v of visited states to avoid non-termination. In the outer call, $Q_v = \varnothing$. The set $\mathcal{C}$ of candidates contains all transition rules ending in the given state α such that the states in the left-hand side of the rule were not visited before. Furthermore, in the outermost call ($Q_v = \{\alpha\}$) rules having a $\bot$ in their list of function symbols are excluded as well, since they do not produce encodings of terms over the original signature $\mathcal{F}$. Then a rule with minimal number of arguments (to obtain small witnesses) is chosen from $\mathcal{C}$ and the function $\mathtt{find\_terms}$ is called recursively for each argument position to get witnesses for the argument states. This might fail in case the automaton was not normalized beforehand and we end up in non-reachable states, in which case we move on to the next candidate rule from $\mathcal{C}$.

In order to apply this algorithm to generate an n-tuple of terms accepted by an RR_n automaton, one has to call the function $\mathtt{find\_terms}$ with a final state of the automaton and decode the resulting term over $\mathcal{F}^{(n)}$.

---

[1] http://cl-informatik.uibk.ac.at/software/FORT.

Input: • RR_n automaton $\mathcal{A} = (\mathcal{F}^{(n)}, Q, Q_f, \Delta)$

   • state $\alpha \in Q$, set of states $Q_v \subseteq Q$

Output: • term accepted in α not using states in Q_v

$Q_v := Q_v \cup \{\alpha\}$;

$\mathcal{C} := \{\, fs(\alpha_1, \ldots, \alpha_m) \to \alpha \in \Delta \mid \alpha_1, \ldots, \alpha_m \notin Q_v \wedge (Q_v = \{\alpha\} \implies \bot \notin fs) \,\}$;

while $\mathcal{C} \neq \varnothing$ **do**

    select $fs(\alpha_1, \ldots, \alpha_m) \to \alpha \in \mathcal{C}$ with minimal m and remove it from $\mathcal{C}$;

    if $(t_i := \mathtt{find\_terms}(\mathcal{A}, \alpha_i, Q_v)$ for $i = 1, \ldots, m)$ does not fail **then**

        return $fs(t_1, \ldots, t_m)$

od;

fail

Fig. 1. Function `find_terms` for witness generation.

Example 1. Consider the signature $\mathcal{F} = \{\mathsf{a}\colon 0, \mathsf{b}\colon 0, \mathsf{g}\colon 1, \mathsf{f}\colon 2, \mathsf{h}\colon 3\}$ and the RR_2 automaton $\mathcal{A}$ over $\mathcal{F}^{(2)}$ with final state $\checkmark$ and transition rules

$$\begin{array}{llllll}
\mathsf{aa} \to \alpha & \bot\mathsf{a} \to \alpha' & \mathsf{fg}(\alpha, \beta') \to \gamma & \mathsf{fh}(\alpha, \beta, \alpha') \to \checkmark & \mathsf{ff}(\alpha, \gamma) \to \checkmark \\
\mathsf{bb} \to \beta & \mathsf{b}\bot \to \beta' & \mathsf{gf}(\beta, \alpha') \to \gamma & \mathsf{gg}(\gamma) \to \gamma & \mathsf{ff}(\checkmark, \alpha) \to \checkmark
\end{array}$$

We compute $\mathtt{find\_terms}(\mathcal{A}, \checkmark, \varnothing)$. We have $\mathcal{C} = \{\mathsf{fh}(\alpha, \beta, \alpha') \to \checkmark, \mathsf{ff}(\alpha, \gamma) \to \checkmark\}$. Note that $\mathsf{ff}(\checkmark, \alpha) \to \checkmark$ does not belong to $\mathcal{C}$. We select the rule $\mathsf{ff}(\alpha, \gamma) \to \checkmark$ having the least number of arguments. The recursive call $\mathtt{find\_terms}(\mathcal{A}, \alpha, \{\checkmark\})$ returns aa, since $\mathsf{aa} \to \alpha$ is the only rule ending in α. Depending on the selected transition rule ending in γ, after further recursive calls we obtain $\mathsf{fg}(\mathsf{aa}, \mathsf{b}\bot)$ or $\mathsf{gf}(\mathsf{bb}, \bot\mathsf{a})$. The latter term gives rise to $\mathsf{ff}(\mathsf{aa}, \mathsf{gf}(\mathsf{bb}, \bot\mathsf{a}))$, which encodes the pair of witnessing terms $\mathsf{f}(\mathsf{a}, \mathsf{g}(\mathsf{b}))$ and $\mathsf{f}(\mathsf{a}, \mathsf{f}(\mathsf{b}, \mathsf{a}))$.

3   Combinations

Several important properties, like (normalization) equivalence, commutation, and relative termination, refer to two or more TRSs. Inspired by Zantema's work on Carpa [10], we added support for combinations of rewrite systems in FORT 2.0. For instance, the commutation property can be written as

```
forall s, t, u ([0] s ->* t & [1] s ->* u =>
      exists v ([1] t ->* v & [0] u ->* v))
```

in FORT syntax. Here the indices 0 and 1 refer to different TRSs (provided by the user in decision mode). Lists of indices (e.g. [0,2,3]) are also supported, indicating that the subsequent (sub)formula is checked for the union of the TRSs corresponding to the listed indices. If no index is specified, the union of all involved TRSs is taken. We return to commutation in Sect. 6. Here we compare FORT with Carpa. The following task is mentioned in the Carpa distribution.[2]

---

[2] https://www.win.tue.nl/~hzantema/carpa.html.

Example 2. If we want to generate two terminating abstract rewrite systems (ARSs) such that their union is non-terminating, the formula ([0] SN & [1] SN & ~ SN) can be used. The additional requirement that the composition of both relations is a subset of the transitive closure of one of them is expressed as

```
forall s, t, u ([0] s -> t & [1] t -> u => [0] s ->+ u | [1] s ->+ u)
```

FORT synthesizes the following two ARSs satisfying the conjunction of these requirements: $\mathcal{A}_0 = \{a \to b, c \to a\}$ and $\mathcal{A}_1 = \{a \to b, b \to c\}$. Using completely different techniques, the same ARSs are generated by Carpa.

Whereas Carpa is restricted to ARSs, its successor Carpa+ can synthesize TRSs that admit rules of the shape $a \to b$, $a \to f(b)$, and $f(a) \to b$ with exactly one unary function symbol f. The properties supported by Carpa+ are restricted to those that can be encoded into the conjunctive fragment of SMT-LRA (linear real arithmetic). For this reason properties like (local) confluence are only approximated. In Carpa these and many others properties were encoded exactly in SAT, which is possible since the number of different terms (constants) is finite in the case of ARSs.

Small ARSs as in Example 2 are easily synthesized by FORT. Checking the examples from the Carpa website for correctness poses no problem for the decision mode of FORT, but the method does not scale very well in synthesis mode.

4   Properties on Open Terms

Since the decision algorithm implemented in FORT is based on tree automata, variables in formulas range over ground terms and hence the properties that FORT is able to decide are restricted to ground terms. Whereas for properties like termination and normalization (restricted to right-ground rewrite systems) this makes no difference, for most properties it does, even for left-linear right-ground rewrite systems, as illustrated by the following example.

Example 3. The TRS $\mathcal{R}$ consisting of the rewrite rules $a \to b$, $f(x, a) \to b$, and $f(b, b) \to b$ is ground confluent since all ground terms rewrite to the normal form b. However, $\mathcal{R}$ is not confluent as $b \leftarrow f(x, a) \to f(x, b)$ with normal forms b and $f(x, b)$.

In this section we consider the following properties of single TRSs:

$$\begin{aligned}
&\text{CR:}   &&\forall s \, \forall t \, \forall u \, (s \to^* t \wedge s \to^* u \implies t \downarrow u) \\
&\text{WCR:} &&\forall s \, \forall t \, \forall u \, (s \to t \wedge s \to u \implies t \downarrow u) \\
&\text{SCR:} &&\forall s \, \forall t \, \forall u \, (s \to t \wedge s \to u \implies \exists v \, (t \to^= v \wedge u \to^* v)) \\
&\text{UN:}   &&\forall s \, \forall t \, \forall u \, (s \to^! t \wedge s \to^! u \implies t = u) \\
&\text{UNC:} &&\forall t \, \forall u \, (t \leftrightarrow^* u \wedge \mathsf{NF}(t) \wedge \mathsf{NF}(u) \implies t = u) \\
&\text{NFP:} &&\forall s \, \forall t \, \forall u \, (s \to t \wedge s \to^! u \implies t \to^! u)
\end{aligned}$$

The results stated for confluence below apply to commutation as well. Let $\mathfrak{P} = \{\mathsf{CR}, \mathsf{WCR}, \mathsf{SCR}, \mathsf{UN}, \mathsf{UNC}, \mathsf{NFP}\}$. For $P \in \mathfrak{P}$ we write GP to denote the property P restricted to ground terms. Let $\mathfrak{R}$ consist of all pairs $(\mathcal{F}, \mathcal{R})$ where $\mathcal{R}$ is a finite left-linear right-ground TRSs over the finite signature $\mathcal{F}$ (containing at least one constant).

For all properties $P \in \mathfrak{P}$, GP does not imply P. Example 3 gives a counterexample to the implication for all properties except SCR. For SCR the TRS consisting of the rules $\mathsf{a} \to \mathsf{b}$, $\mathsf{b} \to \mathsf{f}(\mathsf{a}, \mathsf{a})$, and $\mathsf{f}(\mathsf{a}, x) \to \mathsf{a}$ can be used. The peak $\mathsf{f}(\mathsf{b}, x) \leftarrow \mathsf{f}(\mathsf{a}, x) \to \mathsf{a}$ cannot be joined using $\to^= \cdot {}^* \! \leftarrow$ but any ground instance of $\mathsf{f}(\mathsf{b}, x)$ can be reached from a. Nevertheless, according to the following result (whose proof can be found in [7]), it is possible to check a property $P \in \mathfrak{P}$ using tree automata techniques.

Lemma 4. *If $(\mathcal{F}, \mathcal{R}) \in \mathfrak{R}$ then*

1. $(\mathcal{F}, \mathcal{R}) \vDash P$ $\quad\Longleftrightarrow\quad$ $(\mathcal{F} \cup \{c\}, \mathcal{R}) \vDash GP$ $\qquad$ *for all $P \in \mathfrak{P} \setminus \{\mathsf{UNC}\}$*
2. $(\mathcal{F}, \mathcal{R}) \vDash \mathsf{UNC}$ $\quad\Longleftrightarrow\quad$ $(\mathcal{F} \cup \{c, c'\}, \mathcal{R}) \vDash \mathsf{GUNC}$

with fresh constants c and c'. $\qquad\qquad\qquad\qquad\qquad\qquad\qquad\qquad\qquad$ $\square$

The following example shows that adding a single fresh constant is not sufficient for UNC.

Example 5. The left-linear right-ground TRS $\mathcal{R}$ consisting of the rules

$$\mathsf{a} \to \mathsf{b} \qquad \mathsf{f}(x, \mathsf{a}) \to \mathsf{f}(\mathsf{b}, \mathsf{b}) \qquad \mathsf{f}(\mathsf{b}, x) \to \mathsf{f}(\mathsf{b}, \mathsf{b}) \qquad \mathsf{f}(\mathsf{f}(x, y), z) \to \mathsf{f}(\mathsf{b}, \mathsf{b})$$

does not satisfy UNC since $\mathsf{f}(x, \mathsf{b}) \leftarrow \mathsf{f}(x, \mathsf{a}) \to \mathsf{f}(\mathsf{b}, \mathsf{b}) \leftarrow \mathsf{f}(y, \mathsf{a}) \to \mathsf{f}(y, \mathsf{b})$ is a conversion between distinct normal forms. Adding a single fresh constant c is not enough to violate GUNC as $\mathsf{f}(c, \mathsf{b})$ is the only ground instance of $\mathsf{f}(x, \mathsf{b})$ that is a normal form. The latter is ensured by the last two rewrite rules. Adding another fresh constant c' solves the issue. FORT 2.0 generates the witnessing terms $\mathsf{f}(\$, \mathsf{b})$ and $\mathsf{f}(\%, \mathsf{b})$: $\mathsf{f}(\$, \mathsf{b}) \leftarrow \mathsf{f}(\$, \mathsf{a}) \to \mathsf{f}(\mathsf{b}, \mathsf{b}) \leftarrow \mathsf{f}(\%, \mathsf{a}) \to \mathsf{f}(\%, \mathsf{b})$. Here $\$$ and $\%$ are the fresh constants added by FORT.

Lemma 4 does not generalize to arbitrary properties that are expressible in the first-order theory of rewriting. Consider for example the formula φ:

$$\neg \exists s \, \exists t \, \exists u \, \forall v \, (v \leftrightarrow^* s \lor v \leftrightarrow^* t \lor v \leftrightarrow^* u)$$

which is satisfied on open terms (with respect to any $(\mathcal{F}, \mathcal{R}) \in \mathfrak{R}$). For the TRS consisting of the rule $\mathsf{f}(x) \to \mathsf{a}$ and two additional constants c and c', φ does not hold for ground terms because every ground term is convertible to a, c or c'.

The following result (whose proof can be found in [7]) shows that for properties in $\mathfrak{P}$ it is not always necessary to add fresh constants. Here a *monadic* signature consists of constants and unary function symbols.

Lemma 6. *Let $(\mathcal{F}, \mathcal{R}) \in \mathfrak{R}$ such that $\mathcal{R}$ is ground or $\mathcal{F}$ is monadic. For all $P \in \mathfrak{P}$, $(\mathcal{F}, \mathcal{R}) \vDash P$ if and only if $(\mathcal{F}, \mathcal{R}) \vDash GP$.* $\qquad\qquad\qquad\qquad$ $\square$

FORT indeed benefits from this optimization. For instance, deciding GCR of Cops #506 whose signature is monadic takes 1.73 s if a fresh constant is added, compared to 0.85 s if Lemma 6 is used.

We now report on some synthesis experiments that we performed in FORT, based on the following diagram which summarizes the relationships between properties P and GP for $P \in \mathfrak{P}$:

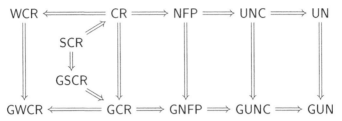

The following TRSs were produced by FORT on the given formulas when restricting the signature (using the command-line option -f "f:2 a:0 b:0") to a binary function symbol f and two constants a and b:

GWCR & $\sim$ WCR & $\sim$ GCR	$a \to b$	$f(x, a) \to a$	$a \to f(a, a)$
GCR & $\sim$ CR & $\sim$ GSCR	$a \to b$	$f(x, a) \to b$	$b \to f(a, a)$
GNFP & $\sim$ NFP & $\sim$ GCR	$a \to b$	$f(x, a) \to f(a, a)$	$f(b, b) \to f(a, a)$
GUNC & $\sim$ UNC & $\sim$ GNFP	$a \to a$	$f(x, a) \to a$	$f(b, x) \to b$

We do not know whether there exist TRSs over the restricted signature that satisfy GUN & $\sim$ UN & $\sim$ GUNC. Human expertise was used to produce a witness over a larger signature, which was subsequently simplified using the decision mode of FORT:

$b \to a$	$c \to c$	$d \to c$	$f(x, a) \to A$	$f(x, A) \to A$
$b \to c$		$d \to e$	$f(x, e) \to A$	$f(c, x) \to A$

FORT produces the following terms as witnesses for the fact that UN is not satisfied: $s = f(d, \$)$, $t = A$, and $u = f(e, \$)$. Indeed both A and $f(e, \$)$ are normal forms reachable from $f(d, \$)$. Moreover, we obtain witnesses $t = a$ and $u = e$ showing that GUNC does not hold. (The rule $c \to c$ is needed to satisfy GUN.)

Since the previous release (1.0) of FORT, many-sorted TRSs are supported. As the set of many-sorted ground terms is accepted by a tree automaton, this extension was mostly straightforward. However, concerning confluence-related properties on non-ground terms, one has to add one (or two for UNC) fresh constant(s) for every sort that variables appearing in the rules can take.

5   Other Extensions

Apart from the extensions detailed before, the efficiency of FORT was improved (in FORT 1.0) using the multithreading features of Java for parallelizing both synthesis and decision. Furthermore, we now also admit variables in right-hand sides of rewrite rules, provided they appear only once in the rule. This extension opens up the possibility of using FORT to compute dependency graphs based on the non-variable approximation for termination analysis [5], check infeasibility of conditional critical pairs for confluence analysis of conditional TRSs [9], and compute needed redexes based on the strong and non-variable approximations for the analysis of optimal normalizing strategies [4]. Even inside FORT this extension was already useful. We previously used $(\twoheadrightarrow_{\mathcal{R}} \cup \twoheadrightarrow_{\mathcal{R}}^{-1})^{+}$ to construct an RR_2 automaton for the conversion relation $(\leftrightarrow^{*})$ but now we can use $\twoheadrightarrow_{\mathcal{R} \cup \mathcal{R}^{-1}}^{+}$ which results in smaller automata for many TRSs. For instance, the conversion relation $\leftrightarrow^{*}$ induced by the TRS

$$\mathsf{g}(\mathsf{f}(\mathsf{a})) \rightarrow \mathsf{f}(\mathsf{g}(\mathsf{f}(\mathsf{a}))) \qquad \mathsf{g}(\mathsf{f}(\mathsf{a})) \rightarrow \mathsf{f}(\mathsf{f}(\mathsf{a})) \qquad \mathsf{f}(\mathsf{f}(\mathsf{a})) \rightarrow \mathsf{f}(\mathsf{a})$$

is modeled as an RR_2 automaton consisting of 118 transitions, down from 427.

6   Experimental Results

In this section we report on the experiments we performed to compare FORT 2.0 with AGCP [1,2] and CoLL [8]. As starting point we consider the 121 left-linear right-ground TRSs in the latest version (765) of the Cops database.[3]

AGCP is a ground confluence tool for many-sorted TRSs based on rewriting induction. In Table 1 we compare FORT and AGCP v0.03 on one-sorted versions of the selected problems. Internally, FORT computes a compatible many-sorted signature with maximal number of sorts, when faced with a ground TRS. This is beneficial to reduce the set of possible ground terms, resulting in smaller automata. We used a 60 s time limit. FORT subsumes AGCP on our collection, with one exception. On Cops #741 AGCP reports "no" whereas FORT does not deliver an answer within 60 s. Increasing the time limit to 150 s enables FORT to report "no" as well.

Table 1. AGCP versus FORT on 121 ground confluence problems.

Tool	Yes	(∅ time)	No	(∅ time)	Maybe	Timeout	Total time
AGCP	24	(0.02 s)	78	(0.06 s)	15	4	276 s
FORT	37	(0.45 s)	81	(0.70 s)	–	3	253 s

Table 2. CoLL versus FORT on 7381 commutation problems.

Tool	Yes	(∅ time)	No	(∅ time)	Maybe	Timeout	Total time
CoLL	623	(0.21 s)	–		276	6482	390682 s
FORT	761	(0.25 s)	6567	(0.60 s)	–	36	6308 s

---

[3] http://cops.uibk.ac.at/.

CoLL is a confluence tool for left-linear TRSs based on commutation and it can establish commutation of multiple TRSs. In Table 2 we compare FORT 2.0 and the latest version[4] of CoLL on $7381 = \binom{121}{2} + 121$ commutation problems stemming from the selected 121 TRSs. To ensure compatibility of the signatures of the separate TRSs, we consistently renamed all function symbols (c_0, c_1, ... for constants, g_0, g_1, ... for unary symbols, etc.). Also for this comparison we used a 60 s time limit. FORT fully subsumes CoLL on our collection.

Detailed results can be obtained from the FORT website. AGCP and CoLL are not restricted to left-linear right-ground TRSs, but we believe that our research can help to make these (and other) tools stronger.

Acknowledgments. We are grateful to Bertram Felgenhauer, Nao Hirokawa, and Julian Nagele for their support with the experiments. The comments by the anonymous reviewers helped to improve the presentation.

References

1. Aoto, T., Toyama, Y.: Ground confluence prover based on rewriting induction. In: Proceedings 1st FSCD. LIPIcs, vol. 52, pp. 33:1–33:12 (2016). https://doi.org/10.4230/LIPIcs.FSCD.2016.33

2. Aoto, T., Toyama, Y., Kimura, Y.: Improving rewriting induction approach for proving ground confluence. In: Proceedings 2nd FSCD. LIPIcs, vol. 84, pp. 7:1–7:18 (2017). https://doi.org/10.4230/LIPIcs.FSCD.2017.7

3. Dauchet, M., Tison, S.: The theory of ground rewrite systems is decidable. In: Proceedings 5th LICS, pp. 242–248 (1990). https://doi.org/10.1109/LICS.1990.113750

4. Durand, I., Middeldorp, A.: Decidable call-by-need computations in term rewriting. Inf. Comput. **196**(2), 95–126 (2005). https://doi.org/10.1016/j.ic.2004.10.003

5. Middeldorp, A.: Approximating dependency graphs using tree automata techniques. In: Proceedings 1st IJCAR. LNAI, vol. 2083, pp. 593–610 (2001). https://doi.org/10.1007/3-540-45744-5_49

6. Rapp, F., Middeldorp, A.: Automating the first-order theory of left-linear right-ground term rewrite systems. In: Proceedings 1st FSCD. LIPIcs, vol. 52, pp. 36:1–36:12 (2016). https://doi.org/10.4230/LIPIcs.FSCD.2016.36

7. Rapp, F., Middeldorp, A.: Confluence properties on open terms in the first-order theory of rewriting. In: Proceedings 5th IWC, pp. 26–30 (2016)

8. Shintani, K., Hirokawa, N.: CoLL: a confluence tool for left-linear term rewrite systems. In: Felty, A.P., Middeldorp, A. (eds.) CADE 2015. LNCS (LNAI), vol. 9195, pp. 127–136. Springer, Cham (2015). https://doi.org/10.1007/978-3-319-21401-6_8

9. Sternagel, C., Sternagel, T.: Certifying confluence of almost orthogonal CTRSs via exact tree automata completion. In: Proceedings 1st FSCD. LIPIcs, vol. 52, pp. 29:1–29:16 (2016). https://doi.org/10.4230/LIPIcs.FSCD.2016.29

10. Zantema, H.: Automatically finding non-confluent examples in term rewriting. In: Proceedings 2nd IWC, pp. 11–15 (2013)

---

[4] Version 1.2 released on January 17, 2018; comparing FORT with version 1.1 of CoLL brought several bugs to light in the latter.

Formalizing Bachmair and Ganzinger's Ordered Resolution Prover

Anders Schlichtkrull[1]([⊠]), Jasmin Christian Blanchette[2,3], Dmitriy Traytel[4], and Uwe Waldmann[3]

[1] DTU Compute, Technical University of Denmark, Kongens Lyngby, Denmark
andschl@dtu.dk
[2] Vrije Universiteit Amsterdam, Amsterdam, The Netherlands
[3] Max-Planck-Institut für Informatik, Saarland Informatics Campus,
Saarbrücken, Germany
[4] Institute of Information Security, Department of Computer Science, ETH Zürich,
Zurich, Switzerland

Abstract. We present a formalization of the first half of Bachmair and Ganzinger's chapter on resolution theorem proving in Isabelle/HOL, culminating with a refutationally complete first-order prover based on ordered resolution with literal selection. We develop general infrastructure and methodology that can form the basis of completeness proofs for related calculi, including superposition. Our work clarifies several of the fine points in the chapter's text, emphasizing the value of formal proofs in the field of automated reasoning.

1 Introduction

Much research in automated reasoning amounts to metatheoretical arguments, typically about the soundness and completeness of logical inference systems or the termination of theorem proving processes. Often the proofs contain more insights than the systems or processes themselves. For example, the superposition calculus rules [2], with their many side conditions, look rather arbitrary, whereas in the completeness proof the side conditions emerge naturally from the model construction. And yet, despite being crucial to our field, today such proofs are usually carried out without tool support beyond TeX.

We believe proof assistants are becoming mature enough to help. In this paper, we present a formalization, developed using the Isabelle/HOL system [16], of a first-order prover based on ordered resolution with literal selection. We follow Bachmair and Ganzinger's account [3] from Chap. 2 of the *Handbook of Automated Reasoning*, which we will simply refer to as "the chapter." Our formal development covers the refutational completeness of two resolution calculi for ground (i.e., variable-free) clauses and general infrastructure for theorem proving processes and redundancy, culminating with a completeness proof for a first-order prover expressed as transition rules operating on triples of clause sets. This material corresponds to the chapter's first four sections.

© Springer International Publishing AG, part of Springer Nature 2018
D. Galmiche et al. (Eds.): IJCAR 2018, LNAI 10900, pp. 89–107, 2018.
https://doi.org/10.1007/978-3-319-94205-6_7

From the perspective of automated reasoning, increased trustworthiness of the results is an obvious benefit of formal proofs. But formalizing also helps clarify arguments, by exposing and explaining difficult steps. Making theorem statements (including definitions and hypotheses) precise can be a huge gain for communicating results. Moreover, a formal proof can tell us exactly where hypotheses and lemmas are used. Once we have created a library of basic results and a methodology, we will be in a good position to study extensions and variants. Given that automatic theorem provers are integrated in modern proof assistants, there is also an undeniable thrill in applying these tools to reason about their own metatheory. From the perspective of interactive theorem proving, formalization work constitutes a case study in the use of a proof assistant. It gives us, as developers and users of such a system, an opportunity to experiment, contribute to lemma libraries, and get inspiration for new features and improvements.

Our motivation for choosing Bachmair and Ganzinger's chapter is manyfold. The text is a standard introduction to superposition-like calculi (together with *Handbook* Chaps. 7 [14] and 27 [26]). It offers perhaps the most detailed treatment of the lifting of a resolution-style calculus's static completeness to a saturation prover's dynamic completeness. It introduces a considerable amount of general infrastructure, including different types of inference systems (sound, reductive, counterexample-reducing, etc.), theorem proving processes, and an abstract notion of redundancy. The resolution calculus, extended with a term order and literal selection, captures most of the insights underlying ordered paramodulation and superposition, but with a simple notion of model.

The chapter's level of rigor is uneven, as shown by the errors and imprecisions revealed by our formalization. We will see that the main completeness result does not hold, due to the improper treatment of self-inferences. Naturally, our objective is not to diminish Bachmair and Ganzinger's outstanding achievements, which include the development of superposition; rather, it is to demonstrate that even the work of some of the most celebrated researchers in our field can benefit from formalization. Our view is that formal proofs can be used to complement and improve their informal counterparts.

This work is part of the IsaFoL (Isabelle Formalization of Logic) project,[1] which aims at developing a library of results about logical calculi. The Isabelle files are available in the *Archive of Formal Proofs* (AFP).[2] They amount to about 8000 lines of source text. Below we provide implicit hyperlinks from theory names. A better way to study the theory files, however, is to open them in Isabelle/jEdit [28]. We used Isabelle version 2017, but the AFP is continuously updated to track Isabelle's evolution. Due to lack of space, we assume the reader has some familiarity with the chapter's content. An extended version of this paper is available as a technical report [21].

---

[1] https://bitbucket.org/isafol/isafol/wiki/Home
[2] https://devel.isa-afp.org/entries/Ordered_Resolution_Prover.html

2   Preliminaries

Ordered resolution depends on little background metatheory. Much of it, concerning partial and total orders, well-foundedness, and finite multisets, is provided by standard Isabelle libraries. We also need literals, clauses, models, terms, and substitutions.

Clauses and Models. We use the same library of clauses (Clausal_Logic.thy) as for the verified SAT solver by Blanchette et al. [6], which is also part of IsaFoL. Atoms are represented by a type variable $'a$, which can be instantiated by arbitrary concrete types—e.g., numbers or first-order terms. A literal, of type $'a$ *literal* (where the type constructor is written in ML-style postfix syntax), can be of the form Pos A or Neg A, where $A :: 'a$ is an atom. The literal order $>$ extends a fixed atom order $>$ by comparing polarities to break ties, with Neg $A >$ Pos A. A clause is a finite multiset of literals, $'a$ *clause* $= 'a$ *literal multiset*, where *multiset* is the Isabelle type constructor of finite multisets. Thus, the clause $A \lor B$, where A and B are atoms, is identified with the multiset $\{A, B\}$; the clause $C \lor D$, where C and D are clauses, is $C \uplus D$; and the empty clause $\perp$ is $\{\}$. The clause order is the multiset extension of the literal order.

A Herbrand interpretation I is a value of type $'a$ *set*, specifying which ground atoms are true (Herbrand_Interpretation.thy). The "models" operator $\vDash$ is defined on atoms, literals, clauses, sets, and multisets of clauses; for example, $I \vDash C \Leftrightarrow \exists L \in C. I \vDash L$. Satisfiability of a set or multiset of clauses N is defined by sat $N \Leftrightarrow \exists I. I \vDash N$.

Multisets are central to our development. Isabelle provides a multiset library, but it is much less developed than those of sets and lists. As part of IsaFoL, we have already extended it considerably and implemented further additions in a separate file (Multiset_More.thy). Some of these, notably a plugin for Isabelle's simplifier to apply cancellation laws, are described in a recent paper [7, Sect. 3].

Terms and Substitutions. The IsaFoR (Isabelle Formalization of Rewriting) library—an inspiration for IsaFoL—contains a definition of first-order terms and results about substitutions and unification [23]. It makes sense to reuse this functionality. A practical issue is that most of IsaFoR is not accessible from the AFP.

Resolution depends only on basic properties of terms and atoms, such as the existence of most general unifiers (MGUs). We exploit this to keep the development parameterized by a type of atoms $'a$ and an abstract type of substitutions $'s$, through Isabelle locales [4] (Abstract_Substitution.thy). A locale represents a module parameterized by types and terms that satisfy some assumptions. Inside the locale, we can refer to the parameters and assumptions in definitions, lemmas, and proofs. The basic operations provided by our locale are application ($\cdot ::$ $'a \Rightarrow 's \Rightarrow 'a$), identity (id $:: 's$), and composition ($\circ :: 's \Rightarrow 's \Rightarrow 's$), about which some assumptions are made (e.g., $A \cdot \text{id} = A$). Substitution is lifted to literals, clauses, sets of clauses, and so on. Many other operations can be defined in terms of the primitives—for example, is_ground $A \Leftrightarrow \forall \sigma. A = A \cdot \sigma$.

To complete our development and ensure that our assumptions are legitimate, we instantiate the locale's parameters with IsaFoR types and operations and discharge its assumptions (IsaFoR_Term.thy). This bridge is currently hosted outside the AFP.

3   Refutational Inference Systems

In their Sect. 2.4, Bachmair and Ganzinger introduce basic conventions for refutational inference systems. In Sect. 3, they present two ground resolution calculi and prove them refutationally complete in Theorems 3.9 and 3.16. In Sect. 4.2, they introduce a notion of counterexample-reducing inference system and state Theorem 4.4 as a generalization of Theorems 3.9 and 3.16 to all such systems. For formalization, two courses of actions suggest themselves: follow the book closely and prove the three theorems separately, or focus on the most general result. We choose the latter, as being more consistent with the goal of providing a well-designed, reusable library.

We collect the abstract hierarchy of inference systems in a single Isabelle theory file (Inference_System.thy). An inference, of type $'a$ $inference$, is a triple $(\mathcal{C}, D, E)$ that consists of a multiset of side premises $\mathcal{C}$, a main premise D, and a conclusion E. An inference system, or calculus, is a possibly infinite set of inferences:

locale $inference\_system$ = **fixes** $\Gamma :: {}'a$ $inference$ set

We use an Isabelle locale to fix, within a named context ($inference\_system$), a set Γ of inferences between clauses over atom type $'a$. Inside the locale, we define a function infers_from that, given a clause set N, returns the subset of Γ inferences whose premises all belong to N. A satisfiability-preserving (or consistency-preserving) inference system enriches the inference system locale with an assumption, whereas sound systems are characterized by a different assumption:

locale $sat\_preserving\_inference\_system$ = $inference\_system$ +
    assumes sat $N \Longrightarrow$ sat $(N \cup$ concl_of ' infers_from $N)$
locale $sound\_inference\_system$ = $inference\_system$ +
    assumes $(\mathcal{C}, D, E) \in \Gamma \Longrightarrow I \vDash \mathcal{C} \cup \{D\} \Longrightarrow I \vDash E$

The notation $f'X$ above stands for the image of the set or multiset X under function f.

Soundness is a stronger requirement than satisfiability preservation. In Isabelle:

sublocale $sound\_inference\_system$ < $sat\_preserving\_inference\_system$

This command emits a proof goal stating that $sound\_inference\_system$'s assumption implies $sat\_preserving\_inference\_system$'s. Afterwards, all the definitions and lemmas about satisfiability-preserving calculi become available about sound ones.

In reductive inference systems ($reductive\_inference\_system$), the conclusion of each inference is smaller than the main premise according to the clause order. A related notion, the counterexample-reducing inference systems, is specified as follows:

locale *counterex_reducing_inference_system* = *inference_system* +
  fixes l_of :: $'a$ *clause set* $\Rightarrow$ $'a$ *set*
  assumes $\{\} \notin N \Longrightarrow D \in N \Longrightarrow$ l_of $N \nvDash D \Longrightarrow$
  $(\forall C \in N.\ $ l_of $N \nvDash C \Longrightarrow D \leq C) \Longrightarrow$
  $\exists C \subseteq N.\ \exists E.\ $ l_of $N \vDash C \wedge (C, D, E) \in \Gamma \wedge$ l_of $N \nvDash E \wedge E < D$

The "model functor" parameter l_of maps clause sets to candidate models. The assumption is that for any set N that does not contain $\{\}$ (i.e., $\bot$), if $D \in N$ is the smallest counterexample—the smallest clause in N falsified by l_of N—we can derive a smaller counterexample E using an inference from clauses in N. This property is useful because if N is saturated (i.e., closed under Γ), we must have $E \in N$, violating D's minimality:

theorem *saturated_model*: saturated $N \Longrightarrow \{\} \notin N \Longrightarrow$ l_of $N \vDash N$
corollary *saturated_complete*: saturated $N \Longrightarrow \neg$ sat $N \Longrightarrow \{\} \in N$

    Bachmair and Ganzinger claim that compactness of clausal logic follows from the refutational completeness of ground resolution (Theorem 3.12), although they give no justification. Our argument relies on an inductive definition of saturation of a set of clauses: saturate :: $'a$ *clause set* $\Rightarrow$ $'a$ *clause set*. Most of the work goes into proving this key lemma, by rule induction on the saturate function:

lemma *saturate_finite*:
  $C \in$ saturate $N \Longrightarrow \exists M \subseteq N.$ finite $M \wedge C \in$ saturate M

The interesting case is when $C = \bot$. We establish compactness in a locale that combines *counterex_reducing_inference_system* and *sound_inference_system*:

theorem *clausal_logic_compact*: $\neg$ sat $N \Leftrightarrow \exists M \subseteq N.$ finite $M \wedge \neg$ sat M

4   Ground Resolution

A useful strategy for establishing properties of first-order calculi is to initially restrict our attention to ground calculi and then to lift the results to first-order formulas containing terms with variables. Accordingly, the chapter's Sect. 3 presents two ground calculi: a simple binary resolution calculus and an ordered resolution calculus with literal selection. Both consist of a single resolution rule, with built-in positive factorization. Most of the explanations and proofs concern the simpler calculus. To avoid duplication, we factor out the candidate model construction (Ground_Resolution_Model.thy). We then define the two calculi and prove that they are sound and reduce counterexamples (Unordered_Ground_Resolution.thy, Ordered_Ground_Resolution.thy).

Candidate Models. Refutational completeness is proved by exhibiting a model for any saturated clause set N that does not contain $\bot$. The model is constructed incrementally, one clause $C \in N$ at a time, starting with an empty Herbrand interpretation. The idea appears to have originated with Brand [10] and Zhang and Kapur [29].

Bachmair and Ganzinger introduce two operators to build the candidate model: I_C denotes the current interpretation before considering C, and ε_C denotes the set of (zero or one) atoms added, or *produced*, to ensure that C is satisfied. The candidate model construction is parameterized by a literal selection function $S :: {}'a\ clause \Rightarrow {}'a\ clause$. We also fix a clause set N. Then we define two operators corresponding to ε_C and I_C:

> **function** production :: ${}'a\ clause \Rightarrow {}'a\ set$ **where**
>   production $C = \{A \mid C \in N \land C \neq \{\} \land \mathsf{Max}\ C = \mathsf{Pos}\ A$
> $\land\ (\bigcup_{D<C} \mathsf{production}\ D) \not\vDash C \land S\ C = \{\}\}$
> **definition** interp :: ${}'a\ clause \Rightarrow {}'a\ set$ **where**
>   interp $C = \bigcup_{D<C} \mathsf{production}\ D$

To ensure monotonicity of the construction, any produced atom must be maximal in its clause. Moreover, productive clauses may not contain selected literals. In the chapter, ε_C and I_C are expressed in terms of each other. We simplified the definition by inlining I_C in ε_C, so that only ε_C is recursive. Since the recursive calls operate on clauses D that are smaller with respect to a well-founded order, the definition is accepted. Bachmair and Ganzinger's I^C and I_N operators are introduced as abbreviations: $\mathsf{Interp}\ C = \mathsf{interp}\ C \cup \mathsf{production}\ C$ and $\mathsf{INTERP} = \bigcup_{C \in N} \mathsf{production}\ C$.

We then prove a host of lemmas about these concepts. Lemma 3.4 amounts to six monotonicity properties, including these:

> **lemma** *interp_imp_Interp*:
> $C \leq D \Longrightarrow D \leq D' \Longrightarrow \mathsf{interp}\ D \vDash C \Longrightarrow \mathsf{Interp}\ D' \vDash C$
> **lemma** *Interp_imp_INTERP*: $C \leq D \Longrightarrow \mathsf{Interp}\ D \vDash C \Longrightarrow \mathsf{INTERP} \vDash C$

Lemma 3.3, whose proof depends on monotonicity, is better proved *after* 3.4:

> **lemma** *productive_imp_INTERP*: production $C \neq \{\} \Longrightarrow \mathsf{INTERP} \vDash C$

A more serious oddity is Lemma 3.7. Using our notations, it can be stated as $D \in N \Longrightarrow C \neq D \Longrightarrow (\forall D' < D.\ \mathsf{Interp}\ D' \vDash C) \Longrightarrow \mathsf{interp}\ D \vDash D'$. However, the last occurrence of D' is clearly wrong—the context suggests C instead. Even after this amendment, we have a counterexample, corresponding to a gap in the proof: $D = \{\}$, $C = \{\mathsf{Pos}\ A\}$, and $N = \{D, C\}$. Since this "lemma" is not actually used, we can simply ignore it.

Unordered Resolution. The unordered ground resolution calculus consists of a single binary inference rule, with the side premise $C \lor A \lor \cdots \lor A$, the main premise $\neg\ A \lor D$, and the conclusion $C \lor D$. Formally, this rule is captured by a predicate:

> **inductive** unord_resolve :: ${}'a\ clause \Rightarrow {}'a\ clause \Rightarrow {}'a\ clause \Rightarrow bool$ **where**
>   unord_resolve $(C \uplus \mathsf{replicate}\ (n+1)\ (\mathsf{Pos}\ A))\ (\{\mathsf{Neg}\ A\} \uplus D)\ (C \uplus D)$

To prove completeness, it suffices to show that the calculus reduces counterexamples (Theorem 3.8). By instantiating the *sound_inference_system* and *counterex_* *reducing_inference_system* locales, we obtain refutational completeness (Theorem 3.9 and Corollary 3.10) and compactness of clausal logic (Theorem 3.12).

Ordered Resolution with Selection. Ordered ground resolution consists of a single rule, ord_resolve. Like unord_resolve, it is sound and counterexample-reducing (Theorem 3.15). Moreover, it is reductive (Lemma 3.13): the conclusion is always smaller than the main premise according to the clause order. The rule is given as

$$\frac{C_1 \lor A_1 \lor \cdots \lor A_1 \quad \cdots \quad C_n \lor A_n \lor \cdots \lor A_n \quad \neg A_1 \lor \cdots \lor \neg A_n \lor D}{C_1 \lor \cdots \lor C_n \lor D}$$

with multiple side conditions whose role is to prune the search space and to make the rule reductive. In Isabelle, we represent the n side premises by three parallel lists of length n: CAs gives the entire clauses, whereas Cs and $\mathcal{A}s$ store the C_i and the $\mathcal{A}_i = A_i \lor \cdots \lor A_i$ parts separately. In addition, As is the list $[A_1, \ldots, A_n]$. The following inductive definition captures the rule formally:

> **inductive** ord_resolve :: *'a clause list* $\Rightarrow$ *'a clause* $\Rightarrow$ *'a clause* $\Rightarrow$ *bool*
> **where**
> $|CAs| = n \Longrightarrow |Cs| = n \Longrightarrow |\mathcal{A}s| = n \Longrightarrow |As| = n \Longrightarrow n \neq 0 \Longrightarrow$
> $(\forall i < n.\ CAs\,!\,i = Cs\,!\,i \uplus \mathsf{Pos}\,'\,\mathcal{A}s\,!\,i) \Longrightarrow (\forall i < n.\ \mathcal{A}s\,!\,i \neq \{\}) \Longrightarrow$
> $(\forall i < n.\ \forall A \in \mathcal{A}s\,!\,i.\ A = As\,!\,i) \Longrightarrow$ eligible $As\ (D \uplus \mathsf{Neg}\,'\, \mathsf{mset}\ As) \Longrightarrow$
> $(\forall i < n.\ \mathsf{strict\_max\_in}\ (As\,!\,i)\ (Cs\,!\,i)) \Longrightarrow (\forall i < n.\ S\ (CAs\,!\,i) = \{\}) \Longrightarrow$
> ord_resolve $CAs\ (D \uplus \mathsf{Neg}\,'\,\mathsf{mset}\ As)\ ((\bigcup \mathsf{mset}\ Cs) \uplus D)$

The $xs\,!\,i$ operator returns the $(i + 1)$st element of xs, and mset converts a list to a multiset. Initially, we tried storing the n premises in a multiset, since their order is irrelevant. However, due to the permutative nature of multisets, there can be no such things as "parallel multisets"; the alternative, a single multiset of tuples, is very unwieldy.

Formalization revealed an error and a few ambiguities in the rule's statement. References to $S(D)$ in the side conditions should have been to $S(\neg A_1 \lor \cdots \lor \neg A_n \lor D)$. The ambiguities are discussed in our technical report [21, Appendix A].

5 Theorem Proving Processes

In their Sect. 4, Bachmair and Ganzinger switch to a dynamic view of saturation: from clause sets closed under inferences to theorem proving processes that start with a clause set N_0 and keep deriving new clauses until no inferences are possible. Redundant clauses can be deleted at any point, and redundant inferences need not be performed.

A derivation performed by a proving process is a possibly infinite sequence $N_0 \vartriangleright N_1 \vartriangleright N_2 \vartriangleright \cdots$, where $\vartriangleright$ relates clause sets (Proving_Process.thy). In Isabelle, such sequences are captured by lazy lists, a codatatype [5] generated by LNil :: *'a llist* and LCons :: *'a* $\Rightarrow$ *'a llist* $\Rightarrow$ *'a llist*, and equipped with lhd ("head") and ltl ("tail") selectors that extract LCons's arguments. The coinductive predicate chain checks that its argument is a nonempty lazy list whose elements are linked by a binary predicate R:

> **coinductive** chain :: $('a \Rightarrow 'a \Rightarrow bool) \Rightarrow 'a \; llist \Rightarrow bool$ **where**
>   chain R (LCons x LNil)
> | chain $R \; xs \Longrightarrow R \; x$ (lhd xs) $\Longrightarrow$ chain R (LCons $x \; xs$)

A derivation is a lazy list Ns of clause sets satisfying the chain predicate with $R = \rhd$. Derivations depend on a redundancy criterion presented as two functions, $\mathcal{R}_{\mathcal{F}}$ and $\mathcal{R}_{\mathcal{I}}$:

> **locale** $redundancy\_criterion = inference\_system +$
>   **fixes** $\mathcal{R}_{\mathcal{F}} :: 'a \; clause \; set \Rightarrow 'a \; clause \; set$ **and**
> $\mathcal{R}_{\mathcal{I}} :: 'a \; clause \; set \Rightarrow 'a \; inference \; set$
>   **assumes** $\mathcal{R}_{\mathcal{I}} \; N \subseteq \Gamma$ **and** sat $(N \setminus \mathcal{R}_{\mathcal{F}} \; N) \Longrightarrow$ sat N **and**
> $N \subseteq N' \Longrightarrow \mathcal{R}_{\mathcal{F}} \; N \subseteq \mathcal{R}_{\mathcal{F}} \; N' \wedge \mathcal{R}_{\mathcal{I}} \; N \subseteq \mathcal{R}_{\mathcal{I}} \; N'$ **and**
> $N' \subseteq \mathcal{R}_{\mathcal{F}} \; N \Longrightarrow \mathcal{R}_{\mathcal{F}} \; N \subseteq \mathcal{R}_{\mathcal{F}} \; (N \setminus N') \wedge \mathcal{R}_{\mathcal{I}} \; N \subseteq \mathcal{R}_{\mathcal{I}} \; (N \setminus N')$

By definition, a transition from M to N is possible if the only new clauses added are conclusions of inferences from M and any deleted clauses would be redundant in N:

$$M \rhd N \Leftrightarrow N \setminus M \subseteq \text{concl\_of `infers\_from } M \wedge M \setminus N \subseteq \mathcal{R}_{\mathcal{F}} \; N$$

This rule combines deduction (the addition of inferred clauses) and deletion (the removal of redundant clauses) in a single transition. The chapter keeps the two operations separated, but this is problematic, as we will see in Sect. 7.

A key concept to connect static and dynamic completeness is that of the set of persistent clauses, or limit: $N_\infty = \bigcup_i \bigcap_{j \geq i} N_j$. These are the clauses that belong to all clause sets except for at most a finite prefix of the sequence N_i. We also need the supremum of a sequence, $\bigcup_i N_i$. We introduce these missing functions (Lazy_List_Liminf.thy):

$$\text{Liminf } xs = \bigcup_{i < |xs|} \bigcap_{j : i \leq j < |xs|} xs \; ! \; j \qquad \text{Sup } xs = \bigcup_{i < |xs|} xs \; ! \; i$$

When interpreting the notation $\bigcup_i \bigcap_{j \geq i} N_j$ for the case of a finite sequence of length n, it is crucial to use the right upper bounds, namely $i, j < n$. For i, the danger is subtle: if $i \geq n$, then $\bigcap_{j : i \leq j < n} N_j$ collapses to the trivial infimum $\bigcap_{j \in \{\}} N_j$, i.e., the set of all clauses.

Lemma 4.2 connects the redundant clauses and inferences at the limit to those of the supremum, and the satisfiability of the limit to that of the initial clause set. Formally:

> **lemma** Rf_limit_Sup: chain $(\rhd)$ $Ns \Longrightarrow \mathcal{R}_{\mathcal{F}}$ (Liminf Ns) $= \mathcal{R}_{\mathcal{F}}$ (Sup Ns)
> **lemma** Ri_limit_Sup: chain $(\rhd)$ $Ns \Longrightarrow \mathcal{R}_{\mathcal{I}}$ (Liminf Ns) $= \mathcal{R}_{\mathcal{I}}$ (Sup Ns)
> **lemma** sat_limit_iff: chain $(\rhd)$ $Ns \Longrightarrow$ (sat (Liminf Ns) $\Leftrightarrow$ sat (lhd Ns))

In the chapter, the proof relies on "the soundness of the inference system," contradicting the claim that "we will only consider consistency-preserving inference systems" [3, Sect. 2.4]. Thanks to Isabelle, we now know that soundness is unnecessary.

Next, we show that the limit is saturated, under some assumptions and for a relaxed notion of saturation. A clause set N is saturated up to redundancy if all inferences from nonredundant clauses in N are redundant:

$$\mathsf{saturated\_upto}\ N \iff \mathsf{infers\_from}\ (N \setminus \mathcal{R}_\mathcal{F}\ N) \subseteq \mathcal{R}_\mathcal{I}\ N$$

The limit is saturated for fair derivations, defined by $\mathsf{fair\_clss\_seq}\ Ns \iff$ $\mathsf{concl\_of}\ '\ \mathsf{infers\_from}\ N' \setminus \mathcal{R}_\mathcal{I}\ N' \subseteq \mathsf{Sup}\ Ns \cup \mathcal{R}_\mathcal{F}\ (\mathsf{Sup}\ Ns)$ with $N' = \mathsf{Liminf}\ Ns$ $\setminus\ \mathcal{R}_\mathcal{F}\ (\mathsf{Liminf}\ Ns)$. The criterion must also be effective, meaning $\gamma \in \Gamma \implies$ $\mathsf{concl\_of}\ \gamma \in N \cup \mathcal{R}_\mathcal{F}\ N \implies \gamma \in \mathcal{R}_\mathcal{I}\ N$. Under these assumptions, we have Theorem 4.3:

theorem *fair_derive_saturated_upto*:
  $\mathsf{chain}\ (\triangleright)\ Ns \implies \mathsf{fair\_clss\_seq}\ Ns \implies \mathsf{saturated\_upto}\ (\mathsf{Liminf}\ Ns)$

The standard redundancy criterion is an instance of the framework. It relies on a counterexample-reducing inference system Γ (Standard_Redundancy.thy):

$$\mathcal{R}_\mathcal{F}\ N = \{C \mid \exists \mathcal{D} \subseteq N.\ (\forall I.\ I \vDash \mathcal{D} \implies I \vDash C) \wedge \forall D' \in \mathcal{D}.\ D' < C\}$$
$$\mathcal{R}_\mathcal{I}\ N = \{(\mathcal{C}, D, E) \in \Gamma \mid \exists \mathcal{D} \subseteq N.\ (\forall I.\ I \vDash \mathcal{D} \uplus \mathcal{C} \implies I \vDash E) \wedge \forall D' \in \mathcal{D}.\ D' < D\}$$

Standard redundancy qualifies as *effective_redundancy_criterion*. In the chapter, this is stated as Theorems 4.7 and 4.8, which depend on two auxiliary properties, Lemmas 4.5 and 4.6. The main result, Theorem 4.9, is that counterexample-reducing calculi are refutationally complete also under the application of standard redundancy:

theorem *saturated_upto_complete*:
  $\mathsf{saturated\_upto}\ N \implies (\neg\ \mathsf{sat}\ N \iff \{\} \in N)$

The informal proof of Lemma 4.6 applies Lemma 4.5 in a seemingly impossible way, confusing redundant clauses and redundant inferences and exploiting properties that appear only in the first lemma's proof. Our solution is to generalize the core argument into a lemma and apply it to prove Lemmas 4.5 and 4.6. Incidentally, the informal proof of Theorem 4.9 also needlessly invokes Lemma 4.5.

Finally, given a redundancy criterion $(\mathcal{R}_\mathcal{F}, \mathcal{R}_\mathcal{I})$ for Γ, its standard extension for $\Gamma' \supseteq \Gamma$ is defined as $(\mathcal{R}_\mathcal{F}, \mathcal{R}'_\mathcal{I})$, where $\mathcal{R}'_\mathcal{I}\ N = \mathcal{R}_\mathcal{I}\ N \cup (\Gamma' \setminus \Gamma)$ (Proving_Process.thy). The standard extension preserves effectiveness, saturation up to redundancy, and fairness.

6 First-Order Resolution

The chapter's Sect. 4.3 presents a first-order version of the ordered resolution rule and a first-order prover, RP, based on that rule. The first step towards lifting the completeness of ground resolution is to show that we can lift individual ground resolution inferences (FO_Ordered_Resolution.thy).

Inference Rule. First-order ordered resolution consists of the single rule

$$\frac{C_1 \vee A_{11} \vee \cdots \vee A_{1k1} \quad \cdots \quad C_n \vee A_{n1} \vee \cdots \vee A_{nkn} \quad \neg A_1 \vee \cdots \vee \neg A_n \vee D}{C_1 \cdot \sigma \vee \cdots \vee C_n \cdot \sigma \vee D \cdot \sigma}$$

where σ is the (canonical) MGU that solves all unification problems $A_{i1} \overset{?}{=} \cdots \overset{?}{=} A_{ik_i} \overset{?}{=} A_i$, for $1 \leq i \leq n$. As expected, the rule has several side conditions. The Isabelle representation of this rule is based on that of its ground counterpart, generalized to apply σ:

> **inductive** ord_resolve :: $'a$ *clause list* $\Rightarrow$ $'a$ *clause* $\Rightarrow$ $'s$ $\Rightarrow$ $'a$ *clause* $\Rightarrow$ *bool*
> **where**
> $|CAs| = n \Longrightarrow |Cs| = n \Longrightarrow |\mathcal{A}s| = n \Longrightarrow |As| = n \Longrightarrow n \neq 0 \Longrightarrow$
> $(\forall i < n.\ CAs\,!\,i = Cs\,!\,i \uplus \mathsf{Pos}\,'\,\mathcal{A}s\,!\,i) \Longrightarrow (\forall i < n.\ \mathcal{A}s\,!\,i \neq \{\}) \Longrightarrow$
>   Some σ = mgu (set_mset $'$ set (map2 add_mset $As\ \mathcal{A}s$)) $\Longrightarrow$
>   eligible $\sigma\ As$ $(D \uplus \mathsf{Neg}\,'$ mset $As) \Longrightarrow$
> $(\forall i < n.\ \mathsf{strict\_max\_in}\ (As\,!\,i \cdot \sigma)\ (Cs\,!\,i \cdot \sigma)) \Longrightarrow (\forall i < n.\ S\ (CAs\,!\,i) = \{\}) \Longrightarrow$
>   ord_resolve CAs $(D \uplus \mathsf{Neg}\,'$ mset $As)\ \sigma\ (((\bigcup$ mset $Cs) \uplus D) \cdot \sigma)$

The rule as stated is incomplete; for example, $\mathsf{p}(x)$ and $\neg\,\mathsf{p}(\mathsf{f}(x))$ cannot be resolved because x and $\mathsf{f}(x)$ are not unifiable. In the chapter, the authors circumvent this issue by stating, "We also implicitly assume that different premises and the conclusion have no variables in common; variables are renamed if necessary." For the formalization, we first considered enforcing the invariant that all derived clauses use mutually disjoint variables, but this does not help when a clause is repeated in an inference's premises. Instead, we rely on a predicate ord_resolve_rename, based on ord_resolve, that standardizes the premises apart. The renaming is performed by a function called renamings_apart :: $'a$ *clause list* $\Rightarrow$ $'s$ *list* that, given a list of clauses, returns a list of corresponding substitutions to apply. This function is part of the abstract interface for terms and substitutions (which we presented in Sect. 2) and is implemented using IsaFoR.

Lifting Lemma. To lift ground inferences to the first-order level, we consider a set of clauses M and introduce an adjusted version S_M of the selection function S. This new selection function depends on both S and M and works in such a way that any ground instance inherits the selection of at least one of the nonground clauses of which it is an instance. This property is captured formally as

> **lemma** S_M_grounding_of_clss:
> $C \in \mathsf{grounding\_of}\ M \Longrightarrow \exists D \in M.\ \exists \sigma.\ C = D \cdot \sigma \wedge S_M\ C = S\ D \cdot \sigma$

where grounding_of M is the set of ground instances of a set of clauses M.

The lifting lemma, Lemma 4.12, states that whenever there exists a ground inference of E from clauses belonging to grounding_of M, there exists a (possibly) more general inference from clauses belonging to M:

lemma *ord_resolve_rename_lifting*:
$(\forall \rho\ C.\ \text{is\_renaming}\ \rho \implies S\ (C \cdot \rho) = S\ C \cdot \rho) \implies$
$\text{ord\_resolve}\ S_M\ CAs\ DA\ \mathcal{A}s\ As\ \sigma\ E \implies$
$\{DA\} \cup \text{set}\ CAs \subseteq \text{grounding\_of}\ M \implies$
$\exists \eta s\ \eta\ \theta\ CAs_0\ DA_0\ \mathcal{A}s_0\ As_0\ E_0\ \tau.$
  $\text{ord\_resolve\_rename}\ S\ CAs_0\ DA_0\ \mathcal{A}s_0\ As_0\ \tau\ E_0\ \wedge$
  $CAs_0 \cdot \eta s = CAs \wedge DA_0 \cdot \eta = DA \wedge E_0 \cdot \theta = E \wedge \{DA_0\} \cup \text{set}\ CAs_0 \subseteq M$

The informal proof of this lemma consists of two sentences spanning four lines of text. In Isabelle, these two sentences translate to 250 lines and 400 lines, respectively, excluding auxiliary lemmas. Our proof involves six steps:

1. Obtain a list of first-order clauses CAs_0 and a first-order clause DA_0 that belong to M and that generalize CAs and DA with substitutions ηs and η, respectively.
2. Choose atoms $\mathcal{A}s_0$ and As_0 in the first-order clauses on which to resolve.
3. Standardize CAs_0 and DA_0 apart, yielding CAs_0' and DA_0'.
4. Obtain the MGU τ of the literals on which to resolve.
5. Show that ordered resolution on CAs_0' and DA_0' with τ as MGU is applicable.
6. Show that the resulting resolvent E_0 generalizes E with substitution θ.

In step 1, suitable clauses must be chosen so that $S\ (CAs_0\ !\ i)$ generalizes $S_M\ (CAs\ !\ i)$, for $0 \leq i < n$, and $S\ DA_0$ generalizes $S_M\ DA$. By the definition of S_M, this is always possible. In step 2, we choose the literals to resolve upon in the first-order inference depending on the selection on the ground inference. If some literals are selected in DA, we define As_0 as the selected literals in DA_0, such that $(As_0\ !\ i) \cdot \eta = As\ !\ i$ for each i. Otherwise, As must be a singleton list containing some atom A, and we define As_0 as the singleton list consisting of an arbitrary $A_0 \in DA_0$ such that $A_0 \cdot \eta = A$. Step 3 may seem straightforward until one realizes that renaming variables can in principle influence selection. To rule this out, our lemma assumes stability under renaming: $S\ (C \cdot \rho) = S\ C \cdot \rho$ for any renaming substitution ρ and clause C. This requirement seems natural, but it is not mentioned in the chapter.

The above choices allow us to perform steps 4 to 6. In the chapter, the authors assume that the obtained CAs_0 and DA_0 are standardized apart from each other as well as their conclusion E_0. This means that they can obtain a single ground substitution μ that connect CAs_0, DA_0, E_0 to CAs, DA, E. By contrast, we provide separate substitutions ηs, η, θ for the different side premises, the main premise, and the conclusion.

7   A First-Order Prover

Modern resolution provers interleave inference steps with steps that delete or reduce (simplify) clauses. In their Sect. 4.3, Bachmair and Ganzinger introduce the nondeterministic abstract prover RP that works on triples of clause sets and that generalizes the Otter-style and DISCOUNT-style loops. RP's core

rule, called inference computation, performs first-order ordered resolution as
described above; the other rules delete or reduce clauses or move them between
clause sets. We formalize RP and prove it complete assuming a fair strategy
(FO_Ordered_Resolution_Prover.thy).

Abstract First-Order Prover. The RP prover is a relation $\rightsquigarrow$ on states of
the form $(\mathcal{N}, \mathcal{P}, \mathcal{O})$, where $\mathcal{N}$ is the set of *new clauses*, $\mathcal{P}$ is the set of *processed
clauses*, and $\mathcal{O}$ is the set of *old clauses*. RP's formal definition is very close to
the original formulation:

> **inductive** $\rightsquigarrow$ :: $'a\ state \Rightarrow 'a\ state \Rightarrow bool$ **where**
> $\quad$ Neg $A \in C \Longrightarrow$ Pos $A \in C \Longrightarrow (\mathcal{N} \cup \{C\}, \mathcal{P}, \mathcal{O}) \rightsquigarrow (\mathcal{N}, \mathcal{P}, \mathcal{O})$
> $\quad | \ D \in \mathcal{P} \cup \mathcal{O} \Longrightarrow$ subsumes $D\ C \Longrightarrow (\mathcal{N} \cup \{C\}, \mathcal{P}, \mathcal{O}) \rightsquigarrow (\mathcal{N}, \mathcal{P}, \mathcal{O})$
> $\quad | \ D \in \mathcal{N} \Longrightarrow$ strictly_subsumes $D\ C \Longrightarrow (\mathcal{N}, \mathcal{P} \cup \{C\}, \mathcal{O}) \rightsquigarrow (\mathcal{N}, \mathcal{P}, \mathcal{O})$
> $\quad | \ D \in \mathcal{N} \Longrightarrow$ strictly_subsumes $D\ C \Longrightarrow (\mathcal{N}, \mathcal{P}, \mathcal{O} \cup \{C\}) \rightsquigarrow (\mathcal{N}, \mathcal{P}, \mathcal{O})$
> $\quad | \ D \in \mathcal{P} \cup \mathcal{O} \Longrightarrow$ reduces $D\ C\ L \Longrightarrow (\mathcal{N} \cup \{C \uplus \{L\}\}, \mathcal{P}, \mathcal{O}) \rightsquigarrow (\mathcal{N} \cup \{C\}, \mathcal{P}, \mathcal{O})$
> $\quad | \ D \in \mathcal{N} \Longrightarrow$ reduces $D\ C\ L \Longrightarrow (\mathcal{N}, \mathcal{P} \cup \{C \uplus \{L\}\}, \mathcal{O}) \rightsquigarrow (\mathcal{N}, \mathcal{P} \cup \{C\}, \mathcal{O})$
> $\quad | \ D \in \mathcal{N} \Longrightarrow$ reduces $D\ C\ L \Longrightarrow (\mathcal{N}, \mathcal{P}, \mathcal{O} \cup \{C \uplus \{L\}\}) \rightsquigarrow (\mathcal{N}, \mathcal{P} \cup \{C\}, \mathcal{O})$
> $\quad | \ (\mathcal{N} \cup \{C\}, \mathcal{P}, \mathcal{O}) \rightsquigarrow (\mathcal{N}, \mathcal{P} \cup \{C\}, \mathcal{O})$
> $\quad | \ (\{\}, \mathcal{P} \cup \{C\}, \mathcal{O}) \rightsquigarrow$ (concl_of ' infers_between $\mathcal{O}\ C, \mathcal{P}, \mathcal{O} \cup \{C\}$)

The rules correspond, respectively, to tautology deletion, forward subsumption,
backward subsumption in $\mathcal{P}$ and $\mathcal{O}$, forward reduction, backward reduction in
$\mathcal{P}$ and $\mathcal{O}$, clause processing, and inference computation.

$\quad$ Initially, $\mathcal{N}$ consists of the problem clauses and the other two sets are empty.
Clauses in $\mathcal{N}$ are reduced using $\mathcal{P} \cup \mathcal{O}$, or even deleted if they are tautological or
subsumed by $\mathcal{P} \cup \mathcal{O}$; conversely, $\mathcal{N}$ can be used for reducing or subsuming clauses
in $\mathcal{P} \cup \mathcal{O}$. Clauses eventually move from $\mathcal{N}$ to $\mathcal{P}$, one at a time. As soon as $\mathcal{N}$
is empty, a clause from $\mathcal{P}$ is selected to move to $\mathcal{O}$. Then all possible resolution
inferences between this given clause and the clauses in $\mathcal{O}$ are computed and put
in $\mathcal{N}$, closing the loop.

$\quad$ The subsumption and reduction rules depend on the following predicates:

$$\text{subsumes } D\ C \Leftrightarrow \exists \sigma.\ D \cdot \sigma \subseteq C$$
$$\text{strictly\_subsumes } D\ C \Leftrightarrow \text{subsumes } D\ C \wedge \neg \text{ subsumes } C\ D$$
$$\text{reduces } D\ C\ L \Leftrightarrow \exists D'\ L'\ \sigma.\ D = D' \uplus \{L'\} \wedge -L = L' \cdot \sigma \wedge D' \cdot \sigma \subseteq C$$

The definition of the set infers_between $\mathcal{O}\ C$, on which inference computation
depends, is more subtle. In the chapter, the set of inferences between C and
$\mathcal{O}$ consists of all inferences from $\mathcal{O} \cup \{C\}$ that have C as *exactly one* of their
premises. This, however, leads to an incomplete prover, because it ignores infer-
ences that need multiple copies of C. For example, assuming a maximal selection
function, the resolution inference

$$\frac{\mathsf{p} \quad \mathsf{p} \quad \neg \mathsf{p} \vee \neg \mathsf{p}}{\bot}$$

is possible. Yet if the clause $\neg \mathsf{p} \vee \neg \mathsf{p}$ reaches $\mathcal{O}$ earlier than p, the inference would not be performed. This counterexample requires ternary resolution,

but there also exists a more complicated one for binary resolution, where both premises are the same clause. Consider the clause set containing

(1) $q(a, c, b)$ (2) $\neg q(x, y, z) \vee q(y, z, x)$ (3) $\neg q(b, a, c)$

and an order $>$ on atoms such that $q(c, b, a) > q(b, a, c) > q(a, c, b)$. Inferences between (1) and (2) or between (2) and (3) are impossible due to order restrictions. The only possible inference involves two copies of (2):

$$\frac{\neg q(x, y, z) \vee q(y, z, x) \quad \neg q(x', y', z') \vee q(y', z', x')}{\neg q(x, y, z) \vee q(z, x, y)}$$

From the conclusion, we derive $\neg q(a, c, b)$ by (3) and $\perp$ by (1). This incompleteness is a severe flaw, although it is probably just an oversight.

Projection to Theorem Proving Process. On the first-order level, a derivation can be expressed as a lazy list Ss of states, or as three parallel lazy lists Ns, Ps, Os. The limit state of a derivation Ss is defined as Liminf Ss = (Liminf Ns, Liminf Ps, Liminf Os), where Liminf on the right-hand side is as in Sect. 5.

Bachmair and Ganzinger use the completeness of ground resolution to prove RP complete. The first step is to show that first-order derivations can be projected down to theorem proving processes on the ground level. This corresponds to Lemma 4.10. Adapted to our conventions, its statement is as follows:

> If $S \rightsquigarrow S'$, then grounding_of $S \triangleright^*$ grounding_of S', with $\triangleright$ based on some extension of ordered resolution with selection function S and the standard redundancy criterion $(\mathcal{R}_{\mathcal{F}}, \mathcal{R}_{\mathcal{I}})$.

This raises some questions: (1) Exactly which instance of the calculus are we extending? (2) Which calculus extension should we use? (3) How can we repair the mismatch between $\triangleright^*$ in the lemma statement and $\triangleright$ where the lemma is invoked?

Regarding question (1), it is not clear which selection function to use. Is the function the same S as in the definition of RP or is it arbitrary? It takes a close inspection of the proof of Lemma 4.13, where Lemma 4.10 is invoked, to find out that the selection function used there is $S_{\text{Liminf } Os}$.

Regarding question (2), the phrase "some extension" is cryptic. It suggests an existential reading, and from the context it would appear that a standard extension (Sect. 5) is meant. However, neither the lemma's proof nor the context where it is invoked supplies the desired existential witness. A further subtlety is that the witness should be independent of S and S', so that transitions can be joined to form a single theorem proving derivation. Our approach is to let $\triangleright$ be the extension consisting of all *sound* derivations: $\Gamma = \{(\mathcal{C}, D, E) \mid \forall I. \; I \models \mathcal{C} \cup \{D\} \Longrightarrow I \models E\}$. This also eliminates the need for Bachmair and Ganzinger's subsumption resolution rule, a special calculus rule that is, from what we understand, implicitly used in the proof of Lemma 4.10 for the subcases associated with RP's reduction rules.

As for question (3), the need for $\rhd^*$ instead of $\rhd$ arises because one of the cases requires a combination of deduction and deletion, which Bachmair and Ganzinger model as separate transitions. By merging the two transitions (Sect. 5), we avoid the issue altogether and can use $\rhd$ in the formal counterpart of Lemma 4.10.

With these issues resolved, we can prove Lemma 4.10 for single steps and extend it to entire derivations:

lemma *RP_ground_derive*: $\mathcal{S} \rightsquigarrow \mathcal{S}' \implies$ grounding_of $\mathcal{S} \rhd$ grounding_of $\mathcal{S}'$
lemma *RP_ground_derive_chain*:
  chain $(\rightsquigarrow)$ $\mathcal{S}s \implies$ chain $(\rhd)$ (lmap grounding_of $\mathcal{S}s$)

The lmap function applies its first argument elementwise to its second argument.

Fairness and Clause Movement. From a given initial state $(\mathcal{N}_0, \{\}, \{\})$, many derivations are possible, reflecting RP's nondeterminism. In some derivations, we could leave a crucial clause in $\mathcal{N}$ or $\mathcal{P}$ without ever reducing it or moving it to $\mathcal{O}$, and then fail to derive $\bot$ even if $\mathcal{N}_0$ is unsatisfiable. For this reason, refutational completeness is guaranteed only for fair derivations. These are defined as derivations such that Liminf $\mathcal{N}s$ = Liminf $\mathcal{P}s$ = $\{\}$, guaranteeing that no clause will stay forever in $\mathcal{N}$ or $\mathcal{P}$.

Fairness is expressed by the fair_state_seq predicate, which is distinct from the fair_clss_seq predicate presented in Sect. 5. In particular, Theorem 4.3 is used in neither the informal nor the formal proof, and appears to play a purely pedagogic role in the chapter. For the rest of this section, we fix a lazy list of states $\mathcal{S}s$, and its projections $\mathcal{N}s$, $\mathcal{P}s$, and $\mathcal{O}s$, such that chain $(\rightsquigarrow)$ $\mathcal{S}s$, fair_state_seq $\mathcal{S}s$, and lhd $\mathcal{O}s = \{\}$.

Thanks to fairness, any nonredundant clause C in $\mathcal{S}s$'s projection to the ground level eventually ends up in $\mathcal{O}$ and stays there. This is proved informally as Lemma 4.11, but again there are some difficulties. The vagueness concerning the selection function can be resolved as for Lemma 4.10, but there is another, deeper flaw.

Bachmair and Ganzinger's proof idea is as follows. By hypothesis, the ground clause C must be an instance of a first-order clause D in $\mathcal{N}s \,!\, j \cup \mathcal{P}s \,!\, j \cup \mathcal{O}s \,!\, j$ for some index j. If $C \in \mathcal{N}s \,!\, j$, then by nonredundancy of C, fairness of the derivation, and Lemma 4.10, there must exist a clause D' that generalizes C in $\mathcal{P}s \,!\, l \cup \mathcal{O}s \,!\, l$ for some $l > j$. By a similar argument, if D' belongs to $\mathcal{P}s \,!\, l$, it will be in $\mathcal{O}s \,!\, l'$ for some $l' > l$, and finally in all $\mathcal{O}s \,!\, k$ with $k \geq l'$. The flaw is that backward subsumption can delete D' without moving it to $\mathcal{O}$. The subsumer clause would then be a strictly more general version of D' (and of the ground clause C).

Our solution is to choose D, and consequently D', such that it is minimal, with respect to subsumption, among the clauses that generalize C in the derivation. This works because strict subsumption is well founded—which we also proved, by reduction to a well-foundedness result about the strict generalization relation on first-order terms, included in IsaFoR [13, Sect. 2]. By minimality, D' cannot be deleted by backward subsumption. This line of reasoning allows us to prove Lemma 4.11, where $\mathcal{O}$_of extracts the $\mathcal{O}$ component of a state:

lemma *fair_imp_Liminf_minus_Rf_subset_ground_Liminf_state*:
   $Gs =$ lmap grounding_of $Ss \implies$
   Liminf $Gs - \mathcal{R}_{\mathcal{F}}$ (Liminf Gs) $\subseteq$ grounding_of ($\mathcal{O}$_of (Liminf Ss))

Completeness. Once we have brought Lemmas 4.10, 4.11, and 4.12 into a suitable shape, the main completeness result, Theorem 4.13, is not difficult to formalize:

theorem *RP_saturated_if_fair*:
   saturated_upto (Liminf (lmap grounding_of Ss))
corollary *RP_complete_if_fair*:
   $\neg$ sat (grounding_of (lhd Ss)) $\implies$ {} $\in \mathcal{O}$_of (Liminf Ss)

A crucial point that is not clear from the text is that we must always use the selection function S on the first-order level and $S_{\text{Liminf } \mathcal{O}s}$ on the ground level. Another noteworthy part of the proof is the passage "Liminf Gs (and hence Liminf Ss) contains the empty clause" (using our notations). Obviously, if grounding_of (Liminf Ss) contains $\bot$, then Liminf Ss must as well. However, the authors do not explain the step from Liminf Gs, the limit of the grounding, to grounding_of (Liminf Ss), the grounding of the limit. Fortunately, by Lemma 4.11, the latter contains all the nonredundant clauses of the former, and the empty clause is nonredundant. Hence the informal argument is fundamentally correct.

8   Discussion and Related Work

Bachmair and Ganzinger cover a lot of ground in a few pages. We found much of the material straightforward to formalize: it took us about two weeks to reach their Sect. 4.3, which introduces the RP prover. By contrast, we needed months to fully understand and formalize that section. While the *Handbook* chapter succeeds at conveying the key ideas at the propositional level, the lack of rigor makes it difficult to develop a deep understanding of ordered resolution proving on first-order clauses.

There are several reasons why Sect. 4.3 did not lend itself easily to a formalization. The proofs often depend on lemmas and theorems from previous sections without explicitly mentioning them. The lemmas and proofs do not quite fit together. And while the general idea of the proofs stands up, they have many confusing flaws that must be repaired. Our methodology involved the following steps: (1) rewrite the informal proofs to a handwritten pseudo-Isabelle; (2) fill in the gaps, emphasizing which lemmas are used where; (3) turn the pseudo-Isabelle into real Isabelle, but with **sorry** placeholders for the proofs; and (4) replace the **sorry**s with proofs. Progress was not always linear. As we worked on each step, more than once we discovered an earlier mistake.

The formalization helps us answer questions such as, "Is effectiveness of ordered resolution (Lemma 3.13) actually needed, and if so, where?" It also allows us to track definitions and hypotheses precisely, so that we always know the scope and meaning of every definition, lemma, or theorem. If a hypothesis appears too strong or superfluous, we can try to rephrase or eliminate it; the proof assistant tells us where the proof breaks.

Starting from RP, we could refine it to obtain an efficient imperative implementation, following the lines of Fleury, Blanchette, and Lammich's verified SAT solver with the two-watched-literals optimization [12]. However, this would probably involve a huge amount of work. To increase provers' trustworthiness, a more practical approach is to have them generate detailed proofs. Such output can be independently reconstructed using a proof assistant's inference kernel. This is the approach implemented in Sledgehammer [8], which integrates automatic provers in Isabelle. Formalized metatheory could in principle be used to deduce a formula's satisfiability from a finite saturation.

We found Isabelle/HOL eminently suitable to this kind of formalization work. Its logic—based on classical simple type theory—balances expressiveness and automatability. We benefited from many features of the system, including codatatypes [5], Isabelle/jEdit [28], the Isar proof language [27], locales [4], and Sledgehammer [8]. It is perhaps indicative of the maturity of theorem proving technology that most of the issues we encountered were unrelated to Isabelle. The main challenge was to understand the informal proof well enough to design suitable locale hierarchies and state the definitions and lemmas precisely, and correctly.

Formalizing the metatheory of logic and deduction is an enticing proposition for many researchers. Two recent, independent developments are particularly pertinent. Peltier [17] proved static completeness of a variant of the superposition calculus in Isabelle/HOL. Since superposition generalizes ordered resolution, his result subsumes our static completeness theorem. It would be interesting to extend his formal development to obtain a verified superposition prover. We could also consider calculus extensions such as polymorphism [11,25], type classes [25], and AVATAR [24]. Hirokawa et al. [13] formalized, also in Isabelle/HOL, an abstract Knuth–Bendix completion procedure as well as ordered (unfailing) completion [1]. Superposition combines ordered resolution (to reason about clauses) and ordered completion (to reason about equality).

The literature contains many other formalized completeness proofs. Early work was carried out by Shankar [22] and Persson [18]. Some of our own efforts are also related: completeness of unordered resolution using semantic trees by Schlichtkrull [20]; completeness of a Gentzen system by Blanchette, Popescu, and Traytel [9]; and completeness of CDCL by Blanchette, Fleury, Lammich, and Weidenbach [6]. We refer to our earlier papers for further discussions of related work.

9   Conclusion

We presented a formal proof that captures the core of Bachmair and Ganzinger's *Handbook* chapter on resolution theorem proving. For all its idiosyncrasies, the chapter withstood the test of formalization, once we had added self-inferences to the RP prover. Given that the text is a basic building block of automated reasoning, we believe there is value in clarifying its mathematical content for the next generations of researchers. We hope that our work will be useful to the editors of a future revision of the *Handbook*.

Formalization of the metatheory of logical calculi is one of the many connections between automatic and interactive theorem proving. We expect to see wider adoption of proof assistants by researchers in automated reasoning, as a convenient way to develop metatheory. By building formal libraries of standard results, we aim to make it easier to formalize state-of-the-art research as it emerges. We also see potential uses of formal proofs in teaching automated reasoning, inspired by the use of proof assistants in courses on the semantics of programming languages [15,19].

Acknowledgment. Christoph Weidenbach discussed Bachmair and Ganzinger's chapter with us on many occasions and hosted Schlichtkrull at the Max-Planck-Institut in Saarbrücken. Christian Sternagel and René Thiemann answered our questions about IsaFoR. Mathias Fleury, Florian Haftmann, and Tobias Nipkow helped enrich and reorganize Isabelle's multiset library. Mathias Fleury, Robert Lewis, Mark Summerfield, Sophie Tourret, and the anonymous reviewers suggested many textual improvements.

Blanchette was partly supported by the Deutsche Forschungsgemeinschaft (DFG) project Hardening the Hammer (grant NI 491/14-1). He also received funding from the European Research Council (ERC) under the European Union's Horizon 2020 research and innovation program (grant agreement No. 713999, Matryoshka). Traytel was partly supported by the DFG program Program and Model Analysis (PUMA, doctorate program 1480).

References

1. Bachmair, L., Dershowitz, N., Plaisted, D.A.: Completion without failure. In: Aït-Kaci, H., Nivat, M. (eds.) Rewriting Techniques-Resolution of Equations in Algebraic Structures, vol. 2, pp. 1–30. Academic Press (1989)
2. Bachmair, L., Ganzinger, H.: Rewrite-based equational theorem proving with selection and simplification. J. Log. Comput. **4**(3), 217–247 (1994)
3. Bachmair, L., Ganzinger, H.: Resolution theorem proving. In: Robinson, A., Voronkov, A. (eds.) Handbook of Automated Reasoning, vol. I, pp. 19–99. Elsevier and MIT Press (2001)
4. Ballarin, C.: Locales: a module system for mathematical theories. J. Autom. Reason. **52**(2), 123–153 (2014)
5. Biendarra, J., et al.: Foundational (co)datatypes and (co)recursion for higher-order logic. In: Dixon, C., Finger, M. (eds.) FroCoS 2017. LNCS (LNAI), vol. 10483, pp. 3–21. Springer, Cham (2017). https://doi.org/10.1007/978-3-319-66167-4_1
6. Blanchette, J.C., Fleury, M., Lammich, P., Weidenbach, C.: A verified SAT solver framework with learn, forget, restart, and incrementality. J. Autom. Reason. **61**(3), 333–366
7. Blanchette, J.C., Fleury, M., Traytel, D.: Nested multisets, hereditary multisets, and syntactic ordinals in Isabelle/HOL. In: Miller, D. (ed.) FSCD 2017. LIPIcs, vol. 84, pp. 11:1–11:18. Schloss Dagstuhl—Leibniz-Zentrum für Informatik (2017)

8. Blanchette, J.C., Kaliszyk, C., Paulson, L.C., Urban, J.: Hammering towards QED. J. Formal. Reason. **9**(1), 101–148 (2016)
9. Blanchette, J.C., Popescu, A., Traytel, D.: Soundness and completeness proofs by coinductive methods. J. Autom. Reason. **58**(1), 149–179 (2017)
10. Brand, D.: Proving theorems with the modifiction method. SIAM J. Comput. **4**(4), 412–430 (1975)
11. Cruanes, S.: Logtk: a logic toolkit for automated reasoning and its implementation. In: Schulz, S., de Moura, L., Konev, B. (eds.) PAAR-2014. EPiC Series in Computing, vol. 31, pp. 39–49. EasyChair (2014)
12. Fleury, M., Blanchette, J.C., Lammich, P.: A verified SAT solver with watched literals using Imperative HOL. In: Andronick, J., Felty, A.P. (eds.) CPP 2018, pp. 158–171. ACM (2018)
13. Hirokawa, N., Middeldorp, A., Sternagel, C., Winkler, S.: Infinite runs in abstract completion. In: Miller, D. (ed.) FSCD 2017. LIPIcs, vol. 84, pp. 19:1–19:16. Schloss Dagstuhl—Leibniz-Zentrum für Informatik (2017)
14. Nieuwenhuis, R., Rubio, A.: Paramodulation-based theorem proving. In: Robinson, A., Voronkov, A. (eds.) Handbook of Automated Reasoning, vol. I, pp. 371–443. Elsevier and MIT Press (2001)
15. Nipkow, T.: Teaching Semantics with a proof assistant: no more LSD trip proofs. In: Kuncak, V., Rybalchenko, A. (eds.) VMCAI 2012. LNCS, vol. 7148, pp. 24–38. Springer, Heidelberg (2012). https://doi.org/10.1007/978-3-642-27940-9_3
16. Nipkow, T., Wenzel, M., Paulson, L.C. (eds.): Isabelle/HOL: A Proof Assistant for Higher-Order Logic. LNCS, vol. 2283. Springer, Heidelberg (2002). https://doi.org/10.1007/3-540-45949-9
17. Peltier, N.: A variant of the superposition calculus. Archive of Formal Proofs 2016 (2016). https://www.isa-afp.org/entries/SuperCalc.shtml
18. Persson, H.: Constructive Completeness of Intuitionistic Predicate Logic—a Formalisation in Type Theory. Licentiate thesis, Chalmers tekniska högskola and Göteborgs universitet (1996)
19. Pierce, B.C.: Lambda, the ultimate TA: using a proof assistant to teach programming language foundations. In: Hutton, G., Tolmach, A.P. (eds.) ICFP 2009, pp. 121–122. ACM (2009)
20. Schlichtkrull, A.: Formalization of the resolution calculus for first-order logic. J. Autom. Reason **61**(4), 455–484
21. Schlichtkrull, A., Blanchette, J.C., Traytel, D., Waldmann, U.: Formalizing Bachmair and Ganzinger's ordered resolution prover (technical report). Technical report (2018). http://matryoshka.gforge.inria.fr/pubs/rp_report.pdf
22. Shankar, N.: Towards mechanical metamathematics. J. Autom. Reason. **1**(4), 407–434 (1985)
23. Thiemann, R., Sternagel, C.: Certification of termination proofs using CeTA. In: Berghofer, S., Nipkow, T., Urban, C., Wenzel, M. (eds.) TPHOLs 2009. LNCS, vol. 5674, pp. 452–468. Springer, Heidelberg (2009). https://doi.org/10.1007/978-3-642-03359-9_31
24. Voronkov, A.: AVATAR: the architecture for first-order theorem provers. In: Biere, A., Bloem, R. (eds.) CAV 2014. LNCS, vol. 8559, pp. 696–710. Springer, Cham (2014). https://doi.org/10.1007/978-3-319-08867-9_46
25. Wand, D.: Polymorphic + typeclass superposition. In: Schulz, S., de Moura, L., Konev, B. (eds.) PAAR-2014. EPiC Series in Computing, vol. 31, pp. 105–119. EasyChair (2014)

26. Weidenbach, C.: Combining superposition, sorts and splitting. In: Robinson, A., Voronkov, A. (eds.) Handbook of Automated Reasoning, vol. II, pp. 1965–2013. Elsevier and MIT Press (2001)

27. Wenzel, M.: Isabelle/Isar—a generic framework for human-readable proof documents. In: Matuszewski, R., Zalewska, A. (eds.) From Insight to Proof: Festschrift in Honour of Andrzej Trybulec, Studies in Logic, Grammar, and Rhetoric, vol. 10, no. 23, University of Białystok (2007)

28. Wenzel, M.: Isabelle/jEdit—a prover IDE within the PIDE framework. In: Jeuring, J., Campbell, J.A., Carette, J., Dos Reis, G., Sojka, P., Wenzel, M., Sorge, V. (eds.) CICM 2012. LNCS (LNAI), vol. 7362, pp. 468–471. Springer, Heidelberg (2012). https://doi.org/10.1007/978-3-642-31374-5_38

29. Zhang, H., Kapur, D.: First-order theorem proving using conditional rewrite rules. In: Lusk, E., Overbeek, R. (eds.) CADE 1988. LNCS, vol. 310, pp. 1–20. Springer, Heidelberg (1988). https://doi.org/10.1007/BFb0012820

The Higher-Order Prover Leo-III

Alexander Steen[1(✉)] and Christoph Benzmüller[1,2]

[1] Institute of Computer Science, Freie Universität Berlin, Berlin, Germany
{a.steen,c.benzmueller}@fu-berlin.de
[2] Computer Science and Communications, University of Luxembourg,
Esch-sur-Alzette, Luxembourg

Abstract. The automated theorem prover Leo-III for classical higher-order logic with Henkin semantics and choice is presented. Leo-III is based on extensional higher-order paramodulation and accepts every common TPTP dialect (FOF, TFF, THF), including their recent extensions to rank-1 polymorphism (TF1, TH1). In addition, the prover natively supports almost every normal higher-order modal logic. Leo-III cooperates with first-order reasoning tools using translations to many-sorted first-order logic and produces verifiable proof certificates. The prover is evaluated on heterogeneous benchmark sets.

1 Introduction

Leo-III is an automated theorem prover (ATP) for classical higher-order logic (HOL) with Henkin semantics and choice.[1] It is the successor of the well-known LEO-II prover [1], whose development significantly influenced the build-up of the TPTP THF infrastructure [2]. Leo-III exemplarily utilizes and instantiates the associated LEOPARD system platform [3] for higher-order (HO) deduction systems implemented in Scala.

In the tradition of the cooperative nature of the LEO prover family, Leo-III collaborates with external theorem provers during proof search, in particular, with first-order (FO) ATPs such as E [4]. Unlike LEO-II, which translated proof obligations into untyped FO languages, Leo-III, by default, translates its HO clauses to (monomorphic or polymorphic) many-sorted FO formulas. That way clutter is reduced during translation, resulting in a more effective cooperation.

Leo-III supports all common TPTP [2,5] dialects (CNF, FOF, TFF, THF) as well as the polymorphic variants TF1 and TH1 [6,7]. The prover returns results according to the standardized SZS ontology and additionally produces a TSTP-compatible (refutation) proof certificate, if a proof is found. Furthermore, Leo-III natively supports reasoning for almost every normal HO modal logic, including (but not limited to) logics **K**, **D**, **T**, **S4** and **S5** with constant, cumulative or

---

The work was supported by the German National Research Foundation (DFG) under grant BE 2501/11-1. For an extended version of this paper see arXiv:1802.02732.

[1] Leo-III is freely available (BSD license) at http://github.com/leoprover/Leo-III.

D. Galmiche et al. (Eds.): IJCAR 2018, LNAI 10900, pp. 108–116, 2018.
https://doi.org/10.1007/978-3-319-94205-6_8

varying domain quantifiers [8]. These hybrid logic competencies make Leo-III, up to the authors' knowledge, the most widely applicable ATP available to date.

The most current release of Leo-III (version 1.2) comes with several novel features, including specialized calculus rules for function synthesis, injective functions and equality-based simplification. An evaluation of Leo-III 1.2 confirms that it is on a par with other current state-of-the-art HO ATP systems.

This paper outlines the base calculus of Leo-III and highlights the features of version 1.2. As a pioneering contribution this also includes Leo-III's native support for reasoning in HO modal logics (which, of course, includes propositional and FO modal logics). Finally, an evaluation of Leo-III is presented for all monomorphic and polymorphic THF problems from the TPTP problem library and for all mono-modal logic problems from the QMLTP library [9].

Related ATP systems. These include TPS, Satallax, cocATP and agsyHOL. Also, some interactive proof assistants such as Isabelle/HOL can be used for automated reasoning in HOL. More weakly related systems include the various recent attempts to lift FO ATPs to the HO domain, e.g. Zipperposition.

Higher-Order Logic. HOL as addressed here has been proposed by Church, and further studied by Henkin, Andrews and others, cf. [10,11] and the references therein. It provides lambda-notation, as an elegant and useful means to denote unnamed functions, predicates and sets (by their characteristic functions). In the remainder a notion of HOL with Henkin semantics and choice is assumed.

2   Higher-Order Paramodulation

Leo-III extends a complete, paramodulation based calculus for HOL with practically motivated, heuristic inference rules, cf. Fig. 1. They are grouped as follows:

Clause normalization. Leo-III employs *definitional clausification* to reduce the number of clauses. Moreover, *miniscoping* is employed prior to clausification. Further normalization rules are straightforward.

Primary inferences. The primary inference rules of Leo-III are *paramodulation* (Para), *equality factoring* (EqFac) and *primitive substitution* (PS) as displayed in Fig. 1. The first two introduce *unification constraints* that are encoded as negative literals. Note that these rules are unordered and produce numerous redundant clauses. Leo-III uses several heuristics to restrict the number of inferences, including a *HO term ordering*. While these restrictions sacrifice completeness in general, recent evaluations confirm practicality of this approach (cf. evaluation in §4); complete search may be retained though. PS instantiates free variables at top-level with approximations of predicate formulas using so-called *general bindings* $\mathcal{GB}_\tau^C$ [1, §2].

Unification. Unification in Leo-III uses a variant of *Huet's pre-unification rules*. Negative equality literals are interpreted as unification constraints and are attempted to be *solved eagerly* by unification. In contrast to LEO-II, Leo-III uses *pattern unification* whenever possible. In order to ensure termination, the pre-unification *search is limited* to a configurable depth.

PRIMARY INFERENCES

$$\frac{\mathcal{C} \vee [l \simeq r]^{\text{tt}} \qquad \mathcal{D} \vee [s \simeq t]^{\alpha}}{\mathcal{C} \vee \mathcal{D} \vee [s[r]_{\pi} \simeq t]^{\alpha} \vee [s|_{\pi} \simeq l]^{\text{ff}}} \text{ (Para)} \qquad \frac{\mathcal{C} \vee [l \simeq r]^{\alpha} \vee [s \simeq t]^{\alpha}}{\mathcal{C} \vee [l \simeq r]^{\alpha} \vee [l \simeq s]^{\text{ff}} \vee [r \simeq t]^{\text{ff}}} \text{ (EqFac)}$$

$$\frac{\mathcal{C} \vee [X_{\overline{\tau_i} \to o} \, \overline{t^i_{\tau_i}}]^{\alpha} \qquad p \in \mathcal{GB}^{\{\neg, \vee\} \cup \{\Pi^{\tau}, =^{\tau} | \tau \in T\}}_{\overline{\tau_i} \to o}}{\left(\mathcal{C} \vee [X_{\overline{\tau_i} \to o} \, \overline{t^i_{\tau_i}}]^{\alpha}\right)\{X/p\}} \text{ (PS)}$$

FURTHER RULES (Choice): E is a choice operator or a free variable; X is a fresh variable.
(INJ): sk is a fresh constant symbol of appropriate type.

$$\frac{\mathcal{C} \vee [s[E \, t]]^{\alpha}}{[t \, X]^{\text{ff}} \vee [t \, (\epsilon \, t)]^{\text{tt}}} \text{ (Choice)} \qquad \frac{[f \, X \simeq f \, Y]^{\text{tt}} \vee [X \simeq Y]^{\text{ff}}}{[\text{sk} \, (f \, X) \simeq X]^{\text{tt}}} \text{ (INJ)}$$

$$\frac{\mathcal{C} := \mathcal{C}' \vee [F_{\overline{\tau_j} \to \tau} \, \overline{s^{1,j}}^{1 \leq j \leq n} \simeq t^1_{\tau}]^{\text{ff}} \vee \cdots \vee [F \, \overline{s^{m,j}}^{1 \leq j \leq n} \simeq t^m_{\tau}]^{\text{ff}}}{\mathcal{C}\left\{F/\lambda \overline{X^j_{\tau_j}}.\epsilon Z_{\tau}. \bigwedge_{k=1}^{m} \left(\left(\bigwedge_{j=1}^{n} X^j = s^{k,j} \right) \longrightarrow Z = t^k \right) \right\}} \text{ (FS)}$$

Fig. 1. Examples of Leo-III's calculus rules. *Technical preliminaries*: $s \simeq t$ denotes an equation of HOL terms, where $\simeq$ is assumed to be symmetric. A literal ℓ is a signed equation, written $[s \simeq t]^{\alpha}$ where $\alpha \in \{\text{tt}, \text{ff}\}$ is the polarity of ℓ. Literals of form $[s_o]^{\alpha}$ are a shorthand for $[s_o \simeq \top]^{\alpha}$. A clause $\mathcal{C}$ is a multiset of literals, denoting its disjunction. For brevity, if $\mathcal{C}, \mathcal{D}$ are clauses and ℓ is a literal, $\mathcal{C} \vee \ell$ and $\mathcal{C} \vee \mathcal{D}$ denote the multi-union $\mathcal{C} \cup \{\ell\}$ and $\mathcal{C} \cup \mathcal{D}$, respectively. $s|_{\pi}$ is the subterm of s at position π, and $s[r]_{\pi}$ denotes the term that is created by replacing the subterm of s at position π by r.

Extensionality rules. Dedicated *extensionality rules* are used in order to eliminate the need for extensionality axioms in the search space. The rules are similar to those of LEO-II [1].

Clause contraction. In addition to standard simplification routines, Leo-III implements are variety of (equational) *simplification procedures*, including *subsumption*, destructive *equality resolution*, heuristic *rewriting* and contextual *unit cutting* (simplify-reflect).

Defined Equalities. Leo-III scans for common definitions of equality predicates and heuristically instantiates (or replaces) them with *primitive equality*.

Choice. Leo-III supports HOL with choice ($\epsilon^{\tau}_{(\tau \to o) \to \tau}$ being a choice operator for type τ). Rule (Choice) *instantiates choice predicates* for subterms that represent either concrete choice operator applications (if $E \equiv \epsilon$) or potential applications of choice (if E is a free variable of the clause).

Function synthesis. If plain unification fails for a set of unification constraints, Leo-III may try to *synthesise function specifications* by rule (FS) using special choice instances that simulate suitable if-then-else terms. In general, this rule tremendously increases the search space. However, it also enables Leo-III to solve some hard problems (with TPTP rating 1.0). Also, Leo-III supports improved *reasoning with injective functions* by postulating the existence of left-inverses, cf. rule (INJ).

Heuristic instantiation. Prior to clause normalization, Leo-III might *instantiate universally quantified* variables. This include exhaustive instantiation of finite types as well as partial instantiation for otherwise interesting types.

3   Modal Logic Reasoning

Modal logics have many relevant applications in computer science, artificial intelligence, mathematics and computational linguistics. They also play an important role in many areas of philosophy, including ontology, ethics, philosophy of mind and philosophy of science. Many challenging applications, as recently explored in metaphysics, require FO or HO modal logics (HOMLs). The development of ATPs for these logics, however, is still in its infancy.

Leo-III is addressing this gap. In addition to its HOL reasoning capabilities, it is the first ATP that natively supports a very wide range of normal HOMLs. To achieve this, Leo-III internally implements a shallow semantical embeddings approach [12,13]. The key idea in this approach is to provide and exploit faithful mappings for HOML input problems to HOL that encode its Kripke-style semantics. An example is as follows:

A The user inputs a HOML problem in a suitably adapted TPTP syntax, e.g.
   ```
   thf(1,conjecture,( ! [P:$i>$o,F:$i>$i, X:$i]: (? [G:$i>$i]:
       (($dia @ ($box @ (P @ (F @ X)))) => ($box @ (P @ (G @ X)))))))).
   ```
   which encodes $\forall P_{\iota \to o} \forall F_{\iota \to \iota} \forall X_\iota \exists G_{\iota \to \iota}(\Diamond \Box P(F(X)) \Rightarrow \Box P(G(X)))$, with `$box` and `$dia` representing the (mono-)modal operators. This example formula (an instance of a corollary of Becker's postulate) is valid in **S5**.

B In the header of the input file the user specifies the logic of interest, say **S5** with rigid constants, constant domain quantifiers and a global consequence relation. For this purpose the TPTP language has been suitably extended:[2]
   ```
   thf(simple_s5, logic, ($modal := [
       $constants := $rigid, $quantification := $constant,
       $consequence := $global, $modalities := $modal_system_S5 ])).
   ```

C When being called with this input file, Leo-III parses and analyses it, automatically selects and unfolds the corresponding definitions of the semantical embedding approach, adds appropriate axioms and then starts reasoning in (meta-logic) HOL. Subsequently, it returns SZS compliant result information and, if successful, also a proof object just as for standard HOL problems. Leo-III's proof for the embedded example is verified by GDV [5] in 356 s.

As of version 1.2, Leo-III supports (but is not limited to) FO and HO extensions of the well known modal logic cube. When taking the different parameter combinations into account (constant/cumulative/varying domain semantics, rigid/non-rigid constants, local/global consequence relation, and further semantical parameters) this amounts to more than 120 supported HOMLs.[3] The exact

---

[2] Cf. http://www.cs.miami.edu/~tptp/TPTP/Proposals/LogicSpecification.html.
[3] Cf. [13, §2.2]; we refer to the literature [8] for more details on HOML.

number of supported logics is in fact much higher, since Leo-III also supports multi-modal logics with independent modal system specification for each modality. Also, user-defined combinations of rigid and non-rigid constants and different quantification semantics per type domain are possible. *Related provers* are in contrast limited to propositional logics or support a small range of FO modal logics only [14–16]. In the restricted logic settings of the related systems, the embedding approach used by Leo-III is still competitive; cf. evaluation below.

4   Evaluation

In order to quantify the performance of Leo-III, an evaluation based on various benchmarks was conducted. Three benchmark data sets were used:

- *TPTP TH0* (2463 problems) is the set of all monomorphic HOL (TH0) problems from the TPTP library v7.0.0 [5] that are annotated as theorems. The TPTP library is a de-facto standard for the evaluation of ATP systems.
- *TPTP TH1* (442 problems) is the subset of all 666 polymorphic HOL (TH1) problems from TPTP v7.0.0 that are annotated as theorems and do not contain arithmetic. The problems mainly consist of HOL Light core exports and Sledgehammer translations of various Isabelle theories.
- *QMLTP* (580 problems) is the subset of all mono-modal benchmarks from the QMLTP library 1.1 [9]. The QMLTP library only contains propositional and FO modal logic problems. Since each problem may have a different validity status for each semantics of modal logic, all problems (and not only those marked as theorem) are selected. The total number of tested problems thus is 580 (raw problems) × 5 (logics) × 3 (domain conditions). QMLTP assumes rigid constant symbols and a local consequence relation.

The evaluation measurements were taken on the StarExec cluster in which each compute node is a 64 bit Red Hat Linux (kernel 3.10.0) machine featuring 2.40 GHz Intel Xeon quad-core processors and a main memory of 128 GB. For each problem, every prover was given a CPU time limit of 240 s. The following theorem provers were employed in one or more of the experiments: Leo-III 1.2 (TH0/TH1/QMLTP) used in conjuction with E, CVC4 and iProver, Isabelle/HOL 2016 (TH0/TH1) [17], Satallax 3.0 (TH0) [18], Satallax 3.2 (TH0), LEO-II 1.7.0 (TH0), Zipperposition 1.1 (TH0) and MleanCoP 1.3 [19] (QMLTP).

The experimental results are discussed next:

TPTP TH0. Table 1 (a) displays each system's performance on the TPTP TH0 data set. For each system the absolute number (Abs.) and relative share (Rel.) of solved problems is displayed. Solved here means that a system is able to establish the SZS status **Theorem** and also emits a proof certificate that substantiates this claim. All results of the system, whether successful or not, are counted and categorized as THM (**Theorem**), CAX (**ContradictoryAxioms**), GUP (**GaveUp**) and TMO (**TimeOut**) for the respective SZS status of the returned

Table 1. Detailed result of the benchmark measurements

(a) *TPTP TH0 data set (2463 problems)*

Systems	Solved		SZS results				Avg. time [s]		Σ time [s]	
	Abs.	Rel.	THM	CAX	GUP	TMO	CPU	WC	CPU	WC
Satallax 3.2	2140	86.89	2140	0	2	321	12.26	12.31	26238	26339
Leo-III	2053	83.39	2045	8	16	394	15.39	5.61	31490	11508
Satallax 3.0	1972	80.06	2028	0	2	433	17.83	17.89	36149	36289
LEO-II	1788	72.63	1789	0	43	631	5.84	5.96	10452	10661
Zipperposition	1318	53.51	1318	0	360	785	2.60	2.73	3421	3592
Isabelle/HOL	0	0.00	2022	0	1	440	46.46	33.44	93933	67610

(b) *TPTP TH1 data set (442 problems)*

Systems	Solved		SZS results				Avg. Time [s]		Σ time [s]	
	Abs.	Rel.	THM	CAX	GUP	TMO	CPU	WC	CPU	WC
Leo-III	185	41.86	183	2	8	249	49.18	24.93	9099	4613
Isabelle/HOL	0	0.00	237	0	23	182	93.53	81.44	22404	19300

result.[4] Additionally, the average and sum of all CPU times and wall clock (WC) times over all solved problems is presented.

Leo-III successfully solves 2053 of 2463 problems (roughly 83.39%) from the TPTP TH0 data set. This is 735 (35.8%) more than Zipperposition, 264 (12.86%) more than LEO-II and 81 (3.95%) more than Satallax 3.0. The only ATP system that solves more problems is the most recent version of Satallax (3.2) that successfully solves 2140 problems, which is approximately 4.24% more than Leo-III. Isabelle currently does not emit proof certificates (hence zero solutions). Even if results without explicit proofs are counted, Leo-III would still have a slightly higher number of problems solved than Satallax 3.0 and Isabelle/HOL with 25 (1.22%) and 31 (1.51%) additional solutions, respectively. Leo-III, Satallax (3.2), Zipperposition and LEO-II produce 18, 17, 15 and 3 unique solutions, respectively. Evidently, Leo-III currently produces more unique solutions than any other ATP system in this setting. Leo-III solves twelve problems that are currently not solved by any other system indexed by TPTP.[5]

Satallax, LEO-II and Zipperposition show only small differences between their individual CPU and WC time on average and sum. A more precise measure for a system's utilization of multiple cores is the so-called *core usage*. It is given by the average of the ratios of used CPU time to used wall clock time over

---

[4] Remark on CAX: In this special case of THM (theorem) the given axioms are inconsistent, so that anything follows, including the given conjecture. Unlike most other provers, Leo-III checks for this special situation.

[5] This information is extracted from the TPTP problem rating information that is attached to each problem. The unsolved problems are NLP004^7, SET013^7, SEU558^1, SEU683^1, SEV143^5, SY0037^1, SY0062^4.004, SY0065^4.001, SY0066^4.004, MSC007^1.003.004, SEU938^5 and SEV106^5.

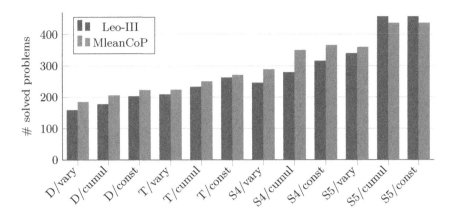

Fig. 2. Comparison of Leo-III and MleanCoP on the QMLTP data set (580 problems)

all solved problems. The core usage of Leo-III for the TPTP TH0 data set is roughly 2.52. This means that, on average, two to three CPU cores are used during proof search by Leo-III. Satallax (3.2), LEO-II and Zipperposition show a quite opposite behavior with core usages of 0.64, 0.56 and 0.47, respectively.

TPTP TH1. Currently, there exist only few ATP systems that are capable of reasoning within polymorphic HOL as specified by TPTP TH1. The only exceptions are HOL(y)Hammer and Isabelle/HOL that schedule proof tactics within HOL Light and Isabelle/HOL, respectively. Unfortunately, only Isabelle/HOL was available for instrumentation in a reasonably recent and stable version. Table 1 (b) displays the measurement results for the TPTP TH1 data set. When disregarding proof certificates, Isabelle/HOL finds 237 theorems (53.62%) which is roughly 28.1% more than the number of solutions founds by Leo-III. Leo-III and Isabelle/HOL produce 35 and 69 unique solutions, respectively.

QMLTP. For each semantical setting supported by MleanCoP, which is the strongest FO modal logic prover available to date [16], the number of theorems found by both Leo-III and MleanCoP in the QMLTP data set is presented in Fig. 2. Leo-III is fairly competitive with MleanCoP (weaker by maximal 14.05%, minimal 2.95% and 8.90% on average) for all **D** and **T** variants. For all **S4** variants, the gap between both systems increases (weaker by maximal 20.00%, minimal 13.66% and 16.18% on average). For **S5** variants, Leo-III is very effective (stronger by 1.36% on average) and it is ahead of MleanCoP for **S5**/const and **S5**/cumul (which coincide). This is due to the encoding of the **S5** accessibility relation in Leo-III 1.2 as the universal relation between possible worlds as opposed to its prior encoding (cf. [13,16]) as an equivalence relation. Leo-III contributes 199 solutions to previously unsolved problems.

For HOML there exist no competitor systems with which Leo-III could be compared. However, Leo-III can e.g. prove all problems from [20].

5 Summary

Leo-III is a state-of-the-art higher-order reasoning system offering many relevant features and capabilities. Due to its wide range of natively supported classical and non-classical logics, which include polymorphic HO logic and numerous FO and HO modal logics, the system has many topical applications in computer science, AI, maths and philosophy. Additionally, an evaluation on heterogeneous benchmark sets shows that Leo-III is also one of the most effective HO ATP systems to date. Leo-III complies with existing TPTP/TSTP standards, gives detailed proof certificates and plays a pivotal role in the ongoing extension of the TPTP library and infrastructure to support modal logic reasoning.

References

1. Benzmüller, C., Sultana, N., Paulson, L.C., Theiß, F.: The higher-order prover LEO-II. J. Autom. Reason. **55**(4), 389–404 (2015)
2. Sutcliffe, G., Benzmüller, C.: Automated reasoning in higher-order logic using the TPTP THF infrastructure. J. Formaliz. Reason. **3**(1), 1–27 (2010)
3. Wisniewski, M., Steen, A., Benzmüller, C.: LEoPARD — a generic platform for the implementation of higher-order reasoners. In: Kerber, M., Carette, J., Kaliszyk, C., Rabe, F., Sorge, V. (eds.) CICM 2015. LNCS (LNAI), vol. 9150, pp. 325–330. Springer, Cham (2015). https://doi.org/10.1007/978-3-319-20615-8_22
4. Schulz, S.: E - a brainiac theorem prover. AI Commun. **15**(2–3), 111–126 (2002)
5. Sutcliffe, G.: The TPTP problem library and associated infrastructure - from CNF to TH0, TPTP v6.4.0. J. Autom. Reason. **59**(4), 483–502 (2017)
6. Blanchette, J.C., Paskevich, A.: TFF1: the TPTP typed first-order form with rank-1 polymorphism. In: Bonacina, M.P. (ed.) CADE 2013. LNCS (LNAI), vol. 7898, pp. 414–420. Springer, Heidelberg (2013). https://doi.org/10.1007/978-3-642-38574-2_29
7. Kaliszyk, C., Sutcliffe, G., Rabe, F.: TH1: the TPTP typed higher-order form with rank-1 polymorphism. In: Fontaine, P., et al. (eds.) 5th PAAR Workshop. CEUR Workshop Proceedings, vol. 1635, pp. 41–55. CEUR-WS.org (2016)
8. Blackburn, P., van Benthem, J.F., Wolter, F.: Handbook of modal logic, vol. 3. Elsevier, Amsterdam (2006)
9. Raths, T., Otten, J.: The QMLTP problem library for first-order modal logics. In: Gramlich, B., Miller, D., Sattler, U. (eds.) IJCAR 2012. LNCS (LNAI), vol. 7364, pp. 454–461. Springer, Heidelberg (2012). https://doi.org/10.1007/978-3-642-31365-3_35
10. Andrews, P.: Church's type theory. In: Zalta, E.N. (ed.) The Stanford Encyclopedia of Philosophy, Metaphysics Research Lab. Stanford University (2014)
11. Benzmüller, C., Miller, D.: Automation of higher-order logic. In: Gabbay, D.M., Siekmann, J.H., Woods, J. (eds.) Handbook of the History of Logic. Computational Logic, vol. 9, pp. 215–254. North Holland/Elsevier, Amsterdam (2014)
12. Benzmüller, C., Paulson, L.: Multimodal and intuitionistic logics in simple type theory. Log. J. IGPL **18**(6), 881–892 (2010)
13. Gleißner, T., Steen, A., Benzmüller, C.: Theorem provers for every normal modal logic. In: Eiter, T., Sands, D. (eds.) LPAR-21. EPiC Series in Computing, Maun, Botswana, vol. 46, pp. 14–30. EasyChair (2017)

14. Hustadt, U., Schmidt, R.A.: MSPASS: modal reasoning by translation and first-order resolution. In: Dyckhoff, R. (ed.) TABLEAUX 2000. LNCS (LNAI), vol. 1847, pp. 67–71. Springer, Heidelberg (2000). https://doi.org/10.1007/10722086_7

15. Tishkovsky, D., Schmidt, R.A., Khodadadi, M.: The tableau prover generator Met-TeL2. In: del Cerro, L.F., Herzig, A., Mengin, J. (eds.) JELIA 2012. LNCS (LNAI), vol. 7519, pp. 492–495. Springer, Heidelberg (2012). https://doi.org/10.1007/978-3-642-33353-8_41

16. Benzmüller, C., Otten, J., Raths, T.: Implementing and evaluating provers for first-order modal logics. In: Raedt, L.D., et al. (eds.) ECAI 2012 of Frontiers in AI and Applications, Montpellier, France, vol. 242, pp. 163–168. IOS Press (2012)

17. Nipkow, T., Wenzel, M., Paulson, L.C. (eds.): Isabelle/HOL: A Proof Assistant for Higher-Order Logic. LNCS, vol. 2283. Springer, Heidelberg (2002). https://doi.org/10.1007/3-540-45949-9

18. Brown, C.E.: Satallax: an automatic higher-order prover. In: Gramlich, B., Miller, D., Sattler, U. (eds.) IJCAR 2012. LNCS (LNAI), vol. 7364, pp. 111–117. Springer, Heidelberg (2012). https://doi.org/10.1007/978-3-642-31365-3_11

19. Otten, J.: MleanCoP: a connection prover for first-order modal logic. In: Demri, S., Kapur, D., Weidenbach, C. (eds.) IJCAR 2014. LNCS (LNAI), vol. 8562, pp. 269–276. Springer, Cham (2014). https://doi.org/10.1007/978-3-319-08587-6_20

20. Benzmüller, C., Woltzenlogel Paleo, B.: The inconsistency in Gödel's ontological argument: a success story for AI in metaphysics. In: IJCAI 2016, vol. 1–3, pp. 936–942. AAAI Press (2016)

Well-Founded Unions

Jeremy Dawson[1], Nachum Dershowitz[2](✉), and Rajeev Goré[1]

[1] Research School of Computer Science,
Australian National University, Canberra, Australia
[2] School of Computer Science, Tel Aviv University, Tel Aviv, Israel
nachum@tau.ac.il

Abstract. Given two or more well-founded (terminating) binary relations, when can one be sure that their union is likewise well-founded? We suggest new conditions for an arbitrary number of relations, generalising known conditions for two relations. We also provide counterexamples to several potential weakenings. All proofs have been machine checked.

1 Introduction

A binary relation R (which need not be an ordering) over some underlying set is *well-founded* (or *terminating*) if there is no infinite *descending* chain $x_0 R x_1 R \cdots R x_{n-1} R x_n R \cdots$.[1] Given well-founded binary relations $R_0, R_1, \ldots, R_n$ over some common (fixed) underlying set X, under what conditions is their union $R_0 \cup R_1 \cup \cdots \cup R_n$ also well-founded?

For two well-founded relations A and B, their union $A \cup B$ is well-founded (Corollary 6 below) if the following relatively powerful condition holds [11]: see also [12]. It is called *Jumping* in [7]:

$$BA \subseteq A(A \cup B)^* \cup B .\tag{*}$$

Juxtaposition is being used for composition ($xBAz$ iff there's a y such that xBy and yAz) and the asterisk for the reflexive-transitive closure (xB^*z iff there are $y_0, y_1, \ldots, y_n$, $n \geq 0$, such that $x = y_0 B y_1 B \cdots B y_n = z$).

Jumping (*) generalises simpler ways of showing well-foundedness of the union of two relations. Eliding the rightmost B possibility gives *quasi-commutation* [2], which is relevant to many rewriting situations (e.g. [2,5,6,15]):

$$BA \subseteq A(A \cup B)^* .\tag{1}$$

Likewise, the simple A option also suffices for the well-foundedness of the union:

$$BA \subseteq A \cup B .\tag{2}$$

Based on preliminary work reported in [4,8].

J. Dawson—Supported by Australian Research Council Discovery Project DP140101540.

[1] We choose to view the forward direction as descent.

D. Galmiche et al. (Eds.): IJCAR 2018, LNAI 10900, pp. 117–133, 2018.
https://doi.org/10.1007/978-3-319-94205-6_9

To gain purchase on the manner of reasoning, let $R = A \cup B$ and imagine a minimal infinite descending chain in R: $x_0 R x_1 R \cdots R x_{n-1} R x_n R \cdots$. By "minimal" we mean that its elements are as small as possible vis-à-vis A, which – as it is well-founded – always enjoys minimal elements. Thus x_0 is the smallest element in the underlying set from which an infinite chain in R ensues. By the same token, x_1 is the smallest possible y such that $x_0 R y R \cdots$. And so on. By the well-foundedness of both A and B, any such chain must have (indeed, must have infinitely many) adjacent BA-steps: $\cdots x B x' A x'' R \cdots$. Now, if (2) holds, we could have taken a giant step $x R x''$, instead, before continuing down the infinite path from x''. But this would imply that the chain is not actually minimal because x'' is less than x' with respect to A, and should have been next after x.

Similarly, to show that (1) suffices, we choose a "preferred" infinite counterexample, in the sense that an A-step is always better than a B-step, given the choice. Again, an infinite chain containing a pair of steps $x B x' A x''$ could not be right since there is a preferred alternative, $x A y R \cdots R x'' R \cdots$, dictated by (1).

Combining these two arguments gives the sufficiency of the combined jumping condition (∗). Among preferred counterexamples, always choose B-steps, $x B x'$, having minimal x' with respect to A. Preference precludes taking an A-first detour instead of a BA pair $x B x' A x''$, while minimality precludes a B-shortcut $x B x''$.

To garner further insight, we first tackle – in the next two sections – the easier case of just three relations. Then, in Sect. 4, we extend the tripartite results and describe the general pattern for an arbitrary number of relations. We also show in Sect. 5 that under the same conditions any chain in the union can be rearranged so that the individual relations appear contiguously. This is followed in Sect. 6 by an example of the use of Preferential Commutation for four relations involved in the dependency-pair method [1].

Letting $R_{i:n} = \bigcup_{j=i}^{n} R_j$ be the union of well-founded relations $R_i, R_{i+1} .. R_n$, and letting R_i^+ be the transitive closure of R_i, our efforts culminate in Sect. 7 with the following sufficient condition (Theorem 28) for the well-foundedness and rearrangeability of $R_{0:n}$: There is some k, $0 \leq k \leq n$, such that

$$R_{i+1:n} R_i \subseteq R_0 R_{0:n}^* \cup R_i^+ \cup R_{i+1:n} \qquad \text{for } i = 0 .. k - 1 \qquad (**)$$

$$R_{i+1:n} R_i \subseteq R_i R_{i:n}^* \cup R_{i+1:n} \qquad \text{for } i = k .. n - 1. \qquad (***)$$

In the quadripartite case ($n = 3$), with $k = 2$, this amounts to the following:

$$(B \cup C \cup D) A \subseteq A(A \cup B \cup C \cup D)^* \cup B \cup C \cup D \qquad (3a)$$

$$(C \cup D) B \subseteq A(A \cup B \cup C \cup D)^* \cup B^+ \cup C \cup D \qquad (3b)$$

$$DC \subseteq C(C \cup D)^* \cup D . \qquad (3c)$$

All proofs have been machine-checked using Isabelle/HOL; see Sect. 8. We conclude with an open quadripartite problem and ideas for future work.

2   Tricolour Unions

We now study the three-relation case $n = 2$. We will refer to the relations A, B, and C as "colours". Ramsey's Theorem may be applied directly:

Theorem 1 (d'après Ramsey). *The union $A \cup B \cup C$ of well-founded relations A, B, and C is well-founded if it is transitive:*

$$(A \cup B \cup C)(A \cup B \cup C) \subseteq A \cup B \cup C . \tag{4}$$

Proof. The infinite version of Ramsey's Theorem applies when the union is transitive, so that every two (distinct) nodes within an infinite chain in the union of the colours have a (directed) edge that is coloured in one of the three colours. Then, there must lie an infinite monochromatic subchain within any infinite chain, contradicting the well-foundedness of each colour alone.   □

The suggestion to use Ramsey's Theorem for such a purpose is due to Franz Baader in 1989 [16, items 38–41]; see [13, Sect. 3.1]. Its use in a termination prover was pioneered in the TermiLog system [10]. Other uses followed; see [3].

Only three of the nine cases implicit in the left-hand side of (4) are actually needed for the limited outcome that we are seeking, an infinite monochromatic path, rather than a clique as in Ramsey's Theorem – as we observe next.

Theorem 2. *The union $A \cup B \cup C$ of well-founded relations A, B, and C is well-founded if*

$$BA \cup CA \cup CB \subseteq A \cup B \cup C . \tag{5}$$

Proof. When the union is not well-founded, there are infinite chains $Y = \{x_i\}_i$ with each x_i being connected to its neighbor x_{i+1} by one of the relations A, B, or C. Extract a maximal (noncontiguous) subsequence $Z = \{x_{i_j}\}_j$ of Y that consists of "hops" $x_{i_j} A x_{i_{j+1}}$ via A for each j. If it's finite and ends at some x_k, then at the *first* opportunity in the tail of Y beginning $x_{k+1}, x_{k+2}, \ldots$ extract another such sequence $\{x_{i'_j}\}_j$. Tack on to Z the intervening steps from x_{k+1} to the start $x_{i'_1}$ of the second maximal subsequence, followed by the rest, $x_{i'_2}, x_{i'_3}$, etc. Repeat and repeat. If any such subsequence turns out to be infinite, we have a contradiction to well-foundedness of A. If they're all finite, then consider $x_{i'_1 - 1}(B \cup C)x_{i'_1} A x_{i'_2}$ in Z. Since we could not take an A-step from $x_{i'_1 - 1}$ or else we would have, condition (5) tell us that $x_{i'_1 - 1}(B \cup C)x_{i'_2}$. Swallowing up all such (non-initial) A-steps in this way, we are left with an infinite chain in $B \cup C$, for which we also know that no A-hops are possible anywhere. Now extract maximal B-chains in the same fashion and then erase them, replacing $xCyBz$ with xCz (A- and B-steps having been precluded), leaving an infinite chain coloured purely C.   □

Condition (5) above is better than what we get by just iterating the simple condition (2) as shown below, with the difference being the option $BA \subseteq C$:

$$BA \subseteq A \cup B$$
$$CA \cup CB \subseteq A \cup B \cup C .$$

To guarantee an infinite clique, not just well-foundedness, instead of (4), one can insist on the three transitivity cases ($AA \subseteq A$, $BB \subseteq B$, $CC \subseteq C$), too:

Corollary 3. *If A, B, and C are transitive relations satisfying (5) and there is an infinite path in $A \cup B \cup C$, then there is an infinite monochromatic clique.*

Proof. By Theorem 2, (at least) one of A, B, C is not well-founded. By transitivity, the elements of any infinite chain in that non-well-founded colour form an infinite clique in the underlying undirected graph. □

Let's refer to the elements in any infinite descending chain in the union $A \cup B \cup C$ as *immortal*. We can do considerably better than the previous theorem:

Theorem 4 (Tripartite). *The union $A \cup B \cup C$ of well-founded relations A, B, and C is well-founded if*

$$(B \cup C)A \subseteq A(A \cup B \cup C)^* \cup B \cup C \tag{6a}$$

$$CB \subseteq A(A \cup B \cup C)^* \cup B^+ \cup C . \tag{6b}$$

Proof (sketch). First construct an infinite chain $Y = \{x_i\}_i$, in which an A-step is always preferred over B or C, as long as immortality is maintained. To do this, start with an immortal element x_0 in the underlying set. At each stage, if the chain so far ends in x_i, check if there is any y such that $x_i A y$ and from which proceeds some infinite chain in the union, in which case y is chosen to be x_{i+1}. Otherwise, x_{i+1} is any immortal element z such that $x_i B z$ or $x_i C z$.

If there are infinitely many B's and/or C's in Y, use them – by means of the first condition – to remove all subsequent A-steps, leaving only B- and C-steps, which go out of points from which A leads of necessity to mortality. From what remains, if there is any C-step at a point where one could take one or more B-steps to any place later in the chain, take the latter route instead. What remains now are C-steps at points where B^+ detours are also precluded. If there are infinitely many such C-steps, then applying the condition for CB will result in a pure C-chain, because neither $A(A \cup B \cup C)^*$ nor B^+ are options. □

Dropping C from the conditions of the previous theorem, one gets the Jumping criterion, which we explored in the introduction:

Definition 5 (Jumping criterion [11,12]**).** *Binary relation A jumps over binary relation B if*

$$BA \subseteq A(A \cup B)^* \cup B . \tag{*}$$

Corollary 6 (Jumping [11,12]**).** *The union $A \cup B$ of well-founded relations A and B is well-founded whenever A jumps over B.*

Applying this Jumping criterion twice, one gets somewhat different (incomparable) conditions for well-foundedness of the union of three relations.

Theorem 7 (Jumping I). *The union $A \cup B \cup C$ of well-founded relations A, B, and C is well-founded if*

$$BA \subseteq A(A \cup B)^* \cup B \tag{7a}$$
$$C(A \cup B) \subseteq (A \cup B)(A \cup B \cup C)^* \cup C . \tag{7b}$$

Proof. The first inequality is the Jumping criterion $(*)$. The second is the same with C for B and $A \cup B$ in place of A. □

For two relations, Jumping provides a substantially weaker well-foundedness criterion than does Ramsey. For three, whereas Jumping allows more than one step in lieu of BA (in essence, AA^*B^*), it doesn't allow for C, as does Ramsey.

Switching rôles, start with Jumping for $B \cup C$ before combining with A, we get slightly different conditions yet:

Theorem 8 (Jumping II). *The union $A \cup B \cup C$ of well-founded relations A, B, and C is well-founded if*

$$(B \cup C)A \subseteq A(A \cup B \cup C)^* \cup B \cup C \tag{8a}$$
$$CB \subseteq B(B \cup C)^* \cup C . \tag{8b}$$

Both this version of Jumping and our tripartite condition allow

$$(B \cup C)A \subseteq A(A \cup B \cup C)^* \cup B \cup C \tag{6a, 8a}$$
$$CB \subseteq B^+ \cup C . \tag{cf. 6b, 8b}$$

They differ in that Jumping also allows the condition shown below on the left whereas tripartite has the one shown on the right instead:

Jumping allows   Tripartite allows
$CB \subseteq B(B \cup C)^* .$ $CB \subseteq A(A \cup B \cup C)^* .$

Example 9. (a) Sadly, we cannot have the best of both worlds. Let's colour edges A, B, and C with (solid) azure, (dashed) black, and (dotted) crimson ink, respectively. The graph below only has multicoloured loops despite satisfying the inclusions below.

$(B \cup C)A \subseteq C$
$\qquad CB \subseteq A \cup B(B \cup C)^* .$

(b) Even the conditions shown below are insufficient since the double loop in the graph below harbours no monochromatic subchain:

$(B \cup C)A \subseteq C$
$\qquad CB \subseteq B(A \cup B)^*$

(c) By the same token, the putative hypothesis that the conditions shown below suffice is countered by the graph at its side:

$BA \cup CB \subseteq C$

$CA \subseteq BA^*$

3   Tripartite Proof

In preparation for the general case, we decompose the proof of the Tripartite Theorem 4 of the previous section into a sequence of notions and lemmata.

Definition 10 (Immortality [7]). *Let $R \subseteq X \times X$ be a binary relation over some underlying set X. The set $R^\infty \subseteq X$ of R-immortal elements are those elements $x_0 \in X$ that head infinite (descending) R-chains, $x_0 R x_1 R \cdots$.*

So, a relation R is well-founded if and only if every element of the underlying set is mortal ($R^\infty = \varnothing$).

Two trivial observations, first.

Proposition 11. *If $R \subseteq S^+$, for binary relations R and S, then perforce $R^\infty \subseteq S^\infty$, that is, every R-immortal is also S-immortal.*

It follows that

Proposition 12. *Binary relation R is well-founded if it is contained in a well-founded relation S, and, more generally, if $R \subseteq S^+$.*

As usual, the (forward) *image* $Q[Y]$ of a set Y under relation Q consists of those z such that yQz for some $y \in Y$, and the *inverse* (or *pre-*) *image* $Q^{-1}[Y]$ of Y under Q are those y such that yQz for some $z \in Y$.

If yRz for (R-)immortal z, then y is also immortal:

Proposition 13. *The inverse image of immortals is immortal: $R^{-1}[R^\infty] = R^\infty$.*

We will make repeated use of the Jumping criterion $(*)$, $BA \subseteq A(A \cup B)^* \cup B$. By induction (on the number of A's), Jumping extends to the transitive closure:

Lemma 14. *If binary relation A jumps over relation B, then*

$$BA^* \subseteq A(A \cup B)^* \cup B . \tag{9}$$

A central tool will be the following concept:

Definition 15 (Constriction). *The* constriction *$B^\sharp$ of binary relation B over X (with respect to relation A) excludes from B all steps of the form zBw for which there is an $A \cup B$-immortal y such that zAy:*

$$B^\sharp = B \setminus \{(z, w) \mid z \in A^{-1}[(A \cup B)^\infty], w \in X\} .$$

The idea of constriction is inspired by its use by Plaisted [17] for subterms.

Lemma 16. *The union $A \cup B$ of binary relations A and B is well-founded whenever $A \cup B^\sharp$ is.*

Proof. Construct an infinite descending $A \cup B$-chain, using A when it leads to immortality, and using B only when needed (making it a constricted step). □

Lemma 17. *If binary relation A jumps over relation B and both A and $B^\sharp$ are well-founded, then $A \cup B^\sharp$ is well-founded.*

Proof. Consider any infinite descending $A \cup B^\sharp$-chain. As A is well-founded, it must contain infinitely many $B^\sharp$-steps. As A jumps over B, Lemma 14 tells us that $B^\sharp A^* \subseteq A(A \cup B)^* \cup B^\sharp$. We have $B^\sharp$ on the right, because that position is constricting on the left. But in any infinite $A \cup B^\sharp$-chain, we cannot replace $B^\sharp A^*$ by $A(A \cup B)^*$ since that would mean that A leads to immortality, violating constriction. Hence, all (non-initial) A-steps may be removed from the chain, leaving an impossible infinite $B^\sharp$-chain. □

Combining the previous two lemmata, we can improve on Corollary 6.

Corollary 18. *If binary relation A jumps over relation B and both A and $B^\sharp$ are well-founded, then $A \cup B$ is well-founded.*

When there are more than three relations, as in the next section, we will need to revise the following lemma with a more flexible notion of constriction. For now, let $C^\flat$ be like $C^\sharp$ except that B-steps may be needed for immortality. Thus $C^\flat$ excludes all C-steps zCw with an $A \cup B \cup C$-immortal y such that zAy.

Lemma 19. *The union $B \cup C^\flat$ is well-founded if well-founded binary relations B and $C^\flat$ satisfy*

$$CB \subseteq A(A \cup B \cup C)^* \cup B^+ \cup C . \tag{6b}$$

Proof. Suppose that B and $C^\flat$ are well-founded, but $B \cup C^\flat$ is not. So there exist $B \cup C^\flat$-immortal elements. Choose z to be a B-minimal such element, and also to be $C^\flat$-minimal among all possible B-minimal choices.

As z is B-minimal, the first step of an infinite $B \cup C^\flat$-chain must be $zC^\flat y$, for some y. Since B is well-founded, let y be B-minimal among possible choices for such a y. By the aforementioned $C^\flat$-minimality of z, although y is $B \cup C^\flat$-immortal, it is not B-minimal among $B \cup C^\flat$-immortals. So we have yBx, where x is $B \cup C^\flat$-immortal.

Relying on (6b), we could replace $zCyBx$ in the putative infinite chain by any one of the following:

- $zAy'(A \cup B \cup C)^*x$, for some y' – but x heads an infinite descending $B \cup C$-chain, contradicting the constriction of $zC^\flat y$; or
- zB^+x, which would contradict our choice of z to be B-minimal; or

– zCx, and so $zC^\flat x$, which would contradict our choice of y to be B-minimal, since yBx and x could have been chosen in place of y.

Since each alternative leads to a contradiction, $B \cup C^\flat$ must be well-founded. $\square$

Everything is in place now for a modular proof of the Tripartite Theorem.

Proof (of Theorem 4). Since A jumps over $B \cup C$ (6a), by Corollary 18, it is enough to show that $(B \cup C)^\sharp$ is well-founded. Given (6b), by Lemma 19, we have that $B \cup C^\flat$ is well-founded. Clearly $(B \cup C)^\sharp \subseteq B \cup C^\flat$, because constricted B is in B and C is constricted to the same degree in both $(B \cup C)^\sharp$ and $B \cup C^\flat$, namely that A does not lead to immortality in the full union. By Proposition 12, the required well-foundedness of $(B \cup C)^\sharp$ follows. $\square$

4   Preferential Commutation

The two three-relation conditions, Jumping I and Jumping II, can each be straightforwardly extended by induction to arbitrarily many relations.

Corollary 20 (Jumping I). *The union $R_{0:n}$ of well-founded relations $R_0, R_1, \ldots, R_n$ is well-founded if*

$$R_{i+1}R_{0:i} \subseteq R_{0:i}R_{0:i+1}^* \cup R_{i+1} \qquad \text{for all } i = 0 \mathinner{..} n - 1 .$$

Proof. Since $B = R_{i+1}$ is well-founded, assume $A = R_{0:i}$ is well-founded by induction. Jumping (Corollary 18) then implies that so is $A \cup B = R_{0:i+1}$. $\square$

Corollary 21 (Jumping II). *The union $R_{0:n}$ of well-founded relations $R_0, R_1, \ldots, R_n$ is well-founded if*

$$R_{i+1:n}R_i \subseteq R_i R_{i:n}^* \cup R_{i+1:n} \qquad \text{for all } i = 0 \mathinner{..} n - 1 . \tag{10}$$

Proof. Let $A = R_i$ and $B = R_{i+1:n}$ in Corollary 18, and reason by induction. $\square$

We next extend Theorem 4 to an arbitrary number of relations and show the sufficiency of what we call *Preferential Commutation*.

Theorem 22 (Preferential Commutation). *The union $R_{0:n}$ of well-founded relations $R_0, R_1, \ldots, R_n$ is well-founded if it satisfies the following* Preferential Commutation Condition:

$$R_{i+1:n}R_i \subseteq R_0 R_{0:n}^* \cup R_i^+ \cup R_{i+1:n} \qquad \text{for all } i = 0 \mathinner{..} n - 1 . \tag{11}$$

Preferential Commutation (11) specializes to the two conditions (6a, 6b) of Theorem 4 in the tripartite case. In the quadripartite case ($n = 3$), it asserts that $A \cup B \cup C \cup D$ is well-founded if

$$(B \cup C \cup D)A \subseteq A(A \cup B \cup C \cup D)^* \cup B \cup C \cup D \tag{11a}$$

$$(C \cup D)B \subseteq A(A \cup B \cup C \cup D)^* \cup B^+ \cup C \cup D \tag{11b}$$

$$DC \subseteq A(A \cup B \cup C \cup D)^* \cup C^+ \cup D . \tag{11c}$$

Notice the inclusion of the options B^+ and C^+ in (11b) and (11c), respectively, when compared with the Jumping criteria. The A^+ has been omitted from (11a) on account of its inclusion in $A(A \cup \cdots)^*$.

We apply Preferential Commutation to four relations in Sect. 6.

Foremost to the argument for the above theorem is a general "detour" condition given below: replacing R_0 in Preferential Commutation with arbitrary P and $R_{0:n}$ with any S. Consider conditions (11a, 11b, 11c) on A, B, C, D in the case of four relations. The point is that we require the union of B, C, D to be well-founded so as to apply Jumping in conjunction with A, but were we to simply use the same method of jumping to establish that, we would not be allowed to introduce any A-steps in the inclusions for compositions of pairs from B, C, D.

We first generalise the notion of constriction (Definition 15).

Definition 23 (Constriction). *For arbitrary binary relation S, the S-constriction $B^{Q\sharp S}$ of binary relation B over X, with respect to Q, excludes from B all steps of the form zBw where there exists some S-immortal element in the Q-image of z:*

$$B^{Q\sharp S} = B \setminus \left(Q^{-1}[S^\infty] \times X \right) .$$

Think of this as B minus cases where Q could have granted S-immortality.

The basic constriction $B^\sharp$ of Definition 15 in the previous section is $B^{A\sharp A \cup B}$, while $C^\flat$ of Lemma 19 is $C^{A\sharp A \cup B \cup C}$.

We note that $B \subseteq C$, $Q \subseteq PR^*$ and $S \subseteq R^+$ imply $B^{P\sharp R} \subseteq C^{Q\sharp S}$.

Definition 24 (Detour). *Binary relations A, B, P, S satisfy the detour condition $\Delta_{B;A}^{P\sharp S}$ if*

$$BA \subseteq PS^* \cup A^+ \cup B . \tag{12}$$

Our central lemma is next; it generalises Lemma 19 of the previous section. Though it does have a proof very similar to the latter, we give here an alternative, distinct argument, one we find quite interesting. Contrary to earlier proofs, here we modify relations to include *only immortal* points – if any!

Lemma 25. *For all binary relations A, B, P, S, such that $A \cup B \subseteq S^+$ and both A and $B^{P\sharp S}$ are well-founded, if the detour condition $\Delta_{B;A}^{P\sharp S}$ holds, then the union $A \cup B^{P\sharp S}$ is well-founded, as is the more constricted union $(A \cup B)^{P\sharp S}$.*

Proof. Let $\underline{A}$ and $\underline{B}$ be relations A and B, respectively, restricted to those pairs (x, y) for which y is an $A \cup B$-immortal element (of X). Assuming A and $B^{P\sharp S}$ are well-founded, so are $\underline{A}$ and $\underline{B}^{P\sharp S}$. Consider any pair of adjacent steps

$$x \, \underline{B}^{P\sharp S} \, y \, \underline{A} \, z .$$

By constriction, the detour $xPy'S^*z$ allowed by (12) in place of $x\underline{B}Az$ is not a viable option, since z is immortal in $A \cup B \subseteq S^+$, and hence y' is immortal in S.

Thus, xBy would not actually be constricting with respect to P. So, we always have the following special case of the Jumping criterion $(*)$:

$$\underline{B}^{P\sharp S}\underline{A} \subseteq \underline{A}^+ \cup \underline{B}^{P\sharp S}\ .$$

Note that the B-step on the right is constricting because it is on the left. By Corollary 6, $\underline{A} \cup \underline{B}^{P\sharp S}$ is well-founded, and so is $A \cup B^{P\sharp S}$, as claimed, since it surely terminates for the excluded mortal elements of $A \cup B$.

Finally, $(A \cup B)^{P\sharp S} = A^{P\sharp S} \cup B^{P\sharp S} \subseteq A \cup B^{P\sharp S}$, so, a fortiori, $(A \cup B)^{P\sharp S}$ is well-founded (per Proposition 12).   □

For binary relations $R_0, \ldots, R_n$, let $R = R_{0:n}$, and let $\Delta_j = \Delta_{R_{j+1:n};R_j}^{R_0\sharp R}$ be the detour $R_{j+1:n}R_j \subseteq R_0 R^* \cup R_j^+ \cup R_{j+1:n}$. Preferential Commutation (11) is:

$$\Delta_0 \wedge \Delta_1 \wedge \cdots \wedge \Delta_{n-1}\ . \tag{11}$$

Lemma 26. *The constricted unions $R_{j:n}^{R_0\sharp R}$, $j = 0 \mathrel{..} n$, of preferentially-commuting well-founded binary relations $R_0, \ldots, R_n$ are all well-founded.*

Proof. By induction, starting with $j = n$ (when the conclusion holds by assumption) and working our way to $j = 0$. For the inductive step, given Δ_j and the well-foundedness of $R_{j+1:n}^{R_0\sharp R}$, and substituting $A = R_j$, $B = R_{j+1:n}$, $P = R_0$, and $S = R$ in the previous lemma, we obtain that $(R_j \cup R_{j+1:n})^{R_0\sharp R} = R_{j:n}^{R_0\sharp R}$ is likewise well-founded. The side condition $A \cup B \subseteq S^+$ of the lemma is satisfied by all the detours, as $R_j \cup R_{j+1:n} \subseteq R^+$ for all j.   □

We are ready for our main result, namely that the union R of well-founded $R_0, R_1, \ldots, R_n$ is well-founded when the detour conditions (11) hold for them.

Proof (of Theorem 22). Let $A = R_0$ and $B = R_{1:n}$. Lemma 26 tells us in particular $(j = 1)$ that $R_{1:n}^{R_0\sharp R} = B^\sharp$ is well-founded. Since Δ_0 means precisely that A jumps over B, Corollary 18 gives the well-foundedness of $A \cup B = R$.   □

5   Preferential Rearrangement

As with Jumping with two relations [7, Thm. 54], Preferential Commutation also means that any chain in the union can be rearranged from "a to z", so to speak.

Theorem 27 (Preferential Rearrangement). *If well-founded relations $R_0, R_1, \ldots, R_n$ satisfy*

$$R_{i+1:n}R_i \subseteq R_0 R_{0:n}^* \cup R_i^+ \cup R_{i+1:n} \qquad \text{for all } i = 0 \mathrel{..} n-1\ . \tag{11}$$

then finite chains can always be rearranged:

$$R_{0:n}^* \subseteq R_0^* R_1^* \cdots R_n^*\ . \tag{13}$$

Proof. By our main theorem (Theorem 22), the union $R = R_{0:n}$ is well-founded, so we can argue by induction on it. Were this theorem false, there would be counterexamples xR^+t containing an inversion $xR^*R_jR_iR^*t$ with $j > i$, and such that no alternative properly ordered chain $xR_0^*R_1^* \cdots R_n^*t$ would be possible. Now, let $xRyR^*t$ be a *minimal* counterexample, in the sense that each element in the chain is minimal vis-à-vis R: there is no smaller head than x for any counterexample; y is minimal among chains with irreparable inversions that begin with x; and so on.

The counterexample must possess an inversion, hence must comprise at least two steps xR_jyR^+t. By minimality, x being larger than y, the lesser chain yR^+t must be rearrangeable, so there is a chain $xR_jyR_izR_i^*R_{i+1}^* \cdots R_n^*t$ for some particular i and j ($y \neq t$ for sure). If $j \leq i$, all is fine and dandy, meaning that the example was not in fact a counterexample. Otherwise, $j > i$, and by (11) one of the following should hold true: (i) xR_0vR^*z; (ii) xR_i^+z; or (iii) $xR_{i+1:n}z$. But (i) is impossible, because vR^*zR^*t itself would of necessity be a better counterexample, as were it resolvable, so too would be the original example, starting with xR_0v. Also, (ii) is impossible, because $xR_i^+zR_{i:n}^*t$ is a perfectly good rearrangement. Lastly, (iii) is impossible, since goodness (no inversions) of $xR_{i+1:n}zR_i^* \cdots R_n^*t$ would provide a viable rearrangement, while badness of $xR_{i+1:n}zR_i^* \cdots R_n^*t$ would make it a smaller counterexample than xR_jyR^*t, z being less than y in R. □

Well-foundedness is necessary [7, Note 43].

It follows that Preferential Commuting (11) of well founded-relations is equivalent to an ordered version of the condition:

$$R_{i+1:n}R_i \subseteq R_0^+R_1^* \cdots R_n^* \cup R_i^+ \cup R_{i+1:n} \qquad \text{for all } i = 0 \mathbin{..} n - 1 . \qquad (11')$$

6 Example: Preferential Dependencies

Term-rewriting systems (see [9]) compute by applying equations to terms left-to-right, replacing arbitrary subterms when they match the left-hand side of an equation. If $\ell = r$ is such an oriented equation, it is used to rewrite a term s by replacing a subterm of s that is an instance $\ell\sigma$ of ℓ with the corresponding right-hand side $r\sigma$, resulting in some new term t. We write $s \to t$.

Termination of the rewriting relation $\to$ of a given system is normally established by showing that each such rewrite results in a decrease in some well-founded term ordering, or, in other words, that $\to\ \subseteq\ >$ for some well-founded relation $>$. A popular method [1] extends the given system of equations with additional replacement rules, called *dependency pairs*, in a way that can make the overall proof easier. Strict decrease in $>$ is only required for top-level applications of rules or their extensions. But only a "quasi-decrease" is needed for applications of (original) rules at *proper* subterms, for some quasi-ordering $\gtrsim$ that is *compatible* with $>$ in the sense that $> \gtrsim\ \subseteq >$ and $\gtrsim >\ \subseteq >$. (See the version in [14].) We very briefly sketch the use of Preferential Commuting to justify variants of this approach.

Given a rewriting system, we deal with four relations: instances $D = \{(\ell\sigma, r\sigma) \mid \ell = r$ is a rule, σ is a substitution$\}$ of original and extended rules; the immediate subterm relation $\rhd$ $[f(\ldots s \ldots) \rhd s]$; the intersection $>$ of $\rightarrow^+$ (for the original rules) with some well-founded partial order (think of it as "decreasing rewriting"); and inner-rewriting $\Rightarrow$, which is $\rightarrow$ applied to a *proper* (not necessarily immediate) subterm $[f(\ldots s \ldots) \Rightarrow f(\ldots t \ldots)$ if $s \rightarrow t]$. It can be seen without difficulty that if there exists any infinite rewrite chain with $\rightarrow$, then there also is an infinite $(D \cup \rhd \cup \Rightarrow)$-chain, wherein D occurs infinitely often.

Preferred Commutation for the union E of $A{:}\Rightarrow$, $B{:}>$, $C{:}\rhd$, and D is achieved by ensuring the following detours:

$$> \Rightarrow \; \subseteq \; \Rightarrow^+ >^* \cup > \qquad \rhd \Rightarrow \; \subseteq \; \Rightarrow \rhd \qquad D \Rightarrow \; \subseteq \; \Rightarrow E^* \cup > \qquad \text{(ba/ca/da)}$$

$$\rhd > \; \subseteq \; \Rightarrow^+ \rhd \qquad D > \; \subseteq \; > \qquad\qquad\quad \text{(cb/db)}$$

$$D \rhd \; \subseteq \; \Rightarrow^* \rhd^+ \cup D \; . \qquad\qquad\quad \text{(dc)}$$

Conditions (ba) and (db) are usually guaranteed by showing that applying a rule can, if anything, only cause a decrease with respect to $>$ (because $D \subseteq \gtrsim$ and $\Rightarrow \subseteq \gtrsim$, for example). Condition (ba) holds automatically when $>$ is on account of an inner step. Conditions (ca) and (cb) hold by the nature of rewriting (recalling that $> \subseteq \rightarrow^+$). For (da) we require a strict decrease in $>$ for each extended rule $(D \subseteq >)$, except perhaps for some instances that allow inner rewriting) and then rely on (ba) for an overall decrease. Condition (dc) is what guides the addition of rules: if there is a directed equation $\ell = r \in D$ and $r \rhd s$, but s is not also a subterm of ℓ ($\ell \not\rhd^+ s$), or of an inner reduct of ℓ (after some $\Rightarrow$ steps), then include $\ell = s$ as an extended rule in D. For each extended rule, a strict decrease in the ordering $>$ ensures (da) and (db). Extended rules may engender additional extended rules, per (dc).

Since $>$ and $\rhd$ are well-founded, and $\Rightarrow$ may be assumed well-founded by an inductive argument, all that remains to be shown is that D on its own is well-founded, which it is if $D \subseteq >$.

7   Preferential Jumping

Preferential Commutation (11) generalises the conjunction of conditions (6a, 6b) of the Tripartite Theorem 4. Its beauty lies in that it allows initial "preferred" R_0-steps and multiple R_i-steps. It does not, however, generalise condition condition (8b) of Jumping II (Theorem 8).

We can, however, extend Theorem 22 to allow a mix of Preferential Commutation and Jumping, with Jumping taking over from Commuting at some point.

Theorem 28 (Preferential Jumping). *The union $R_{0:n}$ of well-founded relations $R_0, R_1, \ldots, R_n$ is well-founded if, for some k, $0 \leq k \leq n$,*

$$R_{i+1:n} R_i \subseteq R_0 R_{0:n}^* \cup R_i^+ \cup R_{i+1:n} \qquad for\ i = 0 \, .. \, k-1 \qquad (**)$$
$$R_{i+1:n} R_i \subseteq R_i R_{i:n}^* \cup R_{i+1:n} \qquad\qquad for\ i = k \, .. \, n-1 \, . \qquad (***)$$

When $k = n$ this leaves only $(**)$, which is pure Preferential Commutation (Theorem 22); when $k = 0$ this leaves $(***)$, which is Jumping II (Corollary 21).

Proof. By $(***)$ and Jumping II, $R_{k:n}$ is well-founded. Taking that into account, by $(**)$ and Preferential Commuting, $R_{1:n}$ is.   □

By iterating the similar result for Jumping for two relations [7, Theorem. 54], we get a rearrangement theorem with conditions as for Corollary 21, that is, $(***)$ for $k = 0$. Interestingly, this seems to require an inductive proof, unlike Theorem 27. Then, combining that with Theorem 27, we get the following result.

Theorem 29. *Well-founded relations $R_0, R_1, \ldots, R_n$ satisfying the Preferential Jumping conditions $(**, ***)$ for some k have re-arrangeable finite chains:*

$$R_{0:n}^* \subseteq R_0^* R_1^* \cdots R_n^* . \tag{13}$$

8   Formalising the Proof

All the results of the preceding sections have been verified using Isabelle/HOL 2005.[2] When formalising this work in Isabelle, we faced a problem in defining "well-foundedness" and "relational composition" since these are defined in exactly opposite ways in the term-rewriting and interactive theorem-proving communities. Fortunately, the two notions are almost always used together, meaning that the two effects cancel each other out, as we explain next.

In Isabelle, the well-foundedness and composition of relations are as follows: *Relation R is* well-founded *if there is no infinite descending chain* where $x <_R y$ means $(x, y) \in R$, and descent goes to the left:

$$\cdots <_R x_n <_R x_{n-1} <_R \cdots <_R x_1 <_R x_0 .$$

The Isabelle definition below is the positive form: a relation R is well-founded iff the principle of well-founded induction over R holds for all properties P:

```
wf ?R == ALL P.
   (ALL x. (ALL y. (y, x) : ?R --> P y) --> P x)
   --> (ALL x. P x)
```

We display Isabelle code explicitly so that readers can make a visual connection with our repository. In this definition, the question mark symbol ? indicates implicit universal quantification and so ?R is a free variable (parameter) that is instantiated. The explicit quantifiers are ALL and EX.

Next, we give its equivalent, which says that a relation R is well-founded if every non-empty set Y has an R-minimal member:

---

[2] After 2005, it became too onerous to keep pace with changes in Isabelle. This does not detract from our verification in any way since Isabelle 2005 is a trusted system. Instructions on running the proofs are at http://users.cecs.anu.edu.au/~jeremy/ isabelle/2005/gen/tripartite-README.

```
wf ?R = (ALL Y x. x : Y --> (EX z:Y. ALL y. (y, z) : ?R --> y ~: Y))
```

Then the Isabelle expression that states precisely that wf R iff there are no infinite descending chains is as follows, where Suc signifies successor in the naturals:

```
wf ?R = (~ (EX f. ALL i. (f (Suc i), f i) : ?R))
```

The symbol ~ encodes classical negation and infix : encodes $\in$, so ~: encodes $\notin$.

In Isabelle, the *composition* of relations R and S (denoted O) is defined by

```
?R O ?S == {(x, z). EX y. (x, y) : ?S & (y, z) : ?R}
```

$$R \circ S = \{(x, z) \mid \exists y. (x, y) \in S \ \& \ (y, z) \in R\} .$$

Our notation RS from Sect. 2 and the Isabelle notation $R \circ S$ for "relational composition" are inverses, obeying $RS = (R^{-1} \circ S^{-1})^{-1}$:

$$RS = S \circ R = \{(a, c) \mid \exists b. (a, b) \in R \ \& \ (b, c) \in S\}$$
$$(RS)^{-1} = R^{-1} \circ S^{-1} = \{(c, a) \mid \exists b. (c, b) \in S^{-1} \ \& \ (b, a) \in R^{-1}\} .$$

Since the Isabelle definitions of composition O and wf of well-founded are *both* mirror images of those from this paper, our Isabelle theorems and the theorems in this paper correspond exactly: if only one were different, we would have to reverse the order of relation composition to make the two notions coincide. For example, the Jumping theorem for two relations [11] appears in our repository as below, where binary Un encodes union ($\cup$) and ^* encodes transitive closure:

```
[| ?S O ?R <= (?R O (?R Un ?S)^*) Un ?S; wf ?R; wf ?S |]
 ==> wf (?R Un ?S)
```

Using the positive definition of well-foundedness leads to Isabelle proofs rather different from our original pen-and-paper proofs. Consider, on the one hand, the arguments given in Sect. 1 and in the proof of Theorem 4, involving infinite sequences in which A is preferred, and in which members are A-minimal, with – on the other hand – the argument in the proof of Lemma 19, which chooses a C^b-minimal B-minimal immortal element. This latter proof reflects much better the flavour of the arguments used in the Isabelle proofs.

We have formulated two distinct Isabelle proofs of the crucial Lemma 25, one along the lines of that of Lemma 19 and another that follows the proof in Sect. 4, with relations restricted to immortal elements.

9   Conclusion and Prospects

Previous work provided sufficient conditions for the union of two well-founded orderings to be well-founded. We discovered a corresponding result for the union of three well-founded orderings and discussed how our sufficient conditions differ from those (viz. Jumping) that result from repeatedly applying the result for two orderings.

We then repackaged the proof of this result for three orderings to extend it to the union of any number of well-founded orderings – in a condition called Preferential Commutation. We showed that whenever there is a finite chain in the union, then there is also one between its two endpoints that takes steps from the relations one after the other, in order. We also gave an example of its use in proofs of termination of rewriting. Finally, we combined Jumping with Preferential Commutation. We expect these results to have significant and varied applications, concomitant with the versatility of binary Jumping.

Usually, when formalising a result, the pen-and-paper proofs have been completed, but in our case, the situation was the opposite. We actually found some proofs using Isabelle and have reworded them for presentation here. The proofs in Isabelle all use "positive" notions (`wf`) rather than "negative" notions ("no infinite chains"). In this case, formalising Theorem 4 involved splitting the proof up into lemmas, which in fact led us to Lemmas 19 and 25, and thence to formulating and proving Theorem 22. As always, formalising a proof confirms that no details have been overlooked or other errors made.

The answer to the question whether the following conditions suffice in the quadripartite case has so far eluded us:

$$(B \cup C \cup D)A \subseteq A(A \cup B \cup C \cup D)^* \cup B \cup C \cup D \qquad (3a, 11a)$$

$$(C \cup D)B \subseteq A(A \cup B \cup C \cup D)^* \cup B^+ \cup C \cup D \qquad (3b, 11b)$$

$$DC \subseteq B(B \cup C \cup D)^* \cup C^+ \cup D . \qquad (cf.\ 3c, 11c)$$

All we can say is the following about any counterexample (where these conditions hold, the individual relations are well-founded, but their union is not):

– $A \cup B$ is not well-founded; for, were it, then Theorem 4 (for relations $A \cup B$, C and D) would give us well-foundedness of the union.
– $(C \cup D)^{A \sharp A \cup B \cup (C \cup D)}$ is not well-founded; for, if it were, then $(B \cup (C \cup D))^{A \sharp A \cup B \cup (C \cup D)} = (B \cup C \cup D)^\sharp$ would also be well-founded by Lemma 25, whence the union would also be by Corollary 18.

Unfortunately, these considerations have not yielded a counterexample.

Further matters worth exploring include:

– What effect would transitivity of the individual relations have on the conditions for well-foundedness? It is known to allow weakening of the Jumping criterion [7]. This suggests a weakening of the first Tripartite condition (6a):

$$(B \cup C)A \subseteq A(A \cup B \cup C)^* \cup (B \cup C)^+ . \qquad (6a^+)$$

– Can we obtain a better understanding of the detour condition Δ that might allow the results reported here to be extended even further? For example, can we exploit the fact that the proof of the crucial Lemma 25 holds with

$A(A \cup B^{P\sharp S})^*$ on the right of Eq. (12), not just A^+? This has the effect of weakening the second Tripartite condition (6b) to

$$CB \subseteq A(A \cup B \cup C)^* \cup B(B \cup C^\flat)^* \cup C , \qquad (6b^\flat)$$

which is why, in Example 9(a), there had to be an immortalising A-step out of what would otherwise have been a perfectly nice $BC^\flat BB$ detour in place of the offending CB cycle, and not the unacceptable $BCBB$ cycling detour.

– Can one "extract" any code (semi-) automatically? For example, if we express results in the contrapositive, then given an infinite descending chain in one relation, can we derive an infinite descending chain in another, as in the manual proof of Lemma 16?
– Focussing on the infinite descending chains, do these results have applications in terms of liveness?
– One of the motivations for this work is the search for novel termination orderings, particularly for term rewriting. The conditions herein may be applicable to a path ordering based on Takeuti's ordinal diagrams [18], for which ramified jumping conditions play a rôle.

References

1. Arts, T., Giesl, J.: Termination of term rewriting using dependency pairs. Theor. Comput. Sci. **236**, 133–178 (2000). Preliminary version. http://verify.rwth-aachen.de/giesl/papers/ibn-97-46.ps
2. Bachmair, L., Dershowitz, N.: Commutation, transformation, and termination. In: Siekmann, J.H. (ed.) CADE 1986. LNCS, vol. 230, pp. 5–20. Springer, Heidelberg (1986). https://doi.org/10.1007/3-540-16780-3_76
3. Bruynooghe, M., Codish, M., Gallagher, J.P., Genaim, S., Vanhoof, W.: Termination analysis of logic programs through combination of type-based norms. ACM Trans. Program. Lang. Syst. **29**, 10 (2007)
4. Dawson, J., Goré, R.: Proving results from the paper Nachum Dershowitz: tripartite unions. Technical report, Australian National University (2016). http://users.cecs.anu.edu.au/~rpg/tripartite-unions.pdf
5. Dershowitz, N.: Termination of linear rewriting systems. In: Even, S., Kariv, O. (eds.) ICALP 1981. LNCS, vol. 115, pp. 448–458. Springer, Heidelberg (1981). https://doi.org/10.1007/3-540-10843-2_36
6. Dershowitz, N.: Hierarchical termination. In: Dershowitz, N., Lindenstrauss, N. (eds.) CTRS 1994. LNCS, vol. 968, pp. 89–105. Springer, Heidelberg (1995). https://doi.org/10.1007/3-540-60381-6_6
7. Dershowitz, N.: Jumping and escaping: modular termination and the abstract path ordering. Theor. Comput. Sci. **464**, 35–47 (2012). http://nachum.org/papers/Toyama.pdf
8. Dershowitz, N.: Tripartite unions. arXiv:1606.01148 [cs.LO], informal (2016)
9. Dershowitz, N., Jouannaud, J.P.: Rewrite systems. In: Handbook of Theoretical Computer Science, Formal Methods and Semantics, vol. B, pp. 243–320. North-Holland, Amsterdam (1990). http://nachum.org/papers/survey-draft.pdf
10. Dershowitz, N., Lindenstrauss, N., Sagiv, Y., Serebrenik, A.: A general framework for automatic termination analysis of logic programs. Appl. Algebra Eng. Commun. Comput. **12**, 117–156 (2001). http://nachum.org/papers/aaecc.pdf

11. Doornbos, H., Backhouse, R., van der Woude, J.: A calculational approach to mathematical induction. Theor. Comput. Sci. **179**, 103–135 (1997). http://www. cs.nott.ac.uk/~rcb/MPC/MadeCalculational.ps.gz

12. Doornbos, H., von Karger, B.: On the union of well-founded relations. Logic J. IGPL **6**, 195–201 (1998)

13. Geser, A.: Relative termination. Ph.D. thesis, Fakultät für Mathematik und Informatik, Universität Passau, Germany (1990). http://homepage.cs.uiowa.edu/ ~astump/papers/geser_dissertation.pdf

14. Giesl, J., Kapur, D.: Dependency pairs for equational rewriting. In: Middeldorp, A. (ed.) RTA 2001. LNCS, vol. 2051, pp. 93–107. Springer, Heidelberg (2001). https://doi.org/10.1007/3-540-45127-7_9

15. Klop, J.W.: Combinatory Reduction Systems. CWI Math. Centre Tract. **127** (1980)

16. Lescanne, P.: Rewriting list contributions (1989). http://www.ens-lyon.fr/LIP/ REWRITING/CONTRIBUTIONS

17. Plaisted, D.A.: Polynomial time termination and constraint satisfaction tests. In: Kirchner, C. (ed.) RTA 1993. LNCS, vol. 690, pp. 405–420. Springer, Heidelberg (1993). https://doi.org/10.1007/978-3-662-21551-7_30

18. Takeuti, G.: Ordinal diagrams. II. J. Math. Soc. Jpn. **12**, 385–391 (1960)

Implicit Hitting Set Algorithms for Maximum Satisfiability Modulo Theories

Katalin Fazekas[1(✉)], Fahiem Bacchus[2], and Armin Biere[1]

[1] Johannes Kepler University, Linz, Austria
katalin.fazekas@jku.at
[2] University of Toronto, Toronto, Canada

Abstract. Solving optimization problems with SAT has a long tradition in the form of MaxSAT, which maximizes the weight of satisfied clauses in a propositional formula. The extension to maximum satisfiability modulo theories (MaxSMT) is less mature but allows problems to be formulated in a higher-level language closer to actual applications. In this paper we describe a new approach for solving MaxSMT based on lifting one of the currently most successful approaches for MaxSAT, the implicit hitting set approach, from the propositional level to SMT. We also provide a unifying view of how optimization, propositional reasoning, and theory reasoning can be combined in a MaxSMT solver. This leads to a generic framework that can be instantiated in different ways, subsuming existing work and supporting new approaches. Experiments with two instantiations clearly show the benefit of our generic framework.

1 Introduction

SMT solvers have become indispensable tools for solving a wide range of problems in many areas. Such solvers provide either a satisfying assignment (e.g., a witness for a bug) or a proof of unsatisfiability (e.g., proving that a particular abstraction does not display a bug). However, in many applications the problem to be solved is more naturally cast as an optimization problem: find an assignment that minimizes some cost function. Li et al. [1], for instance, give a range of applications where optimization is critical. The need to solve such applications has led to a range of work addressing optimization in SMT (e.g., [1–8]).

Work on SMT optimization varies in the generality of the objective functions that can be modeled. For example, [1,4] address optimizing objective functions stated in the theory of linear real arithmetic, while [7] can deal with linear objective functions in which some variables are restricted to be integer. MaxSMT [2] is a restricted but important sub-problem in which the objective functions are linear expressions over Boolean variables (Pseudo Boolean expressions).

In this paper we focus on MaxSMT. Although MaxSMT is not as general as some other optimization approaches, MaxSMT specific solvers are often more efficient on problems where Pseudo Boolean objectives suffice [8], and recent rapid progress in the efficiency of MaxSAT solvers [9] indicates that this special

© Springer International Publishing AG, part of Springer Nature 2018
D. Galmiche et al. (Eds.): IJCAR 2018, LNAI 10900, pp. 134–151, 2018.
https://doi.org/10.1007/978-3-319-94205-6_10

case may more likely scale to practical problems than more general optimization approaches. Furthermore, MaxSAT already has a wide and growing range of applications including planning, fault localization in C code, design debugging, and a variety of problems in data analysis (see [10]). This indicates that Pseudo Boolean objectives are sufficient in a range of applications, and hence MaxSMT, with its addition of theories, is likely to have even greater applicability.

The implicit hitting set (IHS) approach [11] for solving MaxSAT has seen considerable recent progress and is now one of the most effective ways of solving MaxSAT. For example, IHS solvers have been the top performing solvers on weighted problems in the most recent 2016 and 2017 evaluations of MaxSAT solvers [9]. One of the key benefits of the IHS approach is that it provides a clear separation between optimization and propositional reasoning. In particular, in IHS solvers optimization is performed by a separate minimum cost hitting set solver, while the SAT solver is used solely for propositional reasoning. This separation of concerns supports the observed improved performance by allowing the exploitation of more efficient specialized solvers for each component.

Since MaxSAT and MaxSMT are quite similar problems, this naturally leads to the question of how MaxSMT can be similarly separated into optimization, propositional reasoning, and theory reasoning. In this paper we provide a general view of how these separate components can be combined to solve MaxSMT by providing a formal reasoning calculus [12] for MaxSMT solvers that achieves a clear separation of these different components. The calculus formalizes a notion of state that abstracts the more complex notions of state used in implemented solvers, and a set of inference rules for transforming the state that abstracts the operations performed by implemented solvers. The power of the calculus is that almost any scheme for scheduling the application of these rules leads to a solution. Hence, it supports the design of a wide range of different implementations of the basic inferences and of control structures for scheduling their application. It also provides a formal framework for effective harvesting of advances in MaxSAT for improving MaxSMT and vice versa.

2   Preliminaries

We consider formulas F in conjunctive normal form (CNF) consisting of a set of clauses, where each clause C is a disjunction of literals, which are first-order atoms or propositional variables, or their negation. MaxSAT problems are specified by a purely propositional CNF F, without first-order atoms, partitioned into **hard** and **soft** clauses, $hard(F)$ and $soft(F)$. A **feasible solution** to the MaxSAT problem is a truth assignment that satisfies all of the hard clauses. A **core** in MaxSAT solving is a *set of soft clauses* (a subset of $soft(F)$) that when combined with the hard clauses forms an unsatisfiable set of clauses.

Each soft clause C has a positive weight, denoted by $cost(C)$, which specifies the cost of falsifying it. The cost of a set of soft clauses S is the sum of the costs of the soft clauses it contains: $cost(S) = \sum_{C \in S} cost(C)$. The cost of a feasible solution π is the sum of the costs of the soft clauses it falsifies:

$cost(\pi) = cost(\{C \in S \mid \pi \not\models C\})$. An **optimal solution** for a MaxSAT problem is a feasible solution with *minimum cost* among all feasible solutions. Solving a MaxSAT problem is the task of finding an optimal solution.

We can restrict our attention, w.l.o.g, to formulas F in which all soft clauses are unit. In particular, any non-unit soft clause C can be converted to a unit soft clause by (i) adding a new (*relaxed*) hard clause $C \vee v$ where v is a *new* propositional variable (called a *relaxing* or *selector* variable), and (ii) replacing the soft clause C with a new unit soft clause $(\neg v)$ with $cost((\neg v)) = cost(C)$. This transformation is sound since any optimal solution satisfies $C \leftrightarrow \neg v$.

Considering ground first-order atoms generalizes MaxSAT to MaxSMT [2], as SMT [13] generalizes SAT. As in MaxSAT, a MaxSMT problem consists of a set of hard and soft clauses with each soft clause having a weight. However, in MaxSMT literals can be formed from theory atoms as well as from propositional variables. For example, over the theory of linear real arithmetic (LRA) we could form clauses like $(p \vee \neg(1 \leq y) \vee (x + y \geq 2))$, with a propositional variable p and LRA theory atoms $(1 \leq y)$ and $(x + y \geq 2)$.

Let $atoms(F)$ be the set of atoms in F, which range over propositional variables as well as theory atoms. We extend this notion to literals, clauses, and sequences of literals accordingly. A (partial truth) **assignment** over $atoms(F)$ is a sequence of literals from $atoms(F)$ that (i) does not contain both x and $\neg x$ for any $x \in atoms(F)$ and (ii) has no repeated literals. If A and M are two sequences of literals we write AM to indicate their concatenation. An assignment π over $atoms(F)$ is called a **propositional model** of F, denoted by $\pi \models F$, if it satisfies the Boolean abstraction of F in which theory atoms are treated simply as new independent propositional variables. A propositional model of F, π, is also a **theory consistent model** of F if the conjunction of theory literals made true by π is consistent with all theory axioms, denoted $\pi \models_T F$.

A feasible solution π for a MaxSMT formula F is required to be theory consistent model of $hard(F)$ ($\pi \models_T hard(F)$). The cost of π is defined as in MaxSAT. Accordingly, solving MaxSMT means finding an optimal (minimum cost) feasible solution. Again, w.l.o.g., we can assume that all soft clauses are unit clauses. Similarly, a **core** in MaxSMT is a subset of *soft clauses* that when combined with the hard clauses does not have a theory consistent model.

Let K be a set of cores, i.e., a set of sets of soft clauses. A **hitting set** η of K is a set of soft clauses that has a non-empty intersection with every set in K: $\forall \kappa \in K. \eta \cap \kappa \neq \emptyset$. As defined above $cost(\eta) = \sum_{C \in \eta} cost(C)$.

3   Abstract Hitting Set Based MaxSMT Solving

The main contribution of our paper is to introduce and formalize a calculus for the implicit hitting set (IHS) approach for MaxSMT, which at the same time provides the first formal calculus for the IHS approach to MaxSAT [11]. Our calculus captures a flexible separation between optimization, propositional reasoning, and theory reasoning, supporting a number of different implementation strategies. The separation between optimization and propositional reasoning is

Table 1. Transition rules for solving SAT under assumptions (**A-Sat**)

UnitProp $A \mid M \mid F \Longrightarrow A \mid M \ell \mid F$	**if** $\begin{cases} \text{There is a clause } (C \vee \ell) \in F \text{ s.t.} \\ AM \models \neg C \text{ and } atom(\ell) \notin atoms(AM) \end{cases}$
Decide $A \mid M \mid F \Longrightarrow A \mid M \ell^d \mid F$	**if** $atom(\ell) \in \big(atoms(F) \setminus atoms(AM)\big)$
Backjump $A \mid M\ell^d N \mid F \Longrightarrow A \mid M \ell' \mid F$	**if** $\begin{cases} \text{There is a clause } C \in F \text{ s.t. } AM\ell^d N \models \neg C \\ \text{and a clause } C' \vee \ell' \text{ s.t. } F \models C' \vee \ell', \\ \quad AM \models \neg C' \text{ and } atom(\ell') \in atoms(\ell^d N) \end{cases}$
Learn $A \mid M \mid F \Longrightarrow A \mid M \mid F, C$	**if** $\begin{cases} F \models C \text{ and } C \notin F \\ atoms(C) \subseteq \big(atoms(F) \cup atoms(AM)\big) \end{cases}$
Forget $A \mid M \mid F, C \Longrightarrow A \mid M \mid F$	**if** $F \models C$
SatModel $A \mid M \mid F \Longrightarrow SAT(AM, F)$	**if** $AM \models F$
UnSat $A \mid M \mid F \Longrightarrow conflict(F, C)$	**if** $\begin{cases} \text{There is a clause } D \in F \text{ s.t. } AM \models \neg D \\ M \text{ contains no decision literals} \\ \text{and } C \text{ is a clause s.t. } F \models C \text{ and } A \models \neg C \end{cases}$

achieved by exploiting the IHS approach for solving MaxSAT/MaxSMT. Other approaches to MaxSAT solving, e.g., [14–16], employ exclusively propositional reasoning, doing optimization by solving a sequence of SAT decision problems.

Our calculus can be modified to model such approaches by combining the optimization and propositional reasoning components into a single "MaxSAT" component. This would provide a formal model of MaxSMT approaches like [3]. However, as we will demonstrate below, even without such a formal model our calculus still provides a framework for understanding the approach of [3].

IHS and the above cited approaches to solving MaxSAT use propositional reasoning to find cores by exploiting SAT solving under assumptions [17]. In particular, for any subset of soft clauses S we can determine if S and the hard clauses are satisfiable by assuming that the literal of each (unit) soft clause in S is true. If the conjunction of these literals with the hard clauses is unsatisfiable, the SAT solver assumption mechanism returns a clause falsified by the assumptions. Hence, this clause contains only negations of assumed literals, identifying a subset of S that, with the hard clauses, is unsatisfiable (i.e., a core).

Hence, as a first step towards a formal calculus for IHS MaxSMT solving, we provide a calculus for assumption based SAT and SMT reasoning. To the best of our knowledge, such a calculus has not been specified before. This contribution should be useful independent of MaxSMT since assumption based reasoning is used in many different applications besides optimization.

3.1   SAT/SMT Solving Under Assumptions

To formalize assumption based incremental SAT solving [17] and lift it to SMT, we extend the DPLL(T) calculus originally presented in [12]. As above let F be a first-order quantifier-free CNF formula over theory T. The states of our calculus are specified by a triple $A \mid M \mid F$, where F is a CNF formula (initially

Table 2. Additional rules for solving SMT under assumptions (**A-Smt**)

T-Backjump $A \mid M\ell^d N \mid F \Longrightarrow A \mid M\,\ell' \mid F$	**if** $\begin{cases} \text{There is a clause } C \in F \text{ s.t. } AM\ell^d N \models \neg C \\ \text{and a clause } C' \vee \ell' \text{ s.t. } F \models_T C' \vee \ell', \\ \quad AM \models \neg C' \text{ and } atom(\ell') \in atoms(\ell^d N) \end{cases}$
T-Learn $A \mid M \mid F \Longrightarrow A \mid M \mid F,C$	**if** $\begin{cases} F \models_T C \text{ and } C \notin F \\ atoms(C) \subseteq \big(atoms(F) \cup atoms(AM)\big) \end{cases}$
T-Forget $A \mid M \mid F,C \Longrightarrow A \mid M \mid F$	**if** $F \models_T C$
T-Model $A \mid M \mid F \Longrightarrow T\text{-}SAT(AM,F)$	**if** $AM \models_T F$

the input formula), A and M are non-overlapping assignments over $atoms(F)$. A is the given set of assumptions, and M is the solver's current set of implied and decided (noted by a superscript d, e.g., ℓ^d) literals.

The transition rules given in Table 1 specify an abstract assumption based SAT solver (**A-Sat**). These rules follow [12] but are adapted to handle assumptions; the main changes are as follows. First, the abstract states and rules have been extended with a (possibly empty) set of assumption literals A over $atoms(F)$. For example, **Learn** is the same, but **UnitProp** requires $AM \models \neg C$, instead of $M \models \neg C$. Second, we modified the rule Fail to obtain a new rule **UnSat** that transitions into a *conflict(F,C)* state when M has no decision literals and $AM \models \neg D$ for some $D \in F$. In that case, $F \wedge A$ must be unsatisfiable, and we can always find a clause C implied by F and falsified by A (e.g., by resolving all literals negated by M from the clause D). And third, we introduce a transition rule that leads to an explicit $SAT(AM,F)$ state when $AM \models F$ holds. This facilitates combining the assumption based transitions with a MaxSAT or MaxSMT transition system. It can be noted that our calculus captures the technique of [17] which uses one particular control scheme to derive the clauses D and C used in the **UnSat** rule (it is irrelevant that [17] intermixes A and M).

Abstract assumption based SMT solving (**A-Smt**) is specified by the rules of Table 2 along with the rules **UnitProp**, **Decide** and **UnSat** of Table 1. Note that T-entailment subsumes propositional entailment, i.e., $F \models C$ implies $F \models_T C$. Hence, T-**Learn** can learn any clauses that **Learn** can, and T-**Learn** need not always employ theory reasoning (it can also use propositional reasoning to perform learning). This can be important in practice if reasoning in T is expensive. The same remark holds for T-**Backjump** and T-**Forget**.

It can also be noted that **UnSat** requires a falsified clause D to be in F. Hence, when $F \wedge A$ is propositionally satisfiable but T-unsatisfiable our calculus requires sufficient theory lemmas from T-**Learn** so as to obtain a falsified clause in F and to derive a clause C falsified by A.

We say that a state S in a transition system is **final** when no rules are applicable to it. Given a set of assumed literals A and a formula F, the initial state of assumption based SAT/SMT solving is $A \mid \emptyset \mid F$. Deciding

Table 3. Transition rules for optimization ($*$ is any SAT/SMT state) (**A-MaxSMT**)

SAT/SMT-Transition	$\begin{cases} *' \text{ is reachable from } * \text{ by} \\ \text{a single } \textbf{A-Sat/A-Smt} \text{ transition step} \\ (\text{see Table 1 and Table 2}) \end{cases}$
$(LB, UB, \mu) \mid K \mid \langle * \rangle \Longrightarrow$	if
$\quad (LB, UB, \mu) \mid K \mid \langle *' \rangle$	
Core	$\begin{cases} \kappa = \{(\neg \ell) \mid \ell \in C\} \text{ and } \kappa \notin K \\ (\kappa \text{ is set of soft clauses}) \end{cases}$
$(LB, UB, \mu) \mid K \mid \langle \mathit{conflict}(F, C) \rangle \Longrightarrow$	if
$\quad (LB, UB, \mu) \mid K, \kappa \mid \langle \mathit{conflict}(F, C) \rangle$	
HS	$\begin{cases} \eta = HS(K) \\ A' = \{\ell \mid (\ell) \in (\mathit{soft}(F) - \eta)\} \end{cases}$
$(LB, UB, \mu) \mid K \mid \langle * \rangle \Longrightarrow$	if
$\quad (LB, UB, \mu) \mid K \mid \langle A' \mid \emptyset \mid F \rangle$	
MinHS	$\begin{cases} \eta = minHS(K) \\ A' = \{\ell \mid (\ell) \in (\mathit{soft}(F) - \eta)\} \\ LB' = \max(LB, \mathit{cost}(\eta)) \end{cases}$
$(LB, UB, \mu) \mid K \mid \langle * \rangle \Longrightarrow$	if
$\quad (LB', UB, \mu) \mid K \mid \langle A' \mid \emptyset \mid F \rangle$	
ImprovedSolution	if $\mathit{cost}(AM) < UB$
$(LB, UB, \mu) \mid K \mid \langle T\text{-}SAT(AM, F) \rangle \Longrightarrow$	
$\quad (LB, \mathit{cost}(AM), AM) \mid K \mid \langle T\text{-}SAT(AM, F) \rangle$	
OptimalSolution	if $LB \geq UB$
$(LB, UB, \mu) \mid K \mid \langle * \rangle \Longrightarrow \mathit{optSoln}(\mu)$	

the satisfiability/T-satisfiability of F assuming A is a derivation of the form $A \mid \emptyset \mid F \Longrightarrow \cdots \Longrightarrow S_n$, where S_n is a final state in the **A-Sat/A-Smt** system.

Theorem 1 (Termination). *Any sequence of transitions $A \mid \emptyset \mid F \Longrightarrow \cdots$ in A-Sat (A-Smt) that contains no infinite subsequence consisting only of rules from the set $\{$Learn, Forget$\}$ ($\{T$-Learn, T-Forget$\}$), is finite.*

Theorem 2 (Soundness). *For any derivation $A \mid \emptyset \mid F \Longrightarrow \cdots \Longrightarrow S$ in A-Sat (A-Smt) where S is final with respect to A-Sat (A-Smt) we have*

1. $S = \mathit{conflict}(F', C)$ with $F' \models C$, $A \models \neg C$ iff $F \wedge A$ is (T-)unsatisfiable.
2. $S = (T\text{-})SAT(AM, F')$ with $AM \models_{(T)} F'$ iff $F \wedge A$ is (T-)satisfiable.

We can treat A as a prefix of decision literals of M that can not be changed by backjumping. Under this interpretation the results of [12] can be extended to obtain proofs for Theorems 1 and 2. We omit the details due to space constraints.

3.2 IHS MaxSAT/MaxSMT Solving

To obtain an abstract IHS based MaxSMT solver we add the rules given in Table 3. These rules extend the states of **A-Smt** by adding K and the triple (LB, UB, μ), where K is a set of cores, LB and UB are lower and upper bounds on the cost of an optimal solution to the input CNF F, and μ is a feasible solution, represented as a sequence of literals over $atoms(F)$, with $\mathit{cost}(\mu) = UB$. Let **A-MaxSMT** be the transition system defined by the rules in Table 3 along with the rules **A-Smt**.[1] The initial state of **A-MaxSMT** is always the state $IS = (0, \infty, undef) \mid \emptyset \mid \langle \{\ell \mid (\ell) \in soft(F)\} \mid \emptyset \mid hard(F) \rangle$, i.e., we start with valid

---

[1] IHS MaxSAT solvers can be obtained by using the **A-Sat** rules and replacing $T\text{-}SAT(AM, F)$ in **ImprovedSolution** with $SAT(AM, F)$.

lower and upper bounds, an empty set of cores, an initial assumption that all soft clauses are satisfied, and all of the hard clauses of F.

The calculus computes a growing set of cores K, each obtained from assumption based SMT solving, and uses the two subroutines, $minHS(K)$ which returns a minimum cost hitting set of K, and $HS(K)$ which returns an arbitrary hitting set of K. It can be noted that the assumptions (initially and after the rules **HS** or **MinHS**) are always asserting that some subset of the soft clauses along with the hard clauses are satisfied. Hence, as explained above, the subset of $soft(F)$ identified by the returned conflict and added to K by rule **Core** must be a core. Furthermore, the assumptions always specify that all soft clauses *except* those in some hitting set η of K are true. Thus, the returned conflict must identify a *new* core κ that cannot already be in K. In particular, κ is a subset of $soft(F) - \eta$ (it is a subset of the assumed true soft clauses) but no $s \in K$ is a subset of $soft(F) - \eta$ since s contains a non-empty subset $s \cap \eta$ not in $soft(F) - \eta$.

We say that $S_1 \Longrightarrow \cdots \Longrightarrow S_n$ is a **progressing subsequence** if (a) S_1 is the result of applying the **MinHS** rule, (b) all transitions in the sequence arise from applying one of the **A-Smt** rules (i.e. are **SAT/SMT-Transition** steps), and (c) S_n is final with respect to the rules of **A-Smt** (i.e., no **SAT/SMT-Transition** is applicable).

Theorem 3 (Termination). *If $hard(F)$ is T-satisfiable then any derivation $IS \Rightarrow S_1 \Rightarrow \cdots$ of **A-MaxSMT** is finite if it satisfies the following conditions:*

1. *contains no infinite subsequence of rules from the set $\{T\text{-}\textbf{Learn}, T\text{-}\textbf{Forget}\}$*
2. *contains no infinite subsequence not containing a progressing subsequence*
3. *always applies the transitions **OptimalSolution**, **ImprovedSolution** and **Core** whenever they are applicable (with **OptimalSolution** being applied first).*

Theorem 4 (Soundness). *If $hard(F)$ is T-satisfiable, $IS \Rightarrow \cdots \Rightarrow S_n$ is a finite sequence of transitions in **A-MaxSMT**, and S_n is final in **A-MaxSMT**, then S_n is $optSoln(\mu)$ and μ is an optimal solution of F.*

Theorem 4 is immediate from the fact that (a) $optSoln(\mu)$ is the only final state in **A-MaxSMT**, (b) LB and UB are always valid bounds, and (c) $cost(\mu) = UB$.

Hence, the main result is that the calculus terminates under the conditions of Theorem 3. A sketch of the proof follows. First, from Theorem 1 it can be seen that all progressing subsequences must be finite, and thus any infinite sequence of transitions must contain an infinite number of progressing subsequences. Theorem 2 shows that every progressing subsequence must reach either a $T\text{-}SAT$ or a *conflict* final state. If a *conflict* state is reached, then **Core** must be applied next. As explained above this must add a *new* core to K. Each core is a subset of $soft(F)$ so only a finite number of cores exist. Hence, only a finite number of progressing subsequences can end in *conflict*. Otherwise, the progressing subsequence reaches $T\text{-}SAT$. But this can happen only once since the feasible solution found, AM, must be an optimal solution. AM satisfies all clauses except those in a minimum cost hitting set η of K (obtained from

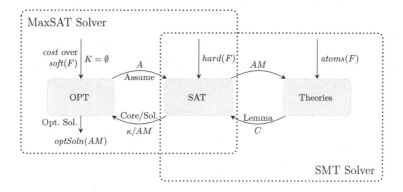

Fig. 1. General architecture for an IHS based MaxSMT solver

MinHS), and hence $cost(AM) \leq cost(\eta)$. Every feasible solution π satisfies $hard(F)$ and every core is unsatisfiable when added to $hard(F)$. Hence, π must falsify at least one soft clause in every core; i.e., the set of clauses falsified by π is a hitting set of K. So by the definition of $cost$ and the minimality of η, $cost(\eta) \leq cost(\pi)$, and thus $cost(AM) \leq cost(\pi)$ for every feasible solution π. Once AM is found, **ImprovedSolution** must be applicable and the condition $cost(AM) = UB \leq cost(\eta) \leq LB$ is achieved (**MinHS** ensures $cost(\eta) \leq LB$). Then **OptimalSolution** must be applied and the derivation terminates. In sum, under the stated conditions only a finite number of progressing subsequences can be executed and so the derivation must be finite.

4 Generic Hitting Set Based MaxSMT

Here we present a general framework that realizes the previously introduced ideas for IHS based MaxSMT solving. Following the desiderata presented in our introduction, we decompose the problem of MaxSMT into three sub-problems: optimization (over Boolean atoms), Boolean satisfiability, and theory reasoning. Although modern SMT solvers are equipped with efficient engines for arithmetic reasoning, in MaxSMT the optimization problem depends purely on the Boolean abstraction of the formula and thus delegating the task of optimization to a specialized solver can be more efficient [8]. Figure 1 shows a general architecture to solve MaxSMT as an implicit hitting set problem [18,19]. The method combines three components that are responsible for our three subtasks: OPT, an optimizer for hitting set computation; SAT, a SAT solver for Boolean reasoning; and Theories, a set of theory solvers to perform theory reasoning. The framework expects as input a MaxSMT formula (F) with a satisfiable set of hard clauses. The SAT and Theory solvers consider only the hard clauses, while the soft (unit) clauses and their costs are only considered by the optimizer. Note that we can initially check the hard clauses for satisfiability. If they are unsatisfiable there is no optimal solution.

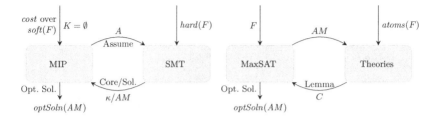

Fig. 2. Example merged (i.e. single-SAT) instantiations of our framework

The evaluation starts with OPT, which computes a (potentially optimal) hitting set η of the current set of unsatisfiable cores (K). This is translated into a set of assumptions (noted as A in Fig. 1) that requires the satisfaction of all soft clauses not in η (see **HS** and **MinHS** steps of **A-MaxSMT**).

The SAT solver can then decide if there exists a feasible solution satisfying $hard(F)$ and the assumptions. Theory solvers can be invoked at various points to check the T-consistency of the SAT solver's current partial assignment AM and to perform T-learning. As in [12] there are a range of flexible (e.g., more eager or more lazy) strategies for deciding when theory reasoning should be invoked.

For a given conjunction of theory literals, a theory solver might return a subset that is T-unsatisfiable forming a **conflict clause** after negation, or additional theory clauses for T-learning. In both cases the returned clauses are valid **lemmas** of the theory (C in Fig. 1). The SAT and theory solvers continue their collaboration under the assumption of A until either a theory consistent model of $hard(F) \wedge A$ is found (i.e. state $T\text{-}SAT(AM, F)$ is reached), or $hard(F) \wedge A$ is found to be unsatisfiable (i.e. state $conflict(F, C)$ is reached). In the latter case, the SAT solver constructs an unsatisfiable core (κ in Fig. 1) that consists of a subset of the soft clauses assumed to be satisfied in A. After that, the optimizer can compute a new hitting set that hits κ as well. Note that the new hitting set need not be of minimum cost. From the new hitting set, a new A is constructed and a new iteration starts. Any theory consistent model that is found for $hard(F) \wedge A$ is a feasible solution of the MaxSMT problem. The optimality of these solutions can be decided by the optimizer component based on their costs. In case the found solution is not optimal, a new hitting set is computed in order to find a better solution. Otherwise, the model is returned as a final optimal solution.

4.1   Possible Instantiations

A practical tool following our proposed general architecture can be achieved in various ways. Based on Fig. 1, one could combine a hitting set calculator with a SAT solver and a set of theory solvers. However, this implementation would not automatically benefit from the advanced techniques implemented in MaxSAT and SMT solvers nor from any future improvements to such solvers. So a more practical question is how to combine already existing tools to obtain a MaxSMT

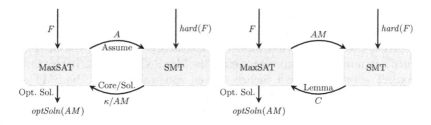

Fig. 3. Example combined (i.e. double-SAT) instantiations of our framework

solver. Here we consider mixed integer programming solvers (MIP), for example CPLEX, for solving the minimum hitting set problem since they are widely available and display state of the art performance on a range of instances.

As Fig. 1 hints, some MaxSAT solvers already implement efficient collaboration between MIP and SAT solvers, while SMT solvers combine SAT and theory solvers. Combining these solvers as black-boxes results in an engine that contains two SAT solvers, while merging these engines results in a tool with a single SAT solver. Figure 2 shows two possible instances of the latter case. On the left, we keep an SMT solver as a black-box and combine it with a MIP solver that is responsible for the hitting sets (and so the assumptions) in each iteration. A benefit of this instance is efficient SMT solving and the ability to use the full power of the MIP solver to express complex objective functions (e.g., multi-objective optimization). One disadvantage is the lack of MaxSAT preprocessing and simplifications. A tighter combination could replace the SAT solver in an IHS based MaxSAT solver with an SMT solver. However, in IHS MaxSAT solving SAT calls are considered relatively cheap compared to MIP calls [20], but SMT calls can be more expensive, so the tradeoffs of some techniques would have to be reevaluated. The instance on the right side of Fig. 2 considers a (not necessarily IHS based) MaxSAT solver as a black-box to find an optimal solution for the abstraction of the problem, and forms a lazy lemmas on demand structure with a set of theory solvers for theory consistency checks. The benefit here is efficient optimization solving, but the disadvantage is delayed theory support.

Instantiations in Fig. 3 present possibilities for combining black-box (i.e. not necessarily IHS based) MaxSAT and SMT solvers, providing the advantage of efficient optimization and SMT solving at the same time. These combinations contain multiple SAT solvers where the connecting interface determines the work distribution among them. On the left side, the solvers communicate via assumptions and cores or solutions. Whenever the MaxSAT solver finds an optimal propositional model for its current problem, the SMT solver has to verify that the soft clauses satisfied in that model are also T-satisfiable (via a set of assumptions that forces their satisfaction). If not, it returns a new core to refine the MaxSAT problem. In practice, the effectiveness of this instance would be compromised if many iterations are needed to refine the MaxSAT model. Another possible disadvantage of this instance is that the SMT solver could learn lemmas that would be useful to the MaxSAT engine but are never passed to it.

An alternate instance (right side in Fig. 3) involves the MaxSAT solver giving to the SMT solver the complete propositional model it found (the optimal model for its current problem). If that model is not theory consistent the SMT solver can return any number of lemmas to refine the MaxSAT problem. In this instance the MaxSAT solver can learn theory related constraints from the SMT solver beyond unsatisfiable cores. The approach introduced in [3] can be seen as a combination of the instances in Fig. 3. There the MaxSAT optimal model is used to provide assumptions to the SMT solver (as in the left-hand instance), but the SMT solver can return many lemmas to the MaxSAT solver not just cores (as in the right-hand instance). A potential drawback of that approach is that the returned lemmas might or might not be useful to the MaxSAT solver, and there is a risk of overloading the MaxSAT solver.

Based on these instances, it appears that support of assumption based incremental solving and efficient extraction of small cores are important features of the involved tools. Thus techniques that improve these aspects of solvers (e.g., [21]) have the potential to improve modular MaxSMT solvers as well. Further, note that our calculus allows the interruption of SMT calls in certain cases (see conditions in Theorem 3), which may be worth considering in practice.

5 Related Work

As argued in the introduction the focus of this paper is on the important class of MaxSMT solvers. Thus this section will concentrate on the closest related approaches. Additional experimental results are provided in the next section.

We modify and extend a general DPLL(T) framework introduced in [12] to formalize our MaxSMT solving approach. Another extension of DPLL(T) by Nieuwenhuis and Oliveras in [2] represents the optimization task explicitly as a set of theory constraints and progressively strengthens this theory by deriving tighter bounds. Our extension of DPLL(T) focuses only on MaxSMT problems and separates the optimization task from theory reasoning.

A modular approach was proposed by Cimatti et al. in [3] where MaxSAT and SMT solvers are employed as black-boxes for MaxSMT solving. As we showed in Sect. 4.1, our framework includes this approach. In Sect. 6 we present some empirical results comparing their approach with other instantiations.

In the context of core-guided MaxSAT solving, SMT solvers have been used instead of SAT, e.g., [22], to handle cardinality constraints more efficiently. We focus on IHS based MaxSMT solving in which no cardinality constraints are introduced into the SMT sub-problems.

Manolios et al. introduced the theoretical underpinnings of a Branch and Cut Modulo Theories framework and developed an optimization procedure where integer linear programming (ILP) and stably-infinite theories are combined [7]. Our approach delegates Boolean reasoning to a SAT solver, while in their construction this is done by the ILP solver.

6   Experimental Evaluation

We implemented two instantiations of our framework. Both use MathSAT5 [23] version 5.5.1 as the SMT component. Our first implementation **maxhs-msat** follows the architecture proposed on the left side of Fig. 3. It combines maxHS 3.0 as the optimizer with MathSAT5. To evaluate the potential of lifting theory lemmas to the MaxSAT level, as proposed in [3] and described in Sect. 4.1 as a combination of the instances in Fig. 3, the configuration **maxhs-msat-ll** lifts and adds all used theory lemmas to the MaxSAT solver in addition to unsatisfiable cores. Our second solver **cplex-msat** implements the architecture shown on the left of Fig. 2 which combines MathSAT5 directly with a hitting set solver (CPLEX 12.7 as in maxHS 3.0) as the optimizer. In this implementation the components interface only with assumptions and unsatisfiable cores.

In both solvers the optimizers compute an optimal hitting set η. In **maxhs-msat** maxHS computes an optimal solution to its current Boolean abstraction, but the clauses falsified by that solution form an optimal hitting set. The SMT solver then tests if the other soft clauses $(soft(F) - \eta)$ are T-satisfiable. If not, a new core is added to the optimizer (along with additional theory lemmas in **maxhs-msat-ll**). Following [20], rather than calling the optimizer in each iteration we allow non-optimal hitting sets. In particular, the new SMT core can be added to the previous hitting set (**-djnt**), or a single minimum weight clause from the new core can be added to the hitting set (**-min**). In both cases we obtain a new (non-minimum) hitting set covering the new core. For **cplex-msat** only, we can also use CPLEX to compute a linear programming solution of the hitting set problem which when rounded up yields a new hitting set (**-lp**). In these cases we continue to use non-minimum hitting sets η' until $soft(F) - \eta'$ becomes T-satisfiable, and then we again use the optimizer to compute a hitting set with minimum cost.

We compare against two state-of-the-art MaxSMT solvers. OptiMathSAT (version 1.4.5) [6] is a general purpose Optimization Modulo Theories (OMT) solver that we use in two different configurations. The default configuration is denoted by **optimathsat-omt**, while **optimathsat-maxres** employs the maximum resolution approach of [14]. We also compare against z3 (version 4.6.0) with two different MaxSAT engine configurations (**z3-maxres** and **z3-wmax**). Note, that the hitting set based engine in z3 has been deprecated and was removed.

We considered three sets of benchmarks from three different sources. The *LL-benchmark* set consists of all 398 quantifier free MaxSMT benchmarks used in [3] with annotations replaced by soft assertions, split into 212 benchmarks over the theory of linear integer arithmetic and 186 benchmarks over linear real arithmetic. For each theory, half the instances have **U**nit weight for soft assertions, while the other half contains **R**andom weights in the interval of 1 and 100. The runtime limit on these instances was set to 20 min.

Our second benchmark set *LEX-benchmark*, consisting of equalities over propositional atoms, are lexicographically-optimum realization problems used in [8]. We only considered the 6098 instances where three groups of soft assertions (**T**ime, **C**ost and **W**eight) have different priorities and the objective is to

Table 4. Results of various solvers and configurations on LL-benchmarks from [3].

Solver	LIA (212)		LRA (186)		Total	SMT%	OPT%
	U	R	U	R			
cplex-msat	82	90	85	85	342	99.22%	0.13%
cplex-msat-djnt	85	**91**	85	85	346	98.83%	0.33%
cplex-msat-min	83	86	85	85	339	99.22%	0.04%
cplex-msat-lp	84	89	85	85	343	98.26%	0.97%
maxhs-msat	85	87	85	85	342	88.15%	11.20%
maxhs-msat-djnt	86	89	85	85	345	83.85%	15.36%
maxhs-msat-min	84	89	85	85	343	92.31%	7.04%
maxhs-msat-ll	80	84	83	78	325	82.57%	15.45%
maxhs-msat-ll-djnt	78	84	83	77	322	87.97%	10.37%
maxhs-msatll-min	79	86	82	85	332	80.13%	17.03%
optimathsat-maxres	**87**	90	85	86	**348**	–	–
optimathsat-omt	75	72	85	85	317	–	–
z3-maxres	73	79	86	85	323	–	–
z3-wmax	69	77	**88**	**88**	322	–	–

lexicographically minimize the sum of the falsified assertions with respect to a given priority order of **T**, **C**, **W**). The time limit was set to 100 s.

Finally, in order to further exercise the strengths of the different approaches, we generated a set of scaled problems from one (arbitrarily chosen) QF-LIA SMT-LIB benchmark family (Bofill-scheduling waste water treatment scheduling problems from [24]). The original family contained 156 randomly generated (referred as *rand-wwtp*) and 251 industrial (*ind-wwtp*) satisfiable SMT problems. We derived instances from these SMT problems by adding randomly chosen theory atoms with random polarity as unit soft clauses. The four groups of derived instances introduced four different percentages (10%, 25%, 50% and 100%) of the atoms in the original problem as soft assertions. All instances were generated once with unit weights and once more with random weights between 1 and the total number of atoms. Due to space constraints, we only present results on instances derived from *rand-wwtp* problems, where we observed an interesting pattern. The time limit was set to 5 min.

The experiments were performed on a cluster in which each computing node consisted of two Intel(R) Xeon(R) E5-2620 v4 @ 2.10 GHz CPUs and 128 GB of main memory. We limited memory usage of each tool to 7 GB on each instance and used different time limits as described above.

Table 4 presents results on the LL-benchmarks. For each solver configuration the first two columns list the number of solved instances with linear integer arithmetic as background theory, where the soft assertions have **U**nit weights in the first column and **R**andom weights in the second. Analogously, the next

Table 5. Results of various solvers and configurations on LEX-benchmarks from [8].

Solver	CTW	Time[s]	WTC	Time[s]	Cores	Opt. HS
cplex-msat	3499	27825	2399	1942	1610031	1615150
cplex-msat-djnt	3687	5936	2399	1455	920387	137339
cplex-msat-min	3699	2479	2399	1391	909828	27245
cplex-msat-lp	3699	4564	2399	1493	1260683	19056
maxhs-msat	3699	2401	2399	1367	0	5319
maxhs-msat-djnt	3699	**2224**	2399	**1359**	0	5319
maxhs-msat-min	3699	2451	2399	1409	0	5319
maxhs-msat-ll	3699	2302	2399	1518	0	5319
maxhs-msat-ll-djnt	3699	2394	2399	1406	0	5319
maxhs-msatll-min	3699	2441	2399	1437	0	5319
optimathsat-maxres	3410	13851	1850	10209	–	–
optimathsat-omt	3481	9710	2068	10483	–	–
z3-maxres	3699	4555	2399	2231	–	–
z3-wmax	3651	5566	2295	9513	–	–

two columns present results in linear real arithmetic. The fifth column contains the total sum of solved instances in the previous four columns. The last two columns show the percentage of time spent in the SMT and in the optimization component (considering only solved instances). The optimization component in **cplex-msat** is CPLEX, while in **maxhs-msat** it is maxHS.

It turns out that **optimathsat-maxres** outperforms the other tools and configurations on these instances, but our implementations remain competitive. Furthermore, lemma lifting (**maxhs-msat-ll** and its different configurations) reduces the percentage time spent in SMT solving, but has a negative effect with respect to the number of solved instances compared to **maxhs-msat** and its different configurations. None of the involved tools appears to be sensitive to the type of weights (**Uniform** vs. **Random**). Although **cplex-msat** does not contain any MaxSAT preprocessing or simplification technique, the results of that tool in this experiment are similar to **maxhs-msat**.

Results on the LEX-benchmark are shown in Table 5. The 6098 problems contained two groups of problems. The first group of 3699 instances used the lexicographic preference ordering **C**ost, **T**ime and then **W**eight, and are shown in the first two columns which list the number of solved instances and the total run time used to solve them. The second group of 2399 instances used the reversed lexicographic preference and are shown in the next two columns. For our tools we also give the total number of unsatisfiable cores and of optimal hitting set calculations (considering again only solved instances) in the last two columns.

On these instances most versions of our approach solve at least as many problems as the state-of-the-art tools and in significantly less time. Due to the background theory of these instances it is enough to find a propositional model,

Table 6. Results of considered solvers and configurations on rand-wwtp family with 10%-100% random unit soft clauses. Each percentage group consists of 312 problems.

Solver	10%	25%	50%	100%	Total	SMT%	OPT%
cplex-msat	289	271	**203**	4	**767**	60.85%	38.46%
cplex-msat-djnt	286	247	114	2	649	97.35%	1.96%
cplex-msat-min	282	244	142	**16**	684	91.46%	7.68%
cplex-msat-lp	287	262	184	13	746	83.4%	15.27%
maxhs-msat	288	270	179	0	737	42.28%	57.31%
maxhs-msat-djnt	289	249	112	1	651	93.91%	5.69%
maxhs-msat-min	281	242	132	15	670	87.99%	11.59%
maxhs-msat-ll	266	166	16	0	448	7.69%	84.93%
maxhs-msat-ll-djnt	266	161	9	0	436	11.30%	77.59%
maxhs-msatll-min	263	166	27	0	456	11.36%	68.11%
optimathsat-maxres	291	258	123	0	672	–	–
optimathsat-omt	240	130	0	0	370	–	–
z3-maxres	280	224	103	0	607	–	–
z3-wmax	**304**	**288**	4	0	596	–	–

i.e., solve a MaxSAT problem, since every propositional solution also happens to be T-satisfiable. This is reflected in the last two columns, where the two instantiations of our framework show different behaviour. For **maxhs-msat**, which combines maxHS with the SMT solver, the number of iterations is always one (in all 5319 satisfiable instances). In this case maxHS finds an optimal Boolean model (through several iterations of its *internal* SAT solver), which the SMT solver then verifies to be theory consistent in one call. In case of **cplex-msat** there is no additional SAT solver between the SMT and the optimization components. Therefore it has to learn all the necessary transitivity properties of the equalities in form of cores from the SMT solver. Thus the number of unsatisfiable cores is higher for **cplex-msat**, which can significantly increase solving time depending on the type of hitting sets used.

These benchmarks in essence allow us to compare the effectiveness of the optimization components independently of the SMT component. This benefits our hitting set based methods, while other solvers rely on alternative approaches. Another important difference is that our prototypes solve lexicographic problems as single objective functions in one run by aggregating the cost functions [25].

The last table (Table 6) presents results for the randomized rand-wwtp benchmarks on which **cplex-msat** performs better than **maxhs-msat**. Using nonminimum hitting sets measurably reduces the performance of both implementations on these instances. From the last two columns we can deduce that the best performing methods are those where more time was spent within the optimization component. Although lemma lifting does result in significant more time spent in maxHS calls, some part of it is spent in the SAT solver, and not in actual optimization. This might explain its bad performance.

To summarize, the experiments support the need for a generic framework for MaxSMT. More concretely we make the following three observations. First, there is no overall best configuration. Performance depends on the distribution of the workload among the involved components, since in general the difficulty of the optimization and SMT problems differ. For instance, improved MaxSAT performance does not necessarily translate into improved MaxSMT performance, simply because of different relative costs between SMT calls and SAT calls. Accordingly, non-minimum hitting sets (like disjoint cores or LP relaxation) usually reduce the workload of the optimizer but put more stress on the SMT solver.

Second, the number of extracted unsatisfiable cores or calculated optimal hitting sets is not always an expedient metric to measure the performance of MaxSMT. Finally, most of the time, lemma lifting does not improve but actually seems to reduce performance of a modular MaxSMT solver, particularly with an implicit hitting set based approach.

All of our experimental results as well as the evaluated benchmarks are available at http://fmv.jku.at/maxsmt/.

7   Conclusion

We have proposed an abstract framework to gain a unifying view of how optimization, propositional reasoning, and theory reasoning can be combined in IHS based MaxSMT solving. Our framework is very flexible supporting a rich space of possible implementation architectures all of which are provably sound. Our empirical results show that different architectures yield quite different performance on different problems sets. This implies that there is considerable potential in more fully exploiting the flexibility of our framework to obtain improved and more robust performance in MaxSMT solvers.

Acknowledgments. This research has been supported by the Austrian Science Fund (FWF) under projects W1255-N23 and S11408-N23.

References

1. Li, Y., Albarghouthi, A., Kincaid, Z., Gurfinkel, A., Chechik, M.: Symbolic optimization with SMT solvers. In: POPL, pp. 607–618. ACM (2014)
2. Nieuwenhuis, R., Oliveras, A.: On SAT modulo theories and optimization problems. In: Biere, A., Gomes, C.P. (eds.) SAT 2006. LNCS, vol. 4121, pp. 156–169. Springer, Heidelberg (2006). https://doi.org/10.1007/11814948_18
3. Cimatti, A., Griggio, A., Schaafsma, B.J., Sebastiani, R.: A modular approach to MaxSAT modulo theories. In: Järvisalo, M., Van Gelder, A. (eds.) SAT 2013. LNCS, vol. 7962, pp. 150–165. Springer, Heidelberg (2013). https://doi.org/10.1007/978-3-642-39071-5_12

4. Sebastiani, R., Tomasi, S.: Optimization modulo theories with linear rational costs. ACM Trans. Comput. Log. **16**(2), 12:1–12:43 (2015)
5. Bjørner, N., Phan, A.-D., Fleckenstein, L.: νZ - an optimizing SMT solver. In: Baier, C., Tinelli, C. (eds.) TACAS 2015. LNCS, vol. 9035, pp. 194–199. Springer, Heidelberg (2015). https://doi.org/10.1007/978-3-662-46681-0_14
6. Sebastiani, R., Trentin, P.: OptiMathSAT: a tool for optimization modulo theories. In: Kroening, D., Păsăreanu, C.S. (eds.) CAV 2015. LNCS, vol. 9206, pp. 447–454. Springer, Cham (2015). https://doi.org/10.1007/978-3-319-21690-4_27
7. Manolios, P., Pais, J., Papavasileiou, V.: The Inez mathematical programming modulo theories framework. In: Kroening, D., Păsăreanu, C.S. (eds.) CAV 2015. LNCS, vol. 9207, pp. 53–69. Springer, Cham (2015). https://doi.org/10.1007/978-3-319-21668-3_4
8. Sebastiani, R., Trentin, P.: On optimization modulo theories, MaxSMT and sorting networks. In: Legay, A., Margaria, T. (eds.) TACAS 2017. LNCS, vol. 10206, pp. 231–248. Springer, Heidelberg (2017). https://doi.org/10.1007/978-3-662-54580-5_14
9. Ansótegui, C., Bacchus, F., Järvisalo, M., Martins, R.: MaxSAT evaluation 2017 (2017). http://mse17.cs.helsinki.fi/
10. Bacchus, F., Järvisalo, M.: Algorithms for maximum satisfiability with applications to AI. In: AAAI-2016 Tutoral (2016). https://www.cs.helsinki.fi/group/coreo/aaai16-tutorial/
11. Davies, J., Bacchus, F.: Solving MAXSAT by solving a sequence of simpler SAT instances. In: Lee, J. (ed.) CP 2011. LNCS, vol. 6876, pp. 225–239. Springer, Heidelberg (2011). https://doi.org/10.1007/978-3-642-23786-7_19
12. Nieuwenhuis, R., Oliveras, A., Tinelli, C.: Solving SAT and SAT modulo theories: from an abstract Davis-Putnam-Logemann-Lovel and procedure to DPLL(T). J. ACM **53**(6), 937–977 (2006)
13. Sebastiani, R.: Lazy satisability modulo theories. JSAT **3**(3–4), 141–224 (2007)
14. Narodytska, N., Bacchus, F.: Maximum satisfiability using core-guided MaxSAT resolution. In: AAAI, pp. 2717–2723. AAAI Press (2014)
15. Martins, R., Joshi, S., Manquinho, V., Lynce, I.: Incremental cardinality constraints for MaxSAT. In: O'Sullivan, B. (ed.) CP 2014. LNCS, vol. 8656, pp. 531–548. Springer, Cham (2014). https://doi.org/10.1007/978-3-319-10428-7_39
16. Ansótegui, C., Bonet, M.L., Levy, J.: SAT-based MaxSAT algorithms. Artif. Intell. **196**, 77–105 (2013)
17. Eén, N., Sörensson, N.: An extensible SAT-solver. In: Giunchiglia, E., Tacchella, A. (eds.) SAT 2003. LNCS, vol. 2919, pp. 502–518. Springer, Heidelberg (2004). https://doi.org/10.1007/978-3-540-24605-3_37
18. Chandrasekaran, K., Karp, R.M., Moreno-Centeno, E., Vempala, S.: Algorithms for implicit hitting set problems. In: SODA, SIAM, pp. 614–629 (2011)
19. Saikko, P., Wallner, J.P., Järvisalo, M.: Implicit hitting set algorithms for reasoning beyond NP. In: KR, pp. 104–113. AAAI Press (2016)
20. Davies, J., Bacchus, F.: Postponing optimization to speed up MAXSAT solving. In: Schulte, C. (ed.) CP 2013. LNCS, vol. 8124, pp. 247–262. Springer, Heidelberg (2013). https://doi.org/10.1007/978-3-642-40627-0_21
21. Lagniez, J.-M., Biere, A.: Factoring out assumptions to speed Up MUS extraction. In: Järvisalo, M., Van Gelder, A. (eds.) SAT 2013. LNCS, vol. 7962, pp. 276–292. Springer, Heidelberg (2013). https://doi.org/10.1007/978-3-642-39071-5_21
22. Ansótegui, C., Gabàs, J., Levy, J.: Exploiting subproblem optimization in SAT-based MaxSAT algorithms. J. Heuristics **22**(1), 1–53 (2016)

23. Cimatti, A., Griggio, A., Schaafsma, B.J., Sebastiani, R.: The MathSAT5 SMT solver. In: Piterman, N., Smolka, S.A. (eds.) TACAS 2013. LNCS, vol. 7795, pp. 93–107. Springer, Heidelberg (2013). https://doi.org/10.1007/978-3-642-36742-7_7

24. Bofill, M., Muñoz, V., Murillo, J.: Solving the wastewater treatment plant problem with SMT. CoRR abs/1609.05367 (2016)

25. Marques-Silva, J., Argelich, J., Graça, A., Lynce, I.: Boolean lexicographic optimization: algorithms & applications. Ann. Math. AI **62**(3–4), 317–343 (2011)

Cubicle-$\mathcal{W}$: Parameterized Model Checking on Weak Memory

Sylvain Conchon[1,2], David Declerck[1,2(✉)], and Fatiha Zaïdi[1]

[1] LRI (CNRS & University Paris-Sud), Université Paris-Saclay,
F-91405 Orsay, France
david.declerck@u-psud.fr
[2] Inria, Université Paris-Saclay, F-91120 Palaiseau, France

Abstract. We present Cubicle-$\mathcal{W}$, a new version of the Cubicle model checker to verify parameterized systems under weak memory models. Its main originality is to implement a backward reachability algorithm modulo weak memory reasoning using SMT. Our experiments show that Cubicle-$\mathcal{W}$ is expressive and efficient enough to automatically prove safety of concurrent algorithms, for an arbitrary number of processes, ranging from mutual exclusion to synchronization barriers.

Keywords: Parameterized model checking · MCMT · SMT
Weak memory

1 Introduction

Concurrent algorithms are usually designed under the sequential consistency (SC) memory model [20] which enforces a global-time linear ordering of (read or write) accesses to shared memories. However, modern multiprocessor architectures do not follow this SC semantics. Instead, they implement several optimizations which lead to relaxed consistency models on shared memory where read and write operations may be reordered. For instance, in x86-TSO [21,22] writes can be delayed after reads due to a store buffering mechanism. Other relaxed models (PowerPC [6], ARM) allow even more types of reorderings.

The new behaviors induced by these models may make out-of-the-shelf algorithms incorrect for subtle reasons mixing interleaving and reordering of events. In this context, finding bugs or proving the correctness of concurrent algorithms is very challenging. The challenge is even more difficult if we consider that most algorithms are *parameterized*, that is designed to be run on architectures containing an arbitrary (large) number of processors.

One of the most efficient technique for verifying concurrent systems is model checking. While this technique has been used to verify parameterized algorithms [2,4,5,9,12,16] and systems under some relaxed memory assumptions [2,3,7,10,11], hardly any state-of-the-art model checker support *both* parameterized verification and weak memory models [2].

Work supported by the French ANR project PARDI (ANR-16-CE25-0006).

D. Galmiche et al. (Eds.): IJCAR 2018, LNAI 10900, pp. 152–160, 2018.
https://doi.org/10.1007/978-3-319-94205-6_11

In this paper, we present Cubicle-$\mathcal{W}$ [1], the new version of the Cubicle [13–15] model checker for verifying safety properties of parameterized array-based transition systems on weak memory. Cubicle-$\mathcal{W}$ is a conservative extension which allows the user to manipulate both SC and weak variables. Its relaxed consistency model is similar to x86-TSO : each process has a FIFO buffer of pending store operations whose side effect is to delay the outcome of its memory writes to all processes.

Like Cubicle, Cubicle-$\mathcal{W}$ is based on the MCMT framework of Ghilardi and Ranise [17]. Its core extends the SMT-based backward reachability procedure with a new pre-image computation which takes into account the delays between write and read operations. In order to consider only coherent read/write pairs, Cubicle-$\mathcal{W}$ relies on a buffer-free memory model inspired by the logical framework of [8] which is implemented as a new theory in its SMT solver. Cubicle-$\mathcal{W}$ is an open-source software freely available at http://cubicle.lri.fr/cubiclew.

2   Tool Presentation

The syntax of Cubicle-$\mathcal{W}$ extends Cubicle's with new constructs for manipulating weak memories. The reader can refer to [13] for the description of Cubicle's input language.

Variable and array declarations can now be prefixed by the keyword **weak** for defining weak memories.

```
weak var X : int
weak array A[proc] : bool
```

Transitions in Cubicle-$\mathcal{W}$ have the same syntactic guard/action form as in Cubicle and they are also supposed to be executed atomically. The new feature is that they must now have *at least* one parameter which represents the process that performs the operations. This parameter is identified using the [.] notation. For instance, in the following example, the parameter [i] of transition t1 represents the process performing all read/write operations on X, A[i] and A[j] when t1 is triggered.

```
transition t1 ([i] j)
requires { X = 42 && A[i] = False }
{ A[j] := False }
```

Even if there is no use of parameter [i] in transitions' guards and actions, this parameter is still mandatory, as in the transition t2 below, to indicate which process performs the operations.

```
transition t2 ([i]) { X := 42 }
```

Note that, as Cubicle-$\mathcal{W}$'s transitions are atomic, having several processes performing reads or writes operations in the same transition would require an unrealistic powerful synchronization mechanism between processes.

The main aspect of our relaxed memory semantics is that, from a global viewpoint, the effect of a write operation on a `weak` memory is not immediately visible to all processes. It is only locally visible to the process that performs it. For instance, if some process `i` executes the transition `t2` above, then `X = 42` is true for `i` after the transition (as the effect of the assignment is immediately locally visible), while all other processes can still read a different value for `X`.

To enforce the global visibility of a write operation, one has to use a *memory barrier*. In Cubicle-$\mathcal{W}$, barriers are provided as a new built-in predicate `fence()`. When used in the guard of a transition, `fence` is true only when the FIFO buffer of the parameter `[i]` of the transition is empty. For instance, if a process executes `t2` then the following transition `t3`:

transition `t3` `([i])` **requires** `{ fence() }{ ... }`

The `fence` predicate in `t3`'s guard ensures that the effect of all previous assignments done by `i` are visible to all processes after `t3`. Note that `fence` is not an action: it does not force buffers to be flushed on memory, but just *waits* for a buffer to be empty. As a consequence, it can only be used in a guard.

Implicit memory barriers are also activated when a transition contains both a read and a write to weak variables (not necessarily the same). For instance, the execution of the following transition `t4` guarantees that the buffer of process `i` is empty *before* and *after* `t4`.

transition `t4` `([i])`
requires `{ A[i] = False }`
`{ X := 1 }`

Because there is no unique view of the contents of weak variables, one can not talk about *the value* of `X`, but rather the value of `X` from the *point of view* of a process `i`, denoted `i@X` in Cubicle-$\mathcal{W}$. This notation is used when describing unsafe states. For instance, in the following formula, a state is defined as unsafe when there exist two (distinct) processes `i` and `j` reading respectively `42` and `0` in the weak variable `X`:

`unsafe (i j) { i@X = 42 && j@X = 0 }`

This notation is not used for describing initial states as Cubicle-$\mathcal{W}$ implicitly assumes that *all* processes have the same view of each weak variable in those states. For instance, the following formula defines initial states where, for all processes, `X` equals `0` and all cells of array `A` contain `False`.

`init (i) { X = 0 && A[i] = False }`

Finally, it is important to note that non weak arrays are restricted to be used only locally by processes: given a non weak array `T`, only `i` can read or write to `T[i]`.

3   Backward Reachability Modulo Weak Memory

The core of Cubicle-$\mathcal{W}$ is an extension of Cubicle's symbolic backward reachability algorithm [13,14]. We first briefly recall how the original Cubicle works, then we give details about our new algorithm.

States in Cubicle are represented by cubes, i.e., formulas of the form $\exists \bar{i}.(\Delta \wedge F)$, where $\bar{i}$ is a set of process variables, Δ is the conjunction of all disequations between the variables in $\bar{i}$ and F is a conjunction of literals. Each literal in F is a comparison ($=, \neq, <, \leq$) between two terms. A term can be a constant (integer, boolean, real, constructor), a process variable (i), a variable (X) or an array access ($A[i]$, where i is a process variable). All process variables in a state are implicitly existentially quantified. Initial states are represented by a universally quantified formula I of the form $\forall \bar{i}.(\Delta \wedge F)$, where Δ and F are as described above.

The core of Cubicle is a symbolic backward reachability loop that maintains two collections of states: $\mathcal{Q}$ contains the states to visit (it is initialized with the states declared as **unsafe**), and $\mathcal{V}$ is filled with the visited states (initially empty). Each iteration of the loop performs the following operations:

1. (*pop*) retrieve and remove a formula φ from $\mathcal{Q}$
2. (*safety test*) check the satisfiability of $\varphi \wedge I$, i.e. determine if the states described by φ intersect with the initial states I. If so, the system is declared as *unsafe*
3. (*fixpoint test*) check if $\varphi \models \mathcal{V}$ is valid, i.e. determine if the states described by φ have already been visited. If so, discard φ and go back to 1
4. (*pre-image computation*) compute the pre-image $pre(\varphi, t)$ of φ for all instances of transitions t, i.e. determine the set of states that can reach φ in one step by applying t with the processes identifiers $\#1, \ldots, \#n$ as parameters, add these states to $\mathcal{Q}$ and add φ to $\mathcal{V}$.

If $\mathcal{Q}$ is empty at step 1, then all the states space has been explored and the system is declared *safe*. Note that the (non-trivial) fixpoint and safety tests are discharged to an embedded SMT solver.

Cubicle-$\mathcal{W}$ uses the same procedure but some operations have been extended to reason modulo an axiomatic description of our weak memory model. This axiomatization uses the notion of *events* to describe weak memory accesses and a global-happens-before (ghb) relation defined as a partial order relation over these events. This relation is used to determine if an execution is valid.

Our logic is extended with new literals to represent read and write operations on weak memories. We assume given a (countable) set of events $\mathcal{E}$. A literal of the form $e{:}\mathtt{Rd}_X(i)$ denotes a read access on variable X by a process i labeled with an event identifier $e \in \mathcal{E}$. Similarly, literals of the form $e{:}\mathtt{Wr}_X(i)$ represent write accesses. The value returned by a read (resp. assigned by a write) is given by the term $val(e)$, where e is the event identifier associated to the operation. Operations on weak arrays are represented by literals of the form $e{:}\mathtt{Rd}_A(i,j)$ and $e{:}\mathtt{Wr}_A(i,j)$, which represent an access by a process i to the cell j of an array A. Last, there is also literals of the form $e{:}\mathtt{fence}(i)$ which indicate that a process

i has a memory barrier on the event e, where e is an event identifier associated to a read by the same process.

The reachability loop of Cubicle-$\mathcal{W}$ implements a new pre-image computation. At step 4, $pre(\varphi, t)$ is modified so that read and write operations from t give rise to Rd and Wr literals labeled with fresh event identifiers. These new events are ordered w.r.t the older ones in the ghb relation expressed by predicates of the form $ghb(e_1, e_2)$, indicating that event e_1 is $ghb\text{-}before$ (i.e., occurs before) event e_2. The ghb-ordering of events is built w.r.t. the following rules:

- New read events are $ghb\text{-}before$ old read and write events from the same process.
- New write events are $ghb\text{-}before$ old write events from the same process, however they are $ghb\text{-}before$ old reads events from the same process only if there is a fence on these reads.
- New write events are $ghb\text{-}before$ all the old write events to the same variable.
- New read events are $ghb\text{-}before$ all the old write events to the same variable.

Finally, when a memory fence is encountered, a literal e:fence(i) is added on all old reads events e which belong to the process i executing the transition.

The treatment of write events is also specific when we have to consider the delays introduced by store buffers: when a new write event e is produced, all possible combinations of e with older compatible reads are considered (unlike in SC), as a write operation *may or may not* satisfy subsequent reads. By *compatible read*, we mean a read on the same variable or array cell as the write, though we may also consider the constant values associated to these events in order to obtain a more accurate set of compatible reads. The connection between a write and an older read obeys the following rules:

- When the write event satisfies an old read event from a different process, the write is $ghb\text{-}before$ the read.
- When the write event does not satisfy an old read event from a different process, the read is $ghb\text{-}before$ the write.
- When the write and the read events belong to the same process, none of them is considered $ghb\text{-}before$ the other (unless there is a fence on the read event).

In order to show how our reachability procedure works, we consider the simple parameterized mutual exclusion algorithm and the exploration graph given below. Cubicle-$\mathcal{W}$ starts with the unsafe formula in node 1. Then, each node represents the result of a pre-image computation by an instance of a transition (denoted by the label of the edge). Remark that formulas in the graph's nodes are implicitly existentially quantified and that a process identifier i is written #$_i$.

```
type loc = Idle | Want | Crit

weak array X[proc] : bool
array PC[proc] : loc

init (i) {PC[i] = Idle && X[i] = False}

unsafe (i j) {PC[i] = Crit && PC[j] = Crit}
```

```
transition t_req ([i])
requires { PC[i] = Idle }
{ X[i] := True ; PC[i] := Want }

transition t_enter ([i])
requires { PC[i] = Want && fence() &&
                forall_other k. X[k] = False }
{ PC[i] := Crit }

transition t_exit ([i])
requires { PC[i] = Crit }
{ X[i] := False ; PC[i] := Idle }
```

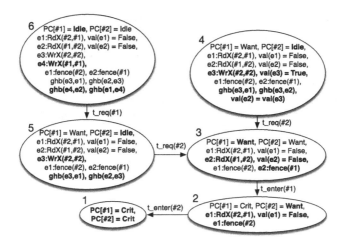

We focus on node 3 which results from the pre-image of node 1 by t_enter(#2) then t_enter(#1). In this state, both processes have read False in X (events e_1 and e_2). Also, since there is a memory barrier in t_enter, both reads are associated to a fence literal. The pre-image of node 3 by t_req(#2) introduces a new write event e3:WrX(#2,#2) with an associated value val(e3) = True. Since there is a memory barrier e1:fence(#2) on e1 by the same process #2, we add ghb(e3,e1) in the formula. Now, this new write event may or may not satisfy the read e2, so we must consider both cases (node 4 and 5).

In node 4, event e3 satisfies e2. The equality val(e2) = val(e3) is then added to the formula which obviously makes it inconsistent. In node 5, the write e3 does not satisfy the read e2, then the value val(e3) is discarded and ghb(e2,e3) is added to the formula. Similarly, the pre-image of node 5 by t_req(#1) yields the formula in state 6 where the new write e4 does not satisfy the read e1. Now, the *ghb* relation is not a valid partial order as the sequence ghb(e2, e3), ghb(e3, e1), ghb(e1, e4), ghb(e4, e2) forms a cyclic relation. Therefore, this state is discarded and the program is declared *safe*.

Remark that if we removed the fence predicate in t_enter, then we would only have ghb(e3, e1), ghb(e4, e2) in state 6, which is a valid partial order relation, so the formula would intersect with the initial state and the program would be *unsafe*.

4 Benchmarks and Conclusion

We have evaluated Cubicle-$\mathcal{W}$ on some classical parameterized concurrent algorithms (available on the tool's webpage [1]). Most of these algorithms are abstraction of real world protocols, expressed with up to eight transitions and up to four weak variables or two unbounded weak arrays. The spinlock example is a manual translation of an actual x86 implementation of a spinlock from the Linux 2.6 kernel. We compared Cubicle-$\mathcal{W}$'s performances with state-of-the-art model checkers supporting the TSO weak memory model, since our model is similar.

The model checkers we used are CBMC [7], Trencher [10,11], MEMORAX [3] and Dual-TSO [2]. As most of these tools do not support parameterized systems, we used them on fixed-size instances of our benchmarks and increased the number of processes until we obtained a timeout (or until we reached a high number of processes, *i.e.* 11 in our case). Dual-TSO supports a restricted form of parameterized systems, but does not allow process-indexed arrays, which are often needed to express parameterized programs. When it was possible, we used it on both parameterized and non parameterized versions of our benchmarks.

		Cubicle $\mathcal{W}$	Memorax SB	Memorax PB	Trencher	CBMC Unwind 2	CBMC Unwind 3	Dual TSO
naive mutex	US	0.04s [N]	–	–	–	–	–	NT [N]
			TO [6]	TO [10]	TO [5]	23.6s [11]	5m37s [11]	TO [6]
			7m54s [5]	12m02s [9]	10.1s [4]	14.7s [10]	3m39s [10]	1m12s [5]
naive mutex	S	0.30s [N]	–	–	–	–	–	NT [N]
			TO [5]	TO [11]	TO [6]	TO [5]	TO [3]	TO [5]
			23.3s [4]	2m28s [10]	54.8s [5]	2m24s [4]	19.4s [2]	35.7s [4]
lamport	US	0.10s [N]	–	–	–	–	–	NT [N]
			TO [4]	TO [4]	KO [4]	7m42s [11]	TO [7]	TO [6]
			17.4s [3]	25.4s [3]	1.73s [3]	4m29s [10]	5m12s [6]	13m12s [5]
lamport	S	0.60s [N]	–	–	–	–	–	NT [N]
			TO [3]	TO [4]	KO [5]	TO [4]	TO [3]	TO [4]
			0.14s [2]	3m02s [3]	3.37s [4]	8m39s [3]	1m55s [2]	9.42s [3]
spinlock [22]	S	0.07s [N]	–	–	–	–	–	TO [N]
			TO [5]	TO [7]	TO [7]	TO [3]	TO [3]	TO [6]
			8m51s [4]	9m52s [6]	21.45s [6]	19.58s [2]	5m08s [2]	1m16s [5]
sense reversing barrier [19]	S	0.06s [N]	–	–	–	–	–	NT [N]
			TO [3]	TO [3]	TO [5]	TO [9]	TO [4]	TO [3]
			0.34s [2]	0.09s [2]	1m58s ☠ [4]	12m25s [8]	1m43s [3]	0.09s [2]
arbiter v1 [18]	S	0.18s [N]	–	–	–	–	–	NT [N]
			TO [1+2]	TO [1+2]	KO [1+5]	TO [1+6]	TO [1+3]	TO [1+6]
					4.57s [1+4]	12m02s [1+5]	44.3s [1+2]	2m45s ☠ [1+5]
arbiter v2 [18]	S	13.5s [N]	–	–	–	–	–	NT [N]
			TO [1+2]	TO [1+2]	KO [1+4]	TO [1+4]	TO [1+2]	TO [1+3]
					1.62s [1+3]	2m56s [1+3]		24.2s [1+2]
two phase commit	S	54.1s [N]	–	–	–	–	–	NT [N]
			TO [2]	TO [4]	TO [4]	TO [11]	TO [11]	TO [3]
				39.7s [3]	7.08s ☠ [3]	12m39s [10]	13m41s [10]	12.3s [2]

The table above gives the running time for each benchmark, with the number of processes between square brackets, where N indicates the parametric case. The second column indicates whether the program is expected to be unsafe (US) or safe (S). Unsafe programs have a second version that was fixed by adding fence predicates. ☠ indicates that a tool gave a wrong answer. **KO** means that a tool crashed. **NT** indicates a benchmark that was not translatable to Dual-TSO.

The tests were run on a MacBook Pro with an Intel Core i7 CPU @ 2,9 Ghz and 8GB of RAM, under OSX 10.11.6. The timeout (TO) was set to 15 min.

These results show that in spite of the relatively small size of each benchmark, state-of-the-art model checkers suffer from scalability issues, which justifies the use of parameterized techniques. Cubicle-$\mathcal{W}$ is thus a very promising approach to the verification of concurrent programs that are both parameterized and operating under weak memory. We have yet to tackle larger programs, which can be achieved by adapting Cubicle's invariant generation mechanism to our weak memory model.

References

1. Cubicle-$\mathcal{W}$. http://cubicle.lri.fr/cubiclew/
2. Abdulla, P.A., Atig, M.F., Bouajjani, A., Ngo, T.P.: The benefits of duality in verifying concurrent programs under TSO. In: CONCUR (2016)
3. Abdulla, P.A., Atig, M.F., Chen, Y., Leonardsson, C., Rezine, A.: Memorax, a precise and sound tool for automatic fence insertion under TSO. In: TACAS (2013)
4. Abdulla, P.A., Delzanno, G., Henda, N.B., Rezine, A.: Regular model checking without transducers. In: TACAS (2007)
5. Abdulla, P.A., Delzanno, G., Rezine, A.: Parameterized verification of infinite-state processes with global conditions. In: Damm, W., Hermanns, H. (eds.) CAV 2007. LNCS, vol. 4590, pp. 145–157. Springer, Heidelberg (2007). https://doi.org/10.1007/978-3-540-73368-3_17
6. Alglave, J., Fox, A., Ishtiaq, S., Myreen, M.O., Sarkar, S., Sewell, P., Nardelli, F.Z.: The semantics of power and arm multiprocessor machine code. In: DAMP (2008)
7. Alglave, J., Kroening, D., Nimal, V., Tautschnig, M.: Software Verification for weak memory via program transformation. In: Felleisen, M., Gardner, P. (eds.) ESOP 2013. LNCS, vol. 7792, pp. 512–532. Springer, Heidelberg (2013). https://doi.org/10.1007/978-3-642-37036-6_28
8. Alglave, J., Maranget, L., Tautschnig, M.: Herding cats: modelling, simulation, testing, and data mining for weak memory. In: ACM TPLS (2014)
9. Apt, K.R., Kozen, D.C.: Limits for automatic verification of finite-state concurrent systems. Inf. Process. Lett. **22**(6), 307–309 (1986)
10. Bouajjani, A., Calin, G., Derevenetc, E., Meyer, R.: Lazy tso reachability. In: FASE (2015)
11. Bouajjani, A., Derevenetc, E., Meyer, R.: Checking and enforcing robustness against TSO. In: Felleisen, M., Gardner, P. (eds.) ESOP 2013. LNCS, vol. 7792, pp. 533–553. Springer, Heidelberg (2013). https://doi.org/10.1007/978-3-642-37036-6_29
12. Clarke, E.M., Grumberg, O., Browne, M.C.: Reasoning about networks with many identical finite-state processes. In: PODC (1986)
13. Conchon, S., Goel, A., Krstić, S., Mebsout, A., Zaïdi, F.: Cubicle: a parallel SMT-based model checker for parameterized systems. In: Madhusudan, P., Seshia, S.A. (eds.) CAV 2012. LNCS, vol. 7358, pp. 718–724. Springer, Heidelberg (2012). https://doi.org/10.1007/978-3-642-31424-7_55
14. Conchon, S., Goel, A., Krstic, S., Mebsout, A., Zaidi, F.: Invariants for finite instances and beyond. In: FMCAD (2013)
15. Conchon, S., Mebsout, A., Zaïdi, F.: Certificates for parameterized model checking. In: Bjørner, N., de Boer, F. (eds.) FM 2015. LNCS, vol. 9109, pp. 126–142. Springer, Cham (2015). https://doi.org/10.1007/978-3-319-19249-9_9
16. German, S.M., Sistla, A.P.: Reasoning about systems with many processes. J. ACM **39**(3), 675–735 (1992)
17. Ghilardi, S., Ranise, S.: MCMT: a model checker modulo theories. In: Giesl, J., Hähnle, R. (eds.) IJCAR 2010. LNCS (LNAI), vol. 6173, pp. 22–29. Springer, Heidelberg (2010). https://doi.org/10.1007/978-3-642-14203-1_3
18. Goeman, H.J.M.: The arbiter: an active system component for implementing synchronizing primitives. Fundam. Inform. **4**(3), 517–530 (1981)
19. Herlihy, M., Shavit, N.: The Art of Multiprocessor Programming (2008)
20. Lamport, L.: How to make a multiprocessor computer that correctly executes multiprocess programs. IEEE Trans. Comput. **9**, 690–691 (1979)

21. Owens, S., Sarkar, S., Sewell, P.: A better x86 memory model: x86-TSO. In: Berghofer, S., Nipkow, T., Urban, C., Wenzel, M. (eds.) TPHOLs 2009. LNCS, vol. 5674, pp. 391–407. Springer, Heidelberg (2009). https://doi.org/10.1007/978-3-642-03359-9_27

22. Sewell, P., Sarkar, S., Owens, S., Nardelli, F.Z., Myreen, M.O.: X86-TSO: A rigorous and usable programmer's model for x86 multiprocessors. In: CACM (2010)

QRAT$^+$: Generalizing QRAT by a More Powerful QBF Redundancy Property

Florian Lonsing$^{(\boxtimes)}$ and Uwe Egly

Research Division of Knowledge Based Systems,
Institute of Logic and Computation, TU Wien, Vienna, Austria
{florian.lonsing,uwe.egly}@tuwien.ac.at

Abstract. The QRAT (quantified resolution asymmetric tautology) proof system simulates virtually all inference rules applied in state of the art quantified Boolean formula (QBF) reasoning tools. It consists of rules to rewrite a QBF by adding and deleting clauses and universal literals that have a certain redundancy property. To check for this redundancy property in QRAT, propositional unit propagation (UP) is applied to the quantifier free, i.e., propositional part of the QBF. We generalize the redundancy property in the QRAT system by QBF specific UP (QUP). QUP extends UP by the universal reduction operation to eliminate universal literals from clauses. We apply QUP to an abstraction of the QBF where certain universal quantifiers are converted into existential ones. This way, we obtain a generalization of QRAT we call QRAT$^+$. The redundancy property in QRAT$^+$ based on QUP is more powerful than the one in QRAT based on UP. We report on proof theoretical improvements and experimental results to illustrate the benefits of QRAT$^+$ for QBF preprocessing.

1 Introduction

In practical applications of propositional logic satisfiability (SAT), it is necessary to establish correctness guarantees on the results produced by SAT solvers by proof checking [7]. The DRAT (deletion resolution asymmetric tautology) [23] approach has become state of the art to generate and check propositional proofs.

The logic of quantified Boolean formulas (QBF) extends propositional logic by existential and universal quantification of the propositional variables. Despite the PSPACE-completeness of QBF satisfiability checking, QBF technology is relevant in practice due to the potential succinctness of QBF encodings [4].

DRAT has been lifted to QBF to obtain the QRAT (quantified RAT) proof system [8,10]. QRAT allows to represent and check (un)satisfiability proofs of QBFs and compute Skolem function certificates of satisfiable QBFs. The QRAT system simulates virtually all inference rules applied in state of the art QBF reasoning tools, such as Q-resolution [15] including its variant long-distance Q-resolution [13,25], and expansion of universal variables [3].

Supported by the Austrian Science Fund (FWF) under grant S11409-N23.

A QRAT proof of a QBF in prenex CNF consists of a sequence of inference steps that rewrite the QBF by adding and deleting clauses and universal literals that have the QRAT *redundancy property*. Informally, checking whether a clause C has QRAT amounts to checking whether all possible resolvents of C on a literal $l \in C$ (under certain restrictions) are *propositionally implied* by the quantifier-free CNF part of the QBF. The principle of redundancy checking by inspecting resolvents originates from the RAT property in propositional logic [12] and was generalized to first-order logic in terms of *implication modulo resolution* [14]. Instead of a complete (and thus computationally hard) propositional implication check on a resolvent, the QRAT system relies on an incomplete check by *propositional unit propagation* (UP). Thereby, it is checked whether UP can derive the empty clause from the CNF augmented by the negated resolvent. Hence redundancy checking in QRAT is unaware of the quantifier structure, which is entirely ignored in UP.

We generalize redundancy checking in QRAT by making it aware of the quantifier structure of a QBF. To this end, we check the redundancy of resolvents based on *QBF specific UP* (QUP). It extends UP by the *universal reduction* (UR) operation [15] and is a polynomial-time procedure like UP. UR is central in resolution based QBF calculi [1,15] as it shortens individual clauses by eliminating universal literals depending on the quantifier structure. We apply QUP to *abstractions* of the QBF where certain universal quantifiers are converted into existential ones. The purpose of abstractions is that if a resolvent is found redundant by QUP on the abstraction, then it is also redundant in the original QBF.

Our contributions are as follows: (1) by applying QUP and QBF abstractions instead of UP, we obtain a *generalization of the* QRAT system which we call $QRAT^+$. In contrast to QRAT, redundancy checking in $QRAT^+$ is aware of the quantifier structure of a QBF. We show that (2) the redundancy property in $QRAT^+$ based on QUP is *more powerful* than the one in QRAT based on UP. $QRAT^+$ can detect redundancies which QRAT cannot. As a formal foundation, we introduce (3) a *theory of QBF abstractions* used in $QRAT^+$. Redundancy elimination by $QRAT^+$ or QRAT can lead to (4) *exponentially shorter proofs* in certain resolution based QBF calculi, which we point out by a concrete example. Note that here we do not study the power of QRAT or $QRAT^+$ as proof systems themselves, but the impact of redundancy elimination. Finally, we report on experimental results (5) to illustrate the benefits of redundancy elimination by $QRAT^+$ and QRAT for QBF preprocessing. Our implementation of $QRAT^+$ and QRAT for preprocessing is the first one reported in the literature.

2   Preliminaries

We consider QBFs $\phi := \Pi.\psi$ in *prenex conjunctive normal form (PCNF)* with a *quantifier prefix* $\Pi := Q_1 B_1 \ldots Q_n B_n$ and a quantifier free CNF ψ not containing tautological clauses. The prefix consists of *quantifier blocks* $Q_i B_i$, where B_i are *blocks* (i.e., sets) of propositional variables and $Q_i \in \{\forall, \exists\}$ are *quantifiers*. We have $B_i \cap B_j = \emptyset$, $Q_i \neq Q_{i+1}$ and $Q_n = \exists$. The CNF ψ is defined precisely over the variables $vars(\phi) = vars(\psi) := B_1 \cup \ldots \cup B_n$ in Π so that all variables are quantified, i.e., ϕ is *closed*. The *quantifier* $Q(\Pi, l)$ of literal l is Q_i if the variable $var(l)$ of l appears in B_i. The set of variables in a clause C is $vars(C) := \{x \mid l \in C, var(l) = x\}$. A literal l is *existential* if $Q(\Pi, l) = \exists$ and *universal* if $Q(\Pi, l) = \forall$. If $Q(\Pi, l) = Q_i$ and $Q(\Pi, k) = Q_j$, then $l \leq_\Pi k$ iff $i \leq j$. We extend the ordering $\leq_\Pi$ to an arbitrary but fixed ordering on the variables in every block B_i.

An *assignment* $\tau : vars(\phi) \to \{\top, \bot\}$ maps the variables of a QBF ϕ to truth constants $\top$ (*true*) or $\bot$ (*false*). Assignment τ is *complete* if it assigns every variable in ϕ, otherwise τ is *partial*. By $\tau(\phi)$ we denote ϕ*under* τ, where each occurrence of variable x in ϕ is replaced by $\tau(x)$ and x is removed from the prefix of ϕ, followed by propositional simplifications on $\tau(\phi)$. We consider τ as a set of literals such that, for some variable x, $x \in \tau$ if $\tau(x) = \top$ and $\bar{x} \in \tau$ if $\tau(x) = \bot$.

An *assignment tree* [10] T of a QBF ϕ is a *complete binary tree* of depth $|vars(\phi)| + 1$ where the internal (non-leaf) nodes of each level are associated with a variable of ϕ. An internal node is universal (existential) if it is associated with a universal (existential) variable. The order of variables along every path in T respects the extended order $\leq_\Pi$ of the prefix Π of ϕ. An internal node associated with variable x has two outgoing edges pointing to its children: one labelled with $\bar{x}$ and another one labelled with x, denoting the assignment of x to false and true, respectively. Each path τ in T from the root to an internal node (leaf) represents a partial (complete) assignment. A leaf at the end of τ is labelled by $\tau(\phi)$, i.e., the value of ϕ under τ. An internal node associated with an existential (universal) variable is labelled with $\top$ iff one (both) of its children is (are) labelled with $\top$. The QBF ϕ is *satisfiable* (*unsatisfiable*) iff the root of T is labelled with $\top$ ($\bot$).

Given a QBF ϕ and its assignment tree T, a subtree T' of T is a *pre-model* [10] of ϕ if (1) the root of T is the root of T', (2) for every universal node in T' both children are in T', and (3) for every existential node in T' exactly one of its children is in T'. A pre-model T' of ϕ is a *model* [10] of ϕ, denoted by $T' \models_t \phi$, if each node in T' is labelled with $\top$. A QBF ϕ is *satisfiable* iff it has a model. Given a QBF ϕ and one of its *models* T', T'' is a *rooted subtree* of T' ($T'' \subseteq T'$) if T'' has the same root as T' and the leaves of T'' are a subset of the leaves of T'.

We consider CNFs ψ defined over a set B of variables without an explicit quantifier prefix. A *model* of a CNF ψ is a model τ of the QBF $\exists B.\psi$ which consists only of the single path τ. We write $\tau \models \psi$ if τ is a model of ψ. For CNFs ψ and ψ', ψ' is *implied* by ψ ($\psi \models \psi'$) if, for all τ, it holds that if $\tau \models \psi$ then $\tau \models \psi'$. Two CNFs ψ and ψ' are *equivalent* ($\psi \equiv \psi'$), iff $\psi \models \psi'$ and $\psi' \models \psi$. We define notation to explicitly refer to QBF models. For QBFs ϕ and ϕ', ϕ' is *implied* by ϕ ($\phi \models_t \phi'$) if, for all T, it holds that if $T \models_t \phi$ then $T \models_t \phi'$. QBFs ϕ and ϕ' are *equivalent* ($\phi \equiv_t \phi'$) iff $\phi \models_t \phi'$ and $\phi' \models_t \phi$, and *satisfiability equivalent* ($\phi \equiv_{sat} \phi'$) iff ϕ is satisfiable whenever ϕ' is satisfiable. Satisfiability equivalence of CNFs is defined analogously and denoted by the same symbol '$\equiv_{sat}$'.

3   The Original **QRAT** Proof System

Before we generalize QRAT, we recapitulate the original proof system [10] and emphasize that redundancy checking in QRAT is unaware of quantifier structures.

Definition 1 ([10]). *The* outer clause *of clause C on literal $l \in C$ with respect to prefix Π is the clause* $\mathsf{OC}(\Pi, C, l) := \{k \mid k \in C, k \leq_\Pi l, k \neq l\}$.

The outer clause $\mathsf{OC}(\Pi, C, l) \subset C$ of C on $l \in C$ contains only literals that are smaller than or equal to l in the variable ordering of prefix Π, excluding l.

Definition 2 ([10]). *Let C be a clause with $l \in C$ and D be a clause with $\bar{l} \in D$ occurring in QBF $\Pi.\psi$. The* outer resolvent *of C with D on l with respect to Π is the clause* $\mathsf{OR}(\Pi, C, D, l) := (C \setminus \{l\}) \cup \mathsf{OC}(\Pi, D, \bar{l})$.

Example 1. Given $\phi := \exists x_1 \forall u \exists x_2.(C \wedge D)$ with $C := (x_1 \vee u \vee x_2)$ and $D := (\bar{x}_1 \vee \bar{u} \vee \bar{x}_2)$, we have $\mathsf{OR}(\Pi, C, D, x_1) = (u \vee x_2)$, $\mathsf{OR}(\Pi, C, D, u) = (x_1 \vee \bar{x}_1 \vee x_2)$, $\mathsf{OR}(\Pi, C, D, x_2) = (x_1 \vee u \vee \bar{x}_1 \vee \bar{u})$, and $\mathsf{OR}(\Pi, D, C, \bar{u}) = (x_1 \vee \bar{x}_1 \vee \bar{x}_2)$. Computing outer resolvents is asymmetric since $\mathsf{OR}(\Pi, C, D, u) \neq \mathsf{OR}(\Pi, D, C, \bar{u})$.

Definition 3 ([10]). *Clause C has property* QIOR *(quantified implied outer resolvent) on literal $l \in C$ with respect to QBF $\Pi.\psi$ iff $\psi \models \mathsf{OR}(\Pi, C, D, l)$ for all $D \in \psi$ with $\bar{l} \in D$.*

Property QIOR relies on checking whether every possible outer resolvent OR of some clause C on a literal is redundant by checking if OR is *propositionally implied* by the *quantifier-free CNF* ψ of the given QBF $\Pi.\psi$. If C has QIOR on literal $l \in C$ then, depending on whether l is existential or universal and side conditions, either C is redundant and can be removed from QBF $\Pi.\psi$ or l is redundant and can be removed from C, respectively, resulting in a *satisfiability-equivalent* QBF.

Theorem 1 ([10]). *Given a QBF $\phi := \Pi.\psi$ and a clause $C \in \psi$ with QIOR on an* existential *literal $l \in C$ with respect to QBF $\phi' := \Pi.\psi'$ where $\psi' := \psi \setminus \{C\}$. Then $\phi \equiv_{sat} \phi'$.*

Theorem 2 ([10]). *Given a QBF $\phi_0 := \Pi.\psi$ and $\phi := \Pi.(\psi \cup \{C\})$ where C has* QIOR *on a* universal *literal $l \in C$ with respect to ϕ_0. Let $\phi' := \Pi.(\psi \cup \{C'\})$ with $C' := C \setminus \{l\}$. Then $\phi \equiv_{sat} \phi'$.*

Note that in Theorems 1 and 2 clause C is actually removed from the QBF for the check whether C has QIOR on a literal. Checking propositional implication ($\models$) as in Definition 3 is co-NP hard and hence intractable. Therefore, in practice a polynomial-time incomplete implication check based on propositional unit propagation (UP) is applied. The use of UP is central in the QRAT proof system.

Definition 4 (propositional unit propagation, UP). *For a CNF ψ and clause C, let $\psi \wedge \overline{C} \vdash_1 \emptyset$ denote the fact that* propositional unit propagation (UP) *applied to $\psi \wedge \overline{C}$ produces the empty clause, where $\overline{C}$ is the conjunction of the negation of all the literals in C. If $\psi \wedge \overline{C} \vdash_1 \emptyset$ then we write $\psi \vdash_1 C$ to denote that C can be derived from ψ by UP (since $\psi \models C$).*

Definition 5 ([10]). *Clause C has property* AT (asymmetric tautology) *with respect to a CNF ψ iff $\psi \vdash_1 C$.*

AT is a propositional clause redundancy property that is used in the QRAT proof system to check whether outer resolvents are redundant, thereby replacing propositional implication ($\models$) in Definition 3 by unit propagation ($\vdash_1$) as follows.

Definition 6 ([10]). *Clause C has property* QRAT (quantified resolution asymmetric tautology) *on literal $l \in C$ with respect to QBF $\Pi.\psi$ iff, for all $D \in \psi$ with $\bar{l} \in D$, the outer resolvent $\mathrm{OR}(\Pi, C, D, l)$ has* AT *with respect to CNF ψ.*

Example 2. Consider $\phi := \exists x_1 \forall u \exists x_2.(C \wedge D)$ with $C := (x_1 \vee u \vee x_2)$ and $D := (\bar{x}_1 \vee \bar{u} \vee \bar{x}_2)$ from Example 1. C does not have AT with respect to CNF D, but C has QRAT on x_2 with respect to QBF $\exists x_1 \forall u \exists x_2.(D)$ since $\mathrm{OR}(\Pi, C, D, x_2) = (x_1 \vee u \vee \bar{x}_1 \vee \bar{u})$ has AT with respect to CNF D.

QRAT is a restriction of QIOR, i.e., a clause that has QRAT also has QIOR but not necessarily vice versa. Therefore, the soundness of removing redundant clauses and literals based on QRAT follows right from Theorems 1 and 2.

Based on the QRAT redundancy property, the QRAT proof system [10] consists of rewrite rules to eliminate redundant clauses, denoted by QRATE, to add redundant clauses, denoted by QRATA, and to eliminate redundant universal literals, denoted by QRATU. In a QRAT *satisfaction proof (refutation)*, a QBF is reduced to the empty formula (respectively, to a formula containing the empty clause) by applying the rewrite rules. The QRAT proof systems has an additional rule to eliminate universal literals by *extended universal reduction* (EUR). We do not present EUR because it is not affected by our generalization of QRAT, which we define in the following. Observe that QIOR and AT (and hence also QRAT) are based on *propositional* implication ($\models$) and unit propagation ($\vdash_1$), i.e., the quantifier structure of the given QBF is not exploited.

4   QRAT$^+$: A More Powerful QBF Redundancy Property

We make redundancy checking of outer resolvents in QRAT aware of the quantifier structure of a QBF. To this end, we generalize QIOR and AT by replacing propositional implication ($\models$) and unit propagation ($\vdash_1$) by QBF implication ($\models_t$) and QBF unit propagation, respectively. Thereby, we obtain a more general and more powerful notion of the QRAT redundancy property, which we call QRAT$^+$.

First, in Proposition 2 we point out a property of QIOR (Definition 3) which is due to the following result from related work [21]: if we attach a quantifier prefix Π to equivalent CNFs ψ and ψ', then the resulting QBFs are equivalent.

Proposition 1 ([21]). *Given CNFs ψ and ψ' such that $vars(\psi) = vars(\psi')$ and a quantifier prefix Π defined precisely over $vars(\psi)$. If $\psi \equiv \psi'$ then $\Pi.\psi \equiv_t \Pi.\psi'$.*

Proposition 2. *If clause C has QIOR on literal $l \in C$ with respect to QBF $\Pi.\psi$, then $\Pi.\psi \equiv_t \Pi.(\psi \wedge \mathsf{OR}(\Pi, C, D, l))$ for all $D \in \psi$ with $\bar{l} \in D$.*

Proof. Since C has QIOR on literal $l \in C$ with respect to QBF $\Pi.\psi$, by Definition 3 we have $\psi \models \mathsf{OR}(\Pi, C, D, l)$ for all $D \in \psi$ with $\bar{l} \in D$, and further also $\psi \equiv \psi \wedge \mathsf{OR}(\Pi, C, D, l)$. Then $\Pi.\psi \equiv_t \Pi.(\psi \wedge \mathsf{OR}(\Pi, C, D, l))$ by Proposition 1.   □

By Proposition 2 any outer resolvent OR of some clause C that has QIOR with respect to some QBF $\Pi.\psi$ is redundant in the sense that it can be *added to the QBF $\Pi.\psi$ in an equivalence preserving way* ($\equiv_t$), i.e., OR is *implied by the QBF $\Pi.\psi$* ($\models_t$). This is the central characteristic of our generalization QRAT^+ of QRAT. We develop a redundancy property used in QRAT^+ which allows to, e.g., remove a clause C from a QBF $\Pi.\psi$ in a satisfiability preserving way (like in QRAT, cf. Theorem 1.) if all respective outer resolvents of C are implied by the QBF $\Pi.(\psi \setminus \{C\})$. Since checking QBF implication is intractable just like checking propositional implication in QIOR, in practice we apply a polynomial-time incomplete QBF implication check based on *QBF unit propagation*.

In the following, we develop a theoretical framework of *abstractions* of QBFs that underlies our generalization QRAT^+ of QRAT. Abstractions are crucial for the soundness of checking QBF implication by QBF unit propagation.

Definition 7 (nesting levels, prefix/QBF abstraction). *Let $\phi := \Pi.\psi$ be a QBF with prefix $\Pi := Q_1 B_1 \ldots Q_i B_i Q_{i+1} B_{i+1} \ldots Q_n B_n$. For a clause C, $levels(\Pi, C) := \{i \mid \exists l \in C, Q(\Pi, l) = Q_i\}$ is the set of nesting levels in C.[1] The abstraction of Π with respect to i with $0 \le i \le n$ produces the abstracted prefix $Abs(\Pi, i) := \Pi$ for $i = 0$ and otherwise $Abs(\Pi, i) := \exists(B_1 \cup \ldots \cup B_i) Q_{i+1} B_{i+1} \ldots Q_n B_n$. The abstraction of ϕ with respect to i with $0 \le i \le n$ produces the abstracted QBF $Abs(\phi, i) := Abs(\Pi, i).\psi$ with prefix $Abs(\Pi, i)$.*

Example 3. Given the QBF $\phi := \Pi.\psi$ with prefix $\Pi := \forall B_1 \exists B_2 \forall B_3 \exists B_4$. We have $Abs(\phi, 0) = \phi$, $Abs(\phi, 1) = Abs(\phi, 2) = \exists(B_1 \cup B_2) \forall B_3 \exists B_4.\psi$, $Abs(\phi, 3) = Abs(\phi, 4) = \exists(B_1 \cup B_2 \cup B_3 \cup B_4).\psi$.

In an abstracted QBF $Abs(\phi, i)$ universal variables from blocks smaller than or equal to B_i are converted into existential ones. If the original QBF ϕ has a model T, then *all* nodes in T associated to universal variables must be labelled with $\top$, in particular the universal variables that are existential in $Abs(\phi, i)$. Hence, for *all* models T of ϕ, every model T^A of $Abs(\phi, i)$ is a subtree of T.

Proposition 3.
Given a QBF $\phi := \Pi.\psi$ with prefix $\Pi := Q_1 B_1 \ldots Q_i B_i \ldots Q_n B_n$ and $Abs(\phi, i)$ for some arbitrary i with $0 \le i \le n$. For all T and T^A we have that if $T \models_t \phi$ and $T^A \subseteq T$ is a pre-model of $Abs(\phi, i)$, then $T^A \models_t Abs(\phi, i)$.

---

[1] In general, clauses C are always (implicitly) interpreted under a quantifier prefix Π.

Proof. By induction on i. The base case $i := 0$ is trivial.

As induction hypothesis (IH), assume that the claim holds for some i with $0 \le i < n$, i.e., for all T and T^A we have that if $T \models_t \phi$ and $T^A \subseteq T$ is a pre-model of $Abs(\phi, i)$, then $T^A \models_t Abs(\phi, i)$. Consider $Abs(\phi, j)$ for $j = i+1$, which is an abstraction of $Abs(\phi, i)$. We have to show that, for all T and T^B we have that if $T \models_t \phi$ and $T^B \subseteq T$ is a pre-model of $Abs(\phi, j)$, then $T^B \models_t Abs(\phi, j)$. We distinguish cases by the type of Q_j in the abstracted prefix $Abs(\Pi, i) = \exists(B_1 \cup \ldots \cup B_i)Q_j B_j \ldots Q_n B_n$ of $Abs(\phi, i)$.

If $Q_j = \exists$ then $Abs(\Pi, i) = Abs(\Pi, j) = \exists(B_1 \cup \ldots B_i \cup B_j) \ldots Q_n B_n$. Since $Abs(\phi, i) = Abs(\phi, j)$, the claim holds for $Abs(\phi, j)$ by IH.

If $Q_j = \forall$ then, towards a contradiction, assume that, for some T and T^B, $T \models_t \phi$ and $T^B \subseteq T$ is a pre-model of $Abs(\phi, j)$, but $T^B \not\models_t Abs(\phi, j)$. Then the root of T^B is labelled with $\bot$, and in particular the nodes of all the variables which are existential in B_j with respect to $Abs(\Pi, j)$ are also labelled with $\bot$. These existential variables appear along a single branch τ' in T^B, i.e., τ' is a partial assignment of the variables in B_j. Since $T^B \subseteq T^A$ and $Q_j = \forall$ in $Abs(\Pi, i)$, the root of T^A is labelled with $\bot$ since there is the branch τ' containing the variables in B_j whose nodes are labelled with $\bot$ in T^A. Hence $T^A \not\models_t Abs(\phi, i)$, which is a contradiction to IH. Therefore, we conclude that $T^B \models_t Abs(\phi, j)$. $\square$

If an abstraction $Abs(\phi, i)$ is unsatisfiable then also the original QBF ϕ is unsatisfiable due to Proposition 3. We generalize Proposition 1 from CNFs to QBFs and their abstractions. Note that the full abstraction $Abs(\phi, i)$ for $i := n$ of a QBF ϕ is a CNF, i.e., it does not contain any universal variables.

Lemma 1. *Let $\phi := \Pi.\psi$ and $\phi' := \Pi.\psi'$ be QBFs with the same prefix $\Pi := Q_1 B_1 \ldots Q_i B_i \ldots Q_n B_n$. Then for all i, if $Abs(\phi, i) \equiv_t Abs(\phi', i)$ then $\phi \equiv_t \phi'$.*

Proof. By induction on $i := 0$ up to $i := n$. The base case $i := 0$ is trivial.

As induction hypothesis (IH), assume that the claim holds for some i with $0 \le i < n$, i.e., if $Abs(\phi, i) \equiv_t Abs(\phi', i)$ then $\phi \equiv_t \phi'$. Let $j = i + 1$ and consider $Abs(\phi, j)$ and $Abs(\phi', j)$, which are abstractions of $Abs(\phi, i)$ and $Abs(\phi', i)$. We have $Abs(\Pi, i) = \exists(B_1 \cup \ldots \cup B_i)Q_j B_j \ldots Q_n B_n$ and $Abs(\Pi, j) = \exists(B_1 \cup \ldots \cup B_j) \ldots Q_n B_n$. We show that if $Abs(\phi, j) \equiv_t Abs(\phi', j)$ then $Abs(\phi, i) \equiv_t Abs(\phi', i)$, and hence also $\phi \equiv_t \phi'$ by IH. Assume that $Abs(\phi, j) \equiv_t Abs(\phi', j)$. We distinguish cases by the type of Q_j in $Abs(\Pi, i)$. If $Q_j = \exists$ then $Abs(\Pi, i) = Abs(\Pi, j) = \exists(B_1 \cup \ldots B_i \cup B_j) \ldots Q_n B_n$, and hence $Abs(\phi, i) \equiv_t Abs(\phi', i)$.

If $Q_j = \forall$, then towards a contradiction, assume that $Abs(\phi, j) \equiv_t Abs(\phi', j)$ but $Abs(\phi, i) \not\equiv_t Abs(\phi', i)$. Then there exists T such that $T \models_t Abs(\phi, i)$ but $T \not\models_t Abs(\phi', i)$. Since $T \not\models_t Abs(\phi', i)$ there exists a pre-model $T^A \subseteq T$ of $Abs(\phi', j)$ such that the root of T^A is labelled with $\bot$, and in particular the nodes of all the variables which are existential in B_j with respect to $Abs(\Pi, j)$ (and universal with respect to $Abs(\Pi, i)$) are also labelled with $\bot$. These existential variables appear along a single branch τ' in T^A, i.e., τ' is a partial assignment of the variables in B_j. Therefore we have $T^A \not\models_t Abs(\phi', j)$. Since $T \models_t Abs(\phi, i)$ and $T^A \subseteq T$, we have $T^A \models_t Abs(\phi, j)$ by Proposition 3, which contradicts the assumption that $Abs(\phi, j) \equiv_t Abs(\phi', j)$.   □

The converse of Lemma 1 does not hold. From the equivalence of two QBFs ϕ and ϕ' we cannot conclude that the abstractions $Abs(\phi, i)$ and $Abs(\phi', i)$ are equivalent. In our generalization QRAT$^+$ of the QRAT system we check whether an outer resolvent of some clause C is implied ($\models_t$) by an *abstraction* of the given QBF. If so then by Lemma 1 the outer resolvent is also implied by the original QBF. Below we prove that this condition is sufficient for the soundness of redundancy removal in QRAT$^+$. To check QBF implication in an incomplete way and in polynomial time, in practice we apply *QBF unit propagation*, which is an extension of propositional unit propagation, to abstractions of the given QBF.

Definition 8 (universal reduction, UR [15]). *Given a QBF $\phi := \Pi.\psi$ and a non-tautological clause C, universal reduction (UR) of C produces the clause* $UR(\Pi, C) := C \setminus \{l \in C \mid Q(\Pi, l) = \forall, \forall l' \in C, Q(\Pi, l') = \exists : var(l') \leq_\Pi var(l)\}$.

Definition 9 (QBF unit propagation, QUP). QBF unit propagation (QUP) *extends UP (Definition 4) by applications of UR. For a QBF $\phi := \Pi.\psi$ and a clause C, let $\Pi.(\psi \wedge \overline{C}) \vdash_{uv} \emptyset$ denote the fact that QUP applied to $\Pi.(\psi \wedge \overline{C})$ produces the empty clause, where $\overline{C}$ is the conjunction of the negation of all the literals in C. If $\Pi.(\psi \wedge \overline{C}) \vdash_{uv} \emptyset$ and additionally $\Pi.\psi \models_t \Pi.(\psi \wedge C)$ then we write $\phi \vdash_{uv} C$ to denote that C can be derived from ϕ by QUP.*

In contrast to UP (Definition 4), deriving the empty clause by QUP by propagating $\overline{C}$ on a QBF ϕ is not sufficient to conclude that C is implied by ϕ.

Example 4. Given the QBF $\Pi.\psi$ with prefix $\Pi := \forall u \exists x$ and CNF $\psi := (u \vee \bar{x}) \wedge (\bar{u} \vee x)$ and the clause $C := (x)$. We have $\Pi.((u \vee \bar{x}) \wedge (\bar{u} \vee x) \wedge (\bar{x})) \vdash_{uv} \emptyset$ since propagating $\overline{C} = (\bar{x})$ produces $(\bar{u})$, which is reduced to $\emptyset$ by UR. However, $\Pi.\psi \not\models_t \Pi.(\psi \wedge C)$ since $\Pi.\psi$ is satisfiable whereas $\Pi.(\psi \wedge C)$ is unsatisfiable. Note that $Abs(\Pi.((u \vee \bar{x}) \wedge (\bar{u} \vee x) \wedge \bar{x}), 2) \not\vdash_{uv} \emptyset$. ·

To correctly apply QUP for checking whether some clause C (e.g., an outer resolvent) is implied by a QBF $\phi := \Pi.\psi$ and thus avoid the problem illustrated in Example 4, we carry out QUP on a *suitable abstraction* of ϕ with respect to C. Let $i = \max(levels(\Pi, C))$ be the maximum nesting level of variables that appear in C. We show that if QUP derives the empty clause from the abstraction $Abs(\phi, i)$ augmented by the negated clause $\overline{C}$, i.e., $Abs(\Pi.(\psi \wedge \overline{C}), i) \vdash_{uv} \emptyset$, then we can safely conclude that C is implied by the *original* QBF, i.e., $\Pi.\psi \models_t \Pi.(\psi \wedge C)$. This approach extends failed literal detection for QBF preprocessing [16].

Lemma 2. *Let $\Pi.\psi$ be a QBF with prefix $\Pi := Q_1 B_1 \ldots Q_n B_n$ and C a clause such that $vars(C) \subseteq B_1$. If $\Pi.(\psi \wedge \overline{C}) \vdash_{uv} \emptyset$ then $\Pi.\psi \equiv_t \Pi.(\psi \wedge C)$.*

Proof. By contradiction, assume $T \models_t \Pi.\psi$ but $T \not\models_t \Pi.(\psi \wedge C)$. Then there is a path $\tau \subseteq T$ such that $\tau(C) = \bot$. Since $vars(C) \subseteq B_1$ and $\Pi.(\psi \wedge \overline{C}) \vdash_\forall \emptyset$, the QBF $\Pi.(\psi \wedge \overline{C})$ is unsatisfiable and in particular $T \not\models_t \Pi.(\psi \wedge \overline{C})$. Since $\tau(C) = \bot$, we have $\tau(\overline{C}) = \top$ and hence $T \models_t \Pi.(\psi \wedge \overline{C})$, which is a contradiction.   □

Lemma 3. *Let $\Pi.\psi$ be a QBF, C a clause, and $i = \max(levels(\Pi, C))$. If $Abs(\Pi.(\psi \wedge \overline{C}), i) \vdash_\forall \emptyset$ then $Abs(\Pi.\psi, i) \equiv_t Abs(\Pi.(\psi \wedge C), i)$.*

Proof. The claim follows from Lemma 2 since all variables that appear in C are existentially quantified in $Abs(\Pi.(\psi \wedge \overline{C}), i)$ in the leftmost quantifier block.   □

Lemma 4. *Let $\Pi.\psi$ be a QBF, C a clause, and $i = \max(levels(\Pi, C))$. If $Abs(\Pi.(\psi \wedge \overline{C}), i) \vdash_\forall \emptyset$ then $\Pi.\psi \equiv_t \Pi.(\psi \wedge C)$.*

Proof. By Lemmas 1 and 3.   □

Lemma 4 provides us with the necessary theoretical foundation to lift AT (Definition 5) from UP, which is applied to CNFs, to QUP, which is applied to *suitable abstractions* of QBFs. The abstractions are constructed depending on the maximum nesting level of variables in the clause we want to check.

Definition 10 (QAT). *Let ϕ be a QBF, C a clause, and $i = \max(levels(\Pi, C))$ Clause C has property QAT (quantified asymmetric tautology) with respect to ϕ iff $Abs(\phi, i) \vdash_\forall C$.*

As an immediate consequence from the definition of QUP (Definition 9) and Lemma 3, we can conclude that a clause C has QAT with respect to a QBF $\Pi.\psi$ if QUP derives the empty clause from the suitable abstraction of $\Pi.\psi$ with respect to C (i.e., $Abs(\Pi.(\psi \wedge \overline{C}), i) \vdash_\forall \emptyset$). Further, if C has QAT then we have $\Pi.\psi \equiv_t \Pi.(\psi \wedge C)$ by Lemma 4, i.e., C is implied by the given QBF $\Pi.\psi$.

Example 5. Given the QBF $\phi := \Pi.\psi$ with $\Pi := \forall u_1 \exists x_3 \forall u_2 \exists x_4$ and $\psi := (u_1 \vee \bar{x}_3) \wedge (u_1 \vee \bar{x}_3 \vee x_4) \wedge (\bar{u}_2 \vee \bar{x}_4)$. Clause $(u_1 \vee \bar{x}_3)$ has QAT with respect to $Abs(\phi, 2)$ with $\max(levels(C)) = 2$ since $\forall u_2$ is still universal in the abstraction. By QUP clause $(u_1 \vee \bar{x}_3 \vee x_4)$ becomes unit and clause $(\bar{u}_2 \vee \bar{x}_4)$ becomes empty by UR. However, clause $(u_1 \vee \bar{x}_3)$ does *not* have AT since $\forall u_2$ is treated as an existential variable in UP, hence clause $(\bar{u}_2 \vee \bar{x}_4)$ does not become empty by UR.

In contrast to AT, QAT is aware of quantifier structures in QBFs as shown in Example 5. We now generalize QRAT to QRAT$^+$ by replacing AT by QAT. Similarly, we generalize QIOR to QIOR$^+$ by replacing propositional implication ($\models$) and equivalence (Proposition 1), by *QBF implication and equivalence* (Lemma 4).

Definition 11 (QRAT$^+$). *Clause C has property QRAT$^+$ on literal $l \in C$ with respect to QBF $\Pi.\psi$ iff, for all $D \in \psi$ with $\bar{l} \in D$, the outer resolvent $OR(\Pi, C, D, l)$ has QAT with respect to QBF $\Pi.\psi$.*

Definition 12 (QIOR$^+$). *Clause C has property QIOR$^+$ on literal $l \in C$ with respect to QBF $\Pi.\psi$ iff $\Pi.\psi \equiv_t \Pi.(\psi \wedge OR(\Pi, C, D, l))$ for all $D \in \psi$ with $\bar{l} \in D$.*

If a clause has QRAT then it also has QRAT$^+$. Moreover, due to Proposition 2, if a clause has QIOR then it also has QIOR$^+$. Hence QRAT$^+$ and QIOR$^+$ indeed are generalizations of QRAT and QIOR, which are strict, as we argue below. The soundness of removing redundant clauses and universal literals based on QIOR$^+$ (and on QRAT$^+$) can be proved by the *same* arguments as original QRAT, which we outline in the following. We refer to an appendix for full proofs [18].

Definition 13 (prefix/suffix assignment [10]). *For a QBF $\phi := \Pi.\psi$ and a complete assignment τ in the assignment tree of ϕ, the partial prefix and suffix assignments of τ with respect to variable x, denoted by τ^x and τ_x, respectively, are defined as $\tau^x := \{y \mapsto \tau(y) \mid y \leq_\Pi x, y \neq x\}$ and $\tau_x := \{y \mapsto \tau(y) \mid y \not\leq_\Pi x\}$.*

For a variable x from block B_i of a QBF, Definition 13 allows us to split a complete assignment τ into three parts $\tau^x l \tau_x$, where the prefix assignment τ^x assigns variables (excluding x) from blocks smaller than or equal to B_i, l is a literal of x, and the suffix assignment τ_x assigns variables from blocks larger than B_i.

Prefix and suffix assignments are important for proving the soundness of satisfiability-preserving redundancy removal by QIOR$^+$ (and QIOR). Soundness is proved by showing that certain paths in a model of a QBF can safely be modified based on prefix and suffix assignments, as stated in the following.

Lemma 5 (cf. Lemma 6 in [10]). *Given a clause C with QIOR$^+$ with respect to QBF $\phi := \Pi.\psi$ on literal $l \in C$ with $\mathsf{var}(l) = x$. Let T be a model of ϕ and $\tau \subseteq T$ be a path in T. If $\tau(C \setminus \{l\}) = \bot$ then $\tau^x(D) = \top$ for all $D \in \psi$ with $\bar{l} \in D$.*

Proof (sketch, see appendix [18]). Let $D \in \psi$ be a clause with $\bar{l} \in D$ and $R := \mathsf{OR}(\Pi, C, D, l) = (C \setminus \{l\}) \cup \mathsf{OC}(\Pi, D, \bar{l})$. By Definition 12, we have $\Pi.\psi \equiv_t \Pi.(\psi \wedge \mathsf{OR}(\Pi, C, D, l))$ for all $D \in \psi$ with $\bar{l} \in D$. The rest of the proof considers a path τ in T and works in the same way as the proof of Lemma 6 in [10]. □

Theorem 3. *Given a QBF $\phi := \Pi.\psi$ and a clause $C \in \psi$ with QIOR$^+$ on an existential literal $l \in C$ with respect to QBF $\phi' := \Pi.\psi'$ where $\psi' := \psi \setminus \{C\}$. Then $\phi \equiv_{sat} \phi'$.*

Proof (sketch, see appendix [18]). The proof relies on Lemma 5 and works in the same way as the proof of Theorem 7 in [10]. A model T of ϕ is obtained from a model T' of ϕ' by flipping the assignment of variable $x = \mathsf{var}(l)$ on a path τ in T' to satisfy clause C. All $D \in \psi$ with $\bar{l} \in D$ are satisfied by such modified τ. □

Theorem 4. *Given a QBF $\phi_0 := \Pi.\psi$ and $\phi := \Pi.(\psi \cup \{C\})$ where C has QIOR$^+$ on a universal literal $l \in C$ with respect to ϕ_0. Let $\phi' := \Pi.(\psi \cup \{C'\})$ with $C' := C \setminus \{l\}$. Then $\phi \equiv_{sat} \phi'$.*

Proof (sketch, see appendix [18]). The proof relies on Lemma 5 and works in the same way as the proof of Theorem 8 in [10]. A model T' of ϕ' is obtained from a model T of ϕ by modifying the subtree under the node associated to variable

$x = \text{var}(l)$. Suffix assignments of some paths τ in T are used to construct modified paths in T' under which clause C' is satisfied. All $D \in \psi$ with $\bar{l} \in D$ are still satisfied after such modifications. $\square$

Analogously to the QRAT proof system that is based on the QRAT redundancy property (Definition 6), we obtain the QRAT$^+$ *proof system* based on property QRAT$^+$ (Definition 11). The system consists of rewrite rules QRATE$^+$, QRATA$^+$, and QRATU$^+$ to eliminate or add redundant clauses, and to eliminate redundant universal literals. On a conceptual level, these rules in QRAT$^+$ are similar to their respective counterparts in the QRAT system. The extended universal reduction rule EUR is the same in the QRAT and QRAT$^+$ systems. In contrast to QRAT, QRAT$^+$ is aware of quantifier structures of QBFs because it relies on the QBF specific property QAT and QUP instead of on propositional AT and UP.

The QRAT$^+$ system has the same desirable properties as the original QRAT system. QRAT$^+$ *simulates* virtually all inference rules applied in QBF reasoning tools and it is based on redundancy property QRAT$^+$ that can be checked in *polynomial time* by QUP. Further, QRAT$^+$ allows to represent proofs in the *same proof format* as QRAT. However, proof checking, i.e., checking whether a clause listed in the proof has QRAT$^+$ on a literal, must be adapted to the use of QBF abstractions and QUP. Consequently, the available QRAT proof checker QRATtrim [10] cannot be used out of the box to check QRAT$^+$ proofs.

Notably, *Skolem functions* can be extracted from QRAT$^+$ proofs of satisfiable QBFs in the *same* way as in QRAT (consequence of Theorem 3, cf. Corollaries 26 and 27 in [10]). Hence like QRAT, QRAT$^+$ can be integrated in complete QBF workflows that include preprocessing, solving, and Skolem function extraction [5].

5   Exemplifying the Power of QRAT$^+$

In the following, we point out that the QRAT$^+$ system is more powerful than QRAT in terms of redundancy detection. In particular, we show that the rules QRATE$^+$ and QRATU$^+$ in the QRAT$^+$ system can eliminate certain redundancies that their counterparts QRATE and QRATU cannot eliminate.

Definition 14. *For $n \geq 1$, let $\Phi_C(n) := \Pi_C(n).\psi_C(n)$ be a class of QBFs with prefix $\Pi_C(n)$ and CNF $\psi_C(n)$ defined as follows.*

$$\Pi_C(n) := \exists B_1 \forall B_2 \exists B_3 \forall B_4 \exists B_5:$$

$$B_1 := \{x_{4i+1}, x_{4i+2} \mid 0 \leq i < n\}$$
$$B_2 := \{u_{2i+1} \mid 0 \leq i < n\}$$
$$B_3 := \{x_{4i+3} \mid 0 \leq i < n\}$$
$$B_4 := \{u_{2i+2} \mid 0 \leq i < n\}$$
$$B_5 := \{x_{4i+4} \mid 0 \leq i < n\}$$

$$\psi_C(n) := \bigwedge_{i:=0}^{n-1} \mathcal{C}(i) \quad \text{with} \quad \mathcal{C}(i) := \bigwedge_{j:=0}^{6} C_{i,j}:$$

$$C_{i,0} := (x_{4i+1} \vee u_{2i+1} \vee \neg x_{4i+3})$$
$$C_{i,1} := (x_{4i+2} \vee \neg u_{2i+1} \vee x_{4i+3})$$
$$C_{i,2} := (\neg x_{4i+1} \vee \neg u_{2i+1} \vee \neg x_{4i+3})$$
$$C_{i,3} := (\neg x_{4i+2} \vee u_{2i+1} \vee x_{4i+3})$$
$$C_{i,4} := (u_{2i+1} \vee \neg x_{4i+3} \vee x_{4i+4})$$
$$C_{i,5} := (\neg u_{2i+2} \vee \neg x_{4i+4})$$
$$C_{i,6} := (\neg x_{4i+1} \vee u_{2i+2} \vee \neg x_{4i+4})$$

Example 6. For $n{:=}1$, we have $\Phi_C(n)$ with prefix $\Pi_C(n){:=}\exists x_1, x_2 \forall u_1 \exists x_3 \forall u_2 \exists x_4$ and CNF $\psi_C(n){:=}\mathcal{C}(0)$ with $\mathcal{C}(0){:=}\bigwedge_{j:=0}^{6} C_{0,j}$ as follows.

$C_{0,0}{:=} (x_1 \lor u_1 \lor \neg x_3)$ $C_{0,4}{:=} (u_1 \lor \neg x_3 \lor x_4)$

$C_{0,1}{:=} (x_2 \lor \neg u_1 \lor x_3)$ $C_{0,5}{:=} (\neg u_2 \lor \neg x_4)$

$C_{0,2}{:=} (\neg x_1 \lor \neg u_1 \lor \neg x_3)$ $C_{0,6}{:=} (\neg x_1 \lor u_2 \lor \neg x_4)$

$C_{0,3}{:=} (\neg x_2 \lor u_1 \lor x_3)$

Proposition 4. *For $n \geq 1$, QRATE$^+$ can eliminate all clauses in $\Phi_C(n)$ whereas QRATE cannot eliminate any clause in $\Phi_C(n)$.*

Proof (sketch). For i and k with $i \neq k$, the sets of variables in $\mathcal{C}(i)$ and $\mathcal{C}(k)$ are disjoint. Thus it suffices to prove the claim for an arbitrary $\mathcal{C}(i)$. Clause $C_{i,0}$ has QRAT$^+$ on literal x_{4i+1} and can be removed. The relevant outer resolvents are $\mathsf{OR}_{0,2} = \mathsf{OR}(\Pi_C(n), C_{i,0}, C_{i,2}, x_{4i+1})$ and $\mathsf{OR}_{0,6} = \mathsf{OR}(\Pi_C(n), C_{i,0}, C_{i,6}, x_{4i+1})$, and we have $\mathsf{OR}_{0,2} = \mathsf{OR}_{0,6} = (u_{2i+1} \lor \neg x_{4i+3})$. Since $\max(levels(\mathsf{OR}_{0,2})) = \max(levels(\mathsf{OR}_{0,6})) = 3$, we apply QUP to the abstraction $Abs(\Phi_C(n), 3)$. Note that variable u_{2i+2} from block B_4 still is universal in the prefix of $Abs(\Phi_C(n), 3)$. Propagating $\overline{\mathsf{OR}_{0,2}}$ and $\overline{\mathsf{OR}_{0,6}}$, respectively, in either case makes $C_{i,4}$ unit, finally $C_{i,5}$ becomes empty under the derived assignment x_{4i+4} since UR reduces the literal $\neg u_{2i+2}$. After removing $C_{i,0}$, clauses $C_{i,2}$ and $C_{i,6}$ trivially have QRAT$^+$ on $\neg x_{4i+1}$. Then $C_{i,1}$ has QRAT$^+$ on x_{4i+3}. Finally, the remaining clauses trivially have QRAT$^+$. In contrast to that, QRATE cannot eliminate any clause in $\Phi_C(n)$. Clause $C_{i,5}$ does not become empty by UP since all variables are existential. The claim can be proved by case analysis of all possible outer resolvents. □

Definition 15. *For $n \geq 1$, let $\Phi_L(n){:=}\Pi_L(n).\psi_L(n)$ be a class of QBFs with prefix $\Pi_L(n)$ and CNF $\psi_L(n)$ defined as follows.*

$\Pi_L(n){:=}\forall B_1 \exists B_2 \forall B_3 \exists B_4$:

$B_1{:=}\{u_{3i+1}, u_{3i+2} \mid 0 \leq i < n\}$ $B_3{:=}\{u_{3i+3} \mid 0 \leq i < n\}$

$B_2{:=}\{x_{3i+1}, x_{3i+2} \mid 0 \leq i < n\}$ $B_4{:=}\{x_{3i+3} \mid 0 \leq i < n\}$

$\psi_L(n) := \bigwedge_{i:=0}^{n-1} \mathcal{C}(i)$ with $\mathcal{C}(i) := \bigwedge_{j:=0}^{7} C_{i,j}$:

$C_{i,0} := (\neg u_{3i+1} \lor \neg x_{3i+1} \lor \neg x_{3i+2})$ $C_{i,4} := (\neg x_{3i+1} \lor \neg x_{3i+2} \lor x_{3i+3})$

$C_{i,1} := (\neg u_{3i+1} \lor \neg x_{3i+1} \lor x_{3i+2})$ $C_{i,5} := (u_{3i+3} \lor \neg x_{3i+3})$

$C_{i,2} := (u_{3i+1} \lor x_{3i+1} \lor \neg x_{3i+2})$ $C_{i,6} := (\neg x_{3i+1} \lor x_{3i+2} \lor \neg x_{3i+3})$

$C_{i,3} := (u_{3i+2} \lor x_{3i+1} \lor x_{3i+2})$ $C_{i,7} := (\neg u_{3i+3} \lor x_{3i+3})$

Proposition 5. *For $n \geq 1$, QRATU$^+$ can eliminate the entire quantifier block $\forall B_1$ in $\Phi_L(n)$ whereas QRATU cannot eliminate any universal literals in $\Phi_L(n)$.*

Proof (sketch, see appendix [18]). Formulas $\Phi_L(n)$ are constructed based on a similar principle as $\Phi_C(n)$ in Definition 14. E.g., clauses $C_{i,0}$ and $C_{i,1}$ have QRAT$^+$ but not QRAT on literals $\neg u_{3i+2}$ and $\neg u_{3i+1}$. During QUP, clauses $C_{i,5}$ and $C_{i,7}$ become empty only due to UR, which is not possible when using UP. □

6   Proof Theoretical Impact of **QRAT** and **QRAT**$^+$

As argued in the context of *interference-based proof systems* [6], certain proof steps may become applicable in a proof system only after redundant parts of the formula have been eliminated. We show that redundancy elimination by QRAT$^+$ or QRAT can lead to exponentially shorter proofs in the resolution based LQU$^+$-resolution [1] QBF calculus. Note that we do not compare the power of QRAT or QRAT$^+$ as proof systems themselves, but the impact of redundancy elimination on other proof systems. The following result relies only on QRATU, i.e., it does not require the more powerful redundancy property QRATU$^+$ in QRAT$^+$.

LQU$^+$-resolution is a calculus that generalizes traditional Q-resolution [15]. It allows to generate resolvents on both existential and universal variables and admits tautological resolvents of a certain kind. LQU$^+$-resolution is among the strongest resolution calculi currently known [1,2], yet the following class of QBFs provides an exponential lower bound on the size of LQU$^+$-resolution proofs.

Definition 16 ([2]). *For $n > 1$, let $\Phi_Q(n):=\Pi_Q(n).\psi_Q(n)$ be the* QUPar*ity QBFs with $\Pi_Q(n):=\exists x_1,\ldots,x_n \forall z_1, z_2 \exists t_2,\ldots,t_n$ and $\psi_Q(n):=C_0 \wedge C_1 \wedge \bigwedge_{i=2}^n \mathcal{C}(i)$ where $C_0:=(z_1 \vee z_2 \vee t_n)$, $C_1:=(\bar{z}_1 \vee \bar{z}_2 \vee \bar{t}_n)$, and $\mathcal{C}(i):=\bigwedge_{j=0}^7 C_{i,j}$:*

$C_{2,0}:= (\bar{x}_1 \vee \bar{x}_2 \vee z_1 \vee z_2 \vee t_2)$	$C_{i,0}:= (\bar{t}_{i-1} \vee \bar{x}_i \vee z_1 \vee z_2 \vee \bar{t}_i)$
$C_{2,1}:= (x_1 \vee x_2 \vee z_1 \vee z_2 \vee \bar{t}_2)$	$C_{i,1}:= (t_{i-1} \vee x_i \vee z_1 \vee z_2 \vee \bar{t}_i)$
$C_{2,2}:= (\bar{x}_1 \vee x_2 \vee z_1 \vee z_2 \vee t_2)$	$C_{i,2}:= (\bar{t}_{i-1} \vee x_i \vee z_1 \vee z_2 \vee t_i)$
$C_{2,3}:= (x_1 \vee \bar{x}_2 \vee z_1 \vee z_2 \vee t_2)$	$C_{i,3}:= (t_{i-1} \vee \bar{x}_i \vee z_1 \vee z_2 \vee t_i)$
$C_{2,4}:= (\bar{x}_1 \vee \bar{x}_2 \vee \bar{z}_1 \vee \bar{z}_2 \vee \bar{t}_2)$	$C_{i,4}:= (\bar{t}_{i-1} \vee \bar{x}_i \vee \bar{z}_1 \vee \bar{z}_2 \vee \bar{t}_i)$
$C_{2,5}:= (x_1 \vee x_2 \vee \bar{z}_1 \vee \bar{z}_2 \vee \bar{t}_2)$	$C_{i,5}:= (t_{i-1} \vee x_i \vee \bar{z}_1 \vee \bar{z}_2 \vee \bar{t}_i)$
$C_{2,6}:= (\bar{x}_1 \vee x_2 \vee \bar{z}_1 \vee \bar{z}_2 \vee t_2)$	$C_{i,6}:= (\bar{t}_{i-1} \vee x_i \vee \bar{z}_1 \vee \bar{z}_2 \vee t_i)$
$C_{2,7}:= (x_1 \vee \bar{x}_2 \vee \bar{z}_1 \vee \bar{z}_2 \vee t_2)$	$C_{i,7}:= (t_{i-1} \vee \bar{x}_i \vee \bar{z}_1 \vee \bar{z}_2 \vee t_i)$

Any refutation of $\Phi_Q(n)$ in LQU$^+$-resolution is exponential in n [2]. The QUParity formulas are a modification of the related *LQParity* formulas [2]. An LQParity formula is obtained from a QUParity formula $\Phi_Q(n)$ by replacing $\forall z_1, z_2$ in prefix $\Pi_Q(n)$ by $\forall z$ and by replacing every occurrence of the literal pairs $z_1 \vee z_2$ and $\bar{z}_1 \vee \bar{z}_2$ in the clauses in $\psi_Q(n)$ by the literal z and $\bar{z}$, respectively.

Proposition 6. QRATU *can eliminate either variable z_1 or z_2 from a QUParity formula $\Phi_Q(n)$ to obtain a related LQParity formula in polynomial time.*

Proof. We eliminate z_2 (z_1 can be eliminated alternatively) in a polynomial number of QRATU steps. Every clause C with $z_2 \in C$ has QRAT on z_2 since $\{z_1, \bar{z}_1\} \subseteq$ OR for all outer resolvents OR. UP immediately detects a conflict when propagating $\overline{\text{OR}}$. After eliminating all literals z_2, the clauses containing $\bar{z}_2$ trivially have QRAT on $\bar{z}_2$, which can be eliminated. Finally, z_1 including all of its occurrences is renamed to z.   □

In the proof above the universal literals can be eliminated by QRATU in any order. Hence in this case the non-confluence [9,14] of rewrite rules in the QRAT and QRAT$^+$ systems is not an issue. LQU$^+$-resolution has polynomial proofs for LQParity formulas [2]. Hence the combination of QRATU and LQU+-resolution results in a calculus that is more powerful than LQU+-resolution. A related result [13] was obtained for the combination of QRATU and the weaker QU-resolution calculus [22].

7   Experiments

We implemented QRAT+ redundancy removal in a tool called QRATPre+ for QBF preprocessing.[2] It applies rules QRATE+ and QRATU+ to remove redundant clauses and universal literals. We did not implement clause addition (QRATA+) or extended universal reduction (EUR). QRATPre+ is the *first implementation* of QRAT+ and QRAT for QBF preprocessing. The preprocessors HQSpre [24] and Bloqqer [10] (which generates partial QRAT proofs to trace preprocessing steps) do not apply QRAT to eliminate redundancies. The following experiments were run on a cluster of Intel Xeon CPUs (E5-2650v4, 2.20 GHz) running Ubuntu 16.04.1. We used the benchmarks from the PCNF track of the QBFEVAL'17 competition. In terms of scheduling the non-confluent (cf. [9,14]) rewrite rules QRATE+ and QRATU+, we have not yet optimized QRATPre+. Moreover, in general large numbers of clauses in formulas may cause run time overhead. In this respect, our implementation leaves room for improvements.

We illustrate the impact of QBF preprocessing by QRAT+ and QRAT on the performance of QBF solving. To this end, we applied QRATPre+ in addition to the state of the art QBF preprocessors Bloqqer and HQSpre. In the experiments, first we preprocessed the benchmarks using Bloqqer and HQSpre, respectively, with a generous limit of two hours wall clock time. We considered 39 and 42 formulas where Bloqqer and HQSpre timed out, respectively, in their original form. Then we applied QRATPre+ to the preprocessed formulas with a *soft wall clock time limit* of 600 seconds. When QRATPre+ reaches the limit, it prints the formula with redundancies removed that have been detected so far. These preprocessed formulas are then solved. Table 1 shows the performance of our solver DepQBF [17] in addition to the top-performing solvers[3] RAReQS [11], CAQE [20], and Qute [19] from QBFEVAL'17, using limits of 7 GB and 1800 s wall clock time. The solvers implement different solving paradigms such as expansion or resolution-based QCDCL. The results clearly indicate the benefits of preprocessing by QRATPre+. The number of solved instances increases. Qute is an exception to this trend. We conjecture that QRATPre+ blurs the formula structure in addition to Bloqqer and HQSpre, which may be harmful to Qute.

We emphasize that we hardly observed a difference in the effectiveness of redundancy removal by QRAT+ and QRAT on the considered benchmarks. The benefits of QRATPre+ shown in Table 1 are due to redundancy removal by QRAT already, and not by QRAT+. However, on additional 672 instances from class *Gent-Rowley* (encodings of the Connect Four game) available from QBFLIB, QRATE+ on average removed 54% more clauses than QRATE. We attribute

---

[2] Source code of QRATPre+: https://github.com/lonsing/qratpreplus.

[3] We excluded the top-performing solver AIGSolve due to observed assertion failures.

Table 1. Solved instances (S), solved unsatisfiable ($\bot$) and satisfiable ones ($\top$), and total wall clock time in kiloseconds (K) including time outs on instances from QBFEVAL'17. Different combinations of preprocessing by Bloqqer, HQSpre, and our tool QRATPre+.

(a) Original instances (no prepr.).

Solver	S	$\bot$	$\top$	Time
CAQE	170	128	42	656K
RAReQS	167	133	34	660K
DepQBF	152	108	44	690K
Qute	130	91	39	720K

(b) Prepr. by QRATPre+ only.

Solver	S	$\bot$	$\top$	Time
CAQE	209	141	68	594K
RAReQS	203	152	51	599K
DepQBF	157	109	48	689K
Qute	131	98	33	724K

(c) Prepr. by Bloqqer only.

Solver	S	$\bot$	$\top$	Time
RAReQS	256	180	76	508K
CAQE	251	168	83	522K
DepQBF	187	121	66	630K
Qute	154	109	45	682K

(d) Prepr. by Bloqqer and QRATPre+.

Solver	S	$\bot$	$\top$	Time
RAReQS	262	178	84	492K
CAQE	255	172	83	507K
DepQBF	193	127	66	622K
Qute	148	107	41	688K

(e) Prepr. by HQSpre only.

Solver	S	$\bot$	$\top$	Time
CAQE	306	197	109	415K
RAReQS	294	194	100	429K
DepQBF	260	171	89	494K
Qute	255	171	84	497K

(f) Prepr. by HQSpre and QRATPre+.

Solver	S	$\bot$	$\top$	Time
CAQE	314	200	114	407K
RAReQS	300	195	105	418K
DepQBF	262	177	85	488K
Qute	250	169	81	500K

this effect to larger numbers of quantifier blocks in the Gent-Rowley instances (median 73, average 79) compared to QBFEVAL'17 (median 3, average 27). The advantage of QBF abstractions in the QRAT$^+$ system is more pronounced on instances with many quantifier blocks.

8   Conclusion

We presented QRAT$^+$, a generalization of the QRAT proof system, that is based on a more powerful QBF redundancy property. The key difference between the two systems is the use of QBF specific unit propagation in contrast to propositional unit propagation. Due to this, redundancy checking in QRAT$^+$ is aware of quantifier structures in QBFs, as opposed to QRAT. Propagation in QRAT$^+$ potentially benefits from the presence of universal variables in the underlying formula. This is exploited by the use of abstractions of QBFs, for which we developed a theoretical framework, and from which the soundness of QRAT$^+$ follows. By concrete classes of QBFs we demonstrated that QRAT$^+$ is more powerful than QRAT in terms of redundancy detection. Additionally, we reported on proof theoretical improvements of a certain resolution based QBF calculus made by QRAT (or QRAT$^+$) redundancy removal. A first experimental evaluation illustrated the potential of redundancy elimination by QRAT$^+$.

As future work, we plan to implement a workflow for checking QRAT$^+$ proofs and extracting Skolem functions similar to QRAT proofs [10]. In our QRAT$^+$ preprocessor QRATPre+ we currently do not apply a specific strategy to handle the non-confluence of rewrite rules. We want to further analyze the effects of non-confluence as it may have an impact on the amount of redundancy detected. In our tool QRATPre+ we considered only redundancy removal. However, to get closer to the full power of the QRAT$^+$ system, it may be beneficial to also add redundant clauses or universal literals to a formula.

References

1. Balabanov, V., Widl, M., Jiang, J.-H.R.: QBF resolution systems and their proof complexities. In: Sinz, C., Egly, U. (eds.) SAT 2014. LNCS, vol. 8561, pp. 154–169. Springer, Cham (2014). https://doi.org/10.1007/978-3-319-09284-3_12

2. Beyersdorff, O., Chew, L., Janota, M.: Proof complexity of resolution-based QBF calculi. In: STACS. LIPIcs, vol. 30, pp. 76–89. Schloss Dagstuhl-Leibniz-Zentrum fuer Informatik (2015)

3. Biere, A.: Resolve and expand. In: Hoos, H.H., Mitchell, D.G. (eds.) SAT 2004. LNCS, vol. 3542, pp. 59–70. Springer, Heidelberg (2005). https://doi.org/10.1007/11527695_5

4. Faymonville, P., Finkbeiner, B., Rabe, M.N., Tentrup, L.: Encodings of bounded synthesis. In: Legay, A., Margaria, T. (eds.) TACAS 2017. LNCS, vol. 10205, pp. 354–370. Springer, Heidelberg (2017). https://doi.org/10.1007/978-3-662-54577-5_20

5. Fazekas, K., Heule, M.J.H., Seidl, M., Biere, A.: Skolem function continuation for quantified Boolean formulas. In: Gabmeyer, S., Johnsen, E.B. (eds.) TAP 2017. LNCS, vol. 10375, pp. 129–138. Springer, Cham (2017). https://doi.org/10.1007/978-3-319-61467-0_8

6. Heule, M.J.H., Kiesl, B.: The potential of interference-based proof systems. In: ARCADE. EPiC Series in Computing, vol. 51, pp. 51–54. EasyChair (2017)

7. Heule, M.J.H., Kullmann, O.: The science of brute force. Commun. ACM **60**(8), 70–79 (2017)

8. Heule, M.J.H., Seidl, M., Biere, A.: A unified proof system for QBF preprocessing. In: Demri, S., Kapur, D., Weidenbach, C. (eds.) IJCAR 2014. LNCS (LNAI), vol. 8562, pp. 91–106. Springer, Cham (2014). https://doi.org/10.1007/978-3-319-08587-6_7

9. Heule, M.J.H., Seidl, M., Biere, A.: Blocked literals are universal. In: Havelund, K., Holzmann, G., Joshi, R. (eds.) NFM 2015. LNCS, vol. 9058, pp. 436–442. Springer, Cham (2015). https://doi.org/10.1007/978-3-319-17524-9_33

10. Heule, M.J.H., Seidl, M., Biere, A.: Solution validation and extraction for QBF preprocessing. J. Autom. Reason. **58**(1), 97–125 (2017)

11. Janota, M., Klieber, W., Marques-Silva, J., Clarke, E.: Solving QBF with counterexample guided refinement. Artif. Intell. **234**, 1–25 (2016)

12. Järvisalo, M., Heule, M.J.H., Biere, A.: Inprocessing rules. In: Gramlich, B., Miller, D., Sattler, U. (eds.) IJCAR 2012. LNCS (LNAI), vol. 7364, pp. 355–370. Springer, Heidelberg (2012). https://doi.org/10.1007/978-3-642-31365-3_28

13. Kiesl, B., Heule, M.J.H., Seidl, M.: A little blocked literal goes a long way. In: Gaspers, S., Walsh, T. (eds.) SAT 2017. LNCS, vol. 10491, pp. 281–297. Springer, Cham (2017). https://doi.org/10.1007/978-3-319-66263-3_18

14. Kiesl, B., Suda, M.: A unifying principle for clause elimination in first-order logic. In: de Moura, L. (ed.) CADE 2017. LNCS (LNAI), vol. 10395, pp. 274–290. Springer, Cham (2017). https://doi.org/10.1007/978-3-319-63046-5_17

15. Kleine Büning, H., Karpinski, M., Flögel, A.: Resolution for quantified Boolean formulas. Inf. Comput. **117**(1), 12–18 (1995)

16. Lonsing, F., Biere, A.: Failed literal detection for QBF. In: Sakallah, K.A., Simon, L. (eds.) SAT 2011. LNCS, vol. 6695, pp. 259–272. Springer, Heidelberg (2011). https://doi.org/10.1007/978-3-642-21581-0_21

17. Lonsing, F., Egly, U.: DepQBF 6.0: A search-based QBF solver beyond traditional QCDCL. In: de Moura, L. (ed.) CADE 2017. LNCS (LNAI), vol. 10395, pp. 371–384. Springer, Cham (2017). https://doi.org/10.1007/978-3-319-63046-5_23

18. Lonsing, F., Egly, U.: QRAT$^+$: Generalizing QRAT by a more powerful QBF redundancy property. CoRR abs/1804.02908 (2018). https://arxiv.org/abs/1804.02908, IJCAR 2018 proceedings version with appendix

19. Peitl, T., Slivovsky, F., Szeider, S.: Dependency learning for QBF. In: Gaspers, S., Walsh, T. (eds.) SAT 2017. LNCS, vol. 10491, pp. 298–313. Springer, Cham (2017). https://doi.org/10.1007/978-3-319-66263-3_19

20. Rabe, M.N., Tentrup, L.: CAQE: A certifying QBF solver. In: FMCAD, pp. 136–143. IEEE (2015)

21. Samulowitz, H., Davies, J., Bacchus, F.: Preprocessing QBF. In: Benhamou, F. (ed.) CP 2006. LNCS, vol. 4204, pp. 514–529. Springer, Heidelberg (2006). https://doi.org/10.1007/11889205_37

22. Van Gelder, A.: Contributions to the theory of practical quantified Boolean formula solving. In: Milano, M. (ed.) CP 2012. LNCS, pp. 647–663. Springer, Heidelberg (2012). https://doi.org/10.1007/978-3-642-33558-7_47

23. Wetzler, N., Heule, M.J.H., Hunt, W.A.: DRAT-trim: Efficient checking and trimming using expressive clausal proofs. In: Sinz, C., Egly, U. (eds.) SAT 2014. LNCS, vol. 8561, pp. 422–429. Springer, Cham (2014). https://doi.org/10.1007/978-3-319-09284-3_31

24. Wimmer, R., Reimer, S., Marin, P., Becker, B.: HQSpre – An effective preprocessor for QBF and DQBF. In: Legay, A., Margaria, T. (eds.) TACAS 2017. LNCS, vol. 10205, pp. 373–390. Springer, Heidelberg (2017). https://doi.org/10.1007/978-3-662-54577-5_21

25. Zhang, L., Malik, S.: Conflict driven learning in a quantified Boolean satisfiability solver. In: ICCAD, pp. 442–449. ACM/IEEE Computer Society (2002)

A Why3 Framework for Reflection Proofs and Its Application to GMP's Algorithms

Guillaume Melquiond[1]($\boxtimes$) and Raphaël Rieu-Helft[2,1]

[1] Inria, Université Paris-Saclay, 91120 Palaiseau, France
`guillaume.melquiond@inria.fr`
[2] TrustInSoft, 75014 Paris, France

Abstract. Earlier work showed that automatic verification of GMP's algorithms using Why3 exceeds the current capabilities of automatic solvers. To complete this verification, numerous cut indications had to be supplied by the user, slowing the project to a crawl. This paper shows how we have extended Why3 with a framework for proofs by reflection, with minimal impact on the trusted computing base. This framework makes it easy to write dedicated decision procedures that make full use of Why3's imperative features and are formally verified. We evaluate how much work could have been saved when verifying GMP's algorithms, had this framework been available. This approach opens the way to efficiently tackling the further verification of GMP's algorithms.

Keywords: Decision procedures · Proofs by reflection
Deductive program verification · Nonlinear integer arithmetic

1 Introduction

The Why3 software-verification tool[1] offers an ML-like language (WhyML) that makes it possible to write programs and to specify the functional behavior of these programs using pre- and post-conditions, and loop invariants [7]. The tool then turns programs and specifications into theorem statements that can be sent to external provers, be they automated (e.g., SMT or TPTP solvers) or interactive (e.g., Coq, Isabelle/HOL, PVS). Once these theorems have been proved, and assuming that Why3 and the external provers are sound, the programs are known to satisfy their specification.

In an earlier work, we used Why3 to implement algorithms from the GNU Multi-Precision library,[2] GMP for short, to prove them correct, and to generate a compatible C library [12]. The proofs were done using automated provers only, mostly SMT ones. While some algorithms are extremely intricate (e.g., division [11]), we ended up having to litter the code with many more assertions than

---

[1] http://why3.lri.fr/.
[2] http://gmplib.org/.

we initially envisioned, as exemplified on Fig. 4, line 24. Seemingly trivial theorems were confusing solvers to no end. Indeed, they involved nonlinear integer arithmetic and large proof contexts. For some theorems (e.g., for the naive multiplication algorithm), we had to write several 100-line assertions, which defeats the point of using automated tools rather than an interactive theorem prover. Thus, we decided to put that experiment on hold, until we got a way to make the proof of these theorems straightforward.

When one wants to extends a theorem prover with new capabilities (e.g., an inference rule dedicated to the problem at hand), one way is to "incorporate a reflection principle, so that the user can verify within the existing theorem proving infrastructure that the code implementing a new rule is correct, and to add that code to the system" [9]. This article shows how we have modified Why3 to offer computational reflection. It was especially important to make the user process straightforward, so that reflection can be routinely used whenever external provers get lost. As an illustration, this paper shows how we made use of our approach to design and prove a decision procedure suitable for verifying GMP-like algorithms.

In Sect. 2, we illustrate computational reflection on the correctness of Strassen's algorithm for matrix multiplication in Why3. While straightforward to verify by hand, this algorithm already exceeds the capabilities of SMT solvers [5]. So we perform a reflection-based proof in a traditional way: we represent logical propositions about matrix polynomials by inductive objects, we define functions over these objects in the logical system, we prove some lemmas about them, and we use these functions and lemmas to prove the correctness property of Strassen's algorithm.

This approach does not require any modification to Why3 or to the external provers, but we have not yet explained how to reify logical propositions into inductive objects that can be manipulated inside Why3's logic. Section 3 shows how we have extended Why3 to do so.

Traditionally, computational reflection performs proofs by evaluating some pure terms occurring in logical propositions. Yet, Why3's programming language is much richer: mutable variables, arrays, exceptions, loops, and so on. Section 4.1 shows how the designer of decision procedures can benefit from the whole extent of WhyML. This required us to add a WhyML interpreter to Why3 (Sect. 4.2).

While the reification component does not extend the trusted computing base of Why3 at all, the interpreter does, albeit in a minimal way. We discuss the soundness of our approach in Sect. 5.

Given the ability to write decision procedures in WhyML, to verify them using Why3, and to execute them inside Why3 on reified logical propositions, we have all the tools to design and use a decision procedure dedicated to verifying GMP's algorithms. Section 6 presents this procedure. While it might look like a naive procedure for solving systems of linear equalities, the coefficients it manipulates are not simple rationals, they are products of rationals by powers with symbolic exponents, e.g., $-5/3 \cdot \beta^{i+j-2}$. These powers occur because we are proving the soundness of algorithms manipulating power series $\sum a_i \beta^i$.

This work is part of Why3 and the examples presented in this article are available at http://toccata.lri.fr/gallery/reflection.en.html.

2 Computational Reflection Proofs

When designing a decision procedure by reflection, one first finds an embedding of the propositions P of interest into the logical language of the formal system. Let us denote $\ulcorner P \urcorner$ the resulting term, e.g., the abstract syntax tree of P. Then one proves that, if $\ulcorner P \urcorner$ satisfies some property φ, then P holds. Thus, when one wants to prove that some proposition P holds, one just has to check that $\varphi(\ulcorner P \urcorner)$ does. If φ is designed so that $\varphi(\ulcorner P \urcorner)$ can be validated just by computations, then we have a proof procedure by computational reflection. This approach has been used in various contexts [1,3–5,8,9].

Let us illustrate this process on a toy example: the correctness of Strassen's matrix multiplication algorithm. Among other properties, one has to prove four matrix equalities such as the following one:

$$\mathbf{A}_{1,1}\mathbf{B}_{1,1} + \mathbf{A}_{1,2}\mathbf{B}_{2,1} = \mathbf{M}_{1,1},$$

with

$$\mathbf{M}_{1,1} = (\mathbf{A}_{1,1} + \mathbf{A}_{2,2}) \cdot (\mathbf{B}_{1,1} + \mathbf{B}_{2,2}) + \mathbf{A}_{2,2} \cdot (\mathbf{B}_{2,1} - \mathbf{B}_{1,1})$$
$$- (\mathbf{A}_{1,1} + \mathbf{A}_{1,2}) \cdot \mathbf{B}_{2,2} + (\mathbf{A}_{1,2} - \mathbf{A}_{2,2}) \cdot (\mathbf{B}_{2,1} + \mathbf{B}_{2,2}).$$

By the group laws of matrix addition and by distributivity of matrix multiplication, one easily shows that the right-hand side of the equality can be turned into the left-hand side. Unfortunately, in practice, SMT solvers (Alt-Ergo, CVC4, Z3) and TPTP solvers (Eprover) fail to prove such a proposition. There are two reasons. First, a solver should instantiate the above algebraic laws on the order of one hundred times, assuming they apply them in an optimal way. Second, when verifying programs, the proof context is usually filled with hundreds of other instantiable theorems, which will delay applying the algebraic laws. As a consequence, unless an automated prover implements a dedicated decision procedure for this kind of property, there is no way its proof can be found.

Let us see how to supplement the lack of such a dedicated decision procedure. While this paper presents it in the context of Why3, the exact same process could be followed in any formal system with some computational capabilities.

2.1 Embedding Terms

The first step is to embed $\mathbf{M}_{1,1}$ into the logical language of Why3. We define the following inductive type t to represent its abstract syntax tree:

```
type t = Var int | Add t t | Mul t t | Sub t t | Ext r t
```

```
type vars = int → a
let rec function interp (x: t) (y: vars) : a =
  match x with
  | Var n → y n
  | Add x1 x2 → aplus  (interp x1 y) (interp x2 y)
  | Mul x1 x2 → atimes (interp x1 y) (interp x2 y)
  | Sub x1 x2 → asub   (interp x1 y) (interp x2 y)
  | Ext r x → ($) r (interp x y)
  end
```

Fig. 1. Interpreting the abstract syntax tree of a polynomial

Matrices appearing at the leaves of the expression (e.g., $\mathbf{A}_{2,1}$) are assigned a unique integer identifier and are represented using the `Var` constructor. The sum, product, and differences of two matrices, are represented using the constructors `Add`, `Mul`, and `Sub`. Finally, the `Ext` constructor represents the external product (by a value of type `r`), which is not needed in the case of Strassen's algorithm.

Note that the function $\mathbf{M} \mapsto \ulcorner\mathbf{M}\urcorner$ cannot be expressed in the logical language, but its inverse can. We thus define a function that maps a term of type `t` into a matrix, as shown in Fig. 1. That definition causes Why3 to create a recursive function `interp` inside the logical system, since its termination is visibly guaranteed by the structural decrease of its argument `x`.

When `aplus`, resp. `atimes`, is instantiated using matrix sum, resp. product, one can prove that the Why3 term

```
interp (Mul (Add (Var 0) (Var 1)) (Var 7)) y
```

is equal to $(\mathbf{A}_{1,1} + \mathbf{A}_{1,2}) \cdot \mathbf{B}_{2,2}$, assuming that `y` maps 0 to $\mathbf{A}_{1,1}$, 1 to $\mathbf{A}_{1,2}$, and 7 to $\mathbf{B}_{2,2}$. This proof can be done by unfolding the definition of `interp`, by reducing the `match with` constructs, and by substituting the applications of `y` by the corresponding results. Why3 provides a small rewriting engine that is powerful enough for such a proof, but one could also use an external prover.

2.2   Normalizing Terms

Let us suppose that we now have two concrete expressions `x1` and `x2` of type `t` and a single map `y` of type `vars` and that we want to prove the following equality:

```
goal g: interp x1 y = interp x2 y
```

The actual value of `y` does not matter, but the facts that `aplus` is a group operation and that `amult` is distributive do. In other words, we want to see `x1` and `x2` as non-commutative polynomials and we want to prove that they have the same monomials with the same coefficients. To do so, let us turn them into weighted lists of monomials. Figure 2 shows an excerpt of the code. For example, the term $(\mathbf{A}_{1,1} + \mathbf{A}_{1,2}) \cdot \mathbf{B}_{2,2}$ gets turned into the list

```
Cons (M 1 (Cons 0 (Cons 7 Nil))) (Cons (M 1 (Cons 1 (Cons 7 Nil))) Nil)
```

```
type m = M int (list int)
type t' = list m

let rec function interp' (x: t') (y: vars) : a =
  match x with
  | Nil → azero
  | Cons (M r m) l → aplus (($) r (mon m y)) (interp' l y)
  end

let rec function conv (x:t) : t'
    ensures { forall y. interp x y = interp' result y }
= match x with
  | Var v → Cons (M rone (Cons v Nil)) Nil
  | Add x1 x2 → (conv x1) ++ (conv x2)
  | Mul x1 x2 → ...
  end
```

Fig. 2. Converting a polynomial to a list of monomials

Note that we have introduced a new interpretation function `interp'` and we have stated the postcondition of `conv` accordingly. Why3 requires us to prove that this postcondition holds. The proof is straightforward, even in the multiplication case. Once done, we obtain the following lemma in the context:

```
lemma conv_def: forall x y. interp x y = interp' (conv x) y
```

We define one last function, `norm`, which sorts a weighted list of monomials by insertion using a lexicographic order, merging contiguous monomials along the way. Its postcondition, once proved, leads to

```
lemma norm_def: forall x y. interp' x y = interp' (norm x) y
```

Note that we do not even need to prove that `norm` actually sorts the input list or that it merges monomials, so the proof is again trivial. If there is some bug in `norm`, it only endangers the completeness of the approach, not its soundness. For example, defining `norm` as the identity function would ultimately be fine but pointless.

By composing `norm` and `conv` and equality, we get our decision procedure φ dedicated to verifying Strassen's algorithm. Indeed, to prove the goal **g** above, we just need to prove the following intermediate lemma:

```
lemma g_aux: norm (conv x1) = norm (conv x2)
```

As with `interp` before, `norm` and `conv` are logic functions defined by induction on their argument, so there is no difficulty in proving `g_aux` using the rewriting engine of Why3 or an external automated prover.

2.3   Advantages

There are several advantages to this approach. The most important one is that the user can easily design a decision procedure dedicated to the problem at hand.

Indeed, the inductive type for representing expressions does not have to handle the full extent of the language, but can focus on the constructions that matter (e.g., addition). Moreover, the soundness of the system is not endangered, since the user has to prove the correctness of the procedure (e.g., the lemmas conv_def and norm_def). Finally, since the procedure is ad hoc, performances in the general case do not matter much, so one can write it so that both the code and the proof are straightforward. For instance, in the example above, the sorting algorithm has quadratic complexity and one only has to prove that the interpretation of the list is left unchanged. Thus, SMT solvers quickly discharge all the verification conditions that Why3 generates to guarantee that the implementation of the decision procedure satisfies its specification.

Even if this normalization procedure is dedicated to proving Strassen's algorithm, we took advantage of Why3's module system to make it generic: coefficients are in an arbitrary commutative ring and variables are in a (noncommutative) ring. Both rings are potentially different, as in the case of matrices. The genericity of the presented decision procedure does not extend to supporting variables in a commutative ring, but it is just a matter of duplicating the code of the decision procedure to modify the ordering relation, which we did.

A very similar reflection-based tactic is used by the Coq proof assistant to formally verify equalities in a commutative ring or semi-ring [8]. This tactic was implemented, part as an OCaml plugin for Coq, part in the meta-language Ltac of Coq. Rather than lists of monomials, that work uses Horner's representation of polynomials: $p_0 + x_1 \cdot (p_1 + x_1 \cdot (p_2 + x_1 \cdots))$ with $(p_i)_i$ being polynomials where variable x_1 does not occur.

3   Reification

We have not yet explained how one obtains the inductive objects used to instantiate the decision procedure. Without modifying Why3, it is up to the user to provide them. Even for an algorithm as simple to verify as Strassen's, the user might forfeit before finishing to translate all the terms of the algorithm.

3.1   Possible Approaches

To circumvent this issue, the original Why3 proof of Strassen's algorithm uses a clever approach [5]. The type of matrices has been modified so that a matrix contains not only the values of its cells but also the normalized list of monomials representing all the operations performed to obtain the matrix. In other words, the decision procedure has been split and embedded into all the matrix operations and it is executed symbolically along them. The lists of monomials (and the operations to build them) are declared *ghost*, so they do not interfere with actual matrix computations and can be erased from the final algorithm, which is therefore still fundamentally the same. Nonetheless, this approach forces the user to instrument the matrix operations, and while these modifications are suitable to prove Strassen's algorithm, they might be useless when verifying another matrix algorithm, if not detrimental by polluting the proof context with all the symbolic computations.

Thus, for a reflection-based decision procedure to be useful, we have to provide some ways to automate the *reification* process, that is, the conversion of expressions into their inductive representation.

As mentioned above, one difficulty lies in defining ⌜⌝, which is an inverse function of `interp`. This inverse is usually written using the meta-language of a formal system to parse the term and to produce the corresponding inductive object. Since Why3 can load plugins written in OCaml, one could certainly use OCaml as a meta-language for Why3. This unfortunately requires the user to learn the inner workings of Why3.

Another possibility would be to use WhyML as a meta-language by providing some primitives to visit the abstract syntax trees of expressions and by making Why3 able to interpret it. As is the case for other formal systems [6,13], any WhyML function using such primitives would no longer be meaningful for the remainder of the logical system, so as to avoid inconsistencies. The user would thus no longer need to leave the confines of WhyML, but this is still not completely satisfactory. Indeed, as written before, ⌜⌝ is the inverse of `interp`, so any explicit definition seems superfluous. Instead, we could have some OCaml code, e.g., a Why3 plugin, that inverts `interp` on the fly.

3.2   Inversion of the Interpretation Function

Consider the following function, which is just a variant of the decision procedure for Strassen's algorithm:

```
let norm_f (x1 x2: t) : bool
  ensures { forall y:vars. result = True → interp x1 y = interp x2 y }
= match norm (conv (Sub x1 x2)) with
  | Nil → True (* the difference evaluates to the empty polynomial *)
  | _ → False
  end
```

Whenever the user wants to use this decision procedure to prove a goal, we would like Why3 to automatically find `x1`, `x2`, and `y`, so that the right-hand side of the post-condition matches the goal. This is done by a straightforward recursive walk of the goal. Let us illustrate this walk with `foo a + b = c`. This goal is an equality, and so is the right-hand side of the postcondition of `norm_f`, so Why3 proceeds recursively on each side of the equality. The left-hand side starts with an addition, while there is an application of `interp` in the postcondition, so Why3 assumes that `interp` is an interpretation function.

This function starts with a pattern matching on its first argument, so Why3 looks at all of the branches. The second branch starts with an addition (i.e., `aplus`, which we assume was instantiated with `+`). So Why3 registers that `x1` should start with the constructor `Add`. And so on, recursively. Eventually, Why3 has to match `foo a` against a branch. None of them matches, but the one for the `Var` constructor returns `y n`, with `y` a variable of type *arrow*. So Why3 selects a fresh integer for n, e.g., 0, and remembers that it should choose y so that it maps 0 to foo a.

3.3   Extensions

The previous process works fine when a goal has to be proved in isolation, irrespective of the proof context. To remove this limitation, Why3 also recognizes the presence of an implication inside a branch of an interpretation function. In that case, it tries to match a hypothesis of the proof context against the left-hand side of the implication, and it does so recursively until all the hypotheses of the context have been tried. The following functions illustrate this behavior. They serve as interpretation functions of a decision procedure that needs to consider all the equalities from the proof context. In this example taken from the verification of GMP's algorithms, the fact that the goal also has to be an equality is a coincidence.

```
function interp_eq (g:equality) (y:vars) (z:C.cvars) : bool
= match g with (g1, g2) → interp g1 y z = interp g2 y z end

function interp_ctx (l:list equality) (g:equality) (y:vars) (z:C.cvars)
      : bool
= match l with
  | Nil → interp_eq g y z (* goal *)
  | Cons h t → (interp_eq h y z) → (interp_ctx t g y z)
  end
```

Notice that, since Why3's logical system does not permit functions returning logical propositions, we have defined these interpretation functions as returning Boolean values. But this has no impact on the way reification proceeds.

While the decision procedures presented in this paper ignore quantified formulas, our reification transformation does support them. For example, the excerpt below would handle universal quantifiers in a nameless fashion, using negative indices to store the depth of the quantifier:

```
function interp_fmla (f:fmla) (l:int) (b:vars) : bool
= match f with
  | Forall f' → forall v. interp_fmla f' (l-1) b[l ← v]
  | ...
  end
```

A current limitation of our approach is the purely syntactic nature of the reification step. For example, for an uninterpreted function foo, the terms foo (a+b) and foo (b+a) are mapped to distinct variables, even though they are provably equal. This requires a significant amount of extra work from the user. However, we are optimistic that this can be mitigated either in the reification step itself or by composition with another decision procedure (as shown in Sect. 6.3).

4   Effectful Decision Procedures

Computations in the reflection-based proof from Sect. 2 are all done in logic functions, which are unfolded by automated provers or Why3's rewriting engine. A limitation of this approach is that Why3's language of logic functions is not very expressive, as they must be side effect-free and their termination must be guaranteed by a structurally decreasing argument.

In this section, we show how we can instead write decision procedures as regular WhyML programs, making full use of the language's imperative features such as loops, references, arrays, and exceptions. These decision procedures are proved correct using Why3 and some automated theorem provers. Their contract can then be instantiated by reification of the goal and context, and used as a cut indication.

4.1   Running Example: Systems of Linear Equalities

As an example, let us consider a decision procedure for linear equation systems in an arbitrary field (code excerpts in Fig. 3). Given some assumed-valid linear equalities in the context, the procedure attempts to prove a linear equality by showing that it is a linear combination of the context.

This is done by representing the context and goal by a matrix and performing a Gaussian elimination (function `gauss_jordan`). In case of success, we obtain a vector of coefficients and we check whether the corresponding linear combination of the context is equal to the goal (function `check_combination`). Otherwise, the procedure returns `False` and proves nothing, since its postcondition has `result = True` as premise.

As is done in Coq with the tactics `lia` and `lra` [1], this is a proof by certificate, since we check if the linear combination of the context returned by `gauss_jordan` matches the goal. There is no need to prove the Gaussian elimination algorithm itself, nor to define a semantics for the matrix passed to it as a parameter. In fact, we do not prove anything about the content of any matrix in the program. This makes the proof of the decision procedure very easy in relation to its length and intricacy.

Let us now examine the contract of the decision procedure. The postcondition states that the goal holds if the procedure returns `True`, for any valuations y and z of the variables such that the equalities in the context hold. The `valid_ctx` and `valid_eq` preconditions state that the integers used as variable identifiers (second argument of the `Term` constructor) in the context and goal are all non-negative. This is needed to prove the safety of array accesses. The nature of the reification procedure ensures that these preconditions will always be true in practice, but as reification is not trusted, the user has to verify them explicitly; SMT solvers do this very easily. Finally, the `raises` clause expresses that an exception may escape the procedure (typically an arithmetic error, as we allow the field operations to be partial). In that case, nothing is proven.

Notice that the decision procedure is independent from Why3 (apart from the fact that it is formally verified), in the sense that it does not contain

```
clone LinearEquationsCoeffs as C with type t = coeff

type expr = Term coeff int | Add expr expr | Cst coeff
type equality = (expr, expr)

let linear_decision (l: list equality) (g: equality) : bool
  requires { valid_ctx l }
  requires { valid_eq g }
  ensures { forall y z. result = True → interp_ctx l g y z }
  raises  { C.Unknown → true }
= ...
  fill_ctx l 0;
  let (ex, d) = norm_eq g in
  fill_goal ex;
  let ab = m_append a b in
  let cd = v_append v d in
  match gauss_jordan (m_append (transpose ab) cd) with
    | Some r → check_combination l g (to_list r 0 ll)
    | None → False
  end
```

Fig. 3. Decision procedure for linear equation systems

meta-instructions for reification or anything linked to Why3 internals. One could easily imagine finding the same kind of code in an automatic prover.

4.2   Interpreter

Due to their side effects, functions from WhyML programs only have abstract declarations in the logical world (as opposed to the concrete logic functions used in Sect. 2). Therefore, they cannot be unfolded by automatic provers or by Why3's rewriting engine. In order to compute the results of decision procedures such as the previous one, we have added an interpreter to Why3. It operates on an ML-like intermediate language that corresponds to WhyML programs from which logic terms, assertions, and ghost code, were erased, thereby assuming that the program was proved beforehand and that the preconditions are met. This intermediate code is produced by the existing extraction mechanism, which is used to produce OCaml and C programs from proved WhyML programs.

Our interpreter provides built-in implementations for some axiomatized parts of the Why3 standard library, such as integer arithmetic and arrays. For performances purposes, we also chose to implement references as a builtin rather than interpret their WhyML definition (records with a single mutable field), in order to reduce the number of indirections. To ease debugging decision procedures, we have added to Why3's standard library a print function of type 'a → unit and without effects. It is interpreted as a polymorphic printf function.

There have been few works on computational reflections using effectful decision procedures. One may cite Claret *et al.* [4]. They use a monadic encoding of effectful computations in Coq (e.g., non-termination). Monadic decision procedures are turned into impure programs that are executed outside of Coq. The result of these external computations is used as a "prophecy" to simulate the execution of the decision procedure inside of Coq. Since we are working with Why3, which natively supports impure computations, we sidestep the need for a heavyweight simulation mechanism.

5   Soundness

The implementation of our framework requires two additions to Why3: a reification transformation and an interpreter of WhyML programs. Let us discuss the soundness of our approach.

First, the rather large and intricate code needed for reification is not part of the trusted computing base of Why3. Indeed, the reification merely guesses values for all the relevant variables and asks Why3 to instantiate the contract of the decision procedure with them. Assuming the user has proved the soundness of the decision procedure, this instantiated proposition holds, whether the reification algorithm is correct or not. A reification failure would either prevent a well-typed instantiation of the post-condition, or the resulting cut would be useless for proving the current goal.

Contrarily to the reification code, our interpreter is part of the trusted computing base. Fortunately, it is very simple, since it only manipulates concrete values. There is no need for partial evaluation nor symbolic execution nor polymorphic equality, which makes this new interpreter much simpler than the existing rewriting engine. Another reason for its simplicity is that the intermediate language has relatively few constructions, since program transformations performed by the existing extraction mechanism eliminate potentially confusing behaviors from the surface language such as parallel assignation.

6   Application: GMP

In this section, we briefly present our verified multiprecision library [12] and show how we eliminated a large number of assertions by implementing a dedicated reflection-based decision procedure.

6.1   A GMP Function

In GMP, natural integers are represented as little-endian buffers of unsigned machine integers called *limbs*. We set a radix β (typically 2^{64}). Any natural number N has a unique radix-β decomposition $\sum_{k=0}^{n-1} a_k \beta^k$, which is represented as the buffer $a_0 a_1 \ldots a_{n-1}$.

In the low-level functions, there is almost no memory management; operands are specified as pointers to their least significant limb and a size of type int32.

```
1    (** [add_limbs r x y sz] adds [x[0..sz-1]] and [y[0..sz-1]]
2        and writes the result in [r].
3        Returns the carry, either [0] or [1].
4        Corresponds to the function [mpn_add_n]. *)
5
6    let add_limbs (r x y: ptr limb) (sz: int32) : limb
7      ensures { 0 ≤ result ≤ 1 }
8      ensures { value r sz + (power radix sz) * result =
9               value x sz + value y sz }
10   =
11     let limb_zero = Limb.of_int 0 in
12     let i = ref (Int32.of_int 0) in
13     let lx = ref limb_zero in
14     let ly = ref limb_zero in
15     let c = ref limb_zero in
16     while Int32.(<) !i sz do
17       invariant { value r !i + (power radix !i) * !c =
18                  value x !i + value y !i }
19       lx := get_ofs x !i;
20       ly := get_ofs y !i;
21       let res, carry = add_with_carry !lx !ly !c in
22       set_ofs r !i res;
23       c := carry;
24       assert { value r (!i+1) + (power radix (!i+1)) * !c
25                = value x (!i+1) + value y (!i+1)
26                by value r (!i+1) + (power radix (!i+1)) * !c
27                  = value r !i + (power radix !i) * res
28                  + (power radix !i) * c
29                  = ... (* 10+ subgoals *) };
30       i := Int32.(+) !i (Int32.of_int 1);
31     done;
32     !c
```

Fig. 4. Addition of two integers

Given such a pointer **a** and a size **n**, provided the pointer is valid over the size **n**, we denote

$$\texttt{value a n} = \overline{a_0 \ldots a_{n-1}} = \sum_{k=0}^{n-1} a_k \beta^k.$$

As an example, Fig. 4 shows the function that adds two natural integers of identical limb count. Part of the specification and most invariants and assertions have been omitted for readability. The algorithm is the schoolbook addition: starting from the least significant limb, the input numbers are added limb by limb, keeping track of the carry.

Unfortunately, even such a simple algorithm somewhat stumps the SMT solvers. In order to prove the loop invariant, we needed the assertion at line 24.

Its proof consists in a sequence of about ten rather simple steps (rewrite an equality in the context, use distributivity, etc.) but the large search space prevents the automatic provers from succeeding. Therefore, we had to provide many cut indications by hand using the by construct.

Yet, with a judicious choice of coefficients, this goal (and many others in the proofs of our library) can be seen as a linear combination of the context. Therefore, we should be able to use the decision procedure from Sect. 4 to prove the assertion in one go.

6.2 Coefficients

The following is a simplified version of the context and goal obtained for the assertion of the main loop of add_limbs (Fig. 4, line 24). Variables r1 and c1 denote the values of r and c at the start of the loop (before the modifications that occur at lines 22 and 23).

```
axiom H:   value r1 i + (power radix i) * c1 = value x i + value y i
axiom H1:  res + radix * c = lx + ly + c1
axiom H2:  value r i = value r1 i
axiom H3:  value x (i+1) = value x i + (power radix i) * lx
axiom H4:  value y (i+1) = value y i + (power radix i) * ly
axiom H5:  value r (i+1) = value r i + (power radix i) * res
goal g:    value r (i+1) + power radix (i+1) * c
           = value x (i+1) + value y (i+1)
```

Notice that the linear combination $H5 - H4 - H3 + H2 + \beta^i \cdot H1 + H$ simplifies to an equality equivalent to g. In order to prove this, our decision procedure has to include powers of β (radix in the WhyML code) in its coefficients, and to support symbolic exponents (as i is a variable).

More precisely, the coefficients of our decision procedure are the product of a rational number and a (symbolic) power of β. Figure 5 is an excerpt of the WhyML implementation of the coefficients. The decision procedure of Fig. 3 is instantiated with type coeff = t.

One can define addition, multiplication, and multiplicative inverse over these coefficients. Addition is partial, since one may only add two coefficients with equal exponents. If this is not the case, the addition raises an exception, which is accounted for in the specification of the decision procedure (exception C.Unknown in Fig. 3). Note that exponents do not have to be structurally equal, only to have equal interp_exp interpretations for all values of y, which can be automatically proved within the decision procedure.

6.3 Modular Decision Procedures

The coefficients above are expressive enough to prove assertions such as the one in Fig. 4. However, notice that their interpretation (function interp in Fig. 5) is expressed in terms of real numbers (this is needed because the Gaussian elimination algorithm used in the decision procedure needs to compute the multiplicative

```
type exp = Lit int | Var int | Plus exp exp | Minus exp | Sub exp exp
type rat = (int, int)
type t = (rat, exp)

function qinterp (q:rat) : real
= match q with (n,d) → from_int n /. from_int d end

function interp_exp (e:exp) (y:vars) : int
= match e with
  | Lit n → n
  | Var v → y v
  | Plus e1 e2 → interp_exp e1 y + interp_exp e2 y
  | Sub e1 e2 → interp_exp e1 y - interp_exp e2 y
  | Minus e' → - (interp_exp e' y)
  end

function interp (t:t) (y:vars) : real
= match t with
  (q,e) → qinterp q *. pow radix (from_int (interp_exp e y))
  end
```

Fig. 5. Definition of the coefficients

inverse of some coefficients), while the context and goal consist in equalities over integers. Moreover, the inductive type for expressions that is used in the decision procedure (type `expr` in Fig. 3) is quite restrictive, which avoids repetitions in the code of the decision procedure. However, this is problematic for the user, since a term such as 2 * 3 * x cannot be reified by inversion of `interp`.

These constraints can be lifted thanks to an approach similar to the `conv` function in Sect. 2. We compose the decision procedure `linear_decision` with a function that converts integer-valued coefficients to real-valued coefficients, and a function that converts from a more expressive expression type to the `expr` type (code excerpts in Fig. 6).

The conversion procedure from integer-valued to real-valued coefficients is only sound when the exponents of β are nonnegative. This is always the case for GMP algorithms. Due to the symbolic exponents, it is not yet possible to automatically prove this property within the decision procedure, so we instead add it as an extra precondition (the `pos_*` predicates in `mp_decision`). In practice, SMT solvers prove it easily.

While the final decision procedure is specialized for GMP goals, almost all the reasoning is done in the generic linear decision procedure `linear_decision`, which we did not modify at all. We expect that, for other use cases than GMP, users will also be able to develop their own interpretation and conversion layers and reuse the primary linear decision procedure as is.

```
let decision (l:list equality') (g:equality') : bool
  requires { valid_ctx' l ∧ valid_eq' g }
  ensures { forall y z. result = True → interp_ctx' l g y z }
  raises  { Unknown → true }
= let sl, sg = simp_ctx l g in
  linear_decision sl sg

let mp_decision (l: list equality'') (g: equality'') : bool
  requires { valid_ctx'' l ∧ valid_eq'' g }
  ensures  { forall y z. result = True → pos_ctx'' l z → pos_eq'' g z
             → interp_ctx'' l g y z }
  raises  { Unknown → true }
= decision (m_ctx l) (m_eq g)
```

Fig. 6. Composition of decision procedures

7   Conclusion

This paper presents two contributions. First, we have developed a framework for proofs by reflection that uses effectful WhyML programs as decision procedures. Second, we have implemented and verified a procedure for automatically solving systems of linear equalities with symbolic coefficients. We have used this decision procedure to prove many goals throughout our formalization of GMP algorithms.

As a point of comparison, we have revisited all our existing proofs of addition, subtraction, and multiplication algorithms, which previously required numerous user-supplied assertions. The decision procedure was able to discharge all the large assertions (in the vein of Fig. 4, line 24). This section of our library was previously about 1660 lines long. The 660 lines of program code were obviously left unchanged, but the 1000 lines of specifications and proofs were halved. Moreover, a large part of the remaining 500 lines consists in function contracts and loop invariants, which are essentially incompressible.

The hardest goal we have successfully used our decision procedure on (an assertion in the proof of the generic-case long division) involves Gaussian elimination on a matrix of size about 150×90, and it terminates in about $3\,\mathrm{s}$, which is acceptable from a user-experience standpoint. Should larger matrices become problematic, one option to improve performance would be, instead of using a WhyML interpreter, to extract the decision procedure to OCaml and execute the resulting binary.

Note that while our decision procedure only deals with linear equation systems, we have successfully used it to prove goals in the proofs of multiplication, division, and logical shifts that, at first glance, are completely nonlinear. In these cases, we had to supply one or two cut indications that took care of the nonlinear part of the reasoning, but this is very acceptable considering that many of these goals previously required more than fifty user-supplied cut indications each. We are optimistic that this new tool will allow us to verify new GMP algorithms much more efficiently than we used to.

The approach presented in this article is not limited to Why3 in principle. All that is required to develop a similar framework is the capability to specify and prove the correctness of decision procedures, and the capability to execute verified programs. As such, it would likely not take much work to adapt our framework to verification platforms such as Leon [2] and Dafny [10]. For example, Leon is already able to compile ground terms to Java bytecode and execute them.

References

1. Besson, F.: Fast reflexive arithmetic tactics the linear case and beyond. In: Altenkirch, T., McBride, C. (eds.) TYPES 2006. LNCS, vol. 4502, pp. 48–62. Springer, Heidelberg (2007). https://doi.org/10.1007/978-3-540-74464-1_4
2. Blanc, R.W., Kneuss, E., Kuncak, V., Suter, P.: An overview of the Leon verification system: verification by translation to recursive functions. In: 4th Annual Scala Workshop (2013)
3. Chaieb, A., Nipkow, T.: Proof synthesis and reflection for linear arithmetic. J. Autom. Reason. **41**(1), 33–59 (2008)
4. Claret, G., del Carmen González Huesca, L., Régis-Gianas, Y., Ziliani, B.: Lightweight proof by reflection using a posteriori simulation of effectful computation. In: Blazy, S., Paulin-Mohring, C., Pichardie, D. (eds.) ITP 2013. LNCS, vol. 7998, pp. 67–83. Springer, Heidelberg (2013). https://doi.org/10.1007/978-3-642-39634-2_8
5. Clochard, M., Gondelman, L., Pereira, M.: The matrix reproved. J. Autom. Reason. **60**(3), 365–383 (2017)
6. Ebner, G., Ullrich, S., Roesch, J., Avigad, J., de Moura, L.: A metaprogramming framework for formal verification. In: 22nd ACM SIGPLAN International Conference on Functional Programming, Oxford, UK, pp. 34:1–34:29, September 2017
7. Filliâtre, J.-C., Paskevich, A.: Why3—where programs meet provers. In: Felleisen, M., Gardner, P. (eds.) ESOP 2013. LNCS, vol. 7792, pp. 125–128. Springer, Heidelberg (2013). https://doi.org/10.1007/978-3-642-37036-6_8
8. Grégoire, B., Mahboubi, A.: Proving equalities in a commutative ring done right in Coq. In: Hurd, J., Melham, T. (eds.) 18th International Conference on Theorem Proving in Higher Order Logics, Oxford, UK, pp. 98–113, August 2005
9. Harrison, J.: Metatheory and reflection in theorem proving: a survey and critique. Technical report CRC-053, SRI International Cambridge Computer Science Research Centre (1995)
10. Leino, K.R.M.: Dafny: an automatic program verifier for functional correctness. In: Clarke, E.M., Voronkov, A. (eds.) LPAR 2010. LNCS (LNAI), vol. 6355, pp. 348–370. Springer, Heidelberg (2010). https://doi.org/10.1007/978-3-642-17511-4_20
11. Moller, N., Granlund, T.: Improved division by invariant integers. IEEE Trans. Comput. **60**(2), 165–175 (2011)
12. Rieu-Helft, R., Marché, C., Melquiond, G.: How to get an efficient yet verified arbitrary-precision integer library. In: Paskevich, A., Wies, T. (eds.) VSTTE 2017. LNCS, vol. 10712, pp. 84–101. Springer, Cham (2017). https://doi.org/10.1007/978-3-319-72308-2_6
13. Ziliani, B., Dreyer, D., Krishnaswami, N.R., Nanevski, A., Vafeiadis, V.: Mtac: a monad for typed tactic programming in Coq. J. Funct. Program. **25**, 1–59 (2015)

Probably Half True: Probabilistic Satisfiability over Łukasiewicz Infinitely-Valued Logic

Marcelo Finger[(✉)] and Sandro Preto

Department of Computer Science, University of São Paulo, São Paulo, Brazil
{mfinger,spreto}@ime.usp.br

Abstract. We study probabilistic-logic reasoning in a context that allows for "partial truths", focusing on computational and algorithmic properties of non-classical Łukasiewicz Infinitely-valued Probabilistic Logic. In particular, we study the satisfiability of joint probabilistic assignments, which we call LIPSAT. Although the search space is initially infinite, we provide linear algebraic methods that guarantee polynomial size witnesses, placing LIPSAT complexity in the NP-complete class. An exact satisfiability decision algorithm is presented which employs, as a subroutine, the decision problem for Łukasiewicz Infinitely-valued (non probabilistic) logic, that is also an NP-complete problem. We develop implementations of the algorithms described and discuss the empirical presence of a phase transition behavior for those implementations.

1 Introduction

This paper deals with the problem of determining the consistency of probabilistic assertions allowing for "partial truths" considerations. This means that we depart from the classical probabilistic setting and instead employ a many-valued underlying logic. In this way we enlarge our capacity to model situations in which a gradation of truth may be closer to the perceptions of agents involved. We employ Łukasiewicz Infinitely-valued logic as it is one of the best studied many-valued logics, having interesting properties which lead to amenable computational treatment. Notably, it has been shown that foundational properties of probabilistic theory such as de Finetti coherence criteria also applies to Łukasiewicz Infinitely-valued probabilistic theories [27].

We provide theoretical presentation leading to algorithms that decide the satisfiability of probabilistic assertions in which the underlying logic is Łukasiewicz logic with infinity truth values in the interval $[0, 1]$. For that, we employ techniques from linear programming and many-valued logics. In the latter case we need to solve several instances of the satisfiability problem in Łukasiewicz Infinitely-valued logic. This problem has been shown to be NP-complete [25]

M. Finger—Partially supported by Fapesp projs. 2015/21880-4 and 2014/12236-1 and CNPq grant 306582/2014-7.

D. Galmiche et al. (Eds.): IJCAR 2018, LNAI 10900, pp. 194–210, 2018.
https://doi.org/10.1007/978-3-319-94205-6_14

and there are some implementations discussed in the literature [3], but there are many implementation options with considerable efficiency differences which we also analyze in this work.

To understand the kind of situation in which our techniques can be applicable consider the following example.

Example 1. Three friends have the habit of going to a bar to watch their soccer team's matches. Staff at the bar claims that at every such match at least two of the friends come to the premises, but if you ask them, they will say that each of them comes to watch at most 60% of the games.

In classical terms, the claims of the staff and of the three friends are in contradiction. In fact, if there are always two of the three friends present at matches, someone must attend to least two-thirds of the team's matches.

However, one may allow someone to arrive for the second half of the match, and consider his attendance only "partially true", say, a truth value of 0.5 in that case. Then it may well be the case that staff and customers are both telling the truth, that is, their claims are jointly satisfiable.   □

It turns out that the example above is unsatisfiable in classical probabilistic logic, but it is satisfiable in Łukasiewicz Infinitely-valued Probabilistic logic. In this work we are going to formalize such problems and present techniques and algorithms to solve them.

1.1 Classical and Non-classical Probabilistic Logic

Classical probabilistic logic combines classical propositional inference with classical (discrete) probability theory. The original formulation of such a mix of logic and probability is due to George Boole who, in his seminal work introducing what is now known as Boolean Algebras, already discussed the problem [4]. Among the foundational works on classical probabilistic theory we highlight that provided by de Finetti's notion of coherent probabilities [9,11].

The decision problem over classical probabilistic logic is called Probabilistic Satisfiability (PSAT). PSAT has been extensively discussed in the literature [18, 20,28], and has recently received a lot of attention due to the improvements in SAT solving and linear programming techniques, having generated a variety of algorithms, for which the empirical phenomenon of phase-transition is by now established [14,15].

Łukasiewicz Infinitely-valued Logic is widely used in the literature to model situations that require the notion of "partial truth", seen as a many-valued logic and algebra [8]. A probability theory over such a many-valued context, including a notion of coherent probabilities in line with de Finetti's original work, was developed as a sound basis for non-classical probability theory [27]. The problem of deciding whether a set of probabilistic assignments over Łukasiewicz Infinitely-valued Logic is coherent was shown to be NP-complete by [6]. It is the goal of this paper to explore equivalent formulations and algorithmic ways to solve this problem and study the existence of a phase transition in its empirical behavior.

The rest of this paper is organized as follows. In Sect. 2 we describe the notions pertaining Łukasiewicz Infinitely-valued Logic and Łukasiewicz Infinitely-valued Probabilistic Logic and the notion of coherent probability over such logic. In Sect. 3 we study the theoretical relationship between linear algebraic methods and the solution of the LIPSAT problem. In Sect. 4 we develop a column generation algorithm for LIPSAT solving and show its correctness. Finally, we discuss implementation issues and the phase transition behavior of the solvers in Sect. 5.

Due to space restrictions, proofs of some results have been omitted. Source code of the solvers developed are publicly available.

2   Preliminaries

Łukasiewicz Infinitely-valued Logic (L_∞) is arguably one of the best studied many-valued logics [8]. It has several interesting properties, such as a truth-functional semantics that is continuous, having classical logic as a limit case and possessing well developed proof-theoretical and algebraic presentations. The semantics of L_∞-formulas represent all piecewise linear functions and only those [23, 26].

The basic L_∞-language is built from a countable set of propositional symbols $\mathbb{P}$, and disjunction ($\oplus$) and negation ($\neg$) operators. For the semantics, define a L_∞-valuation $v : \mathbb{P} \to [0, 1]$, which maps propositional symbols to a value in the rational interval $[0, 1]$. Then v is extended to all L_∞-formulas as follows

$$v(\varphi \oplus \psi) = \min(1, v(\varphi) + v(\psi))$$
$$v(\neg\varphi) = 1 - v(\varphi)$$

From those operations one usually derives the following:

Conjunction: $\varphi \odot \psi =_{\text{def}} \neg(\neg\varphi \oplus \neg\psi)$	$v(\varphi \odot \psi) = \max(0, v(\varphi) + v(\psi) - 1)$
Implication: $\varphi \to \psi =_{\text{def}} \neg\varphi \oplus \psi$	$v(\varphi \to \psi) = \min(1, 1 - v(\varphi) + v(\psi))$
Maximum: $\varphi \vee \psi =_{\text{def}} \neg(\neg\varphi \oplus \psi) \oplus \psi$	$v(\varphi \vee \psi) = \max(v(\varphi), v(\psi))$
Minimum: $\varphi \wedge \psi =_{\text{def}} \neg(\neg\varphi \vee \neg\psi)$	$v(\varphi \wedge \psi) = \min(v(\varphi), v(\psi))$
Bi-implication: $\varphi \leftrightarrow \psi =_{\text{def}} (\varphi \to \psi) \wedge (\psi \to \varphi)$	$v(\varphi \leftrightarrow \psi) = 1 - |v(\varphi) - v(\psi)|$

A formula φ is L_∞-*valid* if $v(\varphi) = 1$ for every valuation v. A formula φ is L_∞-*satisfiable* if there exists a v such that $v(\varphi) = 1$; otherwise it is L_∞-*unsatisfiable*. A set of formulas Φ is satisfiable if there exists a v such that $v(\varphi) = 1$ for all $\varphi \in \Phi$. Note that $v(\varphi \to \psi) = 1$ iff $v(\varphi) \leq v(\psi)$; similarly, $v(\varphi \leftrightarrow \psi) = 1$ iff $v(\varphi) = v(\psi)$.

L_∞ also serves as a basis for a well-founded non-classical probability theory [24]. Define a *convex combination* over a finite set of valuations $v_1, \cdots, v_m$ as a function on formulas into $[0, 1]$ such that

$$C(\varphi) = \lambda_1 v_1(\varphi) + \cdots + \lambda_m v_m(\varphi) \tag{1}$$

where $\lambda_i \geq 0$ and $\sum_{i=1}^{m} \lambda_i = 1$. So a L_∞-probability distribution $\lambda = [\lambda_1, \cdots, \lambda_m]$ is a set of coefficients that form the convex combination of L_∞-valuations. To distinguish L_∞-probabilities from classical ones, we use the notation $C(\cdot)$, following [24]; it is important to note that C is defined over *any finite* set of valuations[1]. Note that classical discrete probabilities are also convex combinations of $\{0, 1\}$-valuations.

This notion of probability associates non-zero values only to a finite number of L_∞-valuations; thus the notion of L_∞-probability is intrinsically discrete. As there are infinitely many possible L_∞-valuations, the remaining ones are assumed to be zero. In this work we are interested in deciding the existence of convex combinations of the form (1) given a set of constraints. So, in theory, the search space is infinite.

It follows immediately from this definition that $C(\alpha) = 1$ if there is a convex combination over $v_1, \cdots, v_m$ where $v_i(\alpha) = 1, 1 \leq i \leq m$.

Lemma 1. $C(\alpha \to \beta) = 1$ iff $C(\alpha) \leq C(\beta)$.   □

Lemma 1 is a direct consequence from the fact that $v(\varphi \to \psi) = 1$ iff $v(\varphi) \leq v(\psi)$.

We define a Łukasiewicz Infinitely-valued Probabilistic (LIP) assignment as an expression of the form

$$\Sigma = \Big\{ C(\alpha_i) = q_i \mid q_i \in [0, 1], 1 \leq i \leq k \Big\}.$$

As a foundational view of probabilities, it is possible to define a coherence criterion over LIP-assignments, in analogy to the de Finetti classical notion of coherent assignment of probabilities [10,11]. Thus, define the L_∞-coherence of a LIP-assignment $\{C(\alpha_i) = q_i \mid 1 \leq i \leq k\}$ in terms of a bet between two players, Alice the bookmaker and Bob the bettor. The outcome on which the players bet is a L_∞-valuation describing an actual "possible world". For each formula α_i, Alice states her betting odd $C(\alpha_i) = q_i \in [0, 1]$ and Bob chooses a "stake" $\sigma_i \in \mathbb{Q}$; Bob pays Alice $\sum_{i=1}^{k} \sigma_i \cdot C(\alpha_i)$ with the promise that Alice will pay back $\sum_{i=1}^{k} \sigma_i \cdot v(\alpha_i)$ if the outcome is the possible world (or valuation) v. As in the classical case, the chosen stake σ_i is allowed to be negative, in which case Alice pays Bob $|\sigma_i| \cdot C(\alpha_i)$ and gets back $|\sigma_i| \cdot v(\alpha_i)$ if the world turns out to be v. Alice's total balance in the bet is

$$\sum_{i=1}^{k} \sigma_i(C(\alpha_i) - v(\alpha_i)).$$

We say that there is a *LIP-Dutch Book* against Alice's LIP-assignment if there is a choice of stakes σ_i such that, for every possible outcome v, Alice's total balance is always negative, indicating a bad choice of betting odds made by Alice.

---

[1] Thus C is more restrictive than the full class of states of an MV-algebra, in the sense of [24], which will not be discussed here.

Definition 1. Given a probability assignment to propositional formulas $\{C(\alpha_i) = q_i \mid 1 \leq i \leq k\}$, the LIP-assignment is *coherent* if there are no Dutch Books against it.

While the coherence of an assignment provides a foundational view to deal with L_∞-probabilities, a more computational view is possible, based on the satisfiability of assignments. Such a view will allow a more operational way of dealing with L_∞-probabilistic assignments.

Definition 2. A LIP-assignment is *satisfiable* if there exists a convex combination C and a set of valuations that jointly verifies all restrictions in it.

Example 2. Consider again Example 1, let x_1, x_2, x_3 be variables representing the presence at the bar of each of the three friends. An L_∞-valuation assigns to each variable a value in $[0, 1]$. The probabilistic constraint expressing that each friend comes at most 60% of the games can be expressed as

$$C(x_1) = C(x_2) = C(x_3) \leq 0.6, \tag{*}$$

and the fact that at least two of them are present is expressed by the constraints

$$C(x_1 \oplus x_2) = C(x_1 \oplus x_3) = C(x_2 \oplus x_3) = 1 \tag{**}$$

which means that no two of them are simultaneously absent. There are infinitely many ways of obtaining a convex combination of L_∞-valuations that satisfy all six conditions, the simplest of which is achieved with a single L_∞-valuation v, $v(x_1) = v(x_2) = v(x_3) = 0.6$; in fact, $v(x_1 \oplus x_2) = v(x_1 \oplus x_3) = v(x_2 \oplus x_3) = \min(1, 0.6 + 0.6) = 1$, so we can attribute 100% of probability mass to v.

A similar result can be obtained with three "classical" valuations $v_i(x_i) = 0, v_i(x_j) = v_i(x_k) = 1$, for pair-wise distinct $i, j, k \in \{1, 2, 3\}$ and a fourth valuation $v_4(x_1) = v_4(x_2) = v_4(x_3) = 0.5$. Note all four valuation satisfy the formulas in (**). The convex valuation assigns probability 0.2 to v_1, v_2, v_3 and 0.4 to v_4, satisfying all constraints (*) and (**).   □

The following result is the characterization of coherence for Łukasiewicz Infinitely-valued Probabilistic Logic.

Proposition 1 (Mundici [27]). *Given a LIP-assignment* $\Sigma = \{C(\alpha_i) = q_i \mid 1 \leq i \leq k\}$, *the following are equivalent:*

(a) Σ *is a coherent LIP-assignment.*
(b) Σ *is a satisfiable LIP-assignment.*

Proposition 1 asserts that deciding LIP coherence is the same as determining LIP-assignment satisfiability, which we call *LIPSAT*. This result is the L_∞ analogous to de Finetti's characterization of coherence of classical probabilistic assignment as equivalent to the *probabilistic satisfiability* (PSAT) of the assignment, which was shown to be an NP-complete problem that can be solved using linear algebraic methods [18,28]. It has also been shown by Bova and Flaminio [6] that deciding the coherence of a LIP-assignment is also an NP-complete problem.

Our goal here is to explore efficient ways to decide the coherence of LIP-assignments. In analogy to the algorithms used for deciding PSAT [14,15], we explore a linear algebraic formulation of the problem.

3   Algebraic Formulation of LIPSAT

We consider an *extended* version of LIP-assignments of the form

$$\Sigma = \left\{ C(\alpha_i) \bowtie_i q_i \mid q_i \in [0,1], \bowtie_i \in \{=, \leq, \geq\}, 1 \leq i \leq k \right\}. \qquad (2)$$

Extended LIP-assignments may have both inequalities and equalities. Such an assignment is satisfiable if there is a L_∞-probability distribution λ that verify all inequalities and equalities in it.

Given an extended LIP-assignment $\Sigma = \{C(\alpha_i) \bowtie_i q_i\}$, let $q = (q_1, \ldots, q_k)'$ be the vector of probabilities in Σ, $\bowtie$ the "vector" of (in)equality symbols. Suppose we are given L_∞-valuations $v_1, \ldots, v_m$ and let $\lambda = (\lambda_1, \ldots, \lambda_m)'$ be a vector of convex weights. Consider the $k \times m$ matrix $A = [a_{ij}]$ where $a_{ij} = v_j(\alpha_i)$. Then an extended LIP-assignment of the form (2) is satisfiable if there are $v_1, \ldots, v_m$ and λ such that the set of algebraic constrains (3) has a solution:

$$A \cdot \lambda \bowtie q$$
$$\sum \lambda_j = 1 \qquad (3)$$
$$\lambda \geq 0$$

The condition $\sum \lambda_j = 1$ can be incorporated as an all-1 row $k+1$ in matrix A, $q = (q_1, \ldots, q_k, 1)'$ and $\bowtie_{k+1}$ is "=". Note that the number m of columns in A is in principle unbounded, but the following consequence of Carathéodory's Theorem [13] yields that a if (3) has a solution, than it has a "small" solution.

Proposition 2 (Carathéodory's Theorem for LIP). *If a set of restrictions of the form (3) has a solution, then it has a solution in which at most $k+1$ elements of λ are non-zero.*   □

Given the algebraic formulation in (3), NP-completeness of LIP satisfiability, originally shown by Bova and Flaminio [6], can be seen as a direct corollary of Proposition 2. In fact, that LIPSAT is NP-hard comes from the fact that when all $q_i = 1$, the problem becomes L_∞-satisfiability, which is NP-complete [25]; and Proposition 2 asserts the existence of a polynomial size witness for LIPSAT, hence is in NP; so LIPSAT is NP-complete. See Corollary 1.

However, to apply linear algebraic methods to efficiently solve LIPSAT, first we need to provide a normal form for it.

3.1   A Normal Form for LIP-Assignments

An extended assignment may seem more expressive than regular LIP-assignments, but we show that no expressivity is gained by this extension. In fact, we define a *normal form* LIP assignment as a pair $\langle \Gamma, \Theta \rangle$, where Γ is a set of L_∞-formulas and Θ is a set of LIP restrictions over propositional symbols of the form

$$\Theta = \Big\{ C(p_i) = q_i \mid q_i \in [0,1], p_i \in \mathbb{P}, 1 \le i \le k \Big\}. \tag{4}$$

The formulas $\gamma \in \Gamma$ represent LIP-assignments of the form $C(\gamma) = 1$, that is, a set of hard constrains in the form of L_∞-formulas which must be satisfied by all valuations in the convex combination that compose a L_∞-probability distribution.

A normal form assignment $\langle \Gamma, \Theta \rangle$ is satisfiable if there are L_∞-valuations $v_1, \cdots, v_m$ such that $v_i(\gamma) = 1$ for every $\gamma \in \Gamma$ and there is a L_∞-probability distribution $\lambda_1, \cdots, \lambda_m$, such that for each assignment $C(p_i) = q_i \in \Theta$, $\sum_{j=1}^{m} \lambda_j \cdot v_j(p_i) = q_i$.

The satisfiability of extended LIP-assignments reduces to that of normal form ones, as follows.

Theorem 1 (Atomic Normal Form). *For every extended LIP-assignments Σ there exists a normal form LIP-assignment $\langle \Gamma, \Theta \rangle$ such that Σ is a satisfiable iff $\langle \Gamma, \Theta \rangle$ is; the normal form assignment can be built from Σ in polynomial time.*

Proof. Start with $\Gamma = \Theta = \varnothing$. Given Σ, first transform it into Σ' in which all assignments are of the form $C(\alpha) \le p$; for that, if Σ contains a constraint of the form $C(\alpha) \bowtie 1$, $\bowtie \in \{=, \ge\}$ (resp. $C(\alpha) = 0, C(\alpha) \le 0$) we insert α (resp. $\neg\alpha$) in Γ and do not insert the constraint in Σ'. If $C(\alpha) = q \in \Sigma$ we insert $C(\alpha) \le q$ and $C(\alpha) \ge q$ in Σ'. Then all assignments of the latter form are transformed into $C(\neg\alpha) \le 1 - q$. All transformation steps preserve satisfiability and can be made in linear time, so $\Gamma \cup \Sigma'$ is satisfiable iff Σ is.

For every $C(\alpha_i) \le q_i \in \Sigma'$, $0 < q_i < 1$, consider a *new* symbol y_i; insert $\alpha_i \to y_i$ in Γ and $C(y_i) = q_i$ in Θ. Clearly $\langle \Gamma, \Theta \rangle$ is in normal form and is obtained in linear time. The fact that Σ is satisfiable iff $\langle \Gamma, \Theta \rangle$ is follows from Lemma 1. □

Example 3. Note that the formalization presented in Example 2 is already in normal form, witnessing that this format is quite a natural one to formulate LIP-assignments. □

3.2 Algebraic Methods for Normal Form LIP-Assignments

For the rest of this paper we assume that LIP-assignments are in normal form. Here we explore their algebraic structure as it allows for the interaction between a LIP problem Θ and a L_∞-SAT instance Γ, such that solutions satisfying the normal form assignment can be seen as probabilistic solutions to Θ constrained by the SAT instance Γ.

Furthermore, to construct a convex combination of the form (1) we will *only* consider Γ-satisfiable valuations. Given a LIP-assignment $\langle \Gamma, \Theta = \{C(p_i) = q_i\} \rangle$, a partial assignment v over $p_i, \ldots, p_k$ is Γ-*satisfiable* if it can be extended to a full assignment that satisfies all formulas in Γ. Let q be a $k + 1$ dimensional vector $(q_1, \ldots, q_k, 1)'$. The following is a direct consequence of Theorem 1.

Lemma 2. *A normal form instance $\langle \Gamma, \Theta \rangle$ is satisfiable iff there is a $(k+1) \times (k+1)$-matrix A_Θ, such that all of its columns are Γ-satisfiable, A_Θ last row is all 1's, and $A_\Theta \lambda = q$ has a solution $\lambda \geq 0$.*

Lemma 2 leads to a linear algebraic PSAT solving method as follows. Let V be the set of partial valuations over the symbols in Θ; consider a $|V|$-dimensional vector c such that

$$c_j = \begin{cases} 0, & v_j \in V \text{ is } \Gamma\text{-satisfiable} \\ 1, & \text{otherwise} \end{cases} \tag{5}$$

The vector c is a boolean "cost" associated to each partial valuation $v_j \in V$, such that the cost is 1 iff v_j is Γ-unsatisfiable. Consider a matrix A whose columns are the valuations in V. Now consider linear program (6) which aims at minimizing that cost, weighted by the corresponding probability value λ_j.

$$
\begin{aligned}
\min \quad & c' \cdot \lambda \\
\text{subject to } & A \cdot \lambda = q \\
& \sum \lambda_i = 1 \\
& \lambda \geq 0
\end{aligned} \tag{6}
$$

A's columns are partial valuations in V

Theorem 2. *A normal form instance $\langle \Gamma, \Theta = \{C(p_i) = q_i \mid 1 \leq i \leq k\} \rangle$ is satisfiable iff linear program (6) reaches a minimal solution $c' \cdot \lambda = 0$. Furthermore, if there is a solution, then there is a solution in which at most $k+1$ values of λ are not null.*

Proof. If linear program (6) reaches 0, we obtain $v_1, \ldots, v_m$ by selecting only the Γ-satisfiable columns A_j for which $\lambda_j > 0$, obtaining a convex combination satisfying Θ. So $\langle \Gamma, \Theta \rangle$ is satisfiable. Conversely, if $\langle \Gamma, \Theta \rangle$ is satisfiable, by Lemma 2 there exists a matrix A_Θ such that all of its columns are Γ-satisfiable partial valuations and $A_\Theta \cdot \lambda = q$; clearly A_Θ is a submatrix of A; make $\lambda_j = 0$ when A_j is a A_Θ column and thus $c' \cdot \lambda = 0$. Again by Lemma 2, A_Θ has at most $k+1$ columns so at most $k+1$ values of λ are not null. □

The following consequence of Theorem 2 was originally proven by Bova and Flaminio [6] as the decision of LIP-assignment coherence, which is equivalent to LIP satisfiability by Proposition 1.

Corollary 1 (LIPSAT Complexity). *The problem of deciding the satisfiability of a LIP-assignment is NP-complete.*

Despite the fact that solvable linear programs of the form (6) always have polynomial size solutions, with respect to the size of the corresponding normal form LIP-assignment, the elements of linear program itself (6) may be exponentially large, rendering the explicit representation of matrix A impractical. In the following, we present an algorithmic technique that avoids that exponential explosion.

4   A LIPSAT-Solving Algorithm

Based on the results of the previous section we are going to present an algorithm employing a linear programming technique called *column generation* [21,22], to obtain a decision procedure for Łukasiewicz Infinitely-valued Probabilistic Logic, which we call *LIPSAT solving*. This algorithm solves the potentially large linear program (6) without explicitly representing all columns and making use of an extended solver for L_∞-satisfiability as an auxiliary procedure to generate columns.

To avoid the exponential blow of the size of matrix in (6), the algorithm basic idea is to employ the simplex algorithm [2,29] over a normal form LIP-assignment $\langle \Gamma, \Theta \rangle$, coupled with a strategy that generates cost decreasing columns without explicitly representing the full matrix A. In this process, we start with a *feasible solution*, which may contain several L_∞ Γ-unsatisfiable columns. We minimize the cost function consisting of the sum of the probabilities associated to Γ-unsatisfiable columns, such that when it reaches zero, we know that the problem is satisfiable; if no column can be generated and the minimum achieved is bigger than zero, a negative decision is reached.

The general strategy employed here is similar to that employed to PSAT solving [14,15], but the column generation algorithm is considerably distinct and requires an extension of L_∞ decision procedure.

From the input $\langle \Gamma, \Theta \rangle$, we implicitly obtain an unbounded matrix A and explicit obtain the vector of probabilities q mentioned in (6). The basic idea of the simplex algorithm is to move from one feasible solution to another one with a decreasing cost. The feasible solution consists of a square matrix B, called the basis, whose columns are extracted from the unbounded matrix A. The pair $\langle B, \lambda \rangle$ consisting of the basis B and a LIP probability distribution λ is a *feasible solution* if $B \cdot \lambda = q$ and $\lambda \geq 0$. We assume that $q_{k+1} = 1$ such that the last line of B we will force $\sum_G \lambda_j = 1$, where G is the set of B columns that are Γ-satisfiable. Each step of the algorithm replaces one column of the feasible solution $\langle B^{(s-1)}, \lambda^{(s-1)} \rangle$ at step $s-1$ obtaining a new feasible solution $\langle B^{(s)}, \lambda^{(s)} \rangle$. The cost vector $c^{(s)}$ is a $\{0,1\}$ vector such that $c_j^{(s)} = 1$ iff B_j is Γ-unsatisfiable. The column generation and substitution is designed such that the total cost is never increasing, that is $c^{(s)\prime} \cdot \lambda^{(s)} \leq c^{(s-1)\prime} \cdot \lambda^{(s-1)}$.

Algorithm 4.1 presents the top level LIPSAT decision procedure. Lines 1–3 present the initialization of the algorithm. We assume the vector q is in ascending order. Let the D_{k+1} be a $k+1$ square matrix in which the elements on the diagonal and below are 1 and all the others are 0. At the initial step we make $B^{(0)} = D_{k+1}$, this forces $\lambda_1^{(0)} = q_1 \geq 0$, $\lambda_{j+1}^{(0)} = q_{j+1} - q_j \geq 0, 1 \leq j \leq k$; and $c^{(0)} = [c_1 \cdots c_{k+1}]'$, where $c_k = 0$ if column j in $B^{(0)}$ is Γ-satisfiable; otherwise $c_j = 1$. Thus the initial state $s = 0$ is a feasible solution.

Algorithm 4.1 main loop covers lines 5–12 which contains the column generation strategy described above. Column generation occurs at beginning of the loop (line 5) which we are going to detail bellow. If column generation fails the process ends with failure in line 7. Otherwise a column is removed and the

---

Algorithm 4.1. LIPSAT-CG: a LIPSAT solver via Column Generation

---

Input: A normal form LIPSAT instance $\langle \Gamma, \Theta \rangle$.
Output: No, if $\langle \Gamma, \Theta \rangle$ is unsatisfiable. Or a solution $\langle B, \lambda \rangle$ that minimizes (6).
1: $q := [\{q_i \mid C(p_i) = q_i \in \Theta, 1 \leq i \leq k\} \cup \{1\}]$ in ascending order;
2: $B^{(0)} := D_{k+1}$;
3: $s := 0$, $\lambda^{(s)} = (B^{(0)})^{-1} \cdot q$ and $c^{(s)} = [c_1 \cdots c_{k+1}]'$;
4: **while** $c^{(s)\prime} \cdot \lambda^{(s)} \neq 0$ **do**
5:     $y^{(s)} = GenerateColumn(B^{(s)}, \Gamma, c^{(s)})$;
6:     **if** $y^{(s)}$ column generation failed **then**
7:         **return** No;   \\ LIPSAT instance is unsatisfiable
8:     **else**
9:         $B^{(s+1)} = merge(B^{(s)}, b^{(s)})$
10:        $s\mathrm{++}$, recompute $\lambda^{(s)}$ and $c^{(s)}$;
11:    **end if**
12: **end while**
13: **return** $\langle B^{(s)}, \lambda(s) \rangle$;   \\ LIPSAT instance is satisfiable

---

generated column is inserted in a process we called *merge* at line 9. The loop ends successfully when the objective function (total cost) $c^{(s)\prime} \cdot \lambda^{(s)}$ reaches zero and the algorithm outputs a probability distribution λ and the set of Γ-satisfiable columns in B, at line 13.

The procedure *merge* is part of the simplex method which guarantees that given a $k+1$ column y and a feasible solution $\langle B, \lambda \rangle$ there always exists a column j in B such that if $B[j := y]$ is obtained from B by replacing column j with y, then there is λ' such that $\langle B[j := y], \lambda' \rangle$ is a feasible solution.

Lemma 3. *Let $\langle B, \lambda \rangle$ be a feasible solution of (6), such that B is non-singular, and let y be a column. Then there always exists a column j such that $\langle B[j := y], \lambda' \rangle$ is a non-singular feasible solution.*

Lemma 3 guarantees the existence of a column which may not be unique and further selection heuristic is necessary; in our implementation we give priority to remove columns which are associated to probability zero on a left-to-right order.

We now describe the column generation method, which takes as input the current basis B, the current cost c, and the L_∞ restrictions Γ; the output is a column y, if it exists, otherwise it signals **No**. The basic idea for column generation is the property of the simplex algorithm called the *reduced cost* of inserting a column y with cost c_y in the basis. The reduced cost is given by equation

$$r_y = c_y - c'B^{-1}y \tag{7}$$

and the simplex method guarantees that the objective function is non increasing if $r_y \leq 0$. Furthermore the generation method is such that the column y is Γ-satisfiable so that $c_y = 0$. We thus obtain

$$c'B^{-1}y \geq 0 \tag{8}$$

which is an inequality on the elements of y. To force λ to be a probability distribution, we make $y_{k+1} = 1$, the remaining elements y_i are valuations of the variables in Θ, so that we are searching for solution to (8) such that $0 \leq y_i \leq 1, 1 \leq i \leq k$. To finally obtain column y we must extend a L_∞-solver that generates valuations satisfying Γ so that it also respects the linear restriction (8). In fact this is not an expressive extension of L_∞ as the McNaughton property guarantees that (8) is equivalent to some L_∞-formula on variables $y_1, \ldots, y_k$ [8]. In practice, we tested two ways of obtaining a joint solver for Γ and (8):

- Employ an SMT (SAT modulo theories) solver that can handle linear algebraic equations such as (8) and the linear inequalities generated by the L_∞-semantics. L_∞-solvers based on SMT can be found in the literature, see [3];
- Use a MIP (mixed integer programming) solver that encodes L_∞-semantics. Equation (8) is simply a new linear restriction to be dealt by the MIP solver. L_∞-solvers based on MIP solvers have been proposed by [19].

In both cases, the restrictions posed by Γ-formulas and (8) are jointly handled by the semantics of the underlying solver. Note that both MIP solving and SMT(linear algebra) are NP-complete problems. We have thus the following result.

Lemma 4. *There are algorithmic solutions to the problem of jointly satisfying L_∞-formulas and inequalities with common variables.*

We now deal with the problem of termination. Column generation as above guarantees that the cost is never increasing. The simplex method ensures that a solvable problem always terminates if the costs always decrease, we are left with the problem of guaranteeing that the objective function does not become stationary. This is guaranteed in the implementation by a column selection strategy that respects *Bland's Rule* and also by plateau escaping strategies such as *Tabu search* [2,29].

Lemma 5. *There are column selection strategies that guarantee that the Algorithm 4.1 always terminates.*

We know that there are no column selection heuristics that guarantee that the simplex method terminates in a polynomial number of steps. However, the simplex method performs very well in most practical cases and its average complexity is known to be polynomial [5].

By placing all the results above together we can state the correction of Algorithm 4.1.

Theorem 3. *Consider the output of Algorithm 4.1 with normal form input $\langle \Gamma, \Theta \rangle$. If the algorithm succeeds with solution $\langle B, \lambda \rangle$, then the input problem is satisfiable with distribution λ over the valuations which are columns of B. If the program outputs no, then the input problem is unsatisfiable. Furthermore, there are column selection strategies that guarantee termination.*

Proof. Lemma 3 guarantees that all steps $\langle B^{(s)}, \lambda(s) \rangle$ is a feasible solution to the problem. If Algorithm 4.1 terminates with success, than cost zero has been reached, so by Theorem 2 the input problem is satisfiable. On the other hand, if column generation fails, this fails with a positive cost, this means there are no Γ-satisfiable columns that can reduce the cost. So, the problem in unsatisfiable. Finally, a suitable column selection strategy by Lemma 5 guarantees termination.

Example 4. We show the steps for the solution of Example 2. Initially, we have

$$
q = \begin{bmatrix} 0.6 \\ 0.6 \\ 0.6 \\ 1 \end{bmatrix}, B^{(0)} = \begin{bmatrix} 1 & 0 & 0 & 0 \\ 1 & 1 & 0 & 0 \\ 1 & 1 & 1 & 0 \\ 1 & 1 & 1 & 1 \end{bmatrix}, \lambda^{(0)} = (B^{(0)})^{-1} \cdot q = \begin{bmatrix} 0.6 \\ 0 \\ 0 \\ 0.4 \end{bmatrix}, c^{(0)} = \begin{bmatrix} 0 \\ 0 \\ 1 \\ 1 \end{bmatrix}.
$$

$c^{(0)}$ expresses that the first two columns of $B^{(0)}$ are Γ-satisfiable. The total cost $\text{cost}^{(0)} = c^{(0)\prime} \cdot \lambda^{(0)} = 0.4$. At this point, column $y^{(1)}$ is generated substituting $B^{(0)}$'s column 3 in the *merge* procedure:

$$
y^{(1)} = \begin{bmatrix} 1 \\ 0 \\ 1 \\ 1 \end{bmatrix}, B^{(1)} = \begin{bmatrix} 1 & 0 & 1 & 0 \\ 1 & 1 & 0 & 0 \\ 1 & 1 & 1 & 0 \\ 1 & 1 & 1 & 1 \end{bmatrix}, \lambda^{(1)} = \begin{bmatrix} 0.6 \\ 0 \\ 0 \\ 0.4 \end{bmatrix}, c^{(1)} = \begin{bmatrix} 0 \\ 0 \\ 0 \\ 1 \end{bmatrix}.
$$

$\text{cost}^{(1)} = 0.4$. Again, column generation provides $y^{(2)}$ in place of column 1:

$$
y^{(2)} = \begin{bmatrix} 1 \\ 1 \\ 0 \\ 1 \end{bmatrix}, B^{(2)} = \begin{bmatrix} 1 & 0 & 1 & 0 \\ 1 & 1 & 0 & 0 \\ 0 & 1 & 1 & 0 \\ 1 & 1 & 1 & 1 \end{bmatrix}, \lambda^{(2)} = \begin{bmatrix} 0.3 \\ 0.3 \\ 0.3 \\ 0.1 \end{bmatrix}, c^{(2)} = \begin{bmatrix} 0 \\ 0 \\ 0 \\ 1 \end{bmatrix}.
$$

$\text{cost}^{(2)} = 0.1$. Finally, column generation provides $y^{(3)}$ in place of column 4:

$$
y^{(3)} = \begin{bmatrix} 0.5 \\ 0.5 \\ 0.5 \\ 1 \end{bmatrix}, B^{(3)} = \begin{bmatrix} 1 & 0 & 1 & 0.5 \\ 1 & 1 & 0 & 0.5 \\ 0 & 1 & 1 & 0.5 \\ 1 & 1 & 1 & 1 \end{bmatrix}, \lambda^{(3)} = \begin{bmatrix} 0.2 \\ 0.2 \\ 0.2 \\ 0.4 \end{bmatrix}, c^{(3)} = \begin{bmatrix} 0 \\ 0 \\ 0 \\ 0 \end{bmatrix}.
$$

$\text{cost}^{(3)} = 0$, so that the problem is satisfiable with solution $\langle B^{(3)}, \lambda^{(3)} \rangle$.   □

5   Implementation and Results

The mere development of a solver over a handful of tests is, in our opinion, an insufficient way to assess the quality of an implementation. In this section we explore a qualitative behavior of solvers, called phase transition, over a large class of randomly generated formulas.

A decision problem displays a *phase transition* when there is an ordering of classes of problems that presents a transition from predominantly satisfiable instances (answer "yes") to predominantly unsatisfiable instances (answer "no"), which is called *a first order phase transition*. Furthermore the decision problem displays a peak in average execution time around the middle of that transition in which fifty percent of answers "yes" and fifty percent of answers "no", which is called a second order phase transition, following the terminology of mechanical statistics [7].

It is conjectured that there is a (second order) phase transition for every NP-complete decision problem [7]. Empirical phase transition behavior are well established for classical SAT [17] and PSAT [14], among many others. In fact, the empirical verification of phase transition for solvers of an NP-complete problem can be perceived as a quality test for its implementation. In the following we present our empirical results, searching for a phase transition behavior, for L_∞-solvers and LIPSAT solver.

5.1   Phase Transition for L_∞-Solvers

In a classical setting one usually employs 3-SAT format to obtain a phase transition diagram. The randomly generated formulas are clauses with three literals each, the number of symbols n is fixed and the rate between the number n of clauses and the rate $\frac{m}{n}$ is used as the control parameter, where m is the number of clauses. In classical 3-SAT, the shape of the curve and the phase transition point is maintained when n is changed. Unfortunately for L_∞ Logic there is no clausal normal form. So instead we employ a set of formulas which are used by [3] consisting of

$$l_1 \oplus l_2 \oplus l_3 \tag{9}$$
$$\neg(l_4 \oplus l_5) \oplus l_6 \tag{10}$$

where l_i are literals (negated or non-negated symbols). The generation of the formulas is parametrized by the number n of propositional symbols and the number m of formulas, which define the class of randomly generated formulas. Following [3], formulas are generated as follows: 70% of formulas are of format (9) and 30% of the format (10). Each literal is randomly chosen from the n possible symbols with equal probability, then there is a 50% chance of being a positive or negative literal.

Two implementations were developed using publicly available open source software[2]:

- a C++-implementation using the C++ interface to the YICES SMT(LA) solver [12];
- a C++-implementation using the C++ interface to the SCIP MIP solver [1].

For each implementation, the experiment proceed as follows: with a fixed $n = 100$ we varied the value of m such that the rate $\frac{m}{n}$ varies from 0.2 to 8 in 0.2 steps. For each pair $\langle n, m \rangle$ we construct a set of 100 randomly generated formulas as described above. And for each set we compute the percentage of L_∞-satisfiable formulas and the average decision time (user time).

All the experiments in this section were run on a UNIX machine with a i7-6900K CPU @ 3.20 GHz with 16 processors. The results of the experiments using two L_∞-solvers are shown in Fig. 1. In Fig. 1a we see the results of an SMT(LA)

---

[2] The source code for all experiments under license GPLv3 are publicly available at http://lipsat.sourceforge.net.

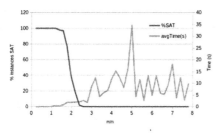
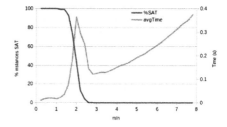

(a) Based on SMT(LA) using YICES   (b) Based on MIP solver using SCIP

Fig. 1. L_∞-solvers performance, randomly gen. instances: $n = 100$, $m = 20$ to 780

L_∞-solver using YICES which presents a first-order phase transition from SAT to unSAT with a middle point occurring at rate $\frac{m}{n} \approx 2$; however the average decision time peak occurs at $\frac{m}{n} \approx 5$, unlike what is expected. Furthermore, the peak time for solving a L_∞ problem is about 35 s. This unexpected behavior may be credited to the fact that YICES converts internally all floating point numbers to pair of integers, which impacts the efficiency of problems whose formulation involves a lot of floating point numbers as is the case of L_∞ decision.

Figure 1b presents an L_∞-solver build with using MIP solver SCIP, in which we can see a phase transition from SAT to unSAT also at $\frac{m}{n} \approx 2$, with an average time peak also around $\frac{m}{n} \approx 2$, as expected. Furthermore, the peak time is 0.35 s, two orders of magnitude more efficient than the YICES solver. Observing the average time, we note an always increasing right tail, which can be credited to the fact that MIP solvers are not implemented with a "fail early" strategy commonly used in logic based solvers, which normally employ what is called restriction learning strategies; furthermore, the size of the matrices used by the MIP solver increases with m. Another possibility to explain such a behavior is the fact that the choice of the family of formulas may be inappropriate, however no such increasing tail was observed in the SMT based method, which reinforces the hypothesis that this behavior is due to the MIP solver. Due to its superior efficiency we only use the SCIP solver as an auxiliary procedure for the LIPSAT solver described next.

5.2   Phase Transition for LIPSAT

The input for the LIPSAT solver is a normal form $\langle \Gamma, \Theta \rangle$. We developed a C++-implementation for Algorithm 4.1 using the C++ interface of the SoPlex linear algebra solver which is part of the SCIP suite of optimizers. We used the L_∞-solver based on SCIP MIP.

The experiments were obtained as follows. The input L_∞-formula Γ was generated in the form we describe above, with a fixed number of symbols n and a varying number of clauses of format (9) and (10) as described above. The probabilistic Θ-restrictions of the form $\{C(y_i) = q_i \mid i \leq i \leq k\}$ were generated

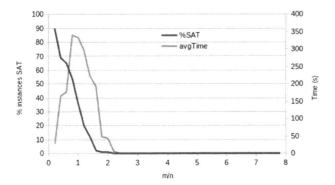

Fig. 2. Phase transition for LIPSAT solver: $k = 20$, $n = 100$ and $m = 20$ to 780

fixing $k \leq n$ and randomly choosing the probabilities q_i uniformly over the interval $[0, 1]$.

The results of the experiment can be seen in Fig. 2. We clearly see a second order phase transition with a peak average time execution that overlaps the decreasing part of the percentage SAT curve. Note that no increasing tail is observed, so that the "fail early" mechanism is achieved in the combination of logic and linear algebra. The peak is near but does not coincide with the fifty percent point of the first order phase transition which may be credited to the increasing shape of the right tail in the L_∞-solver presented in Fig. 2. Also, there is a left shift of the phase transition point $\frac{m}{n} \approx 1$, similar to the shift of PSAT phase transition point with respect to SAT [16]. Overall the phase transition format can be considered satisfactory.

6   Conclusion and the Future

We provided the theoretical basis for the development and implementation of probabilistic reasoning over "partial truth" that respect Łukasiewicz Infinitely-valued Logic restricts. A phase transition behavior could be empirically observed. For the future we hope to develop better solvers for the logics employed having the analysis of the phase transition as a qualitative guideline; and hope to employ the mechanisms developed here to linearly approximate generic functions.

References

1. Achterberg, T.: SCIP: solving constraint integer programs. Math. Program. Comput. **1**(1), 1–41 (2009). http://scip.zib.de/
2. Bertsimas, D., Tsitsiklis, J.N.: Introduction to Linear Optimization. Athena Scientific, Belmont (1997)
3. Bofill, M., Manya, F., Vidal, A., Villaret, M.: Finding hard instances of satisfiability in Łukasiewicz logics. In: ISMVL, pp. 30–35. IEEE (2015)

4. Boole, G.: An Investigation on the Laws of Thought. Macmillan, London (1854). Available on project Gutemberg at http://www.gutenberg.org/etext/15114

5. Borgward, K.H.: The Simplex Method: A Probabilistic Analysis. Algorithms and Combinatorics, vol. 1. Springer, Heidelberg (1986). https://doi.org/10.1007/978-3-642-61578-8

6. Bova, S., Flaminio, T.: The coherence of Łukasiewicz assessments is NP-complete. Int. J. Approx. Reason. **51**(3), 294–304 (2010)

7. Cheeseman, P., Kanefsky, B., Taylor, W.M.: Where the really hard problems are. In: 12th IJCAI, pp. 331–337. Morgan Kaufmann (1991)

8. Cignoli, R., d'Ottaviano, I., Mundici, D.: Algebraic Foundations of Many-Valued Reasoning. Trends in Logic. Springer, Heidelberg (2000). https://doi.org/10.1007/978-94-015-9480-6

9. de Finetti, B.: Sul significato soggettivo della probabilità. Fundamenta Mathematicae **17**(1), 298–329 (1931)

10. de Finetti, B.: La prévision: Ses lois logiques, ses sources subjectives (1937)

11. de Finetti, B.: Theory of Probability: A Critical Introductory Treatment. Wiley, Hoboken (2017). Translated by Antonio Machí and Adrian Smith

12. Dutertre, B.: Yices 2.2. In: Biere, A., Bloem, R. (eds.) CAV 2014. LNCS, vol. 8559, pp. 737–744. Springer, Cham (2014). https://doi.org/10.1007/978-3-319-08867-9_49

13. Eckhoff, J.: Helly, Radon, and Caratheodory type theorems. In: Handbook of Convex Geometry, pp. 389–448. Elsevier Science Publishers (1993)

14. Finger, M., Bona, G.D.: Probabilistic satisfiability: logic-based algorithms and phase transition. In: Walsh, T. (ed.) IJCAI, IJCAI/AAAI, pp. 528–533 (2011)

15. Finger, M., De Bona, G.: Probabilistic satisfiability: algorithms with the presence and absence of a phase transition. AMAI **75**(3), 351–379 (2015)

16. Finger, M., De Bona, G.: Probabilistic satisfiability: algorithms with the presence and absence of a phase transition. Ann. Math. Artif. Intell. **75**(3), 351–379 (2015)

17. Gent, I.P., Walsh, T.: The SAT phase transition. In: Proceedings of the Eleventh European Conference on Artificial Intelligence, ECAI 1994, pp. 105–109. Wiley (1994)

18. Georgakopoulos, G., Kavvadias, D., Papadimitriou, C.H.: Probabilistic satisfiability. J. Complex. **4**(1), 1–11 (1988)

19. Hähnle, R.: Towards an efficient tableau proof procedure for multiple-valued logics. In: Börger, E., Kleine Büning, H., Richter, M.M., Schönfeld, W. (eds.) CSL 1990. LNCS, vol. 533, pp. 248–260. Springer, Heidelberg (1991). https://doi.org/10.1007/3-540-54487-9_62

20. Hansen, P., Jaumard, B.: Probabilistic satisfiability. In: Kohlas, J., Moral, S. (eds.) Handbook of Defeasible Reasoning and Uncertainty Management Systems. HAND, vol. 5, pp. 321–367. Springer, Dordrecht (2000). https://doi.org/10.1007/978-94-017-1737-3_8

21. Hansen, P., Jaumard, B.: Algorithms for the maximum satisfiability problem. Computing **44**, 279–303 (1990). https://doi.org/10.1007/BF02241270

22. Kavvadias, D., Papadimitriou, C.H.: A linear programming approach to reasoning about probabilities. AMAI **1**, 189–205 (1990)

23. McNaughton, R.: A theorem about infinite-valued sentential logic. J. Symb. Log. **16**, 1–13 (1951)

24. Mundici, D.: Advanced Łukasiewicz calculus and MV-algebras. Trends in Logic. Springer, Heidelberg (2011). https://doi.org/10.1007/978-94-007-0840-2

25. Mundici, D.: Satisfiability in many-valued sentential logic is NP-complete. Theor. Comput. Sci. **52**(1–2), 145–153 (1987)

26. Mundici, D.: A constructive proof of McNaughton's theorem in infinite-valued logic. J. Symb. Log. **59**(2), 596–602 (1994)
27. Mundici, D.: Bookmaking over infinite-valued events. Int. J. Approx. Reason. **43**(3), 223–240 (2006)
28. Nilsson, N.: Probabilistic logic. Artif. Intell. **28**(1), 71–87 (1986)
29. Papadimitriou, C., Steiglitz, K.: Combinatorial Optimization: Algorithms and Complexity. Dover, Mineola (1998)

Uniform Substitution for Differential Game Logic

André Platzer[✉]

Computer Science Department, Carnegie Mellon University, Pittsburgh, USA
aplatzer@cs.cmu.edu

Abstract. This paper presents a uniform substitution calculus for *differential game logic* (dGL). Church's *uniform substitutions* substitute a term or formula for a function or predicate symbol everywhere. After generalizing them to differential game logic and allowing for the substitution of hybrid games for game symbols, uniform substitutions make it possible to *only* use axioms instead of axiom schemata, thereby substantially simplifying implementations. Instead of subtle schema variables and soundness-critical side conditions on the occurrence patterns of logical variables to restrict infinitely many axiom schema instances to sound ones, the resulting axiomatization adopts only a finite number of ordinary dGL formulas as axioms, which uniform substitutions instantiate soundly. This paper proves soundness and completeness of uniform substitutions for the monotone modal logic dGL. The resulting axiomatization admits a straightforward modular implementation of dGL in theorem provers.

1 Introduction

Church's *uniform substitution* is a classical proof rule for first-order logic [2, §35/40]. Uniform substitutions uniformly instantiate function and predicate symbols with terms and formulas, respectively, as functions of their arguments. If ϕ is valid, then so is any admissible instance $\sigma\phi$ for any uniform substitution σ:

$$(\text{US}) \quad \frac{\phi}{\sigma\phi}$$

Uniform substitution $\sigma = \{p(\cdot) \mapsto x + \cdot^2 \geq \cdot\}$, e.g. turns $\phi \equiv (p(4y) \rightarrow \exists y\, p(x^2+y))$ into $\sigma\phi \equiv (x + (4y)^2 \geq 4y \rightarrow \exists y\, x + (x^2 + y)^2 \geq x^2 + y)$. The introduction of x is sound, but introducing variable y via $\sigma = \{p(\cdot) \mapsto y + \cdot^2 \geq \cdot\}$ would not be. The occurrence of the variable y of the argument $x^2 + y$ that was already present previously, however, can correctly continue to be used in the instantiation.

Differential game logic (dGL), which is the specification and verification logic for *hybrid games* [5], originally adopted uniform substitution for predicates, because they streamline and simplify completeness proofs. A subsequent investigation of uniform substitutions for differential *dynamic* logic (dL)

This material is based upon work supported by the National Science Foundation under NSF CAREER Award CNS-1054246.

for hybrid *systems* [6] confirmed how impressively Church's original motivation for uniform substitutions manifests in significantly simplifying prover implementations.

Church developed uniform substitutions to relate the study of (object-level) axioms to that of (meta-level) axiom schemata (which stand for an infinite family of axioms). Beyond their philosophical considerations, uniform substitutions significantly impact prover designs by eliminating the usual gap between a logic and its prover. After implementing the recursive application of uniform substitutions, the soundness-critical part of a theorem prover reduces to providing a copy of each concrete logical formula that the logic adopts as axioms. Uniform substitutions provide a modular interface to the static semantics of the logic, because they are the only soundness-critical part of the prover that needs to know free or bound variables of an expression. This simplicity is to be contrasted with the subtle soundness-critical side conditions that usually infest axiom schema and proof rule schema implementations, especially for the more involved binding structures of program logics. The beneficial impact of uniform substitutions on provers made it possible to reduce the size of the soundness-critical core of the differential dynamic logic prover KeYmaera X [3] down to 2% compared to the previous prover KeYmaera [9] and formally verify dL in Isabelle and Coq [1].

This paper generalizes uniform substitution to the significantly more expressive differential game logic for hybrid *games* [5]. The modular structure of the soundness argument for dL is sufficiently robust to work for dGL: (i) prove correctness of the static semantics, (ii) relate syntactic effect of uniform substitution to semantic effect of its adjoint interpretation, (iii) conclude soundness of rule US, and (iv) separately establish soundness of each axiom. The biggest challenge is that hybrid game semantics cannot use state reachability, so correctness notions and their uses for the static semantics need to be phrased as functions of winning condition projections. The interaction of game operators with repetitions causes transfinite fixpoints instead of the arbitrary finite iterations in hybrid systems. Relative completeness follows from previous results, but exploits the new game symbols to simplify the proof. After new soundness justifications, the resulting uniform substitution mechanism and axioms for dGL end up close to those for hybrid systems [6] (apart from the ones that are unsound for hybrid games [5]). The modularity caused by uniform substitutions explains why it was possible to generalize the KeYmaera X prover kernel from hybrid systems to hybrid games with about 10 lines of code.[1] All proofs are inline or in the report [8].

2   Preliminaries: Differential Game Logic

This section reviews differential game logic (dGL), a specification and verification logic for hybrid games [5,7]. Hybrid games support the discrete, continuous, and

---

[1] The addition of games to the previous KeYmaera prover was more complex [10], with an implementation effort measured in months not minutes. Unfortunately, this is not quite comparable, because both provers implement markedly different flavors of games for hybrid systems. The game logic for KeYmaera [10] was specifically tuned as an exterior extension to be more easily implementable than dGL in KeYmaera.

adversarial dynamics of two-player games in hybrid systems between players Angel and Demon. Compared to previous work [5], the logic is augmented to form *(differential-form) differential game logic* with differentials and function symbols [6] and with game symbols a that can be substituted with hybrid games.

2.1 Syntax

Differential game logic has three syntactic categories. Its terms θ are polynomial terms, function symbols interpreted over $\mathbb{R}$, and differential terms $(\theta)'$. Its hybrid games α describe the permitted player actions during the game in program notation. Its formulas ϕ include first-order logic of real arithmetic and, for each hybrid game α, a modal formula $\langle\alpha\rangle\phi$, which expresses that player Angel has a winning strategy in the hybrid game α to reach the region satisfying dGL formula ϕ. In the formula $\langle\alpha\rangle\phi$, the dGL formula ϕ describes Angel's objective while the hybrid game α describes the moves permitted for the two players, respectively.

The set of all *variables* is $\mathcal{V}$. Variables of the form x' for a variable $x \in \mathcal{V}$ are called *differential variables*, which are just independent variables associated to variable x. For any subset $V \subseteq \mathcal{V}$ is $V' \stackrel{\text{def}}{=} \{x' : x \in V\}$ the set of *differential variables* x' for the variables in V. The set of all variables is assumed to contain all its differential variables $\mathcal{V}' \subseteq \mathcal{V}$ (although x'', x''' are not usually used).

Definition 1 (Terms). Terms *are defined by this grammar (with $\theta, \eta, \theta_1, \ldots, \theta_k$ as terms, $x \in \mathcal{V}$ as variable, and f as function symbol of arity k):*

$$\theta, \eta ::= x \mid f(\theta_1, \ldots, \theta_k) \mid \theta + \eta \mid \theta \cdot \eta \mid (\theta)'$$

As in dL [6], *differentials* $(\theta)'$ of terms θ are exploited for the purpose of axiomatically internalizing reasoning about differential equations. The differential $(\theta)'$ describes how the value of θ changes locally depending on how the values of its variables x change, i.e., as a function of the values of the corresponding differential variables x'. Differentials reduce reasoning about *differential equations* to reasoning about *equations of differentials* [6] with their single-state semantics.

Definition 2 (Hybrid games). *The* hybrid games of differential game logic dGL *are defined by the following grammar (with α, β as hybrid games, a as game symbol, x as variable, θ as term, and ψ as dGL formula):*

$$\alpha, \beta ::= a \mid x := \theta \mid x' = \theta \,\&\, \psi \mid ?\psi \mid \alpha \cup \beta \mid \alpha; \beta \mid \alpha^* \mid \alpha^d$$

Atomic games are the following. *Game symbols* a are uninterpreted. The *discrete assignment game* $x := \theta$ evaluates term θ and assigns it to variable x. The *continuous evolution game* $x' = \theta \,\&\, \psi$ allows Angel to follow differential equation $x' = \theta$ for any real duration during which the evolution domain constraint ψ is true ($x' = \theta$ stands for $x' = \theta \,\&\, true$). If ψ is not true in the current state, then no solution exists and Angel loses the game. *Test game* $?\psi$ has no effect except that Angel loses the game prematurely unless ψ is true in the current state.

Compound games are the following. The *game of choice* $\alpha \cup \beta$ allows Angel to choose whether she wants to play game α or, instead, play game β. The *sequential game* $\alpha; \beta$ first plays α and then plays β (unless a player lost prematurely during α). The *repeated game* α^* allows Angel to decide how often to repeat game α by inspecting the state reached after the respective α game to decide whether she wants to play another round. The *dual game* α^d makes the players switch sides: all of Angel's decisions are now Demon's and all of Demon's decisions are now Angel's. Where Angel would have lost prematurely in α (for failing a test or evolution domain) now Demon does in α^d, and vice versa. This makes game play interactive but semantically quite rich [5]. All other operations are definable, e.g., the game where Demon chooses between α and β as $(\alpha^d \cup \beta^d)^d$.

Definition 3 (dGL formulas). *The formulas of differential game logic dGL are defined by the following grammar (with ϕ, ψ as dGL formulas, p as predicate symbol of arity k, θ, η, θ_i as terms, x as variable, and α as hybrid game):*

$$\phi, \psi ::= \theta \geq \eta \mid p(\theta_1, \ldots, \theta_k) \mid \neg \phi \mid \phi \wedge \psi \mid \exists x\, \phi \mid \langle \alpha \rangle \phi$$

The box modality $[\alpha]$ in formula $[\alpha]\phi$ describes that the player Demon has a winning strategy to achieve ϕ in hybrid game α. But dGL satisfies the determinacy duality $[\alpha]\phi \leftrightarrow \neg\langle\alpha\rangle\neg\phi$ [5, Theorem 3.1], which we now take as its definition to simplify matters. Other operators are definable as usual, e.g., $\forall x\, \phi$ as $\neg\exists x\, \neg\phi$. The following dGL formula, for example, expresses that Angel has a winning strategy to follow the differential equation $x' = v$ to a state where $x > 0$ even after Demon chooses $v := 2$ or $v := x^2 + 1$ first: $\langle (v := 2 \cup v := x^2 + 1)^d; x' = v \rangle\, x > 0$.

2.2   Semantics

While the syntax of dGL is close to that of dL (with the only change being the addition of the duality operator ^{d}), its semantics is significantly more involved, because it needs to recursively support *interactive* game play, instead of mere reachability. Variables may have different values in different states of the game. A *state* ω is a mapping from the set of all variables $\mathcal{V}$ to the reals $\mathbb{R}$. Also, ω_x^r is the state that agrees with state ω except for variable x whose value is $r \in \mathbb{R}$. The set of all states is denoted $\mathcal{S}$. The set of all subsets of $\mathcal{S}$ is denoted $\wp(\mathcal{S})$.

The semantics of function, predicate, and game symbols is independent from the state. They are interpreted by an *interpretation* I that maps each arity k function symbol f to a k-ary smooth function $I(f) : \mathbb{R}^k \to \mathbb{R}$, and each arity k predicate symbol p to a k-ary relation $I(p) \subseteq \mathbb{R}^k$. The semantics of differential game logic in interpretation I defines, for each formula ϕ, the set of all states $I[\![\phi]\!]$, in which ϕ is true. Since hybrid games appear in dGL formulas and vice versa, the semantics $I[\![\alpha]\!](X)$ of hybrid game α in interpretation I is defined by simultaneous induction (Definition 5) as the set of all states from which Angel has a winning strategy in hybrid game α to achieve X. The real value of term

θ in state ω for interpretation I is denoted $I\omega[\![\theta]\!]$ and defined as usual.[2] An interpretation I maps each game symbol a to a function $I(a) : \wp(\mathcal{S}) \to \wp(\mathcal{S})$, where $I(a)(X) \subseteq \mathcal{S}$ are the states from which Angel has a winning strategy to achieve $X \subseteq \mathcal{S}$.

Definition 4 (dGL semantics). *The* semantics of a dGL *formula ϕ for each interpretation I with a corresponding set of states $\mathcal{S}$ is the subset $I[\![\phi]\!] \subseteq \mathcal{S}$ of states in which ϕ is true. It is defined inductively as follows*

1. $I[\![\theta \geq \eta]\!] = \{\omega \in \mathcal{S} : I\omega[\![\theta]\!] \geq I\omega[\![\eta]\!]\}$
2. $I[\![p(\theta_1, \ldots, \theta_k)]\!] = \{\omega \in \mathcal{S} : (I\omega[\![\theta_1]\!], \ldots, I\omega[\![\theta_k]\!]) \in I(p)\}$
3. $I[\![\neg\phi]\!] = (I[\![\phi]\!])^\complement = \mathcal{S} \setminus I[\![\phi]\!]$ *is the complement of $I[\![\phi]\!]$*
4. $I[\![\phi \wedge \psi]\!] = I[\![\phi]\!] \cap I[\![\psi]\!]$
5. $I[\![\exists x\, \phi]\!] = \{\omega \in \mathcal{S} : \omega_x^r \in I[\![\phi]\!] \text{ for some } r \in \mathbb{R}\}$
6. $I[\![\langle\alpha\rangle\phi]\!] = I[\![\alpha]\!](I[\![\phi]\!])$

A dGL *formula ϕ is* valid in I, *written $I \models \phi$, iff it is true in all states, i.e., $I[\![\phi]\!] = \mathcal{S}$. Formula ϕ is* valid, *written $\models \phi$, iff $I \models \phi$ for all interpretations I.*

Definition 5 (Semantics of hybrid games). *The* semantics of a hybrid game α *for each interpretation I is a function $I[\![\alpha]\!](\cdot)$ that, for each set of Angel's winning states $X \subseteq \mathcal{S}$, gives the* winning region, *i.e., the set of states $I[\![\alpha]\!](X) \subseteq \mathcal{S}$ from which Angel has a winning strategy to achieve X in α (whatever strategy Demon chooses). It is defined inductively as follows*

1. $I[\![a]\!](X) = I(a)(X)$
2. $I[\![x := \theta]\!](X) = \{\omega \in \mathcal{S} : \omega_x^{I\omega[\![\theta]\!]} \in X\}$
3. $I[\![x' = \theta \,\&\, \psi]\!](X) = \{\omega \in \mathcal{S} : \omega = \varphi(0) \text{ on } \{x'\}^\complement \text{ and } \varphi(r) \in X \text{ for some function } \varphi : [0, r] \to \mathcal{S} \text{ of some duration } r \text{ satisfying } I, \varphi \models x' = \theta \wedge \psi\}$ *where $I, \varphi \models x' = \theta \wedge \psi$ iff $\varphi(\zeta) \in I[\![x' = \theta \wedge \psi]\!]$ and $\varphi(0) = \varphi(\zeta)$ on $\{x, x'\}^\complement$ for all $0 \leq \zeta \leq r$ and $\frac{d\varphi(t)(x)}{dt}(\zeta)$ exists and equals $\varphi(\zeta)(x')$ for all $0 \leq \zeta \leq r$ if $r > 0$.*
4. $I[\![?\psi]\!](X) = I[\![\psi]\!] \cap X$
5. $I[\![\alpha \cup \beta]\!](X) = I[\![\alpha]\!](X) \cup I[\![\beta]\!](X)$
6. $I[\![\alpha; \beta]\!](X) = I[\![\alpha]\!](I[\![\beta]\!](X))$
7. $I[\![\alpha^*]\!](X) = \bigcap\{Z \subseteq \mathcal{S} : X \cup I[\![\alpha]\!](Z) \subseteq Z\}$
8. $I[\![\alpha^d]\!](X) = (I[\![\alpha]\!](X^\complement))^\complement$

The semantics $I[\![x' = \theta \,\&\, \psi]\!](X)$ is the set of all states from which there is a solution of the differential equation $x' = \theta$ of some duration that reaches a state in X without ever leaving the set of all states $I[\![\psi]\!]$ where evolution domain constraint ψ is true. The initial value of x' in state ω is ignored for that solution. It is crucial that $I[\![\alpha^*]\!](X)$ gives a least fixpoint semantics to repetition [5].

Lemma 6 (Monotonicity [5, Lemma 2.7]). *The semantics is* monotone, *i.e., $I[\![\alpha]\!](X) \subseteq I[\![\alpha]\!](Y)$ for all $X \subseteq Y$.*

---

[2] Even if not critical here, differentials have a differential-form semantics [6] as the sum of all partial derivatives by $x \in \mathcal{V}$ multiplied by the corresponding values of x':
$I\omega[\![(\theta)']\!] = \sum_{x \in \mathcal{V}} \omega(x') \frac{\partial I\omega[\![\theta]\!]}{\partial x}(\omega) = \sum_{x \in \mathcal{V}} \omega(x') \frac{\partial I\omega[\![\theta]\!]}{\partial x}.$

3   Static Semantics

The central bridge between a logic and its uniform substitutions is the definition of its static semantics via its free and bound variables. The static semantics captures static variable relationships that are more tractable than the full nuances of the dynamic semantics. It will be used in crucial ways to ensure that no variable is introduced free into a context within which it is bound during the uniform substitution application. It is imperative for the soundness of uniform substitution that the static semantics be sound, so expressions only depend on their free variables and only their bound variables change during hybrid games.

The most tricky part for the soundness justification for dGL is that the semantics of hybrid games is not a reachability relation, such that the usual semantic characterizations of free and bound variables from programs do not work for hybrid games. Hybrid games have a more involved winning region semantics.

The first step is to define *upward projections* $X{\uparrow}V$ that increase the winning region $X \subseteq \mathcal{S}$ from the variables $V \subseteq \mathcal{V}$ to all states that are "on V like X", i.e., similar on V to states in X (and arbitrary on complement $V^{\complement}$). The *downward projection* $X{\downarrow}\omega(V)$ shrinks the winning region X and selects the values of state ω on variables $V \subseteq \mathcal{V}$ to keep just those states of X that agree with ω on V.

Definition 7. *The set* $X{\uparrow}V = \{\nu \in \mathcal{S} : \exists \omega \in X \; \omega = \nu \text{ on } V\} \supseteq X$ *extends* $X \subseteq \mathcal{S}$ *to the states that agree on* $V \subseteq \mathcal{V}$ *with some state in* X *(written* $\exists$*). The set* $X{\downarrow}\omega(V) = \{\nu \in X : \omega = \nu \text{ on } V\} \subseteq X$ *selects state* ω *on* $V \subseteq \mathcal{V}$ *in* $X \subseteq \mathcal{S}$*.*

Remark 8. It is easy to check these properties of up and down projections:

1. Composition: $X{\uparrow}V{\uparrow}W = X{\uparrow}(V \cap W)$
2. Antimonotone: $X{\uparrow}W \subseteq X{\uparrow}V$ for all $W \supseteq V$
3. $X{\uparrow}\emptyset = \mathcal{S}$ (unless $X = \emptyset$) and $X{\uparrow}\mathcal{V} = X$, where $\mathcal{V}$ is the set of all variables
4. Composition: $X{\downarrow}\omega(V){\downarrow}\omega(W) = X{\downarrow}\omega(V \cup W)$
5. Antimonotone: $X{\downarrow}\omega(W) \subseteq X{\downarrow}\omega(V)$ for all $W \supseteq V$
6. $X{\downarrow}\omega(\emptyset) = X$ and $X{\downarrow}\omega(\mathcal{V}) = X \cap \{\omega\}$. Thus, $\omega \in X{\downarrow}\omega(V)$ for any V iff $\omega \in X$.

Projections make it possible to define (*semantic!*) free and bound variables of hybrid games by expressing suitable variable dependence and ignorance. Variable x is free iff two states that only differ in the value of x have different membership in the winning region for hybrid game α for some winning region $X{\uparrow}\{x\}^{\complement}$ that is insensitive to the value of x. Variable x is bound iff it is in the winning region for hybrid game α for some winning condition X but not for the winning condition $X{\downarrow}\omega(\{x\})$ that limits the new value of x to stay at its initial value $\omega(x)$.

Definition 9 (Static semantics). *The* static semantics *defines the* free variables, *which are all variables that the value of an expression depends on, as well as* bound variables, $\mathsf{BV}(\alpha)$, *which can change their value during game* α, *as:*

$$\mathsf{FV}(\theta) = \left\{x \in \mathcal{V} : \exists I, \omega, \tilde{\omega} \text{ such that } \omega = \tilde{\omega} \text{ on } \{x\}^{\complement} \text{ and } I\omega[\![\theta]\!] \neq I\tilde{\omega}[\![\theta]\!]\right\}$$

$$\mathsf{FV}(\phi) = \left\{x \in \mathcal{V} : \exists I, \omega, \tilde{\omega} \text{ such that } \omega = \tilde{\omega} \text{ on } \{x\}^{\complement} \text{ and } \omega \in I[\![\phi]\!] \not\ni \tilde{\omega}\right\}$$

$$\mathsf{FV}(\alpha) = \big\{ x \in \mathcal{V} : \exists I, \omega, \tilde{\omega}, X \text{ with } \omega = \tilde{\omega} \text{ on } \{x\}^{\complement} \text{ and } \omega \in I[\![\alpha]\!](X{\uparrow}\{x\}^{\complement}) \not\ni \tilde{\omega} \big\}$$

$$\mathsf{BV}(\alpha) = \big\{ x \in \mathcal{V} : \exists I, \omega, X \text{ such that } I[\![\alpha]\!](X) \ni \omega \notin I[\![\alpha]\!](X{\downarrow}\omega(\{x\})) \big\}$$

The signature, *i.e., set of function, predicate, and game symbols in ϕ is denoted $\Sigma(\phi)$; accordingly $\Sigma(\theta)$ for term θ and $\Sigma(\alpha)$ for hybrid game α.*

The static semantics from Definition 9 satisfies the coincidence property (the value of an expression only depends on the values of its free variables) and bound effect property (a hybrid game only changes the values of its bound variables).

Lemma 10 (Coincidence for terms). $\mathsf{FV}(\theta)$ *is the smallest set with the coincidence property for θ: If $\omega = \tilde{\omega}$ on $\mathsf{FV}(\theta)$ and $I = J$ on $\Sigma(\theta)$ then $I\omega[\![\theta]\!] = J\tilde{\omega}[\![\theta]\!]$.*

Lemma 11 (Coincidence for formulas). $\mathsf{FV}(\phi)$ *is the smallest set with the coincidence property for ϕ: If $\omega = \tilde{\omega}$ on $\mathsf{FV}(\phi)$ and $I = J$ on $\Sigma(\phi)$, then $\omega \in I[\![\phi]\!]$ iff $\tilde{\omega} \in J[\![\phi]\!]$.*

From which states a hybrid game α can be won only depends on α, the winning region, and the values of its free variables, as $X{\uparrow}\mathsf{FV}(\alpha)$ is only sensitive to $\mathsf{FV}(\alpha)$.

Lemma 12 (Coincidence for games). *The set $\mathsf{FV}(\alpha)$ is the smallest set with the coincidence property for α: If $\omega = \tilde{\omega}$ on $V \supseteq \mathsf{FV}(\alpha)$ and $I = J$ on $\Sigma(\alpha)$, then $\omega \in I[\![\alpha]\!](X{\uparrow}V)$ iff $\tilde{\omega} \in J[\![\alpha]\!](X{\uparrow}V)$.*

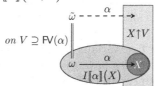

Proof. Let $\mathcal{M}$ be the set of all sets $M \subseteq \mathcal{V}$ satisfying for all $I, \omega, \tilde{\omega}, X$ that $\omega = \tilde{\omega}$ on $M^{\complement}$ implies: $\omega \in I[\![\alpha]\!](X{\uparrow}V)$ iff $\tilde{\omega} \in I[\![\alpha]\!](V)$. One implication suffices.

1. If $x \notin V$, then $\{x\} \in \mathcal{M}$: Assume $\omega = \tilde{\omega}$ on $\{x\}^{\complement}$ and $\omega \in I[\![\alpha]\!](X{\uparrow}V) \subseteq I[\![\alpha]\!](X{\uparrow}V{\uparrow}\{x\})$ by Lem. 6, Def. 7. Then, as $x \notin \mathsf{FV}(\alpha)$, $\tilde{\omega} \in I[\![\alpha]\!](X{\uparrow}V{\uparrow}\{x\}) = I[\![\alpha]\!](X{\uparrow}(V \cap \{x\}^{\complement}))$ by Rem. 8(1). Finally, $X{\uparrow}(V \cap \{x\}^{\complement}) = X{\uparrow}V$ as $x \notin V$.
2. If $M_i \in \mathcal{M}$ is a sequence of sets in $\mathcal{M}$, then $\bigcup_{i \in \mathbb{N}} M_i \in \mathcal{M}$: Assume $\omega = \tilde{\omega}$ on $(\bigcup_i M_i)^{\complement}$ and $\omega \in I[\![\alpha]\!](X{\uparrow}V)$. The state ω_n defined as $\tilde{\omega}$ on $\bigcup_{i<n} M_i$ and as ω on $(\bigcup_{i<n} M_i)^{\complement}$ satisfies $\omega_n \in I[\![\alpha]\!](X{\uparrow}V)$ by induction on n. For $n = 0$, $\omega_0 = \omega$. Since $\omega_n = \omega_{n+1}$ on $M_n^{\complement}$ and $M_n \in \mathcal{M}$, $\omega_n \in I[\![\alpha]\!](X{\uparrow}V)$ implies $\omega_{n+1} \in I[\![\alpha]\!](X{\uparrow}V)$. Finally, $\omega = \tilde{\omega} = \omega_n$ on $(\bigcup_i M_i)^{\complement}$ already.

This argument succeeds for any $V \supseteq \mathsf{FV}(\alpha)$, so $\mathsf{FV}(\alpha)^{\complement} \in \mathcal{M}$ as a (countable) union of $\{x\}$ for all $x \notin \mathsf{FV}(\alpha)$. Finally, if $I = J$ on $\Sigma(\alpha)$ then also $\tilde{\omega} \in J[\![\alpha]\!](X{\uparrow}V)$ by a simple induction, since I gives meaning to function, predicate, and game symbols, but only those that occur in α are relevant.

No set $W \not\supseteq \mathsf{FV}(\alpha)$ has the coincidence property for α, because there, then, is a variable $x \in \mathsf{FV}(\alpha) \setminus W$, which implies there are $I, X, \omega = \tilde{\omega}$ on $\{x\}^{\complement} \supseteq W$ such that $\omega \in I[\![\alpha]\!](X{\uparrow}\{x\}^{\complement}) \not\ni \tilde{\omega}$. But for the set $V \stackrel{\text{def}}{=} \{x\}^{\complement} \supseteq W$ it is, then, the case that $\omega \in I[\![\alpha]\!](X{\uparrow}V)$ but $\tilde{\omega} \notin I[\![\alpha]\!](X{\uparrow}V)$.   □

By Definition 7 and Lemma 6, $\omega \in I[\![\alpha]\!](X)$ implies $\omega \in I[\![\alpha]\!](X{\uparrow}V)$ for all $V \subseteq \mathcal{V}$. All supersets of $\mathsf{FV}(\theta)$ or $\mathsf{FV}(\phi)$ or $\mathsf{FV}(\alpha)$ have the respective coincidence property.

Only its bound variables $\mathsf{BV}(\alpha)$ change their values during hybrid game α, because from any state from which α can be won to achieve X, one can already win α to achieve $X{\downarrow}\omega(\mathsf{BV}(\alpha)^{\complement})$, which stays at ω except for the values of $\mathsf{BV}(\alpha)$.

Lemma 13 (Bound effect). *The set $\mathsf{BV}(\alpha)$ is the smallest set with the bound effect property: $\omega \in I[\![\alpha]\!](X)$ iff $\omega \in I[\![\alpha]\!](X{\downarrow}\omega(\mathsf{BV}(\alpha)^{\complement}))$.*

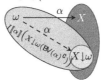

All supersets $V \supseteq \mathsf{BV}(\alpha)$ have the bound effect property, as $I[\![\alpha]\!](X{\downarrow}\omega(V^{\complement})) \supseteq I[\![\alpha]\!](X{\downarrow}\omega(\mathsf{BV}(\alpha)^{\complement}))$ by Remark 8(5) because $V^{\complement} \subseteq \mathsf{BV}(\alpha)^{\complement}$. Other states that agree except on the bound variables share the same selection of the winning region: if $\omega = \tilde{\omega}$ on $\mathsf{BV}(\alpha)^{\complement}$, then $\tilde{\omega} \in I[\![\alpha]\!](X)$ iff $\tilde{\omega} \in I[\![\alpha]\!](X{\downarrow}\omega(\mathsf{BV}(\alpha)^{\complement}))$.

Since all supersets of the free variables have the coincidence property and all supersets of the bound variables have the bound effect property, algorithms that *syntactically compute* supersets FV and BV of free and bound variables [6, Lemma 17] can be soundly augmented by $\mathrm{FV}(\alpha^d) = \mathrm{FV}(\alpha)$ and $\mathrm{BV}(\alpha^d) = \mathrm{BV}(\alpha)$.

4   Uniform Substitution

The static semantics provides, in a modular way, what is needed to define the application $\sigma\phi$ of uniform substitution σ to dGL formula ϕ. The dGL axiomatization uses uniform substitutions that affect terms, formulas, and games, whose application $\sigma\phi$ will be defined in Definition 14 using Fig. 1. A *uniform substitution* σ is a mapping from expressions of the form $f(\cdot)$ to terms $\sigma f(\cdot)$, from $p(\cdot)$ to formulas $\sigma p(\cdot)$, and from game symbols a to hybrid games σa. Vectorial extensions are accordingly for other arities $k \geq 0$. Here $\cdot$ is a reserved function symbol of arity 0, marking the position where the respective argument, e.g., argument θ to $p(\cdot)$ in formula $p(\theta)$, will end up in the replacement $\sigma p(\cdot)$ used for $p(\theta)$.

Definition 14 (Admissible uniform substitution). *A uniform substitution σ is U-admissible for ϕ (or θ or α, respectively) with respect to the variables*

$$\sigma(x) = x \qquad\qquad \text{for variable } x \in \mathcal{V}$$
$$\sigma(f(\theta)) = (\sigma f)(\sigma\theta) \stackrel{\text{def}}{=} \{\cdot \mapsto \sigma\theta\}\sigma f(\cdot) \text{ for function symbol } f$$
$$\sigma(\theta + \eta) = \sigma\theta + \sigma\eta$$
$$\sigma(\theta \cdot \eta) = \sigma\theta \cdot \sigma\eta$$
$$\sigma((\theta)') = (\sigma\theta)' \qquad\qquad \text{if } \sigma \text{ is } \mathcal{V}\text{-admissible for } \theta$$
$$\overline{\sigma(\theta \geq \eta) = \sigma\theta \geq \sigma\eta}$$
$$\sigma(p(\theta)) = (\sigma p)(\sigma\theta) \stackrel{\text{def}}{=} \{\cdot \mapsto \sigma\theta\}\sigma p(\cdot) \text{ for predicate symbol } p$$
$$\sigma(\neg\phi) = \neg\sigma\phi$$
$$\sigma(\phi \wedge \psi) = \sigma\phi \wedge \sigma\psi$$
$$\sigma(\exists x\,\phi) = \exists x\,\sigma\phi \qquad\qquad \text{if } \sigma \text{ is } \{x\}\text{-admissible for } \phi$$
$$\sigma(\langle\alpha\rangle\phi) = \langle\sigma\alpha\rangle\sigma\phi \qquad\qquad \text{if } \sigma \text{ is } \mathsf{BV}(\sigma\alpha)\text{-admissible for } \phi$$
$$\overline{\sigma(a) = \sigma a} \qquad\qquad\qquad\qquad \text{for game symbol } a$$
$$\sigma(x := \theta) = x := \sigma\theta$$
$$\sigma(x' = \theta\,\&\,\psi) = (x' = \sigma\theta\,\&\,\sigma\psi) \qquad \text{if } \sigma \text{ is } \{x, x'\}\text{-admissible for } \theta, \psi$$
$$\sigma(?\psi) = ?\sigma\psi$$
$$\sigma(\alpha \cup \beta) = \sigma\alpha \cup \sigma\beta$$
$$\sigma(\alpha; \beta) = \sigma\alpha; \sigma\beta \qquad\qquad \text{if } \sigma \text{ is } \mathsf{BV}(\sigma\alpha)\text{-admissible for } \beta$$
$$\sigma(\alpha^*) = (\sigma\alpha)^* \qquad\qquad \text{if } \sigma \text{ is } \mathsf{BV}(\sigma\alpha)\text{-admissible for } \alpha$$
$$\sigma(\alpha^d) = (\sigma\alpha)^d$$

Fig. 1. Recursive application of uniform substitution σ

$U \subseteq \mathcal{V}$ iff $\mathsf{FV}(\sigma|_{\Sigma(\phi)}) \cap U = \emptyset$, where $\sigma|_{\Sigma(\phi)}$ is the restriction of σ that only replaces symbols that occur in ϕ, and $\mathsf{FV}(\sigma) = \bigcup_f \mathsf{FV}(\sigma f(\cdot)) \cup \bigcup_p \mathsf{FV}(\sigma p(\cdot))$ are the free variables that σ introduces. A uniform substitution σ is admissible for ϕ (θ or α, respectively) iff the bound variables U of each operator of ϕ are not free in the substitution on its arguments, i.e., σ is U-admissible. These admissibility conditions are listed in Fig. 1, which defines the result $\sigma\phi$ of applying σ to ϕ.

The remainder of this section proves soundness of uniform substitution for dGL. All subsequent uses of uniform substitutions are required to be admissible.

4.1   Uniform Substitution Lemmas

Uniform substitution lemmas equate the syntactic effect that a uniform substitution σ has on a syntactic expression in a state ω and interpretation I with the semantic effect that the switch to the adjoint interpretation $\sigma_\omega^* I$ has on the original expression. Adjoints make it possible to capture in semantics the effect that a uniform substitution has on the syntax.

Let $I_\cdot^d$ denote the interpretation that agrees with interpretation I except for the interpretation of arity 0 function symbol $\cdot$ which is changed to $d \in \mathbb{R}$.

Definition 15 (Substitution adjoints). *The adjoint to substitution σ is the operation that maps I, ω to the adjoint interpretation $\sigma_\omega^* I$ in which the interpretation of each function symbol f, predicate symbol p, and game symbol a are modified according to σ (it is enough to consider those that σ changes):*

$$\sigma_\omega^* I(f) : \mathbb{R} \to \mathbb{R}; \; d \mapsto I_\cdot^d \omega [\![\sigma f(\cdot)]\!]$$
$$\sigma_\omega^* I(p) = \{ d \in \mathbb{R} : \omega \in I_\cdot^d [\![\sigma p(\cdot)]\!] \}$$
$$\sigma_\omega^* I(a) : \wp(\mathcal{S}) \to \wp(\mathcal{S}); \; X \mapsto I[\![\sigma a]\!](X)$$

Corollary 16 (Admissible adjoints). *If $\omega = \nu$ on $\mathsf{FV}(\sigma)$, then $\sigma_\omega^* I = \sigma_\nu^* I$. If $\omega = \nu$ on $U^\complement$ and σ is U-admissible for θ (or ϕ or α, respectively), then*

$$\sigma_\omega^* I[\![\theta]\!] = \sigma_\nu^* I[\![\theta]\!] \; i.e., \; \sigma_\omega^* I\mu[\![\theta]\!] = \sigma_\nu^* I\mu[\![\theta]\!] \; \text{for all states } \mu \in \mathcal{S}$$
$$\sigma_\omega^* I[\![\phi]\!] = \sigma_\nu^* I[\![\phi]\!]$$
$$\sigma_\omega^* I[\![\alpha]\!] = \sigma_\nu^* I[\![\alpha]\!] \; i.e., \; \sigma_\omega^* I[\![\alpha]\!](X) = \sigma_\nu^* I[\![\alpha]\!](X) \; \text{for all sets } X \subseteq \mathcal{S}$$

Substituting equals for equals is sound by the compositional semantics of dL. The more general uniform substitutions are still sound, because the semantics of uniform substitutes of expressions agrees with the semantics of the expressions themselves in the adjoint interpretations. The semantic modification of adjoint interpretations has the same effect as the syntactic uniform substitution.

Lemma 17 (Uniform substitution for terms). *The uniform substitution σ and its adjoint interpretation $\sigma_\omega^* I, \omega$ for I, ω have the same semantics for all terms θ:*

$$I\omega[\![\sigma\theta]\!] = \sigma_\omega^* I\omega[\![\theta]\!]$$

The uniform substitute of a formula is true in an interpretation iff the formula itself is true in its adjoint interpretation. Uniform substitution lemmas are proved by simultaneous induction, since formulas and games are mutually recursive.

Lemma 18 (Uniform substitution for formulas). *The uniform substitution σ and its adjoint interpretation $\sigma_\omega^* I, \omega$ for I, ω have the same semantics for all formulas ϕ:*

$$\omega \in I[\![\sigma\phi]\!] \; \text{iff} \; \omega \in \sigma_\omega^* I[\![\phi]\!]$$

Proof. The proof is by structural induction on ϕ and the structure of σ, simultaneously with Lemma 19. It is in [8] with this case for modalities:

6. $\omega \in I[\![\sigma(\langle \alpha \rangle \phi)]\!]$ iff $\omega \in I[\![\langle \sigma\alpha \rangle \sigma\phi]\!] = I[\![\sigma\alpha]\!](I[\![\sigma\phi]\!])$ (provided σ is $\mathsf{BV}(\sigma\alpha)$-admissible for ϕ) iff (by Lemma 13) $\omega \in I[\![\sigma\alpha]\!](I[\![\sigma\phi]\!] {\downarrow} \omega(\mathsf{BV}(\sigma\alpha)^\complement))$.
   Starting conversely: $\omega \in \sigma_\omega^* I[\![\langle \alpha \rangle \phi]\!] = \sigma_\omega^* I[\![\alpha]\!](\sigma_\omega^* I[\![\phi]\!])$ iff (by Lemma 19) $\omega \in I[\![\sigma\alpha]\!](\sigma_\omega^* I[\![\phi]\!])$ iff (by Lemma 13) $\omega \in I[\![\sigma\alpha]\!](\sigma_\omega^* I[\![\phi]\!] {\downarrow} \omega(\mathsf{BV}(\sigma\alpha)^\complement))$.
   Consequently, it suffices to show that both winning conditions are equal:

$$I[\![\sigma\phi]\!] {\downarrow} \omega(\mathsf{BV}(\sigma\alpha)^\complement) = \sigma_\omega^* I[\![\phi]\!] {\downarrow} \omega(\mathsf{BV}(\sigma\alpha)^\complement)$$

For this, consider any $\nu = \omega$ on $\mathsf{BV}(\sigma\alpha)^\complement$ and show: $\nu \in I[\![\sigma\phi]\!]$ iff $\nu \in \sigma_\omega^* I[\![\phi]\!]$. By induction hypothesis, $\nu \in I[\![\sigma\phi]\!]$ iff $\nu \in \sigma_\nu^* I[\![\phi]\!]$ iff $\nu \in \sigma_\omega^* I[\![\phi]\!]$ by Corollary 16, because $\nu = \omega$ on $\mathsf{BV}(\sigma\alpha)^\complement$ and σ is $\mathsf{BV}(\sigma\alpha)$-admissible for ϕ. $\qquad \square$

The uniform substitute of a game can be won into X from state ω in an interpretation iff the game itself can be won into X from ω in its adjoint interpretation. The most complicated part of the uniform substitution lemma proofs is the case of repetition α^*, because it has a least fixpoint semantics. The proof needs to be set up carefully by transfinite induction (instead of induction along the number of program loop iterations, which is finite for hybrid systems).

Lemma 19 (Uniform substitution for games). *The uniform substitution σ and its adjoint interpretation $\sigma_\omega^* I, \omega$ for I, ω have the same semantics for all games α:*

$$\omega \in I[\![\sigma\alpha]\!](X) \ \text{iff} \ \omega \in \sigma_\omega^* I[\![\alpha]\!](X)$$

Proof. The proof is by structural induction on α, simultaneously with Lemma 18, simultaneously for all ω and X.

1. $\omega \in I[\![\sigma(a)]\!](X) = I[\![\sigma a]\!](X) = \sigma_\omega^* I(a)(X) = \sigma_\omega^* I[\![a]\!](X)$ for game symbol a
2. $\omega \in I[\![\sigma(x := \theta)]\!](X) = I[\![x := \sigma\theta]\!](X)$ iff $X \ni \omega_x^{I\omega[\![\sigma\theta]\!]} = \omega_x^{\sigma_\omega^* I\omega[\![\theta]\!]}$ by using Lemma 17, which is, thus, equivalent to $\omega \in \sigma_\omega^* I[\![x := \theta]\!](X)$.
3. $\omega \in I[\![\sigma(x' = \theta \,\&\, \psi)]\!](X) = I[\![x' = \sigma\theta \,\&\, \sigma\psi]\!](X)$ (provided that σ is $\{x, x'\}$-admissible for θ, ψ) iff $\exists \varphi : [0, T] \to \mathcal{S}$ with $\varphi(0) = \omega$ on $\{x'\}^\complement$, $\varphi(T) \in X$ and for all $t \geq 0$: $\frac{d\varphi(s)}{ds}(t) = I\varphi(t)[\![\sigma\theta]\!] = \sigma_{\varphi(t)}^* I\varphi(t)[\![\theta]\!]$ by Lemma 17 and $\varphi(t) \in I[\![\sigma\psi]\!]$, which, by Lemma 18, holds iff $\varphi(t) \in \sigma_{\varphi(t)}^* I[\![\psi]\!]$.
   Conversely, $\omega \in \sigma_\omega^* I[\![x' = \theta \,\&\, \psi]\!](X)$ iff $\exists \varphi : [0, T] \to \mathcal{S}$ with $\varphi(0) = \omega$ on $\{x'\}^\complement$ and $\varphi(T) \in X$ and for all $t \geq 0$: $\frac{d\varphi(s)}{ds}(t) = \sigma_\omega^* I\varphi(t)[\![\theta]\!]$ and $\varphi(t) \in \sigma_\omega^* I[\![\psi]\!]$. Both sides agree since $\sigma_\omega^* I[\![\theta]\!] = \sigma_{\varphi(t)}^* I[\![\theta]\!]$ and $\sigma_{\varphi(t)}^* I[\![\psi]\!] = \sigma_\omega^* I[\![\psi]\!]$ by Corollary 16 as σ is $\{x, x'\}$-admissible for θ and ψ and $\omega = \varphi(t)$ on $\mathsf{BV}(x' = \theta \,\&\, \psi)^\complement \supseteq \{x, x'\}^\complement$ by Lemma 13.
4. $\omega \in I[\![\sigma(?\psi)]\!](X) = I[\![?\sigma\psi]\!](X) = I[\![\sigma\psi]\!] \cap X$ iff, by Lemma 18, it is the case that $\omega \in \sigma_\omega^* I[\![\psi]\!] \cap X = \sigma_\omega^* I[\![?\psi]\!](X)$.
5. $\omega \in I[\![\sigma(\alpha \cup \beta)]\!](X) = I[\![\sigma\alpha \cup \sigma\beta]\!](X) = I[\![\sigma\alpha]\!](X) \cup I[\![\sigma\beta]\!](X)$, which, by induction hypothesis, is equivalent to $\omega \in \sigma_\omega^* I[\![\alpha]\!](X)$ or $\omega \in \sigma_\omega^* I[\![\beta]\!](X)$, which is $\omega \in \sigma_\omega^* I[\![\alpha]\!](X) \cup \sigma_\omega^* I[\![\beta]\!](X) = \sigma_\omega^* I[\![\alpha \cup \beta]\!](X)$.
6. $\omega \in I[\![\sigma(\alpha; \beta)]\!](X) = I[\![\sigma\alpha; \sigma\beta]\!](X) = I[\![\sigma\alpha]\!](I[\![\sigma\beta]\!](X))$ (provided σ is $\mathsf{BV}(\sigma\alpha)$-admissible for β), which holds iff $\omega \in I[\![\sigma\alpha]\!](I[\![\sigma\beta]\!](X) \downarrow_{\omega(\mathsf{BV}(\sigma\alpha)^\complement)})$ by Lemma 13.
   Starting conversely: $\omega \in \sigma_\omega^* I[\![\alpha; \beta]\!](X) = \sigma_\omega^* I[\![\alpha]\!](\sigma_\omega^* I[\![\beta]\!](X))$, iff, by IH, $\omega \in I[\![\sigma\alpha]\!](\sigma_\omega^* I[\![\beta]\!](X))$ iff, by Lemma 13, $\omega \in I[\![\sigma\alpha]\!](\sigma_\omega^* I[\![\beta]\!](X) \downarrow_{\omega(\mathsf{BV}(\sigma\alpha)^\complement)})$. Consequently, it suffices to show that both winning conditions are equal:

$$I[\![\sigma\beta]\!](X) \downarrow_{\omega(\mathsf{BV}(\sigma\alpha)^\complement)} = \sigma_\omega^* I[\![\beta]\!](X) \downarrow_{\omega(\mathsf{BV}(\sigma\alpha)^\complement)}$$

Consider any $\nu = \omega$ on $\mathsf{BV}(\sigma\alpha)^\complement$ to show: $\nu \in I[\![\sigma\beta]\!](X)$ iff $\nu \in \sigma_\omega^* I[\![\beta]\!](X)$. By IH, $\nu \in I[\![\sigma\beta]\!](X)$ iff $\nu \in \sigma_\nu^* I[\![\beta]\!](X)$ iff $\nu \in \sigma_\omega^* I[\![\beta]\!](X)$ by Corollary 16, because $\nu = \omega$ on $\mathsf{BV}(\sigma\alpha)^\complement$ and σ is $\mathsf{BV}(\sigma\alpha)$-admissible for β.

7. The case $\omega \in I[\![\sigma(\alpha^*)]\!](X) = I[\![(\sigma\alpha)^*]\!](X)$ (provided σ is $\mathsf{BV}(\sigma\alpha)$-admissible for α) uses an equivalent inflationary fixpoint formulation [5, Theorem 3.5]:

$$\tau^0(X) \overset{\text{def}}{=} X$$

$$\tau^{\kappa+1}(X) \overset{\text{def}}{=} X \cup I[\![\sigma\alpha]\!]\big(\tau^\kappa(X)\big) \qquad \kappa+1 \text{ a successor ordinal}$$

$$\tau^\lambda(X) \overset{\text{def}}{=} \bigcup_{\kappa<\lambda} \tau^\kappa(X) \qquad \lambda \neq 0 \text{ a limit ordinal}$$

where the union $\tau^\infty(X) = \bigcup_{\kappa<\infty} \tau^\kappa(X)$ over all ordinals is $I[\![(\sigma\alpha)^*]\!](X)$. Define a similar fixpoint formulation for the other side $\sigma^*_\omega I[\![\alpha^*]\!](X) = \varrho^\infty(X)$:

$$\varrho^0(X) \overset{\text{def}}{=} X$$

$$\varrho^{\kappa+1}(X) \overset{\text{def}}{=} X \cup \sigma^*_\omega I[\![\alpha]\!]\big(\varrho^\kappa(X)\big) \qquad \kappa+1 \text{ a successor ordinal}$$

$$\varrho^\lambda(X) \overset{\text{def}}{=} \bigcup_{\kappa<\lambda} \varrho^\kappa(X) \qquad \lambda \neq 0 \text{ a limit ordinal}$$

The equivalence $\omega \in I[\![\sigma(\alpha^*)]\!](X) = \tau^\infty(X)$ iff $\omega \in \sigma^*_\omega I[\![\alpha^*]\!](X) = \varrho^\infty(X)$ follows from a proof that:

for all κ and all X and all $\nu = \omega$ on $\mathsf{BV}(\sigma\alpha)^\complement$: $\nu \in \tau^\kappa(X)$ iff $\nu \in \varrho^\kappa(X)$

This is proved by induction on ordinal κ, which is either 0, a limit ordinal $\lambda \neq 0$, or a successor ordinal.

$\kappa = 0$: $\nu \in \tau^0(X)$ iff $\nu \in \varrho^0(X)$, because both sets equal X.

λ: $\nu \in \tau^\lambda(X) = \bigcup_{\kappa<\lambda} \tau^\kappa(X)$ iff there is a $\kappa < \lambda$ such that $\nu \in \tau^\kappa(X)$ iff, by IH, $\nu \in \varrho^\kappa(X)$ for some $\kappa < \lambda$, iff $\nu \in \bigcup_{\kappa<\lambda} \varrho^\kappa(X) = \varrho^\lambda(X)$.

$\kappa + 1$: $\nu \in \tau^{\kappa+1}(X) = X \cup I[\![\sigma\alpha]\!]\big(\tau^\kappa(X)\big)$, which, by Lemma 13, is equivalent to $\nu \in X \cup I[\![\sigma\alpha]\!]\big(\tau^\kappa(X){\downarrow}\nu_{(\mathsf{BV}(\sigma\alpha)^\complement)}\big)$.

Starting from the other end, $\nu \in \varrho^{\kappa+1}(X) = X \cup \sigma^*_\omega I[\![\alpha]\!]\big(\varrho^\kappa(X)\big)$ iff, by Corollary 16 using $\nu = \omega$ on $\mathsf{BV}(\sigma\alpha)^\complement \supseteq \mathsf{BV}(\alpha)^\complement$, $\nu \in X \cup \sigma^*_\nu I[\![\alpha]\!]\big(\varrho^\kappa(X)\big)$ iff, by induction hypothesis on α, $\nu \in X \cup I[\![\sigma\alpha]\!]\big(\varrho^\kappa(X)\big)$ iff, by Lemma 13, $\nu \in X \cup I[\![\sigma\alpha]\!]\big(\varrho^\kappa(X){\downarrow}\nu_{(\mathsf{BV}(\sigma\alpha)^\complement)}\big)$ Consequently, it suffices to show that both winning conditions are equal: $\tau^\kappa(X){\downarrow}\nu_{(\mathsf{BV}(\sigma\alpha)^\complement)}=\varrho^\kappa(X){\downarrow}\nu_{(\mathsf{BV}(\sigma\alpha)^\complement)}$. Consider any state $\mu = \omega$ on $\mathsf{BV}(\sigma\alpha)^\complement$, then $\mu \in \tau^\kappa(X)$ iff $\mu \in \varrho^\kappa(X)$ by induction hypothesis on $\kappa < \kappa + 1$.

8. $\omega \in I[\![\sigma(\alpha^d)]\!](X) = I[\![(\sigma\alpha)^d]\!](X) = \big(I[\![\sigma\alpha]\!](X^\complement)\big)^\complement$ iff $\omega \notin I[\![\sigma\alpha]\!](X^\complement)$, which, by IH, is equivalent to $\omega \notin \sigma^*_\omega I[\![\alpha]\!](X^\complement)$, which is, in turn, equivalent to $\omega \in \big(\sigma^*_\omega I[\![\alpha]\!](X^\complement)\big)^\complement = \sigma^*_\omega I[\![\alpha^d]\!](X)$. □

4.2   Soundness

Soundness of uniform substitution for dGL now follows from the above uniform substitution lemmas with the same proof that it had from corresponding lemmas

$$[\cdot] \quad [a]p(\bar{x}) \leftrightarrow \neg\langle a\rangle\neg p(\bar{x})$$

$$\langle := \rangle \quad \langle x := f\rangle p(x) \leftrightarrow p(f)$$

$$\text{DS} \quad \langle x' = f\rangle p(x) \leftrightarrow \exists t{\geq}0 \, \langle x := x + ft\rangle p(x)$$

$$\langle ?\rangle \quad \langle ?q\rangle p \leftrightarrow q \wedge p$$

$$\langle \cup\rangle \quad \langle a \cup b\rangle p(\bar{x}) \leftrightarrow \langle a\rangle p(\bar{x}) \vee \langle b\rangle p(\bar{x})$$

$$\langle ;\rangle \quad \langle a;b\rangle p(\bar{x}) \leftrightarrow \langle a\rangle\langle b\rangle p(\bar{x})$$

$$\langle *\rangle \quad \langle a^*\rangle p(\bar{x}) \leftrightarrow p(\bar{x}) \vee \langle a\rangle\langle a^*\rangle p(\bar{x})$$

$$\langle {}^d\rangle \quad \langle a^d\rangle p(\bar{x}) \leftrightarrow \neg\langle a\rangle\neg p(\bar{x})$$

$$\text{M} \quad \frac{p(\bar{x}) \to q(\bar{x})}{\langle a\rangle p(\bar{x}) \to \langle a\rangle q(\bar{x})}$$

$$\text{FP} \quad \frac{p(\bar{x}) \vee \langle a\rangle q(\bar{x}) \to q(\bar{x})}{\langle a^*\rangle p(\bar{x}) \to q(\bar{x})}$$

$$\text{MP} \quad \frac{p \quad p \to q}{q}$$

$$\forall \quad \frac{p(x)}{\forall x \, p(x)}$$

Fig. 2. Differential game logic axioms and axiomatic proof rules

in dL [6] (see [8]). Due to the modular setup of uniform substitutions, the change from dL to dGL is reflected in how the uniform substitution lemmas are proved, not in how they are used for the soundness of proof rule US. A proof rule is *sound* iff validity of all its premises implies validity of its conclusion.

Theorem 20 (Soundness of uniform substitution). *Proof rule US is sound.*

$$(\text{US}) \quad \frac{\phi}{\sigma\phi}$$

As in dL, uniform substitutions can soundly instantiate locally sound proof rules or proofs [6] just like proof rule US soundly instantiates axioms or other valid formulas (Theorem 20). An inference or proof rule is *locally sound* iff its conclusion is valid in any interpretation I in which all its premises are valid. All locally sound proof rules are sound. The use of Theorem 21 in a proof is marked USR.

Theorem 21 (Soundness of uniform substitution of rules). *If* $\text{FV}(\sigma) = \emptyset$, *all uniform substitution instances of locally sound inferences are locally sound:*

$$\frac{\phi_1 \quad \cdots \quad \phi_n}{\psi} \text{ locally sound} \quad implies \quad \frac{\sigma\phi_1 \quad \cdots \quad \sigma\phi_n}{\sigma\psi} \text{ locally sound}$$

5 Axioms

Axioms and axiomatic proof rules for differential game logic are listed in Fig. 2, where $\bar{x}$ is the (finite-dimensional) vector of all relevant variables. The axioms are concrete dGL formulas that are valid. The axiomatic proof rules are concrete formulas for the premises and concrete formulas for the conclusion that are locally sound. This makes Fig. 2 straightforward to implement by copy-and-paste. Theorem 20 can be used to instantiate axioms to other dGL formulas. Theorem 21 can be used to instantiate axiomatic proof rules to other concrete dGL inferences. Complete axioms for first-order logic from elsewhere [6] and a proof rule (written ℝ) for decidable real arithmetic [11] are assumed as a basis.

The axiom $\langle;\rangle$, for example, expresses that Angel has a winning strategy in game $a; b$ to achieve $p(\bar{x})$ if and only if she has a winning strategy in game a to achieve $\langle b\rangle p(\bar{x})$, i.e., to reach the region from which she has a winning strategy in game b to achieve $p(\bar{x})$. Rule US can instantiate axiom $\langle;\rangle$, for example, with $\sigma = \{a \mapsto (v := 2 \cup v := x+1)^d, b \mapsto x' = v, p(\bar{x}) \mapsto x > 0\}$ to prove

$$\langle(v := 2 \cup v := x+1)^d; x' = v\rangle x > 0 \leftrightarrow \langle(v := 2 \cup v := x+1)^d\rangle\langle x' = v\rangle x > 0$$

The right-hand formula can be simplified when using US again to instantiate axiom $\langle^d\rangle$ with $\sigma = \{a \mapsto v := 2 \cup v := x+1, p(\bar{x}) \mapsto \langle x' = v\rangle x > 0\}$ to prove

$$\langle(v := 2 \cup v := x+1)^d\rangle\langle x' = v\rangle x > 0 \leftrightarrow \neg\langle v := 2 \cup v := x+1\rangle\neg\langle x' = v\rangle x > 0$$

When eliding the equivalences and writing down the resulting formula along with the axiom that was uniformly substituted to obtain it, this yields a proof:

$$
\begin{array}{ll}
& j(x) \to \neg(\neg\exists t{\geq}0\, x + 2t > 0 \lor \neg\exists t{\geq}0\, x + (x{+}1)t > 0) \\
\hline
^{\langle:=\rangle} & j(x) \to \neg(\neg\exists t{\geq}0\, \langle x := x + 2t\rangle x > 0 \lor \langle v := x{+}1\rangle\neg\exists t{\geq}0\, \langle x := x + vt\rangle x > 0) \\
\hline
^{DS} & j(x) \to \neg(\neg\langle x' = 2\rangle x > 0 \lor \langle v := x{+}1\rangle\neg\langle x' = v\rangle x > 0) \\
\hline
^{\langle:=\rangle} & j(x) \to \neg(\langle v := 2\rangle\neg\langle x' = v\rangle x > 0 \lor \langle v := x{+}1\rangle\neg\langle x' = v\rangle x > 0) \\
\hline
^{\langle\cup\rangle} & j(x) \to \neg\langle v := 2 \cup v := x{+}1\rangle\neg\langle x' = v\rangle x > 0 \\
\hline
^{\langle d\rangle} & j(x) \to \langle(v := 2 \cup v := x{+}1)^d\rangle\langle x' = v\rangle x > 0 \\
\hline
^{\langle;\rangle} & j(x) \to \langle(v := 2 \cup v := x{+}1)^d; x' = v\rangle x > 0
\end{array}
$$

It is soundness-critical that US checks velocity v is not bound in the ODE when substituting it for f in DS, since $x + vt$ is not, otherwise, the correct solution of $x' = v$. Likewise, the velocity assignment $v := x+1$ cannot soundly be substituted into the differential equation via $\langle:=\rangle$, which US prevents as x is bound in $x' = v$. Instead, axiom $\langle:=\rangle$ for $v := x+1$ needs to be delayed until after solving by DS. If it were $v := x^2+1$ instead of $v := x+1$, then rule ℝ would finish the proof. But for the above proof with $v := x+1$ to finish, extra assumptions need to be identified.

With $\sigma = \{a \mapsto (v := 2 \cup v := x+1)^d; x' = v, p(\bar{x}) \mapsto x{>}0, q(\bar{x}) \mapsto x^2{>}0\}$, USR instantiates axiomatic rule M to prove an inference continuing the proof:

$$
\begin{array}{l}
x{>}0 \to x^2{>}0 \\
\hline
^{USR,M}\ \langle(v := 2 \cup v := x+1)^d; x' = v\rangle x{>}0 \to \langle(v := 2 \cup v := x+1)^d; x' = v\rangle x^2{>}0
\end{array}
$$

Variable x can be used in the postconditions despite being bound in the game. Likewise, rule USR can instantiate the above proof with $\sigma = \{j(\cdot) \mapsto \cdot{>}{-}1\}$ to:

$$
\begin{array}{l}
^{\mathbb{R}}\ x > -1 \to \neg(\neg\exists t{\geq}0\, x + 2t > 0 \lor \neg\exists t{\geq}0\, x + (x{+}1)t > 0) \\
\hline
^{USR}\ x > -1 \to \langle(v := 2 \cup v := x+1)^d; x' = v\rangle x > 0
\end{array}
$$

USR soundly instantiates the inference from premise to conclusion of the proof without having to change or repeat any part of the proof. Uniform substitutions enable flexible but sound reasoning forwards, backwards, on proofs, or mixed [6]. Without USR, these features would complicate soundness-critical prover cores.

Since the axioms and axiomatic proof rules in Fig. 2 are themselves instances of axiom schemata and proof rule schemata that axiomatize dGL [5], they are (even locally!) *sound*. Axiom DS stems from dL [6] and is for solving constant differential equations. Now that differentials are available, all differential axioms such as the Leibniz axiom $(f(\bar{x}) \cdot g(\bar{x}))' = (f(\bar{x}))' \cdot g(\bar{x}) + f(\bar{x}) \cdot (g(\bar{x}))'$ and all other axioms for differential equations [6] can be added to dGL. Furthermore, hybrid games make it possible to equivalently replace differential equations with evolution domains by hybrid games without domain constraints [5, Lemma 3.4].

The converse challenge for *completeness* is to prove that uniform substitutions are flexible enough to prove all required instances of dGL axioms and axiomatic proof rules. A dGL formula ϕ is called *surjective* iff rule US can instantiate ϕ to any of its axiom schema instances, which are those formulas that are obtained by just replacing game symbols a uniformly by any hybrid game etc. An axiomatic rule is called *surjective* iff USR can instantiate it to any of its proof rule schema instances. The axiom $\langle ? \rangle$ is surjective, as it does not have any bound variables, so its instances are admissible. Similarly rules MP and rule $\forall$ become surjective [6]. The proof of the following lemma transfers from prior work [6, Lemma 39], since any hybrid game can be substituted for a game symbol.

Lemma 22 (Surjective axioms). *If ϕ is a dGL formula that is built only from game symbols but no function or predicate symbols, then ϕ is surjective. Axiomatic rules consisting of surjective dGL formulas are surjective.*

Unfortunately, none of the axioms from Fig. 2 satisfy the assumptions of Lemma 22. While the argument from previous work would succeed [6], the trick to simplify the proof is to consider $p(\bar{x})$ to be $\langle c \rangle true$ for some game symbol c. Then any formula φ can be instantiated for $p(\bar{x})$ alias $\langle c \rangle true$ by substituting the game symbol c with the game $?\varphi$ and subsequently using the surjective axiom $\langle ? \rangle$ to replace the resulting $\langle ?\varphi \rangle true$ by $\varphi \wedge true$ or its equivalent φ as intended. This makes axioms $[\cdot], \langle ? \rangle, \langle \cup \rangle, \langle ; \rangle, \langle * \rangle, \langle^d\rangle$ and all axiomatic rules in Fig. 2 surjective.

With Lemma 22 to show that all schema instantiations required for completeness are provable by US, USR from axioms or axiomatic rules, relative completeness of dGL follows immediately from a previous schematic completeness result for dGL [5] and relative completeness of uniform substitution for dL [6].

Theorem 23 (Relative completeness). *The dGL calculus is a sound and complete axiomatization of hybrid games relative to any differentially expressive logic[3] L, i.e., every valid dGL formula is provable in dGL from L tautologies.*

---

[3] A logic L closed under first-order connectives is *differentially expressive* (for dGL) if every dGL formula ϕ has an equivalent $\phi^\flat$ in L and all differential equation equivalences of the form $\langle x' = \theta \rangle G \leftrightarrow (\langle x' = \theta \rangle G)^\flat$ for G in L are provable in its calculus.

6   Related Work

Since the primary impact of uniform substitution is on conceptual simplicity and a significantly simpler prover implementation, this related work discussion focuses on hybrid games theorem proving. A broader discussion of both hybrid games and uniform substitution themselves is provided in the literature [5,6]. The approach presented here also helps discrete game logic [4], but that is only challenging after a suitable generalization beyond the propositional case.

Prior approaches to hybrid games theorem proving are either based on differential game logic [5,7] or on an exterior game embedding of differential dynamic logic [10]. This paper is based on prior findings on differential game logic [5] that it complements by giving an *explicit construction* for uniform substitution. This enables a purely axiomatic version of dGL that does not need the axiom schemata or proof rule schemata from previous approaches [5,7]. This change makes it substantially simpler to implement dGL soundly in a theorem prover. The exterior game embedding of differential dynamic logic [10] was implemented with proof rule schemata in KeYmaera and was, thus, significantly more complex.

The primary and significant challenge of this paper compared to previous uniform substitution approaches [1,2,6] arose from the semantics of hybrid games, which need a significantly different set-valued winning region style. The root-cause is that, unlike the normal modal logic dL, dGL is a subregular modal logic [5]. Especially, Kripke's axiom $[\alpha](\phi \to \psi) \to ([\alpha]\phi \to [\alpha]\psi)$ is unsound for dGL.

7   Conclusion and Future Work

This paper provides an explicit construction of uniform substitutions and proves it sound for differential game logic. It also indicates that uniform substitutions are flexible when a logic is changed. The modularity principles of uniform substitution hold what they promise, making an implementation in a theorem prover exceedingly straightforward. The biggest challenge was the semantic generalization of the soundness proofs to the subtle interactions caused by hybrid games.

In future work it could be interesting to devise a framework for the general construction of uniform substitutions for arbitrary logics from a certain family. The challenge is that such an approach partially goes against the spirit of uniform substitution, which is built for flexibility (straightforward and easy to change), not necessarily generality (already preequipped to reconfigure for all possible future changes). Such generality seems to require a schematic understanding, possibly self-defeating for the simplicity advantages of uniform substitutions.

References

1. Bohrer, B., Rahli, V., Vukotic, I., Völp, M., Platzer, A.: Formally verified differential dynamic logic. In: Bertot, Y., Vafeiadis, V. (eds.) CPP. ACM (2017). https://doi.org/10.1145/3018610.3018616

2. Church, A.: Introduction to Mathematical Logic. Princeton University Press, Princeton (1956)
3. Fulton, N., Mitsch, S., Quesel, J.D., Völp, M., Platzer, A.: KeYmaera X: an axiomatic tactical theorem prover for hybrid systems. In: Felty, A., Middeldorp, A. (eds.) CADE. LNCS, vol. 9195, pp. 527–538. Springer, Berlin (2015). https://doi.org/10.1007/978-3-319-21401-6_36
4. Parikh, R.: Propositional game logic. In: FOCS, pp. 195–200. IEEE (1983). https://doi.org/10.1109/SFCS.1983.47
5. Platzer, A.: Differential game logic. ACM Trans. Comput. Log. **17**(1), 1:1–1:52 (2015). https://doi.org/10.1145/2817824
6. Platzer, A.: A complete uniform substitution calculus for differential dynamic logic. J. Autom. Reas. **59**(2), 219–265 (2017). https://doi.org/10.1007/s10817-016-9385-1
7. Platzer, A.: Differential hybrid games. ACM Trans. Comput. Log. **18**(3), 19:1–19:44 (2017)
8. Platzer, A.: Uniform substitution for differential game logic. CoRR abs/1804.05880 (2018)
9. Platzer, A., Quesel, J.D.: KeYmaera: a hybrid theorem prover for hybrid systems. In: Armando, A., Baumgartner, P., Dowek, G. (eds.) IJCAR. LNCS, vol. 5195, pp. 171–178. Springer, Berlin (2008). https://doi.org/10.1007/978-3-540-71070-7_15
10. Quesel, J.-D., Platzer, A.: Playing hybrid games with KeYmaera. In: Gramlich, B., Miller, D., Sattler, U. (eds.) IJCAR 2012. LNCS (LNAI), vol. 7364, pp. 439–453. Springer, Heidelberg (2012). https://doi.org/10.1007/978-3-642-31365-3_34
11. Tarski, A.: A Decision Method for Elementary Algebra and Geometry, 2nd edn. University of California Press, Berkeley (1951)

A Logical Framework with Commutative and Non-commutative Subexponentials

Max Kanovich[1], Stepan Kuznetsov[1,2], Vivek Nigam[3,4(✉)], and Andre Scedrov[1,5]

[1] National Research University Higher School of Economics, Moscow, Russia
mkanovich@hse.ru
[2] Steklov Mathematical Institute of RAS, Moscow, Russia
sk@mi.ras.ru
[3] Federal University of Paraíba, João Pessoa, Brazil
[4] fortiss, Munich, Germany
nigam@fortiss.org
[5] University of Pennsylvania, Philadelphia, USA
scedrov@math.upenn.edu

Abstract. Logical frameworks allow the specification of deductive systems using the same logical machinery. Linear logical frameworks have been successfully used for the specification of a number of computational, logics and proof systems. Its success relies on the fact that formulas can be distinguished as linear, which behave intuitively as resources, and unbounded, which behave intuitionistically. Commutative subexponentials enhance the expressiveness of linear logic frameworks by allowing the distinction of multiple contexts. These contexts may behave as multisets of formulas or sets of formulas. Motivated by applications in distributed systems and in type-logical grammar, we propose a linear logical framework containing both commutative and non-commutative subexponentials. Non-commutative subexponentials can be used to specify contexts which behave as lists, not multisets, of formulas. In addition, motivated by our applications in type-logical grammar, where the weakenening rule is disallowed, we investigate the proof theory of formulas that can only contract, but not weaken. In fact, our contraction is non-local. We demonstrate that under some conditions such formulas may be treated as unbounded formulas, which behave intuitionistically.

1 Introduction

Logical frameworks [7,8,13,23,33] have been proposed to specify deductive systems, such as proof systems [7,13,24,26,33], logics [7,22] and operational semantics [25,27,29,33]. The systems that can be encoded depend on the expressive power of the logical framework. Linear logical frameworks, based on Linear Logic [6], allow the encoding of, for example, stateful systems [22,33]. Logical Frameworks with subexponentials allow the encoding of, for example, distributed systems [25,27], authorization logics [22]. Ordered Logical Frameworks [29] allow the specification of systems whose behavior respects some order, for example, evaluation strategies.

© Springer International Publishing AG, part of Springer Nature 2018
D. Galmiche et al. (Eds.): IJCAR 2018, LNAI 10900, pp. 228–245, 2018.
https://doi.org/10.1007/978-3-319-94205-6_16

One key idea [2] of logical frameworks is to distinguish formulas according to the structural rules (weakening, contraction and exchange rules) that are applicable. For example, linear logical frameworks distinguish two types of formulas: *Unbounded Formulas* which behave intuitionistically, that is, can be considered as a set of formulas and *Linear Formulas* which behave linearly, that is, should be considered as a multiset of formulas. Ordered logical frameworks also consider *Ordered Formulas* which are non-commutative, that is, can be considered as a list, not multiset, of formulas. This distinction is reflected in the syntax. Linear logical frameworks have two contexts $\Theta : \Gamma$, where Θ is a set of unbounded formulas and Γ a multiset of linear[1] formulas. Ordered linear logic, on the other hand, has three contexts $\Theta : \Gamma : \Delta$ where Δ is a list of ordered formulas.

Logical Frameworks with Subexponentials refine Linear Logical Frameworks by distinguishing different types of unbounded and linear formulas. They work, therefore, on sequents with multiple contexts. This increased expressiveness allows for the specification of a greater number of proof systems [26] and distributed systems [27] when compared to logical frameworks without subexponentials. However, existing logical frameworks with subexponentials do not allow ordered formulas.

Our main contribution is the logical framework SNILLF which has the following two innovations:

1. **Non-commutative Subexponentials:** SNILLF allows both commutative and non-commutative subexponentials [10]. This means that SNILLF works not only with multiple contexts for unbounded and linear formulas, but also multiple ordered contexts. As an illustration of the power of this system, we encode a distributed system where machines have FIFO buffers storing messages received from the network;

2. **Proof Search with formulas that can contract, but not weaken:** Motivated by applications in type-logical grammar, where weakening of formulas is not allowed, SNILLF allows formulas to be marked with subexponentials that can contract, but not weaken. We classify such formulas as relevant. Relevant formulas lead to complications for proof search because contracting a formula implies that it should be necessarily used in the proof. Thus the contraction of relevant formulas involves a "don't know" non-determinism. This paper investigates the proof theory of relevant formulas. We demonstrate that in some situations it is safe (sound and complete) to consider relevant formulas as unbounded, that is, formulas that can both weaken and contract. We illustrate the use relevant formulas by using SNILLF in type-logical grammar applications.

In Sect. 2, we review the basic proof theory of non-commutative proof systems, namely Lambek Calculus, and subexponentials. Then in Sect. 3 we motivate the use of non-commutatitive subexponentials and relevant formulas with some concrete examples. Section 4 investigates the proof theory of relevant

---

[1] Or affine which can be weakened.

formulas. The Logical Framework SNILLF is introduced in Sect. 5 as a focused proof system. We revisit our main examples in Sect. 6. Finally, we comment on related and future work in Sects. 7 and 8.

2   Lambek Calculus with Subexponentials

While we assume some familiarity with Lambek Calculus [12], we review some of its proof theory. Its rules are depicted in Fig. 1 contaning atomic formulas, the unit constant $\mathbf{1}$, universal quantifier $\forall$, and binary connectives: $\cdot$ (product), $\backslash$ (left division) and $/$ (right division). The formulas in the sequent should be seen as lists, not multisets, of formulas. For example, the $\Gamma, F_1, F_2, \Delta \longrightarrow G$ and $\Gamma, F_2, F_1, \Delta \longrightarrow G$ are not equivalent in general as there may be a proof for one, but not for the other.

$$\frac{}{F \to F} \, I \quad \frac{\Gamma_1, \Gamma_2 \to C}{\Gamma_1, \mathbf{1}, \Gamma_2 \to C} \, 1_L \quad \frac{}{\to \mathbf{1}} \, 1_R \quad \frac{\Pi \to G \quad \Gamma_1, F, \Gamma_2 \to C}{\Gamma_1, F/G, \Pi, \Gamma_2 \to C} \, /_L \quad \frac{\Pi, F \to G}{\Pi \to G/F} \, /_R$$

$$\frac{\Gamma_1, F, G, \Gamma_2 \to C}{\Gamma_1, F \cdot G, \Gamma_2 \to C} \, \cdot_L \quad \frac{\Gamma_1 \to F \quad \Gamma_2 \to G}{\Gamma_1, \Gamma_2 \to F \cdot G} \, \cdot_R \quad \frac{\Pi \to F \quad \Gamma_1, G, \Gamma_2 \to C}{\Gamma_1, \Pi, F \backslash G, \Gamma_2 \to C} \, \backslash_L \quad \frac{F, \Pi \to G}{\Pi \to F \backslash G} \, \backslash_R$$

$$\frac{\Pi \to F\{e/x\}}{\Pi \to \forall x.F} \, \forall_R \quad \frac{\Gamma_1, F\{t/x\}, \Gamma_2 \to C}{\Gamma_1, \forall x.F, \Gamma_2 \to C} \, \forall_L$$

Fig. 1. Cut-free proof system Lambek proof system. Here $\{t/x\}$ denotes the capture avoiding substitution of x by t. Moreover, e is a fresh eigenvariable, that is, not appearing in Π and F.

In our previous work [10], we proposed the proof system SNILL$_\Sigma$ (Subexponential Non-Commutative Intuitionistic Linear Logic)[2] which extends propositional Lambek Calculus with subexponentials. Subexponentials derive from an observation from Linear Logic [5,6,23]. Namely, the linear logic exponentials, !, are non-canonical. That is, LL allows for an unbounded number of subexponentials, !^{s}, indexed by elements in a set of indexes $s \in \mathcal{I}$.

Formally, SNILL$_\Sigma$ contains all rules in Fig. 1. Furthermore, it is parametrized by a subexponential signature $\Sigma = \langle \mathcal{I}, \preceq, \mathcal{W}, \mathcal{C}, \mathcal{E} \rangle$, where $\mathcal{W}, \mathcal{C}, \mathcal{E} \subseteq \mathcal{I}$ and $\preceq$ is a pre-order over the elements of $\mathcal{I}$ upwardly closed with respect to $\mathcal{W}, \mathcal{C}, \mathcal{E}$, that is, if $s_1 \in \mathcal{W}$ and $s_1 \preceq s_2$, then $s_2 \in \mathcal{W}$ and similar for $\mathcal{C}, \mathcal{E}$. SNILL$_\Sigma$ contains the following rules:

– For each $s \in \mathcal{I}$, SNILL$_\Sigma$ contains the dereliction and promotion rules:

$$\frac{\Gamma_1, F, \Gamma_2 \to G}{\Gamma_1, !^s F, \Gamma_2 \to G} \, Der \quad \frac{!^{s_1} F_1, \dots, !^{s_n} F_n \longrightarrow F}{!^{s_1} F_1, \dots, !^{s_n} F_n \longrightarrow !^s F} \, !^s_R, \text{provided, } s \preceq s_i, 1 \leq i \leq n$$

---

[2] In that paper, the system was called SMALC.

– For each $w \in \mathcal{W}$ and $c \in \mathcal{C}$, SNILL$_\Sigma$ contains the rules:

$$\frac{\Gamma, \Delta \longrightarrow G}{\Gamma, !^w F, \Delta \longrightarrow G} \; W \qquad \frac{\Gamma_1, !^c F, \Delta, !^c F, \Gamma_2 \to G}{\Gamma_1, !^c F, \Delta, \Gamma_2 \to G} \; C_1 \qquad \frac{\Gamma_1, !^c F, \Delta, !^c F, \Gamma_2 \to G}{\Gamma_1, \Delta, !^c F, \Gamma_2 \to G} \; C_2$$

– For each $e \in \mathcal{E}$, SNILL$_\Sigma$ contains the rules:

$$\frac{\Gamma_1, \Delta, !^e F, \Gamma_2 \to C}{\Gamma_1, !^e F, \Delta, \Gamma_2 \to C} \; E_1 \qquad \frac{\Gamma_1, !^e F, \Delta, \Gamma_2 \to C}{\Gamma_1, \Delta, !^e F, \Gamma_2 \to C} \; E_2$$

Intuitively, the set $\mathcal{I}$ specifies the subexponential names, $\mathcal{W}$ the subexponentials that are allowed to weaken, $\mathcal{C}$ the subexponentials that allow to contract, and $\mathcal{E}$ the subexponentials that allow to exchange.

Notice additionally that contraction is non-local, that is, the contracted formula can appear anywhere in left hand side of the premise.

In [10], we proved that the propositional fragment of SNILL$_\Sigma$ (with additive connectives), admits cut-elimination. The following extends this result to first-order SNILL$_\Sigma$.

Theorem 1. *For any subexponential signature Σ, SNILL$_\Sigma$ admits cut-elimination.*

The proof is essentially the same as in [10], since in the interesting cases a formula of the form $\forall x.F$ is never the active one, and the $\forall$ rules just permute with the mix rule.

For our applications, we will consider subexponential signatures $\Sigma = \langle \mathcal{I}, \preceq, \mathcal{W}, \mathcal{C}, \mathcal{E} \rangle$ with the following restrictions:

$$\mathcal{W} \subseteq \mathcal{E} \qquad \text{and} \qquad \mathcal{C} \subseteq \mathcal{E}$$

That is, all subexponentials that can be weakened or contracted can also be exchanged. This restriction on subexponentials will be used to establish conditions for reducing "don't know" non-determinism as we describe in Sect. 4. Moreover, they are enough to specify our intended applications as described in Sect. 6.

In the remainder of this paper, we will elide the subexponential signature Σ whenever it is clear from the context.

Given the restriction above on subexponential signtures, we can classify formulas of the form $!^s F$ according to the structural rules that are applicable to s:

– **Linear Formulas:** These formulas are not allowed to be contracted nor weakened, that is, subexponentials $s \notin \mathcal{W} \cup \mathcal{C}$. Linear subexponentials range over $l, l_1, l_2, \ldots$. They can be commutative when $l \in \mathcal{E}$ or non-commutative otherwise;

– **Unbounded Formulas:** These formulas can be both weakened and contracted, that is, subexponentials $s \in \mathcal{W} \cap \mathcal{C}$. Unbounded subexponentials range over $u, u_1, u_2, \ldots$. As $\mathcal{W} \subseteq \mathcal{E}$, these formulas are always commutative that is $u \in \mathcal{E}$;

- **Affine Formulas:** These formulas can only be weakened and not contracted, that is, subexponentials $s \in \mathcal{W}$ and $s \notin \mathcal{C}$; Affine subexponentials range over $a, a_1, a_2, \ldots$. As $\mathcal{W} \subseteq \mathcal{E}$, these formulas are always commutative that is $a \in \mathcal{E}$;
- **Relevant Formulas:** These formulas cannot be weakened but can be contracted, that is, subexponentials $s \in \mathcal{C}, s \notin \mathcal{W}$. Relevant subexponentials range over $r, r_1, r_2, \ldots$. As $\mathcal{C} \subseteq \mathcal{E}$, these formulas are always commutative that is $r \in \mathcal{E}$.

Logical frameworks have been proposed with unbounded, linear and affine formulas, but without relevant formulas. To illustrate the difficulty involving relevant formulas, consider the following derivations with an instance of the dot rule and contraction rules. In the derivation to the left, only the formula $!^u F$ is contracted, while in the right the formula $!^r H$ is also contracted.

$$\frac{\dfrac{!^u F, !^r H, \Gamma \longrightarrow G_1 \quad !^u F, \Delta \longrightarrow G_2}{!^u F, !^r H, \Gamma, !^u F, \Delta \longrightarrow G_1 \cdot G_2} \otimes R}{!^u F, !^r H, \Gamma, \Delta \longrightarrow G_1 \cdot G_2} C \qquad \frac{\dfrac{!^u F, !^r H, \Gamma \longrightarrow G_1 \quad !^u F, !^r H, \Delta \longrightarrow G_2}{!^u F, !^r H, \Gamma, !^u F, !^r H, \Delta \longrightarrow G_1 \cdot G_2} \otimes R}{!^u F, !^r H, \Gamma, \Delta \longrightarrow G_1 \cdot G_2} 2 \times C$$

As unbounded formulas can always be weakened, it is always safe to contract them. If the contracted formula is needed then it can be used and if it turns out not to be needed, the unbounded formula can be weakened before applying the initial rule. Thus, a collection of unbounded formulas can be safely treated as a set of formulas. *This means that the non-determinism due to unbounded formulas is a don't care non-determinism.*

The same is not the case for relevant formulas. As these formulas cannot be weakened, provability may depend on whether one contracts a relevant formula or not. For example, in the derivation to the right, the formula $!^r H$ has to be necessarily used in both premises, while in the derivation to the left, the formula $!^r H$ can only be used in the left premise. *This means that the choice of contracting a relevant formula or not involves a don't know non-determinism.*

3   Examples

We detail two different domain applications for which SNILLF can be applied. The first is on the specification of distributed systems. The second is on type-logical grammar.

3.1   Distributed Systems Semantics

Computer systems work with data structures which behave as sets, multisets and as lists. As an example, consider a system with n machines called $m_1, \ldots, m_n$. Assume that each machine has an input FIFO buffer. Whenever a machine receives a message, it is stored at the beginning of the buffer, and the message at the end of the buffer is processed first by a machine.

A buffer at machine m_i with elements Γ_i is specified as the list of formulas where start and end mark the start and end of the list $[\mathsf{start}, \Gamma_i, \mathsf{end}]_{mi}$. Thus a system with n machines is specified as the collection of contexts of the form which are associated to non-commutative subexponentials $m1, \ldots, mn$, respectively:

$$[\mathsf{start}, \Gamma_1, \mathsf{end}]_{m1} \; [\mathsf{start}, \Gamma_2, \mathsf{end}]_{m2} \; \cdots \; [\mathsf{start}, \Gamma_n, \mathsf{end}]_{mn}$$

As we describe in detail in Sect. 6, since these contexts behave as lists, the order of the elements of the buffers allows to specify the correct FIFO behavior of such buffers.

3.2   Type-Logical Grammar

The Lambek calculus was initially designed by Joachim Lambek [12] as a basic logic in a framework for describing natural language syntax. The idea of such frameworks goes back to works of Ajdukiewicz [1] and Bar-Hillel [3]; nowadays formal grammars of such sort are called *type-logical,* or *categorial* grammars.

The idea of a type-logical grammar is simple: the central part of the grammar is the *lexicon,* a finite binary correspondence $\triangleright$ between words of the language and formulae of the basic logic (such as Lambek Calculus). These formulae are also called *syntactic categories,* or *types.* Thus, in this framework the grammar is fully *lexicalised, i.e.,* all syntactic information is kept in the types associated to words, and one does not need to formulate "global" syntactic rules like "a sentence is a combination of a noun phrase and a verb phrase." The second component of a type-logical grammar is the *goal type.* Usually it is a designated variable (*primitive type*) S (meaning "sentence").

A sentence $w = a_1 \, a_2 \ldots a_n$ is accepted by the grammar, if there exist such formulae $F_1, F_2, \ldots, F_n$ that $a_i \triangleright F_i$ for $1 \leq i \leq n$ and the sequent $F_1, F_2, \ldots, F_n \to S$ is derivable. The language generated by the grammar is defined as the set of all accepted sentences.

As shown by Pentus [28], grammars based on the Lambek calculus can generate only context-free languages. It is known, however, that certain natural language structures are beyond the context-free formalism (as discussed, for example, by Shieber [31] on Swiss German material). This also served as motivation for extending the Lambek calculus with extra connectives, in particular, subexponential modalities.

In order to show how a subexponential connective can be useful in type-logical grammar, let us consider the following series of examples. The syntactic analysis shown in these examples is due to Morrill and Valentín [19]. In our toy grammar for a small fragment of English we associate the following types to words:

John, Mary $\triangleright N$	(noun phrase)
loves, signed $\triangleright N \backslash S / N$	(transitive verb)
girl, paper $\triangleright CN$	(common noun)
the $\triangleright N / CN$	(article: transforms a common noun into a noun phrase)
without $\triangleright (N \backslash S) \backslash (N \backslash S) / GC$	
reading $\triangleright GC / N$	("*reading the paper*" is a gerund clause, GC)
that, whom $\triangleright (CN \backslash CN) / (S / !^{s}N)$	(dependent clause coordinator)

The simplest example, *"John loves Mary,"* is justified as a correct sentence (of type S) by the following derivation in Lambek calculus:

$$\frac{N \to N \quad \dfrac{N \to N \quad S \to S}{N, N \backslash S \to S}}{N, N \backslash S / N, N \to S}$$

There are more sophisticated syntactic constructions for which the *contraction* rule is used. First consider the following sentence: *"John signed the paper without reading it"* (of type S), supported by the following Lambek derivation:

$$\frac{CN \to CN \quad \dfrac{N \to N \quad \dfrac{GC / N, N \to GC \quad N, N \backslash S, (N \backslash S) \backslash (N \backslash S) \to S}{N, N \backslash S, (N \backslash S) \backslash (N \backslash S) / GC, GC / N, N \to S}}{N, N \backslash S / N, N, (N \backslash S) \backslash (N \backslash S) / GC, GC / N, N \to S}}{N, N \backslash S / N, N / CN, CN, (N \backslash S) \backslash (N \backslash S) / GC, GC / N, N \to S}$$

Now let us transform this sentence into a dependent clause: *"the paper that John signed without reading"* (this phrase should be of type N, noun phrase). Notice that here we removed not only *"the paper,"* but also *"it,"* forming two gaps which should be filled with the same $!^{\mathsf{s}}N$. This phenomenon is called *parasitic extraction* and can be handled using dereliction, exchange and contraction:

$$\frac{\dfrac{\dfrac{\dfrac{N, N \backslash S / N, N, (N \backslash S) \backslash (N \backslash S) / GC, GC / N, N \to S}{N, N \backslash S / N, !^{\mathsf{s}}N, (N \backslash S) \backslash (N \backslash S) / GC, GC / N, !^{\mathsf{s}}N \to S}\,Der}{N, N \backslash S / N, (N \backslash S) \backslash (N \backslash S) / GC, GC / N, !^{\mathsf{s}}N \to S}\,C_L}{N, N \backslash S / N, (N \backslash S) \backslash (N \backslash S) / GC, GC / N \to S / !^{\mathsf{s}}N} \quad N / CN, CN, CN \backslash CN \to N}{N / CN, CN, (CN \backslash CN) / (S / !^{\mathsf{s}}N), N, N \backslash S / N, (N \backslash S) \backslash (N \backslash S) / GC, GC / N \to N}$$

Contraction can be used several times, generating examples like *"the paper that the editor of received, but left in the office without reading."*

Finally, the last example shows that *weakening* should not be allowed. Consider *"the girl whom John loves Mary."* This should not be a legal noun phrase, but can be derived using weakening:

$$\frac{\dfrac{\dfrac{N, N \backslash S / N, N \to S}{N, N \backslash S / N, N, !^{\mathsf{s}}N \to S}\,W_L}{N, N \backslash S / N, N \to S / !^{\mathsf{s}}N} \quad N / CN, CN, CN \backslash CN \to N}{N / CN, CN, (CN \backslash CN) / (S / !^{\mathsf{s}}N), N, N \backslash S / N, N \to N}$$

Thus, the subexponential used for type-logical grammar is a *relevant* one; in other words, $\mathsf{s} \in \mathcal{E}$, $\mathsf{s} \in \mathcal{C}$, $\mathsf{s} \notin \mathcal{W}$.

4 Treating Relevant Formulas as Unbounded Formulas

Given that contraction of relevant formulas involves "don't know nondeterminism", during proof search, we would like to postpone (from a bottom-up perspective) as much as possible the application of contraction of relevant formulas. The following lemma provides us with insight on which rules are problematic:

Lemma 1. *Contraction rules permute over all rules except rules* $\cdot_R, \backslash_L, /_L$ *and Der.*

For proof search, this means that for rules R other than $\cdot_R, \backslash_L, /_L$ and Der, it is safe to not contract relevant formulas. This is because from the lemma above, if there is a proof where a formula is contracted before the application of R, then there is also a proof where the formula is contracted after R.

However, the same is not the case for $\cdot_R, \backslash_L, /_L$ and Der. For example, it is not possible to permute contraction over $\backslash_L$ in the following derivation as the occurrences of $!^r F$ are split among the premises:

$$\frac{\dfrac{\Pi_1, !^r F, \Pi_2 \longrightarrow F_1 \qquad \Gamma_1, !^r F, \Gamma_2, F_2, \Gamma_3 \longrightarrow G}{\Gamma_1, !^r F, \Gamma_2, \Pi_1, !^r F, \Pi_2, F_1 \backslash F_2, \Gamma_3 \longrightarrow G} \backslash_L}{\Gamma_1, \Gamma_2, \Pi_1, !^r F, \Pi_2, F_1 \backslash F_2, \Gamma_3 \longrightarrow G} C_L$$

We analyse the rules $\cdot_R, \backslash_L, /_L$ and Der individually and investigate how to reduce don't know non-determinism.

Consider the following derivation to the left containing an instance of $\cdot_R$ rule where r is a relevant formula and the relevant formula $!^r H$ is moved to the right premise. The symmetric reasoning applies if $!^r H$ is moved to the left premise.

$$\frac{\Gamma_1 \to F \qquad \Gamma_2, !^r H, \Gamma_3 \to G}{\Gamma_1, \Gamma_2, !^r H, \Gamma_3 \to F \cdot G} \cdot_R \qquad\qquad \frac{\dfrac{\Gamma_1' \to F \qquad \Gamma_2, !^r H, \Gamma_3 \to G}{\Gamma_1', \Gamma_2, !^r H, \Gamma_3 \to F \cdot G} \cdot_R}{\Gamma_1, \Gamma_2, !^r H, \Gamma_3 \to F \cdot G} n \times C_L$$

As $!^r H$ cannot be weakened, it should be necessarily used in the right premise. That is, it behaves as a linear formula. How about the left premise? Since contraction is not local, it is possible to contract $!^r H$ as many times such that the contracted formulas are moved to the left premise. This means that during proof search, it is safe to consider the formula H unbounded in the left premise. If n copies of H are used in the proof of the left premise, where $n \geq 0$, we can contract it as illustrated by the derivation above to the right where Γ_1' contains the contracted occurrences of the formula $!^r H$.

Similarly, consider the following instance of $\backslash_L$ to the left where the relevant formula $!^r H$ is moved to the left premise. A symmetric observation can be carried out for $/_L$.

$$\frac{\Pi_1, !^r H, \Pi_2 \to F \qquad \Gamma_1, G, \Gamma_2 \to C}{\Gamma_1, \Pi_1, !^r H, \Pi_2 F \backslash G, \Gamma_2 \to C} \backslash_L \qquad\qquad \frac{\dfrac{\Pi_1, !^r H, \Pi_2, \to F \qquad \Gamma_1', G, \Gamma_2' \to C}{\Gamma_1', \Pi_1, !^r H, \Pi_2, F \backslash G, \Gamma_2' \to C} \backslash_L}{\Gamma_1, \Pi_1, !^r H, \Pi_2, F \backslash G, \Gamma_2 \to C} n \times C_L$$

As before, since $!^r H$ cannot be weakened, it should be necessarily used in the left premise. That is, it behaves like a linear non-commutative formula. By similar reasoning as for $\cdot$, we can treat this formula as unbounded in the right premise. Since contractions are non-local, we can copy $!^r H$ so that they are moved to the right premise as illustrated by the derivation above to the right where Γ_1', Γ_2' contain the contracted occurrences of the formula $!^r H$.

The same reasoning applies for relevant formulas moved to the right premise. It is safe to consider the formula H as unbounded in the left premise.

The leads to the our first key observation:

Key Observation 1: *During proof search, any relevant formula moved to one premise of $\cdot_R, \backslash_L, /_L$ can be considered unbounded in the other premise.*

Finally, consider the following instance of Der_L on a relevant formula:

$$\frac{\Gamma_1, H, \Gamma_2 \longrightarrow G}{\Gamma_1, !^r H, \Gamma_2 \longrightarrow G} \; Der$$

Applying the same reasoning as above, the formula $!^r H$ can be treated as unbounded as one can make as many copies as needed before the dereliction. This leads to the following key observation:

Key Observation 2: *During proof search, any relevant formula derelicted by Der can be considered unbounded in its premise.*

Example 1. Consider the derivation below left with the relevant formula $!^r A$:

$$\frac{\dfrac{\overline{!^r A \longrightarrow A}}{!^r A, A \backslash A' \longrightarrow A \cdot A' \cdot A} \; Der, I \quad A' \longrightarrow A \cdot A' \cdot A}{!^r A, A \backslash A' \longrightarrow A \cdot A' \cdot A} \; \backslash_L \qquad \frac{\dfrac{\overline{!^r A \longrightarrow A}}{!^r A, A \backslash A' \longrightarrow A \cdot A' \cdot A} \; Der, I \quad \underline{!^r A}, A' \longrightarrow A \cdot A' \cdot A}{!^r A, A \backslash A' \longrightarrow A \cdot A' \cdot A} \; \backslash_L$$

Following the Key Observation 1 above, as $!^r A$ is moved to the left premise, we can treat $!^r A$ as unbounded in the right premise. This is denoted by the formula $\underline{!^r A}$ as shown in the derivation to the right. We can now prove the right premise using $\underline{!^r A}$ as illustrated by the derivation Ξ below. (Recall unbounded formulas can be contracted safely):

$$\Xi = \frac{\dfrac{\overline{\underline{!^r A} \longrightarrow A}}{} \; Der, I \quad \dfrac{\dfrac{\overline{A' \longrightarrow A'}}{\underline{!^r A}, A' \longrightarrow A'} \; I}{\underline{!^r A}, A' \longrightarrow A'} \; W_L \quad \dfrac{\overline{\underline{!^r A} \longrightarrow A}}{} \; Der, I}{\underline{!^r A}, A' \longrightarrow A \cdot A' \cdot A} \; 2 \times \cdot_R$$

Notice that it may seem unsound to weaken $\underline{!^r A}$ in the middle branch. However, as we can control the number of times $!^r A$ is contracted, we can transform this derivation into a SNILL proof: In particular, we can infer from Ξ that we require two copies of $!^r A$. Thus the corresponding SNILL proof starts with two contractions:

$$\frac{\dfrac{\dfrac{\overline{!^r A \longrightarrow A}}{} \; Der, I \quad !^r A, A', !^r A \longrightarrow A \cdot A' \cdot A}{!^r A, !^r A, A \backslash A', !^r A \longrightarrow A \cdot A' \cdot A} \; \backslash_L}{!^r A, A \backslash A' \longrightarrow A \cdot A' \cdot A} \; 2 \times C_L$$

It remains to construct a proof based on Ξ.

Example 2. Given that we allow non-local contractions, one could expect that Key Observation 1 would also work for non-commutative relevant subexponentials s such that $s \in \mathcal{C}$ and $s \notin \mathcal{E} \cup \mathcal{W}$. However this is not true in general. Consider the following derivation where we attempt to use Key Observation 1, that is, where $!^s A$ is treated as an unbounded formula:

$$
\cfrac{
\cfrac{
\cfrac{!^s A, A_1, A_2 \longrightarrow A_1 \cdot A \cdot A_2}{!^s A, A_1 \cdot A_2 \longrightarrow A_1 \cdot A \cdot A_2}
}{!^s A \longrightarrow A \quad \cfrac{!^s A \longrightarrow (A_1 \cdot A_2 \,/\, A_1 \cdot A \cdot A_2)}{}}
}{!^s A \longrightarrow A \cdot (A_1 \cdot A_2 \,/\, A_1 \cdot A \cdot A_2)}
$$

In the open premise, it would be tempting to move $!^s A$ to the place between A_1 and A_2 and finish the "proof". However, the resulting derivation would not correspond to a valid SNILL proof as it is not possible to contract the original $!^s A$ so that it is placed exactly between A_1 and A_2. While we conjecture that this could be solved by also recalling the places where relevant formulas can be contracted, we leave this investigation for future work. Moreover, such non-commutative relevant formulas are not needed for our applications here.

5   Focused Proof System for SNILL

Logical frameworks are defined proof theoretically by a focused proof system. This section introduces the focused proof system SNILLF for SNILL. We prove that SNILLF is sound and complete with respect to SNILL.

First proposed by Andreoli [2] for Linear Logic, focused proof systems reduce proof search space by distinguishing rules which have don't know non-determinism, classified as *positive*, from rules which have don't care non-determinism, classified as *negative*. For SNILL, the rules $\cdot_R, \backslash_L, /_L, \forall_L$ are positive rules and the rules $\cdot_L, \backslash_R, /_R, \forall_R$ are negative. Formulas of the form $F \cdot G$ and $!^s F$ and 1 are classified as positive while the remaining formulas as negative.

SNILLF sequents are constructed using the following four types of contexts:

- **Commutative Contexts ($\mathcal{K}$):** A commutative context $\mathcal{K}$ maps a commutative subexponentials $s \in \mathcal{E}$ to a set of formulas if $s \in \mathcal{W} \cap \mathcal{C}$, that is, it is unbounded, and to a multiset of formula otherwise. Intutively, such a context $\mathcal{K}$ denotes the formulas: $\mathcal{K}[s_1], \mathcal{K}[s_2], \ldots, \mathcal{K}[s_n]$ where $\{s_1, \ldots, s_n\} = \mathcal{E}$;
- **Unrestricted Relevant Context ($\mathcal{R}^u$):** An unrestricted context $\mathcal{R}^u$ maps relevant subexponentials $r \in \mathcal{C}$ and $r \notin \mathcal{W}$ to sets of formulas. Intuitively, this context stores the relevant formulas which can be treated as unbounded. Using the notation in Sect. 4, $\mathcal{R}^u$ represents the formulas $\mathcal{R}^u[r_1], \ldots, \mathcal{R}^u[r_n]$, where $\{r_1, \ldots, r_n\}$ is the set of all relevant subexponentials;
- **Subexponential Boxes:** $[F_1, \ldots, F_k]_s$ where $s \notin \mathcal{E}$ and $F_1, \ldots, F_k$ is a list, not a multiset, of formulas. This box should be interpreted as the list of formulas $!^s F_1, \ldots, !^s F_k$;

- **Unmarked Boxes:** $[F_1, \ldots, F_k \Uparrow G_1, \ldots, G_m]$, where $F_1, \ldots, F_k$ and $G_1, \ldots, G_m$ are both lists, not multisets, of formulas. This box should be interpreted as the list of formulas $F_1, \ldots, F_k, G_1, \ldots, G_m$. When $m = 0$, we write such box as $[F_1, \ldots, F_k]_\star$.

We use $\mathcal{NC}$ and its variants to denote a sequence of boxed formulas (Subexponential Boxes and Unmarked Boxes). We write $\mathcal{NC}^\star$ whenever all unmarked boxes are of the form $[F_1, \ldots, F_k]_\star$. We define the set $\mathcal{NC}[\mathsf{s}] = \{F \mid [\Gamma_1, F, \Gamma_2]_\mathsf{s} \in \mathcal{NC}\}$. Also, if $\mathcal{NC}_1 = [\Gamma_1]_{\mathsf{s}_1} \cdots [\Gamma]_{\mathsf{s}_i}$ and $\mathcal{NC}_2 = [\Delta]_{\mathsf{s}_i} \cdots [\Gamma]_{\mathsf{s}_n}$, then $\mathcal{NC}_1 \cdot \mathcal{NC}_2$ is defined to be $[\Gamma_1]_{\mathsf{s}_1} \cdots [\Gamma_i, \Delta]_{\mathsf{s}_i} \cdots [\Gamma_n]_{\mathsf{s}_n}$. Empty boxes $[\cdot]_\mathsf{s}, [\cdot]_\star$ are always elided. These also act as identity elements, that is $[F_1, \ldots, F_n]_\mathsf{s} \cdot []_\mathsf{s} = [F_1, \ldots, F_n]_\mathsf{s}$ and similarly for unmarked boxes. Finally, we define the following auxiliary operations on commutative contexts:

$$\mathcal{K}[\mathcal{S}] = \bigcup_{\mathsf{s} \in \mathcal{S}} \mathcal{K}[\mathsf{s}] \qquad (\mathcal{K} +_\mathsf{s} F)[\mathsf{s}'] = \begin{cases} \mathcal{K}[\mathsf{s}'] \uplus \{F\} & \text{if } \mathsf{s}' = \mathsf{s} \\ \mathcal{K}[\mathsf{s}'] & \text{otherwise} \end{cases}$$

$$(\mathcal{K}_1 \otimes \mathcal{K}_2)[\mathsf{s}] = \begin{cases} \mathcal{K}_1[\mathsf{s}] \uplus \mathcal{K}_2[\mathsf{s}] & \text{if } \mathsf{s} \notin \mathcal{W} \cap \mathcal{C} \\ \mathcal{K}_1[\mathsf{s}] & \text{otherwise} \end{cases} \qquad \mathcal{K} \leq_\mathsf{s} = \begin{cases} \mathcal{K}[\mathsf{s}_1] & \text{if } \mathsf{s} \preceq \mathsf{s}_1 \\ \emptyset & \text{otherwise} \end{cases}$$

$(\mathcal{K}_1 \star \mathcal{K}_2) \mid_{\mathcal{S}}$ is true if and only if for all $\mathsf{s} \in \mathcal{S}, \mathcal{K}_1[\mathsf{s}] \star \mathcal{K}_2[\mathsf{s}]$, for $\star \in \{\subset, \subseteq, =\}$

Similar operations are also defined *(mutatis mutandis)* for Unrestricted Relevant Contexts ($\mathcal{R}^u$). These operations are similar to the ones proposed in [23] used in the formalization of the side conditions of the rules for proof systems with subexponentials.

The rules for the focused proof system SNILLF for SNILL are depicted in Fig. 2. They contain the following types of sequents:

- **Negative:** $\mathcal{K} : \mathcal{R}^u : \mathcal{NC}_1, [\Delta \Uparrow \Gamma], \mathcal{NC}_2 \longrightarrow \mathcal{G}$ and $\mathcal{K} : \mathcal{R}^u : \mathcal{NC} \longrightarrow [\Uparrow F]$. Here $\mathcal{G}$ can be either $[\Uparrow F]$ or $[F]$. Moreover, Γ, Δ are lists of formulas.
- **Positive:** $\mathcal{K} : \mathcal{R}^u : \mathcal{NC}^\star \longrightarrow [\Downarrow F]$ and $\mathcal{K} : \mathcal{R}^u : \mathcal{NC}_1^\star [\Downarrow F] \mathcal{NC}_2^\star \longrightarrow [G]_\mathsf{s}$. In the former, the formula F on the r.h.s. is focused on and the latter on the l.h.s.;
- **Decision:** $\mathcal{K} : \mathcal{R}^u : \mathcal{NC}^\star \longrightarrow [G]$: Sequents at the border of negative and positive phases.

During the negative phase, formulas (Δ) to the right of Unmarked Boxes ($[\Gamma \Uparrow \Delta]$) are introduced or moved to the left (Γ) or to other contexts using the Reaction rules $\Uparrow_L, \Uparrow_R$. Notice the negative rule $!^{ne}$. There since the formulas Δ are all not marked with subexponentials, the rule creates a new box $[\Delta]_\star$.

Once a negative phase ends, that is, all unmarked boxes are of the form $[\Gamma]_\star$, one should decide in a formula to focus on using one of the Decide Rules. Decide rules implicitly apply the Dereliction rule whenever applicable. The rules $D_\mathsf{u}, D_{\mathsf{nc}}, D_\mathsf{r}$ choose a formula marked with a subexponential for which exchange rule applies. Therefore, one can place F any where in the context. This D_{nc} which forces the formula F to be where it is. It also causes the box where the formula is to be split. Finally, notice that if an unbounded formula is focused on then it is contracted (as in Andreoli's original system). Moreover following

Negative Phase

$$\frac{\mathcal{K} : \mathcal{R}^u : NC_1, [\Delta \Uparrow F_1, F_2, \Gamma], NC_2 \longrightarrow \mathcal{G}}{\mathcal{K} : \mathcal{R}^u : NC_1, [\Delta \Uparrow F_1 \cdot F_2, \Gamma], NC_2 \longrightarrow \mathcal{G}} \cdot_L \qquad \frac{\mathcal{K} : \mathcal{R}^u : NC_1, [\Delta \Uparrow \Gamma], NC_2 \longrightarrow \mathcal{G}}{\mathcal{K} : \mathcal{R}^u : NC_1, [\Delta \Uparrow 1, \Gamma], NC_2 \longrightarrow \mathcal{G}} 1_L$$

$$\frac{\mathcal{K} : \mathcal{R}^u : NC [\cdot \Uparrow F] \longrightarrow [\Uparrow G]}{\mathcal{K} : \mathcal{R}^u : NC \longrightarrow [\Uparrow G / F]} /_R \qquad \frac{\mathcal{K} : \mathcal{R}^u : [\cdot \Uparrow F] NC \longrightarrow [\Uparrow G]}{\mathcal{K} : \mathcal{R}^u : NC \longrightarrow [\Uparrow F / G]} \backslash_R \qquad \frac{\mathcal{K} : \mathcal{R}^u : NC \longrightarrow [\Uparrow F\{x/e\}]}{\mathcal{K} : \mathcal{R}^u : NC \longrightarrow [\Uparrow \forall x.F]} \forall_R$$

$$\frac{\mathcal{K} +_e F : \mathcal{R}^u : NC_1, [\Delta \Uparrow \Gamma], NC_2 \longrightarrow \mathcal{G}}{\mathcal{K} : \mathcal{R}^u : NC_1, [\Delta \Uparrow !^e F, \Gamma], NC_2 \longrightarrow \mathcal{G}} !^e \qquad \frac{\mathcal{K} : \mathcal{R}^u : NC_1, [\Delta]_\star [F]_{\mathsf{ne}} [\Uparrow \Gamma], NC_2 \longrightarrow \mathcal{G}}{\mathcal{K} : \mathcal{R}^u : NC_1, [\Delta \Uparrow !^{\mathsf{ne}} F, \Gamma], NC_2 \longrightarrow \mathcal{G}} !^{\mathsf{ne}}$$

Positive Phase

$$\frac{\mathcal{K}_1 : \mathcal{R}^u \otimes \mathcal{R}_1 : NC_2^\star \longrightarrow [\Downarrow F] \quad \mathcal{K}_2 : \mathcal{R}^u \otimes \mathcal{R}_2 : NC_1^\star [\Downarrow G] NC_3^\star \longrightarrow [H]}{\mathcal{K}_1 \otimes \mathcal{K}_2 : \mathcal{R}^u : NC_1^\star \cdot NC_2^\star [\Downarrow F \backslash G] NC_3^\star \longrightarrow [H]} \backslash_L$$

$$\frac{\mathcal{K}_1 : \mathcal{R}^u \otimes \mathcal{R}_1 : NC_2^\star \longrightarrow [\Downarrow G] \quad \mathcal{K}_2 : \mathcal{R}^u \otimes \mathcal{R}_2 : NC_1^\star [\Downarrow F] NC_3^\star \longrightarrow [H]}{\mathcal{K}_1 \otimes \mathcal{K}_2 : \mathcal{R}^u : NC_1^\star [\Downarrow F / G] NC_2^\star \cdot NC_3^\star \longrightarrow [H]} /_L$$

where $\mathcal{R}_1[r] = \mathcal{K}_2[r]$ and $\mathcal{R}_2[r] = \mathcal{K}_1[r]$ for all $r \in C$ and $r \notin \mathcal{W}$.

$$\frac{\mathcal{K}_1 : \mathcal{R}^u \otimes \mathcal{R}_1 : NC_1^\star \longrightarrow [\Downarrow F] \quad \mathcal{K}_2 : \mathcal{R}^u \otimes \mathcal{R}_2 : NC_2^\star \longrightarrow [\Downarrow G]}{\mathcal{K}_1 \otimes \mathcal{K}_2 : \mathcal{R}^u : NC_1^\star \cdot NC_2^\star \longrightarrow [\Downarrow F \cdot G]} \cdot_R$$

where $\mathcal{R}_1[r] = \mathcal{K}_2[r]$ and $\mathcal{R}_2[r] = \mathcal{K}_1[r]$ for all $r \in C$ and $r \notin \mathcal{W}$.

$$\frac{}{\mathcal{K} : \mathcal{R}^u : \cdot \longrightarrow [\Downarrow 1]} 1_R \qquad \frac{}{\mathcal{K} : \mathcal{R}^u : [\Downarrow A] \longrightarrow [A]} I \quad \text{where } \mathcal{K}[\mathsf{s}] = \emptyset \text{ for all } \mathsf{s} \notin \mathcal{W}$$

$$\frac{\mathcal{K} : \mathcal{R}^u : NC_1^\star [F\{t/x\}] NC_2^\star \longrightarrow [H]}{\mathcal{K} : \mathcal{R}^u : NC_1^\star [\Downarrow \forall x.F] NC_2^\star \longrightarrow [H]}$$

$$\frac{\mathcal{K} \leq_\mathsf{s} : \mathcal{R}^u \leq_\mathsf{s} : NC^\star \longrightarrow [\Uparrow F]}{\mathcal{K} : \mathcal{R}^u : NC^\star \longrightarrow [\Downarrow !^\mathsf{s} F]} !^\mathsf{s}{}_R, \text{ if } \mathcal{K}[x] = \emptyset = NC^\star[x] \text{ for all } \mathsf{s} \not\leq x$$

Decide Rules

$$\frac{\mathcal{K} +_u F : \mathcal{R}^u : NC^\star [\Gamma_1]_\mathsf{s} [\Downarrow F] [\Gamma_2]_\mathsf{s} NC_2^\star \longrightarrow [G]}{\mathcal{K} +_u F : \mathcal{R}^u : NC^\star [\Gamma_1, \Gamma_2]_\mathsf{s} NC_2^\star \longrightarrow [G]} D_u \qquad \frac{\mathcal{K} : \mathcal{R}^u : NC^\star [\Gamma_1]_\mathsf{s} [\Downarrow F] [\Gamma_2]_\mathsf{s} NC_2^\star \longrightarrow [G]}{\mathcal{K} +_{\mathsf{nc}} F : \mathcal{R}^u : NC^\star [\Gamma_1, \Gamma_2]_\mathsf{s} NC_2^\star \longrightarrow [G]} D_{\mathsf{nc}}$$

$$\frac{\mathcal{K} : \mathcal{R}^u +_r F : NC^\star [\Gamma_1]_\mathsf{s} [\Downarrow F] [\Gamma_2]_\mathsf{s} NC_2^\star \longrightarrow [G]}{\mathcal{K} +_r F : \mathcal{R}^u : NC^\star [\Gamma_1, \Gamma_2]_\mathsf{s} NC_2^\star \longrightarrow [G]} D_r$$

$$\frac{\mathcal{K} : \mathcal{R}^u : NC^\star [\Gamma_1]_\mathsf{s} [\Downarrow F] [\Gamma_2]_\mathsf{s} NC_2^\star \longrightarrow [G]}{\mathcal{K} : \mathcal{R}^u : NC^\star [\Gamma_1, F, \Gamma_2]_\mathsf{s} NC_2^\star \longrightarrow [G]} D_\mathsf{s} \qquad \frac{\mathcal{K} : \mathcal{R}^u : NC^\star \longrightarrow [\Downarrow G]}{\mathcal{K} : \mathcal{R}^u : NC^\star \longrightarrow [G]} D_R$$

Reaction Rules

$$\frac{\mathcal{K} : \mathcal{R}^u : NC^\star [\cdot \Uparrow P] NC_2^\star \longrightarrow [G]}{\mathcal{K} : \mathcal{R}^u : NC^\star [\Downarrow P] NC_2^\star \longrightarrow [G]} R_L \qquad \frac{\mathcal{K} : \mathcal{R}^u : NC^\star \longrightarrow [\Uparrow N_a]}{\mathcal{K} : \mathcal{R}^u : NC^\star \longrightarrow [\Downarrow N_a]} R_R$$

$$\frac{\mathcal{K} : \mathcal{R}^u : NC [\Delta, P_a :\Uparrow \Gamma] NC_2 \longrightarrow \mathcal{G}}{\mathcal{K} : \mathcal{R}^u : NC [\Delta :\Uparrow P_a, \Gamma] NC_2 \longrightarrow \mathcal{G}} \Uparrow_L \qquad \frac{\mathcal{K} : \mathcal{R}^u : NC \longrightarrow [P_a]}{\mathcal{K} : \mathcal{R}^u : NC \longrightarrow [\Uparrow P_a]} \Uparrow_R$$

Fig. 2. SNILLF: focused proof system for SNILL. Here P is a positive formula; N_a is a negative or atomic formula; P_a is a positive or atomic formula; e is a fresh eigenvariable, not appearing in $\mathcal{K}, \mathcal{R}^u, NC, F$; $e \in \mathcal{E}$; $\mathsf{ne} \notin \mathcal{E}$; $u \in \mathcal{W} \cap \mathcal{C} \cap \mathcal{E}$; $\mathsf{nc} \notin \mathcal{C}$; $r \in \mathcal{C}$ and $r \notin \mathcal{W}$.

Key Observation 2 described Sect. 4, whenever a relevant formula is added to the context $\mathcal{R}^u$ and is treated as an unbouded formula.

In the positive phase, one can only introduce the formula that is focused on. The rules $\backslash_L, /_L, \cdot_R$ implement the Key Observation 1 described in Sect. 4. That is, all relevant formula moved to one premise are added to the $\mathcal{R}^u$ context of the other premise and treated as unbounded formulas in that premise. This is specified by the side conditions of that rule.

For soundness of SNILLF with respect to SNILL, we rely on the transformations described in Sect. 4, namely, that is sound to consider relevant formulas as unbounded in some premises. Given this result, soundness just amounts to erasing the focusing annotations and replacing contexts by formulas. For completeness of focusing, we use the modular technique proposed in [14] based on the following permutation lemmas. Lemma 2 justifies the eager application of negative rules (negative phase). Lemma 3 justifies the preservation of focusing in the positive phase.

Lemma 2. *All positive rules permute over all negative rules.*

Lemma 3. *All positive rules permute over all positive rules.*

Theorem 2. *Let $\Sigma = \langle \mathcal{I}, \preceq, \mathcal{W}, \mathcal{C}, \mathcal{E} \rangle$ be a subexponential signature with $\mathcal{C}, \mathcal{W} \subseteq \mathcal{E}$. Let $\mathcal{K}_\emptyset$ and $\mathcal{R}^u_\emptyset$ be the empty contexts, that is, $\mathcal{K}[\mathsf{s}] = \mathcal{R}^u[\mathsf{s}] = \emptyset$ for all s. For any subexponential signature, the sequent $\Gamma \longrightarrow G$ is provable in SNILL$_\Sigma$ if and only if the sequent $\mathcal{K}_\emptyset : \mathcal{R}^u_\emptyset : [\cdot \Uparrow \Gamma] \longrightarrow [\Uparrow G]$ is provable in SNILLF$_\Sigma$.*

6 Applications

We illustrate the power of SNILLF by revisiting the examples described in Sect. 3.

6.1 Distributed Systems

Assume a subexponential signature $\Sigma = \langle \mathcal{I}, \preceq, \mathcal{W}, \mathcal{C}, \mathcal{E} \rangle$ where $\mathcal{I} = \{\mathsf{u}, \mathsf{N}, \mathsf{m}_1, \ldots, \mathsf{m}_n\}$, $\preceq$ is the reflexive relation, that is $i \preceq j$, then $i = j, \mathcal{E} = \{u, \mathsf{N}\}$ and $\mathcal{C} = \mathcal{W} = \{u\}$. Intuitively, we use the subexponential m_i to specify machine m_i's buffer, N to specify the messages sent on the network and u to specify the behavior of the system. Notice that as there are no relevant formulas $\mathcal{R}^u$ is always empty and therefore elided.

A buffer at machine m_i with elements Γ_i is specified as the list of formulas where start and end mark the start and end of the list $[\mathsf{start}, \Gamma_i, \mathsf{end}]_{\mathsf{m}i}$. Thus a system with n machines is specified as the collection of formulas:

$$\mathcal{NC} = [\mathsf{start}, \Gamma_1, \mathsf{end}]_{\mathsf{m}1} \, [\mathsf{start}, \Gamma_2, \mathsf{end}]_{\mathsf{m}2} \, \cdots \, [\mathsf{start}, \Gamma_n, \mathsf{end}]_{\mathsf{m}n}$$

For a better presentation, instead of using the context $\mathcal{K}$, we show the formulas in the sequent explicitly where $\mathcal{K}[\mathsf{u}] = \mathcal{U}$ and $\mathcal{K}[\mathsf{N}] = \mathcal{N}$:

$$\mathcal{U} : \mathcal{N} : \mathcal{NC} \longrightarrow \mathcal{G}$$

Notice that since buffers are lists of formulas, we use non-commutative subexponentials to specify them. However, messages on the network are not necessarily delivered in a particular order. Moreover, messages should be consumed exactly once. Therefore, we use the commutative subexponential N to mark these messages.

We now describe how to specify the transmission of messages between machines. For our example, assume two collections of messages syn_{mj}, ack_{mj} specifying, respectively, a synchronization message from mj and an acknowledgement message to mj. Whenever a machine mi processes the message syn_{mj}, it sends the message ack_{mj} to mj.

The following two clauses specifies this behavior:

$$Deq(i,j) = !^{mi}syn_{mj} \cdot !^{mi}end \setminus !^{mi}end \cdot !^{N} ack_{mj}$$
$$Enq(i,j) = !^{mj}start \cdot !^{mj}ack_{mj} / !^{N}ack_{mj} \cdot !^{mj}start$$

$Deq(i,j)$ specifies the processing of syn_{mj} sending ack_{mj} to the network and $Enq(i,j)$ the receival of ack_{mj}.

The correctness of this encoding can be easily visualized using focusing. Consider two machines $1, 2$. The focused derivation introducing $Deq = Deq(1,2)$ is necessarily of the following form where $\mathcal{M}_2 = [start, \Gamma_2, end]_{m2}$ and $\Theta = Deq(1,2), Enq(1,2)$:

$$\cfrac{\cfrac{\Theta : \mathcal{N}, ack_{m2} : [start, \Gamma_1, end]_{m1} \; \mathcal{M}_2 \longrightarrow [G]}{\Theta : \mathcal{N} : [start, \Gamma_1]_{m1}, [\Uparrow !^{m1}end \cdot !^{N} \; ack_{m2}] \; \mathcal{M}_2 \longrightarrow [G]}}{\cfrac{\Theta : \cdot : [syn_{m2}, end]_{m1} \longrightarrow [\Downarrow !^{m1}syn_{m2} \cdot !^{m1}end] \quad \Theta : \mathcal{N} : [start, \Gamma_1]_{m1}, [\Downarrow !^{m1}end \cdot !^{N} \; ack_{m2}] \; \mathcal{M}_2 \longrightarrow [G]}{\cfrac{\Theta : \mathcal{N} : [start, \Gamma_1, syn_{m2}, end]_{m1}, [\Downarrow Deq] \; \mathcal{M}_2 \longrightarrow [G]}{\Theta : \mathcal{N} : [start, \Gamma_1, syn_{m2}, end]_{m1}, \mathcal{M}_2 \longrightarrow [G]}}}$$

Notice that the messages in the network $\mathcal{N}$ are necessarily moved to the right premise, i.e., no message is lost. Otherwise, the introduction of $!^{m1}$ to the left would fail since N does not allow weakening and $m1 \not\preceq N$. Moreover, notice that Deq can only be focused on at the location shown above (to the left of $\mathcal{M}_2$). Otherwise, the formula $!^{m1}end$ would not be provable: if it is focused not adjacent to a end atom then it would not be provable, and if it is focused to the right of $\mathcal{M}_2$, then one could not introduce $!^{m1}$. Finally, the message syn_{m2} should necessarily appear at the end m1's buffer.

A similar exercise can be carried out when focusing on $Enq = Enq(1,2)$. In this case, the message ack_{m2} should be necessarily in $\mathcal{N}$ and moreover, an element is added to the beginning of the buffer of m2. The corresponding derivation is elided.

6.2   Type-Logical Grammar

We return to the sentence *"the paper that John signed without reading"* described in Sect. 3. The focused proof system SNILLF considerably reduces the proof search space for validating this sentence. Assume just a single relevant subexponential r. The corresponding focused proof is as follows where $\Gamma = CN, (CN \setminus CN)/(S/!^r N), \Gamma_1$ and $\Gamma_1 = N, N \setminus S/N, (N \setminus S)\setminus(N \setminus S)/GC, GC/N$. Moreover, we write explicitly the elements of $\mathcal{K}$ and $\mathcal{R}^u$ as in the previous section.

$$
\cfrac{
\cfrac{}{\cdots : [\Downarrow N] \to [N]} \, I \quad
\cfrac{
\cfrac{
\cdots : [\Gamma_1]_\star \to [\Downarrow S/!^r N] \quad \overline{\cdots : [CN]_\star \, [\Downarrow CN \setminus CN] \to [CN]}
}{
\cdots : [CN]_\star \, [\Downarrow (CN \setminus CN)/(S/!^s N)] \, [\Gamma_1]_\star \to [CN]
} \, D_L
}{
\cfrac{\cdots : [\Gamma]_\star \to [CN]}{\cdots : [\Downarrow N \setminus CN] \, [\Gamma]_\star \to [N]} \, /_L
} \, /_L
}{
\cdots : [\Uparrow N \setminus CN, \Gamma] \to [N]
} \, 7\times \Uparrow_L, D_L
$$

Continuing the left premise, we obtain the following derivation, we release focus and apply $/_R$. At this point, the relevant formula $!^r N$ is moved to the commutative context:

$$
\cfrac{
\cfrac{
\cfrac{}{N : \cdots : [GC/N] \to [GC]} \quad
\cfrac{
\overline{\cdot : \underline{N} : [N \setminus S/N]_\star \to [\Downarrow (N \setminus S)]} \quad \overline{\cdot : \underline{N} : [N]_\star [\Downarrow N \setminus S] \to [S]}
}{
\cdot : \underline{N} : [N, N \setminus S/N]_\star \, [\Downarrow (N \setminus S)\setminus(N \setminus S)] \to [S]
} \, 2\times/_L
}{
\cfrac{
N : \cdots : [N, N \setminus S/N]_\star \, [\Downarrow (N \setminus S)\setminus(N \setminus S)/GC] \, [GC/N]_\star \to [S]
}{
N : \cdots : [\Gamma_1]_\star \to [S]
} \, D_L
}{
\cdots : [\Gamma_1]_\star \, [\Uparrow !^r N] \to [S]
} \, !^r_L
$$

When compared to the derivation in Sect. 3, focusing reduces proof search in two different ways. First, the proof follows a "back-chaining" strategy [8]. This means that one decides on a formula that can immediately prove the goal. For example, decide on the formula $N \setminus GC$. Search fails immediately if one decides on other formulas. The second way is on deciding when to contract the formula $!^r N$. Indeed, in the derivation above, when the formula N is moved to the leftmost branch, it is treated as unbounded in the remaining two branches. This means that one can freely use it as in the middle branch or not as in the right branch.

7 Related Work

Logical Frameworks. When compared to existing logical frameworks, SNILLF has an increased expressiveness. When compared to Intuitionistic Linear Logical (ILL) Frameworks [8,33], SNILLF also allows ordered and relevant formulas. It also seem possible to encode Ordered Logical Frameworks [29,30] in SNILLF. In particular, one should only consider three subexponentials, one unbounded, one linear (or affine) and another non-commutative. The resulting system behaves similarly to Ordered Logical Frameworks. Moreover, ILL frameworks with subexponentials do not consider relevant formulas. It seems possible to apply the ideas here for reducing "don't know non-determinism" in the same way as done here. A proof of the focusing completeness theorem for the ordered logic [29] is detailed in the technical report [32]. We believe that work could also be extended to prove the completeness of SNILLF.

Finally, as SNILLF is intuitionistic, it cannot be directly compared to classical logical frameworks such as Forum [13] and Classical Linear Logic with Subexponentials [21]. We leave the proposal of a classical version of SNILLF to future work.

Type-Logical Grammar. A structural modality closely related to the relevant subexponential discussed above is used in the *CatLog* theorem prover and type-logical grammar parser, which is an ongoing project of Glyn Morrill and his group in Barcelona [17,18]. The difference of the calculus used in CatLog in comparison to our system is the use of *bracket modalities* that introduce controlled non-associativity and also interact with the relevant subexponential in a non-trivial fashion (see [19] for more details). Bracket modalities are used to block unwanted derivations like *"the girl whom John loves Mary and Pete loves"* or *"the paper that John signed the article without reading."* (Both examples are incorrect from the point of view of English grammar, but accepted by the grammar discussed above.) As shown by Kanovich *et al.* [9], the derivability problem for the Lambek calculus with bracket and subexponential modalities is undecidable. There exists, however, a natural decidable fragment, which is actually used in CatLog. This fragment belongs to the NP class, and CatLog utilises several techniques and heuristics in order to speed-up the parsing procedure. In particular, it uses count-invariants for pruning proof search [11] (which generalise multiplicative count-invariants by van Benthem [4]) and focusing for reducing spurious ambiguity. For the multiplicative-additive fragment focusing for the system used in CatLog is discussed in detail in [20]; completeness of focusing for the full set of connectives used in CatLog, including subexponential, is left by Morrill as a topic for further research [18].

There also exist other type-logical grammar frameworks based on different variants of the Lambek calculus. A notable one is the *Grail* system developed by Moot [16] on the basis of Moortgat's *multi-modal* extension of the non-associative Lambek calculus [15]. Like the subexponential extension of the Lambek calculus discussed in this paper, Moortgat's system uses an indexed family of structural connectives.

8 Conclusions

This paper introduced the logical framework SNILLF which allows for both commutative and non-commutative subexponentials. We demonstrate the power of SNILLF by specifying the structural semantics of distributed systems with buffers and specifying type-logical grammars. For the latter, SNILLF uses commutative relevant formulas, that is, formulas $!^s F$ that can contract, but not weaken. We investigate the proof theory of such formulas in order to reduce "don't know non-determinism" involved demonstrating that under some conditions, these formulas can be treated as unbouded. We believe that this paper lays the foundations for the development of concrete systems for, *e.g.*, type-logical grammars.

We are currently investigating a number of future work directions. We intend to investigate through prototype implementations the impact of SNILLF for categorial parsers. Such an implementation will help us investigate possible further uses of subexponentials for capturing other grammatical constructions. From the proof theory, we are investigating how to reduce the "don't know non-determism" of non-commutative relevant formulas. We are also investigating classical versions for SNILLF following our previous work [10].

Acknowledgements. We are grateful to Glyn Morrill, Frank Pfenning, and the anonymous referees.

Financial Support: The work of Max Kanovich and Andre Scedrov was supported by the Russian Science Foundation under grant 17-11-01294 and performed at National Research University Higher School of Economics, Moscow, Russia. The work of Stepan Kuznetsov was supported by the Young Russian Mathematics award, by the Program of the Presidium of the Russian Academy of Sciences No. 01 'Fundamental Mathematics and Its Applications' under grant PRAS-18-01, and by the Russian Foundation for Basic Research grant 18-01-00822. The work of Vivek Nigam was supported by CNPq grant number 304193/2015-1. Sections 1, 2, 3, 7 and 8 were contributed jointly and equally by all co-authors; Sect. 4 was contributed by Scedrov and Kanovich. Section 5 was contributed by Nigam. Section 6 was contributed by Kuznetsov.

References

1. Ajdukiewicz, K.: Die syntaktische Konnexität. Studia Philosophica **1**, 1–27 (1935)
2. Andreoli, J.-M.: Logic programming with focusing proofs in linear logic. J. Logic Comput. **2**(3), 297–347 (1992)
3. Bar-Hillel, Y.: A quasi-arithmetical notation for syntactic description. Language **29**, 47–58 (1953)
4. van Benthem, J.: Language in Action: Categories, Lambdas and Dynamic Logic. Elsevier, North Holland (1991)
5. Danos, V., Joinet, J.-B., Schellinx, H.: The structure of exponentials: uncovering the dynamics of linear logic proofs. In: Gödel, K. (ed.) Colloquium, pp. 159–171 (1993)
6. Girard, J.-Y.: Linear logic. Theor. Comput. Sci. **50**, 1–102 (1987)
7. Harper, R., Honsell, F., Plotkin, G.D.: A framework for defining logics. In: LICS (1987)
8. Hodas, J.S., Miller, D.: Logic programming in a fragment of intuitionistic linear logic: extended abstract. In: LICS (1991)
9. Kanovich, M., Kuznetsov, S., Scedrov, A.: Undecidability of the Lambek calculus with subexponential and bracket modalities. In: Klasing, R., Zeitoun, M. (eds.) FCT 2017. LNCS, vol. 10472, pp. 326–340. Springer, Heidelberg (2017). https://doi.org/10.1007/978-3-662-55751-8_26
10. Kanovich, M., Kuznetsov, S., Nigam, V., Scedrov, A.: Subexponentials in noncommutative linear logic. Math. Struct. Comput. Sci., FirstView, 1–33 (2018). https://doi.org/10.1017/S0960129518000117
11. Kuznetsov, S., Morrill, G., Valentín, O.: Count-invariance including exponentials. In: Proceedings of MoL 2017, volume W17–3413 of ACL Anthology, pp. 128–139 (2017)

12. Lambek, J.: The mathematics of sentence structure. Amer. Math. Mon. **65**, 154–170 (1958)
13. Miller, D.: Forum: a multiple-conclusion specification logic. Theor. Comput. Sci. **165**(1), 201–232 (1996)
14. Miller, D., Saurin, A.: From proofs to focused proofs: a modular proof of focalization in linear logic. In: CSL, pp. 405–419 (2007)
15. Moortgat, M.: Multimodal linguistic inference. J. Logic Lang. Inf. **5**(3–4), 349–385 (1996)
16. Moot, R.: The grail theorem prover: type theory for syntax and semantics. In: Chatzikyriakidis, S., Luo, Z. (eds.) Modern Perspectives in Type-Theoretical Semantics. SLP, vol. 98, pp. 247–277. Springer, Cham (2017). https://doi.org/10.1007/978-3-319-50422-3_10
17. Morrill, G.: CatLog: a categorial parser/theorem-prover. In: LACL System demostration (2012)
18. Morrill, G.: Parsing logical grammar: CatLog3. In: Proceedings of LACompLing (2017)
19. Morrill, G., Valentín, O.: Computation coverage of TLG: nonlinearity. In: NLCS (2015)
20. Morrill, G., Valentín, O.: Multiplicative-additive focusing for parsing as deduction. In: First International Workshop on Focusing (2015)
21. Nigam, V.: Exploiting non-canonicity in the sequent calculus. Ph.D. thesis (2009)
22. Nigam, V.: A framework for linear authorization logics. TCS **536**, 21–41 (2014)
23. Nigam, V., Miller, D.: Algorithmic specifications in linear logic with subexponentials. In: PPDP, pp. 129–140 (2009)
24. Nigam, V., Miller, D.: A framework for proof systems. J. Autom. Reasoning **45**(2), 157–188 (2010)
25. Nigam, V., Olarte, C., Pimentel, E.: A general proof system for modalities in concurrent constraint programming. In: CONCUR (2013)
26. Nigam, V., Pimentel, E., Reis, G.: An extended framework for specifying and reasoning about proof systems. J. Logic Comput. **26**(2), 539–576 (2016)
27. Olarte, C., Pimentel, E., Nigam, V.: Subexponential concurrent constraint programming. Theor. Comput. Sci. **606**, 98–120 (2015)
28. Pentus, M.: Lambek grammars are context-free. In: LICS, pp. 429–433 (1993)
29. Pfenning, F., Simmons, R.J.: Substructural operational semantics as ordered logic programming. In: LICS, pp. 101–110 (2009)
30. Polakow, J.: Linear logic programming with an ordered context. In: PPDP (2000)
31. Shieber, S.M.: Evidence against the context-freeness of natural languages. Linguist. Philos. **8**, 333–343 (1985)
32. Simmons, R.J., Pfenning, F.: Weak focusing for ordered linear logic. Technical report CMU-CS-10-147 (2011)
33. Watkins, K., Cervesato, I., Pfenning, F., Walker, D.: A concurrent logical framework: the propositional fragment. In: Berardi, S., Coppo, M., Damiani, F. (eds.) TYPES 2003. LNCS, vol. 3085, pp. 355–377. Springer, Heidelberg (2004). https://doi.org/10.1007/978-3-540-24849-1_23

Exploring Approximations for Floating-Point Arithmetic Using **UppSAT**

Aleksandar Zeljić[1]([✉]), Peter Backeman[1], Christoph M. Wintersteiger[2], and Philipp Rümmer[1]

[1] Uppsala University, Uppsala, Sweden
aleksandar.zeljic@it.uu.se
[2] Microsoft Research, Cambridge, UK

Abstract. We consider the problem of solving floating-point constraints obtained from software verification. We present UppSAT—an new implementation of a systematic approximation refinement framework [21] as an abstract SMT solver. Provided with an approximation and a decision procedure (implemented in an off-the-shelf SMT solver), UppSAT yields an approximating SMT solver. Additionally, UppSAT includes a library of predefined approximation components which can be combined and extended to define new encodings, orderings and solving strategies. We propose that UppSAT can be used as a sandbox for easy and flexible exploration of new approximations. To substantiate this, we explore encodings of floating-point arithmetic into reduced precision floating-point arithmetic, real-arithmetic, and fixed-point arithmetic (encoded into the theory of bit-vectors in practice). In an experimental evaluation we compare the advantages and disadvantages of approximating solvers obtained by combining various encodings and decision procedures.

1 Introduction

The construction of satisfying assignments of a formula, or showing that no such assignments exist, is one of the most central tasks in automated reasoning. Although this problem has been addressed extensively in research fields including constraint programming, and more recently in Satisfiability Modulo Theories (SMT), there are still constraint languages and background theories where effective model construction is challenging. Such theories are, in particular, arithmetic domains such as bit-vectors, nonlinear real arithmetic (or real-closed fields), and floating-point arithmetic; even when decidable, the high computational complexity of such problems turns model construction into a bottleneck in applications such as model checking, test-case generation, or hybrid systems analysis.

In several recent papers, the notion of *approximation* has been proposed as a means to speed up the construction of (precise) satisfying assignments. Generally speaking, approximation-based solvers follow a two-tier strategy to find a satisfying assignment of a formula ϕ. First, a simplified or *approximated* version $\hat{\phi}$ of ϕ is solved, resulting in an approximate solution $\hat{m}$ that (hopefully)

© Springer International Publishing AG, part of Springer Nature 2018
D. Galmiche et al. (Eds.): IJCAR 2018, LNAI 10900, pp. 246–262, 2018.
https://doi.org/10.1007/978-3-319-94205-6_17

lies close to a precise solution. Second, a *reconstruction* procedure is applied to check whether $\hat{m}$ can be turned into a precise solution m of the original formula ϕ. If no precise solution m close to $\hat{m}$ can be found, *refinement* can be used to successively obtain better, more precise, approximations.

This high-level approach opens up a large number of design choices, some of which have been discussed in the literature. The approximations considered have different properties; for instance, they might be over- or under-approximations (in which case they are commonly called *abstractions*), or be non-conservative and exhibit neither of those properties. The approximated formula $\hat{\phi}$ can be formulated in the same logic as ϕ, or in some *proxy* theory that enables more efficient reasoning. The reconstruction of m from $\hat{m}$ can follow various strategies, including simple re-evaluation, precise constraint solving on partially evaluated formulas, or randomised optimisation. Refinement can be performed with the help of approximate assignments $\hat{m}$, using proofs or unsatisfiable cores, or be independent of the actual reason for failure. The only requirement is that approximations are improved in such a way that finally a most precise approximation is reached (a "non-approximation" so to speak), in which case UppSAT will fall back on a back-end, thus guaranteeing that the final result is correct.

In this paper we focus on the case of (quantifier-free) floating-point arithmetic (FPA) constraints, a particularly challenging domain that has been studied extensively in the SMT context over the past few years [4,13,14,19–21]. To enable uniform exploration of approximation, reconstruction, and refinement methods, as well as simple prototyping and comparative studies, we present UppSAT[1] as a general framework for building approximating solvers. UppSAT is implemented in Scala, open-sourced under the GPL license, and allows the implementation of approximation schemes in a modular and high-level fashion, such that different components can easily be combined with various back-ends. At this point, we exclusively focus on satisfiable benchmarks, and note that in the current version of UppSAT unsatisfiable benchmarks will never be solved faster than by the chosen back-end. This is because a definite statement about unsatisfiability can only be made after reaching the most precise approximation, which means that the back-end has to show unsatisfiability of the original, non-approximated formula. Techniques for unsatisfiable problems are given in [21].

With the help of the UppSAT framework we explore several ways of approximating SMT reasoning for FPA. The main contributions of this paper are:

- a conceptual framework for defining approximations in a modular way, with the help of a library of approximation components that can easily be instantiated and combined, and which are implemented in the UppSAT tool.
- detailed definition of three concrete FPA approximations within the UppSAT framework: reduced-precision approximation [21]; fixed-point approximation; and real arithmetic approximation.
- an extensive experimental evaluation of all three approximations, considering as back-end solvers the decision procedures available in Z3 [11] and Math-SAT5 [7]. This evaluation confirms that approximations can significantly

---

[1] https://github.com/uuverifiers/uppsat/releases/tag/v0.5-alpha.

boost the performance of bit-blasting-based FPA solvers, but interestingly do not help much in combination with the ACDCL solver of MathSAT5.

1.1 Related Work

The SMT solvers MathSAT5 [7], Z3 [11], and Sonolar [17] feature bit-precise conversions from FPA to bit-vector constraints, known as bit-blasting, and represent the currently most commonly used solvers in program verification. As we show in our experiments, the performance of bit-blasting can be boosted significantly with the help of our approximation approach. An alternative, constraint programming-based approach to solve FPA constraints is implemented in COLIBRI [1]. We became aware of this solver only late and have thus not been able to make a thorough experimental comparison, but note that it does display competitive performance. As future work, it would in particular be interesting to experiment with COLIBRI as a back-end solver in UppSAT.

A general framework for decision procedures is Abstract CDCL, introduced by D'Silva et al. [12], which was also instantiated for FPA [3,13]. This approach relies on the definition of suitable abstract domains (as defined for abstract interpretation [8]) for constraint propagation and learning.

The work presented in this paper builds on previous research on the use of approximations for solving FPA constraints [20,21]. UppSAT is also close in spirit to the framework presented by Ramachandran and Wahl [19] for efficiently solving FPA constraints based on the notion of 'proxy' theories, which correspond to our 'output theories'. This framework applies a sophisticated method of reconstruction, by applying a fall-back FPA solver to a version of the input constraint in which all but one variables have been substituted by their value in a failing candidate model. Such reconstruction could also be realized in UppSAT, and an implementation in UppSAT is planned as future work.

A further recent approximation-based solver for FPA is XSat [14]. In XSat, reconstruction of models is implemented with the help of randomized optimization, which results in good performance, but does not give rise to a decision procedure (incorrect sat/unsat results can be produced).

Specific instantiations of abstraction schemes in related areas also include the bit-vector abstractions by Bryant et al. [6] and Brummayer and Biere [5], as well as the (mixed) floating-point abstractions by Brillout et al. [4].

There is a long history of formalization and analysis of FPA concerns using proof assistants, among others in Coq by Melquiond [18] and in HOL Light by Harrison [15]. Coq has also been integrated with a dedicated FPA prover called Gappa by Boldo et al. [2], which is based on interval reasoning and forward error propagation to determine bounds on arithmetic expressions in programs [10]. The ASTRÉE static analyzer [9] features abstract interpretation-based analyses for FPA overflow and division-by-zero problems in ANSI-C programs.

2   Reduced-Precision FPA by Example

We begin by illustrating key notions of the UppSAT framework using the reduced-precision floating-point approximation (RPFP). This approximation

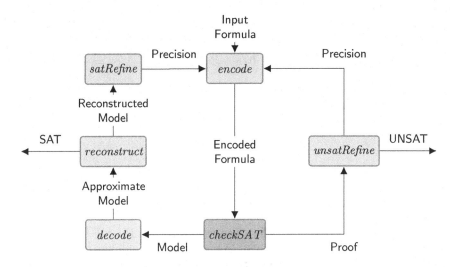

Fig. 1. The approximation refinement algorithm implemented by UppSAT.

uses floating-point operations of reduced precision, i.e., with fewer bits for the exponent and significand. Approximations of this kind have previously been studied in [20,21], and found to be an effective way to boost the performance of bit-blasting-based SMT solvers, since the size of FPA circuits tends to grow quickly with the bit-width. The approximation encodes the same floating-point constraints, but over smaller floating-point domains, resulting in a smaller propositional formula.

The UppSAT framework implements an abstract approximating SMT solver with the solving algorithm shown in Fig. 1. The framework relies on a background solver providing the **checkSAT** routine, reasoning about approximated formulas, while the other (green) boxes have to be implemented in order to specify an approximation. We showcase these elements using an example on the RPFP approximation with the following floating-point formula ϕ over two single-precision floating-point variables x and y:

$$y = x + 1.75 \wedge y \geq 0 \wedge (x = 2.0 \vee x = -4.0) \tag{1}$$

The rounding mode of the addition operation is omitted and assumed to be *RoundTowardZero* in this example. The formula can be satisfied by the model $m = \{x \mapsto 2.0_{8,24}, y \mapsto 3.75_{8,24}\}$, mapping to single-precision values which use 8 bits to represent the exponent and 24 bits for the significand, denoted $FP_{8,24}$.

The RPFP approximation initially **encodes** the formula in the $FP_{3,3}$ floating-point format, i.e., the format using 3 bits for the exponent, and 3 bits for the significand. The approximate formula $\hat{\phi}_{3,3}$ is obtained by replacing the single-precision variables x and y with re-typed variants $x_{3,3}, y_{3,3}$, casting all floating-point literals to the new format, and replacing the addition operator $+$ and comparison predicates $=$ and $\leq$ with the operator $+_{3,3}$ and the predicates $=_{3,3}$

and $\geq_{3,3}$ for reduced-precision arguments (we omit the subscripts for the operators and predicates for aesthetic reasons, except where relevant):

$$y_{3,3} = x_{3,3} + 1.75_{3,3} \wedge y_{3,3} \geq 0_{3,3} \wedge (x_{3,3} = 2.0_{3,3} \vee x_{3,3} = -4.0_{3,3}) \quad (2)$$

Though $\hat{\phi}_{3,3}$ is satisfiable, its models might not be models of the original formula. The models might satisfy the reduced-precision formula only because of over/under-flows and rounding errors in the $FP_{3,3}$ domain, e.g.:

$$\hat{m} = \{x \mapsto 2.0, y \mapsto 3.5\} \quad (3)$$

satisfies $\hat{\phi}_{3,3}$ because $2.0_{3,3} + 1.75_{3,3} = 3.5_{3,3}$ when the rounding mode is *RoundTowardZero*.

To determine whether the approximate solution is indeed a solution for the original formula, we **decode** the model $\hat{m}$ into a candidate model m, by casting the model values from the $FP_{3,3}$ representation to their $FP_{8,24}$ representation. The represented values do not change, but the number of bits used to represent them does. **Model reconstruction** checks whether the original constraints are satisfied by the decoded model and can even make adjustments to the model. A naïve model reconstruction strategy would determine that the candidate model m based on $\hat{m}$ does not satisfy formula ϕ, because $2.0 + 1.75 \neq 3.5$ in single-precision floating-point arithmetic, and would not attempt to correct the failed model. Therefore we need to **refine** the approximation, and a simple strategy is to increase the precision of every node by the same amount, yielding for instance (after encoding):

$$y_{5,5} = x_{5,5} + 1.75_{5,5} \wedge y_{5,5} \geq 0_{5,5} \wedge (x_{5,5} = 2.0_{5,5} \vee x_{5,5} = -4.0_{5,5}) \quad (4)$$

This formula has sufficient bit-width to avoid rounding errors, and the model:

$$\hat{m}_2 = \{x \mapsto 2.0, y \mapsto 3.75\} \quad (5)$$

which is also a model for the original formula. As a side remark, another possibility would be to identify that the cause of the imprecision is that the value y is not correctly represented. Thus it would be necessary only to increase the precision of y (along with predicates and operators involving y):

$$y_{5,5} = x_{3,3} +_{5,5} 1.75_{3,3} \wedge y_{5,5} \geq 0_{3,3} \wedge (x_{3,3} = 2.0_{3,3} \vee x_{3,3} = -4.0_{3,3}) \quad (6)$$

This example shows how the solving proceeds when an approximate solution is found, depicted by the left cycle in Fig. 1, and exiting with a SAT answer. The right cycle in Fig. 1 corresponds to the case when the approximation does not have a model. The **satRefine** and **unsatRefine** can implement different refinement strategies, based on models and proofs/unsatisfiable cores, respectively. In general, the algorithm might take a number of iterations before finding a model (or concluding that the problem is unsatisfiable).

Theorem 1 (Correctness, paraphrased from [21]). *The framework preserves termination, soundness, and completeness of the back-end procedure, provided that: 1. maximal precision $\top$ is reached within a finite amount of steps; and 2. no approximation takes place at maximal precision.*

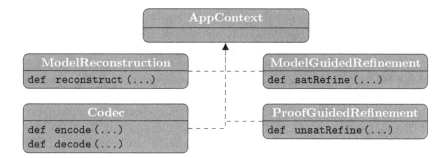

Fig. 2. The basic traits necessary to specify an approximation in UppSAT

```
0   object RPFPApp
1   extends RPFPContext
2   with RPFPCodec
3   with EAAReconstruction
4   with RPFPModelRefinement
5   with RPFPProofRefinement
```

```
0   trait RPFPContext extends AppContext {
1       val   inTheory  = FPTheory
2       val   outTheory = FPTheory
3       type  Prec      = Int
4       val   pOrdering = new IntPOrder(0,5)
5   }
```

Fig. 3. RPFP as a Scala object.   **Fig. 4.** Approximation context for RPFP.

3   Specifying Approximations in UppSAT

In this section we show how to specify approximations in UppSAT, using the example of the RPFP approximation from Sect. 2 and [20,21]. It should be remarked that one of the design goals of UppSAT is the ability to define approximations in a convenient, high-level way; the code we show in this section is mostly identical to the actual implementation in UppSAT, modulo a small number of simplifications for the purpose of presentation.

Reduced-Precision FPA Approximation in UppSAT. An approximation consist of: an approximation context, a codec, a model reconstruction strategy, and a refinement strategy for model- and proof-guided refinement. In UppSAT these components are implemented using several Scala *mix-in traits* that agree on the signature of the approximation, represented by the shared AppContext trait in Fig. 2. The traits are simply combined into an approximation object which will be used by the UppSAT solver. The Fig. 3 shows the object RPFPApp implementing the reduced-precision floating-point approximation by combining instances of the traits shown in Fig. 2 that all extend the RPFPAppContext. Using traits enables the modular mix-and-match approximation design. In the following paragraphs, we present the key points of reduced precision floating-point approximation through its component traits.

Approximation Context. An approximation context specifies input and output theory, a precision domain and a precision ordering. Figure 4 shows the specification of RPFPContext, the approximation context object for the reduced precision floating-point (RPFP) approximation, which approximates floating-point constraints by scaling them down to a smaller floating-point sort, as presented in Sect. 2. Therefore, both the input and the output theories are the quantifier-free theory of FPA (FPTheory). The precision is associated with each node in the formula tree and uniformly affects both the exponent and the significand, so a scalar data type Prec = Int is sufficient to represent precision. In particular, we choose integers in the range $[0, 5]$ with the usual ordering as the precision domain, thus yielding a linear sort scaling which consists of 6 sorts, starting with $FP_{3,3}$ and scaling up to (and including) the original sort (implemented by scaleSort in Fig. 5). In general, a precision domain can range over tuples of any size, but in order to preserve completeness and termination, for the case of decidable theories, we assume that every precision domain contains a top element $\top$, and that precision domains satisfy the ascending chain condition (every ascending chain is finite) [21].

Codec. The RPFPCodec trait implements the encoding of the formula and the decoding of the approximate model. UppSAT provides general traits that implement the *encode* and *decode* methods using a post-order visitor pattern over formulas (PostOrderCodec). This allows the codec to be implemented by implementing the two hook functions that work over nodes in the formula tree.

The function encodeNode, shown in Fig. 5 shows how the approximation scales-down the sort of floating-point variables and operations, while keeping the high-level structure of the formula. Scaling is performed based on precision values, with the exception of predicates, which are scaled dynamically based on the maximum sort of its arguments. Constant literals and rounding modes remain unaffected by this encoding. There is no guarantee that the sorts of nodes of different precisions will match, so cast operations are used to ensure well-sortedness. To ensure consistency of the approximate models, all occurrences of a variable share the same precision.

After the back-end solver returns a model of the approximate constraints, the decodeNode function casts variable assignments to their sort in the original formula. For example, the formula ϕ from Sect. 2 over single-precision floating-point variables is encoded as the formula $\hat{\phi}_{3,3}$. A *checkSAT* call returns a model $\hat{m} = \{x \mapsto 2.0_{3,3}; y \mapsto 3.5_{3,3}\}$. Decoding will cast the values of the approximate model to their original sort (the values will not change, only their sorts), resulting in $m = \{x \mapsto 2.0_{8,24}; y \mapsto 3.5_{8,24}\}$.

Model Reconstruction Strategy. A model reconstruction strategy specifies how to obtain a model of the input constraints starting from the decoded model. Since the RPFP approximation retains the Boolean structure of the original formula, a simple strategy to obtain a reconstructed model is by ensuring that the same atomic constraints are satisfied. Reconstruction chooses a subset of the atoms occurring in the formula, called the *critical atoms*, which if evaluated identically as in the approximate model guarantee that the formula is satisfied; this means, the conjunction of the critical atoms is an *implicant* of the formula.

```
0  trait RPFPCodec extends RPFPContext with PostOrderCodec {
1    def scaleSort(node : AST, p : Int, children : List[AST]) = {
2      node.symbol match {
3        case _ : FloatingPointPredicateSymbol => {
4          val sorts = children.filterNot(_.isLiteral).map(_.symbol.sort)
5          sorts.foldLeft(sorts.head)(fpsortMaximum(_,_))
6        }
7        case _ : FloatingPointFunSymbol => {
8          val FPSort(eBitWidth, sBitWidth) = sort
9          val eBits = 3 + ((eBitWidth - 3) * p)/pOrder.maxPrecision
10         val sBits = 3 + ((sBitWidth - 3) * p)/pOrder.maxPrecision
11         FPSort(eBits, sBits)
12       }
13       case _ => sort
14     }
15   }
16   def encodeNode(node : AST, children : List[AST], p : Int) = {
17     val sort = scaleSort(node, p, children)
18     val castChildren = children.map(cast(_, sort))
19     val symbol = encodeSymbol(node.symbol, sort, castChildren)
20     AST(symbol, node.label, castChildren)
21   }
22   def decodeNode(args : (Model, PrecMap[Prec]), decodedModel : Model,
23                  node : AST) = {
24     val (appModel, pmap) = args
25     val AST(symbol, label, _) = node
26     val decodedValue = decodeFPValue(symbol, appModel(node), pmap(label))
27     decodedModel.set(ast, Leaf(decodedValue))
28     decodedModel
29   }
30 }
```

Fig. 5. Reduced-precision encoding and decoding.

Due to the difference in semantics (e.g., rounding error), when evaluating the original formula, errors accumulate. This can result in critical atoms changing values under the original semantics. Therefore, evaluation of critical atoms under the original semantics is necessary to ensure that the model satisfies the original formula. UppSAT provides a bottom-up reconstruction strategy, which is specified on a node-by-node basis and applied using a post-order visitor. To specify this reconstruction strategy only the reconstructNode hook function needs to be implemented, shown in Fig. 6.

Equality as Assignment. An important heuristic used in the RPFP model reconstruction is equality-as-assignment. The idea is that given an equality constraint $y = f(x_1, \ldots, x_n)$ in which the arguments $x_1, \ldots, x_n$ are *fixed* (see below), but y is not, we can calculate the value of $f(x_1, \ldots, x_n)$ and use it as the value of y in the reconstructed model; this is indeed the only way to satisfy the equality constraint. To put this observation to use, variables are not fixed to a value in the reconstructed model until they are used to evaluate an expression or atom. When a predicate is evaluated, if some its arguments are not fixed, it means that they have not been used yet and can be safely modified at this point. To ensure maximal utilisation of this heuristic, the atoms are topologically sorted to process implicating atoms, such as equalities, before the other critical atoms.

```
0  def reconstructNode(decodedM : Model, candidateM : Model, node : AST) = {
1    val AST(symbol, label, children) = node
2    if (children.length > 0)
3      if (equalityAsAssignment(ast, decodedM, candidateM)) {
4        return candidateM
5      } else {
6        val args =
7          for (c <- children) yield getCurrentValue(c, decodedM, candidateM)
8        val expr = AST(symbol, label, args.toList)
9        val value = ModelEvaluator.evalAST(expr, inputTheory)
10       candidateM.set(node, value)
11     }
12   }
13   candidateM
14 }
```

Fig. 6. Post-order reconstruction using equality-as-assignment

Example 1. Consider the reconstruction outlined in Sect. 2. It reconstructed the model $\hat{m} = \{x \mapsto 2.0_{3,3}, y \mapsto 3.5_{3,3}\}$ by just up-casting the values, yielding $m = \{x \mapsto 2.0_{8,24}, y \mapsto 3.5_{8,24}\}$, which did not satisfy the original formula. Here `equalityAsAssignment` can be applied to the critical atoms $x = 2.0$, $y = x + 1.75$ and $y \geq 0$. Processing them from left to right, the first atom ($x = 2.0$) is satisfied by m, but not the second one ($y = x + 1.75$). This is an equality constraint with an unfixed variable on the left-hand side and the right-hand side is fixed ($x + 1.75 = 3.75$). Therefore, the model is updated m with $y \mapsto 3.75_{8,24}$ (ignoring the value of y in the candidate model), yielding $m_e = \{x \mapsto 2.0_{8,24}, y \mapsto 3.75_{8,24}\}$ which is a model for the original formula.

Model-guided refinement strategy. A model-guided refinement strategy increases the precision of the formula, based on the decoded model and a failed model. When an approximate model can not be reconstructed to a solution, the refinement strategy increases the precision of certain operations, to refine parts of the approximate formula that were too coarse.

Comparing the evaluation of the formula under the decoded and the failed models identifies the critical atoms to be refined. These atoms evaluate as true in the approximate model and as false in the candidate model. Since FPA is a numerical domain, it is possible to apply some notion of *error* to determine which nodes contribute the most to the discrepancies in evaluation and use them to rank the sub-expressions. After ranking, only a portion of them is refined, in our case 30%. Refinement is achieved by increasing the precision by one (in the range $[0, 5]$, as described above). In general, one could use the error to determine by how much to increase the precision. Since error-based refinement can be applied to any numerical domain, UppSAT implements an abstract *error-based refinement strategy*, which allows us to specify refinement by only instantiating the `nodeError` hook function, shown in Fig. 7.

Proof-guided refinement strategy. If we fail to find an approximate model, the proof-guided refinement strategy can use unsatisfiable cores to refine the

```
0  trait RPFPMGRefinementStrategy extends RPFPContext
1                              with ErrorBasedRefinementStrategy {
2   def nodeError(decodedM : Model, failedM : Model
3                 acc : Map[AST, Double], node : AST) = {
4     node.symbol match {
5       case literal : FloatingPointLiteral => acc
6       case fpfs : FloatingPointFunSymbol => {
7         val Some(outErr) = relativeError(node, decodedM, failedM)
8         val argErrors =
9                node.children.map{relativeError(_, decodedM, failedM)}
10        val inErrors = argErrors.collect{case Some(x) => x}
11        val sumInErrors = inErrors.fold(0.0){(x,y) => x + y}
12        val avgInErr = sumInErrors /  inErrors.length
13        acc + (ast -> outErr / (1 + avgInErr))
14      }
15      case _ => acc
16    }
17  }
18 }
```

Fig. 7. Model-guided refinement strategy based on relative errors

formula [21]. At the moment UppSAT has no support for obtaining proofs from the back-end solvers. Instead, a naïve refinement strategy is used, which increases all the precisions by a constant.

4   Other Approximations of FPA

We have shown in detail the RPFP approximation of FPA, and discussed different components that can be used in general. In this section we outline two further approximations of FPA that have been implemented in UppSAT: the fixed-point approximation BV, encoded as bit-vectors, and the real-arithmetic approximation RA. Both approximations are currently implemented as a proof-of-concept for cross-theory approximations. Despite their lack of maturity these approximations show promising results (see Sect. 5).

BV—the Fixed-Point Approximation of FPA. The idea behind the BV approximation is to avoid the overhead of the rounding semantics and special values of FPA, by encoding all the FPA values and operations as values and operations in fixed-point arithmetic.

The BV Context. The input theory is the theory of FPA, and the intended output theory is the theory of fixed-point arithmetic. However, since fixed-point arithmetic is not commonly supported by SMT solvers, we encode fixed-point constraints in the theory of fixed-width bit-vectors. The precision determines the number of integer and fractional binary digits in the fixed-point representation of a number. For simplicity, at this point we do not mix multiple fixed-point formats in one formula, but instead apply uniform precision in the BV approximation; as a result, all operations in a constraint are encoded using the same fixed-point sort. As a proof of concept, the precision domain is two-dimensional, with the first component p_i in a pair (p_i, p_f) denoting the number of integral, and the second component p_f the number of fractional bits in the encoding, respectively. The precision domain ranges from $(5, 5)$ to $(25, 25)$, with the maximum element $(25, 25) = \top$ being interpreted as sending the original, unapproximated FPA constraint to Z3 as a fall-back solver. As an example, given a variable of precision $(5, 5)$, we will have a domain of numbers between 10000.00000_2 and 01111.11111_2, which when interpreted in two's-complement notation are numbers between -16 and 15.96875. Returning to the formula ϕ in Sect. 2, it would be encoded with a precision of $(5, 5)$ into the formula $\hat{\phi}_{5,5}^F$:

$$y_{10} = x_{10} \oplus_{10} 00001\,11000_2 \wedge y_{10} \geq_s 00000\,00000_2 \wedge$$
$$(x_{10} = 00010\,00000_2 \vee x_{10} = 11100\,00000_2)$$

We can note that fixed-point $(5, 5)$-addition is exactly implemented by bit-vector addition $\oplus_{10}$ over 10 bits, and fixed-point comparison $\geq$ by signed bit-vector comparison $\geq_s$ over 10 bits, so that the translation becomes relatively straightforward.

Constants are interpreted as 2's complement numbers with 5 fractional and 5 integral bits, e.g., $11100\,00000_2$ represents the binary number -00100.00000_2, which is -4.0 in decimal notation. It can be seen that the constraint $\hat{\phi}_{5,5}^F$ is satisfied by the model $\hat{m} = \{x_{10} \mapsto 00010\,00000_2, y_{10} \mapsto 00011\,11000_2\}$, which corresponds to the fixed-point solution $x = 2.0$ and $y = 3.75$, which is equal to the floating point model found earlier.

BV Reconstruction and Refinement. The model reconstruction strategy in the BV approximation is the same as in the RPFP approximation. The refinement strategy is very simple: it increases precision along both dimensions by 4, adding 4 more bits to both the integral and fractional bits in the encoding.

RA—the Real Arithmetic Approximation of FPA. The third approximation of FPA we consider, is by encoding it into real arithmetic constraints. We briefly present a simple implementation of this approximation.

Ramachandran and Wahl [19] describe a topological notion of refinement, that requires a back-end solver that handles the combined theory of real arithmetic and FPA. However, solving constraints over this combination of theories is challenging in itself, and efficient SMT solvers are not publicly available, to the best of our knowledge. Therefore, in this paper we only us a binary precision domain of $\{\bot, \top\}$, where either the entire formula is translated into real arithmetic, or the original formula is solved.

The encoding is fairly straightforward: the FPA operations are translated as their real counter-parts, omitting the rounding modes in the process. While the special values can be encoded, currently they are not supported by the RA approximation. Decoding will translate a real number to the closest FPA numeral under the given rounding mode. As discussed above, the refinement is trivial and the reconstruction is the same as in the the RPFP approximation.

5 Experimental Evaluation

In this section we evaluate the effectiveness of the discussed approximations. We instantiate the framework for the three presented approximations. The RPFP approximation is instantiated with three back-ends: Z3, MathSAT5 and Math-SAT5 using ACDCL. The BV approximation is instantiated with the bit-vector solver of Z3 as a back-end, and the RA approximation uses Z3's nlsat tactic [16].

Experimental Setup. We evaluate UppSAT on the *satisfiable* benchmarks of the QF_FP category of the SMT-LIB[2]. Currently, none of the approximations have a meaningful proof-based refinement strategy, so the performance on unsatisfiable problems is left for future work. All experiments were performed on an AMD Opteron 2220 SE machine, running 64-bit Linux, with memory limited to 1.0 GB, and with a timeout of one hour.

Table 1. Comparison of the three back-ends and five instantiations of UppSAT, showing # of benchmarks solved within 1 hour, # of timeouts, # of instances for which the solver was fastest, average # of refinement iterations on solved problems, # of benchmarks where refinement reached maximum precision, average time to process all benchmarks (excluding timeouts), and # of instances only solved by the respective solver.

	ACDCL	MathSAT	Z3 (Z3)	BV (Z3)	RPFP (ACDCL)	RPFP (MathSAT)	RPFP (Z3)	RA (nlsat)
Solved	86	99	97	91	78	**101**	**101**	90
Timeouts	44	31	33	39	52	29	29	40
Best	65	4	6	9	3	9	9	4
Avg. iterations	-	-	-	2.69	3.59	3.16	3.02	1.85
Max precision	-	-	-	23	2	1	2	110
Avg. time (s)	117.10	169.17	355.94	131.64	108.30	81.97	148.43	301.87
Only solver	1	0	2	0	0	1	0	0

We compare the performance of the back-ends and the UppSAT instances on 130 non-trivial satisfiable benchmarks. The results are summarized in Table 1, and a more detailed view of this data is provided by the cactus plot in Fig. 8.

Table and Cactus Plot. Looking at Table 1, we observe that the RPFP approximation combined with bit-blasting, either in Z3 or MathSAT, solves the largest number of instances. When comparing average runtime, MathSAT comes out as the marginally better choice of back-end. This is expected, based on the performance on the back-ends themselves. All the configurations shine on at least a few benchmarks, indicating that the approximations do offer an improvement. Furthermore, the ACDCL algorithm outperforms all the other solvers on 65

---

[2] The regression tests in the **wintersteiger** family were ignored for the evaluation.

benchmarks, but it solves fewer benchmarks that the bit-blasting approaches in total. This is corroborated in the cactus plot, where in the left part of the graph ACDCL is solving many benchmarks, however, eventually it gets overtaken by the other solvers. Looking more closely at the RPFP approximation, we can conclude that it improves performance of bit-blasting considerably, regardless of the implementation (MathSAT or Z3). On the other hand, RPFP seems to hinder, rather than help, the already very efficient ACDCL algorithm.[3]

Looking only at the approximations, we can see that on average the benchmarks are solved using around three iterations (the RA always performs at most two iterations, the RA approximation and the FPA semantics). This indicates that for many of the benchmarks, full-precision encoding is not really necessary, since the RPFP approximation rarely reaches maximum precision.

Virtual Portfolios. In Table 2, we compare the virtual best portfolio over all approximating solvers against the baseline of the virtual best portfolio over back-end solvers. Inclusion of UppSAT instances in the portfolio cuts the average solving time in half.

Table 2. Virtual portfolio performance.

	VP (Back-ends)	VP (All)
Solved	110	112
T/O	20	18
Total time (s)	25135	12516
Avg. time (s)	228.50	111.75

Scatter Plots. Figure 9 shows the runtime comparison of the RPFP and BV approximations against the bit-blasting back-end Z3. The x-axis denotes the runtime of UppSAT instances, while the y-axis denotes the runtime of Z3. Maximum value along either axes denotes a timeout. Data points above the diagonal indicate that UppSAT takes less time and below the diagonal that Z3 takes less time on an instance. The left plot shows a comparison of the RPFP(Z3) instance against the bit-blasting approach in Z3. The majority of benchmarks are solved faster by the UppSAT instance, and the plot is in line with previously published results, but the trend suggests a super-linear speedup in performance which was not as pronounced before. The right plot comparing the runtime of BV(Z3) to that of Z3 is similar to that of RPFP(Z3), with the difference that gains and losses in runtime are even greater with the RPFP approximation. The greater speed-ups are due to even simpler propositional encodings, since the exponent is implicit and fixed upfront. The losses in solving time are due to the fact the BV approximation is not yet mature, since it lacks a fine-tuned precision order, tailor-made refinement and simply re-uses the strategies used by the RPFP approximation. With this in mind, we believe that these results are very promising.

[3] Earlier experiments using the stable version 5.4.1 of MathSAT have shown similar effects of the RPFP approximation to those on the bit-blasting methods. However, overall the performance results were not consistent with performance of MathSAT in previous publications, and indicated a bug. We thank Alberto Griggio for promptly providing us with a corrected version of MathSAT, which we use in the evaluation.

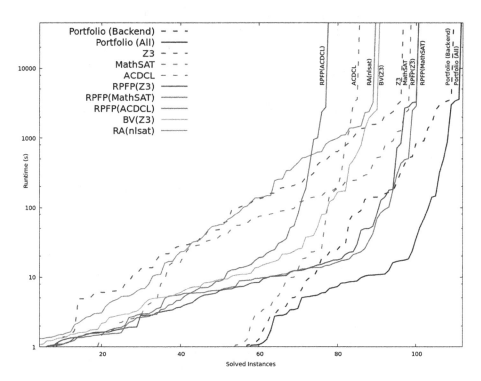

Fig. 8. The X axis shows how many instances can be solved in the amount of time shown on the Y axis, by each of the solvers and the portfolios. The UppSAT instances are shown using full lines, while the back-ends are presented using dashed lines. The colors denote the same back-end, e.g., MathSAT and RPFP(MathSAT) are both colored green.

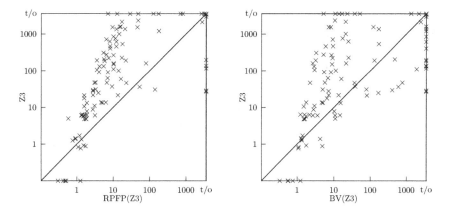

Fig. 9. Runtime comparison of RPFP(Z3) with Z3 (left) and BV(Z3) with Z3.

We omit scatter plots for other UppSAT instances[4], but offer a brief summary of the results. The comparison of RPFP(MathSAT) instance against MathSAT is very similar to that of RPFP(Z3) against Z3. The RPFP(ACDCL) did not improve on the runtime of the ACDCL solver. This appears to be due to the fact that RPFP approximation does not make formulas significantly easier to solve for ACDCL, in contrast to the situation with bit-blasting. The RA(nlsat) instance does currently not show satisfactory results; the approximation is a proof of concept, and is an *on-off* approximation, since there is no space for refinement in the absence of a back-end that would support the combination of non-linear real arithmetic and floating-point arithmetic.

Overall, these results show that the RPFP and BV approximations can indeed speed up the performance of the bit-blasting back-ends, and in case of the BV approximation with not much effort.

6   Conclusion and Future Work

We have presented a methodology and new framework, UppSAT, for implementing approximating SMT solvers. UppSAT enables simple and high-level definition of approximations, can be combined with different back-ends (at the moment Z3 and MathSAT, but further back-ends can be added with little effort), and is useful both for rapid prototyping and for tailoring solvers to particular use-cases.

The experimental evaluation demonstrates the efficacy of approximations. The approximation instances presented here (RPFP(z3), RPFP(MathSAT)) are shown to be state-of-the art in handling formulas in FPA, where they improve their performance of the respective back-end to a even greater extent than previous work. For ACDCL this is not the case, indicating that perhaps a different method of approximation should be utilized.

The fixed point and real arithmetic approximations are presented here as a proof of concept. They are simple and not much effort went into instantiating the framework for these approximations. However, the results shows that even uncomplicated approaches can be competitive; this opens up the line of future work to design tailored refinement and reconstruction strategies.

The clear direction for improving UppSAT is to extend the general framework with more abstract strategies, e.g., retrieve multiple models from an approximate formula and/or apply multiple different reconstruction strategies on approximate models. Currently, much time is spent on looking for models which means there is plenty of room to make more sophisticated strategies in the framework. UppSAT could also be extended to allow approximations to be written in a high-level domain specific language, and allow them to be loaded as dynamic libraries.

Another big challenge is to extend UppSAT to be able to handle unsatisfiable formulas efficiently. Currently, the proof refinement is naive uniform refinement, but there is potential to do much more intelligent refinement.

---

[4] Detailed plots of all approximations and back-ends can be found at https://github.com/uuverifiers/uppsat/wiki/Scatter-Plots---IJCAR-2018.

References

1. Bobot, F., Chihani, Z., Marre, B.: Real behavior of floating point. In: 15th International Workshop on Satisfiability Modulo Theories (2017)
2. Boldo, S., Filliâtre, J.-C., Melquiond, G.: Combining Coq and Gappa for certifying floating-point programs. In: Carette, J., Dixon, L., Coen, C.S., Watt, S.M. (eds.) CICM 2009. LNCS (LNAI), vol. 5625, pp. 59–74. Springer, Heidelberg (2009). https://doi.org/10.1007/978-3-642-02614-0_10
3. Brain, M., D'Silva, V., Griggio, A., Haller, L., Kroening, D.: Deciding floating-point logic with abstract conflict driven clause learning. FMSD **45**, 213–245 (2013)
4. Brillout, A., Kroening, D., Wahl, T.: Mixed abstractions for floating-point arithmetic. In: FMCAD. IEEE (2009)
5. Brummayer, R., Biere, A.: Effective bit-width and under-approximation. In: Moreno-Díaz, R., Pichler, F., Quesada-Arencibia, A. (eds.) EUROCAST 2009. LNCS, vol. 5717, pp. 304–311. Springer, Heidelberg (2009). https://doi.org/10.1007/978-3-642-04772-5_40
6. Bryant, R.E., Kroening, D., Ouaknine, J., Seshia, S.A., Strichman, O., Brady, B.: Deciding bit-vector arithmetic with abstraction. In: Grumberg, O., Huth, M. (eds.) TACAS 2007. LNCS, vol. 4424, pp. 358–372. Springer, Heidelberg (2007). https://doi.org/10.1007/978-3-540-71209-1_28
7. Cimatti, A., Griggio, A., Schaafsma, B.J., Sebastiani, R.: The MathSAT5 SMT solver. In: Piterman, N., Smolka, S.A. (eds.) TACAS 2013. LNCS, vol. 7795, pp. 93–107. Springer, Heidelberg (2013). https://doi.org/10.1007/978-3-642-36742-7_7
8. Cousot, P., Cousot, R.: Abstract interpretation: a unified lattice model for static analysis of programs by construction or approximation of fixpoints. In: POPL, pp. 238–252. ACM Press (1977)
9. Cousot, P., Cousot, R., Feret, J., Mauborgne, L., Monniaux, D., Rival, X.: The ASTREÉ analyzer. In: ESOP, Antoine Miné (2005)
10. Daumas, M., Melquiond, G.: Certification of bounds on expressions involving rounded operators. ACM Trans. Math. Softw. **37**(1), 2 (2010)
11. de Moura, L., Bjørner, N.: Z3: an efficient SMT solver. In: Ramakrishnan, C.R., Rehof, J. (eds.) TACAS 2008. LNCS, vol. 4963, pp. 337–340. Springer, Heidelberg (2008). https://doi.org/10.1007/978-3-540-78800-3_24
12. D'Silva, V., Haller, L., Kroening, D.: Abstract conflict driven learning. In: POPL, pp. 143–154. ACM (2013)
13. D'Silva, V., Haller, L., Kroening, D., Tautschnig, M.: Numeric bounds analysis with conflict-driven learning. In: Flanagan, C., König, B. (eds.) TACAS 2012. LNCS, vol. 7214, pp. 48–63. Springer, Heidelberg (2012). https://doi.org/10.1007/978-3-642-28756-5_5
14. Fu, Z., Su, Z.: XSat: a fast floating-point satisfiability solver. In: Chaudhuri, S., Farzan, A. (eds.) CAV 2016. LNCS, vol. 9780, pp. 187–209. Springer, Cham (2016). https://doi.org/10.1007/978-3-319-41540-6_11
15. Harrison, J.: Floating point verification in HOL Light: the exponential function. TR 428, University of Cambridge Computer Laboratory (1997)
16. Jovanovic, D., de Moura, L.: Solving non-linear arithmetic. ACM Comm. Comput. Algebra **46**(3/4), 104–105 (2012)
17. Lapschies, F., Peleska, J., Gorbachuk, E., Mangels, T.: SONOLAR SMT-solver. In: SMT-COMP system description (2012)
18. Melquiond, G.: Floating-point arithmetic in the Coq system. In: Conference on Real Numbers and Computers, volume 216 of Information & Computation. Elsevier (2012)

19. Ramachandran, J., Wahl, T.: Integrating proxy theories and numeric model lifting for floating-point arithmetic. In: FMCAD. IEEE (2016)
20. Zeljić, A., Wintersteiger, C.M., Rümmer, P.: Approximations for model construction. In: Demri, S., Kapur, D., Weidenbach, C. (eds.) IJCAR 2014. LNCS (LNAI), vol. 8562, pp. 344–359. Springer, Cham (2014). https://doi.org/10.1007/978-3-319-08587-6_26
21. Zeljić, A., Wintersteiger, C.M., Rümmer, P.: An approximation framework for solvers and decision procedures. JAR **58**(1), 127–147 (2017)

Complexity of Combinations of Qualitative Constraint Satisfaction Problems

Manuel Bodirsky[(⊠)] and Johannes Greiner[(⊠)]

Institut für Algebra, TU Dresden, Dresden, Germany
{manuel.bodirsky,johannes.greiner}@tu-dresden.de

Abstract. The CSP of a first-order theory T is the problem of deciding for a given finite set S of atomic formulas whether $T \cup S$ is satisfiable. Let T_1 and T_2 be two theories with countably infinite models and disjoint signatures. Nelson and Oppen presented conditions that imply decidability (or polynomial-time decidability) of $\mathrm{CSP}(T_1 \cup T_2)$ under the assumption that $\mathrm{CSP}(T_1)$ and $\mathrm{CSP}(T_2)$ are decidable (or polynomial-time decidable). We show that for a large class of ω-categorical theories T_1, T_2 the Nelson-Oppen conditions are not only sufficient, but also necessary for polynomial-time tractability of $\mathrm{CSP}(T_1 \cup T_2)$ (unless P = NP).

1 Introduction

Two independent proofs of the finite-domain constraint satisfaction tractability conjecture have recently been published by Bulatov and Zhuk [20,31], settling the Feder-Vardi dichotomy conjecture. In contrast, the computational complexity of constraint satisfaction problems over infinite domains cannot be classified in general [8]. However, for a restricted class of constraint satisfaction problems that strictly all finite-domain CSPs and captures the vast majority of the problems studied in *qualitative reasoning* (see the survey article [9]) there also is a tractability conjecture (see [3–5,17]). The situation is similar to the situation for finite-domain CSPs before Bulatov and Zhuk: there is a formal condition which provably implies NP-hardness, and the conjecture is that every other CSP in the class is in P.

For finite domain CSPs, it turned out that only few fundamentally different algorithms were needed to complete the classification; the key in both the solution of Bulatov and the solution of Zhuk was a clever combination of the existing algorithmic ideas. An intensively studied method for obtaining (polynomial-time) decision procedures for infinite-domain CSPs is the Nelson-Oppen combination method; see, e.g., [2,30]. The method did not play any role for the classification of finite-domain CSPs, but is extremely powerful for combining algorithms for infinite-domain CSPs.

M. Bodirsky and J. Greiner have received funding from the European Research Council (ERC Grant Agreement no. 681988), the German Research Foundation (DFG, project number 622397), and the DFG Graduiertenkolleg 1763 (QuantLA).

© Springer International Publishing AG, part of Springer Nature 2018
D. Galmiche et al. (Eds.): IJCAR 2018, LNAI 10900, pp. 263–278, 2018.
https://doi.org/10.1007/978-3-319-94205-6_18

In order to conveniently state what type of combinations of CSPs can be studied with the Nelson-Oppen method, we slightly generalise the notion of a CSP. The classical definition is to fix an infinite structure $\mathfrak{B}$ with finite relational signature τ; then CSP($\mathfrak{B}$) is the computational problem of deciding whether a given finite set of *atomic τ-formulas* (i.e., formulas of the form $x_1 = x_2$ or of the form $R(x_1, \ldots, x_n)$ for $R \in \tau$ and variables $x_1, \ldots, x_n$) is satisfiable in $\mathfrak{B}$. Instead of fixing a τ-structure $\mathfrak{B}$, we fix a *τ-theory T* (i.e., a set of first-order τ-sentences). Then CSP(T) is the computational problem of deciding for a given finite set S of atomic τ-formulas whether $T \cup S$ has a model. Clearly, this is a generalisation of the classical definition since CSP($\mathfrak{B}$) is the same as CSP(Th($\mathfrak{B}$)) where Th($\mathfrak{B}$) is the *first-order theory of* $\mathfrak{B}$, i.e., the set of all first-order sentences that hold in $\mathfrak{B}$. The definition for theories is *strictly* more expressive (we give an example in Sect. 2 that shows this).

Let T_1 and T_2 be two theories with disjoint finite relational signatures τ_1 and τ_2. We are interested in the question when CSP($T_1 \cup T_2$) can be solved in polynomial time; we refer to this problem as the *combined CSP* for T_1 and T_2. Clearly, if CSP(T_1) or CSP(T_2) is NP-hard, the CSP($T_1 \cup T_2$) is NP-hard, too. Suppose now that CSP(T_1) and CSP(T_2) can be solved in polynomial-time. In this case, there are examples where CSP($T_1 \cup T_2$) is in P, and examples where CSP($T_1 \cup T_2$) is NP-hard. Even if we know the complexity of CSP(T_1) and of CSP(T_2), a classification of the complexity of CSP($T_1 \cup T_2$) for arbitrary theories T_1 and T_2 is too ambitious (see Sect. 4 for a formal justification). But such a classification should be feasible at least for the mentioned class of infinite-domain CSPs for which the tractability conjecture applies.

1.1 Qualitative CSPs

The idea of *qualitative formalisms* is that reasoning tasks (e.g. about space and time) is not performed with absolute numerical values, but rather with *qualitative* predicates (such as *within, before*, etc.). There is no universally accepted definition in the literature that defines what a *qualitative CSP* is, but a proposal has been made in [9]; the central mathematical property for this proposal is *ω-categoricity*. A theory is called *ω-categorical* if it has up to isomorphism only one countable model. A structure is called *ω-categorical* if and only if its first-order theory is ω-categorical. Examples are $(\mathbb{Q}; <)$, Allen's Interval Algebra, and more generally all homogeneous structures with a finite relational signature (a structure $\mathfrak{B}$ is called *homogeneous* if all isomorphisms between finite substructures can be extended to an automorphism; see [6, 25]). The class of CSPs for ω-categorical theories arguably coincides with the class of CSPs for *qualitative formalisms* studied e.g. in temporal and spatial reasoning; see [9].

For an ω-categorical theory T, the complexity of CSP(T) can be studied using the universal-algebraic approach that led to the proof of the Feder-Vardi dichotomy conjecture. One of the central concepts for this approach is the concept of a *polymorphism* of a structure $\mathfrak{B}$, i.e., a homomorphism from $\mathfrak{B}^k$ to $\mathfrak{B}$ for $k \in \mathbb{N}$. It is known that the polymorphisms of a finite structure $\mathfrak{B}$ fully capture the complexity of CSP($\mathfrak{B}$) up to P-time reductions (in fact, up to Log-space reductions; see [24] for a collection of survey articles about the complexity of CSPs), and the same is true for structures $\mathfrak{B}$ with an ω-categorical theory. For an ω-categorical relational structure Γ, the relations that are primitive positive definable in Γ are uniquely determined by the polymorphisms of Γ and vice versa [15]. The possibility to use relations and polymorphisms exchangeably, to study their interplay and to combine known solutions with polymorphisms make the universal algebraic approach a versatile tool. In order to understand when we can apply the universal-algebraic approach to study the complexity of $\mathrm{CSP}(T_1 \cup T_2)$, we need to understand the following fundamental question.

Question 1: Suppose that T_1 and T_2 are theories with disjoint finite relational signatures τ_1 and τ_2. When is there an ω-categorical $(\tau_1 \cup \tau_2)$-theory T such that $\mathrm{CSP}(T)$ equals[1] $\mathrm{CSP}(T_1 \cup T_2)$?

Note that ω-categorical theories are *complete*, i.e., for every first-order sentence ϕ either T implies ϕ or T implies $\neg\phi$. In general, it is not true that $\mathrm{CSP}(T_1 \cup T_2)$ equals $\mathrm{CSP}(T)$ for a complete theory T (we present an example in Sect. 2).

Question 1 appears to be very difficult. However, we present a broadly applicable condition for ω-categorical theories T_1 and T_2 with infinite models that implies the existence of an ω-categorical theory T such that $\mathrm{CSP}(T_1 \cup T_2)$ equals $\mathrm{CSP}(T)$ (Proposition 1 below). The theory T that we construct has many utile properties, in particular:

1. $T_1 \cup T_2 \subseteq T$;
2. if $\phi_1(\bar{x})$ is a τ_1-formula and $\phi_2(\bar{x})$ is a τ_2-formula, both with free variables $\bar{x} = (x_1, \ldots, x_n)$, then $T \models \exists \bar{x}(\phi_1(\bar{x}) \wedge \phi_2(\bar{x}) \wedge \bigwedge_{i<j} x_i \neq x_j)$ if and only if $T_1 \models \exists \bar{x}(\phi_1(\bar{x}) \wedge \bigwedge_{i<j} x_i \neq x_j)$ and $T_2 \models \exists \bar{x}(\phi_2(\bar{x}) \wedge \bigwedge_{i<j} x_i \neq x_j)$;
3. For every $\tau_1 \cup \tau_2$ formula ϕ there exists a Boolean combination of τ_1 and τ_2 formulas that is equivalent to ϕ modulo T.

In fact, T is uniquely given by these three properties (up to equivalence of theories; see Lemma 2) and again ω-categorical, and we call it the *generic combination of T_1 and T_2*. Let $\mathfrak{B}_1$ and $\mathfrak{B}_2$ be two ω-categorical structures whose first-order theories have a generic combination T; then we call the (up to isomorphism unique) countably infinite model of T the *generic combination of $\mathfrak{B}_1$ and $\mathfrak{B}_2$*.

1.2 The Nelson-Oppen Criterion

Let T_1, T_2 be theories with disjoint finite relational signatures τ_1, τ_2 and suppose that $\mathrm{CSP}(T_1)$ is in P and $\mathrm{CSP}(T_2)$ is in P. Nelson and Oppen gave sufficient conditions for $\mathrm{CSP}(T_1 \cup T_2)$ to be solvable in polynomial time, too. Their conditions are:

1. Both T_1 and T_2 are *stably infinite*: a τ-theory T is called *stably infinite* if for every quantifier-free τ-formula $\phi(x_1, \ldots, x_n)$, if ϕ is satisfiable over T,

---

[1] In other words: for all sets S of atomic $(\tau_1 \cup \tau_2)$-formulas, we have that $S \cup T$ is satisfiable if and only if $S \cup (T_1 \cup T_2)$ is satisfiable.

then there also exists an *infinite* model $\mathfrak{A}$ and elements $a_1, \ldots, a_n$ such that $\mathfrak{A} \models \phi(a_1, \ldots, a_n)$.

2. for $i = 1$ and $i = 2$, the signature τ_i contains a binary relation symbol $\neq_i$ that denotes the inequality relation, i.e., T_i implies the sentence
$\forall x, y \, (x \neq_i y \Leftrightarrow \neg(x = y))$;

3. Both T_1 and T_2 are *convex* (here we follow established terminology). A τ-theory T is called *convex* if for every finite set S of atomic τ-formulas the set $T \cup S \cup \{x_1 \neq y_1, \ldots, x_m \neq y_m\}$ is satisfiable whenever $T \cup S \cup \{x_j \neq y_j\}$ is satisfiable for each $j \leq m$.

The assumption that a relation symbol denoting the inequality relation is part of the signatures τ_1 and τ_2 is often implicit in the literature treating the Nelson-Oppen method. It would be interesting to explore when it can be dropped, but we will not pursue this here. The central question of this article is the following.

Question 2. In which settings are the Nelson-Oppen conditions (and in particular, the convexity condition) not only sufficient, but also necessary for polynomial-time tractability of the combined CSP?

Again, for general theories T_1 and T_2, this is a too ambitious research goal; but we will study it for generic combinations of ω-categorical theories T_1, T_2 with infinite models. In this setting, the first condition that both T_1 and T_2 are stably infinite is trivially satisfied. The third condition on T_i, convexity, is equivalent to the existence of a binary injective polymorphism of the (up to isomorphism unique) countably infinite model of T_i (see Sect. 5). We mention that binary injective polymorphisms played an important role in several recent infinite-domain complexity classifications [10,11,27].

1.3   Results

To state our results concerning Question 1 and Question 2 we need basic terminology for permutation groups. A permutation group G on a set A is called

- *n-transitive* if for all tuples $\bar{b}, \bar{c} \in A^n$ having pairwise distinct entries there exists a permutation $g \in G$ such that $g(\bar{b}) = \bar{c}$ (where permutations are applied to tuples componentwise). G is called *transitive* if it is 1-transitive.
- *n-set-transitive* if for all subsets B, C of A with $|B| = |C| = n$ there exists a permutation $g \in G$ such that $g(B) := \{g(b) \mid b \in B\} = C$.

A structure is called *n*-transitive (or *n*-set-transitive) if its automorphism group is. The existence of generic combinations can be characterised as follows (see Sect. 3 for the proof).

Proposition 1. *Let $\mathfrak{B}_1$ and $\mathfrak{B}_2$ be countably infinite ω-categorical structures with disjoint relational signatures. Then $\mathfrak{B}_1$ and $\mathfrak{B}_2$ have a generic combination if and only if both $\mathfrak{B}_1$ and $\mathfrak{B}_2$ do not have algebraicity (in the model-theoretic sense; see Sect. 3) or at least one of $\mathfrak{B}_1$ and $\mathfrak{B}_2$ has an automorphism group which is n-transitive for all $n \in \mathbb{N}$.*

Our main result concerns Question 2 for generic combinations $\mathfrak{B}$ of countably infinite ω-categorical structures $\mathfrak{B}_1$ and $\mathfrak{B}_2$; as we mentioned before, if the generic combination exists, it is up to isomorphism unique, and again ω-categorical. Note that a structure $\mathfrak{A}$ that is 2-set-transitive gives rise to a directed graph $(A; E)$: fix two distinct elements b_1, b_2 of $\mathfrak{A}$; then two vertices c_1, c_2 are joined by a directed edge iff there exists an automorphism α with $\alpha(b_1) = c_1$ and $\alpha(b_2) = c_2$. Note that by 2-set-transitivity, it does not matter which elements b_1 and b_2 we choose, there are at most two resulting graphs and they are always isomorphic. Also note that if the structure is 2-set-transitive and not 2-transitive, then the resulting directed graph is a *tournament*, i.e., it is without loops and for any two distinct vertices a, b either $(a, b) \in E$ or $(b, a) \in E$, but not both. Examples of 2-set-transitive tournaments are the order of the rationals $(\mathbb{Q}; <)$, the countable random tournament (see, e.g., Lachlan [28]), and the countable homogeneous local order $S(2)$ (also see [23]). If $\bar{a} = (a_1, \ldots, a_n) \in B^n$ and G is a permutation group on B then $G\bar{a} := \{(\alpha(a_1), \ldots, \alpha(a_n)) \mid \alpha \in G\}$ is called the *orbit* of $\bar{a}$ (with respect to G); orbits of pairs (i.e., $n = 2$) are also called *orbitals*. Orbitals of pairs of equal elements are called *trivial*. To simplify the presentation, we introduce the following shortcut.

Definition 1. *A structure has property* J *if it is a countably infinite ω-categorical structure which is 2-set-transitive, but not 2-transitive, and contains binary symbols for the inequality relation and for one of the two non-trivial orbitals.*

We give some examples of structures with property J.

Example 1. *The structure* $(\mathbb{Q}; \neq, <, R_{\mathrm{mi}})$ *where*

$$R_{\mathrm{mi}} := \{(x, y, z) \in \mathbb{Q}^3 \mid x \geq y \vee x > z\}.$$

Polynomial-time tractability of the CSP of this structure has been shown in [12].

Example 2. *The structure* $(\mathbb{Q}; \neq, <, R_{ll})$ *where*

$$R_{ll} := \{(x, y, z) \in \mathbb{Q}^3 \mid x < y \vee x < z \vee x = y = z\}.$$

Polynomial-time tractability of the CSP of this structure has been shown in [13].

Further examples of structure with property J come from expansions of the countable random tournament and the countable homogeneous local order mentioned above. The proof of the following theorem can be found in Sect. 5.

Theorem 3. *Let $\mathfrak{B}$ be the generic combination of two structures $\mathfrak{B}_1$ and $\mathfrak{B}_2$ with property J such that $\mathrm{CSP}(\mathfrak{B}_1)$ and $\mathrm{CSP}(\mathfrak{B}_2)$ are in P. Then one of the following applies:*

- *$\mathrm{Th}(\mathfrak{B}_1)$ or $\mathrm{Th}(\mathfrak{B}_2)$ is not convex; in this case, $\mathrm{CSP}(\mathfrak{B})$ is NP-hard.*
- *Each of $\mathrm{Th}(\mathfrak{B}_1)$ and $\mathrm{Th}(\mathfrak{B}_2)$ is convex, and $\mathrm{CSP}(\mathfrak{B})$ is in P.*

In other words, either the Nelson-Oppen conditions apply, and CSP($\mathfrak{B}$) is in P, or otherwise CSP($\mathfrak{B}$) is NP-complete.

Example 4. *Let $\mathfrak{B}_1$ be the relational structure $(\mathbb{Q}; <, \neq, R_{\mathrm{mi}})$ where R_{mi} is defined as above. Let $\mathfrak{B}_2 := (\mathbb{Q}; \prec, \not\prec)$ where $\prec$ also denotes the strict order of the rationals, and $\not\prec$ also denotes the inequality relation (we chose different symbols than $<$ and $\neq$ to make the signatures disjoint). It is easy to see that $\mathfrak{B}_1$ and $\mathfrak{B}_2$ satisfy the assumptions of Proposition 1, so they have a generic combination $\mathfrak{B}$. It is also easy to see that $\mathfrak{B}_1$ and $\mathfrak{B}_2$ are 2-set-transitive, but not 2-transitive. We have already mentioned that they also have polynomial-time tractable CSPs. However, $\mathfrak{B}_1$ does not have a convex theory, and hence our result implies that the CSP of the combined structure is NP-complete (we invite the reader to find an NP-hardness proof without using our theorem!).*

A structure $\mathfrak{B}_1$ is called a *reduct* of a structure $\mathfrak{B}_2$, and $\mathfrak{B}_2$ is called an *expansion* of $\mathfrak{B}_1$, if $\mathfrak{B}_1$ is obtained from $\mathfrak{B}_2$ by dropping some of the relations of $\mathfrak{B}_1$. If $\mathfrak{B}_1$ is a reduct of $\mathfrak{B}_2$ with the signature τ then we write $\mathfrak{B}_2^\tau$ for $\mathfrak{B}_1$. An expansion $\mathfrak{B}_2$ of $\mathfrak{B}_1$ is called a *first-order expansion* if all additional relations in $\mathfrak{B}_2$ have a first-order definition in $\mathfrak{B}_1$. A structure $\mathfrak{B}_1$ is called a *first-order reduct* if $\mathfrak{B}_1$ is a reduct of a first-order expansion of $\mathfrak{B}_2$. Note that if a structure $\mathfrak{B}$ is 2-set-transitive then so is every first-order reduct of $\mathfrak{B}$ (since its automorphism group contains the automorphisms of $\mathfrak{B}$).

The CSPs for first-order reducts of $\mathbb{Q}$ have been called *temporal CSPs*; their computational complexity has been classified completely [12]. There are many interesting polynomial-time tractable temporal CSPs that have non-convex theories, which makes temporal CSPs a particularly interesting class for understanding the situation where the Nelson-Oppen conditions do not apply. Generic combinations of temporal CSPs are isomorphic to first-order reducts of the *countable random permutation* introduced in [22] and studied in [29]; a complexity classification of the CSPs of all reducts of the random permutation (as e.g. in [10,12,27] for simpler structures than the random permutation) is out of reach for the current methods (in particular, the classification method via a reduction to the finite-domain CSP dichotomy from [14] cannot be applied).

Examples of ω-categorical structures with 2-transitive automorphism groups can be found in phylogenetic analysis; see [10]. A generic combination of a structure with a 2-transitive automorphism with $(\mathbb{Q}; <)$ is no longer 2-transitive, but still 2-set-transitive (this will become obvious from the results in Sect. 3). So any 2-transitive structure without algebraicity can be used to produce further interesting examples that satisfy the conditions of Theorem 3.

2   Combinations of CSPs

We already mentioned that our definition of CSPs for theories is a strict generalisation of the notion of CSPs for structures, and this will be clarified by the following proposition which is an immediate consequence of Proposition 2.4.6 in [6].

Proposition 2. *Let T be a first-order theory with finite relational signature. Then there exists a structure $\mathfrak{B}$ such that $\mathrm{CSP}(\mathfrak{B}) = \mathrm{CSP}(T)$ if and only if T has the* Joint Homomorphism Property (JHP), *that is, for any two models $\mathfrak{A}, \mathfrak{B}$ of T there exists a model $\mathfrak{C}$ of T such that both $\mathfrak{A}$ and $\mathfrak{B}$ homomorphically map to $\mathfrak{C}$.*

Example 5. *A simple example of two theories T_1, T_2 with the JHP such that $T_1 \cup T_2$ does not have the JHP is given by*

$$T_1 := \{\forall x, y \, ((O(x) \wedge O(y)) \Rightarrow x = y)\}$$
$$T_2 := \{\forall x. \, \neg(P(x) \wedge Q(x))\}$$

Suppose for contradiction that $T_1 \cup T_2$ has the JHP. Note that

$$T_1 \cup T_2 \cup \{\exists x(O(x) \wedge P(x))\} \quad and \quad T_1 \cup T_2 \cup \{\exists y(O(y) \wedge Q(y))\}$$

are satisfiable. The JHP implies that

$$T_1 \cup T_2 \cup \{\exists x(O(x) \wedge P(x)), \exists y(O(y) \wedge Q(y))\}$$

has a model $\mathfrak{A}$, so $\mathfrak{A}$ has elements u, v satisfying $O(u) \wedge O(v) \wedge P(u) \wedge Q(v)$. Since $\mathfrak{A} \models T_1$ we must have $u = v$, and so $\mathfrak{A}$ does not satisfy the sentence $\forall x. \, \neg(P(x) \wedge Q(x))$ from T_2, a contradiction.

3   Generic Combinations

For general theories T_1, T_2 even the question whether $T_1 \cup T_2$ has the JHP might be a difficult question. But if both T_1 and T_2 are ω-categorical with a countably infinite model that *does not have algebraicity*, then $T_1 \cup T_2$ always has the JHP (a consequence of Lemma 1 below). A structure $\mathfrak{B}$ (and its first-order theory) *does not have algebraicity* if for all first-order formulas $\phi(x_0, x_1, \ldots, x_n)$ and all elements $a_1, \ldots, a_n \in B$ the set $\{a_0 \in B \mid \mathfrak{B} \models \phi(a_0, a_1, \ldots, a_n)\}$ is either infinite or contained in $\{a_1, \ldots, a_n\}$; otherwise, we say that the structure *has algebraicity*.

It is a well-known fact from model theory that the concept of having no algebraicity is closely related to the concept of *strong amalgamation* (see [25], p. 138f). The *age* of a relational τ-structure $\mathfrak{B}$ is the class of all finite τ-structures that embed into $\mathfrak{B}$. A class $\mathcal{K}$ of structures has the *amalgamation property* if for all $\mathfrak{A}, \mathfrak{B}_1, \mathfrak{B}_2 \in \mathcal{K}$ and embeddings $f_i \colon \mathfrak{A} \to \mathfrak{B}_i$, for $i = 1$ and $i = 2$, there exist $\mathfrak{C} \in \mathcal{K}$ and embeddings $g_i \colon \mathfrak{B}_i \to \mathfrak{C}$ such that $g_1 \circ f_1 = g_2 \circ f_2$. It has the *strong amalgamation property* if additionally $g_1(B_1) \cap g_2(B_2) = g_1(f_1(A)) = g_2(f_2(A))$. If $\mathcal{K}$ is a class of structures with finite relational signature which is closed under isomorphism, substructures, and has the amalgamation property, then there exists an (up to isomorphism unique) countable homogeneous structure $\mathfrak{B}$ whose age is $\mathcal{K}$ (see [26]). Moreover, in this case $\mathfrak{B}$ has no algebraicity if and only if $\mathcal{K}$ has the strong amalgamation property (see, e.g., [21]). The significance of strong amalgamation in the theory of combining decision procedures has already been pointed out by Bruttomesso et al. [19]. By the theorem of Ryll-Nardzewski, Engeler, and Svenonius (see [25]) a homogeneous structure with finite relational signature is ω-categorical, and the expansion of an ω-categorial structure by all first-order definable relations is homogeneous.

When $\mathcal{K}$ is a class of structures, we write $I(\mathcal{K})$ for the class of all structures isomorphic to a structure in $\mathcal{K}$. Let τ_1 and τ_2 be disjoint relational signatures, and let $\mathcal{K}_i$ be a class of finite τ_i-structures, for $i \in \{1,2\}$. Then $\mathcal{K}_1 * \mathcal{K}_2$ denotes the class of $(\tau_1 \cup \tau_2)$-structures given by $\{\mathfrak{A} \mid \mathfrak{A}^{\tau_1} \in I(\mathcal{K}_1) \text{ and } \mathfrak{A}^{\tau_2} \in I(\mathcal{K}_2)\}$. If B is a set and $n \in \mathbb{N}$, we write $B^{(n)}$ for the set of tuples from B^n with pairwise distinct entries.

Lemma 1. *Let T_1 and T_2 be ω-categorical theories with disjoint relational signatures τ_1 and τ_2, with infinite models without algebraicity. Then there exists an ω-categorical model $\mathfrak{B}$ of $T_1 \cup T_2$ without algebraicity such that*

$$\text{for all } k \in \mathbb{N}, \bar{a}, \bar{b} \in B^{(k)}: \quad \mathrm{Aut}(\mathfrak{B}^{\tau_1})\bar{a} \cap \mathrm{Aut}(\mathfrak{B}^{\tau_2})\bar{b} \neq \emptyset \qquad (1)$$

$$\text{and for all } k \in \mathbb{N}, \bar{a} \in B^{(k)}: \quad \mathrm{Aut}(\mathfrak{B}^{\tau_1})\bar{a} \cap \mathrm{Aut}(\mathfrak{B}^{\tau_2})\bar{a} = \mathrm{Aut}(\mathfrak{B})\bar{a}. \qquad (2)$$

The proof works via expansion with all first-order definable relations and a Fraïssé-limit. It can be found in the extended version [7].

Note that by the facts on ω-categorical structures mentioned above, the Properties (1) and (2) for $\mathfrak{B}$, $\mathfrak{B}^{\tau_1}$, $\mathfrak{B}^{\tau_2}$ are equivalent to items (2) and (3) in Sect. 1.1 for $T = \mathrm{Th}(\mathfrak{B})$, $T_1 = \mathrm{Th}(\mathfrak{B}^{\tau_1})$, $T_2 = \mathrm{Th}(\mathfrak{B}^{\tau_2})$, respectively. Lemma 1 motivates the following definition.

Definition 2 (Generic Combination). *Let $\mathfrak{B}_1$ and $\mathfrak{B}_2$ be countably infinite ω-categorical structures with disjoint relational signatures τ_1 and τ_2, and let $\mathfrak{B}$ be a model of $\mathrm{Th}(\mathfrak{B}_1) \cup \mathrm{Th}(\mathfrak{B}_2)$. If $\mathfrak{B}$ satisfies item (1) then we say that $\mathfrak{B}$ is a* free combination *of $\mathfrak{B}_1$ and $\mathfrak{B}_2$. If $\mathfrak{B}$ satisfies both item (1) and item (2) then we say that $\mathfrak{B}$ is a* generic combination *(or* random combination; see [1]*) of $\mathfrak{B}_1$ and $\mathfrak{B}_2$.*

The following can be shown via a back-and-forth argument.

Lemma 2. *Let $\mathfrak{B}_1$ and $\mathfrak{B}_2$ be countable ω-categorical structures. Then up to isomorphism, there is at most one generic combination of $\mathfrak{B}_1$ and $\mathfrak{B}_2$.*

In later proofs we need the following lemma.

Lemma 3 (Extension Lemma). *For $i = 1$ and $i = 2$, let $\mathfrak{B}_i$ be an ω-categorical structure with signature τ_i such that $\mathfrak{B}_1$ and $\mathfrak{B}_2$ have a generic combination. Let $\bar{a}, \bar{b}_1, \bar{b}_2$ be tuples such that the tuples $(\bar{a}, \bar{b}_1)$ and $(\bar{a}, \bar{b}_2)$ have pairwise distinct entries and equal length. Then there exist $\alpha_i \in \mathrm{Aut}(\mathfrak{B}_i, \bar{a})$ such that $\alpha_2(\alpha_1(\bar{b}_1)) = \bar{b}_2$.*

Proof. By the definition of free combinations (Property (1)) there exist $\alpha \in$ Aut$(\mathfrak{B}_1)$ and $\beta \in$ Aut$(\mathfrak{B}_2)$ such that $\beta(\alpha(\bar{a}, \bar{b}_1)) = (\bar{a}, \bar{b}_2)$. Note that $\alpha(\bar{a})$ lies in the same orbit as $\bar{a}$ both with respect to $\mathfrak{B}_1$ and with respect to $\mathfrak{B}_2$, so by Property (2) of generic combinations there exists an automorphism $\delta \in$ Aut$(\mathfrak{B})$ that maps $\alpha(\bar{a})$ to $\bar{a}$. Then $\alpha_1 := \delta \circ \alpha$ and $\alpha_2 := \beta \circ \delta^{-1}$ have the desired properties.

We now prove Proposition 1 that we already stated in the introduction, and which states that two countably infinite ω-categorical structures with disjoint relational signatures have a generic combination if and only if both have no algebraicity, or at least one of the structures has an automorphism group which is n-transitive for all $n \in \mathbb{N}$. Note that the countably infinite structures whose automorphism group is n-transitive for all $n \in \mathbb{N}$ are precisely the structures that are isomorphic to a first-order reduct of $(\mathbb{N}; =)$.

Proof. If both $\mathfrak{B}_1$ and $\mathfrak{B}_2$ do not have algebraicity then the existence of an ω-categorical generic combination follows from Lemma 1. If on the other hand $\mathfrak{B}_1$ is n-transitive for all n then an ω-categorical generic combination trivially exists (it will be a first-order expansion of $\mathfrak{B}_2$). The case that $\mathfrak{B}_2$ is n-transitive for all n is analogous.

For the converse direction, let $\mathfrak{B}$ be the generic combination of the τ_1-structure $\mathfrak{B}_1$ and the τ_2-structure $\mathfrak{B}_2$. Recall that $\mathfrak{B}^{\tau_i}$ is isomorphic to $\mathfrak{B}_i$, for $i \in \{1, 2\}$. By symmetry between $\mathfrak{B}_1$ and $\mathfrak{B}_2$, we will assume towards a contradiction that $\mathfrak{B}^{\tau_1}$ has algebraicity and Aut$(\mathfrak{B}^{\tau_2})$ is not n-transitive for some $n \in \mathbb{N}$. Choose n to be smallest possible, so that Aut$(\mathfrak{B}^{\tau_2})$ is not n-transitive. Therefore there exist tuples $(b_0, \ldots, b_{n-1})$ and $(c_0, \ldots, c_{n-1})$, each with pairwise distinct entries, that are in different orbits with respect to Aut$(\mathfrak{B}^{\tau_2})$. By the minimality of n, there exists $\alpha \in$ Aut$(\mathfrak{B}^{\tau_2})$ such that $\alpha(b_1, \ldots, b_{n-1}) = (c_1, \ldots, c_{n-1})$. Algebraicity of $\mathfrak{B}^{\tau_1}$ implies that there exists a first-order τ_1-formula $\phi(x_0, x_1, \ldots, x_m)$ and pairwise distinct elements $a_1, \ldots, a_m$ of $\mathfrak{B}$ such that $\phi(x, a_1, \ldots, a_m)$ holds for precisely one element $x = a_0$ other than $a_1, \ldots, a_m$ in $\mathfrak{B}$. By adding unused extra variables to ϕ we can assume that $m \geq n - 1$. Choose elements $b_n, \ldots, b_m$ such that the entries of $(b_0, \ldots, b_{n-1}, b_n, \ldots, b_m)$ are pairwise distinct and define $c_i := \alpha(b_i)$ for $i \in \{n, \ldots, m\}$. Since $\mathfrak{B}$ is a free combination, there exist tuples $(b'_0, \ldots, b'_m), (c'_0, \ldots, c'_m)$ and $\beta_1, \gamma_1 \in$ Aut$(\mathfrak{B}^{\tau_1})$ and $\beta_2, \gamma_2 \in$ Aut$(\mathfrak{B}^{\tau_2})$ such that

$$\beta_2(b_0, \ldots, b_m) = (b'_0, \ldots, b'_m), \qquad \beta_1(b'_0, \ldots, b'_m) = (a_0, \ldots, a_m),$$
$$\gamma_2(c_0, \ldots, c_m) = (c'_0, \ldots, c'_m), \qquad \gamma_1(c'_0, \ldots, c'_m) = (a_0, \ldots, a_m).$$

Because $\gamma_1^{-1} \circ \beta_1 \in$ Aut$(\mathfrak{B}^{\tau_1})$ and $\gamma_2 \circ \alpha \circ \beta_2^{-1} \in$ Aut$(\mathfrak{B}^{\tau_2})$ both map $(b'_1, \ldots, b'_m)$ to $(c'_1, \ldots, c'_m)$, and due to the second condition for generic combinations, there exists $\mu \in$ Aut$(\mathfrak{B})$ such that $\mu(b'_1, \ldots, b'_m) = (c'_1, \ldots, c'_m)$. Since any operation in Aut$(\mathfrak{B}^{\tau_1})$ preserves ϕ, we have $\gamma_1 \circ \mu \circ \beta_1^{-1}(a_0, \ldots, a_m) = (a_0, \ldots, a_m)$. Therefore μ must map b'_0 to c'_0. Hence, $\gamma_2^{-1} \circ \mu \circ \beta_2 \in$ Aut$(\mathfrak{B}^{\tau_2})$ maps $(b_0, \ldots, b_{n-1})$ to $(c_0, \ldots, c_{n-1})$, contradicting our assumption that they lie in different orbits with respect to Aut$(\mathfrak{B}^{\tau_2})$.

4   Difficulties for a General Complexity Classification

Let T_1 and T_2 be ω-categorical theories with disjoint finite relational signatures such that $\mathrm{CSP}(T_1)$ is in P and $\mathrm{CSP}(T_2)$ is in P. The results in this section suggest that in general we cannot hope to get a classification of the complexity of $\mathrm{CSP}(T_1 \cup T_2)$. We use the result from [8] that there are homogeneous directed graphs $\mathfrak{B}$ such that $\mathrm{CSP}(\mathfrak{B})$ is undecidable. There are even homogeneous directed graphs $\mathfrak{B}$ such that $\mathrm{CSP}(\mathfrak{B})$ is *coNP-intermediate*, i.e., in coNP, but neither coNP-hard nor in P [8] (unless P = coNP). All of the homogeneous graphs $\mathfrak{B}$ used in [8] can be described by specifying a set of finite tournaments $\mathcal{T}$. Let $\mathcal{C}$ be the class of all finite directed loopless graphs $\mathfrak{A}$ such that no tournament from $\mathcal{T}$ embeds into $\mathfrak{A}$. It can be checked that $\mathcal{C}$ is a strong amalgamation class; the Fraïssé-limits of those classes are called the *Henson digraphs*.

Proposition 3. *For every Henson digraph $\mathfrak{B}$ there exist ω-categorical convex theories T_1 and T_2 with disjoint finite relational signatures such that $\mathrm{CSP}(T_1)$ is in P, $\mathrm{CSP}(T_2)$ is in P, and $\mathrm{CSP}(T_1 \cup T_2)$ is polynomial-time Turing equivalent to $\mathrm{CSP}(\mathfrak{B})$.*

The proof is omitted for reasons of space, but can be found in [7]. Note that the Nelson-Oppen conditions do not apply here because it is crucial for our construction that T_1 does not contain a symbol for inequality. We mention that another example of two theories such that $\mathrm{CSP}(T_1)$ and $\mathrm{CSP}(T_2)$ are decidable but $\mathrm{CSP}(T_1 \cup T_2)$ is not can be found in [18].

5   On the Necessity of the Nelson-Oppen Conditions

In this section we introduce a large class of ω-categorical theories where the condition of Nelson and Oppen (the existence of binary injective polymorphisms) is not only a sufficient, but also a necessary condition for the polynomial-time tractability of generic combinations (unless P = NP); in particular, we prove Theorem 3 from the introduction. We need the following characterisation of convexity of ω-categorical theories.

Theorem 6 (Lemma 6.1.3 in [6]).
Let $\mathfrak{B}$ be an ω-categorical structure and let T be its first-order theory. Then the following are equivalent.

- *T is convex;*
- *$\mathfrak{B}$ has a binary injective polymorphism.*

Moreover, if $\mathfrak{B}$ contains the relation $\neq$, these conditions are also equivalent to the following.

- *for every finite set S of atomic τ-formulas such that $S \cup T \cup \{x_1 \neq y_1\}$ is satisfiable and $S \cup T \cup \{x_2 \neq y_2\}$ is satisfiable, then $T \cup S \cup \{x_1 \neq y_1, x_2 \neq y_2\}$ is satisfiable, too.*

The following well-known fact easily follows from many published results, e.g., from the results in [16]. An operation $f\colon B^k \to B$ is called *essentially unary* if there exists an $i \le k$ and a function $g\colon B \to B$ such that $f(x_1, \ldots, x_k) = g(x_i)$ for all $x_1, \ldots, x_k \in B$. The operation f is called *essential* if it is not essentially unary.

Proposition 4 (see [16]). *Let $\mathfrak{B}$ be an infinite ω-categorical structure with finite relational signature containing the relation $\neq$ and such that all polymorphisms of $\mathfrak{B}$ are essentially unary. Then $\mathrm{CSP}(\mathfrak{B})$ is NP-hard.*

Hence, we want to show that the existence of an essential polymorphism of the generic combination of two countably infinite ω-categorical structures $\mathfrak{B}_1$ and $\mathfrak{B}_2$ implies the existence of a binary injective polymorphism. The key technical result, which we prove at the end of this section, is the following proposition.

Proposition 5. *Let $\mathfrak{B}_1, \mathfrak{B}_2$ be ω-categorical structures with generic combination $\mathfrak{B}$ so that*

- *each of $\mathfrak{B}_1$ and $\mathfrak{B}_2$ has a relation symbol that denotes the relation $\neq$;*
- *$\mathfrak{B}$ has a binary essential polymorphism;*
- *$\mathfrak{B}_1$ is 2-set-transitive; and*
- *$\mathfrak{B}_2$ is 1-transitive and contains a binary antisymmetric irreflexive relation.*

Then $\mathfrak{B}_1$ must have a binary injective polymorphism.

To apply Proposition 5, we therefore need to prove the existence of binary essential polymorphisms of generic combinations $\mathfrak{B}$. For this, we use an idea that first appeared in [12] and was later generalized in [6], based on the following concept. A permutation group G on a set B has the *orbital extension property (OEP)* if there is an orbital O such that for all $b_1, b_2 \in B$ there is an element $c \in B$ where $(b_1, c) \in O$ and $(b_2, c) \in O$. The relevance of this property comes from the following lemma.

Lemma 4 (Kára's Lemma; see [6], Lemma 5.3.10). *Let $\mathfrak{B}$ be a structure with an essential polymorphism and an automorphism group with the OEP. Then $\mathfrak{B}$ must have a binary essential polymorphism.*

To apply this lemma to the generic combination $\mathfrak{B}$ of $\mathfrak{B}_1$ and $\mathfrak{B}_2$, we have to verify that $\mathrm{Aut}(\mathfrak{B})$ has the OEP.

Lemma 5. *Any 2-set-transitive permutation group action on a set with at least 3 elements has the OEP.*

Lemma 6. *Let $\mathfrak{B}$ be a generic combination of two ω-categorical structures $\mathfrak{B}_1$ and $\mathfrak{B}_2$ with the OEP. Then $\mathfrak{B}$ has the OEP.*

Now, we proof Theorem 3. Property J from Definition 1 is needed in order to apply Proposition 5 twice.

Proof (Proof of Theorem 3). If all polymorphisms of $\mathfrak{B}$ are essentially unary then Proposition 4 shows that $\mathrm{CSP}(\mathfrak{B})$ is NP-hard. Otherwise, $\mathfrak{B}$ has a binary essential polymorphism by Lemma 4, because $\mathfrak{B}$ has the OEP by Lemmas 5 and 6. Property J implies that $\mathfrak{B}_i$, for $i = 1$ and $i = 2$, is 2-set-transitive and contains a binary relation symbol that denotes $\neq$ and a binary relation symbol that denotes the orbital of $\mathfrak{B}_i$, which is a binary antisymmetric irreflexive relation. Thus, $\mathfrak{B}_1$ and $\mathfrak{B}_2$ satisfy the assumptions of Proposition 5. It follows that $\mathfrak{B}_1$ has a binary injective polymorphism. By Theorem 6, this shows that $\mathrm{Th}(\mathfrak{B}_1)$ is convex. Since we have the same assumptions on $\mathfrak{B}_1$ and on $\mathfrak{B}_2$, we can use Proposition 5 again to show that also $\mathrm{Th}(\mathfrak{B}_2)$ is convex. Now, the Nelson-Oppen combination procedure implies that $\mathrm{CSP}(\mathfrak{B})$ is in P.

Proof (Proof of Proposition 5). Since $\mathfrak{B}_1$ and $\mathfrak{B}_2$ have a generic combination, by Proposition 1 either both $\mathfrak{B}_1$ and $\mathfrak{B}_2$ have no algebraicity or at least one of $\mathfrak{B}_1, \mathfrak{B}_2$ is n-transitive for all $n \in \mathbb{N}$. The structure $\mathfrak{B}_2$ is not 2-transitive. Suppose that $\mathfrak{B}_1$ is n-transitive for all $n \in \mathbb{N}$. Since $\mathfrak{B}$ has a binary essential polymorphism, so has $\mathfrak{B}_1$. Since $\mathfrak{B}_1$ also contains a symbol that denotes the relation $\neq$, it must also have a binary injective polymorphism (see [11]) and we are done. So we assume in the following that both $\mathfrak{B}_1$ and $\mathfrak{B}_2$ do not have algebraicity. Since $\mathfrak{B}_1$ and $\mathfrak{B}_2$ are isomorphic to reducts of $\mathfrak{B}$, we may assume that they actually are reducts of $\mathfrak{B}$. Let ϕ be a primitive positive formula over the signature of $\mathfrak{B}_1$ and suppose that $\phi \wedge x_1 \neq y_1$ has a satisfying assignment s_1 over $\mathfrak{B}_1$ and $\phi \wedge x_2 \neq y_2$ has a satisfying assignment s_2 over $\mathfrak{B}_2$. By Theorem 6 it suffices to show that in $\mathfrak{B}_1$ there exists a satisfying assignment to

$$\phi \wedge x_1 \neq y_1 \wedge x_2 \neq y_2. \tag{3}$$

If $s_1(x_2) \neq s_1(y_2)$ or if $s_2(x_1) \neq s_2(y_1)$ then there is nothing to be shown, so we assume that this is not the case. Let f be the binary essential polymorphism of $\mathfrak{B}$. Then there are $a_1, a_2, a_3, b_1, b_2, b_3 \in B$ such that $f(a_2, b_1) \neq f(a_3, b_1)$ and $f(a_1, b_2) \neq f(a_1, b_3)$. It is easy to see that then there also exist elements $u_1, u_2, v_1, v_2 \in B$ such that $f(u_1, v_1) \neq f(u_2, v_1)$ and $f(u_1, v_1) \neq f(u_1, v_2)$ (choose $u_1 = a_1$, $v_1 = b_1$ and suitable $u_2 \in \{a_2, a_3\}$, $v_2 \in \{b_2, b_3\}$). Note that in particular $u_1 \neq u_2$ and $v_1 \neq v_2$. By the 2-set-transitivity of $\mathfrak{B}_1$, there exist $\alpha_1, \alpha_2 \in \mathrm{Aut}(\mathfrak{B}_1)$ such that

$$\alpha_1(\{s_1(x_1), s_1(y_1)\}) = \{u_1, u_2\} \quad \text{and} \quad \alpha_2(\{s_2(x_2), s_2(y_2)\}) = \{v_1, v_2\}.$$

By renaming variables if necessary we may assume that $\alpha_1(s_1(x_1), s_1(y_1)) = (u_1, u_2)$ and $\alpha_2(s_2(x_2), s_2(y_2)) = (v_1, v_2)$.
Note that $|s_1(\{x_1, y_1, x_2, y_2\})|, |s_2(\{x_1, y_1, x_2, y_2\})| \in \{2, 3\}$.

Case 1. $|s_1(\{x_1, y_1, x_2, y_2\})| = |s_2(\{x_1, y_1, x_2, y_2\})| = 3$. In other words, $s_1(x_2) = s_1(y_2) \notin \{s_1(x_1), s_1(y_1)\}$ and $s_2(x_1) = s_2(y_1) \notin \{s_2(x_2), s_2(y_2)\}$.
   By the transitivity of $\mathrm{Aut}(\mathfrak{B}_1)$ there exist $\beta_1, \beta_2 \in \mathrm{Aut}(\mathfrak{B}_1)$ such that $\beta_1(s_1(x_2)) = u_1$ and $\beta_2(s_2(y_1)) = v_1$. We can choose $\beta_1 \in \mathrm{Aut}(\mathfrak{B}_1)$ such that

$\beta_1(s_1(x_1)), \beta_1(s_1(y_1))$ are distinct from $\alpha_1(s_1(x_2))$ and u_2: to see this, note that $\mathrm{Aut}(\mathfrak{B}_1, u_1)$ has no finite orbits other than $\{u_1\}$ because $\mathfrak{B}_1$ has no algebraicity, and by Neumann's lemma (see e.g. [25], p. 141, Corollary 4.2.2) there exists a $g \in \mathrm{Aut}(\mathfrak{B}_1, u_1)$ such that

$$g(\{\beta_1(s_1(x_1)), \beta_1(s_1(x_1))\}) \cap \{\alpha_1(s_1(x_2)), u_2\} = \emptyset.$$

We can thus replace β by $g \circ \beta$. Hence, $u_1, u_2, \beta_1(s_1(x_1)), \alpha_1(s_1(x_2)), \beta_1(s_1(y_1))$ are pairwise distinct. Likewise, we can choose $\beta_2 \in \mathrm{Aut}(\mathfrak{B}_2)$ such that $v_1, v_2,$ $\beta_2(s_2(x_2)), \alpha_2(s_2(x_1)), \beta_2(s_2(y_2))$ are pairwise distinct.

Let R be the binary antisymmetric irreflexive relation of $\mathfrak{B}_2$, choose any $(a, b) \in R$, and let $\alpha \in \mathrm{Aut}(\mathfrak{B}_2)$ be such that $\alpha(a) = b$. Define $c := \alpha(b)$ and note that $c \neq a$ since otherwise $(a, b), (b, a) \in R$ contrary to our assumptions. Since $\mathfrak{B}$ is a generic combination and $\mathfrak{B}_1, \mathfrak{B}_2$ are transitive, $\mathfrak{B}$ is transitive as well and we can choose a, b, c disjoint from u_1, u_2, v_1, v_2 by Neumanns Lemma as above. Then the Extension Lemma (Lemma 3) asserts the existence of elements u_3, u_4, u_5 and automorphisms $\delta_{i,1} \in \mathrm{Aut}(\mathfrak{B}_i)$, for $i \in \{1, 2\}$, such that

$$\delta_{1,1}(u_1, u_2, u_3, u_4, u_5) = (u_1, u_2, \beta_1(s_1(x_1)), \alpha_1(s_1(x_2)), \beta_1(s_1(y_1)))$$
$$\text{and } \delta_{2,1}(u_1, u_2, u_3, u_4, u_5) = (u_1, u_2, a, b, c).$$

Similarly, there are elements v_3, v_4, v_5 and $\delta_{i,2} \in \mathrm{Aut}(\mathfrak{B}_i)$, for $i \in \{1, 2\}$, such that

$$\delta_{1,2}(v_1, v_2, v_3, v_4, v_5) = (v_1, v_2, \beta_2(s_2(x_2)), \alpha_2(s_2(x_1)), \beta_2(s_2(y_2)))$$
$$\text{and } \delta_{2,2}(v_1, v_2, v_3, v_4, v_5) = (v_1, v_2, a, b, c).$$

See Fig. 1. If $f(u_4, v_3) \neq f(u_4, v_5)$, then

$$s := f(\delta_{1,1}^{-1}\alpha_1 s_1, \delta_{1,2}^{-1}\beta_2 s_2)$$

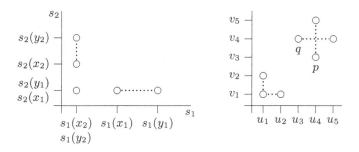

Fig. 1. An illustration of the first case in the proof of Proposition 5. Dashed edges indicate (potential) inequalities between function values.

is a solution to (3):

$$s(x_1) = f(\delta_{1,1}^{-1}\alpha_1 s_1(x_1), \delta_{1,2}^{-1}\beta_2 s_2(x_1)) = f(\delta_{1,1}^{-1}(u_1), \delta_{1,2}^{-1}(v_1)) = f(u_1, v_1)$$
$$\neq f(u_2, v_1) = f(\delta_{1,1}^{-1}(u_2), \delta_{1,2}^{-1}(v_1)) = f(\delta_{1,1}^{-1}\alpha_1 s_1(y_1), \delta_{1,2}^{-1}\beta_2 s_2(y_1)) = s(y_1)$$
$$s(x_2) = f(\delta_{1,1}^{-1}\alpha_1 s_1(x_2), \delta_{1,2}^{-1}\beta_2 s_2(x_2)) = f(u_4, v_3)$$
$$\neq f(u_4, v_5) = f(\delta_{1,1}^{-1}\alpha_1 s_1(x_2), v_5) = f(\delta_{1,1}^{-1}\alpha_1 s_1(y_2), \delta_{1,2}^{-1}\beta_2 s_2(y_2)) = s(y_2).$$

So let us assume that $p := f(u_4, v_3) = f(u_4, v_5)$. If $f(u_3, v_4) \neq f(u_5, v_4)$, then

$$s := f(\delta_{1,1}^{-1}\beta_1 s_1, \delta_{1,2}^{-1}\alpha_2 s_2)$$

is a solution to (3), by similar reasoning as above. Thus, we also assume that $q := f(u_3, v_4) = f(u_5, v_4)$. As $(a, b) \in R$, $(b, c) \in R$, $\delta_{2,1}, \delta_{2,2} \in \mathrm{Aut}(\mathfrak{B}_2)$, and f preserves R,

$$(p, q) = (f(u_4, v_3), f(u_5, v_4)) = (f(\delta_{2,1}^{-1}(b), \delta_{2,2}^{-1}(a)), f(\delta_{2,1}^{-1}(c), \delta_{2,2}^{-1}(b))) \in R.$$

Similarly,

$$(q, p) = (f(u_3, v_4), f(u_4, v_5)) = (f(\delta_{2,1}^{-1}(a), \delta_{2,2}^{-1}(b)), f(\delta_{2,1}^{-1}(b), \delta_{2,2}^{-1}(c))) \in R.$$

Hence, both $(p, q) \in R$ and $(q, p) \in R$, contradicting our assumptions.

Case 2. $|s_1(\{x_1, y_1, x_2, y_2\})| = |s_2(\{x_1, y_1, x_2, y_2\})| = 2$.

Case 2a. $s_1(x_2) = s_1(y_2) = s_1(x_1)$ and $s_2(x_1) = s_2(y_1) = s_2(x_2)$.
In this case, it is easy to verify that

$$s := f(\alpha_1(s_1), \alpha_2(s_2))$$

is a solution to (3).

Case 2b. $s_1(x_2) = s_1(y_2) = s_1(y_1)$ and $s_2(x_1) = s_2(y_1) = s_2(x_2)$. This case can be proven similarly to Case 1 and is written out in [7].

Case 2c. $s_1(x_2) = s_1(y_2) = s_1(x_1)$ and $s_2(x_1) = s_2(y_1) = s_2(y_2)$. This case can be shown analogously to case 2b (swap x and y).

Case 2d. $s_1(x_2) = s_1(y_2) = s_1(y_1)$ and $s_2(x_1) = s_2(y_1) = s_2(y_2)$. This case can be shown analogously to case 2a (swap x and y).

Case 3. $|s_1(\{x_1, y_1, x_2, y_2\})| = 3$ and $|s_2(\{x_1, y_1, x_2, y_2\})| = 2$.

Case 3a. $s_1(x_2) = s_1(y_2) \notin \{s_1(x_1), s_1(y_1)\}$ and $s_2(x_1) = s_2(y_1) = s_2(x_2)$. The proof is similar to the first and to the second case.

Case 3b. $s_1(x_2) = s_1(y_2) \notin \{s_1(x_1), s_1(y_1)\}$ and $s_2(x_1) = s_2(y_1) = s_2(y_2)$. The proof is analogous to Case 3a.

Case 4. $|s_1(\{x_1, y_1, x_2, y_2\})| = 2$ and $|s_2(\{x_1, y_1, x_2, y_2\})| = 3$. This case is symmetric to Case 3 (swap s_1 and s_2).

6   Conclusion and Future Work

For many theories T_1 and T_2 we have shown that the Nelson-Oppen conditions are not only a sufficient, but also a necessary condition for the polynomial-time tractability of the combined constraint satisfaction problem $\mathrm{CSP}(T_1 \cup T_2)$. Our results imply for example the following complexity classification for combinations of temporal CSPs.

Corollary 1. *Let $\mathfrak{B}_1$ and $\mathfrak{B}_2$ be two first-order expansions of $(\mathbb{Q}; <, \neq)$; rename the relations of $\mathfrak{B}_1$ and $\mathfrak{B}_2$ so that $\mathfrak{B}_1$ and $\mathfrak{B}_2$ have disjoint signatures. Then $\mathrm{CSP}(\mathrm{Th}(\mathfrak{B}_1) \cup \mathrm{Th}(\mathfrak{B}_2))$ is in P if $\mathrm{CSP}(\mathfrak{B}_1)$ and $\mathrm{CSP}(\mathfrak{B}_2)$ are in P and if both $\mathrm{Th}(\mathfrak{B}_1)$ and $\mathrm{Th}(\mathfrak{B}_2)$ are convex. Otherwise, $\mathrm{CSP}(\mathrm{Th}(\mathfrak{B}_1) \cup \mathrm{Th}(\mathfrak{B}_2))$ is NP-hard.*

This follows from Proposition 1 which characterises the existence of a generic combination of T_1 and T_2, and from Theorem 3 which classifies the computational complexity of the generic combination.

It would be interesting to show our complexity result for even larger classes of ω-categorical theories T_1 and T_2. It would also be interesting to drop the assumption that the signatures of T_1 and T_2 contain a symbol for the inequality relation.

References

1. Ackerman, N., Freer, C., Patel, R.: Invariant measures concentrated on countable structures. In: Forum of Mathematics Sigma, vol. 4 (2016)
2. Baader, F., Schulz, K.U.: Combining constraint solving. In: Goos, G., Hartmanis, J., van Leeuwen, J., Comon, H., Marché, C., Treinen, R. (eds.) CCL 1999. LNCS, vol. 2002, pp. 104–158. Springer, Heidelberg (2001). https://doi.org/10.1007/3-540-45406-3_3
3. Barto, L., Kompatscher, M., Olšák, M., Pinsker, M., Pham, T.V.: The equivalence of two dichotomy conjectures for infinite domain constraint satisfaction problems. In: Proceedings of the 32nd Annual ACM/IEEE Symposium on Logic in Computer Science, LICS 2017 (2017). Preprint arXiv:1612.07551
4. Barto, L., Opršal, J., Pinsker, M.: The wonderland of reflections. Israel J. Math. **223**, 363–398 (2017). Preprint arXiv:1510.04521
5. Barto, L., Pinsker, M.: The algebraic dichotomy conjecture for infinite domain constraint satisfaction problems. In: Proceedings of the 31st Annual IEEE Symposium on Logic in Computer Science, LICS 2016, pp. 615–622 (2016). Preprint arXiv:1602.04353
6. Bodirsky, M.: Complexity classification in infinite-domain constraint satisfaction. Mémoire d'habilitation à diriger des recherches, Université Diderot - Paris 7. arXiv:1201.0856 (2012)
7. Bodirsky, M., Greiner, J.: Complexity of combinations of qualitative constraint satisfaction problems. Preprint arXiv:1801.05965 (2018)
8. Bodirsky, M., Grohe, M.: Non-dichotomies in constraint satisfaction complexity. In: Aceto, L., Damgård, I., Goldberg, L.A., Halldórsson, M.M., Ingólfsdóttir, A., Walukiewicz, I. (eds.) ICALP 2008. LNCS, vol. 5126, pp. 184–196. Springer, Heidelberg (2008). https://doi.org/10.1007/978-3-540-70583-3_16
9. Bodirsky, M., Jonsson, P.: A model-theoretic view on qualitative constraint reasoning. J. Artif. Intell. Res. **58**, 339–385 (2017)

10. Bodirsky, M., Jonsson, P., Pham, T.V.: The complexity of phylogeny constraint satisfaction problems. ACM Trans. Comput. Log. (TOCL) **18**(3), 23 (2017). An extended abstract appeared in the conference STACS 2016
11. Bodirsky, M., Kára, J.: The complexity of equality constraint languages. Theory Comput. Syst. **3**(2), 136–158 (2008). A conference version appeared in the proceedings of Computer Science Russia, CSR 2006
12. Bodirsky, M., Kára, J.: The complexity of temporal constraint satisfaction problems. J. ACM **57**(2), 1–41 (2009). An extended abstract appeared in the Proceedings of the Symposium on Theory of Computing, STOC
13. Bodirsky, M., Kára, J.: A fast algorithm and Datalog inexpressibility for temporal reasoning. ACM Trans. Comput. Log. **11**(3), 15 (2010)
14. Bodirsky, M., Mottet, A.: Reducts of finitely bounded homogeneous structures, and lifting tractability from finite-domain constraint satisfaction. In: Proceedings of the 31st Annual IEEE Symposium on Logic in Computer Science, LICS 2016, pp. 623–632 (2016). Preprint ArXiv:1601.04520
15. Bodirsky, M., Nešetřil, J.: Constraint satisfaction with countable homogeneous templates. J. Log. Comput. **16**(3), 359–373 (2006)
16. Bodirsky, M., Pinsker, M.: Topological Birkhoff. Trans. Am. Math. Soc. **367**, 2527–2549 (2015)
17. Bodirsky, M., Pinsker, M., Pongrácz, A.: Projective clone homomorphisms. J. Symb. Log. (2014, accepted for publication). Preprint arXiv:1409.4601
18. Bonacina, M.P., Ghilardi, S., Nicolini, E., Ranise, S., Zucchelli, D.: Decidability and undecidability results for Nelson-Oppen and rewrite-based decision procedures. In: Furbach, U., Shankar, N. (eds.) IJCAR 2006. LNCS, vol. 4130, pp. 513–527. Springer, Heidelberg (2006). https://doi.org/10.1007/11814771_42
19. Bruttomesso, R., Ghilardi, S., Ranise, S.: Quantifier-free interpolation in combinations of equality interpolating theories. ACM Trans. Comput. Log. **15**(1), 5 (2014)
20. Bulatov, A.A.: A dichotomy theorem for nonuniform CSPs. FOCS 2017 (2017, accepted for publication). arXiv:1703.03021
21. Cameron, P.J.: Oligomorphic Permutation Groups. Cambridge University Press, Cambridge (1990)
22. Cameron, P.J.: Homogeneous permutations. Electron. J. Comb. **9**(2), 2 (2002)
23. Cherlin, G.: The Classification of Countable Homogeneous Directed Graphs and Countable Homogeneous n-Tournaments, vol. 131, no. 621. AMS Memoirs, New York, January 1998
24. Creignou, N., Kolaitis, P.G., Vollmer, H. (eds.): Complexity of Constraints. LNCS, vol. 5250. Springer, Heidelberg (2008). https://doi.org/10.1007/978-3-540-92800-3
25. Hodges, W.: Model Theory. Cambridge University Press, Cambridge (1993)
26. Hodges, W.: A Shorter Model Theory. Cambridge University Press, Cambridge (1997)
27. Kompatscher, M., Pham, T.V.: A complexity dichotomy for poset constraint satisfaction. In: 34th Symposium on Theoretical Aspects of Computer Science, STACS 2017. Leibniz International Proceedings in Informatics, LIPIcs, vol. 66, pp. 47:1–47:12 (2017)
28. Lachlan, A.H.: Countable homogeneous tournaments. TAMS **284**, 431–461 (1984)
29. Linman, J., Pinsker, M.: Permutations on the random permutation. Electron. J. Comb. **22**(2), 1–22 (2015)
30. Nelson, G., Oppen, D.C.: Fast decision procedures based on congruence closure. J. ACM **27**(2), 356–364 (1980)
31. Zhuk, D.: The proof of CSP dichotomy conjecture. FOCS 2017 (2017, accepted for publication). arXiv:1704.01914

A Generic Framework for Implicate Generation Modulo Theories

Mnacho Echenim, Nicolas Peltier$^{(\boxtimes)}$, and Yanis Sellami

Univ. Grenoble Alpes, CNRS, LIG, 38000 Grenoble, France
{Mnacho.Echenim,Nicolas.Peltier,Yanis.Sellami}@univ-grenoble-alpes.fr

Abstract. The clausal logical consequences of a formula are called its implicates. The generation of these implicates has several applications, such as the identification of missing hypotheses in a logical specification. We present a procedure that generates the implicates of a quantifier-free formula modulo a theory. No assumption is made on the considered theory, other than the existence of a decision procedure. The algorithm has been implemented (using the solvers MINISAT, CVC4 and Z3) and experimental results show evidence of the practical relevance of the proposed approach.

1 Introduction

We present a novel approach based on the usage of a generic SMT solver as a black box to generate ground implicates of a formula modulo a theory. Formally, the implicates of a formula ϕ modulo a theory $\mathcal{T}$ are the ground clauses C such that every model of $\mathcal{T}$ that satisfies ϕ also satisfies C; in other words, these are the *clausal $\mathcal{T}$-consequences of ϕ*. The problem of generating such implicates (up to logical entailment) is of great practical relevance, since for any implicate $\bigvee_{i=1}^{n} l_i$, the formula $\bigwedge_{i=1}^{n} \neg l_i \wedge \phi$ is $\mathcal{T}$-unsatisfiable. The set $\{\neg l_i \mid i \in [1,n]\}$ can thus be viewed as a set of hypotheses under which ϕ is $\mathcal{T}$-unsatisfiable or, dually, $\neg\phi$ is provable. This means that generating implicates can permit to identify missing hypothesis in a theorem, such as omitted lemmata or side conditions. Such hypotheses are useful to correct mistakes in specifications, but also to quickly spot why a given statement is not provable. They can be far more informative than counter-examples in this respect, since the latter are hard to analyze and can be clouded with superfluous information.

Consider for example the simple program over an array defined in Algorithm 1. It turns out that the postcondition of the program is not verified. This can be evidenced by translating the preconditions, the algorithm and the negation of the post-condition into a conjunction of logical formulas, and using an SMT solver to construct a model for this conjunction; this model can then be analyzed to determine what precondition is missing. The obtained model, however, will generally contain a hard to read array definition, and the missing precondition will not be explicitly returned. For instance, the model returned by the Z3 SMT solver [6] is (using our notations):

© Springer International Publishing AG, part of Springer Nature 2018
D. Galmiche et al. (Eds.): IJCAR 2018, LNAI 10900, pp. 279–294, 2018.
https://doi.org/10.1007/978-3-319-94205-6_19

Algorithm 1. Example(Array[Int] T, Int a, Int b)
---
1 **requires** $\forall x, y \in [a,b], x \le y \implies T[x] \le T[y]$;
2 **requires** $T[a] \ge 0$;
3 **let** $T[b+1] = T[b-1] + T[b]$;
4 **ensures** $\forall x, y \in [a, b+1], x \le y \implies T[x] \le T[y]$;

$a : 533, \ b : 533,$
$f : x \mapsto x \ge 533 \ ? \ (x \ge 534 \ ? \ 534 : 533) : 532,$
$g : x \mapsto x = 533 \ ? \ 535 : (x = 534 \ ? \ 19 : -516),$
$T : x \mapsto g(f(x)).$

Implicate generation on the other hand permits to identify the missing precondition in a more efficient manner. The first step consists in selecting the literals that can be used to generate potential explanations; these are called *abducible literals*. In this example, the natural literals to consider are all the (negations of) equalities and inequalities constructed using constants a and b, along with additional predefined constants such as 0 and 1. The second step simply consists in invoking our system, GPiD, to generate the potential missing preconditions. For this example, GPiD plugged with Z3 generates the missing precondition $a \ne b$ in less than 0.2 s. If abducible literals can be constructed using also the function symbol T, then our tool generates the other potential precondition $T[b-1] \ge 0$ in the same amount of time.

In previous work [8–10,12], we devised refinements of the superposition calculus specially tuned to derive such implicates for quantifier-free formula modulo equality with uninterpreted function symbols. We proved the soundness and deductive-completeness of the obtained procedures, i.e., we showed that the procedure derives all implicates up to redundancy. In the present work, we investigate a different approach. We provide a generic algorithm for generating such implicates, relying only on the existence of a decision procedure for the underlying theory, possibly augmented with counter-example generation capabilities to further restrict the search space. The main advantage of this approach is that it is possible to use efficient SMT solvers as black boxes, instead of having to develop specific systems for the purpose of implicate generation. Our method is based on decomposition, in the spirit of the DPLL approach. The generated implicates are constructed on a given set of candidate literals, called *abducible literals*, which is assumed to be fixed before the beginning of the search, e.g., by a human user. As far as flexibility is concerned, the algorithm also permits to only generate implicates satisfying so-called $\subseteq$-*closed* predicates without any post-processing step. We show that the algorithm is sound and complete, and we provide experimental results showing that the obtained system is much more efficient than the previous one based on superposition. We also devise generic approaches to store sets of implicates efficiently, while removing implicates that are redundant modulo the considered theory. Again, the proposed procedure relies only on the possibility of deciding validity in the underlying theory.

Related Work. The implicate generation problem has been thoroughly investigated in the context of propositional logic (see for instance [19]). Earlier approaches are based mainly on refinements of the Resolution rule [15,16,26,30], and they focus on the definition of efficient strategies to generate saturated clause sets and of compact data structures for storing the generated sets of implicates [5,14,23,29]. Other approaches use decomposition-based methods, in the style of the DPLL procedure, for generating trie-based representations of sets of prime implicates [20,21]. Recently [25], a new approach that outperforms previous algorithms has been proposed, based on max-satisfiability solving and problem reformulation. Our algorithm can be used for propositional implicate generation but it is not competitive with this new approach. Our aim with this work was rather to extend the scope of implicate generation to more expressive logics. Indeed, there have been only very few approaches dealing with logics other than propositional. Some extensions have been considered in modal logics [3,4], and algorithms have been proposed for first-order formulas, based on first-order resolution [17,18] or tableaux [22,24]. However, none of these approaches is capable of handling equality efficiently. More recently, algorithms were devised to generate sets of implicants of formulas interpreted in decidable theories [7], by combining quantifier-elimination for discarding useless variables, with model building to construct sufficient conditions for satisfiability.

The rest of the paper is structured as follows. In Sect. 2, basic definitions and notations are introduced. Section 3 contains the definition of the algorithm for generating implicates, starting with a straightforward, naive algorithm and refining it to make it more efficient. In Sect. 4 data-structures and algorithms are presented to store implicates efficiently modulo redundancy. Section 5 contains the description of the implementation and experimental results, and Sect. 6 concludes the paper. Due to space restrictions, some of the proofs are omitted. The full version is available on arXiv.

2 Preliminary Notions

Ground terms and non-quantified formulas are built inductively as usual on a sorted signature Σ. The notions of validity, models, satisfiability, etc. are defined as usual. The set of literals built on Σ is denoted by $\mathcal{L}$. Let $\mathcal{T}$ be a theory. A set of formulas S is $\mathcal{T}$-*satisfiable* if there exists an interpretation I such that $I \models S$ and $I \models \mathcal{T}$. We assume that the $\mathcal{T}$-satisfiability problem is decidable, i.e., that there exists an SMT solver that, given a formula ϕ with no quantifier, can decide whether ϕ is $\mathcal{T}$-satisfiable.

We consider clauses as unordered disjunctions of literals with no repetition. Thus, when we write $C \vee D$, we implicitly assume that C and D share no literal. We also identify unit clauses with the literal they contain. For every literal l, $\bar{l}$ denotes the literal complementary of l. The empty clause is denoted by `false`. If $Q = \{l_1, \dots, l_n\}$ is a set of literals, then we denote by $\overline{Q}$ the clause $\overline{l_1} \vee \dots \vee \overline{l_n}$. Conversely, given a clause $C = l_1 \vee \dots \vee l_n$, we denote by $\overline{C}$ the set of literals (or unit clauses) $\{\overline{l_1}, \dots, \overline{l_n}\}$.

We consider a finite set of *abducible literals* $\mathcal{A}$. We assume that each of these literals is $\mathcal{T}$-satisfiable. Given a set of clauses S, we call a clause C a $(\mathcal{T}, \mathcal{A})$-implicate *of* S if $\overline{C} \subseteq \mathcal{A}$ and $S \models_{\mathcal{T}} C$. We say that C is a *prime* $(\mathcal{T}, \mathcal{A})$-implicate *of* S if C is a $(\mathcal{T}, \mathcal{A})$-implicate of S and for every $(\mathcal{T}, \mathcal{A})$-implicate D of S, if $D \models_{\mathcal{T}} C$ then $C \models_{\mathcal{T}} D$. The set of $(\mathcal{T}, \mathcal{A})$-implicates of S is denoted by $\mathfrak{I}_{\mathcal{A}}(S)$, and the set of prime $(\mathcal{T}, \mathcal{A})$-implicates of S is denoted by $\mathfrak{P}_{\mathcal{A}}(S)$.

Given a set of clauses S and a clause C, we write $S \trianglelefteq_{\mathcal{T}} C$ if there is a clause $D \in S$ such that $D \models_{\mathcal{T}} C$. If S' is a set of clauses, then we write $S \trianglelefteq_{\mathcal{T}} S'$ if for all $C \in S'$, we have $S \trianglelefteq_{\mathcal{T}} C$. We write $S \sim S'$ if $S \trianglelefteq_{\mathcal{T}} S'$ and $S' \trianglelefteq_{\mathcal{T}} S$ (i.e., S and S' are identical modulo $\mathcal{T}$-equivalence).

Proposition 1. *Let l be a literal and let C, D be clauses. The following statements hold:*

1. *$l \vee C \models_{\mathcal{T}} D$ iff $l \models_{\mathcal{T}} D$ and $C \models_{\mathcal{T}} D$.*
2. *$C \models_{\mathcal{T}} l \vee D$ iff $C \wedge \bar{l} \models_{\mathcal{T}} D$.*

We assume an order $\prec$ is given on clauses built on $\mathcal{A}$ that agrees with inclusion, i.e., such that $C \subsetneq D \Rightarrow C \prec D$.

Definition 2. *A $\mathcal{T}$-tautology is a clause that is satisfied by every model of $\mathcal{T}$. Given a set of clauses S, we denote by $\mathrm{SubMin}(S)$ the set obtained by deleting from S all clauses D such that either D is a $\mathcal{T}$-tautology, or there exists $C \in S$ such that $C \models_{\mathcal{T}} D$ and $(D \not\models_{\mathcal{T}} C$ or $C \prec D)$.*

Note that in particular, we have $\mathfrak{P}_{\mathcal{A}}(S) \sim \mathrm{SubMin}(\mathfrak{I}_{\mathcal{A}}(S))$.

3   On the Generation of Prime $(\mathcal{T}, \mathcal{A})$-Implicates

3.1   A Basic Algorithm

We present a simple and intuitive algorithm that permits to generate the $(\mathcal{T}, \mathcal{A})$-implicates of a set of formulas S. This algorithm is based on the fact that a clause C is a $(\mathcal{T}, \mathcal{A})$-implicate of S if and only if $\overline{C} \subseteq \mathcal{A}$ and $S \cup \overline{C} \models_{\mathcal{T}}$ false. It will thus basically consist in enumerating the subsets of $\mathcal{A}$ and searching for those whose union with S is $\mathcal{T}$-unsatisfiable. This may be done by starting with an empty set of hypotheses M and repeatedly and nondeterministically adding new abducible literals to M until $S \cup M$ is $\mathcal{T}$-unsatisfiable. This algorithm is naive, as the same clauses will be produced multiple times, but it forms the basis of the more efficient algorithm in Sect. 3.2.

Definition 3. *Let S be a set of formulas. Let M, A be sets of literals such that $M \cup A \subseteq \mathcal{A}$. We define*

$$\mathfrak{I}_{M,A}(S) = \left\{ C \in \mathfrak{I}_{\mathcal{A}}(S) \mid \exists Q \subseteq A, \ C = \overline{M} \vee \overline{Q} \right\},$$
$$\mathfrak{P}_{M,A}(S) = \mathrm{SubMin}(\mathfrak{I}_{M,A}(S)).$$

Intuitively, a clause $\overline{M} \vee \overline{Q}$ thus belongs to $\mathfrak{I}_{M,A}(S)$ if and only if $\overline{Q}$ is a $(\mathcal{T}, \mathcal{A})$-implicate of $S \cup M$.

Proposition 4. *Let S be a set of formulas, M and A be sets of literals such that $M \cup A \subseteq \mathcal{A}$. If M is $\mathcal{T}$-satisfiable, then $S \cup M$ is $\mathcal{T}$-unsatisfiable iff $\mathfrak{P}_{M,A}(S) = \{\overline{M}\}$. If M is $\mathcal{T}$-unsatisfiable, then $\mathfrak{P}_{M,A}(S) = \emptyset$.*

It is clear that it is useless to add a new hypothesis l into M both if $M \cup \{l\}$ is $\mathcal{T}$-unsatisfiable (because the obtained $(\mathcal{T}, \mathcal{A})$-implicate would be a $\mathcal{T}$-tautology), or if this set is equivalent to M (because the $(\mathcal{T}, \mathcal{A})$-implicate would not be minimal). This motivates the following definition:

Definition 5. *Let S be a set of formulas and let M, A be two sets of literals. We denote by $\mathtt{fix}(S, M, A)$ a set obtained by deleting from A some literals l such that either $M \cup S \models_{\mathcal{T}} l$ or $M \models_{\mathcal{T}} \bar{l}$.*

The use of this definition aims to reduce the number of abducible hypotheses to try, and thus the search space of the algorithm. Still, we do not assume that all the literals l satisfying the condition above are deleted because, in practice, such literals may be hard to detect. However, we assume that no element from M is in $\mathtt{fix}(S, M, A)$.

Proposition 6. *Consider a set of formulas S and two sets of literals M, A such that $\mathfrak{P}_{M,A}(S) \neq \{\overline{M}\}$. The following equalities hold:*

1. $\mathfrak{P}_{M,A}(S) = \mathrm{SubMin}(\bigcup_{l \in A} \mathfrak{P}_{M \cup \{l\}, A}(S))$.
2. $\mathfrak{P}_{M,A}(S) = \mathfrak{P}_{M, \mathtt{fix}(S,M,A)}(S)$.

The results above lead to a basic algorithm for generating $(\mathcal{T}, \mathcal{A})$-implicates which is described in Algorithm 2. As explained above, the algorithm works by adding literals from A as hypotheses until a contradiction can be derived. The **return** statement at Line 5 avoids enumerating the subsets that contain M, once it is known that $S \cup M$ is $\mathcal{T}$-unsatisfiable.

Algorithm 2. BP(S, M, A)

---

1 if M *is $\mathcal{T}$-unsatisfiable* then
2 $\quad$ return $\emptyset$;

3 else
4 $\quad$ if $S \cup M$ *is $\mathcal{T}$-unsatisfiable* then
5 $\quad\quad$ return $\{\overline{M}\}$;

6 $\quad$ else
7 $\quad\quad$ $B = \mathtt{fix}(S, M, A)$;
8 $\quad\quad$ foreach $l \in B$ do
9 $\quad\quad\quad$ let $P_l = \mathrm{BP}(S, M \cup \{l\}, B)$;
10 $\quad\quad$ return $\mathrm{SubMin}(\bigcup_{l \in B} P_l)$;

Lemma 7. $\mathfrak{P}_{M,A}(S) = \text{BP}(S, M, A)$.

Proof. The result is proved by a straightforward induction on $\text{card}(A \setminus M)$. By Proposition 4, $\mathfrak{P}_{M,A}(S) = \emptyset$ if M is $\mathcal{T}$-unsatisfiable, and $\mathfrak{P}_{M,A}(S) = \{\overline{M}\}$ if M is $\mathcal{T}$-satisfiable and $S \cup M$ is $\mathcal{T}$-unsatisfiable. Otherwise, by Proposition 6(2), we have $\mathfrak{P}_{M,A}(S) = \mathfrak{P}_{M,\text{fix}(S,M,A)}(S) = \mathfrak{P}_{M,B}(S)$. By the induction hypothesis, for each $l \in A$, $P_l = \mathfrak{P}_{M \cup \{l\}, B}(S)$, and by Proposition 6(1), $P_l = \mathfrak{P}_{M \cup \{l\}, A}(S)$; we deduce that $\mathfrak{P}_{M,A}(S) = \text{SubMin}(\bigcup_{l \in A} P_l)$. Note that at each recursive call, a new element is added to M, since $\text{fix}(S, M, A)$ is assumed not to contain any element from M.

Theorem 8. *If S is a set of formulas then $\mathfrak{P}_A(S) = \text{BP}(S, \emptyset, A)$.*

Although the algorithm described above computes all the prime $(\mathcal{T}, A)$-implicates of any clause set as required, it is very inefficient, in particular because of the large number of useless and redundant recursive calls that are made. In what follows we present several improvements to the algorithm in order to generate implicates as efficiently as possible.

3.2   Restricting the Set of Candidate Hypotheses

It is obvious that the algorithm BP makes a lot of redundant calls: for example, if $l_1 \vee l_2$ is a prime $(\mathcal{T}, A)$-implicate of a clause set S, then this $(\mathcal{T}, A)$-implicate will be generated twice, first as $l_1 \vee l_2$, and then as $l_2 \vee l_1$. Such redundant calls are quite straightforward to avoid by ensuring that every invocation of the algorithm contains a distinct set of literals M. This can be done by fixing an ordering $<$ among literals in A, and by assuming that hypotheses are always added in this order. Another way of restricting the set of candidate hypotheses is to exploit information extracted from the previous satisfiability test. For example, if $S \cup \{l_1\}$ is satisfiable for some literal l_1, and that a model of this set satisfies another literal l_2, then $S \cup \{l_2\}$ is also satisfiable and it is not necessary to consider l_2 as a hypothesis. In particular, if a model of $S \cup \{l_1\}$ validates all the literals in A, then $\mathfrak{P}_{M,A}(S)$ is necessarily empty and no literal should be selected. We can thus take advantage of the existence of a model of $S \cup M$ in order to guide the choice of the next literals in A. However, observe that this refinement interferes with the previous one based on the order $<$. Indeed, non-minimal hypotheses will have to be considered if all the smaller hypotheses are dismissed because they are true in the model. We formalize these principles below.

Definition 9. *In what follows, we consider a total ordering[1] $<$ on the elements of A. For $A \subseteq \mathcal{A}$ and $l \in A$, we define $A[l] \overset{def}{=} \{l' \in A \mid l < l'\}$. If I is a set of literals then we denote by $A^I[l]$ the set $\{l' \in A \mid l' < l \wedge \overline{l'} \notin I\} \cup A[l]$.*

Example 10. Assume that $A = \{p_i, \neg p_i \mid i = 1, \ldots, 6\}$ and that for all literals $l \in \{p_i, \neg p_i\}$ and $l' \in \{p_j, \neg p_j\}$, $l < l'$ if and only if either $i < j$ or $(i = j,$ $l = p_i$ and $l' = \neg p_i)$. Then $A[p_4] = \{\neg p_4, p_5, \neg p_5, p_6, \neg p_6\}$. If $I = \{p_1, \neg p_2\}$, then $A^I[p_4] = \{p_1, \neg p_2, p_3, \neg p_3, \neg p_4, p_5, \neg p_5, p_6, \neg p_6\}$.

---

[1] Note that this ordering is not necessarily related to the ordering $\prec$ on clauses.

Definition 11. *Let S be a set of clauses. A set of literals I is S-compatible with respect to $\mathcal{A}$ (or simply S-compatible) if every prime $(\mathcal{T}, \mathcal{A})$-implicate of S contains a literal l such that $l \in I$.*

Intuitively, an S-compatible set I consists of literals l such that $\bar{l}$ will be allowed to be added as a hypothesis to generate $(\mathcal{T}, \mathcal{A})$-implicates of S (see Lemma 16 below). The set I can always be defined by taking the negations of all the abducible literals from $\mathcal{A}$. In this case, all literals will remain possible hypotheses. It is possible, however, to restrict the size of I when a model of S is known, as evidenced by the following proposition:

Proposition 12. *If S is a set of clauses and J is a model of S, then the set $I \overset{def}{=} \{l \in \mathcal{L} \mid J \models l\}$ is S-compatible.*

Proof. Let Q be a set of literals such that $\overline{Q}$ is a prime $(\mathcal{T}, \mathcal{A})$-implicate of S, and assume that for all $l \in Q$, $\bar{l} \notin I$, i.e., that for all $l \in Q$, $J \not\models \bar{l}$. Then $J \models l$ holds for every $l \in Q$, hence $J \models S \cup Q$ and $\overline{Q}$ cannot be a $(\mathcal{T}, \mathcal{A})$-implicate of S.

Note that the condition of having a model of S was not added to Definition 11 because in practice, such a model cannot always be constructed efficiently.

Being able to derive unit consequences of the set of axioms (for instance by using unit propagation), can pay off if this additional information can be used to simplify the formula at hand. This motivates the following definition.

Definition 13. *Let S be a set of formulas and $M \subseteq \mathcal{A}$. We denote by $\mathrm{U}_M(S)$ the set of unit clauses logically entailed by $S \cup M$ modulo $\mathcal{T}$, i.e., $\mathrm{U}_M(S) \overset{def}{=} \{l \in \mathcal{L} \mid S \cup M \models_{\mathcal{T}} l\}$. Given a set U such that $M \subseteq U \subseteq \mathrm{U}_M(S)$, we denote by $S_{U,M}$ the formula obtained from S by replacing some (arbitrarily chosen) literals l' by false if $U \models_{\mathcal{T}} \bar{l'}$ and by true if $M \models_{\mathcal{T}} l'$.*

Note that U is not necessarily identical to $\mathrm{U}_M(S)$, because in practice the latter set is hard to generate. Similarly we do not assume that all literals l' are replaced in Definition 13 since testing logical entailment may be costly. Lemma 14 shows that the $(\mathcal{T}, \mathcal{A})$-implicates of a set S and those of $S_{U,M}$ are identical.

Lemma 14. *Let S be a set of formulas and $M \subseteq \mathcal{A}$. Consider a set of literals U such that $M \subseteq U \subseteq \mathrm{U}_M(S)$. Then $\mathfrak{I}_{M,\mathcal{A}}(S) = \mathfrak{I}_{M,\mathcal{A}}(S_{U,M})$.*

Definition 15. *Let $U, M, \mathcal{A}$ be sets of literals. We define: $\mathrm{G}_{U,\mathcal{A},M}(S) \overset{def}{=} \{\overline{M} \vee \bar{l} \mid l \in \mathcal{A} \wedge \bar{l} \in U\}$.*

The lemma below can be viewed as a refinement of Proposition 6. It is based on the previous results, according to which, when adding a new hypothesis l, it is possible to remove from the set of abducible literals $\mathcal{A}$ every literal that is strictly smaller than l, provided its complementary is in I (because we can always assume that the smallest available hypothesis is considered first). This is why the recursive call is on $\mathcal{A}^I[l]$ instead of $\mathcal{A}$. Note also that the use of semantic guidance interferes with the use of the ordering $<$: the smaller the set I, the larger $\mathcal{A}^I[l]$.

Lemma 16. *Assume that $S \cup M$ is $\mathcal{T}$-satisfiable and let I be an $(S \cup M)$-compatible set of literals. Let U be a set of literals such that $M \subseteq U \subseteq \mathrm{U}_M(S)$. We have*

$$\mathfrak{P}_{M,A}(S) = \mathrm{SubMin}\left(\mathrm{G}_{U,A,M}(S) \cup \bigcup_{l \in A, \bar{l} \in I} \mathfrak{P}_{M \cup \{l\}, A^I[l]}(S)\right).$$

Proof. First note that $\mathfrak{P}_{M,A}(S) \neq \{\overline{M}\}$, since $S \cup M$ is $\mathcal{T}$-satisfiable. We first prove that $\mathfrak{P}_{M,A}(S) \subseteq \mathrm{G}_{U,A,M}(S) \cup \bigcup_{l \in A, \bar{l} \in I} \mathfrak{P}_{M \cup \{l\}, A^I[l]}(S)$. Let $C \in \mathfrak{P}_{M,A}(S)$. By hypothesis, C is of the form $\overline{M} \vee \overline{Q}$, where $\emptyset \neq Q \subseteq A$. Since I is $(S \cup M)$-compatible, Q necessarily contains a literal $l \in A$ such that $\bar{l} \in I$. Assume that l is the smallest literal in Q satisfying this property. We distinguish the following cases.

Assume that Q contains a literal l' such that $\overline{l'} \in U$. In this case, since $U \subseteq \mathrm{U}_M(S)$, $S \cup M \models_{\mathcal{T}} \overline{l'}$. Since $Q \subseteq A$, we also have $l' \in A$, and since $\mathfrak{P}_{M,A}(S) \neq \{\overline{M}\}$, we deduce that $\overline{M} \vee \overline{l'} \in \mathfrak{P}_{M,A}(S)$. Since $\overline{M} \vee \overline{l'} \models_{\mathcal{T}} C$ and $C \in \mathfrak{P}_{M,A}(S)$, C must be smaller or equal to $\overline{M} \vee \overline{l'}$, which is possible only if $C = \overline{M} \vee \overline{l'}$. We deduce that $C \in \mathrm{G}_{U,A,M}(S)$.

Otherwise, we show that $Q \setminus \{l\} \subseteq A^I[l]$. By Definition 9, we have $A[l] = \{l' \in A \mid l < l'\}$ and $A^I[l] = \{l' \in A \mid l' < l \wedge l' \notin I\} \cup A[l]$. Let $l' \in Q$, with $l' \neq l$. If $l' > l$ then $l' \in A[l] \subseteq A^I[l]$. If $l' \not> l$, then since $>$ is total and $l \neq l'$, necessarily $l > l'$. Since l is the smallest literal in Q such that $\bar{l} \in I$, we deduce that $\overline{l'} \notin I$. Thus $l' < l$ and $\overline{l'} \notin I$, which entails that $l' \in A^I[l]$. Consequently, $Q \setminus \{l\} \subseteq A^I[l]$. Since $C = \overline{M \cup \{l\}} \vee \overline{Q \setminus \{l\}}$, this entails that $C \in \mathfrak{P}_{M \cup \{l\}, A^I[l]}(S)$.

We now prove that $\mathrm{G}_{U,A,M}(S) \cup \bigcup_{l \in B, \bar{l} \in I} \mathfrak{P}_{M \cup \{l\}, B^I[l]}(S) \subseteq \mathfrak{I}_{M,A}(S)$.

Let $C \in \mathrm{G}_{U,A,M}(S)$. By definition, C is of the form $\overline{M} \cup l$ with $l \in \overline{A} \cap U$. Since $U \subseteq \mathrm{U}_M(S)$, we deduce that $S \cup M \models_{\mathcal{T}} l$, i.e., that $S \models_{\mathcal{T}} \overline{M} \vee l$. Since $\bar{l} \in A$, this entails that $\overline{M} \vee l \in \mathfrak{I}_{M,A}(S)$, hence $C \in \mathfrak{I}_{M,A}(S)$.

Let $C \in \mathfrak{P}_{M \cup \{l\}, A^I[l]}(S)$ with $l \in A$, $\bar{l} \in I$. By definition, $C = \overline{M} \vee \bar{l} \vee \overline{Q}$, with $Q \subseteq A^I[l]$ and $C \in \mathfrak{I}_A(S)$. But $A^I[l] \subseteq A$ by definition, thus $Q \cup \{l\} \subseteq A$ and $C = M \vee (\overline{Q} \vee \bar{l}) \in \mathfrak{I}_{M,A}(S)$.

Similarly to cSP (see [12, Sect. 4.2]), we parameterize our algorithm by a predicate in order to filter the implicates that are generated. The goal of this parametrization is to allow the user to restrict the form of the generated implicates. Typically, one could want to generate implicates only up to a given size limit, or only those satisfying some specific semantic constraints.

Definition 17. *A predicate $\mathcal{P}$ on sets of literals is $\subseteq$-closed if for all sets of literals A such that $\mathcal{P}(A)$ holds, if $B \subseteq A$ then $\mathcal{P}(B)$ also holds.*

Examples of $\subseteq$-closed predicates include cardinality constraints: $\mathcal{P}_k \stackrel{def}{=} \lambda A.\ \mathrm{card}(A) \leq k$, where $k \in \mathbb{N}$, or implicant constraints: $\mathcal{P}_\phi \stackrel{def}{=} \lambda A.\ \phi \models A$, where ϕ is a formula. Note that $\subseteq$-closed predicates can safely be combined by the conjunction and disjunction operators.

An important feature of $\subseteq$-closed predicates is that implicates verifying such predicates can be generated on the fly without any post-processing step, thanks to the following result:

Proposition 18. *If $\mathcal{P}$ is $\subseteq$-closed and $\mathcal{P}(M)$ does not hold, then for all sets of literals A, $\mathcal{P}(M \cup A)$ does not hold either.*

The inclusion of these improvements to the original algorithm results in the one described in Algorithm 3.

Algorithm 3. IMP$(S, M, A, \mathcal{P})$

---

1 **if** M *is* $\mathcal{T}$-*unsatisfiable* **or** $\neg \mathcal{P}(M)$ **then**
2 $\quad \lfloor$ **return** $\emptyset$;
3 **if** $S \cup M$ *is* $\mathcal{T}$-*unsatisfiable* **then**
4 $\quad \lfloor$ **return** $\{\overline{M}\}$;
5 **let** $U \subseteq \mathrm{U}_M(S)$ *such that* $M \subseteq U$;
6 **let** $S = S_{U,M}$;
7 **let** $A = \mathtt{fix}(S, M, A)$;
8 **let** I *be an* $(S \cup M)$-*compatible set of literals* ;
9 **foreach** $l \in A$ *such that* $\bar{l} \in I$ **do**
10 $\quad \lfloor$ **let** $P_l = $ IMP$(S, M \cup \{l\}, A^I[l], \mathcal{P})$;
11 **return** SubMin$(\mathrm{G}_{U,A,M}(S) \cup \bigcup_{l \in A} P_l)$;

Lemma 19. *If $\mathcal{P}$ is $\subseteq$-closed then* IMP$(S, M, A, \mathcal{P}) = \mathfrak{P}_{M,A}(S) \cap \{\overline{A} \mid A \in \mathcal{P}\}$.

Proof. If one of M or $S \cup M$ is $\mathcal{T}$-unsatisfiable, or $\mathcal{P}(M)$ does not hold, then the result follows from Propositions 4 and 18. Otherwise the result is proved by induction on card$(\mathcal{A} \setminus M)$, using Proposition 6 and Lemmata 16 and 14.

Theorem 20. *If $\mathcal{P}$ is $\subseteq$-closed then* $\mathfrak{P}_{\mathcal{A}}(S) \cap \{\overline{A} \mid A \in \mathcal{P}\} = $ IMP$(S, \emptyset, \mathcal{A}, \mathcal{P})$.

4   On the Storage of $(\mathcal{T}, \mathcal{A})$-Implicates

The number of implicates of a given formula may be huge, hence it is essential in practice to have appropriate data structures to store them in a compact way and efficient algorithms to check that a newly generated implicate C is not redundant (forward subsumption modulo $\mathcal{T}$), and if so, to delete all the already generated implicates that are less general than C (backward subsumption modulo $\mathcal{T}$), before inserting C into the stored implicates. In this section, we devise a trie-like data-structure to perform these tasks. As in the previous section, we only rely on the existence of a decision procedure for testing $\mathcal{T}$-satisfiability.

Definition 21. *Let $<_t$ be an order on the literals in $\mathcal{A}$, possibly, but not necessarily, equal to the order $<$ used for literal ordering in the implicate generation algorithm. An $\mathcal{A}$-tree is inductively defined as $\bot$ or a possibly empty set of pairs $\{l_1 : \tau_1, \ldots, l_n : \tau_n\}$, where $l_1, \ldots, l_n$ are pairwise distinct literals in $\mathcal{A}$ and τ_i (for $i = 1, \ldots, n$) is an $\mathcal{A}$-tree only containing literals that are strictly $<_t$-greater than l_i. An $\mathcal{A}$-tree is associated with a set of $\mathcal{A}$-clauses inductively defined as follows:*

$$\mathcal{S}(\bot) \overset{def}{=} \{\texttt{false}\},$$
$$\mathcal{S}(\{l_1 : \tau_1, \ldots, l_n : \tau_n\}) \overset{def}{=} \bigcup_{i=1}^n \{l_i \vee C \mid C \in \mathcal{S}(\tau_i)\}.$$

In particular, $\mathcal{S}(\emptyset) = \emptyset$. Intuitively an $\mathcal{A}$-tree may be seen as a tree in which the edges are labeled by literals and the leaves are labeled by $\emptyset$ or $\bot$, and represents a set of clauses corresponding to paths from the root to $\bot$. We introduce the following simplification rule (which may be applied at any depth inside a tree, not only at the root level):

$$\texttt{Simp} : \tau \cup \{l : \emptyset\} \rightarrow \tau$$

Informally, the rule deletes all leaves labeled by $\emptyset$ except for the root. It may be applied recursively, for instance $\{l : \{l_1 : \emptyset, \ldots, l_n : \emptyset\}\} \rightarrow_{\texttt{Simp}}^{n+1} \emptyset$. Termination is immediate since the size of the tree is strictly decreasing.

Proposition 22. *If $\tau \rightarrow_{\texttt{Simp}} \tau'$ then τ' is an $\mathcal{A}$-tree and $\mathcal{S}(\tau) = \mathcal{S}(\tau')$.*

The algorithm permitting the insertion of a clause in an $\mathcal{A}$-tree is straightforward and thus omitted. The following lemma provides a simple algorithm to check whether a clause is a logical consequence modulo $\mathcal{T}$ of some clause in $\mathcal{S}(\tau)$ (forward subsumption). The algorithm proceeds by induction on the $\mathcal{A}$-tree.

Lemma 23. *Let C be a clause and let τ be an $\mathcal{A}$-tree. We have $\mathcal{S}(\tau) \unlhd_{\mathcal{T}} C$ iff one of the following conditions hold:*

- *$\tau = \bot$.*
- *$\tau = \{l_1 : \tau_1, \ldots, l_n : \tau_n\}$ and there exists $i \in [1, n]$ such that $l_i \models_{\mathcal{T}} C$ and $\mathcal{S}(\tau_i) \unlhd_{\mathcal{T}} C$.*

Proof. If $\tau = \bot$ then $\mathcal{S}(\tau) = \{\texttt{false}\} \unlhd_{\mathcal{T}} C$ hence the equivalence holds. Otherwise, let $\tau = \{l_1 : \tau_1, \ldots, l_n : \tau_n\}$. By definition, $\mathcal{S}(\tau) \unlhd_{\mathcal{T}} C$ holds iff there exists a clause $D \in \mathcal{S}(\tau)$ such that $D \models_{\mathcal{T}} C$. Since $\mathcal{S}(\{l_1 : \tau_1, \ldots, l_n : \tau_n\}) = \bigcup_{i=1}^n \{l_i \vee E \mid E \in \mathcal{S}(\tau_i)\}$, the previously property holds iff there exists $i \in [1, n]$ and $E \in \mathcal{S}(\tau_i)$ such that $l_i \vee E \models_{\mathcal{T}} C$, i.e., such that $l_i \models_{\mathcal{T}} C$ and $E \models_{\mathcal{T}} C$ by Proposition 1(1). By definition, $\exists E\,(E \models_{\mathcal{T}} C \wedge E \in \mathcal{S}(\tau_i))$ iff $(\mathcal{S}(\tau_i) \unlhd_{\mathcal{T}} C)$. Furthermore, $l_i \unlhd_{\mathcal{T}} C$ holds iff $\overline{C} \cup \{l_i\}$ is $\mathcal{T}$-unsatisfiable, hence the result.

The following definition provides an algorithm to remove, in a given $\mathcal{A}$-tree, all branches corresponding to clauses that are logical consequences of a given formula modulo $\mathcal{T}$ (backward subsumption).

Definition 24. *Let ϕ be a formula and let τ be an $\mathcal{A}$-tree. $\mathrm{rm}(\tau, \phi)$ denotes the $\mathcal{A}$-tree defined as follows:*

- *If ϕ is $\mathcal{T}$-unsatisfiable, then $\mathrm{rm}(\tau, \phi) \overset{def}{=} \emptyset$.*
- *If ϕ is $\mathcal{T}$-satisfiable, then:*
    - $\mathrm{rm}(\bot, \phi) \overset{def}{=} \bot$,
    - $\mathrm{rm}(\{l_1 : \tau_1, \ldots, l_n : \tau_n\}, \phi) \overset{def}{=} \bigcup_{i=1}^{n} \{l_i : \mathrm{rm}(\tau_i, \phi \wedge \overline{l_i})\}$.

Intuitively, starting with some clause C, the algorithm incrementally adds literals $\overline{l_1}, \ldots, \overline{l_n}$ occurring in the clauses $D = l_1 \vee \cdots \vee l_n \in \mathcal{S}(\tau)$ and invokes the SMT solver after each addition. If a contradiction is found then this means that $C \models_{\mathcal{T}} D$, hence the branch corresponding to D can be removed. The calls are shared among all common prefixes. Of course, this algorithm is interesting mainly if the SMT solver is able to perform incremental satisfiability testing, with "push" and "pop" commands to add and remove formulas from the set of axioms (which is usually the case).

Lemma 25. *Let ϕ be a formula and let τ be an $\mathcal{A}$-tree. Then $\mathrm{rm}(\tau, \phi)$ is an $\mathcal{A}$-tree, and $\mathcal{S}(\mathrm{rm}(\tau, \phi)) = \{C \in \mathcal{S}(\tau) \mid \phi \not\models_{\mathcal{T}} C\}$.*

Remark 26. The $\mathcal{A}$-trees may be represented as dags instead of trees. In this case, it is clear that the complexity, defined as the number of satisfiability tests of forward subsumption (as defined in Lemma 23) is of the same order as the size of the dag, since the recursive calls only depend on the considered subtree. For backward subsumption (see Definition 24) the situation is different since the recursive calls have an additional parameter that is the formula ϕ, which depends on the path in the $\mathcal{A}$-tree. The maximal number of satisfiability tests is therefore equal to the size of underlying tree, and not that of the dag. Note that it would be necessary to make copies of some of the subtrees, if two pruning operations are applied on the same (shared) subtree with different formulas.

5   Experimental Evaluation

Algorithm 3 has been implemented in a C++ framework called GPID. The SMT solver is used as a black box and GPID can thus be plugged with any tool serving this purpose, provided an interface is written for it. As a consequence, the handled theory is only restricted by the SMT solver. Three interfaces were implemented, respectively for MINISAT [13], CVC4 [1] and Z3 [6]. The implicate generator used in the reported experiments is the one based on Z3, which turned out to be more efficient on the considered benchmarks. All the tests were run on one core of an Intel(R) Core(TM) i5-4250U machine running at 1.9 GHz with 1 GiB of RAM. The benchmarks are extracted from the SMTLib [2] library, the considered theories are quantifier-free uninterpreted functions (QF_UF) and quantifier-free linear integer arithmetic with uninterpreted functions (QF_UFLIA). For obvious reasons, only satisfiable examples have been kept

Table 1. Number of problems for which at least one $(\mathcal{T}, \mathcal{A})$-implicate of a given maximal size can be generated in a given amount of time (in seconds), for the QF_UF SMTLib benchmark (2549 examples).

Size	Time							
	$[0, 0.5[$	$[0.5, 1[$	$[1, 1.5[$	$[1.5, 2[$	$[2, 5[$	$[5, 10[$	$[10, 35[$	None
1	2235	75	28	16	33	32	61	69
2	2236	81	27	16	30	23	67	69
3	2236	79	27	16	34	23	65	69
4	2230	84	23	18	33	24	68	69
5	2231	79	27	12	36	22	73	69
6	2234	73	29	15	30	24	75	69
7	2231	81	23	15	33	22	75	69
8	2233	78	23	16	33	21	76	69

Table 2. Number of problems for which at least one $(\mathcal{T}, \mathcal{A})$-implicate of a given maximal size can be generated in a given amount of time (in seconds), for the QF_UFLIA SMTLib benchmark (400 examples).

Size	Time							
	$[0, 0.5[$	$[0.5, 1[$	$[1, 1.5[$	$[1.5, 2[$	$[2, 5[$	$[5, 10[$	$[10, 35[$	None
1	120	23	46	76	100	6	25	4
2	120	23	6	0	0	0	247	4
3	120	23	6	0	96	4	147	4
4	120	23	6	0	0	0	247	4
5	120	23	6	0	0	0	247	4
6	120	22	7	0	0	0	247	4
7	121	22	6	0	0	0	247	4
8	116	24	6	3	0	0	247	4

for analysis. Abducible literals are part of the problem input, they are generated by considering all ground equalities and disequalities with a maximal depth provided by the user; all the experiments were conducted using a maximal depth of 1 and the average number of abducible literals is around 13397 (min. 1741, max. 17.10^6). We chose not to apply unit propagation simplifications to the considered sets of clauses. More precisely, this means that we let $U = M$ at line 5 of Algorithm 3 and delegate the simplifications that could occur in the following line to the satisfiability checker. The reason for this decision is that efficiently performing such simplifications can be difficult and strongly depends on the theory. We also define $\texttt{fix}(,)$ as the complementation on literals and $\mathcal{P}$ as either $\texttt{true}$ or a predicate ensuring $\text{card}(M) \leq n$ to generate $(\mathcal{T}, \mathcal{A})$-implicates of size at most n. In all the experiments, the prime implicates filter (SubMin) was not active, so that implicates can be generated on the fly. Finally, if available, we recover models of $S \cup M$ from the SMT solver in order to further prune the set of abducibles (see Line 8 of Algorithm 3). Tables 1 and 2 show the number of examples for which our tool generates at least one $(\mathcal{T}, \mathcal{A})$-implicate for a given timespan, for the QF_UF and QF_UFLIA benchmarks respectively. The results show that our tool is quite efficient, since it fails to generate any $(\mathcal{T}, \mathcal{A})$-implicate within 35 s for only 2% (resp. 1%) of the QF_UF (resp. QF_UFLIA) benchmarks. Figure 1 shows the proportion of the QF_UFLIA set for which GPID generates an implicate in less than 15 s, depending on the maximal size constraint. For the QF_UF benchmark, the proportion decreases from 97% for a maximal size constraint of 1 to 95% when there are no size restrictions. We also point out that for 57% of the QF_UF benchmark, we are actually able to generate all the $(\mathcal{T}, \mathcal{A})$-implicates of size 1 in less than 15 s.

We ran additional experiments to compare this approach with a previous one based on a superposition-based approach [11,12] and implemented in the CSP tool. As far as we are aware, CSP is the only other available tool for implicate generation in the theory of equality with uninterpreted function symbols. Previous experiments (see, e.g., [11,12]) showed that CSP is already more efficient than approaches based on a reduction to propositional logic for generating implicates of ground equational formulas, which is why we did not run comparisons against tools for propositional implicate generation. CSP is based on a constrained calculus defined by

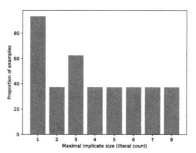

Fig. 1. Proportion (out of 100) of examples of the QF_UFLIA benchmark where GPID generates at least one implicate under 15 s.

the usual inference rules of the superposition calculus together with additional rules to dynamically assert new abducible hypotheses on demand into the search space. The asserted hypotheses are attached to the clauses as constraints and, when an empty clause is generated, the negation of these hypotheses yields a $(\mathcal{T}, \mathcal{A})$-implicate. We chose to compare the tools by focusing on their ability to generate at least one $(\mathcal{T}, \mathcal{A})$-implicate of a given size. Indeed, generating all (prime) $(\mathcal{T}, \mathcal{A})$-implicates is unfeasible within a reasonable amount of time except for very simple formulas, and comparing the raw number of $(\mathcal{T}, \mathcal{A})$-implicates generated is not relevant because some of these may actually be redundant w.r.t. non-generated ones[2]. We believe in practice, being able to efficiently compute a small number of $(\mathcal{T}, \mathcal{A})$-implicates for a complex problem is more useful than computing huge sets of $(\mathcal{T}, \mathcal{A})$-implicates but only for simple formulas. The following experiments are only based on benchmarks that can be solved by both prototypes, as CSP is not capable of handling integer arithmetics. We represented on Fig. 2 the number of examples for which both tools can generate

---

[2] A refined comparison of the set of generated $(\mathcal{T}, \mathcal{A})$-implicates modulo theory entailment is left for future work.

(a)   (b)

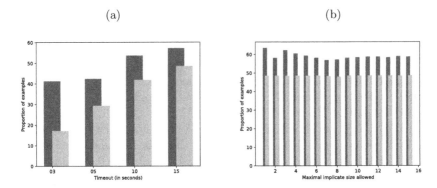

Fig. 2. Number of examples from the QF_UF benchmark set for which GPiD (on the left, darker color) and CSP (on the right, lighter color) generate at least one $(\mathcal{T}, \mathcal{A})$-implicate within a given time (a) and generate at least one implicate of a given maximal size under 15 s (b) (Color figure online)

at least one $(\mathcal{T}, \mathcal{A})$-implicate with a given maximal size constraint for various timeouts (a) and generate at least one $(\mathcal{T}, \mathcal{A})$-implicate within a given time limit for various maximal size constraints (b).

6   Conclusion

We devised a generic algorithm to generate implicates modulo theories and showed that the corresponding implementation is more efficient than a previous attempt based on superposition. This result was to be expected since the DPLL($\mathcal{T}$) approach is more efficient than engines based on the Superposition Calculus for testing the satisfiability of quantifier-free formulas with a large combinatorial structure. Furthermore, the used superposition engine had to be specifically tuned for implicate generation, and it is far less efficient than state-of-the-art systems such as Vampire [27], E [28] or Spass [31] (this is of course the advantage of having a generic algorithm using decision procedures as black boxes). While our aim was to be completely generic, it is clear than the efficiency of the procedure could be improved in practice by integrating theory-specific algorithms for deriving consequences and normalizing formulas. For instance, in the case of the theory of equality with uninterpreted function symbols, the implicates could be normalized by replacing each term by its minimal representative, as is done in [12]. Efficient, theory-dependent simplification procedures will also be explored in future work. A combination between the superposition-based approach [12], in which the assertion of hypotheses is guided by the proof procedure could also be beneficial. Our approach could also be combined with that of [7], which is based on model building and quantifier-elimination.

References

1. Barrett, C., Conway, C.L., Deters, M., Hadarean, L., Jovanović, D., King, T., Reynolds, A., Tinelli, C.: CVC4. In: Gopalakrishnan, G., Qadeer, S. (eds.) CAV 2011. LNCS, vol. 6806, pp. 171–177. Springer, Heidelberg (2011). https://doi.org/10.1007/978-3-642-22110-1_14
2. Barrett, C., Stump, A., Tinelli, C.: The Satisfiability Modulo Theories Library (SMT-LIB) (2010). www.SMT-LIB.org
3. Bienvenu, M.: Prime implicates and prime implicants in modal logic. In: Proceedings of the National Conference on Artificial Intelligence, vol. 22, p. 379. AAAI Press/MIT Press, Menlo Park, Cambridge, London (1999, 2007)
4. Blackburn, P., Van Benthem, J., Wolter, F.: Handbook of Modal Logic. Studies in Logic and Practical Reasoning, vol. 3. Elsevier, Amsterdam (2007). ISSN 1570–2464
5. De Kleer, J.: An improved incremental algorithm for generating prime implicates. In: Proceedings of the National Conference on Artificial Intelligence, p. 780. Wiley (1992)
6. de Moura, L., Bjørner, N.: Z3: an efficient SMT solver. In: Ramakrishnan, C.R., Rehof, J. (eds.) TACAS 2008. LNCS, vol. 4963, pp. 337–340. Springer, Heidelberg (2008). https://doi.org/10.1007/978-3-540-78800-3_24
7. Dillig, I., Dillig, T., McMillan, K.L., Aiken, A.: Minimum satisfying assignments for SMT. In: Madhusudan, P., Seshia, S.A. (eds.) CAV 2012. LNCS, vol. 7358, pp. 394–409. Springer, Heidelberg (2012). https://doi.org/10.1007/978-3-642-31424-7_30
8. Echenim, M., Peltier, N.: A superposition calculus for abductive reasoning. J. Autom. Reason. **57**(2), 97–134 (2016)
9. Echenim, M., Peltier, N., Tourret, S.: An approach to abductive reasoning in equational logic. In: Proceedings of International Conference on Artificial Intelligence, IJCAI 2013, pp. 3–9. AAAI (2013)
10. Echenim, M., Peltier, N., Tourret, S.: A rewriting strategy to generate prime implicates in equational logic. In: Demri, S., Kapur, D., Weidenbach, C. (eds.) IJCAR 2014. LNCS, vol. 8562, pp. 137–151. Springer, Cham (2014). https://doi.org/10.1007/978-3-319-08587-6_10
11. Echenim, M., Peltier, N., Tourret, S.: Quantifier-free equational logic and prime implicate generation. In: Felty, A.P., Middeldorp, A. (eds.) CADE 2015. LNCS, vol. 9195, pp. 311–325. Springer, Cham (2015). https://doi.org/10.1007/978-3-319-21401-6_21
12. Echenim, M., Peltier, N., Tourret, S.: Prime implicate generation in equational logic. J. Artif. Intell. Res. **60**, 827–880 (2017)
13. Eén, N., Sörensson, N.: An extensible SAT-solver. In: Giunchiglia, E., Tacchella, A. (eds.) SAT 2003. LNCS, vol. 2919, pp. 502–518. Springer, Heidelberg (2004). https://doi.org/10.1007/978-3-540-24605-3_37
14. Fredkin, E.: Trie memory. Commun. ACM **3**(9), 490–499 (1960)
15. Jackson, P.: Computing prime implicates incrementally. In: Kapur, D. (ed.) CADE 1992. LNCS, vol. 607, pp. 253–267. Springer, Heidelberg (1992). https://doi.org/10.1007/3-540-55602-8_170
16. Kean, A., Tsiknis, G.: An incremental method for generating prime implicants/implicates. J. Symb. Comput. **9**(2), 185–206 (1990)
17. Knill, E., Cox, P.T., Pietrzykowski, T.: Equality and abductive residua for Horn clauses. Theoret. Comput. Sci. **120**(1), 1–44 (1993)

18. Marquis, P.: Extending abduction from propositional to first-order logic. In: Jorrand, P., Kelemen, J. (eds.) FAIR 1991. LNCS, vol. 535, pp. 141–155. Springer, Heidelberg (1991). https://doi.org/10.1007/3-540-54507-7_12

19. Marquis, P.: Consequence finding algorithms. In: Kohlas, J., Moral, S. (eds.) Handbook of Defeasible Reasoning and Uncertainty Management Systems. HAND, vol. 5, pp. 41–145. Springer, Dordrecht (2000). https://doi.org/10.1007/978-94-017-1737-3_3

20. Matusiewicz, A., Murray, N.V., Rosenthal, E.: Prime implicate tries. In: Giese, M., Waaler, A. (eds.) TABLEAUX 2009. LNCS, vol. 5607, pp. 250–264. Springer, Heidelberg (2009). https://doi.org/10.1007/978-3-642-02716-1_19

21. Matusiewicz, A., Murray, N.V., Rosenthal, E.: Tri-based set operations and selective computation of prime implicates. In: Kryszkiewicz, M., Rybinski, H., Skowron, A., Raś, Z.W. (eds.) ISMIS 2011. LNCS, vol. 6804, pp. 203–213. Springer, Heidelberg (2011). https://doi.org/10.1007/978-3-642-21916-0_23

22. Mayer, M.C., Pirri, F.: First order abduction via tableau and sequent calculi. Log. J. IGPL **1**(1), 99–117 (1993)

23. Mishchenko, A.: An introduction to zero-suppressed binary decision diagrams. Technical report, Proceedings of the 12th Symposium on the Integration of Symbolic Computation and Mechanized Reasoning (2001)

24. Nabeshima, H., Iwanuma, K., Inoue, K., Ray, O.: SOLAR: an automated deduction system for consequence finding. AI Commun. **23**(2), 183–203 (2010)

25. Previti, A., Ignatiev, A., Morgado, A., Marques-Silva, J.: Prime compilation of non-clausal formulae. In: Proceedings of the 24th International Conference on Artificial Intelligence, pp. 1980–1987. AAAI Press (2015)

26. Quine, W.: A way to simplify truth functions. Am. Math. Mon. **62**(9), 627–631 (1955)

27. Riazanov, A., Voronkov, A.: Vampire 1.1 (system description). In: Goré, R., Leitsch, A., Nipkow, T. (eds.) IJCAR 2001. LNCS, vol. 2083, pp. 376–380. Springer, Heidelberg (2001). https://doi.org/10.1007/3-540-45744-5_29

28. Schulz, S.: System description: E 1.8. In: McMillan, K., Middeldorp, A., Voronkov, A. (eds.) LPAR 2013. LNCS, vol. 8312, pp. 735–743. Springer, Heidelberg (2013). https://doi.org/10.1007/978-3-642-45221-5_49

29. Simon, L., Del Val, A.: Efficient consequence finding. In: Proceedings of the 17th International Joint Conference on Artificial Intelligence, pp. 359–370 (2001)

30. Tison, P.: Generalization of consensus theory and application to the minimization of boolean functions. IEEE Trans. Electron. Comput. **4**, 446–456 (1967)

31. Weidenbach, C., Afshordel, B., Brahm, U., Cohrs, C., Engel, T., Keen, E., Theobalt, C., Topić, D.: System description: Spass version 1.0.0. In: CADE 1999. LNCS, vol. 1632, pp. 378–382. Springer, Heidelberg (1999). https://doi.org/10.1007/3-540-48660-7_34

A Coinductive Approach to Proving Reachability Properties in Logically Constrained Term Rewriting Systems

Ştefan Ciobâcă[(✉)] and Dorel Lucanu

Alexandru Ioan Cuza University, Iaşi, Romania
{stefan.ciobaca,dlucanu}@info.uaic.ro

Abstract. We introduce a sound and complete coinductive proof system for reachability properties in transition systems generated by logically constrained term rewriting rules over an order-sorted signature modulo builtins. A key feature of the calculus is a circularity proof rule, which allows to obtain finite representations of the infinite coinductive proofs.

1 Introduction

We propose a framework for specifying and proving reachability properties of systems whose behaviour is modelled using transition systems described by logically constrained term rewriting systems (LCTRSs). By reachability properties we mean that a set of target states are reached in all terminating system computations starting from a given set of initial states. We assume transition systems are generated by constrained term rewriting rules of the form

$$l \twoheadrightarrow r \text{ if } \phi,$$

where l and r are terms and ϕ is a logical constraint. The terms l, r may contain both uninterpreted function symbols and function symbols interpreted in a builtin model, e.g., the model of booleans and integers. The constraint ϕ is a first-order formula that limits the application of the rule and which may contain predicate symbols interpreted in the builtin model. The intuitive meaning of a constrained rule $l \twoheadrightarrow r \text{ if } \phi$ is that any instance of l that satisfies ϕ transitions in one step into a corresponding instance of r.

Example 1. The following set of constrained rewrite rules specifies a procedure for compositeness:

$$init(n) \twoheadrightarrow loop(n, 2) \text{ if } \top,$$
$$loop(i \times k, i) \twoheadrightarrow comp \text{ if } k > 1,$$
$$loop(n, i) \twoheadrightarrow loop(n, i + 1) \text{ if } \neg(\exists k.k > 1 \wedge n = i \times k).$$

If n is not composite, the computation of the procedure is infinite.

© Springer International Publishing AG, part of Springer Nature 2018
D. Galmiche et al. (Eds.): IJCAR 2018, LNAI 10900, pp. 295–311, 2018.
https://doi.org/10.1007/978-3-319-94205-6_20

Given a LCTRS, which serves as a specification for a transition system, it is natural to define the notion of constrained term $\langle t \mid \phi \rangle$, where t is an ordinary term (with variables) and ϕ is a logical constraint. The intuitive meaning of such a term is the set of ground instances of t that satisfy ϕ.

Example 2. The constrained term $\langle init(n) \mid \exists u.1 < u < n \wedge n \bmod u = 0 \rangle$ defines exactly the instances of $init(n)$ where n is composite.

A *reachability formula* is a pair of constrained terms $\langle t \mid \phi \rangle \Rightarrow \langle t' \mid \phi' \rangle$. The intuitive meaning of a reachability formula is that any instance of $\langle t \mid \phi \rangle$ reaches, along all terminating paths of the transition system, an instance of $\langle t' \mid \phi' \rangle$ that agrees with $\langle t \mid \phi \rangle$ on the set of shared variables.

Example 3. The reachability formula

$$\langle init(n) \mid \exists u.1 < u < n \wedge n \bmod u = 0 \rangle \Rightarrow \langle comp \mid \top \rangle$$

captures a functional specification for the algorithm described in Example 1: each terminating computation starting from a state in which n is composite reaches the state *comp*. Computations that start with a negative number (composite or not) are infinite and therefore vacuously covered by the specification above.

We propose an effective proof system that, given a LCTRS, proves valid reachability formulas such as the one above, assuming an oracle that solves logical constraints. In practice, we use an SMT solver instead of the oracle.

Contributions. 1. As computations can be finite or infinite, an inductive approach for reachability is not practically possible. In Sect. 2, we propose a coinductive approach for specifying transition systems, which is an elegant way to look at reachability, but also essential in handling both finite and infinite executions. 2. We formalize the semantics of LCTRSs as a reduction relation over a particular model that combines order-sorted terms with builtin elements such as integers, booleans, arrays, etc. The new approach, introduced in Sect. 3, is simpler than the usual semantics for constrained term rewriting systems [14,18–20], but it also lifts several technical restrictions that are important for our case studies. 3. We introduce a sound and complete coinductive proof system for deriving valid reachability formulas for transition systems specified by a LCTRS. We present our proof system in two steps: in the first step, we provide a three-rule proof system (Fig. 1) for *symbolic execution of constrained terms*. When interpreting the proof system coinductively, its proof trees can be finite or infinite. The finite proof trees correspond to reachability formulas $\langle t \mid \phi \rangle \Rightarrow \langle t' \mid \phi' \rangle$ where there is a bounded number of symbolic steps between $\langle t \mid \phi \rangle$ and $\langle t' \mid \phi' \rangle$. The infinite proof trees correspond to proofs of reachability formulas $\langle t \mid \phi \rangle \Rightarrow \langle t' \mid \phi' \rangle$ that hold for an unbounded number of symbolic steps between $\langle t \mid \phi \rangle$ and $\langle t' \mid \phi' \rangle$ (obtained, e.g., by unrolling loops). Symbolic execution has similarities to narrowing, but unlike narrowing, where each step computes a possible successor, symbolic execution must consider *all* successors of a state at the same time. 4. The infinite proof trees above cannot be obtained in finite time in practice.

In order to derive reachability formulas that require an unbounded number of symbolic steps in finite time, we introduce a fourth proof rule to the system that we call *circularity*. The circularity proof rule can be used to compress infinite proof trees into finite proof trees. The intuition is to use as axioms the goals that are to be proven, when they satisfy a *guardedness* condition. This compression of infinite coinductive trees into finite proof trees via the guardedness condition nicely complements our coinductive approach. This separation between symbolic execution and circularity answers an open question in [21]. 5. We introduce the RMT tool, an implementation of the proof system that validates our approach on a number of examples. RMT uses an SMT solver to discharge logical constraints. The tool is expressive enough for specifying various transition systems, including operational semantics of programming languages, and proving reachability properties of practical interest and is intended to be the starting point of a library for rewriting modulo builtins, which could have more applications.

Related Work. A number of approaches [1,2,14,25,29] to combining rewriting and SMT solving have appeared lately. The rewrite tool Maude [12] has been extended with SMT solving in [25] in order to enable the analysis of open systems. A method for proving invariants based on an encoding into reachability properties is presented in [29]. Both approaches above are restricted to *topmost rewrite theories*. While almost any theory can be written as a topmost theory [22], the encoding can significantly increase the number of transitions, which raises performance concerns. Our definition for constrained term is a generalization of that of constructor constrained pattern used in [29]. In particular [29] does not allow for quantifiers in constraints, but quantifiers are critical to obtaining a complete proof system, as witnessed by their use in the subsumption rule in our proof system ([subs], Fig. 1). The approach without quantifiers is therefore not sufficient to prove reachabilities in a general setting.

A calculus for reachability properties in a formalism similar to LCTRSs is given in [1]. However, the notion of reachability in [1] is different from ours: while we show reachability along *all terminating paths of the computation*, [1] solves reachability properties of the form $\exists \widetilde{x}.t(\widetilde{x}) \rightarrow^* t'(\widetilde{x})$ (i.e. does there exists an instance of t that reaches, along some path, an instance of t').

Work on constrained term rewriting systems appeared in [14,18–20]. In contrast to this approach to constrained rewriting, our semantics is simpler (it does not require two reduction relations), it does not have restrictions on the terms l, r in a rule $l \rightarrow r$ if ϕ and the constraint is an arbitrary first-order formula ϕ, possibly with quantifiers, which are crucial to obtain symbolic execution in its full generality. Constrained terms are generalized to guarded terms in [2], in order to reduce the state space.

Reachability in rewriting is explored in depth in [13]. The work by Kirchner and others [17] is the first to propose the use of rewriting with symbolic constraints for deduction. Subsequent work [14,20,25] extends and unifies previous approaches to rewriting with constraints. The related work section in [25] includes a comprehensive account of literature related to rewriting modulo constraints.

Our previous work [10,21] on proving program correctness was in the context of the K framework [27]. K, developed by Roşu and others, implements semantics-based program verifiers [11] for any language that can be specified by a rewriting-based operational semantics, such as C [15], Java [4] and JavaScript [23]. Our formalism is not more expressive than that of reachability logic [10] for proving partial correctness of programs in a language-independent manner, but it does have several advantages. Firstly, we make a clear separation between *rewrite rules* (used to define transition systems), for which it makes no sense to have constraints on both the lhs and the rhs, and *reachability formulas* (used to specify reachability properties), for which there can be constraints on both the lhs and the rhs. We provide clear semantics of both syntactic constructs above, which makes it unnecessary to check well-definedness of the underlying rewrite system, as required in [10]. Additionally, this separation, which we see as a contribution, makes it easy to get rid of the top-most restriction in previous approaches. Another advantage is that the proposed proof system is very easy to automate, while being sufficiently expressive to specify real-world applications. Additionally, we work in the more general setting of LCTRSs, not just language semantics, which enlarges the possible set of applications of the technique. We also have several major technical improvements compared to [21], where the proof system is restricted to the cases where unification can be reduced to matching and topmost rewriting. The totality property required for languages specifications, which was quite restrictive, was replaced by a local property in proof rules and all restrictions needed to reduce unification to matching were removed.

In contrast to the work on partial correctness in [11], the approach on reachability discussed here is meant for any LCTRS, not just operational semantics. The algorithm in [11] contains a small source of incompleteness, as when proving a reachability property it is either discharged completely through implication or through circularities/rewrite rules. We allow a reachability rule to be discharged partially by subsumption and partially by other means. Constrained terms are a fragment of Matching Logic (see [26]), where no distinction is made between terms and constraints. Coinduction and circular or cyclic proofs have been proposed in other contexts. For example, circular proof systems have been proposed for first-order logic with inductive predicates in [6] and for separation logic in [5]. In the context of interactive theorem provers, circular coinduction has been proposed as an incremental proof method for bisimulation in process calculi (see [24]). A compositional and incremental approach to coinduction that uses a semantic guardedness check instead of a syntactic check is given in [16].

Paper Structure. We present coinductive definitions for execution paths and reachability predicates in Sect. 2. In Sect. 3, we introduce logically constrained term rewriting with builtins in an order-sorted setting. In Sect. 4, we propose a sound and complete coinductive calculus for reachability and a circularity rule for compressing infinite proof trees into finite proof trees. Section 6 discusses the implementation before concluding. The proofs can be found in [8].

2 Reachability Properties: Coinductive Definition

In this section we introduce a class of reachability properties, defined coinductively. A *state predicate* is a subset of states. A *reachability property* is a pair $P \Rightarrow Q$ of state predicates. Such a reachability property is *demonically valid* iff each execution path starting from a state in P eventually reaches a state in Q, or if it is infinite. Since the set of finite and infinite executions is coinductively defined, the set of valid predicates can be defined coinductively as well. Formally, consider a transition system $(M, \rightsquigarrow)$, with $\rightsquigarrow \subseteq M \times M$. We write $\gamma \rightsquigarrow \gamma'$ for $(\gamma, \gamma') \in \rightsquigarrow$. An element $\gamma \in M$ is *irreducible* if $\gamma \not\rightsquigarrow \gamma'$ for any $\gamma' \in M$.

Definition 1 (Execution Path). *The set of* (complete) execution paths *is coinductively defined by the following rules:*

$$\frac{}{\gamma} \gamma \in M, \gamma \text{ irreducible} \qquad \frac{\tau}{\gamma_0 \circ \tau} \gamma_0 \rightsquigarrow hd(\tau)$$

where the function hd is defined by $hd(\gamma) = \gamma$ *and* $hd(\gamma_0 \circ \tau) = \gamma_0$.

The above definition includes both the finite execution paths ending in a irreducible state and the infinite execution paths, defined as the greatest fixed point of the associated functional (see [8]).

Definition 2 (State and Reachability Predicates). *A state predicate is a subset* $P \subseteq M$. *A reachability predicate is a pair of state predicates* $P \Rightarrow Q$. *The predicate* P *is* runnable *if* $P \neq \emptyset$ *and for all* $\gamma \in P$ *there is* $\gamma' \in M$ *s.t.* $\gamma \rightsquigarrow \gamma'$.

A *derivative* measures the sensitivity to change of a quantity. For the case of transition systems, the change of states is determined by the transition relation.

Definition 3 (Derivative of a State Predicate). *The derivative of a state predicate* P *is the state predicate* $\partial(P) = \{\gamma' \mid \gamma \rightsquigarrow \gamma' \text{ for some } \gamma \in P\}$.

As a reachability predicate specifies reachability property of execution paths, we define when a particular execution path satisfies a reachability predicate.

Definition 4 (Satisfaction of a Reachability Predicate). *An execution path* τ *satisfies a reachability predicate* $P \Rightarrow Q$, *written* $\tau \vDash^\forall P \Rightarrow Q$, *iff* $\langle \tau, P \Rightarrow Q \rangle \in \nu \widehat{\mathsf{EPSRP}}$, *where* EPSRP *consists of the following rules:*

$$\frac{}{\langle \tau, P \Rightarrow Q \rangle} hd(\tau) \in P \cap Q \qquad \frac{\langle \tau, \partial(P) \Rightarrow Q \rangle}{\langle \gamma_0 \circ \tau, P \Rightarrow Q \rangle} \gamma_0 \in P, \gamma_0 \rightsquigarrow hd(\tau).$$

The notation $\widehat{\mathsf{EPSRP}}$ stands for the functional of EPSRP and $\nu\widehat{\mathsf{EPSRP}}$ stands for its greatest fixed point (see [8]). We coinductively define the set of *demonically valid reachability predicates* over $(M, \rightsquigarrow)$. This allows to use coinductive proof techniques to prove validity of reachability predicates.

Definition 5 (Valid Reachability Predicates, Coinductively). *We say that $P \Rightarrow Q$ is* demonically valid, *and we write*

$$(M, \rightsquigarrow) \models^\forall P \Rightarrow Q,$$

iff $P \Rightarrow Q \in \nu \widehat{\mathsf{DVP}}$, where DVP consists of the following rules:

$$[\![\mathsf{Subsumption}]\!] \quad \frac{}{P \Rightarrow Q} \; P \subseteq Q \qquad [\![\mathsf{Step}]\!] \quad \frac{\partial(P \setminus Q) \Rightarrow Q}{P \Rightarrow Q} \; P \setminus Q \text{ runnable.}$$

The condition $P \setminus Q$ *runnable* in the second rule is essential to avoid the cases where execution is stuck. These blocking states have no successor in $\partial(P \setminus Q)$ and, in the absense of the condition, we would wrongly conclude that they satisfy $P \Rightarrow Q$. The terminating executions are captured by $[\![\mathsf{Subsumption}]\!]$.

The following proposition justifies our definition of demonically valid reachability predicates.

Proposition 1. *Let $P \Rightarrow Q$ be a reachability predicate. We have $(M, \rightsquigarrow) \models^\forall P \Rightarrow Q$ iff any execution path τ starting from P $(hd(\tau) \in P)$ satisfies $P \Rightarrow Q$.*

3   Logically Constrained Term Rewriting Systems

In this section we introduce our formalism for LCTRSs. We interpret LCTRSs in a model combining order-sorted terms with builtins such as integers, booleans, etc. Logical constraints are first-order formulas interpreted over the fixed model.

We assume a *builtin model* M^b for a many-sorted *builtin signature* $\Sigma^b = (S^b, F^b)$, where S^b is a set of *builtin sorts* that includes at least the sort *Bool* and F^b is the S^b-sorted set of *builtin function symbols*. We assume that the set interpreting the sort *Bool* in the model M^b is $M^b_{Bool} = \{\top, \bot\}$. We use the standard notation M_o for the interpretation of the sort/symbol o in the model M. The set CF^b, defined as the set of (many-sorted) first-order formulas with equality over the signature Σ^b, is the set of *builtin constraint formulas*. Functions returning *Bool* play the role of predicates and terms of sort *Bool* are atomic formulas. We will assume that the builtin constraint formulas can be decided by an oracle (implemented as an SMT solver).

A *signature modulo builtins* is an order-sorted signature $\Sigma = (S, \leq, F)$ that includes Σ^b as a subsignature and such that the only builtin constants in Σ are elements of the builtin model ($\{c \mid c \in F_{\varepsilon,s}, s \in S^b\} = M^b_s$) – therefore the signature might be infinite. By $F_{w,s}$ we denoted the set of function symbols of arity w and result sort s. Σ^b is called the *builtin subsignature* of Σ and $\Sigma^c = (S, \leq, (F \setminus F^b) \cup \bigcup_{s \in S^b} F_{\varepsilon,s})$ the *constructor subsignature* of Σ. We let $\mathcal{X}$ be an S-sorted set of variables.

We extend the builtin model M^b to an $(S, \leq, F)$-model M^Σ defined as follows: • $M^\Sigma_s = T_{\Sigma^c,s}$, for each $s \in S \setminus S^b$ (M^Σ_s is the set of ground constructor terms of sort s, i.e. terms built from constructors applied to builtin elements);

- $M_f^{\Sigma} = M_f^{b}$ for each builtin function symbol $f \in F^{b}$; • M_f^{Σ} is the term constructor $M_f^{\Sigma}(t_1, \ldots, t_n) = f(t_1, \ldots, t_n)$, for each non-builtin function symbol $f \in F \setminus F^{b}$. By fixing the interpretation of the non-builtin function symbols, we can reduce constraint formulas to built-in constraint formulas by relying on an unification algorithm described in detail in [7]. We also make the standard assumption that $M_s \neq \emptyset$ for any $s \in S$.

Example 4. Let $\Sigma^{b} = (S^{b}, F^{b})$, where $S^{b} = \{Int, Bool\}$ and F^{b} include the usual operators over *Bool*eans $(\vee, \wedge, \ldots)$ and over the *Int*egers $(+, -, \times, \ldots)$. The builtin model M^{b} interprets the above sorts and operations as expected.

We consider the signature modulo builtins $\Sigma = (S, \leq, F)$, where the set of sorts $S = \{Cfg, Int, Bool\}$ consists of the builtin sorts and an additional sort *Cfg*, where the subsorting relation $\leq \subseteq S \times S = \emptyset$ is empty, and where the set of function symbols F includes, in addition to the builtin symbols in F^{b}, the following function symbols: $init : Int \to Cfg, loop : Int \times Int \to Cfg, comp : Cfg$. We have that $M_{Cfg}^{\Sigma} = \{init(i) \mid i \in \mathbb{Z}\} \cup \{loop(i, j) \mid i, j \in \mathbb{Z}\} \cup \{comp\}$.

The set CF of *constraint formulas* is the set of first-order formulas with equality over the signature Σ. The subset of the builtin constraint formulas is denoted by CF^{b}. Let $var(\phi)$ denote the set of variables freely occurring in ϕ. We write $M^{\Sigma}, \alpha \vDash \phi$ when the formula ϕ is satisfied by the model M^{Σ} with a valuation $\alpha : X \to M^{\Sigma}$.

Example 5. The constraint formula $\phi \triangleq \exists u.1 < u < n \wedge n \bmod u = 0$ is satisfied by the model M^{Σ} defined in Example 4 and any valuation α such that $\alpha(n)$ is a composite number.

Definition 6 (Constrained Terms). *A constrained term φ of sort $s \in S$ is a pair $\langle t \mid \phi \rangle$, where $t \in T_{\Sigma, s}(\mathcal{X})$ and $\phi \in$ CF.*

Example 6. Continuing the previous example, the following is a constrained term: $\langle init(n) \mid \exists u.1 < u < n \wedge n \bmod u = 0 \rangle$.

We consistently use φ for constrained terms and ϕ for constraint formulas.

Definition 7 (Valuation Semantics of Constraints). *The valuation semantics of a constraint ϕ is the set $\lfloor \phi \rfloor \triangleq \{\alpha : X \to M^{\Sigma} \mid M^{\Sigma}, \alpha \vDash \phi\}$.*

Example 7. Continuing the previous example, we have that

$$\lfloor \exists u.1 < u < n \wedge n \bmod u = 0 \rfloor = \{\alpha : X \to M^{\Sigma} \mid \alpha(n) \text{ is composite}\}.$$

Definition 8 (State Predicate Semantics of Constrained Terms). *The state predicate semantics of a constrained term $\langle t \mid \phi \rangle$ is the set*

$$\llbracket \langle t \mid \phi \rangle \rrbracket \triangleq \{\alpha(t) \mid \alpha \in \lfloor \phi \rfloor\}.$$

Example 8. Continuing the previous example, we have that

$$\llbracket \langle init(n) \mid \exists u.1 < u < n \wedge n \bmod u = 0 \rangle \rrbracket = \{init(n) \mid n \text{ is composite}\}.$$

We now introduce our formalism for logically constrained term rewriting systems. Syntactically, a rewrite rule consists of two terms (the left hand side and respectively the right hand side), together with a constraint formula. As the two terms could *share some variables*, these shared variables should be instantiated consistently in the semantics:

Definition 9 (LCTRS). *A logically constrained rewrite rule is a tuple* (l, r, ϕ), *often written as* $l \rightarrow r$ if ϕ, *where* l, r *are terms in* $T_\Sigma(\mathcal{X})$ *having the same sort, and* $\phi \in CF$. *A logically constrained term rewriting system* $\mathcal{R}$ *is a set of logically constrained rewrite rules.* $\mathcal{R}$ *defines an order-sorted transition relation* $\leadsto_\mathcal{R}$ *on* M^Σ *as follows:* $t \leadsto_\mathcal{R} t'$ *iff there exist a rule* $l \rightarrow r$ if ϕ *in* $\mathcal{R}$, *a context* $c[\cdot]$, *and a valuation* $\alpha : X \rightarrow M^\Sigma$ *such that* $t = \alpha(c[l])$, $t' = \alpha(c[r])$ *and* $M^\Sigma, \alpha \vDash \phi$.

Example 9. We recall the LCTRS given in the introduction:

$$\mathcal{R} = \left\{ \begin{array}{l} init(n) \rightarrow loop(n, 2) \text{ if } \top, \\ loop(i \times k, i) \rightarrow comp \text{ if } k > 1, \\ loop(n, i) \rightarrow loop(n, i + 1) \text{ if } \neg(\exists k. k > 1 \wedge n = i \times k) \end{array} \right\}.$$

A LCTRS $\mathcal{R}$ defines a sort-indexed transition system $(M^\Sigma, \leadsto_\mathcal{R})$. As each constrained term φ defines a state predicate $[\![\varphi]\!]$, it is natural to specify reachability predicates as pairs of constrained terms sharing a subset of variables. The shared variables must be instantiated in the same way by the execution paths connecting states specified by the two constrained terms.

Definition 10 (Reachability Properties of LCTRSs). *A reachability formula* $\varphi \Rightarrow \varphi'$ *is a pair of constrained terms, which may share variables. We say that a LCTRS* $\mathcal{R}$ *demonically satisfies* $\varphi \Rightarrow \varphi'$, *written*

$$\mathcal{R} \vDash^\forall \varphi \Rightarrow \varphi',$$

iff $(M^\Sigma, \leadsto_\mathcal{R}) \vDash^\forall [\![\sigma(\varphi)]\!] \Rightarrow [\![\sigma(\varphi')]\!]$ *for each* $\sigma : var(\varphi) \cap var(\varphi') \rightarrow M^\Sigma$.

Since the carriers sets of M^Σ consist of ground terms, σ is both a substitution and a valuation in the definition above. Its role is critical: to ensure that the shared variables of φ and φ' are instantiated by the same values.

Example 10. Continuing the previous example, we have that the reachability formula $\langle init(n) \, | \, \exists u.1 < u < n \wedge n \bmod u = 0 \rangle \Rightarrow \langle comp \, | \, \top \rangle$ is demonically satisfied by the constrained rule system $\mathcal{R}$ defined in Example 9:

$$\mathcal{R} \vDash^\forall \langle init(n) \, | \, \exists u.1 < u < n \wedge n \bmod u = 0 \rangle \Rightarrow \langle comp \, | \, \top \rangle.$$

We have checked the above reachability formula against $\mathcal{R}$ mechanically, using an implementation of the approach described in this paper.

4   Proving Reachability Properties of LCTRSs

We introduce two proof systems for proving reachability properties in transition systems specified by LCTRSs. The first proof system formalizes *symbolic execution* in a LCTRS, in the following sense: a reachability formula $\varphi \Rightarrow \varphi'$ can be proven if any execution path starting with φ either reaches a state that is an instance of φ', or is divergent. Note that this intuition holds when the proof system is interpreted coinductively, where infinite proof trees are allowed. Unfortunately, these infinite proof trees have a limited practical use because they cannot be obtained in finite time.

In order to solve this limitation, we introduce a second proof system, which contains an additional inference rule, called *circularity*. The *circularity* rule allows to use the reachability formula to be proved as an axiom. This allows to fold infinite proof trees into finite proof trees, which can be obtained in finite time. Adding the reachability formulas that are to be proved as axioms seems at first to be unsound, but it corresponds to a natural intuition: when reaching a proof obligation that we have handled before, there is no need to prove it again, because the previous reasoning can be reused (possibly leading to an infinite proof branch). However, the circularity rule must be used in a guarded fashion in order to preserve soundness. We introduce a simple criterion to select the sound proof trees.

4.1   Derivatives of Constrained Terms

Our proof system relies on the notion of derivative at the syntactic level:

Definition 11 (Derivatives of Constrained Terms). *The* set of derivatives *of a constrained term* $\varphi \triangleq \langle t \mid \phi \rangle$ *w.r.t. a rule* $l \twoheadrightarrow r$ if ϕ_{lr} *is*

$$\Delta_{l,r,\phi_{lr}}(\varphi) \triangleq \{ \langle c[r] \mid \phi' \rangle \mid \phi' \triangleq \phi \wedge t = c[l] \wedge \phi_{lr},$$

$$c[\cdot] \text{ an appropriate context and } \phi' \text{ is satisfiable} \}, \quad (1)$$

where the variables in $l \twoheadrightarrow r$ if ϕ_{lr} *are renamed such that* $var(l, r, \phi_{lr})$ *and* $var(\varphi)$ *are disjoint. If* $\mathcal{R}$ *is a set of rules, then* $\Delta_{\mathcal{R}}(\varphi) = \bigcup_{(l,r,\phi_{lr}) \in \mathcal{R}} \Delta_{l,r,\phi_{lr}}(\varphi)$. *A constrained term* φ *is* $\mathcal{R}$-derivable *if* $\Delta_{\mathcal{R}}(\varphi) \neq \emptyset$.

Example 11. Continuing the previous examples, we have that

$$\Delta_{\mathcal{R}}(\langle init(n) \mid \exists u.1 < u < n \wedge n \ mod \ u = 0 \rangle) =$$
$$\{ \langle loop(n, 2) \mid \exists u.1 < u < n \wedge n \ mod \ u = 0 \rangle \}.$$

In the above case, $\Delta_{\mathcal{R}}$ includes only the derivative computed w.r.t. the first rule in $\mathcal{R}$, because the constraints of the ones computed w.r.t. the other rules are unsatisfiable. Intuitively, the derivatives of a constrained term denote all its possible successor configurations in the transition system generated by $\mathcal{R}$.

The symbolic derivatives and the concrete ones are related as expected:

[axiom] $\dfrac{}{\langle t_l \mid \bot \rangle \Rightarrow \langle t_r \mid \phi_r \rangle}$

[subs] $\dfrac{\langle t_l \mid \phi_l \wedge \neg(\exists \tilde{x}.t_l = t_r \wedge \phi_r) \rangle \Rightarrow \langle t_r \mid \phi_r \rangle}{\langle t_l \mid \phi_l \rangle \Rightarrow \langle t_r \mid \phi_r \rangle}$ $\begin{array}{l}\tilde{x} \triangleq var(t_r, \phi_r) \setminus var(t_l, \phi_l) \\ \exists \tilde{x}.t_l = t_r \wedge \phi_r \text{ satisfiable}\end{array}$

[der$^\forall$] $\dfrac{\langle t^j \mid \phi^j \rangle \Rightarrow \langle t_r \mid \phi_r \rangle, j \in \{1, \ldots, n\}}{\langle t_l \mid \phi_l \rangle \Rightarrow \langle t_r \mid \phi_r \rangle}$ $\begin{array}{l}\langle t_l \mid \phi_l \rangle \text{ is } \mathcal{R}-\text{derivable and} \\ \phi_l \to \bigvee_{j \in \{1,\ldots,n\}} \exists \tilde{y}^j.\phi^j \text{ is valid}\end{array}$

where $\Delta_{\mathcal{R}}(\langle t_l \mid \phi_l \rangle) = \{\langle t^1 \mid \phi^1 \rangle, \ldots, \langle t^n \mid \phi^n \rangle\}$ and
$\tilde{y}^j = var(t^j, \phi^j) \setminus var(t_l, \phi_l)$

Fig. 1. The DSTEP($\mathcal{R}$) Proof System

Theorem 1. *Let $\varphi \triangleq \langle t \mid \phi \rangle$ be a constrained term, $\mathcal{R}$ a constrained rule system, and $(M^\Sigma, \leadsto_{\mathcal{R}})$ the transition system defined by $\mathcal{R}$. Then $[\![\Delta_{\mathcal{R}}(\varphi)]\!] = \partial([\![\varphi]\!])$.*

Our proof systems allow to replace any reachability formula by an equivalent one. Two reachability formulas, $\varphi_1 \Rightarrow \varphi_1'$ and $\varphi_2 \Rightarrow \varphi_2'$, are *equivalent*, written $\varphi_1 \Rightarrow \varphi_1' \equiv \varphi_2 \Rightarrow \varphi_2'$, if, for all LCTRSs $\mathcal{R}$,

$$\mathcal{R} \vDash^\forall \varphi_1 \Rightarrow \varphi_1' \text{ iff } \mathcal{R} \vDash^\forall \varphi_2 \Rightarrow \varphi_2'.$$

We write $[\![\varphi]\!] \subseteq_{shared} [\![\varphi']\!]$ iff for each $\sigma : var(\varphi) \cap var(\varphi') \to M^\Sigma$, we have $[\![\sigma(\varphi)]\!] \subseteq [\![\sigma(\varphi')]\!]$. The next result, used in our proof system, shows that inclusion of the state predicate semantics of two constrained terms can be expressed as a constraint formula, when the shared variables are instantiated consistently.

Proposition 2. *The inclusion $[\![\langle t \mid \phi \rangle]\!] \subseteq_{shared} [\![\langle t' \mid \phi' \rangle]\!]$ holds if and only if $M^\Sigma \vDash \phi \to (\exists \tilde{x})(t = t' \wedge \phi')$, where $\tilde{x} \triangleq var(t', \phi') \setminus var(t, \phi)$.*

4.2 Proof System for Symbolic Execution

The first proof system, DSTEP, derives sequents of the form $\langle t_l \mid \phi_l \rangle \Rightarrow \langle t_r \mid \phi_r \rangle$. The proof system consists of three proof rules presented in Fig. 1 and an implicit structural rule that allows to replace reachability formulas by equivalent reachability formulas. The instances of this implicit structural rule are not included in the proof trees. We explain the three rules in the proof system.

- The [axiom] rule discharges goals where the left hand side of the goal does not match any state. As our structural rule identifies equivalent reachability formulas, this rule can be applied to any left-hand side where the constraint is unsatisfiable (equivalent to $\bot$). This rule discharges reachability formulas where there are no execution paths starting from the left-hand side, and therefore there is no need to continue the proof process along this branch.
- The [subs] rule discharges the cases where the left-hand side is an instance of the right-hand side. The constraint $\exists \tilde{x}.t_l = t_r \wedge \phi_r$ is true exactly when the left-hand side is an instance of the right-hand side, which is ensured by Proposition 2. The proof of the current goal continues only for the cases where

the negation of this constraint holds (i.e., the cases where the left-hand side is not included in the right-hand side).

- The [der$^\forall$] rule allows to take a symbolic step in the left-hand side of the current goal. It computes all derivatives of the left-hand side; the proof process must continue with each such derivative. Let $\psi \triangleq \phi_l \rightarrow \bigvee\{\exists \widetilde{y}^j . \phi^j\}$ be the logical constraint that occurs in the condition of [der$^\forall$]. The formula ψ is valid iff, for any instance of $\langle t_l \,|\, \phi_l \rangle$, there is at least one rule of $\mathcal{R}$ that can be applied to it, meaning that $\mathcal{R}$ is *total for* $\langle t_l \,|\, \phi_l \rangle$. Summarising, the condition of [der$^\forall$] says that $[\![\langle t_l \,|\, \phi_l \rangle]\!]$ must have at least one successor and furthermore that any instance $\gamma \in [\![\langle t_l \,|\, \phi_l \rangle]\!]$ has a $\leadsto_{\mathcal{R}}$-successor.

The following result shows that $\mathsf{DSTEP}(\mathcal{R})$ is sound and complete, modulo an oracle for solving logical constraints.

Theorem 2. *Let $\mathcal{R}$ be a LCTRS. For any reachability formula $\varphi \Rightarrow \varphi'$, we have*

$$\mathcal{R} \models^\forall \varphi \Rightarrow \varphi' \text{ iff } \varphi \Rightarrow \varphi' \in \nu \,\widehat{\mathsf{DSTEP}}(\mathcal{R}).$$

Example 12. Consider the LCTRS $\mathcal{R}$ defined in Example 9. The proof tree for the reachability formula $\langle init(n) \,|\, \psi \rangle \Rightarrow \varphi_r$, where $\psi \triangleq \exists u.1 < u < n \wedge n \bmod u = 0$ denotes the fact that n is composite and $\varphi_r \triangleq \langle comp \,|\, \top \rangle$, is infinite:

$$
\cfrac{
\cfrac{\langle comp \,|\, \bot \rangle \Rightarrow \varphi_r}{\langle comp \,|\, \psi \wedge \phi_a \rangle \Rightarrow \varphi_r}\text{[axiom]}\text{[subs]}
\qquad
\cfrac{
\cfrac{
\cfrac{\langle comp \,|\, \bot \rangle \Rightarrow \varphi_r}{\langle comp \,|\, \psi \wedge \phi_2 \wedge \phi_b \rangle \Rightarrow \varphi_r}\text{[axiom]}\text{[subs]}
}{\langle loop(n,3) \,|\, \psi \wedge \phi_2 \rangle \Rightarrow \varphi_r} \quad \vdots
}{\text{[der}^\forall\text{]}}
}{
\cfrac{\langle loop(n,2) \,|\, \psi \rangle \Rightarrow \varphi_r}{\langle init(n) \,|\, \psi \rangle \Rightarrow \varphi_r}\text{[der}^\forall\text{]}
}
$$

The right branch of the above proof tree is infinite, and:

$$\phi_2 \triangleq \neg \exists k.k > 1 \wedge n = 2 \times k \qquad \phi_a \triangleq loop(n,2) = loop(i' \times k', i') \wedge k' > 1$$
$$\phi_3 \triangleq \neg \exists k.k > 1 \wedge n = 3 \times k \qquad \phi_b \triangleq loop(n,3) = loop(i' \times k', i') \wedge k' > 1$$

$$\dots$$

Note that in the presentation of the tree above, we used the structural rule to replace reachability formulas by equivalent reachability formulas as follows:

$$\langle comp \,|\, \psi \wedge \phi_a \wedge \neg(comp = comp \wedge \top) \rangle \Rightarrow \varphi_r \qquad \equiv \quad \langle comp \,|\, \bot \rangle \Rightarrow \varphi_r,$$
$$\langle comp \,|\, \psi \wedge \phi_2 \wedge \phi_b \wedge \neg(comp = comp \wedge \top) \rangle \Rightarrow \varphi_r \qquad \equiv \quad \langle comp \,|\, \bot \rangle \Rightarrow \varphi_r,$$
$$\langle loop(n',2) \,|\, \top \wedge init(n') = init(n) \wedge \psi \rangle \Rightarrow \varphi_r \qquad \equiv \quad \langle loop(n,2) \,|\, \psi \rangle \Rightarrow \varphi_r,$$
$$\langle loop(n',i'+1) \,|\, \psi \wedge \phi_2' \rangle \Rightarrow \varphi_r \qquad \equiv \quad \langle loop(n,3) \,|\, \psi \wedge \phi_2 \rangle \Rightarrow \varphi_r,$$

where $\phi_2' \triangleq loop(n,2) = loop(n',i') \wedge \neg \exists k.k > 1 \wedge n' = i' \times k$. The ticks appear in the formulas above because, to compute derivatives, we used the following fresh instance of $\mathcal{R}$:

$$\mathcal{R} = \left\{ \begin{array}{l} init(n') \rightarrow\!\!\!\rightarrow loop(n', 2) \text{ if } \top, \\ loop(i' \times k', i') \rightarrow\!\!\!\rightarrow comp \text{ if } k' > 1, \\ loop(n', i') \rightarrow\!\!\!\rightarrow loop(n', i' + 1) \text{ if } \neg(\exists k.k > 1 \wedge n' = i' \times k) \end{array} \right\}.$$

4.3 Extending the Proof System with a Circularity Rule

As we said at the beginning of the section, the use of DSTEP is limited because of the infinite proof trees. The next inference rule is intended to use the initial goals as axioms to fold infinite DSTEP-proof trees into sound finite proof trees.

Definition 12 (Demonic circular coinduction). *Let G be a finite set of reachability formulas. Then the set of rules $\mathsf{DCC}(\mathcal{R}, G)$ consists of $\mathsf{DSTEP}(\mathcal{R})$, together with*

$$[\text{circ}] \quad \frac{\langle t_r^c \mid \phi_l \wedge \phi \wedge \phi_r^c \rangle \Rightarrow \varphi_r, \quad \langle t_l \mid \phi_l \wedge \neg\phi \rangle \Rightarrow \varphi_r}{\langle t_l \mid \phi_l \rangle \Rightarrow \varphi_r} \quad \begin{array}{l} \phi \text{ is } \exists var(t_l^c, \phi_l^c).t_l = t_l^c \wedge \phi_l^c, \\ \langle t_l^c \mid \phi_l^c \rangle \Rightarrow \langle t_r^c \mid \phi_r^c \rangle \in G \end{array}$$

where $\langle t_l^c \mid \phi_l^c \rangle \Rightarrow \langle t_r^c \mid \phi_r^c \rangle$ is a rule in G whose variables have been renamed with fresh names.

The idea is that G should be chosen conveniently so that $\mathsf{DCC}(\mathcal{R}, G)$ proves G itself. We call such goals G (that are used to prove themselves) *circularities*. The intuition behind the rule is that the formula ϕ defined in the rule holds when a circularity can be applied. In that case, it is sufficient to continue the current proof obligation from the rhs of the circularity $\langle t_r^c \mid \phi_r^c \wedge \phi_l \wedge \phi \rangle$. The cases when ϕ does not hold (the circularity cannot be applied) are captured by the proof obligation $\langle t_l \mid \phi_l \wedge \neg\phi \rangle \Rightarrow \varphi_r$.

Of course, not all proof trees under $\mathsf{DCC}(\mathcal{R}, G)$ are sound. The next two definitions identify a class of sound proof trees (cf. Theorem 3).

Definition 13. *Let PT be a proof tree of $\varphi \Rightarrow \varphi'$ under $\mathsf{DCC}(\mathcal{R}, G)$. A $[\text{circ}]$ node in PT is guarded iff it has as ancestor a $[\text{der}^\forall]$ node. PT is guarded iff all its $[\text{circ}]$ nodes are guarded.*

Definition 14. *We write $(\mathcal{R}, G) \vdash^\forall \varphi \Rightarrow \varphi'$ iff there is a proof tree of $\varphi \Rightarrow \varphi'$ under $\mathsf{DCC}(\mathcal{R}, G)$ that is guarded. If F is a set of reachability formulas, we write $(\mathcal{R}, G) \vdash^\forall F$ iff $(\mathcal{R}, G) \vdash^\forall \varphi \Rightarrow \varphi'$ for all $\varphi \Rightarrow \varphi' \in F$.*

The criterion stated by Definition 13 can be easily checked in practice. The following theorem states that the guarded proof trees under DCC are sound.

Theorem 3 (Circularity Principle). *Let $\mathcal{R}$ be a constrained rule system and G a set of goals. If $(\mathcal{R}, G) \vdash^\forall G$ then $\mathcal{R} \models^\forall G$.*

Theorem 3 can be used by finding a set of circularities and using them in a guarded fashion to prove themselves. Then the circularity principle states that such circularities hold.

Example 13. In order to prove $\langle init(n) \mid \psi \rangle \Rightarrow \langle comp \mid \top \rangle$, we choose the following set of circularities

$$G = \left\{ \begin{array}{l} \langle init(n) \mid \psi \rangle \Rightarrow \langle comp \mid \top \rangle, \\ \langle loop(n, i) \mid 2 \leq i \wedge \exists u.i \leq u < n \wedge n \bmod u = 0 \rangle \Rightarrow \langle comp \mid \top \rangle \end{array} \right\}.$$

The second circularity is inspired by the infinite branch of the proof tree under DSTEP. We will show that $(\mathcal{R}, G) \vdash^{\forall} G$, and by Theorem 3, it follows that all reachability formulas in G hold in $\mathcal{R}$.

First Circularity. To obtain a proof of the circularity $\langle init(n) \mid \psi \rangle \Rightarrow \langle comp \mid \top \rangle$, we replace the infinite subtree rooted at $\langle loop(n, 2) \mid \psi \rangle \Rightarrow \varphi_r$ in Example 12 by the following finite proof tree (that uses [circ]):

$$\cfrac{\cfrac{\cfrac{}{\langle comp \mid \bot \rangle \Rightarrow \varphi_r} \; [\text{axiom}]}{\langle comp \mid \psi \wedge \phi \wedge \top \rangle \Rightarrow \varphi_r} \; [\text{subs}] \quad \cfrac{\langle loop(n, 2) \mid \psi \wedge \neg\phi \rangle \Rightarrow \varphi_r}{} \; \begin{array}{l} [\text{axiom}] \\ \hline [\text{circ}] \end{array}}{\langle loop(n, 2) \mid \psi \rangle \Rightarrow \varphi_r}$$

where $\phi \triangleq \exists n', i'.loop(n, 2) = loop(n', i') \wedge 2 \leq i' \wedge \exists u.i' \leq u < n' \wedge n' \bmod u = 0$.

Second Circularity. To complete the proof of G, we have to find a finite proof tree for

$$\langle loop(n, i) \mid 2 \leq i \wedge \exists u.i \leq u < n \wedge n \bmod u = 0 \rangle \Rightarrow \langle comp \mid \top \rangle$$

as well. This is also obtained using [circ] as follows:

$$\cfrac{\cfrac{\cfrac{}{\langle comp \mid \bot \rangle \Rightarrow \varphi_r} \; [\text{axiom}]}{\langle comp \mid \psi_i \wedge \psi_a \rangle \Rightarrow \varphi_r} \; [\text{subs}] \quad \cfrac{T_1 \qquad T_2}{\langle loop(n, i+1) \mid \psi_i \wedge \psi_b \rangle \Rightarrow \varphi_r} \; \begin{array}{l} [\text{circ}] \\ \hline [\text{der}^{\forall}] \end{array}}{\langle loop(n, i) \mid \psi_i \rangle \Rightarrow \langle comp \mid \top \rangle}$$

where

$$\psi_a \triangleq k' > 1 \wedge loop(n, i) = loop(i' \times k', i'),$$
$$\psi_b \triangleq \neg\exists k.k > 1 \wedge n = i \times k,$$
$$\psi_i \triangleq 2 \leq i \wedge \exists u.i \leq u < n \wedge n \bmod u = 0.$$

The subtree

$$\cfrac{T_1 \qquad T_2}{\langle loop(n, i+1) \mid \psi_i \wedge \psi_b \rangle \Rightarrow \varphi_r} \; [\text{circ}]$$

is:

$$\cfrac{\cfrac{\cfrac{}{\langle comp \mid \bot \rangle \Rightarrow \varphi_r} \; [\text{axiom}]}{\langle comp \mid \psi_i \wedge \psi_b \wedge \psi_c \rangle \Rightarrow \varphi_r} \; [\text{subs}] \quad \cfrac{\langle loop(n, i+1) \mid \psi_i \wedge \psi_b \wedge \neg\psi_c \rangle \Rightarrow \varphi_r}{} \; \begin{array}{l} [\text{axiom}] \\ \hline [\text{circ}] \end{array}}{\langle loop(n, i+1) \mid \psi_i \wedge \psi_b \rangle \Rightarrow \varphi_r,}$$

where

$$\psi_c \triangleq \exists n', i'.loop(n, i+1) = loop(n', i') \land 2 \leq i' \land \exists u.i' \leq u < n' \land n' \bmod u = 0.$$

The constraint ψ_c holds when the circularity can be applied and therefore this branch is discharged immediately by [subs] and [axiom]. The other branch, when the circularity cannot be applied, is discharged directly by [axiom], as $\psi_i \land \psi_b \land \neg\psi_c$ is unsatisfiable (ψ_i says that n has a divisor between i and n, ψ_b says that i is not a divisor of n, and ψ_c that n has a divisor between $i+1$ and n).

Note that in both proof trees of the two circularities in G, in order to apply the [circ] rule, we used the following fresh instance of the second circularity:

$$\langle loop(n', i') \mid 2 \leq i' \land \exists u.i' \leq u < n' \land n' \bmod u = 0 \rangle \Rightarrow \langle comp \mid \top \rangle.$$

The proof trees for both goals (circularities) in G are guarded. We have shown therefore that $(\mathcal{R}, G) \vdash^\forall G$. By the Circularity Principle (Theorem 3), we obtain that $\mathcal{R} \vDash^\forall G$ and therefore

$$\mathcal{R} \vDash^\forall \left\{ \begin{array}{l} \langle init(n) \mid \exists u.1 < u < n \land n \bmod u = 0) \rangle \Rightarrow \langle comp \mid \top \rangle, \\ \langle loop(n, i) \mid 2 \leq i \land \exists u.i \leq u < n \land n \bmod u = 0 \rangle \Rightarrow \langle comp \mid \top \rangle \end{array} \right\}$$

which includes what we wanted to show of our transition system defined $\mathcal{R}$ in the running example.

5 Implementation

We have implemented the proof system for reachability in a tool called RMT (for rewriting modulo theories). RMT is open source and can be obtained from http://github.com/ciobaca/rmt/.

To prove a reachability property, the RMT tool performs a bounded search in the proof system given above. The bounds can be set by the user. We have also tested the tool on reachability problems where we do not use strong enough circularities. In these cases, the tool will not find proofs. A difficulty that appears when a proof fails, difficulty shared by all deductive approaches to correctness, is that it is not known if the specification is wrong or if the circularities are not strong enough. Often, by analyzing the failed proof attempt, the user may have the chance to find a hint for the missing circularities, if any. In addition, proofs might also fail because of the incompleteness of the SMT solver. In addition to the running example, we have used RMT on a number of examples, summarized in the table below:

LCTRS	Reachability Property
Computation of $1 + \ldots + n$	Result is $n * (n+1)/2$
Comp. of $gcd(u, v)$ by *rptd. subtractions*	Result matches builtin gcd function
Comp. of $gcd(u, v)$ by *rptd. divisions*	Result matches builtin gcd function
Mult. of two naturals by *rptd. additions*	Result matches builtin $\times$ function
Comp. of $1^2 + \ldots + n^2$	Result is $n(n+1)(2n+1)/6$
Comp. of $1^2 + \ldots + n^2$ *w/out multiplications*	Result is $n(n+1)(2n+1)/6$
Semantics of an IMPerative language	Program computing $1 + \ldots + n$ is correct
Semantics of a FUNctional language	Program computing $1 + \ldots + n$ is correct
Semantics of a FUNctional language	Program computing $1^2 + \ldots + n^2$ is correct

Implementation Details. RMT contains roughly 5000 lines of code, including comments and blank lines. RMT depends only on the standard C++ libraries and it can be compiled by any relatively modern C++ compiler out of the box. At the heart of RMT is a hierarchy of classes for representing variables, function symbols and terms. Terms are stored in DAG format, with maximum structure sharing. The RMT tool relies on an external SMT solver to check satisfiability of constraints. By default, the only dependency is the Z3 SMT solver, which should be installed and its binary should be in the system path. A compile time switch allows to use any other SMT solver that supports the SMTLIB interface, such as CVC4 [3]. In order to reduce constraints over the full signature to constraints over the builtin signature, RMT uses a unification modulo builtins algorithm (see [7]), which transforms any predicate $t_1 = t_2$ (where the terms t_1, t_2 can possibly contain constructor symbols) into a set of builtin constraints.

6   Conclusion and Future Work

We introduced a coinduction based method for proving reachability properties of logically constrained term rewriting systems. We use a coinductive definition of transition systems that unifies the handling of finite and infinite executions. We propose two proof systems for the problem above. The first one formalizes symbolic execution in LCTRSs coinductively, with possibly infinite proof trees. This proof system is complete, but its infinite proof trees cannot be used in practice as proofs. In the second proof system we add to symbolic execution a circularity proof rule, which allows to transform infinite proof trees into finite trees. It is not always possible to find finite proof trees, and we conjecture that establishing a given reachability property is higher up in the arithmetic hierarchy.

We also proposed a semantics for logically constrained term rewriting systems as transition systems over a model combining order-sorted terms with builtin elements such as booleans, integers, etc. The proposed semantics has the advantage of being simpler than the usual semantics of LCTRSs defined in [20], which requires two reduction relations (one for rewriting and one for computing). The approach proposed here also removes some technical constraints such as variable inclusion of the rhs in the lhs, which is important in modelling open systems, where the result of a transition is non-deterministically chosen by the environment. In addition, working in an order-sorted setting is indispensable in order to model easily the semantics of programming languages.

In fact, proving program properties, like correctness and equivalence, is one application of our method. A tool such as C2LCTRS (http://www.trs.cm.is. nagoya-u.ac.jp/c2lctrs/) can be used to convert the semantics of a C program into a LCTRS and then RMT can prove reachability properties of the C program. Additionally, the operational semantics of any language can be encoded as a LCTRS [28] and then program correctness is reducible to a particular reachability formula. But our approach is not limited to programs, as any system that can be modelled as a LCTRS is also amenable to our approach. We define reachability in the sense of partial correctness (i.e., nonterminating executions vacuously satisfy any reachability property). Termination should be established in some other way [18], as it is an orthogonal concern. Our approach to reachability and LCTRSs extends to working modulo AC (or more generally, modulo any set of equations E), but we have not formally presented this to preserve brevity and simplicity. For future work, we would like to test our approach on other interesting problems that arrise in various domains. In particular, it would be interesting to extend our approach to reachability in the context of program equivalence [9]. An interesting challenge is to add defined operations to the algebra underlying the constrained term rewriting systems, which would allow a user to define their own functions, which are not necessarily builtin.

Acknowledgements. We thank the anonymous reviewers for their valuable suggestions. This work was supported by a grant of the Romanian National Authority for Scientific Research and Innovation, CNCS/CCCDI - UEFISICDI, project number PN-III-P2-2.1-BG-2016-0394, within PNCDI III.

References

1. Aguirre, L., Martí-Oliet, N., Palomino, M., Pita, I.: Conditional narrowing modulo SMT and Axioms. In: PPDP 2017, pp. 17–28 (2017)
2. Bae, K., Rocha, C.: Guarded terms for rewriting modulo SMT. In: Proença, J., Lumpe, M. (eds.) FACS 2017. LNCS, vol. 10487, pp. 78–97. Springer, Cham (2017). https://doi.org/10.1007/978-3-319-68034-7_5
3. Barrett, C., Conway, C.L., Deters, M., Hadarean, L., Jovanovic, D., King, T., Reynolds, A., Tinelli, C.: CVC4. In: CAV 2011, pp. 171–177 (2011)
4. Bogdănaş, D., Roşu, G.: K-Java: a complete semantics of Java. In: POPL 2015, pp. 445–456 (2015)
5. Brotherston, J., Gorogiannis, N., Petersen, R.L.: A generic cyclic theorem prover. In: Jhala, R., Igarashi, A. (eds.) APLAS 2012. LNCS, vol. 7705, pp. 350–367. Springer, Heidelberg (2012). https://doi.org/10.1007/978-3-642-35182-2_25
6. Brotherston, J., Simpson, A.: Sequent calculi for induction and infinite descent. J. Log. Comput. **21**(6), 1177–1216 (2011)
7. Ciobâcă, Ş., Arusoaie, A., Lucanu, D.: Unification modulo builtins. In: WoLLIC 2018 (2018, to appear)
8. Ciobâcă, Ş., Lucanu, D.: A coinductive approach to proving reachability properties in logically constrained term rewriting systems (2018). arXiv:1804.08308
9. Ciobâcă, Ş., Lucanu, D., Rusu, V., Roşu, G.: A language-independent proof system for full program equivalence. Formal Asp. Comput. **28**(3), 469–497 (2016)
10. Ştefănescu, A., Ciobâcă, Ş., Mereuta, R., Moore, B.M., Şerbănuță, T.F., Roşu, G.: All-path reachability logic. In: Dowek, G. (ed.) RTA 2014. LNCS, vol. 8560, pp. 425–440. Springer, Cham (2014). https://doi.org/10.1007/978-3-319-08918-8_29
11. Ştefănescu, A., Park, D., Yuwen, S., Li, Y., Roşu, G.: Semantics-based program verifiers for all languages. In: OOPSLA 2016, pp. 74–91 (2016)
12. Durán, F., Eker, S., Escobar, S., Martí-Oliet, N., Meseguer, J., Talcott, C.: Built-in variant generation and unification, and their applications in Maude 2.7. In: Olivetti, N., Tiwari, A. (eds.) IJCAR 2016. LNCS (LNAI), vol. 9706, pp. 183–192. Springer, Cham (2016). https://doi.org/10.1007/978-3-319-40229-1_13
13. Escobar, S., Meseguer, J., Thati, P.: Narrowing and rewriting logic: from foundations to applications. ENTCS **177**, 5–33 (2007)

14. Fuhs, C., Kop, C., Nishida, N.: Verifying procedural programs via constrained rewriting induction. ACM TOCL **18**(2), 14:1–14:50 (2017)
15. Hathhorn, C., Ellison, C., Roşu, G.: Defining the undefinedness of C. In: PLDI 2015, pp. 336–345 (2015)
16. Hur, C.-K., Neis, G., Dreyer, D., Vafeiadis, V.: The power of parameterization in coinductive proof. In: POPL 2013, pp. 193–206 (2013)
17. Kirchner, C., Kirchner, H., Rusinowitch, M.: Deduction with symbolic constraints. Technical report RR-1358, INRIA (1990)
18. Kop, C.: Termination of LCTRSs. CoRR abs/1601.03206 (2016)
19. Kop, C., Nishida, N.: Constrained term rewriting tool. In: Davis, M., Fehnker, A., McIver, A., Voronkov, A. (eds.) LPAR 2015. LNCS, vol. 9450, pp. 549–557. Springer, Heidelberg (2015). https://doi.org/10.1007/978-3-662-48899-7_38
20. Kop, C., Nishida, N.: Term rewriting with logical constraints. In: Fontaine, P., Ringeissen, C., Schmidt, R.A. (eds.) FroCoS 2013. LNCS (LNAI), vol. 8152, pp. 343–358. Springer, Heidelberg (2013). https://doi.org/10.1007/978-3-642-40885-4_24
21. Lucanu, D., Rusu, V., Arusoaie, A.: A generic framework for symbolic execution: a coinductive approach. J. Symb. Comput. **80**, 125–163 (2017)
22. Meseguer, J., Thati, P.: Symbolic reachability analysis using narrowing and its application to verification of cryptographic protocols. High.-Order Symb. Comput. **20**(1–2), 123–160 (2007)
23. Park, D., Ştefănescu, A., Roşu, G.: KJS: a complete formal semantics of JavaScript. PLDI **2015**, 346–356 (2015)
24. Popescu, A., Gunter, E.L.: Incremental pattern-based coinduction for process algebra and its isabelle formalization. In: Ong, L. (ed.) FoSSaCS 2010. LNCS, vol. 6014, pp. 109–127. Springer, Heidelberg (2010). https://doi.org/10.1007/978-3-642-12032-9_9
25. Rocha, C., Meseguer, J., Muñoz, C.A.: Rewriting modulo SMT and open system analysis. J. Log. Algebr. Meth. Program. **86**(1), 269–297 (2017)
26. Roşu, G.: Matching logic. Log. Methods Comp. Sci. **13**(4), 1–61 (2017)
27. Roşu, G., Şerbănuţă, T.F.: An overview of the K semantic framework. J. Log. Algebr. Program. **79**(6), 397–434 (2010)
28. Şerbănuţă, T.-F., Roşu, G., Meseguer, J.: A rewriting logic approach to operational semantics. Inf. and Comp. **207**(2), 305–340 (2009)
29. Skeirik, S., Ştefănescu, A., Meseguer, J.: A constructor-based reachability logic for rewrite theories. TR. http://hdl.handle.net/2142/95770

A New Probabilistic Algorithm
for Approximate Model Counting

Cunjing Ge[1,3], Feifei Ma[1,2,3(✉)], Tian Liu[4], Jian Zhang[1,3], and Xutong Ma[3,5]

[1] State Key Laboratory of Computer Science, Institute of Software,
Chinese Academy of Sciences, Beijing, China
{gecj,maff,zj}@ios.ac.cn
[2] Laboratory of Parallel Software and Computational Science, Institute of Software,
Chinese Academy of Sciences, Beijing, China
[3] University of Chinese Academy of Sciences, Beijing, China
[4] School of Electronics Engineering and Computer Science, Peking University,
Beijing, China
[5] Technology Center of Software Engineering, Institute of Software,
Chinese Academy of Sciences, Beijing, China

Abstract. Constrained counting is important in domains ranging from artificial intelligence to software analysis. There are already a few approaches for counting models over various types of constraints. Recently, hashing-based approaches achieve success but still rely on solution enumeration. In this paper, a new probabilistic approximate model counter is proposed, which is also a hashing-based universal framework, but with only satisfiability queries. A dynamic stopping criteria, for the new algorithm, is presented, which has not been studied yet in previous works of hashing-based approaches. Although the new algorithm lacks theoretical guarantee, it works well in practice. Empirical evaluation over benchmarks on propositional logic formulas and SMT(BV) formulas shows that the approach is promising.

1 Introduction

Constrained counting, the problem of counting the number of solutions for a set of constraints, is important in theoretical computer science and artificial intelligence. Its interesting applications in several fields include program analysis [17,18,20,21,28,30], probabilistic inference [12,31], planning [14] and privacy/confidentiality verification [19]. Constrained counting for propositional formulas is also called model counting, to which probabilistic inference is easily reducible. However, model counting is a canonical #P-complete problem, even

This work has been supported by the National 973 Program of China under Grant 2014CB340701, Key Research Program of Frontier Sciences, Chinese Academy of Sciences (CAS) under Grant QYZDJ-SSW-JSC036, and the National Natural Science Foundation of China under Grant 61100064. Feifei Ma is also supported by the Youth Innovation Promotion Association, CAS.

D. Galmiche et al. (Eds.): IJCAR 2018, LNAI 10900, pp. 312–328, 2018.
https://doi.org/10.1007/978-3-319-94205-6_21

for polynomial-time solvable problems like 2-SAT [37], thus it presents fascinating challenges for both theoreticians and practitioners.

There are already a few approaches for counting solutions over propositional logic formulas and SMT(BV) formulas. Recently, hashing-based approximate counting achi-eves both strong theoretical guarantees and good scalability [29]. The use of universal hash functions in counting problems began in [34, 36], but the resulting algorithm scaled poorly in practice. A scalable approximate counter ApproxMC in [10] scales to large problem instances, while preserving rigorous approximation guarantees. ApproxMC has been extended to finite-domain discrete integration, with applications to probabilistic inference [4, 7, 15]. It was improved by designing efficient universal hash functions [8, 25] and reducing the use of NP-oracle calls from linear to logarithmic [11].

The basic idea in ApproxMC is to estimate the model count by randomly and iteratively cutting the whole space down to a small "enough" cell, using hash functions, and sampling it. The total model count is estimated by a multiplication of the number of solutions in this cell and the ratio of the whole space to the small cell. To determine the size of the small cell, which is essentially a small-scale model counting problem with the model counts bounded by some thresholds, a model enumeration in the cell is adopted. In previous works, the enumeration query was handled by transforming it into a series of satisfiability queries, which is much more time-consuming than a single satisfiability query. An algorithm called MBound [23] only invokes satisfiability query once for each cut. Its model count is determined with high precision by the number of cuts down to the boundary of being unsatisfiable. However, this property is not strong enough to give rigorous guarantees, and MBound only returns an approximation of upper or lower bound of the model count.

In this paper, a new hashing-based approximate counting algorithm, with only satisfiability query, is proposed. Dynamic stopping criterion for the algorithm to terminate, once meeting the criterion of accuracy, is presented, which has not been proposed yet in previous works of hashing-based approaches. Theoretical insights over the efficiency of a prevalent heuristic strategy called leapfrogging are also provided. The new algorithm works well in practice but does not provide theoretical guarantees, since it builds on an assumption of a correlation between the model count and the probability of the hashed formula being unsatisfiable, which has not been proved yet.

The proposed approach is a general framework easy to handle various types of constraints. Prototype tools for propositional logic formulas and SMT(BV) formulas are implemented. An extensive evaluation on a suite of benchmarks demonstrates that (i) the approach significantly outperforms the state-of-the-art approximate model counters, including a counter designed for SMT(BV) formulas, and (ii) the dynamic stopping criterion is promising.

The rest of this paper is organized as follows. Preliminary material is in Sect. 2, related works in Sect. 3, the algorithm in Sect. 4, analysis in Sect. 5, experimental results in Sect. 6, and finally, concluding remarks in Sect. 7.

2   Preliminaries

Let $F(x)$ denote a propositional logic formula on n variables $x = (x_1, \ldots, x_n)$. Let S and S_F denote the whole space (the space of assignments) and the solution space of F, respectively. Let $\#F$ denote the cardinality of S_F, i.e. the number of solutions of F.

(ϵ, δ)-**bound** To count $\#F$, an (ϵ, δ) approximation algorithm, $\epsilon > 0$ and $\delta > 0$, is an algorithm which on every input formula F, outputs a number $\tilde{Y}$ such that $\Pr[(1 + \epsilon)^{-1}\#F \leq \tilde{Y} \leq (1 + \epsilon)\#F] \geq 1 - \delta$. Such an algorithm is called a (ϵ, δ)-counter and the bound is called a (ϵ, δ)-bound [26].

Hash Function. Let $\mathcal{H}_F$ be a family of XOR-based bit-level hash functions on the variables of a formula F. Each hash function $H \in \mathcal{H}_F$ is of the form $H(x) = a_0 \bigoplus_{i=1}^{n} a_i x_i$, where $a_0, \ldots, a_n$ are Boolean constants. In the hashing procedure `Hashing(F)`, a function $H \in \mathcal{H}_F$ is generated by independently and randomly choosing a_is from a uniform distribution. Thus for an assignment α, it holds that $\Pr_{H \in \mathcal{H}_F}(H(\alpha) = true) = \frac{1}{2}$. Given a formula F, let F_i denote a hashed formula $F \wedge H_1 \wedge \cdots \wedge H_i$, where $H_1, \ldots, H_i$ are independently generated by the hashing procedure.

Satisfiability Query. Let `Solving(F)` denote the satisfiability query of a formula F. With a target formula F as input, the satisfiability of F is returned by `Solving(F)`.

Enumeration Query. Let `Counting(F, p)` denote the bounded solution enumeration query. With a constraint formula F and a threshold p ($p \geq 2$) as inputs, a number s is returned such that $s = \min(p - 1, \#F)$. Specifically, 0 is returned for unsatisfiable F, or $p = 1$ which is meaningless.

SMT(BV) Formula. SMT(BV) formulas are quantifier-free and fixed-size that combine propositional logic formulas with constraints of bit-vector theory. For example, $\neg(x + y = 0) \vee (x = y << 1)$, where x and y are bit-vector variables, $<<$ is the shift-left operator. It can be regarded as a propositional logic formula $\neg A \vee B$ that combines bit-vector constraints $A \equiv (x+y = 0)$ and $B \equiv (x = y << 1)$. To apply hash functions to an SMT(BV) formula, a bit-vector is bit-blasted to a set of Boolean variables.

3   Related Works

[3] showed that almost uniform sampling from propositional constraints, a closely related problem to constrained counting, is solvable in probabilistic polynomial time with an NP oracle. Building on this, [10] proposed the first scalable approximate model counting algorithm `ApproxMC` for propositional formulas. `ApproxMC` is based on a family of 2-universal bit-level hash functions that compute XOR of randomly chosen propositional variables. In the current work, this family of hash functions is adopted, which was shown to be 3-independent in [24], and is

Algorithm 1

1: **function** APPROXMC(F, T, *pivot*)
2:     **for** 1 to T **do**
3:         $c \leftarrow$ ApproxMCCore(F, *pivot*)
4:         **if** ($c \neq 0$) **then** AddToList(C, c)
5:     **end for**
6:     **return** FindMedian(C)
7: **end function**
8: **function** APPROXMCCORE(F, *pivot*)
9:     $F_0 \leftarrow F$
10:     **for** $i \leftarrow 0$ **to** ∞ **do**
11:         $s \leftarrow$ Counting(F_i, *pivot* + 1)
12:         **if** ($0 \leq s \leq pivot$) **then return** $2^i s$
13:         $H_{i+1} \leftarrow$ Hashing(F)
14:         $F_{i+1} \leftarrow F_i \wedge H_{i+1}$
15:     **end for**
16: **end function**

revealed to potentially possess better properties than expected by the experimental results and the theoretical analysis in the current work.

The sketch framework of `ApproxMC` [10,13] is listed as Algorithm 1. Its inputs are a formula F and two accuracy parameters T and *pivot*, where T determines the number of times `ApproxMCCore` is invoked, and *pivot* determines the threshold of the enumeration query. The function `ApproxMCCore` starts from the formula F_0, iteratively calls `Counting` and `Hashing` over each F_i, to cut the space (cell) of all models of F_0 using random hash functions, until the count of F_i is no larger than *pivot*, then breaks out of the loop and adds the approximation $2^i s$ into list C. The main procedure `ApproxMC` repeatedly invokes `ApproxMCCore` and collects the returned values, at last returning the median number of list C. The general algorithm in [8] is similar to Algorithm 1, but cuts the cell with dynamically determined proportion instead of the constant $\frac{1}{2}$, due to the word-level hash functions. [11] improves `ApproxMCCore` via binary search to reduce the number of enumeration queries from linear to logarithmic. This binary search improvement is orthogonal to our approach.

A recent work [1] considered a special family of shorter XOR-constraints to improve the efficiency of SAT solving while preserving rigorous guarantee. This improvement of hash functions is also orthogonal to our approach as we use hash functions and SAT solving as black boxes. However, it is unknown whether there exist similar theoretical results like [1].

4   Algorithm

In this section, a new hashing-based algorithm for approximate model counting, with only satisfiability queries, will be proposed, building on an assumption of a probabilistic approximate correlation between the model count and the probability of the hashed formula being unsatisfiable.

Let $F_d = F \wedge H_1 \wedge \cdots \wedge H_d$ be a hashed formula resulted by iteratively hashing d times independently over a formula F. F_d is unsatisfiable if and only if no solution of F satisfies F_d, thus $\Pr_{F_d}(F_d \text{ is unsat}) = \Pr_{F_d}(F_d(\alpha) = false, \alpha \in S_F)$. Assume we have

$$\Pr_{F_d}(F_d \text{ is unsat}) \approx (1 - 2^{-d})^{\#F}. \tag{1}$$

Then based on Eq. (1), an approximation of $\#F$ is achieved by taking logarithm on the value of $\Pr_{F_d}(F_d \text{ is unsat})$, which is estimated in turn by sampling F_d. This is the general idea of our approach. The pseudo-code is presented in Algorithm 2. The inputs are the target formula F and a constant T which determines the number of times GetDepth invoked. GetDepth calls Solving and Hashing repeatedly until an unsatisfiable formula F_{depth} is encountered, and returns the $depth$. Every time GetDepth returns a $depth$, the value of $C[i]$ is increased, for all $i < depth$. At line 9, the algorithm picks a number d such that $C[d]$ is close to $T/2$, since the error estimation fails when $C[d]/T$ is close to 0 or 1. The final result is returned by the formula $\log_{1-(1/2)^d} \frac{counter}{T}$ at line 11.

Algorithm 2. Satisfiability Testing based Approximate Counter (STAC)

1: **function** STAC(F, T)
2:     initialize $C[i]$s with zeros
3:     **for** $t \leftarrow 1$ **to** T **do**
4:         $depth \leftarrow$ GetDepth(F)
5:         **for** $i \leftarrow 0$ **to** $depth - 1$ **do**
6:             $C[i] \leftarrow C[i] + 1$
7:         **end for**
8:     **end for**
9:     pick a number d such that $C[d]$ is closest to $T/2$
10:     $counter \leftarrow T - C[d]$
11:     **return** $\log_{1-2^{-d}} \frac{counter}{T}$ /* return 0 when $d = 0$ */
12: **end function**
13: **function** GETDEPTH(F)
14:     $F_0 \leftarrow F$
15:     **for** $i \leftarrow 0$ **to** ∞ **do**
16:         $b \leftarrow$ Solving(F_i)
17:         **if** (b is false) **then return** i
18:         $H_{i+1} \leftarrow$ Hashing(F_i)
19:         $F_{i+1} \leftarrow F_i \wedge H_{i+1}$
20:     **end for**
21: **end function**

Note that our approach is based on Eq. (1) which is only an assumption. In Sect. 5, we provide theoretical analysis, including the bound of the approximation and the correctness of algorithm, based on the hypothesis. Then in Sect. 6, experimental results on an extensive set of benchmarks show that the approximation given by our approach fits the bound well. It indicates that Eq. (1) is probably true as it is a reasonable explanation to the positive results.

Dynamic Stopping Criterion. The essence of Algorithm 2 is a randomized sampler over a binomial distribution. The number of samples is determined by the value of T, which is pre-computed for a given (ϵ, δ)-bound, and we loosen the value of T to meet the guarantee in theoretical analysis. However, it usually does not loop T times in practice. A variation with dynamic stopping criterion is presented in Algorithm 3.

Algorithm 3. STAC with Dynamic Stopping Criterion

1: **function** STAC_DSC(F, T, ϵ, δ)
2:     initialize $C[i]$s with zeros
3:     **for** $t \leftarrow 1$ **to** T **do**
4:         $depth \leftarrow$ GetDepth(F)
5:         **for** $i \leftarrow 0$ **to** $depth - 1$ **do**
6:             $C[i] \leftarrow C[i] + 1$
7:         **end for**
8:         **for each** d **that** $C[d] > 0$ **do**
9:             $q \leftarrow \frac{t - C[d]}{t}$
10:            $M \leftarrow \log_{1-2^{-d}} q$
11:            $U \leftarrow \log_{1-2^{-d}}(q - z_{1-\delta}\sqrt{\frac{q(1-q)}{t}})$
12:            $L \leftarrow \log_{1-2^{-d}}(q + z_{1-\delta}\sqrt{\frac{q(1-q)}{t}})$
13:            **if** $U < (1 + \epsilon)M$ **and** $L > (1 + \epsilon)^{-1}M$ **then**
14:                **return** M
15:            **end if**
16:        **end for**
17:    **end for**
18: **end function**

Lines 2 to 7 is the same as Algorithm 2, still setting T as a stopping rule and terminating whenever $t = T$. Line 8 to 16 is the key part of this variation, calculating the binomial proportion confidence interval $[L, U]$ of an intermediate result M for each cycle. A commonly used formula $q \pm z_{1-\delta}\sqrt{\frac{q(1-q)}{t}}$ [5,38] is adopted, which is justified by the central limit theorem to compute the $1 - \delta$ confidence interval. However, it becomes invalid for small sample size or proportion close to 0 or 1. In practice, we also considered some improvements, e.g., Wilson score interval [40]. The exact count $\#F$ lies in the interval $[L, U]$ with probability $1 - \delta$. Combining the inequalities presented in line 13, the interval $[(1+\epsilon)^{-1}M, (1+\epsilon)M]$ is the (ϵ, δ)-bound (if the assumption of Formula 1 holds). So the algorithm terminates when the condition in line 13 comes true. The time complexity of Algorithm 3 is still the same as the original algorithm, though it usually terminates earlier.

Satisfiability and Enumeration Query. The bounded counting can be done by negating solution and calling SAT oracle repeatedly, which is employed by

ApproxMC. In practice, enumerating solutions in this way is not very efficient. In evaluation section, experimental results show that the average number of SAT calls of ApproxMC is usually 20 to 30 times to STAC. It may also cause problems while extending to other kinds of formulas. For example, for linear integer arithmetic formula, inserting solution negation clauses will exponentially increase the number of calls of LIP solver.

Leap-frogging Strategy. Recall that GetDepth is invoked T times with the same arguments, and the loop of line 15 to 20 in the pseudo-code of GetDepth in Algorithm 2 is time consuming for large i. A heuristic called leap-frogging to overcome this bottleneck was proposed in [9,10]. Their experiments indicate that this strategy is extremely efficient in practice. The average depth $\bar{d}$ of each invocation of GetDepth is recorded. In all subsequent invocations, the loop starts by initializing i to $\bar{d} - k \cdot$ offset, where $k \geq 1$. Note that if F_i is unsatisfiable, the algorithm repeatedly decreases i by increasing k and check the satisfiability of the new F_i, until a proper initialization i is found for satisfiable F_i. In practice, the constant offset is usually set to 5. Theorem 3 in Sect. 5 shows that the *depth* computed by GetDepth lies in an interval $[d, d+7]$ with probability over 90%. So it suffices to invoke Solving in constant time since the second iteration.

5   Analysis

In this section, we assume Eq. (1) holds. Based on this assumption, theoretical results on the error estimation of our approach are presented. For lack of space, we omit proofs in this section.

Recall that in Algorithm 2, $\#F$ is approximated by a value $\log_{1-2^{-d}} \frac{counter}{T}$. Let q_d denote the value of $(1 - 2^{-d})^{\#F}$. We obtain that $\Pr(F_d \ is \ unsat) = q_d$ for a randomly generated formula F_d. This is justified by Eq. (1). Since the ratio $\frac{counter}{T}$ in Algorithm 2 is a proportion of successes in a Bernoulli trial process, which is used to estimate the value of q_d. Then *counter* is a random variable following a binomial distribution $\mathbb{B}(T, q_d)$.

Theorem 1. *Let $z_{1-\delta}$ be the $1 - \delta$ quantile of $\mathbb{N}(0, 1)$ and*

$$T = max \left(\lceil (\frac{z_{1-\delta}}{2q_d(1 - q_d^\epsilon)})^2 \rceil, \lceil (\frac{z_{1-\delta}}{2(q_d^{(1+\epsilon)^{-1}} - q_d)})^2 \rceil \right). \tag{2}$$

Then $\Pr[\frac{\#F}{1+\epsilon} \leq \log_{1-2^{-d}} \frac{counter}{T} \leq (1 + \epsilon)\#F] \geq 1 - \delta$.

Proof. By above discussions, the ratio $\frac{counter}{T}$ is the proportion of successes in a Ber-noulli trial process which follows the distribution $\mathbb{B}(T, q_d)$. Then we use the approximate formula of a binomial proportion confidence interval $q_d \pm z_{1-\delta}\sqrt{\frac{q_d(1-q_d)}{T}}$, i.e., $\Pr[q_d - z_{1-\delta}\sqrt{\frac{q_d(1-q_d)}{T}} \leq \frac{counter}{T} \leq q_d + z_{1-\delta}\sqrt{\frac{q_d(1-q_d)}{T}}] \geq$

$1 - \delta$. The log function is monotone, so we only have to consider the following two inequalities:

$$\log_{1-2^{-d}}(q_d - z_{1-\delta}\sqrt{\frac{q_d(1 - q_d)}{T}}) \leq (1 + \epsilon)\#F, \tag{3}$$

$$(1 + \epsilon)^{-1}\#F \leq \log_{1-2^{-d}}(q_d + z_{1-\delta}\sqrt{\frac{q_d(1 - q_d)}{T}}). \tag{4}$$

We first consider Eq. (3). By substituting $\log_{1-2^{-d}} q_d$ for $\#F$, we have

$$\log_{1-2^{-d}}(q_d - z_{1-\delta}\sqrt{\frac{q_d(1 - q_d)}{T}}) \leq (1 + \epsilon)\log_{1-2^{-d}} q_d$$

$$\Leftrightarrow q_d - z_{1-\delta}\sqrt{\frac{q_d(1 - q_d)}{T}} \geq q_d^{(1+\epsilon)}$$

$$\Leftrightarrow q_d(1 - q_d^\epsilon) \geq z_{1-\delta}\sqrt{\frac{q_d(1 - q_d)}{T}}$$

$$\Leftrightarrow T \geq (\frac{z_{1-\delta}}{q_d(1 - q_d^\epsilon)})^2 q_d(1 - q_d).$$

Since $0 \leq q_d \leq 1$, we have $\sqrt{q_d(1 - q_d)} \leq \frac{1}{2}$. Therefore, $T = \lceil(\frac{z_{1-\delta}}{2q_d(1-q_d^\epsilon)})^2\rceil \geq (\frac{z_{1-\delta}}{q_d(1-q_d^\epsilon)})^2 q_d(1 - q_d)$.

We next consider Eq. (4). Similarly, we have

$$\log_{1-2^{-d}}(q_d + z_{1-\delta}\sqrt{\frac{q_d(1 - q_d)}{T}}) \geq (1 + \epsilon)^{-1}\log_{1-2^{-d}} q_d$$

$$\Leftrightarrow q_d + z_{1-\delta}\sqrt{\frac{q_d(1 - q_d)}{T}} \leq q_d^{1/(1+\epsilon)}$$

$$\Leftrightarrow T \geq (\frac{z_{1-\delta}}{q_d^{1/(1+\epsilon)} - q_d})^2 q_d(1 - q_d).$$

So Eq. (2) implies Eqs. (3) and (4).

Theorem 1 shows that the result of Algorithm 2 lies in the interval $[(1 + \epsilon)^{-1}\#F, (1 + \epsilon)\#F]$ with probability at least $1 - \delta$ when T is set to a proper value. So we focus on the possible smallest value of T in subsequent analysis.

The next two lemmas are easy to show by derivations.

Lemma 1. $\frac{z_{1-\delta}}{2x(1-x^\epsilon)}$ is monotone increasing and monotone decreasing in $[(1 + \epsilon)^{-\frac{1}{\epsilon}}, 1]$ and $[0, (1 + \epsilon)^{-\frac{1}{\epsilon}}]$ respectively.

Lemma 2. $\frac{z_{1-\delta}}{2(x^{1/(1+\epsilon)}-x)}$ is monotone increasing and monotone decreasing in $[(1 + \epsilon)^{-\frac{1+\epsilon}{\epsilon}}, 1]$ and $[0, (1 + \epsilon)^{-\frac{1+\epsilon}{\epsilon}}]$ respectively.

Theorem 2. If $\#F > 5$, then there exists a proper integer value of d such that $q_d \in [0.4, 0.65]$.

Proof. Let x denote the value of $q_d = (1 - \frac{1}{2^d})^{\#F}$, then we have $(1 - \frac{1}{2^{d+1}})^{\#F} = (\frac{1}{2} + \frac{x^{\frac{1}{\#F}}}{2})^{\#F}$. Consider the derivation

$$\frac{d}{d\#F}(\frac{1}{2} + \frac{x^{\frac{1}{\#F}}}{2})^{\#F} = (\frac{1}{2} + \frac{x^{\frac{1}{\#F}}}{2})^{\#F} \ln(\frac{1}{2} + \frac{x^{\frac{1}{\#F}}}{2}) \frac{x^{\frac{1}{\#F}}}{2} \ln x \frac{d}{d\#F}(\#F^{-1}).$$

Note that $(\frac{1}{2} + \frac{x^{\frac{1}{\#F}}}{2})^{\#F}$ and $\frac{x^{\frac{1}{\#F}}}{2}$ are the positive terms and $\ln(\frac{1}{2} + \frac{x^{\frac{1}{\#F}}}{2})$, $\ln x$ and $\frac{d}{d\#F}(\#F^{-1})$ are the negative terms. Therefore, the derivation is negative, i.e., $(\frac{1}{2} + \frac{x^{\frac{1}{\#F}}}{2})^{\#F}$ is monotone decreasing with respect to $\#F$. In addition, $(\frac{1}{2} + \frac{x^{\frac{1}{5}}}{2})^5$ is the upper bound when $\#F \geq 5$.

Let $x = 0.4$, then $(1 - \frac{1}{2^{d+1}})^{\#F} \leq (\frac{1}{2} + \frac{0.4^{\frac{1}{5}}}{2})^5 \approx 0.65$. Since $(1 - \frac{1}{2^0})^{\#F} = 0$ and $\lim_{d \to +\infty}(1 - \frac{1}{2^d})^{\#F} = 1$ and $(1 - \frac{1}{2^d})^{\#F}$ is continuous with respect to d, we consider the circumstances close to the interval $[0.4, 0.65]$. Assume there exists an integer σ such that $(1 - \frac{1}{2^\sigma})^{\#F} < 0.4$ and $(1 - \frac{1}{2^{\sigma+1}})^{\#F} > 0.65$. According to the intermediate value theorem, we can find a value $e > 0$ such that $(1 - \frac{1}{2^{\sigma+e}})^{\#F} = 0.4$. Obviously, we have $(1 - \frac{1}{2^{\sigma+e+1}})^{\#F} \leq 0.65$ which is contrary with the monotone decreasing property.

From Theorem 2 and Lemmas 1 and 2, it suffices to consider the results of Eq. (2) when $q_d = 0.4$ and $q_d = 0.65$. For example, $T = 22$ for $\epsilon = 0.8$ and $\delta = 0.2$, $T = 998$ for $\epsilon = 0.1$ and $\delta = 0.1$, etc. We therefore pre-computed a table of the value of T. The proof of next theorem is omitted.

Theorem 3. *There exists an integer d such that $q_d < 0.05$ and $q_{d+7} > 0.95$.*

Let *depth* denote the result of GetDepth in Algorithm 2. Then F_d is unsatisfiable only if $d \geq depth$. Theorem 3 shows that there exists an integer d such that $\Pr(depth < d) < 0.05$ and $\Pr(depth < d + 7) > 0.95$, i.e., $\Pr(d \leq depth \leq d + 7) > 0.9$. So in most cases, the value of *depth* lies in an interval $[d, d + 7]$. Also, it is easy to see that $\log_2 \#F$ lies in this interval as well. The following theorem is obvious now.

Theorem 4. *Algorithm 2 runs in time linear in $\log_2 \#F$ relative to an NP-oracle.*

6   Evaluation

To evaluate the performance and effectiveness of our approach, two prototype implementations STAC_CNF and STAC_BV with dynamic stopping criterion for propositional logic formulas and SMT(BV) formulas are built respectively. We considered a wide range of benchmarks from different domains: grid networks, plan recognition, DQMR networks, Langford sequences, circuit synthesis, random 3-CNF, logistics problems and program synthesis [8,10,27,33][1]. For lack of

---

[1] Our tools STAC_CNF and STAC_BV and the suite of benchmarks are available at https://github.com/bearben/STAC.

space, we only list a part of results here. All our experiments were conducted on a single core of an Intel Xeon 2.40 GHz (16 cores) machine with 32 GB memory and CentOS6.5 operating system.

6.1 Quality of Approximation

Recall that our approach is based on Eq. (1) which has not been proved. So we would like to see whether the approximation fits the bound. We experimented 100 times on each instance.

Table 1. Statistical results of 100-times experiments on STAC_CNF ($\epsilon = 0.8, \delta = 0.2$)

Instance	n	$\#F$	$[1.8^{-1}\#F, 1.8\#F]$	Freq.	$\bar{t}$ (s)	$\bar{T}$	$\bar{Q}$
special-1	20	1.0×10^6	$[5.8 \times 10^5, 1.9 \times 10^6]$	82	0.3	12.2	86.7
special-2	20	1	$[0.6, 1.8]$	86	0.6	12.6	37.6
special-3	25	3.4×10^7	$[1.9 \times 10^7, 6.0 \times 10^7]$	82	11.2	11.8	90.1
5step	177	8.1×10^4	$[4.5 \times 10^4, 1.5 \times 10^5]$	90	0.1	11.9	80.5
blockmap_05_01	1411	6.4×10^2	$[3.6 \times 10^2, 1.2 \times 10^3]$	84	1.1	12.0	73.8
blockmap_05_02	1738	9.4×10^6	$[5.2 \times 10^6, 1.7 \times 10^7]$	89	12.7	11.8	87.7
blockmap_10_01	11328	2.9×10^6	$[1.6 \times 10^6, 5.2 \times 10^6]$	83	80.3	12.0	85.0
fs-01	32	7.7×10^2	$[4.3 \times 10^2, 1.4 \times 10^3]$	80	0.02	12.6	76.2
or-50-10-10-UC-20	100	3.7×10^6	$[2.0 \times 10^6, 6.6 \times 10^6]$	77	7.7	12.0	86.1
or-60-10-10-UC-40	120	3.4×10^6	$[1.9 \times 10^6, 6.1 \times 10^6]$	91	3.5	12.1	86.0

In Table 1, column 1 gives the instance name, column 2 the number of Boolean variables n, column 3 the exact counts $\#F$, and column 4 the interval $[1.8^{-1}\#F, 1.8\#F]$. The frequencies of approximations that lie in the interval $[1.8^{-1}\#F, 1.8\#F]$ in 100 times of experiments are presented in column 5. The average time consumptions, average number of iterations, and average number of SAT query invocations are presented in columns 6, 7 and 8 respectively, which also indicate the advantages of our approach.

Under the dynamic stopping criterion, the counts returned by our approach should lie in an interval $[1.8^{-1}\#F, 1.8\#F]$ with probability 80% for $\epsilon = 0.8$ and $\delta = 0.2$. The statistical results in Table 1 show that the frequencies are around 80 for 100-times experiments which fit the 80% probability. The average number of iterations $\bar{T}$ listed in Table 1 is smaller than the loop termination criterion $T = 22$ which is obtained via Formula 2, indicating that the dynamic stopping technique significantly improves the efficiency. In addition, the values of $\bar{T}$ appear to be stable for different instances, hinting that there exists a constant upper bound on T which is irrelevant to instances.

Intuitively, our approach may start to fail on "loose" formulas, i.e., with an "infinitesimal" fraction of non-models. Instance *special-1* and *special-3* are such "loose" formulas where *special-1* has 2^{20} models with only 20 variables and *special-3* has $2^{25} - 1$ models with 25 variables. Instance *special-2* is another extreme case which only has one model. The results in Table 1 demonstrate that STAC_CNF also works fine on these extreme cases.

Table 2. Statistical results of 100-times experiments on STAC_CNF ($\epsilon = 0.2, \delta = 0.1$)

Instance	n	$\#F$	$[1.2^{-1}\#F, 1.2\#F]$	Freq.	$\bar{t}$ (s)	$\bar{T}$	$\bar{Q}$
special-1	20	1.0×10^6	$[8.7 \times 10^5, 1.3 \times 10^6]$	86	4.0	179	1023
special-2	20	1	$[0.8, 1.2]$	91	0.1	179	540
special-3	25	3.4×10^7	$[2.8 \times 10^7, 4.0 \times 10^7]$	91	138	178	1029
5step	177	8.1×10^4	$[6.8 \times 10^4, 9.8 \times 10^5]$	96	1.9	190	1078
blockmap_05_01	1411	6.4×10^2	$[5.3 \times 10^2, 7.7 \times 10^2]$	94	17.1	190	1069
blockmap_05_02	1738	9.4×10^6	$[7.9 \times 10^6, 1.1 \times 10^7]$	87	281	193	1088
blockmap_10_01	11328	2.9×10^6	$[2.4 \times 10^6, 3.5 \times 10^6]$	93	1371	180	1034
fs-01	32	7.7×10^2	$[6.4 \times 10^2, 9.2 \times 10^2]$	91	0.1	172	975
or-50-10-10-UC-20	100	3.7×10^6	$[3.1 \times 10^6, 4.4 \times 10^6]$	90	140	166	925
or-60-10-10-UC-40	120	3.4×10^6	$[2.8 \times 10^6, 4.1 \times 10^6]$	92	66	167	949

Table 3. Statistical results of 100-times experiments on STAC_BV ($\epsilon = 0.8, \delta = 0.2$)

Instance	TB	$\#F$	$[1.8^{-1}\#F, 1.8\#F]$	Freq	$\bar{t}$ (s)	$\bar{T}$	$\bar{Q}$
FINDpath1	32	4.1×10^6	$[2.3 \times 10^6, 7.3 \times 10^6]$	83	27.5	12.4	88.0
queue	16	8.4×10	$[4.7 \times 10, 1.5 \times 10^2]$	75	1.7	12.0	70.6
getopPath2	24	8.1×10^3	$[4.5 \times 10^3, 1.5 \times 10^4]$	88	2.7	12.2	79.5
coloring_4	32	1.8×10^9	$[1.0 \times 10^9, 3.3 \times 10^9]$	76	51.9	12.0	96.1
FISCHER2-7-fair	240	3.0×10^4	$[1.7 \times 10^4, 5.4 \times 10^4]$	79	149	11.8	79.8
case2	24	4.2×10^6	$[2.3 \times 10^6, 7.6 \times 10^6]$	79	16.5	12.4	89.3
case4	16	3.3×10^4	$[1.8 \times 10^4, 5.9 \times 10^4]$	87	2.2	12.5	85.2
case7	18	1.3×10^5	$[7.3 \times 10^4, 2.4 \times 10^5]$	83	2.9	12.4	84.1
case8	24	8.4×10^6	$[4.7 \times 10^6, 1.5 \times 10^7]$	82	14.4	12.1	91.1
case11	15	1.6×10^4	$[9.1 \times 10^3, 2.9 \times 10^4]$	76	2.1	12.0	81.2

We considered another pair of parameters $\epsilon = 0.2, \delta = 0.1$. Then the interval should be $[1.2^{-1}\#F, 1.2\#F]$ and the probability should be 90%. Table 2 shows the results on such parameter setting. The frequencies that the approximation lies in interval $[1.2^{-1}\#F, 1.2\#F]$ are all around or over 90 which fits the 90% probability.

We also conducted 100-times experiments on SMT(BV) problems and the results show that STAC_BV is also promising. Table 3 similarly shows the results of 100-times experiments on STAC_BV. Its column 2 gives the sum of widths of all bit-vector variables (Boolean variable is counted as a bit-vector of width 1) instead. The statistical results demonstrate that the dynamic stopping criterion is also promising on SMT(BV) problems.

6.2   Performance Comparison with (ϵ, δ)-counters

We compared our tools with ApproxMC2 [11] and SMTApproxMC [8] which are hashing-based (ϵ, δ)-counters. Both STAC_CNF and ApproxMC2 use CryptoMini-SAT [35], an efficient SAT solver designed for XOR clauses. STAC_BV and SMT-ApproxMC use the state-of-the-art SMT(BV) solver Boolector [6].

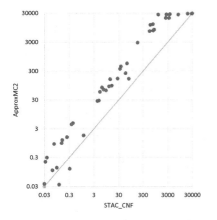

Fig. 1.   Performance   comparison   between STAC_CNF and ApproxMC2

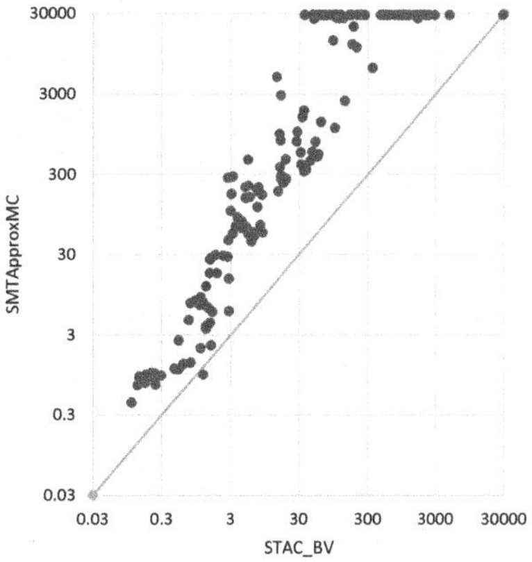

Fig. 2.   Performance   comparison   between STAC_BV and SMTApproxMC

We first conducted experiments with $\epsilon = 0.8, \delta = 0.2$ and 8 hours timeout which are also used in evaluation in previous works [8,11]. Figure 1 presents a comparison on performance between STAC_CNF and ApproxMC2. Each point represents an instance, whose x-coordinate and y-coordinate are the running times of STAC_CNF and ApproxMC2 on this instance, respectively. The figure is in logarithmic coordinates and demonstrates that STAC_CNF outperforms ApprxMC2 by about one order of magnitude. Figure 2 presents a similar comparison on performance between STAC_BV and SMTApproxMC, showing that STAC_BV outperforms SMTApproxMC by one or two orders of magnitude. Furthermore, the advantage enlarges as the scale grows.

Table 4 presents more experimental results with (ϵ, δ) parameters other than $(\epsilon = 0.8, \delta = 0.2)$. Nine pairs of parameters were experimented. "Time Ratio" represents the ratio of the running times of ApproxMC2 to STAC_CNF. "#Calls Ratio" represents the ratio of the number of SAT calls of ApproxMC2 to STAC_CNF. The results show that ApproxMC2 gains advantage as ϵ decreases and STAC_CNF gains advantage as δ decreases. On the whole, ApproxMC2 gains advantage when

Table 4. Performance comparison between STAC_CNF and ApproxMC2 with different pairs of (ϵ, δ) parameters

(ϵ, δ)		Instance							
		blockmap			fs-01	5step	ran5	ran6	ran7
		05_01	05_02	10_01					
$(0.8, 0.3)$	Time Ratio	1.11	3.99	1.22	3.00	3.83	6.53	8.24	5.57
	#Calls Ratio	22.60	39.02	17.91	19.12	23.11	22.53	21.28	23.68
$(0.8, 0.2)$	Time Ratio	1.84	6.16	2.44	2.80	6.05	9.61	15.41	7.37
	#Calls Ratio	26.70	34.68	25.16	33.46	27.24	33.35	38.22	30.94
$(0.8, 0.1)$	Time Ratio	2.27	7.36	3.72	5.25	12.62	9.60	9.54	8.19
	#Calls Ratio	44.88	48.26	40.01	49.40	43.03	46.12	44.84	52.63
$(0.4, 0.3)$	Time Ratio	0.75	1.37	0.42	3.00	5.04	1.97	2.31	2.74
	#Calls Ratio	17.75	36.20	14.69	16.40	27.63	21.07	27.34	21.63
$(0.4, 0.2)$	Time Ratio	0.77	1.44	0.86	4.50	7.70	2.82	1.77	3.02
	#Calls Ratio	20.91	26.35	29.16	26.72	40.66	26.49	27.82	28.94
$(0.4, 0.1)$	Time Ratio	1.08	2.57	1.29	4.90	7.09	3.84	3.43	3.11
	#Calls Ratio	37.16	46.28	39.40	31.99	39.36	41.02	35.88	34.11
$(0.2, 0.3)$	Time Ratio	0.42	0.47	0.23	5.08	3.79	1.26	1.14	1.81
	#Calls Ratio	13.75	20.82	24.35	13.37	19.74	25.20	19.19	20.06
$(0.2, 0.2)$	Time Ratio	0.57	0.92	0.26	8.42	3.37	2.07	1.50	2.45
	#Calls Ratio	21.80	29.62	25.60	21.83	21.59	25.88	22.72	22.98
$(0.2, 0.1)$	Time Ratio	0.87	0.92	0.44	16.69	3.17	3.61	2.27	2.60
	#Calls Ratio	27.86	29.91	33.36	34.17	31.58	40.81	29.01	29.90

ϵ and δ both decrease. Note that the numbers of SAT calls represent the complexity of both algorithms. In Table 4, #Calls Ratio is more stable than Time Ratio among different pairs of parameters and also different instances. It indicates that the difficulty of NP-oracle is also an important factor of running time performance.

6.3 Performance Comparison with Bounding and Guarantee-Less Counters

Since our approach is not a (ϵ, δ)-counter in theory, we also compared STAC_CNF with bounding counters (SampleCount [22], MBound [23]) and guarantee-less counters (ApproxCount [39], SampleTreeSearch [16]). Table 5 shows the experimental results.

For SampleCount, we used $\alpha = 2$ and $t = 3.5$ so that $\alpha t = 7$, giving a correctness confidence of $1 - 2^{-7} = 99\%$. The number of samples per variable setting, z, was chosen to be 20. Our results show that the lower-bound approximated by SampleCount is smaller than exact count $\#F$ by one or more orders of magnitude. We tried larger z, such as $z = 100$ and $z = 1000$, but still failed to obtain

Table 5. Performance comparison of STAC_CNF with existing bounding counters and guarantee-less counters

Instance	n	#F (if known)	STAC_CNF ($\epsilon = 0.8, \delta = 0.2$)		SampleCount (99% confidence)		MBound (99% confidence)		ApproxCount		SampleTreeSearch	
			Models	Time	L-bound	Time	L-bound	Time	Models	Time	Models	Time
blockmap_05_01	1411	640	≈ 807	1 s	≥ 22	116 s	> 64	9 s	$= 640$	7 s	≈ 646	71 s
blockmap_05_02	1738	9.4×10^6	$\approx 7.1 \times 10^6$	16 s	3.6×10^4	5 m	$> 2.1 \times 10^6$	5 m	$= 9.4 \times 10^6$	13 s	$\approx 9.4 \times 10^6$	114 s
blockmap_10_01	11328	2.9×10^6	$\approx 2.6 \times 10^6$	96 s	—	$\geq$8 h	$> 5.2 \times 10^5$	9 m	—	$\geq$8 h	$\approx 3.0 \times 10^6$	62 m
blockmap_15_01	33035	—	$\approx 2.0 \times 10^9$	41 m	—	$\geq$8 h	—	$\geq$8 h	—	$\geq$8 h	—	$\geq$8 h
fs-01	32	768	≈ 709	0.1 s	≥ 68	0.2 s	> 64	2 s	≈ 925	17 s	≈ 769	0.1 s
PLAN RECOGNITION												
5step	177	8.1×10^4	$\approx 8.1 \times 10^4$	0.2 s	$\geq 2.8 \times 10^3$	4 s	$> 8.2 \times 10^3$	6 s	$= 8.1 \times 10^4$	18 s	$\approx 7.5 \times 10^4$	1 s
tire-1	352	7.3×10^8	$\approx 9.8 \times 10^8$	64 m	$\geq 7.0 \times 10^5$	14 s	$> 6.7 \times 10^7$	8 h	$= 7.3 \times 10^8$	48 s	$\approx 7.6 \times 10^8$	5 s
tire-3	577	2.2×10^{11}	—	$\geq$8 h	$\geq 1.3 \times 10^6$	49 s	—	$\geq$8 h	$\approx 2.1 \times 10^{11}$	63 s	$\approx 1.2 \times 10^{11}$	5 s
LANGFORD PROBS.												
lang12	576	2.2×10^5	$\approx 3.2 \times 10^5$	80 m	$\geq 3.6 \times 10^3$	3 m	$> 1.6 \times 10^4$	4 h	≈ 0	111 s	$\approx 1.9 \times 10^5$	101 m
lang15	1024	—	—	$\geq$8 h	$\geq 4.7 \times 10^5$	4 m	—	$\geq$8 h	≈ 0	122 s	—	$\geq$8 h
DQMR NETWORKS												
or-100-20-6-UC-60	200	2.8×10^7	$\approx 3.4 \times 10^7$	14 m	$\geq 1.1 \times 10^{29}$	15 s	$> 4.2 \times 10^6$	6 m	≈ 0	17 s	$\approx 2.8 \times 10^7$	0.8 s
or-50-10-10-UC-40	100	3.1×10^3	$\approx 3.2 \times 10^3$	0.1 s	$\geq 2.7 \times 10^2$	0.1 s	$> 5.1 \times 10^2$	4 s	$\approx 2.1 \times 10^{16}$	68 s	$\approx 3.1 \times 10^3$	0.5 s
or-50-20-10-UC-30	100	6.8×10^8	$\approx 7.4 \times 10^8$	35 m	$\geq 6.0 \times 10^7$	0.2 s	$> 1.3 \times 10^8$	3 h	$\approx 2.7 \times 10^{16}$	62 s	$\approx 7.9 \times 10^8$	0.7 s
or-60-10-10-UC-30	120	6.8×10^7	$\approx 6.2 \times 10^7$	4 m	$\geq 1.0 \times 10^{17}$	9 s	$> 1.7 \times 10^7$	38 m	$\approx 2.4 \times 10^{19}$	83 s	$\approx 6.5 \times 10^7$	1 s
or-60-5-2-UC-40	120	2.1×10^6	$\approx 1.9 \times 10^6$	6 s	$\geq 1.0 \times 10^{17}$	16 s	$> 5.2 \times 10^5$	199 s	$\approx 2.3 \times 10^{19}$	89 s	$\approx 2.1 \times 10^6$	0.9 s
or-70-10-6-UC-40	140	1.2×10^4	$\approx 7.2 \times 10^3$	0.1 s	$\geq 1.0 \times 10^{20}$	7 s	$> 2.0 \times 10^3$	4 s	≈ 0	165 s	$\approx 1.2 \times 10^4$	0.5 s
or-70-5-2-UC-30	140	1.7×10^7	$\approx 4.7 \times 10^7$	51 s	$\geq 1.0 \times 10^{20}$	10 s	$> 2.1 \times 10^6$	11 m	≈ 0	165 s	$\approx 1.7 \times 10^7$	0.8 s
RANDOM 3-CNF												
ran6	30	1.2×10^6	$\approx 1.9 \times 10^6$	0.6 s	$\geq 1.1 \times 10^5$	0.2 s	$> 1.3 \times 10^5$	11 s	$\approx 8.3 \times 10^5$	23 s	$\approx 1.3 \times 10^6$	0.2 s
ran12	40	3.5×10^8	$\approx 4.2 \times 10^8$	10 m	$\geq 3.1 \times 10^7$	0.5 s	$> 6.7 \times 10^7$	15 m	$\approx 2.6 \times 10^8$	23 s	$\approx 3.9 \times 10^8$	0.3 s
ran27	50	1.5×10^8	$\approx 1.2 \times 10^8$	8 m	$\geq 1.3 \times 10^7$	18 s	$> 1.7 \times 10^7$	28 m	$\approx 5.5 \times 10^7$	59 s	$\approx 1.1 \times 10^8$	0.3 s
ran44	60	1.1×10^6	$\approx 1.9 \times 10^6$	6 s	$\geq 9.5 \times 10^4$	8 s	$> 1.3 \times 10^5$	52 s	$\approx 3.3 \times 10^5$	139 s	$\approx 9.1 \times 10^5$	0.5 s

a lower-bound larger than $\#F/10$. Moreover, there are some wrong approximations on DQMR networks problems, e.g., `or-100-20-6-UC-60` only has 2.8×10^7 models but `SampleCount` returns a lower-bound $\geq 1.1 \times 10^{29}$. `SampleCount` is more efficient on Langford problems and random 3-CNF problems, but weaker on problems with a large number of variables, such as `blockmap` problems.

For `MBound`, we used $\alpha = 1$ and $t = 7$ so that $\alpha t = 7$, also giving a correctness confidence of $1 - 2^{-7} = 99\%$. `MBound` also employs a family of XOR hashing functions which is similar to the function used by our approach. The size of XOR constraints k should be no more than half of the number of variables n, i.e., $k \leq n/2$. We found that XOR constraints start to fail as $k << n/2$. So in our experiments, k was chosen to be close to $n/2$. Since `MBound` can only check the bound and may return failure as the bound is too close to the exact count, we implemented a binary search to find the best lower-bound verified by `MBound`. The results in Table 5 are the best lower-bounds and the running times of the whole binary search procedure. Though the lower-bounds are better than `SampleCount`, they are still around $\#F/10$. Similar to our approach, the running times of `MBound` are also quite relevant to the size of $\#F$.

For `ApproxCount`, we manually increased the value of "cutoff" as `Approx-Count` requires. Note that `ApproxCount` calls exact model counter `Cachet` [32] and `Relsat` [2] after formula simplifications, so it sometimes returns the exact counts, such as `blockmap_05_01`, `blockmap_05_02`, `5step` and `tire-1`. On Langford problems and DQMR networks problems, wrong approximations were provided. On other instances, the results show that `STAC_CNF` usually outperforms `ApproxCount`.

For `SampleTreeSearch`, we used its default setting about the number of samples, which is a constant. The results show that it is very efficient and provides good approximations. Our approach only outperforms `SampleTreeSearch` on `blockmap` problems which consist of a large number of variables. However, there is a lack of analysis on the accuracy of the approximation of `SampleTreeSearch`, i.e., no explicit relation between the number of samples and the accuracy.

7   Conclusion

In this paper, we propose a new hashing-based approximate algorithm with dynamic stopping criterion. Our approach has two key strengths: it requires only one satisfiability query for each cut, and it terminates once meeting the criterion of accuracy. We implemented prototype tools for propositional logic formulas and SMT(BV) formulas. Extensive experiments demonstrate that our approach is efficient and promising. Despite that we are unable to prove the correctness of Eq. (1), the experimental results fit it quite well. This phenomenon might be caused by some hidden properties of the hash functions. To fully understand these functions and their correlation with the model count of the hashed formula might be an interesting problem to the community. In addition, extending the idea in this paper to count solutions of other formulas is also a direction of future research.

References

1. Achlioptas, D., Theodoropoulos, P.: Probabilistic model counting with short XORs. In: Gaspers, S., Walsh, T. (eds.) SAT 2017. LNCS, vol. 10491, pp. 3–19. Springer, Cham (2017). https://doi.org/10.1007/978-3-319-66263-3_1
2. Bayardo, Jr, R.J., Schrag, R.: Using CSP look-back techniques to solve real-world SAT instances. In: Proceedings of AAAI, pp. 203–208 (1997)
3. Bellare, M., Goldreich, O., Petrank, E.: Uniform generation of NP-witnesses using an NP-oracle. Inf. Comput. **163**(2), 510–526 (2000)
4. Belle, V., Broeck, G.V., Passerini, A.: Hashing-based approximate probabilistic inference in hybrid domains. In: Proceedings of UAI, pp. 141–150 (2015)
5. Brown, L.D., Cai, T.T., Dasgupta, A.: Interval estimation for a binomial proportion. Stat. Sci. **16**(2), 101–133 (2001)
6. Brummayer, R., Biere, A.: Boolector: an efficient SMT solver for bit-vectors and arrays. In: Kowalewski, S., Philippou, A. (eds.) TACAS 2009. LNCS, vol. 5505, pp. 174–177. Springer, Heidelberg (2009). https://doi.org/10.1007/978-3-642-00768-2_16
7. Chakraborty, S., Fremont, D.J., Meel, K.S., Seshia, S.A., Vardi, M.Y.: Distribution-aware sampling and weighted model counting for SAT. In: Proceedings of AAAI, pp. 1722–1730 (2014)
8. Chakraborty, S., Meel, K.S., Mistry, R., Vardi, M.Y.: Approximate probabilistic inference via word-level counting. In: Proceedings of AAAI, pp. 3218–3224 (2016)
9. Chakraborty, S., Meel, K.S., Vardi, M.Y.: A scalable and nearly uniform generator of SAT witnesses. In: Sharygina, N., Veith, H. (eds.) CAV 2013. LNCS, vol. 8044, pp. 608–623. Springer, Heidelberg (2013). https://doi.org/10.1007/978-3-642-39799-8_40
10. Chakraborty, S., Meel, K.S., Vardi, M.Y.: A scalable approximate model counter. In: Schulte, C. (ed.) CP 2013. LNCS, vol. 8124, pp. 200–216. Springer, Heidelberg (2013). https://doi.org/10.1007/978-3-642-40627-0_18
11. Chakraborty, S., Meel, K.S., Vardi, M.Y.: Algorithmic improvements in approximate counting for probabilistic inference: from linear to logarithmic SAT calls. In: Proceedings of IJCAI, pp. 3569–3576 (2016)
12. Chavira, M., Darwiche, A.: On probabilistic inference by weighted model counting. Artif. Intell. **172**(6–7), 772–799 (2008)
13. Chistikov, D., Dimitrova, R., Majumdar, R.: Approximate counting in SMT and value estimation for probabilistic programs. In: Baier, C., Tinelli, C. (eds.) TACAS 2015. LNCS, vol. 9035, pp. 320–334. Springer, Heidelberg (2015). https://doi.org/10.1007/978-3-662-46681-0_26
14. Domshlak, C., Hoffmann, J.: Probabilistic planning via heuristic forward search and weighted model counting. J. Artif. Intell. Res. (JAIR) **30**, 565–620 (2007)
15. Ermon, S., Gomes, C.P., Sabharwal, A., Selman, B.: Embed and project: discrete sampling with universal hashing. Adv. Neural Inf. Process. Syst. **26**, 2085–2093 (2013)
16. Ermon, S., Gomes, C.P., Selman, B.: Uniform solution sampling using a constraint solver as an oracle. In: Proceedings of UAI, pp. 255–264 (2012)
17. Filieri, A., Pasareanu, C.S., Visser, W.: Reliability analysis in symbolic pathfinder: a brief summary. In: Proceedings of ICSE, pp. 39–40 (2014)
18. Filieri, A., Pasareanu, C.S., Yang, G.: Quantification of software changes through probabilistic symbolic execution (N). In: Proceedings of ASE, pp. 703–708 (2015)
19. Fredrikson, M., Jha, S.: Satisfiability modulo counting: a new approach for analyzing privacy properties. In: Proceedings of CSL-LICS, pp. 42:1–42:10 (2014)

20. Geldenhuys, J., Dwyer, M.B., Visser, W.: Probabilistic symbolic execution. In: Proceedings of ISSTA, pp. 166–176 (2012)
21. von Gleissenthall, K., Köpf, B., Rybalchenko, A.: Symbolic polytopes for quantitative interpolation and verification. In: Kroening, D., Păsăreanu, C.S. (eds.) CAV 2015. LNCS, vol. 9206, pp. 178–194. Springer, Cham (2015). https://doi.org/10.1007/978-3-319-21690-4_11
22. Gomes, C.P., Hoffmann, J., Sabharwal, A., Selman, B.: From sampling to model counting. In: Proceedings of IJCAI, pp. 2293–2299 (2007)
23. Gomes, C.P., Sabharwal, A., Selman, B.: Model counting: a new strategy for obtaining good bounds. In: Proceedings of AAAI, pp. 54–61 (2006)
24. Gomes, C.P., Sabharwal, A., Selman, B.: Near-uniform sampling of combinatorial spaces using XOR constraints. Adv. Neural Inf. Process. Syst. **19**, 481–488 (2006)
25. Ivrii, A., Malik, S., Meel, K.S., Vardi, M.Y.: On computing minimal independent support and its applications to sampling and counting. Constraints **21**(1), 41–58 (2016)
26. Karp, R.M., Luby, M., Madras, N.: Monte-carlo approximation algorithms for enumeration problems. J. Algorithms **10**(3), 429–448 (1989)
27. Kroc, L., Sabharwal, A., Selman, B.: Leveraging belief propagation, backtrack search, and statistics for model counting. Ann. OR **184**(1), 209–231 (2011)
28. Liu, S., Zhang, J.: Program analysis: from qualitative analysis to quantitative analysis. In: Proceedings of ICSE, pp. 956–959 (2011)
29. Meel, K.S., Vardi, M.Y., Chakraborty, S., Fremont, D.J., Seshia, S.A., Fried, D., Ivrii, A., Malik, S.: Constrained sampling and counting: universal hashing meets SAT solving. In: Proceedings of Workshop on Beyond NP (BNP) (2016)
30. Phan, Q., Malacaria, P., Pasareanu, C.S., d'Amorim, M.: Quantifying information leaks using reliability analysis. In: Proceedings of SPIN, pp. 105–108 (2014)
31. Roth, D.: On the hardness of approximate reasoning. Artif. Intell. **82**(1–2), 273–302 (1996)
32. Sang, T., Bacchus, F., Beame, P., Kautz, H.A., Pitassi, T.: Combining component caching and clause learning for effective model counting. In: Proceedings of SAT (2004)
33. Sang, T., Beame, P., Kautz, H.A.: Performing bayesian inference by weighted model counting. In: Proceedings of AAAI, pp. 475–482 (2005)
34. Sipser, M.: A complexity theoretic approach to randomness. In: Proceedings of the 15th Annual ACM Symposium on Theory of Computing, pp. 330–335 (1983)
35. Soos, M., Nohl, K., Castelluccia, C.: Extending SAT solvers to cryptographic problems. In: Kullmann, O. (ed.) SAT 2009. LNCS, vol. 5584, pp. 244–257. Springer, Heidelberg (2009). https://doi.org/10.1007/978-3-642-02777-2_24
36. Stockmeyer, L.J.: The complexity of approximate counting (preliminary version). In: Proceedings of the 15th Annual ACM Symposium on Theory of Computing, pp. 118–126 (1983)
37. Valiant, L.G.: The complexity of enumeration and reliability problems. SIAM J. Comput. **8**(3), 410–421 (1979)
38. Wallis, S.: Binomial confidence intervals and contingency tests: mathematical fundamentals and the evaluation of alternative methods. J. Quant. Linguist. **20**(3), 178–208 (2013)
39. Wei, W., Selman, B.: A new approach to model counting. In: Bacchus, F., Walsh, T. (eds.) SAT 2005. LNCS, vol. 3569, pp. 324–339. Springer, Heidelberg (2005). https://doi.org/10.1007/11499107_24
40. Wilson, E.B.: Probable inference, the law of succession and statistical inference. J. Am. Stat. Assoc. **22**(158), 209–212 (1927)

A Reduction from Unbounded Linear Mixed Arithmetic Problems into Bounded Problems

Martin Bromberger[1,2(✉)]

[1] Max Planck Institute for Informatics and Saarland University,
Saarland Informatics Campus, Saarbrücken, Germany
`mbromber@mpi-inf.mpg.de`
[2] Graduate School of Computer Science,
Saarland Informatics Campus, Saarbrücken, Germany

Abstract. We present a combination of the Mixed-Echelon-Hermite transformation and the Double-Bounded reduction for systems of linear mixed arithmetic that preserve satisfiability and can be computed in polynomial time. Together, the two transformations turn any system of linear mixed constraints into a bounded system, i.e., a system for which termination can be achieved easily. Existing approaches for linear mixed arithmetic, e.g., branch-and-bound and cuts from proofs, only explore a finite search space after application of our two transformations. Instead of generating *a priori* bounds for the variables, e.g., as suggested by Papadimitriou, unbounded variables are eliminated through the two transformations. The transformations orient themselves on the structure of an input system instead of computing *a priori* (over-)approximations out of the available constants. Experiments provide further evidence to the efficiency of the transformations in practice. We also present a polynomial method for converting certificates of (un)satisfiability from the transformed to the original system.

Keywords: Linear arithmetic · Integer arithmetic
Mixed arithmetic · SMT · Linear Transformations · Constraint solving

1 Introduction

Efficient linear arithmetic decision procedures are important for various independent research lines, e.g., optimization, system modeling, and verification. We are interested in feasibility of linear arithmetic problems in the context of the combination of theories, as they occur, e.g., in SMT solving or theorem proving.

The SMT and theorem proving communities have presented several interesting and efficient approaches for pure linear rational arithmetic [18] as well as linear integer arithmetic [5,8,16,20]. SMT research also starts to extend into linear mixed arithmetic [12,18] because some applications require both rational and integer variables, e.g., planning/scheduling problems and verification of timed automata and hybrid systems.

© Springer International Publishing AG, part of Springer Nature 2018
D. Galmiche et al. (Eds.): IJCAR 2018, LNAI 10900, pp. 329–345, 2018.
https://doi.org/10.1007/978-3-319-94205-6_22

We are interest in decision procedures for mixed arithmetic because of a possible combination with superposition [1,4,19]. In the superposition context, arithmetic constraints are part of the first-order clauses. The problems are typically unbounded due to transformations that turn the input formula into a superposition specific input format. Since these problems are unbounded, the search space becomes infinite, which is the case where termination becomes difficult for most linear arithmetic approaches. Unbounded problems appear also in other areas of automated reasoning. Either because of bad encodings, necessary but complicating transformations, e.g., slacking (see Sect. 5), or the sheer complexity of the verification goal. Hence, efficient techniques for handling unbounded problems are necessary for a generally reliable combined procedure.

It is theoretically very easy to achieve termination for linear integer and mixed arithmetic because of so called *a priori bounds*. For example, the *a priori* bounds presented by Papadimitriou [22] guarantee that a problem has a mixed solution if and only if the problem extended by the bounds $|x_i| \leq 2n(ma)^{2m+1}$ for every variable x_i has a mixed solution. In these *a priori* bounds, n is the number of variables, m the number of inequalities, and a the largest absolute value of any integer coefficient or constant in the problem. By extending a problem with those *a priori* bounds, we reduce the search space for a branch-and-bound solver (and many other mixed arithmetic decision procedures) to a finite search space. So branch-and-bound is guaranteed to terminate.

However, these bounds are so large that the resulting search space cannot be explored in reasonable time for many practical problems. One reason for the impracticability of *a priori* bounds is that they only take parameter sizes into account and not actually the structure of each problem. *A priori* bounds are not integrated in any state-of-the-art SMT solvers [3,13–15,17] since they are no help in practice. As far as we know, none of the state-of-the-art SMT solvers use any method that guarantees termination for linear integer or mixed arithmetic.

In this paper, we present satisfiability preserving transformations that reduce unbounded problems into bounded problems. On these bounded problems, most linear mixed decision procedures become terminating, which we show on the example of branch-and-bound. Our reduction works by eliminating unbounded variables. First, we use the Double-Bounded reduction (Sect. 4) to eliminate all unbounded inequalities from our constraint system. Then we use the Mixed-Echelon-Hermite transformation (Sect. 3) to shift the variables of our system to ones that are either bounded or do not appear in the new inequalities and are, therefore, eliminated. With Corollary 14 and Lemma 22 we explain how to efficiently convert certificates of (un)satisfiability between the transformed and the original system. Our method is efficient because it is fully guided by the structure of the problem. This is confirmed by experiments (Sect. 5). We also show how to efficiently determine when a problem is unbounded (Lemma 19). This prevents our solver from applying our transformations on bounded problems.

An extended version of this paper is available on arXiv [7]. It contains an appendix, where we explain how to implement the presented procedures in an incrementally efficient way. This is relevant for the implementation of an efficient

SMT theory solver. The extended version also contains several new examples as well as additional implementation tricks.

2   Preliminaries

While the difference between matrices, vectors, and their components is always clear in context, we generally use upper case letters for matrices (e.g., A), lower case letters for vectors (e.g., x), and lower case letters with an index i or j (e.g., b_i, x_j) as components of the associated vector at position i or j, respectively. The only exceptions are the row vectors $a_i^T = (a_{i1}, \ldots, a_{in})$ of a matrix $A = (a_1, \ldots, a_m)^T$, which already contain an index i that indicates the row's position inside A. We also abbreviate the n-dimensional origin $(0, \ldots, 0)^T$ as 0^n. Moreover, we denote by $\mathrm{piv}(A, j)$ the row index of the *pivot* of a column j, i.e., the smallest row index i with a non-zero entry a_{ij} or $m + j$ if there are no non-zero entries in column j.

A system of constraints $Ax \leq b$ is just a set of non-strict inequalities[1] $\{a_1^T x \leq b_1, \ldots, a_m^T x \leq b_m\}$ and the *rational solutions* of this system are exactly those points $x \in \mathbb{Q}^n$ that satisfy all inequalities in this set. The row coefficients are given by $A = (a_1, \ldots, a_m)^T \in \mathbb{Q}^{m \times n}$, the variables are given by $x = (x_1, \ldots, x_n)^T$, and the inequality bounds are given by $b = (b_1, \ldots, b_m)^T \in \mathbb{Q}^m$. Moreover, we assume that any constant rows $a_i = 0^n$ were eliminated from our system during an implicit preprocessing step. This is a trivial task and eliminates some unnecessarily complicated corner cases.

In this paper, we consider mixed constraint systems, i.e., variables are assigned a type: either rational or integer. Due to convenience, we assume that the first n_1 variables $(x_1, \ldots, x_{n_1})$ are rational and the remaining n_2 variables $(x_{n_1+1}, \ldots, x_n)$ are integer, where $n = n_1 + n_2$. A *mixed solution* is a point $x \in (\mathbb{Q}^{n_1} \times \mathbb{Z}^{n_2})$ that satisfy all inequalities in $Ax \leq b$ and we denote by $\mathcal{M}(Ax \leq b) = \{x \in (\mathbb{Q}^{n_1} \times \mathbb{Z}^{n_2}) : Ax \leq b\}$ the *set of mixed solutions* to the system of inequalities $Ax \leq b$. We sometimes need to relax the variables to be completely rational. Therefore, we denote by $\mathcal{Q}(Ax \leq b) = \{x \in \mathbb{Q}^n : Ax \leq b\}$ the *set of rational solutions* to the system of inequalities $Ax \leq b$.

Since $Ax \leq b$ and $A'x \leq b'$ are just sets, we can write their combination as $(Ax \leq b) \cup (A'x \leq b')$. A special system of inequalities is a system of equations $Dx = c$, which is equivalent to the combined system of inequalities $(Dx \leq c) \cup (-Dx \leq -c)$. We say that a constraint system implies an inequality $h^T x \leq g$, where $h \in \mathbb{Q}^n$, $h \neq 0^n$, and $g \in \mathbb{Q}$, if $h^T x \leq g$ holds for all $x \in \mathcal{Q}(Ax \leq b)$. In the same manner, a constraint system implies an equality $h^T x = g$, where $h \in \mathbb{Q}^n$, $h \neq 0^n$, and $g \in \mathbb{Q}$, if $h^T x = g$ holds for all $x \in \mathcal{Q}(Ax \leq b)$. A constraint implied by $Ax \leq b$ is *explicit* if it does appear in $Ax \leq b$. Otherwise, it is called *implicit*.

Most deductions on linear inequalities are based on Farkas' Lemma:

---

[1] All techniques discussed in this paper can be extended to strict inequalities with the help of δ-rationals [18]. We will omit the strict inequalities and focus only on non-strict inequalities due to lack of space.

Lemma 1 (Farkas' Lemma [6]). $\mathcal{Q}(Ax \leq b) = \emptyset$ *iff there exists a* $y \in \mathbb{Q}^m$ *with* $y \geq 0^m$ *and* $y^T A = 0^n$ *so that* $y^T b < 0$, *i.e., there exists a non-negative linear combination of inequalities in* $Ax \leq b$ *that results in an inequality* $y^T Ax \leq y^T b$ *that is constant and unsatisfiable. If such a* y *exists, then we call it a* certificate of unsatisfiability.

We also frequently use the following lemma, which is just a reformulation of Farkas' Lemma:

Lemma 2 (Linear Implication Lemma). *Let* $\mathcal{Q}(Ax \leq b) \neq \emptyset$, $h \in \mathbb{Q}^n \setminus \{0^n\}$, *and* $g \in \mathbb{Q}$. *Then,* $Ax \leq b$ *implies* $h^T x \leq g$ *iff there exists a* $y \in \mathbb{Q}^m$ *with* $y \geq 0^m$ *and* $y^T A = h^T$ *so that* $y^T b \leq g$, *i.e., there exists a non-negative linear combination of inequalities in* $Ax \leq b$ *that results in the inequality* $h^T x \leq g$.

As we mentioned in the introduction, this paper describes equisatisfiable transformations for constraint systems. We transform the systems in such a way that most linear mixed decision procedures become terminating and still retain their general efficiency. We even show this on the example of branch-and-bound. Although we do not have the time to discuss all facets of branch-and-bound [23], we still want to give a short summary of the algorithm. Branch-and-bound is a recursive algorithm that computes mixed solutions for constraint systems. In each call of the algorithm, it first computes a rational solution s to a constraint system $Ax \leq b^2$. If there are none, then we know that $Ax \leq b$ has no mixed solution. We are also done in the case that s is a mixed solution. Otherwise, we select one of the integer variables x_i assigned to a fractional value $s_i \notin \mathbb{Z}$ and call branch-and-bound recursively on $(Ax \leq b) \cup (x_i \geq \lceil s_i \rceil)$ and $(Ax \leq b) \cup (x_i \leq \lfloor s_i \rfloor)$. If none of the recursive calls returns a mixed solution, then $Ax \leq b$ also does not have a mixed solution. Likewise, if one of them returns a mixed solution s, then it also is a mixed solution to $Ax \leq b$.

Branch-and-bound alone is already complete on bounded constraint systems, i.e., systems where all directions are bounded:

Definition 3 (Bounded Direction). *A direction/vector* $h \in \mathbb{Q}^n \setminus \{0^n\}$ *is* bounded *in the constraint system* $Ax \leq b$ *if there exist* $l, u \in \mathbb{Q}$ *such that* $Ax \leq b$ *implies* $h^T x \leq u$ *and* $-h^T x \leq -l$. *Otherwise, it is called* unbounded.

Definition 4 (Bounded System). *A constraint system* $Ax \leq b$ *is* bounded *if all directions* $h \in \mathbb{Q}^n \setminus \{0^n\}$ *are bounded. Otherwise, it is called* unbounded.

For bounded systems, branch-and-bound is one of the most popular and efficient algorithms. It may, however, diverge if the system has unbounded directions. Even so, not all unbounded systems are equally difficult. For instance, a system where all directions are unbounded has always a mixed solution:

Lemma 5 (Absolutely Unbounded [10]). *If all directions are unbounded in a constraint system* $Ax \leq b$, *then the constraint system has an integer solution.*

---

[2] A rational solution can be computed in polynomial time [23].

In a previous article, we described two cube tests that detect and solve constraint systems with infinite lattice width (another name for absolutely unbounded systems) in polynomial time [10]. The case of absolutely unbounded systems is, therefore, trivial and branch-and-bound can be easily extended so it also becomes complete for absolutely unbounded systems. The actual difficult case is when some directions are bounded and others unbounded. We call these systems *partially unbounded*. Here, branch-and-bound and most other algorithms diverge or become inefficient in practice. The transformations, which we present, are designed to efficiently handle this subclass of problems.

3 Mixed-Echelon-Hermite Transformation

Our overall goal is to present an equisatisfiable transformation that turns any constraint system into a system that is bounded, i.e., a system on which branch-and-bound and many other arithmetic decision procedures terminate. In this section, we only present such a transformation for a subset of constraint systems, which we call *double-bounded constraint systems*. We then show in the next section that each constraint system can be reduced to an equisatisfiable double-bounded system. We also show how to efficiently transform a mixed solution from the double-bounded reduction to a mixed solution for the original system.

Definition 6 (Double-Bounded Constraint System). *A constraint system $Dx \leq u$ is* double-bounded *if $Dx \leq u$ implies $Dx \geq l$ for $l \in \mathbb{Q}^m$. For such a double-bounded system, we call the bounds u the* upper bounds *of Dx and the bounds l the* lower bounds *of Dx. Moreover, we typically write $l \leq Dx \leq u$ instead of $Dx \leq u$ although the* lower bounds l *are only implicit.*

Note that only the inequalities in a double-bounded constraint system are guaranteed to be bounded. Variables might still be unbounded. For instance, in the constraint system $1 \leq 3x_1 - 3x_2 \leq 2$ both inequalities are bounded but the variables x_1 and x_2 are not. Moreover, the above constraint system is also an example where branch-and-bound diverges. This means that even bounding all inequalities does not yet guarantee termination. So for our purposes, a double-bounded constraint system is still too complex.

This changes, however, if we also require that the coefficient matrix D of our constraint system is a *lower triangular matrix with gaps*:

Definition 7 (Lower Triangular Matrix with Gaps). *A matrix $A \in \mathbb{Q}^{m \times n}$ is* lower triangular with gaps *if it holds for each column j that $piv(A, j) > m$ or that $piv(A, j) < piv(A, k)$ for all columns k with $j < k \leq n$, i.e., column j either has only zero entries or all pivoting entries right of j have a higher row index.*

A matrix is lower triangular if and only if the row indices of its pivots are strictly increasing, i.e., $\mathrm{piv}(A, 1) < \ldots < \mathrm{piv}(A, n)$. If we also allow it to have gaps, only the row indices of pivots with non-zero columns have to be strictly increasing. Now we get termination for free because of our restrictions:

Lemma 8 (Lower Triangular Double-Bounded Systems). *Let $D \in \mathbb{Q}^{m \times n}$ be a lower triangular matrix with gaps and $l \leq Dx \leq u$ be a double-bounded constraint system. Then each variable x_j is either bounded, i.e., $l \leq Dx \leq u$ implies that $l'_j \leq x_j \leq u'_j$ or its column in D has only zero entries.*

Proof. Proof by induction. Assume that the above property already holds for all variables x_k with $k < j$. Let $p = \text{piv}(D, j)$. If $p > m$, then the column j of D is zero and we are done. If $p \leq m$, then the pivoting entry d_{pj} of column j is non-zero. Because of Definition 7 and our induction hypothesis, this also means that each column k with $k < j$ has either a zero entry in row p or the variable x_k is bounded by our induction hypothesis, i.e., $l \leq Dx \leq u$ implies $l'_k \leq x_k \leq u'_k$. Since Definition 7 also implies that row p has only zero entries to the right of d_{pj}, the row p has only one unbounded variable with a non-zero entry, viz., x_j. This means we can transform the row $l_p \leq d_p^T x \leq u_p$ into the following two inequalities: $l_p - \sum_{k=1}^{j-1} d_{pk} x_k \leq d_{pj} x_j$ and $u_p - \sum_{k=1}^{j-1} d_{pk} x_k \geq d_{pj} x_j$, where the variables x_k on the left sides are either bounded or $d_{pk} = 0$. Hence, we can derive an upper and lower bound for x_j via bound propagation/refinement [21].   □

Corollary 9 (BnB-LTDB-Termination). *Branch-and-bound terminates on every double-bounded system $l \leq Dx \leq u$ where D is lower triangular with gaps.*

Our next goal is to efficiently transform every double-bounded system $l \leq Dx \leq u$ into an equisatisfiable system that also has a lower triangular coefficient matrix with gaps. We start by defining a class of transformations that do not only preserve mixed equisatisfiability, but are also very expressive.

Definition 10 (Mixed Column Transformation Matrix [12]). *Given a mixed constraint system. A matrix $V \in \mathbb{Q}^{n \times n}$ is a mixed column transformation matrix if it is invertible and consists of an invertible matrix $V_{(\mathbb{Q})} \in \mathbb{Q}^{n_1 \times n_1}$, a unimodular matrix $V_{(\mathbb{Z})} \in \mathbb{Z}^{n_2 \times n_2}$, and a matrix $V_{(M)} \in \mathbb{Q}^{n_1 \times n_2}$ such that*

$$V = \begin{pmatrix} V_{(\mathbb{Q})} & V_{(M)} \\ 0^{n_2 \times n_1} & V_{(\mathbb{Z})} \end{pmatrix}.$$

The inverse of a mixed column transformation matrix V is also a mixed column transformation matrix and can be used to undo the transformation V:

Lemma 11 (Mixed Column Transformation Inversion [12]). *Given a mixed constraint system. Let $V \in \mathbb{Q}^{n \times n}$ be a mixed column transformation matrix. Then V^{-1} is also a mixed column transformation matrix.*

This means that each mixed column transformation matrix defines a bijection from $(\mathbb{Q}^{n_1} \times \mathbb{Z}^{n_2})$ to $(\mathbb{Q}^{n_1} \times \mathbb{Z}^{n_2})$. Hence, they guarantee mixed equisatisfiability:

Lemma 12 (Mixed Column Transformation Equisatisfiability [12]). *Let $Ax \leq b$ be a mixed constraint system. Let $V \in \mathbb{Q}^{n \times n}$ be a mixed column transformation matrix. Then every solution $y \in \mathcal{M}((AV)y \leq b))$ can be converted into a solution $Vy = x \in \mathcal{M}(Ax \leq b)$ and vice versa.*

Moreover, the mixed column transformation matrix V also establishes a direct relationship between the linear combinations of the original constraint system and the transformed one:

Lemma 13 (Mixed Column Transformation Implications). *Let $Ax \leq b$ be a constraint system. Let $V \in \mathbb{Q}^{n \times n}$ be a mixed column transformation matrix. Let $Ax \leq b$ imply $h^T x \leq g$. Then $AVz \leq b$ implies $h^T Vz \leq g$.*

Proof. By Lemma 2, $Ax \leq b$ implies $h^T x \leq g$ iff there exists a non-negative linear combination $y \in \mathbb{Q}^n$ such that $y \geq 0$, $y^T A = h^T$ and $y^T b \leq g$. Multiplying $y^T A = h^T$ with V results in $y^T AV = h^T V$ and thus y is also the non-negative linear combination of inequalities $AVz \leq b$ that results in $h^T Vz \leq g$.   □

Corollary 14 (Mixed Column Transformation Certificates). *Let $Ax \leq b$ be a constraint system. Let $V \in \mathbb{Q}^{n \times n}$ be a mixed column transformation matrix. Then y is a certificate of unsatisfiability for $Ax \leq b$ iff it is one for $AVz \leq b$.*

Now we only need a mixed column transformation matrix V for every coefficient matrix A such that $H = AV$ is lower triangular with gaps. One such matrix V is the one that transforms A into *Mixed-Echelon-Hermite normal form*:

Definition 15 (Mixed-Echelon-Hermite Normal Form [12]). *A matrix $H \in \mathbb{Q}^{m \times n}$ is in* Mixed-Echelon-Hermite normal form *if*

$$H = \begin{pmatrix} E & 0^{r \times (n_1 - r)} & 0^{r \times n_2} \\ E' & 0^{(m-r) \times (n_1 - r)} & H' \end{pmatrix},$$

where E is an $r \times r$ identity matrix (with $r \leq n_1$), $E' \in \mathbb{Q}^{(m-r) \times r}$, and $H' \in \mathbb{Q}^{(m-r) \times n_2}$ is a matrix in hermite normal form, i.e., a lower triangular matrix without gaps, where each entry $h'_{piv(H',j)k}$ in the row $piv(H',j)$ is non-negative and smaller than $h'_{piv(H',j)j}$.

The following proof for the existence of the Mixed-Echelon-Hermite normal form is constructive and presents the Mixed-Echelon-Hermite transformation.

Lemma 16 (Mixed-Echelon-Hermite Transformation). *Let $A \in \mathbb{Q}^{m \times n}$ be a matrix, where the upper left $r \times n_1$ submatrix has the same rank r as the complete left $m \times n_1$ submatrix. Then there exists a mixed transformation matrix $V \in \mathbb{Q}^{n \times n}$ such that $H = AV$ is in* Mixed-Echelon-Hermite normal form.

Proof. Proof from [12] with slight modifications so it also works for singular matrices. We subdivide A into

$$A = \begin{pmatrix} A_{11} & A_{12} \\ A_{21} & A_{22} \end{pmatrix}$$

such that $A_{11} \in \mathbb{Q}^{r \times n_1}$, $A_{12} \in \mathbb{Q}^{r \times n_2}$, $A_{21} \in \mathbb{Q}^{m-r \times n_1}$, and $A_{21} \in \mathbb{Q}^{m-r \times n_2}$. Then we bring A_{11} with an invertible matrix $V_{11} \in \mathbb{Q}^{n_1 \times n_1}$ into reduced echelon column form $H_{11} = (E \ \ 0^{r \times (n_1 - r)}) = A_{11}V_{11}$, where E is an $r \times r$ identity matrix. We get V_{11} and H_{11} by using Bareiss algorithm instead of the better known Gaussian elimination as it is polynomial in time [2].[3] Note that the last $n_1 - r$ columns of $H_{21} = (H'_{21}\ 0^{(m-r)\times(n_1-r)}) = A_{21}V_{11}$ are also zero because all rows in A_{21} are linear dependent of A_{11} (due to the rank). Next we notice that

$$A_{12} - A_{11}V_{11}\begin{pmatrix} A_{12} \\ 0^{(n_1-r)\times n_2} \end{pmatrix} = A_{12} - (E\ 0^{r\times(n_1-r)})\begin{pmatrix} A_{12} \\ 0^{(n_1-r)\times n_2} \end{pmatrix} = 0^{r\times n_2}$$

so we can reduce the upper right submatrix A_{12} to zero by adding multiples of the n_1 columns with rational variables to the n_2 columns with integer variables. However, this also transforms the lower right submatrix A_{22} into

$$H'_{22} = A_{22} - A_{21}V_{11}\begin{pmatrix} A_{12} \\ 0^{(n_1-r)\times n_2} \end{pmatrix}.$$

Finally, we transform this new submatrix H'_{22} into hermite normal form H_{22} via the algorithm of Kannan and Bachem (or a similar polynomial time algorithm) (see footnote 3). This algorithm also returns a unimodular matrix $V_{22} \in \mathbb{Z}^{n_2 \times n_2}$ such that $H_{22} = H'_{22}V_{22}$. To summarize: our total mixed transformation matrix is

$$V = \begin{pmatrix} V_{11} & -V_{11}\cdot\begin{pmatrix} A_{12} \\ 0^{(n_1-r)\times n_2} \end{pmatrix}\cdot V_{22} \\ 0^{n_2\times n_1} & V_{22} \end{pmatrix} \quad \text{and} \quad H = AV = \begin{pmatrix} H_{11} & 0^{r\times n_2} \\ H_{21} & H_{22} \end{pmatrix}.$$

$\square$

It is not possible to transform every matrix $A \in \mathbb{Q}^{m\times n}$ into Mixed-Echelon-Hermite normal form. We have to restrict ourselves to matrices, where the upper left $r \times n_1$ submatrix has the same rank r as the complete left $m \times n_1$ submatrix. However, this is very easy to accomplish for a system of linear mixed arithmetic constraints $l \leq Ax \leq u$. The reason is that the order of inequalities does not change the set of satisfiable solutions. Hence, we can swap the inequalities and, thereby, the rows of A until its upper left $r \times n_1$ submatrix has the desired form. This also means that there are usually multiple possible inequality orderings that each have their own Mixed-Echelon-Hermite normal form H.

To conclude this section: whenever we have a double-bounded constraint system $l \leq Dx \leq u$, we can transform it (after some row swapping) into an equisatisfiable system $l \leq Hy \leq u$ where $H = DV$ is in Mixed-Echelon-Hermite normal form and $Vy = x$. Since H is also a lower triangular matrix with gaps, branch-and-bound terminates on $l \leq Hy \leq u$ with a mixed solution t or it will return unsatisfiable (Corollary 9). Moreover, we can convert any mixed solution t for $l \leq Hy \leq u$ into a mixed solution s for $l \leq Dx \leq u$ by setting $s := Vt$. Hence, we have a complete algorithm for double-bounded constraint systems.

---

[3] We do actually use less efficient, Gaussian-elimination-based transformations in our own implementation [7]. The reason is that these transformations are incrementally efficient. Our experiments show that the transformation cost still remains negligible in practice.

4 Double-Bounded Reduction

In the previous Section, we have shown how to solve a double-bounded constraint system. Now we show how to reduce any constraint system $A'x \leq b'$ to an equisatisfiable double-bounded system $l \leq Dx \leq u$. Moreover, we explain how to take any solution of $l \leq Dx \leq u$ and turn it into a solution for $A'x \leq b'$.

As the first step of our reduction, we reformulate the constraint system into a so called *split system*:

Definition 17 (Split System). $(Ax \leq b) \cup (l \leq Dx \leq u)$ *is a* split system *if:*
(i) all directions are unbounded in $Ax \leq b$; *(ii) all row vectors* a_i *from* A *are also unbounded in* $(Ax \leq b) \cup (l \leq Dx \leq u)$. *Moreover, we call* $Ax \leq b$ *the unbounded part and* $l \leq Dx \leq u$ *the bounded part of the split system.*

A split system consists of an unbounded part $Ax \leq b$ that is guaranteed to have (infinitely many) integer solutions (see Lemma 5) and a double-bounded part $l \leq Dx \leq u$. Any constraint system can be brought into the above form. We just have to move all unbounded inequalities into the unbounded part and all bounded inequalities into the bounded part.

Lemma 18 (Split Equivalence). *Let* $A'x \leq b'$ *be a constraint system with* $A' \in \mathbb{Q}^{m \times n}$. *Then there exists an equivalent split system* $(Ax \leq b) \cup (l \leq Dx \leq u)$ *where: (i)* $A \in \mathbb{Q}^{m_1 \times n}$ *and* $D \in \mathbb{Q}^{m_2 \times n}$ *so that* $m_1 + m_2 = m$; *(ii) all rows* d_i^T *of* D *and* a_k^T *of* A *appear as rows in* A'; *and (iii)* $Dx \leq u$ *implies* $l \leq Dx$.

Proof. For (i), (ii), and the equivalence, it is enough to move all bounded inequalities $a_i'^T x \leq b_i'$ of $A'x \leq b'$ into $Dx \leq u$ and all unbounded inequalities into $Ax \leq b$. For (iii), we assume for a contradiction that $Dx \leq u$ does not imply $l_i \leq d_i^T x$ but $(Dx \leq u) \cup (Ax \leq b)$ does. By Lemma 2, this means that there exists a $y \in \mathbb{Q}^{m_2}$ with $y \geq 0^{m_2}$ and a $z \in \mathbb{Q}^{m_1}$ with $z \geq 0^{m_1}$ so that $y^T D + z^T A = -d_i^T$ and $y^T u + z^T b \leq -l_i$. We also know that there exists a $z_k > 0$ because $Dx \leq u$ alone does not imply $l_i \leq d_i^T x$. We use this fact to reformulate $y^T D + z^T A = -d_i^T$ into $-a_k^T = \frac{1}{z_k} \left[y^T D + d_i^T + \sum_{j=1, j \neq k}^{m_1} z_j a_j^T \right]$, and use the bounds of the inequalities in $Dx \leq u$ and $Ax \leq b$ to derive a lower bound for $a_k^T x$: $-a_k^T x \leq \frac{1}{z_k} \left[y^T u + u_i + \sum_{j=1, j \neq k}^{m_1} z_j b_j \right]$. Hence, a_k^T is bounded in $A'x \leq b'$ and we have our contradiction. □

The above Lemma also shows that the bounded part of a constraint system is self-contained, i.e., a constraint system implies that a direction is bounded if and only if its bounded part does. The actual difficulty of reformulating a system into a split system is not the transformation per se, but finding out which inequalities are bounded or not. There are many ways to detect whether an inequality is bounded by a constraint system. Most work even in polynomial time. For instance, solving the linear rational optimization problem "minimize $a_i^T x$ such that $Ax \leq b$" returns $-\infty$ if a_i is unbounded, ∞ if $Ax \leq b$ has no rational solution, and the optimal lower bound l_i for $a_i^T x$ otherwise. However, it still requires us to solve m linear optimization problems.

A, in our opinion, more efficient alternative is based on our previously presented algorithm for finding equality bases [9]. This is due to the following relationship between bounded directions and equalities:

Lemma 19 (Bounds and Equalities). *Let $\mathcal{Q}(Ax \leq b) \neq \emptyset$. Then h is bounded in $Ax \leq b$ iff $Ax \leq 0^m$ implies that $h^T x = 0$.*

Proof. By Definition 3, h is bounded in $Ax \leq b$ means that there exists $l, u \in \mathbb{Q}$ such that $Ax \leq b$ implies $h^T x \leq u$ and $-h^T x \leq -l$. By Lemma 2, this is equivalent to: there exist $l, u \in \mathbb{Q}$, $y, z \in \mathbb{Q}^m$ with $y, z \geq 0^m$, and $y^T A = h^T = -z^T A$ so that $y^T b \leq u$ and $z^T b \leq -l$. Symmetrically, $Ax \leq 0$ implies that $h^T x = 0$ is equivalent to: there exist a $y, z \in \mathbb{Q}^m$ with $y, z \geq 0^m$ and $y^T A = h^T = -z^T A$ so that $y^T 0^m \leq 0$ and $z^T 0^m \leq 0$. Since u and l only have to exists, we can trivially choose them as $u := y^T b$ and $l := -z^T b$. This means that $y^T b \leq u$, $z^T b \leq -l$, $y^T 0^m \leq 0$, and $z^T 0^m \leq 0$ are all trivially satisfied by any pair of linear combinations $y, z \in \mathbb{Q}^m$ with $y, z \geq 0^m$ such that $y^T A = h^T = -z^T A$. Hence, the two definitions are equivalent and our lemma holds. □

It is easy and efficient to compute an equality basis for $Ax \leq 0^m$ and to determine with it the inequalities in $Ax \leq b$ that are bounded [9]. The only disadvantage towards the optimization approach is that we do not derive an optimal lower bound l for the inequalities. This is no problem because only the existence of lower bounds is relevant and not the actual bound values.

In a split system $(Ax \leq b) \cup (l \leq Dx \leq u)$, the unbounded part is actually inconsequential to the rational/mixed satisfiability of the system. It may reduce the number of rational/mixed solutions, but it never removes them all. Hence, $(Ax \leq b) \cup (l \leq Dx \leq u)$ is equisatisfiable to just $l \leq Dx \leq u$. We first show this equisatisfiability for the rational case:

Lemma 20 (Rational Extension). *Let $(Ax \leq b) \cup (l \leq Dx \leq u)$ be a split system. Let $s \in \mathbb{Q}^n$ be a rational solution to the bounded part $l \leq Dx \leq u$ such that $Ds = g$, where $g \in \mathbb{Q}^{m_2}$. Then $(Ax \leq b) \cup (Dx = g)$ has a solution s'.*

Proof. Assume for a contradiction that $(Ax \leq b) \cup (Dx = g)$ has no solution. By Lemma 1, this means that there exist a $y \in \mathbb{Q}^{m_1}$ with $y \geq 0^{m_1}$ and $z, z' \in \mathbb{Q}^{m_2}$ with $z, z' \geq 0^{m_2}$ such that $y^T A + z^T D - z'^T D = 0^n$ and $y^T b + z^T g - z'^T g < 0$. Since $Dx = g$ is satisfiable by itself, there must exist a $y_i > 0$. Now we use this fact to reformulate the equation $y^T A + z^T D - z'^T D = 0^n$ into

$$-a_i^T = \frac{1}{y_i} \left[\left(\sum_{j=1 j \neq i}^{m_1} y_j a_j^T \right) + z^T D - z'^T D \right],$$

from which we deduce a lower bound for $a_i^T x$ in $(Ax \leq b) \cup (l \leq Dx \leq u)$:

$$-a_i^T x \leq \frac{1}{y_i} \left[\left(\sum_{j=1 j \neq i}^{m_1} y_j b_j \right) + z^T u - z'^T l \right].$$

Therefore, a_i is bounded in $(Ax \leq b) \cup (l \leq Dx \leq u)$, which is a contradiction. □

Note that the bounded part $l \leq Dx \leq u$ of a split system can still have unbounded directions (not inequalities). Some of these unbounded directions in $l \leq Dx \leq u$ are the orthogonal directions to the row vectors d_i, i.e., vectors $v_j \in \mathbb{Z}^n$ such that $d_i^T v_j = 0$ for all $i \in \{1, \ldots, m_2\}$. This also means that the existence of one mixed solution $s \in (\mathbb{Q}^{n_1} \times \mathbb{Z}^{n_2})$ and one unbounded direction proves the existence of infinitely many mixed solutions. We just need to follow the orthogonal directions, i.e., for all $\lambda \in \mathbb{Z}$, $s' = \lambda \cdot v_j + s$ is also a mixed solution because $d_i^T s' = \lambda \cdot d_i^T v_j + d_i^T s = d_i^T s$. In the next two steps, we prove that $Ax \leq b$ cannot cut off all of these orthogonal solutions because it is completely unbounded. The first step proves that $Ax \leq b$ remains absolutely unbounded even if we settle on one set of orthogonal solutions, i.e., enforce $Dx = Ds$ for some solution s.

Lemma 21 (Persistence of Unboundedness). *Let $(Ax \leq b) \cup (l \leq Dx \leq u)$ be a split system. Let $s \in \mathbb{Q}^n$ be a rational solution for $l \leq Dx \leq u$ such that $Ds = g$ (with $g \in \mathbb{Q}^{m_2}$). Then all row vectors a_i from A are still unbounded in $(Ax \leq b) \cup (Dx = g)$.*

Proof. By Lemma 20, $(Ax \leq b) \cup (Dx = g)$ has at least a rational solution s^*. Moreover, $(Ax \leq 0) \cup (Dx = 0)$ does not imply $a_i^T x = 0$ because of Lemma 19 and the assumption that the row vectors a_i from A are unbounded in $(Ax \leq b) \cup (l \leq Dx \leq u)$. In reverse, $(Ax \leq b) \cup (Dx = g)$ having a real solution, $(Ax \leq 0) \cup (Dx = 0)$ does not imply $a_i^T x = 0$, and Lemma 19 prove together that the row vectors a_i from A are also unbounded in $(Ax \leq b) \cup (Dx = g)$. □

The next step proves how to extend the mixed solution from the bounded part to the complete system with the help of the Mixed-Echelon-Hermite normal form and the absolute unboundedness of $Ax \leq b$.

Lemma 22 (Mixed Extension). *Let $(Ax \leq b) \cup (l \leq Dx \leq u)$ be a split system. Let $s \in (\mathbb{Q}^{n_1} \times \mathbb{Z}^{n_2})$ be a mixed solution for $l \leq Dx \leq u$. Then $(Ax \leq b) \cup (l \leq Dx \leq u)$ has a mixed solution s'.*

Proof. Let $g = Ds$. Without loss of generality we assume that the upper left $r \times n_1$ submatrix of D has the same rank r as the complete left $m_1 \times n_1$ submatrix of D. (Otherwise, we just reorder the rows accordingly.) Therefore, there exists a mixed column transformation matrix V such that $H = DV$ is in Mixed-Echelon-Hermite normal form (see Lemma 16). By Lemma 12, there exists a mixed vector $t \in (\mathbb{Q}^{n_1} \times \mathbb{Z}^{n_2})$ such that $s = Vt$ and t is a mixed-solution to $l \leq Hy \leq u$ as well as $Hy = g$. Let $\mathcal{U}$ be the set of indices with 0 columns in H and $\mathcal{B}$ the column indices with bounded variables. Then the equation system $(Hy = g)$ fixes each variable y_j with $j \in \mathcal{B}$ to the value t_j because H is lower triangular with gaps. Hence, $((AV)y \leq b) \cup (Hy = g)$ is equivalent to

$$A \left[\sum_{j \in \mathcal{U}} \begin{pmatrix} v_{1j} \\ \vdots \\ v_{nj} \end{pmatrix} \cdot y_j \right] \leq b - A \left[\sum_{j \in \mathcal{B}} \begin{pmatrix} v_{1j} \\ \vdots \\ v_{nj} \end{pmatrix} \cdot t_j \right]. \tag{1}$$

Due to Lemmas 13 and 21, all directions are unbounded in (1). This means (1) has an integer solution (Lemma 5) assigning each variable y_j with $j \in \mathcal{U}$ to a $t'_j \in \mathbb{Z}$. (Can be computed via the unit cube test [11]). We extend this solution to all variables y by setting $t'_j := t_j$ for $j \in \mathcal{B}$ and we have a mixed solution $t' \in (\mathbb{Q}^{n_1} \times \mathbb{Z}^{n_1})$ for $((AV)y \leq b) \cup (l \leq Hy \leq u)$. Hence, we have via Lemma 12 a mixed solution $s' \in (\mathbb{Q}^{n_1} \times \mathbb{Z}^{n_2})$ for $(Ax \leq b) \cup (l \leq Dx \leq u)$ with $s' = Vt'$. $\square$

Corollary 23 (Double-Bounded Reduction). *The split system* $(Ax \leq b) \cup (l \leq Dx \leq u)$ *is mixed equisatisfiable to* $(l \leq Dx \leq u)$.

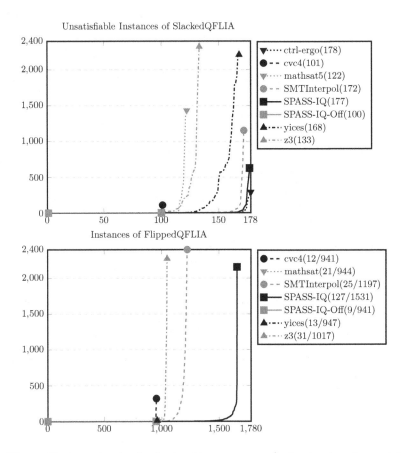

Fig. 1. Horizontal axis: # of solved instances; vertical axis: time (seconds)

5   Experiments

We integrated the Double-Bounded reduction and the Mixed-Echelon-Hermite transformation into our own theory solver *SPASS-IQ v0.2*[4] and ran it on four

---

[4] Available on http://www.spass-prover.org/spass-iq.

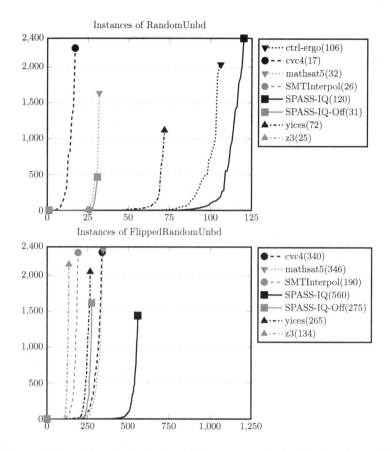

Fig. 2. Horizontal axis: # of solved instances; vertical axis: time (seconds)

families of newly constructed benchmarks (see footnote 4). Once with the transformations turned on (*SPASS-IQ*) and once with the transformations turned off (*SPASS-IQ-Off*). If SPASS-IQ encounters a system $Ax \leq b$ that is not explicitly bounded, i.e., where not all variables have an explicit upper and lower bound, then it computes an equality basis for $Ax \leq 0^m$. This basis is used to determine whether the system is implicitly bounded, absolutely unbounded or partially bounded, as well as which of the inequalities are bounded. Our solver only applies our two transformations if the problem is partially unbounded. The resulting equisatisfiable but bounded problem is then solved via branch-and-bound. The other two cases, absolutely unbounded and implicitly bounded, are solved respectively via the unit cube test [11] and branch-and-bound on the original system. Our solver also converts any mixed solutions from the transformed system into mixed solutions for the original system following the proof of Lemma 22. Rational conflicts are converted between the two systems by using Corollary 14.

We tried to restrict our benchmarks to partially unbounded problems since we only apply our transformations on those problems. We even found some partially unbounded problems in the SMT-LIB benchmarks for QF_LIA (quantifier free linear arithmetic). However, there are not many such benchmarks: only one in *CAV-2009*, five in *cut_lemmas*, and three in *slacks*. So we created in addition four new benchmark families:

SlackedQFLIA: are linear integer benchmarks based on the SMT-LIB classes *CAV-2009* [16], *cut_lemmas* [20], and *dillig* [16]. We simply took all of the unsatisfiable benchmarks and replaced in them all variables x with $x_+ - x_-$ where x_+ and x_- are two new variables such that $x_+, x_- \geq 0$. This transformation, called slacking, is equisatisfiable and the slacked version of the *dillig*-benchmarks, called *slacked* [21], is already in the SMT-LIB. Slacking turns any unsatisfiable problem into a partially unbounded one. Hence, all problems in *SlackedQFLIA* are partially unbounded. Slacking is commonly used to integrate absolute values into linear systems or for solvers that require non-negative variables [23].

RandomUnbd: are linear integer benchmarks that are all partially unbounded and satisfiable with 10, 25, 50, 75, and 100 variables. All problems are randomly created via a sagemath script (see footnote 4).

FlippedQFLIA and *FlippedRandomUnbd*: are linear mixed benchmarks that are all partially unbounded. They are based on *SlackedQFLIA* and *RandomUnbd*. We constructed them by first copying ten versions of the integer benchmarks and then randomly flipping the type of some of the variables to rational (probability of 20%). Some of the flipped instances of *SlackedQFLIA* became satisfiable.

We compared our solver with some of the state-of-the-art SMT solvers currently available for linear mixed arithmetic: *cvc4-1.5* [3], *mathsat5-5.1* [14], *SMTInterpol 2.1-335-g4c543a5* [13], *yices2.5.4* [17], and *z3-4.6.0* [15]. Most of these solvers employ a branch-and-bound approach with an underlying dual simplex solver [18], which is also the basis for our own solver. As far as we are aware, none of them employ any techniques that guarantee termination.

SMTInterpol extends branch-and-bound via the cuts from proofs approach, which uses the Mixed-Echelon-Hermite transformation to find more versatile branches and cuts [12]. Although the procedure is not complete, the similarities to our own approach make an interesting comparison. Actually, the Double-Bounded reduction alone would be sufficient to make SMTInterpol terminating since it already builds branches via a Mixed-Echelon-Hermite transformation.

We also compared our solver with the *ctrl-ergo* solver [5] although it is restricted to pure integer arithmetic. Ctrl-ergo is complete over linear integer arithmetic and uses the most similar approach to our transformations that we found in the literature. It dynamically eliminates one linear independent bounded direction at a time via transformation. The disadvantages of the dynamic approach are that it is very restrictive and does not leave enough freedom to change strategies or to add complementing techniques. Moreover, ctrl-ergo uses this transformation approach for all problems and not only the partially unbounded ones, which sometimes leads to a massive overhead on bounded problems.

For the experiments, we used a Debian Linux cluster and allotted to each problem and solver combination 2 cores of an Intel Xeon E5620 (2.4 GHz) processor, 4 GB RAM, and 40 min. The only solver benefiting from multiple cores is SMTInterpol. The plots in Figs. 1 and 2 depict the results of the different solvers. In the legends of the plots, the numbers behind the solver names are the number of solved instances. For *FlippedQFLIA*, there are two numbers to indicate the number of satisfiable/unsatisfiable instances solved. This is only necessary for *FlippedQFLIA* because it is the only tested benchmark family with satisfiable and unsatisfiable instances. (We verified that the results match if two solvers solved the same problem.)

Although our solver could not solve all problems (due to time and memory limits) it was still able to solve more problems than the other solvers. It was also faster on most instances than the other solvers. In some of the unsatisfiable, partially unbounded benchmarks ctrl-ergo is better than SPASS-IQ. This is due to its conflict focused, dynamic approach. For the same reason, ctrl-ergo is slower on the satisfiable, partially unbounded benchmarks. Only SPASS-IQ, ctrl-ergo, and yices solved all of the ten original SMT-LIB benchmarks that are partially unbounded, though the complete methods were still a lot faster (SPASS-IQ took 23 s, ctrl-ergo took 42 s, and yices took 1273 s). On one of these benchmarks, 20-14.slacks.smt2 from *slacks*, all other solvers seem to diverge. Another interesting result of our experiments is that relaxing some integer variables to rational variables seems to make the problems harder instead of easier. We expected this for our transformations because the resulting systems become more complex and less sparse, but it is also true for the other solvers. The reason might be that bound refinement, a technique used in most branch-and-bound implementations, is less effective on mixed problems.

The time SPASS-IQ needs to detect the bounded inequalities and to apply our transformations is negligible. This is even true for the implicitly bounded problems we tested. As mentioned before, we do not have to apply our transformations to terminate on bounded problems. This is also the only advantage we gain from detecting that a problem is implicitly bounded. Since there is no noticeable difference in the run time, we do not further elaborate the results on bounded problems, e.g. with graphs.

An actual disadvantage of our approach is that the Mixed-Echelon-Hermite transformation increases the density of the coefficient matrix as well as the absolute size of the coefficients. Both are important factors for the efficiency of the underlying simplex solver. Moreover, SPASS-IQ reaches more often the memory limit than the time limit because it needs a (too) large number of branches and bound refinements before terminating.

6   Conclusion

We have presented the Mixed-Echelon-Hermite transformation (Lemma 16) and the Double-Bounded reduction (Lemma 18 and Corollary 23). We have shown that both transformations together turn any constraint system into an equisatisfiable system that is also bounded (Lemma 8). This is sufficient to make branch-and-bound, and many other linear mixed decision procedures, complete and terminating. We have also shown how to convert certificates of (un)satisfiability efficiently between the transformed and original systems (Corollary 14 and Lemma 22). Moreover, experimental results on partially unbounded benchmarks show that our approach is also efficient in practice.

Our approach can be nicely combined with the extensive branch-and-bound framework and its many extensions, where other complete techniques cannot be used in a modular way [5,8]. For future research, we plan to test our transformations in combination with other algorithms, e.g., cuts from proofs, or as a dynamic version similar to the approach used by ctrl-ergo [5]. We also want to test whether our transformations are useful preprocessing steps for select constraint system classes that are bounded.

References

1. Althaus, E., Kruglov, E., Weidenbach, C.: Superposition modulo linear arithmetic SUP(LA). In: Ghilardi, S., Sebastiani, R. (eds.) FroCoS 2009. LNCS (LNAI), vol. 5749, pp. 84–99. Springer, Heidelberg (2009). https://doi.org/10.1007/978-3-642-04222-5_5
2. Bareiss, E.H.: Sylvester's identity and multistep integer-preserving Gaussian elimination. Math. Comput. **22**(103), 565–578 (1968)
3. Barrett, C., Conway, C.L., Deters, M., Hadarean, L., Jovanović, D., King, T., Reynolds, A., Tinelli, C.: CVC4. In: Gopalakrishnan, G., Qadeer, S. (eds.) CAV 2011. LNCS, vol. 6806, pp. 171–177. Springer, Heidelberg (2011). https://doi.org/10.1007/978-3-642-22110-1_14
4. Baumgartner, P., Waldmann, U.: Hierarchic superposition with weak abstraction. In: Bonacina, M.P. (ed.) CADE 2013. LNCS (LNAI), vol. 7898, pp. 39–57. Springer, Heidelberg (2013). https://doi.org/10.1007/978-3-642-38574-2_3
5. Bobot, F., Conchon, S., Contejean, E., Iguernelala, M., Mahboubi, A., Mebsout, A., Melquiond, G.: A simplex-based extension of Fourier-Motzkin for solving linear integer arithmetic. In: Gramlich, B., Miller, D., Sattler, U. (eds.) IJCAR 2012. LNCS (LNAI), vol. 7364, pp. 67–81. Springer, Heidelberg (2012). https://doi.org/10.1007/978-3-642-31365-3_8
6. Boyd, S. Vandenberghe, L.: Convex optimization. In: CUP (2004)
7. Bromberger, M.: A reduction from unbounded linear mixed arithmetic problems into bounded problems. ArXiv e-Prints, abs/1804.07703 (2018)
8. Bromberger, M., Sturm, T., Weidenbach, C.: Linear integer arithmetic revisited. In: Felty, A.P., Middeldorp, A. (eds.) CADE 2015. LNCS (LNAI), vol. 9195, pp. 623–637. Springer, Cham (2015). https://doi.org/10.1007/978-3-319-21401-6_42
9. Bromberger, M., Weidenbach, C.: Computing a complete basis for equalities implied by a system of LRA constraints. In: SMT 2016 (2016)
10. Bromberger, M., Weidenbach, C.: Fast cube tests for LIA constraint solving. In: Olivetti, N., Tiwari, A. (eds.) IJCAR 2016. LNCS (LNAI), vol. 9706, pp. 116–132. Springer, Cham (2016). https://doi.org/10.1007/978-3-319-40229-1_9
11. Bromberger, M., Weidenbach, C.: New techniques for linear arithmetic: cubes and equalities. Form. Methods Syst. Des. **51**(3), 433–461 (2017)

12. Christ, J., Hoenicke, J.: Cutting the mix. In: Kroening, D., Păsăreanu, C.S. (eds.) CAV 2015. LNCS, vol. 9207, pp. 37–52. Springer, Cham (2015). https://doi.org/10.1007/978-3-319-21668-3_3

13. Christ, J., Hoenicke, J., Nutz, A.: SMTInterpol: an interpolating SMT solver. In: Donaldson, A., Parker, D. (eds.) SPIN 2012. LNCS, vol. 7385, pp. 248–254. Springer, Heidelberg (2012). https://doi.org/10.1007/978-3-642-31759-0_19

14. Cimatti, A., Griggio, A., Schaafsma, B.J., Sebastiani, R.: The MathSAT5 SMT solver. In: Piterman, N., Smolka, S.A. (eds.) TACAS 2013. LNCS, vol. 7795, pp. 93–107. Springer, Heidelberg (2013). https://doi.org/10.1007/978-3-642-36742-7_7

15. de Moura, L., Bjørner, N.: Z3: an efficient SMT solver. In: Ramakrishnan, C.R., Rehof, J. (eds.) TACAS 2008. LNCS, vol. 4963, pp. 337–340. Springer, Heidelberg (2008). https://doi.org/10.1007/978-3-540-78800-3_24

16. Dillig, I., Dillig, T., Aiken, A.: Cuts from proofs: a complete and practical technique for solving linear inequalities over integers. In: Bouajjani, A., Maler, O. (eds.) CAV 2009. LNCS, vol. 5643, pp. 233–247. Springer, Heidelberg (2009). https://doi.org/10.1007/978-3-642-02658-4_20

17. Dutertre, B.: Yices 2.2. In: Biere, A., Bloem, R. (eds.) CAV 2014. LNCS, vol. 8559, pp. 737–744. Springer, Cham (2014). https://doi.org/10.1007/978-3-319-08867-9_49

18. Dutertre, B., de Moura, L.: A fast linear-arithmetic solver for DPLL(T). In: Ball, T., Jones, R.B. (eds.) CAV 2006. LNCS, vol. 4144, pp. 81–94. Springer, Heidelberg (2006). https://doi.org/10.1007/11817963_11

19. Fietzke, A., Weidenbach, C.: Superposition as a decision procedure for timed automata. Math. Comput. Sci. 6(4), 409–425 (2012)

20. Griggio, A.: A practical approach to satisfiability modulo linear integer arithmetic. JSAT 8(1/2), 1–27 (2012)

21. Jovanović, D., de Moura, L.: Cutting to the chase. JAR 51(1), 79–108 (2013)

22. Papadimitriou, C.H.: On the complexity of integer programming. J. ACM 28(4), 765–768 (1981)

23. Schrijver, A.: Theory of Linear and Integer Programming. Wiley, New York (1986)

Cops and CoCoWeb:
Infrastructure for Confluence Tools

Nao Hirokawa[1]([✉])[iD], Julian Nagele[2][iD], and Aart Middeldorp[3][iD]

[1] School of Information Science, JAIST, Nomi, Japan
hirokawa@jaist.ac.jp
[2] School of Electronic Engineering and Computer Science,
Queen Mary University of London, London, UK
j.nagele@qmul.ac.uk
[3] Department of Computer Science, University of Innsbruck, Innsbruck, Austria
aart.middeldorp@uibk.ac.at

Abstract. In this paper we describe the infrastructure supporting confluence tools and competitions: Cops, the confluence problems database, and CoCoWeb, a convenient web interface for tools that participate in the annual confluence competition.

1 Introduction

In recent years several tools have been developed to automatically prove confluence and related properties of a variety of rewrite formats. These tools compete annually in the confluence competition [1] (CoCo).[1] Confluence tools run on confluence problems which are organized in the confluence problems (Cops) database. Cops is managed via a web interface

http://cops.uibk.ac.at/

that comes equipped with a useful tagging system. Cops has recently been revamped and we describe its unique features in this paper.

Most of the tools that participate in CoCo can be downloaded, installed, and run on one's local machine, but this can be a painful process.[2] Only few confluence tools—we are aware of CO3 [8], ConCon [11], and CSI [7,14]—provide a convenient web interface to painlessly test the status of a confluence problem that is provided by the user. In this paper we present CoCoWeb, a web interface to execute confluence tools on confluence problems. This provides a single entry point to all tools that participate in CoCo. CoCoWeb is available at

http://cocoweb.uibk.ac.at/

This research is supported by FWF (Austrian Science Fund) project P27528, JSPS Core to Core Program, and JSPS KAKENHI Grant Nos. 25730004 and 17K00011.

[1] http://coco.nue.ie.niigata-u.ac.jp/.
[2] StarExec provides a VM image with their environment, which can be helpful in case a local setup is essential.

D. Galmiche et al. (Eds.): IJCAR 2018, LNAI 10900, pp. 346–353, 2018.
https://doi.org/10.1007/978-3-319-94205-6_23

The typical use of CoCoWeb is to test whether a given confluence problem is known to be confluent or not. This is useful when preparing or reviewing an article, preparing or correcting exams about term rewriting, and when contemplating submitting a challenging problem to Cops. In particular, CoCoWeb is useful when crafting or looking for examples to illustrate a new technique. For instance, in [4] a rewrite system is presented that can be shown to be confluent with the technique introduced in that paper. The authors write "Note that we have tried to show confluence [...] by confluence checker ACP and Saigawa, and both of them failed." Despite having an easy to use web interface, CSI was not tried. CoCoWeb could also be useful for the CoCo steering committee when integrating newly submitted problems into Cops and also when investigating dubious answers of confluence tools.

The remainder of the paper is organized as follows. In the next section we describe the functionality of Cops, and in Sect. 3 we present the web interface of CoCoWeb by means of a number of screenshots. Both sections contain a few implementation details as well. In Sect. 4 we list some possibilities for extending the functionality of Cops and CoCoWeb in the future.

2   Cops: Confluence Problems Database

Cops is an online database for confluence problems in term rewriting. It was created in 2012 to facilitate the organization of CoCo and development of confluence tools. Via its web interface, everyone can retrieve and download confluence problems, and also upload new problems. Uploaded problems are reviewed by the CoCo steering committee and then integrated into Cops. Figure 1 shows the web interface of Cops. Below, we explain basic features of the interface. The interface is designed in a way that novice users can easily learn problem formats, and at the same time experts on confluence can retrieve a problem set for their experiments.

Problems. While there are several variations of rewrite systems, Cops comprises the following five rewriting formats: ordinary term rewrite systems (TRS), extended term rewrite systems (eTRS) that do not impose the variable conditions of TRS, conditional term rewrite systems (CTRS), higher-order term rewrite systems (HRS), and many-sorted term rewrite systems (MSTRS). In the database, confluence problems are maintained as text files, and identification numbers are assigned to them. Currently, Cops contains 765 systems and more than half of them have been collected from the literature.

The main box in Fig. 1 shows confluence problem number 1 (1.trs), which consists of five rewrite rules. To increase readability, Cops supports syntax highlighting for the above five formats. By clicking the hyperlinked number in brackets in the comment field, the source of the problem with a corresponding BibTeX entry is displayed. Typically the comment field also includes the name of the person(s) who submitted the problem. This is to acknowledge the effort involved in locating interesting problems and making these available to the community.

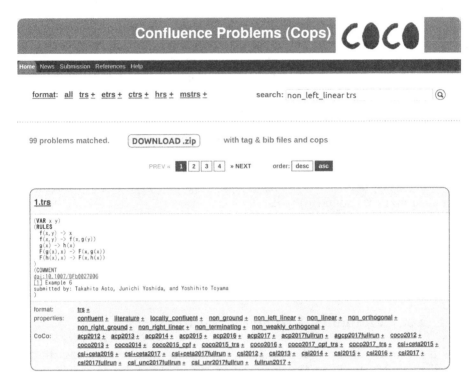

Fig. 1. The web interface of Cops.

Tags. Cops has no directory structure. Instead, *tags*—which can be seen as a multi-dimensional directory structure—are assigned to problems. Different kinds of tags are supported. On the one hand, properties of rewrite systems like left-linearity, groundness, and termination are useful to filter the database for those problems that are supported by a particular tool or technique. These include the tags to distinguish the four different input formats, and they are automatically computed when problems are submitted. A second category of tags refers to tools that could solve the problem (i.e., prove or disprove confluence) in earlier confluence competitions. An example is `acp2017` which is assigned to all problems selected for CoCo 2017 that were solved by ACP [2].

Finally there is the `literature` tag that is assigned to problems that appear in the literature, which includes papers presented at informal workshops like the International Workshop on Confluence and PhD theses. This tag is important since CoCo uses problems from the literature, rather than generated problems that are biased towards one particular tool or technique.

The data of Cops consists of confluence problems and tags. Every tag file is a list of problem numbers in text format. Most of the tag files are generated automatically or updated by a collection of scripts that call external tools. The current collection includes tools to check syntactical properties like left-linearity or right-groundness, ConCon [11] for tags that are specific to CTRSs, and

T⊤T₂ [5] for checking termination and non-termination of TRSs. Since some properties (e.g. termination) are undecidable, tags like **non_terminating** also exist. In addition, Felgenhauer's duplication checker for TRSs (modulo symbol renaming) is included.[3] Duplication is not desirable for fair evaluations. The tag "**duplicate**" is assigned to such a problem.

Queries. Problems can be filtered by typing queries in the search box. Queries are specified by Boolean combinations of tags and problem numbers:

$$\phi ::= tag \mid number \mid \; ! \; \phi \mid \phi \, \phi \mid \phi \; \texttt{OR} \; \phi \mid \{\phi\}$$

Conjunction is denoted by juxtaposition and negation by an exclamation mark. For instance, the query "**left_linear trs**" yields all problems with the two tags **left_linear** and **trs**. In order to search for non-left-linear TRSs whose termination is not known "**!{left_linear OR terminating} trs**" is used. This functionality is also useful for comparing tools. The query "**csi2017 !acp2017**" shows all confluence problems that were solved by CSI but not by ACP in CoCo 2017. It is worth noting that advisory board members of CoCo exploit the tag-based queries (besides random seeds) to compile problem sets used for the live competitions of CoCo. Problems resulting from search queries can be downloaded as a zip file. Optionally, tag files and BibTeX files are included too.

The search engine of Cops consists of only 235 lines of Ruby code. This is implemented as a command-line tool and bundled with a problem set when the aforementioned download option is selected. The tool name is **cops** and one can run it in a Unix environment. For example, the command

```
./cops 'oriented deterministic 3_ctrs'
```

outputs all problem numbers of oriented deterministic 3-CTRSs in the downloaded problem set. The web interface is built on it. The corresponding code is about 5,000 lines of PHP, Ruby, and JavaScript code. Syntax highlighting in the submission page has been implemented on the top of CodeMirror.[4] Finally, BibTeX2HTML[5] is used for generating HTML for the references.

3   CoCoWeb: Web Interface for Confluence Tools

CoCoWeb is a web service to access confluence tools in a web browser. Figure 2 shows a screenshot of the entry page of CoCoWeb. Problems can be entered in three different ways:

1. using the text box,
2. uploading a file,
3. entering the number of a problem in Cops.

---

[3] https://github.com/haskell-rewriting/canonical-trs.

[4] https://codemirror.net/.

[5] https://www.lri.fr/~filliatr/bibtex2html/.

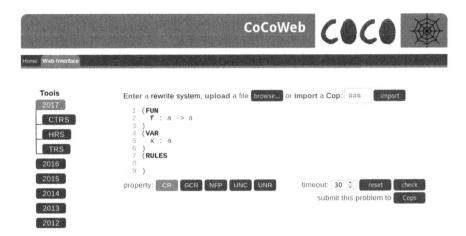

Fig. 2. The entry page of CoCoWeb.

The problem can be submitted to Cops via the submit button. The tools that should be executed can be selected from the tools panel on the left. Tools are grouped into categories, similar to the grouping in CoCo. Multiple tools can be selected. This is illustrated in Fig. 3. Here we selected CR as property, the CoCo 2016 and 2015 versions of ACPH [9] and the CoCo 2015 version of CSI^ho [6], and Cop 500 is chosen is input problem.

The final screenshot (in Fig. 4) shows the output of CoCoWeb after clicking the check button. The output of the selected tools is presented in separate tabs. The colors of these tabs reveal useful information: Green means that the tool answered yes, red (not shown) means that the tool answered no, and a maybe answer or a timeout is shown in blue. By clicking on a tab, the color is made lighter and the output of the tool is presented. The final line of the output is timing information provided by CoCoWeb.

The tools in CoCoWeb run on hardware compatible with a single node of StarExec [12] that is used for CoCo, allowing for a proper comparison of tools. By specializing the service to confluence, CoCoWeb offers easy access to all tools that participated in CoCo without requiring users to register first, and immediate scheduling of executions as well as syntax highlighting.

Most of CoCoWeb is built using PHP. User input in forms, i.e., rewrite systems and tool selections, is sent using the HTTP POST method. The dynamic parts of the website, namely folding and unfolding in the tool selection menu and the tabs used for viewing tool output are implemented using JavaScript. To layout the tool selection menu we made extensive use of CSS3 selectors. For instance, the buttons to select tools are implemented as checkboxes with labels that are styled according to whether the checkbox is ticked or not:

```
.tools input[type="checkbox"]:checked + label
     {  color: white;  background-color: #799BB3;  }
```

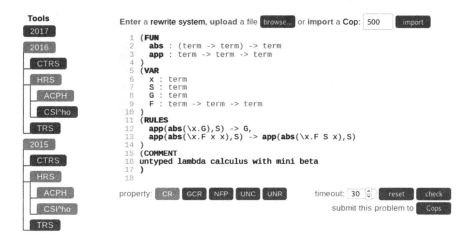

Fig. 3. Problem and tool selection in CoCoWeb.

Drawing the edges of the tree menu is also done using CSS, relying mainly on its `::before` selector.

Since its second edition, CoCo has adopted StarExec as competition platform. Competition participants upload binary executables of their tools together with necessary files to StarExec. Importing and complementing them with missing software, we reproduce the competition versions of tools on the CoCoWeb server. The collection of tools is maintained and associated with the web interface with help of small scripts. The content of the tool menu, i.e., years, the grouping by categories, and the actual tools, is generated automatically from a directory tree that has the structure of the menu in CoCoWeb. The directories contain small configuration files that specify how the tools are to be run, in case they are selected. Two environment variables are set in such a file, for example the one for the 2012 version of Saigawa [3] reads as follows:

```
TOOLDIR="Saigawa-2012/bin"
TOOL="./starexec_run_saigawa -t $TO $FILE"
```

The variable `TOOLDIR` specifies the directory that contains the tool binary, while `TOOL` gives the tool invocation, which in turn refers to `TO`, the timeout, and `FILE`, the input rewrite system. Using such configuration files tools are run using the following script, whose first, second, and third argument are the configuration of the tool, input rewrite system, and timeout respectively:

```
DIR=$(pwd -P)
FILE=$(readlink -f $2)
TO=$3;   TOT=$((TO + 2));   TOK=$((TOT + 2))
source $1
pushd $DIR/bin/$TOOLDIR > /dev/null
/usr/bin/time -f "\\nTook %es" timeout -k $TOK $TOT $TOOL
popd > /dev/null
```

Fig. 4. Testing Example 14 from [4] in CoCoWeb. (Color figure online)

The script uses three different timeouts: `TO` is the timeout passed to the tool itself if supported, while after `TOT` and `TOK` the signals `SIGTERM` and `SIGKILL` are sent to the tool, using GNU coreutils timeout, in case it did not terminate on its own.[6] When multiple tools are selected, CoCoWeb runs them sequentially, in order to avoid interference.

4   Conclusion

In this paper we introduced Cops and CoCoWeb, two convenient systems that provide support for researchers that are interested in (developing tools for) confluence and related properties of rewrite systems. We believe the developed infrastructure could be useful for other competitions besides CoCo.

Both systems can be extended in several ways, which we plan to address in future work. For Cops, we are mainly concerned with two issues. One is about the reorganization of tags. Every year CoCo extends its scope to capture emerging trends, causing some tags to be redefined or renamed. Another is about reproducibility of search queries, which is crucial as Cops has been used as a standard benchmark for confluence techniques. To address these issues, we are seeking for a way to support versioning the database.

For CoCoWeb, preselection of tools based on the input problem would be a nice feature. This is not as trivial as it sounds, since different properties share the same problem format. We plan to investigate the selection method for ATP systems [13]. Supporting pretty-printing for XML output is another future task.

---

[6] To account for timing imprecisions and tools performing internal cleanup, 2 extra seconds are granted to the tool before sending `SIGTERM` and another 2 before `SIGKILL` is sent, which turned out to work well in practice.

While several tools support output of (non-)confluence proofs in the Certification Problem Format [10], the current web interface just displays the raw XML code.

Acknowledgments. We thank Harald Zankl, Christian Nemeth, and Takahito Aoto for their involvement in CoCo and the first release of Cops. Suggestions by the former helped to improve the paper.

References

1. Aoto, T., Hirokawa, N., Nagele, J., Nishida, N., Zankl, H.: Confluence competition 2015. In: Felty, A.P., Middeldorp, A. (eds.) CADE 2015. LNCS (LNAI), vol. 9195, pp. 101–104. Springer, Cham (2015). https://doi.org/10.1007/978-3-319-21401-6_5
2. Aoto, T., Yoshida, J., Toyama, Y.: Proving confluence of term rewriting systems automatically. In: Treinen, R. (ed.) RTA 2009. LNCS, vol. 5595, pp. 93–102. Springer, Heidelberg (2009). https://doi.org/10.1007/978-3-642-02348-4_7
3. Hirokawa, N., Klein, D.: Saigawa: a confluence tool. In: Proceedings of 1st IWC, p. 49 (2012). http://cl-informatik.uibk.ac.at/iwc/iwc2012.pdf
4. Ishizuki, S., Oyamaguchi, M., Sakai, M.: Conditions for confluence of innermost terminating term rewriting systems. In: Proceedings of 5th IWC, pp. 70–74 (2016). http://cl-informatik.uibk.ac.at/iwc/iwc2016.pdf
5. Korp, M., Sternagel, C., Zankl, H., Middeldorp, A.: Tyrolean Termination Tool 2. In: Treinen, R. (ed.) RTA 2009. LNCS, vol. 5595, pp. 295–304. Springer, Heidelberg (2009). https://doi.org/10.1007/978-3-642-02348-4_21
6. Nagele, J.: CoCo 2015 participant: CSI^ho 0.1. In: Proceedings of 4th IWC, p. 47 (2015). http://cl-informatik.uibk.ac.at/iwc/iwc2015.pdf
7. Nagele, J., Felgenhauer, B., Middeldorp, A.: CSI: new evidence – a progress report. In: de Moura, L. (ed.) CADE 2017. LNCS (LNAI), vol. 10395, pp. 385–397. Springer, Cham (2017). https://doi.org/10.1007/978-3-319-63046-5_24
8. Nishida, N., Kuroda, T., Yanagisawa, M., Gmeiner, K.: CO3: a COnverter for proving COnfluence of COnditional TRSs (version 1.2). In: Proceedings of 4th IWC, p. 42 (2015). http://cl-informatik.uibk.ac.at/iwc/iwc2015.pdf
9. Onozawa, K., Kikuchi, K., Aoto, T., Toyama, Y.: ACPH: system description for CoCo 2016. In: Proceedings of 5th IWC, p. 76 (2016). http://cl-informatik.uibk.ac.at/iwc/iwc2016.pdf
10. Sternagel, C., Thiemann, R.: The certification problem format. In: Proceedings of 11th UITP, EPTCS, vol. 167, pp. 61–72 (2014). https://doi.org/10.4204/EPTCS.167.8
11. Sternagel, T., Middeldorp, A.: Conditional confluence (system description). In: Dowek, G. (ed.) RTA 2014, TLCA 2014. LNCS, vol. 8560, pp. 456–465. Springer, Cham (2014). https://doi.org/10.1007/978-3-319-08918-8_31
12. Stump, A., Sutcliffe, G., Tinelli, C.: StarExec: a cross-community infrastructure for logic solving. In: Demri, S., Kapur, D., Weidenbach, C. (eds.) IJCAR 2014. LNCS (LNAI), vol. 8562, pp. 367–373. Springer, Cham (2014). https://doi.org/10.1007/978-3-319-08587-6_28
13. Sutcliffe, G., Suttner, C.: Evaluating general purpose automated theorem proving systems. Artif. Intell. **131**(1), 39–54 (2001). https://doi.org/10.1016/S0004-3702(01)00113-8
14. Zankl, H., Felgenhauer, B., Middeldorp, A.: CSI – a confluence tool. In: Bjørner, N., Sofronie-Stokkermans, V. (eds.) CADE 2011. LNCS (LNAI), vol. 6803, pp. 499–505. Springer, Heidelberg (2011). https://doi.org/10.1007/978-3-642-22438-6_38

Investigating the Existence of Large Sets of Idempotent Quasigroups via Satisfiability Testing

Pei Huang[1,3], Feifei Ma[1,2,3(✉)], Cunjing Ge[1,3], Jian Zhang[1,3(✉)], and Hantao Zhang[4]

[1] State Key Laboratory of Computer Science, Institute of Software, Chinese Academy of Sciences, Beijing, China
{huangpei,maff,gecj,zj}@ios.ac.cn
[2] Laboratory of Parallel Software and Computational Science, Institute of Software, Chinese Academy of Sciences, Beijing, China
[3] University of Chinese Academy of Sciences, Beijing, China
[4] Department of Computer Science, The University of Iowa, Iowa City, IA 52242, USA
hantao-zhang@uiowa.edu

Abstract. In this paper, we describe a method for solving some open problems in design theory based on SAT solvers. Modern SAT solvers are efficient and can produce unsatisfiability proofs. However, the state-of-the-art SAT solvers cannot solve the so-called *large set* problem of idempotent quasigroups. Two idempotent quasigroups over the same set of elements are said to be *disjoint* if at any position other than the main diagonal, the two elements from the two idempotent quasigroups at the same position are different. A collection of $n-2$ idempotent quasigroups of order n is called a *large set* if all idempotent quasigroups are mutually disjoint, denoted by $LIQ(n)$. The existence of $LIQ(n)$ satisfying certain identities has been a challenge for modern SAT solvers even if $n = 9$. We will use a finite-model generator to help the SAT solver avoiding symmetric search spaces, and take advantages of both first order logic and the SAT techniques. Furthermore, we use an incremental search strategy to find a maximum number of disjoint idempotent quasigroups, thus deciding the non-existence of large sets. The experimental results show that our method is highly efficient. The use of symmetry breaking is crucial to allow us to solve some instances in reasonable time.

1 Introduction

In recent decades, automated reasoning tools have been applied to some combinatorial problems which are difficult for conventional mathematical methods. For example, Heule et al. solved the boolean pythagorean triples problem via a parallelized SAT solver with 800 cores in about 2 days [10]. Generally, these combinatorial problems are hard to solve. The quasigroup problem is among these

© Springer International Publishing AG, part of Springer Nature 2018
D. Galmiche et al. (Eds.): IJCAR 2018, LNAI 10900, pp. 354–369, 2018.
https://doi.org/10.1007/978-3-319-94205-6_24

problems and it has attracted much focus by researchers in the field of combinatorics and automated reasoning. The improvement of automated reasoning techniques made computer search play an important role in the study of quasigroups [24]. For example, many open problems of the type from QG2 to QG9 have been solved by some finite-model generators such as *MGTP, FINDER, SEM, MACE4* and propositional satisfiability provers *SATO, DDPP*, respectively [6, 19, 23, 25, 27].

The large set problem, which seeks to find a set of combinatorial objects rather than one, is a classic and challenging research topic in combinatorial design theory. Sylvester first proposed the existence of large sets of Kirkman triple systems in the 1850s [3]. In the late last century, the large set of idempotent quasigroups were proposed. Due to its difficulty in construction, the research progress is quite slow in the mathematics field. So, any progress of the large set is something expected [2, 13, 22].

Investigating the existence of large sets of moderate order via computer can provide support for mathematicians to further explore general issues. Sometimes they can use mathematical construction to produce large objects from smaller ones recursively. Besides, many hard combinatorial problems related to quasigroups also have potential value in the field of cryptography [7, 12].

A collection of $n - 2$ idempotent quasigroups (IQs) of order n is called a *large set* if any two of them are disjoint, denoted by $LIQ(n)$. Ma et al. applied *SEM* [27] to some open cases about large sets of idempotent quasigroups with certain identities summarized by Zhu [28, 29], and solved $LIQ(n)$ ($n \leq 8$) [14].

In this paper, we attempt to further study the open cases of $LIQ(n)$ for $n \geq 9$. Unlike $LIQ(n)$ ($n \leq 8$), it is difficult to solve these cases via encoding them directly as SAT, SMT, CSP or first order logic formulae, even though $n = 9$ is just one more than $n = 8$. We tested direct encoding ways and used SAT solvers like *MiniSat, Glucose, Treengeling* (1st in SAT 2016 competition in parallel track), SMT solver like *Z3* [5], CSP solver like *Minizinc* [17] and finite-model generators like *SEM, SEMD, MACE4* [15] and they all failed to produce a result in a week for many instances. However, $LIQ(n)$ ($n \leq 8$) can be solved in seconds. These challenging problems, $LIQ(n)$ ($n \geq 9$), promote us to seek for more powerful search strategies.

There have been much progress in SAT solving during the past 20 years; and the state-of-the-art SAT solvers can make very efficient low-level inferences. They have become the core search engine in many tools used for combinational [8, 16] and sequential equivalence checking [1, 11]. Apart from high efficiency, the state-of-the-art SAT solvers can make a claim that the formula is unsatisfiable with a *formal proof*. One can verify the proof emitted by a SAT solver and ensure the result is correct [9, 21]. Yet when other kinds of solvers claim that a formula is unsatisfiable, one has to trust that the solver fully exhausted the search space for the problem. However, SAT solvers are weak in dynamic symmetry breaking, hence may revisit a lot of redundant search space. Usually, when a problem is encoded as Boolean formulae, its structural characteristics may be hidden.

On the other hand, when a problem is encoded as first order logic formulae, the structural information and some properties are well preserved. The finite-model generator can make use of this information to break symmetries. Finite-model generators such as *SEM, SEMD* and *MACE4* are good at dynamic symmetry breaking and enumerating all solutions. A benefit of using SEM-style finite-model generators is that the search process can exploit high-level structural information in the formulas (e.g., symmetries) to reduce the search space. The core dynamic symmetry breaking method is a heuristic called least number heuristic (LNH) [27]. The basic idea is that many element names, which have not yet been used in the search, are essentially the same. Furthermore, *MACE4* can eliminate isomorphic solutions statically.

A question naturally arises: would it be helpful to combine these two paradigms? In 2001, Zhang proposed to combine automatic symmetry breaking with SAT [26]. The experimental results in [26] showed the advantage of this method. We employ a similar search strategy in solving the *LIQ* problem: Use the first order model generator to generate some asymmetric partial potential solutions; with the help of these partial solutions as candidates, a SAT solver can avoid a lot of symmetric search spaces. This simple combination can take advantages of their respective strengths. In addition to that, we also statically add some symmetry breaking constraints. Adding constraints to the basic model has been most used historically by constraint programmers [18]. The experimental results show that this combination can greatly improve the solving efficiency. We found some instances, which cannot be solved with a single solver in a week before, can be solved in minutes now. Due to these strategies, a number of open cases of $LIQ(n)$ have been solved. We not only establish the non-existence of these cases, but also find the maximum number of disjoint $IQ(n)$s, and some other interesting mathematical results.

This paper is organized as follows: in Sect. 2, we introduce some preliminaries about *LIQ*; In Sects. 3 and 4, we present the encoding method and how to break symmetry and speed up the search process; In Sects. 5 and 6, we present the results about *LIQ* and the experiments. Furthermore, we evaluate and discuss the experimental results; In the final section, conclusions are drawn.

2   Preliminaries

2.1   Definitions

Let us recall some notations.

A *quasigroup* is denoted as an ordered pair $(Q, \oplus)$, where Q is a set and $\oplus$ is a binary operation on Q. For all constants $a, b \in Q$ and variables $x, y \in Q$ equations $a \oplus x = b$ and $y \oplus a = b$ are uniquely solvable. $|Q|$ is said be the *order* of $(Q, \oplus)$.

It is well-known that the multiplication table of quasigroup $(Q, \oplus)$ is a Latin square. Thus, Latin square and quasigroup are often treated as synonyms. Figure 1 shows the multiplication table of a quasigroup $(Q, \oplus)$ where $Q = \{0, 1, 2, 3\}$.

```
L |  0  1  2  3
----+------------
0 |  0  2  3  1
1 |  3  1  0  2
2 |  1  3  2  0
3 |  2  0  1  3
```

Fig. 1. A quasigroup of order 4

For all $x \in Q$, if $x \oplus x = x$ (briefly $x^2 = x$), the quasigroup $(Q, \oplus)$ is *idempotent*. We denote an idempotent quasigroup of order n as $IQ(n)$.

Two quasigroups $(Q, \oplus)$ and $(Q, \cdot)$ are said to be *disjoint* if for all $x, y \in Q$, $x \oplus y \neq x \cdot y$ whenever $x \neq y$.

Definition 1 (Large Set). *A collection of idempotent quasigroups* $(Q, \oplus_1)$, $(Q, \oplus_2)$, ..., $(Q, \oplus_{n-2})$, *where* $n = |Q|$, *is called a* large set, *if any two of the idempotent quasigroups are disjoint.*

A large set of idempotent quasigroups of order n is denoted by $LIQ(n)$. Figure 2 shows a large set of idempotent quasigroups of order 4, i.e., $LIQ(4)$, which consists of two disjoint $IQ(4)$s.

```
L1 |  0  1  2  3          L2 |  0  1  2  3
----+------------         ----+------------
0 |  0  2  3  1           0 |  0  3  1  2
1 |  3  1  0  2           1 |  2  1  3  0
2 |  1  3  2  0           2 |  3  0  2  1
3 |  2  0  1  3           3 |  1  2  0  3
```

Fig. 2. Two disjoint $IQ(4)$s in $LIQ(4)$

Besides the existence of $LIQ(n)$, the maximum number of disjoint $IQ(n)$s for non-existent instances is also concerned.

Definition 2 (Orthogonal). *Two quasigroups* $(Q, \oplus)$ *and* $(Q, \cdot)$ *are said to be* orthogonal, *if for all* $x_1, x_2, y_1, y_2 \in Q$, *the ordered pair* $(x_1 \oplus x_2, y_1 \cdot y_2)$ *is unique.*

$L1$ and $L2$ in Fig. 2 are also *orthogonal*. Ordered pairs are shown in Fig. 3 and every ordered pair appears only once.

A LIQ is said to have orthogonality, if any two quasigroups in the LIQ are orthogonal. In general, for idempotent quasigroups, orthogonality implies disjointness, but the reverse does not hold.

```
(L1,L2) |   0      1      2      3
----+--------------------------
   0 | (0,0)  (2,3)  (3,1)  (1,2)
   1 | (3,2)  (1,1)  (0,3)  (2,0)
   2 | (1,3)  (3,0)  (2,2)  (0,1)
   3 | (2,1)  (0,2)  (1,0)  (3,3)
```

Fig. 3. The ordered pair of L1 and L2

2.2   The Problems

Teirlinck and Lindner [20], and Chang [4] have already established the existence of $LIQ(n)$. In [4], Chang concluded that there exists an $LIQ(n)$ for any $n \geq 3$ with the exception $n = 6$. The spectrums of some large sets in which the IQs satisfy certain identities have not been explored extensively up to now. The existence of idempotent quasigroups satisfying the seven "short identities" has been studied systematically. These identities are:

1. $xy \oplus yx = x$ Schröder quasigroup
2. $yx \oplus xy = x$ Stein's third law
3. $(xy \oplus y)y = x$ C_3-quasigroup
4. $x \oplus xy = yx$ Stein's first law; Stein quasigroup
5. $(yx \oplus y)y = x$
6. $yx \oplus y = x \oplus yx$ Stein's second law
7. $xy \oplus y = x \oplus xy$ Schröder's first law

In the above equations, xy is an abbreviation of $(x \oplus y)$. That means xy has higher precedence than $x \oplus y$. Let $LIQ^{(i)}(n)$ denote the large set of idempotent quasigroups of order n satisfying identity (i). The existence of $LIQ^{(i)}(n)$ is still an open problem. In [28], Zhu listed several open cases. Since the search spaces of the problem grow exponentially with order n, we pick out some open cases of moderate orders which may be suitable for computer search. In Table 1, we list these open cases where $n \leq 13$.

Table 1. Open cases for LIQ of moderate sizes

1.	$LIQ^{(1)}(12)$	$LIQ^{(1)}(13)$	2.	$LIQ^{(2)}(9)$	$LIQ^{(2)}(12)$
3.	$LIQ^{(3)}(10)$	$LIQ^{(3)}(13)$	4.	$LIQ^{(4)}(9)$	$LIQ^{(4)}(11)$
5.	$LIQ^{(5)}(11)$		6.	$LIQ^{(6)}(9)$	$LIQ^{(6)}(13)$
7.	$LIQ^{(7)}(9)$	$LIQ^{(7)}(13)$			

In [14], Ma et al. studied $LIQ^{(i)}(n)$s with n no more than 8 in Table 1. They modeled $LIQ^{(i)}(n)$ via first order formulae and used the finite-model generator *SEM*. However, using the direct method is impracticable for $LIQ^{(i)}(n)$s where n is more than 8. So $LIQ^{(i)}(n)$s, where $n \geq 9$, are still open cases.

The search space grows exponentially with the size of the problem. In practice, we encoded $LIQ^{(i)}(n)$s, where $n \geq 9$, as a SAT instance, using uninterpreted functions and first order formulae in a naive way. However, we could not get any result via SAT solvers like *MiniSat, glucose, Treengeling*, the SMT solver like *Z3*, the CSP solver like *Minizinc* and finite-model generators like *SEM, MACE4* in a week for some instances with $n = 9$. However, for $LIQ(n)$s ($n \leq 8$), results could be obtained in a few seconds to minutes.

3   Encoding

In the introduction we have expounded the reason why we choose SAT as core engine. First, a notation $ExactOne(x_1, x_2, ..., x_n)$ will be introduced. $x_1, x_2, ..., x_n$ are Boolean variables and $ExactOne(x_1, x_2, ..., x_n)$ is a formula composed of $x_1, x_2, ..., x_n$. $ExactOne(x_1, x_2, ..., x_n)$ expresses the fact that exactly one of these Boolean variables is true for any satisfying assignment to this formula.

$$ExactOne(x_1, x_2, ..., x_n) = (x_1 \lor x_2 \lor, ..., \lor x_n) \land \underbrace{(\overline{x_1} \lor \overline{x_2}) \land (\overline{x_1} \lor \overline{x_3}), ..., \land (\overline{x_{n-1}} \lor \overline{x_n})}_{\binom{n}{2}}$$

Without loss of generality, we assume the domain Q to be the set $\{0, 1, \ldots, n-1\}$. $\oplus_f$ is actually a function $L_f : Q \times Q \mapsto Q$ satisfying the constraints of idempotent quasigroup and identity (i). $LIQ(n) = \{L_1, L_2, \ldots, L_{n-2}\}$, where $n = |Q|$, denotes a large set. $L_f(x, y)$ denotes $x \oplus_f y$ in quasigroup L_f. In Sect. 2, we mentioned that a quasigroup can be seen as a multiplication table (or a matrix called **Latin square**). Then we can use Boolean variable $P_{f,x,y,v}$ to denote that the position(x, y) of f_{th} IQ in the large set is v.

According to the definition of **quasigroup**, it is easy to know that each element in Q occurs exactly once in each row and exactly once in each column in the matrix. So for each row x of f_{th} IQ, we add formula:

$$\bigwedge_{0 \leq v \leq n-1} ExactOne(P_{f,x,0,v}, P_{f,x,1,v}, ..., P_{f,x,n-1,v})$$

For each column y of f_{th} IQ, we add formula:

$$\bigwedge_{0 \leq v \leq n-1} ExactOne(P_{f,0,y,v}, P_{f,1,y,v}, ..., P_{f,n-1,y,v})$$

For each cell (x, y) of f_{th} IQ, we add formula:

$$P_{f,x,y,1} \lor P_{f,x,y,2} \lor, ..., \lor P_{f,x,y,n-1}$$

For f_{th} IQ in the large set, The idempotent property $(x^2 = x)$ can be encoded as:

$$\bigwedge_{0 \leq x \leq n-1} P_{f,x,x,x}$$

The disjoint property depicts that for any two latin squares L_j and L_k ($1 \leq j, k \leq n-2$), $L_j(x,y) \neq L_k(x,y)$ except for $x = y$. So the encoding for m disjoint latin squares is:

$$\bigwedge_{1 \leq j < k \leq m} \bigwedge_{x \neq y} \bigwedge_{0 \leq v \leq n-1} \overline{P_{j,x,y,v}} \vee \overline{P_{k,x,y,v}}$$

The encoding for the seven identities:

(1) For $(xy \oplus yx = x)$:

$$\bigwedge_{0 \leq x,y,v_1,v_2 \leq n-1} \overline{P_{f,x,y,v_1}} \vee \overline{P_{f,y,x,v_2}} \vee P_{f,v_1,v_2,x}$$

(2) For $(yx \oplus xy = x)$:

$$\bigwedge_{0 \leq x,y,v_1,v_2 \leq n-1} \overline{P_{f,y,x,v_1}} \vee \overline{P_{f,x,y,v_2}} \vee P_{f,v_1,v_2,x}$$

(3) For $(xy \oplus y)y = x$:

$$\bigwedge_{0 \leq x,y,v_1,v_2 \leq n-1} \overline{P_{f,x,y,v_1}} \vee \overline{P_{f,v_1,y,v_2}} \vee P_{f,v_2,y,x}$$

(4) For $(x \oplus xy = yx)$:

$$\bigwedge_{0 \leq x,y,v_1,v_2 \leq n-1} \overline{P_{f,x,y,v_1}} \vee \overline{P_{f,y,x,v_2}} \vee P_{f,x,v_1,v_2}$$

(5) For $(yx \oplus y)y = x$:

$$\bigwedge_{0 \leq x,y,v_1,v_2 \leq n-1} \overline{P_{f,y,x,v_1}} \vee \overline{P_{f,v_1,y,v_2}} \vee P_{f,v_2,y,x}$$

(6) For $yx \oplus y = x \oplus yx$:

$$\bigwedge_{0 \leq x,y,v_1,v_2 \leq n-1} \overline{P_{f,y,x,v_1}} \vee \overline{P_{f,v_1,y,v_2}} \vee P_{f,x,v_1,v_2}$$

(7) For $xy \oplus y = x \oplus xy$:

$$\bigwedge_{0 \leq x,y,v_1,v_2 \leq n-1} \overline{P_{f,x,y,v_1}} \vee \overline{P_{f,v_1,y,v_2}} \vee P_{f,x,v_1,v_2}$$

Actually, there is a lot of redundancy in the encoding. For example, $P_{f,x,x,x}$ must be assigned true and $P_{f,x,x,v}(v \neq x)$ must be assigned false. So there are many redundant clauses like $\overline{P_{f,1,1,2}} \vee \overline{P_{f,1,1,3}} \vee P_{f,2,3,1}$. These clauses can be eliminated in the encoding phase. Besides redundant clauses, many clauses can be simplified. For instance, known $P_{f,0,1,2}$(we will explain it in the next section), $\overline{P_{f,0,1,2}} \vee \overline{P_{f,1,0,v_2}} \vee P_{f,2,v_2,0}$ can be simplified as $\overline{P_{f,1,0,v_2}} \vee P_{f,2,v_2,0}$. Although the state-of-the-art SAT solvers can simplify these during the preprocessing phase, it is better to remove them in the encoding phase.

4   Search Strategies

Arguably, many hard combinatorial problems allow isomorphic solutions, and we say these problems have symmetries. The search may revisit equivalent states over and over again. Exploiting symmetry can reduce the search time to solve the problem. It is common to identify three main approaches to break symmetry. The first method is to reformulate the problem so it has a reduced number of symmetries. The second is to add symmetry breaking constraints before search starts, thereby making some symmetric solutions unacceptable while leaving at least one solution in each symmetric equivalence class. The final approach is to break symmetry dynamically during search, adapting the search procedure appropriately. Although symmetry breaking technique and automatic identification of the symmetry have been concerned by researchers in the past, state-of-the-art SAT solvers do not have the ability to identify and break symmetry automatically. So, it is vital for us to handle the symmetries of the problem at hand.

4.1 Symmetry Breaking via Adding Constraints

First, we examine the structure of the problem to identify symmetries.

Lemma 1. *If there is a large set $LIQ(n) = \{L_1, L_2, \ldots, L_{n-2}\}$, then there exists a large set such that $L_j(0, 1) = j + 1$.*

Proof. According to $x^2 = x$, we know that $L_j(0, 0) = 0$ and $L_j(1, 1) = 1$. Since $a \oplus_j x = b$ and $y \oplus_j a = b$ are uniquely solvable, $L_i(0, 1)$ can not be 0 or 1. All candidates for it include $2, 3, \ldots, n - 1(n - 2$ elements). Due to the property of *disjoint*, for any $1 \le j_1, j_2 \le n - 2$, $L_{j_1}(0, 1) \ne L_{j_2}(0, 1)$ but the cardinality of collection of $L_j(0, 1)$ is $n - 2$. So they must be exactly $\{2, 3, \ldots n - 1\}$.   □

Obviously, if $\{L_1, L_2, \ldots, L_{n-2}\}$ is a large set then any permutation of L_1, L_2, ..., L_{n-2} is also a large set. The permutation may not affect search time when a large set exists. However, when the large set does not exist, the solver will nearly enumerate all permutations and conclude that it is unsatisfiable. In other words, $\forall f_1, f_2 \in \{1, 2, \ldots n - 2\}$ there is no essential difference between Boolean variable $P_{f_1,x,y,v}$ and $P_{f_2,x,y,v}$.

We can take advantage of *Lemma* 1 to fix the order of $\{L_1, L_2, \ldots, L_{n-2}\}$. Since $L_j(0, 1)$ must be exactly $\{2, 3, \ldots n - 1\}$, we specify its order by assigning

$$L_1(0, 1) = 2 < L_2(0, 1) = 3 < \ldots < L_{n-2}(0, 1) = n - 1$$

So, by *Lemma* 1, we can add the following unit clauses:

$$\bigwedge_{1 \le f \le n-2} P_{f,0,1,f+1}$$

then Boolean variable $P_{f_1,x,y,v}$ and $P_{f_2,x,y,v}$ are different in a way and $(n-3)! - 1$ isomorphic cases eliminated.

Now that the sequence of $IQ(n)$s is fixed, we use $DIQ^{(i)}_{(n)}(l)$ to denote the first l disjoint $IQ(s)$s satisfying identity (i).

$$DIQ^{(i)}_{(n)}(l) = \{L_1, L_2, \ldots, L_l \mid L_j(0, 1) = j + 1\}(1 \le l \le n - 2)$$

Proposition 1. *If $DIQ^{(i)}_{(n)}(l)$ does not exist and $DIQ^{(i)}_{(n)}(l - 1)$ exists, then the maximum number of disjoint $IQ^{(i)}(n)$s is $l - 1$.*

Proof. Suppose there exist m $(m \geq l)$ disjoint IQs when $DIQ^{(i)}_{(n)}(l)$ does not exist and $DIQ^{(i)}_{(n)}(l-1)$ exists. Then we know that l disjoint IQs also exist. We assume they are $\{L_{j_1}, L_{j_2}, ..., L_{j_l}\}$ where $j_1 < j_2 <, ..., < j_l$. According to Lemma 1 we know $L_{j_1}(0,1) = j_1 + 1$, $L_{j_2}(0,1) = j_2 + 1, ..., L_{j_l}(0,1) = j_l + 1$. We can define a permutation in Cauchy form:

$$\sigma : \begin{pmatrix} 0 & 1 & 2 & 3 & ... & l & l+1 ... n-1 \\ 0 & 1 & j_1+1 & j_2+1 & ... & j_l+1 & * & ... & * \end{pmatrix}$$

The '*' can be any legitimate number. We can perform σ^{-1} on $\{L_{j_1}, L_{j_2}, ..., L_{j_l}\}$. It is easy to know that all constraints still hold under permutation σ^{-1} and $\{L_{j_1}, L_{j_2}, ..., L_{j_l}\}^{\sigma^{-1}} = \{L_1, L_2, ..., L_l\} = DIQ^{(i)}_{(n)}(l)$. Thus $DIQ^{(i)}_{(n)}(l)$ must exist however it contradicts with the known conditions. So the assumptions can not be true.   □

Proposition 1 reveals that any l disjoint $IQ^{(i)}(n)$s must be isomorphic to some $DIQ^{(i)}_{(n)}(l)$. When $l = n - 2$, $DIQ^{(i)}_{(n)}(l)$ is exactly $LIQ^{(i)}(n)$. So we can search for $DIQ^{(i)}_{(n)}(l)$s successively, increasing l step by step. Once for some l $DIQ^{(i)}_{(n)}(l)$ is unsatisfiable we can conclude that $LIQ^{(i)}(n)$ doesn't exist and the maximum number of disjoint IQs is $l - 1$. The reason why $DIQ^{(i)}_{(n)}(l)$s are searched by increasing l instead of decreasing l is that it usually takes much more time to solve an unsatisfiable case than a satisfiable one.

The framework of incremental search for $LIQ^{(i)}(n)$ is shown in Algorithm 1. At each step, $DIQ^{(i)}(n)(l)$ is encoded as a set of Boolean formulas, denoted by $Encode(DIQ^{(i)}_{(n)}(l))$, and solved by a SAT solver.

Algorithm 1. Incremental search for $LIQ^{(i)}(n)$

Input: n is order, i is identity (i)
Output: The existence of LIQ and the maximum number of disjoint IQs
1 **for** $l \leftarrow 2$ **to** $n - 2$ **do**
2 | $result \leftarrow$ Solve $Encode(DIQ^{(i)}_{(n)}(l))$;
3 | **if** $result$ is $UNSAT$ **then**
4 | | **return** $NONEXISTENT$ AND $max = l - 1$;
5 | **end**
6 **end**
7 **return** $EXISTENT$

4.2   Combine Solvers to Break Symmetry

Due to adding constraints, some apparent symmetries have been eliminated but there still remains a lot. Although the state-of-the-art SAT solvers are very efficient for constraint propagation and conflict analysis, they are not good at breaking symmetry dynamically during search. Thus, we can use other solvers, which can break symmetry dynamically during search, to enumerate a small subspace and then add these candidate partial solutions to the clauses set to help a SAT solver to eliminate a lot of symmetric states.

It is easy to know that if there is a large set $\{L_1', L_2', ..., L_{n-2}'\}$ then there exists an isomorphic $DIQ_{(n)}^{(i)}(n-2) = \{L_1, L_2, ..., L_{n-2}\}$ where $L_1(0,1) = L_1'(0,1) = 2$. In Sect. 4.1, the search process can be seen as expanding L_1 to L_{n-2} step by step. So, if we find all non-isomorphic L_1s as a candidate set and test whether they can be expanded to a large set, then a lot of isomorphic search spaces can be avoided. In this case, all non-isomorphic L_1s can be seen as candidate partial solutions. So, we use the first order logic solver to generate all non-isomorphic L_1s. $C^{(i)}(n)$ is used to denote the candidate set formed by all non-isomorphic L_1s which satisfy identity (i) of order n. We use $\Sigma_{C^{(i)}(n)}$ to denote the first order logic formula encoding.

$$\Sigma_{C^{(i)}(n)} = \{\forall x \forall y \forall z (y = z \bigvee P(x,y) \neq P(x,z)),$$
$$\forall x \forall y \forall z (x = z \bigvee P(x,y) \neq P(z,y)),$$
$$\forall x P(x,x) = x,$$
$$Identity(i)$$
$$P(0,1) = 2,$$
$$\forall x \forall y (x = y \bigvee P(x,y) \neq P(x,y))\}$$

The (un)satisfiability of the problem is preserved. We summarize this process as Algorithm 2.

We only implemented the sequential version of Algorithm 2; it is easy to implement a parallel version. This also can be seen as an example of the divide and conquer method with symmetry breaking.

The search framework of Algorithm 2 can help us get the result about a $LIQ(n)$ quickly but it may fail in getting the maximum number of disjoint IQs. A small modification can make it capable of getting the maximum number of disjoint IQs. One just needs to delete the symmetry breaking constraints which are introduced in Sect. 4.1. However, this will be slower than the original version. In addition, $C^{(i)}(n)$ can also be extended to denote more non-isomorphic disjoint IQs. We use *Hybrid search* to denote the original version, *Hybrid search1* to denote the version without symmetry breaking constraints and *Hybrid search2* to denote the version extending the concept of $C^{(i)}(n)$.

5   New Results

Table 2 lists the results about investigating the existence of large sets of idempotent quasigroups. The column of '*Maximum*' presents the maximum number of

---

Algorithm 2. Hybrid search for $LIQ^{(i)}(n)$

Input: n is order, i is identity (i)
Output: The existence of LIQ and the maximum number of disjoint IQs
1   $max \leftarrow 1$;
2   Generate $C^{(i)}(n)$ by $\Sigma_{C^{(i)}(n)}$ with a finite model generator which can break symmetry.;
3   **for** *all* $L_1 \in C^{(i)}(n)$ **do**
4   **for** $l \leftarrow 2$ **to** $n-2$ **do**
5         $result \leftarrow$ Solve $\{Encode(DIQ^{(i)}_{(n)}(l)) + Encode(L_1)\}$ with SAT solver ;
6   **if** $result$ *is* $UNSAT$ **then** Break ;
7   **else if** $l > max$ **then** $max \leftarrow l$;
8   **end**
9   **if** $max == n-2$ **then return** $EXISTENT$;
10 **end**
11 **if** $max < n-3$ **then** $max = unknown$;
12 **return** $NONEXISTENT$ AND max

---

disjoint $IQ(n)$s for a non-existent instance. And the fifth column marks whether the second strategy, hybrid search, was used. The maximum number of disjoint $IQ(n)$s for $LIQ^{(2)}(9)$, $LIQ^{(4)}(11)$ and $LIQ^{(5)}(11)$ are obtained by *Hybrid search1* which is a modified version of Algorithm 2 without symmetry breaking constraints.

Table 2. The result about LIQ

Order n	Identity i	Existence of LIQ	Maximum	Hybrid
9	2	NO	6	✓
	4	NO	6	-
	6	NO	6	-
	7	NO	6	-
11	4	NO	4	✓
	5	NO	2	✓

The current generation of SAT solvers support emission of unsatisfiability proofs. And standards for such proofs exist, as well as checkers. When the hybrid method was applied we should check the proof file and $C^{(i)}(n)$, although $C^{(i)}(n)$ may be hard to be verified by a formal method. $C^{(i)}(n)$ is a small fraction of the whole problem which can be double checked by different solvers. If one wants to verify the result, one just need to check $DIQ^{(i)}_{(n)}(max + 1)$, where max denotes the maximum number of disjoint IQs.

In addition to investigating the existence of large sets, we also check the orthogonality of some LIQs. The definition of the orthogonality (Definition 2) of

LIQs has been introduced in Sect. 2.1. While LIQ cannot imply orthogonality, all small order LIQs we found so far do have the orthogonality. This is also true for all $LIQ^{(i)}(n)$, where $n \leq 8$, found by Ma et al. Only $LIQ^{(1)}(8)$, $LIQ^{(3)}(4)$, $LIQ^{(4)}(4)$, $LIQ^{(7)}(8)$ exist according to [14]. $LIQ^{(3)}(4)$ and $LIQ^{(4)}(4)$ can be constructed simply by hand, and they have been checked by mathematicians. Our work is to check $LIQ^{(1)}(8)$ and $LIQ^{(7)}(8)$. We enumerate all large sets of $LIQ^{(1)}(8)$ and $LIQ^{(7)}(8)$. The number of $LIQ^{(1)}(8)$s and $LIQ^{(7)}(8)$s are 240 (some isomorphic solutions are eliminated). We examined all the $LIQ^{(1)}(8)$s and $LIQ^{(7)}(8)$s and concluded that all large sets of $LIQ^{(1)}(8)$s and $LIQ^{(7)}(8)$s have orthogonality.

6 Experimental Evaluation

In this section, we will use some experimental data to show the efficiency of our method. The experiment is performed on a Dell laptop with Intel(R) Core(TM) i7-6500U CPU(2.50 GHz), operating system Ubuntu 16.04 and 16 GB memory.

The first strategy of adding symmetry breaking constraints will not be discussed in this section, because it is a universal method to save search time. The search framework also can help us avoid some isomorphic search spaces when getting the maximum number of disjoint IQs for nonexistent instance according to *Proposition* 1. It may prove that the large set does not exist when l is small. Due to the first search strategy, we can use the SAT solver *Glucose 4.1* to prove that $LIQ^{(4)}(9)$, $LIQ^{(6)}(9)$ and $LIQ^{(7)}(9)$ do not exist.

However, $LIQ^{(2)}(9)$, $LIQ^{(3)}(10)$, $LIQ^{(4)}(11)$ and $LIQ^{(5)}(11)$ are still hard for the first search strategy. We used *Treengeling* and parallel version of *Glucose 4.1* to solve these hard instances on a computer server with 100 cores (Intel Xeon CPU E7-8870 @ 2.40GHz 32M Cache). However, these instances exhausted a week without any results. Owing to the hybrid method we prove that they do not exist within several minutes except for $LIQ^{(3)}(10)$.

In order to evaluate the efficiency of the hybrid method, we compared the running times of using the hybrid method against just using a single SAT solver or a finite-model generator. The results are shown in Table 3. The first three columns show the search time of only using *Algorithm* 1 and the fourth column shows search time of using the hybrid method. *SEM* and *MACE4* did almost the same in our experiment. We used *Glucose 4.1* and *MACE4* in *Hybrid search*. All the implementations are in github[1].

From Table 3, we know that combining SAT and first order logic can significantly improve the efficiency of search for LIQ. In particular, for some hard instances that finite-model generators and SAT solvers cannot solve in a week, the hybrid method can solve it in minutes.

Table 4 shows the results of different versions of the hybrid method which have been introduced in Sect. 4.2. *Hybrid search* is the original version and *Hybrid search1* is the version that can find the maximum number of disjoint $IQ(n)$s.

---

[1] https://github.com/huangdiudiu/LIQ-search.

Table 3. The running times of different methods in solving $LIQ^{(i)}(n)$

Instance	Glucose 4.1		Glucose(Parallel)		MACE 4		Hybrid search	
	Time	Result	Time	Result	Time	Result	Time	Result
$LIQ^{(2)}(9)$	>1 week	-	>1 week	-	>1 week	-	51.88 s	UNSAT
$LIQ^{(4)}(9)$	14.372 s	UNSAT	10.60 s	UNSAT	>24 h	-	0.25 s	UNSAT
$LIQ^{(6)}(9)$	889.23 s	UNSAT	616.735 s	UNSAT	>24 h	-	0.94 s	UNSAT
$LIQ^{(7)}(9)$	1020.74 s	UNSAT	560.07 s	UNSAT	>24 h	-	0.32 s	UNSAT
$LIQ^{(4)}(11)$	>1 week	-	>1 week	-	>1 week	-	8.19 s	UNSAT
$LIQ^{(5)}(11)$	>1 week	-	>1 week	-	>1 week	-	12.88 s	UNSAT

Hybrid search2 extends the concept of $C^{(i)}(n)$, which is trying to find all non-isomorphic $\{L_1, L_2\}$ by the finite-model generator.

Table 4. Comparison of different versions of the hybrid method

Instance	Hybrid search			Hybrid search1			Hybrid search2		
	MACE	SAT	Total(s)	MACE	SAT	Total(s)	MACE	SAT	Total(s)
$LIQ^{(2)}(9)$	4.79	12.18	16.97	4.79	282.52	241.46	921.33	21.00	942.33
$LIQ^{(4)}(9)$	0.01	0.28	0.29	0.01	1.06	0.39	1.45	1.19	2.03
$LIQ^{(6)}(9)$	0.08	0.55	0.63	0.08	20.78	20.86	>3000	-	>3000
$LIQ^{(7)}(9)$	0.03	0.50	0.53	0.03	2.15	2.18	32.84	1.97	34.81
$LIQ^{(4)}(11)$	0.01	10.83	10.84	0.01	211.17	211.27	731.11	13.62	744.73
$LIQ^{(5)}(11)$	1.01	23.06	24.07	1.01	95.84	96.85	>3000	-	>3000

It is easy to know that fixing the sequence of IQs can improve the search process from the comparison between *Hybrid search* and *Hybrid search1*. However, it will sacrifice the ability of getting the maximum number of disjoint $IQ(n)$s. From the comparison of *Hybrid search* and *Hybrid search2*, we know that finding all non-isomorphic L_1s as candidate set $C^{(i)}(n)$ is more efficient than finding two disjoint IQs as candidate set.

The hybrid method divides the problem into two parts. One part, $C^{(i)}(n)$, is solved by a finite-model generator (first order logic) and the other part is solved by a SAT solver. So the hybrid ratio of first order logic formulae and SAT encoding will affect the performance. Actually, we observe that finding all non-isomorphic L_1s as candidate set $C^{(i)}(n)$ is the most efficient in our experiment. We take $LIQ^{(7)}(9)$ and $LIQ^{(4)}(11)$ as examples to show that how the hybrid ratio affects the performance. If m denotes the number of $IQ(n)$s that are encoded as first order logic formulae, then we use $m/(n-2)$ to denote the hybrid ratio of first order logic formulae. Figure 4 shows the relationship between the hybrid ratio $m/(n-2)$ and search time. We find that $m/(n-2) = 1/(n-2)$ is the best hybrid ratio for almost all instances. When $m/(n-2) = 0$, that means all of the problems are solved by a SAT solver. When $m/(n-2) = 1$, that means all of the problems are solved by a finite-model generator.

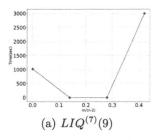

(a) $LIQ^{(7)}(9)$

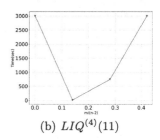

(b) $LIQ^{(4)}(11)$

Fig. 4. The relationship between the hybrid ratio $m/(n-2)$ and run time. The abscissa axis displays the value of $m/(n-2)$.

7 Conclusion

This paper describes an application of automated reasoning techniques and tools to an interesting problem in combinatorics: the large set of idempotent quasigroups ($LIQs$) satisfying the short identities. The $LIQs$ of moderate orders which are difficult for mathematical methods can also be quite challenging for computer search. We present some effective search strategies for this problem, and the core idea is symmetry breaking. We find that combining the power of SAT solving and finite model generation is far more efficient than using a single solver. As a result, a number of open cases have been solved. Besides, we find that all $LIQ^{(1)}(8)$s and $LIQ^{(7)}(8)$s have orthogonality.

Acknowledgements. This work has been supported by the National 973 Program of China under Grant 2014CB340701, the National Natural Science Foundation of China under Grant 61100064, and the CAS/SAFEA International Partnership Program for Creative Research Teams. Feifei Ma is also supported by the Youth Innovation Promotion Association, CAS. We thank Lie Zhu and Yanxun Chang for suggesting these open problems and help.

References

1. Baumgartner, J., Mony, H., Paruthi, V., Kanzelman, R., Janssen, G.: Scalable sequential equivalence checking across arbitrary design transformations. In: International Conference on Computer Design, pp. 259–266 (2006)
2. Cao, H., Ji, L., Zhu, L.: Large sets of disjoint packings on 6k + 5 points. J. Comb. Theory Ser. A **108**(2), 169–183 (2004)
3. Cayley, A.: On the triadic arrangements of seven and fifteen things. Philos. Mag. **37**(247), 50–53 (1946)

4. Chang, Y.: The spectrum for large sets of idempotent quasigroups. J. Comb. Des. **8**(2), 79–82 (2015)

5. de Moura, L., Bjørner, N.: Z3: an efficient SMT solver. In: Ramakrishnan, C.R., Rehof, J. (eds.) TACAS 2008. LNCS, vol. 4963, pp. 337–340. Springer, Heidelberg (2008). https://doi.org/10.1007/978-3-540-78800-3_24

6. Fujita, M., Slaney, J.K., Bennett, F.: Automatic generation of some results in finite algebra. In: International Joint Conference on Artificial Intelligence, pp. 52–57. Morgan Kaufmann (1993)

7. Gligoroski, D., Markovski, S., Knapskog, S.J.: A public key block cipher based on multivariate quadratic quasigroups. CoRR, abs/0808.0247 (2008)

8. Goldberg, E.I., Prasad, M.R., Brayton, R.K.: Using SAT for combinational equivalence checking. In: Proceedings of the Conference on Design, Automation and Test in Europe, pp. 114–121 (2001)

9. Heule, M.J.H., Hunt, W.A., Wetzler, N.: Verifying refutations with extended resolution. In: Bonacina, M.P. (ed.) CADE 2013. LNCS (LNAI), vol. 7898, pp. 345–359. Springer, Heidelberg (2013). https://doi.org/10.1007/978-3-642-38574-2_24

10. Heule, M.J.H., Kullmann, O., Marek, V.W.: Solving and verifying the Boolean pythagorean triples problem via cube-and-conquer. In: Creignou, N., Le Berre, D. (eds.) SAT 2016. LNCS, vol. 9710, pp. 228–245. Springer, Cham (2016). https://doi.org/10.1007/978-3-319-40970-2_15

11. Kaiss, D., Skaba, M., Hanna, Z., Khasidashvili, Z.: Industrial strength SAT-based alignability algorithm for hardware equivalence verification. In: Formal Methods in Computer-Aided Design, pp. 20–26 (2007)

12. Koscielny, C.: Generating quasigroups for cryptographic applications. Int. J. Appl. Math. Comput. Sci. **12**, 559–569 (2002)

13. Jiaxi, L.: On large sets of disjoint steiner triple systems II. J. Comb. Theory **37**(2), 147–155 (1983)

14. Ma, F., Zhang, J.: Computer search for large sets of idempotent quasigroups. In: Kapur, D. (ed.) ASCM 2007. LNCS (LNAI), vol. 5081, pp. 349–358. Springer, Heidelberg (2008). https://doi.org/10.1007/978-3-540-87827-8_30

15. McCune, W.: Mace4 reference manual and guide. CoRR, cs.SC/0310055 (2003)

16. Mishchenko, A., Chatterjee, S., Brayton, R.K., Eén, N.: Improvements to combinational equivalence checking. In: International Conference on Computer-Aided Design, pp. 836–843 (2006)

17. Nethercote, N., Stuckey, P.J., Becket, R., Brand, S., Duck, G.J., Tack, G.: MiniZinc: towards a standard CP modelling language. In: Bessière, C. (ed.) CP 2007. LNCS, vol. 4741, pp. 529–543. Springer, Heidelberg (2007). https://doi.org/10.1007/978-3-540-74970-7_38

18. Rossi, F., van Beek, P., Walsh, T.: Handbook of Constraint Programming. Elsevier, New York (2006)

19. Slaney, J.K., Fujita, M., Stickel, M.: Automated reasoning and exhaustive search: quasigroup existence problems. Comput. Math. Appl. **29**(2), 115–132 (1995)

20. Teirlinck, L., Lindner, C.C.: The construction of large sets of idempotent quasigroups. Eur. J. Comb. **9**(1), 83–89 (1988)

21. Wetzler, N., Heule, M.J.H., Hunt, W.A.: DRAT-trim: efficient checking and trimming using expressive clausal proofs. In: Sinz, C., Egly, U. (eds.) SAT 2014. LNCS, vol. 8561, pp. 422–429. Springer, Cham (2014). https://doi.org/10.1007/978-3-319-09284-3_31

22. Yuan, L., Kang, Q.: Some infinite families of large sets of Kirkman triple systems. J. Comb. Des. **16**(3), 202–212 (2008)

23. Zhang, H.: SATO: an efficient prepositional prover. In: McCune, W. (ed.) CADE 1997. LNCS, vol. 1249, pp. 272–275. Springer, Heidelberg (1997). https://doi.org/10.1007/3-540-63104-6_28
24. Zhang, H.: Combinatorial designs by SAT solvers. In: Handbook of Satisfiability, pp. 533–568. IOS Press (2009)
25. Zhang, H., Stickel, M.: Implementing the Davis-Putnam method. J. Autom. Reason. **24**(1–2), 277–296 (2000)
26. Zhang, J.: Automatic symmetry breaking method combined with SAT. In: ACM Symposium on Applied, Computing, pp. 17–21 (2001)
27. Zhang, J., Zhang, H.: SEM: a system for enumerating models. In: International Joint Conference on Artificial Intelligence, vol. 2, pp. 298–303 (1995)
28. Zhu, L.: Personal communication, September 2007
29. Zhu, L.: Large set problems for various idempotent quasigroups (2014)

Superposition with Datatypes
and Codatatypes

Jasmin Christian Blanchette[1,2], Nicolas Peltier[3], and Simon Robillard[4(✉)]

[1] Vrije Universiteit Amsterdam, Amsterdam, The Netherlands
[2] Max-Planck-Institut für Informatik, Saarland Informatics Campus,
Saarbrücken, Germany
[3] Univ. Grenoble Alpes, CNRS, Grenoble INP, LIG, 38000 Grenoble, France
[4] Chalmers University of Technology, Gothenburg, Sweden
simon.robillard@chalmers.se

Abstract. The absence of a finite axiomatization of the first-order theory of datatypes and codatatypes represents a challenge for automatic theorem provers. We propose two approaches to reason by saturation in this theory: one is a conservative theory extension with a finite number of axioms; the other is an extension of the superposition calculus, in conjunction with axioms. Both techniques are refutationally complete with respect to nonstandard models of datatypes and nonbranching codatatypes. They take into account the acyclicity of datatype values and the existence and uniqueness of cyclic codatatype values. We implemented them in the first-order prover Vampire and compare them experimentally.

1 Introduction

The ability to reason about inductive and coinductive datatypes has many applications in program verification, formalization of the metatheory of programming languages, and even formalization of mathematics. Inductive datatypes, or simply *datatypes*, consist of finite values freely generated from constructors. Coinductive datatypes, or *codatatypes*, additionally support infinite values. Non-freely generated (co)datatypes are also useful. All of these variants can be seen as members of a single unifying framework (Sect. 2).

It is well known that the first-order theory of datatypes cannot be finitely axiomatized. Distinctness, injectivity, and exhaustiveness of constructors are easy to axiomatize, but acyclicity is more subtle, and for induction we would need an axiom schema or a second-order axiom. Codatatypes are also problematic: Besides a coinduction principle that is dual to induction, they are characterized by the existence of all possible infinite values, corresponding intuitively to infinite ground terms. Both datatypes and codatatypes represent a challenge for automatic theorem provers.

Superposition [2] is a highly successful calculus for reasoning about first-order clauses and equality. There has been some work on extending superposition with induction [10,24], including by Kersani and Peltier [11], and on the

© Springer International Publishing AG, part of Springer Nature 2018
D. Galmiche et al. (Eds.): IJCAR 2018, LNAI 10900, pp. 370–387, 2018.
https://doi.org/10.1007/978-3-319-94205-6_25

axiomatization of datatypes, including by Kovács, Robillard, and Voronkov [12]. In this paper, we propose both axiomatizations and extensions of the superposition calculus to support freely and non-freely generated datatypes as well as codatatypes.

We first focus on a conservative extension of the theory with a finite number of first-order axioms that capture the basic properties of constructors, acyclicity of datatype values, uniqueness of cyclic (ω-regular) codatatype values, and existence of all codatatype cyclic values (Sect. 3). These axioms admit nonstandard models; for example, for the Peano-style natural numbers freely generated by zero : nat and suc : $nat \rightarrow nat$, we cannot exclude the familiar nonstandard models of arithmetic, in which arbitrarily many copies of $\mathbb{Z}$ may appear besides $\mathbb{N}$. Similarly, the domains interpreting codatatypes are not guaranteed to contain all infinite acyclic values.

The axiomatization of codatatypes up to a suitable notion of nonstandard models constitutes the first theoretical contribution of this paper. Our second theoretical contribution is an extension of superposition with inference rules to reason about datatypes and codatatypes (Sect. 4). This is inspired by an acyclicity rule that Robillard presented at the Vampire 2017 workshop [22]. The main distinguishing feature of our rules is that they are (in combination with a few axioms) refutationally complete and their side conditions have some new order restrictions, helping prune the search space. On the other hand, our approach also requires a relaxation of the side conditions of the superposition rule: For clauses of the form $c(\bar{s}) \approx t \vee \mathcal{C}$, where c is a constructor and the first literal is maximal, inferences onto t must be performed, as in ordered paramodulation [1]. In addition, we propose calculus extensions to reason about codatatypes.

Both the theory extension and the calculus extension are designed to be refutationally complete with respect to nonstandard models of datatypes and *nonbranching* codatatypes—codatatypes whose constructors have at most one corecursive argument (Sect. 5). Due to space restrictions, we can only briefly sketch the proof in this paper. We refer to our technical report [8] for detailed justifications and further explanations.

The calculus extension can be integrated into the given clause algorithm that forms the core of a prover's saturation loop (Sect. 6). The inference partners for the acyclicity and uniqueness rules can be located efficiently. We implemented both the axiomatic and the calculus approaches in the first-order prover Vampire [13] and compare them empirically on Isabelle/HOL [17] benchmarks and on crafted benchmarks (Sect. 7).

2   Syntax and Semantics

Our setting is a many-sorted first-order logic. We let τ, υ range over simple types (sorts), s, t, u, v range over terms, $\mathsf{a}, \mathsf{b}, \mathsf{c}, \dots$ range over function symbols, x, y, z range over variables, and $\mathcal{C}, \mathcal{D}, \mathcal{E}$ range over clauses. Literals are atoms of the form $s \approx t$ or $\neg\, s \approx t$, also written $s \not\approx t$. Clauses are finite disjunctions of literals, viewed as multisets. Substitutions are written in postfix notation, with

$s\sigma\theta = (s\sigma)\theta$. The notation $\bar{x}$ represents a tuple $(x_1, \ldots, x_m)$, and $[m, n]$ denotes the set $\{m, m + 1, \ldots, n\}$, where $m \leq n + 1$.

A *position p of type τ in t* is a position in t such that $t|_p$ is of type τ. If s, t are terms and P is a set of positions of the same type as s in t, then $t[s]_P$ denotes the term obtained from t by replacing the subterms occurring at a position in P by s: $t[s]_P := s$ if $\varepsilon \in P$; $t[s]_P := t$ if $P = \varnothing$; and $\mathsf{f}(t_1, \ldots, t_n)[s]_P := \mathsf{f}(t_i[s]_{P_i})_{i \in [1,n]}$, with $P_i = \{q \mid i.q \in P\}$ otherwise. Given two positions p and q, we write $p < q$ if p is a proper prefix of q. Let $\mathcal{C}tr$ be a distinguished finite set of function symbols, called *constructors*. We reserve the letters $\mathsf{c}, \mathsf{d}, \mathsf{e}$ for constructors. A *constructor position* in t is a position q in t such that for every $p < q$, the head symbol of $t|_p$ is a constructor.

Definition 1. The set of *constructor contexts* of profile $\tau \to \upsilon$ is defined inductively as follows: (1) if t is a term of type υ, then t is a constructor context of profile $\tau \to \upsilon$; (2) if $\Gamma_1, \ldots, \Gamma_n$ are constructor contexts of profile $\tau \to \tau_i$ and $\mathsf{c} : \tau_1 \times \cdots \times \tau_n \to \upsilon$ is a constructor, then $\mathsf{c}(\Gamma_1, \ldots, \Gamma_n)$ is a constructor context of profile $\tau \to \upsilon$; (3) the hole $\bullet$ is a constructor context of profile $\upsilon \to \upsilon$.

Every constructor context can be written as $\Gamma[\bullet]_P$, where P is a set of constructor positions of the same type in Γ, denoting the positions of $\bullet$ in Γ. It is *empty* if $\varepsilon \in P$, and *constant* if $P = \varnothing$. We write $\Gamma[\bullet]_p$ as an abbreviation for $\Gamma[\bullet]_{\{p\}}$, and we write $\Gamma[t]_P$ to denote the term obtained by replacing every position of P by the term t in the context $\Gamma[\bullet]_P$. Moreover, we write $\tau \rhd \upsilon$ ("υ depends on τ") if there exists a constructor of profile $\tau_1 \times \cdots \times \tau_n \to \upsilon$, with $\tau = \tau_i$ for some $i \in [1, n]$, and $\tau \sim \upsilon$ if $\tau \rhd^* \upsilon$ and $\upsilon \rhd^* \tau$.

The axioms and rules in this paper are parameterized by the following sets. Let $\mathcal{T}_{\mathsf{ind}}$ and $\mathcal{T}_{\mathsf{coind}}$ be disjoint sets of types, intended to model datatypes and codatatypes, respectively, and assume that the codomain of every constructor is in $\mathcal{T}_{\mathsf{ind}} \cup \mathcal{T}_{\mathsf{coind}}$. Let $\mathcal{C}tr_{\mathsf{inj}} \subseteq \mathcal{C}tr$ be a set of constructors, denoting injective constructors. Let $\bowtie$ be a binary symmetric and irreflexive relation among constructors; $\mathsf{c} \bowtie \mathsf{d}$ indicates that terms with head symbol c are always distinct from terms with head symbol d.

We introduce some properties of interpretations that are intended to capture some of the properties of (co)datatypes. An interpretation $\mathcal{I}$ satisfies

- **Exh** (exhaustiveness) iff, for every type $\tau \in \mathcal{T}_{\mathsf{ind}} \cup \mathcal{T}_{\mathsf{coind}}$, $\mathcal{I} \models \bigvee_{i=1}^m \exists \bar{x}_i.\ x \approx \mathsf{c}_i(\bar{x}_i)$, where x is a variable of type τ, $\{\mathsf{c}_1, \ldots, \mathsf{c}_m\}$ is the set of constructors of codomain τ, and $\bar{x}_i$ is a vector of pairwise distinct variables of the appropriate length and types;
- **Inf** (infiniteness) iff, for every type $\tau \in \mathcal{T}_{\mathsf{ind}} \cup \mathcal{T}_{\mathsf{coind}}$, the domain of τ is infinite;
- **Acy** (acyclicity, for datatypes) iff, for every type $\tau \in \mathcal{T}_{\mathsf{ind}}$ and for every nonempty constructor context $\Gamma[\bullet]_p$ of profile $\tau \to \tau$, where p is a position, we have $\mathcal{I} \models \Gamma[x]_p \not\approx x$, where x is a variable of type τ not occurring in Γ;
- **FP** (existence and uniqueness of fixpoints, for codatatypes) iff, for every type $\tau \in \mathcal{T}_{\mathsf{coind}}$, for every nonempty constructor context $\Gamma[\bullet]_P : \tau \to \tau$, $\mathcal{I} \models (\exists x.\ \Gamma[x]_P \approx x) \wedge (\Gamma[x]_P \approx x \wedge \Gamma[y]_P \approx y \Rightarrow x \approx y)$, where x, y are fresh variables of type τ;

- **Dst** (distinctness of constructors) iff, for every pair of constructors c, d of the same codomain such that $c \bowtie d$, $\mathcal{I} \models c(\bar{x}) \not\approx d(\bar{y})$ where $\bar{x}$ and $\bar{y}$ are disjoint vectors of pairwise distinct variables of the appropriate length and types;
- **Inj** (injectivity) iff, for every n-ary constructor $c \in \mathcal{C}tr_{\mathsf{inj}}$ and pairwise distinct variables $x_1, \ldots, x_n, y_1, \ldots, y_n$ of the appropriate types, $\mathcal{I} \models c(x_1, \ldots, x_n) \approx c(y_1, \ldots, y_n) \Longrightarrow \bigwedge_{i=1}^{n} x_i \approx y_i$.

Most datatypes occurring in practice are recursive, so condition **Inf** is usually satisfied. In particular, it is the case for any nonempty freely generated (co)datatype τ such that $\tau \rhd^+ \tau$. Conditions **Dst** and **Inj** are defined by finite sets of axioms, but not conditions **Acy** and **FP**. In Sect. 3, we introduce conservative extensions of the considered formula so that conditions **Acy** and **FP** are satisfied. Then in Sect. 4, we replace some of these axioms by inference rules.

We assume that $\tau \not\sim \upsilon$ whenever $\tau \in \mathcal{T}_{\mathsf{ind}}$ and $\upsilon \in \mathcal{T}_{\mathsf{coind}}$. Intuitively, this condition means that a datatype cannot be defined by mutual recursion with a codatatype, which is a very natural restriction [7]. If this condition does not hold, it is easy to see that there is no interpretation that satisfies both **Acy** and **FP**. On the other hand, we may have $\tau \rhd^+ \upsilon$ or $\upsilon \rhd^+ \tau$ with $\tau \in \mathcal{T}_{\mathsf{ind}}$ and $\upsilon \in \mathcal{T}_{\mathsf{coind}}$. There may also exist types not belonging to $\mathcal{T}_{\mathsf{ind}} \cup \mathcal{T}_{\mathsf{coind}}$, and the types in $\mathcal{T}_{\mathsf{ind}} \cup \mathcal{T}_{\mathsf{coind}}$ may depend on them. Finally, we assume that for each type τ, there exists a ground term t of type τ.

3   Axioms

The axioms Exhaust for exhaustiveness, Dist for distinctness, and Inj for injectivity correspond to the formulas used to express the properties **Exh**, **Dst**, and **Inj** in Sect. 2. The other axioms are introduced below.

For all types $\tau \sim \upsilon$, we introduce a predicate symbol $\mathsf{sub}_\upsilon^\tau$ on $\tau \times \upsilon$ together with the following axioms, where $\tau \sim \upsilon \sim \upsilon'$ and $c : \cdots \times \upsilon \times \cdots \to \upsilon'$ is a constructor:

$$\text{Sub}_1:\ \mathsf{sub}_\tau^\tau(x,x) \qquad \text{Sub}_2:\ \neg\,\mathsf{sub}_\upsilon^\tau(x,y) \vee \mathsf{sub}_{\upsilon'}^\tau(x, c(\bar{z}, y, \bar{z}'))$$
$$\text{NSub}:\ \neg\,\mathsf{sub}_\tau^{\upsilon'}(c(\bar{z}, x, \bar{z}'), x) \quad \text{if } \tau \in \mathcal{T}_{\mathsf{ind}}$$

Let $\mathsf{Sub} = \mathsf{Sub}_1 \wedge \mathsf{Sub}_2$. The least fixpoint model of Sub is the usual subterm relation for constructor terms. The axiom NSub states that no term of a type in $\mathcal{T}_{\mathsf{ind}}$ may occur at a nonempty constructor position in itself.

Definition 2. An interpretation $\mathcal{I}$ is sub-*minimal* if, for all $\tau \sim \upsilon$, it satisfies the equivalence $\mathsf{sub}_\upsilon^\tau(x,y) \Leftrightarrow \bigvee\{\exists \bar{z}.\ y \approx \Gamma[x]_p \mid \Gamma[\bullet]_p$ is a constructor context of profile $\tau \to \upsilon\}$, where $\bar{z}$ denotes the vector of variables in Γ that are distinct from x, y.

For every pair of types $\tau, \upsilon \in \mathcal{T}_{\mathsf{coind}}$ with $\tau \sim \upsilon$, we introduce a type $\boxed{\tau}_\upsilon$ to denote contexts $\Gamma[\bullet]_P$ of profile $\upsilon \to \tau$. Let hole_υ be a constant of type $\boxed{\upsilon}_\upsilon$, denoting an empty context. All constructors $c : \tau_1 \times \cdots \times \tau_n \to \tau$ and types υ such that $\exists i \, \upsilon \triangleright^* \tau_i$ are associated with new n-ary constructors $\boxed{\mathsf{c}}_\upsilon : \upsilon_1 \times \cdots \times \upsilon_n \to \boxed{\tau}_\upsilon$, where for every $i \in [1, n]$, $\upsilon_i = \boxed{\tau_i}_\upsilon$ if $\upsilon \triangleright^* \tau_i$ and $\upsilon_i = \tau_i$ otherwise. Let $\mathsf{app}_\upsilon^\tau : \boxed{\tau}_\upsilon \times \upsilon \to \tau$, $\mathsf{cyc}_\upsilon : \boxed{\upsilon}_\upsilon \to \upsilon$, and $\mathsf{cst}_\upsilon^\tau : \tau \to \boxed{\tau}_\upsilon$ be new function symbols. Intuitively, if y denotes the context $\Gamma[\bullet]_P$, then $\mathsf{app}(y, x)$ denotes the term $\Gamma[x]_P$, $\mathsf{cyc}(y)$ denotes the fixpoint of $\Gamma[\bullet]_P$, and $\mathsf{cst}_\upsilon^\tau$ denotes a constant context (i.e., a context $\Gamma[\bullet]_P$ with $P = \varnothing$).

We consider the following axioms, where $\upsilon \in \mathcal{T}_{\mathsf{coind}}$ and x, y, x_i, z_i are pairwise distinct variables of the appropriate types:

App_1: $\mathsf{app}_\upsilon^\tau(\mathsf{cst}_\upsilon^\tau(x), y) \approx x$ App_2: $\mathsf{app}_\upsilon^\upsilon(\mathsf{hole}_\upsilon, y) \approx y$

App_3: $\mathsf{app}_\upsilon^\tau(\boxed{\mathsf{c}}_\upsilon(x_1, \ldots, x_n), y) \approx \mathsf{c}(t_1, \ldots, t_n)$
    if $\mathsf{c} : \tau_1 \times \cdots \times \tau_n \to \tau$ is a constructor and $\exists i \, \upsilon \triangleright^* \tau_i$
    with $t_i = \mathsf{app}_\upsilon^{\tau_i}(x_i, y)$ if $\upsilon \triangleright^* \tau_i$ and $t_i = x_i$ otherwise

Uniq: $x \approx \mathsf{hole}_\upsilon \lor y \not\approx \mathsf{app}_\upsilon^\upsilon(x, y) \lor z \not\approx \mathsf{app}_\upsilon^\upsilon(x, z) \lor y \approx z$

Cycl: $\mathsf{cyc}_\upsilon(x) \approx \mathsf{app}_\upsilon^\upsilon(x, \mathsf{cyc}_\upsilon(x))$

Hole_1: $\mathsf{hole}_\upsilon \not\approx \mathsf{cst}_\upsilon^\upsilon(x)$ Hole_2: $\mathsf{hole}_\upsilon \not\approx \boxed{\mathsf{c}}_\upsilon(x_1, \ldots, x_n)$ if $\mathsf{c} : \cdots \to \upsilon$

Let $\mathsf{App} = \mathsf{App}_1 \land \mathsf{App}_2 \land \mathsf{App}_3$ and $\mathsf{Hole} = \mathsf{Hole}_1 \land \mathsf{Hole}_2$.

Example 3. Let $\mathsf{c} : \tau_0 \times \upsilon \to \tau$ be a constructor, with $\upsilon \triangleright^* \tau_0$. Then the profile of $\boxed{\mathsf{c}}_\upsilon$ is $\boxed{\tau_0}_\upsilon \times \boxed{\upsilon}_\upsilon \to \boxed{\tau}_\upsilon$. The term $t := \boxed{\mathsf{c}}_\upsilon(\mathsf{cst}_\upsilon^{\tau_0}(x), \mathsf{hole}_\upsilon)$ encodes the constructor context $\mathsf{c}(x, \bullet)$. If $\mathsf{a} : \upsilon$, then $\mathsf{app}_\upsilon^\tau(t, \mathsf{a}) =_{\mathsf{App}} \mathsf{c}(x, \mathsf{a})$, where $=_{\mathsf{App}}$ denotes equality modulo App (i.e., $s =_{\mathsf{App}} t \Leftrightarrow \mathsf{App} \models s \approx t$).

Lemma 4 (Soundness of the Axioms). *If interpretation $\mathcal{I}$ satisfies* **Acy** *and* **FP**, *there exists a* sub-*minimal extension of $\mathcal{I}$ validating* Sub, NSub, App, Uniq, Cycl, *and* Hole.

Lemma 5 (Completeness of the Axioms). *Any model of the set of axioms* $\{\mathsf{Sub}, \mathsf{NSub}, \mathsf{App}, \mathsf{Uniq}, \mathsf{Cycl}, \mathsf{Hole}\}$ *fulfills* **Acy** *and* **FP**.

Lemma 6 (Completeness of the Theory). *Let $\mathcal{T}$ be the theory of free constructors, as defined by the properties* **Exh**, **Inf**, **Acy**, **FP**, **Dst**, *and* **Inj**, *with $Ctr_{\mathsf{inj}} = Ctr$ and $\mathsf{c} \bowtie \mathsf{d}$ for all distinct constructors c and d. If S is a first-order sentence in which the only symbols occurring are constructors and equality ($\approx$), then either $\mathcal{T} \models S$ or $\mathcal{T} \models \neg S$.*

Comon and Lescanne [9] provide a decision procedure for equational formulas over finite and infinite trees, which correspond respectively to freely generated datatypes and codatatypes. It is based on a collection of equivalence-preserving transformation rules for eliminating quantifiers and normalizing the formulas. The set of formulas $T = \{\mathsf{Dist}, \mathsf{Inj}, \mathsf{Exhaust}, \mathsf{Sub}, \mathsf{NSub}, \mathsf{App}, \mathsf{Uniq}, \mathsf{Cycl}, \mathsf{Hole}\}$ forms the axiomatization of a conservative extension of the theory of (co)datatypes. We can thus derive a decision procedure for testing satisfiability of first-order sentences S containing only constructors symbols and the equality predicate in the above theory. By interleaving the steps of two fair saturation procedures of the superposition calculus, the first over $S \cup T$ and the second over $\neg S \cup T$, one of the two attempts is guaranteed to derive a refutation in finite time.

4   Inference Rules

As an alternative to the above axiomatization, we propose an extension of the superposition calculus [2] with dedicated rules. Unless otherwise noted, the usual conventions of superposition apply. The standard notion of redundancy is used, with respect to the theory of equality. The notation $s \not\approx t$ indicates that the literal is selected, or that it is maximal in its clause, after the substitution σ has been applied, and nothing is selected, whereas $s \approx t$ indicates that the literal is strictly maximal in its clause, after σ, and no literal is selected. We let $[\neg]\, s \approx t$ stand for either $s \approx t$ or $s \not\approx t$.

Superposition. We denote by $\mathcal{SP}$ the usual rules of the superposition calculus with a slight relaxation: Superposition inside the nonmaximal term of an equation is allowed if the head of the maximal term is a constructor. This ensures that in the rewrite system built from saturated clause sets for defining a model, the right-hand side of every rule is irreducible if the head of the left-hand side is a constructor. Thus, our superposition rule is as follows:

$$\frac{u \approx v \lor \mathcal{D} \qquad [\neg]\, s[u'] \approx t \lor \mathcal{C}}{\sigma([\neg]\, s[v] \approx t \lor \mathcal{D} \lor \mathcal{C})} \ \text{Sup}$$

where $\sigma = \text{mgu}\,\{u \stackrel{?}{=} u'\}$, u' is not a variable, and $\sigma(u) \not\preceq \sigma(v)$; moreover, $\sigma(s[u']) \not\preceq \sigma(t)$ if $[\neg]$ is $\neg$ or if the head symbol of t is not a constructor. The equality resolution and equality factoring rules are the standard ones.

Infiniteness. The next rule captures infiniteness of (co)datatypes:

$$\frac{\left(\bigvee_{i=1}^{n} x \approx t_i\right) \lor \mathcal{C}}{\mathcal{C}} \ \text{Inf}$$

if x is a variable of a type $\tau \in \mathcal{T}_{\text{ind}} \cup \mathcal{T}_{\text{coind}}$ and does not occur in $\mathcal{C}$ or $t_1, \ldots, t_n$.

Lemma 7 (Soundness of Inf). *Let N be a clause set, and let $\mathcal{I}$ be a model of N satisfying **Inf**. If $\mathcal{C}$ is derived from N by Inf, then $\mathcal{I} \models \mathcal{C}$.*

Distinctness. The distinctness property of constructors takes the form of two rules:

$$\frac{\mathsf{c}(\bar{s}) \approx t \lor \mathcal{C}}{\mathcal{C}\sigma} \ \text{Dist}_1$$

if $\sigma = \text{mgu}\,\{t \stackrel{?}{=} \mathsf{d}(\bar{x})\}$, where $\mathsf{c} \bowtie \mathsf{d}$ and $\bar{x}$ are fresh pairwise distinct variables; and

$$\frac{\mathsf{d}(\bar{t}) \approx u' \lor \mathcal{D} \qquad \mathsf{c}(\bar{s}) \approx u \lor \mathcal{C}}{(\mathcal{C} \lor \mathcal{D})\sigma} \ \text{Dist}_2$$

if $\mathsf{c} \bowtie \mathsf{d}$, $\sigma = \text{mgu}\,\{u \stackrel{?}{=} u'\}$, $\mathsf{c}(\bar{s})\sigma \not\preceq u\sigma$, and $\mathsf{d}(\bar{t})\sigma \not\preceq u'\sigma$.

Proposition 8 (Soundness of Dist_1 and Dist_2). *Let N be a clause set, and let $\mathcal{I}$ be a model of N satisfying* **Dst**. *If a clause $\mathcal{C}$ is derived from N by Dist_1 or Dist_2, then $\mathcal{I} \models \mathcal{C}$.*

Remark 9. If t is not a variable, the premise of Dist_1 is redundant after the rule is applied. Unifying t with $\mathsf{c}(\bar{x})$ can be useful when t is a variable. For example, from the clause $\mathsf{c}(x) \approx x$, we can derive $\square$ by unifying x with $\mathsf{d}(\bar{y})$, where $\mathsf{d} \bowtie \mathsf{c}$.

Injectivity. The injectivity property of constructors is also captured by two rules:

$$\frac{\mathsf{c}(s_1, \ldots, s_m) \approx t \vee \mathcal{C}}{(s_i \approx x_i \vee \mathcal{C})\sigma} \; \mathsf{Inj}_1$$

if $i \in [1, m]$, $\mathsf{c} \in \mathcal{C}tr_{\mathsf{inj}}$, $\sigma = \operatorname{mgu}\{t \stackrel{?}{=} \mathsf{c}(x_1, \ldots, x_m)\}$, and $x_1, \ldots, x_m$ are fresh; and

$$\frac{\mathsf{c}(s_1, \ldots, s_m) \approx u' \vee \mathcal{D} \quad \mathsf{c}(t_1, \ldots, t_m) \approx u \vee \mathcal{C}}{(s_i \approx t_i \vee \mathcal{C} \vee \mathcal{D})\sigma} \; \mathsf{Inj}_2$$

if $i \in [1, m]$, $\mathsf{c} \in \mathcal{C}tr_{\mathsf{inj}}$, $\sigma = \operatorname{mgu}\{u \stackrel{?}{=} u'\}$, $u\sigma \not\succeq \mathsf{c}(\bar{s})\sigma$, and $u'\sigma \not\succeq \mathsf{c}(\bar{t})\sigma$.

Proposition 10 (Soundness of Inj_1 and Inj_2). *Let N be a clause set, and let $\mathcal{I}$ be a model of N satisfying* **Inj**. *If a clause $\mathcal{C}$ is derived from N by Inj_1 or Inj_2, then $\mathcal{I} \models \mathcal{C}$.*

Remark 11. If Inj_1 is applied on every argument $i \in [1, m]$ and t is not a variable, the premise becomes redundant and can be removed. Unifying t with the term $\mathsf{c}(x_1, \ldots, x_m)$ is useful when t is a variable. For example, given the clause $\mathsf{c}(x, \mathsf{a}) \approx x$, we can derive $\mathsf{a} \approx x_2$ by Inj_1, from which $\square$ can be derived by Inf.

Acyclicity. The acyclicity rule attempts to detect constraints that would force a datatype value to be cyclic. The simplest example is a clause of the form $\Gamma[s] \approx s$, where Γ is a nonempty constructor context. More generally, the clauses

$$s_1 \approx \Gamma_1[s_2] \qquad s_2 \approx \Gamma_2[s_3] \qquad \cdots \qquad s_{n-1} \approx \Gamma_{n-1}[s_n] \qquad s_n \approx \Gamma_n[s_1]$$

entail a constraint $s_1 \approx \Gamma_1[\Gamma_2[\cdots [\Gamma_{n-1}[\Gamma_n[s_1]]] \cdots]]$. Moreover, the rule must support variables and nonunit clauses, and it should be finitely branching if we want to incorporate it in saturation-based provers—i.e., the set of clauses derivable from a given finite set of premises by a single rule should be finite. Finally, clauses of the form $\Gamma[x] \approx s \vee \mathcal{C}$, where x occurs in $\mathcal{C}$, are problematic, because there are infinitely many instantiations of x that can result in a cyclic constraint: s, $\mathsf{c}(s)$, $\mathsf{c}(\mathsf{c}(s))$, etc. To cope with all these subtleties, we first need to develop a considerable theoretical apparatus.

Definition 12. A *chain* built on a nonempty sequence of clauses $(\mathcal{C}_1, \ldots, \mathcal{C}_n)$ under condition $\mathcal{D}$ is a sequence $(t_1, \ldots, t_{n+1})$ of terms satisfying the following conditions:

1. for every $i \in [1, n]$, C_i is of the form $s_i \approx \Gamma_i[s'_{i+1}]_{p_i} \vee C'_i$, where p_i is a nonempty constructor position in Γ_i;
2. there exists a substitution σ such that either σ is an mgu of $E = \{s'_i \stackrel{?}{=} s_i \mid i \in [2, n]\}$ or σ is an mgu of $\{s'_{n+1} \stackrel{?}{=} s_1\} \cup E$;
3. $t_i = s_i\sigma$ for $i \in [1, n]$ and $t_{n+1} = s'_{n+1}\sigma$;
4. $\mathcal{D} = \bigvee_{i=1}^{n} C'_i\sigma$;
5. $\mathrm{type}(t_1) \sim \cdots \sim \mathrm{type}(t_{n+1})$;
6. $(\Gamma_i[s'_{i+1}]_{p_i} \approx s_i)\sigma$ is strictly maximal in $C_i\sigma$, and no literal is selected in $C_i\sigma$;
7. $s_i\sigma \not\succ \Gamma_i[s'_{i+1}]_{p_1}\sigma$, for $i \in [1, n]$;
8. for every $i \in [2, n]$, s'_i is not a variable.

The expression $\Gamma_1[\cdots[\Gamma_n[\bullet]_{p_n}]\cdots]_{p_1}\sigma$ is the chain's constructor context, σ is its mgu, and $p_1.\cdots.p_n$ is its constructor position. If $t_1 = t_{n+1}$, the sequence is called a *cycle*. A chain is *direct* if $t_i \neq t_j$ for all $i, j \in [1, n+1]$ with $i \neq j$ and $\{i, j\} \neq \{1, n+1\}$, and *variable-ended* if s'_{n+1} is a variable.

Remark 13. Conditions 5 to 8 are optional. They help prune the search space.

Definition 14. A chain $(t_1, \ldots, t_{n+1})$ built on a clause sequence $(C_1, \ldots, C_n)$ is an *extension* of an acyclic chain $(s_1, \ldots, s_{m+1})$ if $n \geq m$, the latter chain is built on $(C_1, \ldots, C_m)$, and the same literals and positions are considered in each clause C_i in both chains.

Since chains can be arbitrarily long, we need to impose some additional conditions to prune them and ensure that the rules are finitely branching. Let **Keep** be a property of chains that fulfills the following requirements:

(i) if a chain $\bar{t}$ does not satisfy **Keep**, no extension of $\bar{t}$ satisfies **Keep**;
(ii) for every finite clause set N, the set of chains built on a sequence of renamings of clauses in N and satisfying **Keep** is finite;
(iii) for every cycle $(t_1, \ldots, t_n, t_1)$, there exists a chain $(s_1, \ldots, s_m)$ with $m \leq n$ satisfying **Keep** such that for some k, the cycle $(t_{1+k}, \ldots, t_{n+k}, t_{1+k})$ (with $t_i := t_{i-n}$ if $i > n$) is an extension of $(s_1, \ldots, s_m)$.

For example, **Keep** can be defined as the set of chains built on clauses C_i that are pairwise distinct modulo renaming and such that C_1 is the most recently processed clause. This is the definition we use in our description of the extended saturation loop (Sect. 6) and in the implementation in Vampire.

Remark 15. Condition (i) is essential in practice, to ensure that the chains can be incrementally constructed in an efficient way, because it ensures that the construction can be stopped when a prefix not satisfying **Keep** is obtained. Condition (ii) ensures that the rule is finitely branching. Condition (iii) is essential for completeness.

Definition 16. A chain of length n is *eligible* if it is variable-ended and $n = 1$, or if it is not variable-ended, it satisfies **Keep**, and either it is a cycle or there exists an extension of length $n + 1$ that does not satisfy **Keep**.

Remark 17. The conditions on eligible chains are the strongest ones preserving completeness, but they are not necessary for soundness.

The acyclicity rule follows:

$$\frac{\mathcal{C}_1 \quad \cdots \quad \mathcal{C}_n}{\mathcal{D} \vee \mathcal{E}} \text{ Acycl}$$

if there exists a direct, eligible chain $(t_1, \ldots, t_{n+1})$ built on $(\mathcal{C}_1, \ldots, \mathcal{C}_n)$ under condition $\mathcal{D}$ and either $t_1 = t_{n+1}$ and $\mathcal{E} = \varnothing$ or $t_1 \neq t_{n+1}$ and $\mathcal{E} = \neg \mathsf{sub}(t_1, t_{n+1})$.

Intuitively, the existence of the chain guarantees (if $\mathcal{D}$ is false) that there exists a nonempty constructor context $\Gamma[\bullet]_p$ such that $t_1 \approx \Gamma[t_{n+1}]_p$ holds. If $t_1 = t_{n+1}$, this contradicts acyclicity. Otherwise, we deduce that t_1 cannot occur at a constructor position inside the constructor term corresponding to t_{n+1}; hence $\mathsf{sub}(t_1, t_{n+1})$ is false.

Lemma 18 (Soundness of Acycl). *Let N be a clause set, and let $\mathcal{I}$ be a sub-minimal model of N satisfying* **Acy**. *If $\mathcal{C}$ is derived from N by* Acycl, *then $\mathcal{I} \models \mathcal{C}$.*

Uniqueness of Fixpoints. The uniqueness rule also depends on the notion of chain:

$$\frac{\mathcal{C}_1 \quad \cdots \quad \mathcal{C}_n}{\mathcal{D} \vee \left(\bigvee_{p \in P} u|_p \not\approx \mathsf{app}(s_p, t_1)\right) \vee u' \not\approx z \vee z \approx t_1} \text{ Uniq}$$

if there exists an eligible chain $(t_1, \ldots, t_{n+1})$ of constructor context $\Gamma[\bullet]_q$ built on $(\mathcal{C}_1, \ldots, \mathcal{C}_n)$ under condition $\mathcal{D}$ and the following requirements are met:

1. $u = \Gamma[t_{n+1}]_q$;
2. P is the set of prefix-minimal positions p of some type $\tau \sim \mathsf{type}(t_1)$ in u with $p \not< q$;
3. for every $p \in P$, s_p is a fresh variable of the appropriate context type;
4. u' is obtained from u by replacing all terms at a position $p \in P$ by $\mathsf{app}(s_p, z)$.

Intuitively, the existence of the chain ensures (if $\mathcal{D}$ is false) that $t_1 \approx \Gamma[t_{n+1}]_q$. If $t_1 = t_{n+1}$, we could derive $y \not\approx \Gamma[y]_q \vee y \approx t_1$ by uniqueness. However, this would not be sufficient for completeness. First, t_1 may be distinct from t_{n+1}, but we may have $t_{n+1} = \Delta[t_1]_Q$, for some constructor context Δ, in which case we should derive $y \not\approx \Gamma[\Delta[y]_Q]_q \vee y \approx t_1$ instead. Second, t_1 may also occur at other positions in Γ. To capture all these cases using a finitely branching rule, we introduce new variables s_p whose purpose is to denote the context Γ_p such that $\Gamma_p[t_1] = u|_p$. (If t_1 does not occur inside $u|_p$, then Γ_p is constant.)

Example 19. From the clause $\mathsf{a} \approx \mathsf{c}(\mathsf{b}, x)$, using the chain (a, x), with the constructor context $\mathsf{c}(\mathsf{b}, \bullet)$, we derive

$$\mathsf{b} \not\approx \mathsf{app}(x_1, \mathsf{a}) \vee x \not\approx \mathsf{app}(x_2, \mathsf{a}) \vee z \not\approx \mathsf{c}(\mathsf{app}(x_1, z), \mathsf{app}(x_2, z)) \vee z \approx \mathsf{a}$$

Then $u = \mathsf{c}(\mathsf{b}, x)$ and $P = \{1, 2\}$.

From the clauses $a \approx c(b, a)$ and $b \approx d(a, a)$, using the chain (a, b, a), with the constructor context $c(d(a, \bullet), a)$, we derive

$$a \not\approx \mathsf{app}(x_{1.1}, a) \vee a \not\approx \mathsf{app}(x_{1.2}, a) \vee a \not\approx \mathsf{app}(x_2, a)$$
$$\vee \, z \not\approx c(c(\mathsf{app}(x_{1.1}, z), \mathsf{app}(x_{1.2}, z)), \mathsf{app}(x_2, z)) \vee z \approx a$$

In this case, $u = c(d(a, a), x)$ and $P = \{1.1, 1.2, 2\}$.

Lemma 20 (Soundness of Uniq). *Let N be a clause set, and let $\mathcal{I}$ be a model of $N \cup \{\mathsf{App}, \mathsf{Hole}\}$ satisfying* **FP***. If C is derived from N by* Uniq*, then $\mathcal{I} \models C$.*

We also introduce the following optional simplification rule:

$$\frac{\Gamma_n[\cdots[\Gamma_1[s']_P]\cdots]_P \approx s \vee C}{(\Gamma_1[s]_P \approx s \vee C)\sigma} \; \text{Compr}$$

where s and s' are terms of the same type $\tau \in \mathcal{T}_{\mathsf{coind}}$ and P is a nonempty set of constructor positions in Γ_i, for $i \in [1, n]$, such that $\varepsilon \notin P$, and $\sigma = \mathrm{mgu} \{s \stackrel{?}{=} s', \Gamma_1 \stackrel{?}{=} \cdots \stackrel{?}{=} \Gamma_n\}$.

Proposition 21 (Soundness of Compr). *Let N be a clause set, and let $\mathcal{I}$ be a model of N satisfying* **FP***. If $\mathcal{D}$ is derived from N by* Compr*, then $\mathcal{I} \models \mathcal{D}$.*

5 Refutational Completeness

We establish the refutational completeness of the calculus presented in Sect. 4. This result ensures that the axioms for distinctness, injectivity, and acyclicity (NSub) may be omitted. The axiom Uniq may also be omitted in some cases, formally defined below. The axiom Sub is still needed since it is used in the completeness proof for Acycl.

If $N \not\ni \square$ is a clause set saturated under $\mathcal{SP}$, then R_N denotes the set of rewrite rules constructed as usual from N and $\rightarrow_{R_N}$ denotes the (one-step) reduction relation. We refer to the literature [2,16] for details about the construction of R_N. The notation $\mathcal{M}_N$ denotes the model of N defined by the congruence $\stackrel{*}{\leftrightarrow}_{R_N}$ on ground terms.

We first establish some results about the form of the rules in R_N.

Proposition 22. *Let N be a clause set saturated under $\mathcal{SP}$ and* Inf*. Let $u \approx v \vee C \in N$, and let θ be a substitution such that $u\theta \succ v\theta$, $(u \approx v)\theta \succ C\theta$, and $\mathcal{M}_N \not\models C\theta$. If $\mathrm{type}(u) \in \mathcal{T}_{\mathsf{ind}} \cup \mathcal{T}_{\mathsf{coind}}$, then u is not a variable.*

Corollary 23. *Let N be a clause set saturated under $\mathcal{SP}$ and* Inf*. For every rule $c(\bar{t}) \rightarrow_{R_N} s$ in R_N, where c is a constructor, s is R_N-irreducible.*

Lemma 24 (Infiniteness). *Let N be a clause set saturated under $\mathcal{SP}$ and* Inf*. If $\square \notin N$, then $\mathcal{M}_N$ satisfies* **Inf***.*

Lemma 25 (Distinctness). *Let N be a clause set saturated under $\mathcal{SP}$, $Dist_1$, $Dist_2$, and Inf. For all ground terms $a = \mathsf{c}(\bar{a})$ and $b = \mathsf{d}(\bar{b})$ such that $\mathsf{c} \bowtie \mathsf{d}$, we have $a \not\overset{*}{\rightarrow}_{R_N} b$.*

Lemma 26 (Injectivity). *Let N be a clause set saturated under $\mathcal{SP}$, Inf, Inj_1, and Inj_2. For all ground terms $a = \mathsf{c}(a_1, \ldots, a_n)$ and $b = \mathsf{c}(b_1, \ldots, b_n)$ with $\mathsf{c} \in \mathcal{C}tr_{inj}$ and such that $a_i \not\overset{*}{\rightarrow}_{R_N} b_i$ for some $i \in [1, n]$, we have $a \not\overset{*}{\rightarrow}_{R_N} b$.*

The completeness proof for acyclicity requires further definitions and results.

Definition 27. Let $\mathcal{I}$ be an interpretation and t be a term. A constructor context $\Gamma[\bullet]_p$ is a *minimal cyclicity witness* for t and $\mathcal{I}$ if it is of the same type as t, p is a position of the same type as t in Γ, $\mathcal{I} \models t \approx \Gamma[t]_p$, and $|q| \geq |p|$ for every position $q \neq \varepsilon$ and constructor context $\Delta[\bullet]_q$ such that $\mathcal{I} \models t \approx \Delta[t]_q$.

Proposition 28. *Let $(t_1, \ldots, t_n, t_1)$ be a cycle of constructor context $\Gamma[\bullet]_p$ for a clause set N under condition $\mathcal{D}$. If $\mathcal{I} \models N \cup \{\neg \mathcal{D}\sigma\}$, and $\Gamma[\bullet]_p$ is a minimal cyclicity witness for $t_1\sigma$ and $\mathcal{I}$, then $(t_1, \ldots, t_n, t_1)$ is direct.*

Lemma 29. *Let $t : \tau$ and $s : \upsilon$ be ground terms with $\tau \sim \upsilon$. Let $\Gamma[\bullet]_p$ be a ground constructor context of type τ, where p is a position of type υ in Γ. Let N be a clause set saturated under $\mathcal{SP}$ and Inf. Assume that t, s, and $\Gamma|_{p'}$ are R_N-irreducible, for every position $p' \not\leq p$. If $\mathcal{M}_N \models \Gamma[s]_p \approx t$, then R_N contains n rules $\Gamma_i[a_{i+1}]_{p_i} \rightarrow_{R_N} a_i$, for $i \in [1, n]$, with $\Gamma[s]_p = \Gamma_0[\Gamma_1[\cdots [\Gamma_n[a_{n+1}]_{p_n}] \cdots]_{p_1}]_{p_0}$, $p_0.p_1.\cdots.p_n = p$, $a_{n+1} = s$, and $t = \Gamma_0[a_1]_{p_0}$.*

Lemma 30 (Acyclicity). *If $\mathsf{Sub} \subseteq N$ and $N \not\ni \square$ is saturated under $\mathcal{SP}$, Acycl, and Inf, then $\mathcal{M}_N$ satisfies condition $\mathbf{Acy}$.*

Remark 31. The Inf rule is needed for completeness. For example, it is clear that the clause $x \approx \mathsf{a} \vee x \approx \mathsf{b}$ contradicts acyclicity, but no contradiction can be derived without using Inf. The relaxation of the application conditions of Sup is also essential. Consider the set $N = \{\mathsf{a}_1 \approx \mathsf{c}(\mathsf{a}_2), \mathsf{a}_2 \approx \mathsf{a}_3, \mathsf{a}_3 \approx \mathsf{c}(\mathsf{a}_1)\}$, with $\mathsf{c}(\ldots) \succ \mathsf{a}_{i+1} \succ \mathsf{a}_i$. It is clear that N is saturated without the relaxation, and N contradicts acyclicity, since $N \models \mathsf{a}_1 \approx \mathsf{c}(\mathsf{c}(\mathsf{a}_1))$. With the relaxation, Sup derives the clause $\mathsf{a}_2 \approx \mathsf{c}(\mathsf{a}_1)$; then Acycl exploits the cycle $(\mathsf{a}_1, \mathsf{a}_2, \mathsf{a}_1)$ to derive $\square$.

For the Uniq rule, we provide a restricted completeness result, under the assumption that the considered constructor context contains at most one occurrence of $\bullet$.

Lemma 32 (Uniqueness of Fixpoints). *If $\mathsf{App} \subseteq N$ and $N \not\ni \square$ is saturated under $\mathcal{SP}$, Uniq, and Inf, then $\mathcal{M}_N \models x \approx \Gamma[x]_r \wedge y \approx \Gamma[y]_r \Rightarrow x \approx y$ for every constructor context of the form $\Gamma[\bullet]_r$ of type $\tau \in \mathcal{T}_{coind}$, where r is a nonempty position of type τ in Γ.*

Definition 33. A signature is *coinductively nonbranching* if for every constructor $\mathsf{c} : \tau_1 \times \cdots \times \tau_n \rightarrow \tau$ such that $\tau \in \mathcal{T}_{coind}$, there exists at most one $i \in [1, n]$ such that $\tau_i \sim \tau$.

For example, the signature is coinductively nonbranching for infinite streams and possibly infinite lists, but not for infinite binary trees.

Corollary 34 (Fixpoints). *Assume that the signature is coinductively non-branching. If* $\mathsf{Cycl} \cup \mathsf{App} \subseteq N$ *and* $N \not\ni \square$ *is saturated under* $\mathcal{SP}$, Uniq, *and* Inf, *then* $\mathcal{M}_N$ *satisfies condition* **FP**.

Example 35. Corollary 34 does not hold for arbitrary signatures. The clause set $\{\mathsf{a} \approx \mathsf{c}(\mathsf{d}(\mathsf{a},\mathsf{b})),\ \mathsf{b} \approx \mathsf{e}(\mathsf{d}(\mathsf{a},\mathsf{b})),\ \mathsf{a}' \approx \mathsf{c}(\mathsf{d}(\mathsf{a}',\mathsf{b}')),\ \mathsf{b}' \approx \mathsf{e}(\mathsf{d}(\mathsf{a}',\mathsf{b}')),\ \mathsf{d}(\mathsf{a},\mathsf{b}) \not\approx \mathsf{d}(\mathsf{a}',\mathsf{b}')\}$ contradicts **FP**, because $\mathsf{d}(\mathsf{a},\mathsf{b})$ and $\mathsf{d}(\mathsf{a}',\mathsf{b}')$ are both solutions of $x \approx \mathsf{d}(\mathsf{c}(x),\mathsf{e}(x))$. However, the Uniq rule applies only with constructor contexts of head symbol c (if the chain starts with a or a') or e (if it starts with b or b').

6 Saturation Procedure

The inference rules of the calculus presented in Sect. 4 are all finitely branching, provided that the eligibility criterion is applied for the Acycl and Uniq rules. As a result, saturation of a clause set can be carried out using standard saturation procedures. These generally work by maintaining a set of *passive clauses* that initially contains all the clauses to saturate and a set of *active clauses* that is initially empty. The algorithm heuristically chooses a passive clause that becomes the *given clause*, moves it to the active clauses, and performs all possible inferences between it and the active clauses. Conclusions are added to the set of passive clauses, and the procedure is iterated until $\square$ is derived, or until the set of passive clauses is empty.

To improve search, it is useful to distinguish between *simplifying rules* and *generating rules*. In simplifying rules, at least one of the premises is redundant with respect to the conclusion. The Inf rule is simplifying, as well as the Dist_1 and Inj_1 rules when the term t is not a variable, and the Acycl rule when there is only one premise and $t_1 = t_n$.

In addition to the calculus, we propose the following simplifying rules to eliminate theory tautologies:

$$\frac{\mathsf{c}(\bar{s}) \not\approx \mathsf{d}(\bar{t}) \vee \mathcal{C}}{\varnothing}\ \mathsf{Dist}^- \qquad\qquad \frac{s \not\approx \Gamma[s] \vee \mathcal{C}}{\varnothing}\ \mathsf{Acycl}^-$$

where $\mathsf{c} \bowtie \mathsf{d}$, $\Gamma[\bullet]$ is a nonempty constructor context, and $\mathrm{type}(s) \in \mathcal{T}_{\mathsf{ind}}$. Moreover, the following rule applies injectivity of $\mathsf{c} \in \mathcal{C}tr_{\mathsf{inj}}$ to simplify literals:

$$\frac{\mathsf{c}(s_1,\ldots,s_n) \not\approx \mathsf{c}(t_1,\ldots,t_n) \vee \mathcal{C}}{\left(\bigvee_{i=1}^{n} s_i \not\approx t_i\right) \vee \mathcal{C}}\ \mathsf{Inj}^-$$

The soundness of Inj^- follows from c's being a function symbol, but since it is also injective, the premise is redundant with respect to the theory. We conjecture that the addition of these simplification rules preserves refutational completeness.

If all constructors are free (i.e., $Ctr_{inj} = Ctr$ and $c \bowtie d$ holds for all distinct constructors c and d), by applying the above rules eagerly, we also guarantee that in any literal $[\neg]s \approx t$ in an active clause, at most one of s or t has a constructor for head symbol, as (dis)equalities between constructor terms will have been simplified directly after clause generation. This invariant enables a few optimizations in the implementation of the generating rules, notably during the detection of chains.

The relaxation of the application conditions of the Sup rule increases the number of clauses it must generate and may hence be detrimental to the search. We can reduce the incidence of this scenario by choosing a term order that considers constructors as smaller than non-constructors. For path orders, we can choose a symbol precedence $\succ$ such that $f \succ c$ for all non-constructor symbols f and constructors c.

To implement the Acycl and Uniq rules, we must be able to efficiently detect eligible chains among the set of active clauses. Testing all subsets of the active clauses is impractical, and the detection of a chain requires the computation of an mgu over a set of equations, instead of a single equation. We present a procedure that takes the given clause $\mathcal{C}_1$ as input and applies the two rules to all subsets of clauses containing $\mathcal{C}_1$ and upon which an eligible chain can be built. There are three cases in which the rules must be applied: when the chain is a cycle, when it is variable-ended and has length 1, and when there exists an extension of the chain that violates **Keep**. The procedure relies on a data structure that provides a $\mathsf{nextLinks}(s')$ operation, where s' is a term. For each literal $s \approx t$ in an active clause $\mathcal{C}$ such that s is unifiable with s' under an mgu σ and $s\sigma \not\succeq t\sigma$, the operation returns the tuple $(\mathcal{C}, \sigma, T)$, where T is the set of terms under nonempty constructor positions in t. This operation can be implemented using term indexing techniques already found in state-of-the-art provers [22, Sect. 5.1].

The procedure $\mathsf{considerGiven}(\mathcal{C}_1)$ applies the rule Acycl or Uniq to all subsets of actives clauses that contain the given clause $\mathcal{C}_1$ and form an eligible chain:

Procedure $\mathsf{considerGiven}(\mathcal{C}_1)$ **is**
    for s'_2 such that $\mathcal{C}_1 = s_1 \approx \Gamma[s'_2] \vee \mathcal{D}_1$ **do**
        $\mathsf{extendChain}(s_1, s'_2, \{\}, \{\mathcal{C}_1\})$

Procedure $\mathsf{extendChain}(s_1, s'_i, \theta, Ch)$ **is**
    if $s_1\theta = s'_i\theta$ **then**
        apply rule Acycl or Uniq to chain Ch under mgu θ
    else if s'_i is a variable **then**
        if $|Ch| = 1$ **then**
            apply rule Acycl or Uniq to chain Ch under mgu θ
    else if exists $(\mathcal{C}_i, \sigma, T) \in \mathsf{nextLinks}(s'_i\theta)$ such that $\mathcal{C}_i \in Ch$ **then**
        apply rule Acycl or Uniq to chain Ch under mgu θ
    else
        for $(\mathcal{C}_i, \sigma, T) \in \mathsf{nextLinks}(s'_i\theta)$ **do**
            for $s'_{i+1} \in T$ **do**
                $\mathsf{extendChain}(s_1, s'_{i+1}, \sigma\theta, Ch \uplus \{\mathcal{C}_i\})$

7   Evaluation

We implemented the calculus presented above in the first-order theorem prover Vampire [13]. Our source code is publicly available.[1] The new rules are added to the existing calculus, which includes other sound rules and a sophisticated redundancy elimination mechanism. Vampire can process input files in SMT-LIB [4] format and recognizes both the `declare-datatypes` command and the non-standard `declare-codatatypes` command. These commands trigger the addition of relevant axioms or the activation of inference rules, according to user-specified options. This implementation is an extension of previous work done in Vampire [12]. The behavior of this older implementation can be replicated by enabling only the simplification rules of the calculus and adding the axioms Dist, Inj, Exhaust, Sub, and NSub to the initial clause set.

We evaluated the implementation on 4170 problems that were used previously by Reynolds and Blanchette [20] to evaluate CVC4. These were generated by translating Isabelle problems to SMT-LIB using the Sledgehammer bridge [18]. We also used synthetic problems that exercise the properties of cyclic values. Both benchmark sets and detailed results are available online.[2]

All the experiments in this section were carried out on a cluster on which each node is equipped with two quad-core Intel processors running at 2.4 GHz, with 24 GiB of memory. A 60 s time limit per problem was enforced. We used a single basic saturation strategy relying on the DISCOUNT saturation algorithm. The calculus was parameterized by a Knuth–Bendix term order, unless otherwise noted. This simple approach provides a homogeneous basis on which to compare the performance of the different procedures. It typically solves fewer problems than the portfolio approach commonly used with Vampire, in which several different strategies are tried in short time slices.

We first compare the performance of three configurations of the prover on the Isabelle problems. The first configuration corresponds to the axiomatic approach presented in Sect. 3: the axioms Dist, Inj, Exhaust, Sub, NSub, App, Uniq, Cycl, and Hole are added to the set of clauses to saturate, and only standard inferences rules are used by the prover. Superposition need not rewrite the nonmaximal side of an equation.

The second configuration implements part of the calculus presented in Sect. 4. Only the axioms Exhaust, Sub, NSub, App, Uniq, Cycl, and Hole are added to the clauses, and the rules $Dist_1$, $Dist_2$, Inj_1, and Inj_2 are used during the search, in addition to the simplification rules described in Sect. 6. The side conditions of Sup are also relaxed. The rules Acycl and Uniq are not used; instead, reasoning on the properties of cyclic terms is based on axioms.

The third configuration uses all the rules described in Sect. 4. Only the axioms Sub and App are added, on which the Acycl and Uniq rules depend, and the axioms Cycl and Exhaust. This configuration is the only one which does not

---

[1] http://github.com/vprover/vampire/releases/tag/ijcar2018-data.
[2] http://matryoshka.gforge.inria.fr/pubs/supdata_data.tar.gz.

ensure refutational completeness, since Uniq is incomplete with respect to the uniqueness of fixpoints for branching codatatypes.

The first two configurations both solved 1114 problems and the third one solved 1113 problems; 1116 problems are solved by at least one configuration. These homogeneous results do not reveal significant differences between the approaches. To assess the role of the acyclicity property of datatypes and the properties of codatatype fixpoints in the benchmarks, we also tested a system that did not include any axioms and rules related to these properties. With such an incomplete system, we found that 12 problems could not be solved. This is roughly in line with the results of Reynolds and Blanchette using CVC4 on the same problems [20]. No new problems were solved by this configuration, suggesting that reasoning about properties of cyclic terms does not lead to worse performance even when these properties are not needed for refutation.

We also tested variants of the last two configurations in which the calculus was parameterized by a lexicographic path order, to assess whether this term order could improve the performance when used with the relaxed superposition rule. These configurations solved a total of 1104 problems, including 5 new problems.

Since properties of cyclic values are seldom used in the Isabelle benchmarks, we crafted a set of (refutable) problems to assess the performance of the rules Acycl and Uniq. For a term s and a nonconstant context $\Gamma[\bullet]$, let $\mathsf{exchain}(s, \Gamma[\bullet])$ denote any sentence $\exists s_2, \ldots s_n \forall t_1, \ldots t_m.\ s \approx \Gamma_1[s_2] \wedge \ldots \wedge s_n \approx \Gamma_n[s]$, where $t_1, \ldots, t_m$ all occur in Γ and such that $\Gamma_1[\ldots [\Gamma_n[\bullet]] \ldots] = \Gamma[\bullet]$. The formula $\exists s.\, \mathsf{exchain}(s, \Gamma[\bullet])$, where $\mathrm{type}(s) \in \mathcal{T}_{\mathsf{ind}}$, forms an acyclicity problem. The set of acyclicity problems used in our experiments is denoted AC. If $m = 0$, the clausified form of this problem is ground (ACG). The formula $\exists s_1, s_2.\, \mathsf{exchain}(s_1, \Gamma[\bullet]) \wedge \mathsf{exchain}(s_2, \Gamma[\bullet]) \wedge s_1 \not\approx s_2$, where $\mathrm{type}(s_1) \in \mathcal{T}_{\mathsf{coind}}$, forms a uniqueness problem (U). Note that in such a problem, the two chains may not be formed upon the same equalities, although they build the same constructor context. Similarly, if $m = 0$, we obtain a ground uniqueness problem (UG). Finally, the sentence $\forall s.\, \neg\, \mathsf{exchain}(s, \Gamma[\bullet])$, for $\mathrm{type}(s) \in \mathcal{T}_{\mathsf{coind}}$, forms an existence problem (EX).

We generated 100 instances of each type of problem. The number of problems solved by Vampire (V) on these problems are presented in the following table, along with the results obtained using CVC4's [3] and Z3's [15] native support for datatypes and, in CVC4's case, for codatatypes:

	AC			ACG			U		UG		EX	
	V	CVC4	Z3	V	CVC4	Z3	V	CVC4	V	CVC4	V	CVC4
Axioms	65	–	–	100	–	–	14	–	10	–	40	–
Calculus	82	100	59	100	100	100	14	12	13	100	35	0

The number of problems solved shows that the Acycl rule performs better than the axioms for acyclicity problems with variables. Only one of these problems could be solved by the axiomatic approach and not by the Acycl rule. Both approaches managed to solve all of the ground acyclicity problems. Z3 solved all of the ground problems, performing slightly less well on those featuring universal quantifiers. CVC4 was able to solve all of the acyclicity problems, including those with universal quantifiers, a notable improvement over previous results obtained on similar problems [22, Sect. 6].

On uniqueness problems, the Uniq rule solved a superset of the ground problems solved by the axiomatic approach, whereas on nonground problems each approach uniquely solved 3 problems, for a total of 17 problems solved. Again, CVC4 performed remarkably well on ground problems, while the presence of variables in the problem led to a marked degradation of its performance. Finally, for existence problems, the refutation relies mostly on the Cycl axiom, which is included in the clause set in both Vampire configurations. Yet, the purely axiomatic approach was able to solve 6 problems that could not be solved when the Uniq rule was activated, indicating that the rule might lead the search in a suboptimal direction. The theory solver in CVC4 does not take into account the existence of fixpoints for codatatypes, which is a nonground property. Consequently, none of the existence problems were solved by CVC4.

From the results, it appears that the calculus supersedes the axiomatic approach for problems with datatypes. For codatatypes, both approaches solve different problems, suggesting that they should both be included in a strategy portfolio. However, the conceptual simplicity and easy implementation of the axiomatic approach may outweigh these differences in performance.

8   Related Work

The potential of (co)datatypes for automated reasoning has been studied mostly in the context of satisfiability modulo theories (SMT). Datatypes are parts of the SMT-LIB 2.6 standard [4]. They were implemented in CVC3 by Barrett et al. [5], in Z3 [15] by de Moura, and in CVC4 by Reynolds and Blanchette [20]. The CVC4 work also includes a decision procedure for the ground theory of codatatypes. Moreover, CVC4 supports automatic structural induction [21] and dedicated reasoning support for selectors.

Structural induction has also been added to superposition by Kersani and Peltier [11], Cruanes [10], and Wand [24]. In unpublished work, Wand implemented incomplete inference rules for datatypes, including acyclicity, in his superposition prover Pirate. Robillard's earlier Acycl rule [22] has inspired our Acycl rule, but it suffered from many forms of incompleteness. For example, given the unsatisfiable clause set $\{\, \mathsf{a} \approx \mathsf{c}(x) \lor \mathsf{p}(x),\, \lnot\,\mathsf{p}(\mathsf{c}(\mathsf{a}))\}$, the old Acycl rule derived only $\mathsf{p}(\mathsf{a})$ before reaching saturation. Another issue concerned cycles built from multiple copies of the same premise.

In the context of program verification, Bjørner [6] introduced a decision procedure for (co)datatypes in STeP, the Stanford Temporal Prover. The program verification tool Dafny provides both a syntax for defining (co)datatypes and some support for automatic (co)induction proofs [14]. Other verification tools such as Leon [23] and RADA [19] also include (semi-)decision procedures for datatypes. We refer to Barrett et al. [5] and Reynolds and Blanchette [20] for further discussions of related work.

9   Conclusion

We presented two approaches to reason about datatypes and codatatypes in first-order logic: an axiomatization and an extension of the superposition calculus. We established completeness results about both. We also showed how to integrate the new inference rules in a saturation prover's main loop and implemented them in the Vampire prover. The empirical results look promising, although it is not clear from our benchmarks how often the most difficult properties—acyclicity for datatypes, existence and uniqueness of fixpoints for codatatypes—are useful in practice.

This work is part of a wider research program that aims at bridging the gap between automatic theorem provers and their applications to program verification and interactive theorem proving. In future work, we want to reconstruct the new proof rules in Isabelle, to make it possible to enable datatype reasoning in Sledgehammer. We also believe that further tuning and evaluations could help improve the calculus and the heuristics.

Acknowledgment. We thank Alexander Bentkamp, Simon Cruanes, Uwe Waldmann, Daniel Wand, and Christoph Weidenbach for fruitful discussions that led to this work. We also thank Mark Summerfield and the anonymous reviewers for suggesting textual improvements.

Blanchette has received funding from the European Research Council (ERC) under the European Union's Horizon 2020 research and innovation program (grant agreement No. 713999, Matryoshka). Robillard has received funding from the ERC Starting Grant 2014 SYMCAR 639270, the Wallenberg Academy Fellowship 2014 TheProSE, the Swedish Research Council grant GenPro D0497701, and the Austrian FWF research project RiSE S11409-N23.

References

1. Bachmair, L., Dershowitz, N., Hsiang, J.: Orderings for equational proofs. In: LICS 1986, pp. 346–357. IEEE Computer Society (1986)
2. Bachmair, L., Ganzinger, H.: Rewrite-based equational theorem proving with selection and simplification. J. Log. Comput. **4**(3), 217–247 (1994)
3. Barrett, C., Conway, C.L., Deters, M., Hadarean, L., Jovanović, D., King, T., Reynolds, A., Tinelli, C.: CVC4. In: Gopalakrishnan, G., Qadeer, S. (eds.) CAV 2011. LNCS, vol. 6806, pp. 171–177. Springer, Heidelberg (2011). https://doi.org/10.1007/978-3-642-22110-1_14
4. Barrett, C., Fontaine, P., Tinelli, C.: The SMT-LIB standard: version 2.7. Technical report, University of Iowa (2017). http://smt-lib.org/
5. Barrett, C., Shikanian, I., Tinelli, C.: An abstract decision procedure for satisfiability in the theory of inductive data types. J. Satisf. Boolean Model. Comput. **3**, 21–46 (2007)
6. Bjørner, N.S.: Integrating decision procedures for temporal verification. Ph.D. thesis, Stanford University (1998)
7. Blanchette, J.C., Hölzl, J., Lochbihler, A., Panny, L., Popescu, A., Traytel, D.: Truly modular (co)datatypes for Isabelle/HOL. In: Klein, G., Gamboa, R. (eds.) ITP 2014. LNCS, vol. 8558, pp. 93–110. Springer, Cham (2014). https://doi.org/10.1007/978-3-319-08970-6_7

8. Blanchette, J.C., Peltier, N., Robillard, S.: Superposition with datatypes and codatatypes. Technical report (2018). http://matryoshka.gforge.inria.fr/pubs/supdata_report.pdf

9. Comon, H., Lescanne, P.: Equational problems and disunification. J. Symb. Comput. **7**(3–4), 371–425 (1989)

10. Cruanes, S.: Superposition with structural induction. In: Dixon, C., Finger, M. (eds.) FroCoS 2017. LNCS (LNAI), vol. 10483, pp. 172–188. Springer, Cham (2017). https://doi.org/10.1007/978-3-319-66167-4_10

11. Kersani, A., Peltier, N.: Combining superposition and induction: a practical realization. In: Fontaine, P., Ringeissen, C., Schmidt, R.A. (eds.) FroCoS 2013. LNCS (LNAI), vol. 8152, pp. 7–22. Springer, Heidelberg (2013). https://doi.org/10.1007/978-3-642-40885-4_2

12. Kovács, L., Robillard, S., Voronkov, A.: Coming to terms with quantified reasoning. In: Castagna, G., Gordon, A.D. (eds.) POPL 2017, pp. 260–270. ACM (2017)

13. Kovács, L., Voronkov, A.: First-order theorem proving and VAMPIRE. In: Sharygina, N., Veith, H. (eds.) CAV 2013. LNCS, vol. 8044, pp. 1–35. Springer, Heidelberg (2013). https://doi.org/10.1007/978-3-642-39799-8_1

14. Leino, K.R.M., Moskal, M.: Co-induction simply. In: Jones, C., Pihlajasaari, P., Sun, J. (eds.) FM 2014. LNCS, vol. 8442, pp. 382–398. Springer, Cham (2014). https://doi.org/10.1007/978-3-319-06410-9_27

15. de Moura, L., Bjørner, N.: Z3: an efficient SMT solver. In: Ramakrishnan, C.R., Rehof, J. (eds.) TACAS 2008. LNCS, vol. 4963, pp. 337–340. Springer, Heidelberg (2008). https://doi.org/10.1007/978-3-540-78800-3_24

16. Nieuwenhuis, R., Rubio, A.: Paramodulation-based theorem proving. In: Robinson, J.A., Voronkov, A. (eds.) Handbook of Automated Reasoning, vol. I, pp. 371–443. Elsevier and MIT Press (2001)

17. Nipkow, T., Wenzel, M., Paulson, L.C. (eds.): Isabelle/HOL. LNCS, vol. 2283. Springer, Heidelberg (2002). https://doi.org/10.1007/3-540-45949-9

18. Paulson, L.C., Blanchette, J.C.: Three years of experience with Sledgehammer, a practical link between automatic and interactive theorem provers. In: Sutcliffe, G., Schulz, S., Ternovska, E. (eds.) IWIL 2010. EPiC, vol. 2, pp. 1–11. EasyChair (2012)

19. Pham, T., Whalen, M.W.: RADA: a tool for reasoning about algebraic data types with abstractions. In: Meyer, B., Baresi, L., Mezini, M. (eds.) ESEC/FSE 2013, pp. 611–614. ACM (2013)

20. Reynolds, A., Blanchette, J.C.: A decision procedure for (co)datatypes in SMT solvers. J. Autom. Reason. **58**(3), 341–362 (2017)

21. Reynolds, A., Kuncak, V.: Induction for SMT Solvers. In: D'Souza, D., Lal, A., Larsen, K.G. (eds.) VMCAI 2015. LNCS, vol. 8931, pp. 80–98. Springer, Heidelberg (2015). https://doi.org/10.1007/978-3-662-46081-8_5

22. Robillard, S.: An inference rule for the acyclicity property of term algebras. In: Kovács, L., Voronkov, A. (eds.) Vampire 2017. EPiC, EasyChair, to appear

23. Suter, P., Köksal, A.S., Kuncak, V.: Satisfiability modulo recursive programs. In: Yahav, E. (ed.) SAS 2011. LNCS, vol. 6887, pp. 298–315. Springer, Heidelberg (2011). https://doi.org/10.1007/978-3-642-23702-7_23

24. Wand, D.: Superposition: types and Polymorphism. Ph.D. thesis, Universität des Saarlandes (2017)

Efficient Encodings of First-Order Horn Formulas in Equational Logic

Koen Claessen and Nicholas Smallbone[(✉)]

Chalmers University of Technology, Gothenburg, Sweden
{koen,nicsma}@chalmers.se

Abstract. We present several translations from first-order Horn formulas to equational logic. The goal of these translations is to allow equational theorem provers to efficiently reason about non-equational problems. Using these translations we were able to solve 37 problems of rating 1.0 (i.e. which had not previously been automatically solved) from the TPTP.

1 Introduction

Equational theorem provers such as Waldmeister [14] and Twee [20] are highly effective on equational problems, and often outperform first-order theorem provers. But they are also quite limited: while many problems require heavy use of equational reasoning, few problems consist purely of unit equations.

Take, for example, problem `LAT224-1` from the TPTP [21]. In many ways this problem is perfect for equational theorem provers. It is about lattice theory, and includes all the usual lattice axioms such as associativity, commutativity, idempotence and absorption. It also has the rather juicy-looking axiom

$$x \sqcap (y \sqcup (x \sqcap z)) = (x \sqcap y) \sqcup (x \sqcap (y \sqcup (z \sqcap (x \sqcup (y \sqcap z)))))$$

which any equational prover would love to reason about. Unfortunately, it has exactly one non-unit axiom,

$$x \sqcap y = \bot \wedge x \sqcup y = \top \rightarrow \bar{x} = y, \tag{1}$$

so we can not use an equational prover. No theorem prover is able to automatically prove `LAT224-1`: it has always had rating 1.0 on the TPTP.

It is possible to prove `LAT224-1` if we *encode* the non-unit axiom as a unit equation. Suppose we add a new function ifeq together with the axiom

$$\mathsf{ifeq}(x, x, y, z) = y.$$

The idea is that $\mathsf{ifeq}(x, y, z, w)$ represents the expression "if $x = y$ then z else w". We can then reformulate axiom (1) as the equation

$$\mathsf{ifeq}(x \sqcap y, \bot, \mathsf{ifeq}(x \sqcup y, \top, \bar{x}, y), y) = y$$

© Springer International Publishing AG, part of Springer Nature 2018
D. Galmiche et al. (Eds.): IJCAR 2018, LNAI 10900, pp. 388–404, 2018.
https://doi.org/10.1007/978-3-319-94205-6_26

and now we have a unit equality problem. When we present this transformed problem to the equational prover Twee, it solves it in 5 s.

This encoding works for any Horn formula, it is easy to automate, and it does not alter unit clauses. Thus we can use it to cheaply add Horn clause reasoning to an equational prover without weakening its equational reasoning powers.

The idea of using ifeq to encode clauses as equations is not new: it originated as a way of encoding *full* first-order logic as equations [7,16]. The fact that all first-order formulas can be encoded as equations is remarkable, but the cited encoding needs many axioms for ifeq, as well as congruence axioms for each function in the input problem, so it is not a practical way of proving theorems.

On the other hand, as LAT224-1 shows, if we take advantage of the simple nature of Horn formulas, we can come up with encodings that are simple and practical for theorem proving. These encodings let an equational theorem prover reason *efficiently* about Horn formulas, and turn it into a powerful theorem prover for mostly-equational problems.

This paper introduces several such encodings, and shows that they work in theory and in practice.

Contributions. We describe and prove correct four efficient encodings of Horn formulas into equational logic. The first two encodings are inspired by existing (but impractical) encodings for full first-order logic [7,16], and the last two are our own invention. We evaluate our encodings on the TPTP and are able to solve 37 problems of rating 1.0, in other words problems that no existing prover could automatically solve.

Notation and Definitions. A Horn clause is a clause with at most one positive literal, for example $\neg A \vee \neg B \vee C$ or $\neg A$, where A, B and C are atomic formulas; a Horn formula is a set of Horn clauses. As in first-order logic, an atomic formula is either a predicate or an equation. A Horn clause with a positive literal is a *definite clause*; a Horn clause with no positive literal is a *goal clause*. We freely write Horn clauses as implications; for example, instead of $\neg A \vee \neg B \vee C$, we often write $A \wedge B \rightarrow C$, and we also write goal clauses as $A \wedge B \rightarrow \textbf{false}$. When given a Horn formula, as is usual in theorem proving, the problem is to prove the conjunction of the clauses unsatisfiable.

When writing formulas, we adopt the convention that x, y, z and w are variables and s, t, u and v stand for terms.

2   Encoding Equational Horn Clauses as Equations

In this section we present four encodings from *equational* Horn formulas to unit equations. The encodings take as input a Horn formula with no predicate symbols (other than equality), and produce a set of unit equalities plus a set of goal clauses. In fact, the goal clauses from the input formula are passed through unchanged. In Sect. 3 we discuss how to handle predicate symbols, and in Sect. 4 we discuss different ways to handle the goal clauses.

Each of the encodings in this section consists of:

– A set of axioms which are unconditionally added to the input formula.
– A rule which eliminates *one* negative literal from a clause, by replacing a clause of the form $C \wedge s = t \rightarrow u = v$ with a clause of the form $C \rightarrow u' = v'$. The rule can also add new unit clauses to the problem.

To apply the encoding, we must add the axioms specified by the encoding, and then repeatedly apply the prescribed rule until no negative literals remain, except those that are in goal clauses.

We demonstrate all the encodings on the following example clauses:

$$f(x) = f(y) \rightarrow x = y$$
$$f(a) = b \wedge f(c) = d \rightarrow a = c$$

2.1   Encoding 1: if-then-else

We start with the encoding described in the introduction. To recap, the idea is to have a function $\mathsf{ifeq}(x, y, z, w)$ which is supposed to mean "if $x = y$ then z else w", and to encode Horn clauses using if-then-else.

First we add to the input formula the axiom

$$\mathsf{ifeq}(x, x, y, z) = y$$

where ifeq must of course be a fresh symbol.

The rule we use to eliminate a negative literal is: given a clause of the form $C \wedge s = t \rightarrow u = v$, replace it with the clause

$$C \rightarrow \mathsf{ifeq}(s, t, u, v) = v.$$

Since the term v is duplicated, if v has a greater size than u, we swap u and v before applying the encoding rule, in order to reduce formula size.

Example. The example formula above is encoded as the following three equations:

$$\mathsf{ifeq}(x, x, y, z) = y$$
$$\mathsf{ifeq}(f(x), f(y), x, y) = y$$
$$\mathsf{ifeq}(f(a), b, \mathsf{ifeq}(f(c), d, a, c), c) = c$$

The third clause above is derived as follows:

$$f(a) = b \wedge f(c) = d \rightarrow a = c$$
$$\equiv \quad f(a) = b \rightarrow \mathsf{ifeq}(f(c), d, a, c) = c$$
$$\equiv \quad \mathsf{ifeq}(f(a), b, \mathsf{ifeq}(f(c), d, a, c), c) = c$$

Efficiency. An efficient encoding should have several characteristics:

1. It should not alter unit equations, so that the prover can deal efficiently with the equational part of the problem.
2. It should increase the size of the problem as little as possible.
3. Discharging a condition should not require lots of book-keeping inferences. In other words, from the encoded versions of $s = t$ and $s = t \to u = v$ it should be easy to deduce $u = v$.
4. It should not increase the search space by allowing needless or unproductive inferences. If possible, any valid inference in the encoded problem should correspond to a reasonable inference in the original Horn problem.

The if-then-else encoding does well on most of those fronts:

1. It does not alter unit equations.
2. The term v is duplicated during encoding, so the problem may blow up. In practice, the fact that we pick v to be smaller than u helps.
3. Discharging a condition takes at most two inferences: from $s = t$ we can deduce $\mathsf{ifeq}(s, s, u, v) = v$ and from that and $\mathsf{ifeq}(x, x, y, z) = y$ we get $u = v$.
4. Assuming that the prover uses a simplification ordering, in the equation $\mathsf{ifeq}(s, t, u, v)$, the only inferences allowed will be paramodulations into s, t, u and v. The first two are useful but the last two are not; we fix this in encoding 3.

Proof of Correctness. Suppose that we are encoding the formula ϕ. We start by adding the function ifeq and the axiom $\mathsf{ifeq}(x, x, y, z) = y$ to obtain the formula ϕ_0. We then eliminate negative literals one at a time to obtain a sequence of formulas $\phi_1, \ldots, \phi_n$. That is, we obtain ϕ_{i+1} from ϕ_i by replacing one clause $C \wedge s = t \to u = v$ with $C \to \mathsf{ifeq}(s, t, u, v) = v$. Our goal is to show that ϕ and ϕ_n are equisatisfiable. A note on notation: in this paper, given a model $\mathcal{M}$ and a variable assignment σ, we write $\mathcal{M}, \sigma \models \phi$ for the valuation of a formula, and $\mathcal{M}^\sigma(t)$ for the valuation of a term.

Soundness. Given a model $\mathcal{M}$ of ϕ, we extend it to a model $\mathcal{M}_0$ of ϕ_0 by interpreting ifeq as follows:

$$\mathsf{ifeq}(x, y, z, w) = \begin{cases} z, & \text{if } x = y \\ w, & \text{otherwise} \end{cases}$$

This definition clearly satisfies the axiom $\mathsf{ifeq}(x, x, y, z) = z$, so we have $\mathcal{M}_0 \models \phi_0$. From the following lemma, it follows immediately by induction that $\mathcal{M}_0 \models \phi_i$ for all i, so in particular $\mathcal{M}_0 \models \phi_n$.

Lemma 1 (Single step soundness). *If* $\mathcal{M}_0 \models s = t \to u = v$ *then* $\mathcal{M}_0 \models \mathsf{ifeq}(s, t, u, v) = v$.

Proof. Let σ be a variable assignment such that $\mathcal{M}_0, \sigma \models s = t \rightarrow u = v$. We show that $\mathcal{M}_0, \sigma \models \mathsf{ifeq}(s, t, u, v) = v$ by case analysis on the values of s, t, u and v in M_0^σ:

- If $s \neq t$,[1] then $\mathsf{ifeq}(s, t, u, v) = v$ by definition of ifeq in $\mathcal{M}_0$.
- If $s = t$ and $u = v$, then $\mathsf{ifeq}(s, t, u, v) = \mathsf{ifeq}(s, s, v, v) = v$.   □

Completeness. Since ϕ_0 is stronger than ϕ, any model of ϕ_0 is a model of ϕ. It remains to show that if $\mathcal{M} \models \phi_n$ then $\mathcal{M} \models \phi_0$. This follows immediately by induction from the following lemma and the fact that $\mathsf{ifeq}(x, x, y, z) = y$ is an axiom of ϕ_n:

Lemma 2 (Single step completeness). *Suppose that* $\mathcal{M} \models \mathit{ifeq}(x, x, y, z) = y$. *If* $\mathcal{M} \models \mathit{ifeq}(s, t, u, v) = v$ *then* $\mathcal{M} \models s = t \rightarrow u = v$.

Proof. Given a variable assignment σ, and assuming that $\mathcal{M}, \sigma \models s = t$, we prove that $\mathcal{M}, \sigma \models u = v$. Again we drop the heavy "$\mathcal{M}, \sigma \models$" notation. From $s = t$ and $\mathsf{ifeq}(x, x, y, z) = z$ we get $\mathsf{ifeq}(s, t, u, v) = u$. Combined with the assumption $\mathsf{ifeq}(s, t, u, v) = v$ this gives $u = v$.   □

2.2   Encoding 2: if-then

The if-then-else encoding is asymmetric: when encoding the clause $s = t \rightarrow u = v$, the term v is duplicated but u is not. The if-then encoding is a symmetric variant. The encoding uses a function $\mathsf{ifeq}(x, y, z)$ which is intended to mean "if $x = y$ then z else unspecified". We add the following axiom to the input formula:

$$\mathsf{ifeq}(x, x, y) = y.$$

The rule we use to eliminate a negative literal is: given a clause of the form $C \wedge s = t \rightarrow u = v$, replace it with

$$C \rightarrow \mathsf{ifeq}(s, t, u) = \mathsf{ifeq}(s, t, v).$$

Example. The example clause set becomes:

$$\mathsf{ifeq}(x, x, y) = y$$
$$\mathsf{ifeq}(f(x), f(y), x) = \mathsf{ifeq}(f(x), f(y), y)$$
$$\mathsf{ifeq}(f(a), b, \mathsf{ifeq}(f(c), d, a)) = \mathsf{ifeq}(f(a), b, \mathsf{ifeq}(f(c), d, c))$$

Efficiency. Compared to the if-then-else encoding, the if-then encoding is likely to produce bigger equations, as the equation $\mathsf{ifeq}(s, t, u) = \mathsf{ifeq}(s, t, v)$ duplicates both s and t. It also requires more inference steps to discharge a condition, as both sides of the equation must be rewritten. However, if u and v are large terms, it may produce smaller equations than the if-then-else encoding.

---

[1] Really we mean $\mathcal{M}_0^\sigma(s) \neq \mathcal{M}_0^\sigma(t)$, but we leave out the heavy notation in this proof.

Proof of Correctness. Almost identical to the if-then-else encoding. The only change is that ϕ_0 is now ϕ together with the axiom $\mathsf{ifeq}(x, x, y) = y$, and we construct its model differently. Given a model $\mathcal{M}$ of ϕ, we extend it to a model $\mathcal{M}_0$ of ϕ_0 by interpreting ifeq as follows, where a is an arbitrary domain element of $\mathcal{M}$:

$$\mathsf{ifeq}(x, y, z) = \begin{cases} z, \text{ if } x = y \\ a, \text{ otherwise} \end{cases}$$

2.3 Encoding 3: Specialised if-then-else

The third encoding is designed to work well with Knuth-Bendix completion. The aim is to encode $s = t \rightarrow u = v$ in such a way that the resulting equations become rewrite rules in which u and v only appear on the *right-hand* side. This means that only s and t participate in critical pairs, not u and v.

The rule we use is: given a clause of the form $C \wedge s = t \rightarrow u = v$, replace it with the two clauses

$$\mathsf{fresh}_i(y, y, x_1, \ldots, x_n) = u$$
$$C \rightarrow \mathsf{fresh}_i(s, t, x_1, \ldots, x_n) = v$$

where fresh_i is a fresh function symbol and $x_1, \ldots, x_n$ is the union of the free variables of u and v.[2] We introduce a new symbol fresh_i each time the rule is applied. The idea is that $\mathsf{fresh}_i(x, y, x_1, \ldots, x_n)$ represents the expression "if $x = y$ then u else v".[3] Once the prover derives $s = t$, the two equations can be combined to yield $u = v$. Thus, fresh_i is really the function symbol ifeq from before (which is not used in this encoding), *specialised* to the implication at hand, which removes the need to have u and v as an argument to ifeq.

In our testing, the encoding works best if we always let u be the smaller term and v the bigger term of the positive literal, ordering by weight—that is, the opposite way to the if-then-else encoding. We are not sure why.

Efficiency. If the two equations above are oriented left-to-right, the only inference a prover can make is to paramodulate into s and t in order to make them equal. Once s and t are made equal, the two rules can be combined to derive $u = v$. Thus the search space is reduced: the theorem prover effectively simulates a first-order prover working forward from positive unit literals.

Example. Take the example clause set. The first clause is encoded as

$$\mathsf{fresh}_1(z, z, x, y) = x$$
$$\mathsf{fresh}_1(f(x), f(y), x, y) = y.$$

---

[2] Since we are working with clauses, free variables are universally quantified.

[3] The arguments $x_1, \ldots, x_n$ help to unambiguously identify u and v. Without them, this interpretation of fresh_i would not make sense and the encoding would be unsound.

The second clause is first transformed into

$$\mathsf{fresh}_2(x, x) = a$$
$$f(a) = b \rightarrow \mathsf{fresh}_2(f(c), d) = c$$

and this latter clause becomes

$$\mathsf{fresh}_3(x, x) = \mathsf{fresh}_2(f(c), d)$$
$$\mathsf{fresh}_3(f(a), b) = c.$$

The final result is five equations:

$$\mathsf{fresh}_1(z, z, x, y) = x$$
$$\mathsf{fresh}_1(f(x), f(y), x, y) = y$$
$$\mathsf{fresh}_2(x, x) = a$$
$$\mathsf{fresh}_3(x, x) = \mathsf{fresh}_2(f(c), d)$$
$$\mathsf{fresh}_3(f(a), b) = c$$

We argued above that we would like each of these equations to be oriented left-to-right, but for the fourth equation it is not clear if that will happen. This suggests that the encoding could be improved if we could give the prover an appropriate ordering for the fresh symbols. See Sect. 2.5 for another solution.

Proof of Correctness. We first introduce some notation. We write $\overrightarrow{x}$ for $(x_1, \ldots, x_n)$, the sequence of all free variables of u and v. If σ is a variable assignment, we write $\sigma(\overrightarrow{x})$ for $(\sigma(x_1), \ldots, \sigma(x_n))$. If $\mathcal{M}$ is an interpretation and $(c_1, \ldots, c_n)$ are domain elements then we write $\mathcal{M}^{\overrightarrow{c}}(u)$ or $\mathcal{M}^{\overrightarrow{c}}(v)$ for the value of u or v under the variable assignment $\{x_1 \mapsto c_1, \ldots, x_n \mapsto c_n\}$. Note that if $\sigma(\overrightarrow{x}) = \overrightarrow{c}$ then $\mathcal{M}^{\sigma}(u) = \mathcal{M}^{\overrightarrow{c}}(u)$ and $\mathcal{M}^{\sigma}(v) = \mathcal{M}^{\overrightarrow{c}}(v)$.

It is enough to show that a single application of the encoding rule is sound and complete. In other words, if ϕ is a formula which contains the clause

$$C \wedge s = t \rightarrow u = v,$$

and ϕ_{enc} is ϕ with that clause replaced by the following two clauses:

$$\mathsf{fresh}(y, y, \overrightarrow{x}) = u$$
$$C \rightarrow \mathsf{fresh}(s, t, \overrightarrow{x}) = v,$$

then we must show that ϕ and ϕ_{enc} are equisatisfiable.

Soundness. Suppose $\mathcal{M}$ is a model of ϕ. We extend $\mathcal{M}$ to a model $\mathcal{M}_{\mathsf{enc}}$ of ϕ_{enc} by interpreting fresh as follows:

$$\mathsf{fresh}(a, b, \overrightarrow{c}) = \begin{cases} \mathcal{M}^{\overrightarrow{c}}(u), & \text{if } a = b \\ \mathcal{M}^{\overrightarrow{c}}(v), & \text{otherwise} \end{cases}$$

Note that $\mathcal{M}$ and $\mathcal{M}_{\mathsf{enc}}$ agree on the truth of any formula not involving fresh, and that since fresh was freshly generated, it does not occur in s, t, u, v or C.

We need to check that both of the new clauses of ϕ_{enc} hold. First we show that $\mathcal{M}_{\mathsf{enc}} \models \mathsf{fresh}(y, y, \overrightarrow{x}) = u$: given a variable assignment σ, let $\overrightarrow{c} = \sigma(\overrightarrow{x})$; then $\mathcal{M}_{\mathsf{enc}}^{\sigma}(\mathsf{fresh}(y, y, \overrightarrow{x})) = \mathsf{fresh}(\sigma(y), \sigma(y), \overrightarrow{c}) = \mathcal{M}_{\mathsf{enc}}^{\overrightarrow{c}}(u) = \mathcal{M}_{\mathsf{enc}}^{\sigma}(u)$.

Then we assume that σ is a variable assignment such that $\mathcal{M}_{\mathsf{enc}}, \sigma \models C$, and show that $\mathcal{M}_{\mathsf{enc}}, \sigma \models \mathsf{fresh}(s, t, \overrightarrow{x}) = v$. Let $\overrightarrow{c} = \sigma(\overrightarrow{x})$. Note that since fresh does not occur in C, we must have $\mathcal{M}, \sigma \models C$. Recalling that $\mathcal{M}, \sigma \models C \wedge s = t \rightarrow u = v$, the proof proceeds by case analysis:

- If $\mathcal{M}, \sigma \not\models s = t$, then $\mathcal{M}_{\mathsf{enc}}^{\sigma}(\mathsf{fresh}(s, t, \overrightarrow{x})) = \mathsf{fresh}(\mathcal{M}^{\sigma}(s), \mathcal{M}^{\sigma}(t), \overrightarrow{c}) = \mathcal{M}^{\overrightarrow{c}}(v) = \mathcal{M}_{\mathsf{enc}}^{\sigma}(v)$, since $\mathcal{M}^{\sigma}(s) \neq \mathcal{M}^{\sigma}(t)$.
- If $\mathcal{M}, \sigma \models s = t$ and $\mathcal{M}, \sigma \models u = v$, then $\mathcal{M}_{\mathsf{enc}}^{\sigma}(\mathsf{fresh}(s, t, \overrightarrow{x})) = \mathsf{fresh}(\mathcal{M}^{\sigma}(s), \mathcal{M}^{\sigma}(t), \overrightarrow{c}) = \mathcal{M}^{\overrightarrow{c}}(u) = \mathcal{M}^{\sigma}(u) = \mathcal{M}_{\mathsf{enc}}^{\sigma}(v)$.

Completeness. To go from a model of ϕ_{enc} to a model of ϕ, we show that if (i) $\mathcal{M} \models \mathsf{fresh}(y, y, \overrightarrow{x}) = u$ and (ii) $\mathcal{M}, \sigma \models C \rightarrow \mathsf{fresh}(s, t, \overrightarrow{x}) = v$, then $\mathcal{M}, \sigma \models C \wedge s = t \rightarrow u = v$.

Assume that $\mathcal{M}, \sigma \models C \wedge s = t$. From (i) and $\mathcal{M}^{\sigma}(s) = \mathcal{M}^{\sigma}(t)$ we get $\mathcal{M}^{\sigma}(\mathsf{fresh}(s, t, \overrightarrow{x})) = \mathcal{M}^{\sigma}(u)$. From (ii) and $\mathcal{M}, \sigma \models C$ we get $\mathcal{M}^{\sigma}(\mathsf{fresh}(s, t, \overrightarrow{x})) = \mathcal{M}^{\sigma}(v)$. Therefore $\mathcal{M}^{\sigma}(u) = \mathcal{M}^{\sigma}(\mathsf{fresh}(s, t, \overrightarrow{x})) = \mathcal{M}^{\sigma}(v)$ and $\mathcal{M}, \sigma \models u = v$.

2.4   Encoding 4: Split if

When using Knuth-Bendix completion, the equation $\mathsf{fresh}(s, t, \overrightarrow{x}) = u$ has the disadvantage that the number of critical pairs it creates is the *product* of the number of critical pairs created using s and using t. The fourth encoding solves this problem by having two equations, one for s and one for t.

The rule we use is: replace a clause of the form $C \wedge s = t \rightarrow u = v$ with the two clauses

$$\mathsf{fresh}_i(s, x_1, \ldots, x_n) = u$$
$$C \rightarrow \mathsf{fresh}_i(t, x_1, \ldots, x_n) = v,$$

where $x_1, \ldots, x_n$ consists of the union of the free variables of s, t, u and v (not just u and v as in encoding 3) and fresh_i is a fresh function symbol. Just as in encoding 3, we introduce a new symbol fresh_i each time the rule is applied.

Example. For the example clause set, the first clause is encoded as:

$$\mathsf{fresh}_1(f(x), x, y) = x$$
$$\mathsf{fresh}_1(f(y), x, y) = y$$

The second clause becomes:

$$\mathsf{fresh}_2(f(c)) = a$$
$$f(a) = b \rightarrow \mathsf{fresh}_2(d) = c$$

and the latter clause in turn is encoded as:

$$\mathsf{fresh}_3(f(a)) = \mathsf{fresh}_2(d)$$
$$\mathsf{fresh}_3(b) = c$$

Efficiency. The chief gain compared to encoding 3 is the reduced number of critical pairs. One disadvantage is that we must include the free variables of all four terms, s, t, u and v, which may lead to rather large equations.

Proof of Correctness. We use the same setup and notation, and indeed most of the same proof, as for encoding 3. Suppose that ϕ is a formula which contains the clause

$$C \wedge s = t \rightarrow u = v,$$

and that in ϕ_{enc} that clause has been replaced by the following two:

$$\mathsf{fresh}(s, \overrightarrow{x}) = u$$
$$C \rightarrow \mathsf{fresh}(t, \overrightarrow{x}) = v.$$

We show that ϕ and ϕ_{enc} are equisatisfiable.

Soundness. Suppose that $\mathcal{M}$ is a model of ϕ. We extend $\mathcal{M}$ to a model $\mathcal{M}_{\mathsf{enc}}$ of ϕ_{enc} by interpreting fresh as follows:

$$\mathsf{fresh}(a, \overrightarrow{b}) = \begin{cases} \mathcal{M}^{\overrightarrow{b}}(u), \text{ if } \mathcal{M}^{\overrightarrow{b}}(s) = a \\ \mathcal{M}^{\overrightarrow{b}}(v), \text{ otherwise} \end{cases}$$

As before, $\mathcal{M}$ and $\mathcal{M}_{\mathsf{enc}}$ agree on the truth of any formula not involving fresh, and fresh does not occur in s, t, u, v or C.

First we check that $\mathcal{M}_{\mathsf{enc}} \models \mathsf{fresh}(s, \overrightarrow{x}) = u$. Given a variable assignment σ, let $\overrightarrow{b} = \sigma(\overrightarrow{x})$. Then $\mathcal{M}_{\mathsf{enc}}^{\sigma}(\mathsf{fresh}(s, \overrightarrow{x})) = \mathsf{fresh}(\mathcal{M}^{\sigma}(s), \overrightarrow{b}) = \mathsf{fresh}(\mathcal{M}^{\overrightarrow{b}}(s), \overrightarrow{b}) = \mathcal{M}^{\overrightarrow{b}}(u) = \mathcal{M}_{\mathsf{enc}}^{\sigma}(u)$.

Then we show that if $\mathcal{M}_{\mathsf{enc}} \models C$ then $\mathcal{M}_{\mathsf{enc}} \models \rightarrow \mathsf{fresh}(t, \overrightarrow{x}) = v$. Take a variable assignment σ such that $\mathcal{M}_{\mathsf{enc}} \models C$, which implies that $\mathcal{M} \models C$, and let $\overrightarrow{b} = \sigma(\overrightarrow{x})$. Since $\mathcal{M}, \sigma \models C \wedge s = t \rightarrow u = v$, we can do a case split:

- If $\mathcal{M}, \sigma \not\models s = t$, then $\mathcal{M}_{\mathsf{enc}}^{\sigma}(\mathsf{fresh}(t, \overrightarrow{x})) = \mathsf{fresh}(\mathcal{M}^{\sigma}(t), \sigma(\overrightarrow{x})) = \mathsf{fresh}(\mathcal{M}^{\overrightarrow{b}}(t), \overrightarrow{b}) = \mathcal{M}^{\overrightarrow{b}}(v) = \mathcal{M}_{\mathsf{enc}}^{\sigma}(v)$, since $\mathcal{M}^{\sigma}(t) \neq \mathcal{M}^{\sigma}(s)$.
- If $\mathcal{M}, \sigma \models s = t$ and $\mathcal{M}, \sigma \models u = v$, then $\mathcal{M}_{\mathsf{enc}}^{\sigma}(\mathsf{fresh}(t, \overrightarrow{x})) = \mathsf{fresh}(\mathcal{M}^{\sigma}(t), \sigma(\overrightarrow{x})) = \mathsf{fresh}(\mathcal{M}^{\sigma}(s), \sigma(\overrightarrow{x})) = \mathsf{fresh}(\mathcal{M}^{\overrightarrow{b}}, \overrightarrow{b}) = \mathcal{M}^{\overrightarrow{b}}(v) = \mathcal{M}_{\mathsf{enc}}^{\sigma}(v)$.

Completeness. To go from a model of ϕ_{enc} to a model of ϕ, we show that if (i) $\mathcal{M} \models \mathsf{fresh}(s, \overrightarrow{x}) = u$ and (ii) $\mathcal{M}, \sigma \models C \rightarrow \mathsf{fresh}(t, \overrightarrow{x}) = v$, then $\mathcal{M}, \sigma \models C \wedge s = t \rightarrow u = v$.

Assume that $\mathcal{M}, \sigma \models C \wedge s = t$. From (i) and $\mathcal{M}^{\sigma}(s) = \mathcal{M}^{\sigma}(t)$ we get $\mathcal{M}^{\sigma}(\mathsf{fresh}(t, \overrightarrow{x})) = \mathcal{M}^{\sigma}(u)$. From (ii) and $\mathcal{M}, \sigma \models C$ we get $\mathcal{M}^{\sigma}(\mathsf{fresh}(t, \overrightarrow{x})) = \mathcal{M}^{\sigma}(v)$. Therefore $\mathcal{M}^{\sigma}(u) = \mathcal{M}^{\sigma}(\mathsf{fresh}(t, \overrightarrow{x})) = \mathcal{M}^{\sigma}(v)$ and $\mathcal{M}, \sigma \models u = v$.

2.5   Tupling, an Optional Transformation

In encodings 3 and 4, nested implications become rather messy. For example, in encoding 4, the implication

$$a = b \wedge c = d \rightarrow e = f$$

turns into the following equations:

$$\mathsf{fresh}_1(c) = e$$
$$\mathsf{fresh}_2(a) = \mathsf{fresh}_1(d)$$
$$\mathsf{fresh}_2(b) = f$$

In Sect. 2.3, we argued that for efficiency the encoded equations should always be oriented with fresh_i on the left, but this is not possible for the middle equation, and efficiency may suffer.

If we introduce a function symbol tuple, we can transform the above implication into the following binary clause:

$$\mathsf{tuple}(a, c) = \mathsf{tuple}(b, d) \rightarrow e = f$$

This exploits the fact that the only way to prove $\mathsf{tuple}(a, c) = \mathsf{tuple}(b, d)$ is to prove $a = b$ and $c = d$, because there are no extra axioms about tuple[4].

Applying encoding 4 now gives a much cleaner translation, and the equations will be oriented correctly:

$$\mathsf{fresh}(\mathsf{tuple}(a, c)) = e$$
$$\mathsf{fresh}(\mathsf{tuple}(b, d)) = f$$

In fact, we can fuse the function symbol fresh with tuple, because fresh is never used without tuple, and the result is:

$$\mathsf{fresh}(a, c) = e$$
$$\mathsf{fresh}(b, d) = f$$

In general, tupling transforms the clause $t_1 = u_1 \wedge \ldots \wedge t_n = u_n \rightarrow t = u$ into the clause $\mathsf{tuple}_n(t_1, \ldots, t_n) = \mathsf{tuple}_n(u_1, \ldots, u_n) \rightarrow t = u$, where tuple_n is a function symbol of arity n.

If the input problem is sorted, the result sort of tuple should be a fresh sort, otherwise we risk unsoundness. If the input problem is unsorted, then it is safe for tuple to be unsorted too. To show this, we require two lemmas.

Lemma 3 (Safe sort erasure). *Let ϕ be a sorted formula and ϕ_{erased} be the same formula with all sorts erased. Suppose that ϕ has the property that, if it is satisfiable, it has a model where (i) all domains have the same cardinality, or (ii) all domains are infinite. Then ϕ and ϕ_{erased} are equisatisfiable.*

---

[4] Also, no special support for tuples is needed in the theorem prover.

Proof. For (i), see Lemma 1 of [8]. For (ii), use the Löwenheim–Skolem theorem to get a model where all domains are countably infinite, then use (i).   □

Lemma 4 (The cardinality of Horn formula models). *If an unsorted Horn formula has a model of cardinality at least 2, it also has an infinite model.*

Proof. Suppose that ϕ is such a Horn formula. Since it has a model of cardinality at least 2, we know that $\phi \not\vdash x = y$. We now make use of the fact that every satisfiable Horn formula has a minimal model. Take the signature of ϕ and adjoin a countably infinite set of constant symbols $c_1, c_2, \ldots$; this preserves satisfiability, and we claim that the resulting minimal model $\mathcal{M}$ is infinite.

Since none of the constants c_i occurs in ϕ, any resolution proof of $\phi \vdash c_i = c_j$ (for $i \neq j$) also proves $\phi \vdash x = y$. Since $\phi \not\vdash x = y$, this implies that $\phi \not\vdash c_i = c_j$, and hence in any minimal model $c_i \neq c_j$. Since $\mathcal{M}$ is minimal, none of the c_i are equal in $\mathcal{M}$, and so $\mathcal{M}$ is infinite.   □

After these two lemmas, we are ready to prove the satefy of tupling for unsorted Horn formulas.

Lemma 5 (Safety of unsorted tupling for Horn formulas). *Let ϕ be an unsorted Horn formula and ϕ_{enc} be a transformed version in which we have applied (sorted) tupling. Then ϕ_{enc} and its sort erasure are equisatisfiable, i.e., the sorts can safely be erased.*

Proof. ϕ_{enc} has two sorts; let us call them ι and τ (for tuple). Since τ is a tuple sort it may always be interpreted by a tuple of domain elements of ι; in that case, if the cardinality of ι is κ, the cardinality of τ is κ^n for some n.

We are going to invoke Lemma 3. There are three cases, and in all of them the requirements of Lemma 3 hold:

- ϕ is unsatisfiable. In this case ϕ_{enc} is unsatisfiable too.
- ϕ has a model of cardinality 1. In this case, ϕ_{enc} has a model where ι has cardinality 1 and τ has cardinality $1^n = 1$.
- ϕ only has models of cardinality greater than 1. By Lemma 4, it has an infinite model, so ϕ_{enc} has a model where ι and τ are both infinite.   □

3   Eliminating Predicates

Equational provers do not usually support predicates. Before using the encodings of the previous section, we eliminate predicates in the standard way, by replacing them with functions: we introduce a new sort bool and a constant true : bool, and we replace each atomic formula $p(t_1, \ldots, t_n)$ with the equation $p(t_1, \ldots, t_n) =$ true (p is now a function of result sort bool). We assume without proof that this well-known transformation, which we call *sorted predicate elimination*, is correct.

We would like to avoid introducing sorts if the input problem is unsorted. This suggests that we should do *unsorted* predicate elimination: the same transformation as above, but without introducing the bool sort. Unfortunately, this is not sound: the set of clauses $\{x = y, \neg p(a)\}$ is satisfiable, but $\{x = y, p(a) \neq \mathsf{true}\}$ is unsatisfiable because $x = y$ implies $p(a) = \mathsf{true}$. In fact, unsoundness occurs only when the input problem implies $x = y$, as the following lemma implies:

Lemma 6. *Let ϕ_{enc} be obtained from an unsorted formula ϕ by sorted predicate elimination. If ϕ has no model of size 1 then ϕ_{enc} and its sort erasure are equisatisfiable, i.e., the sorts can safely be erased.*

Proof (rather similar to Lemma 5). The formula ϕ_{enc} has two sorts; let us call them bool and ι. We show that either ϕ_{enc} is unsatisfiable or it has a model where bool and ι are infinite, and then invoke Lemma 3. There are two cases:

- If ϕ is unsatisfiable, then so is ϕ_{enc}.
- If ϕ is satisfiable, then by assumption it has a model of cardinality greater than 1. By Lemma 4, ϕ has an infinite model. Therefore ϕ_{enc} has a model where ι is infinite. This model can be extended to one where bool is also infinite, since the monotonicity test of [8] is satisfied.   □

We can exploit Lemma 6 to eliminate predicates without introducing sorts, if the input problem is unsorted:

- Check if the input problem has a model of size 1. If so, abort the encoding and report that the formula is satisfiable.
- Otherwise, perform unsorted predicate elimination.

If the input problem is sorted, Lemma 6 does not help, so for sorted problems we always introduce the sort bool.

This leaves the problem of checking if the input formula has a model of size 1, which is easy to solve by observing that, in a model of size 1, all terms are equal and all predicates are constant-valued. To check if ϕ has a model of size 1, we replace all equality literals $t = u$ with **true**, and all predicates $p(t_1, \ldots, t_n)$ with a Boolean variable p. We then check if the resulting propositional Horn formula is satisfiable, for example by doing unit propagation or using a SAT solver.

Example. Given the Horn clauses $\{x = y, \neg p(a)\}$, we check if the set of clauses $\{\mathsf{true}, \neg p\}$ is satisfiable. It is, so the original problem has a model of size 1. Given the clauses $\{f(x) = f(y), p(a), \neg p(b)\}$, we check if $\{\mathsf{true}, p, \neg p\}$ is satisfiable. It is not, so we eliminate the predicates to get $\{f(x) = f(y), p(a) = \mathsf{true}, p(b) \neq \mathsf{true}\}$.

4   Encoding Goal Clauses

The goal clauses of a Horn formula are negated conjectures. By having several goal clauses of several literals each, a formula can have a conjecture which is an arbitrary Boolean combination (without negation) of positive literals. Most equational provers do not accept such expressive goals; some require the goal to be a single ground unit equation.

To solve this, we introduce two new constants false, true : bool, where bool is the sort introduced by predicate elimination. We then replace each goal clause

$C \to$ **false** with the clause $C \to$ false = true, and add one goal clause, false $\neq$ true.

Some provers accept slightly more general goals, and the goal encoding can be adapted to the situation. For example, a Twee goal can be a disjunction of ground equations, so the transformation of this section is only used for non-ground conjectures, and tupling (Sect. 2.5) is used for ground conjectures that use conjunction. A Waldmeister goal can be a conjunction of ground equations, and so the transformation of this section must be used for disjunctions.

5   Evaluation

We evaluated our encodings on two equational theorem provers, Waldmeister [14] and Twee [20], and one first-order prover, E 2.0 [19], using the 2159 unsatisfiable Horn problems available in TPTP v7.0.0 [21]. We tried all four encodings, with tupling enabled and disabled. Each prover was allowed to run for five minutes. We ran Waldmeister with the flag `--auto` and E with the flag `--auto-schedule`. We ran Twee with a heuristic designed for Horn clauses, which we describe below.

The results are shown in Table 1, with the best result of each prover marked in bold. To provide a baseline for the comparison, we also gave the original unencoded problems to E, which solved 1972, and SPASS [22], which solved 1370.

We see that Twee solved more problems than Waldmeister, but both provers have respectable performance: as a prover for Horn problems, Waldmeister is about as powerful as SPASS, while Twee lies in between SPASS and E. Nonetheless, E is clearly better than Twee or Waldmeister at solving typical Horn problems. The results also show that the encoding has a cost: E solves about 300 fewer problems when the problems are encoded.

Table 1. Number of problems solved using each encoding.

Prover	Tupling?	Encoding			
		1	2	3	4
Twee	No	1621	1596	1671	**1683**
	Yes	1534	1465	1493	1648
Waldmeister	No	1340	1246	1088	1151
	Yes	**1378**	1281	1176	1173
E	No	1698	**1710**	1672	1701
	Yes	1673	1676	1579	1615

We also see that Twee and Waldmeister prefer entirely different encodings. Since the two provers use a similar proof procedure, the difference may lie in the heuristics used. Special heuristics for encoded Horn formulas are future work.

Rating 1 Problems. Together, the three provers solved the following 37 problems of rating 1.0, i.e., problems which are not currently solved by any automatic prover. All three provers solved several rating 1.0 problems; see Table 2 for details.

ALG212+1	ALG213+1	KLE077+1	KLE156+2	LAT064-1	LAT178-1
LAT180-1	LAT181-1	LAT184-1	LAT185-1	LAT186-1	LAT187-1
LAT188-1	LAT189-1	LAT190-1	LAT191-1	LAT193-1	LAT202-1
LAT203-1	LAT206-1	LAT207-1	LAT221-1	LAT224-1	LAT225-1
LAT226-1	LAT228-1	LAT229-1	LAT231-1	LAT242-1	LAT256-1
LCL147-1	LCL148-1	LCL151-1	REL020+1	REL040+3	REL040-4
REL041+1					

These problems come from several domains of the TPTP, but all consist mostly of equations, and most involve an algebraic structure having a rich equational theory. This suggests that the encodings are most effective on problems where the bulk of the reasoning is equational—as we might expect.

Table 2. Number of rating-1 problems solved using each encoding.

Prover	Tupling?	Encoding			
		1	2	3	4
Twee	No	**29**	19	24	18
	Yes	22	23	5	17
Waldmeister	No	9	10	5	5
	Yes	8	7	**15**	13
E	No	3	4	1	3
	Yes	3	5	1	**7**

A Heuristic for Twee. Twee employed a special heuristic designed to let it eagerly discharge preconditions of equations. Twee decides which critical pair to join next by picking the one with the lowest *score*, which is computed based on, among other things, the size of the critical pair. The heuristic is that, if we are scoring a term like $\mathsf{ifeq}(s, t, u, v) = v$, and s and t happen to be unifiable, we perform the unification but then count s and t as having zero size.

We use this heuristic in encodings 1, 2 and 3. In encoding 4, there is no way to apply it, since s and t are never present in the same term. With the heuristic enabled, Twee solves more rating 1 problems—but slightly fewer problems overall.

Satisfiable Problems. We also evaluated our encodings on the 312 satisfiable, Horn, non-unit equality problems in the TPTP. The results were that, regardless of the prover or the encoding used, we always solved 210–220 problems. The only exception was if-then without tupling, which solved about 190 problems depending on the prover; the problems lost were a large collection of rating-0 problems from NLP, such as NLP002-1.p. As a baseline, E with no encoding solved 195 problems, so we gained about 20 problems.

On closer inspection, it turned out that most of the problems gained had no goal clauses. Any such problem has a model of size 1, and our translation immediately reports it as satisfiable without even running the prover (see Sect. 3). This explains why the choice of encoding had almost no effect.

Twee was able to complete a few problems that E could not (such as GRP112-1.p, of rating 0.62), but all the problems that we inspected could be solved by Paradox [9] in under a second. Thus our translations are not very helpful for showing satisfiability. (We did not evaluate them on Paradox, but as the translations do not alter the size of any model we do not expect any effect.)

6   Discussion and Related Work

The proof search of a superposition-based prover working on a Horn problem, and a completion-based prover working on an encoded Horn problem, is quite similar. Superposition [4,18] incorporates almost all the deduction rules of unfailing completion [3], only having a slightly stricter condition for backwards simplification. Thanks to literal selection, a superposition prover working on a Horn problem can work by forward reasoning from positive unit literals alone, much like a completion prover will do when using the "split if" encoding.

One important difference is that equational provers use more powerful redundancy tests. For example, ordered completion [2] allows a critical pair to be discarded if all of its ground instances can be joined (see e.g. [15]). A superposition prover would typically implement only a small subset of this feature, e.g., for handling commutative functions. Connectedness testing [1] is another powerful technique. Equational logic has a well-developed theory of proof orderings [2,5,6] which can be used to prove the correctness of many redundancy criteria.

There have been many and varied attempts to apply equational reasoning to first-order and Horn logic. As mentioned in the introduction, our first two encodings are inspired by work in universal algebra on encoding first-order logic as equations [7,16]. Completion has been generalised to Horn clauses [3,10,13,17]. Other term rewriting approaches to first-order logic include one using Boolean rings [11] and one using Gröbner bases [12].

7   Conclusion and Future Work

We have demonstrated that by encoding Horn formulas as equations, we can transform an equational theorem prover into a practical prover for Horn formulas. The resulting prover is strong on problems that combine difficult equational reasoning with some Horn clause reasoning. The encodings have a number of possible applications, including reasoning about functional programs, and reasoning about abstract algebra.

The encodings we presented do have some overhead. To eliminate this overhead, we plan to investigate hybrid approaches, where Horn clauses are encoded but the prover's strategy is specially tailored for reasoning about them—for example, by building certain equations into the prover or discarding unnecessary inferences. We hope that equational encodings of first-order logic [7,16] can perhaps be made practical using such a hybrid approach.

Acknowledgements. This work was supported by the Swedish Research Council (VR) grant 2016-06204, *Systematic testing of cyber-physical systems (SyTeC)*.

References

1. Bachmair, L., Dershowitz, N.: Critical pair criteria for completion. J. Symb. Comput. **6**(1), 1–18 (1988). https://doi.org/10.1016/S0747-7171(88)80018-X
2. Bachmair, L., Dershowitz, N.: Equational inference, canonical proofs, and proof orderings. J. ACM **41**(2), 236–276 (1994)
3. Bachmair, L., Dershowitz, N., Plaisted, D.A.: Completion without failure. In: Rewriting Techniques, pp. 1–30. Elsevier (1989)
4. Bachmair, L., Ganzinger, H.: Rewrite-based equational theorem proving with selection and simplification. J. Log. Comput. **4**, 217–247 (1994)
5. Bonacina, M.P., Dershowitz, N.: Abstract canonical inference. ACM Trans. Comput. Logic **8**(1) (2007). http://doi.acm.org/10.1145/1182613.1182619
6. Bonacina, M.P., Hsiang, J.: Towards a foundation of completion procedures as semidecision procedures. Theor. Comput. Sci. **146**(1–2), 199–242 (1995)
7. Burris, S.: Discriminator varieties and symbolic computation. J. Symb. Comput. **13**(2), 175–207 (1992). https://doi.org/10.1016/S0747-7171(08)80089-2
8. Claessen, K., Lillieström, A., Smallbone, N.: Sort it out with monotonicity. In: Bjørner, N., Sofronie-Stokkermans, V. (eds.) CADE 2011. LNCS (LNAI), vol. 6803, pp. 207–221. Springer, Heidelberg (2011). https://doi.org/10.1007/978-3-642-22438-6_17
9. Claessen, K., Sörensson, N.: New techniques that improve MACE-style finite model finding. In: Proceedings of the CADE-19 Workshop: Model Computation-Principles, Algorithms, Applications, pp. 11–27. Citeseer (2003)
10. Dershowitz, N.: A maximal-literal unit strategy for Horn clauses. In: Kaplan, S., Okada, M. (eds.) CTRS 1990. LNCS, vol. 516, pp. 14–25. Springer, Heidelberg (1991). https://doi.org/10.1007/3-540-54317-1_78
11. Hsiang, J.: Rewrite method for theorem proving in first order theory with equality. J. Symb. Comput. **3**, 133–151 (1987)
12. Kapur, D., Narendran, P.: An equational approach to theorem proving in first-order predicate calculus. SIGSOFT Softw. Eng. Notes **10**(4), 63–66 (1985)
13. Kounalis, E., Rusinowitch, M.: On word problems in Horn theories. In: Lusk, E., Overbeek, R. (eds.) CADE 1988. LNCS, vol. 310, pp. 527–537. Springer, Heidelberg (1988). https://doi.org/10.1007/BFb0012854
14. Löchner, B., Hillenbrand, T.: A phytography of WALDMEISTER. AI Commun. **15**(2, 3), 127–133 (2002)
15. Martin, U., Nipkow, T.: Ordered rewriting and confluence. In: Stickel, M.E. (ed.) CADE 1990. LNCS, vol. 449, pp. 366–380. Springer, Heidelberg (1990). https://doi.org/10.1007/3-540-52885-7_100
16. Mckenzie, R.: On spectra, and the negative solution of the decision problem for identities having a finite nontrivial model. J. Symb. Log. **40**(2), 186–196 (1975). https://projecteuclid.org:443/euclid.jsl/1183739380

17. Nieuwenhuis, R., Orejas, F.: Clausal rewriting. In: Kaplan, S., Okada, M. (eds.) CTRS 1990. LNCS, vol. 516, pp. 246–258. Springer, Heidelberg (1991). https://doi.org/10.1007/3-540-54317-1_95

18. Rusinowitch, M.: Theorem-proving with resolution and superposition. J. Symb. Comput. **11**(1–2), 21–49 (1991)

19. Schulz, S.: System description: E 1.8. In: McMillan, K., Middeldorp, A., Voronkov, A. (eds.) LPAR 2013. LNCS, vol. 8312, pp. 735–743. Springer, Heidelberg (2013). https://doi.org/10.1007/978-3-642-45221-5_49

20. Smallbone, N.: Twee, an equational theorem prover (2017). http://nick8325.github.io/twee/

21. Sutcliffe, G.: The TPTP problem library and associated infrastructure. From CNF to TH0, TPTP v6.4.0. J. Autom. Reason. **59**(4), 483–502 (2017)

22. Weidenbach, C., Dimova, D., Fietzke, A., Kumar, R., Suda, M., Wischnewski, P.: SPASS version 3.5. In: Schmidt, R.A. (ed.) CADE 2009. LNCS (LNAI), vol. 5663, pp. 140–145. Springer, Heidelberg (2009). https://doi.org/10.1007/978-3-642-02959-2_10

A FOOLish Encoding of the Next State Relations of Imperative Programs

Evgenii Kotelnikov[1]([✉]), Laura Kovács[1,2], and Andrei Voronkov[3,4]

[1] Chalmers University of Technology, Gothenburg, Sweden
evgenyk@chalmers.se
[2] TU Wien, Vienna, Austria
[3] The University of Manchester, Manchester, UK
[4] EasyChair, Manchester, UK

Abstract. Automated theorem provers are routinely used in program analysis and verification for checking program properties. These properties are translated from program fragments to formulas expressed in the logic supported by the theorem prover. Such translations can be complex and require deep knowledge of how theorem provers work in order for the prover to succeed on the translated formulas. Our previous work introduced FOOL, a modification of first-order logic that extends it with syntactical constructs resembling features of programming languages. One can express program properties directly in FOOL and leave translations to plain first-order logic to the theorem prover. In this paper we present a FOOL encoding of the next state relations of imperative programs. Based on this encoding we implement a translation of imperative programs annotated with their pre- and post-conditions to partial correctness properties of these programs. We present experimental results that demonstrate that program properties translated using our method can be efficiently checked by the first-order theorem prover Vampire.

1 Introduction

Automated program analysis and verification requires discovering and proving program properties ensuring program correctness. These program properties are usually expressed in combined theories of various data structures, such as integers and arrays. SMT solvers and first-order theorem provers that are used to check these properties need efficient handling of both theories and quantifiers. Moreover, formalisation of the program properties in the logic supported by the SMT solver or theorem prover plays a crucial role in making the prover succeed proving program correctness.

The translation of program properties into logical formulas accepted by a theorem prover is not straightforward. The reason for this is a mismatch between the semantics of the programming language constructs and that of the input language of the theorem prover. If program properties are not directly expressible in the input language, one needs to implement a translation of these properties to the language of the theorem prover. Such translations can be complex and error

© Springer International Publishing AG, part of Springer Nature 2018
D. Galmiche et al. (Eds.): IJCAR 2018, LNAI 10900, pp. 405–421, 2018.
https://doi.org/10.1007/978-3-319-94205-6_27

prone. Furthermore, one might need deep knowledge of how theorem provers work to obtain formulas in a form that theorem provers can handle efficiently.

Program verification systems reduce the mismatch between program properties and their formalisation as logical formulas from two ends. On the one hand, intermediate verification languages, such as Boogie [12] and WhyML [5], are designed to represent programs and their properties in a way that is friendly for translations to logic. On the other hand, theorem provers extend their supported logics with syntactic constructs that mirror those of programming languages.

Our previous work introduced FOOL [8], a modification of many-sorted first-order logic (FOL). FOOL extends FOL with syntactical constructs such as if-then-else and let-in expressions. These constructs can be used to naturally express program properties about conditional statements and variable updates. Users of a theorem prover that supports FOOL do not need to invent translations for these features of programming languages and can use features of FOOL directly. It allows the theorem prover to apply its own translation to FOL that it can use efficiently. We extended the Vampire theorem prover [10] to support FOOL [6] and designed an efficient clausification algorithm VCNF [7] for FOOL.

In summary, FOOL extends FOL with the following constructs.

- First-class boolean sort—one can define function and predicate symbols with boolean arguments and use quantifiers over the boolean sort.
- Boolean variables used as formulas.
- Formulas used as arguments to function and predicate symbols.
- Expressions of the form if φ then s else t, where φ is a formula, and s and t are either both terms or formulas.
- Expressions of the form let $D_1; \ldots; D_k$ in t, where $k > 0$, t is either a term or a formula, and $D_1, \ldots, D_k$ are simultaneous definitions, each of the form
  1. $f(x_1 : \sigma_1, \ldots, x_n : \sigma_n) = s$, where $n \geq 0$, f can be a function or a predicate symbol, and s is either a term or a formula;
  2. $(c_1, \ldots, c_n) = s$, where $n > 1$, $c_1, \ldots, c_n$ are constant symbols of the sorts $\sigma_1, \ldots, \sigma_n$, respectively, and s is a tuple expression. A tuple expression is inductively defined to be either
     (a) $(s_1, \ldots, s_n)$, where $s_1, \ldots, s_n$ are terms of the sorts $\sigma_1, \ldots, \sigma_n$, respectively;
     (b) if φ then s_1 else s_2, where φ is a formula, and s_1 and s_2 are tuple expressions; or
     (c) let $D_1; \ldots; D_k$ in s', where $D_1; \ldots; D_k$ are definitions, and s' is a tuple expression.

To our knowledge, no other logic, efficiently implemented in automated theorem provers, contains these constructs. Some constructs of FOOL have been previously implemented in interactive and higher-order theorem provers. However, there was no special emphasis on the efficiency or friendliness of the translation for the following processing by automatic provers.

In this paper, we extend our previous work on FOOL by demonstrating the efficient use of FOOL for program analysis. To this end, we give an efficient encoding of the next state relations of imperative programs in FOOL. Let us

```
if (x > y) {
    t := x;
    x := y;
    y := t;
}
assert x <= y;
```

$$\textbf{let } (x, y, t) = \textbf{ if } x > y$$
$$\textbf{then let } t = x \textbf{ in}$$
$$\textbf{let } x = y \textbf{ in}$$
$$\textbf{let } y = t \textbf{ in}$$
$$(x, y, t)$$
$$\textbf{else } (x, y, t)$$
$$\textbf{in } x \leq y$$

Fig. 1. An imperative program with an **if** statement.

Fig. 2. A FOOL encoding of the program assertion on Fig. 1.

motivate our work with the simple program on Fig. 1. This program contains an **if** statement and assignments to integer variables. The **assert** statement ensures that x is never greater than y after execution of the **if** statement.

To check that the given program assertion holds using an automated theorem prover, one has to express this assertion as a logical formula. For that, one has to express the updated values of x and y after the sequence of assignments. For example, one can compute the updated value of each individual variable separately for each possible execution trace. However, this approach suffers from a bloated resulting formula that will contain duplicating parts of the program. A more common technique is to first convert a program to a static single assignment (SSA) form. This conversion introduces a new intermediate variable for each assignment and creates a smaller translated formula.

Both excessive naming and excessive duplication of program expressions can make the resulting logical formula very hard for a first-order theorem prover. The encoding of the next state relations of imperative programs given in this paper avoids both by using a FOOL formula that closely matches the structure of the original program (Sect. 3). This way the decision between introducing new symbols and duplicating program expressions is left to the theorem prover that is better equipped to make it. The assertion of the program in Fig. 1 is concisely expressed with our encoding as the FOOL formula on Fig. 2.

While FOOL offers a concise representation of some programming constructs, the efficient implementation of FOOL poses a challenge for first-order theorem provers since their performance on various translations to CNF can be hampered by the (unintended) use of constructs interfering with their internal implementation, including the use of orderings, selection and the saturation algorithm. For example, to deal with the boolean sort, it is not uncommon to add an axiom like $(\forall x)(x = 0 \lor x = 1)$ for this sort. Even this simple axiom can cause a considerable growth of the search space, especially when used with certain term orderings. To address the challenge of dealing with full FOOL, one needs experimental comparison of various translations or various implementations of FOOL. Our paper is the first one to make such an experimental comparison.

Our encoding uses tuple expressions and **let-in** expressions with tuple definitions, available in FOOL. We extend and generalise the use of tuples in first-order theorem provers by introducing a polymorphic theory of first class tuples (Sect. 2). In this theory one can define tuple sorts and use tuples as terms.

Our encoding can be efficiently used in automated program analysis and verification. To demonstrate this, we report on our experimental results obtained by running Vampire on program verification problems (Sect. 4). These verification problems are partial correctness properties that we generated from a collection of imperative programs using an implementation of our encoding to FOOL as well as other tools.

Contributions. We summarise the main contributions of this paper below.

1. We define an encoding of the next state relation of imperative programs in FOOL and show that it is sound (Sect. 3). Using this encoding, we define a translation of certain properties of imperative programs to FOOL formulas.
2. We present a polymorphic theory of first class tuples and its implementation in Vampire (Sect. 2). To our knowledge, Vampire is the only superposition-based theorem prover to support this theory.
3. We present experimental results obtained by running Vampire on a collection of benchmarks expressing partial correctness properties of imperative programs (Sect. 4). We generated these benchmarks using an implementation of our encoding to FOOL and other tools. Our results show Vampire is more efficient on the FOOL encoding of partial correctness properties, compared with other translations.

2 Polymorphic Theory of First Class Tuples

The use of tuple expressions in FOOL is limited. They can only occur on the right hand side of a tuple definition in `let-in`. One cannot use a tuple expression elsewhere, for example, as an argument to a function or predicate symbol.

In this section we describe the theory of first class tuples that enables a more generic use of tuples. This theory contains tuple sorts and tuple terms. Both of them are first class—one can define function and predicate symbols with tuple arguments, quantify over the tuple sort, and use tuple terms as arguments to function and predicate symbols. Tuple expressions in FOOL, combined with the polym orphic theory of tuples, are tuple terms.

Definition. The polymorphic theory of tuples is the union of theories of tuples parametrised by tuple arity $n > 0$ and sorts $\tau_1, \ldots, \tau_n$.

A theory of first class tuples is a first-order theory that contains a sort $(\tau_1, \ldots, \tau_n)$, function symbols $t : \tau_1 \times \ldots \times \tau_n \to (\tau_1, \ldots, \tau_n)$, $\pi_1 : (\tau_1, \ldots, \tau_n) \to \tau_1, \ldots, \pi_n : (\tau_1, \ldots, \tau_n) \to \tau_n$, and two axioms. The function symbol t constructs a tuple from given terms, and function symbols $\pi_1, \ldots, \pi_n$ project a tuple to its individual elements. For simplicity we will write $(t_1, \ldots, t_n)$ instead of $t(t_1, \ldots, t_n)$ to mean a tuple of terms $t_1, \ldots, t_n$. The tuple axioms are

1. exhaustiveness

$$(\forall x_1 : \tau_1) \ldots (\forall x_n : \tau_n)(\pi_1((x_1, \ldots, x_n)) \doteq x_1 \wedge \ldots \wedge \pi_n((x_1, \ldots, x_n)) \doteq x_n);$$

2. injectivity

$$(\forall x_1 : \tau_1) \ldots (\forall x_n : \tau_n)(\forall y_1 : \tau_1) \ldots (\forall y_n : \tau_n)$$
$$((x_1, \ldots, x_n) \doteq (y_1, \ldots, y_n) \Rightarrow x_1 \doteq y_1 \wedge \ldots \wedge x_n \doteq y_n).$$

Tuples are ubiquitous in mathematics and programming languages. For example, one can use the tuple sort $(\mathbb{R}, \mathbb{R})$ as the sort of complex numbers. Thus, the term (a, b), where $a : \mathbb{R}$ and $b : \mathbb{R}$ represents a complex number $a + bi$. One can define the addition function $plus : (\mathbb{R}, \mathbb{R}) \times (\mathbb{R}, \mathbb{R}) \to (\mathbb{R}, \mathbb{R})$ for complex numbers with the formula

$$(\forall x : (\mathbb{R}, \mathbb{R}))(\forall y : (\mathbb{R}, \mathbb{R}))$$
$$(plus(x, y) \doteq (\pi_1(x) + \pi_1(y), \pi_2(x) + \pi_2(y))), \tag{1}$$

where $+$ denotes addition for real numbers.

Tuple terms can be used as tuple expressions in FOOL. If $(c_1, \ldots, c_n) = s$ is a tuple definition inside a `let-in`, where $c_1, \ldots, c_n$ are constant symbols of sorts $\tau_1, \ldots, \tau_n$, respectively, then tuple expression s is a term of the sort $(\tau_1, \ldots, \tau_n)$.

It is not hard to extend tuple definitions to allow arbitrary tuple terms of the correct sort on the right hand side of $=$. For example, one can use a variable of the tuple sort. With such extension, Formula 1 can be equivalently expressed using a `let-in` with two simultaneous tuple definitions as follows

$$(\forall x : (\mathbb{R}, \mathbb{R}))(\forall y : (\mathbb{R}, \mathbb{R}))$$
$$(plus(x, y) \doteq \mathtt{let}\ (a, b) = x;\ (c, d) = y\ \mathtt{in}\ (a + c, b + d)). \tag{2}$$

Implementation. Vampire implements reasoning with the polymorphic theory of tuples by adding corresponding tuple axioms when the input uses tuple sorts and/or tuple functions. For each tuple sort $(\tau_1, \ldots, \tau_n)$ used in the input, Vampire defines a term algebra [9] with the single constructor t and n destructors $\pi_1, \ldots, \pi_n$. Then Vampire adds the corresponding term algebra axioms, which coincide with the tuple theory axioms.

Vampire reads formulas written in the TPTP language [16]. The TFX subset[1] of TPTP contains a syntax for tuples and `let-in` expressions with tuple definitions. The sort $(\mathbb{R}, \mathbb{R})$ is represented in TFX as `[$real,$real]` and the term $(a + c, b + d)$ is represented as `[$sum(a,c),$sum(b,d)]`. Formula 2 can be expressed in TPTP as

```
tff(plus,type,plus:([$real,$real]*[$real,$real])>[$real,$real]).
tff(plus_def,axiom,
    ![X:[$real,$real],Y:[$real,$real]]:
    (plus(X,Y)=$let([[a:$real,b:$real],[c:$real,d:$real]],
              [a,b]:=X;[c,d]:=Y,
              [$sum(a,c),$sum(b,d)]))).
```

Vampire translates `let-in` with tuple definitions to clausal normal form of first-order logic using the VCNF clausification algorithm [7].

---

[1] http://www.cs.miami.edu/~tptp/TPTP/Proposals/TFXTHX.html.

3   Imperative Programs to FOOL

In this section we discuss an efficient translation of imperative programs to FOOL. To formalise the translation we define a restricted imperative programming language and its denotational semantics in Sect. 3.1. This language is capable of expressing variable updates, if-then-else, and sequential composition. Then, we define an encoding of the next state relation for programs of this language, and state the soundness property of this encoding in Sect. 3.2. Finally, in Sect. 3.3 we show a translation that converts a program, annotated with its pre-conditions and post-conditions, to a FOOL formula that expresses the partial correctness property of that program.

We give (rather standard) definitions of our programming language and its semantics and use them to define the main contributions of this section: the encoding of the next state relation (Definition 6) and soundness of this encoding (Theorem 1).

3.1   An Imperative Programming Language

We define a programming language with assignments to typed variables, if-then-else, and sequential composition. We omit variable declarations in our language and instead assume for each program a set of program variables V and a type assignment η. η is a function that maps each program variable into a type. Each type is either int, bool, or $\texttt{array}(\sigma, \tau)$, where σ and τ are types of array indexes and array values, respectively. In the sequel we will assume that V and η are arbitrary but fixed.

Programs in our language select and update elements of arrays, including multidimensional arrays. We do not introduce a distinguished type for multidimensional arrays but instead use nested arrays. We write $\texttt{array}(\sigma_1, \ldots, \sigma_n, \tau)$, $n > 1$, to mean the nested array type $\texttt{array}(\sigma_1, \texttt{array}(\ldots, \texttt{array}(\sigma_n, \tau) \ldots))$.

Definition 1. *An* expression *of the type* τ *is defined inductively as follows.*

1. *An integer n is an expression of the type* int.
2. *Symbols* true *and* false *are expressions of the type* bool.
3. *If $\eta(x) = \tau$, then x is an expression of the type τ.*
4. *If $\eta(x) = \texttt{array}(\sigma_1, \ldots, \sigma_n, \tau)$, $n > 0$, $e_1, \ldots, e_n$ are expressions of types σ_1, $\ldots, \sigma_n$, respectively, then $x[e_1, \ldots, e_n]$ is an expression of the type τ.*
5. *If e_1 and e_2 are expressions of the type τ, then $e_1 \doteq e_2$ is an expression of the type* bool.
6. *If e_1 and e_2 are expressions of the type* int, *then $-e_1$, $e_1 + e_2$, $e_1 - e_2$, $e_1 \times e_2$ are expressions of the type* int.
7. *If e_1 and e_2 are expressions of the type* int, *then $e_1 < e_2$ is an expression of the type* bool.
8. *If e_1 and e_2 are expression of the type* bool, *then $\neg e_1$, $e_1 \lor e_2$, $e_1 \land e_2$ are expressions of the type* bool. $\qquad\square$

Definition 2. *A* statement *is defined inductively as follows.*

1. `skip` *is an empty statement.*
2. *If* $\eta(x_1) = \tau_1, \ldots, \eta(x_n) = \tau_n, n \geq 1$ *and* $e_1, \ldots, e_n$ *are expressions of the types* $\tau_1, \ldots, \tau_n$, *respectively, then* $x_1, \ldots, x_n := e_1, \ldots, e_n$ *is a statement.*
3. *If* $\eta(x) = \mathtt{array}(\sigma_1, \ldots, \sigma_n, \tau), n \geq 1$, *and* $e_1, \ldots, e_n, e$ *are expressions of types* $\sigma_1, \ldots, \sigma_n, \tau$, *respectively, then* $x[e_1, \ldots, e_n] := e$ *is a statement.*
4. *If* e *is an expression of the type* `bool`, s_1 *and* s_2 *are statements, and at least one of* s_1, s_2 *is not* `skip`, *then* `if` e `then` s_1 `else` s_2 *is a statement.*
5. *If* s_1 *and* s_2 *are statements and neither of them is* `skip`, *then* $s_1 ; s_2$ *is a statement.* □

We say that $x_1, \ldots, x_n$ in the statement $x_1, \ldots, x_n := e_1, \ldots, e_n$ and x in the statement $x[e_1, \ldots, e_n] := e$ are *assigned program variables.* For each statement s we denote by updates(s) the set of all assigned program variables that occur in s.

We define the semantics of the programming language by an interpretation function $[\![\,-\,]\!]$ for types, expressions and statements. The interpretation of a type is a set: $[\![\,\mathtt{int}\,]\!] = \mathbb{Z}$, $[\![\,\mathtt{bool}\,]\!] = \{0, 1\}$, and $[\![\,\mathtt{array}(\tau, \sigma)\,]\!] = [\![\,\tau\,]\!] \rightarrow [\![\,\sigma\,]\!]$. The interpretation of expressions and statements is defined using *program states*, that is, mappings of program variables $x \in V$, $\eta(x) = \tau$ to elements of $[\![\,\tau\,]\!]$.

Definition 3. *Let* e *be an expression of the type* τ. *The* interpretation $[\![\,e\,]\!]$ *is a mapping from program states to* $[\![\,\tau\,]\!]$ *defined inductively as follows.*

1. $[\![\,n\,]\!]$ *maps each state to* n, *where* n *is an integer.*
2. $[\![\,\mathtt{true}\,]\!]$ *maps each state to* 1.
3. $[\![\,\mathtt{false}\,]\!]$ *maps each state to* 0.
4. $[\![\,x\,]\!]$ *maps each st to* $st(x)$.
5. $[\![\,x[e_1, \ldots, e_n]\,]\!]$ *maps each st to* $st(x)([\![\,e_1\,]\!](st)) \ldots ([\![\,e_n\,]\!](st))$.
6. $[\![\,e_1 \oplus e_2\,]\!]$ *maps each st to* $[\![\,e_1\,]\!](st) \oplus [\![\,e_2\,]\!](st)$, *where* $\oplus \in \{\doteq, +, -, \times, < , \vee, \wedge\}$.
7. $[\![\,\neg e\,]\!]$ *maps each st to* $\neg[\![\,e\,]\!](st)$. □

Definition 4. *Let* s *be a statement. The* interpretation $[\![\,s\,]\!]$ *is a mapping between program states defined inductively as follows.*

1. $[\![\,\mathtt{skip}\,]\!]$ *is the identity mapping.*
2. $[\![\,x_1, \ldots, x_n := e_1, \ldots, e_n\,]\!]$ *maps each st to st' such that* $st'(x_i) = [\![\,e_i\,]\!](st)$ *for each* $1 \leq i \leq n$ *and otherwise coincides with st.*
3. $[\![\,x[e_1, \ldots, e_n] := e\,]\!]$ *maps each st to st' such that*

$$st'(x)([\![\,e_1\,]\!](st)) \ldots ([\![\,e_n\,]\!](st)) = [\![\,e\,]\!](st)$$

and otherwise coincides with st.
4. $[\![\,\mathtt{if}\ e\ \mathtt{then}\ s_1\ \mathtt{else}\ s_2\,]\!]$ *maps each st to* $[\![\,s_1\,]\!](st)$ *if* $[\![\,e\,]\!](st) = 1$ *and to* $[\![\,s_2\,]\!](st)$ *otherwise.*
5. $[\![\,s_1 ; s_2\,]\!]$ *is* $[\![\,s_2\,]\!] \circ [\![\,s_1\,]\!]$. □

3.2 Encoding the Next State Relation

Our setting is FOOL extended with the theory of linear integer arithmetic, the polymorphic theory of arrays [6], and the polymorphic theory of first class tuples (Sect. 2). The theory of linear integer arithmetic includes the sort $\mathbb{Z}$, the predicate symbol $<$, and the function symbols $+$, $-$, and $\times$. The theory of arrays includes the sort $array(\tau, \sigma)$ for all sorts τ and σ, and function symbols $select$ and $store$. The function symbol $select$ represents a binary operation of extracting an array element by its index. The function symbol $store$ represents a ternary operation of updating an array at a given index with a given value. We point out that sorts $bool$, $\mathbb{Z}$, and $array(\sigma, \tau)$ mirror types $\texttt{bool}$, $\texttt{int}$ and $\texttt{array}(\sigma, \tau)$ of our programming language, and have the same interpretations.

We represent multidimensional arrays in FOOL as nested arrays[2]. To this end we (i) inductively define $select(a, i_1, \ldots, i_n)$, where $n > 1$, to be $select(select(a, i_1), i_2, \ldots, i_n)$; and (ii) inductively define $store(a, i_1, \ldots, i_n, e)$, where $n > 1$, to be $store(a, i_1, store(select(a, i_1), i_2, \ldots, i_n, e))$.

Our encoding of the next state relation produces FOOL terms that use program variables as constants and do not use any other uninterpreted function or predicate symbols. In the sequel we will only consider such FOOL terms. For these FOOL terms, η is a type assignment and each program state can be extended to a η-interpretation, the details of this extension are straightforward (we refer to [8] for the semantics of FOOL). We will use program states as η-interpretations for FOOL terms. For example we will write $\mathrm{eval}_{st}(t)$ for the value of t in st, where t is a FOOL term and st is a program state. We will say that a program state st satisfies a FOOL formula φ if $\mathrm{eval}_{st}(\varphi) = 1$.

To define the encoding of the next state relation we first define a translation of expressions to FOOL terms. Our encoding applies this translation to each expression that occurs inside a statement.

Definition 5. *Let e be an expression of the type τ. $\mathcal{T}(e)$ is a FOOL term of the sort τ, defined inductively as follows.*

$$\mathcal{T}(n) = n, \text{ where } n \text{ is an integer.}$$
$$\mathcal{T}(\texttt{true}) = true.$$
$$\mathcal{T}(\texttt{false}) = false.$$
$$\mathcal{T}(x) = x.$$
$$\mathcal{T}(x[e_1, \ldots, e_n]) = select(x, \mathcal{T}(e_1), \ldots, \mathcal{T}(e_n)).$$
$$\mathcal{T}(e_1 \oplus e_2) = \mathcal{T}(e_1) \oplus \mathcal{T}(e_2), \text{ where } \oplus \in \{\doteq, +, -, <, \times, \vee, \wedge\}.$$
$$\mathcal{T}(-e) = -\mathcal{T}(e).$$
$$\mathcal{T}(\neg e) = \neg \mathcal{T}(e).$$

$\square$

---

[2] Multidimensional arrays can be represented in FOOL also as arrays with tuple indexes. We do not discuss such representation in this work.

Lemma 1. $\text{eval}_{st}(\mathcal{T}(e)) = [\![\, e \,]\!](st)$ *for each expression* e *and state* st.   □

Proof. By structural induction on e.   □

Definition 6. *Let* s *be a statement.* $\mathcal{N}(s)$ *is a mapping between FOOL terms of the same sort, defined inductively as follows.*

1. $\mathcal{N}(\texttt{skip})$ *is the identity mapping.*
2. $\mathcal{N}(x_1, \ldots, x_n := e_1, \ldots, e_n)$ *maps* t *to*

$$\texttt{let } (x_1, \ldots, x_n) = (\mathcal{T}(e_1), \ldots, \mathcal{T}(e_n)) \texttt{ in } t.$$

3. $\mathcal{N}(x[e_1, \ldots, e_n] := e)$ *maps* t *to*

$$\texttt{let } x = store(x, \mathcal{T}(e_1), \ldots, \mathcal{T}(e_n), \mathcal{T}(e)) \texttt{ in } t.$$

4. $\mathcal{N}(\texttt{if } e \texttt{ then } s_1 \texttt{ else } s_2)$ *maps* t *to*

$$\texttt{let } (x_1, \ldots, x_n) = \texttt{if } \mathcal{T}(e) \texttt{ then } \mathcal{N}(s_1)((x_1, \ldots, x_n))$$
$$\texttt{else } \mathcal{N}(s_2)((x_1, \ldots, x_n))$$
$$\texttt{in } t,$$

   where $\text{updates}(s_1) \cup \text{updates}(s_2) = \{x_1, \ldots, x_n\}$.
5. $\mathcal{N}(s_1 \,;\, s_2)$ *is* $\mathcal{N}(s_1) \circ \mathcal{N}(s_2)$.   □

The following theorem is the soundness property of translation $\mathcal{N}$. Essentially, it states that $\mathcal{N}$ encodes the semantics of a given statement as a FOOL formula.

Theorem 1. $\text{eval}_{st}(\mathcal{N}(s)(t)) = \text{eval}_{[\![\, s \,]\!](st)}(t)$ *for each statement* s, *state* st *and FOOL term* t.   □

Proof. By structural induction on s.   □

3.3 Encoding the Partial Correctness Property

We use the encoding of the next state relation to generate partial correctness properties of programs annotated with their pre-conditions and post-conditions.

We define an *annotated program* to be a Hoare triple $\{\varphi\}\, s\, \{\psi\}$, where s is a statement, and φ and ψ are formulas in first-order logic. We say that $\{\varphi\}\, s\, \{\psi\}$ is correct if for each program state st that satisfies φ, $[\![\, s \,]\!](st)$ satisfies ψ. We translate each annotated program $\{\varphi\}\, s\, \{\psi\}$ to the FOOL formula $\varphi \Rightarrow \mathcal{N}(s)(\psi)$.

Theorem 2. *Let* $\{\varphi\}\, s\, \{\psi\}$ *be an annotated program. The FOOL formula* $\varphi \Rightarrow \mathcal{N}(s)(\psi)$ *is valid iff* $\{\varphi\}\, s\, \{\psi\}$ *is correct.*   □

Proof. Directly follows from Theorem 1.   □

We point out the following two properties of the encoding $\mathcal{N}$. First, the size of the encoded formula is $O(v \cdot n)$, where v is the number of variables in the program and n is the program size as each program statement is used once with one or two instances of $(x_1, \ldots, x_n)$. Second, the encoding does not introduce any new symbols. When we translate program correctness properties to FOL, both an excessive number of new symbols and an excessive size of the translation might make the encoded formula hard for a theorem prover. Instead of balancing between the two, encoding to FOOL leaves the decision to the theorem prover.

4   Experiments

In this section we describe our experiments on comparing the performance of the Vampire theorem prover [10] on FOOL and on translations of program properties to FOL. We used a collection of 50 programs written in the Boogie verification language [12]. Each of these programs uses only variable assignments, if-then-else statements, and sequential composition and is annotated with its pre-conditions and post-conditions, expressed in first-order logic. From this collection of programs we generated the following three sets of benchmarks.

1. 50 problems in first-order logic written in the SMT-LIB language [2]. We generated these problems by running the front end of the Boogie [1] verifier.
2. 50 FOOL problems with tuples generated by running our implementation of the translation from Sect. 3.3, named Voogie.
3. 50 FOOL problems generated by running the BLT [3] translator.

We point out that in our experiments we do not aim to compare methods of program verification or specific verification tools. Rather, we compare different ways of translating realistic verification problems for theorem provers.

In what follows, we describe the collection of imperative programs used in our experiments (Sect. 4.1) and discuss our set of benchmarks (Sect. 4.2). All properties that we deal with use integers and arrays, as well as universal and existential quantifiers. To verify these properties one has to reason in the combination of theories and quantifiers. We briefly describe how Vampire implements this kind of reasoning in Sect. 4.3. Our experimental results are summarised in Tables 1, 2 and 3 and discussed in Sect. 4.4.

4.1   Examples of Imperative Programs

We demonstrate the work of our translation on a collection of imperative programs that only use variable assignments, if-then-else statements, and sequential composition. Unfortunately, no large collections of such programs are available. There are many benchmarks for software verification tools, but most of them use control flow statements not covered in this work, such as gotos and exceptions. We also cannot use benchmarks from the hardware verification and model checking communities, because they are mostly about boolean values and bit-vectors. For our experiments we generated our own imperative programs in two steps described below.

First, we crafted 10 programs that implement textbook algorithms and solutions to program verification competitions. Each program uses variables of the integer, boolean, and array type. Each program contains a single `while` loop of the form `while e do s`, where e is a boolean expression and s is a statement. In addition, each program contains variable assignments, `if-then-else` statements, and sequential composition. We annotated each program with its pre-condition φ and each loop with its invariant ψ. The formulas φ and ψ are expressed in first-order logic.

Then, we unrolled the loop of each program k times, where k is an integer between 1 and 5. This resulted in 50 loop-free programs that retain the annotated properties. Each program encodes the loop invariant property of the original program. Multiple unrollings provide us with programs with long sequences of variables updates, `if-then-else` statements and compositions, which are convenient for our experiments. Our loop unrolling program transformation consisted of the following steps.

1. Introduce a fresh boolean variable bad that encodes the under-specified state of the program.
2. Construct a guarded loop iteration i as `if` $\neg e$ `then` $bad := true$ `else skip` ; s.
3. Construct a sequence of iterations i ; $\dots$; i, where i is repeated k times.
4. Finally, construct the annotated program $\{\varphi \wedge \psi\} \, i$; $\dots$; $i \, \{\neg bad \Rightarrow \psi\}$.

It is not hard to show that if a program with a loop satisfies its specification, then the Hoare triple resulting in step 4 of the above transformation also holds.

We wrote our example programs with loops as well as their loop-free unrolled versions in the Boogie language. Boogie can unroll loops automatically, but introduces `goto` statements that our translation does not support. For this reason, we used the loop unrolling described above.

An example of our loop unrolling is available at http://www.cse.chalmers. se/~evgenyk/ijcar18/. It shows the `maxarray` program with a loop from our collection and a program generated from `maxarray` by unrolling its loop twice.

4.2   Benchmarks

We used the 50 annotated loop-free programs and generated their partial correctness statements using Boogie, Voogie and BLT. These statements are encoded as unsatisfiable problems in first-order logic and FOOL. Our collection of imperative programs with loops, their loop-free unrollings and benchmarks expressed in the TPTP language [15] is available at http://www.cse.chalmers.se/~evgenyk/ ijcar18/. The TPTP benchmarks are also available, along with other FOOL problems, on the TPTP website http://tptp.org.

The Boogie verifier generates verification conditions as formulas in first-order logic written in the SMT-LIB language and uses the SMT solver Z3 [4] to check these formulas. We ran Boogie with the option `/proverLog` to print the generated formulas on each of our annotated loop-free programs and in this way obtained a collection of 50 SMT-LIB benchmarks.

Voogie is our implementation of the translation described in Sect. 3. It takes as input programs written in a fragment of the Boogie language and generates FOOL formulas written in the TPTP language. The source code of Voogie is available at https://github.com/aztek/voogie.

The fragment of the Boogie language supported by Voogie can be seen as the smallest fragment that is sufficient to represent the loop-free programs in our collection. This fragment consists of (i) top level variable declarations; (ii) a single procedure main annotated with its pre- and post-conditions; (iii) assignments to variables, including parallel assignments, and assignments to array elements; (iv) if-then-else statements; and (v) arithmetic and boolean operations. Running Voogie on each loop-free program in our collection gave us 50 TPTP benchmarks. An example of the TPTP benchmark obtained from running Voogie on the maxarray program with its loops unrolled twice is available at http://www.cse.chalmers.se/~evgenyk/ijcar18/.

BLT (Boogie Less Triggers) [3] is an automatic tool that takes Boogie programs as input and generates their verification conditions in first-order logic written in the TPTP language. BLT has an experimental feature of generating FOOL formulas with tuple let-in and tuple expressions to represent next state values of program variables in a style similar to Voogie. At the time of our experiments, this feature was not stable enough, and we did not enable it. Running BLT with its default configuration on each of the 50 loop-free programs in our collection gave us 50 TPTP benchmarks.

The representation of program expressions coincides in all three translations. All translations use the theory of linear integer arithmetic and the theory of arrays as realised in their respective languages.

4.3   Theories and Quantifiers in Vampire

Vampire's main algorithm is saturation of a set of first-order clauses using the resolution and superposition calculus. Vampire also implements the AVATAR architecture [17] for splitting clauses. The idea behind AVATAR is to use a SAT or an SMT solver to guide proof search. AVATAR selects sub-problems for the saturation-based prover to tackle by making decisions over a propositional abstraction of the clause search space. The -sas option of Vampire selects the SAT solver.

Vampire handles theories by automatically adding theory axioms to the search space whenever an interpreted sort, function, or predicate is found in the input. This approach is incomplete for theories such as linear and non-linear integer and real arithmetic, but shows good results in practice. The -tha option of Vampire with values on and off controls whether theory axioms are added.

A recent work [13] lifted AVATAR to be modulo theories by replacing the SAT solver by an SMT solver, ensuring that the sub-problem is theory-consistent in the ground part. The result is that the saturation prover and the SMT solver deal with the parts of the problem to which they are best suited. Vampire implements AVATAR modulo theories using Z3.

Our experience with running Vampire on theory- and quantifier-intensive problems shows that some of the theory axioms can degrade the performance of Vampire. These axioms make Vampire infer many theory tautologies making the search space larger. We found that, among others, axioms of commutativity, associativity, left and right identity, and left and right inverse of arithmetic operations are in this sense "expensive". Our solution to this problem is a more refined control over which theory axioms Vampire adds to the search space. We added to the -tha option of Vampire a new value named some that makes Vampire only add "cheap" axioms to the search space. some implements our empirical criterion for choosing theory axioms. Designing other criteria for axiom selection is an interesting task for future work.

4.4 Experimental Results

For our experiments, we compared the performance of Vampire on the Boogie, Voogie, and BLT translations of our benchmarks.

We ran Vampire on all three sets of benchmarks with options -tha some and -sas z3. Vampire supports both TPTP and SMT-LIB syntax, the input language is selected by setting the --input_syntax option to tptp and smtlib2, respectively. We performed our experiments on the StarExec compute cluster [14] using the time limit of 5 min per problem. The detailed experimental results are available at http://www.cse.chalmers.se/~evgenyk/ijcar18/.

Table 1. Runtimes in seconds of Vampire on the Boogie translation of the benchmarks.

Benchmark	Number of loop unrollings				
	1	2	3	4	5
binary-search	0.884	2.420	3.364	10.709	27.648
bubble-sort	–	–	–	–	–
dutch-flag	24.789	–	–	–	–
insertion-sort	122.354	–	–	–	–
matrix-transpose	1.311	–	1.078	–	–
maxarray	0.205	0.587	1.197	1.702	1.692
maximum	0.066	0.078	0.082	0.095	0.129
one-duplicate	–	–	–	–	–
select-k	96.993	–	–	–	–
two-way-sort	0.191	0.205	0.647	1.384	1.344

Tables 1 and 2 summarise the results of Vampire on the Boogie and Voogie translations of the benchmarks, respectively. A dash means that Vampire does not solve the problem within the given time limit.

Table 2. Runtimes in seconds of Vampire on the Voogie translation of the benchmarks.

Benchmark	Number of loop unrollings				
	1	2	3	4	5
binary-search	1.979	25.135	6.560	–	163.803
bubble-sort	0.394	53.192	2.073	–	–
dutch-flag	11.384	–	–	–	–
insertion-sort	18.262	38.169	3.369	21.698	11.639
matrix-transpose	0.266	8.362	–	–	–
maxarray	0.170	0.587	0.489	2.635	6.325
maximum	0.062	0.065	0.070	0.087	0.102
one-duplicate	0.125	2.402	2.231	93.746	145.243
select-k	0.216	0.612	203.655	–	–
two-way-sort	0.464	5.360	–	–	–

- Vampire solves 25 of the problems, translated by Boogie, and 36 problems, translated by Voogie.
- For 16 benchmark programs, Vampire solves their Voogie translations, but not the Boogie translations.
- For 5 benchmark programs, Vampire solves their Boogie translations, but not the Voogie translations.
- For 20 benchmark programs, Vampire solves both of their translations, and is faster on the Voogie translations for 12 of them.

Table 3 summarises the results of Vampire on the BLT translations of the benchmarks.

- Vampire solves 19 of the problems, translated by BLT.
- For all benchmark programs whose BLT translation Vampire is able to solve, Vampire also solves their Voogie translations. There are 3 benchmark programs for which Vampire solves their BLT translations but not their Boogie translations.

Based on the results presented in Tables 1, 2 and 3 we make the following observation. The problems translated from our benchmarks by Voogie are easier for Vampire than the problems translated by Boogie and BLT. Vampire is more efficient both in terms of the number of solved problems and runtime on the problems translated by Voogie. This confirms our conjecture that the use of (efficient translations of) FOOL is better for saturation theorem provers than translations to FOL designed for other purposes. It would be interesting to run these experiments for theorem provers other than Vampire, however Vampire is currently the only prover implementing FOOL.

Table 3. Runtimes in seconds of Vampire on the BLT translation of the benchmarks.

Benchmark	Number of loop unrollings				
	1	2	3	4	5
binary-search	0.821	163.790	–	–	–
bubble-sort	3.511	–	–	–	–
dutch-flag	4.049	–	–	–	–
insertion-sort	1.780	–	–	–	–
matrix-transpose	0.465	12.437	–	–	–
maxarray	0.174	1.567	47.724	–	–
maximum	0.069	0.140	0.724	12.234	–
one-duplicate	0.307	10.039	–	–	–
select-k	3.142	–	–	–	–
two-way-sort	0.319	24.622	–	–	–

5   Related Work

Our previous work introduced FOOL [8], its implementation in Vampire [6], and an efficient clausification algorithm for FOOL formulas [7].

In [6] we sketched a tuple extension of FOOL and an algorithm for computing the next state relations of imperative programs that uses this extension. This paper extends and improves the algorithm. In particular, (i) we described an encoding that uses FOOL in its current form, available in Vampire, (ii) we refined the encoding to only use in let-in the variables updated in program statements, (iii) we gave the definition of the encoding formally and in full detail, and (iv) we presented experimental results that confirm the described benefits of the encoding.

Boogie is used as the name of both the intermediate verification language [12] and the automated verification framework [1]. The Boogie verifier encodes the next state relations of imperative programs in first-order logic by naming intermediate states of program variables [11].

BLT [3] is a tool that automatically generates verification conditions of Boogie programs. The aim of the BLT project is to use first-order theorem provers rather than SMT solvers for checking quantified program properties. BLT produces formulas written in the TPTP language and uses if-then-else and let-in constructs of FOOL. BLT has an experimental option that introduces tuples for encoding of the next state relation. This option implements the encoding described in our earlier work [6].

6   Conclusion and Future Work

We presented an encoding of the next state relations of imperative programs in FOOL. Based on this encoding we defined a translation from imperative programs, annotated with their pre- and post-conditions, to FOOL formulas that encode partial correctness properties of these programs. We presented experimental results obtained by running the theorem prover Vampire on such properties. We generated these properties using our translation and verification tools Boogie and BLT. We described a polymorphic theory of first class tuples and its implementation in Vampire.

The formulas produced by our translation can be efficiently checked by automated theorem provers that support FOOL. The structure of our encoding closely resembles the structure of the program. The encoding contains neither new symbols nor duplicated parts of the program. This way, the efficient representation of the problem in plain first-order logic is left to the theorem prover that is better equipped to do it.

Our encoding is useful for automated program analysis and verification. Our experimental results show that Vampire was more efficient in terms of the number of solved problems and runtime on the problems obtained using our translation.

FOOL reduces the gap between programming languages and languages of automated theorem provers. Our encoding relies on tuple expressions and let-in with tuple definitions, available in FOOL. To our knowledge, these constructs are not available in any other logic efficiently implemented in automated theorem provers.

The polymorphic theory of first class tuples is a useful addition to a first-order theorem prover. On the one hand, it generalises and simplifies tuple expressions in FOOL. On the other hand, it is a convenient theory on its own, and can be used for expressing problems of program analysis and computer mathematics.

For future work we are interested in making automated first-order theorem provers friendlier to program analysis and verification. One direction of this work is design of an efficient translation of features of programming languages to languages of automated theorem provers. Another direction is extensions of first-order theorem provers with new theories, such as the theory of bit vectors. Finally, we are interested in further improving automated reasoning in combination of theories and quantifiers.

Acknowledgments. This work has been supported by the ERC Starting Grant 2014 SYMCAR 639270, the Wallenberg Academy Fellowship 2014, the Swedish VR grant D0497701, the Austrian research project FWF S11409-N23 and the EPSRC grant EP/P03408X/1-QuTie.

References

1. Barnett, M., Chang, B.-Y.E., DeLine, R., Jacobs, B., Leino, K.R.M.: Boogie: a modular reusable verifier for object-oriented programs. In: de Boer, F.S., Bonsangue, M.M., Graf, S., de Roever, W.-P. (eds.) FMCO 2005. LNCS, vol. 4111, pp. 364–387. Springer, Heidelberg (2006). https://doi.org/10.1007/11804192_17
2. Barrett, C., Stump, A., Tinelli, C.: The SMT-LIB standard: version 2.0. Technical report, Department of Computer Science, The University of Iowa (2010). www.SMT-LIB.org

3. Chen, Y.T., Furia, C.A.: Triggerless happy. In: Polikarpova, N., Schneider, S. (eds.) IFM 2017. LNCS, vol. 10510, pp. 295–311. Springer, Cham (2017). https://doi.org/10.1007/978-3-319-66845-1_19

4. de Moura, L., Bjørner, N.: Z3: an efficient SMT solver. In: Ramakrishnan, C.R., Rehof, J. (eds.) TACAS 2008. LNCS, vol. 4963, pp. 337–340. Springer, Heidelberg (2008). https://doi.org/10.1007/978-3-540-78800-3_24

5. Filliâtre, J.-C., Paskevich, A.: Why3—where programs meet provers. In: Felleisen, M., Gardner, P. (eds.) ESOP 2013. LNCS, vol. 7792, pp. 125–128. Springer, Heidelberg (2013). https://doi.org/10.1007/978-3-642-37036-6_8

6. Kotelnikov, E., Kovács, L., Reger, G., Voronkov, A.: The vampire and the FOOL. In: Proceedings of the 5th ACM SIGPLAN Conference on Certified Programs, pp. 37–48 (2016)

7. Kotelnikov, E., Kovács, L., Suda, M., Voronkov, A.: A clausal normal form translation for FOOL. In: Benzmüller, C., Sutcliffe, G., Rojas, R. (eds) 2nd Global Conference on Artificial Intelligence, GCAI 2016. EPiC Series in Computing, vol. 41, pp. 53–71. EasyChair (2016)

8. Kotelnikov, E., Kovács, L., Voronkov, A.: A first class Boolean sort in first-order theorem proving and TPTP. In: Kerber, M., Carette, J., Kaliszyk, C., Rabe, F., Sorge, V. (eds.) CICM 2015. LNCS (LNAI), vol. 9150, pp. 71–86. Springer, Cham (2015). https://doi.org/10.1007/978-3-319-20615-8_5

9. Kovács, L., Robillard, S., Voronkov, A.: Coming to terms with quantified reasoning. In: Proceedings of the 44th ACM SIGPLAN Symposium on Principles of Programming Languages, POPL 2017, pp. 260–270 (2017)

10. Kovács, L., Voronkov, A.: First-order theorem proving and VAMPIRE. In: Sharygina, N., Veith, H. (eds.) CAV 2013. LNCS, vol. 8044, pp. 1–35. Springer, Heidelberg (2013). https://doi.org/10.1007/978-3-642-39799-8_1

11. Rustan, K., Leino, M.: Efficient weakest preconditions. Inf. Process. Lett. **93**(6), 281–288 (2005)

12. Rustan, K., Leino, M.: This is Boogie 2. Manuscr. KRML, **178**(131) (2008)

13. Reger, G., Bjørner, N., Suda, M., Voronkov, A.: AVATAR modulo theories. In: 2nd Global Conference on Artificial Intelligence, GCAI 2016, pp. 39–52 (2016)

14. Stump, A., Sutcliffe, G., Tinelli, C.: StarExec: a cross-community infrastructure for logic solving. In: Demri, S., Kapur, D., Weidenbach, C. (eds.) IJCAR 2014. LNCS (LNAI), vol. 8562, pp. 367–373. Springer, Cham (2014). https://doi.org/10.1007/978-3-319-08587-6_28

15. Sutcliffe, G.: The TPTP problem library and associated infrastructure - from CNF to TH0, TPTP v6.4.0. J. Autom. Reason. **59**(4), 483–502 (2017)

16. Sutcliffe, G., Schulz, S., Claessen, K., Baumgartner, P.: The TPTP typed first-order form with arithmetic. In: Bjørner, N., Voronkov, A. (eds.) LPAR 2012. LNCS, vol. 7180, pp. 406–419. Springer, Heidelberg (2012). https://doi.org/10.1007/978-3-642-28717-6_32

17. Voronkov, A.: AVATAR: the architecture for first-order theorem provers. In: Biere, A., Bloem, R. (eds.) CAV 2014. LNCS, vol. 8559, pp. 696–710. Springer, Cham (2014). https://doi.org/10.1007/978-3-319-08867-9_46

Constructive Decision via Redundancy-Free Proof-Search

Dominique Larchey-Wendling$^{(\boxtimes)}$

Université de Lorraine, CNRS, LORIA, Nancy, France
dominique.larchey-wendling@loria.fr

Abstract. We give a constructive account of Kripke-Curry's method which was used to establish the decidability of Implicational Relevance Logic ($\mathbf{R}_{\rightarrow}$). To sustain our approach, we mechanize this method in axiom-free Coq, abstracting away from the specific features of $\mathbf{R}_{\rightarrow}$ to keep only the essential ingredients of the technique. In particular we show how to replace Kripke/Dickson's lemma by a constructive form of Ramsey's theorem based on the notion of almost full relation. We also explain how to replace König's lemma with an inductive form of Brouwer's Fan theorem. We instantiate our abstract proof to get a constructive decision procedure for $\mathbf{R}_{\rightarrow}$ and discuss potential applications to other logical decidability problems.

Keywords: Constructive decidability · Relevance logic
Redundancy-free proof-search · Almost full relations

1 Introduction

In this paper, we give a fully constructive/inductive account of Kripke's decidability proof of implicational relevance logic $\mathbf{R}_{\rightarrow}$, fulfilling the program outlined in [17]. The result is known as Kripke's but it crucially relies on Curry's lemma [18] which states that if a sequent S_2 is redundant over a sequent S_1 and S_2 has a proof, then S_1 has a shorter proof. Our account of Kripke-Curry's method is backed by an axiom-free mechanized proof of the result in the Coq proof assistant. However, their method and our constructivized implementation is in no way limited to that particular logic. As explained in [19], "Kripke's procedure for deciding $\mathbf{R}_{\rightarrow}$ can be seen as a precursor for many later algorithms that rely on the existence of a well quasi ordering (WQO)." From a logical perspective, Kripke-Curry's method has been adapted to *implicational ticket entailment* [2] and *the multiplicative and exponential fragment of linear logic* [1]. However, both of these recent papers are now contested inside the community because of deeply hidden flaws in the arguments [9, footnote 1], [20, footnote 4] and [6, pp 360–362]. This illustrates that the beauty of Kripke-Curry's method should not hide its subtlety and justifies all the more the need to machine-check such proofs.

Work partially supported by the TICAMORE project (ANR grant 16-CE91-0002).

© Springer International Publishing AG, part of Springer Nature 2018
D. Galmiche et al. (Eds.): IJCAR 2018, LNAI 10900, pp. 422–438, 2018.
https://doi.org/10.1007/978-3-319-94205-6_28

From a complexity perspective, S. Schmitz recently gave a 2-ExpTime complexity characterization [19] of the entailment problem for $\mathbf{R}_\rightarrow$, implying a decision procedure. However, the existence of a complexity characterization does not automatically imply a constructive proof of decidability. Indeed, the decision procedure itself or its termination proof might involve non-constructive arguments. In the case of $\mathbf{R}_\rightarrow$, the result of [19] "relies crucially" on the 2-ExpTime-completeness of the coverability problem in branching vector addition systems (BVASS) [7]. Checking the constructive acceptability of such chains of results implies checking that property for every link in the chain, an intimidating task, all the more problematic when considering mechanization.[1]

Our interest in the entailment problem for $\mathbf{R}_\rightarrow$ lies in the inherent simplicity and genericity of Kripke-Curry's argumentation. It is centered around the notion of redundancy avoidance. But compared to e.g. intuitionistic logic (**IL**), the case of $\mathbf{R}_\rightarrow$ is specific because *redundancy is not reduced to repetition*: the redundancy relation is not the identity. The case of repetition is not so interesting: Curry's lemma is trivial for repetition; and the sub-formula property and the pigeon hole principle ensure that Gentzen's sequent system **LJ** has a finite search space.

In the case of $\mathbf{R}_\rightarrow$, the sequent S_2 is redundant over S_1 if they are cognate and S_1 is included into S_2 for multiset inclusion [18]. In [17], Dickson's lemma is identified as the main difficulty for transforming Kripke-Curry's method into a constructive proof. Dickson's lemma[2] is a consequence of Ramsey's theorem which, stated positively, can be viewed as the following result [23]: the intersection of two WQOs is a WQO. The closure of the class of WQOs under direct products follows trivially and so does Dickson's lemma. We think that the use of König's lemma in Kripke's proof is also a potential difficulty w.r.t. constructivity. Admittedly, there are many variants of this lemma and indeed, we will use one which is suited in a constructive argumentation.

Let us now present the content of this paper. In Sect. 2, we propose an overview of Kripke-Curry's argumentation focusing on the two issues of Dickson's lemma and König's lemma. To constructivize that proof, we approached the problem posed by Dickson's lemma by using Coquand's [3] direct intuitionistic proof of Ramsey's theorem through an intuitionistic formulation of WQOs as *almost full relations* (AF) [24]. These results are recalled in Sect. 3. We also give a constructive version of König's lemma based on AF relations.

In Sect. 4, we give an account of what could be called the central ingredients of Kripke-Curry's proof by outlining the essential steps of our constructive mechanization in the inductive type theory on which Coq is based. Our `proof_decider` of Fig. 3 abstracts away from the particular case of $\mathbf{R}_\rightarrow$, by isolating the essential ingredient: an almost full redundancy relation which satisfies Curry's lemma. Finally, in Sect. 4.6, we instantiate the `proof_decider` into a constructive decision procedure for $\mathbf{R}_\rightarrow$. We also discuss the potential applications to other sub-structural logics.

---

[1] As for coverability in BVASS, it seems that the arguments developed in [7] cannot easily be converted to constructive ones (private communication with S. Demri).

[2] Dickson's lemma states that under pointwise order, $\mathbb{N}^k$ is a WQO for any $k \in \mathbb{N}$.

$$\vdash A \supset A$$
$$\vdash (A \supset A \supset B) \supset (A \supset B)$$

$$\vdash (A \supset B) \supset (C \supset A) \supset (C \supset B)$$
$$\vdash (A \supset B \supset C) \supset (B \supset A \supset C)$$

$$\dfrac{\vdash A \supset B \quad \vdash A}{\vdash B}$$

Fig. 1. Hilbert's style proof system for implicational relevance logic $\mathbf{R}_\rightarrow$.

The technical aspects of our proofs are sustained by a Coq v8.6 mechanization which is available under a Free Software license [14]. The size of this development is significant—around 15 000 lines of code,—but most of the code is devoted to libraries and the implementations of the proof systems $\mathbf{R}_\rightarrow$, $\mathbf{LR1}_\rightarrow$ and $\mathbf{LR2}_\rightarrow$ and the links between them: soundness/completeness results, cut-elimination, sub-formula property, finitely branching proof-search, etc. The case of implicational intuitionistic logic $\mathbf{J}_\rightarrow$ is treated as well in this mechanization. The core of our constructivization of Kripke-Curry's proof can be found in the file proof.v and is only around 800 lines long (including comments).

2   Constructive Issues in Kripke's Decidability Proof

In this section, we recall the main aspects of Kripke's decidability proof for the *implicational fragment of relevance logic* $\mathbf{R}_\rightarrow$, described with Hilbert style proof rules in Fig. 1. We sum up the description of [18] while focusing on the aspects of the arguments that were challenging from a constructive perspective. Among the many research directions later suggested by Riche in [17] for solving the missing link—a constructive proof of IDP or of Dickson's lemma,—the use of Coquand's approach [4] to *Bar induction* turned out as a solution.

Notice that in the notation $\mathbf{R}_\rightarrow$, the symbol $\rightarrow$ represents the logic-level implication to stay coherent with [17–19]. But in this paper, we rather use $\supset$ to denote logical implications to avoid conflicting with the Coq notation for function types $T_1 \rightarrow T_2$ (see e.g. the below definition of HR_proof).

2.1   What Is a Constructive Proof of Relevant Decidability?

Let us formalize the high-level question that.we solve in this paper. Before we give a mechanized constructive proof of decidability for $\mathbf{R}_\rightarrow$, we need to formally define provability or proofs, at least for $\mathbf{R}_\rightarrow$. This can easily be done in Coq using the (informative) inductive predicate:

```
Inductive HR_proof : Form → Set :=
  | id   : ∀A,       ⊢ A ⊃ A
  | pfx  : ∀A B C, ⊢ (A ⊃ B) ⊃ (C ⊃ A) ⊃ (C ⊃ B)
  | comm : ∀A B C, ⊢ (A ⊃ B ⊃ C) ⊃ (B ⊃ A ⊃ C)
  | cntr : ∀A B,   ⊢ (A ⊃ A ⊃ B) ⊃ (A ⊃ B)
  | mp   : ∀A B,   ⊢ A ⊃ B → ⊢ A → ⊢ B
where " ⊢ A " := (HR_proof A).
```

$$\frac{}{A \vdash A} \; \langle \text{AX} \rangle \qquad \frac{\Gamma, A, A \vdash B}{\Gamma, A \vdash B} \; \langle {}^*\text{W} \rangle \qquad \frac{\Gamma \vdash A \quad A, \Delta \vdash B}{\Gamma, \Delta \vdash B} \; \langle \text{CUT} \rangle$$

$$\frac{A, \Gamma \vdash B}{\Gamma \vdash A \supset B} \; \langle {}^*\supset \rangle \qquad \frac{\Gamma \vdash A \quad B, \Delta \vdash C}{\Gamma, \Delta, A \supset B \vdash C} \; \langle \supset{}^* \rangle$$

Fig. 2. The **LR1**$_{\rightarrow}$ sequent calculus rules for **R**$_{\rightarrow}$.

i.e. `HR_proof` A or ($\vdash A$ for short) encodes the *type of proofs of the formula* A in the Hilbert system for **R**$_{\rightarrow}$ of Fig. 1. A *constructive decidability proof* for **R**$_{\rightarrow}$ would then be given by a term `HR_decidability` of type:

$$\texttt{HR\_decidability} : \forall A : \texttt{Form}, \{\texttt{inhabited}(\vdash A)\} + \{\neg\texttt{inhabited}(\vdash A)\}$$

i.e. a total computable function which maps every formula A to a `boolean` value which if `true`, ensures that there is a proof of A, and if `false` ensures that there is no proof of A. A *constructive decider* is a stronger result of type:

$$\texttt{HR\_decider} : \forall A : \texttt{Form}, (\vdash A) + (\vdash A \rightarrow \texttt{False})$$

that is a total computable function that maps every formula A to either a proof of A or else ensures that no such proof exists. In other words, a constructive decider is an (always terminating) constructive proof-search algorithm. Obviously, adding axioms to Coq might hinder the computability of its terms (ensured by the normalization property of Coq). Hence, we allow no axiom and we aim at defining `HR_decidability` or `HR_decider` in axiom-free Coq.

2.2 Sequent Calculi for R$_{\rightarrow}$

Hilbert's **R**$_{\rightarrow}$ formulation is (unsurprisingly) not really suited to designing decision procedures based on proof-search. A standard approach is to convert Hilbert's systems into *sequent rules* such as those of **LR1**$_{\rightarrow}$ in Fig. 2 (see also [18]). In this particular system, a sequent $\Gamma \vdash A$ is composed of a multiset Γ of formulæ on the left of the $\vdash$ symbol and exactly one formula A on the right of the $\vdash$ symbol. There are three structural rules: $\langle \text{AX} \rangle$, $\langle {}^*\text{W} \rangle$ and $\langle \text{CUT} \rangle$, and two logical rules: $\langle {}^*\supset \rangle$ and $\langle \supset{}^* \rangle$. The soundness/completeness of this conversion to sequent calculus is ensured by the following result: a formula A has a Hilbert proof $\vdash A$ if and only if the sequent $\emptyset \vdash A$ has a proof in **LR1**$_{\rightarrow}$ (with $\emptyset$ as the empty multiset). That result is mechanized in the file relevant_equiv.v.

Although designed for proof-search, the sequent system **LR1**$_{\rightarrow}$ still suffers two major problems when considering fully automated procedures: one is the $\langle \text{CUT} \rangle$ rule and the other is the more problematic contraction rule $\langle {}^*\text{W} \rangle$. *Cut-elimination* is one of the central questions of proof-theory, partly because $\langle \text{CUT} \rangle$ makes proof-search infinitely branching. Fortunately, the $\langle \text{CUT} \rangle$ rule is admissible in **LR1**$_{\rightarrow}$ so we can safely remove that rule from **LR1**$_{\rightarrow}$. Cut-admissibility is proved using a relational phase semantic in the file sem_cut_adm.v.

On the other hand $\langle {}^*W \rangle$ needs to be handled much more carefully. The trick of Curry, which is well described in [18] is to absorb several instances of $\langle {}^*W \rangle$ in the rule $\langle {}^*\supset \rangle$ but in a tightly controlled way.[3] This is done by replacing both rules $\langle {}^*W \rangle$ and $\langle {}^*\supset \rangle$ with the rule $\langle {}^*\supset_2 \rangle$:

$$\frac{\Gamma \vdash A \quad B, \Delta \vdash C}{\Theta, A \supset B \vdash C} \; \langle {}^*\supset_2 \rangle \quad \text{with } \mathtt{LR2\_condition}(A \supset B, \Gamma, \Delta, \Theta)$$

obtaining the system $\mathbf{LR2}_{\rightarrow}$ composed of $\langle AX \rangle$, $\langle \supset^* \rangle$ and $\langle {}^*\supset_2 \rangle$. The side $\mathtt{LR2\_condition}(A \supset B, \Gamma, \Delta, \Theta)$ is a bit complicated to express formally so we will informally sum up its central idea: while applying $\langle {}^*\supset_2 \rangle$ top-down, some controlled/bounded form of contraction is allowed on every formula: the principal formula $A \supset B$ can be contracted at most twice while side formulæ in Γ, Δ can be contracted at most once. See the definition of $\mathtt{LR2\_condition}$ in file relevant_contract.v for a precise characterization.

2.3   Irredundant Proofs in $\mathbf{LR2}_{\rightarrow}$

Before using $\mathbf{LR2}_{\rightarrow}$ for deciding $\mathbf{R}_{\rightarrow}$, $\mathbf{LR2}_{\rightarrow}$ must of course be proved *equivalent* to $\mathbf{LR1}_{\rightarrow}$ and this is not a trivial task (see file relevant_equiv.v for the technical details). The cornerstone of the equivalence between $\mathbf{LR2}_{\rightarrow}$ and $\mathbf{LR1}_{\rightarrow}$ lies in a critical property of $\mathbf{LR2}_{\rightarrow}$ called *Curry's lemma*. It ensures both:

– the *admissibility of the contraction rule* $\langle {}^*W \rangle$ in $\mathbf{LR2}_{\rightarrow}$;
– the *completeness of irredundant proof-search* in $\mathbf{LR2}_{\rightarrow}$.

We say that a sequent $\Delta \vdash B$ *is redundant over* a sequent $\Gamma \vdash A$ and we denote $\Gamma \vdash A \prec_r \Delta \vdash B$ if $\Gamma \vdash A$ can be obtained from $\Delta \vdash B$ by repeated top-down applications of the contraction rule $\langle {}^*W \rangle$. We also characterize redundancy using the number of *occurrences* $|\Gamma|_X$ of the formula X in the multiset Γ:

$$\Gamma \vdash A \prec_r \Delta \vdash B \iff A = B \wedge \forall X, |\Gamma|_X \prec_r^{\mathbb{N}} |\Delta|_X \qquad (\prec_r)$$

where the binary relation $n \prec_r^{\mathbb{N}} m$ on $\mathbb{N}$ is defined by $n \leqslant m \wedge (n = 0 \Leftrightarrow m = 0)$. Now we can state Curry's lemma.

Lemma 1 (Curry [5], 1950). *Consider two sequents such that $\Delta \vdash B$ is redundant over $\Gamma \vdash A$, i.e. $\Gamma \vdash A \prec_r \Delta \vdash B$. Then any $\mathbf{LR2}_{\rightarrow}$-proof of $\Delta \vdash B$ can be contracted into a $\mathbf{LR2}_{\rightarrow}$-proof of $\Gamma \vdash A$, that is, a proof of lesser height.*

The Coq proof term is $\mathtt{LR2\_Curry}$ in the file relevant_LR2.v. Admissibility of contraction follows trivially from Curry's lemma and hence the completeness of $\mathbf{LR2}_{\rightarrow}$ w.r.t. $\langle CUT \rangle$-free $\mathbf{LR1}_{\rightarrow}$. Another critical consequence of Curry's lemma is related to irredundant proofs.

---

[3] Unrestricted contraction would generate infinitely *branching* proof-search.

Definition 1 (Irredundant proof). *A proof is* redundant *if there is a redundant pair in one of its branches, i.e.* $\Gamma \vdash A \prec_r \Delta \vdash B$ *where* $\Delta \vdash B$ *occurs in the sub-proof of* $\Gamma \vdash A$. *A proof is* irredundant *if none of its branches contain a redundant pair.*

By Curry's lemma, any sequent provable in $\mathbf{LR2}_\rightarrow$ has an irredundant proof in $\mathbf{LR2}_\rightarrow$ (the argument is not completely trivial, see Sect. 4). Hence, while searching for proofs in $\mathbf{LR2}_\rightarrow$, one can safely stop at redundancies.

2.4 Kripke's Decidability Proof

Building on Curry's lemma, the key insight of Kripke's proof of decidability is the following result. As explained in [8,17], it was discovered many times in different fields of mathematics, as e.g. Hilbert's finite basis theorem, the infinite division principle (IDP by Meyer [15]), Dickson's lemma, etc. We express Kripke's lemma with a concept that was not clearly spotted at that time but was popularized later on, that of well quasi order.

Definition 2 (Well Quasi Order). *A binary relation* $\leqslant$ *over a set* X *is a* well quasi order *(WQO) if it is reflexive, transitive and any infinite sequence* $x : \mathbb{N} \to X$ *contains a good pair* (i,j), *which means both* $i < j$ *and* $x_i \leqslant x_j$.

Lemma 2 (Kripke [12], 1959). *Given a finite set of (sub-)formulæ* $\mathcal{S}$, *the redundancy relation* $\prec_r$ *is a WQO when it is restricted to sequents composed exclusively of formulæ in* $\mathcal{S}$.

Proof. We give a "modernized" account of the proof. By Ramsey's theorem, the product (or intersection) of two WQOs is WQO.[4] Hence, the relation $\prec_r^{\mathbb{N}}$ over $\mathbb{N}$ is a WQO as the intersection of two WQOs. By finiteness of $\mathcal{S}$, the identity relation $=_\mathcal{S}$ on $\mathcal{S}$ is also a WQO (this is an instance of *the pigeon hole principle*).

Denoting $\prec_r^\mathcal{S}$ as the restriction of $\prec_r$ to the sequents composed of formulæ in the finite set $\mathcal{S}$, we can derive the equivalence

$$\Gamma \vdash A \prec_r^\mathcal{S} \Delta \vdash B \iff A =_\mathcal{S} B \wedge \bigwedge_{X \in \mathcal{S}} |\Gamma|_X \prec_r^\mathbb{N} |\Delta|_X$$

hence $\prec_r^\mathcal{S}$ is a WQO as a finite intersection of WQOs. □

Kripke's proof of decidability of $\mathbf{LR2}_\rightarrow$ (and hence $\mathbf{R}_\rightarrow$) can be summarized in the following steps:

- consider a start sequent $\Gamma \vdash A$ and let $\mathcal{S}$ be its finite set of sub-formulæ;
- launch backward proof-search for irredundant proofs of $\Gamma \vdash A$ in $\mathbf{LR2}_\rightarrow$, i.e. search stops when no rule is applicable or at a redundancy. We denote by $\mathcal{T}$ the corresponding (potentially infinite) proof-search tree;

---

[4] This result is known as Dickson's lemma when restricted to $\mathbb{N}^k$ with the point-wise product order. The inclusion relation between multisets built from the finite set $\mathcal{S}$ is a particular case of the product order $\mathbb{N}^k$ where k is the cardinal of $\mathcal{S}$.

- by the sub-formula property, no formula outside of $\mathcal{S}$ can occur in $\mathcal{T}$;
- $\mathcal{T}$ is finitely branching (critically relies on the side condition of rule $\langle {}^\star \supset_2 \rangle$);
- $\mathcal{T}$ cannot have an infinite branch (by Kripke's lemma);
- hence by König's lemma, the proof-search tree $\mathcal{T}$ is finite.

In [17], Riche focuses on Kripke/Dickson's lemma as the main difficulty to get an argumentation that could be accepted from a constructive point of view. We think that König's lemma is also a potentially non-constructive result [21], depending on its precise formulation. In Sect. 4, we explain how to overcome these two difficulties and transform this method into a mechanized HR_decider.

3   Inductive Well Quasi Orders

In this section we describe an inductive formulation of the notion of WQO. We are now going to use the language of Inductive Type Theory instead of Set theoretical language. Type theoretically, Definition 2 becomes: a WQO is a reflexive and transitive predicate $\prec_r : X \to X \to \texttt{Prop}$ such that for any $f : \texttt{nat} \to X$, there exists $i, j : \texttt{nat}$ s.t. $i < j$ and $f_i \prec_r f_j$ (good pair). We can say that any infinite sequence is bound to be redundant. We recall the inductive characterization of WQO due to Fridlender and Coquand [10] and the *constructive Ramsey theorem* [24], from which we derive a constructive proof of Kripke's lemma. Using the inductive *Fan theorem* [10], we derive a *constructive König's lemma*. The corresponding Coq code can be found in the library file almost_full.v.

3.1   Good Lists, Almost Full Relations and Bar Inductive Predicates

Much like *well founded relations* can be defined inductively by *accessibility predicates* (see module Wf of Coq standard library), WQOs can inductively be defined either by the *almost full inductive predicate* (AF) or by *bar inductive predicates*. Notice that these equivalent inductive characterizations are constructively stronger than the usual classical definition (like in the case of well-foundedness).

Let us consider a type $X : \texttt{Type}$ and a redundancy relation $\prec_r : X \to X \to \texttt{Prop}$. We define the good $\prec_r : \texttt{list}\ X \to \texttt{Prop}$ predicate that characterizes the (finite) lists which contain a good pair:

$$\texttt{good}\ \prec_r\ [x_{n-1}; \ldots; x_0] \iff \exists i\, j,\ i < j < n \land x_i \prec_r x_j \qquad \text{(good)}$$

Hence the list $[\ldots; b; \ldots; a; \ldots]$ is good when $a \prec_r b$. The list is read from right to left because we represent *the n-prefix of a sequence* $f : \texttt{nat} \to X$ by $[f_{n-1}; \ldots; f_0]$.

Definition 3 (Ir/redundant). *Given a relation* $\prec_r : X \to X \to \texttt{Prop}$ *called redundancy, a list of values* $l : \texttt{list}\ X$ *is redundant if* good $\prec_r l$ *holds and is irredundant if* $\neg(\texttt{good}\ \prec_r l)$ *holds.*

We write $\mathsf{bad} \prec_{\mathrm{r}} l$ when the list l satisfies $\neg(\mathsf{good} \prec_{\mathrm{r}} l)$. The *lifting of a relation* $R : X \to X \to \mathsf{Prop}$ by $x : X$ is denoted $R \uparrow x$ and characterized by:

$$u\,(R \uparrow x)\,v \iff u\,R\,v \lor x\,R\,u \qquad \text{for any } u, v : X$$

The disjunct $x\,R\,u$ prohibits any u which is R-greater than x to occur in $(R \uparrow x)$-bad sequences. AF relations are defined as those satisfying the af_t predicate:[5]

```
Inductive af_t {X : Type} (R : X → X → Prop) : Type :=
 | in_af_t0 : (∀x y, x R y)      → af_t R
 | in_af_t1 : (∀x, af_t (R ↑ x)) → af_t R.
```

Hence any full relation (i.e. $\forall x\,y,\ x\,R\,y$) is AF, and if every lifting of R is AF, then so is R. Notice that the predicate af_t is *informative*: it contains a well-founded tree of liftings until the relation becomes full (see [24]). This information is important to compute bounds for proof-search. Reflexive and transitive relations which satisfy af_t are WQOs in the classical interpretation. But constructively speaking, they are stronger in the following sense: any sequence $f : \mathsf{nat} \to X$ can *effectively* be transformed into an upper-bound n under which there exists a good pair, upper-bound obtained by finite inspection of the prefixes of f:

```
Lemma af_t_inf_chain (X : Type) (≺r : X → X → Prop) :
   af_t(≺r) → ∀(f : nat → X), {n : nat | ∃i j, i < j < n ∧ fi ≺r fj}.
```

The constructive Ramsey theorem [24] states that almost full relations are closed under (binary) intersection, and as a consequence, under direct products:

```
Theorem af_t_prod (X Y : Type) (R : X → X → Prop) (S : Y → Y → Prop) :
   af_t R → af_t S → af_t (R × S).
```

Notice that reflexivity and transitivity of WQOs are completely orthogonal to almost fullness in these results. They play no important role in our development.

The $\mathsf{af}_t(\prec_{\mathrm{r}})$ property can alternatively be defined by bar inductive predicates [10] as $\mathsf{bar}_t\,(\mathsf{good} \prec_{\mathrm{r}})\,[\,]$ with the following inductive definition:[6]

```
Inductive bar_t {X : Type} (P : list X → Prop) (l : list X) : Type :=
 | in_bar_t0 : P l                     → bar_t P l
 | in_bar_t1 : (∀x, bar_t P (x :: l))  → bar_t P l.
```

Hence, $\mathsf{bar}_t\,P\,l$ means that regardless of the repeated extensions of the list l by adding elements at its head, the predicate P is bound to be reached at some point. With this definition, we can derive the (informative) equivalence:

```
Theorem bar_t_af_t_eq X ≺r l : af_t(≺r ↑↑ l) ⟺ bar_t (good ≺r) l.
```

where $R \uparrow\uparrow [x_1, \ldots, x_n] := R \uparrow x_n \ldots \uparrow x_1$ (see file af_bar_t.v for details). And we deduce the equivalence $\mathsf{af}_t(\prec_{\mathrm{r}})$ iff $\mathsf{bar}_t\,(\mathsf{good} \prec_{\mathrm{r}})\,[\,]$ as the particular case.

---

[5] The braces around $\{X : \mathsf{Type}\}$ specify an *implicit argument*.

[6] $[\,]$ and $\_ :: \_$ are shorthand notations for the two list constructors.

3.2 A Constructive Form of König's lemma

Brouwer's Fan theorem can be proved equivalent to the binary form of König's lemma [21]. So one could wrongfully be led to the conclusion that both of these results cannot be constructively established. Here we explain that using suitable inductive definitions, such results can perfectly be established constructively.

For the rest of this section, we assume a type X : Type. We recall the inductive interpretation of the Fan theorem [10]. Given a list of lists ll : list (list X), we define the list of choice sequences (or fan) of ll denoted list_fan ll

Definition list_fan : list (list X) $\rightarrow$ list (list X).

The precise definition of list_fan uses auxiliary functions (see list_fan.v) but this is unimportant here. Only the following specification which characterizes the elements of list_fan $[l_1; \ldots; l_n]$ as choices sequences for $[l_1; \ldots; l_n]$ matters.[7]

$$[x_1; \ldots; x_n] \in_1 \text{list\_fan } [l_1; \ldots; l_n] \iff x_1 \in_1 l_1 \land \cdots \land x_n \in_1 l_n \qquad \text{(FAN)}$$

These are the lists composed of one element of l_1, then one element of l_2, ... and one element of l_n. The following result [10] states that if P is monotonic and bound to be reached by successive extensions starting from $[\,]$, then P is bound to be reached *uniformly* over the finitary fan represented by choices sequences.

Theorem fan_t_on_list $(P : \text{list } X \rightarrow \text{Prop})$ $\big(H_P : \forall x\, l, P\, l \rightarrow P(x :: l)\big)$:
    $\text{bar}_t\, P\, [\,] \rightarrow \text{bar}_t\, (\text{fun } ll \mapsto \forall l, l \in_1 \text{list\_fan } ll \rightarrow P\, l)\, [\,].$

The proof of this result is recalled in bar_t.v. Combining fan_t_on_list with bar_t_af_t_eq and af_t_inf_chain, we derive the following strong form of König's lemma. Given an almost full redundancy relation $\prec_r : X \rightarrow X \rightarrow \text{Prop}$ and a sequence of finitary choices f : nat $\rightarrow$ list X, beyond an effective lower-bound n, every finite prefix of a choice sequence for f is redundant (see koenig.v).

Theorem Constructive_Koenigs_lemma $\prec_r (f : \text{nat} \rightarrow \text{list } X)$:
    $\text{af}_t(\prec_r) \rightarrow \{n \mid \forall m\, l,\ n \leqslant m \rightarrow l \in_1 \text{list\_fan } [f_{m-1}; \ldots; f_0] \rightarrow \text{good } \prec_r l\}.$

In Sect. 4.5, we over-approximate the branches of the proof-search tree as the choice sequences of f : nat $\rightarrow$ list stm which collects in $f\, n$ the finitely many sequents that occur at height n in the proof-search tree. Thus we get a uniform upper-bound of the length of irredundant (i.e. bad $\prec_r$) proof-search branches.

4   Decision via Redundancy-Free Proof-Search

In this section, we describe the mechanization of a generic constructive decider based on redundancy-avoiding proof-search. We first give an informal account of the main arguments, then we proceed with a more formal description of these steps in the language of Coq. Except for the tree.v and almost_full.v libraries, all the following development is contained in the file proof.v.

---

[7] The notation $x \in_1 l$ is a shortcut for In $x\, l$, the (finitary) membership predicate.

4.1   Overview of the Assumptions and Main Arguments

Let us consider a type stm of *statements* representing logical propositions. These statements could, depending on the intended application, be formulæ like in Hilbert style proof systems, or sequents in sequent proof systems or in some versions of natural deduction, or more generally structures like nested sequents.

Statements are items to be proved or refuted (by showing the impossibility of a proof as a term of type has_proof $s \to$ False). For this, we describe a proof system as a set of valid

$$\frac{H_1 \ \dots \ H_n}{C}$$

rule instances. These instances are generally represented as in the right figure where C : stm is the *conclusion* of the instance and $[H_1; \dots; H_n]$: list stm is the list of *premises*. We collect the set of valid rule instances into a binary relation rules : stm $\to$ list stm $\to$ Prop between individual statements (stm viewed as conclusions) and list of statements (list stm viewed as premises). Hence, the validity of the above rule instance in the proof system is expressed by the predicate rules C $[H_1; \dots; H_n]$. When dealing with proof-search based decision, infinite horizontal branching of proof-search is usually forbidden. Hence, for a given C : stm, only finitely many rule instances exist with C as conclusion. Moreover, that finite set of instances must be computable to be able to enumerate the next steps of backward proof-search. We denote this property by rules_fin and we say that rules *has finite inverse images*.[8]

Valid rules instances are combined to form proof trees. A proof tree is a finite tree of statements where each node is a valid rule instance. Proofs are proof trees of their root node and n-bounded proofs are proofs of height bounded by n. Because of the finite inverse images property rules_fin, the set of n-bounded proofs of a given statement s_0 is finite and computable. We define the notion of *minimal proof*, which is a proof of minimal height among the proofs of the same statement. Every proof can effectively be transformed into a minimal proof by searching among the finitely many proofs of lesser height. An *everywhere minimal proof* is such that each of its sub-proof is minimal. Every proof can effectively be transformed into an everywhere minimal proof.

Our generic constructive technique assumes a binary redundancy relation $\prec_r$ between statements which satisfies Curry's lemma 1: every proof containing a redundancy can be contracted into a lesser proof. As a consequence, everywhere minimal proofs are redundancy free. If we moreover assume that the binary relation $\prec_r$ is almost full (i.e. a *constructive WQO*), then every infinite sequence of statements contains a redundancy. However, remember that in Kripke's lemma 2 for **LR2**$_\to$, only the restriction of $\prec_r$ to finitely generated sequents is a WQO. Hence we only assume $\prec_r$ to be almost full on the set of sub-statements of an initial statement s_0.[9] By using the constructive version of König's lemma

---

[8] Typically, systems which include a *cut-rule* do not satisfy the rules_fin property which is why *cut-elimination* is viewed as a critical requisite to design sequent-based decision procedures. The same remark holds for the *modus-ponens* rule of Hilbert systems, usually making them unsuited for decision procedures.

[9] For this, we need a notion of sub-statement that is reflexive, transitive and such that valid rules instances possess the sub-statement property.

of Sect. 3.2, we show that every sequence of sub-statements of s_0 longer than a bound n_0 contains a redundancy. The bound n_0 can be computed from s_0 only. As a consequence, every irredundant proof of s_0 is a n_0-bounded proof. And deciding the provability of s_0 is reduced to testing whether the set of n_0-bounded proofs of s_0 is empty or not.

4.2   Finiteness, Trees and Branches

In this section, we consider a fixed X : Type. The predicate $\mathtt{finite}_t$ P expresses that a sub-type $P : X \to \mathtt{Prop}$ of X is finite and computable into a list:

Definition $\mathtt{finite}_t$ $(P : X \to \mathtt{Prop}) := \{l : \mathtt{list}\ X \mid \forall x, x \in_1 l \Longleftrightarrow P\ x\}$.

We will use this predicate to encode the `rules_fin` property. We will also need a type of finitely branching (oriented) trees:

Inductive tree $X := \mathtt{in\_tree} : X \to \mathtt{list}\,(\mathtt{tree}\ X) \to \mathtt{tree}\ X$.

where we denote $\langle x|l \rangle$ for $(\mathtt{in\_tree}\ x\ l)$. The root $\mathtt{root} : \mathtt{tree}\ X \to X$ of a tree verifies $\mathtt{root}\ \langle x|l \rangle = x$ and the height $\mathtt{ht} : \mathtt{tree}\ X \to \mathtt{nat}$ of a tree verifies $\mathtt{ht}\ \langle \_|l \rangle = 1 + \max\,(\mathtt{map}\ \mathtt{ht}\ l)$. Branches of trees are represented as specific lists of elements of X. We inductively define a `branch` predicate:

Inductive branch : tree $X \to$ list $X \to$ Prop :=
 | $\mathtt{in\_tb0}$: $\forall t$, branch t $[\,]$
 | $\mathtt{in\_tb1}$: $\forall x$, branch $\langle x|[\,] \rangle$ $(x :: [\,])$
 | $\mathtt{in\_tb2}$: $\forall b\ x\ l\ t,\ t \in_1 l \to$ branch $t\ b \to$ branch $\langle x|l \rangle$ $(x :: b)$.

s.t. the lists b which satisfy branch $t\ b$ collect all the nodes encountered on paths from the root of t to one of its internal nodes. The empty list $[\,]$ is among them.

4.3   Proofs, Minimal Proofs and Everywhere Minimal Proofs

We consider a type `stm` of logical statements and a collection of valid rule instances `rules` which has the finite inverse image property.

Variables   (stm : Type) (rules : stm $\to$ list stm $\to$ Prop).
Hypothesis   (rules_fin : $\forall c$: stm, $\mathtt{finite}_t$ (rules c)).

Hence, not only are there finitely many rule instances for a given conclusion c but the predicate `rules_fin` c contains ll_c : list (list stm), an *effective list* of those instances which verifies the property:

$$[h_1; \ldots; h_n] \in_1 ll_c \Longleftrightarrow \mathtt{rules}\ c\ [h_1; \ldots; h_n] \quad \text{for any } [h_1; \ldots; h_n] : \mathtt{list}\ \mathtt{stm}$$

This effective aspect of finite branching is often implicit in studies on proof-search, because if one cannot even compute the valid instances for a given conclusion, then there is no way to implement backward proof-search.

The notion of proof is based on that of *proof tree*. We define a predicate `proof_tree` that satisfies the below (recursive) characteristic property:

Definition proof_tree : tree stm → Prop.
$\forall s\, l$, proof_tree $\langle s|l\rangle \Longleftrightarrow$ rules s (map root l) $\wedge \forall t,\, t \in_1 l \to$ proof_tree t.

i.e. trees of statements where each node is a valid rule instance, the conclusion being the node itself and the premises being the children of the node. Given a statement s, a *proof* of s is a proof tree t with root s, and a *n-bounded proof* is a proof of height bounded by n:

Definition proof $(s : \text{stm})\ (t : \text{tree stm}) :=$ proof_tree $t \wedge$ root $t = s$.
Definition bproof $(n : \text{nat})\ (s : \text{stm})\ (t : \text{tree stm}) :=$ proof $s\, t \wedge$ ht $t \leqslant n$.

Proofs of a given statement s are not necessarily finitely many but because of the finite inverse image property `rules_fin`, n-bounded proofs are:

Proposition bproof_finite_t $(n : \text{nat})\ (s : \text{stm})$: finite$_\text{t}$ (bproof $n\, s$).

We introduce the notion of *minimal proof*, that is a proof of minimal height among the proofs with a given root s. We show that every proof t can be transformed into a minimal proof by a simple search of the shortest among the (ht t)-bounded proofs of s, of which a list can be computed using `bproof_finite_t`.

Definition   min_proof $s\, t :=$ proof $s\, t \wedge \forall t'$, proof $s\, t' \to$ ht $t \leqslant$ ht t'.
Proposition proof_minimize $s\, t$: proof $s\, t \to \{t_{\min} \mid$ min_proof $s\, t_{\min}\}$.

But to exploit Curry's lemma, we need a much stronger minimality property: this is the notion of *everywhere minimal proof tree*, where every sub-tree is a minimal proof of its own root. We show that every proof can effectively be transformed into an everywhere minimal proof.

Definition   emin_ptree : tree stm → Prop.
$\forall s\, l$, emin_ptree $\langle s|l\rangle \Longleftrightarrow$ min_proof $s\, \langle s|l\rangle \wedge \forall t,\, t \in_1 l \to$ emin_ptree t.
Definition   emin_proof $s\, t :=$ proof $s\, t \wedge$ emin_ptree t.
Proposition proof_eminimize $s\, t$: proof $s\, t \to \{t_{\text{em}} \mid$ emin_proof $s\, t_{\text{em}}\}$.

Proof. The argument proceeds by induction on the height ht t of the proof tree t. It uses `proof_minimize` to compute a minimal proof t_1 for s and then proceeds inductively on every immediate sub-proof of t_1.   □

4.4   The Completeness of Irredundant Proofs via Curry's Lemma

We assume a notion of redundancy on statements, that is a binary relation $\prec_\text{r} : \text{stm} \to \text{stm} \to \text{Prop}$. A list of statements $l : \text{list stm}$ is redundant if it contains a good pair for $\prec_\text{r}$, which is denoted by good $\prec_\text{r} l$ (see Sect. 3.1).

The list l is *irredundant* if it contains no good pair, i.e. bad $\prec_r l$. A tree t : tree stm is an *irredundant proof* if it is a proof and every branch of the tree is irredundant.[10]

Definition irred_proof s t := proof s t $\wedge$ $\forall b$, branch t b $\rightarrow$ bad $\prec_r$ (rev b).

We now state the assumption Curry abstracting Curry's lemma:

Hypothesis Curry : $\forall s_1 s_2 t$, proof $s_2 t \rightarrow s_1 \prec_r s_2 \rightarrow \exists t'$, $\wedge \begin{cases} \text{proof } s_1 \text{ } t' \\ \text{ht } t' \leqslant \text{ht } t. \end{cases}$

assumption under which everywhere minimal proofs become irredundant:

Lemma proof_emin_irred s t : emin_proof s t $\rightarrow$ irred_proof s t.

Proof. Given any branch b of an everywhere minimal proof tree t, we show that b cannot contain a redundancy. Let $s_1 \prec_r s_2$ be a good pair in b and let t_1/t_2 be the sub (proof) trees of roots s_1/s_2. As s_2 occurs after s_1 in b, t_2 is a strict sub-tree of t_1 and thus ht $t_2 <$ ht t_1. Using Curry, we get a proof t_1' of s_1 with ht $t_1' \leqslant$ ht t_2. We derive ht $t_1' <$ ht t_1, and thus t_1 is not a minimal proof of s_1, contradicting the everywhere minimality of t.   □

As a consequence, every proof can be transformed into an irredundant one by direct combination of proof_eminimize and proof_emin_irred.

Theorem proof_reduce s t : proof s t $\rightarrow$ $\{t_{\text{irr}} \mid$ irred_proof s $t_{\text{irr}}\}$.

4.5   Bounding the Height of Irredundant Proofs

Kripke used König's lemma to prove the finiteness of the finitely branching irredundant proof-search tree by showing that it cannot have infinite branches. The constructive argument works positively by showing that one can compute a uniform upper-bound over the length of irredundant proof-search branches. We use the constructive version of König's lemma of Sect. 3.2.

To capture the *sub-formula property* in our setting, we assume an abstract notion of sub-statement denoted by $s_1 \supseteq_{\text{sf}} s_2$ and which intuitively reads as the sub-formulæ of s_2 are also sub-formulæ of s_1. We postulate that $\supseteq_{\text{sf}}$ is both reflexive (sf_refl) and transitive (sf_trans) and more importantly, that every rule instance preserves sub-statements bottom-up (sf_rules):

Variables $(\supseteq_{\text{sf}} : \text{stm} \rightarrow \text{stm} \rightarrow \text{Prop})$ $(\text{sf\_refl} : \forall s, s \supseteq_{\text{sf}} s)$
$(\text{sf\_trans} : \forall s_1 s_2 s_3, s_1 \supseteq_{\text{sf}} s_2 \rightarrow s_2 \supseteq_{\text{sf}} s_3 \rightarrow s_1 \supseteq_{\text{sf}} s_3)$
$(\text{sf\_rules} : \forall c \, l, \text{rules } c \, l \rightarrow \forall s, s \in_1 l \rightarrow c \supseteq_{\text{sf}} s).$

---

[10] Branches are read from the root to leaves, hence the use of rev to reverse lists.

Starting from an initial statement s_0 : stm, we build the *proof-search sequence from* s_0 as the sequence of iterations fun $n \mapsto$ rules_next^n $[s_0]$ of the operator

Let rules_next : list stm $\rightarrow$ list stm.
$\forall l\, h,\ h \in_1$ rules_next $l \iff \exists c\, m,\ c \in_1 l \land h \in_1 m \land$ rules $c\, m$.

i.e. rules_next l is the (finite) inverse image of l by valid rules instances. By sf_rules, the proof-search sequence is composed of sub-statements of s_0:

Proposition proof_search_sf s_0 n s : $s \in_1$ rules_next^n $[s_0] \rightarrow s_0 \supseteq_{\mathrm{sf}} s$.

We can cover all the proof-search branches of length n starting from s_0 using the choices sequences over the proof-search sequence fun $n \mapsto$ rules_next^n $[s_0]$. Indeed, we establish the following covering property:

Let   FAN n s_0 := list_fan $[$rules_next^{n-1} $[s_0];\ldots;$ rules_next^0 $[s_0]]$.
Lemma ptree_proof_search (t : tree stm) (b : list stm) :
     branch t $b \rightarrow$ proof_tree $t \rightarrow$ rev $b \in_1$ FAN (length b) (root t).

Beware that this fan is a strict upper-approximation of proof-search branches.
     We postulate our *redundancy hypothesis* denoted Kripke which states that the relation $\prec_r$ is almost full when restricted to sub-statements of the initial statement s_0 (of which the provability is tested). Using constructive König's lemma of Sect. 3.2 (Constructive_Koenigs_lemma), we derive:

Hypothesis   Kripke : $\forall s_0$: stm, af$_t$ $\left(\prec_r$ restr (fun $s \mapsto s_0 \supseteq_{\mathrm{sf}} s)\right)$.
Proposition irredundant_max_length (s_0 : stm) :
     $\{n_0$: nat $\mid \forall m\, l,\ n_0 \leqslant m \rightarrow l \in_1$ FAN m $s_0 \rightarrow$ good $\prec_r l\}$.

Notice that we need the informative predicate af$_t$ to effectively compute the upper-bound. We conclude that irredundant proofs are bounded proofs:

Lemma proof_irred_bounded s_0 : $\{n_0 \mid$ irred_proof $s_0 \subseteq$ bproof n_0 $s_0\}$.

Hence, given a starting statement s_0, we can compute (from s_0 only) an upper-bound n_0 such that every irredundant proof of s_0 is n_0-bounded.

4.6   A Constructive Decider Based on Redundancy-Free Proof-Search

The proof decider follows trivially. Indeed, the corresponding algorithm uses proof_irred_bounded to first compute a bound n_0 such that every irredundant proof of s_0 has height bounded by n_0. Second, the algorithm computes the list of n_0-bounded proofs of s_0 using bproof_finite_t. If that list is non-empty, then s_0 has a proof. Otherwise, there is no n_0-bounded proofs for s_0, thus there is no irredundant proof for s_0 (this is the property of the upper-bound n_0), and then there is no proof for s_0 at all using proof_reduce. The full abstract result proof_decider is displayed in Fig. 3 and established in the file proof.v.

Theorem proof_decider (stm : Type) (rules : stm → list stm → Prop)
                      (rules_fin : $\big(\forall c,\ \text{finite}_t\ (\text{rules}\ c)\big)$
                      $\big(\supseteq_{\text{sf}} : \text{stm} \to \text{stm} \to \text{Prop}\big)$ (sf_refl : $\forall s, s \supseteq_{\text{sf}} s$)
                      (sf_trans : $\forall r\ s\ t,\ r \supseteq_{\text{sf}} s \to s \supseteq_{\text{sf}} t \to r \supseteq_{\text{sf}} t$)
                      (sf_rules : $\forall c\ l,\ \text{rules}\ c\ l \to \forall h,\ h \in_1 l \to c \supseteq_{\text{sf}} h$)
                      $(\prec_{\text{r}} : \text{stm} \to \text{stm} \to \text{Prop})$
                      (Curry : $\forall s\ t\ p,\ \text{pf}\ t\ p \to s \prec_{\text{r}} t \to \exists q,\ \text{pf}\ s\ q \land \text{ht}\ q \leqslant \text{ht}\ p$)
                      $\big($Kripke : $\forall s_0, \text{af}_t\ (\prec_{\text{r}} \text{restr}\ (\text{fun}\ s \mapsto s_0 \supseteq_{\text{sf}} s))\big)$:

$$\forall s_0 : \text{stm},\ \big\{p : \text{tree stm} \mid \text{pf}\ s_0\ p\big\} + \big\{\forall p : \text{tree stm},\ \neg(\text{pf}\ s_0\ p)\big\}.$$

Fig. 3. Constructive decider by redundancy-free proof-search (pf := proof rules).

We instantiate the proof_decider on the **LR2**$_\to$ sequent calculus in the file relevant_LR2_dec.v:

Theorem LR2_decider $(s : \text{Seq})$:
    $\big\{t \mid \text{proof LR2\_rules}\ s\ t\big\} + \big\{\forall t,\ \neg(\text{proof LR2\_rules}\ s\ t)\big\}.$

Using soundness and completeness results between **R**$_\to$ ⤳ **LR1**$_\to$ ⤳ **LR2**$_\to$ (see the summary file relevant_equiv.v), we get the constructive decider for **R**$_\to$ specified in Sect. 2.1, the proof of which can be found in logical_deciders.v:

Theorem HR_decider : $\forall A : \text{Form},\ \text{HR\_proof}\ A + (\text{HR\_proof}\ A \to \text{False})$.

5   Conclusion and Perspectives

We present an abstract and constructive view of Kripke-Curry's method for deciding Implicational Relevance Logic **R**$_\to$. We get an axiom-free Coq implementation [14] that we instantiate on **LR2**$_\to$ to derive a constructive decider for **R**$_\to$. Although not presented in this paper, our implementation includes a constructive decider for implicational intuitionistic logic **J**$_\to$ which shares the same language for formulæ as **R**$_\to$. It is based on a variant of Gentzen's sequent calculus **LJ**. Unlike what happens which richer fragments of Relevance Logic [22], extensions of this method to full propositional **IL** would present no difficulty.

From a complexity perspective, Kripke's decidability proof for **R**$_\to$ based on Dickson's lemma might be analyzed using control functions as in [8] to classify its complexity in the Fast Growing Hierarchy. Notice however that these techniques involve classical formulations of WQOs and their conversion to a constructive setting is far from evident. Furthermore, the 2-EXPTIME complexity characterization of [19] was not obtained that via control functions nor Dickson's lemma.

Kripke-Curry's method has a potential use well beyond **R**$_\to$ or Dickson's lemma and might be able to provide decidability for logics of still unknown and presumably high complexities. A very difficult case would be to get a constructive proof of decidability for the logic of Bunched Implications **BI** [11] based on Kripke-Curry's method. Indeed, as is the case for **LR1$_\to$**, contraction (and weakening) cannot be completely removed from the bunched sequent calculus **LBI**. It is not obvious what notion of redundancy should be used in that case.

Analyzing the "glitches" in the decidability proof of ticket entailment [6] is another obvious perspective of this work. Indeed, the attempt of [2] is also based on Kripke-Curry's method. This decidability result was independently obtained by Padovani [16] with seemingly much more involved techniques such as the use of Kruskal's tree theorem. Still, Kruskal's tree theorem is also a result about WQOs of which we do already have a mechanized constructive proof in Coq [13]. The mechanization of ticket entailment might not be completely out of reach.

References

1. Bimbó, K.: The decidability of the intensional fragment of classical linear logic. Theoret. Comput. Sci. **597**, 1–17 (2015)
2. Bimbó, K., Dunn, J.M.: On the decidability of implicational ticket entailment. J. Symb. Log. **78**(1), 214–236 (2013)
3. Coquand, T.: A direct proof of the intuitionistic Ramsey theorem. In: Pitt, D.H., Curien, P.-L., Abramsky, S., Pitts, A.M., Poigné, A., Rydeheard, D.E. (eds.) CTCS 1991. LNCS, vol. 530, pp. 164–172. Springer, Heidelberg (1991). https://doi.org/10.1007/BFb0013465
4. Coquand, T.: Constructive topology and combinatorics. In: Myers, J.P., O'Donnell, M.J. (eds.) Constructivity in CS 1991. LNCS, vol. 613, pp. 159–164. Springer, Heidelberg (1992). https://doi.org/10.1007/BFb0021089
5. Curry, H.B.: A Theory of Formal Deductibility. Notre Dame mathematical lectures. University of Notre Dame, Dame (1957)
6. Dawson, J.E., Goré, R.: Issues in machine-checking the decidability of implicational ticket entailment. In: Schmidt, R.A., Nalon, C. (eds.) TABLEAUX 2017. LNCS, vol. 10501, pp. 347–363. Springer, Cham (2017). https://doi.org/10.1007/978-3-319-66902-1_21
7. Demri, S., Jurdziński, M., Lachish, O., Lazić, R.: The covering and boundedness problems for branching vector addition systems. J. Comput. System Sci. **79**(1), 23–38 (2012)
8. Figueira, D., Figueira, S., Schmitz, S., Schnoebelen, P.: Ackermannian and primitive-recursive bounds with Dickson's lemma. In: LICS 2011, pp. 269–278 (2011)
9. Figueira, D., Lazic, R., Leroux, J., Mazowiecki, F., Sutre, G.: Polynomial-space completeness of reachability for succinct branching VASS in dimension one. In: ICALP 2017, LIPIcs, vol. 80, pp. 119:1–14. Schloss Dagstuhl (2017)
10. Fridlender, D.: An interpretation of the fan theorem in type theory. In: Altenkirch, T., Reus, B., Naraschewski, W. (eds.) TYPES 1998. LNCS, vol. 1657, pp. 93–105. Springer, Heidelberg (1999). https://doi.org/10.1007/3-540-48167-2_7
11. Galmiche, D., Méry, D., Pym, D.: The semantics of BI and resource tableaux. Math. Struct. Comput. Sci. **15**(6), 1033–1088 (2005)
12. Kripke, S.: The problem of entailment (abstract). J. Symb. Log. **24**, 324 (1959)
13. Larchey-Wendling, D.: A mechanized inductive proof of Kruskal's tree theorem (2015). http://www.loria.fr/~larchey/Kruskal

14. Larchey-Wendling, D.: A constructive mechanization of Kripke-Curry's method for the decidability of implicational relevance logic (2018). https://github.com/DmxLarchey/Relevant-decidability

15. Meyer, R.K.: Improved decision procedures for pure relevant logic. In: Anderson, C.A., Zelëny, M. (eds.) Logic, Meaning and Computation. Synthese Library (Studies in Epistemology, Logic, Methodology, and Philosophy of Science), pp. 191–217. Springer, Dordrecht (2001). https://doi.org/10.1007/978-94-010-0526-5_9

16. Padovani, V.: Ticket entailment is decidable. Math. Struct. Comput. Sci. **23**(3), 568–607 (2013)

17. Riche, J.: Decision procedure of some relevant logics: a constructive perspective. J. Appl. Non-Class. Log. **15**(1), 9–23 (2005)

18. Riche, J., Meyer, R.K.: Kripke, Belnap, Urquhart and relevant decidability and complexity. In: Gottlob, G., Grandjean, E., Seyr, K. (eds.) CSL 1998. LNCS, vol. 1584, pp. 224–240. Springer, Heidelberg (1999). https://doi.org/10.1007/10703163_16

19. Schmitz, S.: Implicational relevance logic is 2-ExpTime-complete. J. Symb. Log. **81**(2), 641–661 (2016)

20. Schmitz, S.: The complexity of reachability in vector addition systems. ACM SIGLOG News **3**(1), 4–21 (2016)

21. Schwichtenberg, H.: A direct proof of the equivalence between Brouwer's Fan theorem and König's lemma with a uniqueness hypothesis. J. UCS **11**(12), 2086–2095 (2005)

22. Urquhart, A.: The undecidability of entailment and relevant implication. J. Symb. Log. **49**(4), 1059–1073 (1984)

23. Veldman, W., Bezem, M.: Ramsey's theorem and the pigeonhole principle in intuitionistic mathematics. J. Lond. Math. Soc. **s2–47**(2), 193–211 (1993)

24. Vytiniotis, D., Coquand, T., Wahlstedt, D.: Stop when you are almost-full. In: Beringer, L., Felty, A. (eds.) ITP 2012. LNCS, vol. 7406, pp. 250–265. Springer, Heidelberg (2012). https://doi.org/10.1007/978-3-642-32347-8_17

Deciding the First-Order Theory of an Algebra of Feature Trees with Updates

Nicolas Jeannerod[(⊠)] and Ralf Treinen

Univ. Paris Diderot, Sorbonne Paris Cité, IRIF, UMR 8243 CNRS, Paris, France
{nicolas.jeannerod,ralf.treinen}@irif.fr

Abstract. We investigate a logic of an algebra of trees including the update operation, which expresses that a tree is obtained from an input tree by replacing a particular direct subtree of the input tree, while leaving the rest unchanged. This operation improves on the expressivity of existing logics of tree algebras, in our case of feature trees. These allow for an unbounded number of children of a node in a tree.

We show that the first-order theory of this algebra is decidable *via* a weak quantifier elimination procedure which is allowed to swap existential quantifiers for universal quantifiers. This study is motivated by the logical modeling of transformations on UNIX file system trees expressed in a simple programming language.

1 Introduction

Feature trees are trees where nodes have an unbounded number of children, and where edges from nodes to their children carry names such that no node has two different outgoing edges with the same name. Hence, the names on the edges can be used to select the different children of a node. Feature trees have been used in constraint-based formalisms in the field of computational linguistics (e.g. [14]) and constrained logic programming [1,15]. This work is motivated by a different application of feature trees: they are a quite accurate model of UNIX file system trees. The most important abstraction in viewing a file structure as a tree is that we ignore multiple hard links to files. Our mid-term goal is to derive, using symbolic execution techniques, from a shell script a logical formula that describes the semantics of this script as a relation between the initial file tree and the one that results from execution of the script.

Feature tree logics have at their core basic constraints like $x[f]y$, expressing that y is a subtree of x accessible from the root of x via feature f, and $x[f] \uparrow$, expressing that the tree x does not have a feature f at its root node. This is already sufficient to describe some tree languages that are useful in our context. For instance, the script consisting of the single command `mkdir /home/john`, which creates a directory `john` under the directory `home`, succeeds on a tree if the

This work has been partially supported by the ANR project CoLiS, contract number ANR-15-CE25-0001.

tree satisfies the formula $\exists d.(r[\text{home}]d \wedge d[\text{john}] \uparrow)$, which expresses that home is a subdirectory of the root, which does itself *not* have a subdirectory john. We ignore here the difference between directories and regular files, as well as file permissions.

Update Constraints. In order to describe the effect of executing the above script we need more expressivity. A first idea is to introduce an update constraint $y \doteq x[f \mapsto z]$, which states that the tree y is obtained from the tree x by setting its child f to z, and creating the child when it does not exist. Using this, the semantics of mkdir /home/john could be described by

$$\exists d, d', e. \, (in[\text{home}]d \wedge d[\text{john}] \uparrow \wedge out \doteq in[\text{home} \mapsto d'] \wedge d' \doteq d[\text{john} \mapsto e] \wedge e[\emptyset])$$

Here, $e[\emptyset]$ expresses that e is an empty directory. Note that this formula, by virtue of the update constraint, expresses that any existing directories under home different from john are not touched.

Programming constructs translate to combinations of logical formula. For instance, if $\phi_p(in, out)$, resp. $\phi_q(in, out)$ describe the semantics of script fragments p and q, then their composition is described by $\exists t.(\phi_p(in, t) \wedge \phi_q(t, out))$. The reality of our use case is more complex than that due to the hairy details of error handling in shell scripts [10], and is up to future work.

Formulas with more complex quantification structure occur when we express interesting properties of scripts. For instance, p and q are equivalent if

$$\forall in, out. \, (\phi_p(in, out) \leftrightarrow \phi_q(in, out))$$

Debian requires in its policy [7] so-called maintainer scripts to be idempotent, which can be expressed for a script p as

$$\forall in, out. \, (\phi_p(in, out) \leftrightarrow \exists t.(\phi_p(in, t) \wedge \phi_p(t, out)))$$

Since we are interested in verifying these kinds of properties on scripts we need a logic of feature trees including update constraints, and which enjoys a decidable first-order logic.

Related Work. The first decidability result of a full first-order theory of *Herbrand trees* (*i.e.*, based on equations $x = f(x_1, \ldots, x_n)$) is due to Malc'ev [13], this result has later been extended by [6,12]. A first decidability result for the first-order theory of *feature trees* was given for the logic FT [1], which comprises the predicates $x[f]y$ and $x[f] \uparrow$, by [4]. This was later extended to the logic CFT [15], which in addition to FT has an *arity constraint* $x[F]$ for any finite set F of feature symbols, expressing that the root of x has precisely the features F, in [3,5]. Note that in these logics one can only quantify over trees, not over feature symbols. The generalization to a two-sorted logic which allows for quantification over features is undecidable [16], but decidability can be recovered if one restricts the use of feature variables to talk about existence of features only [17]. All these decidable logics of trees have a non-elementary lower bound [18]. The case of a feature logic with update constraints was open up to now.

Choosing the Right Predicates. The difficulty in solving update constraints stems from the fact that an update constraint involves three trees: the original tree, the final tree and the sub-tree that gets grafted on the original tree.

There are no symmetries between these three arguments, and a conjunction of several update constraints may become quite involved. Our approach to handle this rather complex update constraint is to replace it by a more elementary constraint system which is based on the classical $x[f]y$, and the new *similarity constraint* $x \sim_f y$. The latter constraint expresses that x and y have the same children with the same names, except for the name f where they may differ. This system has the same expressive power as update constraints since on the one hand $z \doteq x[f \mapsto y]$ is equivalent to $x \sim_f y \wedge z[f]y$, and on the other hand $x \sim_f y$ is equivalent to $\exists z, v.(z \doteq x[f \mapsto v] \wedge z \doteq y[f \mapsto v])$. In order to simplify these constraints one needs the generalization $x \sim_F y$ where F is a finite set of features. For each set of features F, similarities $\sim_F$ are equivalence relations, which is very useful when designing simplification rules, and these relations have useful properties, like $(x \sim_F y \wedge x \sim_G y) \leftrightarrow x \sim_{F \cap G} y$ and $(x \sim_F y \wedge y \sim_G z) \rightarrow x \sim_{F \cup G} z$.

Eliminating Quantifiers. Our theory of feature trees does not have the property of quantifier elimination in the strict sense [9]. This is already the case without the update (or similarity) constraints, as we can see in the following example: $\exists x.(y[f]x \wedge x[g] \uparrow)$. This formula means that there is a tree denoted by x such that y points to x through the feature f, and that x does not have the feature g. A quantifier elimination procedure would have to conserve this information about the global variable y. This situation is not unusual when designing decision procedures. There are basically two possible remedies: the first one is to extend the logical language by new predicates which express properties which otherwise would need existential quantifiers to express. This approach of achieving the property of quantifier elimination by extension of the logical language is well-known from Presburger arithmetic, it was also used in [3,4].

However, in the case of feature tree logics, the needed extension of the language is substantial and requires the introduction of *path constraints*. For instance, the above formula would be equivalent to the path constraint $y[f][g] \uparrow$ stating that the variable y has a feature f pointing towards a tree where there is no feature g. Unfortunately, this extension entails the need of quite complex simplification rules for these new predicates.

The alternative solution is to our knowledge due to [13] and consists in exploiting the fact that certain predicates of the logic behave like functions. This solution was also used in [6] for Herbrand trees. When switching to feature trees this solution becomes quite elegant [17], the above formula would be replaced by $\neg y[f] \uparrow \wedge \forall x.(y[f]x \rightarrow x[g] \uparrow)$ stating that y has a feature f (by $\neg y[f] \uparrow$) and that for each variable x such that y points towards x via f (in fact, there is only one), x has no feature g. The price is that existential quantifiers are not completely eliminated but swapped for universal ones. This is, however, sufficient, since one can now apply this transformation to a formula in prenex normal form, and successively reduce the number of quantifier eliminations.

Structure of this Paper. We summarize some notions from logic that will be used in the rest of the paper in Sect. 2. Our model of trees as well as the syntax and semantics of our logic are defined formally in Sect. 3. The quantifier elimination procedure in given in Sect. 4. We conclude in Sect. 5. Proofs are only sketched, full proofs are to be found in the companion technical report [11].

2   Preliminaries

We assume logical conjunction and disjunction to be associative and commutative, and equality to be symmetric. For instance, we identify the formula $x \doteq y \wedge (x[f] \uparrow \vee x[g]z)$ with $(x[g]z \vee x[f] \uparrow) \wedge y \doteq x$.

The set of free variables of a formula ϕ is written $\mathcal{V}(\phi)$. We write $\phi\{x \mapsto y\}$ for the formula obtained by replacing in ϕ all free occurrences of x by y. We write $\tilde{\exists}\phi$ for the existential closure $\exists \mathcal{V}(\phi).\phi$, and similarly $\tilde{\forall}\phi$ for $\forall \mathcal{V}(\phi).\phi$.

A *conjunctive clause with existential quantifiers*, or in short *clause*, is either $\perp$, or a finite set of literals prefixed by a string of existential quantifiers. Note that such a clause may still contain free variables, that is we do *not* require all its variables to be quantified. If $\exists X.(a_1 \wedge \ldots \wedge a_n)$ is such a clause, then we can partition its set of literals $c = g_c \cup l_c$ such that g_c contains all the literals of c that contain no variable of X, and l_c the set of literals of c that contain at least one variable of X. We have the following logical equivalence:

$$\models (\exists X.c) \leftrightarrow (g_c \wedge \exists X.l_c)$$

We call (g_c, l_c) the *decomposition* of $\exists X.c$. g_c is the *global part* and l_c the *local part* of c, X is the set of *local variables* and $\mathcal{V}(\exists X.c) \supseteq \mathcal{V}(g_c)$ the set of *global variables*.

A *disjunctive normal form* (dnf) is a finite set of clauses, all of which are different from $\perp$.

A formula is in *prenex normal form* (pnf) if it is of the form $Q_1 x_1 \ldots Q_n x_n.\phi$ where ϕ is quantifier-free, and where the Q_i are existential or universal quantifiers. If all Q_i are $\exists$ (resp. $\forall$) then the formula is called a Σ_1-formula (resp. Π_1-formula).

$A \rightsquigarrow B$ denotes the set of partial functions from the set A to the set B with a finite domain. The domain of a partial function f is written $\mathrm{dom}(f)$. The complement of a set is written X^c. We write $X \setminus Y$ for $\{x \in X \mid x \notin Y\}$.

3   A Logic for an Algebra of Trees with Similarities

3.1   Decorations

In addition to what has been said in the introduction, our model of feature trees also has information attached to the *nodes* of the trees. In our application to UNIX filesystems, these could be records containing the usual file attributes like various timestamps and access permission bits, owner and group, and so on. This work abstracts from the details of the information attached to tree nodes: we take the definition of node decorations, and the pertaining logic as a parameter. We hence assume given an arbitrary set $\mathcal{D}$ of *decorations*.

We assume given a set D of predicate symbols for decorations, and an interpretation $\mathcal{D}$ for D with universe $\mathcal{D}$. These predicate symbols are of course assumed to be disjoint from the predicate symbols that will be introduced in Subsect. 3.3. We also require that D contains a binary predicate $x \not\doteq y$ expressing the *disequality* of two information items.

We also assume that we have a quantifier elimination procedure for $\mathcal{D}$: we can compute for any Σ_1 formula ψ over the language D, possibly with free variables, a quantifier-free formula D-elim(ψ) that is equivalent in $\mathcal{D}$ to ψ and has the same free variables. Furthermore, we can decide for any closed and quantifier-free D-formula whether in holds in $\mathcal{D}$.

3.2 Feature Trees

We assume given an infinite set $\mathcal{F}$ of *features*. The letters f, g, h will always denote features.

The set $\mathcal{FT}$ of *feature trees* is inductively defined as

$$\mathcal{FT} = \mathcal{D} \times (\mathcal{F} \rightsquigarrow \mathcal{FT})$$

Here, the case of a partial function with empty domain serves as base case of the induction. Hence, this amounts to saying that a feature tree is a finite unordered tree where nodes are labeled by decorations, and edges are labeled by features. Each node in a feature tree has a finite number of outgoing edges, and all outgoing edges of a node carry different names. We write $\hat{t}$ for the decoration of the root node of t and we write $\vec{t}$ for its mapping at the root, i.e. $t = (\hat{t}, \vec{t})$. Our notion of equality on trees is *structural equality*, i.e. $t = s$ iff $\hat{t} = \hat{s}$ and $\vec{t} = \vec{s}$, that is $\mathsf{dom}(\vec{t}) = \mathsf{dom}(\vec{s})$ and $\vec{t}(f) = \vec{s}(f)$ for every $f \in \mathsf{dom}(\vec{t})$. Examples of feature trees are given in Fig. 1.

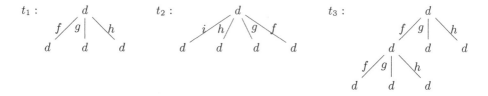

Fig. 1. Examples of feature trees. $d \in \mathcal{D}$ is some arbitrary node decoration.

For the reasons explained in the introduction, our logical language does not contain $y \doteq z[x \mapsto f]$ but the simpler $x \sim_F y$ for any *finite* set $F \subseteq \mathcal{F}$. If $F \subseteq \mathcal{F}$ then we say that *t is similar to s outside F*, written $t \sim_F s$, if for all $f \in F^c = \mathcal{F} \setminus F$ we have that

- either $f \notin \text{dom}(\vec{t}\,) \cup \text{dom}(\vec{s}\,)$
- or $f \in \text{dom}(\vec{t}\,) \cap \text{dom}(\vec{s}\,)$, and $\vec{t}\,(f) = \vec{s}\,(f)$.

In other words, t and s are similar outside F if they have precisely the same children except maybe for the features in F.

3.3   Constraints and Their Interpretation

The set of predicate symbols (or atomic constraints) of our logic is

$x \doteq y$	Equality	$A(x_1 \dots x_n)$	Decoration predicate $A \in D$
$x[f]y$	Feature f from x to y	$x[f] \uparrow$	Absence of feature f from x
$x[F]$	Fence constraint	$x \sim_F y$	Similarity outside F

In fences and similarities, the sets F are finite. We will use the usual syntactic sugar and write $x \neq y$ for $\neg(x \doteq y)$, and $x \not\sim_F y$ for $\neg(x \sim_F y)$. As with equality, we consider similarity predicates to be symmetric, that is we identify $x \sim_F y$ with $y \sim_F x$.

We have one model which has as universe the set $\mathcal{FT}$. As usual, we use the same symbol $\mathcal{FT}$ for the model and for its universe. The predicate symbols are interpreted as follows, where ρ is a valuation of the free variables of the formula in the model $\mathcal{FT}$:

$$\mathcal{FT}, \rho \models x \doteq y \qquad \text{iff } \rho(x) = \rho(y)$$
$$\mathcal{FT}, \rho \models x[f]y \qquad \text{iff } f \in \text{dom}(\overrightarrow{\rho(x)}) \text{ and } \overrightarrow{\rho(x)}(f) = \rho(y)$$
$$\mathcal{FT}, \rho \models x[f] \uparrow \qquad \text{iff } f \notin \text{dom}(\overrightarrow{\rho(x)})$$
$$\mathcal{FT}, \rho \models x[F] \qquad \text{iff } \text{dom}(\vec{x}) \subseteq F$$
$$\mathcal{FT}, \rho \models x \sim_F y \qquad \text{iff } \rho(x) \sim_F \rho(y)$$
$$\mathcal{FT}, \rho \models A(x_1, \dots, x_n) \text{ iff } \mathcal{D}, (\lambda x_i.\overrightarrow{\rho(x_i)}) \models A(x_1, \dots, x_n)$$

Example 1. Let ρ be the valuation $[x \to t_1, y \to t_2, z \to t_3]$ for the trees defined in Fig. 1. The following formulas are satisfied in $\mathcal{FT}, \rho$:

$$z[f]x, \quad x[i] \uparrow, \quad x[\{f, g, h, i\}], \quad x \sim_{\{i\}} y$$

Similarity constraints are actually only of interest in case of an infinite set of features. In case of a finite set $\mathcal{F}$, the similarity constraint could already be expressed in the logic FT that was mentioned in Sect. 1:

$$x \sim_G y \Leftrightarrow \bigwedge_{f \in \mathcal{F} \setminus G} ((x[f] \uparrow \wedge y[f] \uparrow) \vee \exists z(x[f]z \wedge y[f]z))$$

Note the difference between our *fence* constraint, which states an upper bound on the root features of a tree, and the *arity* constraint of [3,15] which states a precise set of root features of a tree. Both are equivalent, since one can express a fence F as a disjunction of all the arities that are subsets of F. Reciprocally, in our logic, we can express that x has arity F as $x[F] \wedge \bigwedge_{f \in F} \neg x[f] \uparrow$.

Also note that decoration predicates behave in $\mathcal{FT}$ as in $\mathcal{D}$:

Proposition 1. *If ψ is a formula using only symbols of D then*

$$\mathcal{FT}, \alpha \models \psi \qquad \Leftrightarrow \qquad \mathcal{D}, \lambda x.\widehat{\alpha(x)} \models \psi.$$

4 Quantifier Elimination

4.1 Clashing Clauses

We say that a clause c that is not $\bot$ *clashes* if one of the patterns of Fig. 2 matches (modulo associativity and commutativity of $\wedge$) a sub-clause $c' \subseteq c$.

C-CYCLE	$x_1[f_1]x_2 \wedge \ldots \wedge x_n[f_n]x_1$	$(n \geq 1)$
C-FEAT-ABS	$x[f]y \wedge x[f] \uparrow$	
C-FEAT-FEN	$x[f]y \wedge x[F]$	$(f \notin F)$
C-NEQ-REFL	$x \neq x$	
C-NSIM-REFL	$x \not\sim_F x$	

Fig. 2. Clash patterns

Remark that C-CYCLE is a clash since our model allows for finite feature trees only, the other clash cases should be obvious.

Lemma 1. *If a clause c clashes then $\mathcal{FT} \models (c \to \bot)$.*

4.2 Positive Clauses with Local Variables

As a preparation for the general case we first consider only one single clause $\exists X.(a_1 \wedge \ldots \wedge a_n)$ containing only positive atoms, prefixed by some existential quantifiers.

S-EQ	$\exists X, x.(x \doteq y \wedge c)$	$\Rightarrow$	$\exists X.c\{x \mapsto y\}$	$(x \neq y)$
S-FEATS	$\exists X, z.(x[f]y \wedge x[f]z \wedge c)$	$\Rightarrow$	$\exists X.(x[f]y \wedge c\{z \mapsto y\})$	
			$(y \neq z$, and if $z \in \mathcal{V}_o$ then $y \in \mathcal{V}_o)$	
S-FEATS-GLOB	$\exists X, x.(x[f]y \wedge x[f]z \wedge c)$	$\Rightarrow$	$\exists X, x.(x[f]y \wedge y \doteq z \wedge c)$	$(y, z \notin X)$
S-SIMS	$x \sim_F y \wedge x \sim_G y \wedge c$	$\Rightarrow$	$x \sim_{F \cap G} y \wedge c$	
P-FEAT	$x \sim_F y \wedge x[f]z \wedge c$	$\Rightarrow$	$x \sim_F y \wedge x[f]z \wedge y[f]z \wedge c$	$(f \notin F)$
P-ABS	$x \sim_F y \wedge x[f] \uparrow \wedge c$	$\Rightarrow$	$x \sim_F y \wedge x[f] \uparrow \wedge y[f] \uparrow \wedge c$	$(f \notin F)$
P-FEN	$x \sim_F y \wedge x[G] \wedge c$	$\Rightarrow$	$x \sim_F y \wedge x[G] \wedge y[F \cup G] \wedge c$	
P-SIM	$x \sim_F y \wedge x \sim_G z \wedge c$	$\Rightarrow$	$x \sim_F y \wedge x \sim_G z \wedge y \sim_{F \cup G} z \wedge c$	
			$(\text{if } \bigcap_{(y \sim_H z) \in c} H \not\subseteq F \cup G)$	

Fig. 3. Transformation rules for the positive case. Existential quantifiers are only written were relevant. Rule S-FEATS is parameterized by a set $\mathcal{V}_o$ of variables.

In this subsection and the following, we will use transformation rules as the ones in Fig. 3. These rules describe transformations that map a clause to a formula (in this subsection the resulting formula is also a clause, but that will no longer be the case in the next subsection). We say that such a rule *left $\Rightarrow$ right* applies to a clause c if:

1. The pattern *left* matches the complete clause c modulo associativity and commutativity of conjunction.
2. The side conditions of the rule, if any, are met.
3. The transformation yields a formula which is *different* from c.

If c is a clause and r a transformation rule then we write $r(c)$ for the formula obtained by applying r to c.

Each of the rules of Fig. 3 describes an equivalence transformation in the model $\mathcal{FT}$. Equation elimination (S-EQ) is a logical equivalence. S-FEATS implements the fact that features are functional. This rule is parameterized by a set $\mathcal{V}_o$ of variables that will be the set of variables (local or global) of the input clause. The variable replacement is $\mathcal{V}_o$-oriented in the sense that we never replace a variable in $\mathcal{V}_o$ by a variable outside $\mathcal{V}_o$. S-FEAT-GLOB is similar to S-FEAT for the case that y and z are both global variables. S-SIMS allows us to contract multiple similarities between the same pair of variables into one. P-FEATS, P-ABS and P-FEN propagate constraints along a similarity, taking into account the index of the similarity. Finally, P-SIM is a kind of transitivity of similarity, where we take care not to add a similarity which is subsumed by already existing similarities.

The propagations play two important roles in that system. First, they move information, possibly leading to a clash. This is the case in the following example where a fence moves through similarities to clash with a feature constraint:

$$\begin{array}{llll}
 & x[f]v & \wedge\ x \sim_{\{g\}} y & \wedge\ y \sim_{\{h\}} z \wedge z[\varnothing] \\
\text{P-FEN} & x[f]v & \wedge\ x \sim_{\{g\}} y \wedge y[\{h\}] \wedge\ y \sim_{\{h\}} z \wedge z[\varnothing] \\
\text{P-FEN} & x[f]v \wedge x[\{g,h\}] \wedge\ x \sim_{\{g\}} y \wedge y[\{h\}] \wedge\ y \sim_{\{h\}} z \wedge z[\varnothing]
\end{array}$$

Second, they take information from local variables and move it to global variables. This mechanism is at the core of the elimination of existential quantifications, the idea being that once all the propagations took place, all interesting information is explicit in the global part, and we can hence drop the local part.

$$\begin{array}{ll}
 & y[h] \uparrow \qquad\qquad \wedge\ \exists z.(x \sim_{\{f\}} z \wedge z \sim_{\{g\}} y) \\
\text{P-SIM} & y[h] \uparrow \wedge\ x \sim_{\{f,g\}} y \wedge \exists z.(x \sim_{\{f\}} z \wedge z \sim_{\{g\}} y) \\
\text{P-ABS} & x[h] \uparrow \wedge\ y[h] \uparrow \wedge\ x \sim_{\{f,g\}} y \wedge \exists z.(x \sim_{\{f\}} z \wedge z \sim_{\{g\}} y)
\end{array}$$

The following function computes a normal form with respect to the rules of Fig. 3:

```
function normalize-positive(c: positive clause)
  V_o := V(c_1) where c = ∃X.c_1
  while c does not clash and some rule r of Fig. 3 applies to c:
    c := r(c)
  return(c)
```

Lemma 2. *For a positive clause c, the function* `normalize-positive` *terminates and yields a positive clause that is equivalent in $\mathcal{FT}$ to c.*

Given a quantifier-free clause c, we define D-part(c) as the conjunction of all D-literals of c.

Lemma 3. *Let the function* `normalize-positive` *return a clause* $\exists X.c$ *that does not clash and* (g_c, l_c) *be its decomposition. Let* $d = D\text{-}elim(\exists X.D\text{-}part(c))$. *If c contains no atom $x[f]y$ with $x \notin X$ and $y \in X$ then*

$$\mathcal{FT} \models \tilde{\forall}((\exists X.c) \leftrightarrow (g_c \wedge d))$$

Actually, both lemmas are special cases of the forthcoming Lemmas 5 and 6 of Sect. 4.3.

Lemma 3 can serve for quantifier elimination in the positive case, at least when there is no feature constraint from a global variable to a local one. We will see in Sect. 4.4 what can be done if this is not the case.

4.3   General Clauses with Local Variables

In case of clauses containing both positive and negative literals we have to consider transformation rules that introduce negations or disjunctions. However, our rules will continue to take a single clause as input. As a consequence, we have to transform the result obtained by a transformation into disjunctive normal form. We assume given a function **dnf** that takes a formula without universal quantifiers and containing only positive occurrences of existential quantifiers, and returns an equivalent dnf that does not contain any clashing clauses. This can be achieved by using a standard dnf transformation and then purging all clashing clauses, or alternatively by applying the clash rules on the fly.

Syntactic Sugar. In the transformation rules to be presented below we will use several abbreviations that allow us to write the rules more concisely. First we have

$$x\langle F \rangle := \bigvee_{f \in F} \exists z.x[f]z$$

where $F \subset \mathcal{F}$ is a finite set. This formula states that x has *at least one* feature in the set F, it can be seen as a dual to the fence constraint $x[F]$ which states that x has *at most* the features in the set F. Note that $x\langle F \rangle$ introduces a disjunction, so introducing such a formula requires the result to be put into dnf.

The formula $x \neq_f y$ states that x and y differ at feature f, that is either one of them has f and the other one does not, or their children at f are different. The formula $x \neq_F y$ generalizes this to a finite set $F \subset \mathcal{F}$, stating that x and y differ at at least one of the features in F.

$$x \neq_f y := \exists z'.(x[f]\uparrow \wedge y[f]z') \vee \exists z.(x[f]z \wedge y[f]\uparrow)$$
$$\vee \exists z, z'.(x[f]z \wedge y[f]z' \wedge (z \not\equiv z' \vee z \not\sim_\varnothing z'))$$
$$x \neq_F y := \bigvee_{f \in F} x \neq_f y$$

We use $(z \not\cong z' \vee z \not\sim_\varnothing z')$ instead of $(z \neq z)$ to denote a difference between two variables in order to avoid problems with the termination. These formulas introduce disjunctions. They also introduce negated similarities at some newly created children of x and y, so we have to take care in the termination proof when these formulas are introduced by a transformation.

R-NEq	$x \neq y \wedge c$	$\Rightarrow$	$(x \not\cong y \vee x \not\sim_\varnothing y) \wedge c$
R-NFeat	$\neg x[f]y \wedge c$	$\Rightarrow$	$(x[f] \uparrow \vee \exists z.(x[f]z \wedge (y \not\cong z \vee y \not\sim_\varnothing z))) \wedge c$
R-NAbs	$\neg x[f] \uparrow \wedge c$	$\Rightarrow$	$\exists z.x[f]z \wedge c$
R-NFen-Fen	$x[F] \wedge \neg x[G] \wedge c$	$\Rightarrow$	$x[F] \wedge x\langle F \setminus G\rangle \wedge c$
R-NSim-Sim	$x \sim_F y \wedge x \not\sim_G y \wedge c$	$\Rightarrow$	$x \sim_F y \wedge x \neq_{F \setminus G} y \wedge c$
R-NSim-Fen	$x[F] \wedge x \not\sim_G y \wedge c$	$\Rightarrow$	$x[F] \wedge \left(\neg y[F \cup G] \vee x \neq_{F \setminus G} y\right) \wedge c$
E-NFen	$x \sim_F y \wedge \neg x[G] \wedge c$	$\Rightarrow$	$x \sim_F y \wedge (\neg x[F \cup G] \vee x\langle F \setminus G\rangle) \wedge c$ $(F \not\subseteq G)$
E-NSim	$x \sim_F y \wedge x \not\sim_G z \wedge c$	$\Rightarrow$	$x \sim_F y \wedge \left(x \not\sim_{F \cup G} z \vee x \neq_{F \setminus G} z\right) \wedge c$ $(F \not\subseteq G)$

Fig. 4. Replacement and enlargement rules for the general case. $\not\cong$ is the disequality of decorations.

New Rules. Figure 4 extends the previously defined set of rules by adding several replacement rules and two enlargement rules. First, we have R-NEq, R-NFeat and R-NAbs that eliminate occurrences of the negated constraints $x \neq y$, $\neg x[f]y$ and $\neg x[f] \uparrow$ respectively. Since no other rule introduces any of these negated constraints we can ignore these two negated constraints in the rest of the section.

Then we have three rules that combine a positive with a negative constraint. R-NFen-Fen applies to the case where we have both a positive fence F and a negated fence G for x. We simplify this by keeping the positive fence F, and replacing the negative fence by saying that x must have a feature that is in F (since that is all it can have), but not in G. Similarly, R-NSim-Sim applies when we have between x and y both a positive similarity except in F, and a negated similarity except in G. We simplify this by keeping the positive similarity, and replacing the negated similarity by stating that x and y differ at a feature that is in F (since these are the only features where they may differ) but not in G. Finally, R-NSim-Fen applies when we have a fence F for x, and a negated similarity with y except in G. Note that for any F and G, $G^c = (F \cup G)^c \cup (F \setminus G)$. Hence, the negated similarity is equivalent to saying that either y has a feature outside $F \cup G$, which is the only possibility to have a difference with x outside $F \cup G$ since x has already fence F, or the difference is in the finite set $F \setminus G$.

Finally, we have the two enlargement rules E-NFen and E-NSim. Their sole purpose is to ensure (by enlarging the negated fence or the index of a negated similarity) that the rules in Fig. 5 can be applied when we have a similarity in conjunction with a negated fence or a negated similarity. The correctness proof of these rules is similar to the three previous rules. In fact, the similarity between x

and y is not needed for the correctness of these two rules and serves only for the termination proof since the requirement of a context $x \sim_F y$ excludes arbitrary enlargements.

$$P\text{-}NF_{EN} \quad x \sim_F y \wedge \neg x[G] \wedge c \;\Rightarrow\; x \sim_F y \wedge \neg x[G] \wedge \neg y[G] \wedge c \quad (F \subseteq G)$$
$$P\text{-}NS_{IM} \quad x \sim_F y \wedge x \not\sim_G z \wedge c \;\Rightarrow\; x \sim_F y \wedge x \not\sim_G z \wedge y \not\sim_G z \wedge c \quad (F \subseteq G)$$

Fig. 5. Propagation rules for the general case.

The two rules in Fig. 5 may propagate a negated fence or a negated similarity through a similarity. In fact, if x and y coincide outside F and $F \subseteq G$, then x and y also coincide outside G. Hence, if x has a feature outside G then so does y (P-NF$_{EN}$), and if x differs from z at some feature outside G then so does y (P-NS$_{IM}$).

We define the set of rules R_1 as the union of all the transformation rules of Figs. 3 and 4, and R_2 as the set of the two transformation rules of Fig. 5.

```
function normalize (c: clause)
    d := {c}
    Vₒ := V(c₁) where c = ∃X.c₁
    while exists c ∈ d to which some rule r ∈ R₁ applies
        d := (d \ {c}) ∪ dnf (r(c))
    while exists c ∈ d to which r ∈ R₂ applies
        d := (d \ {c}) ∪ {r(c)}
    return (d)
```

The function **normalize** normalizes first by rule set R_1, and then by rule set R_2. This decomposition is necessary to ensure termination. It also makes sense since application of rules R_2 conserves normal forms with respect to R_1.

Lemma 4. *The output of* **normalize** *is a dnf where each conjunction is in normal form for* $R_1 \cup R_2$.

Proof (sketch). We have to prove that the application of one of the rules in R_2 to a normal form with respect to R_1 does not produce a redex for any of the rules in R_1. Assume, for instance that the application of P-NF$_{EN}$ to c introduces a redex of R-NF$_{EN}$-F$_{EN}$. This means that the negative fence constraint introduced for y will react with a positive fence constraint (for y) that was already present in c. Since c is in normal form with respect to P-F$_{EN}$, x must have a fence constraint in c. This yields a contradiction since then c is not in normal form with respect to R-NF$_{EN}$-F$_{EN}$. The other cases are similar (details can be found in [11]).

Lemma 5. *The function* **normalize**, *when applied to a clause* c, *terminates and yields a dnf* d *such that* $\mathcal{FT} \models \tilde{\forall}(c \leftrightarrow d)$.

Proof (sketch). Equivalence of c and d follows from the fact that each transformation rule is an equivalence in $\mathcal{FT}$. Termination is shown by defining a well-founded order on clauses such that each rule transforms a clause into a set of stricter smaller clauses. The termination order on dnf formulas is the multiset extension [8] of this order.

This order is a lexicographic order over twelve different measures that decrease with the applications of the rules. We can for instance handle the rules R-NEQ, R-NFEAT and R-NABS first by saying that they decrease the number of negated equalities, feature constraints or absences. Since nothing introduces those literals, this is already a good start.

The first main difficulty in finding that order comes from the fact that all the propagation rules are trying to saturate the clause. A good measure that decreases with them is then the set of all possible atoms that are not in the formula. For P-FEAT, for instance: $\{(x[f]y) \mid x, y \in \mathcal{V}(c); f \in \mathcal{F}(c); (x[f]y) \notin c\}$. That would make a good measure if $\mathcal{V}(c)$ could not increase with the application of other rules such as R-NSIM-FEN. We have thus to handle these other rules first, which leads us to another main difficulty.

The second main difficulty comes from the negated similarities. Indeed, while all other literals may only move "horizontally" following the similarities, negated similarities may "descend" in the constraint, creating variables and feature constraints if needed. It is not obvious when it will stop, and in particular to find a bound on the number of variables introduced.

Let us consider the following example constraint and one of its reduction paths (that is, the reduction may create several branches in the dnf, and we take only the one we are interested in):

$$x_0[f]x_1 \wedge x_1[f]y_0 \qquad\qquad \wedge x_0[\{f\}] \wedge x_1[\{f\}] \wedge x_0 \nsim_\varnothing y_0$$

By R-NSIM-FEN:

$$\exists y_1, \underline{z_1}.\ x_0[f]x_1 \wedge x_1[f]y_0 \wedge \underline{x_0[f]z_1} \wedge y_0[f]y_1 \wedge x_0[\{f\}] \wedge x_1[\{f\}] \wedge z_1 \nsim_\varnothing y_1$$

By S-FEATS:

$$\exists y_1.\quad x_0[f]x_1 \wedge x_1[f]y_0 \wedge y_0[f]y_1 \qquad \wedge x_0[\{f\}] \wedge x_1[\{f\}] \wedge x_1 \nsim_\varnothing y_1$$

In two rules, we created a new variable y_1, and removed a negated similarity just to put it again somewhere else. Note in particular that R-NSIM-FEN can still apply, because x_1 has now a fence and a negative similarity. In fact, if, instead of two, we take a number n of variables x_i, we can extend that example into one that always doubles the number of variables.

The key to our solution to this problem is that rules that make negative similarities descend, thus introducing feature constraints and new variables, need some "fuel", which is the presence of positive fences or similarities. We define the *original variables* as the variables that were in the clause at the beginning of `normalize`. Then, we show that

1. the number of original variables cannot grow;
2. there are never feature constraints from non-original variables towards original ones;
3. the positive fences and similarities can only be present on original variables.

It remains the problem that negative similarities can descend. At some point, they will necessarily go too deep and leave the area where the original variables may live. By doing so, they loose the positive fences and similarities that they need to keep descending, and the process stops.

The full proof, including the lemmas corresponding to the points (1), (2) and (3), the definition of the measures and the technical details can be found in [11].

This is also where we make use of the quantifier elimination procedure for D-formulas. Given a quantifier-free clause c, we define D-part(c) as the conjunction of all D-literals of c.

Lemma 6. *Let the function* `normalize` *return a dnf which contains a clause* $\exists X.c$. *Let* (g_c, l_c) *be the decomposition of* c, *and* $d = D\text{-}elim(\exists X.D\text{-}part(c))$. *If* c *contains no atom* $x[f]y$ *with* $x \notin X$ *and* $y \in X$ *then*

$$\mathcal{FT} \models \tilde{\forall}((\exists X.c) \leftrightarrow g_c \wedge d)$$

The proof can be found in [11].

We call a clause *normalized* when it is an element of a dnf returned as result of function `normalize`.

4.4 Quantifier Elimination

In order to eliminate a block of existential quantifiers from a clause we apply iteratively the following rule:

$$\text{FEAT-FUN} \quad \exists X, x.(y[f]x \wedge c) \quad \Rightarrow \quad \neg y[f]\uparrow \wedge \forall x.\, (y[f]x \rightarrow \exists X.\, (y[f]x \wedge c))$$
$$(y \notin X, y \neq x)$$

This rule follows the idea of [13], and was already applied to feature constraints in [17]. The correctness of this transformation is shown by the following chain of equivalences in the model $\mathcal{FT}$:

$$\begin{array}{ll}
\exists X, x.(y[f]x \wedge c) & \\
\exists x.(y[f]x \wedge \exists X.c) & \text{since } x, y \notin X \\
\neg y[f]\uparrow \wedge \forall x.\, (y[f]x \rightarrow \exists X.c) & \text{since features are functional} \\
\neg y[f]\uparrow \wedge \forall x.\, (y[f]x \rightarrow \exists X.\, (y[f]x \wedge c))
\end{array}$$

The last step is very important, because it ensures that, if $y[f]x \wedge c$ is in normal form, then the right part of the implication is also in normal form. This will be important for the function defined below.

The function `switch` defined below iterates this replacement for all local variables x that occur in the form $y[f]x$ where y is not local: the function applies the transformation FEAT-FUN, and then recursively applies itself on the result. When there remains no more feature constraint $y[f]x$ from a global variable to a local variable in the normalized clause c, we meet the hypotheses of Lemma 6. We then return the conjunction of the global part g_c of the normalized clause, and of the D-part of the local part l_c from which we have eliminated the block of existential quantifiers.

```
recursive function switch(c: normalized clause)
  if ∃X,x.(y[f]x ∧ c') matches c and y ∉ X:
    return(¬y[f] ↑ ∧ ∀x.(y[f]x → switch(∃X.y[f]x ∧ c')))
  else:
    (g_c,l_c) := decomposition(c)
    d := D-elim(∃X.D-part(l_c))
    return(g_c ∧ d)
```

Example 2. When given the following formula

$$\exists v, w.(y[f]v \wedge v[f]w \wedge w[f]z \wedge w[\{f,g\}] \wedge y \sim_\varnothing z)$$

the function `switch` returns

$$\neg y[f] \uparrow \wedge \forall v.(y[f]v \to (\neg v[f] \uparrow \wedge \forall w.(v[f]w \to (w[f]z \wedge y \sim_\varnothing z))))$$

Lemma 7. *Given a normalized clause c, `switch(c)` terminates and yields a formula ψ such that*

1. $\mathcal{FT} \models \tilde{\forall}(c \leftrightarrow \psi)$;
2. $\mathcal{V}(\psi) \subseteq \mathcal{V}(c)$;
3. *ψ contains no existential quantifiers and only positive occurrences of universal quantifiers;*
4. *If $\mathcal{V}(c) = \emptyset$ then ψ is quantifier-free.*

We can now write a function that transforms a Σ_1 formula into an equivalent Π_1 formula. For this we assume given a function `pnf` that transforms any formula into its prenex normal form.

```
function solve(p: Σ₁ formula)
  let ∃X.q = p where q is quantifier-free
  d := dnf(q)
  dt := ⋁_{c∈d} normalize(∃X.c)
  u := ⋁_{c∈dt} switch(c)
  return(pnf(u))
```

Finally, the function `decide` takes a formula in prenex normal form and returns an equivalent (in $\mathcal{FT}$) formula without any quantifiers. If Q is a string of quantifiers, then $\overline{Q}$ is the string of quantifiers obtained from Q by changing $\exists$ into $\forall$ and vice-versa. For instance, $\overline{\exists x \forall y \exists z} = \forall x \exists y \forall z$.

```
recursive function decide(p: pnf)
  if p is quantifier-free:
    return(p)
  else if p is Q.∃X.q
    where q quantifier-free, Q does not end on ∃:
    return(decide(Q. solve(∃X.q)))
  else if p is Q.∀X.q
    where q quantifier-free, Q does not end on ∀:
    return(¬ decide(Q̄. solve(∃X.¬q)))
```

Theorem 1. *Given a formula p in prenex normal form,* `decide(p)` *terminates and yields a formula q such that*

- $\mathcal{FT} \models \tilde{\forall}(p \leftrightarrow q)$
- $\mathcal{V}(q) \subseteq \mathcal{V}(p)$
- *q is a* Π_1 *formula, and quantifier-free in case* $\mathcal{V}(p) = \emptyset$

Proof (sketch). Termination follows from the fact that at each call to `decide`, the number of quantifier *alternations* in the pnf decreases.

If we apply `decide` to a closed formula, we hence obtain an equivalent (in $\mathcal{FT}$) formula that contains no free variables and no quantifiers. Since the only tree-terms are variables, we have obtained formula of the language D, for which we can decide by assumption validity in $\mathcal{D}$.

Corollary 1. *The first order theory of* $\mathcal{FT}$ *is decidable.*

5   Conclusion

We have presented a quantifier elimination procedure for a first-order theory of feature trees with similarity constraints. Since update constraints can be expressed by similarity and feature constraints, this implies in particular that the first-order theory of feature trees with update constraints is decidable.

Our model of feature trees is in several respects an abstraction of UNIX file systems [2]. First, real file systems make a distinction between different kinds of files (directories, regular files, various kinds of device files). This distinction is omitted here just for the sake of presentation. More importantly, real file systems are not really trees as they allow for multiple paths from the root to regular files (which must be sinks), and they provide for symbolic links. Since extending the model by any of these may lead to undecidability of the full first-order theory we might have to look for smaller fragments which are sufficient for our application to the symbolic execution of scripts.

Acknowledgments. The idea of investigating update constraints on feature trees originates from discussions with Gert Smolka a long time ago. We would like to thank the anonymous reviewers for their useful remarks and suggestions, and the members of the CoLiS project for numerous discussions on tree constraints and their use in modeling tree operations, in particular Claude Marché, Kim Nguyen, Joachim Niehren, Yann Régis-Gianas, Sylvain Salvati, and Mihaela Sighireanu.

References

1. Aït-Kaci, H., Podelski, A., Smolka, G.: A feature-based constraint system for logic programming with entailment. Theor. Comput. Sci. **122**(1–2), 263–283 (1994)
2. Bach, M.: The Design of the UNIX Operating System. Prentice-Hall, Upper Saddle River (1986)

3. Backofen, R.: A complete axiomatization of a theory with feature and arity constraints. J. Log. Program. **24**(1&2), 37–71 (1995)
4. Backofen, R., Smolka, G.: A complete and recursive feature theory. Theor. Comput. Sci. **146**(1–2), 243–268 (1995)
5. Backofen, R., Treinen, R.: How to win a game with features. Inf. Comput. **142**(1), 76–101 (1998)
6. Comon, H., Lescanne, P.: Equational problems and disunification. J. Symb. Comput. **7**, 371–425 (1989)
7. Debian Policy Mailing List: Debian Policy Manual, Version 4.1.3. Debian, December 2017. https://www.debian.org/doc/debian-policy/
8. Dershowitz, N., Manna, Z.: Proving termination with multiset orderings. Commun. ACM **22**(8), 465–476 (1979)
9. Hodges, W.: Model Theory, Encyclopedia of Mathematics and Its Applications, vol. 42. Cambridge University Press, Cambridge (1993)
10. Jeannerod, N., Marché, C., Treinen, R.: A formally verified interpreter for a shell-like programming language. In: Paskevich, A., Wies, T. (eds.) VSTTE 2017. LNCS, vol. 10712, pp. 1–18. Springer, Cham (2017). https://doi.org/10.1007/978-3-319-72308-2_1
11. Jeannerod, N., Treinen, R.: Deciding the first-order theory of an algebra of feature trees with updates (extended version), January 2018. https://hal.archives-ouvertes.fr/hal-01760575
12. Maher, M.J.: Complete axiomatizations of the algebras of finite, rational and infinite trees. In: LICS, pp. 348–357. IEEE, Edinburgh, July 1988
13. Malc'ev, A.I.: Axiomatizable classes of locally free algebras of various type (Chap. 23). In: Wells I, B.F. (ed.) The Metamathematics of Algebraic Systems: Collected Papers 1936–1967, pp. 262–281. North Holland, Amsterdam (1971)
14. Smolka, G.: Feature constraint logics for unification grammars. J. Log. Program. **12**, 51–87 (1992)
15. Smolka, G., Treinen, R.: Records for logic programming. J. Log. Program. **18**(3), 229–258 (1994)
16. Treinen, R.: Feature constraints with first-class features. In: Borzyszkowski, A.M., Sokołowski, S. (eds.) MFCS 1993. LNCS, vol. 711, pp. 734–743. Springer, Heidelberg (1993). https://doi.org/10.1007/3-540-57182-5_64
17. Treinen, R.: Feature trees over arbitrary structures (Chap. 7). In: Blackburn, P., de Rijke, M. (eds.) Specifying Syntactic Structures, pp. 185–211. CSLI Publications and FoLLI, Stanford (1997)
18. Vorobyov, S.: An improved lower bound for the elementary theories of trees. In: McRobbie, M.A., Slaney, J.K. (eds.) CADE 1996. LNCS, vol. 1104, pp. 275–287. Springer, Heidelberg (1996). https://doi.org/10.1007/3-540-61511-3_91

A Separation Logic with Data: Small Models and Automation

Jens Katelaan[1]([✉]), Dejan Jovanović[2], and Georg Weissenbacher[1]

[1] TU Wien, Vienna, Austria
jkatelaan@forsyte.at
[2] SRI International, New York, USA

Abstract. Separation logic has become a stock formalism for reasoning about programs with dynamic memory allocation. We introduce a variant of separation logic that supports lists and trees as well as inductive constraints on the data stored in these structures. We prove that this logic has the small model property, meaning that for each satisfiable formula there is a small domain in which the formula is satisfiable. As a consequence, the satisfiability and entailment problems for our fragment are in NP and CONP, respectively. Leveraging this result, we describe a polynomial SMT encoding that allows us to decide satisfiability and entailment for our separation logic.

1 Introduction

Separation logic is a popular formalism to describe the state and shape of dynamically allocated data structures and is used for the verification of programs that manipulate the heap. The formalism prominently features in Facebook's static analyzer INFER [6], which is successfully deployed on an industrial scale to analyze the memory safety of millions of lines of imperative code. This impressive scalability is facilitated by the logic's separating conjunction operator ($*$), which allows the decomposition of the program heap into disjoint regions and thus enables compositional reasoning by isolating data structures modified by a given code fragment from unaffected portions of the memory (the frame). Moreover, separation logic provides recursive predicates which describe the shape of dynamically allocated data structures such as linked lists and trees and thus enable reasoning about programs with unbounded heap.

The high expressiveness of separation logic, however, comes at the cost of undecidability [7]. Consequently, the use of separation logic in deductive verification [19,20] and symbolic execution [4] is often restricted to decidable fragments. To obtain decidability, most of these fragments adhere to at least one of the following restrictions: they only support lists [1,3,8,12,18]; they can only express structural constraints but not constraints on data stored in structures

The research presented in this paper was supported by the Austrian Science Fund (FWF) under project W1255-N23, the Austrian National Research Network S11403-N23 (RiSE), and the National Science Foundation (NSF) grant 1528153.

© Springer International Publishing AG, part of Springer Nature 2018
D. Galmiche et al. (Eds.): IJCAR 2018, LNAI 10900, pp. 455–471, 2018.
https://doi.org/10.1007/978-3-319-94205-6_30

[3,5,10,11]; or they are not closed under Boolean operators, but only under separating conjunction [3,5,10,13]. Yet, the computational complexity is still daunting in many cases: deciding satisfiability is EXPTIME-hard for the fragments in [5,13], for instance, and [10] and the STRAND logic [14] rely on a reduction of structural constraints to monadic second-order logic. Other separation logics and related formalisms are undecidable altogether [22].

We present a decidable separation logic that aims to strike a balance between expressiveness and computational complexity. Our logic supports list and tree segments as well as arbitrary data constraints, allowing us to describe common data structures such as binary search trees and max-heaps. Notably, the spatial formulas of our fragment are closed under classical Boolean operators including negation, allowing us to decide satisfiability and entailment.

This decidability result is established by showing that the structural part of our separation logic has the *small-model property*. In particular, we provide a bound that is linear in the number of variables in the formula. As a consequence, the satisfiability problem of our fragment is in NP and entailment in coNP (exploiting closure under negation). Moreover, the explicit bound provided by our small-model theorem enables a range of SAT/SMT-encodings of our separation logic. We characterize the properties that such encodings must satisfy and provide a complete polynomial-size encoding of our logic.

While we are not the first to propose an encoding of separation logic into SMT ([12] implements a reachability theory in SMT, lists with length constraints are encoded in [18], and an encoding supporting the magic wand operator but not recursive predicates is described in [23]), our approach lifts a number of requirements of encodings of comparable expressiveness and complexity [19,20].

First, the encodings in [19,20] are based on theories for reachability in function graphs [12,25] as well as a theory of finite sets, whereas we rely only on theories supported by off-the-shelf SMT solvers. Second, in [20], reasoning about trees relies on ghost variables representing pointers to parent nodes. Reachability predicates are restricted to these parent fields. In our approach, we support reasoning about left and right descendants. Third, our encoding can be easily combined with arbitrary data theories, whereas the data constraints in [19,20] are harder to generalize, since they depend on local theory extensions. Fourth, our logic supports reasoning about tree segments via a notion of stop points, thus generalizing the structural properties about tree-like structures that can be expressed in the logic compared to [20].

The paper is structured as follows. In Sect. 2, we introduce $\mathbf{SL}_{\mathsf{data}}$, a separation logic with support for lists, trees, and data, but without constraints on the data stored inside the lists and trees. We prove the small-model property for $\mathbf{SL}_{\mathsf{data}}$ in Sect. 3. In Sect. 4, we introduce $\mathbf{SL}_{\mathsf{data}}^*$, an extension of $\mathbf{SL}_{\mathsf{data}}$ in which the data inside of list and tree structures can be constrained by formulas from the data theory. We show how to lift the small-model property to this extended setting. In Sect. 5, we present a polynomial encoding of $\mathbf{SL}^*_{\mathsf{data}}$ into SMT. We conclude in Sect. 6.[1]

2   Separation Logic with Lists, Trees, and Data

In this section we introduce the core fragment of our separation logic of lists, trees, and data. Our approach is parametric with respect to a background theory $\mathcal{T}_{\mathsf{data}}$ of the data domain, and a background theory $\mathcal{T}_{\mathsf{loc}}$ of the location domain. We denote with $\mathcal{F}_{\mathsf{data}}$ and $\mathcal{F}_{\mathsf{loc}}$ the sets of all quantifier-free $\mathcal{T}_{\mathsf{data}}$-formulas and $\mathcal{T}_{\mathsf{loc}}$-formulas, respectively. The background theories can be instantiated with any first-order theory with equality, as usual in satisfiability modulo theories (see, e.g., [2]). We denote this logic with $\mathbf{SL}_{\mathsf{data}}$.

Syntax. We work in a many-sorted logic with equality. The signature of $\mathbf{SL}_{\mathsf{data}}$ contains the sorts $\mathcal{S} = \{\mathsf{loc}, \mathsf{data}\}$, representing locations and data, respectively. We assume a countable infinite set of (sorted) variables $\mathcal{X}$ and a dedicated constant **null** of sort loc. We denote with $\mathbf{s}$ a vector $\langle s_1, \ldots, s_n \rangle$ of variables from $\mathcal{X}$, and write ϵ for an empty vector, and $\mathbf{s}_1 \cdot \mathbf{s}_2$ for the concatenation of two vectors.

Let $\mathsf{Fld} = \{\mathsf{n}, \mathsf{l}, \mathsf{r}, \mathsf{d}\}$ be the set of field identifiers corresponding to the next element of a list node, the left and the right child of a binary tree node, and the data field. To each field $f \in \mathsf{Fld}$ we associate a binary points-to predicate $\rightarrow_f$ with the following signatures:

$$\rightarrow_{\mathsf{n}} : \mathsf{loc} \times \mathsf{loc} \;\mapsto\; \mathsf{Bool}, \qquad\qquad \rightarrow_{\mathsf{l}} : \mathsf{loc} \times \mathsf{loc} \;\mapsto\; \mathsf{Bool},$$
$$\rightarrow_{\mathsf{r}} : \mathsf{loc} \times \mathsf{loc} \;\mapsto\; \mathsf{Bool}, \qquad\qquad \rightarrow_{\mathsf{d}} : \mathsf{loc} \times \mathsf{data} \mapsto \mathsf{Bool}.$$

The logic includes two inductive predicates list and tree with signatures

$$\mathsf{list} : \mathsf{loc} \times \mathsf{loc}^* \mapsto \mathsf{Bool}, \qquad\qquad \mathsf{tree} : \mathsf{loc} \times \mathsf{loc}^* \mapsto \mathsf{Bool}.$$

The syntax of $\mathbf{SL}_{\mathsf{data}}$ is presented in Fig. 1. A formula in $\mathbf{SL}_{\mathsf{data}}$ is a well-sorted Boolean combination of spatial formulas ($F_{Spatial}$). Spatial formulas are constructed by applying the separating conjunction $*$ to the *spatial atoms*. The spatial atoms are $\mathcal{T}_{\mathsf{loc}}$ and $\mathcal{T}_{\mathsf{data}}$ formulas, the points-to predicate $x \rightarrow_f y$, the list predicate $\mathsf{list}(x, \mathbf{s})$ and the tree predicate $\mathsf{tree}(x, \mathbf{s})$. To ease notation, we denote a separating conjunction of several points-to predicates over the same variable x with $x \rightarrow_{\mathsf{p}_1, \ldots, \mathsf{p}_n} (y_1, \ldots, y_n)$. The vector $\mathbf{s}$ is a vector of structural *stop points* delineating the data structure. By abuse of notation, we omit $\mathbf{s}$ when it is empty.

Our logic departs from standard presentations of separation logic (see, e.g., [24]) in several details. First, we do not have **emp**, the empty heap. It can be introduced as syntactic sugar, e.g. **emp** := (**null** = **null**). Second, we include an independent points-to predicate for each field of lists and trees to facilitate extensions of the logic to doubly-linked and/or overlaid data structures, see

---

[1] Due to lack of space some proofs and additional material are omitted and can be found in the extended version.

$$t := \mathbf{null} \mid x \in \mathcal{X}$$

$$
\begin{array}{lll}
A_{Spatial} ::= t \rightarrow_f t \mid \mathsf{list}(t, \mathbf{s}) \mid \mathsf{tree}(t, \mathbf{s}) \mid \mathcal{F}_{\mathsf{loc}} \mid \mathcal{F}_{\mathsf{data}} & \text{Spatial atoms} \\
F_{Spatial} ::= A_{Spatial} \mid F_{Spatial} * F_{Spatial} & \text{Spatial formulas} \\
F ::= F_{Spatial} \mid \neg F \mid F \vee F \mid F \wedge F & \mathbf{SL}_{\mathsf{data}} \text{ formulas}
\end{array}
$$

Fig. 1. Syntax of the core separation logic $\mathbf{SL}_{\mathsf{data}}$ with lists, trees, and data.

e.g. [9]. Third, our lists and tree fragments represent data structures that start from x and end in stop points $\mathbf{s}$ in an ordered fashion. Additionally—unlike in many decidable separation logics [3,10]—we allow arbitrary Boolean structure outside of the spatial conjunction.

Example 1 (Syntax). Let x, y, z be variables of sort loc, and w be of sort data.

- $\mathsf{list}(x, \langle y \rangle) * \mathsf{list}(y)$ are disjoint list segments from x to y and from y to $\mathbf{null}$.
- $\mathsf{tree}(x, \langle y, z \rangle) * \mathsf{tree}(y) * \mathsf{tree}(z)$ represents a binary tree rooted in x that contains two subtrees y and z ordered from left to right, as specified by $\langle y, z \rangle$.
- $(x \rightarrow_{\mathsf{n,d}} (y, w)) * \mathsf{list}(y) * (w > 0)$ (where $\mathcal{T}_{\mathsf{data}}$ is an arithmetic theory) states that x is a list node with data $w > 0$ pointing to a list with head y.

Semantics. We denote with $f = \{x_1, \ldots, x_n \mapsto y_1, \ldots, y_n\}$ a partial function that maps x_i to y_i and is otherwise undefined, and write $f = \emptyset$ if f is undefined everywhere. We write $\mathrm{dom}(f)$ and $\mathrm{img}(f)$ for the domain and image of f.

The semantics of $\mathbf{SL}_{\mathsf{data}}$ formulas are defined in terms of heap interpretations. Let $X \subseteq \mathcal{X}$ be a set of variables. A *heap interpretation* $\mathcal{M}$ over X is a map that interprets each sort $\sigma \in \mathcal{S}$ as a non-empty domain $\sigma^{\mathcal{M}}$, each $x \in X \cup \{\mathbf{null}\}$ of sort σ as an element $x^{\mathcal{M}} \in \sigma^{\mathcal{M}}$, and each points-to predicate $\rightarrow_f$ of sort $\sigma_1 \times \sigma_2 \mapsto \mathsf{Bool}$ is interpreted as a partial function $f^{\mathcal{M}} : \sigma_1 \rightharpoonup \sigma_2$ with finite domain such that $f^{\mathcal{M}}(\mathbf{null})$ is undefined (i.e., null may never be allocated) and such that $\mathrm{dom}(\mathsf{n}) \cap (\mathrm{dom}(\mathsf{l}) \cup \mathrm{dom}(\mathsf{r})) = \emptyset$, i.e., a location cannot be both a list and a tree location.

We denote with $\mathcal{M}[x_1, \ldots, x_n \mapsto v_1, \ldots, v_n]$ a heap interpretation over $X \cup \{x_1, \ldots, x_n\}$ that differs from $\mathcal{M}$ only by interpreting the variables x_i as values v_i. Let $\ell_1, \ell_2 \in \mathsf{loc}^{\mathcal{M}}$. We write $\ell_1 \rightarrow_{\mathcal{M}} \ell_2$ if $\ell_2 = f^{\mathcal{M}}(\ell_1)$ for some $f \in \mathsf{Fld}$. We extend this notation to variables and write $x \rightarrow_{\mathcal{M}} y$ if for two variables $x, y \in X$ it holds that $x^{\mathcal{M}} \rightarrow_{\mathcal{M}} y^{\mathcal{M}}$. We denote with $\rightarrow_{\mathcal{M}}^*$ and $\rightarrow_{\mathcal{M}}^+$ the usual Kleene closures of $\rightarrow_{\mathcal{M}}$ and say that ℓ_2 is *reachable from* ℓ_1 if $\ell_1 \rightarrow_{\mathcal{M}}^* \ell_2$.

A location $\ell \in \mathsf{loc}^{\mathcal{M}}$ is an *allocated location* in $\mathcal{M}$ if there exists an $f \in \{\mathsf{n}, \mathsf{l}, \mathsf{r}\}$ such that $\ell \in \mathrm{dom}(f^{\mathcal{M}})$. We define $\mathsf{loc}_{\mathsf{list}}^{\mathcal{M}} := \{\ell \in \mathsf{loc}^{\mathcal{M}} \mid \ell \in \mathrm{dom}(\mathsf{n})\}$ and $\mathsf{loc}_{\mathsf{tree}}^{\mathcal{M}} := \{\ell \in \mathsf{loc}^{\mathcal{M}} \mid \ell \in \mathrm{dom}(\mathsf{l}) \cup \mathrm{dom}(\mathsf{r})\}$. Location ℓ is *fully allocated* in $\mathcal{M}$ if it allocates data and either the next pointer or both the left and right pointer, i.e., if $\ell \in (\mathrm{dom}(\mathsf{n}^{\mathcal{M}}) \cup (\mathrm{dom}(\mathsf{l}^{\mathcal{M}}) \cap \mathrm{dom}(\mathsf{r}^{\mathcal{M}}))) \cap \mathrm{dom}(\mathsf{d}^{\mathcal{M}})$. Location ℓ is *labeled* in $\mathcal{M}$ if there exists an $x \in X \cup \{\mathbf{null}\}$ with $x^{\mathcal{M}} = \ell$. Otherwise, ℓ is unlabeled.

The *size of* $\mathcal{M}$, denoted $|M|$, is the number of allocated locations in $\mathcal{M}$.[2] The *size of a formula* F, denoted $|F|$, is defined as the numbers of terms, atomic formulas, and operators in F (including symbols from $\mathcal{T}_{\mathsf{loc}}$ and $\mathcal{T}_{\mathsf{data}}$).

Let $\mathcal{M}_1$ and $\mathcal{M}_2$ be heap interpretations over X that agree on the interpretation of all sorts, variables, and constants. We say that $\mathcal{M}_1$ is a *sub-interpretation* of $\mathcal{M}_2$, written $\mathcal{M}_1 \subseteq \mathcal{M}_2$, if $f^{\mathcal{M}_1} \subseteq f^{\mathcal{M}_2}$ for all fields $f \in \mathsf{Fld}$. $\mathcal{M}_1$ and $\mathcal{M}_2$ are *disjoint interpretations* if for all $f \in \mathsf{Fld}$, $\mathrm{dom}(f^{\mathcal{M}_1}) \cap \mathrm{dom}(f^{\mathcal{M}_2}) = \emptyset$. If $\mathcal{M}_1$ and $\mathcal{M}_2$ are disjoint, we denote with $M_1 \oplus M_2$ the composition of $\mathcal{M}_1$ and $\mathcal{M}_2$ defined by taking the point-wise union of the functions f for each field $f \in \mathsf{Fld}$.

$$
\begin{aligned}
\mathcal{M} &\models x \rightarrow_f y && \text{iff } f^{\mathcal{M}} = \{x^{\mathcal{M}} \mapsto y^{\mathcal{M}}\} \wedge \forall h \in \mathsf{Fld} . h \neq f \implies h = \emptyset \\
\mathcal{M} &\models F_{\mathsf{loc}} && \text{iff } (\mathcal{M} \models_{\mathsf{loc}} F_{\mathsf{loc}}) \wedge \forall h \in \mathsf{Fld} . h = \emptyset \\
\mathcal{M} &\models F_{\mathsf{data}} && \text{iff } (\mathcal{M} \models_{\mathsf{data}} \mathcal{F}_{\mathsf{loc}}) \wedge \forall h \in \mathsf{Fld} . h = \emptyset \\
\mathcal{M} &\models F * G && \text{iff } \mathcal{M} = \mathcal{M}_1 \oplus \mathcal{M}_2 \wedge \mathcal{M}_1 \models F \wedge \mathcal{M}_2 \models G \\
\mathcal{M} &\models F \wedge G && \text{iff } \mathcal{M} \models F \wedge \mathcal{M} \models G \\
\mathcal{M} &\models F \vee G && \text{iff } \mathcal{M} \models F \vee \mathcal{M} \models G \\
\mathcal{M} &\models \neg F && \text{iff not } \mathcal{M} \models F \\
\mathcal{M} &\models x \notin \mathbf{s} && \text{iff } \bigwedge_{y \in \mathbf{s}} \mathcal{M} \models x \neq y \\
\mathcal{M} &\models \mathsf{pred}(x, \mathbf{s}) && \text{iff } \exists i . \mathcal{M} \models \mathsf{pred}_s^i(x, \mathbf{s}) \wedge \bigwedge_{y_1 \neq y_2 \in \mathbf{s}} \mathcal{M} \models y_1 \neq y_2 \\
\mathcal{M} &\models \mathsf{pred}_t^0(x, \epsilon) && \text{iff } \mathcal{M} \models x = \mathbf{null} \\
\mathcal{M} &\models \mathsf{pred}_t^0(x, \langle y \rangle) && \text{iff } \mathcal{M} \models x = y \\
\mathcal{M} &\models \mathsf{list}_t^i(x, \mathbf{s}) && \text{iff } \exists \ell \in \mathsf{loc}^{\mathcal{M}}, d \in \mathsf{data}^{\mathcal{M}} . \\
& && \quad \mathcal{M}[z, w \mapsto \ell, d] \models x \notin \mathbf{t} * x \rightarrow_{n,d} (z, w) * \mathsf{list}_t^{i-1}(z, \mathbf{s}) \\
\mathcal{M} &\models \mathsf{tree}_t^i(x, \mathbf{s}) && \text{iff } \mathbf{s} = \mathbf{s}_1 \cdot \mathbf{s}_2 \wedge i = i_1 + i_2 + 1 \\
& && \quad \wedge \exists \ell_1, \ell_2 \in \mathsf{loc}^{\mathcal{M}}, d \in \mathsf{data}^{\mathcal{M}} . \mathcal{M}' = \mathcal{M}[z_1, z_2, w \mapsto \ell_1, \ell_2, d] \\
& && \quad \wedge \mathcal{M}' \models x \notin \mathbf{t} * x \rightarrow_{l,r,d} (z_1, z_2, w) * \mathsf{tree}_t^{i_1}(z_1, \mathbf{s}_1) * \mathsf{tree}_t^{i_2}(z_2, \mathbf{s}_2)
\end{aligned}
$$

Fig. 2. Semantics of the core separation logic $\mathbf{SL}_{\mathsf{data}}$. Variable z is a fresh variable of sort $\mathsf{loc}_{\mathsf{list}}$, z_1 and z_2 are fresh variables of sort $\mathsf{loc}_{\mathsf{tree}}$, and w is a fresh variable of sort data. For brevity we denote with pred either list or tree.

The semantics of a formula $F \in \mathbf{SL}_{\mathsf{data}}$ with respect to a heap interpretation $\mathcal{M}$ is defined inductively over the structure of F, as presented in Fig. 2. The semantics of location and data formulas $F_{\mathsf{loc}} \in \mathcal{F}_{\mathsf{loc}}$ and $F_{\mathsf{data}} \in \mathcal{F}_{\mathsf{data}}$ is defined by their interpretation in $\mathcal{T}_{\mathsf{loc}}$ and $\mathcal{T}_{\mathsf{data}}$ (denoted with $\mathcal{M} \models_{\mathsf{loc}} F_{\mathsf{loc}}$ and $\mathcal{M} \models_{\mathsf{data}} F_{\mathsf{data}}$), respectively. Our semantics is *precise* in the usual separation-logic sense (see e.g. [3]), meaning $\mathcal{M} \models x_1 \rightarrow_f x_2$ implies that $x_1 \rightarrow_f x_2$ is the only pointer that is defined in $\mathcal{M}$.

Example 2 (Semantics). Consider the following graphical representations of three heap interpretations.

---

[2] This size notion captures the amount of allocated memory rather than the amount of addressable memory (which is determined by the interpretation of the location domains).

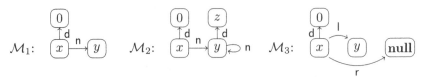

Each field interpretation corresponds directly to the edges of the graph labeled with that field. For example, in $\mathcal{M}_2$, the next fields of lists are interpreted as $\mathsf{n}^{\mathcal{M}_2} = \{x^{\mathcal{M}_2}, y^{\mathcal{M}_2} \mapsto y^{\mathcal{M}_2}, y^{\mathcal{M}_2}\}$. In these interpretations we have that

$$\mathcal{M}_1 \models \mathsf{list}(x, \langle y \rangle), \quad \mathcal{M}_2 \not\models \mathsf{list}(x, \langle y \rangle), \quad \mathcal{M}_2 \not\models \mathsf{list}(x, \langle y \rangle) * \mathsf{list}(y, \langle y \rangle),$$
$$\mathcal{M}_2 \models \mathsf{list}(x, \langle y \rangle) * (y \rightarrow_{\mathsf{n,d}} (y, z)), \quad \mathcal{M}_3 \models \mathsf{tree}(x, \langle y \rangle) * (x \neq y).$$

3   A Small Model Property for $\mathbf{SL_{data}}$

In this section, we show that every satisfiable $\mathbf{SL_{data}}$ formula F is satisfiable by a model with a small interpretation of the loc domain. More precisely, we will derive a size bound that is linear in the number of variables in F. This result has two implications. First, the satisfiability problem for $\mathbf{SL_{data}}$ is in NP if satisfiability for $\mathcal{T}_{\mathsf{loc}}$ and $\mathcal{T}_{\mathsf{data}}$ is in NP, as we can guess and check a polynomially-sized model.[3] Second, as we will argue in Sect. 5, the size bound enables encodings of $\mathbf{SL_{data}}$ into SMT without the need to reason about unbounded reachability.

To derive a tight bound, we distinguish between the *list variables* and the *tree variables* in F. A variable is a list variable if it appears in at least one atom of the form $x_1 \rightarrow_{\mathsf{n}} x_2$ or $\mathsf{list}(x_1, \langle x_2, \ldots, x_k \rangle)$; a variable is a tree variable if it appears in an atom of the form $x_1 \rightarrow_{\mathsf{l}} x_2$, $x_1 \rightarrow_{\mathsf{r}} x_2$ or $\mathsf{tree}(x_1, \langle x_2, \ldots, x_k \rangle)$. The main result in the current section is the following:

Theorem 1 (Small-model property for $\mathbf{SL_{data}}$). *Let F be a satisfiable $\mathbf{SL_{data}}$ formula with n_{list} list variables, n_{tree} tree variables, and at most $k \geq 1$ stop locations per* tree *predicate. Then there is a heap interpretation $\mathcal{M}$ that satisfies F such that $|\mathcal{M}| \leq \max(4, 2n_{\mathsf{list}} + (2 + k)n_{\mathsf{tree}})$.*

Example 3. To illustrate the bound of Theorem 1, let $x_1, \ldots, x_n$ be tree variables, $\mathbf{s} = \langle s_1, \ldots, s_n \rangle$, for $n = 2^k$. Consider the formula $\mathsf{tree}(x_1, \mathbf{s}) * \cdots * \mathsf{tree}(x_n, \mathbf{s})$. A heap that satisfies this formula needs to accommodate n separate trees that all end in n stop points. The smallest such heap $\mathcal{M}$ would therefore include n full binary trees with n leaves and have the allocated size $|\mathcal{M}| = n(n-1)$. Although this size is quadratic in n, this is only because we are using n stop points. In practice, the number of (program) variables pointing into a tree structure will be an upper bound for the number of stop points. This number is generally low—for example at most 2 for many typical tree traversal and tree update algorithms. □

To prove Theorem 1, we take an arbitrary model $\mathcal{M} \models F$ and transform it into a small model $\mathcal{M}'$ such that $\mathcal{M}' \models F$. We define separate transformations

---

[3] This is the case, e.g., in the common case when $\mathcal{T}_{\mathsf{loc}}$ is the theory of equality and $\mathcal{T}_{\mathsf{data}}$ is the theory of linear arithmetic.

for positive and negative heap interpretations. A positive heap interpretation is one that satisfies a positive spatial formula. More precisely, a heap interpretation $\mathcal{M}$ over X is *positive* if there exists a formula $F = A_1 * \cdots * A_k$ such that $\mathcal{M} \models F$. A heap interpretation that is not positive is *negative*.

Positive heap interpretations are well behaved. In particular, every unlabeled location in a positive interpretation $\mathcal{M}$ is contained in exactly one list or tree in $\mathcal{M}$, as the separating conjunction precludes sharing of unlabeled allocated locations between multiple data structures. The semantics of the list and tree predicates additionally enforce acyclicity within each data structure and full allocation of all unlabeled locations. We formalize these and related observations in the following lemma.

Lemma 1. *In a positive heap interpretation $\mathcal{M}$ the following holds:*

1. *Every unlabeled allocated location is fully allocated.*
2. *If $\mathcal{M} = \mathcal{M}_1 \oplus \mathcal{M}_2$, with both $\mathcal{M}_1$ and $\mathcal{M}_2$ positive, then every unlabeled and allocated location of $\mathcal{M}$ is allocated in exactly one of $\mathcal{M}_1$ and $\mathcal{M}_2$.*
3. *For every unlabeled allocated location ℓ, there is exactly one x such that ℓ is reachable from $x^{\mathcal{M}}$ without going through another labeled location, i.e., such that $x \to_{\mathcal{M}}^{+} \ell' \to_{\mathcal{M}}^{*} \ell$ implies that ℓ' is unlabeled.*
4. *Every loop in $\mathcal{M}$ must contain a labeled location: if $\ell_1 \to_{\mathcal{M}}^{+} \ell_1$, then every such closed path can be split into $\ell_1 \to_{\mathcal{M}}^{*} x^{\mathcal{M}} \to_{\mathcal{M}}^{*} \ell_1$ for some $x \in X$.*

Note that by definition, all negative heap interpretations falsify *all* spatial formulas. So if $\mathcal{M}$ is negative and $\mathcal{M} \models F$, we know that $\mathcal{M}$ falsifies all spatial subformulas of F, as would *any* negative heap interpretation. For example, we can replace $\mathcal{M}$ by a small model that contains a loop, because such a model is negative by Lemma 1. Intuitively, this is why all formulas that are satisfied by negative heap interpretations have the small model property.

Lemma 2. *There exists a heap interpretation $\mathcal{M}_0$, with $|\mathcal{M}_0| = 4$, that falsifies all spatial formulas $A_1 * \cdots * A_n$. Moreover, for any negative heap interpretation $\mathcal{M}$ and formula F, if $\mathcal{M} \models F$ then $\mathcal{M}_0 \models F$.*

For the remainder of this section, we assume that F is a formula that is satisfiable in a positive interpretation $\mathcal{M}$ over variables X. First, we define a transformation of heap interpretations that removes a single location from the field interpretations. We then show that, using this transformation, we can minimize $\mathcal{M}$ to a model that is still positive and satisfies F. Finally, we show that the size of this positive minimal model is bounded as in Theorem 1.

Location Removal. Let $\ell \in \mathsf{loc}^{\mathcal{M}}$ be an allocated location, i.e., there is at least one field $f \in \{\mathsf{n}, \mathsf{l}, \mathsf{r}\}$ with $\ell \in \mathrm{dom}(f^{\mathcal{M}})$. We say that such a location ℓ_0 is *removable through its field f* if the field f is defined (allocated) at ℓ_0 and for all other fields $g \neq \mathsf{d}$, $g^{\mathcal{M}}(\ell)$ is either **null** or undefined. (Note that this does not preclude that also $f^{\mathcal{M}}(\ell) = \mathbf{null}$.) If ℓ is a location removable through its field

f, we write $\mathcal{M}' = \mathcal{M} \setminus \{\ell\}$ for the interpretation that mimics $\mathcal{M}$ apart from avoiding the location ℓ, i.e., for all $g \in \{n, l, r\}$ and for all locations ℓ', we define

$$g^{\mathcal{M}'}(\ell') = \begin{cases} g^{\mathcal{M}}(\ell') & \text{if } \ell' \neq \ell \text{ and } g^{\mathcal{M}}(\ell') \neq \ell, \\ g^{\mathcal{M}}(\ell) & \text{if } \ell' \neq \ell \text{ and } g^{\mathcal{M}}(\ell') = \ell \\ \bot & \text{if } \ell' = \ell \end{cases}$$

In addition, location ℓ is removed from the data interpretation, i.e., we set $\mathsf{d}^{\mathcal{M}'} = \mathsf{d}^{\mathcal{M}} \setminus \{\ell \mapsto d\}$. Figure 3 illustrates location removal for lists and trees.

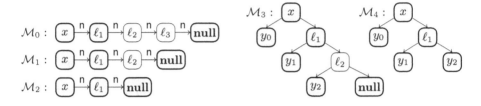

Fig. 3. Subsequent removal of removable non-essential list locations ℓ_3 and ℓ_2 and tree location ℓ_2, transforming $\mathcal{M}_0$ via $\mathcal{M}_1$ into $\mathcal{M}_2$ and $\mathcal{M}_3$ into $\mathcal{M}_4$. Essential locations are displayed in green. (Color figure online)

Essential Locations. Location removal reduces the allocated size of $\mathcal{M}$. We now characterize the locations that can safely be removed from $\mathcal{M}$ without falsifying the formula F. Assume two distinct labeled locations ℓ_1 and ℓ_2 from $\mathrm{loc}^{\mathcal{M}}$. We call a location $\ell \in \mathrm{loc}^{\mathcal{M}}$ an *induction indicator* for (ℓ_1, ℓ_2) if $\ell_1 \to_{\mathcal{M}} \ell \to_{\mathcal{M}}^{+} \ell_2$. Intuitively, a location is an induction indicator if it is a potential witness of a first step of a longer unrolling of an inductive predicate. An induction indicator cannot be removed as such a removal might change the interpretation of a predicate $x \to_f y$ from false to true. An allocated location ℓ is an *essential location* iff ℓ is a labeled location or ℓ is an induction indicator.

Model Minimization. We now proceed to show that if $\mathcal{M}' = \mathcal{M} \setminus \{\ell\}$ for a non-essential, removable location ℓ, then $\mathcal{M} \models F$ if and only if $\mathcal{M}' \models F$.

Lemma 3. *Let $\mathcal{M}$ be a positive heap interpretation over X and let A be a spatial atom. Let ℓ be a non-essential, removable location and let $\mathcal{M}' = \mathcal{M} \setminus \{\ell\}$. Then $A \models \mathcal{M}$ if and only if $\mathcal{M}' \models A$.*

If $F = A_1 * \cdots * A_n$ is a separating conjunction, we use that $\mathcal{M} = \mathcal{M}_1 \oplus \cdots \oplus \mathcal{M}_k$ for some $\mathcal{M}_i$ such that $\mathcal{M}_i \models A_i$. By Lemma 1, ℓ is fully allocated in exactly one $\mathcal{M}_i$, has no direct predecessors outside of $\mathcal{M}_i$, and is removable in $\mathcal{M}_i$. This lets us reduce the case for the separating conjunction to Lemma 3 and conclude:

Lemma 4. *Let $F = A_1 * \cdots * A_n$ be a conjunction of spatial atoms A_i, and let $\mathcal{M}$ be a positive heap interpretation. Let ℓ be a non-essential removable location of $\mathcal{M}$, and let $\mathcal{M}' = \mathcal{M} \setminus \{\ell\}$. Then $\mathcal{M} \models F$ if and only if $\mathcal{M}' \models F$.*

Finally, by a simple induction proof over the Boolean structure of F that relies on Lemma 4, we conclude:

Lemma 5. *Let $\mathcal{M}$ be a positive heap interpretation over X, ℓ be a non-essential removable location, and let $\mathcal{M}' = \mathcal{M} \setminus \{\ell\}$. Then $\mathcal{M} \models F$ if and only if $\mathcal{M}' \models F$.*

It remains to be shown that by iterating location removal, we eventually terminate with a model of a size that satisfies the bound in Theorem 1.

Proof (of Theorem 1). Let $\mathcal{M} \models F$. If F is also satisfied in the small model $\mathcal{M}_0$ from Lemma 2, we are done. Otherwise, consider the DNF form of F. Since $\mathcal{M} \models F$, there is at least one conjunct $L_1 \wedge \ldots \wedge L_m$ of the DNF such that $\mathcal{M} \models L_i$, for all i. If all L_i are negated, then $\mathcal{M}_0$ satisfies them, and therefore $\mathcal{M}_0 \models F$, contrary to assumption. Therefore, there is at least one positive $L_i = A_1 * \cdots * A_n$, with $\mathcal{M} \models A_1 * \cdots * A_n$. We iterate location removal until we end in a positive model $\mathcal{M}' \models F$ and $\mathcal{M}' \models A_1 * \cdots * A_n$, that has no more removable non-essential locations. We estimate the size of $\mathcal{M}'$. An allocated location from $\mathcal{M}'$ is either essential or non-essential.

There are at most $N_1 = 2n_{\mathsf{list}} + 3n_{\mathsf{tree}}$ allocated essential locations in $\mathcal{M}'$: In $\mathcal{M}'$, every list location has at most one successor location, and every tree location has at most two direct successors (with respect to $\to_{\mathcal{M}}$). These are potential induction indicators which, taken together with labeled locations, give a total of at most N_1 essential locations.

Now, let ℓ be an allocated but non-essential location in $\mathcal{M}'$. Since $\mathcal{M}' \models A_1 * \cdots * A_n$. Location ℓ must be allocated by one of the A_i atoms. A_i cannot be a $\to_f$ predicate as ℓ would otherwise be labeled and therefore essential. A_i cannot be a list predicate either, as ℓ would be removable. A_i must therefore be a $\mathsf{tree}(x, \mathbf{s})$ predicate. We claim that both left and right subtree of ℓ must contain (distinct) stop variables s_1 and s_2. If not, then one of the descendants of ℓ would be removable. The location ℓ is therefore the lowest common ancestor of s_1 and s_2. Assuming that the number of stop points in $\mathbf{s}$ is at most k, for a tree starting at x, there are at most $k - 1$ such common ancestors. Since there can be at most n_{tree} non-empty tree predicates among A_i, there are at most $N_2 = n_{\mathsf{tree}}(k - 1)$ allocated non-essential locations in $\mathcal{M}'$. We can thus bound the size of $\mathcal{M}'$ with $N_1 + N_2 = 2n_{\mathsf{list}} + (2 + k)n_{\mathsf{tree}}$. $\qquad\square$

4   Extending $\mathbf{SL_{data}}$ with Data Constraints

In this section we add to $\mathbf{SL_{data}}$ the possibility to constrain the data values in lists and trees by means of passing $\mathcal{T}_{\mathsf{data}}$ formulas as additional parameter to the list and tree predicates. We call the extended logic $\mathbf{SL^*_{data}}$. Our goal is to reason about data properties of inductive structures that appear frequently in practice, e.g., a list being sorted, or a tree being a binary search tree.

We assume two dedicated fresh variables α and β from $\mathcal{X}$ of sort data to be used exclusively in data predicates. We call a formula $P(\alpha)$ a *unary data predicate*, and a pair $(f, P(\alpha, \beta))$, for $f \in \{n, l, r\}$, a *binary data predicate*. Both types of predicates may also contain other variables from $\mathcal{X} \setminus \{\alpha, \beta\}$. We pass a set of data predicates $\mathcal{P}$ as additional parameter to the predicates, obtaining ternary predicates $\mathsf{list}(x, \mathsf{s}, \mathcal{P})$ and $\mathsf{tree}(x, \mathsf{s}, \mathcal{P})$. As before, for brevity, if either of s or $\mathcal{P}$ is empty, we omit them. Semantics of an inductive data predicate $\mathsf{pred}(x, \mathsf{s}, \mathcal{P})$, in a heap interpretation $\mathcal{M}$, are as follows:

1. The predicate holds in $\mathcal{M}$ only if it holds without the data constraints, i.e., $\mathsf{pred}(x, \mathsf{s})$ must be true in $\mathcal{M}$ and therefore $\mathcal{M}$ describes a pred structure.
2. For each unary data predicate $P(\alpha) \in \mathcal{P}$, all allocated data in $\mathcal{M}$ must satisfy P. More precisely, for all $(\ell, d) \in \mathsf{d}^{\mathcal{M}}$, we have that $\mathcal{M}[\alpha \mapsto d] \models_{\mathsf{data}} P$ holds.
3. For each binary predicate $(f, P(\alpha, \beta)) \in \mathcal{P}$, all allocated data must be related with all of its f-descendants through P. More precisely, for all $(\ell_1, d_1), (\ell_2, d_2) \in \mathsf{d}^{\mathcal{M}}$ such that $f^{\mathcal{M}}(\ell_1) \rightarrow^*_{\mathcal{M}} \ell_2$, $\mathcal{M}[\alpha, \beta \mapsto d_1, d_2] \models_{\mathsf{data}} P$ holds.

Example 4 (Data Predicates). We illustrate the inductive predicates through representative examples of data predicates over lists and trees. The predicates

$$\mathsf{list}(x, \{(\alpha = 0)\}), \qquad \mathsf{list}(x, \{(n, \alpha \neq \beta)\}), \qquad \mathsf{list}(x, \{(n, \alpha < \beta)\}),$$

describe a list with all data values equal to 0, a list with all data values distinct, and a list with data values increasing. The predicates

$$\mathsf{tree}(x, \{(l, \beta < \alpha), (r, \beta > \alpha)\}), \qquad \mathsf{tree}(x, \{(l, \beta < \alpha), (r, \beta < \alpha)\}),$$

describe a binary search tree, and a max-heap. Formula $\mathsf{list}(x, \langle m \rangle, \{(\alpha < M)\}) * m \rightarrow_{n,d} (y, M) * \mathsf{list}(y, \{(\alpha > M)\})$ describes a partitioned list where the left partition contains elements smaller than the pivot m, and the right partition contains elements larger than the pivot m. Formula $\mathsf{list}(x) * \mathsf{list}(y, \{(\alpha \neq a)\}) \wedge \neg(\mathsf{list}(x, \{(\alpha \neq a)\}) * \mathsf{list}(y))$ describes a list x that contains a data value a, and a list y that does not contain a, meaning that the sets of values in the lists x and y are different. □

We now lift the small-model property to full $\mathbf{SL}^*_{\mathsf{data}}$.

Theorem 2 (Small-model property for $\mathbf{SL}^*_{\mathsf{data}}$). *Let F be a satisfiable $\mathbf{SL}^*_{\mathsf{data}}$ formula with n_{list} list variables, n_{tree} tree variables, m_{list} list predicates with data constraints, m_{tree} tree predicates with data constraints, and at most $k \geq 1$ stop locations per tree predicate. Then there is a heap interpretation $\mathcal{M}$ that satisfies F such that $|\mathcal{M}| \leq \max(4, 2n_{\mathsf{list}} + (3 + k)n_{\mathsf{tree}} + 2m_{\mathsf{list}} + 2m_{\mathsf{tree}})$.*

Intuitively, the changes to the reasoning are minimal since the data predicates are universal, and the location removal does not invalidate predicates that are true. We only must be careful to ensure that location removal does not change the value of a data predicate from false to true.

Example 5. Let $F = \mathsf{list}(x_1, \langle x_2 \rangle, \{(\mathsf{n}, P_1)\}) \wedge \neg\mathsf{list}(x_1, \langle x_2 \rangle, \{(\mathsf{n}, P_2)\})$, where $P_1 = (\alpha < \beta)$ and $P_2 = (2\alpha \leq \beta)$, and consider the following model $\mathcal{M}_0$.

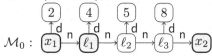

We have that $\mathcal{M}_0 \models F$. In addition, locations ℓ_2 and ℓ_3 are not essential and can be removed. But, in models $\mathcal{M}_1 = \mathcal{M}_0 \setminus \{\ell_2\}$ and $\mathcal{M}_2 = \mathcal{M}_0 \setminus \{\ell_3\}$, we have that $\mathcal{M}_1 \not\models F$ but $\mathcal{M}_2 \models F$.   □

To avoid the situation described above, we must ensure that for each data predicate $\mathsf{pred}(x, \mathbf{s}, \mathcal{P})$ that is falsified due to some $P \in \mathcal{P}$ being false, we have a designated pair of locations that witness the reason why P is false.

Proof (of Theorem 2). As in the proof of Theorem 1, assume that F has a positive model $\mathcal{M}$, i.e., $\mathcal{M} \models A_1 * \cdots * A_n$ for some A_i from F. In this model, for each falsified data predicate, we designate at most 2 additional locations as essential and proceed with reducing $\mathcal{M}$ to $\mathcal{M}'$ by removing all removable non-essential locations. The model $\mathcal{M}'$ therefore contains at most $N_1 = 2n_{\mathsf{list}} + 3n_{\mathsf{tree}} + 2m_{\mathsf{list}} + 2m_{\mathsf{list}}$ locations marked as essential. We now count the number of allocated non-essential locations ℓ in $\mathcal{M}'$. As in the proof of Theorem 1, location ℓ must be allocated as part of a predicate $\mathsf{pred}(x, \mathbf{s})$, whose interpretation can also contain two associated witness locations w_1 and w_2 with $w_1 \rightarrow^+_{\mathcal{M}'} w_2$. Location ℓ is then either the lowest common ancestor of two stop location $s_i^{\mathcal{M}'}$ and $s_j^{\mathcal{M}'}$, or a lowest common ancestor of some stop locations $s_i^{\mathcal{M}'}$ and w_2. Assuming that the number of stop points in $\mathbf{s}$ is at most k, for a tree starting at x there are at most k such common ancestors. Since there can be at most n_{tree} non-empty tree predicates among A_i, there are at most $N_2 = kn_{\mathsf{tree}}$ allocated non-essential locations in $\mathcal{M}'$. We can thus bound the size of $\mathcal{M}'$ with $|\mathcal{M}'| \leq N_1 + N_2 = 2n_{\mathsf{list}} + (3 + k)n_{\mathsf{tree}} + 2m_{\mathsf{list}} + 2m_{\mathsf{tree}}$.   □

As opposed to the symbolic-heap family of separation logics (e.g. [3,5]), the logic $\mathbf{SL}^*_{\mathsf{data}}$ is closed under negation, and we can solve the entailment problem $F \models G$ by checking whether $F \wedge \neg G$ is unsatisfiable.

Corollary 1. *If the satisfiability problem for $\mathcal{T}_{\mathsf{data}}$ is in* NP *then the satisfiability problem for $\mathbf{SL}^*_{\mathsf{data}}$ is in* NP, *and the entailment problem for $\mathbf{SL}^*_{\mathsf{data}}$ is in* CONP.

5 Encoding $\mathbf{SL}^*_{\mathsf{data}}$ into SMT

We now present an encoding of $\mathbf{SL}^*_{\mathsf{data}}$ formulas into SMT. We show that every formula $F \in \mathbf{SL}^*_{\mathsf{data}}$ can be encoded in polynomial time (and size) as a formula in the SMT theory of arrays that is satisfiable iff F is satisfiable. Our approach relies on the theory of arrays extended with combinators that can express constant arrays and express point-wise array operations [16]. We denote this theory by $\mathcal{T}_{\mathsf{array}}$. In $\mathcal{T}_{\mathsf{array}}$, it is possible to express universal statements about array elements without relying on quantifiers. Moreover, the satisfiability of generalized array formulas is decidable in NP with effective decision procedures implemented in popular SMT solvers such as Z3 [17] and Boolector [21].

The basic theory of arrays defines functions store and $\cdot[\cdot]$, as usual (see, e.g. [15]). The generalized theory adds a constant combinator $\mathbf{K}$ and a map combinator map such that $\mathbf{K}(c)[i] = c$, for a constant c, and $\mathsf{map}_f(A_1, \ldots, A_n)[i] = f(A_1[i], \ldots, A_n[i])$, for a function f. The array combinators are expressive enough to express basic set-theoretic operations. For example, we can view a set of locations as an array mapping loc to Bool and define set operations as follows

$$\{x\} = \mathsf{store}(\mathbf{K}(\bot), x, \top) \qquad x \in X = X[x] \qquad \mathsf{empty}(X) = (X = \mathbf{K}(\bot))$$
$$X \subseteq Y = \mathsf{map}_{\Rightarrow}(X, Y) \qquad X \cup Y = \mathsf{map}_{\vee}(X, Y) \qquad X \cap Y = \mathsf{map}_{\wedge}(X, Y)$$

In the following, we use the set notation as a shorthand for the equivalent array encoding. We denote array variables that represent sets in capital letters (e.g., X), and vectors of array variables in boldface (e.g., $\mathbf{X} = \langle X_1, \ldots X_n \rangle$). To ease notation we overload predicates over sets to predicates over vectors of sets in a point-wise manner and write, e.g., $\mathsf{empty}(\mathbf{X})$ for $\bigwedge \mathsf{empty}(X_i)$, and $\mathbf{X} = \mathbf{Y} \cup \mathbf{Z}$ for $\bigwedge X_i = Y_i \cup Z_i$.

To each $\mathbf{SL}^*_{\mathsf{data}}$ interpretation $\mathcal{M}_{\mathrm{SL}}$, of size N, we associate an equivalent first-order model $\mathcal{M}_{\mathrm{SMT}}$ in the theory $\mathcal{T}_{\mathsf{array}} \oplus \mathcal{T}_{\mathsf{data}} \oplus \mathcal{T}_{\mathsf{loc}}$ as follows. $\mathcal{M}_{\mathrm{SMT}}$ interprets each sort from $\mathcal{M}_{\mathrm{SL}}$ as the same sort; each partial function $f \in \mathsf{Fld}$ in $\mathcal{M}_{\mathrm{SL}}$ as an array f of the same sort; and the domain of each partial function $f \in \mathsf{Fld}$ in $\mathcal{M}_{\mathrm{SL}}$ as a dedicated set variable X_f. The interpretation $f^{\mathcal{M}_{\mathrm{SMT}}}$ of a field is an array mapping each $\ell \in \mathsf{dom}(f^{\mathcal{M}_{\mathrm{SL}}})$ to $f^{\mathcal{M}_{\mathrm{SL}}}(\ell)$, and to an arbitrary well-sorted value otherwise. The interpretation of $X_f^{\mathcal{M}_{\mathrm{SMT}}}$ is an array representing the set $\mathsf{dom}(f^{\mathcal{M}_{\mathrm{SL}}})$. The interpretation $\mathcal{M}_{\mathrm{SMT}}$ also includes N dedicated location variables $x_1, \ldots, x_N$, and a set of locations X interpreted so that $X = X_{\mathsf{n}} \cup X_{\mathsf{l}} \cup X_{\mathsf{r}}$ and $X \subseteq \{x_1, \ldots, x_N\}$ holds. Other variables and constants are interpreted in $\mathcal{M}_{\mathrm{SMT}}$ as they are in $\mathcal{M}_{\mathrm{SL}}$.

The following SMT formula $\Delta^N_{\mathbf{SL}}$ defines $\mathbf{SL}^*_{\mathsf{data}}$ heap interpretations of size at most N.

$$\Delta^N_{\mathbf{SL}} \overset{\mathrm{def}}{=} X = \bigcup_{f \in \mathsf{Fld}} X_f \wedge X \subseteq \{x_1, \ldots, x_N\} \wedge \mathbf{null} \notin X \wedge \mathsf{empty}(X_{\mathsf{n}} \cap (X_{\mathsf{l}} \cup X_{\mathsf{r}}))$$

Formula $\Delta^N_{\mathbf{SL}}$ makes sure that the allocated heap size is at most N, that **null** is not allocated, and that no variable is treated as both a list and a tree location. In the following we always denote with $\mathbf{X} = \langle X_{\mathsf{n}}, X_{\mathsf{l}}, X_{\mathsf{r}}, X_{\mathsf{d}} \rangle$ the vector of dedicated set variables denoting field footprints.

SMT Translation. The encoding function T_N that translates basic $\mathbf{SL}^*_{\mathsf{data}}$ formulas to SMT is shown in Fig. 4. We start without inductive predicates, following the approach from [19]. The function T_N takes an $\mathbf{SL}^*_{\mathsf{data}}$ formula F and translates F into an SMT formula $F' = \mathsf{T}_N(F)$ so that F' is satisfiable if and only if F is satisfiable in a model of size at most N. The translation relies on two auxiliary functions: $\mathsf{T}^b_N(F)$, that translates the Boolean structure of F recursively;

$$\mathsf{T}_N^s(F_{\mathsf{loc}}, \mathbf{Y}) = \langle F_{\mathsf{loc}}, \mathsf{empty}(\mathbf{Y}), \emptyset \rangle$$

$$\mathsf{T}_N^s(F_{\mathsf{data}}, \mathbf{Y}) = \langle F_{\mathsf{data}}, \mathsf{empty}(\mathbf{Y}), \emptyset \rangle$$

$$\mathsf{T}_N^s(x \rightarrow_f y, \mathbf{Y}) = \langle f(x) = y, Y_f = \{x\} \wedge \mathsf{empty}(\mathbf{Y} \setminus Y_f), \emptyset \rangle$$

$$\mathsf{T}_N^s(F_1 * F_2, \mathbf{Y}) = \text{let } \mathbf{Y}_1, \mathbf{Y}_2 \text{ be fresh in}$$
$$\text{let } \langle A_1, B_1, Z_1 \rangle = \mathsf{T}_N^s(F_1, \mathbf{Y}_1), \langle A_2, B_2, Z_2 \rangle = \mathsf{T}_N^s(F_2, \mathbf{Y}_2) \text{ in}$$
$$\text{let } Z = Z_1 \cup Z_2 \cup \mathbf{Y}_1 \cup \mathbf{Y_2} \text{ in}$$
$$\langle A_1 \wedge A_2 \wedge \mathsf{empty}(\mathbf{Y}_1 \cap \mathbf{Y}_2), B_1 \wedge B_2 \wedge \mathbf{Y} = \mathbf{Y}_1 \cup \mathbf{Y}_2, Z \rangle$$

$$\mathsf{T}_N^b(F) = \text{let } \mathbf{Y} \text{ be fresh}, \langle A, B, Z \rangle = \mathsf{T}_N^s(F, \mathbf{Y}) \text{ in } \langle A \wedge \mathbf{X} = \mathbf{Y}, B, Z \cup \mathbf{Y} \rangle$$

$$\mathsf{T}_N^b(\neg F) = \text{let } \langle A, B, Z \rangle = \mathsf{T}_N^b(F) \text{ in } \langle \neg A, B, Z \rangle$$

$$\mathsf{T}_N^b(F_1 \wedge F_2) = \text{let } \langle A_1, B_1, Z_1 \rangle = \mathsf{T}_N^b(F_1), \langle A_2, B_2, Z_2 \rangle = \mathsf{T}_N^b(F_2) \text{ in}$$
$$\langle A_1 \wedge A_2, B_1 \wedge B_2, Z_1 \cup Z_2 \rangle$$

$$\mathsf{T}_N^b(F_1 \vee F_2) = \text{let } \langle A_1, B_1, Z_1 \rangle = \mathsf{T}_N^b(F_1), \langle A_2, B_2, Z_2 \rangle = \mathsf{T}_N^b(F_2) \text{ in}$$
$$\langle A_1 \vee A_2, B_1 \wedge B_2, Z_1 \cup Z_2 \rangle$$

$$\mathsf{T}_N(F) = \text{let } \mathsf{T}_N^b(F) = \langle A, B, Z \rangle \text{ in } A \wedge B \wedge \Delta_{\mathbf{SL}}^N$$

Fig. 4. SMT encoding for the core fragment of $\mathbf{SL}_{\mathsf{data}}$ without inductive predicates.

and $\mathsf{T}_N^s(F, \mathbf{Y})$, that translates spatial formulas. Both functions take as input a formula (and a footprint $\mathbf{Y}$ to define) and return a triple $\langle A, B, Z \rangle$, where A and B together define the semantics of F and Z is the set of all fresh variables introduced by the translation. The encoding is straightforward, with the exception of negation. Let $\mathsf{T}_N^b(F) = \langle A, B, Z \rangle$. In order for our encoding to be correct, we make sure that the following properties hold.

Correctness: $\mathcal{M}_{\mathrm{SL}}^N \models F$ iff $\mathcal{M}_{\mathrm{SMT}}^N \models \exists Z \,.\, A \wedge B$;
Z-Existence: $\exists Z \,.\, B$ is valid; and
Z-Equivalence: $B(Z_1) \wedge B(Z_2) \Rightarrow A(Z_1) = A(Z_2)$ is valid.

The correctness property ensures that the encoding correctly encodes the $\mathbf{SL}_{\mathsf{data}}^*$ semantics: F is true in a heap interpretation of size N iff it is true in the corresponding SMT model. Z-Existence and Z-Equivalence make sure that the encoding can accommodate negation: the B part of the translation is a "definition" of the fresh variables Z: variables Z can be assigned in each model to satisfy B, in a way that the A part cannot distinguish. These properties allow us to ensure correctness of the translation of negation $\neg F$. Assuming that $\mathsf{T}_N^b(F) = \langle A, B, Z \rangle$, we can derive the encoding of negation as

$$\mathcal{M}_{\mathrm{SL}}^N \models \neg F \qquad \text{iff} \qquad \mathcal{M}_{\mathrm{SL}}^N \not\models F \qquad \text{iff} \qquad (1)$$

$$\mathcal{M}_{\mathrm{SMT}}^N \not\models \exists Z.A \wedge B \qquad \text{iff} \qquad \mathcal{M}_{\mathrm{SMT}}^N \models \neg \exists Z.A \wedge B \qquad \text{iff} \qquad (2)$$

$$\mathcal{M}_{\mathrm{SMT}}^N \models \forall Z.B \Rightarrow \neg A \qquad \text{iff} \qquad \mathcal{M}_{\mathrm{SMT}}^N \models \exists Z.B \wedge \neg A, \qquad (3)$$

where the equivalence (3) follows from Z-existence and Z-equivalence.

Lists and Trees. The translation of an inductive predicate $\mathsf{T}_N^s(\mathsf{pred}(x, \mathbf{s}, \mathcal{P}), \mathbf{Y})$, with $\mathbf{s} = \langle s_1, \ldots, s_k \rangle$, for model sizes of at most N, introduces fresh binary predicates $r_1^Z, \ldots, r_N^Z$, and a fresh location set Z. These fresh predicates are meant to represent reachability in up to N steps within the set Z. Location set Z will represent all nodes reachable from x allocated within the predicate. Throughout the remainder of the section, we assume that x, $\mathbf{s}$, $\mathcal{P}$, $\mathbf{Y}$, Z, r_i^Z and N are fixed and in scope of all definitions. We also assume sets of fields F_{pred} and $F_{\mathsf{pred}}^{\mathsf{d}}$, defined as $F_{\mathsf{list}} = \{\mathsf{n}\}$ and $F_{\mathsf{list}}^{\mathsf{d}} = \{\mathsf{n}, \mathsf{d}\}$, or $F_{\mathsf{tree}} = \{\mathsf{l}, \mathsf{r}\}$ and $F_{\mathsf{tree}}^{\mathsf{d}} = \{\mathsf{l}, \mathsf{r}, \mathsf{d}\}$.

To fully define translation function T_N^s, we will define auxiliary helper formulas. To start with, we define the following functions for convenience: $\mathsf{isstop}(x) \stackrel{\mathrm{def}}{=} x = \mathbf{null} \vee \bigvee_{s \in \mathbf{s}} x = s$, defining stop nodes; $S(x, y) \stackrel{\mathrm{def}}{=} \bigvee_{f \in F_{\mathsf{pred}}} f[x] = y$, defining a successor node; and $\mathsf{define}\mathbf{Y} \stackrel{\mathrm{def}}{=} \bigwedge_{f \in F_{\mathsf{pred}}^{\mathsf{d}}} Y_f = Z \wedge \bigwedge_{f \in \mathsf{Fld} \setminus F_{\mathsf{pred}}^{\mathsf{d}}} Y_f = \emptyset$, defining pred-relevant elements of $\mathbf{Y}$ in terms of the footprint Z.

Although reachability is not expressible in first-order logic, since we are only interested in finite reachability with respect to the model elements $x_1, \ldots, x_N$, we can define the reachability predicates r_K^Z, for $1 \leq K \leq N$, as follows.

$$R_1 \stackrel{\mathrm{def}}{=} \bigwedge_{1 \leq i, j \leq N} r_1^Z(x_i, x_j) \Leftrightarrow (x_i \in Z \wedge \neg\mathsf{isstop}(x_j) \wedge S(x_i, x_j))$$

$$R_K \stackrel{\mathrm{def}}{=} \bigwedge_{1 \leq i, j \leq N} r_K^Z(x_i, x_j) \Leftrightarrow \left(r_{K-1}^Z(x_i, x_j) \vee \bigvee_{1 \leq k \leq n} (r_{K-1}^Z(x_i, x_k) \wedge r_1^Z(x_k, x_j)) \right)$$

$$\mathsf{reachability} \stackrel{\mathrm{def}}{=} R_1 \wedge R_2 \wedge \cdots \wedge R_N$$

In addition, we define the function $r_N^Z(x, y, f) \stackrel{\mathrm{def}}{=} f[x] = y \vee (f[x] \in Z \wedge r_N^Z(f[x_i], x_j))$ to denote that y is reachable from x through f as the first step. We can now define the formula $\mathsf{footprint}$ that asserts that the set Z (the footprint of pred) is defined as the set of locations reachable from x.

$$\mathsf{emptyZ} \stackrel{\mathrm{def}}{=} \mathsf{isstop}(x) \vee \left(\bigwedge_{1 \leq i \leq N} x \neq x_i \right)$$

$$\mathsf{footprint} \stackrel{\mathrm{def}}{=} Z \subseteq \{x_1, \ldots, x_N\} \wedge (\mathsf{emptyZ} \Rightarrow Z = \emptyset) \wedge$$
$$\wedge \left(\neg\mathsf{emptyZ} \Rightarrow \bigwedge_{1 \leq i \leq N} ((x_i \in Z) \Leftrightarrow ((x_i = x) \vee r_N^Z(x, x_i))) \right)$$

Next, the formula $\mathsf{structure}$ ensures that the elements of the pred are part of an acyclic data structure, starting at x, with no sharing of non-$\mathbf{null}$ nodes.

$$\mathsf{oneparent} \stackrel{\mathrm{def}}{=} \bigwedge_{1 \leq i \leq N} x_i \in Z \Rightarrow \bigwedge_{f \neq g \in F_{\mathsf{pred}}} (f[x_i] = g[x_i] \Rightarrow f[x_i] = \mathbf{null})$$

$$\wedge \bigwedge_{1 \leq j \leq N} x_j \in Z \wedge x_i \neq x_j \Rightarrow \bigwedge_{f, g \in F_{\mathsf{pred}}} (f[x_i] = g[x_j] \Rightarrow f[x_i] = \mathbf{null})$$

$$\mathsf{structure} \stackrel{\mathrm{def}}{=} (\neg\mathsf{isstop}(x) \Rightarrow x \in Z) \wedge \mathsf{oneparent} \wedge \neg r_N^Z(x, x)$$

For ensuring stop node properties, we assert that the stop nodes of pred are pairwise different, occur exactly once, are the only leaves of the structure, and, for trees, are ordered the same way as prescribed by the vector $\mathbf{s} = \langle s_1, \ldots, s_k \rangle$.

$$\mathsf{stopseq} \stackrel{\text{def}}{=} (\mathsf{isstop}(x) \Rightarrow \bigwedge_{s \in \mathbf{s}} x = s) \wedge \bigwedge_{1 \le i < j \le k} s_i \ne s_j$$

$$\mathsf{stopsoccur} \stackrel{\text{def}}{=} \neg\mathsf{isstop}(x) \implies \bigwedge_{s \in \mathbf{s}} \bigvee_{1 \le p \le N} (x_p \in Z \wedge S(x_p, s))$$

$$\mathsf{stopleaves} \stackrel{\text{def}}{=} \bigwedge_{1 \le i \le N} \bigwedge_{f \in F_{\mathsf{pred}}} (x_i \in Z \wedge f[x_i] \notin Z) \Rightarrow \mathsf{isstop}(f[x_i])$$

$$\mathsf{fstop}(x_p, f, s) \stackrel{\text{def}}{=} f[x_p] = s \vee \bigvee_{1 \le c \le N} r_N^Z(x_p, x_c, f) \wedge x_c \in Z \wedge S(x_c, s)$$

$$\mathsf{ordered} \stackrel{\text{def}}{=} \bigwedge_{1 \le i < k} \bigvee_{1 \le p \le N} x_p \in Z \wedge \mathsf{fstop}(x_p, \mathsf{l}, s_i) \wedge \mathsf{fstop}(x_p, \mathsf{r}, s_{i+1})$$

We combine the above constraints into $\mathsf{stops}^{\mathsf{list}} \stackrel{\text{def}}{=} \mathsf{stopsoccur} \wedge \mathsf{stopseq} \wedge \mathsf{stopleaves}$ and $\mathsf{stops}^{\mathsf{tree}} \stackrel{\text{def}}{=} \mathsf{stopsoccur} \wedge \mathsf{stopseq} \wedge \mathsf{stopleaves} \wedge \mathsf{ordered}$. Finally, we define the data formula that ensures that the data allocated in the predicate respects the given (unary and binary) data predicates.

$$\mathsf{udata}(P) \stackrel{\text{def}}{=} \mathsf{map}_\Rightarrow(Z, \mathsf{map}_P(\mathsf{d})) = \mathbf{K}(\top)$$

$$\mathsf{bdata}(f, P) \stackrel{\text{def}}{=} \bigwedge_{1 \le i, j \le N} x_i, x_j \in Z \wedge r_N^Z(x_i, x_j, f) \Rightarrow P(x_i, x_j)$$

$$\mathsf{data} \stackrel{\text{def}}{=} \bigwedge_{P \in \mathcal{P}} \mathsf{udata}(P) \wedge \bigwedge_{(f, P) \in \mathcal{P}} \mathsf{bdata}_N(f, P)$$

Putting all the auxiliary formulas together, we define the translation of inductive predicates $\mathsf{pred} \in \{\mathsf{list}, \mathsf{tree}\}$ to SMT as follows.

$$\begin{aligned} \mathsf{T}_N^s(\mathsf{pred}(x, \mathbf{s}, \mathcal{P}), \mathbf{Y}) &= \text{let } r_1^Z, \dots, r_N^Z, Z \text{ be fresh} \\ &\quad \text{let } A = \mathsf{structure} \wedge \mathsf{stops}^{\mathsf{pred}} \wedge \mathsf{data} \\ &\quad \text{let } B = \mathsf{reachability} \wedge \mathsf{footprint} \wedge \mathsf{defineY} \text{ in} \\ &\quad \langle A, B, \{r_1^Z, \dots, r_N^Z, Z\} \rangle \end{aligned}$$

It is important to note that the formulas R_K only ensure that the predicates r_K^Z are fully defined on the set $\{x_1, \dots, x_N\}$ and can be interpreted arbitrarily elsewhere. Nevertheless, this is sufficient for the translation to be correct. By inspection, it can be seen that the A part of the translation cannot distinguish two interpretations of r_K^Z that differ only outside of $\{x_1, \dots, x_N\}$. This is crucial for the correctness of the encoding as it supports the Z-Equivalence property of the translation.

Theorem 3. *Let F be a $\mathbf{SL}_{\mathsf{data}}^*$ formula and N be the bound given by Theorem 2. Then F is $\mathbf{SL}_{\mathsf{data}}^*$-satisfiable if and only if the SMT translation $F' = T_N(F)$ is satisfiable. Moreover, the translation F' is polynomial in the size of F.*

As $\mathcal{T}_{\mathsf{array}}$ is in NP, this yields an NP decision procedure for $\mathbf{SL}_{\mathsf{data}}^*$ if $\mathcal{T}_{\mathsf{data}}$ is in NP, matching the complexity result from Sect. 4.

6   Conclusion

We defined a new fragment of separation logic, $\mathbf{SL}^*_{\mathsf{data}}$, which supports lists, trees, and data constraints. $\mathbf{SL}^*_{\mathsf{data}}$ allows us to formalize common data structures such as max-heaps and binary search trees. Despite this expressiveness, satisfiability and entailment of $\mathbf{SL}^*_{\mathsf{data}}$ formulas are decidable in NP and CONP, respectively. This follows from the logic's small-model property: Every model of an $\mathbf{SL}^*_{\mathsf{data}}$ formula can be converted into a small model by removing unnecessary locations. We derived a bound that is linear in the number of variables and thus enables a polynomial encoding into SMT. An implementation, which remains future work, can be based on off-the-shelf SMT solvers. In addition, we plan to extend our approach to doubly-linked and nested data structures, as well as to abduction.

References

1. Bansal, K., Brochenin, R., Lozes, E.: Beyond shapes: lists with ordered data. In: de Alfaro, L. (ed.) FoSSaCS 2009. LNCS, vol. 5504, pp. 425–439. Springer, Heidelberg (2009). https://doi.org/10.1007/978-3-642-00596-1_30

2. Barrett, C.W., Sebastiani, R., Seshia, S.A., Tinelli, C.: Satisfiability modulo theories. In: Handbook of Satisfiability. IOS Press, Amsterdam (2009)

3. Berdine, J., Calcagno, C., O'Hearn, P.W.: A decidable fragment of separation logic. In: Lodaya, K., Mahajan, M. (eds.) FSTTCS 2004. LNCS, vol. 3328, pp. 97–109. Springer, Heidelberg (2004). https://doi.org/10.1007/978-3-540-30538-5_9

4. Berdine, J., Calcagno, C., O'Hearn, P.W.: Symbolic execution with separation logic. In: Yi, K. (ed.) APLAS 2005. LNCS, vol. 3780, pp. 52–68. Springer, Heidelberg (2005). https://doi.org/10.1007/11575467_5

5. Brotherston, J., Fuhs, C., Pérez, J.A.N., Gorogiannis, N.: A decision procedure for satisfiability in separation logic with inductive predicates. In: CSL-LICS (2014). https://doi.org/10.1145/2603088.2603091

6. Calcagno, C., Distefano, D., Dubreil, J., Gabi, D., Hooimeijer, P., Luca, M., O'Hearn, P., Papakonstantinou, I., Purbrick, J., Rodriguez, D.: Moving fast with software verification. In: Havelund, K., Holzmann, G., Joshi, R. (eds.) NFM 2015. LNCS, vol. 9058, pp. 3–11. Springer, Cham (2015). https://doi.org/10.1007/978-3-319-17524-9_1

7. Calcagno, C., Yang, H., O'Hearn, P.W.: Computability and complexity results for a spatial assertion language for data structures. In: Hariharan, R., Vinay, V., Mukund, M. (eds.) FSTTCS 2001. LNCS, vol. 2245, pp. 108–119. Springer, Heidelberg (2001). https://doi.org/10.1007/3-540-45294-X_10

8. Cook, B., Haase, C., Ouaknine, J., Parkinson, M., Worrell, J.: Tractable reasoning in a fragment of separation logic. In: Katoen, J.-P., König, B. (eds.) CONCUR 2011. LNCS, vol. 6901, pp. 235–249. Springer, Heidelberg (2011). https://doi.org/10.1007/978-3-642-23217-6_16

9. Drăgoi, C., Enea, C., Sighireanu, M.: Local shape analysis for overlaid data structures. In: Logozzo, F., Fähndrich, M. (eds.) SAS 2013. LNCS, vol. 7935, pp. 150–171. Springer, Heidelberg (2013). https://doi.org/10.1007/978-3-642-38856-9_10

10. Iosif, R., Rogalewicz, A., Simacek, J.: The tree width of separation logic with recursive definitions. In: Bonacina, M.P. (ed.) CADE 2013. LNCS (LNAI), vol. 7898, pp. 21–38. Springer, Heidelberg (2013). https://doi.org/10.1007/978-3-642-38574-2_2

11. Itzhaky, S., Banerjee, A., Immerman, N., Nanevski, A., Sagiv, M.: Effectively-propositional reasoning about reachability in linked data structures. In: Sharygina, N., Veith, H. (eds.) CAV 2013. LNCS, vol. 8044, pp. 756–772. Springer, Heidelberg (2013). https://doi.org/10.1007/978-3-642-39799-8_53

12. Lahiri, S., Qadeer, S.: Back to the future: revisiting precise program verification using SMT solvers. In: POPL, pp. 171–182 (2008). https://doi.org/10.1145/1328438.1328461

13. Le, Q.L., Tatsuta, M., Sun, J., Chin, W.-N.: A decidable fragment in separation logic with inductive predicates and arithmetic. In: Majumdar, R., Kunčak, V. (eds.) CAV 2017. LNCS, vol. 10427, pp. 495–517. Springer, Cham (2017). https://doi.org/10.1007/978-3-319-63390-9_26

14. Madhusudan, P., Parlato, G., Qiu, X.: Decidable logics combining heap structures and data. In: POPL, pp. 611–622 (2011). https://doi.org/10.1145/1926385.1926455

15. McCarthy, J.: Towards a mathematical science of computation. In: IFIP Congress (1962)

16. de Moura, L., Bjørner, N.: Generalized, efficient array decision procedures. In: FMCAD, pp. 45–52 (2009). https://doi.org/10.1109/FMCAD.2009.5351142

17. de Moura, L., Bjørner, N.: Z3: an efficient SMT solver. In: Ramakrishnan, C.R., Rehof, J. (eds.) TACAS 2008. LNCS, vol. 4963, pp. 337–340. Springer, Heidelberg (2008). https://doi.org/10.1007/978-3-540-78800-3_24

18. Navarro Pérez, J.A., Rybalchenko, A.: Separation logic modulo theories. In: Shan, C. (ed.) APLAS 2013. LNCS, vol. 8301, pp. 90–106. Springer, Cham (2013). https://doi.org/10.1007/978-3-319-03542-0_7

19. Piskac, R., Wies, T., Zufferey, D.: Automating separation logic using SMT. In: Sharygina, N., Veith, H. (eds.) CAV 2013. LNCS, vol. 8044, pp. 773–789. Springer, Heidelberg (2013). https://doi.org/10.1007/978-3-642-39799-8_54

20. Piskac, R., Wies, T., Zufferey, D.: Automating separation logic with trees and data. In: Biere, A., Bloem, R. (eds.) CAV 2014. LNCS, vol. 8559, pp. 711–728. Springer, Cham (2014). https://doi.org/10.1007/978-3-319-08867-9_47

21. Preiner, M., Niemetz, A., Biere, A.: Lemmas on demand for lambdas. In: DIFTS@FMCAD. CEUR Workshop Proceedings, vol. 1130. CEUR-WS.org (2013)

22. Qiu, X., Garg, P., Ştefănescu, A., Madhusudan, P.: Natural proofs for structure, data, and separation. In: PLDI (2013)

23. Reynolds, A., Iosif, R., Serban, C., King, T.: A decision procedure for separation logic in SMT. In: Artho, C., Legay, A., Peled, D. (eds.) ATVA 2016. LNCS, vol. 9938, pp. 244–261. Springer, Cham (2016). https://doi.org/10.1007/978-3-319-46520-3_16

24. Reynolds, J.C.: Separation logic: a logic for shared mutable data structures. In: LICS, pp. 55–74 (2002). https://doi.org/10.1109/LICS.2002.1029817

25. Totla, N., Wies, T.: Complete instantiation-based interpolation. JAR **57**(1), 37–65 (2016). https://doi.org/10.1007/s10817-016-9371-7

MædMax: A Maximal Ordered Completion Tool

Sarah Winkler$^{(\boxtimes)}$ and Georg Moser

University of Innsbruck, Innsbruck, Austria
{sarah.winkler,georg.moser}@uibk.ac.at

Abstract. The equational reasoning tool MædMax implements maximal ordered completion. This new approach extends the maxSMT-based method for standard completion developed by Klein and Hirokawa (2011) to ordered completion and equational theorem proving. MædMax incorporates powerful ground completeness checks and supports certification of its proofs by an Isabelle-based certifier. It also provides an order generation mode which can be used to synthesize term orderings for other tools. Experiments show the potential of our approach.

1 Introduction

Equational reasoning has been one of the main research areas of theorem proving endeavors ever since Knuth and Bendix proposed completion [8]. To remedy the fact that completion may fail if unorientable equations are encountered, *ordered* completion was developed [3]. The ideas of this method have since been pervasive in automated deduction whenever equations are involved. Completion and paramodulation procedures are typically based on a given-clause-algorithm [9,14], which implies that facts are processed one at a time. The reduction order—a notoriously critical parameter—is typically fixed once and for all.

Maximal completion follows a very different approach. The idea is to maintain a single pool of equations. One then tries to orient as many equations as possible, by solving a maxSMT problem. If a terminating rewrite system $\mathcal{R}$ obtained in this way joins all its critical pairs as well as the input equalities, it is complete. Otherwise the critical pairs of $\mathcal{R}$ are added to $\mathcal{E}$ and the procedure is reiterated, as sketched in Fig. 1(a). In this way the proof search is guided by a maxSMT solver and steered towards systems with desirable properties. Maximal completion gave rise to the simple yet efficient and powerful completion tool Maxcomp [7]. It was later shown that the tool's search process can be significantly improved by using more complex objective functions, instead of merely maximizing the number of oriented equations [11].

The tool MædMax is an ordered completion and equational theorem proving tool based on a similar approach. As input it takes a set of equalities $\mathcal{E}_0$ and a

S. Winkler—Supported by FWF project T789.

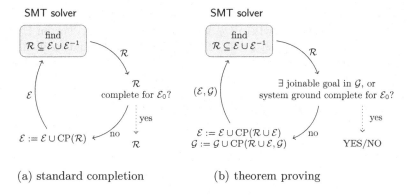

(a) standard completion   (b) theorem proving

Fig. 1. Maximal completion.

goal equality, and tries to decide whether the goal follows from $\mathcal{E}_0$. To this end, it attempts to derive the goal from $\mathcal{E}_0$ or generate an equivalent ground-complete system. It is known that such a system can in particular be used to disprove the goal [1].

Figure 1(b) visualizes our approach (for the common case of a goal without variables). We maintain a set of equalities $\mathcal{E}$ and a set of goals $\mathcal{G}$, which is considered a disjunction. By orienting equations in $\mathcal{E}$ we find a terminating rewrite system $\mathcal{R}$. If a goal in $\mathcal{G}$ can be joined using $\mathcal{R}$, or the system is ground complete then the goal can be decided. Otherwise, critical pairs are added to both $\mathcal{E}$ and $\mathcal{G}$ and the procedure is repeated. Thus in contrast to the given clause algorithm many equations are processed at once, and the proof search is steered towards systems that have desirable properties.

Our experiments show that MædMax is particularly suited to prove (ground) completeness and satisfiability, due to sophisticated joinability criteria. If a proof is found then MædMax can output an equational proof that is checkable by the Isabelle-based verifier CeTA, thus offering a high degree of reliability. The tool also provides an order generation mode, where the reduction order deemed most suitable according to the optimized criteria is displayed. Finally, we illustrate practical relevance by means of examples in recent applications to data integration [12].

The remainder of this paper is organized as follows. In Sect. 2 we recall some relevant concepts and notations. Ordered maximal completion is presented in Sect. 3. Implementation details are highlighted in Sect. 4. In Sect. 5 we report on experimental results and conclude.

2   Preliminaries

In the sequel standard notation from term rewriting is used [2]. We consider the set of terms $\mathcal{T}(\mathcal{F}, \mathcal{V})$ over a signature $\mathcal{F}$ and a set of variables $\mathcal{V}$, while $\mathcal{T}(\mathcal{F})$ denotes the set of all ground terms. An equational system (ES) $\mathcal{E}$ is a set of equations $\ell \approx r$ over $\mathcal{T}(\mathcal{F}, \mathcal{V})$, and a term rewrite system (TRS) $\mathcal{R}$ is a set of equations denoted as $\ell \to r$, where $\ell \notin \mathcal{V}$ and $\text{Var}(r) \subseteq \text{Var}(\ell)$. For an ES $\mathcal{E}$ we write $\mathcal{E}^{\pm}$ to denote $\mathcal{E} \cup \{t \approx s \mid s \approx t \in \mathcal{E}\}$. A TRS $\mathcal{R}$ is terminating if there are no infinite rewrite sequences $t_0 \to_{\mathcal{R}} t_1 \to_{\mathcal{R}} t_2 \to_{\mathcal{R}} \ldots$, and (ground) confluent if $s \;{}_{\mathcal{R}}^{*}{\leftarrow} \cdot \to_{\mathcal{R}}^{*} t$ implies $s \to_{\mathcal{R}}^{*} \cdot {}_{\mathcal{R}}^{*}{\leftarrow} t$ for all (ground) terms s and t. A TRS which is terminating and (ground) confluent is (ground) complete.

One common method to establish termination of a TRS $\mathcal{R}$ is to find a reduction order $>$ which is compatible with $\mathcal{R}$. A reduction order $>$ is ground total if $s > t$, $t > s$, or $s = t$ holds for all ground terms s and t. It is known that LPO and KBO are reduction orders enjoying this property whenever they are based on a total precedence, and polynomial interpretations can be extended to such an order. For a reduction order $>$ and an ES $\mathcal{E}$, the TRS $\mathcal{E}^{>}$ consists of all rules $s\sigma \to t\sigma$ such that $s \approx t \in \mathcal{E}^{\pm}$ and $s\sigma > t\sigma$ [3]. A TRS $\mathcal{R}$ is moreover ground complete for an ES $\mathcal{E}_0$ if $\mathcal{R}$ is ground complete and the relations $\leftrightarrow_{\mathcal{R}}^{*}$ and $\leftrightarrow_{\mathcal{E}_0}^{*}$ coincide when restricted to ground terms. Given a terminating TRS $\mathcal{R}$ and a term t, we write $t\downarrow_{\mathcal{R}}$ to denote some normal form of t. For an ES $\mathcal{E}$, we write $\mathcal{E}\downarrow_{\mathcal{R}}$ for the set of all equations $s\downarrow_{\mathcal{R}} \approx t\downarrow_{\mathcal{R}}$ such that $s \approx t \in \mathcal{E}$ and $s\downarrow_{\mathcal{R}} \neq t\downarrow_{\mathcal{R}}$. An equation $s \approx t$ is ground joinable in $\mathcal{R}$ if $s\sigma \downarrow_{\mathcal{R}} t\sigma$ for all grounding substitutions σ, where $\downarrow_{\mathcal{R}}$ abbreviates $\to_{\mathcal{R}}^{*} \cdot {}_{\mathcal{R}}^{*}{\leftarrow}$.

We use the following notion of *extended critical pairs* [3]: Given a reduction order $>$ and $\ell_1 \approx r_1$ and $\ell_2 \approx r_2$ in $\mathcal{E}^{\pm}$, the equation $\ell_2\sigma[r_1\sigma]_p \approx r_2\sigma$ is an extended critical pair if p is a function symbol position in ℓ_2, the terms $\ell_2|_p$ and ℓ_1 are unifiable with most general unifier σ, and neither $r_1\sigma > \ell_1\sigma$ nor $r_2\sigma > \ell_2\sigma$ hold. The set of extended critical pairs of an ES $\mathcal{E}$ with respect to $>$ is denoted by $\text{CP}_{>}(\mathcal{E})$.

3   Maximal Ordered Completion

We now formalize the approach of maximal ordered completion and theorem proving sketched in Fig. 1(b).

Let $\mathfrak{R}$ be a function mapping an ES $\mathcal{E}$ to a set of terminating TRSs such that for all $\mathcal{R} \in \mathfrak{R}(\mathcal{E})$ we have (1) $\mathcal{R} \subseteq \mathcal{E}^{\pm}$ and (2) there is a ground total reduction order $>$ extending $\to_{\mathcal{R}}$. Moreover, let S be a function from ESs to ESs such that $S(\mathcal{E}) \subseteq \leftrightarrow_{\mathcal{E}}^{*}$ for every ES $\mathcal{E}$. We consider define maximal ordered completion without goals. Our procedure is defined via the following relation φ which maps an ES $\mathcal{E}$ to a tuple $(\mathcal{R}, \mathcal{E}', >)$ consisting of a TRS $\mathcal{R}$, an ES $\mathcal{E}'$, and a reduction order $>$.

Definition 1. *Given a set of input equalities $\mathcal{E}_0$ and an ES $\mathcal{E}$, let*

$$\varphi(\mathcal{E}) = \begin{cases} (\mathcal{R}, \mathcal{E}\downarrow_{\mathcal{R}}, >) & \text{if } \mathcal{R} \cup (\mathcal{E}\downarrow_{\mathcal{R}})^{>} \text{ is ground complete for } \mathcal{E}_0 \\ & \text{for some } \mathcal{R} \in \mathfrak{R}(\mathcal{E}), \text{ and} \\ \varphi(\mathcal{E} \cup S(\mathcal{E})) & \text{otherwise.} \end{cases}$$

The idea is to recursively apply Definition 1 to a set of initial equations $\mathcal{E}_0$. Note that in general φ may be neither defined nor unique. In MædMax the set $S(\mathcal{E})$ is chosen such that $S(\mathcal{E}) \subseteq \bigcup_{\mathcal{R} \in \mathfrak{R}(\mathcal{E})} \text{CP}_{>}(\mathcal{R} \cup \mathcal{E}\downarrow_{\mathcal{R}})\downarrow_{\mathcal{R}}$.

Example 1. Consider the following ES $\mathcal{E}_0$ axiomatizing a Boolean ring, where multiplication is denoted by concatenation.

(1) $(x + y) + z \approx x + (y + z)$ (2) $x + y \approx y + x$ (3) $0 + x \approx x$

(4) $x(y + z) \approx xy + xz$ (5) $(xy)z \approx x(yz)$ (6) $xy \approx yx$

(7) $(x + y)z \approx xz + yz$ (8) $xx \approx x$ (9) $x + x \approx 0$

(10) $1x \approx x$

Let $\mathcal{R}_1$ be the TRS $\{(1), (3), (\overline{4}), (5), (\overline{7}), (8), (9), (10)\}$ obtained by orienting distributivity from right to left and all other equations (except for commutativity) from left to right, and $\mathfrak{R}(\mathcal{E}_0)$ the singleton set containing $\mathcal{R}_1$. This choice orients a maximal number of equations. Now the set $S(\mathcal{E}_0)$ may consist of the following extended critical pairs of rules among $\mathcal{R}_1$ and the unorientable commutativity equations:

(11) $x + (y + z) \approx y + (x + z)$ (12) $x(yz) \approx y(xz)$ (13) $x + 0 \approx x$

(14) $y + (x + y) \approx x$ (15) $x(yx) \approx xy$ (16) $x1 \approx x$

(17) $y + (y + x) \approx x$ (18) $x(xy) \approx xy$ (19) $0x \approx 0$

Note that $\mathcal{R}_1$-joinable critical pairs such as $x + (x + 0) \approx 0$ or $x0 \approx y0$ are not included. We have $\varphi(\mathcal{E}_0) = \varphi(\mathcal{E}_1)$ for $\mathcal{E}_1 = \mathcal{E}_0 \cup S(\mathcal{E}_0)$. Now $\mathfrak{R}(\mathcal{E}_1)$ may contain $\mathcal{R}_2 = \{(1), (3), (4), (5), (7), \ldots, (10), (13), \ldots, (19)\}$. This TRS is LPO-terminating, so there is a ground-total reduction order $>$ that contains $\rightarrow_{\mathcal{R}_2}$. We have $\mathcal{E}_1 \downarrow_{\mathcal{R}_2} = \{(2), (6), (11), (12)\}$, and it can be shown that for $\mathcal{E} = \mathcal{E}_1 \downarrow_{\mathcal{R}_2}$ the system $\mathcal{R}_2 \cup \mathcal{E}^>$ is ground complete. Despite its simplicity, neither WM [1] nor E [14] or Vampire [9] can show satisfiability of this example (considering $\mathcal{E}_0$ as a set of axioms in first-order logic with equality, in the case of the latter).

We next extend our approach to theorem proving, akin to Fig. 1(b). Let S_G be a binary function on ESs such that $S_G(\mathcal{G}, \mathcal{E}) \subseteq \leftrightarrow^*_{\mathcal{E} \cup \mathcal{G}} \setminus \leftrightarrow^*_{\mathcal{E}}$ for all ESs $\mathcal{E}$ and $\mathcal{G}$. In our implementation, $S_G(\mathcal{G}, \mathcal{E})$ contains extended critical pairs between an equation in $\mathcal{G}$ and an equation in $\mathcal{E}$. The following relation ψ maps a pair of ESs $\mathcal{E}$ and $\mathcal{G}$ to YES or NO.

Definition 2. *Given a set of input equalities $\mathcal{E}_0$, an initial ground goal $s_0 \approx t_0$ and ESs $\mathcal{E}$ and $\mathcal{G}$, let*

$$\psi(\mathcal{E}, \mathcal{G}) = \begin{cases} \text{YES} & \text{if } s \downarrow_{\mathcal{R} \cup \mathcal{E}^>} t \text{ for some } s \approx t \in \mathcal{G} \text{ and } \mathcal{R} \in \mathfrak{R}(\mathcal{E}), \\ \text{NO} & \text{if } \mathcal{R} \cup \mathcal{E} \downarrow_{\mathcal{R}}^> \text{ is ground complete for } \mathcal{E}_0 \\ & \text{but } s_0 \not\downarrow_{\mathcal{R} \cup \mathcal{E}^>} t_0, \text{ for some } \mathcal{R} \in \mathfrak{R}(\mathcal{E}), \text{ and} \\ \psi(\mathcal{E}', \mathcal{G}') & \text{for } \mathcal{G}' = S_G(\mathcal{G}, \mathcal{R} \cup \mathcal{E}) \text{ and } \mathcal{E}' = S(\mathcal{E}). \end{cases}$$

For a set of input equations $\mathcal{E}_0$ and an initial goal $s_0 \approx t_0$, the procedure is started with the initial call $\psi(\mathcal{E}_0, \{s_0 \approx t_0\})$. Note that the parameter $\mathcal{G}$ of ψ denotes a disjunction of goals, not a conjunction. Due to the declarative nature of the completion and theorem proving procedures described by Definitions 1 and 2 the following correctness result is straightforward.

Theorem 1. *Let $\mathcal{E}_0$ be an ES and $s_0 \approx t_0$ be a ground goal.*

1. *If $\varphi(\mathcal{E}_0) = (\mathcal{R}, \mathcal{E}, >)$ then $\mathcal{R} \cup \mathcal{E}^>$ is ground complete for $\mathcal{E}_0$.*
2. *If $\psi(\mathcal{E}_0, \{s_0 \approx t_0\})$ is defined then $\psi(\mathcal{E}_0, \{s_0 \approx t_0\}) = \text{YES}$ if and only if $s_0 \leftrightarrow^*_{\mathcal{E}_0} t_0$.*   □

4   Implementation

MædMax is available as a command-line tool and via a web interface.[1] It is implemented in OCaml and accepts input problems in the TPTP [17] as well as the trs format.[2] We describe how the three main phases of our approach are implemented (see Definitions 1 and 2): (1) finding the terminating TRSs $\mathfrak{R}(\mathcal{E})$, (2) success checks, and (3) selection of new equations and goals, i.e., computation of $S(\mathcal{E})$ and $S_G(\mathcal{G}, \mathcal{E})$. Also some further particular features of the tool get highlighted. Many settings can be controlled via a command line option, we refer to the website for details. In the default *auto* mode the settings are determined heuristically.

Finding Rewrite Systems. In phase (1), MædMax computes the set of TRSs $\mathfrak{R}(\mathcal{E})$ for a given ES $\mathcal{E}$ by solving an optimization problem whose objective function can be controlled via a strategy. Assuming we want to find a TRS $\mathcal{R} \in \mathfrak{R}(\mathcal{E})$ the search may involve the following criteria, as well as their (possibly weighted) sums:

(a) maximize the oriented equations in $\mathcal{E}$ (i.e., the size of $\mathcal{R}$),
(b) maximize the equations in $\mathcal{E}$ that are reducible by $\mathcal{R}$,
(c) minimize critical pairs among rules in $\mathcal{R}$, or
(d) maximize reducible critical pairs among rules in $\mathcal{R}$.

Maxcomp relied on criterion (a), and later a combination of (a), (b), and (c) was found most suitable for standard completion [11]. Our tool uses by default criterion (b), which was most effective in experiments, but switches to (a) in cases where the proof search is considered stuck.

In practice the optimization problem is solved by encoding the optimization constraints in SMT and solving a maxSMT problem using Yices [5]. In order to orient equations, MædMax uses (SMT encodings of) LPO, KBO, and linear polynomial interpretations, as well as a dynamic choice among these at runtime depending on which satisfies the criteria best.

Success Checks. In phase (2), MædMax succeeds if (a) a goal can be joined or (b) a ground complete system was found. In the latter case, a ground goal is decided by checking syntactic equality of the two term's normal forms. For non-ground goals basic and normalizing narrowing is supported. MædMax establishes

---

[1] http://cl-informatik.uibk.ac.at/software/maedmax/.
[2] https://www.lri.fr/~marche/tpdb/format.html.

ground confluence by verifying ground joinability of extended critical pairs. However, depending on the signature this may be nontrivial: while the property is decidable for some orders when enriching the signature with infinitely many constants, it is undecidable for a finite given signature [4]. Our tool supports the criteria of [10, 18] to that end, and both kinds of signatures.

Selection. In the selection phase (3), MædMax picks new equations and goals, given the current set of equations $\mathcal{E}$ and a TRS $\mathcal{R} \in \mathfrak{R}(\mathcal{E})$. To that end, it computes the set $S(\mathcal{E})$ containing n new equations and $S_G(\mathcal{G}, \mathcal{E})$ containing m new goals, where by default $n = 12$ and $m = 2$. In the *auto* mode the number n gets adjusted depending on the current state to avoid dealing with too many equations. The selection heuristic prefers small equations and old, but not yet reducible equations.

Order Generation Mode. MædMax can also be run for a couple of iterations and output the term ordering that is deemed best according to the criteria mentioned above (maximal number of oriented equations, etc.).

Certification. MædMax can output proofs in an XML format (CPF) that are checkable by the Isabelle-based certifier CeTA. For the case where a goal is proven (answer YES), certification follows the approach of [16]: The XML certificate gives a stepwise derivation of the goal from the input equations [16] which is checked by CeTA. Due to recent work [6] also some NO answers are checkable, though ground joinability support in CeTA is limited since the criterion from [18] is not supported.

Optimizations. Fingerprint indexing [13] is used to speed up both rewriting and overlap computation. In order to deal with associative and commutative symbols, the approach of [1] is incorporated. In the *auto* mode the tool also triggers restarts: if a current state is considered stuck, the procedure is restarted but where the input problem is extended by a number of small lemmas found so far.

5   Evaluation

All details of the following experiments can be obtained from the website. The tests were run single threaded on an Intel® Core™ i7-5930K CPU at 3.50 GHz with 12 cores, with varying timeouts as indicated below.

Table 1 compares MædMax with Waldmeister (WM) [1], E [14], and Vampire [9] on different problem sets. The first two lines refer to satisfiable/unsatisfiable problems in TPTP's unit equality division [17]. The third row refers to the 23 problems for which a ground complete system is given in [10] (which are hence all satisfiable, in the TPTP terminology). The fourth row refers to 731 problems generated by the conditional confluence tool ConCon to check infeasibility of critical pairs [15, Sect. 7.5], which are partially satisfiable and partially unsatisfiable. The last row refers to 139 satisfiable problems for standard completion [11].

Table 1. Experimental results.

	MædMax	WM	E	Vampire
TPTP UEQ SAT (600 s)	14	9	12	11
TPTP UEQ UNSAT (600 s)	621	832	692	724
Examples in [10] (60 s)	7	4	4	3
ConCon examples (60 s)	704	657	705	704
KB examples (60 s)	91	45	84	48

For TPTP problems the timeout was set to 600 s; for the latter two data sets 60s were chosen since larger timeouts did not induce any changes. Table 1 shows that MædMax outperforms other tools on satisfiable examples.[3] On unsatisfiable examples MædMax does not prevail, but all CPF proofs of unsatisfiability produced by MædMax have been certified by CeTA, they can be found on-line. For 14 problems the proof output for CeTA cannot be accomplished within a timeout of 1200 s, though.

On average, MædMax spends most of its running time on finding the TRSs $\mathfrak{R}(\mathcal{E})$ (20%), critical pair computation (33%), and overlap computation (33%). Only about 1% of the time is actually spent in the SMT solver.

We tested the order generation mode of MædMax with E since it accepts precedence and weight parameters for LPO and KBO as command line options. To that end MædMax was run for 10 s, and the devised reduction order was passed to E. In this way, E solved 605 unsatisfiable and 10 satisfiable TPTP UEQ problems. Though the number of solved examples is lowered wrt. Table 1 the average time is reduced, and different problems could be solved.

We conclude with a practical application example. The tool AQL[4] performs functorial data integration by means of a category-theoretic approach [12], taking advantage of (ground) completion. The following example problem was communicated by the authors.

Example 2. We consider database tables ylsAL and ylsAW relating amphibians to land and water animals, respectively. They are described by 400 ground equations over symbols ylsAL, ylsALL, ylsAW, ylsAWW and 449 constants of the form a_i, w_i, l_i representing data entities. We give six example equations to convey an impression:

$$ylsAW(a_1) \approx w_{29} \qquad ylsAW(a_{78}) \approx w_{16} \qquad ylsAW(a_{61}) \approx w_{30}$$
$$ylsAL(a_{37}) \approx l_{80} \qquad ylsAL(a_{84}) \approx l_6 \qquad ylsAL(a_{29}) \approx l_{47}$$

In addition, the equation $ylsALL(ylsAL(x)) \approx ylsAWW(ylsAW(x))$ describes a mapping to a second database schema. A ground complete presentation of the

---

[3] The Maxcomp version presented in [11] solves 91 KB examples within 60 s, too, but 98 problems in 600 s. For the other tools the numbers hardly change with a larger timeout. Maxcomp is not applicable to the other problem sets though.

[4] http://categoricaldata.net/aql.html.

entire system thus constitutes a representation of the data, translated to the second schema. MædMax discovers a complete presentation of 889 rules in less then 20 s, while AQL's internal completion prover fails. MædMax' automatic mode switches to linear polynomials for such systems with many symbols, which turned out to be faster than LPO or KBO in this situation.

For Example 2 even a complete system can be found. In general ground completeness is achieved by MædMax for those problems encountered by AQL, as required. Further details can be found on the website.

6   Conclusion

We presented the tool MædMax implementing maximal ordered completion, a novel approach to ordered completion and equational theorem proving. Our experiments show that this approach outperforms other tools on satisfiable problems. For unsatisfiable problems MædMax can produce output verifiable by the trusted proof checker CeTA, thus offering a very high degree of reliability. We believe that MædMax is particularly suited to problems with large signatures like Example 2, due to its ability to search for a reduction order which induces a small number of critical pairs and hence fewer steps to completion.

Acknowledgements. The authors thank Ryan Wisnesky for sharing AQL problems, and the anonymous referees for their helpful comments.

References

1. Avenhaus, J., Hillenbrand, T., Löchner, B.: On using ground joinable equations in equational theorem proving. JSC **36**(1–2), 217–233 (2003). https://doi.org/10.1016/S0747-7171(03)00024-5
2. Baader, F., Nipkow, T.: Term Rewriting and All That. Cambridge University Press, Cambridge (1998). https://doi.org/10.1017/CBO9781139172752
3. Bachmair, L., Dershowitz, N., Plaisted, D.A.: Completion without failure. In: Aït Kaci, H., Nivat, M. (eds.) Resolution of Equations in Algebraic Structures, Rewriting Techniques of Progress in Theoretical Computer Science, vol. 2, pp. 1–30. Academic Press, Cambridge (1989)
4. Comon, H., Narendran, P., Nieuwenhuis, R., Rusinowitch, M.: Deciding the confluence of ordered term rewrite systems. ACM TOCL **4**(1), 33–55 (2003). https://doi.org/10.1145/601775.601777
5. Dutertre, B., de Moura, L.: A fast linear-arithmetic solver for DPLL(T). In: Ball, T., Jones, R.B. (eds.) CAV 2006. LNCS, vol. 4144, pp. 81–94. Springer, Heidelberg (2006). https://doi.org/10.1007/11817963_11
6. Hirokawa, N., Middeldorp, A., Sternagel, C., Winkler, S.: Infinite runs in abstract completion. In: Proceedings of the 2nd FSCD. LIPIcs, vol. 84, pp. 19:1–19:16 (2017). https://doi.org/10.4230/LIPIcs.FSCD.2017.19
7. Klein, D., Hirokawa, N.: Maximal completion. In: Proceedings of the 22nd RTA. LIPIcs, vol. 10, pp. 71–80 (2011). https://doi.org/10.4230/LIPIcs.RTA.2011.71

8. Knuth, D.E., Bendix, P.: Simple word problems in universal algebras. In: Leech, J. (ed.) Computational Problems in Abstract Algebra, pp. 263–297. Pergamon Press, Oxford (1970). https://doi.org/10.1016/B978-0-08-012975-4

9. Kovács, L., Voronkov, A.: First-order theorem proving and VAMPIRE. In: Sharygina, N., Veith, H. (eds.) CAV 2013. LNCS, vol. 8044, pp. 1–35. Springer, Heidelberg (2013). https://doi.org/10.1007/978-3-642-39799-8_1

10. Martin, U., Nipkow, T.: Ordered rewriting and confluence. In: Stickel, M.E. (ed.) CADE 1990. LNCS, vol. 449, pp. 366–380. Springer, Heidelberg (1990). https://doi.org/10.1007/3-540-52885-7_100

11. Sato, H., Winkler, S.: Encoding dependency pair techniques and control strategies for maximal completion. In: Felty, A.P., Middeldorp, A. (eds.) CADE 2015. LNCS, vol. 9195, pp. 152–162. Springer, Cham (2015). https://doi.org/10.1007/978-3-319-21401-6_10

12. Schultz, P., Wisnesky, R.: Algebraic data integration. JFP **27**(e24), 51 (2017). https://doi.org/10.1017/S0956796817000168

13. Schulz, S.: Fingerprint indexing for paramodulation and rewriting. In: Gramlich, B., Miller, D., Sattler, U. (eds.) IJCAR 2012. LNCS, vol. 7364, pp. 477–483. Springer, Heidelberg (2012). https://doi.org/10.1007/978-3-642-31365-3_37

14. Schulz, S.: System description: E 1.8. In: McMillan, K., Middeldorp, A., Voronkov, A. (eds.) LPAR 2013. LNCS, vol. 8312, pp. 735–743. Springer, Heidelberg (2013). https://doi.org/10.1007/978-3-642-45221-5_49

15. Sternagel, T.: Reliable confluence analysis of conditional term rewrite systems. Ph.D. thesis. University of Innsbruck (2017)

16. Sternagel, T., Winkler, S., Zankl, H.: Recording completion for certificates in equational reasoning. In: Proceedings of CPP 2015, pp. 41–47 (2015). https://doi.org/10.1145/2676724.2693171

17. Sutcliffe, G.: The TPTP problem library and associated infrastructure: the FOF and CNF parts. JAR **43**(4), 337–362 (2009). https://doi.org/10.1007/s10817-009-9143-8

18. Winkler, S.: A ground joinability criterion for ordered completion. In: Proceedings of 6th IWC, pp. 45–49 (2017)

From Syntactic Proofs to Combinatorial Proofs

Matteo Acclavio and Lutz Straßburger[(✉)]

Inria Saclay & LIX, Ecole Polytechnique, Palaiseau, France
http://www.lix.polytechnique.fr/Labo/Matteo.Acclavio/,
http://www.lix.polytechnique.fr/Labo/Lutz.Strassburger/

Abstract. In this paper we investigate Hughes' combinatorial proofs as a notion of proof identity for classical logic. We show for various syntactic formalisms including sequent calculus, analytic tableaux, and resolution, how they can be translated into combinatorial proofs, and which notion of identity they enforce. This allows the comparison of proofs that are given in different formalisms.

1 Introduction

Proof theory plays an important role in many areas of computer science. However, unlike many other mathematical fields, it is not able to identify its objects. We do not have a clear understanding of when two proofs are the same. The standard proof theoretical answer to this question is *normalization*: two proofs are the same, if they have the same normal form. This certainly makes perfect sense from the viewpoint of functional programming and the Curry-Howard-correspondence, where proofs are programs and the proof normalization is the execution of the program. However, from the viewpoint of logic programming and proof search, this only makes little sense, since all considered proofs are already in normal form.

An alternative approach to the question of proof identity is based on rule permutations. Two proofs are considered the same if they can be transformed into each other by a series of simple rule permutation steps. The fundamental problem with this approach is that both proofs have to be presented in the same proof system. In fact, one can say that proof theory, in its current form, *is not the theory of proofs but the theory of proof systems.* The question of comparing two proofs that are given in two different proof systems (for example, analytic tableaux and resolution) does not even make sense. And most of the important theorems of proof theory, like soundness, completeness, cut admissibility, proof complexity, or focusing, are not about proofs but about proof systems.

Combinatorial proofs [10,11] have been introduced by Hughes to address this problem. They are graphical presentations of proofs, independent from the syntactic restrictions of proof formalisms. Nonetheless, combinatorial proofs form a proof system in the sense of Cook and Reckhow [3], the correctness of a combinatorial proof can be checked in polynomial time in the size of the proof. However, the precise relation between combinatorial proofs and syntactic proofs has so

© Springer International Publishing AG, part of Springer Nature 2018
D. Galmiche et al. (Eds.): IJCAR 2018, LNAI 10900, pp. 481–497, 2018.
https://doi.org/10.1007/978-3-319-94205-6_32

far been been discussed only on a superficial level. In [11], Hughes shows the relation between combinatorial proofs and a nonstandard version of the sequent calculus LK, and in [18], the relation to the deep inference system SKS is shown.

In this paper we explore the relation between combinatorial proofs and syntactic proofs in various formalisms, in particular, we look at a one-sided variant of LK [8] which has an explicit contraction and weakening rule and in which the conjunction rule is multiplicative, and at G3p [19] in which the conjunction rule is additive and there are no contraction and weakening rules. Then we also look at analytic tableaux [15], and resolution. We will show how a syntactic proof in each of these formalisms is translated into a combinatorial proof, and when a combinatorial proof can be translated back. Note that this is not always possible. Even though all systems are semantically complete, i.e., can prove all theorems, they do not see all proofs.

We will also define for analytic tableaux and for resolution a syntactic equivalence on proofs, and then show that this equivalence coincides with the one imposed by combinatorial proofs. This justifies the use of combinatorial proofs for proof identity: two proofs are the same if they are mapped to the same combinatorial proof.[1] To our knowledge, this is the first proposal for a notion of proof identity that allows us to compare syntactic proofs in different formalisms.

The paper is organized as follows: We first give some preliminaries on combinatorial proofs in Sect. 2. Then we discuss the relation between combinatorial proofs and sequent calculus in Sect. 3, and finally, in Sects. 4 and 5, we investigate the translation from tableaux and resolution into combinatorial proofs.

2   Preliminaries on Combinatorial Proofs

For simplicity, we consider formulas (denoted by capital Latin letters $A, B, C, \ldots$) in negation normal form[2], generated from a countable set $\mathcal{V} = \{a, b, c, \ldots\}$ of propositional variables by the following grammar: $A, B ::= a \mid \bar{a} \mid A \wedge B \mid A \vee B$, where $\bar{a}$ is the negation of a. The negation can then be defined for all formulas using the De Morgan laws $\bar{\bar{A}} = A$, and $\overline{A \wedge B} = \bar{A} \vee \bar{B}$ and $\overline{A \vee B} = \bar{A} \wedge \bar{B}$. An *atom* is a variable or its negation. We use $\mathcal{A}$ to denote the set of all atoms. A *sequent* Γ is a multiset of formulas, written as a list separated by comma: $\Gamma = A_1, A_2, \ldots, A_n$. We write $\bar{\Gamma}$ to denote the sequent $\bar{A}_1, \bar{A}_2, \ldots, \bar{A}_n$. We define the *size* of a sequent Γ, denoted by $|\Gamma|$, to be the number of atom occurrences in it. We write $\wedge\Gamma$ (resp. $\vee\Gamma$) for the conjunction (res. disjunction) of the formulas in Γ and $F^{\vee k} := \underbrace{F \vee \cdots \vee F}_{k}$ ($F^{\wedge k} := \underbrace{F \wedge \cdots \wedge F}_{k}$) the disjunction (conjunction) of k copies of F.

A *graph* $\mathfrak{G} = \langle V_{\mathfrak{G}}, E_{\mathfrak{G}} \rangle$ consists of a set of *vertices* $V_{\mathfrak{G}}$ and a set of *edges* $E_{\mathfrak{G}}$ which are two-element subsets of $V_{\mathfrak{G}}$. We omit the index $\mathfrak{G}$ when it is clear from

---

[1] However, this paper does not speak about normalization of combinatorial proofs. For this topic, the reader is referred to [11,17,18].

[2] Note that this is only a cosmetic limitation. The theory of combinatorial proofs can easily be extended to the full language including implication and general negation.

context. For $v, w \in V$ we write vw for $\{v, w\}$. For two graphs $\mathfrak{G} = \langle V, E \rangle$ and $\mathfrak{G}' = \langle V', E' \rangle$, we define the operations *union* $\mathfrak{G} \vee \mathfrak{G}' = \langle V \cup V', E \cup E' \rangle$ and *join* $\mathfrak{G} \wedge \mathfrak{G}' = \langle V \cup V', E \cup E' \cup \{vv' \mid v \in V, v' \in V'\}\rangle$. For a set L, a graph $\mathfrak{G}$ is *L-labeled* if every vertex of $\mathfrak{G}$ is associated with an element L, called its *label*.

If we associate to each atom a a single vertex labeled with a then every formula A uniquely determines an ($\mathcal{A}$-labeled) graph $\mathfrak{G}(A)$ that is constructed via the operations $\wedge$ and $\vee$. We define $\mathfrak{G}(\Gamma) = \mathfrak{G}(\vee\Gamma)$. It is easy to see that for two formulas A and B, we have $\mathfrak{G}(A) = \mathfrak{G}(B)$ iff A and B are equivalent modulo associativity and commutativity of $\wedge$ and $\vee$.

Example 2.1. Let $A = (a \wedge (b \vee \bar{c})) \vee (c \wedge \bar{d})$ then $\bar{A} = (\bar{a} \vee (\bar{b} \wedge c)) \wedge (\bar{c} \vee d)$. The two graphs $\mathfrak{G}(A)$ and $\mathfrak{G}(\bar{A}) = \overline{\mathfrak{G}(A)}$ are:

A graph $\langle V, E \rangle$ is called a *cograph* if V does not contain four distinct vertices u, v, w, z with $uv, vw, wz \in E$ and $vz, zu, uw \notin E$. We have the following well-known proposition, which can already be found in [6].

Proposition 2.2. *A graph is equal to* $\mathfrak{G}(A)$ *for some A iff it is a cograph.*

The following definitions are due to Retoré [14]. An *R&B-graph* $\mathfrak{G} = \langle V, R, B \rangle$ is a triple such that $\langle V, R \rangle$ and $\langle V, B \rangle$ are graphs and such that B is a perfect matching on V, i.e., no two edges in B are adjacent and every vertex $v \in V_{\mathfrak{G}}$ is incident to an edge in B. We write $\mathfrak{G}^{\downarrow}$ for $\langle V, R \rangle$. An *R&B-cograph* is an R&B-graph $\mathfrak{G} = \langle V, R, B \rangle$ where $\mathfrak{G}^{\downarrow} = \langle V, R \rangle$ is a cograph.

A *cordless æ-cycle* in $\mathfrak{G} = \langle V, R, B \rangle$ is a set $\{v_1, \ldots, v_{2n}\} \subseteq V$ of vertices such that $v_{2n}v_1, v_2v_3, \ldots, v_{2n-2}v_{2n-1} \in B$ and $v_iv_j \in R$ if and only if $i = 2k+1$ an $j = 2k+2$ for some $0 \leq k \leq n-1$. We say an R&B-graph is *æ-acyclic* if it has no cordless æ-cycle. Following [14] we will draw B-edges in blue/bold, and R-edges in red/regular. Below are four examples:

The first one is not an R&B-cograph, the other three are. The second one has a chordless æ-cycle, and the last two are æ-acyclic.

A *homomorphism* $f \colon \mathfrak{G} \to \mathfrak{G}'$ is a function from $V_{\mathfrak{G}}$ to $V_{\mathfrak{G}'}$ such that $vw \in E_{\mathfrak{G}}$ implies $f(v)f(w) \in E_{\mathfrak{G}'}$. A *skew fibration*, denoted as $f \colon \mathfrak{G} \rightarrowtail \mathfrak{G}'$, is a graph homomorphism such that for every $v \in V_{\mathfrak{G}}$ and $w' \in V_{\mathfrak{G}'}$ with $f(v)w' \in E'_{\mathfrak{G}}$ there is a $w \in V_{\mathfrak{G}}$ with $vw \in E_{\mathfrak{G}}$ and $f(w)w' \notin E'_{\mathfrak{G}}$.

Let $\mathfrak{C} = \langle V, R, B \rangle$ be an R&B-graph and $f \colon \mathfrak{C}^{\downarrow} \to \mathfrak{G}$ be a homomorphism and let $\mathfrak{G}$ be $\mathcal{A}$-labeled (where $\mathcal{A}$ is the set of atoms). We say f is *axiom preserving* iff $xy \in B$ implies that the labels of $f(w)$ and $f(v)$ are dual to each other. We are now ready to give the definition of a combinatorial proof.

Definition 2.3. A *combinatorial proof* of a sequent Γ consists of a non-empty æ-acyclic R&B-cograph $\mathfrak{C}$ and an axiom preserving skew fibration $f \colon \mathfrak{C}^{\downarrow} \rightarrowtail \mathfrak{G}(\Gamma)$.

In [10], Hughes has shown that combinatorial proofs form a proof system in the sense of Cook and Reckhow [3], i.e., correctness can be checked in polynomial time, that is, given a formula A and a R&B-graph $\langle V, R, B \rangle$ and a map $f \colon \langle V, R \rangle \to \mathfrak{G}(A)$, it can be checked in polynomial time in the size of the input whether (1) $\langle V, R \rangle$ is a cograph, (2) $\langle V, R, B \rangle$ is æ-acyclic, and (3) f is axiom preserving and a skew fibration.

We follow the notational convention of [17,18] and write $\phi \colon \Gamma \vdash \Delta$, to denote a combinatorial proof for the sequent $\bar{\Gamma}, \Delta$, and say that Γ is its *premise* and Δ its *conclusion*. We write $\phi \colon \circ \vdash \Delta$ (resp. $\phi \colon \Gamma \vdash \circ$) if Γ (resp. Δ) is empty.[3] Note that if $\phi \colon \Gamma \vdash \Sigma, \Delta$ is a combinatorial proof then so is $\phi = \phi' \colon \bar{\Sigma}, \Gamma \vdash \Delta$.

Following [18], we draw combinatorial proofs as follow: let $\phi \colon \Gamma \vdash \Delta$ be a combinatorial proof with skew fibration $f \colon \overline{\mathfrak{C}_\Gamma} \vee \mathfrak{C}_\Delta \rightarrowtail \mathfrak{G}(\Gamma) \vee \mathfrak{G}(\Delta)$, and let $F(\overline{\mathfrak{C}_\Gamma})$ and $F(\mathfrak{C}_\Delta)$ be the formula trees corresponding to the cographs $\overline{\mathfrak{C}_\Gamma}$ and $\mathfrak{C}_\Delta$ respectively. We write Γ, $F(\overline{\mathfrak{C}_\Gamma})$, $F(\mathfrak{C}_\Delta)$, and Δ above each other, and we draw the B-edges in bold/blue and the map f by thin/purple arrows (see Fig. 1).

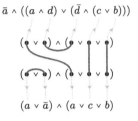

$\bar{a} \wedge ((a \wedge d) \vee (\bar{d} \wedge (c \vee b)))$

$(a \vee \bar{a}) \wedge (a \vee c \vee b)$

Fig. 1. A combinatorial proof

The relation between combinatorial proofs and deep inference proofs in system SKS [2] has been detailed out in [17,18]. We will not go into details here, but we will make heavy use of the following theorem, originally shown in [11,16]:

Theorem 2.4. *Let A and B be formulas. Then the following are equivalent:*

- *There is a skew fibration $f \colon \mathfrak{G}(A) \rightarrowtail \mathfrak{G}(B)$;*
- *There is a derivation Φ from A to B using only deep contraction $\mathsf{c}\downarrow \colon A \vee A \to A$ and deep weakening $\mathsf{w}\downarrow \colon A \to A \vee B$, modulo associativity and commutativity of $\wedge$ and $\vee$, denoted as $A \vDash_{\mathsf{w}\downarrow,\mathsf{c}\downarrow} B$;*
- *There is a derivation $\bar{\Phi}$ from $\bar{B}$ to $\bar{A}$ using only deep cocontraction $A \wedge A$ and deep coweakening $\mathsf{w}\uparrow \colon A \wedge B \to A$, modulo associativity and commutativity of $\wedge$ and $\vee$, denoted as $\bar{B} \vDash_{\mathsf{w}\uparrow,\mathsf{c}\uparrow} \bar{A}$.*

This suggests the following definition:

Definition 2.5. *A formula F' is a* skew *of a formula F iff $F \vDash_{\mathsf{c}\uparrow,\mathsf{w}\uparrow} F'$.*

3   Sequent Calculus

We recall in Fig. 2 the formulation of LK cut-free sequent calculus that we use in this paper. Moreover, we refer to MLL as the fragment of LK calculus consisting of the rules $\wedge, \vee, \mathsf{AX}$ only. We speak of MLL + mix if we additionally allow mix.

Theorem 3.1. ([14]) *Let A and B be formulas. There is an æ-acyclic R&B-cograph $\mathfrak{C}$ with $\mathfrak{C}^\downarrow = \mathfrak{G}(\Gamma)$ iff there is a proof of Γ in MLL + mix.*

---

[3] It cannot happen that both Γ and Δ are empty.

$$\cfrac{}{\vdash A,\bar{A}}\ \text{AX} \qquad \cfrac{\vdash \Gamma,A,B}{\vdash \Gamma,A\vee B}\ \vee \qquad \cfrac{\vdash \Gamma,A \quad \vdash B,\Delta}{\vdash \Gamma,A\wedge B,\Delta}\ \wedge \qquad \cfrac{\vdash \Gamma}{\vdash \Gamma,A}\ \text{W} \qquad \cfrac{\vdash \Gamma,A,A}{\vdash \Gamma,A}\ \text{C} \quad \Big| \quad \cfrac{\Gamma \quad \Delta}{\Gamma,\Delta}\ \text{mix}$$

Fig. 2. Left: sequent system LK (cut free) for classical logic, **Right:** the mix rule

$$\cfrac{}{\vdash A,\bar{A},\Gamma}\ \text{AX}_{\text{G3p}} \qquad \cfrac{\vdash A,B,\Gamma}{\vdash A\vee B,\Gamma}\ \vee_{\text{G3p}} \qquad \cfrac{\vdash A,\Gamma \quad \vdash B,\Gamma}{\vdash A\wedge B,\Gamma}\ \wedge_{\text{G3p}}$$

Fig. 3. Rules of cut-free G3p sequent calculus

$$\cfrac{\cfrac{}{\vdash a,\bar{a},c,\bar{b}}\ \text{AX} \quad \cfrac{}{\vdash a,\bar{a},b,\bar{b}}\ \text{AX}}{\vdash a,\bar{a},c\wedge b,\bar{b}}\ \wedge_{\text{G3p}} \quad \leadsto \quad \cfrac{\cfrac{\cfrac{}{\vdash a,\bar{a}}\ \text{AX}}{\vdash a,\bar{a},c}\ \text{C} \quad \cfrac{}{\vdash b,\bar{b}}\ \text{AX}}{\vdash a,\bar{a},c\wedge b,\bar{b}}\ \wedge_{\text{LK}} \quad \leadsto \quad (\overset{\frown}{\bullet \vee \bullet})\wedge(\overset{\frown}{\bullet \vee \bullet}) \qquad \overset{\frown}{\overset{\frown}{\bullet \vee \bullet}}$$
$$a,\bar{a},(c\wedge b),\bar{b} \qquad a,\bar{a},(c\wedge b),\bar{b}$$

Fig. 4. From left to right: a G3p proof, the corresponding LK proof, the naive (incorrect) translation, and finally, the correct combinatorial proof

$$\cfrac{\cfrac{\vdash A,B,C,D,\Gamma}{\vdash A\vee B,C,D,\Gamma}\ \vee}{\vdash A\vee B,C\vee D,\Gamma}\ \vee \quad \sim \quad \cfrac{\cfrac{\vdash A,B,C,D,\Gamma}{\vdash A,B,C\vee D,\Gamma}\ \vee}{\vdash A\vee B,C\vee D,\Gamma}\ \vee \qquad \cfrac{\vdash A,C,D,\Gamma \quad \vdash B,C,D,\Gamma}{\vdash A\wedge B,C,D,\Gamma}\ \wedge}{\cfrac{}{\vdash A\wedge B,C\vee D,\Gamma}}\ \vee \quad \sim \quad \cfrac{\cfrac{\vdash A,C,D,\Gamma}{\vdash A,C\vee D,\Gamma}\ \vee \quad \cfrac{\vdash B,C,D,\Gamma}{\vdash B,C\vee D,\Gamma}\ \vee}{\vdash A\wedge B,C\vee D,\Gamma}\ \wedge$$

$$\cfrac{\cfrac{\vdash A,C,\Gamma \quad \vdash A,D,\Gamma}{\vdash A,C\wedge D,\Gamma}\ \wedge \quad \cfrac{\vdash B,C,\Gamma \quad \vdash B,D,\Gamma}{\vdash B,C\wedge D,\Gamma}\ \wedge}{\vdash A\wedge B,C\wedge D,\Gamma}\ \wedge \quad \sim \quad \cfrac{\cfrac{\vdash A,C,\Gamma \quad \vdash B,C,\Gamma}{\vdash A\wedge B,C,\Gamma}\ \wedge \quad \cfrac{\vdash A,D,\Gamma \quad \vdash B,D,\Gamma}{\vdash A\wedge B,D,\Gamma}\ \wedge}{\vdash A\wedge B,C\wedge D,\Gamma}\ \wedge$$

Fig. 5. G3p proof equivalence

In [11], Hughes has shown how to translate an LK-proof into a combinatorial proof. In this paper we also consider the (cut-free) sequent calculus G3p [19], shown in Fig. 3, which has no explicit contraction and weakening rules.

Theorem 3.2. *If $d(F)$ is a derivation of the formula F in G3p, then there is a combinatorial proof $\phi_{d(F)}: \circ \vdash F$, such that every B-edges in $\phi_{d(F)}$ correspond to an instance of the AX-rule in $d(F)$.*

Proof. This follows from the result on LK in [11] and the observation that any G3p derivation can be simulated in LK by making heavy use of C and W, but without changing the AX-instances in the proof. □

Remark 3.3. Observe that the relation between B-edges in $\phi_{d(F)}$ and AX-instances in $d(F)$ is not a bijection, as can be seen by the example (due to Hughes) shown in Fig. 4 where the naive translation is not a combinatorial proof because the induced mapping is not a skew fibration. In the correct combinatorial proof the B-edge coming from the AX-instance on $b,\bar{b}$ is deleted.

In sequent calculus, the standard notion of proof identity is defined via rule permutations. The generating permutation we use for G3p are shown in Fig. 5.

Theorem 3.4. *If $d(\Gamma)$ and $d'(\Gamma)$ are two G3p derivations that are equivalent modulo the rule permutations in Fig. 5, then $\phi_{d(\Gamma)} = \phi_{d'(\Gamma)}$.*

To prove this theorem, we will prove a stronger result for analytic tableaux in the next section and then reflect it back to the sequent calculus. For LK, such a statement is less trivial, since due to the presence of weakening and contraction, there are permutations that delete or duplicate subproofs, and such operations are not preserved by combinatorial proofs (see Remark 3.3 above).

4   Analytic Tableaux

Analytic tableaux are a formalism for refutations based on the decomposition of the negation $\bar{F}$ of a formula in order to find contradictions between its subformulas and conclude, by completeness, the provability of F. This is done by expanding a formula $\bar{F}$ over a tree of its subformulas via *expansion rules* until all branches contain a formula and its negation. The resulting tree with root $\bar{F}$ is related to the disjunctive normal form $\mathrm{DNF}(\bar{F})$ as follows: each branch represents the conjunction of the formulas appearing in its nodes and the tree represents the disjunction of its branches.

We work here with a non-cumulative formulation of the tableaux formalism.

Definition 4.1 (Tableau). A *tableau* is a rooted binary tree with nodes labeled by sets of occurrences of formulas according with the following conditions:
- The tree consisting of a single node with formula set $\{\bar{F}\}$ is a tableau of F;
- If T_F is a tableau of F, then the tree obtained by the application of one of the following *tableau expansion rules* is a tableau of F:
    - If ℓ is a leaf of T_F with formula set L containing a conjunction $A \wedge B$, then the tree obtained extending T_F with a leaf ℓ_1 attached to ℓ with fomula set $L \cup \{A, B\} \setminus \{A \wedge B\}$ is a tableau of F;
    - If ℓ is a leaf of T_F with formula set L containing a disjunction $A \vee B$, then the tree obtained by extending T_F with two leaves ℓ_1 and ℓ_2 attached to ℓ with respective formula sets $L \cup \{A\} \setminus \{A \wedge B\}$ and $L \cup \{B\} \setminus \{A \wedge B\}$ is a tableau of F.

A branch of a tableau is *closed* if its leaf contains a formula and its negation, otherwise it is *open*. A tableau is *closed* if all its branches are. A branch is *atomic closed* if the closing formulas are atoms. A tableau is *full* if no expansion rule can be applied to its open branches, *non-expandable* if no expansion rule can be applied to any branch.

If T_Γ is a tableau of Γ, we denote T_Γ, A the tableau obtained by adding to T_Γ the formula A to each node. A *redundant* tableau T_F is a full tableau of F such that there is a closed tableau T_Γ such that T_F has two closed branches T_Γ, A and $T_{\Gamma, \bar{B}}$. Figures 8 and 9 show some examples.

There is a one-to-one correspondence between atoms in the leaves of a full tableau of with root labeled by a formula $\bar{F}$ and occurrences of atoms in its disjunctive normal form (denoted DNF) clauses due to the correspondence between tableau expansion rules and dual clause form algorithm [7].

$$(a \vee b) \wedge (\bar{a} \vee c)$$
$$(a \vee b), (\bar{a} \vee c)$$
$$(a \vee b), \bar{a}, c$$

$$\overrightarrow{Flip_{\mathsf{G3p}}}$$

$$\cfrac{\cfrac{\overline{\vdash a, \bar{a}, \bar{c}} \; \mathsf{AX} \quad \overline{\vdash \bar{b}, a, \bar{c}} \; \mathsf{AX}^T}{\vdash \bar{a} \wedge \bar{b}, \bar{c}, a} \wedge}{\cfrac{\vdash \bar{a} \wedge \bar{b}, \bar{c} \vee a}{\vdash (\bar{a} \wedge \bar{b}) \vee (\bar{c} \vee a)} \vee} \vee$$

$$\diagup \quad \diagdown$$
$$\boxed{a}, \boxed{\bar{a}}, c \quad b, \bar{a}, c$$

Fig. 6. A tableau of $(\bar{a} \wedge \bar{b}) \vee (\bar{c} \vee a)$ and its associate G3p derivation

Proposition 4.2. *If T_F is a full tableau of F with a non-closed branch, then*

$$\mathrm{DNF}(\bar{F}) = \bigvee_{L_i \; leaves \; of \; T_F} \left(\bigwedge_{A_{ij} \; formulas \; in \; L_i} A_{ij} \right).$$

There is a close correspondence between G3p proofs and tableaux, which has been established in [7], and which allows to define a *flipping translation* (here denoted as the function $Flip_{\mathsf{G3p}}$), which associates to any full tableau T_F a G3p derivation tree $d_T(F)$ "by tuning it upside-down and negating everything" and viceversa. In particular, we associate an axiom-rule $\mathsf{AX}_{L_i}^T$ with conclusions $\vdash A_1, \ldots A_n$ whenever there is a non-closed leaf L_i (with formulas $\bar{A}_1, \cdots, \bar{A}_n$); we define the *theory of T_F* (denoted $\mathcal{T}_{T_F}$) to be the set of such axioms. An example is shown in Fig. 6. This translation suggests the definition of an equivalence relation, shown in Fig. 7, over tableaux derived from the G3p proof equivalence (Fig. 5). The last equation does not occur in the G3p equivalence because it cannot be written as a rule permutation. Its interpretation is the following: if $T_{\bar{\Gamma}}$ is closed, then so is any tableau with root $\Gamma, A \vee B$. Hence, any full tableau $T_{\bar{\Gamma}, \bar{B}} \neq T_{\bar{\Gamma}}, B$ is closed, and we consider the contribution given by the formula $\bar{B}$ to be superfluous to the closure of $T_{\bar{\Gamma}}, A \vee B$. However, if B is not a closing formula, we can not discard the corresponding tableau branch because otherwise we lose the information about the branch leaves (see Fig. 8).

In order to keep track of the information about branching and avoid mismatching in translation, we consider different occurrences $a_1, \ldots, a_n$ of the same atom a in F as different atoms for the following definitions.

Definition 4.3 (Tableaux oversaturation, Sprout). If T_F is a tableau of F, its *oversaturation* is a tree T_F^* obtained by updating the formula sets of each vertex of T_F inductively from the leaves to the root as follows:

- If a leaf is closed, the formulas of this vertex in T_F^* are only the two closing formulas;
- If a vertex of T_F has two children then its formulas are $A \vee B, F_1, \ldots, F_n$. If both its children have been updated then:
  - If one child contains A and the other contains B then they contain two skews F_i' and F_i'' (it can be $F_i' = F_i''$) of F_i, then we replace F_i by $F_i' \wedge F_i''$ in T_F^*. If only one of the children contains a skew F_i' of F_i, then we replace F_i by F_i';
  - If no child contains A (similarly if no child contains B), we replace the vertex and the subtree having this vertex as root with the child containing neither A nor B and its corresponding subtree;

$$
\begin{array}{c}
A \wedge B, C \wedge D \\
A \wedge B, C, D \\
A, B, C, D
\end{array}
\sim
\begin{array}{c}
A \wedge B, C \wedge D \\
A, B, C \wedge D \\
A, B, C, D
\end{array}
\qquad
\begin{array}{c}
A \wedge B, C \vee D \\
\diagup \quad \diagdown \\
A \wedge B, C \qquad A \wedge B, D \\
A, B, C \qquad A, B, D
\end{array}
\sim
\begin{array}{c}
A \wedge B, C \vee D \\
A, B, C \vee D \\
\diagup \quad \diagdown \\
A, B, C \qquad A, B, D
\end{array}
$$

$$
\begin{array}{c}
A \vee B, C \vee D \\
\diagup \quad \diagdown \\
A \vee B, C \qquad\qquad A \vee B, D \\
\diagup \; \diagdown \quad\quad \diagup \; \diagdown \\
A, C \quad B, C \quad A, B, D \quad B, D
\end{array}
\sim
\begin{array}{c}
A \vee B, C \vee D \\
\diagup \quad \diagdown \\
A, B, C \vee D \qquad\qquad A, B, C \vee D \\
\diagup \; \diagdown \quad\quad \diagup \; \diagdown \\
A, C \quad B, C \quad A, B, D \quad B, D
\end{array}
$$

$$
\begin{array}{c}
\Gamma, A \vee B \\
\diagup \quad \diagdown \\
T_{\bar{F}}, A \qquad T_{\bar{F}, \bar{B}}
\end{array}
\sim
\begin{array}{c}
\Gamma, A \vee B \\
T_{\bar{F}}, A \vee B
\end{array}
$$
where $T_{\bar{F}}$ is a closed tableau of Γ and $\bar{B}$ is a closing formula of a branch of $T_{\bar{F}, \bar{B}}$.

Fig. 7. Expansion rules permutation generating tableaux standard equivalence

$$
\begin{array}{c}
a, \bar{a}, c \vee b, \bar{b} \\
\diagup \quad \diagdown \\
\boxed{a}, \boxed{\bar{a}}, c, \bar{b} \quad a, \bar{a}, \boxed{b}, \boxed{\bar{b}}
\end{array}
\sim
\begin{array}{c}
\boxed{a}, \boxed{\bar{a}}, c \vee b, \bar{b}
\end{array}
\qquad
\begin{array}{c}
a, \bar{a}, c \vee b \\
\diagup \quad \diagdown \\
\boxed{a}, \boxed{\bar{a}}, c \quad \boxed{a}, \boxed{\bar{a}}, b
\end{array}
\not\sim
\begin{array}{c}
\boxed{a}, \boxed{\bar{a}}, c \vee b
\end{array}
$$

Fig. 8. Tableaux equivalences in case of redundant tableaux

$$
\begin{array}{c}
(a \vee b) \wedge (c \vee d) \wedge \bar{c} \wedge \bar{d} \\
(a \vee b), (c \vee d), \bar{c}, \bar{d} \\
\diagup \qquad\qquad \diagdown \\
a, c \vee d, \bar{c}, \bar{d} \qquad\qquad b, c \vee d, \bar{c}, \bar{d} \\
\diagup \; \diagdown \quad\quad \diagup \; \diagdown \\
a, \boxed{c}, \boxed{\bar{c}}, \bar{d} \quad a, \boxed{d}, \bar{c}, \boxed{\bar{d}} \quad b, \boxed{c}, \boxed{\bar{c}}, \bar{d} \quad b, \boxed{d}, \bar{c}, \boxed{\bar{d}}
\end{array}
\qquad
\begin{array}{c}
(a \vee b) \wedge (c \vee d) \wedge \bar{c} \wedge \bar{d} \\
a \vee b, \boxed{c \vee d}, \boxed{\bar{c} \wedge \bar{d}}
\end{array}
$$

Fig. 9. A redundant tableau and a non-redundant one of the same formula

$$
\begin{array}{c}
(a \vee b) \wedge ((\bar{a} \wedge \bar{a}) \vee c) \\
(a \vee b), ((\bar{a} \wedge \bar{a}) \vee c) \\
(a \vee b), \bar{a} \wedge \bar{a}, c \\
\diagup \quad \diagdown \\
\boxed{a}, \boxed{\bar{a}} \quad b, \bar{a}, c
\end{array}
\qquad
\xrightarrow{Flip_{\mathsf{MLL}}}
\qquad
\cfrac{
\cfrac{
\cfrac{
\cfrac{
\cfrac{
\dfrac{}{\vdash a, \bar{a}} \mathsf{AX} \quad \dfrac{}{\vdash \bar{b}, a, \bar{c}} \mathsf{AX}^T
}{\vdash \bar{a} \wedge \bar{b}, \bar{c}, a, a} \wedge
}{\vdash \bar{a} \wedge \bar{b}, \bar{c}, (a \vee a)} \vee
}{\vdash \bar{a} \wedge \bar{b}, \bar{c} \vee (a \vee a)} \vee
}{\vdash (\bar{a} \wedge \bar{b}) \vee (\bar{c} \vee (a \vee a))} \vee
$$

Fig. 10. The oversaturation of tableau in Fig. 6 and the associated MLL derivation.

- If a vertex of T_F has one child then its formulas are $F_1 \wedge F_2, F_3, \ldots, F_n$. If the child contains a skew F_i' of F_i, then we replace F_i by F_i' in T_F^*.

The root formula of T^* is called the *sprout* of T_F, denoted $spr_T(F)$.

Lemma 4.4. *If T_F is a tableau of F, then $spr_T(F)$ is a skew of $\bar{F}$.*

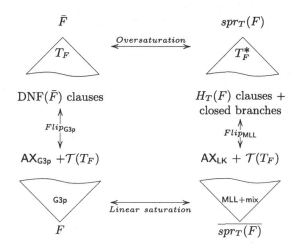

Fig. 11. Rosetta stone of tableaux translation

Proof. By induction over the structure of T_F. If T_F has no branching then $\bar{F}$ is either a conjunction of formulas or $\bar{F} = (A \wedge \bar{A}) \wedge \bar{F}'$; then $\bar{F}$ is respectively $spr_T^*(F)$ or $\bar{F} \vDash_{\mathsf{w}\uparrow} (A \vee \bar{A})$. If T_F has branchings we can assume without loosing generalities T_F expandable with root $\bar{F} = (A \vee B) \wedge C$ and leaves $\{A, C\}$ and $\{B, C\}$. Then $spr_T(F) = (A \wedge B) \vee (C \wedge C)$, and therefore $\bar{F} \vDash_{\mathsf{c}\uparrow} spr_T(F)$. We conclude by composition that $\bar{F} \vDash_{\mathsf{c}\uparrow,\mathsf{w}\uparrow} spr_T(F)$.   □

Definition 4.5 (Harvest of T). We define the *harvest* of a tableau T_F, denoted $H_T(F)$, as as the conjunction over the non-closed leaves of T_F^* of the disjunction of the formulas in each of these leaves:

$$H_T(F) = \bigvee_{L_i \text{ non-closed leaves of } T_F^*} \left(\bigwedge_{A_{ij} \text{ formulas in } L_i} A_{ij} \right).$$

If T_F is a tableau and we define the *theory of T_F* (denoted $\mathcal{T}(T_F)$) as a set of additional axiom-rules $\mathsf{AX}_{L_i}^T$ with conclusions $\vdash A_1, \ldots A_n$ where $\bar{A}_1, \cdots, \bar{A}_n$ are the formulas in the non-closed leaf L_i, we have the following:

Lemma 4.6. *For everty tableau T_F, the formula $\overline{spr_T(F)}$ is derivable in* MLL + $\mathcal{T}(T_F)$. *Furthermore, if T_F is closed then $\overline{spr_T(F)}$ is provable in* MLL + mix, *and finally, if T_F is non-redundant and closed then $\overline{spr_T(F)}$ is provable in* MLL.

Proof. By the $Flip_{\mathsf{G3p}}$ operation we have a G3p proof that we translate inductively via a procedure that we call *linear saturation* into an proof in MLL + mix + $\mathcal{T}(T_F)$ of $\overline{spr_T(F)}$:

- a G3p axiom with conclusion $\vdash A, \bar{A}, \Gamma$ is translated into an axiom $\vdash A, \bar{A}$;
- a $\mathcal{T}(T_F)$-axiom in G3p remains the same $\mathcal{T}(T_F)$-axiom (in LK);
- a $\vee_{\mathsf{G3p}}$ instance is translated into a $\vee_{\mathsf{LK}}$ and formulas are replaces with their relative skews if both principal formulas occurs. If one or both principal formulas are missed, this inference disappears during the translation.

– a $\wedge_{\mathsf{G3p}}$ instance is translated into a $\wedge_{\mathsf{LK}}$ instance (respectively mix instance) on the relative formulas skews if both principal formulas occur (if none of its principal formulas occur) followed by $\vee_{\mathsf{LK}}$ instances on whenever two skews of a same formula F_i appear in both premises. If one of the two principal formula is missed, we consider the $\wedge_{\mathsf{G3p}}$ inference premises in which this should belong, we keep this branch in our derivation, we discard the other one and the inference disappears.

The other statements follow immediately by case inspection.   □

Figure 10 shows an example for Lemma 4.6. Its proof suggests that analogously to the operation $Flip_{\mathsf{G3p}}$ that relates tableaux and $\mathsf{G3p}$ derivations, we can define and operation $Flip_{\mathsf{MLL}}$ that relates oversaturated tableaux and derivations $\mathsf{MLL} + \mathsf{mix} + \mathcal{T}(T_F)$ "by flipping it upside-down and negating everything". The interactions between the two flipping translation, the oversaturation and the sprouting procedure can be summarized by the diagram in Fig. 11.

We can now give a polynomial translation from tableaux into combinatorial proofs.

Theorem 4.7. *Let T_F be a full tableau for F. If T_F is closed then there is a combinatorial proof $\phi_{T_F} : \bar{F} \vdash \circ$. Otherwise, $\phi_{T_F} : \bar{F} \vdash H_T(F)$. In either case, if T_F is non-redundant the atoms pairs in the formula pairs that close the branches are mapped to the B-edges in ϕ_{T_F}. If T_F is closed, then this is a bijection.*

Proof. We define $\phi_{T_F} : \bar{F} \vdash H_T(F)$ as follows:

– The R&B-cograph $\mathfrak{C}_{T_F}$ of ϕ_{T_F} is given by the cograph $\mathfrak{C}^{\downarrow}_{T_F} = \overline{\mathfrak{G}(spr_T(F))} \vee \mathfrak{G}(H_T(F))$ enriched with a matching B_{T_F} constructed as follows:
   - For each closed branch of T_F we consider its closing pair of formulas $(G, \overline{G})$ in its leaf label. For each atom a_i in G and $\overline{a}_i$ in $\overline{G}$, we define an edge between their corresponding vertices in $\overline{\mathfrak{G}(spr(T))}$;
   - For each non-closed branch of T_F, the associate clause of $spr_T(F)$ occurs in its harvest $H_T(F)$. We define an edge between the vertices corresponding the associated atoms in $\mathfrak{G}(H_T(F))$ and $\overline{\mathfrak{G}(spr(T))}$.
– The skew fibration $f : \mathfrak{C}^{\downarrow} \rightarrowtail \mathfrak{G}_{T_F}$ where $\mathfrak{G}_{T_F} = \mathfrak{C}(\bar{F}) \vee \mathfrak{C}(H_T(F))$ is given by the disjunction $f = f^{\uparrow} \vee 1_{H_T(F)}$ of the identity skew fibration $1_{H_T(F)}$ over $\mathfrak{G}(H_T(F))$ and the skew fibration $f^{\uparrow}$ defined by the sprouting derivation $\bar{F} \vDash spr_T(F)$.

Similarly if T_F is closed, then the R&B-cograph $\mathfrak{C}_{T_F}$ of ϕ_{T_F} is the cograph $\mathfrak{C}^{\downarrow}_{T_F} = \overline{\mathfrak{G}(spr_T(F))}$ enriched with the corresponding matching B_{T_F} while the skew fibration $f : \mathfrak{C}^{\downarrow} \rightarrowtail \mathfrak{G}_{T_F}$ where $\mathfrak{G}_{T_F} = \overline{\mathfrak{C}(\bar{F})}$ is given by the sprouting derivation from $\bar{F}$ to $spr_T(F)$.   □

Examples for this construction are shown in Fig. 12.

Theorem 4.8. *If T_F and T'_F are two equivalent tableaux of F, $\phi_{T_F} = \phi_{T'_F}$.*

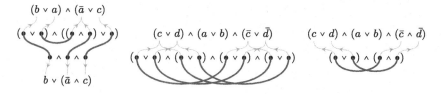

Fig. 12. The combinatorial proofs associate to the tableaux in Figs. 6 and 9

Proof. The two tableaux T_F and T'_F are equivalent if and only if their corresponding derivations $d^*_T(F)$ and $d^*_{T'}(F)$ in MLL + mix + $\mathcal{T}(T_F)$ are. By the canonicity of MLL proof nets [1,14] we conclude that ϕ_{T_F} and $\phi_{T'_F}$ have the same R&B-cograph. Furthermore, from the work in [5,11,16], we know that skew fibrations are canonical. Hence, $\phi_{T_F} = \phi_{T'_F}$.   □

Via the flipping operation $Flip_{\mathsf{G3p}}$ we can now immediately obtain the proof of Theorem 3.4.

In order to translate a combinatorial proofs back into a tableau, we associate to a CP $\phi : F \vdash \circ$ with skew fibration $f : \mathfrak{C}^{\downarrow} \rightarrowtail \mathfrak{G}_{T_F}$ the MLL derivation $d_{T^*}(F')$ represented by the cograph $\mathfrak{C}^{\downarrow}$ where $F' = \overline{spr_T(F)}$ is obtained by labeling the vertices of $\mathfrak{C}^{\downarrow}$ with their images under f. Then, by $Flip_{\mathsf{MLL}}$ we have an oversaturated tableau T^*_F. If mix is absent, T^*_F contains all the information to invert the oversaturation and reconstruct T_F. However, in the presence of mix, even if we can translate any instance of rule mix-rule into a ∧-expansion, we can not recover the structure of T_F in general.

5   Resolution

Resolution is a refutation system related to conjunctive normal forms. A resolution proof consist of applying *resolution rule* on clauses of conjunctive normal form of a formula F in order to produce an empty clause. However, the resolution technique does not require the full conversion of a formula to its conjunctive normal form, in the same way tableaux can be closed before a complete expansion. A general resolution proof consist of a sequence of *expansion rules* intercutted by resolution rules terminating with the production of an empty clause or a clause form formula.

In resolution we also denote formulas in *Davis-Putnam's block notation* used in [2,7,9] which interprets lists $(X_1, \ldots, X_n)$ and $[X_1, \ldots, X_n]$ as respectively the conjunction $X_1 \wedge \cdots \wedge X_n$ and the disjunction $X_1 \vee \cdots \vee X_n$ of their elements.

Definition 5.1. If $F = (C_1, \ldots, C_n, C)$ is a formula with $C = [X_1, \ldots, X_n]$, we define the following *(resolution) expansion rules*:

- if $X_i = A \vee B$, then $F \rightarrow^{A \vee B}_{\vee} (C_1, \ldots, C_n, C')$ and we say that the clause C' is generated by the clause C where $C' = [X_1, \ldots, X_{i-1}, A, B, X_{1+1}, \ldots, X_n]$;

$$\cfrac{\cfrac{\cfrac{[(a \vee b) \wedge (\bar{a} \vee c)]}{[a \vee b][\bar{a} \vee c]}}{\cfrac{[a,b][\bar{a} \vee c]}{[a,b][\bar{a},c]}}\vee}{[b,c]}Res^a \quad \cfrac{\cfrac{\cfrac{\cfrac{\cfrac{\cfrac{[\bar{a} \wedge (a \vee d) \wedge (\bar{d} \vee (b \vee c))]}{[\bar{a} \wedge (a \vee d)][\bar{d} \vee (b \wedge c)]}\wedge}{[\bar{a}][a \vee d][\bar{d} \vee (b \wedge c)]}\wedge}{[\bar{a}][a \vee d][\bar{d},b \wedge c]}\vee}{[\bar{a}][a,d][\bar{d},b \wedge c]}Res^d}{[\bar{a}][a,b \wedge c]}\wedge}{\cfrac{[\bar{a}][a,b][a,c]}{[b][a,c]}}Res^a \quad \cfrac{\cfrac{\cfrac{[(a \vee b) \wedge (c \vee d) \wedge \bar{c} \wedge \bar{d}]}{[a \vee b][(c \vee d) \wedge \bar{c} \wedge \bar{d}]}\wedge}{[a \vee b][c \vee d][\bar{c} \wedge \bar{d}]}\wedge}{[a \vee b][\ ]}Res^{c \vee d}$$

Fig. 13. A resolution expansion of $(a \vee b) \wedge (\bar{a} \vee c)$, one of $\bar{a} \wedge (a \vee b) \wedge (\bar{d} \vee (b \vee c))$ and a resolution proof of $(\bar{a} \wedge \bar{b}) \vee (\bar{c} \wedge \bar{d}) \vee c \vee d$.

- if $X_i = A \wedge B$ then $F \rightarrow_{\wedge}^{A \wedge B} (C_1, \ldots, C_n, C', C'')$ and we say that the clauses C' and C'' are generated by the clause C where C' and C'' are respectively $[X_1, \ldots, X_{i-1}, A, X_{1+1}, \ldots, X_n]$ and $[X_1, \ldots, X_{i-1}, B, X_{1+1}, \ldots, X_n]$;

An *expansion of F* is the last formula F' produced by a sequence of application of expansion rules starting from F.

Definition 5.2 (Resolution Rule). If $F = (C_1, C_2, \Sigma)$, $C_1 = [\Gamma, X_1, \ldots, X_n]$ and $C_2 = [\Delta, \overline{X}_1, \ldots, \overline{X}_m]$ where X_i and $\bar{X}_i$ are respectively occurrences of X and $\bar{X}$, we say $F' = ([\Gamma, \Delta], \Sigma)$ is the result of resolving C_1 and C_2 on the *resolving formula* X (denoted $F \rightarrow_X^R F'$) and that the clause $[\Gamma, \Delta]$ is generated by the clauses C_1 and C_2.

Definition 5.3 (Resolution Proof) A *resolution expansion* R_F of a formula F is a sequence of formulas $F = F_0 \rightarrow \cdots \rightarrow F_n$ such that F_{i+1} is obtained by F_i by applying an expansion rule or a resolution rule. We call F_n the *result* of the resolution expansion R_F.

A resolution expansion is *closed* if F_n contains an empty clause $[\ ]$, *non-expandable* if no expansion rules can be applied to F_n and *full* if is non-expandable and no resolution rule can be applied to F_n.

A resolution proof of $\overline{F}$ is a closed resolution expansion of F. Figure 13 shows some examples. If a clause C generates after a certain number of expansions and resolution a clause C' we say that C' is *derived* by C. As for the other proof and refutation systems in this paper, we want to define a notion of equivalence on resolution expansions.

Definition 5.4. We define the *resolution derivation equivalence* to be the smallest equivalence relation over resolution expansions generated by the following relations for any formulas F and G:

- Resolution rule inferences commute with resolution rule and $\vee$-expansions inferences;
- $\vee$-expansions inferences commute with both resolution rule and $\vee$-expansions inferences;
- If F and G do not belong to the same clause then $\rightarrow_F^\vee \rightarrow_G^\wedge = \rightarrow_G^\wedge \rightarrow_F^\vee$;

- If F and G and also $\bar{F}$ and $\bar{G}$ do not belong to the same clause then $\to_F^{\vee}\to_G^R=\to_G^R\to_F^{\vee}$;
- If G does not belong to the same clause of the resolving formula F or its corresponding $\bar{F}$ then $\to_F^R\to_G^{\wedge}=\to_G^{\wedge}\to_F^R$;
- If F and G belong to the same clause then $\to_F^{\vee}\to_G^{\wedge}=\to_G^{\wedge}\to_{F_1}^{\vee}\to_{F_2}^{\vee}$ where F_1 and F_2 are the two copies of F belonging in the two clauses produced by the $\wedge$-expansion of G.

Moreover, we ask the following condition: if $F_0 \to \dots \to F_{k-1} \to F_k \to F_{k+1} \dots \to F_n$ is a resolution expansion such that $F_{k-1} \to F_k$ is a resolution inference which generate an empty clause, then

$$F_0 \to \dots \to F_{k-1} \to F_k \to F_{k+1} \dots \to F_n = F_0 \to \dots \to F_{k-1} \to F_k.$$

Lemma 5.5. *If a formula F' is obtained by applying an expansion rule to a formula F, then F' is a skew of F.*

Proof. The first transformation given in Definition 5.1 corresponds to the associativity of $\vee$, and the second transformation corresponds to $(A \wedge B) \vee \Delta \overset{c\uparrow}{\to} (A \wedge B) \vee (\Delta \wedge \Delta) \overset{c\uparrow}{\to} ((A \wedge B) \vee (\Delta \wedge \Delta)) \wedge ((A \wedge B) \vee (\Delta \wedge \Delta)) \overset{4 \cdot w\uparrow}{\to} (A \vee \Delta) \wedge (B \vee \Delta)$ where $A \wedge B = X_i$ and $\Delta = \bigwedge_{j \neq i} X_j$. $\qquad \square$

Definition 5.6 (Pseudo-resolution expansion). If $F = (\Gamma, [\Delta])$ is a formula, we define the following *(resolution) pseudo-expansion rules*:

- *clause duplication*: $F \to_{[\Delta]}^{\delta} (\Gamma, [\Delta], [\Delta])$;
- *clause erasing*: $F \to_{[\Delta]}^{\epsilon} (\Gamma)$.

A *pseudo-resolution expansion* of F is a sequence of formulas $F = F_0 \to \dots \to F_n$ such that F_{i+1} is obtained by F_i by applying an expansion rule, a resolution rule or a pseudo-expansion rule.

As for tableaux, we associate to any resolution expansion R_F a formula spr_R which admits a linear derivation from spr_R to the result of the resolution R_F. We introduce an oversaturation procedure to give a pseudo-resolution expansion in which all the resolution rule inference are applied after the expansions and no superfluous information such as non-empty clauses in a closed resolution is kept. We observe that during the oversaturation some clauses may be duplicated if some $\wedge$-expansion and resolution inferences are permuted.

Definition 5.7 (Oversaturation of R_F). We define the *oversaturation of R_F* as a pseudo-resolution R_F^* obtained by the following procedure:

- $R_F^* = R_F$ and we say that all its resolution rule inference are *active*;
- We proceed by induction over the number of active resolution rule inferences in R_F^*. We start from the last resolution rule inferences $F_k \to_R^X F_{k+1}$ in the pseudo-expansion R_F^* and we deactivate it as follows:
  - if it generates the empty clause, then we apply to any clause in F_k which do not contain the resolving formulas a clause erasing rule;

- if no rule inference is applied to the clause generated by an active resolution inference, then move the application of this inference at the end the pseudo-expansion and we deactivate it;
- if a ∨-expansion is applied to the clause $[X_1, \ldots, A \vee B, \ldots X_n]$ of F_{k+1} generated by an active resolution inference resolving on a formula Y, then we permute them: we apply a ∨-expansion to the unique clause C_1 in F_k containing $A \vee B$ and then resolve on Y;
- if a ∧-expansion is applied to the clause $[X_1, \ldots, A \wedge B, \ldots X_n]$ of F_{k+1} generated by an active resolution inference, then we permute them: we apply the ∧-expansion to the unique clause C_1 in F_k containing $A \wedge B$ and the resolving formula X, we apply a clause duplication on the clause C_2 containing the corresponding resolving formula $\bar{X}$ and then we apply two resolution inferences to the corresponding pairs of clause. These two resolution rule inferences are active.

Figure 14 shows two examples.

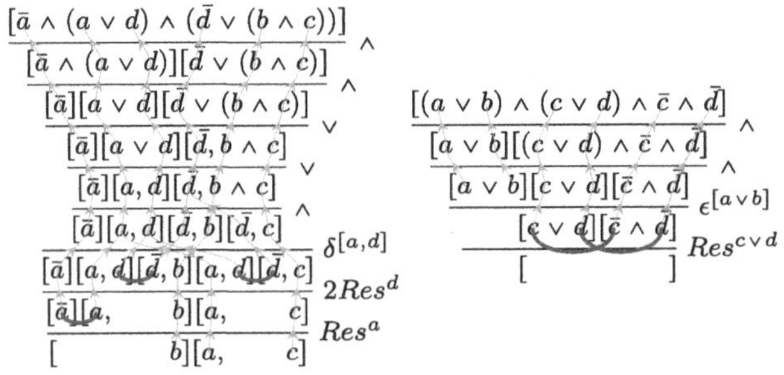

Fig. 14. The oversaturation of the resolution expansions of $\bar{a} \wedge (a \vee b) \wedge (\bar{d} \vee (b \vee c))$ and $\mathfrak{G} \vee \mathfrak{G}' = \langle V \cup V', E \cup E' \rangle$ of in Fig. 13 with their relative combinatorial proofs

Lemma 5.8. *The oversaturation procedure terminates.*

Proof. We define the *weight* of a resolution rule generating a clause C in a resolution expansion R_F as the number of all the ∨- and ∧-expansion inferences applied to any clause derived by C. The weight of a resolution expansion is the sum of the weight of its resolution rules. This decreases at each step.   □

Remark 5.9. Any oversaturatation of R_F is a sequence of the form $F \to^*_{exp} F_{n-k} \to^*_R F_n$ where the sequence $F \to^*_{exp} F_{n-k}$ is made only of expansion and pseudo-expansion rules and $F_{n-k} \to^*_R F_n$ is made of k resolution rules.

Definition 5.10 (Sprout of R_F). If $R_F^* = F \to^* F_n$ is an oversaturation of R_F, the *sprout of* R_F is the formula $spr_R(F)$ obtained by deeply apply w↓ to F_n for each resolution inference in R_F^* in the following way: if C is the clause of F_n generated (directly or inderectly) by resolving k copies of X and h copies of $\bar{X}$, then we weak C with $(X \wedge \bar{X})^{\vee kh}$.

$$\cfrac{[A \wedge B, \Gamma]}{[A, \Gamma][B, \Gamma]} \wedge\text{-exp} \quad \cfrac{[A \wedge B]}{[A][B]} \wedge\text{-exp} \quad \cfrac{[A \vee B, \Gamma]}{[A, B, \Gamma]} \vee\text{-exp} \quad \cfrac{[\Gamma], [\Delta]}{[\Gamma]} \epsilon_{[\Delta]} \quad \cfrac{\Gamma}{\Gamma, \Gamma} \delta_{[\Gamma]}$$

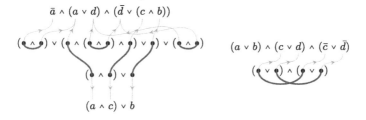

Fig. 15. Interpretation of rules inferences in pseudo-resolution

$$\bar{a} \wedge (a \vee d) \wedge (\bar{d} \vee (c \wedge b))$$

$$(\bullet \wedge \bullet) \vee (\bullet \wedge (\bullet \wedge \bullet) \wedge \bullet) \vee \bullet) \vee (\bullet \wedge \bullet) \qquad (a \vee b) \wedge (c \vee d) \wedge (\bar{c} \vee \bar{d})$$

$$(\bullet \wedge \bullet) \vee \bullet \qquad (\bullet \vee \bullet) \wedge (\bullet \vee \bullet)$$

$$(a \wedge c) \vee b$$

Fig. 16. Combinatorial proofs for the oversaturated resolutions in Fig. 14

Lemma 5.11 (Sprouting fibration). *If R_F is a resolution expansion and $spr_R(F)$ its sprouting, then there is a skew fibration $f: \mathfrak{G}(spr_R(T)) \rightarrowtail \mathfrak{G}(F)$.*

Proof. By composition of the interpretations of expansions, pseudo-expansion and resolution rules (see Fig. 15). □

Theorem 5.12. *If R_F is a resolution expansion of F with result $R_F(F)$, then there is a combinatorial proof representing R_F of the form $\phi_{R_F}: F \vdash R_F(F)$. In particular, we have $\phi_{R_F}: F \vdash \circ$ if R_F is closed.*

Proof. We define $\phi_{R_F}: F \vdash R_F(F)$ as follows:
- The R&B-cograph $\mathfrak{C}_{R_F}$ of ϕ_{R_F} is given by the cograph $\mathfrak{C}_{R_F}^{\downarrow} = \overline{\mathfrak{G}(spr_R(F))} \vee \mathfrak{G}(R_F(F))$ enriched with a matching B_{R_F} constructed as follows:
  - For each resolution rule application in R_F we consider its resolving formula X. For each atom a_i in X and $\bar{a}_i$ in $\overline{X}$, we define an edge between their corresponding vertices in $\mathfrak{G}(spr_R(F))$;
  - We define an edge between each vertex in $\overline{\mathfrak{G}(spr_R(F))}$ and the corresponding vertex in $\mathfrak{G}(R_F(F))$;
- The skew fibration $f: \mathfrak{C}^{\downarrow} \rightarrowtail \mathfrak{G}_{R_F}$ where $\mathfrak{G}_{T_F} = \overline{\mathfrak{C}(F)} \vee \mathfrak{C}(R_F(F))$ is given by the disjunction $f = f^{\uparrow} \vee 1_{\mathfrak{G}(R_F(F))}$ where $f^{\uparrow}$ is the sprouting fibration of $spr_R(F)$ and $1_{\mathfrak{G}(R_F(F))}$ is the identity over $\mathfrak{G}(R_F(F))$.

Similarly, if R_F is closed, then the R&B-cograph $\mathfrak{C}_{R_F}$ of ϕ_{R_F} is the cograph $\mathfrak{C}_{R_F}^{\downarrow} = \overline{\mathfrak{G}(spr(T))}$ enriched with the corresponding matching B_{T_F} and the skew fibration $f: \mathfrak{C}^{\downarrow} \rightarrowtail \mathfrak{G}_{T_F}$ where $\mathfrak{G}_{T_F} = \overline{\mathfrak{C}(F)}$ is given by the sprouting fibration from F to $spr(T)$ (see Fig. 16). □

Remark 5.13. The number of B-edges in ϕ_{R_F} is equal to the sum of the number of atoms occurring in resolved formulas, each of which is counted as many times as the product of positive and negative occurrences of the resolved formula to which the atom belongs. This size explosion can be avoided in the special case where resolution inferences resolve the same number of formulas occurrences. Then we can adapt the translation suggested by Das in [4], by replacing the interpretation of resolution rule in Fig. 15 with the following:

$$[\Gamma, \boxed{\frac{X^{\vee n}][\bar{X}^{\wedge n}}{(\underbrace{\qquad}_{\wedge}\underbrace{\qquad}_{\wedge})} Res^X}, \Delta]$$

Theorem 5.14. *If R_F and R'_F are equivalent, then $\phi_{R_F} = \phi_{R'_F}$.*

Proof. The permutation of expansion rules does not change the number of formula occurrences in the resolution rule instances. Hence, we can conclude by the same reasoning as in the proof of Theorem 4.8.   □

It is possible to associate to $\phi : F \vdash \circ$ with skew fibration $f\colon \mathfrak{C}^{\downarrow} \rightarrowtail \mathfrak{G}_F$ a resolution expansion as follows. If F' is the formula associated to $\mathfrak{C}^{\downarrow}$ by labeling its vertices according to f, then we can transform F' to its conjunctive normal form by applying $c{\uparrow}$ and $w{\uparrow}$. By Lemma 5.5 the composition of this expansion with f represents the expansion part of the pseudo-resolution, while the perfect matching of the R&B-cograph $\mathfrak{C}^{\downarrow}$ takes track of the resolution rule instances. If multiple matchings connect atoms in the same clause this correspond to a unique resolution rule inference, otherwise a clause duplication has been performed during the resolution expansion oversaturation, which means that the some ∧-expansions have been performed after the corresponding resolution rule.

6   Conclusions

In this paper we tried to make a case that *combinatorial proofs* can serve as canonical representation for proofs in classical propositional logic, by showing how natural notions of proof identity in various syntactic formalisms are reflected by combinatorial proofs. We extended the investigation by Hughes [11] from sequent calculus to other formalisms that are employed in automatic reasoning.

There is ongoing research investigating the possible structure for combinatorial proofs for intuitionistic logic, and in future research we plan to investigate combinatorial proofs for first-order logic, based on Hughes' *unification nets* [12], and for modal logics, based on the recent development on *nested sequents* [13].

References

1. Bellin, G., van de Wiele, J.: Subnets of proof-nets in MLL⁻. In: Advances in Linear Logic, pp. 249–270. Cambridge University Press (1995)
2. Brünnler, K., Tiu, A.F.: A local system for classical logic. In: Nieuwenhuis, R., Voronkov, A. (eds.) LPAR 2001. LNCS (LNAI), vol. 2250, pp. 347–361. Springer, Heidelberg (2001). https://doi.org/10.1007/3-540-45653-8_24

3. Cook, S.A., Reckhow, R.A.: The relative efficiency of propositional proof systems. J. Symb. Logic **44**(1), 36–50 (1979)
4. Das, A.: On the relative proof complexity of deep inference via atomic flows. Log. Methods Comput. Sci. **11**(1), 1–23 (2015)
5. Das, A., Straßburger, L.: On linear rewriting systems for Boolean logic and some applications to proof theory. Log. Methods Comput. Sci. **12**(4), 1–27 (2016)
6. Duffin, R.: Topology of series-parallel networks. J. Math. Anal. Appl. **10**(2), 303–318 (1965)
7. Fitting, M.: First-Order Logic and Automated Theorem Proving. Springer Science & Business Media, Heidelberg (2012)
8. Gentzen, G.: Untersuchungen über das logische Schließen. I. Mathematische Zeitschrift **39**, 176–210 (1935)
9. Guglielmi, A., Straßburger, L.: Non-commutativity and MELL in the calculus of structures. In: Fribourg, L. (ed.) CSL 2001. LNCS, vol. 2142, pp. 54–68. Springer, Heidelberg (2001). https://doi.org/10.1007/3-540-44802-0_5
10. Hughes, D.: Proofs Without Syntax. Ann. Math. **164**(3), 1065–1076 (2006)
11. Hughes, D.: Towards Hilbert's 24th problem: combinatorial proof invariants: (preliminary version). Electr. Notes Theor. Comput. Sci. **165**, 37–63 (2006)
12. Hughes, D.J.: Unification nets: canonical proof net quantifiers. In: LICS 2018 (2018)
13. Marin, S., Straßburger, L.: Label-free modular systems for classical and intuitionistic modal logics. In: Advances in Modal Logic 10 (2014)
14. Retoré, C.: Handsome proof-nets: perfect matchings and cographs. Theor. Comput. Sci. **294**(3), 473–488 (2003)
15. Smullyan, R.M.: First-Order Logic. Courier Corporation, Massachusetts (1995)
16. Straßburger, L.: A characterization of medial as rewriting rule. In: Baader, F. (ed.) RTA 2007. LNCS, vol. 4533, pp. 344–358. Springer, Heidelberg (2007). https://doi.org/10.1007/978-3-540-73449-9_26
17. Straßburger, L.: Combinatorial flows and proof compression. Research report RR-9048, Inria Saclay (2017). https://hal.inria.fr/hal-01498468
18. Straßburger, L.: Combinatorial flows and their normalisation. In: Miller, D. (ed.) FSCD 2017. LIPIcs, vol. 84, pp. 31:1–31:17. Schloss Dagstuhl (2017)
19. Troelstra, A.S., Schwichtenberg, H.: Basic Proof Theory, vol. 43. Cambridge University Press, Cambridge (2000)

A Resolution-Based Calculus
for Preferential Logics

Cláudia Nalon[1](✉) and Dirk Pattinson[2]

[1] Department of Computer Science, University of Brasília,
Brasília, DF 70910-090, Brazil
nalon@unb.br
[2] Research School of Computer Science, Australian National University,
Acton ACT, Canberra 2601, Australia
Dirk.Pattinson@anu.edu.au

Abstract. The vast majority of modal theorem provers implement modal tableau, or backwards proof search in (cut-free) sequent calculi. The design of suitable calculi is highly non-trivial, and employs nested sequents, labelled sequents and/or specifically designated transitional formulae. Theorem provers for first-order logic, on the other hand, are by and large based on resolution. In this paper, we present a resolution system for preference-based modal logics, specifically Burgess' system S. Our main technical results are soundness and completeness. Conceptually, we argue that resolution-based systems are not more difficult to design than cut-free sequent calculi but their purely syntactic nature makes them much better suited for implementation in automated reasoning systems.

1 Introduction

Theorem-provers for First-Order logic, such as E [20], Vampire [17] and SPASS [22] are typically based on resolution, often augmented with elements of the superposition calculus [1] to deal with equality. This is in sharp contrast with Modal (or Description Logic) reasoners which are typically based on variants of analytic tableau. Examples are the FACT++ reasoner [21], LoTREC [7], LeanTAP [2] and Racer [11]. The situation is similar for non-normal modal logics, such as Alternating Temporal Logic [5] and various forms of conditional logics [14,15] as well as various logics that can be subsumed under co-algebraic semantics [10]. Modal theorem provers based on resolution, on the other hand, are thin on the ground, but compare favourably with Tableau-based approaches in terms of efficiency [12,13].

As part of an ongoing investigation into resolution theorem-proving for modal logics, this paper presents a resolution system for Burgess' system S [3], a conditional logic that extends classical propositional logic with a binary modal connective, written ⇒, and read as 'if ... then typically ... '. The binary

---

The first author was partially supported by FWF START Y544-N23.

connective $\Rightarrow$ is interpreted over models having a set W of possible worlds, and a preorder relation $\leq_w$ at each world $w \in W$. The preorder relation can be interpreted as local plausibility relation, where $w' \leq_w w''$ is interpreted as w' being as plausible as w'' (from the perspective of w). In finite models, the modal formula $\phi \Rightarrow \psi$ can then be interpreted at w by stipulating that every $\leq_w$-minimal world that satisfies ϕ must also satisfy ψ. The dual of $\Rightarrow$ is denoted by $\not\Rightarrow$, that is, $\varphi \not\Rightarrow \psi$ is defined as $\neg\varphi \Rightarrow \neg\psi$. The interpretation of $\varphi \not\Rightarrow \psi$ is, thus, that there exists a minimal $\neg\varphi$-world which satisfies ψ.

The ensuing logic is part of a family of conditional logics [4] for which sequent, or tableau calculi are notoriously hard to construct, and often require additional syntactic structure. Various conditional logics require nested sequents [14], labelled sequents [8,9] or special transition formulae [15], together with non-trivial proofs of either semantic completeness or cut elimination. Again, this is in sharp contrast to modal calculi based on resolution, where the only extra machinery needed is a global modality.

Our main technical contribution is the design of a resolution calculus for Burgess' system S, together with proofs of soundness and completeness. As with other resolution-based systems found in the literature (including First-Order resolution calculi), our procedure consists of two phases. In the first phase, an input formula is translated into an equisatisfiable set of clauses. Then a set of inference rules is applied to the clause set. There are two types of rules: one corresponding to the usual modal propagation, as seen in modal tableaux calculi; and a set of resolution-based rules. Although the method presented here is essentially clausal, the formula φ in a modal formula of the form $\varphi \not\Rightarrow \psi$ partially retains the structure of the original problem on the left-hand side of the modal operator. This allows for a simpler set of rules for modal propagation based on the set of axioms for S. Besides the resolution-based rules for dealing with the propositional fragment of the logic, the resolution rules operate on modal formulae and propagate potential inconsistencies between modal formulae to the propositional level.

Conceptually, we argue that resolution-based systems are not more difficult to design than cut-free sequent calculi but their purely syntactic nature makes them much better suited for implementation in automated reasoning systems.

The paper is organised as follows. In the next section, the language of S is given, following the presentation in [6]. The resolution-based calculus for S, named RES$_S$, is detailed in Sect. 3: we present the transformation rules for translating a formula into the normal form and the inference rules of RES$_S$, together with a non-trivial example involving nested conditional formulae. In Sect. 4, we show that RES$_S$ is sound, complete, and terminating. We summarise and discuss our results in Sect. 5.

2   Language

In this section we introduce the language of S, following closely the presentation in [6]. Let $\mathsf{P} = \{p, q, r, \ldots, p', q', r', \ldots\}$ be a denumerable set of propositional symbols. Formulae are built from P, the usual classical connectives for negation ($\neg$) and conjunction ($\wedge$), and the conditional implication ($\Rightarrow$). The set of well-formed formulae of S, denoted by $\mathsf{WFF_S}$, is inductively defined as follows:

- for all $p \in \mathsf{P}$, $p \in \mathsf{WFF_S}$;
- if φ_i, $0 \leq i \leq n$, $n \in \mathbb{N}$, are in $\mathsf{WFF_S}$, then so are $\neg\varphi_1$, $(\varphi_1 \wedge \ldots \wedge \varphi_n)$, and $\varphi_1 \Rightarrow \varphi_2$.

The empty conjunction is denoted by **true** (*verum*). Let φ_i, $0 \leq i \leq n$, $n \in \mathbb{N}$, be formulae in $\mathsf{WFF_S}$. The following connectives are introduced as abbreviations: **false** $= \neg\textbf{true}$ (*falsum*), $(\varphi_1 \vee \ldots \vee \varphi_n) = \neg(\neg\varphi_1 \wedge \ldots \wedge \neg\varphi_n)$ (disjunction), $(\varphi_1 \to \varphi_2) = (\neg\varphi_1 \vee \varphi_2)$ (implication), and $(\varphi_1 \leftrightarrow \varphi_2) = (\varphi_1 \to \varphi_2) \wedge (\varphi_2 \to \varphi_1)$ (double implication). We denote the dual of $\Rightarrow$ by $\not\Rightarrow$, that is, $\varphi_1 \not\Rightarrow \varphi_2$ is defined as $\neg(\neg\varphi_1 \Rightarrow \neg\varphi_2)$. Parentheses are omitted if the reading is not ambiguous. We set the precedence order of operators as $\neg < \{\wedge, \vee\} < \to\, < \leftrightarrow\, < \{\Rightarrow, \not\Rightarrow\}$, that is $\neg$ binds stronger than $\wedge$ and $\vee$, which bind stronger than $\to$, as usual.

A *literal* is a proposition or its negation. We denote the set of all literals by L. The set of subformulae of a formula is defined in the usual way. As we take the conjunction as an *n*-ary operator, for a formula φ of the form $\varphi_1 \wedge \ldots \wedge \varphi_n$, any conjunction formed by the subformulae occurring in φ is a subformula of φ. For instance, p, q, r, $p \wedge q$, $p \wedge r$, $q \wedge r$, and $p \wedge q \wedge r$ are all the subformulae of $p \wedge q \wedge r$.

A complete axiomatisation for S is given in [3] and comprises the following axiom schemata (where $\varphi, \psi, \chi \in \mathsf{WFF_S}$):

A0 all propositional tautologies;
A1 $\varphi \Rightarrow \varphi$ (reflexivity);
A2 $((\varphi \Rightarrow \psi) \wedge (\varphi \Rightarrow \chi)) \to (\varphi \Rightarrow (\psi \wedge \chi))$;
A3 $(\varphi \Rightarrow (\psi \wedge \chi)) \to (\varphi \Rightarrow \psi)$ (monotonicity on the right-hand side of $\Rightarrow$);
A4 $((\varphi \Rightarrow \psi) \wedge (\varphi \Rightarrow \chi)) \to ((\varphi \wedge \psi) \Rightarrow \chi)$ (cautious monotonicity);
A5 $((\varphi \Rightarrow \chi) \wedge (\psi \Rightarrow \chi)) \to ((\varphi \vee \psi) \Rightarrow \chi)$ (or);

together with uniform substitution and the following inference rules: *modus ponens* [MP] if $\vdash \varphi$ and $\vdash (\varphi \to \psi)$, then $\vdash \psi$; and *replacement of provable equivalents* [RPE] if $\vdash (\varphi_1 \leftrightarrow \varphi_2)$ and $\vdash \psi$, then $\vdash \psi'$, where ψ' only differs from ψ by replacing some subformulae of ψ of the form φ_1 by φ_2.

The semantics of S is given in terms of Kripke structures with a ternary relation over worlds. Let (W, w_0, π, R) be a Kripke structure where $W \neq \emptyset$ is a set (of *worlds*) with a distinguished world w_0; $\pi : W \longrightarrow (\mathsf{P} \longrightarrow \{true, false\})$ is an evaluation function which maps every world to a truth assignment over P; and R is a ternary relation over W, where $\leq_w = \{(w', w'') \mid (w, w', w'') \in R\}$, for which $\leq_w$ is a preorder (i.e. a reflexive and transitive relation). We say that $\leq_w$ is a *preferential order* over the worlds in W from the point of view of w. We define W_w, for $w \in W$, to be the set $\{w' \mid (w', w'') \in \leq_w$, for some $w''\}$, that is, W_w is the set of worlds considered at least as plausible as some world in W according to the preferential order given by $\leq_w$.

Let $M = (W, w_0, \pi, R)$ be a Kripke structure. Truth of a formula at a world $w \in W$ in M, denoted by $\models$, is defined as follows (where $\varphi_i \in \mathsf{WFF_S}$, for all $0 \leq i \leq n$, $n \in \mathbb{N}$):

- $\langle M, w \rangle \models p$ if, and only if, $\pi(w)(p) = true$, for all $p \in \mathsf{P}$;
- $\langle M, w \rangle \models \neg\varphi_1$ if, and only if, $\langle M, w \rangle \not\models \varphi_1$;
- $\langle M, w \rangle \models (\varphi_1 \wedge \ldots \wedge \varphi_n)$ if, and only if, $\langle M, w \rangle \models \varphi_i$, for all $1 \leq i \leq n$;
- $\langle M, w \rangle \models \varphi_1 \Rightarrow \varphi_2$ if, and only if, for all $w' \in W_w$, if $\langle M, w' \rangle \models \varphi_1$, then there is $w'' \in W$ such that $w'' \leq_w w'$ and $\langle M, w'' \rangle \models \varphi_1 \wedge \varphi_2$; and there is no world $w''' \in W_w$ such that $w''' \leq_w w''$ and $\langle M, w''' \rangle \models \varphi_1 \wedge \neg\varphi_2$.

Satisfiability of a formula is given with respect to the distinguished world w_0 in a structure (W, w_0, π, R). A formula φ is *satisfied in a structure* M if $\langle M, w_0 \rangle \models \varphi$. In this case, we say that $\langle M, w_0 \rangle$ is a *model* for φ. A formula φ is *satisfiable* if there is a model for φ. A formula φ is *valid* if it is satisfiable in all structures. A set of formulae $\Gamma = \{\gamma_1, \ldots, \gamma_n\}$, $n \in \mathbb{N}$, is satisfiable if, and only if, $\bigwedge_{i=1}^{n} \gamma_i$ is satisfiable.

Some further conditions can be imposed on the class of Kripke structures that characterise the semantics of S without affecting the set of valid formulae. For instance, only finite structures need to be considered, as the finite model property for preferential logics holds [3,6]. For finite structures, the interpretation of $\Rightarrow$ is much simpler: $\varphi \Rightarrow \psi$ is satisfiable at a world w if, and only if, all minimal φ-worlds in W_w satisfy $\varphi \wedge \psi$. Let S_n be the sublanguage of S with bounded nesting of at most n preferential operators. We recall the following lemma:

Lemma 1 [6, Lemma 3.1]. *Let φ be a formula of the form $(\psi_0 \not\Rightarrow \psi_0') \wedge \bigwedge_{i=1}^{k} (\psi_i \Rightarrow \psi_i')$ where $\psi_i, \psi_i' \in \mathsf{S}_0$, for all i, $0 \leq i \leq k$, $k \in \mathbb{N}$. If φ is satisfiable, then φ is satisfiable in a Kripke structure with at most $k+1$ worlds which are totally ordered by $\leq$.*

The proof of Lemma 1, as given in [6], shows that only total orderings over the set of worlds need to be considered when checking the satisfiability of a formula with only one occurrence of the dual of the conditional implication. For formulae with more occurrences of the dual operator, a disjoint set of total orderings over the set of worlds needs to be considered, one for each negated conditional. Still, a structure satisfying such a formula is polynomially bounded in the size of the formula [6, Proposition 3.2]. For the language of S_n, with bounded nesting of at most n preferential operators, $n > 1$, testing the satisfiability of a formula can be restricted to structures which are polynomial in the size of the formula, where the degree of the polynomial is bounded by $2 \times n$ [6, Proposition 3.6]. As given in [6], the satisfiability problem for S is PSPACE-complete.

3   Calculus

In this section, we present the clausal resolution-based calculus $\mathsf{RES_S}$ for checking the satisfiability of formulae in the language of S. A *clause* is a disjunction of

Table 1. NFF transformation rules

$$\begin{aligned}
\mathsf{nnf}(l) &= l \text{ for } l \in \mathsf{L} & \mathsf{nnf}(\neg\neg\varphi) &= \mathsf{nnf}(\varphi) \\
\mathsf{nnf}(\bigwedge \varphi_i) &= \bigwedge \mathsf{nnf}(\varphi_i) & \mathsf{nnf}(\neg(\bigwedge \varphi_i)) &= \bigvee \mathsf{nnf}(\neg\varphi_i) \\
\mathsf{nnf}(\bigvee \varphi_i) &= \bigvee \mathsf{nnf}(\varphi_i) & \mathsf{nnf}(\neg(\bigvee \varphi_i)) &= \bigwedge \mathsf{nnf}(\neg\varphi_i) \\
\mathsf{nnf}(\varphi \rightarrow \psi) &= \mathsf{nnf}(\neg\varphi) \vee \mathsf{nnf}(\psi) & \mathsf{nnf}(\neg(\varphi \rightarrow \psi)) &= \mathsf{nnf}(\varphi) \wedge \mathsf{nnf}(\neg\psi) \\
\mathsf{nnf}(\varphi \Rightarrow \psi) &= \mathsf{nnf}(\varphi) \Rightarrow \mathsf{nnf}(\psi) & \mathsf{nnf}(\neg(\varphi \Rightarrow \psi)) &= \mathsf{nnf}(\neg\varphi \not\Rightarrow \neg\psi) \\
\mathsf{nnf}(\varphi \not\Rightarrow \psi) &= \mathsf{nnf}(\varphi) \not\Rightarrow \mathsf{nnf}(\psi) & \mathsf{nnf}(\neg(\varphi \not\Rightarrow \psi)) &= \mathsf{nnf}(\neg\varphi \Rightarrow \neg\psi) \\
\mathsf{nnf}(\varphi \leftrightarrow \psi) &= \mathsf{nnf}(\varphi \rightarrow \psi) \wedge \mathsf{nnf}(\psi \rightarrow \varphi) \\
\mathsf{nnf}(\neg(\varphi \leftrightarrow \psi)) &= \mathsf{nnf}(\varphi \wedge \neg\psi) \vee \mathsf{nnf}(\psi \wedge \neg\varphi)
\end{aligned}$$

literals or modal formulae of the form $(\varphi \Rightarrow \varphi')$ or $(\varphi \not\Rightarrow \varphi')$, where φ and φ' have no subformulae whose main operator is $\Rightarrow$ or $\not\Rightarrow$. A *literal clause* is a clause with no occurrences of modal operators, that is, it is a disjunction of literals. A formula is in Conjunctive Normal Form (CNF) if, and only if, it is a conjunction of initial and global clauses, defined as follows:

initial clause: $\bigvee_{a=1}^{n} l_a$
global clause: $\boxed{*}(\bigvee_{b=1}^{m_1} l'_b \vee \bigvee_{c=1}^{m_2}(\varphi_c \Rightarrow \psi_c) \vee \bigvee_{d=1}^{m_3}(\varphi'_d \not\Rightarrow \psi'_d))$

where $n, m_1, m_2, m_3 \in \mathbb{N}$, $l_a, l'_b \in \mathsf{L}$, $\varphi_c, \psi_c, \psi'_d$ are literal clauses, φ'_d is in Negation Normal Form (NNF), no formulae contains nested modal operators, and $\boxed{*}$ is the universal operator. We introduce the universal operator because the translation into the normal form requires that the definition of formulae being renamed is available throughout the whole model. The universal operator is interpreted as usual: if $M = (W, w_0, \pi, R)$ is a Kripke structure and $w \in W$, then $\langle M, w \rangle \models \boxed{*}\varphi$ if, and only if, for all $w' \in W$, $\langle M, w' \rangle \models \varphi$. The empty clause is denoted by **false**. The transformation into the normal form uses rewriting and renaming, where the renaming technique is used to replace complex formulae in the scope of the disjunctions (except for the left-hand side of the dual operator) and the nesting of conditional operators by new propositional symbols. Clauses and formulae within the scope of modal operators are required to be in simplified form, that is, $\varphi \vee \varphi$, $\varphi \wedge \varphi$, $\varphi \vee$ **false**, and $\varphi \wedge$ **true** simplify to φ; $\varphi \vee \neg\varphi$, $\varphi \vee$ **true**, and $\boxed{*}$ **true** simplify to **true**; and $\varphi \wedge \neg\varphi$, $\varphi \wedge$ **false**, and $\boxed{*}$ **false** simplify to **false**.

The transformation of a formula φ in the language of S into CNF is given as follows. We denote the NNF of φ by $\mathsf{nnf}(\varphi)$, which is obtained by applying the function $\mathsf{nnf} : \mathsf{WFF_S} \longrightarrow \mathsf{WFF_S}$ to φ, whose definition is given in Table 1, where l is a literal, $\varphi, \varphi_i, \psi \in \mathsf{WFF_S}$ and $i \in \mathbb{N}$. Let φ be a well-formed formula in NNF. The translation of φ into Conjunctive Normal Form is defined as

$$\mathsf{cnf}(\varphi) = t_0 \wedge \tau(\boxed{*}(t_0 \rightarrow \varphi))$$

where t_0 is a new propositional symbol and the transformation function $\tau :$ $\mathsf{WFF_S} \longrightarrow \mathsf{WFF_S}$ is defined as follows (where $\varphi, \varphi_i, \psi, \chi \in \mathsf{WFF_S}$, $i \in \mathbb{N}$, t is a literal, and t' is a new propositional symbol). For the base case, the right-hand side of the implication is a disjunction where each disjunct is a literal, or it is of the form $(\varphi' \Rightarrow \varphi'')$ or $(\psi' \not\Rightarrow \psi'')$, where ψ' is a propositional formula (i.e. with no occurrences of subformulae whose main operator is either $\Rightarrow$ or $\not\Rightarrow$), and the formulae φ', φ'' and ψ'' are literal clauses:

$$\tau(\boxed{*}(t \to \varphi)) = \boxed{*}(\neg t \vee \varphi)$$

If the right-hand side of the implication or the right-hand side of a conditional is a conjunction, then rewriting is applied:

$$\tau(\boxed{\Box}(t \to \bigwedge \varphi_i)) = \bigwedge \tau(\boxed{\Box}(t \to \varphi_i))$$
$$\tau(\boxed{\Box}(t \to (\varphi \Rightarrow \bigwedge \varphi_i))) = \bigwedge \tau(\boxed{\Box}(t \to (\varphi \Rightarrow \varphi_i)))$$

If any of the disjuncts on the right-hand side of the implication is not a literal, then renaming is applied:

$$\tau(\boxed{*}(t \to \bigvee \varphi_i \vee \psi)) = \tau(\boxed{*}(t \to \bigvee \varphi_i \vee t')) \wedge \tau(\boxed{*}(\mathsf{nnf}(t' \leftrightarrow \psi)))$$

Conjunctions on the left-hand side of conditionals and on the right-hand side of the dual operator are renamed as follows:

$$\tau(\boxed{*}(t \to (\bigwedge \varphi_i \Rightarrow \chi))) = \tau(\boxed{*}(t \to (t' \Rightarrow \chi))) \wedge \tau(\boxed{*}(\mathsf{nnf}(t' \leftrightarrow \bigwedge \varphi_i)))$$
$$\tau(\boxed{*}(t \to (\varphi \not\Rightarrow \bigwedge \varphi_i))) = \tau(\boxed{*}(t \to (\varphi \not\Rightarrow t'))) \wedge \tau(\boxed{*}(\mathsf{nnf}(t' \leftrightarrow \bigwedge \varphi_i)))$$

If any of the disjuncts on the left-hand side of the conditional or on the right-hand side of a (negated) conditional is not a literal, that is, if ψ in the following is not a literal, then renaming is also applied.

$$\tau(\boxed{*}(t \to ((\bigvee \varphi_i \vee \psi) \Rightarrow \chi))) = \tau(\boxed{*}(t \to ((\bigvee \varphi_i \vee t') \Rightarrow \chi))) \wedge \tau(\boxed{*}(\mathsf{nnf}(t' \leftrightarrow \psi)))$$
$$\tau(\boxed{*}(t \to (\varphi \Rightarrow (\psi \vee \bigvee \varphi_i)))) = \tau(\boxed{*}(t \to (\varphi \Rightarrow (t' \vee \bigvee \varphi_i)))) \wedge \tau(\boxed{*}(\mathsf{nnf}(t' \leftrightarrow \psi)))$$
$$\tau(\boxed{*}(t \to (\varphi \not\Rightarrow (\psi \vee \bigvee \varphi_i)))) = \tau(\boxed{*}(t \to (\varphi \not\Rightarrow (t' \vee \bigvee \varphi_i)))) \wedge \tau(\boxed{*}(\mathsf{nnf}(t' \leftrightarrow \psi)))$$

If the left hand-side of the negated conditional is not a propositional formula, that is, if ψ is of the form $(\psi' \Rightarrow \psi'')$ or $(\psi' \not\Rightarrow \psi'')$, then renaming is also applied. Let $\varphi[\psi \mapsto t']$ denote the result of replacing some occurrences of the subformula ψ in φ by t':

$$\tau(\boxed{*}(t \to (\varphi \not\Rightarrow \chi))) = \tau(\boxed{*}(t \to (\varphi[\psi \mapsto t'] \not\Rightarrow \chi))) \wedge \tau(\boxed{*}(\mathsf{nnf}(t' \leftrightarrow \psi)))$$

Although the resolution-based method for S is essentially clausal, note that formulae on the left-hand side of the modal operator $\not\Rightarrow$ are not required to be literal clauses. This helps preserving more of the structure of the original formula and, as so, to identifying the cases where the axioms **A4** and **A5** should be propagated. We also note that, for a formula φ being renamed by a propositional symbol t, if φ occurs only with positive (resp. negative) polarity, then only the implication $t \to \varphi$ (resp. $(\varphi \to t)$) is needed [16]. As formulae on the left-hand side of conditionals occur with both polarities, in order to simplify our presentation, we have chosen to introduce both sides of the definition of φ, i.e. $(t \to \varphi)$ and $(\varphi \to t)$, into the clause set.

As checking the satisfiability of a conjunction $\bigwedge_{i=1}^{n} \varphi_i$, $n \in \mathbb{N}$, is equivalent to checking the satisfiability of the set $\{\varphi_1, \ldots, \varphi_n\}$, we refer to a formula into CNF as a *set of clauses*. Given a set of clauses in CNF, the resolution procedure is applied until a contradiction, in the form of **false**, is found or no new clauses can be derived. The inference rules can be divided into two sets: a set of rules for propagation of formulae whose main operator are either the conditional or its dual; and a set of resolution-based rules.

Table 2. Inference Rules

[L-AND-2]
$$\frac{\boxed{*}(D \vee ((\varphi \wedge \psi) \not\Rightarrow \chi))}{\boxed{*}(D \vee (\varphi \not\Rightarrow \chi) \vee (\psi \not\Rightarrow \chi))}$$

[L-OR-2]
$$\frac{\boxed{*}(D \vee ((\varphi \vee \psi) \not\Rightarrow \chi))}{\boxed{*}(D \vee (\varphi \not\Rightarrow \psi) \vee (\varphi \not\Rightarrow \chi))}$$

[REF-1]
$$\frac{\boxed{*}(D \vee (\varphi \Rightarrow \psi))}{\boxed{*}(\varphi \Rightarrow \varphi)}$$

[REF-2]
$$\frac{\boxed{*}(D \vee (\varphi \not\Rightarrow \psi))}{\boxed{*}(\neg\varphi \Rightarrow \neg\varphi)}$$

[SIMP-1]
$$\frac{\boxed{*}(D \vee (\varphi \not\Rightarrow \mathbf{false}))}{\boxed{*}(D)}$$

[I-RES-1]
$$\frac{D \vee l \quad D' \vee \neg l}{D \vee D'}$$

[I-RES-2]
$$\frac{D \vee l \quad \boxed{*}(D' \vee \neg l)}{D \vee D'}$$

where D' is a literal clause

[RES]
$$\frac{\boxed{*}(D \vee l) \quad \boxed{*}(D' \vee \neg l)}{\boxed{*}(D \vee D')}$$

[R-RES-$\Rightarrow$-1]
$$\frac{\boxed{*}(D \vee (\varphi \Rightarrow (\psi \vee l))) \quad \boxed{*}(D' \vee (\varphi' \Rightarrow (\chi \vee \neg l)))}{\boxed{*}(D \vee D' \vee \neg(\varphi \leftrightarrow \varphi') \vee (\varphi \Rightarrow (\psi \vee \chi)))}$$

[R-RES-$\Rightarrow$-2]
$$\frac{\boxed{*}(D \vee (\varphi \Rightarrow (\psi \vee l))) \quad \boxed{*}(D' \vee \neg l)}{\boxed{*}(D \vee (\varphi \Rightarrow (\psi \vee D')))}$$

where D' is a literal clause

[L-RES-$\not\Rightarrow$]
$$\frac{\boxed{*}(D \vee (\varphi \Rightarrow \mathbf{false})) \quad \boxed{*}(D' \vee (\varphi' \not\Rightarrow \psi))}{\boxed{*}(D \vee D' \vee \neg(\neg\varphi \leftrightarrow \varphi'))}$$

[R-RES-$\not\Rightarrow$-1]
$$\frac{\boxed{*}(D \vee (\varphi \Rightarrow (\psi \vee l))) \quad \boxed{*}(D' \vee (\varphi' \not\Rightarrow (\chi \vee \neg l)))}{\boxed{*}(D \vee D' \vee \neg(\neg\varphi \leftrightarrow \varphi') \vee (\varphi' \not\Rightarrow (\psi \vee \chi)))}$$

[R-RES-$\not\Rightarrow$-2]
$$\frac{\boxed{*}(D \vee (\varphi \not\Rightarrow (\psi \vee l))) \quad \boxed{*}(D' \vee \neg l)}{\boxed{*}(D \vee (\varphi \not\Rightarrow (\psi \vee D')))}$$

where D' is a literal clause

The inference rules used for propagation given in Table 2 are closely related to the axioms of S. The inference rule [L-OR-2] is related to cautious monotonicity (axiom **A4**). The inference rule [L-AND-2] is related to the disjunction property on the left-hand side of a conditional (axiom **A5**). The inference rules [REF-1] and [REF-2] correspond to reflexivity, that is, the axiom **A1**. Finally, the [SIMP-1] corresponds to simplification, as $\varphi \not\Rightarrow \mathbf{false}$ is unsatisfiable.

The resolution-based inference rules are also given in Table 2, where [I-RES-1] and [I-RES-2], the resolution rules related to *initial* clauses, and [RES] are syntactical variations of the classical binary resolution rule given in [18]. The remaining resolution-based inference rules are justified by the axioms **A2** and **A3**. The resolution-based rule [R-RES-$\Rightarrow$-1] says that when the left-hand side of the conditionals in the premises are equivalent, then the standard binary resolution rule can be applied to the right-hand side of those conditionals. Note that in the case of [R-RES-$\not\Rightarrow$-1], which is similar, the negated conditional in the premises is of the form $(\varphi' \not\Rightarrow \chi')$ and, from the definition of the dual, we have that this conditional is equivalent to $\neg(\neg\varphi' \Rightarrow \neg\chi')$. The disjunct $\neg(\neg\varphi \leftrightarrow \varphi')$ in the conclusion then states that either φ' is not equivalent to the negation of φ (from the other premise) or resolution can be applied to the right-hand side of those conditionals. The inference rule [L-RES-$\not\Rightarrow$] says that if a formula φ cannot be satisfied in any ordering, as given by the premise $\varphi \Rightarrow \mathbf{false}$, then any negated conditional whose left-hand side is equivalent to $\neg\varphi$ cannot be satisfied either. The inference rules [R-RES-$\Rightarrow$-2] and [R-RES-$\nRightarrow$-2] apply resolution to a literal occurring in a global literal clause with its complement occurring on the right-hand side of conditionals.

The inference rules in Table 2 are presented in simplified form, as some of their conclusions are not transformed into the normal form. For the inference rules [R-RES-$\Rightarrow$-1], [R-RES-$\nRightarrow$-1], and [L-RES-$\nRightarrow$] the resolvent should be rewritten into the normal form. In these cases, distribution can be used to avoid further renaming: a formula as $D\vee(\varphi\vee(\psi\wedge\chi))$ can be rewritten as the clauses $(D\vee\varphi\vee\psi)$ and $(D\vee\varphi\vee\chi)$, for a disjunction D, and formulae φ, ψ, and χ. However, for the resolvents of [REF-1] and [REF-2], further renaming may need to be applied. For instance, if $(\varphi\vee\psi)\nRightarrow\chi$ is a subformula in the clause set, then an application of [REF-2] would generate $(\neg\varphi\wedge\neg\psi)\Rightarrow(\neg\varphi\wedge\neg\psi)$. However, as from the definition of the normal form, conjunctions are not allowed on the left-hand side of conditional clauses. Instead of $(\neg\varphi\wedge\neg\psi)\Rightarrow(\neg\varphi\wedge\neg\psi)$, the clauses corresponding to $t\Rightarrow t$ and $t\leftrightarrow(\neg\varphi\wedge\neg\psi)$, where t is a new propositional symbol, are introduced in the clause set. This is not problematic from the point of view of termination, as [REF-1] and [REF-2] are only applied to formulae which can possibly occur in the clause set. As we show later, because the number of such formulae is finite, so it is the number of new propositional symbols that can be introduced as a result of the application of either inference rule.

The soundness of all inference rules follows almost immediately from the axiomatisation of S, as shown in Sect. 4. The following is the formal definition of a derivation.

Definition 1. *Let Φ be a set of clauses. A derivation in $\mathsf{RES_S}$ for Φ is a sequence of clause sets $\Phi_0, \Phi_1, \ldots$ where $\Phi_0 = \Phi$ and, for each $i > 0$, $\Phi_{i+1} = \Phi_i \cup \{D\}$, where D is the conclusion obtained from Φ_i by an application of one of the inference rules given in Table 2 to premises in Φ_i. We require that D is in simplified form, $D \notin \Phi_i$, and that D is not a propositional tautology.*

Note that the inference rules [REF-1] and [REF-2] introduce tautologies of the form $\varphi \Rightarrow \varphi$, where φ or $\neg\varphi$ occurs on the left-hand side of (negated) conditionals. Those tautologies are needed for completeness. Thus, the constraint for including the resolvent D into the clause set is restricted to classical tautologies, that is, of the form $\varphi \vee \neg\varphi$, for a formula φ.

Definition 2. *Let Φ be a set of clauses. A refutation in $\mathsf{RES_S}$ for Φ is a finite derivation $\Phi_0, \Phi_1, \ldots, \Phi_k$, $k \in \mathbb{N}$, where **false** in Φ_k. We write $\Phi \vdash_{\mathsf{RES_S}}$ **false**, if there is a refutation from Φ in $\mathsf{RES_S}$.*

Definition 3. *Let Φ be a set of clauses. We say that Φ is saturated if any further application of the inference rules given in Table 2 to clauses in Φ only generates a clause already in Φ.*

As derivations require progress, a saturated set is a point where a derivation cannot progress any further.

Definition 4. *Let Φ be a set of clauses. A derivation $\Phi_0, \Phi_1, \dots$ in $\mathsf{RES_S}$ for Φ terminates if there is $k \in \mathbb{N}$ such that Φ_k is saturated or $\mathbf{false} \in \Phi_k$.*

Before showing the correctness results concerning our calculus, we present an example of a refutation involving a validity with nested conditionals.

Example 1. We show that $\varphi = ((a \Rightarrow b) \wedge (a \Rightarrow c) \wedge (d \Rightarrow c)) \Rightarrow (((a \wedge b) \vee d) \Rightarrow c)$ is a valid formula in S. The negation normal form of the $\neg\varphi$ is $((\neg a \not\Rightarrow \neg b) \vee (\neg a \not\Rightarrow \neg c) \vee (\neg d \not\Rightarrow \neg c)) \not\Rightarrow (((\neg a \vee \neg b) \wedge \neg d) \not\Rightarrow \neg c)$. Clauses 1 to 6 correspond to the normal form of $\neg\varphi$, noting that Clauses 4 to 6 only show the side of the definitions of the propositional symbols introduced by renaming that are needed in the proof.

1. t_0
2. $\boxed{*}(\neg t_0 \vee ((t_1 \vee t_2 \vee t_3) \not\Rightarrow t_4))$
3. $\boxed{*}(t_1 \vee (a \Rightarrow b))$
4. $\boxed{*}(t_2 \vee (a \Rightarrow c))$
5. $\boxed{*}(t_3 \vee (d \Rightarrow c))$
6. $\boxed{*}(\neg t_4 \vee (((\neg a \vee \neg b) \wedge \neg d) \not\Rightarrow \neg c))$

The following refutation follows from the above set of clauses:

7. $\boxed{*}(\neg t_4 \vee ((\neg a \vee \neg b) \not\Rightarrow \neg c) \vee (\neg d \not\Rightarrow \neg c))$ [L-AND-2,6]
8. $\boxed{*}(t_3 \vee \neg t_4 \vee ((\neg a \vee \neg b) \not\Rightarrow \neg c))$ [R-RES-$\not\Rightarrow$-1,7,5]
9. $\boxed{*}(t_3 \vee \neg t_4 \vee (\neg a \not\Rightarrow \neg b) \vee (\neg a \not\Rightarrow \neg c))$ [L-OR-2,8]
10. $\boxed{*}(t_2 \vee t_3 \vee \neg t_4 \vee (\neg a \not\Rightarrow \neg b))$ [R-RES-$\not\Rightarrow$-1,9,4]
11. $\boxed{*}(t_1 \vee t_2 \vee t_3 \vee \neg t_4)$ [SIMP-1,R-RES-$\not\Rightarrow$-1,10,3]
12. $\boxed{*}(t_5 \Rightarrow t_5)$ [REF-2,2, where $t_5 \leftrightarrow \neg(t_1 \vee t_2 \vee t_3)$]
13. $\boxed{*}(t_5 \vee t_1 \vee t_2 \vee t_3)$ [REF-2,2]
14. $\boxed{*}(\neg t_5 \vee \neg t_1)$ [REF-2,2]
15. $\boxed{*}(\neg t_5 \vee \neg t_2)$ [REF-2,2]
16. $\boxed{*}(\neg t_5 \vee \neg t_3)$ [REF-2,2]
17. $\boxed{*}(\neg t_5 \vee t_2 \vee t_3 \vee \neg t_4)$ [RES,11,14]
18. $\boxed{*}(\neg t_5 \vee t_3 \vee \neg t_4)$ [RES,17,15]
19. $\boxed{*}(\neg t_5 \vee \neg t_4)$ [RES,18,16]
20. $\boxed{*}(\neg t_0 \vee ((t_1 \vee t_2 \vee t_3) \not\Rightarrow \neg t_5))$ [R-RES-$\not\Rightarrow$-2,19,2]
21. $\boxed{*}(\neg t_0 \vee \neg(\neg t_5 \leftrightarrow (t_1 \vee t_2 \vee t_3)))$ [SIMP-1,R-RES-$\not\Rightarrow$-1,20,12]
22. $\boxed{*}\,\neg t_0$ [RES,21,13,14,15,16]
23. **false** [I-RES-2,22,1]

We note that Clause 8 (resp. Clauses 10 and 11) is in simplified form, as $\mathsf{nnf}(\neg(\neg d \leftrightarrow \neg d))$ (resp. $\mathsf{nnf}(\neg(\neg a \leftrightarrow \neg a))$) simplifies to **false**. The justification of Clause 22 abbreviates several applications of the inference rule [RES] between Clause 21 and the clauses corresponding to the definition of t_5, i.e. Clauses 13 to 16.

4   Correctness

In this section we provide the proofs that $\mathsf{RES_S}$ is a sound, complete, and terminating calculus for S. First, we show that given a formula φ, the transformation into CNF is satisfiability preserving.

Theorem 1. *Let φ be a formula in $\mathsf{WFF_S}$. Then, φ is satisfiable if, and only if, $t_0 \wedge \tau(\boxed{*}(t_0 \rightarrow \varphi))$ is satisfiable.*

Proof (sketch). The proof is very standard. We first show that φ is satisfiable if, and only if, $t_0 \wedge \boxed{*}(t_0 \to \varphi)$ is satisfiable, as the evaluation of φ does not depend on the evaluation of t_0 and that the operator $\boxed{*}$ does not occur in φ. Then, we show that each of the transformation rules is satisfiability preserving, that is, a formula of the form $\boxed{*}(t \to \varphi')$ is satisfiable if, and only if, $\tau(\boxed{*}(t \to \varphi'))$ is satisfiable. Rewriting is justified by equivalences. For transformation steps which require renaming, let ψ be the subformula of φ' which is being renamed by the transformation function and t' a new propositional symbol. Given the satisfiability of $\boxed{*}(t \to \varphi')$, then there is a model $M = (W, w_0, \pi, R)$ such that $\langle M, w \rangle \models (t \to \varphi')$, for all $w \in W$. We then build a model $M' = (W, w_0, \pi', R)$, which is exactly as M except by the evaluation function. We define $\pi'(w)(p) = \pi(w)(p)$ for all worlds $w \in W$ and propositional symbols p, such that $p \neq t'$; and $\pi'(w)(t') = true$ if, and only if, $\langle M, w \rangle \models \psi$. We then show that, for all worlds $w \in W$, we have that $\langle M', w \rangle \models (t \to \varphi'[\psi \mapsto t']) \wedge (t' \leftrightarrow \psi)$, where $\varphi'[\psi \mapsto t']$ is the result of replacing some occurrences of the subformula ψ in φ' by t'. The if part follows easily by taking into account that, by construction, t' and ψ are satisfied at the same worlds in a model; hence $(t \to \varphi'[t' \mapsto \psi])$ is satisfiable in all worlds $w \in W$. It follows that $\tau(\boxed{*}(t \to \varphi'))$ is satisfiable. Finally, by induction on the number of steps of a transformation, we obtain that φ and $t_0 \wedge \tau(\boxed{*}(t_0 \to \varphi))$ are equisatisfiable. □

We note that the transformation into the normal form results in a formula which is polynomial in the size of the original formula, as the number of subformulae of a formula is linear in the size of the original formula and also because the renaming procedure introduces at most two copies of every subformula (plus a constant number of connectives). The procedure is also terminating, as only complex subformulae of a formulae are either rewritten or renamed.

Lemma 2. *Let Φ be a set of clauses. Then, any derivation in $\mathsf{RES_S}$ from Φ terminates.*

Proof. Let P_Φ^+ be the set of propositional symbols occurring in Φ. We define $\mathsf{P}_\Phi^- = \{\neg p \mid p \in \mathsf{P}_\Phi\}$ and $\mathsf{L}_\Phi = \mathsf{P}_\Phi^+ \cup \mathsf{P}_\Phi^-$. Let Sub_Φ^+ be the set of all propositional subformulae occurring in Φ, $Sub_\Phi^- = \{\mathsf{nnf}(\neg\varphi) \mid \varphi \in Sub_\Phi^+\}$, and $Sub_\Phi^\pm = Sub_\Phi^+ \cup Sub_\Phi^-$. As only propositional formulae can occur on the left-hand side of the conditional implications and its dual, then the number of additional literals that might be introduced during the application of [REF-1] and [REF-2] is bounded by $|Sub_\Phi^\pm|$. As P_Φ and $Sub_\Phi^\pm$ are finite, so it is $\mathscr{P}(\mathsf{L}_\Phi \cup Sub_\Phi^\pm \bigcup_{\varphi \in Sub_\Phi^\pm}\{t_\varphi \Rightarrow t_\varphi, t \leftrightarrow \varphi\})$, where t_φ is a new propositional symbol. Let C_Φ be the largest set of clauses that can be constructed from L_Φ, $Sub_\Phi^\pm$, and the conditionals introduced by [REF-1] and [REF-2] together with the double-implications introduced for renaming of formulae in $Sub_\Phi^\pm$. From Definition 1, for all derivations $\Phi_0, \Phi_1, \dots$ from Φ, we have that $\Phi_i \subset C_\Phi$ and also that $\Phi_i \subset \Phi_{i+1}$, for all $i > 0$. Thus, every derivation must terminate. □

For soundness of $\mathsf{RES_S}$, we need to show that, for each of the inference rules given in Table 2, if the premises of the inference rules are satisfiable, so it is their conclusion. We omit most of the easy cases, but note that soundness of [L-AND-2] and [L-OR-2] follow almost immediately from the contrapositive forms of the axioms **A5** and **A4**, that is, $((\varphi \wedge \psi) \not\Rightarrow \chi) \rightarrow (\varphi \not\Rightarrow \chi) \vee (\psi \not\Rightarrow \chi))$ and $((\varphi \vee \psi) \not\Rightarrow \chi) \rightarrow (\varphi \not\Rightarrow \psi) \vee (\varphi \not\Rightarrow \chi))$, respectively. The inference rules [REF-1] and [REF-2] are obviously sound, as they introduce instances of the axiom **A1** into the clause set. It is also very easy to see that [I-RES-1], [I-RES-2], and [RES] are only variations of the classical binary resolution: the fact they are sound follows also almost immediately from the results in [18]. The next lemmas show the soundness of [R-RES-$\not\Rightarrow$-1] and [L-RES-$\not\Rightarrow$].

Lemma 3. *Let Φ be a set of clauses with $\{ \boxed{*}(D \vee (\varphi \Rightarrow (\psi \vee l))), \boxed{*}(D' \vee (\varphi' \not\Rightarrow (\chi \vee \neg l)))\} \subseteq \Phi$. If Φ is satisfiable, then $\Phi \cup \{ \boxed{*}(D \vee D' \vee \neg(\neg\varphi \leftrightarrow \varphi') \vee (\varphi' \not\Rightarrow \psi \vee \chi)\}$ is satisfiable.*

Proof. If Φ is satisfiable, as $\{ \boxed{*}(D \vee (\varphi \Rightarrow (\psi \vee l))), \boxed{*}(D' \vee (\varphi' \not\Rightarrow (\chi \vee \neg l)))\} \subseteq \Phi$, from the definition of satisfiability of sets, there is a model $M = (W, w_0, \pi, R)$ such that (1) $\langle M, w_0 \rangle \models \boxed{*}(D \vee (\varphi \Rightarrow (\psi \vee l)))$ and (2) $\langle M, w_0 \rangle \models \boxed{*}(D' \vee (\varphi' \not\Rightarrow (\chi \vee \neg l)))$. From (1) and the semantics of the universal operator, for all $w \in W$, we have that (3) $\langle M, w \rangle \models (D \vee (\varphi \Rightarrow (\psi \vee l)))$. Analogously, from (2), for all $w \in W$, we have that (4) $\langle M, w \rangle \models (D' \vee (\varphi' \not\Rightarrow (\chi \vee \neg l)))$. Let w be any world in W. From (3) and (4), by distribution, there are four cases: (i) $\langle M, w \rangle \models (D \wedge D')$; (ii) $\langle M, w \rangle \models (D \wedge (\varphi' \not\Rightarrow (\chi \vee \neg l)))$; (iii) $\langle M, w \rangle \models (D' \wedge (\varphi \Rightarrow (\psi \vee l)))$; or (iv) $\langle M, w \rangle \models (\varphi \Rightarrow (\psi \vee l)) \wedge (\varphi' \not\Rightarrow (\chi \vee \neg l))$. It is easy to see that if Cases (i), (ii), or (iii) hold, then we have that (5) $\langle M, w \rangle \models (D \vee D')$. For the fourth case, there are two possibilities: either (6) $\langle M, w \rangle \models \neg(\neg\varphi \leftrightarrow \varphi')$; or (7) $\langle M, w \rangle \models (\neg\varphi \leftrightarrow \varphi')$. From (7) and from the fact that $\langle M, w \rangle \models (\varphi' \not\Rightarrow (\chi \vee \neg l))$, by soundness of [RPE], we obtain that (8) $\langle M, w \rangle \models (\neg\varphi \not\Rightarrow (\chi \vee \neg l))$. From (8), the fact that $\langle M, w \rangle \models (\varphi \Rightarrow (\psi \vee l))$, the semantics of conjunctions, and the soundness of **A2**, by the soundness of [MP], we obtain that $\langle M, w \rangle \models (\neg\varphi \not\Rightarrow (\psi \vee l) \wedge (\chi \vee \neg l))$. By the soundness of resolution, applied on the right-hand side of the preferential conditional, we obtain that (9) $\langle M, w \rangle \models (\neg\varphi \not\Rightarrow (\psi \vee \chi))$. From (9) and (7), we obtain that (10) $\langle M, w \rangle \models (\varphi' \not\Rightarrow (\psi \vee \chi))$. From (5), (6), (10), and from the semantics of disjunction, we finally have that $\langle M, w \rangle \models (D \vee D' \vee \neg(\neg\varphi \leftrightarrow \varphi') \vee (\varphi' \not\Rightarrow \psi \vee \chi))$. As this holds for any world w, from the semantics of the universal operator, it follows that $\langle M, w_0 \rangle \models \boxed{*}(D \vee D' \vee \neg(\neg\varphi \leftrightarrow \varphi') \vee (\varphi' \not\Rightarrow \psi \vee \chi))$. We conclude that $\Phi \cup \{ \boxed{*}(D \vee D' \vee \neg(\neg\varphi \leftrightarrow \varphi') \vee (\varphi' \not\Rightarrow \psi \vee \chi))\}$ is satisfiable. ☐

Lemma 4. *Let Φ be a set of clauses with $\{ \boxed{*}(D \vee (\varphi \Rightarrow \mathbf{false})), \boxed{*}(D' \vee (\varphi' \not\Rightarrow \psi))\} \subseteq \Phi$. If Φ is satisfiable, then $\Phi \cup \{ \boxed{*}(D \vee D' \vee \neg(\neg\varphi \leftrightarrow \varphi'))\}$ is satisfiable.*

Proof (Sketch). The proof follows from the fact that $\boxed{*}(D \vee (\varphi \Rightarrow \mathbf{false}))$ is semantically equivalent to $\boxed{*}(D \vee (\varphi \Rightarrow \psi \wedge \neg\psi))$ and, from **A3**, this implies that $\boxed{*}(D \vee (\varphi \Rightarrow \neg\psi))$ is satisfiable. By Lemma 3 taking Φ with $\{ \boxed{*}(D \vee (\varphi \Rightarrow \neg\psi)), \boxed{*}(D' \vee (\varphi' \not\Rightarrow \psi))\} \subseteq \Phi$, together with the soundness of [SIMP-1], we obtain that $\Phi \cup \{ \boxed{*}(D \vee D' \vee \neg(\neg\varphi \leftrightarrow \varphi'))\}$ is satisfiable. ☐

The proof that [R-RES-$\Rightarrow$-1] is sound is pretty similar to that of Lemma 3. Soundness of [R-RES-$\Rightarrow$-2] and [R-RES-$\not\Rightarrow$-2] follow easily from the fact that the right-hand side of the operators $\Rightarrow$ and $\not\Rightarrow$ are monotonic. Thus resolution can be applied on the the right-hand side of modal formulae, taking into account that the other premise is also in the scope of the universal operator. The next theorem shows that $\mathsf{RES_S}$ is sound, the proof of which follows from our argumentation, as above, and Lemmas 3 and 4.

Theorem 2. *Let Φ be a set of clauses and $\Phi_0, \Phi_1, \ldots$ be a derivation in $\mathsf{RES_S}$ for Φ. If Φ is satisfiable, then every Φ_i, $i \geq 0$, is satisfiable.*

The soundness proof, given above, shows that if Φ is satisfiable, then there is no refutation from Φ, that is: if there is a structure M such that $M \models \Phi$, then $\Phi \not\vdash_{\mathsf{RES_S}} \mathbf{false}$. In the following, we prove the completeness of $\mathsf{RES_S}$: if $\Phi \not\vdash_{\mathsf{RES_S}} \mathbf{false}$, then there is a structure M such that $M \models \Phi$. The proof follows the standard construction of canonical models for modal logics and is heavily based on that given in [6].

Given a set of clauses Φ, we construct a structure (W, S), where W is a set (of worlds) and S is a binary relation over W, as follows. Let I and G denote the set of initial and global clauses in Φ, respectively. Let $G' = \{\varphi \mid \boxed{*}\, \varphi \in G\}$. Let φ_I, φ_G, and $\varphi_{G'}$ denote the conjunction of formulae in I, G and G', respectively. Let $Cl(\Phi)$ be the closure of Φ under subformulae and simple negation. That is, $Cl(\Phi)$ is the least set such that:

- $\varphi_I \wedge \varphi_G \wedge \varphi_{G'} \in Cl(\Phi)$;
- If $\varphi \in Cl(\Phi)$ and $\varphi' \in \mathsf{Subf}(\varphi)$, then $\varphi' \in Cl(\Phi)$;
- If $\varphi \in Cl(\Phi)$ and then $\mathsf{nnf}(\neg\varphi) \in Cl(\Phi)$;

where $\mathsf{Subf}(\varphi)$ denotes the set of subformulae of φ. (Recall that we consider sub-conjunctions of subformulae as subformulae.) Let $\mathscr{A}, \mathscr{B} \in \mathscr{P}(Cl(\Phi))$ be sets of formulae in the powerset of the closure of Φ. A set of formulae $\mathscr{A}$ is $\mathsf{RES_S}$-consistent if, and only if, (i) for all $\varphi \in \mathscr{A}$, $\neg\,\boxed{*}\,\varphi \notin \mathscr{A}$; and (ii) $\mathscr{A} \not\vdash_{\mathsf{RES_S}} \mathbf{false}$. A consistent set of formulae $\mathscr{A}$ is maximal with respect to $Cl(\Phi)$ if, and only if, (i) $G \subseteq \mathscr{A}$ (in order to ensure that all global clauses are in all sets); and (ii) there is no consistent $\mathscr{B} \in \mathscr{P}(Cl(\Phi))$ such that $\mathscr{A} \subset \mathscr{B}$. Although there is no specific inference rules for dealing with formulae of the form $\neg\,\boxed{*}\,\varphi$, as they do not occur in the normal form, a set containing such a formula cannot be maximal consistent, as G is a subset of all maximal sets.

An *atom* is a maximal consistent set in $\mathscr{P}(Cl(\Phi))$, the powerset of $Cl(\Phi)$. Let Atoms_Φ be the set of all atoms constructed from Φ. In the following, we denote atoms by a, b, c, d, and set of atoms by A, B, C, D. For an atom a, we write $\bigwedge a$ (resp. $\bigvee a$) as an abbreviation for the conjunction (resp. disjunction) of the formulae in a. A *world* is defined as a pair (a, A), where a is an atom and A a set of atoms. Given two atoms a and b, and a set of atoms B, we define that $\mathsf{Prefer}(a, b, B)$ holds if, and only if, $\bigwedge a \wedge \neg((\bigwedge b \vee \bigvee \bigwedge_{b' \in B} b') \Rightarrow \bigvee \bigwedge_{b' \in B} b')$ is $\mathsf{RES_S}$-consistent. We define the structure $M = (W, w_0, \pi, R)$ as follows:

- $W = \{(a, A) \mid a \in Atoms_\Phi, A \subseteq (Atoms_\Phi \setminus \{a\})\}$;
- $w_0 = (a, \emptyset) \in W$, with $\Phi \subseteq a$;
- For all propositional symbols $p \in \mathsf{P}$, let $\pi((a, A))(p) = true$ if, and only if, $p \in a$;
- For all worlds $(a, A) \in W$, let $W_{(a,A)} = \{(b, B) \in W \mid \mathsf{Prefer}(a, b, B)\}$ and set $(b, B) <_{(a,A)} (c, C)$, if $C \cup \{c\} \subseteq B$. For all worlds $w', w'' \in W_w$, set $\leq_w$ such that, $w' \leq_w w'$; and, if $w' <_w w''$, then $w' \leq_w w''$.

Intuitively, a in (a, A) is a world satisfying $\bigwedge a$ which is strictly preferred to all worlds in A. The evaluation function assigns truth values to propositional symbols according to their value in a. The set $W_{(a,A)}$ contains all worlds (b, B) such that it is consistent with a that b is strictly preferred to worlds satisfying atoms in B. Note that the construction of $W_{(a,A)}$ depends on the set of conditionals in a, as defined by the predicate $\mathsf{Prefer}(\cdot)$. Thus, if two atoms a and b share the same set of conditional formulae, then $W_{(a,A)}$ and $W_{(b,A')}$ are exactly the same. It is easy to check that the relation $\leq_w$ is indeed a preorder. Given those definitions, we establish the completeness of $\mathsf{RES_S}$. First, we note that the construction of a model from a saturated set of clauses is closed under the inference rules of $\mathsf{RES_S}$ and also that the following two properties hold.

Lemma 5. *Let Φ be a saturated set of clauses, G be the set of global clauses in Φ, $Cl(\Phi)$ be the closure of Φ, and a be an atom in $\mathsf{Atoms_\Phi}$, the set of all atoms constructed from Φ. For any formula $\varphi \in Cl(\Phi) \cup G$, $\varphi \in a$ if, and only if, $\mathsf{nnf}(\neg\varphi) \notin a$.*

Lemma 6. *Let Φ be a saturated set of clauses, G be the set of global clauses in Φ, and a be an atom in $\mathsf{Atoms_\Phi}$, the set of all atoms constructed from Φ. For any formula $\circledast\, \varphi \in G$, $\circledast\, \varphi \in a$ if, and only if, $\varphi \in a$.*

The proof of the truth lemma depends on the following two results. For $\varphi \Rightarrow \psi \in (a, A)$, as an additional (induction) hypotheses, we assume that for all subformulae φ' of $\varphi \Rightarrow \psi$ and all worlds (a', A'), we have that $\varphi' \in a'$ if, and only if, $\langle M, (a', A') \rangle \models \varphi'$.

Lemma 7. *Let Φ be a saturated set of clauses, $M = (W, w_0, \pi, R)$ be the structure constructed as above for Φ, $(a, A) \in W$ be a world in M, and φ, ψ be formulae in $Cl(\Phi)$. If $\varphi \Rightarrow \psi \in a$ in (a, A) then $\langle M, (a, A) \rangle \models \varphi \Rightarrow \psi$.*

Proof. For the purpose of contradiction, assume $\varphi \Rightarrow \psi \in (a, A)$, but that $\langle M, (a, A) \rangle \not\models \varphi \Rightarrow \psi$. If $\langle M, (a, A) \rangle \not\models \varphi \Rightarrow \psi$, then, from the semantics of $\Rightarrow$, there is a world $(b, B) \in W_{(a,A)}$ such that (b, B) is a minimal φ-world and $\langle M, (b, B) \rangle \models \varphi \wedge \neg\psi$. It follows that (i) $\langle M, (b, B) \rangle \models \varphi$ and (ii) $\langle M, (b, B) \rangle \models \neg\psi$. By induction hypotheses, from (i), we have that $\varphi \in b$ and, from (ii), that $\psi \notin b$ (or, equivalently, that $\mathsf{nnf}(\neg\psi) \in b$). If $(b, B) \in W_{(a,A)}$, then, from the definition of $W_{(a,A)}$, we have that $\mathsf{Prefer}(a, b, B)$ holds, that is, $a \wedge \neg((b \vee \bigvee B) \Rightarrow \bigvee B)$ is $\mathsf{RES_S}$-consistent. However, we can show that for $\varphi, \neg\psi \in b$, we have that $\mathsf{Prefer}(a, b, B)$ and $\varphi \Rightarrow \psi \in (a, A)$ is not $\mathsf{RES_S}$-consistent (see Appendix A for the detailed proof), which contradicts with having $(b, B) \in W_{(a,A)}$. Thus, it cannot be the case that $\langle M, (a, A) \rangle \not\models \varphi \Rightarrow \psi$. Hence, $\langle M, (a, A) \rangle \models \varphi \Rightarrow \psi$. $\square$

Lemma 8. *Let Φ be a saturated set of clauses, $M = (W, w_0, \pi, R)$ be the structure constructed as above for Φ, $(a, A) \in W$ be a world in M, and φ, ψ be formulae in $Cl(\Phi)$. If $\langle M, (a, A)\rangle \models \varphi \Rightarrow \psi$, then $\varphi \Rightarrow \psi \in a$ in (a, A).*

Proof. We show the contrapositive, i.e. if $\varphi \Rightarrow \psi \notin a$ in (a, A), then $\langle M, (a, A)\rangle \not\models \varphi \Rightarrow \psi$. If $\varphi \Rightarrow \psi \notin a$, then $a \wedge (\varphi \Rightarrow \psi)$ is not $\mathsf{RES_S}$-consistent. By Lemma 5, $\mathsf{nnf}((\neg\varphi \not\Rightarrow \neg\psi)) \in (a, A)$. For the purpose of contradiction, assume that $\langle M, (a, A)\rangle \models \varphi \Rightarrow \psi$. Thus, for all (b, B) in $W_{(a,A)}$ such that (b, B) is a minimal φ-world, $\langle M, (b, B)\rangle \models \varphi \wedge \psi$. It follows that $\langle M, (b, B)\rangle \models \varphi$ and $\langle M, (b, B)\rangle \models \psi$. By inductive hypothesis, $\varphi \in (b, B)$ and $\psi \in (b, B)$. As $(b, B) \in W_{(a,A)}$, from the definition of $W_{(a,A)}$, $\mathsf{Prefer}(a, b, B)$ holds, i.e. $a \wedge \neg((b \vee \bigvee B) \Rightarrow \bigvee B)$ is $\mathsf{RES_S}$-consistent. We can show however that, for $\varphi, \psi \in b$, we have that $\neg((b \vee \bigvee B) \Rightarrow \bigvee B)$ and $\neg(\varphi \Rightarrow \psi)$ is not $\mathsf{RES_S}$-consistent (see Appendix A for the detailed proof). Thus, it cannot be the case that $\langle M, (a, A)\rangle \models \varphi \Rightarrow \psi$. Hence, $\langle M, (a, A)\rangle \not\models \varphi \Rightarrow \psi$.   □

Lemma 9. *Let Φ be a saturated set of clauses, $M = (W, w_0, \pi, R)$ be the structure constructed as above for Φ, $(a, A) \in W$ be a world in M, and φ a formula in $Cl(\Phi)$. Then, $\varphi \in a$ in (a, A) if, and only if, $\langle M, (a, A)\rangle \models \varphi$.*

The proofs for the classical connectives make use of Lemmas 5 and 6 and is routine. For formulae of the form $(\varphi \Rightarrow \psi)$, the proof follows from Lemmas 7 and 8. Completeness of $\mathsf{RES_S}$ follows immediately from the truth lemma (Lemma 9), as stated above.

5   Discussion and Further Work

We have presented a sound and complete resolution calculus for Burgess' system S. Our main motivation is to present a purely syntactic calculus that is both easy and efficient to implement. The only other calculus for S we are aware of is that of [19] which heavily relies on semantic arguments for the definition of proof rules, and is therefore non-trivial to both implement and optimise. In contrast, the resolution system here is purely given in syntactic terms. The design of the resolution rules, while not generated from the axioms by means of an algorithmic procedure, closely follows the axiomatisation. A different axiomatisation would lead to a different set of inference rules, in particular those related to propagation ([L-AND-2], [L-OR-2], [REF-1] and [REF-2]). The main technical challenge was the completeness proof, for which we have adapted a canonical model construction to the resolution setting, obtaining a direct proof (without translating to other calculi) where the main obstacle in the proof was to integrate the construction with pre-processing of formulae into normal form.

To fully substantiate our claim regarding ease of implementation and efficiency, we plan to implement and compare both our calculus and that of [19].

A   Proofs

The next two proofs were automatically generated by a prototype prover which implements the calculus given in this paper. Only clauses needed in the refutation are shown. Also the inference rule [SIMP-1] is always applied together with [R-RES-$\not\Rightarrow$-1], so clauses are already in simplified form. First, as part of the proof of Lemma 7, we show that for $\varphi, \neg\psi \in b$, we have that $\mathsf{Prefer}(a, b, B)$ and $\varphi \Rightarrow \psi \in (a, A)$ is contradictory.

1.	t_1	[Assumption]
2.	$(\neg t_1 \vee (\varphi \Rightarrow \psi))$	[Assumption, $\varphi \Rightarrow \psi \in a$]
3.	$(\neg t_1 \vee (((\neg\varphi \vee \psi) \wedge \neg B) \not\Rightarrow \neg B))$	[Assumption, $\varphi, \neg\psi \in b$, $(b, B) \in W_{(a,A)}$]
4.	$(\neg t_1 \vee (\neg B \not\Rightarrow \neg B) \vee ((\neg\varphi \vee \psi) \not\Rightarrow \neg B)$	[L-AND-2,3]
8.	$(B \vee \neg t_3 \vee t_2)$	[REF-2,3]
9.	$(\neg t_2 \vee t_3)$	[REF-2,3]
11.	$(\varphi \vee \neg t_2)$	[REF-2,3]
12.	$(\neg\psi \vee \neg t_2)$	[REF-2,3]
14.	$((B \Rightarrow B))$	[REF-2,4]
15.	$((t_2 \Rightarrow t_2))$	[REF-2,4]
25.	$(B \vee \neg\psi \vee \neg t_3)$	[RES,8,12,t_2]
26.	$(B \vee \varphi \vee \neg t_3)$	[RES,8,11,t_2]
38.	$(\neg B \vee \neg t_4)$	[R-RES-$\not\Rightarrow$-1,3,14]
39.	$(B \vee \neg t_4 \vee t_2)$	[R-RES-$\not\Rightarrow$-1,3,14]
40.	$(B \vee t_4 \vee t_5)$	[R-RES-$\not\Rightarrow$-1,3,14]
41.	$(\psi \vee \neg\varphi \vee \neg t_5)$	[R-RES-$\not\Rightarrow$-1,3,14]
42.	$(\neg B \vee \neg t_5)$	[R-RES-$\not\Rightarrow$-1,3,14]
44.	$(\neg t_1 \vee t_4)$	[R-RES-$\not\Rightarrow$-1,3,14,B]
49.	$(\varphi \vee \neg t_3 \vee \neg t_5)$	[RES,42,26,B]
50.	$(\neg\psi \vee \neg t_3 \vee \neg t_5)$	[RES,42,25,B]
254.	$(\psi \vee \neg t_3 \vee \neg t_5)$	[RES,41,49,φ]
276.	$(\neg t_3 \vee \neg t_5)$	[RES,254,50,ψ]
292.	$(\neg t_2 \vee \neg t_5)$	[RES,276,9,$\neg t_3$]
493.	$(B \vee \neg t_2 \vee t_4)$	[RES,40,292,t_5]
540.	$(B \vee \neg\psi \vee \neg t_4)$	[RES,39,12,t_2]
547.	$(\neg t_4 \vee t_2)$	[RES,39,38,B]
557.	$(\varphi \vee \neg t_4)$	[RES,547,11,t_2]
558.	$(\neg t_4 \vee t_3)$	[RES,547,9,t_2]
559.	$(\neg t_1 \vee t_2)$	[RES,547,44,$\neg t_4$]
560.	$(\neg t_4 \vee \neg t_5)$	[RES,547,292,t_2]
814.	$(\neg t_1 \vee (((\neg\varphi \vee \psi) \wedge \neg B) \not\Rightarrow \neg\psi \vee \neg t_4))$	[R-RES-$\not\Rightarrow$-2,540,3,B]
834.	$(\neg\varphi \vee \neg t_6 \vee t_5)$	[R-RES-$\not\Rightarrow$-1,814,2]
837.	$(B \vee \neg t_7 \vee t_2)$	[R-RES-$\not\Rightarrow$-1,814,2]
838.	$(\varphi \vee \neg t_7)$	[R-RES-$\not\Rightarrow$-1,814,2]
839.	$(\neg\varphi \vee \neg t_3 \vee t_7)$	[R-RES-$\not\Rightarrow$-1,814,2]
1333.	$(\neg\varphi \vee \neg t_4 \vee t_7)$	[RES,839,558,$\neg t_3$]
1361.	$(\neg t_4 \vee t_7)$	[RES,1333,557,φ]
1372.	$(\neg t_1 \vee t_7)$	[RES,1361,44,$\neg t_4$]
1374.	$(B \vee \neg t_2 \vee t_7)$	[RES,1361,493,$\neg t_4$]
1885.	$(B \vee \neg t_7 \vee t_4)$	[RES,837,493,t_2]

1911. $(\neg t_5 \vee \neg t_7 \vee t_4)$ [RES,1885,42,B]
1981. $(\neg t_5 \vee \neg t_7)$ [RES,1911,560,t_4]
2915. $(\neg\varphi \vee \neg t_6 \vee \neg t_7)$ [RES,834,1981,t_5]
2942. $(\neg t_6 \vee \neg t_7)$ [RES,2915,838,φ]
2984. $(\neg t_1 \vee \neg t_6)$ [RES,2942,1372,$\neg t_7$]
2985. $(B \vee \neg t_2 \vee \neg t_6)$ [RES,2942,1374,$\neg t_7$]
3123. $(\neg t_1 \vee (((\neg\varphi \vee \psi) \wedge \neg B) \not\Rightarrow \neg t_2 \vee \neg t_6))$ [R-RES-$\not\Rightarrow$-2,2985,3,B]
3901. $(\neg t_2 \vee \neg t_9 \vee t_5)$ [R-RES-$\not\Rightarrow$-1,3123,15]
3904. $(B \vee \neg t_{10} \vee t_2)$ [R-RES-$\not\Rightarrow$-1,3123,15]
3906. $(\neg t_2 \vee t_{10})$ [R-RES-$\not\Rightarrow$-1,3123,15]
6854. $(\neg t_1 \vee (((\neg\varphi \vee \psi) \wedge \neg B) \not\Rightarrow \neg t_{10} \vee t_2))$ [R-RES-$\not\Rightarrow$-2,3904,3,B]
10225. $(\neg t_2 \vee \neg t_9)$ [RES,3901,292,t_5]
10314. $(\neg t_1 \vee \neg t_9)$ [RES,10225,559,$\neg t_2$]
22010. $(\neg t_1 \vee (((\neg\varphi \vee \psi) \wedge \neg B) \not\Rightarrow \neg t_{10} \vee \neg\psi))$ [R-RES-$\not\Rightarrow$-2,6854,12,t_2]
23844. $(\neg t_1 \vee (((\neg\varphi \vee \psi) \wedge \neg B) \not\Rightarrow \neg\psi \vee \neg t_2))$ [R-RES-$\not\Rightarrow$-2,22010,3906,$\neg t_{10}$]
23923. $(\neg t_1 \vee t_6 \vee (((\neg\varphi \vee \psi) \wedge \neg B) \not\Rightarrow \neg t_2))$ [R-RES-$\not\Rightarrow$-1,23844,2,$\neg\psi$]
24001. $(\neg t_1 \vee t_6 \vee t_9)$ [R-RES-$\not\Rightarrow$-1,23923,15,$\neg t_2$]
24126. $(\neg t_1 \vee t_6)$ [RES,24001,10314,t_9]
24194. $(\neg t_1)$ [RES,24126,2984,t_6]
24242. **false**   [I-RES-2,1,24194]

The following refutation is part of the proof of Lemma 8, where we show that, for $\varphi, \psi \in b$, we have that $\neg((b \vee \bigvee B) \Rightarrow \bigvee B)$ and $\neg(\varphi \Rightarrow \psi) \in a$ is not RES_S-consistent

1. t_1
2. $(\neg t_1 \vee (((\neg\varphi \vee \neg\psi) \wedge \neg B) \not\Rightarrow \neg B))$ [Assumption, $b \in W_{(a,A)}$]
3. $(\neg t_1 \vee (\neg\varphi \not\Rightarrow \neg\psi))$ [Assumption, $\neg(\varphi \Rightarrow \psi) \in a$]
4. $(\neg t_1 \vee (\neg B \not\Rightarrow \neg B) \vee ((\neg\varphi \vee \neg\psi) \not\Rightarrow \neg B))$ [L-AND-2,2]
11. $(\varphi \vee \neg t_2)$ [REF-2,2]
12. $(\psi \vee \neg t_2)$ [REF-2,2]
13. $(\neg\varphi \vee \neg\psi \vee t_2)$ [REF-2,2]
15. $(B \Rightarrow B)$ [REF-2,4]
16. $(t_2 \Rightarrow t_2)$ [REF-2,4]
44. $(\neg t_1 \vee (\neg\varphi \not\Rightarrow \neg t_2))$ [R-RES-$\not\Rightarrow$-2,3,12,$\neg\psi$]
55. $(\neg\varphi \vee \neg t_2 \vee \neg t_6)$ [R-RES-$\not\Rightarrow$-1,44,16]
56. $(\varphi \vee \neg t_6 \vee t_2)$ [R-RES-$\not\Rightarrow$-1,44,16]
64. $(\neg t_1 \vee t_6)$ [R-RES-$\not\Rightarrow$-1,44,16,$\neg t_2$]
318. $(\neg\psi \vee \neg t_6 \vee t_2)$ [RES,56,13,φ]
578. $(\neg t_2 \vee \neg t_6)$ [RES,55,11,$\neg\varphi$]
627. $(\neg\psi \vee \neg t_6)$ [RES,578,318,$\neg t_2$]
630. $(\neg\psi \vee \neg t_1)$ [RES,627,64,$\neg t_6$]
2580. $(\neg B \vee \neg t_9 \vee t_{10})$ [R-RES-$\not\Rightarrow$-1,2,15]
2582. $(\neg B \vee \neg t_{10})$ [R-RES-$\not\Rightarrow$-1,2,15]
2584. $(B \vee \neg t_9 \vee t_2)$ [R-RES-$\not\Rightarrow$-1,2,15]
2592. $(\neg t_1 \vee t_9)$ [R-RES-$\not\Rightarrow$-1,2,15,$\neg B$]
3769. $(B \vee \psi \vee \neg t_9)$ [RES,2584,12,t_2]
11984. $(\neg B \vee \neg t_9)$ [RES,2580,2582,t_{10}]
12128. $(\psi \vee \neg t_9)$ [RES,11984,3769,$\neg B$]
12165. $(\psi \vee \neg t_1)$ [RES,12128,2592,$\neg t_9$]

12205. ¬t_1 [RES,12165,630,ψ]
12230. **false**   [I-RES-2,1,12205,t_1]

References

1. Bachmair, L., Ganzinger, H.: Rewrite-based equational theorem proving with selection and simplification. JLC **4**(3), 217–247 (1994)
2. Beckert, B., Goré, R.: System description: leanK 2.0. In: Kirchner, C., Kirchner, H. (eds.) CADE 1998. LNCS, vol. 1421, pp. 51–55. Springer, Heidelberg (1998). https://doi.org/10.1007/BFb0054247
3. Burgess, J.P.: Quick completeness proofs for some logics of conditionals. Notre Dame J. Formal Log. **22**(1), 76–84 (1981)
4. Chellas, B.: Modal Logic. Cambridge University Press, Cambridge (1980)
5. David, A.: Deciding ATL* satisfiability by tableaux. In: Felty, A.P., Middeldorp, A. (eds.) CADE 2015. LNCS (LNAI), vol. 9195, pp. 214–228. Springer, Cham (2015). https://doi.org/10.1007/978-3-319-21401-6_14
6. Friedman, N., Halpern, J.Y.: On the complexity of conditional logics. In: Doyle, J., Sandewall, E., Torasso, P. (eds.) Proceedings of KR 1994, pp. 202–213. M. Kaufmann (1994)
7. Gasquet, O., Herzig, A., Longin, D., Sahade, M.: LoTREC: logical tableaux research engineering companion. In: Beckert, B. (ed.) TABLEAUX 2005. LNCS (LNAI), vol. 3702, pp. 318–322. Springer, Heidelberg (2005). https://doi.org/10.1007/11554554_25
8. Giordano, L., Gliozzi, V., Olivetti, N., Pozzato, G.L.: Analytic tableaux calculi for KLM logics of nonmonotonic reasoning. ACM Trans. Comput. Log. **10**(3), 18:1–18:47 (2009)
9. Giordano, L., Gliozzi, V., Olivetti, N., Schwind, C.: Tableau calculus for preference-based conditional logics: PCL and its extensions. ACM Trans. Comput. Log. **10**(3), 21 (2009)
10. Gorín, D., Pattinson, D., Schröder, L., Widmann, F., Wißmann, T.: CooL – a generic reasoner for coalgebraic hybrid logics (system description). In: Demri, S., Kapur, D., Weidenbach, C. (eds.) IJCAR 2014. LNCS (LNAI), vol. 8562, pp. 396–402. Springer, Cham (2014). https://doi.org/10.1007/978-3-319-08587-6_31
11. Haarslev, V., Möller, R.: RACER system description. In: Goré, R., Leitsch, A., Nipkow, T. (eds.) IJCAR 2001. LNCS, vol. 2083, pp. 701–705. Springer, Heidelberg (2001). https://doi.org/10.1007/3-540-45744-5_59
12. Hustadt, U., Gainer, P., Dixon, C., Nalon, C., Zhang, L.: Ordered resolution for coalition logic. In: De Nivelle, H. (ed.) TABLEAUX 2015. LNCS (LNAI), vol. 9323, pp. 169–184. Springer, Cham (2015). https://doi.org/10.1007/978-3-319-24312-2_12
13. Nalon, C., Hustadt, U., Dixon, C.: K$_S$P: a resolution-based prover for multimodal K. In: Olivetti, N., Tiwari, A. (eds.) IJCAR 2016. LNCS (LNAI), vol. 9706, pp. 406–415. Springer, Cham (2016). https://doi.org/10.1007/978-3-319-40229-1_28
14. Olivetti, N., Pozzato, G.L.: Nested sequent calculi and theorem proving for normal conditional logics: the theorem prover NESCOND. Intell. Artif. **9**(2), 109–125 (2015)
15. Olivetti, N., Pozzato, G.L., Schwind, C.: A sequent calculus and a theorem prover for standard conditional logics. ACM Trans. Comput. Log. **8**(4), 22 (2007)

16. Plaisted, D.A., Greenbaum, S.A.: A structure-preserving clause form translation. JLC **2**, 293–304 (1986)
17. Riazanov, A., Voronkov, A.: The design and implementation of VAMPIRE. AI Commun. **15**(2–3), 91–110 (2002)
18. Robinson, J.A.: A machine-oriented logic based on the resolution principle. J. ACM **12**(1), 23–41 (1965)
19. Schröder, L., Pattinson, D., Hausmann, D.: Optimal tableaux for conditional logics with cautious monotonicity. In: Coelho, H., Studer, R., Wooldridge, M. (eds.) Proceedings of the ECAI 2010 Frontiers in Artificial Intelligence and Applications, vol. 215 , pp. 707–712. IOS Press (2010)
20. Schulz, S.: System description: E 0.81. In: Basin, D., Rusinowitch, M. (eds.) IJCAR 2004. LNCS (LNAI), vol. 3097, pp. 223–228. Springer, Heidelberg (2004). https://doi.org/10.1007/978-3-540-25984-8_15
21. Tsarkov, D., Horrocks, I.: FaCT++ description logic reasoner: system description. In: Furbach, U., Shankar, N. (eds.) IJCAR 2006. LNCS (LNAI), vol. 4130, pp. 292–297. Springer, Heidelberg (2006). https://doi.org/10.1007/11814771_26
22. Weidenbach, C., Dimova, D., Fietzke, A., Kumar, R., Suda, M., Wischnewski, P.: SPASS version 3.5. In: Schmidt, R.A. (ed.) CADE 2009. LNCS (LNAI), vol. 5663, pp. 140–145. Springer, Heidelberg (2009). https://doi.org/10.1007/978-3-642-02959-2_10

Extended Resolution Simulates DRAT

Benjamin Kiesl[1][(✉)], Adrián Rebola-Pardo[1], and Marijn J. H. Heule[2]

[1] Institute of Logic and Computation, TU Wien, Vienna, Austria
`kiesl@kr.tuwien.ac.at`
[2] Department of Computer Science, The University of Texas, Austin, USA

Abstract. We prove that extended resolution—a well-known proof system introduced by Tseitin—polynomially simulates DRAT, the standard proof system in modern SAT solving. Our simulation procedure takes as input a DRAT proof and transforms it into an extended-resolution proof whose size is only polynomial with respect to the original proof. Based on our simulation, we implemented a tool that transforms DRAT proofs into extended-resolution proofs. We ran our tool on several benchmark formulas to estimate the increase in size caused by our simulation in practice. Finally, as a side note, we show how blocked-clause addition—a generalization of the extension rule from extended resolution—can be used to replace the addition of resolution asymmetric tautologies in DRAT without introducing new variables.

1 Introduction

Propositional logic presents us with an intricate problem: Does there exist a polynomially-bounded proof system for the unsatisfiable propositional formulas? In other words, can the unsatisfiability of formulas be certified in a compact way? Although we still don't know the answer, the attempts to solve this problem have led to a variety of interesting results in the area of proof complexity (for an excellent survey, see [18]). Many of these results had, and continue to have, a direct impact on automated reasoning.

Already in 1985, Haken [6] proved that the resolution proof system, which is well-suited for mechanization, does not admit polynomial-size proofs for all unsatisfiable formulas. However, by adding a simple rule that allows the introduction of definitions over new variables, Tseitin [17] turned resolution into an exponentially stronger proof system known as *extended resolution*. Up to this day, there are no known exponential lower-bounds on the size of extended-resolution proofs and so it is seen as one of the most powerful proof systems.

While this might convince a theoretician, it seemingly hasn't impressed the practitioners in SAT solving. These practitioners aim at developing tools that can decide the satisfiability of propositional formulas as efficiently as possible, and for this, they need proof systems that succinctly express the techniques used

This work has been supported by the National Science Foundation under grant CCF-1618574, by the Austrian Science Fund (FWF) under project W1255-N23, and by Microsoft Research through its PhD Scholarship Programme.

by their tools. Skeptical that extended resolution could meet their needs, they came up with several proof systems of which DRAT [21] has become their de-facto standard. For instance, participants in the annual SAT competition must produce DRAT proofs and also recent proofs of open mathematical problems, including the Erdős Discrepancy Conjecture [12], were provided in DRAT.

The DRAT proof system generalizes extended resolution insofar as every extended-resolution proof can be seen as a DRAT proof, but beyond that, DRAT allows additional techniques. While in extended resolution we show the unsatisfiability of a formula by successively deriving more and more consequences, in DRAT we iteratively modify a formula in satisfiability-preserving ways. To keep proof checking practical, DRAT allows only the derivation of specific facts that fulfill an efficiently-checkable syntactic criterion—so-called *resolution asymmetric tautologies* [10] (see Definition 4 on page 520).

Although its additional features make DRAT suitable for SAT solving, it remained unclear whether these features can indeed cause exponential gains in expressivity. In this paper, we show that they do *not*. To this end, we prove in a constructive way that extended resolution simulates DRAT polynomially, i.e., we show how every DRAT proof can be feasibly transformed into an extended-resolution proof. This confirms the expected proof-complexity landscape where all top-tier proof systems—including extended resolution, DRAT, and extended Frege systems [18]—are essentially equivalent.

Rounding off the picture, we show how blocked-clause addition [13]—a generalization of the extension rule from extended resolution—can be used to replace the addition of resolution asymmetric tautologies in DRAT without introducing new variables. In combination with recent simulation results regarding DRAT and newer proof systems [7,8], our paper thus bridges the gap between proof systems from the present and from the past.

The main contributions of this paper are as follows: (1) We prove that extended resolution simulates DRAT polynomially. (2) We implemented our simulation as a tool that transforms DRAT proofs into extended-resolution proofs. (3) We present an empirical evaluation of our simulation tool. (4) We show how blocked-clause addition can be used as an alternative for resolution-asymmetric-tautology addition in DRAT.

2   Preliminaries

Here we present the background required for understanding this paper. We consider propositional formulas in *conjunctive normal form* (CNF), which are defined as follows. A *literal* is either a variable x (a *positive literal*) or the negation $\bar{x}$ of a variable x (a *negative literal*). The *complementary literal* $\bar{a}$ of a literal a is defined as $\bar{a} = \bar{x}$ if $a = x$ and $\bar{a} = x$ if $a = \bar{x}$. For a literal l, we denote the variable of l by $var(l)$. A *clause* is a disjunction of literals; we assume that clauses do not contain repeated literals. A *unit clause* is a clause that contains exactly one literal; a *tautology* contains complementary literals. A *formula* is a conjunction of clauses. We view clauses as sets of literals and formulas as sets of clauses. A clause C *subsumes* a clause D if $C \subseteq D$.

An *assignment* is a function from a set of variables to the truth values 1 (*true*) and 0 (*false*). An assignment is *total* with respect to a formula if it assigns a truth value to every variable occurring in the formula, otherwise it is *partial*. We often denote assignments by the sequences of literals they satisfy. For instance, $x\,\bar{y}$ denotes the assignment that assigns 1 to x and 0 to y. A literal l is *satisfied* by an assignment α if l is positive and $\alpha(var(l)) = 1$ or if it is negative and $\alpha(var(l)) = 0$. A literal is *falsified* by an assignment if its complement is satisfied by the assignment. A clause is satisfied by an assignment α if it contains a literal that is satisfied by α. Finally, a formula is satisfied by an assignment α if all its clauses are satisfied by α. A formula is *satisfiable* if there exists an assignment that satisfies it. Two formulas are *logically equivalent* if they are satisfied by the same total assignments. Two formulas are *satisfiability-equivalent* if they are either both satisfiable or both unsatisfiable.

Given a clause C and an assignment α, we define $C\,|\,\alpha$ as the clause obtained from C by removing all literals that are falsified by α. If F is a formula, we define $F\,|\,\alpha = \{C\,|\,\alpha \mid C \in F \text{ and } \alpha \text{ does not satisfy } C\}$. The result of applying the *unit-clause rule* to a formula F is the formula $F\,|\,a$ with (a) being a unit clause in F. We also refer to applications of the unit-clause rule as *unit-propagation steps*. The iterated application of the unit-clause rule to a formula, until no unit clauses are left, is called *unit propagation*. If unit propagation on F yields the empty clause $\bot$, we say that it *derives a conflict* on F. For example, unit propagation derives a conflict on $F = (\bar{a} \vee b) \wedge (\bar{b}) \wedge (a)$ since $F\,|\,a = (b) \wedge (\bar{b})$ and $F\,|\,ab = \bot$.

We define proof systems and polynomial simulations following Cook and Reckhow [5]:

Definition 1. *A proof system for propositional formulas in CNF is a surjective polynomial-time-computable function $f : \Sigma^* \to \mathcal{F}$ where Σ is some alphabet and $\mathcal{F}$ is the set of all unsatisfiable formulas.*

A proof system can thus be seen as a proof-checking function f that takes a *proof candidate* P (which is a string over Σ) together with an unsatisfiable formula F and checks in polynomial time if P is a correct proof of F. The requirement that f is surjective means that there must exist a proof for *every* unsatisfiable formula. We sometimes use the word *proof system* in a more colloquial way to denote the rules that define what constitutes a correct proof of a certain type. The *size* of a proof is the number of symbols occurring in it.

Definition 2. *A proof system $f_1 : \Sigma_1^* \to \mathcal{F}$ polynomially simulates a proof system $f_2 : \Sigma_2^* \to \mathcal{F}$ if there exists a polynomial-time-computable function $g : \Sigma_2^* \to \Sigma_1^*$ such that $f_1(g(x)) = f_2(x)$.*

In other words, f_1 polynomially simulates f_2 if there exists a polynomial-time-computable function that transforms f_2-proofs into f_1-proofs. We next present the proof systems *extended resolution* and DRAT.

3   Extended Resolution (ER) and DRAT

An extended-resolution proof as well as a DRAT proof of a formula F are sequences of the form $C_1, \ldots, C_m, I_{m+1}, \ldots, I_n$ where $C_1, \ldots, C_m$ are clauses of F and $I_{m+1}, \ldots, I_n$ are *instructions* as defined in the following. There are three different kinds of instructions: addition, deletion, and extension. An *addition* is a pair $\langle \mathsf{a}, C \rangle$ where C is a clause; a *deletion* is a pair $\langle \mathsf{d}, C \rangle$ where C is a clause; and an *extension* (also called a *definition introduction*) is a pair $\langle \mathsf{e}, \varphi \rangle$ where φ is a propositional definition of the form $x \leftrightarrow p \vee (c_1 \wedge \cdots \wedge c_k)$ where x is a variable not occurring in any earlier instructions of the proof and $p, c_1, \ldots, c_k$ are literals where $var(x), var(p), var(c_1), \ldots, var(c_k)$ are pairwise distinct. Converting such a definition to CNF yields the clause set $\mathsf{cnf}(\varphi) = \{ x \vee \bar{p}, \ x \vee \bar{c}_1 \vee \cdots \vee \bar{c}_k,$ $\bar{x} \vee p \vee c_1, \ldots, \bar{x} \vee p \vee c_k \}$; in the particular case $k = 0$ we have $\mathsf{cnf}(\varphi) = \{ x \vee \bar{p}, \ \bar{x} \}$. The sequence $C_1, \ldots, C_m, I_{m+1}, \ldots, I_n$ gives rise to formulas $F_0, F_1, \ldots, F_n$ as follows:

$$F_i = \begin{cases} \{C_1, \ldots, C_i\} & \text{if } i \leq m \\ F_{i-1} \cup \{C\} & \text{if } i > m \text{ and } I_i = \langle \mathsf{a}, C \rangle \\ F_{i-1} \setminus \{C\} & \text{if } i > m \text{ and } I_i = \langle \mathsf{d}, C \rangle \\ F_{i-1} \cup \mathsf{cnf}(\varphi) & \text{if } i > m \text{ and } I_i = \langle \mathsf{e}, \varphi \rangle \end{cases}$$

We call F_i the *accumulated formula* corresponding to the i-th instruction. Based on this, we can now define the details of extended resolution and DRAT. In both proof systems, a correct proof of a formula F must derive the empty clause $\bot$, i.e., $\bot \in F_n$. They differ only in the instructions they permit.

3.1   Extended Resolution

Extended resolution combines resolution with the *extension rule*: A sequence $C_1, \ldots, C_m, I_{m+1}, \ldots, I_n$ is a correct extended-resolution proof of a formula F if every instruction $I_i \in I_{m+1}, \ldots, I_n$ is either (1) an addition $\langle \mathsf{a}, C \vee D \rangle$ where $C \vee D$ is the *resolvent* $(C \vee p) \otimes_p (D \vee \bar{p})$ of two clauses $C \vee p$ and $D \vee \bar{p}$ occurring in F_{i-1}, or (2) an extension $\langle \mathsf{e}, \varphi \rangle$. When Tseitin originally introduced the extension rule [17], he only allowed definitions of the form $x \leftrightarrow \bar{a} \vee \bar{b}$ where a and b are variables. These definitions correspond to the clauses $x \vee a$, $x \vee b$, and $\bar{x} \vee \bar{a} \vee \bar{b}$. However, more general definitions can be derived from these basic definitions in a simple but tedious way. Because of this, more general extension rules are common in the literature, some even allowing definitions $x \leftrightarrow \psi$ where ψ is an arbitrary propositional formula over previous variables (cf. [4,6,16]).

3.2   DRAT

A sequence $C_1, \ldots, C_m, I_{m+1}, \ldots, I_n$ is a correct DRAT proof of a formula F if every instruction $I_i \in I_{m+1}, \ldots, I_n$ is either (1) a deletion $\langle \mathsf{d}, C \rangle$ where C is an arbitrary clause, or (2) an addition $\langle \mathsf{a}, C \rangle$ where C is a RAT or a RUP in F_{i-1}; we now proceed to introduce these notions. We start by defining RUPs (short for *reverse unit propagation*) [20]:

Definition 3. *A clause $C = c_1 \vee \cdots \vee c_k$ is a* RUP *in a formula F if unit propagation derives a conflict on $F \wedge (\bar{c}_1) \wedge \cdots \wedge (\bar{c}_k)$. If C is a* RUP *in F, we say that F implies C via unit propagation.*

As an example, $F = (\bar{a} \vee c) \wedge (\bar{b} \vee \bar{c})$ implies $(\bar{a} \vee \bar{b})$ via unit propagation since unit propagation on $F \wedge (a) \wedge (b)$ derives both (c) and $(\bar{c})$, which leads to a conflict. Observe that if C is a resolvent of two clauses in a formula F, or if F contains a clause D that subsumes C, then C is a RUP in F. Now, a RAT is a clause for which all resolvents upon one of its literals are RUPs [10]:

Definition 4. *A clause $C \vee p$ is a* resolution asymmetric tautology (RAT) *on p in a formula F if for every clause $D \vee \bar{p} \in F$, the resolvent $C \vee D$ is implied by F via unit propagation.*

Example 1. Consider the formula $F = (\bar{p} \vee \bar{a}) \wedge (\bar{p} \vee b) \wedge (b \vee c) \wedge (\bar{c} \vee a)$ and the clause $C = a \vee p$. There are two resolvents of C upon p: The resolvent $a \vee \bar{a}$ (obtained by resolving with $\bar{p} \vee \bar{a}$) is a tautology and thus trivially a RUP in F; the resolvent $a \vee b$ (obtained by resolving with $\bar{p} \vee b$) is a RUP in F since unit propagation derives a conflict on $F \wedge (\bar{a}) \wedge (\bar{b})$. It follows that C is a RAT on p in F.   □

Observe that if C is a non-empty RUP in F, it is a RAT in F on any literal $p \in C$ (the empty clause $\bot$ cannot be a RAT as it contains no literals). In the rest of the paper, we thus call a clause a *proper* RAT if it is a RAT on some literal p but not a RUP. The addition of definition clauses, as with the extension rule, is a special case of blocked-clause addition [9] (see Sect. 6), which itself is a particular case of RAT addition. We thus regard DRAT as a generalization of extended resolution.

4   Simulating **DRAT** with Extended Resolution

We perform the transformation of a DRAT proof into an extended-resolution proof in four stages. In the first stage, we use the extension rule together with RUP addition and clause deletion to eliminate all additions of proper RATs. In the second stage, we get rid of all clause deletions. In the third stage, we then replace all RUP additions by resolution inferences and subsumed-clause additions. Finally, in the fourth stage, we also eliminate the subsumed-clause additions to obtain a correct extended-resolution proof.

4.1   Eliminating Additions of Proper **RAT**s

Given a DRAT proof $C_1, \ldots, C_m, I_{m+1}, \ldots, I_n$, we iterate over the instructions $I_{m+1}, \ldots, I_n$ and replace every addition $I_i = \langle \mathsf{a}, p \vee C \rangle$ of a clause $p \vee C$ that is a proper RAT on p in the accumulated formula F_{i-1} by a sequence π_i of instructions. As illustrated in Fig. 1, such a sequence π_i consists of a single definition introduction followed first by several RUP additions and then by

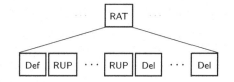

Fig. 1. We transform a RAT addition into a definition introduction (Def), followed by RUP additions and clause deletions (Del).

several clause deletions. In the case where I_i is not the addition of a proper RAT, we simply let π_i be I_i. At the end of this iterative process, we obtain a sequence $C_1, \ldots, C_m, \pi_{m+1}, \ldots, \pi_n$, where every π_i is a sequence of instructions corresponding to the instruction I_i from the original proof. The sequence $C_1, \ldots, C_m, \pi_{m+1}, \ldots, \pi_n$ contains no additions of proper RATs, but instead contains definition introductions.

Each iteration of this process performs the following transformation, where I_i is an addition instruction of a clause $C = p \vee c_1 \vee \cdots \vee c_k$ which is a RAT on literal p in the accumulated formula F_{i-1} before I_i.

$$C_1, \ldots, C_m, \pi_{m+1}, \ldots, \pi_{i-1}, I_i, I_{i+1}, \ldots, I_n$$
$$\updownarrow$$
$$C_1, \ldots, C_m, \pi_{m+1}, \ldots, \pi_{i-1}, \pi_i, I'_{i+1}, \ldots, I'_n$$

We first use the extension rule to introduce a clause $x \vee c_1 \vee \cdots \vee c_k$ as well as some other definition clauses, where x is a *new* variable in the sense that it is not used anywhere else in the proof. Note that $x \vee c_1 \vee \cdots \vee c_k$ differs from C only on the literal p, which is replaced by the variable x. We then use RUP additions and clause deletions to replace all occurrences of p in F_{i-1} by x. Our procedure guarantees that the formula accumulated after π_i in the resulting sequence is exactly $F_i[x/p]$, obtained from $F_i = F_{i-1} \cup \{C\}$ (the accumulated formula after I_i in the original proof) by simultaneously replacing occurrences of p by x and occurrences of $\bar{p}$ by $\bar{x}$.

As a consequence, the correctness of the whole proof is preserved by simply renaming p to x, and $\bar{p}$ to $\bar{x}$, in all later instructions, resulting in the instructions $I'_{i+1}, \ldots, I'_n$. It is thus clear that the size of the accumulated formula after π_i in the new proof is the same as that of F_i in the original proof; this property will be crucial for the complexity analysis in Sect. 4.5. We now explain in detail how the sequence π_i is obtained, and provide an example to illustrate the procedure.

(1) Use the extension rule to introduce the definition $x \leftrightarrow p \vee (\bar{c}_1 \wedge \cdots \wedge \bar{c}_k)$. This adds the clause set $\{x \vee c_1 \vee \cdots \vee c_k, \; x \vee \bar{p}, \; \bar{x} \vee p \vee \bar{c}_1, \ldots, \; \bar{x} \vee p \vee \bar{c}_k\}$. The first clause will be our replacement of the RAT $p \vee c_1 \vee \cdots \vee c_k$. Intuitively, this definition follows the correctness proof of RAT clause addition from [10]: given any interpretation satisfying F_{i-1}, we can construct another interpretation satisfying F_i by conditionally changing the truth value of p, precisely as given by the definition of x. The rest of the transformation simply replaces occurrences of p by x.

(2) Replace the literal p in all clauses of F_{i-1} by the new variable x:

   (a) Add for every clause $D \vee p \in F_{i-1}$ the clause $D \vee x$. This is a correct RUP addition since $D \vee x$ is a resolvent of $D \vee p$ and $x \vee \bar{p}$.

   (b) Add for every clause $D \vee \bar{p} \in F_{i-1}$ the clause $D \vee \bar{x}$. To show that this is a correct RUP addition, we show that unit propagation derives a conflict on $F_{i-1} \wedge \bar{D} \wedge (x)$, where $\bar{D}$ is the conjunction of the negated literals of D. As C is a RAT on p in F_{i-1}, we know that the resolvent $c_1 \vee \cdots \vee c_k \vee D$ of C and $D \vee \bar{p}$ is a RUP in F_{i-1}. Now, by propagating the unit clauses of $\bar{D}$, we derive $(\bar{p})$ because the clause $D \vee \bar{p}$ is in F_{i-1}. After this, we propagate x and $\bar{p}$ to derive all the unit clauses $(\bar{c}_1), \ldots, (\bar{c}_k)$ from the clauses $\bar{x} \vee p \vee \bar{c}_j$ with $j \in 1, \ldots, k$. But then we have derived the negations of all literals in the resolvent $c_1 \vee \cdots \vee c_k \vee D$, and since this resolvent is a RUP in F_{i-1}, unit propagation must eventually derive a conflict.

   (c) Delete all clauses containing p or $\bar{p}$, including those added in step 1. Note that this does not delete the clause $x \vee c_1 \vee \cdots \vee c_k$.

Example 2. Suppose we are given a proof $C_1, \ldots, C_m, I_{m+1}, \ldots, I_i, \ldots, I_n$ and we want to eliminate the addition $I_i = \langle \mathsf{a}, C \rangle$ where $C = p \vee a$ is a proper RAT on p in the accumulated formula $F_{i-1} = \{\bar{p} \vee b, a \vee b \vee c, \bar{c} \vee d, \bar{d}, \bar{a} \vee p\}$. Observe that C is a RAT on p because the resolvent $a \vee b$, obtained by resolving C with $\bar{p} \vee b$ upon p, is a RUP in F_{i-1}.

We first use the extension rule to add the definition $x \leftrightarrow p \vee \bar{a}$. This adds the clauses $x \vee a$, $x \vee \bar{p}$, and $\bar{x} \vee p \vee \bar{a}$. Next, we need to replace the literal p in F_{i-1} by x. To do so, we first resolve $x \vee \bar{p}$ with $\bar{a} \vee p$ to derive $\bar{a} \vee x$. Then, we introduce the RUP $\bar{x} \vee b$ for the existing clause $\bar{p} \vee b$. (It can be easily seen that $\bar{x} \vee b$ is a RUP in $F_{i-1} \cup \{\bar{x} \vee p \vee \bar{a}, x \vee \bar{p}, x \vee a\}$: By propagating $\bar{b}$, we derive $\bar{p}$ from $\bar{p} \vee b$. After this, the propagation of x and $\bar{p}$ derives $\bar{a}$ from $\bar{x} \vee p \vee \bar{a}$. But then further propagation will eventually lead to a conflict because $a \vee b$, which is the resolvent of $p \vee a$ and $\bar{p} \vee b$, is a RUP in F_{i-1}.) Finally, we delete all clauses containing p or $\bar{p}$. We thus obtain the proof $C_1, \ldots, C_m, I_{m+1}, \ldots, I_{i-1}, \pi_i, \ldots, I_n$ where π_i is the sequence $\langle \mathsf{e}, x \leftrightarrow p \vee \bar{a} \rangle$, $\langle \mathsf{a}, \bar{a} \vee x \rangle$, $\langle \mathsf{a}, \bar{x} \vee b \rangle$, $\langle \mathsf{d}, \bar{p} \vee b \rangle$, $\langle \mathsf{d}, \bar{a} \vee p \rangle$, $\langle \mathsf{d}, \bar{x} \vee p \vee \bar{a} \rangle$, $\langle \mathsf{d}, x \vee \bar{p} \rangle$. After the last instruction of π_i, we get the accumulated formula $\{\bar{x} \vee b, a \vee b \vee c, \bar{c} \vee d, \bar{d}, \bar{a} \vee x, x \vee a\}$, which is precisely $F_i[x/p]$. We then just need to replace p by x and $\bar{p}$ by $\bar{x}$ in $I_{i+1}, \ldots, I_n$ to obtain a correct proof $C_1, \ldots, C_m, I_{m+1}, \ldots, I_{i-1}, \pi_i, I'_{i+1}, \ldots, I'_n$. □

4.2   Eliminating Clause Deletions

At this point, our proof is a sequence of (1) clauses from the original formula, (2) definition introductions, (3) RUP additions, and (4) clause deletions. Since no additions of proper RATs remain in the proof, the elimination of a deletion instruction does not affect the correctness of other proof instructions: The addition of RUPs depends only on the existence of clauses in the accumulated formula but not on their non-existence (if C is a RUP in F, it is also a RUP in every

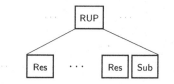

Fig. 2. We transform a RUP addition into a sequence of resolution steps (Res) followed by a single subsumed-clause addition (Sub).

superset of F). Likewise, the extension rule is not affected by additional clauses. By simply eliminating all deletions, we thus end up with a correct proof. Note that this would not work if *proper* RAT additions were still present, because they depend on the non-existence of certain clauses (a clause C is a RAT in a formula F only if F contains *no* resolvents with C that are not RUPs).

4.3 Eliminating RUP Additions

Similar to the first stage of our simulation, we again iterate over the proof from the beginning. In this stage, we now replace all additions of RUPs that are neither resolvents nor subsumed clauses. In the following, we show how the addition of such a RUP can be transformed into a sequence of resolution steps followed by a single subsumed-clause addition. This is illustrated in Fig. 2. We note that this has already been explained on a high level in the literature [15, 19].

Let us first observe that, given a correct proof containing only RUP additions and definition introductions, the RUP additions of tautological clauses can be directly eliminated. To see this, simply observe that definition introductions are never affected by the presence of tautologies. Furthermore, if a clause C is a RUP in F, and F contains a tautology $a \lor \bar{a} \lor D$, the latter never becomes a unit clause in $F|\alpha$ under any assignment α; therefore, C is also a RUP in the formula resulting from removing tautologies from F. In the following, we thus consider only proofs without tautological clauses.

If a non-tautological clause C is a RUP in a formula F, we know that unit propagation derives a conflict on $F \land \bar{C}$ where $\bar{C}$ is the conjunction of the negated literals in C. This is equivalent to saying that unit propagation derives a conflict on $F|\bar{C}$, viewing $\bar{C}$ as the assignment that satisfies $\bar{C}$. Hence, there exists a (possibly empty) sequence of literals $a_1, \ldots, a_n$ such that the unit clause (a_i) occurs in $F|\bar{C}a_1 \ldots a_{i-1}$ for each $1 \leq i \leq n$, and the empty clause $\bot$ occurs in $F|\bar{C}a_1 \ldots a_n$. Intuitively, (a_i) is the unit clause propagated at the i-th propagation step after all unit clauses in $\bar{C}$ have been propagated. These unit clauses and the empty clause stem from clauses $D_1, \ldots, D_{n+1} \in F$ with the following properties: (I) the clause $D_i|\bar{C}a_1 \ldots a_{i-1}$ is the unit clause (a_i) for $1 \leq i \leq n$, (II) D_i is not satisfied by $\bar{C}a_1 \ldots a_{i-1}$ for $1 \leq i \leq n+1$, and (III) the clause $D_{n+1}|\bar{C}a_1 \ldots a_n$ is the empty clause.

Algorithm 1 uses the clauses $D_1, \ldots, D_{n+1}$ as follows: It starts with the last clause, D_{n+1}, and step-by-step resolves it with the clauses $D_n, \ldots, D_1$ until it obtains a clause C_1 that subsumes C. Using C_1, we can then derive C with a subsumed-clause addition. Example 3 illustrates the execution of the algorithm.

$$1 \quad C_{n+1} \leftarrow D_{n+1}$$
$$2 \quad \textbf{for } i = n, \ldots, 1 \textbf{ do}$$
$$3 \quad \quad \textbf{if } \bar{a}_i \in C_{i+1} \textbf{ then } C_i \leftarrow D_i \otimes_{a_i} C_{i+1}$$
$$4 \quad \quad \textbf{else } C_i \leftarrow C_{i+1}$$

Algorithm 1. Given a RUP C, the algorithm derives a clause $C_1 \subseteq C$.

Example 3. Consider the clause $C = a \vee b$ and $F = D_1 \wedge D_2 \wedge D_3 \wedge D_4$ where:

$$D_1 = a \vee c \qquad D_2 = a \vee \bar{c} \vee d \qquad D_3 = \bar{d} \vee e \qquad D_4 = \bar{d} \vee \bar{e}$$

The clause C is a RUP in F because unit propagation derives a conflict on $F \wedge (\bar{a}) \wedge (\bar{b})$, or equivalently, it derives a conflict on $F|\bar{a}\bar{b}$. To illustrate this, we perform the unit propagation:

$$D_1|\bar{a}\bar{b} = (c) \qquad D_2|\bar{a}\bar{b}c = (d) \qquad D_3|\bar{a}\bar{b}cd = (e) \qquad D_4|\bar{a}\bar{b}cde = \bot$$

Our algorithm now performs resolution steps as follows (* marks unit literals):

$$
\begin{array}{c}
\overbrace{a \vee c^*}^{D_1} \qquad \dfrac{\overbrace{a \vee \bar{c} \vee d^*}^{D_2} \qquad \dfrac{\overbrace{\bar{d} \vee e^*}^{D_3} \qquad \overbrace{\bar{d} \vee \bar{e}}^{D_4}}{\bar{d}}}{a \vee \bar{c}} \\
\dfrac{}{\underbrace{a}_{C_1}}
\end{array}
$$

As we can see, the resulting clause $C_1 = (a)$ subsumes $C = a \vee b$.   □

Lemma 1. *If a formula F implies a non-tautological clause C via unit propagation, then the clause C_1, computed by Algorithm 1, subsumes C.*

Proof. We show by induction that, for every $1 \leq i \leq n + 1$, the clause C_i computed by Algorithm 1 satisfies $C_i|\bar{C}a_1 \ldots a_{i-1} = \bot$. The claim then follows from $C_1|\bar{C} = \bot$, which is equivalent to $C_1 \subseteq C$.

BASE CASE $(i = n + 1)$: Follows from $C_{n+1} = D_{n+1}$ and property (III).

INDUCTION STEP $(1 \leq i \leq n)$: Assume the claim holds for $i + 1$. Then, we have $C_{i+1}|\bar{C}a_1 \ldots a_i = \bot$, and from property (I) we know $D_i|\bar{C}a_1 \ldots a_{i-1} = (a_i)$. Now, if C_{i+1} does not contain $\bar{a}_i$, then $C_{i+1}|\bar{C}a_1 \ldots a_{i-1} = \bot$. In this case, the algorithm sets $C_i = C_{i+1}$ and so the claim holds for i. In contrast, if C_{i+1} contains $\bar{a}_i$, then the algorithm sets $C_i = D_i \otimes_{a_i} C_{i+1}$. But then, as C_i contains only literals of D_i and C_{i+1} except for a_i and $\bar{a}_i$, the claim also follows for i. □

The following statement, which is a variant of Theorem 2 in [15] as well as of the Theorem of Lee [14], is a consequence of Lemma 1; it allows us to repeatedly eliminate all additions of RUPs that are not resolvents or subsumed clauses.

Theorem 2. *If a formula F implies a non-tautological clause C via unit propagation using n propagation steps, we can derive C from F via at most n resolution steps followed by one subsumed-clause addition.*

4.4   Eliminating Subsumed-Clause Additions

At this point, every instruction is either a definition introduction or it adds a resolvent or a subsumed clause. Since the extension rule does not depend on previous clauses, we can reorder the instructions of our proof so that all definition introductions occur before all addition instructions.

Now, by a well-known method (e.g., [1]) we can eliminate all subsumed-clause additions from the latter part of our proof. The procedure works by recursively labeling every clause in the proof with a subclause. These labels give a resolution proof, possibly with unnecessary inferences. The labeling proceeds as follows:

1. We label every leaf clause by itself.
2. For each resolvent of two clauses $C_1 \vee x$ and $C_2 \vee \bar{x}$, which are labeled by D_1 and D_2 respectively, we label the resolvent by D_1 if $x \notin D_1$; by D_2 if $\bar{x} \notin D_2$; and by $D_1 \vee D_2$ if $x \in D_1$ and $\bar{x} \in D_2$.
3. For each subsumption inference from a clause C that is labeled by D, we label the subsumed clause by D.

It is straightforward to check that the labels define a resolution derivation without subsumed-clause additions; in fact, a refutation, as the only subclause of $\perp$ is $\perp$ itself. This is polynomial, and can only reduce the size of the input. The resulting derivation may contain redundant parts such as unused subderivations, but these do not affect our analysis and can be easily removed. After eliminating all subsumed-clause additions, we finally obtain an extended-resolution proof.

Example 4. The following proof tree includes the subsumed-clause additions 1 and 2.

$$
\cfrac{\cfrac{a \vee b \ [a \vee b] \quad \cfrac{\bar{b} \ [\bar{b}]}{a \vee \bar{b} \ [\bar{b}]^*}\,{}^1}{a \ [a]} \qquad \cfrac{\cfrac{\bar{a} \vee c \ [\bar{a} \vee c]^* \quad \cfrac{d \ [d]}{\bar{c} \vee d \ [d]^*}\,{}^2}{\bar{a} \vee d \ [d]^*} \qquad \bar{a} \vee \bar{d} \ [\bar{a} \vee \bar{d}]}{\bar{a} \ [\bar{a}]}}{\perp \ [\perp]}
$$

After dropping the marked (*) clauses, the result is the following proof:

$$
\cfrac{\cfrac{a \vee b \qquad \bar{b}}{a} \qquad \cfrac{d \qquad \bar{a} \vee \bar{d}}{\bar{a}}}{\perp}
$$

4.5   Complexity of the Simulation

We show now that our simulation only involves a polynomial blow-up. To simplify the presentation, we use the number of literals (with repetitions) in a proof P as the measure for its size, denoted by $\|P\|$. After we have shown that the size of the resulting extended-resolution proof is polynomial compared to the original DRAT proof, it should be clear that the computation of the simulation is also polynomial, given the simplicity of the used techniques. Let the original DRAT proof be $P = C_1, \ldots, C_m, I_{m+1}, \ldots, I_n$. Note first that for every $m + 1 \leq i \leq n$, the size $\|I_i\|$ of the instruction I_i, and the size $\|F_i\|$ of the accumulated formula F_i are both bounded by $\mathcal{O}(\|P\|)$. Note also that the elimination of clause deletions and subsumed-clause additions shrinks the proof. Hence, out of the four stages in the simulation, we only need to consider the first stage (elimination of RAT additions) and the third stage (elimination of RUP additions) to obtain an upper bound on the proof size.

Elimination of RAT *Additions.* For the following, remark that for $i \in m+1, \ldots, n$, the size of the accumulated formula after the i-th proof fragment π_i (obtained by transforming the instruction I_i) in the new proof is the same as that of F_i in the original DRAT proof (we explained this on page 521). For the elimination of a single RAT addition of a clause $p \vee c_1 \vee \cdots \vee c_k$, we first add the definition $x \leftrightarrow p \vee (\bar{c}_1 \wedge \cdots \wedge \bar{c}_k)$. This step is clearly $\mathcal{O}(\|P\|)$. After this, we add for each clause $D \vee p \in F_{i-1}$ the clause $D \vee x$, and we add for each clause $D \vee \bar{p} \in F_{i-1}$ the clause $D \vee \bar{x}$. This leads to at most $\mathcal{O}(\|F_{i-1}\|) = \mathcal{O}(\|P\|)$ new literals. Finally, we delete all clauses containing p or $\bar{p}$. These deletions together are again of size at most $\mathcal{O}(\|F_{i-1}\|) = \mathcal{O}(\|P\|)$. Overall, the size of the proof generated by eliminating a single RAT addition is thus bounded by $3 \times \mathcal{O}(\|P\|) = \mathcal{O}(\|P\|)$. Finally, as we perform at most n such RAT eliminations and since $n = \mathcal{O}(\|P\|)$, the size of the resulting proof after eliminating all RATs is bounded by $\mathcal{O}(\|P\|^2)$.

Elimination of RUP *Additions.* Before we eliminate RUPs, we have a proof whose size is $\mathcal{O}(\|P\|^2)$. We thus eliminate at most $\mathcal{O}(\|P\|^2)$ RUP additions. It remains to determine a bound for the size of the proof instructions obtained by eliminating a single RUP addition. Theorem 2 tells us that if C is a RUP that is implied via unit propagation using k propagation steps, we can derive C with at most k resolution steps followed by a single subsumed-clause addition. Clearly, the number of unit-propagation steps is bounded by the number of variables occurring in the proof (every variable can be propagated at most once). Now, the number of variables in the original proof P is clearly bounded by $\|P\|$ and since the elimination of RAT additions has introduced at most one new variable for every RAT, we have $\mathcal{O}(\|P\|)$ variables. Hence, a single RUP elimination leads to at most $\mathcal{O}(\|P\|)$ instructions. As the size of a single instruction is bounded by $\mathcal{O}(\|P\|)$ (a clause can contain at most two literals per variable), every RUP elimination results in a proof of size $\mathcal{O}(\|P\|^2)$. We conclude that the size of the resulting extended-resolution proof is $\mathcal{O}(\|P\|^4)$.

Note that our analysis is very conservative. For instance, representing resolvents implicitly (just pointing to their two parent clauses) instead of representing them explicitly shrinks the resulting extended-resolution proof significantly. As we will see in Sect. 5, the increase in size on practical DRAT proofs is way smaller than the theoretical bound we obtain here. Combining our result with the recent result that DRAT polynomially simulates DPR (a generalization of DRAT) [7], we obtain the complexity landscape depicted in Fig. 3.

Fig. 3. A dashed line from X to Y means that X simulates Y polynomially. A solid line from X to Y means that every Y proof can be regarded as an X proof.

5 Experimental Evaluation

We implemented our simulation procedure as an extension of the proof checker DRAT-trim.[1] We then evaluated the simulation tool on existing DRAT proofs for the *pigeon-hole formulas, two-pigeons-per-hole formulas* [2], and *Tseitin formulas* [3,17]. The pigeon-hole formulas (hole*) ask whether $n + 1$ pigeons can be placed into n holes such that each hole contains at most one pigeon. Similarly, the two-pigeons-per-hole formulas (tph*) ask whether $2n + 1$ pigeons can be placed into n holes with at most two pigeons per hole. Finally, the Tseitin formulas (Urquhart*) encode a parity problem on graphs.

We selected the DRAT proofs of these formulas for three reasons. First, out of all DRAT proofs we are aware of, they have the highest ratio of proper-RAT-to-RUP-instructions and so the transformation from DRAT to extended resolution can offer insight into a worst-case scenario regarding existing proofs. Second, the proofs originate from a transformation of DPR proofs to DRAT proofs [7]. We thus also see what happens when we transform the more general DPR proofs, and not only DRAT proofs, to extended resolution. Third, all three formula families are hard for resolution, meaning that they admit only resolution proofs whose size is exponential with respect to the formula [6,18].

Table 1 shows the results of our experiments. Although the extended-resolution proofs are clearly larger than the corresponding DRAT proofs, the blow-up is far from the theoretical worst case. As we already selected proofs with many proper RAT instructions, we imagine that the growth is even smaller on proofs with a modest number of RAT instructions. For a pigeon-hole formula holeX, the increase in size is roughly the factor X. For the two-pigeons-per-hole formulas, the growth is larger. This can be explained by the high clauses-to-variables ratio. Finally, for the Tseitin formulas, the growth lies between a factor of 20 and 30.

As a comparison, Table 2 shows the smallest extended-resolution proofs of the pigeon-hole formulas and of the Tseitin formulas known to us. The proofs of the pigeon-hole formulas were manually constructed by Cook [4] whereas the proofs of the Tseitin formulas were produced using the tool EBDDRES 1.2 [11]. To the best of our knowledge, there is only one tool supporting extended resolution that was able to solve one of the selected two-pigeons-per-hole formulas: EBDDRES 1.1 [16]. It generated an extended-resolution proof with 2 638 385 definitions and 18 848 004 resolution steps for the formula tph8.

---

[1] The simulation tool, checkers, formulas, and proofs discussed in this section are available on http://www.cs.utexas.edu/~marijn/drat2er.

Table 1. A size comparison of DPR, DRAT, and ER proofs of formulas that are hard for resolution. We generated the ER proofs from existing DRAT proofs [7]. Column headers refer to the numbers of variables (#var), clauses (#cls), clause additions (#add), added definitions (#def), and resolution steps (#res).

formula	input		DPR	DRAT	ER	
	#var	#cls	#add	#add	#def	#res
hole20	420	4221	2870	26547	18162	282471
hole30	930	13981	9455	89827	61962	1393411
hole40	1640	32841	22140	213107	147562	4344126
hole50	2550	63801	42925	416387	288962	10517116
tph8	136	5457	1156	25204	13931	1093959
tph12	300	27625	3950	127296	68645	11688956
tph16	528	87329	9416	401004	212847	63391635
tph20	820	213241	18450	976376	512841	236415141
Urquhart-s5-b1	106	714	620	28189	8320	102293
Urquhart-s5-b2	107	742	606	32574	9020	123943
Urquhart-s5-b3	121	1116	692	41230	11404	188875
Urquhart-s5-b4	114	888	636	37978	10497	171576

Table 2. Small existing ER proofs of pigeon-hole formulas and Tseitin formulas.

formula	ER by Cook [4]		formula	ER by EBDDRES [11]	
	#def	#res		#def	#res
hole20	2660	160151	Urquhart-s5-b1	11054	39702
hole30	8990	810161	Urquhart-s5-b2	12684	45389
hole40	21320	2560171	Urquhart-s5-b3	28358	100585
hole50	41650	6250181	Urquhart-s5-b4	16295	58552

6   Replacing **RAT** Addition with Blocked-Clause Addition

In our polynomial simulation, we needed to introduce a new variable for every proper RAT addition. This cannot be avoided because extended resolution without new variables is just ordinary resolution, and ordinary resolution is exponentially weaker than both DRAT and extended resolution [6]. We now show how *blocked-clause addition*, introduced by Kullmann [13] as a generalization of the extension rule from extended resolution, can be used to replace RAT addition *without* introducing new variables. This shows that a simple generalization of the extension rule is essentially as powerful as RAT addition, even when no new variables are introduced. Informally, a clause is blocked if all resolvents upon one of its literals are tautologies [13]:

Definition 5. *A clause C is* blocked *by a literal $p \in C$ in a formula F if all resolvents of C upon p with clauses in F are tautologies.*

Example 5. Consider the formula $F = (\bar{p} \vee \bar{b}) \wedge (\bar{p} \vee \bar{a}) \wedge (p \vee c) \wedge (a \vee c)$ and the clause $a \vee b \vee p$. There are two resolvents of $a \vee b \vee p$ upon p: The clause $a \vee b \vee \bar{b}$, obtained by resolving with $\bar{p} \vee \bar{b}$, and the clause $a \vee b \vee \bar{a}$, obtained by resolving with $\bar{p} \vee \bar{a}$. As both resolvents are tautologies, $a \vee b \vee p$ is blocked by p in F. □

Blocked clauses are thus more restricted than RATs: While the RAT property only requires all the resolvents to be implied via unit propagation, blocked clauses require them to be tautologies, which are trivially implied via unit propagation. Hence, every blocked clause is also a RAT but not vice versa.

We follow an iterative procedure similar to the one presented in Sect. 4. Suppose $C = c_1 \vee \cdots \vee c_k \vee p$ is a proper RAT on p in a formula F. To replace the addition of C to F, we first turn C into a blocked clause by replacing the resolution partners that do not lead to tautological resolvents. We then add the clause with blocked-clause addition and afterwards derive all the original resolution partners again. As illustrated in Fig. 4, this leads to a sequence consisting of RUP additions, clause deletions, and a single blocked-clause addition. Specifically, we perform the following steps:

(1) For every clause $D \vee \bar{p} \in F_{i-1}$ such that the resolvent $R = c_1 \vee \cdots \vee c_k \vee D$ with C upon p is not a tautology, add R with RUP addition. The resolvent R is guaranteed to be a RUP because C is a RAT on p in F_{i-1}.

(2) For every clause $D \vee \bar{p} \in F_{i-1}$ such that the resolvent with C upon p is not a tautology, replace $D \vee \bar{p}$ by the clause set $D_p = \{(\bar{c}_j \vee D \vee \bar{p}) \mid 1 \leq j \leq k\}$. Since all the clauses in D_p are subsumed by $D \vee \bar{p}$, this replacement results in a sequence of deletions and RUP additions. Note that in case C is a unit clause, the set D_p is empty and so all resolution partners are deleted.

(3) Add C with blocked-clause addition. This is a correct addition because after step 2, every clause that contains $\bar{p}$ contains a literal $\bar{c}_j$ with $c_j \in C$. Hence, by resolving such a clause with C we obtain a tautology.

(4) Use RUP addition to add all the clauses $D \vee \bar{p}$, which we replaced in step 2, again. The addition of such a clause $D \vee \bar{p}$ is a correct RUP addition: If C is a unit clause, we have added $R = D$ (which subsumes $D \vee \bar{p}$) in step 1. If C is not a unit clause, then $D_p \cup \{R\}$ implies $D \vee \bar{p}$ via unit propagation: By propagating p and the negated literals of D, we derive the unit clauses $(\bar{c}_1), \ldots, (\bar{c}_k)$ from the clauses in $D_p = \{(\bar{c}_j \vee D \vee \bar{p}) \mid 1 \leq j \leq k\}$. But

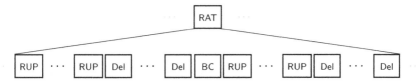

Fig. 4. We transform a RAT addition into a sequence consisting of RUP additions, clause deletions (Del), and a single blocked-clause addition (BC).

these unit clauses lead to a conflict with the clause $c_1 \vee \cdots \vee c_k$, which we derive by propagating the negated literals of D on $R = c_1 \vee \cdots \vee c_k \vee D$.

(5) Delete all the RUPs added in step 1 and the clause sets D_p added in step 2.

Example 6. Consider the formula $F = \{\bar{p}, \, a \vee b \vee c, \, \bar{c} \vee d, \, \bar{d}, \, \bar{a} \vee e, \, \bar{b} \vee e\}$ and the clause $C = a \vee b \vee p$. The clause C is not blocked but it is a RAT on p in F, meaning that F implies the resolvent $a \vee b$ of C and $\bar{p}$ via unit propagation. To turn C into a blocked clause, we first add $a \vee b$ with RUP addition. We next replace $\bar{p}$ by the clauses $\bar{p} \vee \bar{a}$ and $\bar{p} \vee \bar{b}$ (both clauses are subsumed by $\bar{p}$ and thus they are RUPs). Now $\bar{p} \vee \bar{a}$ and $\bar{p} \vee \bar{b}$ contain literals whose complements occur in C. We can thus add C with blocked-clause addition.

After this, we use RUP addition to add the original resolution partner $\bar{p}$ again: This is a correct RUP addition because $a \vee b$, $\bar{p} \vee \bar{a}$, and $\bar{p} \vee \bar{b}$ together imply $\bar{p}$ via unit propagation (to see this, observe that making p true forces $\bar{a}$ and $\bar{b}$ to be true which leads to a conflict with $a \vee b$). This step is actually the reason why we derived $a \vee b$ in the beginning. Finally, we delete the intermediate clauses $a \vee b$, $\bar{p} \vee \bar{a}$, and $\bar{p} \vee \bar{b}$ to obtain the formula $F \cup \{C\}$.   □

7   Conclusion

We showed how every DRAT proof can be feasibly transformed into an extended-resolution proof. To evaluate the increase in size caused by our simulation, we implemented it and performed experiments on existing DRAT proofs for hard formulas. The experiments revealed that the obtained proofs are far smaller than the theoretical worst case and that they are also not much larger than existing extended-resolution proofs of the same formulas. We imagine that the size of the proofs could be reduced even further by performing additional compression steps, which is a direction for future work.

In addition, we showed how blocked-clause addition can be used to simulate the addition of resolution asymmetric tautologies (RATs) without the introduction of new variables. Our results provide us with a better understanding of both DRAT and extended resolution. We now know how extended resolution can mimic the reasoning steps of DRAT. Moreover, our transformations illustrate that the addition of RATs in DRAT combines in an elegant way the benefits of resolution, subsumption, and blocked-clause addition. We thus believe that DRAT is still the preferable proof system for practical SAT solving, even though it offers no exponential gains in expressivity compared to extended resolution.

References

1. Baaz, M., Leitsch, A.: Methods of Cut-Elimination. Trends in Logic, vol. 3. Springer, Heidelberg (2011)
2. Biere, A.: Two pigeons per hole problem. In: Proceedings of SAT Competition 2013: Solver and Benchmark Descriptions, p. 103 (2013)

3. Chatalic, P., Simon, L.: Multi-resolution on compressed sets of clauses. In: Proceedings of the 12th IEEE International Conference on Tools with Artificial Intelligence (ICTAI 2000), pp. 2–10 (2000)
4. Cook, S.A.: A short proof of the pigeon hole principle using extended resolution. SIGACT News **8**(4), 28–32 (1976)
5. Cook, S.A., Reckhow, R.A.: The relative efficiency of propositional proof systems. J. Symb. Log. **44**(1), 36–50 (1979)
6. Haken, A.: The intractability of resolution. Theor. Comput. Sci. **39**, 297–308 (1985)
7. Heule, M.J.H., Biere, A.: What a difference a variable makes. In: Beyer, D., Huisman, M. (eds.) TACAS 2018. LNCS, vol. 10806, pp. 75–92. Springer, Cham (2018)
8. Heule, M.J.H., Kiesl, B., Biere, A.: Short proofs without new variables. In: de Moura, L. (ed.) CADE 2017. LNCS (LNAI), vol. 10395, pp. 130–147. Springer, Cham (2017)
9. Järvisalo, M., Biere, A., Heule, M.J.H.: Blocked clause elimination. In: Esparza, J., Majumdar, R. (eds.) TACAS 2010. LNCS, vol. 6015, pp. 129–144. Springer, Heidelberg (2010)
10. Järvisalo, M., Heule, M.J.H., Biere, A.: Inprocessing rules. In: Gramlich, B., Miller, D., Sattler, U. (eds.) IJCAR 2012. LNCS (LNAI), vol. 7364, pp. 355–370. Springer, Heidelberg (2012)
11. Jussila, T., Sinz, C., Biere, A.: Extended resolution proofs for symbolic SAT solving with quantification. In: Biere, A., Gomes, C.P. (eds.) SAT 2006. LNCS, vol. 4121, pp. 54–60. Springer, Heidelberg (2006)
12. Konev, B., Lisitsa, A.: Computer-aided proof of Erdős discrepancy properties. Artif. Intell. **224**(C), 103–118 (2015)
13. Kullmann, O.: On a generalization of extended resolution. Discret. Appl. Math. **96–97**, 149–176 (1999)
14. Lee, C.T.: A completeness theorem and a computer program for finding theorems derivable from given axioms. Ph.D. thesis (1967)
15. Philipp, T., Rebola-Pardo, A.: Towards a semantics of unsatisfiability proofs with inprocessing. In: Proceedings of the 21st International Conference on Logic for Programming, Artificial Intelligence and Reasoning (LPAR-21). EPiC Series in Computing, vol. 46, pp. 65–84. EasyChair (2017)
16. Sinz, C., Biere, A.: Extended resolution proofs for conjoining BDDs. In: Grigoriev, D., Harrison, J., Hirsch, E.A. (eds.) CSR 2006. LNCS, vol. 3967, pp. 600–611. Springer, Heidelberg (2006)
17. Tseitin, G.S.: On the complexity of derivation in propositional calculus. Stud. Math. Math. Log. **2**, 115–125 (1968)
18. Urquhart, A.: The complexity of propositional proofs. Bull. Symb. Log. **1**(4), 425–467 (1995)
19. Van Gelder, A.: Verifying RUP proofs of propositional unsatisfiability. In: Proceedings of the 10th International Symposium on Artificial Intelligence and Mathematics (ISAIM 2008) (2008)
20. Van Gelder, A.: Producing and verifying extremely large propositional refutations. Ann. Math. Artif. Intell. **65**(4), 329–372 (2012)
21. Wetzler, N., Heule, M.J.H., Hunt Jr., W.A.: DRAT-trim: efficient checking and trimming using expressive clausal proofs. In: Sinz, C., Egly, U. (eds.) SAT 2014. LNCS, vol. 8561, pp. 422–429. Springer, Cham (2014)

Verifying Asymptotic Time Complexity of Imperative Programs in Isabelle

Bohua Zhan and Maximilian P. L. Haslbeck[✉]

Technische Universität München, Munich, Germany
{zhan,haslbema}@in.tum.de
http://www21.in.tum.de/haslbema

Abstract. We present a framework in Isabelle for verifying asymptotic time complexity of imperative programs. We build upon an extension of Imperative HOL and its separation logic to include running time. Our framework is able to handle advanced techniques for time complexity analysis, such as the use of the Akra–Bazzi theorem and amortized analysis. Various automation is built and incorporated into the auto2 prover to reason about separation logic with time credits, and to derive asymptotic behaviour of functions. As case studies, we verify the asymptotic time complexity (in addition to functional correctness) of imperative algorithms and data structures such as median of medians selection, Karatsuba's algorithm, and splay trees.

Keywords: Isabelle · Time complexity analysis · Separation logic
Program verification

1 Introduction

In studies of formal verification of computer programs, most of the focus has been on verifying functional correctness of a program. However, for many algorithms, analysis of its running time can be as difficult, or even more difficult than the proof of its functional correctness. In such cases, it is of interest to verify the run-time analysis, that is, showing that the algorithm, or a given implementation of it, does have the claimed asymptotic time complexity.

Interactive theorem provers are useful tools for performing such a verification, as their soundness is based on a small trusted kernel, hence long derivations can be made with a very high level of confidence. So far, the work of Guéneau et al. [6,12] appears to be the only general framework for asymptotic time complexity analysis of imperative programs in an interactive theorem prover. The framework is built in Coq, based on Charguéraud's CFML package [5] for verifying imperative programs using characteristic formulas.

We present a new framework[1] for asymptotic time complexity analysis in Isabelle/HOL [19]. The framework is an extension of Imperative HOL [2], which

---

[1] Available online at https://github.com/bzhan/Imperative_HOL_Time.

© Springer International Publishing AG, part of Springer Nature 2018
D. Galmiche et al. (Eds.): IJCAR 2018, LNAI 10900, pp. 532–548, 2018.
https://doi.org/10.1007/978-3-319-94205-6_35

represents imperative programs as monads. Compared to [12], we go further in two directions. First, we incorporate the work of Eberl [11] on the Akra–Bazzi theorem to analyze several divide-and-conquer algorithms. Second, we extend the auto2 prover [21] to provide substantial automation in reasoning about separation logic with time credits, as well as deriving asymptotic behaviour of functions.

We also make use of existing work by Nipkow [18] on analysis of amortized complexity for functional programs. Based on this work, we verify the amortized complexity of imperative implementations of two data structures: skew heaps and splay trees.

Throughout our work, we place a great deal of emphasis on modular development of proofs. As the main theorems to be proved are concerned with asymptotic complexity rather than explicit constants, they do not depend on implementation details. In addition, by using an ad-hoc refinement scheme similar to that in [21], the analysis of an imperative program is divided into clearly-separated parts: proof of functional correctness, analysis of asymptotic behaviour of runtime functions, and reasoning about separation logic. Further separation of concerns is used in amortized analysis.

In summary, the main contributions of this paper are as follows:

- We extend Imperative HOL and its separation logic to reason about running time of imperative programs (Sect. 2.1).
- We introduce a methodology to organize the verification so that proofs can be divided cleanly into orthogonal parts (Sect. 3).
- We extend the existing setup of the auto2 prover for separation logic to also work with time credits. We also set up various automation for proving asymptotic behaviour of functions in one or two variables (Sect. 4).
- We demonstrate the broad applicability of our framework with several case studies (Sect. 5), including those involving advanced techniques for runtime analysis such as the use of the Akra–Bazzi theorem (for merge sort, median of medians selection, and Karatsuba's algorithm) and amortized analysis (for dynamic arrays, skew heaps, and splay trees). We also provide an example (Knapsack problem) illustrating asymptotic complexity on two variables.

2   Background

In this section, we review some background material needed in our work. First, we briefly describe the extension of Imperative HOL to reason about running time of imperative programs. Then, we recapitulate the existing theory of asymptotic complexity in Isabelle, and Eberl's formalization of the Akra–Bazzi theorem.

2.1   Imperative HOL with Time

Imperative HOL [2] is a framework for reasoning about imperative programs in Isabelle. Lammich and Meis later constructed a separation logic for this framework [17]. More details on both can be found in [16].

Atkey [1] introduced the idea of including *time credits* in separation logic to enable amortized resource analysis, in particular analysis of the running time of a program. He also provided a formalization of the logic in Coq. In this section, we describe how this idea is implemented by modifying Imperative HOL and its separation logic.

Basic Definitions. In ordinary Imperative HOL, a procedure takes a heap (of type `heap`) as input, and can either fail, or return a pair consisting of a return value and a new heap. In Imperative HOL with time, a procedure returns in addition a natural number when it succeeds, specifying the number of computation steps used. Hence, the type `'a Heap` for a procedure with return type `'a` is given by `heap ⇒ ('a × heap × nat) option`.

In the separation logic for ordinary Imperative HOL, a *partial heap* is defined to be a heap together with a subset of used addresses (type `heap × nat set`). In our case, a partial heap can also contain a number of time credits. Hence, the new type for partial heaps is given by `pheap = (heap × nat set) × nat`.

An assertion (type `assn`) is, as before, a mapping from `pheap` to `bool` that does not depend on values of the heap outside the address set. The notation $((h, as), n) \vDash P$ means the partial heap $((h, as), n)$ satisfies the assertion P. The basic assertions have the same meaning as before, except they also require the partial heap to contain zero time credits. In addition we define the assertion $\$n$, to specify a partial heap with n time credits and nothing else.

The *separating conjunction* of two assertions is defined as follows (differences from original definition are marked in bold):

$$P * Q = \lambda((h, as), \mathbf{n}). \exists u \ v \ \mathbf{n_1} \ \mathbf{n_2}. \begin{cases} u \cup v = as \land u \cap v = \emptyset \land \mathbf{n_1} + \mathbf{n_2} = \mathbf{n} \land \\ ((h, u), \mathbf{n_1}) \vDash P \land ((h, v), \mathbf{n_2}) \vDash Q. \end{cases}$$

That is, time credits can be split in a separation conjunction in the same way as sets of addresses on the heap. In particular $\$(n + m) = \$n * \$m$.

Hoare Triples. A Hoare triple `<P> c <Q>` is a predicate of type

$$\texttt{assn} \Rightarrow \texttt{'a Heap} \Rightarrow (\texttt{'a} \Rightarrow \texttt{assn}) \Rightarrow \texttt{bool},$$

defined as follows: `<P> c <Q>` holds if for any partial heap $((h, as), n)$ satisfying P, the execution of c on h is successful with new heap h', return value r, and time consumption t, such that $n \geq t$, and the new partial heap $((h', as'), n - t)$ satisfies $Q(r)$, where as' is as together with the newly allocated addresses. With this definition of a Hoare triple with time, the frame rule continues to hold.

Most basic commands (e.g. accessing or updating a reference, getting the length of an array) are defined to take one unit of computation time. Commands that operate on an entire array, for example initializing an array, or extracting an array into a functional list, are defined to take $n + 1$ units of computation time, where n is the length of the array. From this, we can prove Hoare triples for the basic commands. We give two examples (here $p \mapsto_a xs$ asserts that p points to the array xs and $\uparrow b$ asserts that b is true):

```
<p ↦ₐ xs * $1> Array.len xs <λr. p ↦ₐ xs * ↑(r = length xs)>

<$(n + 1)> Array.new n x <λr. r ↦ₐ replicate n x>
```

We define the notation `<P> c <Q>ₜ` as a shorthand for `<P> c <Q * true>`. The assertion `true` holds for any partial heap, and in particular can include any number of time credits. Hence, a Hoare triple of the form `<P*$n> c <Q>ₜ` implies that the procedure c costs *at most* n time credits. We very often state Hoare triples in this form, and so only prove upper bounds on the computation time of the program.

2.2 Asymptotic Analysis

Working with asymptotic complexity informally can be particularly error-prone, especially when several variables are involved. Some examples of fallacious reasoning are given in [12, Sect. 2]. In an interactive theorem proving environment, such problems can be avoided, since all notions are defined precisely, and all steps of reasoning must be formally justified.

For the definition of the big-O notation, or more generally Landau symbols, we use the formalization by Eberl [9], where they are defined in a general form in terms of filters, and therefore work also in the case of multiple variables.

In our work, we are primarily interested in functions of type `nat⇒real` (for the single variable case) and `nat × nat⇒real` (for the two-variable case). Given a function g of one of these types, the Landau symbols $O(g)$, $\Omega(g)$ and $\Theta(g)$ are sets of functions of the same type. In the single variable case, using the standard filter (`at_top` for *limit at positive infinity*), the definitions are as follows:

$$f \in O(g) \longleftrightarrow \exists c > 0.\ \exists N.\ \forall n \geq N.\ |f(n)| \leq c \cdot |g(n)|$$
$$f \in \Omega(g) \longleftrightarrow \exists c > 0.\ \exists N.\ \forall n \geq N.\ |f(n)| \geq c \cdot |g(n)|$$
$$f \in \Theta(g) \longleftrightarrow f \in O(g) \wedge f \in \Omega(g)$$

In the two-variable case, we will use the product filter `at_top ×F at_top` throughout. Expanding the definitions, the meaning of the Landau symbols are as expected:

$$f \in O_2(g) \longleftrightarrow \exists c > 0.\ \exists N.\ \forall n, m \geq N.\ |f(n,m)| \leq c \cdot |g(n,m)|$$
$$f \in \Omega_2(g) \longleftrightarrow \exists c > 0.\ \exists N.\ \forall n, m \geq N.\ |f(n,m)| \geq c \cdot |g(n,m)|$$
$$f \in \Theta_2(g) \longleftrightarrow f \in O_2(g) \wedge f \in \Omega_2(g)$$

2.3 Akra–Bazzi Theorem

A well-known technique for analyzing the asymptotic time complexity of divide and conquer algorithms is the Master Theorem (see for example [7, Chap. 4]). The Akra–Bazzi theorem is a generalization of the Master Theorem to a wider range of recurrences. Eberl [11] formalized the Akra–Bazzi theorem in Isabelle, and also wrote tactics for applying this theorem in a semi-automatic manner.

Notably, the automation is able to deal with taking ceiling and floor in recursive calls, an essential ingredient for actual applications but often ignored in informal presentations of the Master theorem.

In this section, we state a slightly simpler version of the result that is sufficient for our applications. Let $f : \mathbb{N} \to \mathbb{R}$ be a non-negative function defined recursively as follows:

$$f(x) = g(x) + \sum_{i=1}^{k} a_i \cdot f(h_i(x)) \qquad \text{for all } x \geq x_0 \tag{1}$$

where $x_0 \in \mathbb{N}$, $g(x) \geq 0$ for all $x \geq x_0$, $a_i \geq 0$ and each $h_i(x) \in \mathbb{N}$ is either $\lceil b_i \cdot x \rceil$ or $\lfloor b_i \cdot x \rfloor$ with $0 < b_i < 1$, and x_0 is large enough that $h_i(x) < x$ for all $x \geq x_0$.

The parameters a_i and b_i determine a single characteristic value p, defined as the solution to the equation

$$\sum_{i=1}^{k} a_i \cdot b_i^p = 1 \tag{2}$$

Depending on the relation between the asymptotic behaviour of g and $\Theta(x^p)$, there are three main cases of the Akra–Bazzi theorem:

Bottom-heavy: if $g \in O(x^q)$ for $q < p$ and $f(x) > 0$ for sufficiently large x, then $f \in \Theta(x^p)$.
Balanced: if $g \in \Theta(x^p \ln^a x)$ with $a \geq 0$, then $f \in \Theta(x^p \ln^{a+1} x)$.
Top-heavy: if $g \in \Theta(x^q)$ for $q > p$, then $f \in \Theta(x^q)$.

All three cases are demonstrated in our examples (in Karatsuba's algorithm, merge sort, and median of medians selection, respectively).

3   Organization of Proofs

In this section, we describe our strategy for organizing the verification of an imperative program together with its time complexity analysis. The strategy is designed to achieve the following goals:

- Proof of functional correctness of the algorithm should be separate from the analysis of memory layout and time credits using separation logic.
- Analysis of time complexity should be separate from proof of correctness.
- Time complexity analysis should work with asymptotic bounds Θ most of the time, rather than with explicit constants.
- Compositionality: verification of an algorithm should result in a small number of theorems, which can be used in the verification of a larger algorithm. The statement of these theorems should not depend on implementation details.

We first consider the general case and then describe the additional layer of organization for proofs involving amortized analysis.

3.1   General Case

For a procedure with name f, we define three Isabelle functions:

f_fun: The functional version of the procedure.
f_impl: The imperative version of the procedure.
f_time: The runtime function of the procedure.

The definition of f_time should be stated in terms of runtime functions of procedures called by f_impl, in a way parallel to the definition of f_impl. If f_impl is defined by recursion, f_time should also be defined by recursion in the corresponding manner.

The theorems to be proved are:

1. The functional program f_fun satisfies the desired correctness property.
2. A Hoare triple stating that f_impl implements f_fun and runs within f_time.
3. The running time f_time satisfies the desired asymptotic behaviour.
4. Combining 1 and 2, a Hoare triple stating that f_impl satisfies the desired correctness property, and runs within f_time.

Here the proof of Theorem 2 is expected to be routine, since the three definitions follow the same structure. Theorem 3 should involve only analysis of asymptotic behaviour of functions, while Theorem 1 should involve only reasoning with functional data structures. In the end, Theorems 3 and 4 present an interface for external use, whose statements do not depend on details of the implementation or of the proofs.

We illustrate this strategy on the final step of merge sort. The definitions of the functional and imperative programs are shown side by side below. Note that the former is working with a functional list, while the latter is working with an imperative array on the heap.

```
merge_sort_fun xs =                merge_sort_impl X = do {
  (let n = length xs in              n ← Array.len X;
   (if n ≤ 1 then xs                 if n ≤ 1 then return ()
    else                             else do {
     let as = take (n div 2) xs;       A ← atake (n div 2) X;
         bs = drop (n div 2) xs;        B ← adrop (n div 2) X;
         as' = merge_sort_fun as;       merge_sort_impl A;
         bs' = merge_sort_fun bs;       merge_sort_impl B;
         r = merge_list as' bs'         mergeinto (n div 2)
     in r                              (n - n div 2) A B X
   )                                 }
  )                                }
```

The runtime function of the procedure is defined as follows:

```
n ≤ 1 ⟹ merge_sort_time n = 2
n > 1 ⟹ merge_sort_time n = 2 + atake_time n + adrop_time n +
    merge_sort_time (n div 2) + merge_sort_time (n - n div 2) +
    mergeinto_time n
```

The theorems to be proved are as follows. First, correctness of the functional algorithm `merge_sort_fun`:

```
merge_sort_fun xs = sort xs
```

Second, a Hoare triple asserting the agreement of the three definitions:

```
<p ↦ₐ xs * $(merge_sort_time (length xs))>
 merge_sort_impl p
<λ_. p ↦ₐ merge_sort_fun xs>ₜ
```

Third, the asymptotic time complexity of `merge_sort_time`:

```
merge_sort_time ∈ Θ(λn. n * ln n)
```

Finally, Theorems 1 and 2 are combined to prove the final Hoare triple for external use, with `merge_sort_fun xs` replaced by `sort xs`.

3.2   Amortized Analysis

In an amortized analysis, we fix some type of data structure and consider a set of primitive operations on it. For simplicity, we assume each operation has exactly one input and output data structure (extension to the general case is straightforward). A potential function P is defined on instances of the data structure and represents time credits that can be used for future operations. Each procedure f is associated an actual runtime f_t and an amortized runtime f_{at}. They are required to satisfy the following inequality: let a be the input data structure of f and let b be its output data structure, then[2]

$$f_{at} + P(a) \geq f_t + P(b). \tag{3}$$

The proof of inequality (3) usually involves arithmetic, and sometimes the correctness of the functional algorithm. For skew heaps and splay trees, the analogous results are already proved in [18], and only slight modifications are necessary to bring them into the right form for our use.

The organization of an amortized analysis in our framework is as follows. We define two assertions: the *raw* assertion `raw_assn t a` stating that the address a points to an imperative data structure refining t, and the *amortized* assertion, defined as

```
amor_assn t a = raw_assn t a * $(P(t)),
```

where P is the potential function.

For each primitive operation implemented by `f`, we define `f_fun`, `f_impl`, and `f_time` as before, where `f_time` is the *actual* runtime. We further define a function `f_atime` to be the proposed amortized runtime. The theorems to be proved are as follows (compare to the list in Sect. 3):

---

[2] In many presentations, the amortized runtime f_{at} is simply *defined* to be $f_t + P(b) - P(a)$. Our approach is more flexible in allowing f_{at} to be defined by a simple formula and isolating the complexity to the proof of (3).

1. The functional program *f_fun* satisfies the desired correctness property.
2. A Hoare triple using the amortized assertion stating that *f_impl* implements *f_fun* and runs within *f_atime*, which is a consequence of the following:
   2a. A Hoare triple using the raw assertion stating that *f_impl* implements *f_fun* and runs within *f_time*.
   2b. The inequality between amortized and actual runtime.
3. The *amortized* runtime *f_atime* satisfies the desired asymptotic behaviour.
4. Combining 1 and 2, a Hoare triple stating that *f_impl* satisfies the desired correctness property and runs within *f_atime*.

In the case of data structures (and unlike merge sort), it is useful to state Theorem 4 in terms of yet another, abstract assertion which hides the concrete reference to the data structure. This follows the technique described in [21, Sect. 5.3]. Theorems 3 and 4 are the final results for external use.

We now illustrate this strategy using splay trees as an example. The raw assertion is called *btree*. The basic operation in a splay tree is the "splay" operation, from which insertion and lookup can be easily defined. For this operation, the functions *splay*, *splay_impl*, and *splay_time* are defined by recursion in a structurally similar manner. Theorem 2a takes the form:

```
<btree t a * $(splay_time x t)>
 splay_impl x a
<btree (splay x t)>ₜ
```

Let *splay_tree_P* be the potential function on splay trees. Then the amortized assertion is defined as:

```
splay_tree t a = btree t a * $(splay_tree_P t)
```

The amortized runtime for splay has a relatively simple expression:

```
splay_atime n = 15 * (⌈3 * log 2 n⌉ + 2)
```

The difficult part is showing the inequality relating actual and amortized runtime (Theorem 2b):

```
bst t ⟹ splay_atime (size1 t) + splay_tree_P t ≥
            splay_time x t + splay_tree_P (splay x t),
```

which follows from the corresponding lemma in [18]. Note the requirement that *t* is a binary search tree. Combining 2a and 2b, we get Theorem 2:

```
bst t ⟹
<splay_tree t a * $(splay_atime (size1 t))>
 splay_impl x a
<splay_tree (splay x t)>ₜ
```

The asymptotic bound on the amortized runtime (Theorem 3) is:

```
splay_atime ∈ Θ(λx. ln x)
```

The functional correctness of `splay` (Theorem 1) states that it maintains sorted-ness of the binary search tree and its set of elements:

```
bst t ⟹ bst (splay a t),     set_tree (splay a t) = set_tree t
```

The abstract assertion hides the concrete tree behind an existential quantifier:

```
splay_tree_set S a = (∃_A t. splay_tree t a * ↑(bst t) * ↑(set_tree t = S))
```

The final Hoare triple takes the form (`card S` denotes the cardinality of `S`):

```
<splay_tree_set S a * $(splay_atime (card S + 1))>
  splay_impl x a
<splay_tree_set S>ₜ
```

4   Setup for Automation

In this section, we describe automation to handle two of the steps mentioned in the previous section: one working with separation logic (for Theorem 2), and the other proving asymptotic behaviour of functions (for Theorem 3).

4.1   Separation Logic with Time Credits

First, we discuss automation for reasoning about separation logic with time credits. This is an extension of the setup discussed in [21] for reasoning about ordinary separation logic. Here, we focus on the additional setup concerning time credits.

The basic step in the proof is as follows: suppose the current heap satisfies the assertion $P * \$T$ and the next command has the Hoare triple

```
<P' * $$T' * ↑ b> c <Q>
```

where b is the pure part of the precondition, apply the Hoare triple to derive the successful execution of c, and some assertion on the next heap. In ordinary separation logic (without $\$T$ and $\$T'$), this involves matching P' with parts of P, proving the pure assertions b, and then applying the frame rule. In the current case, we additionally need to show that $T' \leq T$, so $\$T$ can be rewritten as $\$T = \$(T' + T'') = \$T' * \$T''$.

In general, proving this inequality can involve arbitrarily complex arguments. However, due to the close correspondence in the definitions of `f_time` and `f_impl`, the actual tasks usually lie in a simple case, and we tailor the automation to focus on this case. First, we normalize both T and T' into polynomial form:

$$T = c_1 p_1 + \cdots + c_m p_m, \quad T' = d_1 q_1 + \cdots + d_n q_n, \tag{4}$$

where each c_i and d_j are constants, and each p_i and q_j are non-constant terms or 1. Next, for each term $d_j q_j$ in T', we try to find some term $c_i p_i$ in T such that p_i equals q_j according to the known equalities, and $d_j \leq c_i$. If such a term is found, we subtract $d_j p_i$ from T. This procedure is performed on T in sequence (so d_2q_2 is searched on the remainder of T after subtracting d_1q_1, etc.). If the procedure succeeds with T'' remaining, then we have $T = T' + T''$.

The above procedure suffices in most cases. For example, given the parallel definitions of `merge_sort_impl` and `merge_sort_time` in Sect. 3.1, it is able to show that `merge_sort_impl` runs in time `merge_sort_time`. However, in some special cases, more is needed. The extra reasoning often takes the following form: if s is a term in the normalized form of T, and $s \geq t$ holds for some t (an inequality that must be derived during the proof), then the term s can be replaced by t in T.

In general, we permit the user to provide hints of the form

```
@have "s ≥t t",
```

where the operator $\cdot \geq_t \cdot$ is equivalent to $\cdot \geq \cdot$, used only to remind auto2 that the fact is for modification of time credit only. Given this instruction, auto2 attempts to prove $s \geq t$, and when it succeeds, it replaces the assertion $h_i \vDash P * \$T$ on the current heap with $h_i \vDash P * \$T' * \text{true}$, where the new time credit T' is the normalized form of $T - s + t$. This technique is needed in case studies such as binary search and median of medians selection (see the explanation for the latter in Sect. 5).

4.2 Asymptotic Analysis

The second part of the automation is for analysis of asymptotic behaviour of runtime functions. Eberl [9] already provides automation for Landau symbols in the single variable case. In addition to incorporating it into our framework, we add facilities for dealing with function composition and the two-variable case.

Because side conditions for the Akra–Bazzi theorem are in the Θ form, we mainly deal with Θ and Θ_2, stating the exact asymptotic behaviours of running time functions. However, since running time functions themselves are very often only upper bounds of the actual running times, we are essentially still proving big-O bounds on running times of programs.

In our case, the general problem is as follows: given the definition of `f_time` (n) in terms of some `g_time` $(s(n))$ (runtime of procedures called by `f_impl`), simple terms like $4n$ or 1, or recursive calls to `f_time`, determine the asymptotic behaviour of `f_time`.

To begin with, we maintain a table of the asymptotic behaviour of previously defined runtime functions. The attribute `asym_bound` adds a new theorem to this table. This table can be looked-up by the name of the procedure.

We restrict ourselves to asymptotic bounds of the form

$$\text{polylog}(a, b) = (\lambda n.\ n^a (\ln n)^b),$$

where a and b are natural numbers. In the two-variable case, we work with asymptotic bounds of the form

$$\text{polylog}_2(a, b, c, d) = (\lambda(m, n).\ \text{polylog}(a, b)(m) \cdot \text{polylog}(c, d)(n)).$$

This suffices for our present purposes and can be extended in the future. Note that this restriction does not mean our framework cannot handle other complexity classes, only that they will require more manual proofs (or further setup of automation).

Non-recursive Case. When the runtime function is non-recursive, the analysis proceeds by determining the asymptotic behaviour in a bottom-up manner.

To handle terms of the form $\mathtt{g\_time}\ (s(n))$ where s is linear, we use the following composition rule: if $u \in \Theta(\mathrm{polylog}(a, b))$, and $v \in \Theta(\lambda n.\ n)$, then $u \circ v \in \Theta(\mathrm{polylog}(a, b))$. Composition in general is quite subtle: the analogous rule does not hold if u is the exponential function[3].

The asymptotic behaviour of a sum is determined by the absorption rule: if $g_1 \in O(g_2)$, then $\Theta(g_1 + g_2) = \Theta(g_2)$. Here, we make use of existing automation in [9] for deciding inclusion of big-O classes of polylog functions. The rule for products is straightforward.

The combination of these three rules can solve many examples automatically. E.g. this (artificial) example: if $f_1 \in \Theta(\lambda n.\ n)$ and $f_2 \in \Theta(\lambda n.\ \ln n)$, then

$$(\lambda n.\ f_1(n + 1) + n \cdot f_2(2n) + 3n \cdot f_2(n\ \mathtt{div}\ 3)) \in \Theta(\lambda n.\ n \ln n).$$

Analogous results are proved in the two-variable case (note that unlike in the single variable case, not all pairs of polylog$_2$ functions are comparable. e.g. $O(m^2 n + mn^2)$). For example, the following can be automatically solved: if additionally $f_3 \in \Theta(\lambda(m, n).\ mn)$ and $f_4 \in \Theta(\lambda(m, n).\ m + n)$, then

$$(\lambda(m, n).\ f_1(n) + f_2(m) + mn + f_3(m\ \mathtt{div}\ 3, n + 1)) \in \Theta(\lambda(m, n).\ mn).$$
$$(\lambda(m, n).\ 1 + f_1(n) + f_2(m) + f_4(m + 1, n + 1)) \in \Theta(\lambda(m, n).\ m + n).$$

Recursive Case. There are two main classes of results for analysis of recursively-defined runtime functions: the Akra–Bazzi theorem and results about linear recurrences. For both classes of results, applying the theorem reduces the analysis of a recursive runtime function to the analysis of a non-recursive function, which can be solved using automation described in the previous part.

The Akra–Bazzi theorem is discussed in Sect. 2.3. Theorems about linear recurrences allow us to reason about for-loops written as recursions. They include the following: in the single variable case, if f is defined by induction as

$$f(0) = c, \quad f(n + 1) = f(n) + g(n),$$

where $g \in \Theta(\lambda n.\ n)$, then $f \in \Theta(\lambda n.\ n^2)$.

In the two-variable case, if f satisfies

$$f(0, m) \leq C, \quad f(n + 1, m) = f(n, m) + g(m)$$

---

[3] https://math.stackexchange.com/questions/761006/big-o-and-function-composition.

where $g \in \Theta(\lambda n.\ n)$, then $f \in \Theta_2(\lambda(n, m).\ nm)$.

As an example, consider the problem of showing $\Theta(\lambda n.\ n * \ln n)$ complexity of `merge_sort_time`, defined in Sect. 3.1. This applies the balanced case of the Akra–Bazzi theorem. Using this theorem, the goal is reduced to:

$(\lambda n.\ 2 + $ `atake_time` $n + $ `adrop_time` $n + $ `mergeinto_time` $n) \in \Theta(\lambda n.\ n)$

(the non-recursive calls run in linear time). This can be shown automatically using the method described in the previous section, given that `atake_time`, `adrop_time`, and `mergeinto_time` have already been shown to be linear.

5   Case Studies

In this section, we present the main case studies verified using our framework. The examples can be divided into three classes: divide-and-conquer algorithms (using the Akra–Bazzi theorem), algorithms that are essentially for-loops (using linear recurrences), and amortized analysis.

We measure the complexity of a proof by counting the number of steps in the proof: each lemma statement counts as one step and each hint provided by the user as an additional step. In the table below, #Hoare counts the number of steps for proving the Hoare triples (Theorems 2 and 4). #Time counts the number of steps for reasoning about runtime functions (Theorem 3). We also list the ratio (Ratio) between the sum of #Hoare and #Time to the number of lines of the imperative program (#Imp). This ratio measures the overhead for verifying the imperative program with runtime analysis. In particular this does *not* include verifying the correctness of the functional program (Theorem 1). In addition we list the total lines of code for each case study.

	#Imp	#Time	#Hoare	Ratio	LOC
Binary search	11	10	14	2.18	82
Merge sort	38	11	12	0.61	121
Karatsuba	58	18	28	0.79	250
Select	51	41	31	1.41	447
Insertion sort	15	3	4	0.47	42
Knapsack	27	9	8	0.63	113
Dynamic array	55	19	37	1.02	424
Skew heap	25	38	21	2.36	257
Splay tree	120	51	37	0.73	447

Using our automation the average overhead ratio is slightly over 1. On a dual-core laptop with 2 GHz each, processing all the examples takes around ten minutes. The development of the case studies, together with the framework itself, took about 4 person months.

Next we give details for some of the case studies.

Karatsuba's Algorithm. The functional version of Karatsuba's algorithm for multiplying two polynomials is verified in [8]. To simplify matters, we further restrict us to the case where the two polynomials are of the same degree.

The recursive equation is given by:

$$T(n) = 2 \cdot T(\lceil n/2 \rceil) + T(\lfloor n/2 \rfloor) + g(n). \tag{5}$$

Here $g(n)$ is the sum of the running times corresponding to non-recursive calls, which can be automatically shown to be linear in n. Then the Akra–Bazzi method gives the solution $T(n) \in \Theta(n^{log_2 3})$ (bottom-heavy case).

Median of Medians Selection. Median of medians for quickselect is a worst-case linear-time algorithm for selecting the i-th largest element of an unsorted array [7, Sect. 9.3]. In the first step of the algorithm, it chooses an approximate median p by dividing the array into groups of 5 elements, finding the median of each group, and finding the median of the medians by a recursive call. In the second step, p is used as a pivot to partition the array, and depending on i and the size of the partitions, a recursive call may be made to either the section $x < p$ or the section $x > p$. This algorithm is particularly interesting because its runtime satisfies a special recursive formula:

$$T(n) \leq T(\lceil n/5 \rceil) + T(\lceil 7n/10 \rceil) + g(n), \tag{6}$$

where $g(n)$ is linear in n. The Akra–Bazzi theorem shows that T is linear (top-heavy case).

Eberl verified the correctness of the functional algorithm [10]. There is one special difficulty in verifying the imperative algorithm: the length of the array in the second recursive call is not known in advance, only that it is bounded by $\lceil 7n/10 \rceil$. Hence, we need to prove monotonicity of T, as well as provide the hint $T(\lceil 7n/10 \rceil) \geq_t T(l)$ (where l is the length of the array in the recursive call) during the proof.

Knapsack. The dynamic programming algorithm solving the Knapsack problem is used to test our ability to handle asymptotic complexity with two variables. The time complexity of the algorithm is $\Theta_2(nW)$, where n is the number of items, and W is the capacity of the sack. Correctness of the functional algorithm was proved by Simon Wimmer.

Dynamic Array. Dynamic Arrays [7, Sect. 17.4] are one of the simpler amortized data structures. We verify the version that doubles the size of the array whenever it is full (without automatically shrinking the array).

Skew Heap and Splay Tree. For these two examples, the bulk of the analysis (functional correctness and justification of amortized runtime) is done in [18]. Our work is primarily to define the imperative version of the algorithm and verifying its agreement with the functional version. Some work is also needed to transform the results in [18] into the appropriate form, in particular rounding the real-valued potentials and runtime functions into natural numbers required in our framework.

6 Related Work

We compare our work with recent advances in verification of runtime analysis of programs, starting from those based on interactive theorem provers to the more automatic methods.

The most closely-related is the impressive work by Guéneau et al. [12] for asymptotic time complexity analysis in Coq. We now take a closer look at the similarities and differences:

- Guéneau et al. give a structured overview of different problems that arise when working informally with asymptotic complexity in several variables, then solve this problem by rigorously defining asymptotic domination (which is essentially $f \in O(g)$) with filters and develop automation for reasoning about it. We follow the same idea by building on existing formalization of Landau symbols with filters in Isabelle [9], then extend automation to also handle the two-variable case.
- While they package up the functional correctness together with the complexity claims into one predicate *specO*, we choose to have two separate theorems (the Hoare triple and the asymptotic bound).
- While their automation assists in synthesizing recurrence equations from programs, they leave their solution to the human. In contrast, we write the recurrence relation by hand, which can be highly non-obvious (e.g. in the case of median of medians selection), but focus on solving the recurrences for the asymptotic bounds automatically (e.g. using the Akra–Bazzi theorem).
- Their main examples include binary search, the Bellman–Ford algorithm and union-find, but not those requiring applications of the Master theorem or the Akra–Bazzi method. We present several other advanced examples, including applications of the Akra–Bazzi method, and those involving amortized analysis.

Wang et al. [20] present TiML, a functional programming language which can be annotated by invariants and specifically also with time complexity annotations in types. The type checker extracts verification conditions from these programs, which are handled by an SMT solver. They also make the observation that annotational burden can be lowered by not providing a closed form for a time bound, but only specifying its asymptotic behaviour. For recursive functions, the generated VCs include a recurrence (e.g. $T(n - 1) + 4n \leq T(n)$) and one is left to show that there exists a solution for T which is additionally in some asymptotic bound, e.g. $O(n^2)$. By employing a recurrence solver based on heuristic pattern matching they make use of the Master Theorem in order to discharge such VCs. In that manner they are able to verify the asymptotic complexity of merge sort. Additionally they can handle amortized complexity, giving Dynamic Arrays and Functional Queues as examples. Several parts of their work rely on non-verified components, including the use of SMT solvers and the pattern matching for recurrence relations. In contrast, our work is verified throughout by Isabelle's kernel.

On the other end of the scale we want to mention Automatic Amortized Resource Analysis (AARA). Possibly the first example of a resource analysis logic based on potentials is due to Hofmann and Jost [15]. They pioneer the use of potentials coded into the type system in order to automatically extract bounds in the runtime of functional programs. Hoffmann et al. successfully developed this idea further [13,14]. Carbonneaux et al. [3,4] extend this work to imperative programs and automatically solve extracted inequalities by efficient off-the-shelf LP-solvers. While the potentials involved are restricted to a specific shape, the analysis performs well and at the same time generates Coq proof objects certifying their resulting bounds.

7   Conclusion

In this paper, we presented a framework for verifying asymptotic time complexity of imperative programs. This is done by extending Imperative HOL and its separation logic with time credits. Through the case studies, we demonstrated the ability of our framework to handle complex examples, including those involving advanced techniques of time complexity analysis, such as the Akra–Bazzi theorem and amortized analysis. We also showed that verification of amortized analysis of functional programs [18] can be converted to verification of imperative programs with little additional effort.

One major goal for the future is to extend Imperative HOL with *while* and *for* loops, and add facilities for reasoning about them (both functional correctness and time complexity). Ultimately, we would like to build a single framework in which all deterministic algorithms typically taught in undergraduate study (for example, those contained in [7]) can be verified in a straightforward manner.

The Refinement Framework by Lammich [16] is a framework for stepwise refinement from specifications via deterministic algorithms to programs written in Imperative HOL. It would certainly be interesting to investigate how to combine this stepwise refinement scheme with runtime analysis.

Acknowledgments. This work is funded by DFG Grant NI 491/16-1. We thank Manuel Eberl for his impressive formalization of the Akra–Bazzi method and the functional correctness of the selection algorithm, and Simon Wimmer for the formalization of the DP solution for the Knapsack problem. We thank Manuel Eberl, Tobias Nipkow, and Simon Wimmer for valuable feedback during the project. Finally, we thank Armaël Guéneau and his co-authors for their stimulating paper.

References

1. Atkey, R.: Amortised resource analysis with separation logic. In: Gordon, A.D. (ed.) ESOP 2010. LNCS, vol. 6012, pp. 85–103. Springer, Heidelberg (2010). https://doi.org/10.1007/978-3-642-11957-6_6

2. Bulwahn, L., Krauss, A., Haftmann, F., Erkök, L., Matthews, J.: Imperative functional programming with Isabelle/HOL. In: Mohamed, O.A., Muñoz, C., Tahar, S. (eds.) TPHOLs 2008. LNCS, vol. 5170, pp. 134–149. Springer, Heidelberg (2008). https://doi.org/10.1007/978-3-540-71067-7_14

3. Carbonneaux, Q., Hoffmann, J., Reps, T., Shao, Z.: Automated resource analysis with Coq proof objects. In: Majumdar, R., Kunčak, V. (eds.) CAV 2017. LNCS, vol. 10427, pp. 64–85. Springer, Cham (2017). https://doi.org/10.1007/978-3-319-63390-9_4

4. Carbonneaux, Q., Hoffmann, J., Shao, Z.: Compositional certified resource bounds. In: Grove, D., Blackburn, S. (eds.) PLDI 2015, pp. 467–478. ACM (2015)

5. Charguéraud, A.: Characteristic formulae for the verification of imperative programs. In: Proceedings of the 16th ACM SIGPLAN International Conference on Functional Programming, ICFP 2011, pp. 418–430. ACM, New York (2011). https://doi.org/10.1145/2034773.2034828

6. Charguéraud, A., Pottier, F.: Verifying the correctness and amortized complexity of a union-find implementation in separation logic with time credits. J. Autom. Reason. (2017). https://doi.org/10.1007/s10817-017-9431-7

7. Cormen, T.H., Leiserson, C.E., Rivest, R.L., Stein, C.: Introduction to algorithms, 3rd edn. MIT Press, Cambridge (2009)

8. Divasón, J., Joosten, S., Thiemann, R., Yamada, A.: The factorization algorithm of Berlekamp and Zassenhaus. Archive of formal proofs. Formal proof development, October 2016. http://isa-afp.org/entries/Berlekamp_Zassenhaus.html

9. Eberl, M.: Landau symbols. Archive of formal proofs. Formal proof development, July 2015. http://isa-afp.org/entries/Landau_Symbols.html

10. Eberl, M.: The median-of-medians selection algorithm. Archive of formal proofs. Formal proof development, December 2017. http://isa-afp.org/entries/Median_Of_Medians_Selection.html

11. Eberl, M.: Proving divide and conquer complexities in Isabelle/HOL. J. Autom. Reason. **58**(4), 483–508 (2017)

12. Guéneau, A., Charguéraud, A., Pottier, F.: A fistful of dollars: formalizing asymptotic complexity claims via deductive program verification. In: Ahmed, A. (ed.) ESOP 2018. LNCS, vol. 10801, pp. 533–560. Springer, Cham (2018). https://doi.org/10.1007/978-3-319-89884-1_19

13. Hoffmann, J., Aehlig, K., Hofmann, M.: Multivariate amortized resource analysis. In: ACM SIGPLAN Notices, vol. 46, pp. 357–370. ACM (2011)

14. Hoffmann, J., Das, A., Weng, S.C.: Towards automatic resource bound analysis for OCaml. In: ACM SIGPLAN Notices, vol. 52, pp. 359–373. ACM (2017)

15. Hofmann, M., Jost, S.: Type-based amortised heap-space analysis. In: Sestoft, P. (ed.) ESOP 2006. LNCS, vol. 3924, pp. 22–37. Springer, Heidelberg (2006). https://doi.org/10.1007/11693024_3

16. Lammich, P.: Refinement to imperative/HOL. In: Urban, C., Zhang, X. (eds.) ITP 2015. LNCS, vol. 9236, pp. 253–269. Springer, Cham (2015). https://doi.org/10.1007/978-3-319-22102-1_17

17. Lammich, P., Meis, R.: A separation logic framework for imperative HOL. Archive of formal proofs. Formal proof development, November 2012. http://isa-afp.org/entries/Separation_Logic_Imperative_HOL.html

18. Nipkow, T.: Amortized Complexity Verified. In: Urban, C., Zhang, X. (eds.) ITP 2015. LNCS, vol. 9236, pp. 310–324. Springer, Cham (2015). https://doi.org/10.1007/978-3-319-22102-1_21
19. Nipkow, T., Wenzel, M., Paulson, L.C. (eds.): Isabelle/HOL. LNCS, vol. 2283. Springer, Heidelberg (2002). https://doi.org/10.1007/3-540-45949-9
20. Wang, P., Wang, D., Chlipala, A.: TiML: a functional language for practical complexity analysis with invariants. Proc. ACM Program. Lang. **1**(OOPSLA), 79 (2017)
21. Zhan, B.: Efficient verification of imperative programs using auto2. In: Beyer, D., Huisman, M. (eds.) TACAS 2018. LNCS, vol. 10805, pp. 23–40. Springer, Cham (2018). https://doi.org/10.1007/978-3-319-89960-2_2

Efficient Interpolation
for the Theory of Arrays

Jochen Hoenicke[✉] and Tanja Schindler[✉]

University of Freiburg, Freiburg, Germany
{hoenicke,schindle}@informatik.uni-freiburg.de

Abstract. Existing techniques for Craig interpolation for the quantifier-free fragment of the theory of arrays are inefficient for computing sequence and tree interpolants: the solver needs to run for every partitioning (A, B) of the interpolation problem to avoid creating AB-mixed terms. We present a new approach using Proof Tree Preserving Interpolation and an array solver based on Weak Equivalence on Arrays. We give an interpolation algorithm for the lemmas produced by the array solver. The computed interpolants have worst-case exponential size for extensionality lemmas and worst-case quadratic size otherwise. We show that these bounds are strict in the sense that there are lemmas with no smaller interpolants. We implemented the algorithm and show that the produced interpolants are useful to prove memory safety for C programs.

1   Introduction

Several model-checkers [1,2,8,14,16,17,20,25,26] use interpolants to find candidate invariants to prove the correctness of software. They require efficient tools to check satisfiability of a formula in a decidable theory and to compute interpolants (usually sequence or tree interpolants) for unsatisfiable formulas. Moreover, they often need to combine several theories, e.g., integer or bitvector theory for reasoning about numeric variables and array theory for reasoning about pointers. In this paper we present an interpolation procedure for the quantifier-free fragment of the theory of arrays that allows for the combination with other theories and that reuses an existing unsatisfiability proof to compute interpolants efficiently.

Our method is based on the array solver presented in [10], which fits well into existing Nelson-Oppen frameworks. The solver generates lemmas, valid in the theory of arrays, that explain equalities between terms shared between different theories. The terms do not necessarily belong to the same formula in the interpolation problem and the solver does not need to know the partitioning. Instead, we use the technique of Proof Tree Preserving Interpolation [13], which produces interpolants from existing proofs that can contain propagated equalities between symbols from different parts of the interpolation problem.

J. Hoenicke and T. Schindler—This work is supported by the German Research Council (DFG) under HO 5606/1-1.

© Springer International Publishing AG, part of Springer Nature 2018
D. Galmiche et al. (Eds.): IJCAR 2018, LNAI 10900, pp. 549–565, 2018.
https://doi.org/10.1007/978-3-319-94205-6_36

The contribution of this paper is an algorithm to interpolate the lemmas produced by the solver of the theory of arrays without introducing quantifiers. The solver only generates two types of lemmas, namely a variant of the read-over-write axiom and a variant of the extensionality axiom. However, the lemmas contain array store chains of arbitrary length which need to be handled by the interpolation procedure. The interpolants our algorithm produces summarize array store chains, e.g., they state that two shared arrays at the end of a sub-chain differ at most at m indices, each satisfying a subformula. Bruttomesso et al. [6] showed that adding a diff function to the theory of arrays makes the quantifier-free fragment closed under interpolation, i.e. it ensures the existence of quantifier-free interpolants for quantifier-free problems. We use the diff function to obtain the indices for store chains and give a more efficient algorithm that exploits the special shape of the lemmas provided by the solver.

Nevertheless, the lemma interpolants produced by our algorithm may be exponential in size (with respect to the size of the input lemma). We show that this is unavoidable as there are lemmas that have no small interpolants.

Related Work. The idea of computing interpolants from resolution proofs goes back to Krajíček and Pudlák [22,27]. McMillan [24] extended their work to SMT with a single theory. The theory of arrays can be added by including quantified axioms and can be interpolated using, e.g., the method by Christ and Hoenicke [9] for quantifier instantiation, or the method of Bonacina and Johansson [4] for superposition calculus. Brillout et al [5] apply a similar algorithm to compute interpolants from sequent calculus proofs. In contrast to our approach, using such a procedure generates quantified interpolants.

Equality interpolating theories [7,30] allow for the generation of quantifier-free interpolants in the combination of quantifier-free theories. A theory is equality interpolating if it can express an interpolating term for each equality using only the symbols occurring in both parts of the interpolation problem. The algorithm of Yorsh and Musuvathi [30] only supports convex theories and is not applicable to the theory of arrays. Bruttomesso et al. [7] extended the framework to non-convex theories. They also present a complete interpolation procedure for the quantifier-free theory of arrays that works for theory combination in [6]. However, their solver depends on the partitioning of the interpolation problem. This can lead to exponential blow-up of the solving procedure. Our interpolation procedure works on a proof produced by a more efficient array solver that is independent of the partitioning of the interpolation problem.

Totla and Wies [29] present an interpolation method for arrays based on complete instantiations. It combines the idea of [7] with local theory extension [28]. Given an interpolation problem A and B, they define two sets, each using only symbols from A resp. B, that contain the instantiations of the array axioms needed to prove unsatisfiability. Then an existing solver and interpolation procedure for uninterpreted functions can be used to compute the interpolant. The procedure causes a quadratic blow-up on the input formulas. We also found that their procedure fails for some extensionality lemmas, when we used it to create candidate interpolants. We give an example for this in Sect. 6.

The last two techniques require to know the partitioning at solving time. Thus, when computing sequence [24] or tree interpolants [19], they would require either an adapted interpolation procedure or the solver has to run multiple times. In contrast, our method can easily be extended to tree interpolation [11].

2   Basic Definitions

We assume standard first-order logic. A theory $\mathcal{T}$ is given by a signature Σ and a set of axioms. The theory of arrays $\mathcal{T}_A$ is parameterized by an index theory and an element theory. Its signature Σ_A contains the *select* (or *read*) function $\cdot[\cdot]$ and the *store* (or *write*) function $\cdot\langle\cdot \lhd \cdot\rangle$. In the following, a, b, s, t denote array terms, i, j, k index terms and v, w element terms. For array a, index i and element v, $a[i]$ returns the element stored in a at i, and $a\langle i \lhd v\rangle$ returns a copy of a where the element at index i is replaced by the element v, leaving a unchanged. The functions are defined by the following axioms proposed by McCarthy [23].

$$\forall a \ i \ v. \ a\langle i \lhd v\rangle[i] = v \tag{idx}$$

$$\forall a \ i \ j \ v. \ i \neq j \rightarrow a\langle i \lhd v\rangle[j] = a[j] \tag{read-over-write}$$

We consider the variant of the extensional theory of arrays proposed by Bruttomesso et al. [6] where the signature is extended by the function $\text{diff}(\cdot, \cdot)$. For distinct arrays a and b, it returns an index where a and b differ, and an arbitrary index otherwise. The extensionality axiom then becomes

$$\forall a \ b. \ a[\text{diff}(a,b)] = b[\text{diff}(a,b)] \rightarrow a = b. \tag{ext-diff}$$

The authors of [6] have shown that the quantifier-free fragment of the theory of arrays with diff, $\mathcal{T}_{AxDiff}$, is closed under interpolation. To express the interpolants conveniently, we use the notation from [29] for rewriting arrays. For $m \geq 0$, we define $a \overset{m}{\rightsquigarrow} b$ for two arrays a and b inductively as

$$a \overset{0}{\rightsquigarrow} b := a \qquad a \overset{m+1}{\rightsquigarrow} b := a\langle\text{diff}(a,b) \lhd b[\text{diff}(a,b)]\rangle \overset{m}{\rightsquigarrow} b.$$

Thus, $a \overset{m}{\rightsquigarrow} b$ changes the values in a at m indices to the values stored in b. The equality $a \overset{m}{\rightsquigarrow} b = b$ holds if and only if a and b differ at up to m indices. The indices where they differ are the diff terms occurring in $a \overset{m}{\rightsquigarrow} b$.

An interpolation problem (A, B) is a pair of formulas where $A \wedge B$ is unsatisfiable. A *Craig interpolant* for (A, B) is a formula I such that (i) A implies I in the theory $\mathcal{T}$, (ii) I and B are $\mathcal{T}$-unsatisfiable and (iii) all non-theory symbols occurring in I are shared between A and B. Given an interpolation problem (A, B), the symbols shared between A and B are called *shared*, symbols only occurring in A are called *A-local* and symbols only occurring in B, *B-local*. A literal, e.g. $a = b$, that contains A-local and B-local symbols is called *mixed*.

3   Preliminaries

Our interpolation procedure operates on theory lemmas instantiated from particular variants of the read-over-write and extensionality axioms, and is designed to be used within the proof tree preserving interpolation framework. In the following, we give a short overview of this method and revisit the definitions and results about weakly equivalent arrays.

3.1   Proof Tree Preserving Interpolation

The proof tree preserving interpolation scheme presented by Christ et al. [13] allows to compute interpolants for an unsatisfiable formula using a resolution proof that is unaware of the interpolation problem.

For a partitioning (A, B) of the interpolation problem, two projections $\cdot \downharpoonright A$ and $\cdot \downharpoonright B$ project a literal to its A-part resp. B-part. For a literal ℓ occurring in A, we define $\ell \downharpoonright A \equiv \ell$. If ℓ is A-local, $\ell \downharpoonright B \equiv \mathbf{true}$. For ℓ in B, the projections are defined analogously. These projections are canonically extended to conjunctions of literals. A *partial interpolant* of a clause C occurring in the proof tree is defined as the interpolant of $A \wedge (\neg C) \downharpoonright A$ and $B \wedge (\neg C) \downharpoonright B$. Partial interpolants can be computed inductively over the proof tree and the partial interpolant of the root is the interpolant of A and B. For a theory lemma C, a partial interpolant is computed for the interpolation problem $((\neg C) \downharpoonright A, (\neg C) \downharpoonright B)$.

The core idea of proof tree preserving interpolation is a scheme to handle mixed equalities. For each $a = b$ where a is A-local and b is B-local, a fresh variable x_{ab} is introduced. This allows to define the projections as follows.

$$(a = b) \downharpoonright A \equiv (a = x_{ab}) \qquad (a = b) \downharpoonright B \equiv (x_{ab} = b)$$

Thus, $a = b$ is equivalent to $\exists x_{ab}.(a = b) \downharpoonright A \wedge (a = b) \downharpoonright B$ and x_{ab} is a new shared variable that may occur in partial interpolants. For disequalities we introduce an uninterpreted predicate EQ and define the projections for $a \neq b$ as

$$(a \neq b) \downharpoonright A \equiv \mathrm{EQ}(x_{ab}, a) \qquad (a \neq b) \downharpoonright B \equiv \neg\,\mathrm{EQ}(x_{ab}, b).$$

For an interpolation problem $(A \wedge (\neg C) \downharpoonright A, B \wedge (\neg C) \downharpoonright B)$ where $\neg C$ contains $a \neq b$, we require as additional symbol condition that x_{ab} only occurs as first parameter of an EQ predicate which occurs positively in the interpolant, i.e., the interpolant has the form $I[\mathrm{EQ}(x_{ab}, s_1)] \ldots [\mathrm{EQ}(x_{ab}, s_n)]$[1]. For a resolution step on the mixed pivot literal $a = b$, the following rule combines the partial interpolants of the input clauses to a partial interpolant of the resolvent.

$$\frac{C_1 \vee a = b : I_1[\mathrm{EQ}(x_{ab}, s_1)] \ldots [\mathrm{EQ}(x_{ab}, s_n)] \qquad C_2 \vee a \neq b : I_2(x_{ab})}{C_1 \vee C_2 : I_1[I_2(s_1)] \ldots [I_2(s_n)]}$$

---

[1] One can show that such an interpolant exists for every equality interpolating theory in the sense of Definition 4.1 in [7]. The terms s_i are the terms $\underline{v}$ in that definition.

3.2 Weakly Equivalent Arrays

Proof tree preserving interpolation can handle mixed literals, but it cannot deal with mixed *terms* which can be produced when instantiating (read-over-write) on an A-local store term and a B-local index. The lemmas produced in the decision procedure for the theory of arrays presented by Christ and Hoenicke [10] avoid such mixed terms by exploiting *weak equivalences* between arrays.

For a formula F, let V be the set of terms that contains the array terms in F and in addition the select terms $a[i]$ and their indices i and for each store term $a\langle i \lhd v\rangle$ in F the terms i, v, $a[i]$ and $a\langle i \lhd v\rangle[i]$. Let $\sim$ be the equivalence relation on V representing equality. The *weak equivalence graph* G^W is defined by its vertices, the array-valued terms in V, and its undirected edges of the form (i) $s_1 \leftrightarrow s_2$ if $s_1 \sim s_2$ and (ii) $s_1 \overset{i}{\leftrightarrow} s_2$ if s_1 has the form $s_2\langle i \lhd \cdot\rangle$ or vice versa. If two arrays a and b are connected in G^W by a path P, they are called *weakly equivalent*. This is denoted by $a \overset{P}{\leftrightarrow} b$. Weakly equivalent arrays can differ only at finitely many positions given by $\text{Stores}\,(P) := \{i \mid \exists s_1\, s_2.\, s_1 \overset{i}{\leftrightarrow} s_2 \in P\}$. Two arrays a and b are called *weakly equivalent on* i, denoted by $a \approx_i b$, if they are connected by a path P such that $k \not\sim i$ holds for each $k \in \text{Stores}\,(P)$. Two arrays a and b are called *weakly congruent on* i, $a \sim_i b$, if they are weakly equivalent on i, or if there exist $a'[j], b'[k] \in V$ with $a'[j] \sim b'[k]$ and $j \sim k \sim i$ and $a' \approx_i a$, $b' \approx_i b$. If a and b are weakly congruent on i, they must store the same value at i. For example, if $a\langle i + 1 \lhd v\rangle \sim b$ and $b[i] \sim c[i]$, arrays a and b are weakly equivalent on i while a and c are only weakly congruent on i.

We use $\text{Cond}(a \overset{P}{\leftrightarrow} b), \text{Cond}(a \approx_i b), \text{Cond}(a \sim_i b)$ to denote the conjunction of the literals $v = v'$ (resp. $v \neq v'$), $v, v' \in V$, such that $v \sim v'$ (resp. $v \not\sim v'$) is necessary to show the corresponding property. Instances of array lemmas are generated according to the following rules:

$$\frac{a \approx_i b \qquad i \sim j \qquad a[i], b[j] \in V}{\text{Cond}(a \approx_i b) \wedge i = j \to a[i] = b[j]} \qquad \text{(roweq)}$$

$$\frac{a \overset{P}{\leftrightarrow} b \qquad \forall i \in \text{Stores}\,(P).\, a \sim_i b \qquad a, b \in V}{\text{Cond}(a \overset{P}{\leftrightarrow} b) \wedge \bigwedge_{i \in \text{Stores}(P)} \text{Cond}(a \sim_i b) \to a = b} \qquad \text{(weq-ext)}$$

The first rule, based on (read-over-write), propagates equalities between select terms and the second, based on extensionality, propagates equalities on array terms. These rules are complete for the quantifier-free theory of arrays [10]. In the following, we describe how to derive partial interpolants for these lemmas.

4 Interpolants for Read-Over-Weakeq Lemmas

A lemma generated by (roweq) explains the conflict (negation of the lemma)

$$\text{Cond}(a \approx_i b) \wedge i = j \wedge a[i] \neq b[j].$$

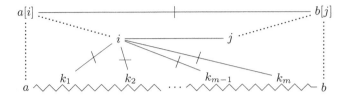

Fig. 1. A read-over-weakeq conflict. Solid lines represent strong (dis-)equalities, dotted lines function-argument relations, and zigzag lines represent weak paths consisting of store steps and array equalities.

The weak equivalence $a \approx_i b$ ensures that a and b are equal at $i = j$ which contradicts $a[i] \neq b[j]$ (see Fig. 1).

The general idea for computing an interpolant for this conflict, similar to [15], is to summarize maximal paths induced by literals of the same part (A or B), relying on the fact that the terms at the ends of these paths are shared. If a shared term is equal to the index i, we can express that the shared arrays at the path ends coincide or must differ at the index. There is a *shared term for $i = j$* if i or j are shared or if $i = j$ is mixed. If there is no shared term for $i = j$, the interpolant can be expressed using diff chains to capture the index. We identify four basic cases: (i) there is a shared term for $i = j$ and $a[i] = b[j]$ is in B or mixed, (ii) there is a shared term for $i = j$ and $a[i] = b[j]$ is A-local, (iii) both i and j are B-local, and (iv) both i and j are A-local.

4.1 Shared Term for $i = j$ and $a[i] = b[j]$ is in B or Mixed

If there exists a shared term x for the index equality $i = j$, the interpolant can contain terms $s[x]$ for shared array terms s occurring on the weak path between a and b. The basic idea is to summarize the weak A-paths by applying rule (roweq) on their end terms.

Example 1. Consider the following read-over-weakeq conflict:

$$a = s_1 \land s_1 \langle k_1 \lhd v_1 \rangle = s_2 \land s_2 \langle k_2 \lhd v_2 \rangle = s_3 \land s_3 = b$$
$$\land\, i \neq k_1 \land i \neq k_2 \land i = j \land a[i] \neq b[j]$$

where a, k_2, v_2, i are A-local, b, k_1, v_1, j are B-local, and s_1, s_2, s_3 are shared. Projecting the mixed literals on A and B as described in Sect. 3.1 yields the interpolation problem

$$A : a = s_1 \land s_2 \langle k_2 \lhd v_2 \rangle = s_3 \land \mathrm{EQ}(x_{ik_1}, i) \land i \neq k_2 \land i = x_{ij} \land \mathrm{EQ}(x_{a[i]b[j]}, a[i])$$
$$B : s_1 \langle k_1 \lhd v_1 \rangle = s_2 \land s_3 = b \land \neg\, \mathrm{EQ}(x_{ik_1}, k_1) \land x_{ij} = j \land \neg\, \mathrm{EQ}(x_{a[i]b[j]}, b[j]).$$

An interpolant is $I \equiv \mathrm{EQ}(x_{a[i]b[j]}, s_1[x_{ij}]) \land s_2[x_{ij}] = s_3[x_{ij}] \land \mathrm{EQ}(x_{ik_1}, x_{ij})$.

Algorithm. The first step is to subdivide the weak path $P : a \approx_i b$ into A- and B-paths. An equality edge $\leftrightarrow$ is assigned to either an A- or B-path depending on whether the corresponding equality is in A or B. A mixed equality $a' = b'$ is split into the A-local equality $a' = x_{a'b'}$ and the B-local equality $x_{a'b'} = b'$. Store edges $\overset{i}{\leftrightarrow}$ are assigned depending on which part contains the store term. If an equality or store term is shared between both parts, the algorithm can assign it to A or B arbitrarily. The whole path from a to b is then an alternation of A- and B-paths, which meet at shared boundary terms.

Let x be the shared term for $i = j$, i.e. x stands for i if i is shared, for j if i is not shared but j is, and for the auxiliary variable x_{ij} if $i = j$ is mixed.

(i) An inner A-path $\pi : s_1 \approx_i s_2$ of P starts and ends with a shared term. The summary is $s_1[x] = s_2[x]$. For a store edge on π with index k, add the disjunct $x = k$ if the corresponding disequality $i \neq k$ is B-local, and the disjunct $\mathrm{EQ}(x_{ik}, x)$ if the disequality is mixed. The interpolant of the subpath is

$$I_\pi \equiv s_1[x] = s_2[x] \vee F_\pi^A(x) \qquad \text{where } F_\pi^A(x) \equiv \bigvee_{\substack{k \in \mathrm{Stores}(\pi) \\ i \neq k \ B\text{-local}}} x = k \ \vee \bigvee_{\substack{k \in \mathrm{Stores}(\pi) \\ i \neq k \ \mathrm{mixed}}} \mathrm{EQ}(x_{ik}, x).$$

(ii) If $a[i] \neq b[j]$ is mixed and $a[i]$ is A-local, the first A-path on P starts with a or a is shared, i.e. $\pi : a \approx_i s_1$ (where s_1 can be a). For the path π, build the term $\mathrm{EQ}(x_{a[i]b[j]}, s_1[x])$ and add $F_\pi^A(x)$ as in case (i).

$$I_\pi \equiv \mathrm{EQ}(x_{a[i]b[j]}, s_1[x]) \vee F_\pi^A(x)$$

(iii) Similarly in the case where $a[i] \neq b[j]$ is mixed and $b[j]$ is A-local, the last A-path on P ends with b or b is shared, $\pi : s_n \approx_i b$. In this case the disjunct $i \neq j$ needs to be added if $i = j$ is B-local and i, j are both shared.

$$I_\pi \equiv \mathrm{EQ}(x_{a[i]b[j]}, s_n[x]) \vee F_\pi^A(x) \ [\vee i \neq j]$$

(iv) For every B-path π, add the conjunct $x \neq k$ for each A-local index disequality $i \neq k$, and the conjunct $\mathrm{EQ}(x_{ik}, x)$ for each mixed index disequality $i \neq k$ on π. We define

$$F_\pi^B(x) \equiv \bigwedge_{\substack{k \in \mathrm{Stores}(\pi) \\ i \neq k \ A\text{-local}}} x \neq k \ \wedge \bigwedge_{\substack{k \in \mathrm{Stores}(\pi) \\ i \neq k \ \mathrm{mixed}}} \mathrm{EQ}(x_{ik}, x).$$

The lemma interpolant is the conjunction of the above path interpolants. If i, j are shared, $b[j]$ is in B, and $i = j$ is A-local, add the conjunct $i = j$.

Lemma 1. *If x is a shared term for $i = j$ and $a[i] = b[j]$ is in B or mixed, a partial interpolant of the lemma* $\mathrm{Cond}(a \approx_i b) \wedge i = j \to a[i] = b[j]$ *is*

$$I \equiv \bigwedge_{\pi \in A\text{-paths}} I_\pi \ \wedge \bigwedge_{\pi \in B\text{-paths}} F_\pi^B(x) \ \ [\wedge i = j].$$

Proof. The interpolant only contains the shared boundary arrays, the shared term x for $i = j$, auxiliary variables for mixed disequalities under an EQ predicate, and shared store indices k where the store term is in a different part than the corresponding index disequality.

$\neg C \mid A$ implies I: For a B-path π, we show that $F_\pi^B(x)$ follows from the A-part. If i is B-local, there are no A-local or mixed index disequalities and $F_\pi^B(x)$ holds trivially. Otherwise $i = x$ follows from A, since either i is shared and x is i, $i = j$ is A-local and x is j, or $i = x$ is the A-projection of the mixed equality $i = j$. Then $F_\pi^B(x)$ follows by replacing i by x in A-local disequalities and A-projections of mixed disequalities on π. For an A-path π, if $F_\pi^A(x)$ does not hold, we get $s_1[x] = s_2[x]$ by applying rule (roweq). Note that $x \neq k$ follows from $i = x$ if $i \neq k$ is A-local, and from $EQ(x_{ik}, k)$ and $\neg F_\pi^A(x)$ in the mixed case. For the outer A-path in case (ii), $a[x] = s_1[x]$ is combined with the A-projection of the mixed disequality $a[i] \neq b[j]$ using $i = x$, which yields the EQ term. Analogously we get the EQ term for (iii), but to derive $j = x$ in the case where both i and j are shared but $i = j$ is B-local, we need to exclude $i \neq j$.

$\neg C \mid B \wedge I$ is unsat: Again if i is in B then $i = x$ follows from B by the choice of x. For a B-path π, we can conclude $s_1[x] = s_2[x]$ by applying rule (roweq) and using the index disequalities in $\neg C \mid B$ and $F_\pi^B(x)$. For an A-path π, $s_1[x] = s_2[x]$ (or, in cases (ii) and (iii), $EQ(x_{a[i]b[j]}, s[x])$) follows from I_π using the B-local index disequalities and $i = x$ to show that $F_\pi^A(x)$ cannot hold. Transitivity and the B-projection of $a[i] \neq b[j]$ lead to a contradiction. If $i = j$ is A-local, i is the shared term, and $b[j]$ is in B, the conjunct $i = j$ in I is needed here.   □

4.2   Shared Term for $i = j$ and $a[i] = b[j]$ is A-local

If there exists a shared index for $i = j$ and $a[i] = b[j]$ is A-local, we build disequalities for the B-paths instead of equalities for the A-paths. This corresponds to obtaining the interpolant of the inverse problem (B, A) by Sect. 4.1 and negating the resulting formula. Only the EQ terms are not negated because of the asymmetry of the projection of mixed disequalities.

Lemma 2. *Using the definitions of F_π^A and F_π^B from the previous section, if x is a shared term for $i = j$ and $a[i] = b[j]$ is A-local, then a partial interpolant of the lemma* $\mathrm{Cond}(a \approx_i b) \wedge i = j \to a[i] = b[j]$ *is*

$$I \equiv \bigvee_{(\pi : s_1 \approx_i s_2) \in B\text{-}paths} (s_1[x] \neq s_2[x] \wedge F_\pi^B(x)) \quad \vee \quad \bigvee_{\pi \in A\text{-}paths} F_\pi^A(x) \quad [\vee\, i \neq j].$$

4.3   Both i and j are B-local

When both i and j are B-local (or both A-local), we may not find a shared term for the index where a and b should be equal. Instead we use the diff function to express all indices where a and b differ. For instance, if $a = b\langle i \triangleleft v\rangle\langle j \triangleleft w\rangle$ for arrays a, b with $a[j] \neq b[j]$, then $\mathrm{diff}(a, b) = j$ or $\mathrm{diff}(a \overset{1}{\leadsto} b, b) = j$ hold.

Example 2. Consider the following conflict:

$$a = s_1 \wedge s_1\langle k \lhd v \rangle = s_2 \wedge s_2 = b \wedge i \neq k \wedge i = j \wedge a[i] \neq b[j]$$

where a, b, i, j are B-local, k, v are A-local, and s_1, s_2 are shared. Splitting the mixed disequality $i \neq k$ as described in Sect. 3.1 yields the interpolation problem

$$A : s_1\langle k \lhd v \rangle = s_2 \wedge \mathrm{EQ}(x_{ik}, k)$$
$$B : a = s_1 \wedge s_2 = b \wedge \neg\,\mathrm{EQ}(x_{ik}, i) \wedge i = j \wedge a[i] \neq b[j].$$

An interpolant should reflect the information that s_1 and s_2 can differ at most at one index satisfying the EQ term. Using diff, we can express the interpolant

$$I \equiv (s_1 = s_2 \vee \mathrm{EQ}(x_{ik}, \mathrm{diff}(s_1, s_2))) \wedge s_1 \overset{1}{\leadsto} s_2 = s_2.$$

To generalize this idea, we define inductively over $m \geq 0$ for the arrays a and b, and a formula $F(\cdot)$ with one free parameter:

$$\mathrm{weq}(a, b, 0, F(\cdot)) \equiv a = b$$
$$\mathrm{weq}(a, b, m + 1, F(\cdot)) \equiv (a = b \vee F(\mathrm{diff}(a, b))) \wedge \mathrm{weq}(a \overset{1}{\leadsto} b, b, m, F(\cdot)).$$

The formula $\mathrm{weq}(a, b, m, F(\cdot))$ states that arrays a and b differ at most at m indices and that each index i where they differ satisfies the formula $F(i)$.

Algorithm. For an A-path $\pi : s_1 \approx_i s_2$, we count the number of stores $|\pi| := |\,\mathrm{Stores}\,(\pi)\,|$. Each index i where s_1 and s_2 differ must satisfy $F_\pi^A(i)$ as defined in Sect. 4.1. There is nothing to do for B-paths.

Lemma 3. *A partial interpolant of the lemma* $\mathrm{Cond}(a \approx_i b) \wedge i = j \to a[i] = b[j]$ *with B-local i and j is*

$$I \equiv \bigwedge_{(\pi:s_1 \approx_i s_2) \in A\text{-}paths} \mathrm{weq}\left(s_1, s_2, |\pi|, F_\pi^A(\cdot)\right).$$

Proof. The symbol condition holds by the same argument as in Lemma 1.

$\neg C \restriction A$ implies I: Let $\pi : s_1 \approx_i s_2$ be an A-path on P. The path π shows that s_1 and s_2 can differ at most at $|\pi|$ indices, hence $s_1 \overset{|\pi|}{\leadsto} s_2 = s_2$ follows from $\neg C \restriction A$. If $s_1 \overset{m}{\leadsto} s_2 \neq s_2$ holds for $m < |\pi|$, then $\mathrm{diff}(s_1 \overset{m}{\leadsto} s_2, s_2) = k$ for some $k \in \mathrm{Stores}\,(\pi)$. If $i \neq k$ is A-local, then $k = k$ holds trivially, if $i \neq k$ is mixed, then $\mathrm{EQ}(x_{ik}, k)$ is part of $\neg C \restriction A$. Hence, $s_1 \overset{m}{\leadsto} s_2 = s_2 \vee F_\pi^A(\mathrm{diff}(s_1 \overset{m}{\leadsto} s_2, s_2))$ holds for all $m < |\pi|$. This shows $\mathrm{weq}(s_1, s_2, |\pi|, F_\pi^A(\cdot))$.

$\neg C \restriction B \wedge I$ is unsat: For every B-path $\pi : s_1 \approx_i s_2$ on P, we get $s_1[i] = s_2[i]$ with (roweq). For every A-path $\pi : s_1 \approx_i s_2$, I implies that s_1 and s_2 differ at finitely many indices which all satisfy $F_\pi^A(\cdot)$. The disequalities and B-projections in B imply that i does not satisfy $F_\pi^A(i)$, and therefore $s_1[i] = s_2[i]$. Then $a[i] = b[i]$ holds by transitivity, in contradiction to $a[i] \neq b[j]$ and $i = j$ in B. $\quad\square$

4.4   Both i and j are A-local

The interpolant is dual to the previous case and we define the dual of weq for arrays a, b, a number $m \geq 0$ and a formula F:

$$\mathrm{nweq}(a, b, 0, F(\cdot)) \equiv a \neq b$$
$$\mathrm{nweq}(a, b, m + 1, F(\cdot)) \equiv (a \neq b \wedge F(\mathrm{diff}(a, b))) \vee \mathrm{nweq}(a \overset{1}{\leadsto} b, b, m, F(\cdot)).$$

The formula $\mathrm{nweq}(a, b, m, F(\cdot))$ expresses that either one of the first m indices i found by stepwise rewriting a to b satisfies the formula $F(i)$, or a and b differ at more than m indices. Like in Sect. 4.2, the lemma interpolant is dual to the one computed in Sect. 4.3.

Lemma 4. *A partial interpolant of the lemma* $\mathrm{Cond}(a \approx_i b) \wedge i = j \rightarrow a[i] = b[j]$ *with A-local i and j is* $I \equiv \bigvee_{(\pi: s_1 \approx_i s_2) \in B\text{-}paths} \mathrm{nweq}(s_1, s_2, |\pi|, F_\pi^B(\cdot))$.

Theorem 1. *For all instantiations of the rule (roweq), quantifier-free interpolants can be computed as described in Sects. 4.1–4.4.*

5   Interpolants for Weakeq-Ext Lemmas

A conflict corresponding to a lemma of type (weq-ext) is of the form

$$\mathrm{Cond}(a \overset{P}{\Leftrightarrow} b) \wedge \bigwedge_{i \in \mathrm{Stores}(P)} \mathrm{Cond}(a \sim_i b) \wedge a \neq b.$$

The main path P shows that a and b differ at most at the indices in Stores (P), and $a \sim_i b$ (called i-path as of now) shows that a and b do not differ at index i.

   To compute an interpolant, we summarize the main path by weq (or nweq) terms to capture the indices where a and b can differ, and include summaries for the i-paths that are similar to the interpolants in Sect. 4. The i-paths can contain a select edge $a' \overset{k_1}{\longleftrightarrow} \overset{k_2}{\longrightarrow} b'$ where $a'[k_1] \sim b'[k_2]$, $i \sim k_1$, and $i \sim k_2$. In the B-local case, Sect. 4.3, B-local select edges make no difference for the construction, as the weq formulas are built over A-paths, and analogously for the A-local case, Sect. 4.4. However, if there are A-local select terms $a'[k]$ in the B-local case or vice versa, then k is shared or the index equality $i = k$ is mixed and we can use k or the auxiliary variable x_{ik} and proceed as in the cases where there is a shared term.

   We have to adapt the interpolation procedures in Sects. 4.1 and 4.2 by adding the index equalities that pertain to a select edge, analogously to the index dise-quality for a store edge. More specifically, we add to $F_\pi^A(x)$ a disjunct $x \neq k$ for each B-local $i = k$ on an A-path, and $x \neq x_{ik}$ for each mixed $i = k$. Here, x is the shared term for the i-path index i. For B-paths we add to $F_\pi^B(x)$ a conjunct $x = k$ for each A-local $i = k$ and $x = x_{ik}$ for each mixed $i = k$. Moreover, if there is a mixed select equality $a'[k_1] = b'[k_2]$ on the i-path, the auxiliary variable $x_{a'[k_1]b'[k_2]}$ is used in the summary for the subpath instead of $s[x]$, i.e., we get a term of the form $s_1[x] = x_{a'[k_1]b'[k_2]}$ in Sect. 4.1, and analogously for Sect. 4.2.

   For (weq-ext) lemmas, we distinguish three cases: (i) $a = b$ is in B, (ii) $a = b$ is A-local, or (iii) $a = b$ is mixed.

5.1 $a = b$ is in B

If the literal $a = b$ is in B, the A-paths both on the main store path and on the weak paths have only shared path ends. Hence, we summarize A-paths similarly to Sects. 4.1 and 4.3.

Algorithm. Divide the main path $a \overset{P}{\leftrightarrow} b$ into A-paths and B-paths. For each $i \in \mathrm{Stores}\,(P)$ on a B-path, summarize the corresponding i-path as in Sects. 4.1 or 4.3. The resulting formula is denoted by I_i. For an A-path $s_1 \overset{\pi}{\leftrightarrow} s_2$ use a weq formula to state that each index where s_1 and s_2 differ satisfies $I_i(\cdot)$ for some $i \in \mathrm{Stores}\,(\pi)$ where I_i is computed as in Sect. 4.1 with the shared term $\cdot$ for $i = j$. If i is also shared we add $i = \cdot$ to the interpolant.

Lemma 5. *The lemma* $\mathrm{Cond}(a \overset{P}{\leftrightarrow} b) \wedge \bigwedge_{i \in \mathrm{Stores}(P)} \mathrm{Cond}(a \sim_i b) \to a = b$ *where $a = b$ is in B has the partial interpolant*

$$I \equiv \bigwedge_{\substack{i \in \mathrm{Stores}(\pi) \\ \pi \in B\text{-paths}}} I_i \quad \wedge \bigwedge_{(s_1 \overset{\pi}{\leftrightarrow} s_2) \in A\text{-paths}} \mathrm{weq}\left(s_1, s_2, |\pi|, \bigvee_{i \in \mathrm{Stores}(\pi)} \left(I_i(\cdot)\,[\,\wedge\, i = \cdot]\right)\right).$$

Proof. The path summaries I_i fulfill the symbol conditions, and the boundary terms s_1, s_2 used in the weq formulas are guaranteed to be shared.

$\neg C \mathop{\llcorner} A$ implies I: By Sects. 4.1 and 4.3, $\mathrm{Cond}(a \sim_i b) \mathop{\llcorner} A$ implies I_i for $i \in \mathrm{Stores}\,(\pi)$ where π is a B-path on P. For an A-path $s_1 \overset{\pi}{\leftrightarrow} s_2$ on P, we know that s_1 and s_2 differ at most at $|\pi|$ positions, namely at the indices $i \in \mathrm{Stores}\,(\pi)$. Each index satisfies the corresponding I_i by Sect. 4.1. Hence, $\mathrm{weq}(s_1, s_2, |\pi|, \bigvee_{i \in \mathrm{Stores}(\pi)} I_i(\cdot)[\wedge\, i = \cdot])$ holds.

$\neg C \mathop{\llcorner} B \wedge I$ is unsat: We first note that if a and b differ at some index i, there must be an A-path or a B-path $s_1 \overset{\pi}{\leftrightarrow} s_2$ on the main path, such that s_1 and s_2 also differ at index i. We show that no such index exists. For a B-path $s_1 \overset{\pi}{\leftrightarrow} s_2$, s_1 and s_2 can only differ at $i \in \mathrm{Stores}\,(\pi)$. But for every $i \in \mathrm{Stores}\,(\pi)$, we get $a[i] = b[i]$ from I_i as in Lemma 1 resp. 3. For an A-path $s_1 \overset{\pi}{\leftrightarrow} s_2$, the interpolant contains $\mathrm{weq}(s_1, s_2, |\pi|, \bigvee_{i \in \mathrm{Stores}(\pi)}(I_i(\cdot)[\wedge\, i = \cdot]))$. Thus, if s_1 and s_2 differ at some index i', the interpolant implies $I_i(i')$ for some index $i \in \mathrm{Stores}\,(\pi)$ and additionally $i = i'$ if i is shared. Together with $\mathrm{Cond}(a \sim_i b) \mathop{\llcorner} B$ this implies $a[i'] = b[i']$ as in the proof of Lemma 1. This shows that there is no index where a and b differ, but this contradicts $a \neq b$ in $\neg C \mathop{\llcorner} B$. $\square$

5.2 $a = b$ is A-local

The case where $a = b$ is A-local is similar with the roles of A and B swapped. For each $i \in \mathrm{Stores}\,(\pi)$ on an A-path π on P, interpolate the corresponding i-path as in Sects. 4.2 or 4.4 and obtain I_i. For each $i \in \mathrm{Stores}\,(\pi)$ on a B-path π on P, interpolate the corresponding i-path as in Sect. 4.2 using $\cdot$ as shared term and obtain $I_i(\cdot)$.

Lemma 6. *The lemma* $\text{Cond}(a \overset{P}{\leftrightarrow} b) \wedge \bigwedge_{i \in \text{Stores}(P)} \text{Cond}(a \sim_i b) \rightarrow a = b$ *where* $a = b$ *is A-local has the partial interpolant*

$$I \equiv \bigvee_{\substack{i \in \text{Stores}(\pi) \\ \pi \in A\text{-paths}}} I_i \quad \vee \quad \bigvee_{(s_1 \overset{\pi}{\leftrightarrow} s_2) \in B\text{-paths}} \text{nweq}\left(s_1, s_2, |\pi|, \bigwedge_{i \in \text{Stores}(\pi)} (I_i(\cdot)\,[\vee\,i \neq \cdot])\right).$$

5.3 $a = b$ is Mixed

If $a = b$ is mixed, where w.l.o.g. a is A-local, the outer A- and B-paths end with A-local or B-local terms respectively. The auxiliary variable x_{ab} may not be used in store or select terms, thus we first need to find a shared term representing a before we can summarize A-paths.

Example 3. Consider the following conflict:

$$
\begin{aligned}
a = s\langle i_1 \lhd v_1\rangle \wedge b = s\langle i_2 \lhd v_2\rangle \wedge a \neq b && \text{(main path)} \\
\wedge\, a[i_1] = s_1[i_1] \wedge b = s_1\langle k_1 \lhd w_1\rangle \wedge i_1 \neq k_1 && (i_1\text{-path}) \\
\wedge\, a = s_2\langle k_2 \lhd w_2\rangle \wedge i_2 \neq k_2 \wedge b[i_2] = s_2[i_2] && (i_2\text{-path})
\end{aligned}
$$

where a, i_1, v_1, k_2, w_2 are A-local, b, i_2, v_2, k_1, w_1 are B-local and s, s_1, s_2 are shared.
Our algorithm below computes the following interpolant for the conflict.

$$I \equiv I_0(s) \vee \text{nweq}\left(s, s_1, 2, I_0(s\langle \cdot \lhd s_1[\cdot]\rangle) \wedge \text{EQ}(x_{i_1 k_1}, \cdot)\right)$$
$$\text{where } I_0(\tilde{s}) = \text{EQ}(x_{ab}, \tilde{s}) \wedge \text{weq}(\tilde{s}, s_2, 1, \text{EQ}(x_{i_2 k_2}, \cdot))$$

Algorithm. Identify in the main path P the first A-path $a \overset{\pi_0}{\leftrightarrow} s_1$ and its store indices $\text{Stores}(\pi_0) = \{i_1, \ldots i_{|\pi_0|}\}$. To build an interpolant, we rewrite s_1 by storing at each index i_m the value $a[i_m]$. We use $\tilde{s}$ to denote the intermediate arrays. We build a formula $I_m(\tilde{s})$ inductively over $m \leq |\pi_0|$. This formula is an interpolant if $\tilde{s}$ is a shared array that differs from a only at the indices $i_1, \ldots, i_m$.
   For $m = 0$, i.e., $a = \tilde{s}$, we modify the lemma by adding the strong edge $\tilde{s} \leftrightarrow a$ in front of all paths and summarize it using the algorithm in Sect. 5.1, but drop the weq formula for the path $\tilde{s} \leftrightarrow a \overset{\pi_0}{\leftrightarrow} s_1$. This yields $I_{5.1}(\tilde{s})$. We define

$$I_0(\tilde{s}) \equiv \text{EQ}(x_{ab}, \tilde{s}) \wedge I_{5.1}(\tilde{s}).$$

   For the induction step we assume that $\tilde{s}$ only differs from a at $i_1, \ldots, i_m, i_{m+1}$. Our goal is to find a shared index term x for i_{m+1} and a shared value v for $a[x]$. We use the i_{m+1}-path to conclude that $\tilde{s}\langle x \lhd v\rangle$ is equal to a at i_{m+1}. Then we can include $I_m(\tilde{s}\langle x \lhd v\rangle)$ computed using the induction hypothesis.
   (i) If there is a select edge on a B-subpath of the i_{m+1}-path or if i_{m+1} is itself shared, we immediately get a shared term x for i_{m+1}. If the last B-path π^{m+1} on the i_{m+1}-path starts with a mixed select equality, then the corresponding auxiliary variable is the shared value v. Otherwise, π^{m+1} starts with a shared

array s^{m+1} and $v := s^{m+1}[x]$. We summarize the i_{m+1}-path from a to the start of π^{m+1} as in Sect. 4.2 and get $I_{4.2}(x)$. Finally, we set

$$I_{m+1}(\tilde{s}) \equiv I_{4.2}(x) \vee (I_m(\tilde{s}\langle x \lhd v\rangle) \wedge F^B_{\pi^{m+1}}(x)).$$

(ii) Otherwise, we split the i_{m+1}-path into a $\sim_{i_{m+1}} s^{m+1}$ and $s^{m+1} \overset{\pi^{m+1}}{\Leftrightarrow} b$, where π^{m+1} is the last B-subpath of the i_{m+1}-path. If s_1 and a are equal at i_{m+1} then also $\tilde{s}$ and a are equal and the interpolant is simply $I_m(\tilde{s})$. If a and s^{m+1} differ at i_{m+1}, we build an interpolant from $a \sim_{i_{m+1}} s^{m+1}$ as in Sect. 4.4 and obtain $I_{4.4}$. Otherwise, s_1 and s^{m+1} differ at i_{m+1}. We build the store path $s_1 \overset{P'}{\Leftrightarrow} s^{m+1}$ by concatenating P and π^{m+1}. Using nweq on the subpaths $s \overset{\pi}{\Leftrightarrow} s'$ of P' we find the shared term x for i_{m+1}. If π is in A we need to add the conjunct $s \overset{|\pi|}{\rightsquigarrow} s' = s'$ to obtain an interpolant. We get

$$I_{m+1}(\tilde{s}) \equiv I_m(\tilde{s}) \vee I_{4.4} \quad [\text{for } a \sim_{i_{m+1}} s^{m+1}] \vee$$
$$\bigvee_{s \overset{\pi}{\Leftrightarrow} s' \text{ in } P'} \text{nweq} \left(s, s', |\pi|, I_m(\tilde{s}\langle \cdot \lhd s^{m+1}[\cdot]\rangle) \wedge F^B_{\pi^{m+1}}(\cdot) \right) [\wedge s \overset{|\pi|}{\rightsquigarrow} s' = s'].$$

Lemma 7. *The lemma* $\text{Cond}(a \overset{P}{\Leftrightarrow} b) \wedge \bigwedge_{i \in \text{Stores}(P)} \text{Cond}(a \sim_i b) \to a = b$ *where* $a = b$ *is mixed has the partial interpolant* $I \equiv I_{|\pi_0|}(s_1)$.

A proof by induction over the length of the path π_0 can be found in [21].

Theorem 2. *Sections 5.1–5.3 give interpolants for all cases of rule* (weq-ext).

6   Complexity

Expanding the definition of an array rewrite term $a \overset{k}{\rightsquigarrow} b$ naïvely already yields a term exponential in k. This is avoided by using let expressions for common subterms. With this optimization the interpolants for read-over-weakeq lemmas are quadratic in the worst case. The interpolants of Sects. 4.1 and 4.2 contain at most one literal for every literal in the lemma, so the interpolant is linear in the size of the lemma. The interpolants of Sects. 4.3 and 4.4 are quadratic, since expanding the definition of weq will copy the formula $F^A_\pi(\cdot)$ resp. $F^B_\pi(\cdot)$, for each local store edge and instantiate it with a different shared term.

Example 4. The following interpolation problem has only quadratic interpolants.

$$A : b = a\langle i_1 \lhd v_1\rangle \cdots \langle i_n \lhd v_n\rangle \wedge p_1(i_1) \wedge \cdots \wedge p_n(i_n)$$
$$B : a[j] \neq b[j] \wedge \neg p_1(j) \wedge \ldots \neg p_n(j)$$
$$I \equiv \text{let } a_0 = a \text{ let } d_1 = \text{diff}(a_0, b) \text{ let } a_1 = a_0\langle d_1 \lhd b[d_1]\rangle$$
$$\ldots \text{let } d_n = \text{diff}(a_{n-1}, b) \text{ let } a_n = a_{n-1}\langle d_n \lhd b[d_n]\rangle$$
$$(p_1(d_1) \vee \cdots \vee p_n(d_1) \vee a_0 = b) \wedge \cdots$$
$$(p_1(d_n) \vee \cdots \vee p_n(d_n) \vee a_{n-1} = b) \wedge a_n = b$$

There is no interpolant that is not quadratic in n. The interpolant has to imply that $p_k(i_k)$ is true for every k. There are no shared index-valued terms in the lemma. Hence, the only way to express the i_k values using shared terms is by applying the diff operator on a and b and constructing diff chains as in the interpolant I. The diff operator returns one of the $i_1, \ldots, i_n$ in every step, but it is not determined which one. Consequently, every combination $p_k(d_l)$ is needed.

The algorithms in Sects. 5.1 and 5.2 produce a worst-case quadratic interpolant as they nest the linear interpolants of Sects. 4.1 and 4.2 in a weq resp. nweq formula, which expands this term a linear number of times. However, the algorithm in Sect. 5.3 is worst-case exponential in the size of the extensionality lemma.

The following example explains why this bound is strict. This example also shows that the method of Totla and Wies [29] is not complete. In particular, for $n = 1$ their preprocessing algorithm produces a satisfiable formula from the original interpolation problem.

Example 5. The following interpolation problem of size $O(n^2)$ has only interpolants of exponential size in n.

$$A : a = s\langle i_1^A \lhd v_1^A \rangle \cdots \langle i_n^A \lhd v_n^A \rangle \wedge p(a) \wedge$$

$$\bigwedge_{j=1}^{n} p_j(i_j^A) \wedge \bigwedge_{j=1}^{n} a[i_j^A] = s_j[i_j^A] \wedge$$

$$\bigwedge_{j=1}^{n} \bigwedge_{l=0, l \neq j}^{n} q_j(i_l^A) \wedge \bigwedge_{j=1}^{n} t_j = a\langle i_0^A \lhd w_{j0}^A \rangle \cdots \cancel{\langle i_j^A \lhd w_{jj}^A \rangle} \cdots \langle i_n^A \lhd w_{jn}^A \rangle$$

$$B : b = s\langle i_1^B \lhd v_1^B \rangle \cdots \langle i_n^B \lhd v_n^B \rangle \wedge \neg p(b) \wedge$$

$$\bigwedge_{j=1}^{n} \bigwedge_{l=0, l \neq j}^{n} \neg p_j(i_l^B) \wedge \bigwedge_{j=1}^{n} s_j = b\langle i_0^B \lhd w_{j0}^B \rangle \cdots \cancel{\langle i_j^B \lhd w_{jj}^B \rangle} \cdots \langle i_n^B \lhd w_{jn}^B \rangle \wedge$$

$$\bigwedge_{j=1}^{n} \neg q_j(i_j^B) \wedge \bigwedge_{j=1}^{n} b[i_j^B] = t_j[i_j^B]$$

The first line of A and the first line of B ensure that there is a store-chain from a over s to b of length $2n$ and $p(a)$ and $\neg p(b)$ are used to derive the contradiction from the extensionality axiom. To prove that a and b are equal, the formulas show that they are equal at the indices i_j^A, $j = 1, \ldots, n$ (second line of A and B). Here p_j is used to ensure that i_j^A is distinct from all i_l^B, $l \neq j$. Analogously the last line of A and B shows that a and b are equal at the indices i_j^B, $j = 1, \ldots, n$.

Since $p(a) \wedge \neg p(b)$ is essential to prove unsatisfiability, the interpolant needs to contain the term $p(\cdot)$ for some shared array term that is equal to a and b. This can only be expressed by store terms of size n, e.g., $p(s\langle i_1 \lhd \cdot \rangle \cdots \langle i_n \lhd \cdot \rangle)$ (alternatively some store term starting on s_j or t_j can be used). As in the previous example, the store indices i_j can only be expressed using diff chains between shared arrays. For each index there is only one shared array that is guaranteed to contain the right value. The diff function returns the indices in arbitrary order. Therefore, the interpolant needs a case for every combination of diff term and value, as it is done by the interpolant computed in Sect. 5.3. This means the interpolant contains exponentially many $p(\cdot)$ terms.

7   Evaluation

We implemented the presented algorithms into SMTINTERPOL [12], an SMT solver computing sequence and tree interpolants. Our implementation verifies at run-time that the returned interpolants are correct. To evaluate the interpolation algorithm we used the ULTIMATE AUTOMIZER software model-checker [17] on the memory safety track of the SV-COMP 2018[2] benchmarks. This track was chosen because ULTIMATE uses arrays to model memory access. We ran our experiments using the open-source benchmarking software benchexec [3] on a machine with a 3.4 GHz Intel i7-4770 CPU and set a 900 s time and a 6 GB memory limit. As comparison, we ran ULTIMATE with Z3[3] and SMTINTERPOL without array interpolation using ULTIMATE's built-in theory-independent interpolation scheme based on unsatisfiable cores and predicate transformers [18].

Table 1 shows the result. From the 326 benchmarks we removed 50 benchmarks which ULTIMATE could not parse. The unknown results come from non-linear arithmetic (SMTINTERPOL), quantifiers (due to incomplete elimination in the setting SMTINTERPOL-NoArrayInterpol), or incomplete interpolation engine (Z3). Our new algorithm solves 12.6% more problems, and both helps to verify safety and guide the counterexample generation for unsafe benchmarks.

Table 1. Evaluation of ULTIMATE AUTOMIZER on the SV-COMP benchmarks for memsafety running with our new interpolation engine, without array interpolation, and Z3.

Setting	Tasks	Safe	Unsafe	Timeout	Unknown
SMTINTERPOL-ArrayInterpol	276	101	96	66	13
SMTINTERPOL-NoArrayInterpol	276	92	83	75	26
Z3	276	32	44	13	187

8   Conclusion

We presented an interpolation algorithm for the quantifier-free fragment of the theory of arrays. Due to the technique of proof tree preserving interpolation, our algorithm also works for the combination with other theories. Our algorithm operates on lemmas produced by an efficient array solver based on weak equivalence on arrays. The interpolants are built by simply iterating over the weak

---

[2] https://sv-comp.sosy-lab.org/2018/.
[3] https://github.com/Z3Prover/z3 in version 4.6.0 (2abc759d0).

equivalence and weak congruence paths found by the solver. We showed that the complexity bound on the size of the produced interpolants is optimal.

In contrast to most existing interpolation algorithms for arrays, the solver does not depend on the partitioning of the interpolation problem. Thus, our technique allows for efficient interpolation especially when several interpolants for different partitionings of the same unsatisfiable formula need to be computed. Although it remains to prove formally that the algorithm produces tree interpolants, during the evaluation all returned tree interpolants were correct.

Acknowledgement. We would like to thank Daniel Dietsch for running the experiments.

References

1. Andrianov, P., Friedberger, K., Mandrykin, M., Mutilin, V., Volkov, A.: CPA-BAM-BnB: block-abstraction memoization and region-based memory models for predicate abstractions. In: Legay, A., Margaria, T. (eds.) TACAS 2017. LNCS, vol. 10206, pp. 355–359. Springer, Heidelberg (2017). https://doi.org/10.1007/978-3-662-54580-5_22
2. Beyer, D., Henzinger, T.A., Jhala, R., Majumdar, R.: The software model checker BLAST. STTT **9**(5–6), 505–525 (2007)
3. Beyer, D., Löwe, S., Wendler, P.: Benchmarking and resource measurement. In: Fischer, B., Geldenhuys, J. (eds.) SPIN 2015. LNCS, vol. 9232, pp. 160–178. Springer, Cham (2015). https://doi.org/10.1007/978-3-319-23404-5_12
4. Bonacina, M., Johansson, M.: On interpolation in automated theorem proving. J. Autom. Reason. **54**(1), 69–97 (2015)
5. Brillout, A., Kroening, D., Rümmer, P., Wahl, T.: Program verification via Craig interpolation for Presburger arithmetic with arrays. In: VERIFY@IJCAR, pp. 31–46. EasyChair (2010)
6. Bruttomesso, R., Ghilardi, S., Ranise, S.: Quantifier-free interpolation of a theory of arrays. Log. Methods Comput. Sci. **8**(2), (2012)
7. Bruttomesso, R., Ghilardi, S., Ranise, S.: Quantifier-free interpolation in combinations of equality interpolating theories. ACM Trans. Comput. Log. **15**(1), 5:1–5:34 (2014)
8. Cassez, F., Sloane, A.M., Roberts, M., Pigram, M., Suvanpong, P., de Aledo, P.G.: Skink: static analysis of programs in LLVM intermediate representation. In: Legay, A., Margaria, T. (eds.) TACAS 2017. LNCS, vol. 10206, pp. 380–384. Springer, Heidelberg (2017). https://doi.org/10.1007/978-3-662-54580-5_27
9. Christ, J., Hoenicke, J.: Instantiation-based interpolation for quantified formulae. In: Decision Procedures in Software, Hardware and Bioware. Dagstuhl Seminar Proceedings, vol. 10161. Schloss Dagstuhl, Germany (2010)
10. Christ, J., Hoenicke, J.: Weakly equivalent arrays. In: Lutz, C., Ranise, S. (eds.) FroCoS 2015. LNCS (LNAI), vol. 9322, pp. 119–134. Springer, Cham (2015). https://doi.org/10.1007/978-3-319-24246-0_8
11. Christ, J., Hoenicke, J.: Proof tree preserving tree interpolation. J. Autom. Reason. **57**(1), 67–95 (2016)
12. Christ, J., Hoenicke, J., Nutz, A.: SMTInterpol: an interpolating SMT solver. In: Donaldson, A., Parker, D. (eds.) SPIN 2012. LNCS, vol. 7385, pp. 248–254. Springer, Heidelberg (2012). https://doi.org/10.1007/978-3-642-31759-0_19

13. Christ, J., Hoenicke, J., Nutz, A.: Proof tree preserving interpolation. In: Piterman, N., Smolka, S.A. (eds.) TACAS 2013. LNCS, vol. 7795, pp. 124–138. Springer, Heidelberg (2013). https://doi.org/10.1007/978-3-642-36742-7_9

14. Dangl, M., Löwe, S., Wendler, P.: CPACHECKER with support for recursive programs and floating-point arithmetic. In: Baier, C., Tinelli, C. (eds.) TACAS 2015. LNCS, vol. 9035, pp. 423–425. Springer, Heidelberg (2015). https://doi.org/10.1007/978-3-662-46681-0_34

15. Fuchs, A., Goel, A., Grundy, J., Krstic, S., Tinelli, C.: Ground interpolation for the theory of equality. Log. Methods Comput. Sci. **8**(1) (2012)

16. Greitschus, M., Dietsch, D., Heizmann, M., Nutz, A., Schätzle, C., Schilling, C., Schüssele, F., Podelski, A.: Ultimate Taipan: trace abstraction and abstract interpretation. In: Legay, A., Margaria, T. (eds.) TACAS 2017. LNCS, vol. 10206, pp. 399–403. Springer, Heidelberg (2017). https://doi.org/10.1007/978-3-662-54580-5_31

17. Heizmann, M., et al.: Ultimate Automizer with an on-demand construction of Floyd-Hoare automata. In: Legay, A., Margaria, T. (eds.) TACAS 2017. LNCS, vol. 10206, pp. 394–398. Springer, Heidelberg (2017). https://doi.org/10.1007/978-3-662-54580-5_30

18. Heizmann, M., Dietsch, D., Leike, J., Musa, B., Podelski, A.: ULTIMATE AUTOMIZER with array interpolation. In: Baier, C., Tinelli, C. (eds.) TACAS 2015. LNCS, vol. 9035, pp. 455–457. Springer, Heidelberg (2015). https://doi.org/10.1007/978-3-662-46681-0_43

19. Heizmann, M., Hoenicke, J., Podelski, A.: Nested interpolants. In: POPL, pp. 471–482. ACM (2010)

20. Henzinger, T., Jhala, R., Majumdar, R., Sutre, G.: Lazy abstraction. In: POPL, pp. 58–70. ACM (2002)

21. Hoenicke, J., Schindler, T.: Efficient interpolation for the theory of arrays. CoRR, abs/1804.07173 (2018)

22. Krajícek, J.: Interpolation theorems, lower bounds for proof systems, and independence results for bounded arithmetic. J. Symb. Log. **62**(2), 457–486 (1997)

23. McCarthy, J.: Towards a mathematical science of computation. In: IFIP Congress, pp. 21–28 (1962)

24. McMillan, K.: An interpolating theorem prover. Theor. Comput. Sci. **345**(1), 101–121 (2005)

25. McMillan, K.L.: Lazy abstraction with interpolants. In: Ball, T., Jones, R.B. (eds.) CAV 2006. LNCS, vol. 4144, pp. 123–136. Springer, Heidelberg (2006). https://doi.org/10.1007/11817963_14

26. Nutz, A., Dietsch, D., Mohamed, M.M., Podelski, A.: ULTIMATE KOJAK with memory safety checks. In: Baier, C., Tinelli, C. (eds.) TACAS 2015. LNCS, vol. 9035, pp. 458–460. Springer, Heidelberg (2015). https://doi.org/10.1007/978-3-662-46681-0_44

27. Pudlák, P.: Lower bounds for resolution and cutting plane proofs and monotone computations. J. Symb. Log. **62**(3), 981–998 (1997)

28. Sofronie-Stokkermans, V.: Hierarchic reasoning in local theory extensions. In: Nieuwenhuis, R. (ed.) CADE 2005. LNCS (LNAI), vol. 3632, pp. 219–234. Springer, Heidelberg (2005). https://doi.org/10.1007/11532231_16

29. Totla, N., Wies, T.: Complete instantiation-based interpolation. J. Autom. Reason. **57**(1), 37–65 (2016)

30. Yorsh, G., Musuvathi, M.: A combination method for generating interpolants. In: Nieuwenhuis, R. (ed.) CADE 2005. LNCS (LNAI), vol. 3632, pp. 353–368. Springer, Heidelberg (2005). https://doi.org/10.1007/11532231_26

ATPBOOST: Learning Premise Selection in Binary Setting with ATP Feedback

Bartosz Piotrowski[1,2(✉)] and Josef Urban[1]

[1] Czech Institute of Informatics, Robotics and Cybernetics, Prague, Czech Republic
[2] Faculty of Mathematics, Informatics and Mechanics, University of Warsaw,
Warsaw, Poland
bartoszpiotrowski@post.pl

Abstract. ATPBOOST is a system for solving sets of large-theory problems by interleaving ATP runs with state-of-the-art machine learning of premise selection from the proofs. Unlike many approaches that use multi-label setting, the learning is implemented as binary classification that estimates the pairwise-relevance of (*theorem, premise*) pairs. ATPBOOST uses for this the fast state-of-the-art XGBoost gradient boosting algorithm. Learning in the binary setting however requires negative examples, which is nontrivial due to many alternative proofs. We discuss and implement several solutions in the context of the ATP/ML feedback loop, and show significant improvement over the multi-label approach.

1 Introduction: Machine Learning for Premise Selection

Assume that c is a conjecture which is a logical consequence of a large set of premises P. The chance of finding a proof of c by an automated theorem prover (ATP) often depends on choosing a small subset of P relevant for proving c. This is known as the *premise selection* task [1]. This task is crucial to make ATPs usable for proof automation over large formal corpora created with systems such as Mizar, Isabelle, HOL, and Coq [4]. Good methods for premise selection typically also transfer to related tasks, such as *internal proof guidance* of ATPs [8, 10,13,17] and *tactical guidance* of ITPs [7].

The most efficient premise selection methods use *data-driven/machine-learning* approaches. Such methods work as follows. Let T be a set of theorems with their proofs. Let C be a set of conjectures without proofs, each associated with a set of available premises that can be used to prove them. We want to learn a (statistical) model from T, which for each conjecture $c \in C$ will rank its available premises according to their relevance for producing an ATP proof of c. Two different machine learning settings can be used for this task:

1. *multilabel classification*: we treat premises used in the proofs as opaque labels and we create a model capable of labeling conjectures based on their features,

---

Supported by the *AI4REASON* ERC grant 649043, Czech project AI&Reasoning CZ.02.1.01/0.0/0.0/15_003/0000466 and European Regional Development Fund.

D. Galmiche et al. (Eds.): IJCAR 2018, LNAI 10900, pp. 566–574, 2018.
https://doi.org/10.1007/978-3-319-94205-6_37

2. *binary classification*: here the aim of the learning model is to recognize pairwise-relevance of the (*conjecture, premise*) pairs, i.e. to decide what is the chance of a premise being relevant for proving the conjecture based on the features of both the conjecture and the premise.

Most of the machine learning methods for premise selection have so far used the first setting [3,9,11]. This includes fast and robust machine learning algorithms such as *naive Bayes* and *k-nearest neighbors* (k-NN) capable of multilabel classification with many examples and labels. This is needed for large formal libraries with many facts and proofs. There are however several reasons why the second approach may be better:

1. Generality: in binary classification it is easier to estimate the relevance of (*conjecture, premise*) pairs where the premise was so far unseen (i.e., not in the training data).
2. State-of-the-art ML algorithms are often capable of learning subtle aspects of complicated problems based on the features. The multilabel approach trades the rich feature representation of the premise for its opaque label.
3. Many state-of-the-art ML algorithms are binary classifiers or they struggle when performing multilabel classification for a large number of labels.

Recently, substantial work [2] has been done in the binary setting. In particular, applying deep learning to premise selection has improved state of the art in the field. There are however modern and efficient learning algorithms such as XGBoost [5] that are much less computationally-intensive then deep learning methods. Also, obtaining negative examples for training the binary classifiers is a very interesting problem in the context of many alternative ATP proofs and a feedback loop between the ATP and the learning system.

1.1 Premise Selection in Binary Setting with Multiple Proofs

The existence of multiple ATP proofs makes premise selection different from conventional machine learning applications. This is evident especially in the binary classification setting. The ML algorithms for recognizing pairwise relevance of (*conjecture, premise*) pairs require good data consisting of two (typically balanced) classes of positive and negative examples. But there is no conventional way how to construct such data in our domain. For every true conjecture there are infinitely many mathematical proofs. The ATP proofs are often based on many different sets of premises. The notions of *useful* or *superfluous premise* are only approximations of their counterparts defined for sets of premises.

As an example, consider the following frequent situation: a conjecture c can be ATP-proved with two sets of axioms: $\{p_1, p_2\}$ and $\{p_3, p_4, p_5\}$. Learning only from one of the sets as positives and presenting the other as negative (*conjecture, premise*) pairs may considerably distort the learned notion of a *useful premise*. This differs from the multilabel setting, where negative data are typically not used by the fast ML algorithms such as naive Bayes and k-NN. They just aggregate different positive examples into the final ranking.

Therefore, to further improve the premise selection algorithms it seems useful to consider learning from multiple proofs and to develop methods producing good negative data. The most suitable way how to do that is to allow multiple interactions of the machine learner with the ATP system. In the following section we present the ATPBOOST system, which implements several such algorithms.

2   ATPBOOST: Setting, Algorithms and Components

ATPBOOST[1] is a system for solving sets of large-theory problems by interleaving ATP runs with learning of premise selection from the proofs using the state-of-the-art XGBoost algorithm. The system implements several algorithms and consists of several components described in the following sections. Its setting is a large theory $\mathcal{T}$, extracted from a large ITP library where facts appear in a chronological order. In more detail, we assume the following inputs and notation:

1. T – names of theorems (and problems) in a large theory $\mathcal{T}$.
2. P – names of all facts (premises) in $\mathcal{T}$. We require $P \supseteq T$.
3. STATEMENTS$_P$ of all $p \in P$ in the TPTP format [15] .
4. FEATURES$_P$ – characterizing each $p \in P$. Here we use the same features as in [11] and write $\boldsymbol{f}_p$ for the (sparse) vector of features of p.
5. ORDER$_P$ ($<_P$) – total order on P; p may be used to prove t iff $p <_P t$. We write A_t for $\{p : p <_P t\}$, i.e. the set of premises allowed for t.
6. PROOFS$_{T'}$ for a subset $T' \subseteq T$. Each $t \in T'$ may have many proofs – denoted by $\mathcal{P}_t$. P_t denotes the premises needed for at least one proof in $\mathcal{P}_t$.

2.1   Algorithms

We first give a high-level overview and pseudocode of the algorithms implemented in ATPBOOST. Section 2.2 then describes the used components in detail.

Algorithm 1 is the simplest setting. Problems are split into the train/test sets, XGBoost learns from the training proofs, and its predictions are ATP-evaluated on the test set. This is used mainly for parameter optimization.

Algorithm 2 evaluates the trained XGBoost also on the training part, possibly finding new proofs that are used to update the training data for the next iteration. The test problems and proofs are never used for training. Negative mining may be used to find the worst misclassified premises and to correspondingly update the training data in the next iteration.

Algorithm 3 begins with no training set, starting with ATP runs on random rankings. XGBoost is trained on the ATP proofs from the previous iteration, producing new ranking for all problems for the next iteration. This is a MaLARea-style [16] feedback loop between the ATP and the learner.

---

[1] The Python package is at https://github.com/BartoszPiotrowski/ATPboost.

2.2   Components

Below we describe the main components of the ATPBoost algorithms and the main ideas behind them. As discussed in Sect. 1, they take into account the binary learning setting, and in particular implement the need to teach the system about multiple proofs by proper choice of examples, continuous interaction with the ATP and intelligent processing of its feedback. The components are available as procedures in our Python package.

Algorithm 1. Simple training/test split.

Require: Set of theorems T, set of premises $P \supseteq T$, PROOFS$_T$, FEATURES$_P$, STATEMENTS$_P$, ORDER$_P$, PARAMS$_{set}$, PARAMS$_{model}$.
1: $T_{train}, T_{test} \leftarrow$ RANDOMLYSPLIT(T)
2: $\mathcal{D} \leftarrow$ CREATETRAININGSET(PROOFS$_{T_{train}}$, FEATURES$_P$, ORDER$_P$, PARAMS$_{set}$)
3: $\mathcal{M} \leftarrow$ TRAINMODEL($\mathcal{D}$, PARAMS$_{model}$)
4: $\mathcal{R} \leftarrow$ CREATERANKINGS($T_{test}, \mathcal{M}$, FEATURES$_P$, ORDER$_P$)
5: $\mathcal{P} \leftarrow$ ATPEVALUATION($\mathcal{R}$, STATEMENTS$_P$)

Algorithm 2. Incremental feedback-loop with training/test split.

Require: Set of theorems T, set of premises $P \supseteq T$, FEATURES$_P$, STATEMENTS$_P$, PROOFS$_T$, ORDER$_P$, PARAMS$_{set}$, PARAMS$_{model}$, PARAMS$_{negmin}$ (optionally).
1: $T_{train}, T_{test} \leftarrow$ RANDOMLYSPLIT(T)
2: $\mathcal{D} \leftarrow$ CREATETRAININGSET(PROOFS$_{T_{train}}$, FEATURES$_P$, ORDER$_P$, PARAMS$_{set}$)
3: **repeat**
4:     $\mathcal{M} \leftarrow$ TRAINMODEL($\mathcal{D}$, PARAMS$_{model}$)
5:     $\mathcal{R}_{train} \leftarrow$ CREATERANKINGS($T_{train}, \mathcal{M}$, FEATURES$_P$, ORDER$_P$)
6:     $\mathcal{R}_{test} \leftarrow$ CREATERANKINGS($T_{test}, \mathcal{M}$, FEATURES$_P$, ORDER$_P$)
7:     $\mathcal{P}_{train} \leftarrow$ ATPEVALUATION($\mathcal{R}_{train}$, STATEMENTS$_P$)
8:     $\mathcal{P}_{test} \leftarrow$ ATPEVALUATION($\mathcal{R}_{test}$, STATEMENTS$_P$)
9:     UPDATE(PROOFS$_{train}, \mathcal{P}_{train}$)
10:     UPDATE(PROOFS$_{test}, \mathcal{P}_{test}$)
11:     **if** PARAMS$_{negmin}$ **then**
12:         $\mathcal{D} \leftarrow$ NEGATIVEMINING($\mathcal{R}$, PROOFS$_{train}$, FEATURES$_P$, ORDER$_P$, PARAMS$_{negmin}$)
13:     **else**
14:         $\mathcal{D} \leftarrow$ CREATETRAININGSET(PROOFS$_{train}$, FEATURES$_P$, ORDER$_P$, PARAMS$_{set}$)
15: **until** Number of PROOFS$_{test}$ increased after UPDATE.

Algorithm 3. Incremental feedback-loop starting with no proofs.

Require: Set of theorems T, set of premises $P \supseteq T$, FEATURES$_P$, STATEMENTS$_P$, ORDER$_P$, PARAMS$_{set}$, PARAMS$_{model}$, PARAMS$_{negmin}$ (optionally).
1: PROOFS$_T \leftarrow \emptyset$
2: $\mathcal{R} \leftarrow$ CREATERANDOMRANKINGS(T)
3: $\mathcal{P} \leftarrow$ ATPEVALUATION($\mathcal{R}$, STATEMENTS$_P$)
4: UPDATE(PROOFS$_T, \mathcal{P}$)
5: $\mathcal{D} \leftarrow$ CREATETRAININGSET(PROOFS$_T$, FEATURES$_P$, ORDER$_P$, PARAMS$_{set}$)
6: **repeat**
7:     $\mathcal{M} \leftarrow$ TRAINMODEL($\mathcal{D}$, PARAMS$_{model}$)
8:     $\mathcal{R} \leftarrow$ CREATERANKINGS($T, \mathcal{M}$, FEATURES$_P$, ORDER$_P$)
9:     $\mathcal{P} \leftarrow$ ATPEVALUATION($\mathcal{R}$, STATEMENTS$_P$)
10:     UPDATE(PROOFS$_T, \mathcal{P}$)
11:     **if** PARAMS$_{negmin}$ **then**
12:         $\mathcal{D} \leftarrow$ NEGATIVEMINING($\mathcal{R}$, PROOFS$_T$, FEATURES$_P$, ORDER$_P$, PARAMS$_{negmin}$)
13:     **else**
14:         $\mathcal{D} \leftarrow$ CREATETRAININGSET(PROOFS$_T$, FEATURES$_P$, ORDER$_P$, PARAMS$_{set}$)
15: **until** Number of PROOFS$_T$ increased after UPDATE.

CREATETRAININGSET (PROOFS$_T$, FEATURES$_P$, ORDER$_P$, PARAMS).
This procedure constructs a TRAININGSET for a binary learning algorithm. This
is a sparse matrix of positive/negative examples and a corresponding vector of
binary labels. The examples (matrix rows) are created from PROOFS$_T$ and FEA-
TURES$_P$, respecting ORDER$_P$. Each example is a concatenation of f_t and f_p,
i.e., the features of a theorem t and a premise p. Positive examples express that
p is relevant for proving t, whereas the negatives mean the opposite.

The default method (SIMPLE) creates positives from all pairs (t,p) where
$p \in P_t$. Another method (SHORT) creates positives only from the *short* proofs of
t. These are the proofs of t with at most $m+1$ premises, where m is the minimal
number of premises used in a proof from $\mathcal{P}_t$. Negative examples for theorem t
are chosen randomly from pairs (t,p) where $p \in A_t \setminus P_t$. The number of such
randomly chosen pairs is RATIO $\cdot N_{\mathrm{pos}}$, where N_{pos} is the number of positives
and RATIO$\in \mathbb{N}$ is a parameter that needs to be optimized experimentally. Since
$|A_t \setminus P_t|$ is usually much larger than $|P_t|$, it seems reasonable to have a large
RATIO. This however increases class imbalance and the probability of presenting
to the learning algorithm a *false negative*. This is a pair (t,p) where $p \notin P_t$, but
there is an ATP proof of t using p that is not yet in our dataset.

TRAINMODEL (TRAININGSET, PARAMS). This procedure trains a binary
learning classifier on the TRAININGSET, creating a MODEL. We use XGBoost [5]
– a state-of-the-art tree-based gradient boosting algorithm performing very well
in machine learning competitions. It is also much faster to train compared to
deep learning methods, performs well with unbalanced training sets, and is optimized for working with sparse data. XGBoost has several important parameters,
such as NUMBEROFTREES, MAXDEPTH (of trees) and ETA (learning rate). These
parameters have significant influence on the performance and require tuning.

CREATERANKINGS (C, MODEL, FEATURES$_P$, ORDER$_P$). This procedure
uses the trained MODEL to construct RANKINGS$_C$ of premises from P for conjectures $c \in C \subseteq T$. Each conjecture c is paired with each premise $p <_P c$ and
concatenations of f_c and f_p are passed to the MODEL. The MODEL outputs a
real number in $[0, 1]$, which is interpreted as the relevance of p for proving c. The
relevances are then used to sort the premises into RANKINGS$_C$.

ATPEVALUATION (RANKINGS, STATEMENTS). Any ATP can be used for
evaluation. By default we use E [14][2]. As usual, we construct the ATP problems
for several top slices (lengths $1, 2, \ldots, 512$) of the RANKINGS. To remove redundant premises we *pseudo-minimize* the proofs: only the premises needed in the
proofs are used as axioms and the ATP is rerun until a fixpoint is reached.

UPDATE (OLDPROOFS, NEWPROOFS). The UPDATE makes a union of the
new and old proofs, followed by a subsumption reduction. I.e., if premises of two
proofs of t are in a superset relation, the proof with the larger set is removed.

---

[2] The default time limit is 10 s and the memory limit is 2 GB. The exact default
command is: `./eprover -auto-schedule -free-numbers -s -R -cpu-limit=10
-memory-limit=2000 -print-statistics -p -tstp-format problem_file`.

NEGATIVEMINING (PROOFS$_T$, FEATURES$_P$, ORDER$_P$, PARAMS). This is used as a more advanced alternative to CREATETRAININGSET. It examines the last RANKINGS$_T$ for the most *misclassified positives*. I.e., for each $t \in T$ we create a set MP_t of those p that were previously ranked high for t, but no ATP proof of t was using p. We define three variants:

1. NEGMIN_ALL: Let m_t be the maximum rank of a t-useful premise ($p \in P_t$) in RANKINGS$_T[t]$. Then $MP_t^1 = \{p : rank_t(p) < m_t \wedge p \notin P_t\}$.
2. NEGMIN_RAND: We randomly choose into MP_t^2 only a half of MP_t^1.
3. NEGMIN_1: $MP_t^3 = \{p : rank_t(p) < |P_t| \wedge p \notin P_t\}$.

The set MP_t^i is then added as negatives to the examples produced by the CREATETRAININGSET procedure. The idea of such negative mining is that the learner takes into account the mistakes it made in the previous iteration.

3 Evaluation

We evaluate[3] the algorithms on a set of 1342 MPTP2078 [1] large (*chainy*) problems that are provable in 60 s using their small (*bushy*) versions.

Parameter Tuning: First we run Algorithm 1 to optimize the parameters. The dataset was randomly split into a train set of 1000 problems and test set of 342. For the train set, we use the proofs obtained by the 60 s run on the bushy versions. We tune the RATIO parameter of CREATETRAININGSET, and the NUMBEROFTREES, MAXDEPTH and ETA parameters of TRAINMODEL. Due to resource constraints we *a priori* assume good defaults: RATIO = 16, NUMBEROFTREES = 2000, MAXDEPTH = 10, ETA = 0.2. Then we observe how changing each parameter separately influences the results. Table 1 shows the ATP results for the RATIO parameter, and Fig. 1 for the model parameters.

Table 1. Influence of the RATIO of randomly generated negatives to positives.

RATIO	1	2	4	8	16	32	64
Proved (%)	74.0	78.4	79.0	78.7	80.1	79.8	80.1

It is clear that a high number of negatives is important. Using RATIO = 16 proves 6% more test problems than the balanced setting (RATIO = 1). It is also clear that a higher number of trees – at least 500 – improves the results. However, too many trees (over 8000) slightly decrease the performance, likely due to overfitting. The ETA parameter gives best results with values between 0.04 and 0.64, and the MAXDEPTH of trees should be around 10.

---

[3] All the scripts we used for the evaluation are available at https://github.com/BartoszPiotrowski/ATPboost/tree/master/experiments.

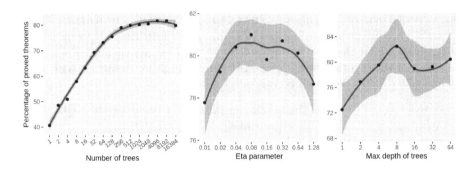

Fig. 1. ATP performance of different parameters of the XGBoost model.

We evaluate Algorithm 1 also on a much bigger ATP-provable part of MML with 29271 theorems in train part and 3253 in test. With parameters RATIO = 20, NUMBEROFTREES = 4000, MAXDEPTH = 10 and ETA = 0.2 we proved 58.78% theorems (1912). This is a 15.7% improvement over k-NN, which proved 50.81% (1653) theorems. For a comparison, the improvement over k-NN obtained (with much higher ATP time limits) with deep learning in [2] was 4.3%.

Incremental Feedback Loop with Train/Test Split: This experiment evaluates Algorithm 2, testing different methods of negative mining. The train/test split and the values of the parameters RATIO, NUMBEROFTREES, MAXDEPTH, ETA are taken from the previous experiment. We test six methods in parallel. Two XGB methods (SIMPLE and SHORT) are the variants of the CREATETRAININGSET procedure, three XGB methods (NEGMIN_ALL, NEGMIN_RAND and NEGMIN_1) are the variants of the NEGATIVEMINING, and the last one is a k-NN learner similar to the one from [11], used here for comparison. The experiment starts with the same proofs for training theorems as in the previous one, and we performed 30 rounds of the feedback loop. Figure 2 shows the results.

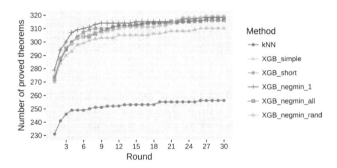

Fig. 2. Number of proved theorems in subsequent iterations of Algorithm 2.

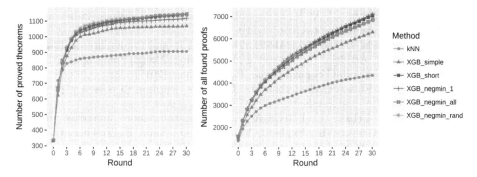

Fig. 3. Number of proved theorems (left) and number of all found proofs (right) in subsequent rounds of the experiment corresponding to Algorithm 3.

All the new methods largely outperform k-NN, and XGB_SHORT is much better than XGB_SIMPLE. I.e., positives from too many proofs seem harmful, as in [12] where this was observed with k-NN. The differences between the XGB variants SHORT, NEGMIN_1, NEGMIN_ALL, and NEGMIN_RAND do not seem significant and all perform well. At the end of the loop (30th round) 315–319 theorems of the 342 (ca 93%) are proved.

Incremental Feedback-loop with no Initial Proofs: The final experiment corresponds to Algorithm 3. There is no train/test split and no initial proofs. The first ATP evaluation is done on random rankings, proving 335 theorems out of the 1342. Then the loop starts running with the same options as in the previous experiment. Figure 3 shows the numbers of theorems that were proved in the subsequent rounds, as well as the growth of the total number of different proofs. This is important, because all these proofs are taken into account by the machine learning. Again, k-NN is the weakest and XGB_SIMPLE is worse than the rest of the methods, which are statistically indistinguishable. In the last round XGB_NEGMIN_RAND proves 1150 (86%) theorems. This is a 26.8% improvement over k-NN (907) and 7.7% more than XGB_SIMPLE (1068).

References

1. Alama, J., Heskes, T., Kühlwein, D., Tsivtsivadze, E., Urban, J.: Premise selection for mathematics by corpus analysis and kernel methods. J. Autom. Reason. **52**(2), 191–213 (2014)
2. Alemi, A.A., Chollet, F., Irving, G., Szegedy, C., Urban, J. (eds.): DeepMath - deep sequence models for premise selection (2016). http://dblp.uni-trier.de/rec/bibtex/conf/nips/IrvingSAECU16
3. Blanchette, J.C., Greenaway, D., Kaliszyk, C., Kühlwein, D., Urban, J.: A learning-based fact selector for Isabelle/HOL. J. Autom. Reason. **57**(3), 219–244 (2016)
4. Blanchette, J.C., Kaliszyk, C., Paulson, L.C., Urban, J.: Hammering towards QED. J. Formal. Reason. **9**(1), 101–148 (2016)

5. Chen, T., Guestrin, C.: XGBoost: a scalable tree boosting system (2016). http://dblp.uni-trier.de/rec/bibtex/conf/kdd/ChenG16

6. Eiter, T., Sands, D. (eds.) LPAR-21, 21st International Conference on Logic for Programming, Artificial Intelligence and Reasoning, Maun, Botswana, 7–12 May 2017. EPiC Series in Computing, vol. 46. EasyChair (2017)

7. Gauthier, T., Kaliszyk, C., Urban, J.: TacticToe: learning to reason with HOL4 tactics. In: Eiter and Sands [6], pp. 125–143 (2017)

8. Jakubův, J., Urban, J.: ENIGMA: efficient learning-based inference guiding machine. In: Geuvers, H., England, M., Hasan, O., Rabe, F., Teschke, O. (eds.) CICM 2017. LNCS (LNAI), vol. 10383, pp. 292–302. Springer, Cham (2017). https://doi.org/10.1007/978-3-319-62075-6_20

9. Kaliszyk, C., Urban, J.: Learning-assisted automated reasoning with flyspeck. J. Autom. Reason. **53**(2), 173–213 (2014)

10. Kaliszyk, C., Urban, J.: FEMaLeCoP: fairly efficient machine learning connection prover. In: Davis, M., Fehnker, A., McIver, A., Voronkov, A. (eds.) LPAR 2015. LNCS, vol. 9450, pp. 88–96. Springer, Heidelberg (2015). https://doi.org/10.1007/978-3-662-48899-7_7

11. Kaliszyk, C., Urban, J.: MizAR 40 for Mizar 40. J. Autom. Reason. **55**(3), 245–256 (2015)

12. Kuehlwein, D., Urban, J.: Learning from multiple proofs: first experiments. In: Fontaine, P., Schmidt, R.A., Schulz, S. (eds.) PAAR-2012, EPiC Series, vol. 21, pp. 82–94. EasyChair (2013)

13. Loos, S.M., Irving, G., Szegedy, C., Kaliszyk, C.: Deep network guided proof search. In: Eiter and Sands [6], pp. 85–105 (2017)

14. Schulz, S.: E - a brainiac theorem prover. AI Commun. **15**(2–3), 111–126 (2002)

15. Sutcliffe, G.: The TPTP problem library and associated infrastructure. J. Autom. Reason. **43**(4), 337–362 (2009)

16. Urban, J., Sutcliffe, G., Pudlák, P., Vyskočil, J.: MaLARea SG1 - machine learner for automated reasoning with semantic guidance. In: Armando, A., Baumgartner, P., Dowek, G. (eds.) IJCAR 2008. LNCS (LNAI), vol. 5195, pp. 441–456. Springer, Heidelberg (2008). https://doi.org/10.1007/978-3-540-71070-7_37

17. Urban, J., Vyskočil, J., Štěpánek, P.: MaLeCoP machine learning connection prover. In: Brünnler, K., Metcalfe, G. (eds.) TABLEAUX 2011. LNCS (LNAI), vol. 6793, pp. 263–277. Springer, Heidelberg (2011). https://doi.org/10.1007/978-3-642-22119-4_21

Theories as Types

Dennis Müller[1]([✉]), Florian Rabe[1,2]([✉]), and Michael Kohlhase[1]([✉])

[1] Computer Science, FAU Erlangen-Nürnberg, Erlangen, Germany
{dennis.mueller,michael.kohlhase}@fau.de, florian.rabe@gmail.com
[2] LRI, Université Paris Sud, Orsay, France

Abstract. Theories are an essential structuring principle that enable modularity, encapsulation, and reuse in formal libraries and programs (called classes there). Similar effects can be achieved by dependent record types. While the former form a separate language layer, the latter are a normal part of the type theory. This overlap in functionality can render different systems non-interoperable and lead to duplication of work.

We present a type-theoretic calculus and implementation of a variant of record types that for a wide class of formal languages naturally corresponds to theories. Moreover, we can now elegantly obtain a contravariant functor that reflects the theory level into the object level: for each theory we obtain the type of its models and for every theory morphism a function between the corresponding types. In particular this allows shallow – and thus structure-preserving – encodings of mathematical knowledge and program specifications while allowing the use of object-level features on models, e.g. equality and quantification.

1 Introduction

In the area of formal systems like type theories, logics, and specification and programming languages, various language features have been studied that allow for inheritance and modularity, e.g., theories, classes, contexts, and records. They all share the motivation of grouping a list of declarations into a new entity such as in $R = [\![x_1 : A_1, \ldots, x_n : A_n]\!]$. The basic intuition behind it is that R behaves like a product type whose values are of the form $\langle x_1 : A_1 := a_1, \ldots, x_n : A_n := a_n \rangle$. Such constructs are indispensable already for elementary applications such as defining the algebraic structure of Semilattices (as in Fig. 1), which we will use as a running example.

$$
\mathsf{Semilattice} = \left\{
\begin{array}{lll}
U & : \mathbf{type} \\
\wedge & : U \to U \to U \\
\mathbf{assoc} & : \vdash \forall x, y, z : U.\ (x \wedge y) \wedge z \doteq x \wedge (y \wedge z) \\
\mathbf{commutative} & : \ldots \\
\mathbf{idempotent} & : \ldots
\end{array}
\right\}
$$

Fig. 1. A grouping of declarations for semilattices

© Springer International Publishing AG, part of Springer Nature 2018
D. Galmiche et al. (Eds.): IJCAR 2018, LNAI 10900, pp. 575–590, 2018.
https://doi.org/10.1007/978-3-319-94205-6_38

Many systems support **stratified grouping** (where the language is divided into a lower level for the base language and a higher level that introduces the grouping constructs) or **integrated grouping** (where the grouping construct is one out of many type-forming operations without distinguished ontological status), or both. The

System	Name of feature	
	Stratified	Integrated
ML	Signature/module	Record
C++	Class	Class, struct
Java	Class	Class
Idris [Bra13]	Module	Record
Coq [Coq15]	Module	Record
HOL Light [Har96]	ML signatures	Records
Isabelle [Wen09]	Theory, locale	Record
Mizar [TB85]	Article	Structure
PVS [ORS92]	Theory	Record
OBJ [Gog+93]	Theory	
FoCaLiZe [Har+12]	Species	Record

names of the grouping constructs vary between systems, and we will call them **theories** and **records** in the sequel. An overview of some representative examples is given in the table on the right. For a discussion of these concepts and a comprehensive review of the related work we refer the reader to [MRK].

The two approaches have different advantages. Stratified grouping permits a separation of concerns between the core language and the module system. It also captures high-level structure well in a way that is easy to manage and discover in large libraries, closely related to the advantages of the little theories approach [FGT92]. But integrated grouping allows applying base language operations (such as quantification or tactics) to the grouping constructs. For this reason, the (relatively simple) stratified Coq module system is disregarded in favor of records in major developments such as [Mat].

Allowing both features can lead to a duplication of work where the same hierarchy is formalized once using theories and once using records. A compromise solution is common in object-oriented programming languages, where classes behave very much like stratified grouping but are at the same time normal types of the type system. We call this **internalizing** the higher level features. While combining advantages of stratified and integrated grouping, internalizing is a very heavyweight type system feature: stratified grouping does not change the type system at all, and integrated grouping can be easily added to or removed from a type system, but internalization adds a very complex type system feature from the get-go. It has not been applied much to logics and similar formal systems: the only example we are aware of is the FoCaLiZe [Har+12] system. A much weaker form of internalization is used in OBJ and related systems based on stratified grouping: here theories may be used as (and only as) the types of parameters of parametric theories. Most similarly to our approach, OCaml's first-class modules internalize the theory (called *module type* in OCaml) M as the type $\mathtt{module}\ M$; contrary to both OO-languages and our approach, this kind of internalization is in addition and unrelated to integrated grouping.

In any case, because theories usually allow for advanced declarations like imports, definitions, and notations, as well as extra-logical declarations, systematically internalizing theories requires a correspondingly expressive integrated grouping construct. Records with defined fields are comparatively rare; e.g., present in [Luo09] and OO-languages. Similarly, imports between record types and/or record terms are featured only sporadically, e.g., in Nuprl [Con+86], maybe even as an afterthought only.

Finally, we point out a closely related trade-off that is orthogonal to our development: even after choosing either a theory or a record to define grouping, many systems still offer a choice whether a declaration becomes a parameter or a field. See [SW11] for a discussion.

Contribution. We present the first formal system that systematically internalizes theories into record types. The central idea is to use an operator Mod that turns the theory T into the type Mod (T), which behaves like a record type. We take special care not to naively compute this record type, which would not scale well to the common situations where theories with hundreds of declarations or more are used. Instead, we introduce record types that allow for defined fields and merging so that Mod (T) preserves the structure of T.

Our approach combines the advantages of stratified and integrated grouping in a lightweight language feature that is orthogonal to and can be easily combined with other foundational language features. Concretely, it is realized as a module in the MMT framework [Rab14], which allows for the modular design of foundational languages. By combining our new modules with existing ones, we obtain many formal systems with internalized theories. In particular, our typing rules conform to the abstractions of MMT so that MMT's type reconstruction [Rab17] is immediately applicable to our features. We showcase the potential in a case study based on this implementation, and which is interesting in its own right: A formal library of elementary mathematical concepts that systematically utilizes Mod $(\cdot)$ throughout for algebraic structures, topological spaces etc.

Overview. We formulate our approach in the setting of a dependently-typed λ-calculus, which we recall in Sect. 2. This section also serves as a gentle primer for defining language features in MMT. Section 3 introduces our notion of record types, based on which we introduce the model-operator in Sect. 4. Section 5 presents our implementation and a major case study on elementary mathematics. This paper is a shortened version of [MRK], which also contains all the proofs.

2   Preliminaries

We introduce the well-known dependently-typed lambda calculus as the starting point of our development. The **grammar** is given in Fig. 2. The only surprise here is that we allow optional definitions in contexts; this is a harmless convenience at this point but will be critical later on when we introduce records with defined fields. As usual, we write $T \rightarrow T'$ instead of $\prod_{x:T} T'$ when possible. We also write $T[x/T']$ for the usual capture-avoiding **substitution** of T' for x in T.

MMT uses a bidirectional type system, i.e., we have two separate judgments for type *inference* and type *checking*. Similarly, we have two equality judgments:

$$\begin{array}{ll}
\Gamma ::= \cdot \mid \Gamma, x[:T][:=T] & \text{contexts} \\
T ::= x \mid \textbf{type} \mid \textbf{kind} & \text{variables and universes} \\
\quad \mid \Pi_{x:T'} T \mid \lambda x : T'.T \mid T_1 T_2 & \text{dependent function types}
\end{array}$$

Fig. 2. Grammar for contexts and expressions

one for checking equality of two given terms and one for reducing a term to another one. Our **judgments** are given in Fig. 3.

Adding record types in Sect. 3 will introduce non-trivial **subtyping**, e.g., $[\![x : T, y : S]\!]$ is a subtype of $[\![x : T]\!]$.[1] Therefore, we already introduce a subtyping judgment here even though it is not needed for dependent function types yet. For our purposes, it is sufficient (and desirable) to consider subtyping to be an abbreviation: $\Gamma \vdash T_1 <: T_2$ iff for all t $\Gamma \vdash t \Leftarrow T_1$ implies $\Gamma \vdash t \Leftarrow T_2$.

Judgment	Intuition
$\vdash \Gamma \ \textbf{ctx}$	Γ is a well-formed context
$\Gamma \vdash t \Leftarrow T$	t checks against type/kind T.
$\Gamma \vdash t \Rightarrow T$	type/kind of term t is inferred to be T
$\Gamma \vdash t_1 \equiv t_2 : T$	t_1 and t_2 are equal at type T
$\Gamma \vdash t_1 \rightsquigarrow t_2$	t_1 computes to t_2
$\Gamma \vdash T_1 <: T_2$	T_1 is a subtype of T_2

Fig. 3. Judgments

The **pre/postconditions** of these judgments are as follows: $\Gamma \vdash t \Leftarrow T$ assumes that T is well-typed and implies that t is well-typed. $\Gamma \vdash t \Rightarrow T$ implies that both t and T are well-typed. $\Gamma \vdash t_1 \rightsquigarrow t_2$ implies that t_2 is well-typed iff t_1 is (which puts additional burden on computation rules that are called on not-yet-type-checked terms). Equality and subtyping are only used for expressions that are assumed to be well-typed, i.e., $\Gamma \vdash t_1 \equiv t_2 : T$ implies $\Gamma \vdash t_i \Leftarrow T$, and $\Gamma \vdash T_1 <: T_2$ implies that T_i is a type/kind.

Remark 1 (Horizontal Subtyping and Equality). The equality judgment could alternatively be formulated as an untyped equality $t \equiv t'$. That would require some technical changes to the rules but would usually not be a huge difference. In our case, however, the use of typed equality is critical.

For example, consider record values $r_1 = \langle a := 1, b := 1 \rangle$ and $r_2 = \langle a := 1, b := 2 \rangle$ as well as record types $R = [\![a : nat]\!]$ and $S = [\![a : nat, b : nat]\!]$. Due to horizontal subtyping, we have $S <: R$ and thus both $r_i \Leftarrow S$ and $r_i \Leftarrow R$. This has the advantage that the function $S \rightarrow R$ that throws away the field

---

[1] This is sometimes called *horizontal* subtyping. In that case, the straightforward covariance rule for record types is called *vertical* subtyping.

b becomes the identity operation. Now our equality at record types behaves accordingly and checks only for the equality of those fields required by the type. Thus, $r_1 \equiv r_2 : R$ is true whereas $r_1 \equiv r_2 : S$ is false, i.e., the equality of two terms may depend on the type at which they are compared. While seemingly dangerous, this makes sense intuitively: r_1 can be replaced with r_2 in any context that expects an object of type R because in such a context the field b, where r_1 and r_2 differ, is inaccessible.

Of course, this treatment of equality precludes downcasts: an operation that casts the equal terms $r_1 : R$ and $r_2 : R$ into the corresponding unequal terms of type S would be inconsistent. But such downcasts are still possible (and valuable) at the meta-level. For example, a tactic $GroupSimp(G, x)$ that simplifies terms x in a group G can check if G is commutative and in that case apply more simplification operations.

The full rules of a lambda calculus can be found in the long version [MRK]. We can now show that the usual variance rule for function types is derivable.

Theorem 1. *The following subtyping rule is derivable:*

$$\frac{\Gamma \vdash A <: A' \quad \Gamma, x : A \vdash B' <: B}{\Gamma \vdash \prod_{x:A'} B' <: \prod_{x:A} B}$$

Moreover, we can show that every well-typed term t has a **principal type** T in the sense that (i) $\Gamma \vdash t \Leftarrow T$ and (ii) whenever $\Gamma \vdash t \Leftarrow T'$, then also $\Gamma \vdash T <: T'$. The principal type is exactly the one inferred by our rules (see Theorem 2).

3   Record Types with Defined Fields

We now introduce record types as an additional module of our framework by extending the grammar and the rules. The basic intuition is that $[\![\Gamma]\!]$ and $\langle\!\langle\Gamma\rangle\!\rangle$ construct record types and terms. We call a context **fully typed** resp. **defined** if all fields have a type resp. a definition. In $[\![\Gamma]\!]$, Γ must be fully typed and may additionally contain defined fields. In $\langle\!\langle\Gamma\rangle\!\rangle$, Γ must be fully defined; the types are optional and usually omitted in practice.

Because we frequently need fully defined contexts, we introduce a notational convention for them: a context denoted by a *lower* case letters like γ is always fully defined. In contrast, a context denoted by an *upper* case letter like Γ may have any number of types or definitions.

3.1   Records

We extend our grammar as in Fig. 4, where the previously existing parts are grayed out.

$$\begin{aligned}
\Gamma &::= \cdot \mid \Gamma, x[: T][:= T] \\
T &::= x \mid \mathsf{type} \mid \mathsf{kind} \\
&\mid \Pi_{x:T'} T \mid \lambda x : T'.T \mid T_1 T_2 \\
&\mid [\![\Gamma]\!] \mid \langle\!\langle \Gamma \rangle\!\rangle \mid T.x \qquad\qquad \text{record types, terms, projections}
\end{aligned}$$

Fig. 4. Grammar for records

Remark 2 (Field Names and Substitution in Records). Note that we use the same identifiers for variables in contexts and fields in records. This allows reusing results about contexts when reasoning about and implementing records. In particular, it immediately makes our records dependent, i.e., both in a record type and — maybe surprisingly — in a record term every variable x may occur in subsequent fields. In some sense, this makes x bound in those fields. However, record types are critically different from Σ-types: we must be able to use x in record projections, i.e., x can *not* be subject to α-renaming.

As a consequence, capture-avoiding substitution is not always possible. This is a well-known problem that is usually remedied by allowing every record to declare a name for itself (e.g., the keyword `this` in many object-oriented languages), which is used to disambiguates between record fields and a variable in the surrounding context (or fields in a surrounding record). We gloss over this complication here and simply make substitution a partial function.

Before stating the rules, we introduce a few critical auxiliary definition:

Definition 1 (Substituting in a Record). We extend substitution $t[x/t']$ to records:

- $[\![x_1 : T_1, \ldots, x_n : T_n]\!] \, [y/t]$
  $$= \begin{cases} [\![x_1 : T_1[y/t], \ldots, x_{i-1} : T_{i-1}[y/t], x_i : T_i, \ldots, x_n : T_n]\!] \text{ if } y = x_i \\ [\![x_1 : (T_1[y/t]), \ldots, x_n : (T_n[y/t])]\!] \text{ else} \end{cases}$$
  if none of the x_i are free in t. Otherwise the substitution is undefined.
- $\langle\!\langle x_1 := t_1, \ldots, x_n := t_n \rangle\!\rangle \, [y/t] = \begin{cases} \langle\!\langle x_1 := t_1, \ldots, x_n := t_n \rangle\!\rangle \text{ if } y \in \{x_1, \ldots, x_n\} \\ \langle\!\langle x_1 := (t_1[y/t]), \ldots, x_n := (t_n[y/t]) \rangle\!\rangle \text{ else} \end{cases}$
  if none of the x_i are free in t. Otherwise the substitution is undefined.
- $(r.x)[y/t] = (r[y/t]).x$.

Definition 2 (Substituting with a Record). We write $t[r/\Delta]$ for the result of substituting any occurrence of a variable x declared in Δ with $r.x$

In the special case where $r = \langle\!\langle \delta \rangle\!\rangle$, we simply write $t[\delta]$ for $t[\langle\!\langle \delta \rangle\!\rangle /\delta]$, i.e., we have $t[x_1 := t_1, \ldots, x_n := t_n] = t[x_n/t_n] \ldots [x_1/t_1]$.

Our rules for records are given in Fig. 5. Their roles are systematically similar to the rules for functions: three inference rules for the three constructors followed by a type and an equality checking rule for record types and the (in this case: two) computation rules. We remark on a few subtleties below.

Formation:

$$\frac{\vdash \Gamma, \Delta \, \mathtt{ctx} \quad \Delta \text{ fully typed} \quad \max \Delta \in \{\mathtt{type}, \mathtt{kind}\}}{\Gamma \vdash [\![\Delta]\!] \Rightarrow \max \Delta}$$

where $\max \Delta$ is the maximal universe of all undefined fields in Δ

Introduction:

$$\frac{\vdash \Gamma, \Delta \, \mathtt{ctx} \quad \delta \text{ fully defined} \quad \Delta \text{ like } \delta \text{ but with all missing types inferred}}{\Gamma \vdash \langle \delta \rangle \Rightarrow [\![\Delta]\!]}$$

Elimination:

$$\frac{\Gamma \vdash r \Rightarrow [\![\Delta_1, x : T[:= t], \Delta_2]\!]}{\Gamma \vdash r.x \Rightarrow T[r/\Delta_1]}$$

Type checking:

$$\frac{\overbrace{\Gamma \vdash r.x \Leftarrow T[r/\Delta]}^{\text{For } x : T[:= t] \,\in\, \Delta} \quad \overbrace{\Gamma \vdash r.x \equiv t[r/\Delta] : T[r/\Delta]}^{\text{additionally for } x : T := t \,\in\, \Delta} \quad \Gamma \vdash r \Rightarrow R}{\Gamma \vdash r \Leftarrow [\![\Delta]\!]}$$

Equality checking (extensionality):

$$\frac{\overbrace{\Gamma \vdash r_1.x \equiv r_2.x : T[r_1/\Delta]}^{\text{For every } (x : T) \,\in\, \Delta}}{\Gamma \vdash r_1 \equiv r_2 : [\![\Delta]\!]}$$

Computation:

$$\frac{\delta = \delta_1, x[: T] := t, \delta_2 \quad \Gamma \vdash \langle \delta \rangle \Rightarrow R}{\Gamma \vdash \langle \delta \rangle.x \rightsquigarrow t[\delta_1]} \qquad \frac{\Gamma \vdash r \Rightarrow [\![\Delta_1, x : T := t, \Delta_2]\!]}{\Gamma \vdash r.x \rightsquigarrow t[r/\Delta_1]}$$

Fig. 5. Rules for records

The formation rule is partial in the sense that not every context defines a record type or kind. This is because the universe of a record type must be as high as the universe of any undefined field to avoid inconsistencies. For example, $\max(a : \mathtt{nat}) = \mathtt{type}$, $\max(a : \mathtt{type}) = \mathtt{kind}$ and $\max(a : \mathtt{kind})$ is not defined. If we switched to a countable hierarchy of universes (which we avoid for simplicity), we could turn every context into a record type.

The introduction rule infers the principal type of every record term. Because we allow record types with defined fields, this is the singleton type containing only that record term. This may seem awkward but does not present a problem in practice, where type checking is preferred over type inference anyway.

The elimination rule is straightforward, but it is worth noting that it is entirely parallel to the second computation rule.[2]

---

[2] Note that it does not matter how the fields of the record are split into Δ_1 and Δ_2 as long as Δ_1 contains all fields that the declaration of x depends on.

The type checking rule has a surprising premise that r must already be well-typed (against some type R). Semantically, this assumption is necessary because we only check the presence of the fields required by $[\![\Delta]\!]$ — without the extra assumption, typing errors in any additional fields that r might have could go undetected. In practice, we implement the rule with an optimization: If r is a variable or a function application, we can efficiently infer some type for it. Otherwise, if $r = \langle\delta\rangle$, some fields of δ have already been checked by the first premise, and we only need to check the remaining fields. The order of premises matters in this case: we want to first use type checking for all fields for which $[\![\Delta]\!]$ provides an expected type before resorting to type inference on the remaining fields.

In the equality checking rule, note that we only have to check equality at undefined fields — the other fields are guaranteed to be equal by the assumption that r_1 and r_2 have type $[\![\Delta]\!]$.

Like the type checking rule, the first computation rule needs the premise that r is well-typed to avoid reducing an ill-typed into a well-typed term. In practice, our framework implements computation with a boolean flag that tracks whether the term to be simplified can be assumed to be well-typed or not; in the former case, this assumption can be skipped.

The second computation rule looks up the definition of a field in the type of the record. Both computation rules can be seen uniformly as definition lookup rules — in the first case the definition is given in the record, in the second case in its type.

Example 1. Figure 6 shows a record type of **Semilattices** (actually, this is a kind because it contains a type field) analogous to the grouping in Fig. 1 (using the usual encoding of axioms via judgments-as-types and higher-order abstract syntax for first-order logic).

$$\text{Semilattice} := \begin{bmatrix} U & : \textbf{type} \\ \wedge & : U \to U \to U \\ \text{assoc} & : \vdash \forall x, y, z : U.\ (x \wedge y) \wedge z \doteq x \wedge (y \wedge z) \\ \text{commutative} & : \ldots \\ \text{idempotent} & : \ldots \end{bmatrix}$$

Fig. 6. The (record-)kind of semilattices

Then, given a record r : Semilattice, we can form the record projection $r.\wedge$, which has type $r.U \to r.U \to r.U$ and $r.\text{assoc}$ yields a proof that $r.\wedge$ is associative. The intersection on sets forms a semilattice so (assuming we have proofs $\cap - \text{assoc}$, $\cap - \text{comm}$, $\cap - \text{idem}$ with the corresponding types) we can give an instance of that type as

$$\text{interSL} : \text{Semilattice} := \langle U := \text{Set}, \wedge := \cap, \text{assoc} := \cap - \text{assoc}, \ldots \rangle$$

Theorem 2 (Principal Types). *Our inference rules infer a principal type for each well-typed normal term.*

In analogy to function types, we can derive the subtyping properties of record types. We introduce context subsumption and then combine horizontal and vertical subtyping in a single statement.

Definition 3 (Context Subsumption). For two fully typed contexts Δ_i we write $\Gamma \vdash \Delta_1 \hookrightarrow \Delta_2$ iff for every declaration $x : T[:= t]$ in Δ_1 there is a declaration $x : T'[:= t']$ in Δ_2 such that

- $\Gamma \vdash T' <: T$ and
- if t is present, then so is t' and $\Gamma \vdash t \equiv t' : T$

Intuitively, $\Delta_1 \hookrightarrow \Delta_2$ means that everything of Δ_1 is also in Δ_2. That yields:

Theorem 3 (Record Subtyping). *The following rule is derivable:*

$$\frac{\Gamma \vdash \Delta_1 \hookrightarrow \Delta_2}{\Gamma \vdash [\![\Delta_2]\!] <: [\![\Delta_1]\!]}$$

3.2   Merging Records

We introduce an advanced operation on records, which proves critical for both convenience and performance: Theories can easily become very large containing hundreds or even thousands of declarations. If we want to treat theories as record types, we need to be able to build big records from smaller ones without exploding them into long lists. Therefore, we introduce an explicit merge operator $+$ on both record types and terms.

In the grammar, this is a single production for terms:

$$T ::= T + T$$

The intended meaning of $+$ is given by the following definition:

Definition 4 (Merging Contexts). Given a context Δ and a (not necessarily well-typed) context E, we define a partial function $\Delta \oplus E$ as follows:

- $\cdot \oplus E = E$
- If $\Delta = d, \Delta_0$ where d is a single declaration for a variable x:
  - if x is not declared in E: $(d, \Delta_0) \oplus E = d, (\Delta_0 \oplus E)$
  - if $E = E_0, e, E_1$ where e is a single declaration for a variable x:
    * if a variable in E_0 is also declared in Δ_0: $\Delta \oplus E$ is undefined,
    * if d and e have unequal types or unequal definitions: $\Delta \oplus E$ is undefined[3],

---

[3] It is possible and important in practice to also define $\Delta \oplus E$ when the types/definitions in d and e are provably equal. We omit that here for simplicity.

* otherwise, $(d, \Delta_0) \oplus (E_0, e, E_1) = E_0, m, (\Delta_0, E_1)$ where m arises by merging d and e.

Note that $\oplus$ is an asymmetric operator: While Δ must be well-typed (relative to some ambient context), E may refer to the names of Δ and is therefore not necessarily well-typed on its own.

We do not define the semantics of $+$ via inference and checking rules. Instead, we give equality rules that directly expand $+$ into $\oplus$ when possible:

$$\frac{\vdash \Gamma, (\Delta_1 \oplus \Delta_2) \; \mathbf{ctx}}{\Gamma \vdash [\![\Delta_1]\!] + [\![\Delta_2]\!] \rightsquigarrow [\![\Delta_1 \oplus \Delta_2]\!]} \qquad \frac{\vdash \Gamma, (\delta_1 \oplus \delta_2) \; \mathbf{ctx}}{\Gamma \vdash \langle\!\langle \delta_1 \rangle\!\rangle + \langle\!\langle \delta_2 \rangle\!\rangle \rightsquigarrow \langle\!\langle \delta_1 \oplus \delta_2 \rangle\!\rangle}$$

$$\frac{\vdash \Gamma, (\Delta \oplus \delta) \; \mathbf{ctx}}{\Gamma \vdash [\![\Delta]\!] + \langle\!\langle \delta \rangle\!\rangle \rightsquigarrow \langle\!\langle \Delta \oplus \delta \rangle\!\rangle}$$

In implementations some straightforward optimizations are needed to verify the premises of these rules efficiently; we omit that here for simplicity. For example, merges of well-typed records with disjoint field names are always well-typed, but e.g., $[\![x : nat]\!] + [\![x : bool]\!]$ is not well-typed even though both arguments are.

In practice, we want to avoid using the computation rules for $+$ whenever possible. Therefore, we prove admissible rules (i.e., rules that can be added without changing the set of derivable judgments) that we use preferentially:

Theorem 4. *If R_1, R_2, and $R_1 + R_2$ are well-typed record types, then $R_1 + R_2$ is the greatest lower bound with respect to subtyping of R_1 and R_2. In particular, $\Gamma \vdash r \Leftarrow R_1 + R_2$ iff $\Gamma \vdash r \Leftarrow R_1$ and $\Gamma \vdash r \Leftarrow R_2$.*
If $\Gamma \vdash r_i \Leftarrow R_i$ and $r_1 + r_2$ is well-typed, then $\Gamma \vdash r_1 + r_2 \Leftarrow R_1 + R_2$.

Inspecting the type checking rule in Fig. 5, we see that a record r of type $[\![\Delta]\!]$ must repeat all defined fields of Δ. This makes sense conceptually but would be a major inconvenience in practice. The merging operator solves this problem elegantly as we see in the following example:

Example 2. Continuing our running example, we can now define a type of semilattices *with order* (and all associated axioms) as in Fig. 7.

$$\mathtt{SemilatticeOrder} := \mathtt{Semilattice} + \left[\!\!\left[\begin{array}{ll} \leq & : U \to U \to U := \lambda x, y : U. \; x \doteq x \wedge y \\ \mathtt{refl} & : \vdash \forall a : U. \; a \leq a := (\mathrm{proof}) \\ \ldots \end{array}\right]\!\!\right]$$

$$\mathtt{interSLO} := \mathtt{SemilatticeOrder} + \mathtt{interSL}$$

Fig. 7. Running example

Now the explicit merging in the type `SemilatticeOrder` allows the projection `interSLO`. $\leq$, which is equal to $\lambda x, y : (\mathtt{interSLO}.U) . (x \doteq x(\mathtt{interSLO}.\wedge)y)$ and `interSLO.refl` yields a proof that this order is reflexive – without needing to define the order or prove the axiom anew for the specific instance `interSL`.

4   Internalizing Theories

4.1   Preliminaries: Theories

We introduce a minimal definition of stratified theories and theory morphisms, which can be seen as a very simple fragment of the MMT language [RK13]. The **grammar** is given in Fig. 8, again graying out the previously introduced parts.

$$
\begin{aligned}
\Theta &::= \cdot \mid \Theta,\, X = \{\Gamma\} \mid \Theta,\, X : X_1 \to X_2 = \{\Gamma\} \quad \text{theory level} \\
\Gamma &::= \cdot \mid \Gamma,x[:T][:=T] \mid \Gamma,\, \text{include}\, X \quad\quad\quad\; \text{includes} \\
T &::= x \mid \text{type} \mid \text{kind} \\
&\quad\; \mid\; \Pi_{x:T'}\, T \mid \lambda x : T'.T \mid T_1 T_2 \\
&\quad\; \mid\; [\![\Gamma]\!] \mid \langle\!\langle \Gamma \rangle\!\rangle \mid T.x \mid T_1 + T_2
\end{aligned}
$$

Fig. 8. A simple stratified language

Each of the two levels has its own context: Firstly, the **theory level** context Θ introduces names X, which can be either theories $X = \{\Gamma\}$ or morphisms $X : P \to Q = \{\Gamma\}$, where P and Q are the names of previously defined theories. Secondly, the **expression level** context Γ is as before but may additionally contain includes $\text{include}\, X$ of other theories resp. morphisms X. We call a context **flat** if it does not contain includes.

All **judgments** are as before except that they acquire a second context, e.g., the typing judgment now becomes $\Theta; \Gamma \vdash t \Leftarrow T$. With this modification, all rules for function and record types remain unchanged. However, we add the restriction that Γ in $[\![\Gamma]\!]$ and $\langle\!\langle \Gamma \rangle\!\rangle$ must be flat.

We omit the **rules** for theories and morphisms for brevity and only sketch their intuitions. We think of theories as named contexts and of morphisms as named substitutions between contexts. Includes allow forming both modularly by copying over the declarations of a previously named object. While theories may contain arbitrary declarations, morphisms are restricted: Let Θ contain $P = \{\Gamma\}$ and $Q = \{\Delta\}$. Then a morphism $V : P \to Q = \{\delta\}$ is well-typed if δ is fully defined (akin to record terms) and contains for each declaration $x : T$ of P a declaration $x = t$ where t may refer to all names declared in Q. V induces a homomorphic extension $\overline{V}$ that maps P-expressions to Q-expressions. The key property of morphisms is that, if V is well-typed, then $\Theta; P \vdash t \Leftarrow T$ implies $\Theta; Q \vdash \overline{V}(t) \Leftarrow \overline{V}(T)$ and accordingly for equality checking and subtyping. Thus, theory morphisms preserve judgments and (via propositions-as-types representations) truth. Moreover, it is straightforward to extend the above with identity and composition so that theories and morphisms form a category. We refer to [Rab14] for details.

4.2   Internalization

We can now add the internalization operator, for which everything so far was preparation. We add one production to the **grammar**:

$$ T ::= \text{Mod}(X) $$

The intended meaning of $\text{Mod}(X)$ is that it turns a theory X into a record type and a morphism $X : P \to Q$ into a function $\text{Mod}(Q) \to \text{Mod}(P)$. For simplicity, we only state the rules for the case where all include declarations are at the beginning of theory/morphism:

$$\frac{P = \{\text{include } P_1, \ldots, \text{include } P_n, \Delta\} \text{ in } \Theta \quad \Delta \text{ flat} \quad \max P \text{ defined}}{\Theta; \Gamma \vdash \text{Mod}(P) \rightsquigarrow \text{Mod}(P_1) + \ldots + \text{Mod}(P_n) + [\![\Delta]\!]}$$

$$\frac{V : P \to Q = \{\text{include } V_1, \ldots, \text{include } V_n, \delta\} \text{ in } \Theta \quad \delta \text{ flat}}{\Theta; \Gamma \vdash \text{Mod}(V) \rightsquigarrow \lambda r : \text{Mod}(Q) . \text{Mod}(P) + (\text{Mod}(V_1)\, r) + \ldots + (\text{Mod}(V_n)\, r) + \langle\!\langle \delta[r] \rangle\!\rangle}$$

where we use the following abbreviations:

- In the rule for theories, $\max P$ is the biggest universe occurring in any declaration transitively included into P, i.e., $\max P = \max\{\max P_1, \ldots, \max P_n, \max \Delta\}$ (undefined if any argument is).
- In the rule for morphisms, $\delta[r]$ is the result of substituting in δ every reference to a declaration of x in Q with $r.x$.

In the rule for morphisms, the occurrence of $\text{Mod}(P)$ may appear redundant; but it is critical to (i) make sure all defined declarations of P are part of the record and (ii) provide the expected types for checking the declarations in δ.

Example 3. Consider the theories in Fig. 9. Applying $\text{Mod}(\cdot)$ to these theories yields exactly the record types of the same name introduced in Sect. 3 (Figs. 6 and 7), i.e., we have $\text{interSL} \Leftarrow \text{Mod}(\text{Semilattice})$ and $\text{interSLO} \Leftarrow \text{Mod}(\text{SemilatticeOrder})$. In particularly, Mod preserves the modular structure of the theory.

$$\text{theory Semilattice} = \begin{cases} U & : \textbf{type} \\ \wedge & : U \to U \to U \\ \text{assoc} & : \vdash \forall x, y, z : U.\, (x \wedge y) \wedge z \doteq x \wedge (y \wedge z) \\ \text{commutative} & : \ldots \\ \text{idempotent} & : \ldots \end{cases}$$

$$\text{theory SemilatticeOrder} = \begin{cases} \text{include Semilattice} \\ \text{order} & : U \to U \to U := \lambda x, y : U.\, x \doteq x \wedge y \\ \text{refl} & : \vdash \forall a : U.\, a \leq a := (\text{proof}) \\ \ldots \end{cases}$$

Fig. 9. A theory of semilattices

The basic properties of $\text{Mod}(X)$ are collected in the following theorem:

Theorem 5 (Functoriality). $\text{Mod}(\cdot)$ *is a monotonic contravariant functor from the category of theories and morphisms ordered by inclusion to the category of types (of any universe) and functions ordered by subtyping. In particular,*

- *if P is a theory in Θ and $\max P \in \{\texttt{type}, \texttt{kind}\}$, then $\Theta; \Gamma \vdash \texttt{Mod}(P) \Leftarrow \max P$*
- *if $V : P \to Q$ is a theory morphism in $\Theta; \Gamma \vdash \texttt{Mod}(V) \Leftarrow \texttt{Mod}(Q) \to \texttt{Mod}(P)$*
- *if P is transitively included into Q, then $\Theta; \Gamma \vdash \texttt{Mod}(Q) <: \texttt{Mod}(P)$.*

An immediate advantage of $\texttt{Mod}(\cdot)$ is that we can now use the expression level to define expression-like theory level operations. As an example, we consider the **intersection** $P \cap P'$ of two theories, i.e., the theory that includes all theories included by both P and P'. Instead of defining it at the theory level, which would begin a slippery slope of adding more and more theory level operations, we can simply build it at the expression level:

$$P \cap P' := \texttt{Mod}(Q_1) + \ldots + \texttt{Mod}(Q_n)$$

where the Q_i are all theories included into both P and P'.[4]

Note that the computation rules for $\texttt{Mod}$ are efficient in the sense that the structure of the theory level is preserved. In particular, we do not flatten theories and morphisms into flat contexts, which would be a huge blow-up for big theories.[5]

However, efficiently *creating* the internalization is not enough. $\texttt{Mod}(X)$ is defined via $+$, which is itself only an abbreviation whose expansion amounts to flattening. Therefore, we establish admissible rules that allow *working with* internalizations efficiently, i.e., without computing the expansion of $+$:

Theorem 6. *Fix well-typed Θ, Γ and $P = \{\texttt{include}\,P_1, \ldots, \texttt{include}\,P_n, \Delta\}$ in Θ. Then the following rules are admissible:*

$$\frac{\overbrace{\Theta; \Gamma \vdash r \Leftarrow \texttt{Mod}(P_i)}^{1 \le i \le n} \quad \overbrace{\Theta; \Gamma \vdash r.x \Leftarrow T[r/P]}^{x:T \in \Delta} \quad \overbrace{\Theta; \Gamma \vdash r.x \equiv t[r/P] : T[r/P]}^{x:T:=t \in \Delta} \quad \Gamma \vdash r \Rightarrow R}{\Theta; \Gamma \vdash r \Leftarrow \texttt{Mod}(P)}$$

$$\frac{\overbrace{\Theta; \Gamma \vdash r_i \Leftarrow \texttt{Mod}(P_i)}^{1 \le i \le n} \quad \overbrace{\Theta; \Gamma \vdash r_i \equiv r_j : P_i \cap P_j}^{1 \le i,j \le n} \quad \Theta; \Gamma \vdash \langle\!\langle \delta \rangle\!\rangle [r/P] \Leftarrow [\![\Delta]\!] \quad \Gamma \vdash r \Rightarrow R}{\Theta; \Gamma \vdash \underbrace{\texttt{Mod}(P) + r_1 + \ldots + r_n + \langle\!\langle \delta \rangle\!\rangle}_{=:r} \Rightarrow \texttt{Mod}(P)}$$

where $[r/P]$ abbreviates the substitution that replaces every x declared in a theory transitively-included into P with $r.x$.[6]

The first rule in Theorem 6 uses the modular structure of P to check r at type $\texttt{Mod}(P)$. If r is of the form $\langle\!\langle \delta \rangle\!\rangle$, this is no faster than flattening $\texttt{Mod}(P)$

---

[4] Note that because $P \cap P'$ depends on the syntactic structure of P and P', it only approximates the least upper bound of $\texttt{Mod}(P)$ and $\texttt{Mod}(P')$ and is not stable under, e.g., flattening of P and P'. But it can still be very useful in certain situations.

[5] The computation of $\max P$ may look like it requires flattening. But it is easy to compute and cache its value for every named theory.

[6] In practice, these substitutions are easy to implement without flattening r because we can cache for every theory which theories it includes and which names it declares.

all the way. But in the typical case where r is also formed modularly using a similar structure as P, this can be much faster. The second rule performs the corresponding type inference for an element of $\mathrm{Mod}\,(P)$ that is formed following the modular structure of P. In both cases, the last premise is again only needed to make sure that r does not contain ill-typed fields not required by $\mathrm{Mod}\,(P)$. Also note that if we think of $\mathrm{Mod}\,(P)$ as a colimit and of elements of $\mathrm{Mod}\,(P)$ as morphisms out of P, then the second rule corresponds to the construction of the universal morphisms out of the colimit.

Example 4. We continue Example 3 and assume we have already checked $\texttt{interSL} \Leftarrow \mathrm{Mod}\,(\texttt{Semilattice})$ (*).

We want to check $\texttt{interSL} + \langle\delta\rangle \Leftarrow \mathrm{Mod}\,(\texttt{SemilatticeOrder})$. Applying the first rule of Theorem 6 reduces this to multiple premises, the first one of which is (*) and can thus be discharged without inspecting $\texttt{interSL}$.

Example 4 is still somewhat artificial because the involved theories are so small. But the effect pays off enormously on larger theories.

5 Implementation and Case Study

We have implemented a variant of the record types and the $\mathrm{Mod}\,(\cdot)$-operator described here in the MMT-system (as part of [LFX]). They are used extensively in the *Math-in-the-Middle* archive (MitM), which forms an integral part in the OpenDreamKit [Deh+16] and MaMoRed [Koh+17] projects. In particular the formalizations of algebra and topology are systematically built on top of the concepts presented in this paper.

$$
\texttt{theory Ring} = \left\{
\begin{array}{ll}
U & : \textbf{type} \\
+ & : U \to U \to U \\
\cdot & : U \to U \to U \\
\texttt{assoc\_plus} & : \vdash \forall x, y, z : U.\ (x+y)+z \doteq x+(y+z) \\
\texttt{commutative\_plus} & : \ldots \\
\ldots &
\end{array}
\right\}
$$

$$
\texttt{theory Field} = \left\{
\begin{array}{ll}
\texttt{include} & \texttt{Ring} \\
\texttt{inverses\_times} : \vdash \forall x : U.\ x \neq 0 \Rightarrow \exists y.\ x \cdot y \doteq 1 \\
\ldots &
\end{array}
\right\}
$$

$$
\texttt{theory Module}(R : \mathrm{Mod}\,(\texttt{Ring})) = \left\{
\begin{array}{l}
\texttt{include} \qquad \texttt{AbelianGroup} \\
\texttt{scalar\_mult} : R.U \to U \to U \\
\ldots
\end{array}
\right\}
$$

$$
\texttt{theory VSpace}(F : \mathrm{Mod}\,(\texttt{Field})) = \left\{
\begin{array}{l}
\texttt{include Module}(F) \\
\ldots
\end{array}
\right\}
$$

Fig. 10. Theories for R-modules and vector spaces

The archive sources can be found at [Mit], and its contents can be inspected and browsed online at https://mmt.mathhub.info under MitM/smglom. Note that the Mod ($\cdot$) operator is called `ModelsOf` here.

For a particularly interesting example that occurs in MitM, consider the theories for modules and vector spaces (over some ring/field) given in Fig. 10, which elegantly follow informal mathematical practice. Going beyond the syntax introduced so far, these use *parametric* theories. Our implementation extends Mod to parametric theories as well, namely in such a way that Mod (`Module`) : $\prod_{R:\text{Mod}(\text{Ring})} \text{Mod}(\text{Module}(R))$ and correspondingly for fields. Thus, we obtain

$$\text{Mod}(\text{VSpace}) = \lambda F : \text{Mod}(\text{Field}).((\text{Mod}(\text{Module})\ F) + \ldots)$$

and, e.g., Mod (`VSpace`) $\mathbb{R}$ $<:$ Mod (`Module`) $\mathbb{R}$. Because of type-level parameters, this requires some kind of parametric polymorphism in the type system. For our approach, the shallow polymorphism module that is available in MMT is sufficient.

6 Conclusion

We have presented a formal system that allows to systematically combine the advantages of stratified and integrated grouping mechanisms found in type theories, logics, and specification/programming languages. Concretely, our system allows internalizing theories into record types in a way that preserves their defined fields and modular structure.

Our MitM case study shows that theory internalization is an important feature of any foundation; especially if it interfaces to differing mathematical software systems. Our experiments have also shown that (predicate) subtyping makes internalization even stronger in practice. But type-inference in the combined system induces non-trivial trade-offs; which we leave to future work.

Acknowledgements. The work reported here has been kicked off by discussions with Jacques Carette and William Farmer who have experimented with theory internalizations into record types in the scope of their MathScheme system. We acknowledge financial support from the OpenDreamKit Horizon 2020 European Research Infrastructures project (#676541).

References

[Bra13] Brady, E.: Idris, a general-purpose dependently typed programming language: design and implementation. J. Funct. Program. **23**(5), 552–593 (2013)

[Con+86] Constable, R., et al.: Implementing Mathematics with the Nuprl Development System. Prentice-Hall, Upper Saddle River (1986)

[Deh+16] Dehaye, P.-O., et al.: Interoperability in the OpenDreamKit project: the math-in-the-middle approach. In: Kohlhase, M., Johansson, M., Miller, B., de Moura, L., Tompa, F. (eds.) CICM 2016. LNCS (LNAI), vol. 9791, pp. 117–131. Springer, Cham (2016). https://doi.org/10.1007/978-3-319-42547-4_9. https://github.com/OpenDreamKit/OpenDreamKit/blob/master/WP6/CICM2016/published.pdf

[FGT92] Farmer, W., Guttman, J., Thayer, F.: Little theories. In: Kapur, D. (ed.) Conference on Automated Deduction, pp. 467–581 (1992)

[Gog+93] Goguen, J., Winkler, T., Meseguer, J., Futatsugi, K., Jouannaud, J.: Introducing OBJ. In: Goguen, J., Coleman, D., Gallimore, R. (eds.) Applications of Algebraic Specification Using OBJ. Cambridge (1993)

[Har+12] Hardin, T., et al.: The FoCaLiZe Essential (2012). http://focalize.inria.fr/

[Har96] Harrison, J.: HOL light: a tutorial introduction. In: Srivas, M., Camilleri, A. (eds.) FMCAD 1996. LNCS, vol. 1166, pp. 265–269. Springer, Heidelberg (1996). https://doi.org/10.1007/BFb0031814

[Koh+17] Kohlhase, M., Koprucki, T., Müller, D., Tabelow, K.: Mathematical models as research data via flexiformal theory graphs. In: Geuvers, H., England, M., Hasan, O., Rabe, F., Teschke, O. (eds.) CICM 2017. LNCS (LNAI), vol. 10383, pp. 224–238. Springer, Cham (2017). https://doi.org/10.1007/978-3-319-62075-6_16

[LFX] MathHub MMT/LFX Git Repository. http://gl.mathhub.info/MMT/LFX. Accessed 15 May 2015

[Luo09] Luo, Z.: Manifest fields and module mechanisms in intensional type theory. In: Berardi, S., Damiani, F., de'Liguoro, U. (eds.) TYPES 2008. LNCS, vol. 5497, pp. 237–255. Springer, Heidelberg (2009). https://doi.org/10.1007/978-3-642-02444-3_15

[Mat] Mathematical Components. http://www.msr-inria.fr/projects/mathematical-components-2/

[Mit] MitM/SMGLoM. https://gl.mathhub.info/MitM/smglom. Accessed 01 Feb 2018

[MRK] Müller, D., Rabe, F., Kohlhase, M.: Theories as Types. http://kwarc.info/kohlhase/submit/tatreport.pdf

[ORS92] Owre, S., Rushby, J.M., Shankar, N.: PVS: a prototype verification system. In: Kapur, D. (ed.) CADE 1992. LNCS, vol. 607, pp. 748–752. Springer, Heidelberg (1992). https://doi.org/10.1007/3-540-55602-8_217

[Rab14] Rabe, F.: How to identify, translate, and combine logics? J. Log. Comput. (2014). https://doi.org/10.1093/logcom/exu079

[Rab17] Rabe, F.: A modular type reconstruction algorithm. ACM Trans. Comput. Log. (2017). Accepted pending minor revision: https://kwarc.info/people/frabe/Research/rabe_recon_17.pdf

[RK13] Rabe, F., Kohlhase, M.: A scalable module system. Inf. Comput. **230**(1), 1–54 (2013)

[SW11] Spitters, B., van der Weegen, E.: Type classes for mathematics in type theory. CoRR abs/1102.1323 (2011). arXiv:1102.1323

[TB85] Trybulec, A., Blair, H.: Computer assisted reasoning with MIZAR. In: Joshi, A. (ed.) Proceedings of the 9th International Joint Conference on Artificial Intelligence, pp. 26–28. Morgan Kaufmann (1985)

[Wen09] Wenzel, M.: The Isabelle/Isar Reference Manual, 3 December 2009. http://isabelle.in.tum.de/documentation.html

[Coq15] Coq Development Team: The Coq Proof Assistant: Reference Manual. Technical report, INRIA (2015)

Datatypes with Shared Selectors

Andrew Reynolds[1], Arjun Viswanathan[1], Haniel Barbosa[1], Cesare Tinelli[1(✉)], and Clark Barrett[2]

[1] University of Iowa, Iowa City, USA
`cesare-tinelli@uiowa.edu`
[2] Department of Computer Science, Stanford University, Stanford, USA

Abstract. We introduce a new theory of algebraic datatypes where selector symbols can be shared between multiple constructors, thereby reducing the number of terms considered by current SMT-based solving approaches. We show that the satisfiability problem for the traditional theory of algebraic datatypes can be reduced to problems where selectors are mapped to shared symbols based on a transformation provided in this paper. The use of shared selectors addresses a key bottleneck for an SMT-based enumerative approach to the Syntax-Guided Synthesis (SyGuS) problem. Our experimental evaluation of an implementation of the new theory in the SMT solver CVC4 on syntax-guided synthesis and other domains provides evidence that the use of shared selectors improves state-of-the-art SMT-based approaches for constraints over algebraic datatypes.

1 Introduction

Algebraic datatypes, also known as inductive or recursive datatypes, are composite types commonly used for expressing finite data structures in computer science applications, such as lists or trees. Reasoning efficiently about (algebraic) datatypes is thus paramount in such fields as program analysis and verification, which has led to numerous approaches for automating solving in this setting. In this paper, we follow the semantic approach introduced by Barrett et al. [10], which is generally the basis for datatype decision procedures in satisfiability modulo theories (SMT) solvers [11].

In semantic presentations of the theory of algebraic datatypes [10,22], a datatype is an absolutely free algebra over a signature of function symbols called *constructors*; the immediate subterms of a datatype value are accessed with function symbols called *selectors*, or *projections*, which are specific for each constructor and its arguments. Datatypes also have *discriminators*, or *testers*, associated with each constructor. They are predicates indicating whether a given datatype value was built with a specific constructor.

The satisfiability of quantifier-free formulas in the theory of algebraic datatypes is decidable. A basic decision procedure for this problem [10,22] used by a number of SMT solvers operates by *progressively unrolling* datatypes: it tries to satisfy constraints by guessing top-level constructors in order to build

© Springer International Publishing AG, part of Springer Nature 2018
D. Galmiche et al. (Eds.): IJCAR 2018, LNAI 10900, pp. 591–608, 2018.
https://doi.org/10.1007/978-3-319-94205-6_39

values for the constraint variables incrementally. Concretely, if x is a datatype variable and c is an n-ary constructor for the datatype, the procedure may guess the equality constraint $x \approx c(x_1, \ldots, x_n)$ where $x_1, \ldots, x_n$ are fresh variables. If such a choice leads to an inconsistency, the procedure backtracks and tries different constructors until it determines that the constraints are satisfiable or no more choices are possible. During this process, lemmas in the form of quantifier-free clauses may be learned by the procedure that prevent the procedure from making guesses already shown to be infeasible. However, these lemmas may include selectors, and because each selector is associated with only a single constructor, the generality and hence the usefulness of such lemmas is limited.

To address this limitation we introduce a new (formulation of the) theory of datatypes that allows certain selectors to be shared by multiple constructors. This way, information previously acquired when reasoning with a constructor, i.e., the learned lemmas on the applications of its selectors, can be reused when an argument of the same type is considered in another constructor. We illustrate this point with the following example.

Example 1. Consider a binary tree whose internal nodes store either one or two integer values, and whose leaves store both a Boolean and an integer value.

A datatype **Tree** modeling this data structure has three constructors: one (N_1) taking an integer and two **Tree** elements as arguments, another (N_2) taking two integers and two **Tree** elements as arguments, and a third (L) taking as arguments a Boolean and an integer element. We write this datatype in the following BNF-style notation:

$$\textbf{Tree} = N_1(\textbf{Int}, \textbf{Tree}, \textbf{Tree}) \mid N_2(\textbf{Int}, \textbf{Int}, \textbf{Tree}, \textbf{Tree}) \mid L(\textbf{Bool}, \textbf{Int})$$

We assume each constructor has selectors associated with them. The *subfields* (i.e., the immediate subterms) of terms constructed by N_1 are accessed, respectively, by the selectors $S^{N_1, 1}$, $S^{N_1, 2}$, and $S^{N_1, 3}$ of type **Tree** $\rightarrow$ **Int**, **Tree** $\rightarrow$ **Tree** and **Tree** $\rightarrow$ **Tree**. The selectors for the other constructors are similar. We also assume each constructor is associated with a tester predicate, i.e. isN_1, isN_2, and isL, each of which takes a **Tree** as an argument. Given term t of type **Tree**, consider the following set of clauses:

$$\{ \neg \mathsf{isN}_1(t) \vee S^{N_1, 1}(t) \geq 0, \ \neg \mathsf{isL}(t) \vee S^{L, 2}(t) \geq 0 \} \tag{1}$$

The first clause states that when t has top symbol N_1, its first subfield (which is of type **Int**) is non-negative. Similarly, the second says that when t has top symbol L, its second subfield is non-negative.

Consider now a different kind of selector symbol $S^{\mathsf{Int}, 1}$ of type **Tree** $\rightarrow$ **Int** which maps each value of type **Tree** to the first (i.e., leftmost) subfield of t of type **Int**, *regardless* of the top constructor symbol of t. We will refer to such selectors as *shared selectors*. While nine selectors in the standard sense are necessary for **Tree**, five shared selectors suffice to access all possible subfields of a value of type **Tree**: two to access the **Tree** subfields, two to access the **Int** subfields, and one to access the **Bool** subfield of L. In particular, clause set (1) can be rewritten as follows using only one shared selector:

$$\{\, \neg\mathsf{isN}_1(t) \vee \mathsf{S}^{\mathsf{Int},\,1}(t) \geq 0,\ \neg\mathsf{isL}(t) \vee \mathsf{S}^{\mathsf{Int},\,1}(t) \geq 0 \,\} \tag{2}$$

stating that when t has top symbol N_1 or L, its first integer child is non-negative.

•

In Example 1, the second set of clauses has one unique arithmetic constraint whereas the first set has two. In practice, reducing the number of unique constraints can substantially improve the performance of SMT solvers. Our experiments show that shared selectors lead to a significant reduction in the number of unique constraints for several classes of benchmarks from real applications, with resulting SMT solver performance improvements that are proportional to the magnitude of this reduction.

Contributions. We introduce a conservative extension of the (generic) theory of algebraic datatypes that features shared selectors. We show how using shared selectors instead of standard (unshared) selectors can improve the performance of current satisfiability procedures for the theory and also, as a result, the performance of procedures for syntax-guided synthesis. Specifically:

1. We formalize the new theory and show that constraints in the original signature can be reduced to equisatisfiable constraints whose selectors are all shared selectors. We present a decision procedure for the satisfiability of quantifier-free formulas in this theory as a natural modification of an earlier procedure for datatypes [22].
2. We provide details on an SMT-based approach for syntax-guided synthesis [24], and demonstrate how it can significantly benefit from native support in the SMT solver for a theory of datatypes with shared selectors.
3. We present an extensive experimental evaluation of our implementation in the SMT solver CVC4 [7] on benchmarks from SMT-LIB [8] and from the most recent edition of SyGuS-COMP [3], the syntax-guided synthesis competition. This evaluation shows that shared selectors can reduce the number of terms introduced during solving, thus leading to more solved problems with respect to the state of the art.

2   Preliminaries

Our setting is a many-sorted classical first-order logic similar in essence to the one adopted by the SMT-LIB standard [9]. A signature $\Sigma = (\mathcal{Y}, \mathcal{F})$ consists of a set $\mathcal{Y}$ of first-order types, or *sorts*, and a set $\mathcal{F}$ of first-order function symbols over these types. Each symbol $\mathsf{f} \in \mathcal{F}$ is associated with a list $\tau_1, \dots, \tau_n$ of argument types and a return type τ, written $\mathsf{f} : \tau_1 \times \cdots \times \tau_n \to \tau$ or just $\mathsf{f} : \tau$ if $n = 0$. The function $\mathsf{arity}(\mathsf{f})$ returns n. We assume that any signature contains a **Bool** type and constants true, false : **Bool**; a family $(\approx\, : \tau \times \tau \to \textbf{Bool})_{\tau \in \mathcal{Y}}$ of equality symbols; a family $(\mathsf{ite} : \textbf{Bool} \times \tau \times \tau \to \tau)_{\tau \in \mathcal{Y}}$ of *if-then-else* symbols; and the

Boolean connectives $\neg, \wedge, \vee$ with their expected types. Function symbols of **Bool** return type play the role of predicate symbols.

Typed terms are built as usual over function symbols from $\mathcal{F}$ and typed variables from a fixed family $(\mathcal{V}_\tau)_{\tau \in \mathcal{Y}}$ of pairwise-disjoint infinite sets. Formulas are terms of type **Bool**. The syntax $t \not\approx u$ is short for $\neg(t \approx u)$. We reserve the names x, y, z for variables; r, s, t, u for terms (which may be formulas); and φ, ψ for formulas. We use the symbol $=$ for equality at the meta-level. The *set of all terms* occurring in a term t is denoted by $\mathbf{T}(t)$. When convenient, we write an enumeration of (meta)symbols $a_1, \ldots, a_n$ as $\bar{a}$. If $b_1, \ldots, b_k$ is another enumeration, $\bar{a}\bar{b}$ denotes the enumeration $a_1, \ldots, a_n, b_1, \ldots, b_k$.

Given a signature $\Sigma = (\mathcal{Y}, \mathcal{F})$, a Σ-interpretation $\mathcal{I}$ maps: each $\tau \in \mathcal{Y}$ to a non-empty set $\tau^{\mathcal{I}}$, the *domain* of τ in $\mathcal{I}$, with $\mathbf{Bool}^{\mathcal{I}} = \{\top, \bot\}$; each $x \in \mathcal{V}_\tau$ to an element of $\tau^{\mathcal{I}}$; each $f \in \mathcal{F}$ s.t. $f : \tau_1 \times \cdots \times \tau_n \to \tau$ to a total function $u^{\mathcal{I}} : \tau_1^{\mathcal{I}} \times \cdots \times \tau_n^{\mathcal{I}} \to \tau^{\mathcal{I}}$ when $n > 0$ and to an element of $\tau^{\mathcal{I}}$ when $n = 0$. A satisfiability relation between Σ-interpretations and Σ-formulas is defined inductively as usual.

A *theory* is a pair $\mathcal{T} = (\Sigma, \mathbf{I})$ where Σ is a signature and $\mathbf{I}$ is a non-empty class of Σ-interpretations, the *models of $\mathcal{T}$*, that is closed under variable reassignment (i.e., every Σ-interpretation that differs from one in $\mathbf{I}$ only in how it interprets the variables is also in $\mathbf{I}$) and isomorphism. A Σ-formula φ is $\mathcal{T}$-satisfiable (respectively $\mathcal{T}$-unsatisfiable) if it is satisfied by some (resp., no) interpretation in $\mathbf{I}$. A satisfying interpretation for φ *models (or is a model of)* φ. A formula φ is *valid in $\mathcal{T}$* (or $\mathcal{T}$-*valid*) if every model of $\mathcal{T}$ is a model of φ.

3 Theory of Datatypes with Shared Selectors

In this section, we consider a theory $\mathcal{D}$ of algebraic datatypes over some signature $\Sigma = (\mathcal{Y}, \mathcal{F})$ and then extend it conservatively to an expanded signature *with shared selectors*. The terms of $\mathcal{D}$ are quantifier-free. As a technical convenience, we treat free variables as constants in a suitable expansion of Σ. The types of $\mathcal{D}$ are partitioned into a set of *datatypes* $\mathcal{Y}_{\mathrm{dt}}$, and a set of other types $\mathcal{Y}_{\mathrm{ord}}$. We use the metavariables δ, ϵ to refer to datatypes and τ, υ for arbitrary first-order types. Each datatype δ is equipped with one or more *constructors*, distinguished function symbols from $\mathcal{F}$ with return type δ. For every argument k of a constructor $\mathsf{C} : \tau_1 \ldots, \tau_n \to \delta$ for δ, we assume $\mathcal{F}$ contains a *(standard) selector* $\mathsf{S}_\delta^{\mathsf{C}, k} : \delta \to \tau_k$. We omit δ from the selector name when it is understood or not important. We refer the reader to the SMT-LIB 2 reference document [9] or Barrett et al. [10] for a formal definition of this theory.[1] We recall salient properties of its symbols as needed.

To start, each model of the theory, when reduced to the constructors of a datatype in the theory, is isomorphic to a term (or *Herbrand*) algebra. Concretely, this means that if $\delta \in \mathcal{Y}_{\mathrm{dt}}$ is a datatype whose constructors

---

[1] The two references differ on how they make selectors (which are naturally partial functions) total. We follow the SMT-LIB 2 standard here.

are $\{C_1, \ldots, C_m\}$, then the following formulas are all $\mathcal{D}$-valid for all distinct $i, j \in \{1, \ldots, m\}$

$$\forall x_1, \ldots, x_{p_i}, z_1, \ldots, z_{q_i}.\ C_i(x_1, \ldots, x_{p_i}) \not\approx C_j(z_1, \ldots, z_{q_i}) \qquad (Distinctness)$$

$$\begin{aligned} \forall x_1, \ldots, x_{p_i}, z_1, \ldots, z_{p_i}. \\ C_i(x_1, \ldots, x_{p_i}) \approx C_i(z_1, \ldots, z_{p_i}) \to x_1 \approx z_1 \wedge \ldots \wedge x_{p_i} \approx z_{p_i} \end{aligned} \qquad (Injectivity)$$

$$\forall x.\ \mathsf{isC}_1(x) \vee \cdots \vee \mathsf{isC}_m(x) \qquad (Exhaustiveness)$$

Above, we write $\mathsf{isC}_i(t)$ to denote the predicate that holds if and only if the top symbol of t is C_i. Strictly speaking, we do not need to extend our signature with the tester symbols isC since a term of the form $\mathsf{isC}(t)$ can be considered an abbreviation for the equality $t \approx C(\mathsf{S}^{C,1}(t), \ldots, \mathsf{S}^{C,n}(t))$ where $n = \mathsf{arity}(C)$.

Interpretations must also respect *acyclicity*, which states that constructor terms cannot be equal to any of their proper subterms.

Since all models of $\mathcal{D}$ interpret a datatype δ in the same way modulo isomorphism, we will say that δ is *finite* if its interpretation is a finite set. For simplicity, we will assume that *every type τ in $\mathcal{D}$ that is not a datatype is interpreted as an infinite set in every model of $\mathcal{D}$*. This is not a strong restriction in practice, since types with some fixed, finite cardinality k can be treated as datatypes with k nullary constructors.

The relationship between an n-ary constructor C and each of its selectors $\mathsf{S}^{C,k}$ with $k = 1, \ldots, n$ is captured by the following $\mathcal{D}$-valid formula:

$$\forall x_1, \ldots, x_{n_i}.\ \mathsf{S}_\delta^{C,k}(C(x_1, \ldots, x_{n_i})) \approx x_k \qquad (Standard\ selection)$$

3.1   Shared Selectors

We extend the signature of $\mathcal{D}$ with additional selectors which we call *shared selectors* and denote as $\mathsf{S}_\delta^{\tau,k}$, for each datatype δ and type τ in $\mathcal{D}$ and each natural number k. Intuitively, a shared selector $\mathsf{S}_\delta^{\tau,k}$ for δ, when applied to a δ-term $C(t_1, \ldots, t_n)$ returns the k-th argument of C that has type τ, if one exists.

Example 2. Consider again the **Tree** datatype introduced in Example 1. For term

$$t = \mathsf{N}_1(1,\ \mathsf{N}_2(2, 3,\ \mathsf{L}(\mathsf{true},\ 4),\ \mathsf{L}(\mathsf{false},\ 5)),\ \mathsf{L}(\mathsf{true},\ 6))$$

the equalities $\mathsf{S}^{\mathsf{Int},1}(t) \approx 1$, $\mathsf{S}^{\mathsf{Int},2}(\mathsf{S}^{\mathsf{Tree},1}(t)) \approx 3$, and $\mathsf{S}^{\mathsf{Int},1}(\mathsf{S}^{\mathsf{Tree},2}(\mathsf{S}^{\mathsf{Tree},1}(t))) \approx 6$ are all valid in our extension of $\mathcal{D}$ to shared selectors. $\qquad \bullet$

To define shared selectors formally, let us first define a partial function stoa (for *selector to argument*) that takes as input a natural number k, a type τ, and a constructor C, and returns the index of the k-th argument of C of type τ. We leave stoa undefined if C has fewer than k arguments of type τ.

Example 3. For the **Tree** datatype, $\mathsf{stoa}(1, \mathbf{Int}, \mathsf{N}_1) = 1$, $\mathsf{stoa}(2, \mathbf{Tree}, \mathsf{N}_1) = 3$ and $\mathsf{stoa}(1, \mathbf{Int}, \mathsf{L}) = 2$, whereas $\mathsf{stoa}(2, \mathbf{Int}, \mathsf{N}_1)$, $\mathsf{stoa}(1, \mathbf{Bool}, \mathsf{N}_2)$, and $\mathsf{stoa}(1, \mathbf{Tree}, \mathsf{L})$ are undefined. $\qquad \bullet$

More formally, in our extension of theory $\mathcal{D}$ with shared selectors, which we also refer to as $\mathcal{D}$ for convenience, the following holds for all datatypes δ, constructors C of δ, and shared selectors $S^{\tau,k}$, whenever $\mathsf{stoa}(k, \tau, C)$ is defined:

$$\forall x_1, \ldots, x_n.\ S_\delta^{\tau,k}(C(x_1, \ldots, x_n)) \approx x_i,\ \text{where } i = \mathsf{stoa}(k, \tau, C) \qquad (\textit{Shared selection})$$

It is not difficult to argue that every Σ-formula φ without shared selectors is valid in the extended theory if and only if it is valid in the original theory.

3.2   From Standard Selectors to Shared Selectors

The satisfiability problem for *constraints*, i.e., finite sets of literals, over the original theory of datatypes (without shared selectors) is decidable [10]. In this section, we introduce a transformation $\mathcal{H}$ that reduces arbitrary constraints in our extended theory $\mathcal{D}$, which may have both standard and shared selectors, to constraints with no standard selectors. Applying this transformation as an initial step allows us to determine the satisfiability of arbitrary Σ-constraints by means of a decision procedure for Σ-constraints without standard selectors.

To define this transformation, let max_Σ denote some natural number that is greater than the arity of all constructors in Σ. We define the dual of the stoa function from Subsect. 3.1 as the partial function atos (for *argument to selector*) that takes as input a type τ, a constructor $C : \tau_1 \times \cdots \times \tau_n \to \delta$, and a natural number $k \leq n$, and returns the number of times τ occurs in $\tau_1, \ldots, \tau_k$.

Figure 1 defines the transformation $\mathcal{H}$, which takes as arguments a Σ-term t and a mapping M. The latter consists of one entry of the form $s \mapsto C$ for each datatype term s in $\mathbf{T}(t)$ where C is one of the constructors for the type of s. Without loss of generality, we assume that all applications of shared selectors $S_\delta^{\tau,k}$ occurring in t are such that $k < \mathsf{max}_\Sigma$. The transformation $\mathcal{H}$ leaves variables unchanged; for terms whose top symbol is a constructor or a shared selector, $\mathcal{H}$ behaves homorphically. For terms t with a standard selector $S_\delta^{C,k} : \delta \to \tau$ as top symbol, we distinguish whether the argument t_1 is mapped to C by M or not. In the first case, we replace $S_\delta^{C,k}$ by the shared selector $S_\delta^{\tau,\mathsf{atos}(\tau,C,k)}$. In the second case, we replace $S_\delta^{C,k}$ by the shared selector $S_\delta^{\tau,\mathsf{err}(C,k)}$, where err is a function that takes as arguments a constructor and a k such that $1 \leq k \leq \mathsf{arity}(C)$, and returns a natural number. Additionally, err has the properties:

1. If $C_1 \neq C_2$ or $k_1 \neq k_2$, then $\mathsf{err}(C_1, k_1) \neq \mathsf{err}(C_2, k_2)$, and
2. $\mathsf{err}(C_1, k) \geq \mathsf{max}_\Sigma$.

We use the function err in this transformation to introduce shared selectors that are unique to the pair (C, k), as guaranteed by Property 1 above, and whose return value is undefined, as guaranteed by Property 2. In either case, $\mathcal{H}$ is applied recursively to t_1.

We extend $\mathcal{H}$ to sets of equalities and disequalities E as follows:

$$\mathcal{H}(E, M) = \{\, \mathcal{H}(t_1, M) \approx \mathcal{H}(t_2, M) \mid t_1 \approx t_2 \in E \,\} \cup$$
$$\{\, \mathcal{H}(t_1, M) \not\approx \mathcal{H}(t_2, M) \mid t_1 \not\approx t_2 \in E \,\} \cup \{\, \mathsf{isC}(t) \mid t \mapsto C \in M \,\}$$

$\mathcal{H}(t, M) = \text{match } t \text{ with}$

$\quad x \to x$

$\quad C(t_1, \ldots, t_n) \to C(\mathcal{H}(t_1, M), \ldots, \mathcal{H}(t_n, M))$

$\quad S_\delta^{\tau,k}(t_1) \to S_\delta^{\tau,k}(\mathcal{H}(t_1, M))$

$\quad S_\delta^{C,k}(t_1) \to \begin{cases} S_\delta^{\tau,\mathsf{atos}(\tau,\,C,\,k)}(\mathcal{H}(t_1, M)) & \text{if } M(t_1) = C \\ S_\delta^{\tau,\mathsf{err}(C,k)}(\mathcal{H}(t_1, M)) & \text{otherwise} \end{cases}$

$\quad\quad \text{where } S_\delta^{C,k} : \delta \to \tau$

Fig. 1. Definition of $\mathcal{H}(t, M)$

In other words, for each (dis)equality, we include the corresponding constraint where the transformation is applied to both its terms. We add to this set an application of the discriminator for C to t for each $t \mapsto C$ in the mapping M.

Example 4. Consider again the **Tree** datatype from Example 1. Let:

$$E = \{x \approx N_1(2, y, S^{N_1,\,2}(x)), S^{N_1,\,1}(x) \approx 2, S^{L,\,2}(x) \not\approx 0\} \text{ and } M = \{x \mapsto N_1, y \mapsto L\}$$

Then, $\mathcal{H}(E, M)$ is the set:

$$\{x \approx N_1(2, y, S^{\mathsf{Tree},\,1}(x)), S^{\mathsf{Int},\,1}(x) \approx 2\} \cup \{S^{\mathsf{Int},\,\mathsf{err}(L,\,2)}(x) \not\approx 0\} \cup \{\mathsf{isN}_1(x), \mathsf{isL}(y)\}$$

Since M maps x to N_1, the standard selector application $S^{N_1,\,2}(x)$ is converted to the shared selector application $S^{\mathsf{Tree},\,1}(x)$, whereas $S^{L,\,2}(x)$ is converted to $S^{\mathsf{Int},\,\mathsf{err}(L,\,2)}(x)$. $\bullet$

The following theorem states the key property of the transformation $\mathcal{H}$, namely that a set of arbitrary Σ-constraints E is satisfiable if and only if there exists some mapping M for which $\mathcal{H}(E, M)$ is satisfiable. The full proof of this statement is available in an extended version of this paper [26].

Theorem 1. *E is $\mathcal{D}$-satisfiable iff $\mathcal{H}(E, M)$ is $\mathcal{D}$-satisfiable for some M.*

Proof (Sketch). We split the statement into its two implications. The proof relies on the construction of a mapping M from a model of E.

"$\Rightarrow$": If E is satisfied by some Σ-model $\mathcal{I}$ of $\mathcal{D}$, there exists a mapping $M_\mathcal{I}$ and Σ-model $\mathcal{J}$ of $\mathcal{D}$ such that $\mathcal{H}(E, M_\mathcal{I})$ is satisfied by $\mathcal{J}$. We show this by a particular construction for $M_\mathcal{I}$ and $\mathcal{J}$. Let the mapping $M_\mathcal{I}$ be $\{t \mapsto C \mid \mathcal{I} \models \mathsf{isC}(t), t \in \mathbf{T}(E)\}$. Construct $\mathcal{J}$ as follows. First, all types τ and constructors are interpreted by $\mathcal{J}$ the same way as in $\mathcal{I}$. Furthermore, we interpret all variables and standard selectors in $\mathcal{J}$ the same as in $\mathcal{I}$. It remains to state how shared selectors are interpreted in $\mathcal{J}$. Notice that our transformation generates shared selectors of the form $S_\delta^{\tau,\,\mathsf{err}(C,\,k)}$. We distinguish these in the following construction.

$$S_\delta^{\tau,\,\mathsf{err}(C,\,k)\mathcal{J}} = S_\delta^{C,\,k\mathcal{I}}, \text{ and } S_\delta^{\tau,\,k\mathcal{J}} = S_\delta^{\tau,\,k\mathcal{I}} \text{ for all other shared selectors.}$$

The above construction is well-defined due to our definition of err. In particular, $\mathsf{err}(\mathsf{C}, k)$ is defined uniquely for each (constructor, natural number) pair. By this construction, it can be shown that $\mathcal{H}(t, M_{\mathcal{I}})^{\mathcal{J}} = t^{\mathcal{I}}$ by structural induction on t for all $t \in \mathbf{T}(E)$. Since $\mathcal{I}$ satisfies E and since $\mathcal{H}(t, M_{\mathcal{I}})^{\mathcal{J}} = t^{\mathcal{I}}$ for all terms $t \in \mathbf{T}(E)$, we have that $\mathcal{J}$ satisfies the equalities and disequalities in $\mathcal{H}(E, M_{\mathcal{I}})$ of the form $(\neg)\mathcal{H}(t_1, M_{\mathcal{I}}) \approx \mathcal{H}(t_2, M_{\mathcal{I}})$. By construction of $M_{\mathcal{I}}$, we have that $\mathcal{J}$ satisfies the constraints in $\mathcal{H}(E, M_{\mathcal{I}})$ of the form $\mathsf{isC}(t)$ where $t \mapsto \mathsf{C} \in M_{\mathcal{I}}$. Hence, $\mathcal{J}$ satisfies $\mathcal{H}(E, M_{\mathcal{I}})$.

"$\Leftarrow$": If $\mathcal{H}(E, M)$ is satisfied by some Σ-model $\mathcal{J}$ of $\mathcal{D}$ for some mapping M, then E is satisfied by some Σ-model $\mathcal{I}$ of $\mathcal{D}$. We show this by constructing $\mathcal{I}$ as follows. First, all types, constructors, variables and have the same interpretation in $\mathcal{I}$ as in $\mathcal{J}$. Furthermore, all shared selectors have the same interpretation in $\mathcal{I}$ as in $\mathcal{J}$. We interpret standard selectors in $\mathcal{I}$ as follows.

$$S_{\delta}^{\mathsf{C}, k}(t)^{\mathcal{I}} = \begin{cases} S_{\delta}^{\tau, \mathsf{atos}(\tau, \mathsf{C}, k)}(t)^{\mathcal{I}} & \text{if } M(t) = \mathsf{C} \\ S_{\delta}^{\tau, \mathsf{err}(\mathsf{C}, k)}(t)^{\mathcal{I}} & \text{otherwise} \end{cases}$$

Similar to the first part, it can be shown that $t^{\mathcal{I}} = \mathcal{H}(t, M)^{\mathcal{J}}$ by structural induction on t for all $t \in \mathbf{T}(E)$. Since $\mathcal{J}$ satisfies $\mathcal{H}(E, M)$ and $t^{\mathcal{I}} = \mathcal{H}(t, M)^{\mathcal{J}}$ for all $t \in \mathbf{T}(E)$, we have that $\mathcal{I}$ satisfies the equalities and disequalities in $\mathcal{H}(E, M_{\mathcal{I}})$ of the form $(\neg)\mathcal{H}(t_1, M) \approx \mathcal{H}(t_2, M)$. Furthermore, since $\mathcal{J}$ satisfies the constraints $\mathsf{isC}(t)$ for all $t \mapsto \mathsf{C} \in M$ and since $t^{\mathcal{I}} = t^{\mathcal{J}}$, we have that $\mathcal{I}$ satisfies these constraints as well. Thus, $\mathcal{I}$ satisfies $\mathcal{H}(E, M)$. $\square$

Corollary 1. *For some index sets I and J, and set E of Σ-literals without standard selectors, let*

$$E_0 = E \cup \{\, S^{\mathsf{C}_{j_i}, k_i}(x_i) \approx y_i \mid i \in I, j \in J \,\} \qquad \text{and}$$
$$E_1 = E \wedge \{\, \mathsf{ite}(\mathsf{isC}_{j_i}(x_i), S^{\tau, \mathsf{atos}(\tau, \mathsf{C}_{j_i}, k_i)}(x_i), S^{\tau, \mathsf{err}(\mathsf{C}_{j_i}, k_i)}(x_i)) \approx y_i \mid i \in I, j \in J \,\} \,.$$

The sets E_0 and E_1 are equisatisfiable in $\mathcal{D}$.

Using this corollary, we can reduce (possibly after some literal flattening) the satisfiability of an arbitrary set of Σ-constraints E_0 to a set of Σ-constraints E_1 not containing standard selectors. In particular, our implementation in CVC4 replaces each application of the form $S^{\mathsf{C}_{j_i}, k_i}(x_i)$ by the term $\mathsf{ite}(\mathsf{isC}_{j_i}(x_i),$ $S^{\tau, \mathsf{atos}(\tau, \mathsf{C}_{j_i}, k_i)}(x_i),$ $S^{\tau, \mathsf{err}(\mathsf{C}_{j_i}, k_i)}(x_i))$ during a preprocessing pass on the input formula.

4   Decision Procedure for Datatypes with Shared Selectors

This section describes a tableau-like calculus for deciding constraint satisfiability in $\mathcal{D}$, with constraint variables interpreted existentially. The calculus is parametrized by the theory's signature Σ. By the results of the previous section, we can restrict with no loss of generality the input language to sets of equalities

$$\frac{t \approx u \in E^* \quad t \not\approx u \in E}{\bot} \text{ CONFLICT}$$

$$\frac{C_1(\bar{t}) \approx C_1(\bar{u}) \in E^*}{E := E, \bar{t} \approx \bar{u}} \text{ DECOMPOSE} \qquad \frac{C_1(\bar{t}) \approx C_2(\bar{u}) \in E^* \quad C_1 \neq C_2}{\bot} \text{ CLASH}$$

$$\frac{C_n(\bar{u}_n u \bar{v}_n) \approx u_{n-1}, \ldots, C_2(\bar{u}_2 u_2 \bar{v}_2) \approx u_1, C_1(\bar{u}_1 u_1 \bar{v}_1) \approx u \in E^* \quad n \geq 1}{\bot} \text{ CYCLE}$$

$$\frac{S_\delta^{\tau,n}(t) \in T(E) \text{ or } \delta \text{ is finite}}{\begin{array}{l} E := E, t \approx C_1(S_\delta^{\tau_{1,1}, \text{atos}(\tau_{1,1}, C_1, 1)}(t), \ldots, S_\delta^{\tau_{1,n_1}, \text{atos}(\tau_{1,n_1}, C_1, n_1)}(t)) \\ \quad\vdots \\ E := E, t \approx C_m(S_\delta^{\tau_{m,1}1, \text{atos}(\tau_{m,1}, C_m, 1)}(t), \ldots, S_\delta^{\tau_{m,n_m}, \text{atos}(\tau_{m,n_m}, C_m, n_m)}(t)) \end{array}} \text{ SPLIT}$$

where δ has constructors $C_1, \ldots, C_m$ and $C_i : \tau_{1,i} \times \cdots \times \tau_{i,n_i} \to \delta, 1 \leq i \leq m$

Fig. 2. Derivation rules.

and disequalities between Σ-terms with no standard selectors and no discriminators. Since our calculus is based on similar calculi for datatypes that have been presented in detail in previous work [10,22], we focus on our modifications to accommodate shared selectors.

The derivation rules of the calculus operate on a current set E of constraints as specified in Fig. 2. A derivation rule can be applied to E if its premises are met. Some of those premises check membership in the *congruence closure* E^* of E, the smallest superset of E that is closed under entailment in the theory of equality.[2] A rule's conclusion either modifies E or replaces it by $\bot$ to indicate unsatisfiability. There, the notation $E, t \approx s$ abbreviates $E \cup \{t \approx s\}$; the notation $\bar{t} \approx \bar{u}$ stands for the set of equalities between the corresponding elements of $\bar{t}$ and $\bar{u}$. The SPLIT rule has multiple alternative conclusions, denoting branching.

A rule application is *redundant* if (one of) its conclusion(s) leaves E unchanged. The rules are applied to build a *derivation tree*, i.e., a tree whose nodes are finite sets of (dis)equalities, with an *initial constraint set* E_0 as its root and child nodes obtained by a non-redundant rule application to their parent. We say that E_0 *has* a derivation tree D if D is a derivation tree with root E_0. A node is *saturated* if it admits only redundant rule applications. A derivation tree is *closed* if all of its leaf nodes are $\bot$. Intuitively, a derivation tree is generated progressively from E_0 by applying a derivation rule to a leaf node. The rules are applied until the derivation tree becomes closed (indicating that the initial set E_0 is $\mathcal{D}$-unsat) or contains a saturated leaf node (indicating that E_0 is $\mathcal{D}$-sat).

---

[2] Such tests are effective by well-known results about the theory of equality [6].

In the calculus, all reasoning based on the general properties of equality is encapsulated in the rule CONFLICT, which detects that congruent terms are forced to be distinct. The remaining rules perform datatype reasoning proper, with DECOMPOSE computing a downward equality closure based on the injectivity of constructors and CLASH detecting failures based on their distinctness. The CYCLE rule recognizes when a constructor term must be equivalent to one of its subterms, which is forbidden in all models of the theory.

The calculus also incrementally unrolls terms by branching on different constructors, with the SPLIT rule performing case distinctions on constructors for various terms occurring in E. The main modification from the previous calculi for the theory of datatypes is that this SPLIT rule operates on shared selectors. Its application can be seen as an on-the-fly transformation from standard to shared selectors as described in Sect. 3.2. Indeed, for each constructor C_i in its conclusion, the following holds with a mapping M such that $M(t) = \mathsf{C}_i$:

$$\mathsf{S}_\delta^{\tau_1,\,\mathsf{atos}(\tau_1,\,\mathsf{C}_i,\,1)}(t) = \mathcal{H}(\mathsf{S}_\delta^{\mathsf{C}_i,\,1}(t),\,M),\ \ldots,\ \mathsf{S}_\delta^{\tau_{n_i},\,\mathsf{atos}(\tau_{n_i},\,\mathsf{C}_i,\,n_i)}(t) = \mathcal{H}(\mathsf{S}_\delta^{\mathsf{C}_i,\,n}(t),\,M)$$

Any derivation strategy for the calculus that does not stop until it generates a closed tree or a saturated node yields a decision procedure for the $\mathcal{D}$-satisfiability of sets of Σ-literals. We prove this similarly to previous work [10,22], but using shared selectors and in the simpler setting obtained by assuming the availability of a congruence closure procedure. The full proofs are available in an extended version of this paper [26].

Proposition 1 (Termination). *All derivation trees in the calculus are finite.*

Proposition 2 (Refutation Soundness). *If a constraint set E_0 has a closed derivation tree, then it is $\mathcal{D}$-unsatisfiable.*

Proposition 3 (Solution Soundness). *If a constraint set E_0 has a derivation tree with a saturated node, then it is $\mathcal{D}$-satisfiable.*

Theorem 2. *Constraint satisfiability in the theory $\mathcal{D}$ of datatypes with (standard and) shared selectors is decidable.*

5   Using Shared Selectors for Syntax-Guided Synthesis

In this section, we show how the theory of datatypes with shared selectors can substantially improve the performance of an approach by Reynolds et al. [24] for performing *syntax-guided synthesis* (SyGuS) [1] directly within an SMT solver.

Syntax-guided synthesis is the problem of automatically synthesizing a function that satisfies a given specification, but with the addition of explicit syntactic restrictions on the solution space. These restrictions specify that the function must be built with selected operators over basic types (such as arithmetic and Boolean operators) and belong to the language generated by a given grammar. Grammars allow users to specify formally a set of candidates for the desired function, thus reducing the search effort of a SyGuS solver.

More technically, a syntax-guided synthesis problem for a function f in a background theory T of the basic types consists of:

1. a set of semantic restrictions, or specification, given by a (second-order) T-formula of the form $\exists f.\, \forall \bar{x}.\, \varphi[f, \bar{x}]$, and
2. a set of syntactic restrictions on the solutions for f, given by a grammar R.

A solution for f is a lambda term $\lambda \bar{y}.\, e$ of the same type as f, such that (i) $\forall \bar{x}.\, \varphi[\lambda \bar{y}.\, e, \bar{x}]$ is valid in T (modulo beta-reductions) and (ii) e is in the language generated by R.

CVC4 incorporates a SyGuS solver that automatically encodes the solution space of a SyGuS problem as a set of algebraic datatypes mirroring the problem's syntactic restrictions [24]. A deep embedding of the datatypes in the problem's background theory T, realized as a set of automatically generated axioms, provides a semantics for datatype values in terms of the semantic values in T.

Example 5. Consider the problem of synthesizing a binary function f over the integers such that f is commutative (i.e., $\exists f\, \forall xy.\, f(x, y) \approx f(y, x)$), and with the solution space for f defined by a context-free grammar R with start symbol A and production rules:

$$A \rightarrow x \mid y \mid 0 \mid 1 \mid A + A \mid A - A \mid \mathsf{ite}(B, A, A) \qquad\qquad B \rightarrow A \geq A \mid A \approx A \mid \neg B$$

The following mutually recursive datatypes capture the grammar R. The datatypes themselves correspond to R's non-terminals (e.g., **a** corresponds to A), their constructors correspond to production rules (e.g., X corresponds to $A \rightarrow x$):

$$\mathbf{a} = \mathsf{X} \mid \mathsf{Y} \mid \mathsf{Zero} \mid \mathsf{One} \mid \mathsf{Plus}(\mathbf{a}, \mathbf{a}) \mid \mathsf{Minus}(\mathbf{a}, \mathbf{a}) \mid \mathsf{Ite}(\mathbf{b}, \mathbf{a}, \mathbf{a})$$
$$\mathbf{b} = \mathsf{Geq}(\mathbf{a}, \mathbf{a}) \mid \mathsf{Eq}(\mathbf{a}, \mathbf{a}) \mid \mathsf{Neg}(\mathbf{b})$$

Datatypes like the ones above are associated with the programs they represent through *evaluation functions* that map datatype values, expressed as variable-free constructor terms, to expressions over the basic types. For example, the evaluation function for **a** is denoted by a function symbol $\mathsf{eval_a} : \mathbf{a} \times \mathsf{Int} \times \mathsf{Int} \rightarrow \mathsf{Int}$, and the specific term $\mathsf{eval_a}(\mathsf{Plus}(\mathsf{X}, \mathsf{X}), 2, 3)$ is interpreted as $(x + x)\{x \mapsto 2, y \mapsto 3\} = 2 + 2 = 4$. The evaluation functions are defined axiomatically by a set of quantified formulas that, in this case, can be handled by any SMT solver that, like CVC4, supports the combined theory of datatypes, linear arithmetic, and uninterpreted functions. The SyGuS problem for f in this example can then be stated as the *first-order* formula:

$$\forall xy.\, \mathsf{eval_a}(d, x, y) \approx \mathsf{eval_a}(d, y, x) \tag{3}$$

where d is a fresh constant of type **a**. This formula has models in which d is interpreted as Zero or Plus(X, Y), which correspond to solutions $f = \lambda xy.\, 0$ and $f = \lambda xy.\, x + y$ for the original problem, respectively.[3]   •

---

[3] For a thorough description of this approach, see [24].

Since CVC4 is a DPLL(T)-based solver [11], for a problem like the one in the example above, it will find a possible solution for d by first guessing its top constructor symbol with an application of the SPLIT rule from Fig. 2. The effect of the rule is achieved in practice with the generation of *splitting lemmas* such as the following, which we write here with discriminators and standard selectors for simplicity:

$$\mathsf{isX}(d) \vee \mathsf{isY}(d) \vee \cdots \vee \mathsf{isIte}(d) \tag{4}$$

$$\mathsf{isX}(\mathsf{S}^{\mathsf{Plus},\,1}(d)) \vee \mathsf{isY}(\mathsf{S}^{\mathsf{Plus},\,1}(d)) \vee \cdots \vee \mathsf{isIte}(\mathsf{S}^{\mathsf{Plus},\,1}(d)) \tag{5}$$

$$\mathsf{isGeq}(\mathsf{S}^{\mathsf{Ite},\,1}(d)) \vee \mathsf{isEq}(\mathsf{S}^{\mathsf{Ite},\,1}(d)) \vee \mathsf{isNeg}(\mathsf{S}^{\mathsf{Ite},\,1}(d)) \tag{6}$$

$$\mathsf{isX}(\mathsf{S}^{\mathsf{Ite},\,2}(d)) \vee \mathsf{isY}(\mathsf{S}^{\mathsf{Ite},\,2}(d)) \vee \cdots \vee \mathsf{isIte}(\mathsf{S}^{\mathsf{Ite},\,2}(d)) \tag{7}$$

The solver will subsequently guess the top constructor for other subterms of d's value. These guesses are represented symbolically by *selector chains*, i.e. zero or more applications of selectors to d; for example, $\mathsf{S}^{\mathsf{Plus},\,1}(d)$ is a selector chain that corresponds to the first *child* of d (if we think of the value of d as a tree) when d is an application of Plus; $\mathsf{S}^{\mathsf{Plus},\,1}(\mathsf{S}^{\mathsf{Plus},\,1}(d))$ is a selector chain that corresponds to the first child of the first child of d when d and its first child are both applications of Plus; and so on.

The bottleneck in solving (3) is the large number of splitting lemmas for selector chains introduced during search which, depending on the datatypes involved, is often highly exponential. Our key observation is that datatypes generated by the SyGuS approach sketched above very often include constructors with arguments of the same type. In Example 5, both **a** and **b** have multiple constructors with arguments of type **a**. Using shared selectors, we can reduce the number of selectors in the example from 7 to 3 for **a** and from 5 to 3 for **b**. Moreover, *using shared selectors in selector chains makes splitting lemmas relevant in multiple contexts*. For example, a splitting lemma for a selector chain $\mathsf{S}^{\mathsf{a},\,1}(d)$ is relevant when d is either Plus, Minus or Ite; likewise $\mathsf{S}^{\mathsf{a},\,1}(\mathsf{S}^{\mathsf{a},\,1}(d))$ is relevant when d *and* its first child of type **a** are applications of either Plus, Minus or Ite. Notice that by using the decision procedure for shared selectors from Sect. 4, lemmas (5) and (7) would be instead both provided to the SAT engine as:

$$\mathsf{isX}(\mathsf{S}^{\mathsf{Int},\,1}(d)) \vee \mathsf{isY}(\mathsf{S}^{\mathsf{Int},\,1}(d)) \vee \cdots \vee \mathsf{isIte}(\mathsf{S}^{\mathsf{Int},\,1}(d))$$

Using shared selectors can lead to a reduction in the number of other kinds of lemmas as well. For instance, during synthesis CVC4 implements *symmetry breaking* techniques to avoid spending time on multiple candidates that are all equivalent in T [24,25]. Redundant candidates are avoided by adding blocking clauses to the SAT engine that are also expressed in terms of discriminators applied to selector chains.

Example 6. Consider again the function f, grammar R, and datatypes **a** and **b** from Example 5. Assume that the solver considers X as a candidate solution for d, and later considers another candidate solution, Plus(X, Zero). Since the corresponding arithmetic terms x and $x + 0$ are equivalent in integer arithmetic, the solver infers a *lemma template* of the form:

$$\neg\mathsf{isPlus}(z) \lor \neg\mathsf{isX}(\mathsf{S}^{\mathsf{Int},\,1}(z)) \lor \neg\mathsf{isZero}(\mathsf{S}^{\mathsf{Int},\,2}(z))$$

to block a redundant candidate solution like (the one corresponding to) $x + 0$. This is achieved by instantiating the template with the substitution $\{z \mapsto d\}$ for variable z. More interestingly, z can be instantiated with other selector chains to rule out *entire families* of redundant candidate solutions. For instance, the lemma obtained with $\{z \mapsto \mathsf{S}^{\mathsf{Int},\,1}(d)\}$ rules out all terms that have $x + 0$ as their first child of type **a**, such as the terms $(x + 0) + y$, $\mathsf{ite}(x \geq y, x + 0, y)$ and $(x + 0) - 1$, which are equivalent to the smaller expressions $x + y$, $\mathsf{ite}(x \geq y, x, y)$ and $x - 1$, respectively, and hence redundant as candidate solutions. Sharing selectors allows the same blocking clause to be reused for the different constructors, whereas standard selectors would require three different clauses in this case, with $z \mapsto \mathsf{S}^{\mathsf{Plus},\,1}(d)$, $z \mapsto \mathsf{S}^{\mathsf{Ite},\,2}(d)$, and $z \mapsto \mathsf{S}^{\mathsf{Minus},\,1}(d)$, respectively. ●

A majority of SyGuS problems can be encoded as datatypes that have significant sharing of selectors across multiple constructors, thus making the use of shared selectors particularly effective in this domain. The next section measures the impact of shared selections when solving SyGuS problems in CVC4.

6   Experiments

We implemented our calculus for the theory of datatypes with shared selectors in CVC4 Version 1.5, together with a preprocessing pass to convert standard selectors in input formulas to shared ones and other modifications to the existing decision procedure for datatypes, as described in Sects. 3.2 and 4. We discuss here our evaluation of two configurations of CVC4, one with and one without support for shared selectors, on two different sets of benchmarks: the SyGuS benchmark suite from the 2017 SyGuS competition [4]; and SMT-LIB [8] benchmarks containing datatypes. Our experiments[4] were performed on the StarExec logic solving service [28].

6.1   Syntax-Guided Synthesis Benchmarks

The benchmarks from the 2017 SyGuS competition are divided into five families across four tracks: (*i*) the General track, with problems over the theories of linear integer arithmetic (LIA) or bit-vectors; (*ii*) the conditional linear integer arithmetic track (CLIA), with problems over LIA; (*iii*) the Invariant synthesis track, also over LIA; and (*iv*) the Programming-by-examples track [17,18], with a family over bit-vectors and another over strings. We measured the impact

---

[4] The data and details on how to reproduce our results are available at https://cvc4.cs.stanford.edu/papers/IJCAR2018-shsel/.

Family	#	Solved: sh / std	Time	SAT Decs	Terms	Sels
General	535	319 / 235 (232)	15.4 / 144.9	67k / 151k	189k / 284k	5.8 / 16.8
CLIA	73	18 / 17 (17)	25.1 / 142.4	158k / 405k	25k / 60k	9.6 / 22.2
Invariant	67	46 / 46 (46)	49.1 / 114.6	374k / 896k	37k / 61k	5.7 / 13.1
PBE_BV	750	665 / 253 (253)	27.4 / 211.9	54k / 3873k	14k / 202k	3.0 / 16.0
PBE_Strings	108	93 / 64 (64)	13.3 / 39.9	90k / 334k	14k / 41k	8.6 / 18.7

Fig. 3. Performance of CVC4 on benchmarks from five families of SyGuS Comp 2017.

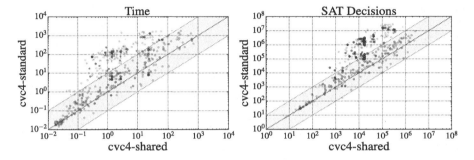

Fig. 4. Impact of shared selectors on solving time and number of SAT decisions.

of shared selectors by comparing for the two configurations of CVC4 the total number of solved problems and the average solving time, number of decisions performed by the SAT engine, quantifier-free terms generated, and number of selectors in the signature. Averages were computed over the set of problems solved by both configurations. We used a timeout of 30 min per benchmark.

A summary of the results is given in Fig. 3. The first two columns show the evaluated family and the number of benchmarks in it, while the other columns present the statistics listed above, with average times expressed in seconds. The number of problems solved by both configurations is given in parentheses in the third column. The results clearly show that sharing selectors reduces the number of selectors in the signature, which generally leads to fewer terms and SAT decisions, with a positive impact on solving speed and number of problems solved. Except for the invariant family, the CVC4 configuration with shared selectors solves more problems than the one without. The impact of shared selectors is particularly significant for the bit-vector benchmark suite (PBE_BV), with a reduction of over 80% in the average number of selectors. In that case, CVC4 is over eight times faster with shared selectors than without, solving 412 more problems, thus reducing the percentage of unsolved problems in this category from over 65% to less than 12%. Significant improvements can also be observed in the PBE_Strings and General families, with the percentages of unsolved problems being reduced from over 40% to almost 13% and from over 55% to almost 40%, respectively.

Family	#	Solved: sh / std	Time	Decs	Terms	Sels
Leon	410	179 / 175 (175)	0.96 / 0.75	9920 / 9925	718 / 929	8.67 / 23.10
Sledgehammer	321	113 / 112 (112)	0.47 / 0.47	6949 / 6942	185 / 185	10.50 / 12.76
Nunchaku	158	67 / 67 (67)	0.49 / 0.44	7149 / 6653	1373 / 1297	6.22 / 7.22

Fig. 5. Performance of CVC4 on benchmarks from three families of SMT LIB.

We present a per-problem comparison in the scatter plots of Fig. 4, which clearly shows that for the vast majority of the benchmarks, sharing selectors reduces the number of SAT decisions and improves the solving time, often by orders of magnitude.

Comparison Against Other SyGuS Solvers. We also compared CVC4's performance with the state-of-the-art SyGuS solver EUSOLVER [2,5]. For fairness, in this comparison we combine the results of the above configurations of CVC4 with its other approach for solving single-invocation synthesis problems (see [24] for details), which impacts the CLIA and General families of benchmarks. We obtained the following results for the problems solved by EUSOLVER and CVC4 with and without shared selectors: 71/73/73 for CLIA, 404/391/334 for General, 42/46/46 for Invariant, 739/665/253 for PBE_BV, and 68/93/64 for PBE_Strings. These numbers show that overall CVC4 is significantly more competitive with shared selectors than without, surpassing EUSOLVER's performance in three of the five families.

6.2   Datatype Benchmarks from SMT LIB

We also considered all SMT-LIB benchmarks containing datatypes. Among these, we excluded from consideration 14, 387 benchmarks that do not have any shareable selectors, as CVC4 with and without shared selectors perform the same on these benchmarks. The remaining 889 benchmarks are divided into three families: (*i*) the Leon set contains benchmarks generated by Leon [12] for verification of Scala programs (AUFBVDTLIA logic); (*ii*) the Sledgehammer set has benchmarks from Isabelle [21] generated by Sledgehammer [14] (UFDT logic); and (*iii*) the Nunchaku set has benchmarks generated for higher order theorem provers by Nunchaku [23] (UFDT logic).

We summarize our results over the two configurations of CVC4, with and without shared selectors, in Fig. 5, following the same schema as in Fig. 3. We used a timeout of 60 s, since in this setting we evaluate SMT solvers as backends of verification and ITP tools, which require fast answers. The configuration with shared selectors solved at least all the benchmarks as the one without. The Leon benchmark set shows the most significant impact of sharing selectors, with a reduction of over 60% in the average number of selectors, and 4 more problems solved.

Comparison Against Other SMT Solvers. To put the shared selector version of CVC4 in context with the state of the art, we also compared it with the only two provers that can reason about datatypes and support the SMT-LIB format: z3 [16] and Vampire [19]. On the Nunchaku and Sledgehammer benchmarks, the number of problems solved by CVC4/z3/Vampire is 67/29/30 and 113/119/138, respectively. The comparison on the Leon set excludes Vampire, since it does not support the theory of bit-vectors; the split between CVC4 and z3 is 179/173 on that set. The results show that CVC4 compares favorably with the other tools.

7   Related Work

The motivation of our work is to reduce the number of terms considered by a decision procedure for the theory of algebraic of datatypes, based on procedures introduced in previous work [10,22]. Thus, our contributions apply to other systems that handle datatypes semantically, such as SMBC [15] and the SMT solver z3 [16]. On systems that reason about datatypes axiomatically, such as the first-order theorem prover and SMT solver Vampire [19], and the higher-order systems Isabelle [21] and Dafny [20], whether to share selectors and how to handle them is simply a matter of axiomatizing the datatypes theory accordingly. For example, the axiomatization in Vampire avoids selectors altogether [19, Sect. 4.3], while in Isabelle users are encouraged to write specifications directly with shared selectors [13, Sect. 3].

Most SyGuS solvers employ a variation of counter-example guided inductive synthesis (CEGIS), introduced by Solar-Lezama [27]. While CVC4 benefits from sharing selectors by representing syntax restrictions with datatypes, other systems use an outer layer with an underlying reasoning engine, for instance using an SMT solver to verify the correctness of candidate solutions, but not for performing the enumerative search [5].

8   Conclusion

We have presented an extension of the theory of algebraic datatypes that adds shared selectors. We have discussed and proved correct a calculus for deciding the constraint satisfiability problem in the new theory. Moreover, we have described how algebraic datatypes can be leveraged in an SMT solver to solve syntax-guided synthesis problems and explained how the use of shared selectors in this setting can lead to significant performance gains. Our experiments demonstrate that an implementation of the new calculus in the CVC4 solver significantly enhances its performance on syntax-guided synthesis problems and is responsible for making CVC4 the best known solver for certain classes of problems.

In future work, we plan to generalize our approach so that distinct selector *chains* can be compressed to a single application of the same selector symbol. This requires more sophisticated criteria for recognizing when two selector chains for a datatype cannot be simultaneously constrained for arbitrary values of that datatype. We believe that this further extension can be done in a manner similar to the one presented here and expect that this will lead to further performance improvements.

References

1. Alur, R., Bodík, R., Juniwal, G., Martin, M.M.K., Raghothaman, M., Seshia, S.A., Singh, R., Solar-Lezama, A., Torlak, E., Udupa, A.: Syntax-guided synthesis. In: FMCAD, pp. 1–8. IEEE (2013)
2. Alur, R., Černý, P., Radhakrishna, A.: Synthesis through unification. In: Kroening, D., Păsăreanu, C.S. (eds.) CAV 2015. LNCS, vol. 9207, pp. 163–179. Springer, Cham (2015). https://doi.org/10.1007/978-3-319-21668-3_10
3. Alur, R., Fisman, D., Singh, R., Solar-Lezama, A.: SyGuS-comp 2017: results and analysis. In: Fisman, D., Jacobs, S. (eds.) Proceedings Sixth Workshop on Synthesis (SYNT). EPTCS, vol. 260, pp. 97–115 (2017)
4. Alur, R., Fisman, D., Singh, R., Solar-Lezama, A.: SyGuS-comp 2017: Results and analysis. CoRR, abs/1711.11438 (2017)
5. Alur, R., Radhakrishna, A., Udupa, A.: Scaling enumerative program synthesis via divide and conquer. In: Legay, A., Margaria, T. (eds.) TACAS 2017. LNCS, vol. 10205, pp. 319–336. Springer, Heidelberg (2017). https://doi.org/10.1007/978-3-662-54577-5_18
6. Baader, F., Nipkow, T.: Term Rewriting and All That. Cambridge University Press, New York (1998)
7. Barrett, C., Conway, C.L., Deters, M., Hadarean, L., Jovanović, D., King, T., Reynolds, A., Tinelli, C.: CVC4. In: Gopalakrishnan, G., Qadeer, S. (eds.) CAV 2011. LNCS, vol. 6806, pp. 171–177. Springer, Heidelberg (2011). https://doi.org/10.1007/978-3-642-22110-1_14
8. Barrett, C., Fontaine, P., Tinelli, C.: The Satisfiability Modulo Theories Library (SMT-LIB) (2016). www.SMT-LIB.org
9. Barrett, C., Fontaine, P., Tinelli, C.: The SMT-LIB Standard: Version 2.6. Technical report, Department of Computer Science, The University of Iowa (2017)
10. Barrett, C., Shikanian, I., Tinelli, C.: An abstract decision procedure for a theory of inductive data types. JSAT 3(1–2), 21–46 (2007)
11. Barrett, C., Tinelli, C.: Satisfiability modulo theories. In: Clarke, E., Henzinger, T., Veith, H., Bloem, R. (eds.) Handbook of Model Checking, pp. 305–343. Springer, Cham (2018). https://doi.org/10.1007/978-3-319-10575-8_11
12. Blanc, R., Kuncak, V., Kneuss, E., Suter, P.: An overview of the Leon verification system: verification by translation to recursive functions. In: Proceedings of the 4th Workshop on Scala, pp. 1:1–1:10. ACM (2013)
13. Blanchette, J.C., Hölzl, J., Lochbihler, A., Panny, L., Popescu, A., Traytel, D.: Truly modular (co)datatypes for Isabelle/HOL. In: Klein, G., Gamboa, R. (eds.) ITP 2014. LNCS, vol. 8558, pp. 93–110. Springer, Cham (2014). https://doi.org/10.1007/978-3-319-08970-6_7
14. Böhme, S., Nipkow, T.: Sledgehammer: judgement day. In: Giesl, J., Hähnle, R. (eds.) IJCAR 2010. LNCS (LNAI), vol. 6173, pp. 107–121. Springer, Heidelberg (2010). https://doi.org/10.1007/978-3-642-14203-1_9
15. Cruanes, S.: Satisfiability modulo bounded checking. In: de Moura, L. (ed.) CADE 2017. LNCS (LNAI), vol. 10395, pp. 114–129. Springer, Cham (2017). https://doi.org/10.1007/978-3-319-63046-5_8
16. de Moura, L., Bjørner, N.: Z3: an efficient SMT solver. In: Ramakrishnan, C.R., Rehof, J. (eds.) TACAS 2008. LNCS, vol. 4963, pp. 337–340. Springer, Heidelberg (2008). https://doi.org/10.1007/978-3-540-78800-3_24
17. Gulwani, S.: Automating string processing in spreadsheets using input-output examples. In: POPL, pp. 317–330. ACM (2011)

18. Gulwani, S.: Programming by examples: applications, algorithms, and ambiguity resolution. In: Olivetti, N., Tiwari, A. (eds.) IJCAR 2016. LNCS (LNAI), vol. 9706, pp. 9–14. Springer, Cham (2016). https://doi.org/10.1007/978-3-319-40229-1_2
19. Kovács, L., Robillard, S., Voronkov, A.: Coming to terms with quantified reasoning. In: POPL, pp. 260–270. ACM (2017)
20. Leino, K.R.M.: Developing verified programs with Dafny. Ada Lett. **32**(3), 9–10 (2012)
21. Nipkow, T., Wenzel, M., Paulson, L.C. (eds.): Isabelle/HOL. LNCS, vol. 2283. Springer, Heidelberg (2002). https://doi.org/10.1007/3-540-45949-9
22. Reynolds, A., Blanchette, J.C.: A decision procedure for (co)datatypes in SMT solvers. J. Autom. Reason. **58**(3), 341–362 (2017)
23. Reynolds, A., Blanchette, J.C., Cruanes, S., Tinelli, C.: Model finding for recursive functions in SMT. In: Olivetti, N., Tiwari, A. (eds.) IJCAR 2016. LNCS (LNAI), vol. 9706, pp. 133–151. Springer, Cham (2016). https://doi.org/10.1007/978-3-319-40229-1_10
24. Reynolds, A., Kuncak, V., Tinelli, C., Barrett, C., Deters, M.: Refutation-based synthesis in SMT. Form. Methods Syst. Des. (2017)
25. Reynolds, A., Tinelli, C.: SyGuS techniques in the core of an SMT solver. arXiv preprint arXiv:1711.10641 (2017)
26. Reynolds, A., Viswanathan, A., Barbosa, H., Tinelli, C., Barrett, C.: Datatypes with shared selectors. Technical report, The University of Iowa (2018). http://cvc4.cs.stanford.edu/papers/IJCAR2018-shsel/
27. Solar-Lezama, A., Tancau, L., Bodík, R., Seshia, S.A., Saraswat, V.A.: Combinatorial sketching for finite programs. In: ASPLOS, pp. 404–415. ACM (2006)
28. Stump, A., Sutcliffe, G., Tinelli, C.: StarExec: a cross-community infrastructure for logic solving. In: Demri, S., Kapur, D., Weidenbach, C. (eds.) IJCAR 2014. LNCS (LNAI), vol. 8562, pp. 367–373. Springer, Cham (2014). https://doi.org/10.1007/978-3-319-08587-6_28

Enumerating Justifications Using Resolution

Yevgeny Kazakov$^{(\boxtimes)}$ and Peter Skočovský

The University of Ulm, Ulm, Germany
yevgeny.kazakov@uni-ulm.de, peter.skoco@gmail.com

Abstract. If a conclusion follows from a set of axioms, then its justification is a minimal subset of axioms for which the entailment holds.
An entailment can have several justifications. Such justifications are commonly used for the purpose of debugging of incorrect entailments in
Description Logic ontologies. Recently a number of SAT-based methods
have been proposed that can enumerate all justifications for entailments
in light-weight ontologies languages, such as $\mathcal{EL}$. These methods work by
encoding $\mathcal{EL}$ inferences in propositional Horn logic, and finding minimal
models that correspond to justifications using SAT solvers. In this paper,
we propose a new procedure for enumeration of justifications that uses
resolution with answer literals instead of SAT solvers. In comparison to
SAT-based methods, our procedure can enumerate justifications in any
user-defined order that extends the set inclusion relation. The procedure is
easy to implement and, like resolution, can be parametrized with ordering
and selection strategies. We have implemented this procedure in PULi—a
new Java-based Proof Utility Library, and performed an empirical comparison of (several strategies of) our procedure and SAT-based tools on
popular $\mathcal{EL}$ ontologies. The experiments show that our procedure provides
a comparable, and often better performance than those highly optimized
tools. For example, using one of the strategies, we were able for the first
time to compute all justifications for all entailed concept subsumptions in
one of the largest commonly used medical ontology Snomed CT.

1 Introduction and Motivation

Axiom pinpointing, or computing justifications—minimal subsets of axioms of
the ontology that entail a given logical consequence—has been a widely studied research topic in ontology engineering [1–12]. Most of the recent methods
focus on the so-called $\mathcal{EL}$ family of Description Logics (DLs), in which logical
consequences can be proved by deriving new axioms from existing ones using
inference rules. The resulting inferences are usually encoded as propositional
(Horn) clauses, and justifications are computed from them using (modifications
of) SAT solvers. To ensure correctness, the input inference set must be complete,
that is, the inferences are enough to derive the consequence from any subset of
the ontology from which it follows.

In this paper, we present a new resolution-based procedure that enumerates
all justifications of an entailment given a complete set of inferences. Apart from

© Springer International Publishing AG, part of Springer Nature 2018
D. Galmiche et al. (Eds.): IJCAR 2018, LNAI 10900, pp. 609–626, 2018.
https://doi.org/10.1007/978-3-319-94205-6_40

requiring completeness, the form of inferences can be arbitrary and does not depend on any logic. For example, our method can be used with the inferences provided by existing consequence-based procedures [13–16]. The procedure can enumerate justifications in any given order, provided it extends the proper subset relation on sets of axioms. Performance of the procedure depends on the strategy it follows while enumerating justifications. We have empirically evaluated three simple strategies and experimentally compared our procedure with other highly optimized justification computation tools.

The paper is organized as follows. In Sect. 2 we describe related work. Section 3 introduces background on DLs, justifications, and resolution. In Sect. 4 we present the new procedure, and in Sect. 5 we describe its implementation and empirical evaluation.

2   Related Work

There are, generally, two kinds of procedures for computing justifications [9] using a DL reasoner. *Black-Box* procedures use a reasoner solely for entailment checking, and thus can be used for any reasoner and DL. *Glass-Box* procedures require additional information from a reasoner, such as inferences that the reasoner has used, and thus can only work with reasoners that can provide such information.

In a nutshell, Black-Box procedures [4,6,7,10] systematically explore subsets of axioms and check using a reasoner, which of these subsets entail the given logical conclusion, and which not. Unnecessary tests are avoided using the monotonicity property of the entailment.

Finding *one* justification is relatively easy. Starting from the set of all axioms that entail the conclusion, one tries to remove axioms one by one. If after the removal the entailment does not hold, the axiom is inserted back. This results in a subset from which no axiom can be further removed without breaking the entailment, i.e., a justification for the entailment. This justification, however, may be not unique as the result depends on the order in which the axioms are considered for removal. In the worst case, there can be exponentially-many different justifications. So, unsurprisingly, there is no polynomial procedure for computing *all* justifications even in languages such as $\mathcal{EL}$, for which entailment checking is polynomially decidable [17]. Further, computing all justifications is even hard in the number of justifications: it is already NP-hard to verify, given a set of justifications, if there exists another justification not in this set [4]. Hence, in practice, one is interested in algorithms for *enumeration* of justifications, i.e., algorithms that can return justifications without necessarily finishing computing all of them.

Most existing algorithms for enumeration of justifications rely, in one way or the other, on the *hitting set duality* that was introduced in the field of Model Based Diagnosis [18,19] and later adapted for DLs [5,7]. A *hitting set* for a collection of sets is a set containing at least one element from each set in the collection. A minimal hitting set of all justifications for an entailment is a *repair*—a minimal set of axioms, removal of which breaks the entailment. Dually, every minimal hitting set of all repairs is a justification. Existing justification enumeration algorithms, in fact, also (implicitly) enumerate repairs in addition to justifications.

Suppose that one has computed some justifications and repairs for an entailment. To find a new justification or a repair, it is sufficient to find a set M of axioms that has at least one axiom from each repair and misses at least one axiom from each justification. I.e., M is a hitting set of the computed repairs, and its complement (within the set of all axioms) is a hitting set of the computed justifications. If no such set M exists, then there are no new justifications or repairs, for otherwise a new justification or the complement of a new repair would satisfy this requirement. Now, if M entails the conclusion, then a justification can be extracted from M by repeatedly removing axioms as described before. This justification will be different from all previously computed justifications because M is not a super-set of any of them. On the other hand, if M does not entail the conclusion, then a new repair can be extracted from the complement of M similarly, by removing axioms until the set is no longer a repair. Likewise, this will be a new repair since the complement of M is not a super-set of any previously computed repairs. By repeating this procedure, one can enumerate all repairs and all justifications.

Finding a suitable set M satisfying the requirements above can be accomplished using a propositional SAT solver. Specifically, for each computed repair, we add a clause consisting of atoms corresponding to the axioms in the repair. Similarly, for each computed justification, we add a clause consisting of the negations of atoms corresponding to the axioms in the justification. Then for every model of these clauses, the set M consisting of the axioms whose atoms are true, satisfies the requirements. SAT solvers can also be used to optimize the entailment tests, which are usually main bottleneck of Black-Box procedures. For example, in the case of $\mathcal{EL}$, all necessary information about entailments from subsets of axioms can be represented by (a polynomial number of) inferences. Every $\mathcal{EL}$ inference can be translated to a propositional (Horn) clause with the negative atoms corresponding to the premises of the inference, and the positive atom corresponding to its conclusion. A conclusion is derivable from axioms using the inferences iff the translation of the inferences entails the (Horn) clause whose negative atoms correspond to axioms and the positive atoms corresponds to the conclusion.

The above Glass-Box procedure was first proposed and implemented in EL+SAT [11,12], and later improved in EL2MUS [3] and SATPin [8]. These tools differ mainly in the way how they enumerate models corresponding to the candidate sets M, and further optimizations employed. EL+SAT and EL2MUS use two instances of a SAT solver—one for enumeration of candidate models, and another for verifying derivability using inferences—whereas SATPin [8] uses one (modified) SAT solver for both of these tasks. The encoding for finding the candidate set M described above is most close to the implementation of EL2MUS. EL+SAT and SATPin do not explicitly enumerate repairs, but each time a model is found that corresponds to a set M that does not entail the conclusion,

Table 1. The syntax and semantics of $\mathcal{EL}$

	Syntax	Semantics
Roles		
Atomic role	R	$R^{\mathcal{I}}$
Concepts		
Atomic concept	A	$A^{\mathcal{I}}$
Top	$\top$	$\Delta^{\mathcal{I}}$
Conjunction	$C \sqcap D$	$C^{\mathcal{I}} \cap D^{\mathcal{I}}$
Existential restriction	$\exists R.C$	$\{x \mid \exists y \in C^{\mathcal{I}} : \langle x, y \rangle \in R^{\mathcal{I}}\}$
Axioms		
Concept inclusion	$C \sqsubseteq D$	$C^{\mathcal{I}} \subseteq D^{\mathcal{I}}$

$$\mathsf{R_0}\ \frac{}{C \sqsubseteq C} \quad \mathsf{R_{\sqcap}^-}\ \frac{C \sqsubseteq D_1 \sqcap D_2}{C \sqsubseteq D_1 \quad C \sqsubseteq D_2} \quad \mathsf{R_{\sqsubseteq}}\ \frac{C \sqsubseteq D}{C \sqsubseteq E} : D \sqsubseteq E \in \mathcal{O}$$

$$\mathsf{R_{\top}}\ \frac{}{C \sqsubseteq \top} \quad \mathsf{R_{\sqcap}^+}\ \frac{C \sqsubseteq D_1 \quad C \sqsubseteq D_2}{C \sqsubseteq D_1 \sqcap D_2} \quad \mathsf{R_{\exists}}\ \frac{C \sqsubseteq \exists R.D \quad D \sqsubseteq E}{C \sqsubseteq \exists R.E}$$

Fig. 1. The inference rules for reasoning in $\mathcal{EL}$

a "blocking" clause is added to ensure that such a model is not returned again. However, the number of such blocking clauses is at least as large as the number of repairs. Further differences are that in EL2MUS the entailment checking solver is specialized in Horn clauses, and that EL+SAT and SATPin extract justifications by a deletion-based procedure (as outlined above), while EL2MUS uses an insertion-based procedure. Another tool EL2MCS [2] uses MaxSAT [20,21] to compute all repairs and extracts justifications from them using the hitting set duality, but it cannot return any justification before all repairs are computed. Further, BEACON [1] is a tool that integrates the justification procedure of EL2MUS.

Up to a few optimizations, the mentioned SAT-based tools use $\mathcal{EL}$ inferences only for the entailment checks. Had they delegated the entailment checks to a separate DL reasoner, they could be regarded as Black-Box. Our approach uses a similar encoding of inferences in propositional logic, however, it relies neither on a SAT solver nor on the hitting set duality. In particular, our method does not enumerate repairs (explicitly or implicitly). As we operate on inferences directly, a Black-Box version of our procedure would not be possible.

3   Preliminaries

3.1   The Description Logic $\mathcal{EL}$

The syntax of $\mathcal{EL}$ is defined using a vocabulary consisting of countably infinite sets of *(atomic) roles* and *atomic concepts*. Complex *concepts* and *axioms* are defined recursively using Table 1. We use letters R, S for roles, C, D, E for concepts, and A, B for atomic concepts. An *ontology* is a finite set of axioms.

An *interpretation* $\mathcal{I} = (\Delta^\mathcal{I}, \cdot^\mathcal{I})$ consists of a nonempty set $\Delta^\mathcal{I}$ called the *domain* of $\mathcal{I}$ and an interpretation function $\cdot^\mathcal{I}$ that assigns to each role R a binary relation $R^\mathcal{I} \subseteq \Delta^\mathcal{I} \times \Delta^\mathcal{I}$, and to each atomic concept A a set $A^\mathcal{I} \subseteq \Delta^\mathcal{I}$. This assignment is extended to complex concepts as shown in Table 1. $\mathcal{I}$ *satisfies* an axiom α (written $\mathcal{I} \models \alpha$) if the corresponding condition in Table 1 holds. $\mathcal{I}$ is a *model* of an ontology $\mathcal{O}$ (written $\mathcal{I} \models \mathcal{O}$) if $\mathcal{I}$ satisfies all axioms in $\mathcal{O}$. We say that $\mathcal{O}$ *entails* an axiom α (written $\mathcal{O} \models \alpha$), if every model of $\mathcal{O}$ satisfies α. A concept C is *subsumed* by D w.r.t. $\mathcal{O}$ if $\mathcal{O} \models C \sqsubseteq D$. The *ontology classification task* requires to compute all entailed subsumptions between atomic concepts occurring in $\mathcal{O}$.

Reasoning in $\mathcal{EL}$ can be performed by applying inference rules that derive subsumptions between concepts [17]. We use a variant of $\mathcal{EL}$ rules shown in Fig. 1 that do not require normalization [14]. As usual, the premises of the rules (if any) are given above the horizontal line, and the conclusions below. Note that rule $\mathsf{R}_\sqsubseteq$ can only use $D \sqsubseteq E$ from the ontology $\mathcal{O}$. This *side condition* should be distinguished from the premises of the rules, where one can use any derived axiom. This restriction has been made for efficiency reasons: if to use $D \sqsubseteq E$ as the second premise of $\mathsf{R}_\sqsubseteq$, like for rule $\mathsf{R}_\exists$, there would be too many unnecessary inferences.

The rules in Fig. 1 are *sound* and *complete* for entailment checking, i.e., the entailment $\mathcal{O} \models \alpha$ holds iff α is derivable from $\mathcal{O}$ using the rules.[1] Furthermore, for deriving α, it is sufficient to use inferences that contain only concepts appearing in $\mathcal{O}$ or α [14]. I.e., it is not necessary to apply $\mathsf{R}_\sqcap^+$ if $D_1 \sqcap D_2$ does not appear in $\mathcal{O}$ or α. This so-called *subformula property* implies that checking the $\mathcal{EL}$ entailment $\mathcal{O} \models \alpha$ can be performed in polynomial time [14,17] since there are at most polynomially-many different rule applications that can use only concepts appearing in $\mathcal{O}$ or α.

3.2   Inferences, Support, and Justifications

Although our experimental evaluation is concerned about $\mathcal{EL}$, our method can be used with a large class of inference systems of which the system in Fig. 1 is just one example. In general, we assume that the rules manipulate with objects that we call *axioms*, and an *ontology* is any finite set of such axioms. An *inference* is an expression *inf* of the form $\langle \alpha_1, \ldots, \alpha_n \vdash \alpha \rangle$ where $\alpha_1, \ldots, \alpha_n$ is a (possibly empty) sequence of axioms called the *premises* of *inf*, and α is an axiom called the *conclusion* of *inf*.

Let I be a set of inferences. An I-*derivation* from $\mathcal{O}$ is a sequence of inferences $d = \langle inf_1, \ldots, inf_k \rangle$ from I such that for every i with $(1 \leq i \leq k)$, and each premise

---

[1] Actually, w.l.o.g., one can also assume that the axioms in $\mathcal{O}$ are only used in the side conditions of rule $\mathsf{R}_\sqsubseteq$. Indeed, any axiom $D \sqsubseteq E \in \mathcal{O}$ can be derived in this way by first deriving a tautology $D \sqsubseteq D$ by R_0 and then deriving $D \sqsubseteq E$ by $\mathsf{R}_\sqsubseteq$ from $D \sqsubseteq D$ using $D \sqsubseteq E \in \mathcal{O}$.

$$\textbf{Resolution } \frac{c \vee a \quad \neg a \vee d}{c \vee d} \qquad \textbf{Factoring } \frac{c \vee a \vee a}{c \vee a}$$

Fig. 2. Propositional resolution and factoring rules

α of inf_i that is not in $\mathcal{O}$, there exists $j < i$ such that α is the conclusion of inf_j. An axiom α is *derivable* from $\mathcal{O}$ using I (notation: $\mathcal{O} \vdash_I \alpha$) if either $\alpha \in \mathcal{O}$ or there exists an I-derivation $d = \langle inf_1, \dots, inf_k \rangle$ from $\mathcal{O}$ such that α is the conclusion of inf_k. A *support* for $\mathcal{O} \vdash_I \alpha$ is a subset of axioms $\mathcal{O}' \subseteq \mathcal{O}$ such that $\mathcal{O}' \vdash_I \alpha$. A *justification* for $\mathcal{O} \vdash_I \alpha$ is a subset-minimal support for $\mathcal{O} \vdash_I \alpha$.

Suppose that $\models$ is an entailment relation between ontologies and axioms. A *justification* for $\mathcal{O} \models \alpha$ (also sometimes called a *minimal axiom set* MinA [4]) is a minimal subset $\mathcal{O}' \subseteq \mathcal{O}$ such that $\mathcal{O}' \models \alpha$. An inference $\langle \alpha_1, \dots, \alpha_n \vdash \alpha \rangle$ is *sound* if $\{\alpha_1, \dots, \alpha_n\} \models \alpha$. A set of inferences I is *complete* for the entailment $\mathcal{O} \models \alpha$ if $\mathcal{O}' \models \alpha$ implies $\mathcal{O}' \vdash_I \alpha$ for every subset $\mathcal{O}' \subseteq \mathcal{O}$. Note that if I is complete for $\mathcal{O} \models \alpha$ then $\mathcal{O}' \models \alpha$ iff $\mathcal{O}' \vdash_I \alpha$ for every $\mathcal{O}' \subseteq \mathcal{O}$. In particular, justifications for $\mathcal{O} \vdash_I \alpha$ coincide with justifications for $\mathcal{O} \models \alpha$.

Example 1. Consider the following applications of rules $\mathbf{R}_\sqcap^-$ and $\mathbf{R}_\sqcap^+$ in Fig. 1: $\mathsf{I}_e = \{\langle A \sqsubseteq B \sqcap C \vdash A \sqsubseteq B \rangle, \langle A \sqsubseteq B \sqcap C \vdash A \sqsubseteq C \rangle, \langle A \sqsubseteq C, A \sqsubseteq B \vdash A \sqsubseteq C \sqcap B \rangle\}$. Thus, I_e is a set of inferences over $\mathcal{EL}$ axioms. Let $\mathcal{O}_e = \{A \sqsubseteq B \sqcap C; A \sqsubseteq B; A \sqsubseteq C\}$ be an $\mathcal{EL}$ ontology and $\alpha_e = A \sqsubseteq C \sqcap B$ an $\mathcal{EL}$ axiom. Note that $\mathcal{O}_e \vdash_{\mathsf{I}_e} \alpha_e$.

It is easy to see that $\mathcal{O}'_e = \{A \sqsubseteq B; A \sqsubseteq C\} \vdash_{\mathsf{I}_e} \alpha_e$ and $\mathcal{O}''_e = \{A \sqsubseteq B \sqcap C\} \vdash_{\mathsf{I}_e} \alpha_e$, but $\{A \sqsubseteq B\} \not\vdash_{\mathsf{I}_e} \alpha_e$ and $\{A \sqsubseteq C\} \not\vdash_{\mathsf{I}_e} \alpha_e$. Hence, $\mathcal{O}'_e$ and $\mathcal{O}''_e$ are justifications for $\mathcal{O}_e \vdash_{\mathsf{I}_e} \alpha_e$. All inferences in I_e are also sound for the $\mathcal{EL}$ entailment relation $\models$. Since $\mathcal{O}'_e$ and $\mathcal{O}''_e$ are the only two justifications for $\mathcal{O}_e \models \alpha_e$, the inference set I_e is complete for the entailment $\mathcal{O}_e \models \alpha_e$.

3.3   Resolution with Answer Literals

Our procedure for enumeration of justifications is based on the *resolution calculus*, which is a popular method for automated theorem proving [22]. We will mainly use resolution for propositional Horn clauses. A (propositional) *literal* is either an atom $l = a$ (*positive literal*) or a negation of atom $l = \neg a$ (*negative literal*). A (propositional) *clause* is a disjunction of literals $c = l_1 \vee \cdots \vee l_n$, $n \geq 0$. As usual, we do not distinguish between the order of literals in clauses, i.e., we associate clauses with multisets of literals. Given two clauses c_1 and c_2, we denote by $c_1 \vee c_2$ the clause consisting of all literals from c_1 plus all literals from c_2. A clause is *Horn* if it has at most one positive literal. The *empty clause* $\square$ is the clause with $n = 0$ literals. The inference rules for the propositional resolution calculus are given in Fig. 2. We say that a set of clauses S is *closed* under the resolution rules if S contains every clause derived by the rules in Fig. 2 from S. The resolution calculus is *refutationally complete*: every set of clauses S closed under the resolution rules is satisfiable if and only if it does not contain the empty clause. This means that for checking satisfiability of the input set of clauses, it is sufficient to deductively close this set under the resolution rules and check if the empty clause is derived in the closure.

To reduce the number of resolution inferences (and hence the size of the closure) several *refinements* of the resolution calculus were proposed. The rules in Fig. 2 can be restricted using orderings and selection functions [22]. In particular, for Horn clauses, it is sufficient to select one (positive or negative) literal in each clause, and require that the resolution inferences are applied only on those (Theorem 7.2 in [22]).[2] This strategy is called *resolution with free selection*. In addition to rule restrictions, one can also use a number of *simplification rules* that can remove or replace clauses in the closure S. We will use two such rules. *Elimination of duplicate literals* removes all duplicate literals from a clause (including duplicate negative literals). *Subsumption deletion* removes a clause c from S if there exists another *sub-clause* c' of c in S, i.e., $c = c' \vee c''$ for some (possibly empty) clause c''. In this case we say that c' *subsumes* c.

Example 2. Consider the set of Horn clauses 1–7 below. We apply resolution with free selection that selects the underlined literals in clauses. Clauses 8–10 are obtained by resolution inferences from clauses shown in angle braces on the right.

1: $\underline{\neg p_1} \vee p_2$	4: $\underline{p_1}$	7: $\underline{\neg p_4}$	8: $\underline{\neg p_3} \vee p_4$ $\langle 3, 5 \rangle$
2: $\underline{\neg p_1} \vee p_3$	5: $\underline{p_2}$		9: $\underline{p_4}$ $\langle 6, 8 \rangle$
3: $\underline{\neg p_2} \vee \neg p_3 \vee p_4$	6: $\underline{p_3}$		10: $\square$ $\langle 7, 9 \rangle$

Note that the resolution rule was not applied, e.g., to clauses 3 and 6 because literal $\neg p_3$ in clause 3 is not selected. Also note that many clauses in the closure above can be derived by several resolution inferences. For example, clause 5 can be obtained by resolving clauses 1 and 4 and clause 6 by resolving 2 and 4. Therefore the empty clause 10 can be derived from several subsets of the original clauses 1–7.

The resolution calculus is mainly used for checking satisfiability of a clause set, and is not directly suitable for finding unsatisfiable subsets of clauses. To solve the latter problem, we use an extension of resolution with so-called answer literals [23]. To determine, which subsets of the input clauses are unsatisfiable, we add to every input clause a fresh positive *answer literal*. Resolution rules can then be applied to the extended clauses on the remaining (ordinary) literals using the usual orderings and selection functions. If some clause with answer literals is derived, then this clause with the answer literals removed, can be derived from the clauses for which the answer literals were introduced. In particular, if a clause containing only answer literals is derived, then the set of clauses that corresponds to these answer literals is unsatisfiable. Completeness of resolution means that all such unsatisfiable sets of clauses can be found in this way. If answer literals are added to some but not all clauses and a clause with only answer literals is derived, then the set of clauses that corresponds to the answer literals plus clauses without answer literals is unsatisfiable.

---

[2] Note that the factoring rule cannot apply to Horn clauses.

Example 3. Consider the clauses 1–7 from Example 2. Let us add answer literals a_1-a_3 to clauses 4–6 and apply the resolution rules on the remaining (underlined) literals like in Example 2, eliminating duplicate literals if they appear.

1:	$\neg p_1 \vee p_2$	8:	$\underline{p_2} \vee a_1$ $\langle 1,4 \rangle$	15:	$\underline{p_4} \vee a_1$ $\langle 9,11 \rangle$
2:	$\neg p_1 \vee p_3$	9:	$\underline{p_3} \vee a_1$ $\langle 2,4 \rangle$	16:	$\boxed{a_2 \vee a_3}$ $\langle 7,12 \rangle$
3:	$\neg p_2 \vee \neg p_3 \vee p_4$	10:	$\neg p_3 \vee \underline{p_4} \vee a_2$ $\langle 3,5 \rangle$	17:	$\boxed{a_1 \vee a_2}$ $\langle 7,13 \rangle$
4:	$\underline{p_1} \vee a_1$	11:	$\neg p_3 \vee \underline{p_4} \vee a_1$ $\langle 3,8 \rangle$	18:	$\boxed{a_1 \vee a_3}$ $\langle 7,14 \rangle$
5:	$\underline{p_2} \vee a_2$	12:	$\underline{p_4} \vee a_2 \vee a_3$ $\langle 6,10 \rangle$	19:	$\boxed{a_1}$ $\langle 7,15 \rangle$
6:	$\underline{p_3} \vee a_3$	13:	$\underline{p_4} \vee a_1 \vee a_2$ $\langle 9,10 \rangle$		
7:	$\neg p_4$	14:	$\underline{p_4} \vee a_1 \vee a_3$ $\langle 6,11 \rangle$		

The framed clauses 16–19 contain only answer literals, so the corresponding sets of clauses are unsatisfiable in conjunction with the input clauses without answer literals. For example, clause 16 means that clauses 1–3, 5–7 are unsatisfiable and clause 19 means that clauses 1–4, 7 are also unsatisfiable. Note that clause 19 subsumes clauses 17–18; if subsumed clauses are deleted, we obtain only clauses with answer literals that correspond to *minimal* subsets of clauses 4–6 that are unsatisfiable in conjunction with the remaining input clauses 1–3, 7.

4   Enumerating Justifications Using Resolution

In this section, we present a new procedure that, given an ontology $\mathcal{O}$, an inference set I and a goal axiom α_g, enumerates justifications for $\mathcal{O} \vdash_I \alpha_g$. It uses the usual reduction of the derivability problem $\mathcal{O} \vdash_I \alpha_g$ to satisfiability of propositional Horn clauses [2,8,11,12] in combination with the resolution procedure with answer literals.

Given a derivability problem $\mathcal{O} \vdash_I \alpha_g$, we assign to each axiom α_i occurring in I a fresh propositional atom p_{α_i}. Each inference $\langle \alpha_1, \ldots, \alpha_n \vdash \alpha \rangle \in I$ is then translated to the Horn clause $\neg p_{\alpha_1} \vee \cdots \vee \neg p_{\alpha_n} \vee p_\alpha$. In addition, for each axiom $\alpha \in \mathcal{O}$ that appears in I, we introduce a (unit) clause p_α. Finally, we add the clause $\neg p_{\alpha_g}$ encoding the assumption that α_g is not derivable. It is easy to see that $\mathcal{O} \vdash_I \alpha_g$ if and only if the resulting set of clauses is unsatisfiable.

We now extend this reduction to find justifications for $\mathcal{O} \vdash_I \alpha_g$. Recall that a subset $\mathcal{O}' \subseteq \mathcal{O}$ is a support for $\mathcal{O} \vdash_I \alpha_g$ if $\mathcal{O}' \vdash_I \alpha_g$. Hence, the subset of clauses p_α for $\alpha \in \mathcal{O}'$ is unsatisfiable in combination with the clauses for the encoding of inferences and $\neg p_{\alpha_g}$. We can find all such minimal subsets (corresponding to justifications) by adding a fresh answer literal to every clause p_α with $\alpha \in \mathcal{O}$, and applying resolution on non-answer literals together with elimination of redundant clauses.

Example 4. Consider the ontology $\mathcal{O}_e$, inferences I_e and axiom α_e from Example 1. To encode the derivability problem $\mathcal{O}_e \vdash_{I_e} \alpha_e$ we assign atoms p_1-p_4 to the axioms occurring in I_e as follows:

$$p_1 : A \sqsubseteq B \sqcap C, \quad p_2 : A \sqsubseteq B, \quad p_3 : A \sqsubseteq C, \quad p_4 : A \sqsubseteq C \sqcap B.$$

Algorithm 1. Enumeration of justifications using resolution

Enumerate($\mathcal{O} \vdash_I \alpha$, $\precsim$): enumerate justifications for $\mathcal{O} \vdash_I \alpha$
input   : $\mathcal{O} \vdash_I \alpha$ – the problem for which to enumerate justifications,
            $\precsim$ – an admissible preorder on clauses

1 $Q \leftarrow$ createEmptyQueue($\precsim$) ; // for unprocessed clauses
2 Q.addAll(encode($\mathcal{O} \vdash_I \alpha$)); // add the clause encoding of the problem
3 $S \leftarrow$ createEmptyList() ; // for processed clauses
4 **while** $Q \neq \emptyset$ **do**
5 $\quad$ $c \leftarrow Q$.remove(); // take one minimal element out of the queue
6 $\quad$ $c \leftarrow$ simplify(c); // remove duplicate literals from c
7 $\quad$ **if** c **is not subsumed by any** $c' \in S$ **then**
8 $\quad\quad$ S.add(c);
9 $\quad\quad$ **if** c **contains only answer literals then**
10 $\quad\quad\quad$ **report** decode(c); // a new justification is found
11 $\quad\quad$ **else** // apply resolution rules to c and clauses in S
12 $\quad\quad\quad$ **for** $c' \in$ resolve(c,S) **do**
13 $\quad\quad\quad\quad$ Q.add(c');

The encoding produces clauses 1–7 from Example 3: the inferences I_e are encoded by clauses 1–3, the axioms in $\mathcal{O}_e$ result in clauses 4–6 with answer literals, and the assumption that α_e is not derivable is encoded by clause 7. The derived clauses 16–19 correspond to supports of $\mathcal{O}_e \vdash_{I_e} \alpha_e$, and by eliminating redundant clauses 17–18, we obtain clauses 16 and 19 that correspond to justifications $\mathcal{O}'_e$ and $\mathcal{O}''_e$ from Example 1.

One disadvantage of the described procedure is that it requires the closure under the resolution rules to be fully computed before any justification can be found. Indeed, since derived clauses may be subsumed by later clauses, one cannot immediately see whether a clause with only answer literals corresponds to a justification. For example, clause 19 in Example 3 subsumes clauses 17–18 derived before, thus 17–18 do not correspond to justifications. We address this problem by using *non-chronological* application of resolution inferences. Intuitively, instead of applying the rules to clauses in the order in which they are derived, we apply the rules to clauses containing fewer answer literals first. Thus, in Example 3, we apply the rules to clause 15 before clauses 12–14.

The improved procedure can *enumerate* justifications, i.e., return justifications one by one without waiting for the algorithm to terminate. This procedure is described in Algorithm 1. It is a minor variation of the standard saturation-based procedure for computing the closure under (resolution) rules, which uses a *priority queue* to store unprocessed clauses instead of an ordinary queue. Let $\precsim$ be a total preorder on clauses (a transitive reflexive relation for which every two clauses are comparable). As usual, we write $c_1 \prec c_2$ if $c_1 \precsim c_2$ but $c_2 \not\precsim c_1$. We say that $\precsim$ is *admissible* if $c_1 \prec c_2$ whenever the set of answer literals of c_1 is a proper subset of the set of answer literals of c_2. For example, it is required that $\neg p_3 \vee p_4 \vee a_1 \prec p_4 \vee a_1 \vee a_2$, but not necessary that $p_4 \vee a_1 \prec p_4 \vee a_2 \vee a_3$. Note that if c is derived by resolution from clauses c_1 and c_2 then $c_1 \precsim c$ and $c_2 \precsim c$ since c contains the answer literals of both c_1 and c_2.

We say that a clause d (not necessarily occurring in Q) is *minimal* w.r.t. Q if there exists no clause $c \in$ Q such that $c \prec d$. A *priority queue* based on $\precsim$ is a queue in which the remove operation returns only a *minimal* element w.r.t. Q.[3] Given such a queue Q, Algorithm 1 initializes it with the translation of the input problem $\mathcal{O} \vdash_l \alpha$ (line 2) and then repeatedly applies resolution between minimal clauses taken out of this queue (loop 4–13) and the clauses in S that were processed before. Specifically, the removed minimal clause c is first simplified by removing duplicate literals (line 6) and then checked if it is subsumed by any previously processed clauses in S (in particular, if c was processed before). If c is subsumed by some $c' \in$ S, it is ignored and the next (minimal) clause is taken from the queue Q. Otherwise, c is added to S (line 8). If c contains only answer literals, then it corresponds to a justification (as we show next), which is then reported by the algorithm (line 10). Otherwise, resolution inferences are then applied on the selected non-answer literal in c (line 12). The new clauses derived by resolution are then added to Q (line 13) and the loop continues until Q is empty.

We now prove that Algorithm 1 in line 10 always returns a (new) justification. It is easy to see that if a clause d was minimal w.r.t. Q in the beginning of the while loop (line 4) then it remains minimal w.r.t. Q at the end of the loop (line 13). Indeed, for the clause c taken from the queue (line 5), we have $c \nprec d$. For all clauses c' obtained by resolving c with clauses from S (line 12) we have $c \precsim c'$. Hence $c' \nprec d$ for all c' added to Q (line 13) (for otherwise, $c \precsim c' \prec d$). This, in particular, implies that each clause in S is always minimal w.r.t. Q and, consequently, if c_1 was added to S before c_2 then $c_1 \precsim c_2$ (for otherwise $c_2 \prec c_1$ and c_1 would not be minimal w.r.t. Q when $c_2 \in$ Q). Hence, there cannot be two clauses c_1 and c_2 in S that contain only answer literals such that c_1 is a proper sub-clause of c_2 since in this case $c_1 \prec c_2$, thus c_2 must have been added to S after c_1, but then c_2 would be subsumed by c_1 (see line 7). Hence each result returned in line 10 is a (new) justification.

Since clauses are added to S in the order defined by $\precsim$, the justifications are also returned according to this order. Hence Algorithm 1 can return justifications in any user-defined order $\precsim$ on subsets of axioms as long as $s_1 \subsetneq s_2$ implies $s_1 \prec s_2$. Indeed, any such an order $\precsim$ can be lifted to an admissible order on clauses by comparing the sets of answer literals of clauses like the corresponding sets of axioms. For example, one can define $s_1 \precsim s_2$ by $\|s_1\| \leq \|s_2\|$ where $\|s\|$ is the cardinality of s. Instead of $\|s\|$ one can use any other measure $m(s)$ that is monotonic over the proper subset relation (i.e., $s_1 \subsetneq s_2$ implies $m(s_1) < m(s_2)$), for example, the *length* of s—the total number of symbols needed to write down all axioms in s.

---

[3] If there are several minimal elements in the queue, one of them is chosen arbitrarily.

5   Implementation and Evaluation

We have implemented Algorithm 1 as a part of the new Java-based Proof Utility Library (PULi).[4] In our implementation, we used the standard Java priority queue for Q, and employed a few optimisations to improve the performance of the algorithm.

First, we have noticed that our implementation spends over 95% of time on checking subsumptions in line 7. To improve subsumption checks, we developed a new datastructure for storing sets of elements and checking if a given set is a superset of some stored set. In a nutshell, we index the sets by 128 *bit vectors*, represented as a pair of 64 bit integers, where each set element assigns 1 to one position of the bit vector based on its hash value. This idea is reminiscent of Bloom filters[5] and can be also seen as a simple version of a feature vector indexing [24]. We store the sets in a trie[6] with the bit vector as the key, and use bitwise operations to determine if one vector has all bits of the other vector, which gives us a necessary condition for set inclusion. Using this datastructure, we were able to significantly improve the subsumption tests.

We have also noticed that the queue Q often contains about 10 times more elements than the closure S. To improve the memory consumption, we do not create the resolvents c' immediately (see line 12), but instead store in the queue Q the pairs of clauses (from S) from which these resolvents were obtained. This does not reduce the number of elements in the queue, but reduces the memory consumed by each element to essentially a few pointers plus an integer for determining the priority of the element.

We have evaluated our implementation on inferences computed for entailed axioms in some large $\mathcal{EL}$ ontologies, and compared performance with SAT-based tools for enumeration of justifications EL2MUS [3], EL2MCS [2] and SATPin [8]. The inferences were extracted using EL+SAT [11] (in the following called sat inferences) and ELK reasoner [25] (in the following called elk inferences). Both are capable of computing small inference sets that derive particular entailed axioms and are complete for these entailments (see Sect. 3.2).

For our evaluation, we chose ontologies GO-PLUS, GALEN and SNOMED, which contain (mostly) $\mathcal{EL}$ axioms. GO-PLUS is a recent version of Gene Ontology,[7] which imports a number of other ontologies. The provided distribution included subsumption axioms that were inferred (annotated with is_inferred), which we have removed. GALEN is the version 7 of OpenGALEN.[8] We did not use the more recent version 8, because the other tools were running out of memory. SNOMED is the 2015-01-31 version of Snomed CT.[9] From the first two ontologies we removed non-$\mathcal{EL}$ axioms, such as functional property axioms, and axioms that

---

[4]https://github.com/liveontologies/puli.

[5]https://en.wikipedia.org/wiki/Bloom_filter.

[6]https://en.wikipedia.org/wiki/Trie.

[7]http://geneontology.org/page/download-ontology.

[8]http://www.opengalen.org/sources/sources.html.

[9]http://www.snomed.org/.

Table 2. Summary of the input ontologies

	GO-PLUS	GALEN	SNOMED
# axioms	105557	44475	315521
# concepts	57173	28482	315510
# roles	157	964	77
# queries	90443	91332	468478

Table 3. Summary of sizes of inference sets

		GO-PLUS	GALEN	SNOMED
sat	Average	470.3	59140.0	997.8
	Median	39.0	110290.0	1.0
	Max	15915.0	152802.0	39381.0
elk	Average	166.9	3602.0	110.3
	Median	43.0	3648.0	8.0
	Max	7919.0	81501.0	1958.0

contain inverse property expressions and disjunctions. We have also adapted the input ontologies, so that they could be processed by (the reasoner of) EL+SAT. We removed disjointness axioms and replaced property equivalences with pairs of property inclusions. Duplicate axioms were removed by loading and saving the ontologies with OWL API.[10] With these ontologies, we have computed justifications for the entailed direct subsumptions between atomic concepts (in the following called *the queries*) using various tools. Table 2 shows the numbers of axioms, atomic concepts, atomic roles, and queries of each input ontology, and Table 3 the statistics about the sizes of inference sets obtained for these queries. All queries were processed by tools in a fixed random order to achieve a fair distribution of easy and hard problems. We used a *global timeout* of one hour for each tool and a *local timeout* of one minute per query.[11] To run the experiments we used a PC with Intel Core i5 2.5 GHz processor and 8 GiB RAM operated under 64-bit OS Ubuntu 16.04. For Java tools, we used OpenJDK v. 1.80_151 with a 7.7 GiB heap space limit.

As an admissible order on clauses for our implementation of Algorithm 1, we chose the relation $\precsim$ that compares the number of different answer literals in clauses. When using this order, cardinality-minimal justifications are found first. To control resolution inferences, we used three different selection strategies (for Horn clauses) that we detail next. For a propositional atom p, let $\#(p)$ be the number of input clauses in which p appears as a (positive) literal. Given a clause c, the *BottomUp* strategy, selects a negative literal $\neg p$ of c whose value $\#(p)$ is minimal; if there are no negative literals, the (only) positive literal of c is selected. The *TopDown* strategy selects a positive literal, if there is one, and otherwise selects a negative literal like in BottomUp. Finally, the *Threshold* strategy selects a negative literal $\neg p$ with the minimal value $\#(p)$ if $\#(p)$ does not exceed a given threshold value or there is no positive literal in c; otherwise the positive literal is selected. In our experiments we used the threshold value of 2.

---

[10]http://owlcs.github.io/owlapi/.

[11]The project for conducting the experiments can be found at https://github.com/liveontologies/pinpointing-experiments; a docker image is available at https://github.com/liveontologies/docker-pinpointing-experiments.

Table 4. Number of queries attempted in 1h/number of 60s timeouts/% of attempted queries in the number of all queries/% of 60s timeouts in the number of queries attempted in 1h

		GO-PLUS	GALEN	SNOMED
sat	BottomUp	2967 / 47 / 3.3 / 1.58	133 / 58 / 0.1 / 43.6	5630 / 26 / 1.2 / 0.46
	TopDown	25025 / 46 / 27.2 / 0.18	5687 / 16 / 6.2 / 0.28	16541 / 28 / 3.5 / 0.17
	Threshold	36236 / 43 / 40.1 / 0.12	3356 / 3 / 3.7 / 0.09	48994 / 16 / 10.5 / 0.03
	EL2MUS	12760 / 48 / 14.1 / 0.38	5077 / 34 / 5.6 / 0.67	14076 / 39 / 3.0 / 0.28
	EL2MCS	6758 / 47 / 7.5 / 0.70	3194 / 27 / 3.5 / 0.85	6275 / 47 / 1.3 / 0.75
	SATPin	4390 / 53 / 4.9 / 1.21	1475 / 39 / 1.6 / 2.64	3490 / 46 / 0.7 / 1.32
elk	BottomUp	3694 / 47 / 4.1 / 1.27	10584 / 28 / 11.6 / 0.26	159820 / 22 / 34.1 / 0.01
	TopDown	20249 / 44 / 22.4 / 0.22	7016 / 35 / 7.7 / 0.50	158992 / 12 / 33.9 / 0.01
	Threshold	35622 / 52 / 39.4 / 0.15	35462 / 16 / 38.8 / 0.05	468478 / 0 / 100 / 0.00
	EL2MUS	13554 / 47 / 15.0 / 0.35	13024 / 38 / 14.3 / 0.29	15708 / 42 / 3.4 / 0.27
	EL2MCS	6758 / 47 / 7.5 / 0.70	8725 / 47 / 9.6 / 0.54	6466 / 48 / 1.4 / 0.74
	SATPin	4625 / 54 / 5.1 / 1.17	3037 / 49 / 3.3 / 1.61	4144 / 50 / 0.9 / 1.21

Intuitively, the BottomUp strategy simulates the *Unit resolution*, the TopDown simulates the *SLD resolution*, and Threshold is some combination thereof.

Table 4 shows for how many queries all justifications were computed within the global and local timeouts.[12] The first six rows correspond to experiments on sat inferences and the other six rows to experiments on elk inferences. Note that, generally, the tools processed more queries and had fewer percentage of timeouts for elk inferences. Also, the Threshold strategy performed best in almost all cases. In particular, it could process all queries of SNOMED without timeouts. None of the SAT-based tools was able to find all justifications for all queries of SNOMED even after running for 24 h. We have then further verified (on a slightly faster PC with more memory) that Threshold without timeouts could process all 5415670 (not necessarily direct) entailed subsumptions of SNOMED in about 21 h using 10 GiB of java heap space. The hardest query took about 17 min and returned 658932 justifications. The largest number of justifications 942658 was returned by the third-hardest query in about 5 min.

To have an idea which strategy was best for which query, we have plotted in Fig. 3 the distributions of the query times for all strategies. Each point $\langle x, y \rangle$ of a plot represents the proportion x of queries that were solved by the method in under the time y. For instance, TopDown solved about 90% of the queries of GO-PLUS for sat inferences in under 0.01 s. Each plot considers only queries attempted by all tools on that plot. Since each plot represents the distribution of times and not a direct comparison of times for each query, even if one line is completely below another one, this does not mean that the corresponding method is faster for *every* query. To get a more detailed comparison, we have also plotted the distribution of minimum query times with a thin black line. For each query, the *minimum time* is the time spent by the tool that was the

---

[12]The raw experimental data is available at https://osf.io/4q6a9/.

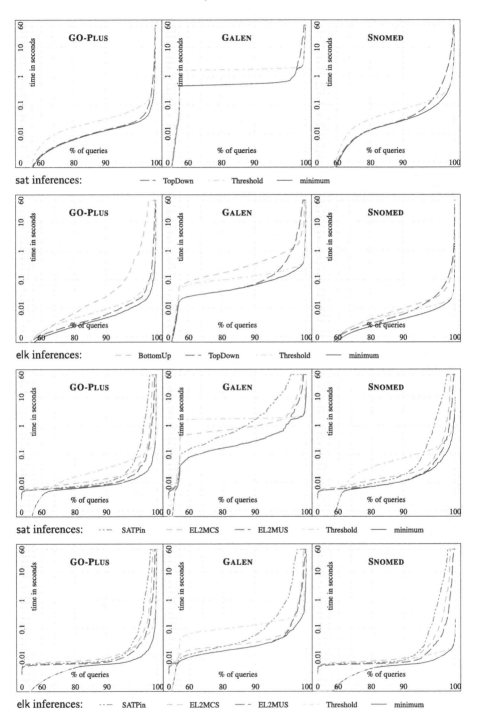

Fig. 3. Distribution of query times for SAT tools and resolution strategies, on sat and elk inference sets. The BottomUp strategy for sat is missing due too few processed queries.

fastest on that query (among the tools on the same plot). If a plot for some tool coincides with this black line at point (x, y), then all queries solved within time y by some tool were also solved within time y by this tool. In particular, this tool is the fastest for all queries with the minimal time y. This analysis shows, for example, that TopDown was often the best resolution tool for easy queries (solved under 0.1 s by some tool), and Threshold was the best tool for hard queries (solved over 1 s by all tools) on all ontologies. The sat-based tool EL2MUS was often the winner on medium-hard queries. Note that the time scale is logarithmic, so the times below 1 ms are not displayed.

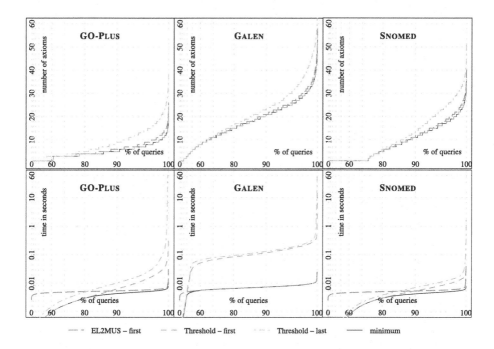

Fig. 4. Distribution of sizes (above) and computation times (below) for first and last justifications computed by Threshold and first justifications computed by EL2MUS on elk inferences

As mentioned before, one important difference between SAT tools and resolution-based tools, is that the latter allow one to enumerate justifications in any admissible order. In particular, it is possible to find cardinality-minimal justifications without computing all justifications, which is useful, e.g., for ontology debugging. Since deciding whether there exists a justification within a given size bound is an NP-complete problem [4], finding a cardinality-minimal justification is not as easy as finding one (arbitrary) justification. To determine whether the difference is significant in practice, we compare computations of the first justification by EL2MUS, the first (cardinality-minimal) justification by Threshold, and the last (before timeout, cardinality-maximal) justification by Threshold. In Fig. 4, we plot distributions of the sizes for these justifications and of times spent on computing them. Interestingly, the sizes of first justifications found by EL2MUS are very close to the minimal sizes. This is probably because small justifications are more likely to be obtained when minimizing sets of axioms for which the entailment holds. Unsurprisingly, EL2MUS can consistently compute the fist justifications in a few milliseconds. Although the times for computing a cardinality-minimal justification can be significantly higher (especially for GALEN, which has large inference sets), they can still be a few orders of magnitude smaller than for computing all (= the last) justifications for the hard cases. In particular, Threshold was able to compute a cardinality-minimal justification for all queries of GO-PLUS, GALEN and SNOMED respectively in about 6 min, 1.5 h, and 25 min.

6   Summary

We presented a new procedure that enumerates justifications using inferences that derive the goal consequence from an ontology. The inferences are encoded as Horn clauses and resolution with answer literals is applied. Our procedure can be parameterized by an ordering in which the justifications should be enumerated (as long as it extends the subset relation) and by a strategy that selects literals for resolution. The algorithm is relatively easy to implement and can be also easily used with non-Horn and non-propositional clauses. Our empirical evaluation shows that the procedure provides comparable, if not better performance than other tools that also use inferences as input. For example, for Snomed CT we were able to compute all justifications for all direct subsumptions in less than 1 h, and for all (possibly indirect) subsumptions in less than 1 day. Currently, we cannot explain the difference in the performance of the evaluated selection strategies. We hope to explore this question in the future.

References

1. Arif, M.F., Mencía, C., Ignatiev, A., Manthey, N., Peñaloza, R., Marques-Silva, J.: BEACON: an efficient SAT-based tool for debugging $\mathcal{EL}^+$ ontologies. In: Creignou, N., Le Berre, D. (eds.) SAT 2016. LNCS, vol. 9710, pp. 521–530. Springer, Cham (2016). https://doi.org/10.1007/978-3-319-40970-2_32
2. Arif, M.F., Mencía, C., Marques-Silva, J.: Efficient axiom pinpointing with EL2MCS. In: Hölldobler, S., Krötzsch, M., Peñaloza, R., Rudolph, S. (eds.) KI 2015. LNCS (LNAI), vol. 9324, pp. 225–233. Springer, Cham (2015). https://doi.org/10.1007/978-3-319-24489-1_17
3. Arif, M.F., Mencía, C., Marques-Silva, J.: Efficient MUS enumeration of horn formulae with applications to axiom pinpointing. CoRR abs/1505.04365 (2015)
4. Baader, F., Peñaloza, R., Suntisrivaraporn, B.: Pinpointing in the description logic $\mathcal{EL}^+$. In: Hertzberg, J., Beetz, M., Englert, R. (eds.) KI 2007. LNCS (LNAI), vol. 4667, pp. 52–67. Springer, Heidelberg (2007). https://doi.org/10.1007/978-3-540-74565-5_7

5. Horridge, M.: Justification based explanation in ontologies. Ph.D. thesis, University of Manchester, UK (2011)
6. Junker, U.: QUICKXPLAIN: preferred explanations and relaxations for over-constrained problems. In: McGuinness, D.L., Ferguson, G. (eds.) Proceedings of the 19th AAAI Conference on Artificial Intelligence (AAAI 2004), pp. 167–172. AAAI Press/The MIT Press (2004)
7. Kalyanpur, A., Parsia, B., Horridge, M., Sirin, E.: Finding all justifications of OWL DL entailments. In: Aberer, K., et al. (eds.) ASWC/ISWC -2007. LNCS, vol. 4825, pp. 267–280. Springer, Heidelberg (2007). https://doi.org/10.1007/978-3-540-76298-0_20
8. Manthey, N., Peñaloza, R., Rudolph, S.: Efficient axiom pinpointing in $\mathcal{EL}$ using SAT technology. In: Lenzerini, M., Peñaloza, R. (eds.) Proceedings of the 29th International Workshop on Description Logics (DL 2016). CEUR Workshop Proceedings, vol. 1577. CEUR-WS.org (2016)
9. Parsia, B., Sirin, E., Kalyanpur, A.: Debugging OWL ontologies. In: Ellis, A., Hagino, T. (eds.) Proceedings of the 14th International Conference on World Wide Web (WWW 2005), pp. 633–640. ACM (2005)
10. Kalyanpur, A.: Debugging and repair of OWL ontologies. Ph.D. thesis, University of Maryland College Park, USA (2006)
11. Sebastiani, R., Vescovi, M.: Axiom pinpointing in lightweight description logics via Horn-SAT encoding and conflict analysis. In: Schmidt, R.A. (ed.) CADE 2009. LNCS (LNAI), vol. 5663, pp. 84–99. Springer, Heidelberg (2009). https://doi.org/10.1007/978-3-642-02959-2_6
12. Vescovi, M.: Exploiting SAT and SMT techniques for automated reasoning and ontology manipulation in description logics. Ph.D. thesis, University of Trento, Italy (2011)
13. Baader, F., Lutz, C., Suntisrivaraporn, B.: Efficient reasoning in $\mathcal{EL}^+$. In: Parsia, B., Sattler, U., Toman, D. (eds.) Proceedings of the 19th International Workshop on Description Logics (DL 2006), CEUR Workshop Proceedings, vol. 189. CEUR-WS.org (2006)
14. Kazakov, Y., Krötzsch, M., Simančík, F.: The incredible ELK: from polynomial procedures to efficient reasoning with $\mathcal{EL}$ ontologies. J. Autom. Reason. **53**(1), 1–61 (2014)
15. Kazakov, Y.: Consequence-driven reasoning for Horn $\mathcal{SHIQ}$ ontologies. In: Boutilier, C. (ed.) Proceedings of the 21st International Joint Conference on Artificial Intelligence (IJCAI 2009), pp. 2040–2045. IJCAI (2009)
16. Simančík, F., Kazakov, Y., Horrocks, I.: Consequence-based reasoning beyond Horn ontologies. In: Walsh, T. (ed.) Proceedings of the 22nd International Joint Conference on Artificial Intelligence (IJCAI 2011), pp. 1093–1098. AAAI Press/IJCAI (2011)
17. Baader, F., Brandt, S., Lutz, C.: Pushing the $\mathcal{EL}$ envelope. In: Kaelbling, L., Saffiotti, A. (eds.) Proceedings 19th International Joint Conference on Artificial Intelligence (IJCAI 2005), pp. 364–369. Professional Book Center (2005)
18. Greiner, R., Smith, B.A., Wilkerson, R.W.: A correction to the algorithm in Reiter's theory of diagnosis. Artif. Intell. **41**(1), 79–88 (1989)
19. Reiter, R.: A theory of diagnosis from first principles. Artif. Intell. **32**(1), 57–95 (1987)
20. Morgado, A., Liffiton, M., Marques-Silva, J.: MaxSAT-based MCS enumeration. In: Biere, A., Nahir, A., Vos, T. (eds.) HVC 2012. LNCS, vol. 7857, pp. 86–101. Springer, Heidelberg (2013). https://doi.org/10.1007/978-3-642-39611-3_13

21. Ansótegui, C., Bonet, M.L., Levy, J.: SAT-based MaxSAT algorithms. Artif. Intell. **196**, 77–105 (2013)
22. Bachmair, L., Ganzinger, H.: Resolution theorem proving. In: Robinson, J.A., Voronkov, A. (eds.) Handbook of Automated Reasoning, pp. 19–99. Elsevier and MIT Press (2001)
23. Green, C.: Theorem proving by resolution as a basis for question-answering systems. Mach. Intell. **4**, 183–205 (1969)
24. Schulz, S.: Simple and efficient clause subsumption with feature vector indexing. In: Bonacina, M.P., Stickel, M.E. (eds.) Automated Reasoning and Mathematics. LNCS (LNAI), vol. 7788, pp. 45–67. Springer, Heidelberg (2013). https://doi.org/10.1007/978-3-642-36675-8_3
25. Kazakov, Y., Klinov, P.: Goal-directed tracing of inferences in EL ontologies. In: Mika, P., et al. (eds.) ISWC 2014. LNCS, vol. 8797, pp. 196–211. Springer, Cham (2014). https://doi.org/10.1007/978-3-319-11915-1_13

A SAT-Based Approach to Learn Explainable Decision Sets

Alexey Ignatiev[1,3]([⊠]), Filipe Pereira[1], Nina Narodytska[2], and Joao Marques-Silva[1]

[1] LASIGE, Faculdade de Ciências, Universidade de Lisboa, Lisbon, Portugal
{aignatiev,jpms}@ciencias.ulisboa.pt, fcpereira@lasige.di.fc.ul.pt
[2] VMWare Research, Palo Alto, CA, USA
nnarodytska@vmware.com
[3] ISDCT SB RAS, Irkutsk, Russia

Abstract. The successes of machine learning in recent years have triggered a fast growing range of applications. In important settings, including safety critical applications and when transparency of decisions is paramount, accurate predictions do not suffice; one expects the machine learning model to also explain the predictions made, in forms understandable by human decision makers. Recent work proposed explainable models based on *decision sets* which can be viewed as unordered sets of rules, respecting some sort of rule non-overlap constraint. This paper investigates existing solutions for computing decision sets and identifies a number of drawbacks, related with rule overlap and succinctness of explanations, the accuracy of achieved results, but also the efficiency of proposed approaches. To address these drawbacks, the paper develops novel SAT-based solutions for learning decision sets. Experimental results on computing decision sets for representative datasets demonstrate that SAT enables solutions that are not only the most efficient, but also offer stronger guarantees in terms of rule non-overlap.

1 Introduction

Machine learning (ML) has witnessed remarkable progress and important successes in recent years [18,22,28]. In some settings, predictions made by machine learning algorithms should provide explanations, preferably explanations that can be interpreted (or understood) by human decision makers. Concrete examples include safety-critical situations, but also when transparency of decisions is paramount. The importance of explainable AI (XAI), i.e. the problem of associating explanations with ML predictions, is underscored by recent research [2,21,42], by ongoing research programs [9], by EU-level legislation which is expected to enforce the automated generation of explanations [11], and also by a number of meetings on computing explainable ML models [16,17,30].

This work was supported by FCT grants SAFETY (SFRH/BPD/120315/2016) and ABSOLV (028986/02/SAICT/2017), and LASIGE Research Unit, ref. UID/CEC/00408/2013.

An often used approach to provide explanations for ML predictions is to resort to some sort of logic-related model, including rule/decision lists, rule/decision sets, and decision trees [2, 21]. These logic-related models can in most cases associate explanations with predictions, represented as conjunctions of literals, that follow from the actual model representation. Clearly, the smaller the model representation, the simpler the explanations are likely to be, and so easier to understand by human decision makers.

Recent approaches include the computation of (smaller or smallest) rule lists [2, 42], the computation of decision sets [21], but also the computation of decision trees [3]. Rule lists impose an order of the rules [35], whereas decision sets do not. Clearly, from an interpretability perspective, decision sets are the most appealing since each prediction depends only on the literals associated with each rule. On the negative side, decision sets can exhibit rule overlap, and so may require decisions to be made when more than one class is predicted. Furthermore, even restricted forms of rule learning are well-known to be hard for NP [35].

This paper analyzes recent work on computing interpretable decision sets [21]. The paper highlights a number of drawbacks of the proposed approach, related with rule overlap, the generation of explanations, but also with the scalability of the approach. The paper then investigates three main topics. The first topic is the proposal of a rigorous definition of rule overlap. The paper relates this new definition with earlier work, and conjectures that solving the problem of overlap when learning optimal (in size) decision sets is hard for the second level of the polynomial hierarchy. The paper then proposes a number of variants of learning decision sets with less demanding constraints on overlap, and shows that these variants are instead hard for NP. The second topic is the issue of generating explanations for predictions. The paper shows that different models for learning decision sets provide different forms of computing explanations, thus enabling the generation of explanations in most settings. The third topic is to develop different propositional models for learning optimal decision sets. The proposed models build on earlier work on inductive inference [19], but introduce a number of variants, allowing for multiple classes, and also accommodating different overlap constraints. Moreover, the paper shows that all these models exhibit symmetries in the problem formulation, and so predicates breaking these symmetries can be used for improving performance.

The paper is organized as follows. Section 2 introduces the definitions and notation used in the remainder of the paper. The issue of overlap and explanation generation is investigated in Sect. 3. Propositional models for learning decision sets subject to different constraints on overlap are proposed in Sect. 4. Section 5 analyzes the performance of the proposed approach on representative datasets, and compares with earlier work [21]. Section 6 concludes the paper.

2   Preliminaries

This section briefly overviews Boolean Satisfiability (SAT), the classification problem in ML, and the learning of decision sets (DS). Throughout the paper, the

Ex.	Vacation (V)	Concert (C)	Meeting (M)	Expo (E)	Hike (H)
e_1	0	0	1	0	0
e_2	1	0	0	0	1
e_3	0	0	1	1	0
e_4	1	0	0	1	1
e_5	0	1	1	0	0
e_6	0	1	1	1	0
e_7	1	1	0	1	1

(a) A classification example

if ¬Meeting **then** Hike
if ¬Vacation **then** ¬Hike

(b) Decision set with some overlap

if Vacation **then** Hike
if ¬Vacation **then** ¬Hike

(c) Decision set with no overlap

Fig. 1. A classification example and its decision set

notation $[R]$ is used to denote the set of natural numbers $\{1, \ldots, R\}$, moreover, for a point $\mathbf{f}$ in some K-dimensional space, the rth coordinate is given by $\mathbf{f}[r]$.

Boolean Satisfiability (SAT). We assume notation and definitions standard in the area of SAT [4]. Formulas are represented in Conjunctive Normal Form (CNF) and defined over a set of variables $X = \{x_1, \ldots, x_n\}$. A formula $\mathcal{F}$ is a conjunction of clauses, a clause is a disjunction of literals, and a literal is a variable x_i or its complement $\neg x_i$. Where appropriate, formulas are viewed as sets of sets of literals. CNF encodings of cardinality constraints have been studied extensively, and will be assumed throughout [4]. Moreover, standard classification techniques are assumed [39].

Classification Problems. We follow the notation used in earlier work [3,21]. We consider a set of features $\mathcal{F} = \{f_1, \ldots, f_K\}$, all of which are assumed to be binary, taking a value in $\{0, 1\}$. When necessary, the fairly standard one-hot-encoding [32] is assumed for handling non-binary categorical features. Numeric features can be handled with standard techniques as well. Since all features are binary, a literal on a feature f_r will be represented as f_r, denoting that the feature takes value 1, i.e. $f_r = 1$, or as $\neg f_r$, denoting that the feature takes value 0, i.e. $f_r = 0$. Hence, the space of features (or *feature space* [14]) is $\mathcal{U} \triangleq \prod_{r=1}^{K} \{f_r, \neg f_r\}$.

To learn a classifier, one starts from given training data (also referred to as examples or samples) $\mathcal{E} = \{e_1, \ldots, e_M\}$. Examples are associated with classes taken from a set of classes $\mathcal{C}$. The paper focuses mostly on binary classification, i.e. $\mathcal{C} = \{c_0, c_1\}$. (We will associate c_0 with 0 and c_1 with 1, for simplicity.) Thus, $\mathcal{E}$ is partitioned into $\mathcal{E}^+$ and $\mathcal{E}^-$, denoting the examples classified as positive ($c_1 = 1$) and as negative ($c_0 = 0$), respectively. Each example $e_q \in \mathcal{E}$ is represented as a 2-tuple (π_q, ς_q), where $\pi_q \in \mathcal{U}$ denotes the literals associated with the example and $\varsigma_q \in \{0, 1\}$ is the class to which the example belongs. We have $\varsigma_q = 1$ if $e_q \in \mathcal{E}^+$ and $\varsigma_q = 0$ if $e_q \in \mathcal{E}^-$. A literal l_r on a feature f_r, $l_r \in \{f_r, \neg f_r\}$, *discriminates* an example e_q iff $\pi_q[r] = \neg l_r$, i.e. the feature takes the value opposite to the value in the set of literals of the example. Moreover, we assume a mapping from feature values to classes, $\mu : \mathcal{U} \to \mathcal{C}$, i.e. we require consistency in the examples. Alternatively, we could allow for possible inconsistencies in the examples, by associating examples with elements of a relation $\rho \subseteq \mathcal{U} \times \mathcal{C}$.

Details on how to handle the extensions to this basic formulation, including non-binary features, the handling of non-binary classes, and allowing for inconsistent examples, are beyond the scope of this paper but are discussed in later sections. Furthermore, in this paper we assume that all features are specified for all examples; the work can be generalized for situations where the value of some features for some examples is left unspecified.

In the remainder of the paper, we will also consider non-conflicting subsets of $\mathcal{L} \triangleq \cup_{r=1}^{K}\{f_r, \neg f_r\}$, such that a subset of $\mathcal{L}$ is *non-conflicting* if for all features f_r, the literals f_r and $\neg f_r$ do not both occur in that subset. When referring to the actual data points representing the examples in $\mathcal{E}$, we use the notation $\mathbf{f}$, with $\mathbf{f} \in \prod_{r=1}^{K}\{f_r, \neg f_r\}$.

Example 1. Figure 1 shows a simple classification example. The set of binary features is $\mathcal{F} = \{f_1, f_2, f_3, f_4\}$ with $f_1 \triangleq V$, $f_2 \triangleq C$, $f_3 \triangleq M$, and $f_4 \triangleq E$. Example e_1 is represented by the 2-tuple (π_1, ς_1), with $\pi_1 = (\neg V, \neg C, M, \neg E)$ and $\varsigma_1 = 0$. Moreover, the literals V, C, $\neg M$ and E discriminate e_1. For this classification example, we have $\mathcal{U} = \{V, \neg V\} \times \{C, \neg C\} \times \{M, \neg M\} \times \{E, \neg E\}$.

The objective of classification is to learn some function $\hat{\phi}$ which matches the actual function ϕ on the training data and generalizes *suitably well* on unseen test data [13,14,27,34]. In this paper, we seek to learn representations of $\hat{\phi}$ corresponding to *decision sets* (DS). Many other representations have been studied, including decision trees [34], rule lists [2], and sums of terms (i.e. DNF) [15,41], among others. These are of interest, including for XAI, but are beyond the scope of this work.

Related Work. Rule learning, as a form of covering problem, can be traced back to the 1960s [26]. Rule learning finds important applications in ML and Data Mining (DM), and it is a standard topic in ML and DM textbooks [13,14,27]. Although rule learning has been investigated at the propositional and predicate levels, in different settings, the focus of this paper is the optimal learning of propositional rules. Rules can be organized as lists, being referred to as rule (or decision) lists, or as sets, being also referred to as rule (or decision) sets. The difference between the two representations is that lists impose an order on the rules, and sets do not. It is well-known that learning optimal rule lists is NP-hard [20]. As a result, most algorithms for learning rule lists or sets are heuristic [7,8,33,34], being in general efficient to run, but providing essentially no guarantees in terms of the quality of the computed rules. Recent work has focused on developing small or optimal rule (or decision) lists [2], but also rule (or decision) sets [21]. The focus of the paper is the learning of decision sets, and so we investigate in more detail the recently proposed IDS (*interpretable decision sets*) approach [21].

3   Learning Explainable Decision Sets

This section introduces, but also generalizes, the definitions proposed in earlier work [21] for the problem of learning decision sets.

Definition 1 (Itemset). Given $\mathcal{F}$, an *itemset* π is an element of $\mathcal{I} \triangleq \prod_{r=1}^{K}\{f_r, \neg f_r, u\}$, where u represents a *don't care* value. Where applicable, an itemset π is also interpreted as the conjunction of the coordinates different from u, i.e. the *specified* literals of π.

Clearly, an itemset represents a cube in the K-dimensional feature space. Moreover, a K-dimensional point in feature space is also a (completely specified) itemset.

Definition 2 (Clashing itemsets). Given two itemsets $\pi_1, \pi_2 \in \mathcal{I}$, the two itemsets *clash*, written $\pi_1 \cap \pi_2 = \emptyset$, if and only if there exists a coordinate r such that $\pi_1[r] = f_r$ and $\pi_2[r] = \neg f_r$, or $\pi_1[r] = \neg f_r$ and $\pi_2[r] = f_r$.

Definition 3 (Rule). A *rule* is a 2-tuple (π, ς), where $\pi \in \mathcal{I}$ is an itemset, and $\varsigma \in \mathcal{C}$ is a class. Moreover, a rule (π, ς) is to be interpreted as follows:

> **IF** the specified literals in π are true, **THEN** pick class ς

Rules can and have been used in different settings [13,14,27]. This paper considers the use of rules as the building block of decision sets.

Definition 4 (Decision Sets). Given a set of (binary) features $\mathcal{F}$, defining a feature space $\mathcal{U}$, and a set of classes $\mathcal{C}$, a *decision set* $\mathbb{S}$ is a finite set of rules.

Given a decision set $\mathbb{S}$, there may exist points in feature space not covered by $\mathbb{S}$. An often used (optional) solution is to consider a *default* rule, which applies whenever the disjunction of the conjunctions of literals associated with each rule of $\mathbb{S}$ takes value 0.

Definition 5 (Default rule $\mathfrak{D}$). A rule of the form $\mathfrak{D} \triangleq (\emptyset, \varsigma)$ denotes the *default* rule of a decision set $\mathbb{S}$, applicable when all the other rules take value 0 on a given point of feature space. The class selected is ς.

Example 2. Referring back to Example 1, Fig. 1b shows an example of a decision set for the dataset of 1a, whereas Fig. 1c shows a different decision set. (The difference between the two relates with the notion of overlap to be introduced below.) Moreover, for the first decision set (see Fig. 1b), a (necessary) default rule could be $(\emptyset, 0)$. For example, for the feature space point (V, C, M, E) we can now say that the class, due to the default rule, is 0.

In contrast with earlier work [21], we consider generalized forms of cover, subject to subsets of the feature space.

Definition 6 (*$\mathcal{X}$-Cover*). Given $\mathcal{X} \subseteq \mathcal{U}$ and an itemset, the $\mathcal{X}$-cover of the itemset is the set of feature space points in $\mathcal{X}$ with a non-empty intersection with the itemset. The cover of the default rule $\mathfrak{D}$ is the set of points in feature space not covered by *any* of the other rules of a decision set.

Earlier work [21] considers a less general definition of cover, where $\mathcal{X}$ corresponds to the training data $\mathcal{E}$. Overlap between two rules assesses whether the set of points covered by two rules intersect. Overlap has been investigated recently in the context of learning decision sets [21]. This earlier work focused on overlap *solely* with respect to the *training data*, i.e. the starting set of examples, providing *no* guarantees on any other point of feature space. As a result, and in contrast with earlier work [21], we consider generalized forms of overlap, subject to subsets of the feature space.

Definition 7 (*$\mathcal{X}$-overlap*). Two rules $r_1 = (\pi_1, \varsigma_1)$ and $r_2 = (\pi_2, \varsigma_2)$ *overlap* in $\mathcal{X} \subseteq \mathcal{U}$ iff,

$$\exists \mathbf{f} \in \mathcal{X}. \mathbf{f} \cap \pi_1 \neq \emptyset \text{ and } \mathbf{f} \cap \pi_2 \neq \emptyset \tag{1}$$

Observe that the simpler definition, requiring $\pi_1 \cap \pi_2 \neq \emptyset$, would not enable restricting overlap to specific subsets of $\mathcal{U}$. Furthermore, the definition of overlap considered in earlier work [21] corresponds to $\mathcal{E}$-overlap.

The above definition can be qualified with $\oplus$ or $\ominus$, depending if we are concerned with overlap where the classification agrees ($\oplus$), i.e. all rules whose bodies are not false predict the same class, or disagrees ($\ominus$), i.e. there exist rules whose bodies are not false that do not predict the same class.

More importantly, the proposed formulation of overlap enables investigating the quality of decision sets in points of feature space *not* covered by the initial set of examples. We will be mostly concerned with $\mathcal{U}^{\ominus}$-overlap between pairs of rules with different classifications, aiming to eliminate such overlap. We will be less concerned with $\mathcal{U}^{\oplus}$-overlap, but this can also be deemed of interest [21].

Example 3. With respect to Example 1 and the decision set shown in Figure 1b there is no $\mathcal{E}$-overlap, but there is overlap in feature space. For the point $(\neg V, \neg C, \neg M, \neg E) \in \mathcal{U}$ we have $\ominus$ overlap. Moreover, the decision set in Figure 1c exhibits no overlap.

Generating Succinct Explanations. For a rule (π, ς), its *explanation* is the conjunction of literals in π. Thus, for *any* point in feature space for which there exists no $\ominus$ overlap, we can simply pick one of the rules consistent with that point as the explanation for the prediction. In this situation, we refer to the explanation as *offline* (or *explicit*). Moreover, assuming there is no $\ominus$ overlap and that all points in feature space are covered by some rule, then the set of rules provides a succinct representation of the explanations of the predictions made. If there exists $\ominus$ overlap, then one can simply pick one of the rules for which the itemset takes value 1, and list the itemset as an explanation. One additional case, is when some point in feature space is not covered by any rule in a decision set. In this case, one resorts to a default rule $\mathfrak{D} = (\emptyset, \varsigma)$, which has no immediate explanation. Nevertheless, it is still possible to provide explanations, albeit the set of justifications can no longer be represented succinctly. We consider a point $\mathbf{f}$ in feature space such that no rule is applicable, and so the default rule is used. For each rule $r_i = (\pi_i, \varsigma_i)$, there must exist one literal $l_{i,k}$ that falsifies the itemset π_i. As a result, an explanation for selecting the default rule can be constructed by picking one falsified literal from each itemset of each rule with a class that is *not consistent* with the class associated with the default rule. We refer to these explanations as *online* (or *implicit*). Clearly, the explanation will depend on each point on feature space not covered by the other rules, but we are still able to produce explanations.

Example 4. We consider again Example 1 and the decision set in Fig. 1b, assuming a default rule $(\emptyset, 0)$. For the point in feature space (V, C, M, E) the prediction will be 0 (i.e. $\neg H$), due to the default rule. Moreover, since the prediction will be 0, then we pick a 0-valued literal from the rules that would predict a different class, M in this case for the first rule. Thus, we can provide the explanation $\{M\}$; i.e. any time there is a `meeting`, then we will *not* take the `hike`.

4   Learning Decision Sets with SAT

This section develops different SAT models for learning decision sets. We can associate a Boolean function E^0 with $\mathcal{E}^-$, which takes value 1 for each point in feature space associated with $\mathcal{E}^-$, i.e. each combination of binary features that represents an example in $\mathcal{E}^-$ is a minterm of E^0. Similarly, we associate a Boolean function E^1 with $\mathcal{E}^+$, which takes value 1 for each point in the feature space associated with $\mathcal{E}^+$. Moreover, each combination of binary features that represents an example in $\mathcal{E}^+$ is a minterm of E^1. Clearly, our working hypothesis is that $E^0 \wedge E^1 \vDash \bot$, i.e. the examples represent a mapping. As shown below, the minimum decision set problem can be formalized in different ways. This paper considers a general formalization of the minimum decision set problem, in terms of computing two sets of terms F^0 and F^1, i.e. two DNF representations, and is defined as follows:

Definition 8 [MINDSET, MINDS$_0$]. Let $\langle \mathcal{E}^-, \mathcal{E}^+ \rangle$ be a 2-tuple of examples associated with two distinct classes, c_0 and c_1, and each represented by Boolean functions E^0 and E^1, respectively. MINDS$_0$ is the problem of finding the smallest DNF representations of Boolean functions F^0 and F^1, measured in the number of *terms*, such that: (i) $E^0 \vDash F^0$; (ii) $E^1 \vDash F^1$; and (iii) $F^1 \leftrightarrow F^0 \vDash \bot$.

Observe that condition (iii) above ensures that a decision set is computed (1) exhibiting no $\mathcal{U}^\ominus$-overlap and (2) covering the complete feature space $\mathcal{U}$. This should be compared with the substantially less demanding constraint of $\mathcal{E}$-overlap investigated in earlier work [21]. Moreover, the cost of the DNF representation could be measured in terms of the number of literals. The paper considers the cost in terms of the number of terms (or rules), but it is straight-forward to extend to considering also literals. Alternatives to take the number of literals into account are investigated in Sect. 5.

Lemma 1. For any decision set respecting Definition 8, it holds that (i) $F^0 \wedge E^1 \models \bot$; and (ii) $F^1 \wedge E^0 \models \bot$.

Proposition 1. The decision version of MINDS_0 is in Σ_2^p.

Proof. (Sketch) Given some size threshold T, simply guess the terms of the two DNFs, F^0 and F^1, using no more than T terms, and then check that, for every assignment, the values of F^0 and F^1 differ. Clearly, this can be encoded as a 2QBF formula.   □

Furthermore, the decision version of the MINDS_0 problem is apparently hard for Σ_2^p. For example, if we minimize E^0 to a DNF F^0, then computing the smallest DNF F^1 subject to F^0 is a well-known Σ_2^p-hard problem [40].

Conjecture 1. MINDS_0 is hard for Σ_2^p.

The proof (or disproof) of this conjecture is left as future work. Given the above, we can envision the following optimization problems, studied in the remainder of the paper, which result from relaxing the constraint $F^1 \leftrightarrow F^0 \models \bot$ of MINDS_0, thus achieving hardness for NP:

1. MINDS_4: Minimize F^0, given $F^1 \equiv E^1$ constant, and such that (i) $E^0 \models F^0$; and (ii) $F^0 \wedge E^1 \models \bot$.
2. MINDS_3: Same as above, but for F^1 given $F^0 \equiv E^0$ constant.
3. MINDS_2: Minimize both F^0 and F^1, such that (i) $E^0 \models F^0$; (ii) $E^1 \models F^1$; (iii) $F^0 \wedge E^1 \models \bot$; and (iv) $F^1 \wedge E^0 \models \bot$.
4. MINDS_1: Minimize F^0 and F^1, such that (i) $E^0 \models F^0$; (ii) $E^1 \models F^1$; and (iii) $F^1 \wedge F^0 \models \bot$.

Observe that all of the above problems are weakened versions of MINDS_0, the main difference being the constraints on the functions associated with E^0 and E^1. Among MINDS_i, $i \neq 0$, MINDS_1 imposes the most severe constraint, ensuring no $\mathcal{U}^\ominus$-overlap takes place, although there may be points for which both F^0 and F^1 take value 0.

Proposition 2. The decision versions of the optimization problems MINDS_1, MINDS_2, MINDS_3 and MINDS_4 above are complete for NP.

Proof (Sketch). The simplest solution is to use earlier results [36,40] to argue that the decision versions of MINDS_3 and MINDS_4 are complete for NP. (Earlier work [19] claims NP-hardness, but citing references that do not actually prove the result.)

It is easy to reduce MINDS_3 or MINDS_4 to MINDS_1 or MINDS_2; i.e. simply ignore the other computed function. Moreover, we show below that the decision versions of MINDS_1 and MINDS_2 are in NP, by reducing these problems to SAT. Thus, completeness of MINDS_1 and MINDS_2 follows.   □

Example 5. With respect to Example 1, the decision set shown in Fig. 1b respects MINDS_2, MINDS_3 and MINDS_4, whereas the decision set of Fig. 1b also respects MINDS_1 and MINDS_0.

In the sections below we investigate SAT-based models for computing decision sets, under one of the relaxed optimization models MINDS_1, MINDS_2, MINDS_3 or MINDS_4. Moreover, we investigate symmetry breaking properties of the problem formulation, which can be used for constraining any of these models.

4.1 SAT Models for MINDS_3 and MINDS_4

This section details SAT models for solving MINDS_3. With minor modifications, similar models can be devised for MINDS_4. The purpose of MINDS_3 is to find a minimum-size representation of F^1, subject to a non-$\mathcal{U}^{\ominus}$-overlap constraint with respect to E^0. To solve this problem, a number of propositional models can be envisioned. We first investigate a model proposed in the literature [19,37,38]. Afterwards, we detail a new model, aiming at better performance when using SAT solvers. Both propositional models encode the decision problem of MINDS_3: can F^1 be represented with N terms? The model considers a grid of N by K entries, each row of K entries denoting the representation of the condition of a rule or, alternatively, a term in the DNF representation of F^1, for a total of N terms. Throughout this section, it holds that $1 \leq j \leq N$ and $1 \leq r \leq K$, with q associated with some example e_q from $\mathcal{E}$, $\mathcal{E}^-$ or $\mathcal{E}^+$.

An Existing SAT Model. One model, proposed by Kamath et al. [19], assumes the representation of a Boolean function in terms of K-dimensional points describing the functions ON-set and the OFF-set, respectively E^1 and E^0 in our case.

The variables used in the propositional representation are:

- $p_{jr} = 1$ iff x_i not included in term j.
- $p'_{jr} = 1$ iff $\neg x_i$ not included in term j.
- sl^q_{jr}: replace either with p'_{jr} if feature f_r occurs positively in $e_q \in \mathcal{E}^+$, or with p_{jr} if feature f_r occurs negatively in $e_q \in \mathcal{E}^+$.
- $cr_{jq} = 1$ iff rule j covers $e_q \in \mathcal{E}^+$.

Furthermore, the constraints proposed in [19] can be translated as follows:

1. One of p_{jr} and p'_{jr} must be true:

$$(p_{jr} \vee p'_{jr}) \qquad j \in [N] \wedge r \in [K] \qquad (2)$$

2. Each negative example $e_q \in \mathcal{E}^-$, with a set of positive features P_q and a set of negative features N_q, must be discriminated by every term:

$$\left(\bigvee_{r \in P_q} \neg p'_{jr} \vee \bigvee_{r \in N_q} \neg p_{jr} \right) \qquad j \in [N] \wedge e_q \in \mathcal{E}^- \qquad (3)$$

3. Each positive example must be covered:
   - Constraint for a term not covering a positive example:

$$(sl^q_{jr} \vee \neg cr_{jq}) \qquad j \in [N] \wedge r \in [K] \wedge e_q \in \mathcal{E}^+ \qquad (4)$$

   - Each positive example must be covered by some term:

$$\left(\bigvee_{j=1}^{N} cr_{jq} \right) \qquad e_q \in \mathcal{E}^+ \qquad (5)$$

Analysis of the constraints yields the following:

Proposition 3. The model uses $\mathcal{O}(N \times M \times K)$ clauses and literals.

An Alternative Model. In contrast with the model of Kamath et al. [19], we propose a model with a different semantics for some of the variables, and a few additional clauses, to elicit propagation. As shown by the experimental results, the motivation has been to devise a model for which the computed solutions are (heuristically) easier to interpret, by specifying fewer literals.

The sets of variables to use are the following:

- s_{jr}: whether for rule j, a literal in feature r is to be skipped.
- l_{jr}: literal on feature r for rule j, in the case the feature is *not* skipped.
- d^0_{jr}: whether feature r of rule j discriminates value 0.
- d^1_{jr}: whether feature r of rule j discriminates value 1.
- cr_{jq}: whether (*used*) rule j covers $e_q \in \mathcal{E}^+$.
  (Observe that this variable is also used in the existing model [19].)

The constraints encoding MINDS$_3$ are:

1. Each term must have some literals:

$$\left(\bigvee_{r=1}^{K} \neg s_{jr} \right) \qquad j \in [N] \qquad (6)$$

2. One must be able to account for which literals are discriminated by which rules:

$$\begin{aligned} d^0_{jr} &\leftrightarrow \neg s_{jr} \wedge l_{jr} \qquad j \in [N] \wedge r \in [K] \\ d^1_{jr} &\leftrightarrow \neg s_{jr} \wedge \neg l_{jr} \qquad j \in [N] \wedge r \in [K] \end{aligned} \qquad (7)$$

3. In addition, one must be able to discriminate all the negative examples in each term. Let $e_q \in \mathcal{E}^-$ be a negative example, and $\sigma(r,q)$ denote the sign of feature f_r for e_q. Then,

$$\left(\bigvee_{r=1}^{K} d^{\sigma(r,q)}_{j,r} \right) \qquad j \in [N] \wedge e_q \in \mathcal{E}^- \qquad (8)$$

4. We must also ensure that each positive example is covered by some rule, associated with its class.

– First, define whether a rule covers some specific positive example:

$$cr_{jq} \leftrightarrow \left(\bigwedge_{r=1}^{K} \neg d_{j,r}^{\sigma(r,q)} \right) \qquad j \in [N] \wedge e_q \in \mathcal{E}^+ \qquad (9)$$

– Second, each $e_q \in \mathcal{E}^+$ must be covered by some rule. This corresponds to (5).

Proposition 4. The propositional encoding uses $\mathcal{O}(N \times M \times K)$ clauses and literals.

4.2 SAT Models for MINDS$_1$ and MINDS$_2$

The models analyzed in the previous section, MINDS$_3$ and MINDS$_4$, learn one function for one class, e.g. F^1 for c_1. For the other class, e.g. c_0, only the original minterms are available, and a default rule that may opt to pick this other class for points of feature space not covered by F^1. It is in general possible to have more accurate representations of the two classes, by considering some of the models described earlier in this paper, concretely MINDS$_2$ and MINDS$_1$. This section develops propositional models for MINDS$_2$ and MINDS$_1$.

The Case of MINDS$_2$. It is immediate to generalize MINDS$_3$ (or MINDS$_4$) to the case of MINDS$_2$. Essentially, the constraints for discriminating classes and for covering classes must be replicated for the target classes[1].

The Case of MINDS$_1$. We consider a grid of N by K entries, each row of K entries denoting the organization of a rule. The (basic) sets of variables to use are the same as for MINDS$_3$, with the addition of c_j, representing a class variable, which is 0 if the class of rule j is false (or negative), and 1 otherwise. Moreover, the constraints encoding MINDS$_1$ are:

1. Every term *must* be used. This constraint corresponds to (6).
2. We must also be able to account for which literals are discriminated by which rules. This constraint corresponds to (7).
3. In addition, we must be able to discriminate positive examples in rules of the negative class and vice-versa. Let $e_q \in \mathcal{E}^+$ be a positive example, and $\sigma(r,q)$ be defined as above. Then,

$$\begin{aligned} \neg c_j \rightarrow \left(\bigvee_{r=1}^{K} d_{j,r}^{\sigma(r,q)} \right) \qquad & j \in [N] \wedge e_q \in \mathcal{E}^+ \\ c_j \rightarrow \left(\bigvee_{r=1}^{K} d_{j,r}^{\sigma(r,q)} \right) \qquad & j \in [N] \wedge e_q \in \mathcal{E}^- \end{aligned} \qquad (10)$$

4. We must also ensure that each example is covered by some rule, associated with its class.

---

[1] The generalization from MINDS$_3$ to MINDS$_2$ is straightforward, and omitted due to space constraints. Moreover, the model for MINDS$_1$ follows a similar approach.

- First, the constraint for a rule to cover some example:

$$cr_{jq} \leftrightarrow \left(\neg c_j \wedge \bigwedge_{r=1}^{K} \neg d_{j,r}^{\sigma_{r,q}} \right) \qquad j \in [N] \wedge e_q \in \mathcal{E}^-$$
$$cr_{jq} \leftrightarrow \left(c_j \wedge \bigwedge_{r=1}^{K} \neg d_{j,r}^{\sigma_{r,q}} \right) \qquad j \in [N] \wedge e_q \in \mathcal{E}^+ \tag{11}$$

- Second, all examples no matter the class must be covered, and so we generalize (5) to get:

$$\left(\bigvee_{j=1}^{N} cr_{jq} \right) \qquad e_q \in \mathcal{E} \tag{12}$$

Thus, every element is covered.

5. Finally, two terms associated with different classes *must not* exhibit $\mathcal{U}^{\ominus}$-overlap:

$$\neg(c_i \leftrightarrow c_j) \rightarrow \left(\bigvee_{r=1}^{K} \neg s_{ir} \wedge \neg s_{jr} \wedge \neg(l_{ir} \leftrightarrow l_{jr}) \right) \qquad i, j \in [N] \wedge i < j \tag{13}$$

4.3   Breaking Symmetries

The propositional models proposed in earlier sections essentially capture (unordered) sets of terms. The lack of order reveals a symmetry. If the number of terms is large, this can impact performance significantly. A standard technique to eliminate such symmetries in the problem formulation is to impose an order in the representation. The approach we take is to sort the terms, such that the number of each feature is inverse to the weight of the feature in the binary representation of the number associated with the term. Unspecified features have the largest weight. Clearly, imposing an order on the terms does *not* affect correctness of the propositional model.

We describe next the constraints for the alternative model proposed in Sect. 4.1. For the other models, a similar solution is used. The additional variables used are the following:

- $eq_{j,r} = 1$ iff term j equals term $j - 1$ until feature r.
- $gt_{j,r} = 1$ iff term j is greater than term $j - 1$ by feature r.

For the constraints below $j \in [N]$ and $r \in [K]$. The constraints for $eq_{j,r}$ are the following, with $eq_{j,0} = 1$:

$$eq_{j,r} \leftrightarrow eq_{j,r-1} \wedge \left(s_{j-1,r} \wedge s_{j,r} \vee d_{j-1,r}^1 \wedge d_{j,r}^1 \vee d_{j-1,r}^0 \wedge d_{j,r}^0 \right) \tag{14}$$

The constraints for gt_{jr}, with $gt_{j0} = 0$, are the following:

$$gt_{j,r} \leftrightarrow gt_{j,r-1} \vee eq_{j,r-1} \wedge \neg s_{j-1,r} \wedge s_{j,r} \vee eq_{j,r-1} \wedge d_{j-1,r}^1 \wedge d_{j,r}^0 \tag{15}$$

Observe that distinguishing a positive literal corresponds to accepting a negative literal. Clearly, additional variables can be introduced to enable clausification [39]. Finally, for the last feature, each term must be greater than the preceeding one:

$$\bigwedge_{j=2}^{N} (gt_{j,K}) \tag{16}$$

5 Experimental Results

This section evaluates the ideas studied in the paper given a variety of datasets.

5.1 Experimental Setup

The proposed models were implemented in a prototype as a Python script instrumenting calls to the MiniSat 2.2 SAT solver [10]. More precisely, the weakened models MinDS_i, $i \in [4]$, were implemented. Although all these models target binary classification, most of the practical benchmark datasets require non-binary classification. Therefore, the implemented prototype supports non-binary classification as well. As a result, we deem interesting for the evaluation to check the performance of models MinDS_2 and MinDS_1 generalized to an arbitrary number of classes, and so we *do not* test models MinDS_3 and MinDS_4, as they are expected to be easier to deal with. Also note that the implementation supports both encodings of MinDS_3 (and, thus, of generalized MinDS_2) studied in the paper: (1) the existing encoding [19] and (2) the alternative encoding proposed above. In the following, the novel encoding of MinDS_2 is simply called MinDS_2 while the encoding from [19] is referred to as MP92. Additionally and for testing how helpful the proposed symmetry breaking predicates (SBPs) are, the basic models were augmented with SBPs resulting in the following configurations: MinDS_2+SBP, MinDS_1+SBP, and MP92+SBP. Finally, IDS[2], a recent approach [21] based on *smooth local search* [12], was also tested in the evaluation. IDS uses the Apriori algorithm [1] for generating candidate itemsets, with the default support threshold[3] equal to 0.2. For simplifying the problem solved by IDS, we increased this value to 0.5, which resulted in two configurations of IDS to run: IDS-supp0.2 and IDS-supp0.5.

The experiments were performed on a subset of datasets of the PMLB repository[4] [31]. The number of samples in the selected datasets varies from 87 to 49621[5] ($\approx$1651.1 on average) while the number of original (i.e. non-binary)

---

[2] https://github.com/lvhimabindu/interpretable_decision_sets/.

[3] A support threshold parameter ϵ in the Apriori algorithm ensures that the candidate itemsets are present in at least ϵ data points.

[4] https://github.com/EpistasisLab/penn-ml-benchmarks/.

[5] Some of the PMLB datasets are inconsistent, i.e. they have multiple occurrences of the same samples marked by different labels. Since the proposed models assume consistent data, the datasets were replaced by their largest consistent subsets. The number of samples shown above corresponds to the size of the resulting consistent datasets.

Table 1. Number of solved instances per model (out of 49 in total).

MP92	MP92+SBP	MinDS$_2$	MinDS$_2$+SBP	MinDS$_1$	MinDS$_1$+SBP	IDS-supp0.2	IDS-supp0.5
42	45	42	45	6	6	0	2

features varies from 4 to 59 ($\approx$15.1 on average). Applying the *one-hot encoding* results in 6 to 2232 binary features ($\approx$353.1 on average). The total number of selected datasets is 49.

All the conducted experiments were performed in Ubuntu Linux on an Intel Xeon E5-2630 2.60 GHz processor with 64 GB of memory. The time limit was set to 600s and the memory limit to 10 GB for each individual process to run. The experimental evaluation was divided into two parts detailed below.

5.2   Testing Scalability

The number of benchmarks solved by each competitor is shown in Table 1. Given the large number of binary features in the datasets, the performance of both MP92 and MinDS$_2$ can be regarded as quite positive. As expected, symmetry breaking improves it further: MP92+SBP and MinDS$_2$+SBP solve all but 4 instances. Observe that MinDS$_1$ and MinDS$_1$+SBP perform significantly worse: these models can solve only 6 instances. However, this is not surprising given that MinDS$_1$ targets computing decision sets exhibiting no $\mathcal{U}^{\ominus}$-overlap, which is in general significantly harder to solve.

Assessing the Performance of IDS [21]. As shown in Table 1, and in contrast to the SAT-based models studied, IDS [21] performs quite poorly in practice. With the default support threshold 0.2, IDS is unable to solve (within 600 s) any instance, and it can solve only 2 instances if the support threshold is increased to 0.5. Moreover, and although IDS aims at maximizing the number of covered training samples and minimizing the rule overlap, the rules produced by IDS exhibit significant overlap, even on examples taken from the training data[6]. Given the poor performance of IDS and the weak guarantees in terms of rule overlap, the rest of this section focuses solely on the SAT-based models.

Performance on Subsampled Datasets. To investigate the performance of the models further, we (1) discarded the 6 instances solved by all models and (2) subsampled the remaining 43 (49 − 6) benchmarks in the following way. For each dataset, we randomly selected 5%, 10%, 20%, and 50% of training samples and repeated this procedure 20 times for each percentage value. This resulted in 80 randomly subsampled datasets for each of the 43 benchmarks. The total number of subsampled benchmarks is 3440.

---

[6] These surprising results motivated in part our detailed analysis of overlap. It should be noted that the authors of IDS [21] have been informed of IDS's poor performance and poor ability to avoid rule overlap, but have been unable to justify the results of IDS.

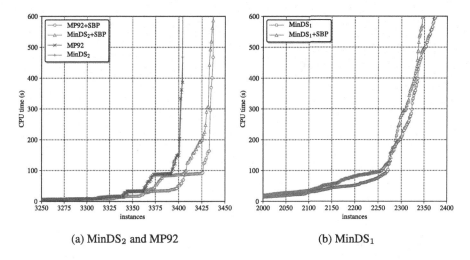

(a) MinDS$_2$ and MP92   (b) MinDS$_1$

Fig. 2. Performance of the considered models on subsampled datasets.

Figure 2 depicts the performance of the proposed models on the subsampled benchmarks. As shown in Fig. 2a, MP92 and MinDS$_2$ demonstrate almost the same performance and solve successfully 3404 benchmarks. Enabling symmetry breaking proves itself helpful allowing them to solve 33 more instances, i.e. 3437 overall. This is, however, not the case for MinDS$_1$, which solves 2374 instances (2346 instances, resp.) if SBPs are disabled (enabled, resp.). This can be explained by the benchmarks' nature as they have a large number of classes and training samples while the solutions are not large enough for SBPs to pay off. Note that although MinDS$_1$ performs significantly worse than MP92 and MinDS$_2$, it can still solve ≈70% of the subsampled benchmarks. These results should be regarded as significant given the size and the properties of the tested datasets, as well as the fact that MinDS$_1$ targets no $\ominus$ overlap on the *complete feature space*. To our best knowledge, these results are far beyond the state of the art [21], and enable solving to optimality a whole new range of challenging datasets.

5.3   Assessing Quality

Observe that the proposed models target minimizing the number of rules in the target decision sets, rather than their total size, i.e. the total number of literals used. Hence, it is of interest to compare the "quality" of solutions reported by MP92 and the novel model MinDS$_2$. One option is to simply compare the number of literals in the decision sets reported by the two models. Alternatively, one can try to minimize the number of literals in the resulting decision sets, by applying Boolean lexicographic optimization (BLO) [23], as soon as a decision set with the smallest number of rules is computed. For this, as soon as the number of rules in the decision set is minimized (i.e. the corresponding CNF formula is satisfiable),

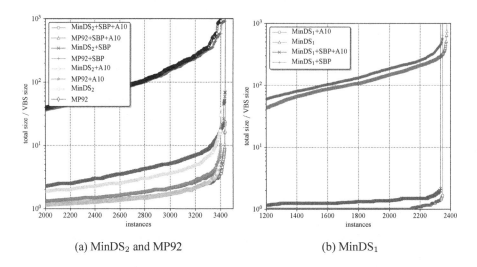

(a) MinDS$_2$ and MP92   (b) MinDS$_1$

Fig. 3. Quality of solutions computed with the considered models.

a simple MaxSAT problem can be devised by augmenting the formula with unit soft clauses, which force all literals of the decision set to be unused. This can be applied to any model MINDS$_i$, $i \in [4]$. Afterwards, the minimum number of literals can be computed by a standalone MaxSAT solver or approximated with the use of an MCS (minimal correction subset) enumerator [24]. While the former approach is exact, it is often outperformed by the latter one. For the purpose of the evaluation, we tried both options with every model considered. The MaxSAT solver used for computing exact solutions was MSCG [29] while the approximation of MaxSAT solutions was done by computing first 10 MCSes with the LBX algorithm [25].

Evaluation of the quality of solutions was done in the following way. Additionally to the tested configurations of MinDS$_2$ and MP92, all of them were ran in the BLO mode with literal minimization done by (1) a MaxSAT solver and (2) an MCS enumerator. Among all configurations, a *virtual best solver* (VBS) was constructed w.r.t. the total number of literals in the solution. Afterwards, we measured how much larger the decision sets for each tested configuration are w.r.t. the VBS, i.e. given the number of literals L in the solution produced by the configuration and the number of literals L^* in the VBS solution, we considered value L/L^*. A similar study was done for MinDS$_1$.

Figure 3 shows the quality of solutions for all tested models. (Y-axis here is scaled logarithmically.) Here, the configurations marked by *+A10* compute 10 MCSes to approximate the solution. All configurations that use a MaxSAT solver represent a constant $f(x) = 1$ and, thus, are omitted in Fig. 3. However, note that they participate in the VBS. In general, our experiments suggest that the MaxSAT-based literal minimization is expensive and results in only ≈85% of the instances solved. One surprising observation is how much worse the quality of

MP92's solutions is when compared to $MinDS_2$ (see Fig. 3a). In some cases, decision sets learned by MP92 have 3 orders of magnitude more literals than the VBS decision sets and 1–2 orders of magnitude more literals than solutions computed by $MinDS_2$. On the other hand, these results indicate that approximate literal number minimization after learning a target decision set is feasible and does not degrade the performance of the overall procedure if done by enumerating a fixed number of MCSes. This is confirmed for the case of $MinDS_1$ (see Fig. 3b). Note that efficient minimization of the total number of literals in the target decision sets is crucial given the requirement that they must be interpretable.

6   Conclusions and Research Directions

Decision (or rule) sets represent a promising approach for providing explanations in different ML settings. This paper shows that learning optimal decision sets raises a number of difficulties, related with overlap of rules, especially when the rules are associated with different classes. The paper conjectures that the exact solution for the learning problem of decision sets, while ensuring no overlap, is hard for the second level of the polynomial hierarchy. Moreover, the paper proposes a number of alternative problem formulations, all of which are shown to be hard for NP, and develops SAT-based solutions, relating with earlier work [19]. The experimental results, obtained on representative datasets, confirm the relevance of the approach, and yield a number of conclusions. Compared with earlier work [21], that exploits a variant of local search, the proposed SAT-based approach is not only far more accurate, but also remarkably more efficient. The results provide evidence that SAT-based learning of optimal decision sets can handle practical datasets of interest, when the goal is to devise ML models that associate explanations with predictions.

The promising results in the paper motivate a number of lines of work, including proving (or disproving) the paper's main conjecture, developing more efficient propositional encodings, but also to consider other approaches that enable finding an optimal solution to the learning problem for decision sets. Also and as mentioned on the paper, the proposed approach can be adapted to study the problem from another perspective, i.e. by minimizing the total number of literals in a decision set instead the number of rules, or alternatively refer to multi-objective optimization. This approach may result in smaller and, thus, better interpretable solutions, in which case it would be appealing to compare it again heuristic rule-based classifiers targeting this same problem, e.g. CN2 [6,7] and PRISM [5] among others. One additional natural line of work will be to extend the work to rule lists [2], but also to more expressive function representation languages, while preserving the ability to provide explanations for predictions.

References

1. Agrawal, R., Srikant, R.: Fast algorithms for mining association rules in large databases. In: VLDB, pp. 487–499 (1994)
2. Angelino, E., Larus-Stone, N., Alabi, D., Seltzer, M., Rudin, C.: Learning certifiably optimal rule lists. In: KDD, pp. 35–44 (2017)
3. Bessiere, C., Hebrard, E., O'Sullivan, B.: Minimising decision tree size as combinatorial optimisation. In: CP, pp. 173–187 (2009)
4. Biere, A., Heule, M., van Maaren, H., Walsh, T. (eds.): Handbook of Satisfiability. IOS Press, Amsterdam (2009)
5. Cendrowska, J.: PRISM: an algorithm for inducing modular rules. Int. J. Man Mach. Stud. **27**(4), 349–370 (1987)
6. Clark, P., Boswell, R.: Rule induction with CN2: some recent improvements. In: Kodratoff, Y. (ed.) EWSL 1991. LNCS, vol. 482, pp. 151–163. Springer, Heidelberg (1991). https://doi.org/10.1007/BFb0017011
7. Clark, P., Niblett, T.: The CN2 induction algorithm. Mach. Learn. **3**, 261–283 (1989)
8. Cohen, W.W.: Fast effective rule induction. In: ICML, pp. 115–123 (1995)
9. DARPA: DARPA explainable Artificial Intelligence (XAI) program (2016). https://www.darpa.mil/program/explainable-artificial-intelligence
10. Eén, N., Sörensson, N.: An extensible SAT-solver. In: SAT, pp. 502–518 (2003)
11. EU Data Protection Regulation: Regulation (EU) 2016/679 of the European Parliament and of the Council (2016). http://eur-lex.europa.eu/legal-content/EN/TXT/PDF/?uri=CELEX:32016R0679&from=en
12. Feige, U., Mirrokni, V.S., Vondrák, J.: Maximizing non-monotone submodular functions. SIAM J. Comput. **40**(4), 1133–1153 (2011)
13. Fürnkranz, J., Gamberger, D., Lavrac, N.: Foundations of Rule Learning. Springer, Heidelberg (2012). https://doi.org/10.1007/978-3-540-75197-7
14. Han, J., Kamber, M., Pei, J.: Data Mining: Concepts and Techniques, 3rd edn. Morgan Kaufmann, Burlington (2012)
15. Hauser, J.R., Toubia, O., Evgeniou, T., Befurt, R., Dzyabura, D.: Disjunctions of conjunctions, cognitive simplicity, and consideration sets. J. Mark. Res. **47**(3), 485–496 (2010)
16. ICML, WHI Workshop: ICML Workshop on Human Interpretability in Machine Learning. https://sites.google.com/view/whi2017/home. Accessed Aug 2017
17. IJCAI, XAI Workshop: IJCAI Workshop on Explainable Artificial Intelligence (XAI). http://home.earthlink.net/~dwaha/research/meetings/ijcai17-xai/. Accessed Aug 2017
18. Jordan, M.I., Mitchell, T.M.: Machine learning: trends, perspectives, and prospects. Science **349**(6245), 255–260 (2015)
19. Kamath, A.P., Karmarkar, N., Ramakrishnan, K.G., Resende, M.G.C.: A continuous approach to inductive inference. Math. Program. **57**, 215–238 (1992)
20. Kearns, M.J., Li, M., Valiant, L.G.: Learning boolean formulas. J. ACM **41**(6), 1298–1328 (1994)
21. Lakkaraju, H., Bach, S.H., Leskovec, J.: Interpretable decision sets: a joint framework for description and prediction. In: KDD, pp. 1675–1684 (2016)
22. LeCun, Y., Bengio, Y., Hinton, G.: Deep learning. Nature **521**(7553), 436 (2015)
23. Marques-Silva, J., Argelich, J., Graça, A., Lynce, I.: Boolean lexicographic optimization: algorithms & applications. Ann. Math. Artif. Intell. **62**(3–4), 317–343 (2011)

24. Marques-Silva, J., Heras, F., Janota, M., Previti, A., Belov, A.: On computing minimal correction subsets. In: IJCAI, pp. 615–622 (2013)
25. Mencía, C., Previti, A., Marques-Silva, J.: Literal-based MCS extraction. In: IJCAI, pp. 1973–1979 (2015)
26. Michalski, R.S.: On the quasi-minimal solution of the general covering problem. In: International Symposium on Information Processing, pp. 125–128 (1969)
27. Mitchell, T.M.: Machine Learning. McGraw-Hill, New York (1997)
28. Mnih, V., Kavukcuoglu, K., Silver, D., Rusu, A.A., Veness, J., Bellemare, M.G., Graves, A., Riedmiller, M., Fidjeland, A.K., Ostrovski, G., et al.: Human-level control through deep reinforcement learning. Nature **518**(7540), 529 (2015)
29. Morgado, A., Ignatiev, A., Marques-Silva, J.: MSCG: robust core-guided MaxSAT solving. JSAT **9**, 129–134 (2015)
30. NIPS, IML Symposium: NIPS Interpretable ML Symposium. http://interpretable. ml/. Accessed Dec 2017
31. Olson, R.S., La Cava, W., Orzechowski, P., Urbanowicz, R.J., Moore, J.H.: PMLB: a large benchmark suite for machine learning evaluation and comparison. BioData Min. **10**(1), 36 (2017)
32. Pedregosa, F., et al.: Scikit-learn: machine learning in Python. J. Mach. Learn. Res. **12**, 2825–2830 (2011)
33. Quinlan, J.R.: Generating production rules from decision trees. In: IJCAI, pp. 304–307 (1987)
34. Quinlan, J.R.: C4.5: Programs for Machine Learning. Morgan Kauffmann, Burlington (1993)
35. Rivest, R.L.: Learning decision lists. Mach. Learn. **2**(3), 229–246 (1987)
36. Schaefer, M., Umans, C.: Completeness in the polynomial-time hierarchy: a compendium. SIGACT news **33**(3), 32–49 (2002)
37. Triantaphyllou, E.: Inference of a minimum size boolean function from examples by using a new efficient branch-and-bound approach. J. Global Optim. **5**(1), 69–94 (1994)
38. Triantaphyllou, E.: Data Mining and Knowledge Discovery via Logic-Based Methods: Theory, Algorithms, and Applications. Springer, Heidelberg (2010). https:// doi.org/10.1007/978-1-4419-1630-3
39. Tseitin, G.S.: On the complexity of derivation in propositional calculus. In: Siekmann, J.H., Wrightson, G. (eds.) Automation of Reasoning. Symbolic Computation (Artificial Intelligence), pp. 466–483. Springer, Heidelberg (1983). https://doi.org/ 10.1007/978-3-642-81955-1_28
40. Umans, C., Villa, T., Sangiovanni-Vincentelli, A.L.: Complexity of two-level logic minimization. IEEE Trans. CAD Integr. Circuits Syst. **25**(7), 1230–1246 (2006)
41. Wang, T., Rudin, C., Doshi-Velez, F., Liu, Y., Klampfl, E., MacNeille, P.: Or's of And's for interpretable classification, with application to context-aware recommender systems. CoRR, abs/1504.07614 (2015)
42. Wang, T., Rudin, C., Doshi-Velez, F., Liu, Y., Klampfl, E., MacNeille, P.: A Bayesian framework for learning rule sets for interpretable classification. J. Mach. Learn. Res. **18**, 1–37 (2017)

Proof-Producing Synthesis of CakeML with I/O and Local State from Monadic HOL Functions

Son Ho[1], Oskar Abrahamsson[2], Ramana Kumar[3], Magnus O. Myreen[2(✉)], Yong Kiam Tan[4], and Michael Norrish[5]

[1] MINES ParisTech, PSL Research University, Paris, France
[2] Chalmers University of Technology, Gothenburg, Sweden
myreen@chalmers.se
[3] Data61, CSIRO, UNSW, Sydney, Australia
[4] Carnegie Mellon University, Pittsburgh, USA
[5] Data61, CSIRO, ANU, Canberra, Australia

Abstract. We introduce an automatic method for producing stateful ML programs together with proofs of correctness from monadic functions in HOL. Our mechanism supports references, exceptions, and I/O operations, and can generate functions manipulating local state, which can then be encapsulated for use in a pure context. We apply this approach to several non-trivial examples, including the type inferencer and register allocator of the otherwise pure CakeML compiler, which now benefits from better runtime performance. This development has been carried out in the HOL4 theorem prover.

1 Introduction

This paper is about bridging the gap between programs verified in logic and verified implementations of those programs in a programming language (and ultimately machine code). As a toy example, consider computing the nth Fibonacci number. Here is a recursion equation for a function, fib, in higher-order logic (HOL) that does the job.

$$\text{fib } n = \text{if } n < 2 \text{ then } n \text{ else fib } (n-1) + \text{fib } (n-2)$$

A hand-written implementation (shown here in CakeML [9], which has similar syntax and semantics to Standard ML) would look something like this:

```
fun fiba i j n = if n = 0 then i else fiba j (i+j) (n-1);
(print (n2s (fiba 0 1 (s2n (hd (CommandLine.arguments()))))));
 print "\n")
handle _ => print_err ("usage:" ^ CommandLine.name() ^ "<n>\n");
```

In moving from mathematics to a real implementation, some issues are apparent:

(1) We use a tail-recursive linear-time algorithm, rather than the exponential-time recursion equation.

© Springer International Publishing AG, part of Springer Nature 2018
D. Galmiche et al. (Eds.): IJCAR 2018, LNAI 10900, pp. 646–662, 2018.
https://doi.org/10.1007/978-3-319-94205-6_42

(2) The whole program is not a pure function: it does I/O, reading its argument from the command line and printing the answer to standard output.

(3) We use exception handling to deal with malformed inputs (if the arguments do not start with a string representing a natural number, hd or s2n may raise an exception).

The first of these issues (1) can easily be handled in the realm of logical functions: We define the tail-recursive version in logic

$$\text{fiba } i\ j\ n = \text{if } n = 0 \text{ then } i \text{ else fiba } j\ (i+j)\ (n-1)$$

then produce a correctness theorem, $\vdash \forall n.\ \text{fiba } 0\ 1\ n = \text{fib } n$, with a simple inductive proof (a 5-line tactic proof in HOL4, not shown).

Now, because fiba is a logical function with an obvious computational counterpart, we can use proof-producing synthesis techniques [13] to automatically synthesise code verified to compute it. We thereby produce something like the first line of the CakeML code above, along with a theorem relating the semantics of the synthesised code back to the function in logic.

But when it comes to handling the other two issues, (2) and (3), and producing and verifying the remaining three lines of CakeML code, our options are less straightforward. The first issue was easy because we were working with a *shallow embedding*, where one writes the program as a function in logic and proves properties about that function directly. Shallow embeddings rely on an analogy between mathematical functions and procedures in a pure functional programming language. Effects, however, like state, I/O, and exceptions, can stretch this analogy too far. The alternative is a *deep embedding*: one writes the program as an input to a formal semantics, which can accurately model computational effects, and proves properties about its execution under those semantics.

Proofs about shallow embeddings are relatively easy since they are in the native language of the theorem prover, whereas proofs about deep embeddings are filled with tedious details because of the indirection through an explicit semantics. Still, the explicit semantics make deep embeddings more realistic. An intermediate option that is suitable for the effects we are interested in—state/references, exceptions, and I/O—is to use *monadic functions*: one writes (shallow) functions that represent computations, aided by a composition operator (monadic bind) for stitching together effects. The monadic approach to writing effectful code in a pure language may be familiar from the Haskell language which made it popular.

For our nth Fibonacci example, we can model the effects of the whole program with a monadic function, fibm, that calls the pure function fiba to do the calculation. Figure 1 shows how fibm can be written using do-notation familiar from Haskell. This is as close as we can get to capturing the effectful behaviour of the desired CakeML program while remaining in a shallow embedding. Now how can we produce real code along with a proof that it has the correct semantics? If we use the proof-producing synthesis techniques mentioned above [13], we produce *pure* CakeML code that exposes the monadic plumbing in an explicit

```
fibm () =
do
  args ← commandline (arguments ());
  a ← hd args;
  n ← s2n a;
  stdio (print (n2s (fiba 0 1 n)));
  stdio (print "\n")
od otherwise
do
  name ← commandline (name ());
  stdio (print_err ("usage: " ^ name ^ " <n>\n"))
od
```

Fig. 1. The Fibonacci program written using do-notation in logic.

state-passing style. But we would prefer verified *effectful* code that uses native features of the target language (CakeML) to implement the monadic effects.

In this paper, we present an automated technique for producing verified effectful code that handles I/O, exceptions, and other issues arising in the move from mathematics to real implementations. Our technique systematically establishes a connection between shallowly embedded functions in HOL with monadic effects and deeply embedded programs in the impure functional language CakeML. The synthesised code is efficient insofar as it uses the native effects of the target language and is close to what a real implementer would write. For example, given the monadic fibm function above, our technique produces essentially the same CakeML program as on the first page (but with a `let` for every monad bind), together with a proof that the synthesised program is a refinement.

Contributions. Our technique for producing verified effectful code from monadic functions builds on a previous limited approach [13]. The new generalised method adds support for the following features:

- global references and exceptions (as before, but generalised),
- mutable arrays (both fixed and variable size),
- input/output (I/O) effects,
- local mutable arrays and references, which can be integrated seamlessly with code synthesis for otherwise pure functions, and,
- composable effects, whereby different state and exception monads can be combined using a lifting operator.

As a result, we can now write *whole programs* as shallow embeddings and obtain real verified code via synthesis. Prior to this paper, whole program verification in CakeML involved manual deep embedding proofs for (at the very least) the I/O wrapper. To exercise our toolchain, we apply it to several examples:

- the nth Fibonacci example already seen (exceptions, I/O)
- the Floyd Warshall algorithm for finding shortest paths (arrays)
- the CakeML compiler's type inferencer (local refs, exceptions)
- the CakeML compiler's register allocator (local refs, arrays)
- the Candle theorem prover's kernel [8] (global refs, exceptions)
- an OpenTheory [7] article checker (global refs, exceptions, I/O).

In Sect. 5, we compare runtimes with the previous non-stateful versions of CakeML's register allocator and type inferencer; and for the OpenTheory reader we compare the amount of code/proof required before and after using our technique.

The HOL4 development is at https://code.cakeml.org; our new synthesis tool is at https://code.cakeml.org/tree/master/translator/monadic.

2   High-Level Ideas

This paper combines the following three concepts in order to deliver the contributions listed above. The main ideas will be described briefly in this section, while subsequent sections will provide details. The three concepts are:

(i) synthesis of stateful ML code as described in our previous work [13],
(ii) separation logic [15] as used by characteristic formulae for CakeML [5], and
(iii) a new abstract synthesis mode for the CakeML synthesis tools [13].

Our previous work on proof-producing synthesis of stateful ML (i) was severely limited by the requirement to have a hard-coded invariant on the program's state. There was no support for I/O and all references had to be declared globally. At the time of developing (i), we did not have a satisfactory way of generalising the hard-coded state invariant.

In this paper we show (in Sect. 3) that the separation logic of CF (ii) can be used to neatly generalise the hard-coded state invariant of our prior work (i). CF-style separation logic easily supports references and arrays, including resizable arrays, and, supports I/O too because it allows us to treat I/O components as if they are heap components. Furthermore, by carefully designing the integration of (i) and (ii), we retain the frame rule from the separation logic. In the context of code synthesis, this frame rule allows us to implement a lifting feature for changing the type of the state-and-exception monads. Being able to change types in the monads allows us to develop *reusable* libraries—e.g. verified file I/O functions—that users can lift into the monad that is appropriate for their application.

The combination of (i) and (ii) does not by itself support synthesis of code with local state due to inherited limitations of (i), wherein the generated code must be produced as a concrete list of global declarations. For example, if monadic functions, say foo and bar, refer to a common reference, say r, the reference r must be defined globally:

```
val r = ref 0;
fun foo n = ...; (* code that uses r *)
fun bar n = ...; (* code that uses r and calls foo *)
```

In this paper (in Sect. 4), we introduce a new *abstract* synthesis mode (iii) which removes the requirement of generating code that only consists of a list of global declarations, and, as a result, we are now able to synthesise code such as the following, where reference r is a local variable.

```
fun pure_bar k n =
  let
    val r = ref k
    fun foo n = ... (* code that uses r *)
    fun bar n = ... (* code that uses r and calls foo *)
  in Success (bar n) end
  handle e => Failure e;
```

In the input to the synthesis tool, this declaration and initialisation of local state corresponds to applying the state-and-exception monad. Expressions that fully apply the state-and-exception monad can subsequently be used in the synthesis of *pure* CakeML code: the monadic synthesis tools can prove a pure specification for such expressions, thereby encapsulating the monadic features.

3 Generalised Approach to Synthesis of Stateful ML Code

This section describes how our previous approach to proof-producing synthesis of stateful ML code [13] has been generalised. In particular, we explain how the separation logic from our previous work on characteristic formulae [5] has been used for the generalisation (Sect. 3.3); and how this new approach adds support for user-defined references, fixed- and variable-length arrays, I/O functions (Sect. 3.4), and a handy feature for reusing state-and-exception monads (Sect. 3.5).

In order to make this paper as self-contained as possible, we start with a brief look at how the semantics of CakeML is defined (Sect. 3.1) and how our previous work on synthesis of pure CakeML code works (Sect. 3.2), since the new synthesis method for stateful code is an evolution of the original approach for pure code.

3.1 Preliminaries: CakeML Semantics

The semantics of the CakeML language is defined in the *functional big-step* style [14], which means that the semantics is an interpreter defined as a functional program in the logic of a theorem prover.

The definition of the semantics is layered. At the top-level the semantics function defines what the observable I/O events are for a given whole program. However, more relevant to the presentation in this paper is the next layer down:

a function called evaluate that describes exactly how expressions evaluate. The type of the evaluate function is shown below. This function takes as arguments a state (with a type variable for the I/O environment), a value environment, and a list of expressions to evaluate. It returns a new state and a value result.

$$\text{evaluate} : \delta \text{ state} \rightarrow$$
$$\text{v sem\_env} \rightarrow \text{exp list} \rightarrow \delta \text{ state} \times (\text{v list, v}) \text{ result}$$

The semantics state is defined as the record type below. The fields relevant for this presentation are: refs, clock and ffi. The refs field is a list of store values that acts as a mapping from reference names (list index) to reference and array values (list element). The clock is a logical clock for the functional big-step style. The clock allows us to prove termination of evaluate and is, at the same time, used for reasoning about divergence. Lastly, ffi is the parametrised oracle model of the foreign function interface, i.e. I/O environment.

$$\delta \text{ state} = <| \text{ clock} : \text{num} ; \text{ refs} : \text{store\_v list} ; \text{ ffi} : \delta \text{ ffi\_state} ; \ldots |>$$

$$\text{where } \text{store\_v} = \text{Refv v} | \text{ W8array (word8 list)} | \text{ Varray (v list)}$$

A call to the function evaluate returns one of two results: Rval *res* for successfully terminating computations, and Rerr *err* for stuck computations.

Successful computations, Rval *res*, return a list *res* of CakeML values. CakeML values are modelled in the semantics using a datatype called v. This datatype includes (among other things) constructors for (mutually recursive) closures (Closure and Recclosure), datatype constructor values (Conv), and literal values (Litv) such as integers, strings, characters etc. These will be explained when needed in the rest of the paper.

Stuck computations, Rerr *err*, carry an error value *err* that is one of the following. For this paper, Rraise *exc* is the most relevant case.

- Rraise *exc* indicates that evaluation results in an uncaught exception *exc*. These exceptions can be caught with a handle in CakeML.
- Rabort Rtimeout_error indicates that evaluation of the expression consumes all of the logical clock. Programs that hit this error for all initial values of the clock are considered diverging.
- Rabort Rtype_error, for other kinds of errors, e.g. when evaluating ill-typed expressions, or attempting to access unbound variables.

3.2   Preliminaries: Synthesis of Pure ML Code

Our previous work [13] describes a *proof-producing* algorithm for synthesising CakeML functions from functions in higher-order logic. Here proof-producing means that each execution proves a theorem (called a certificate theorem) guaranteeing correctness of that execution of the algorithm. In our setting, these theorems relate the CakeML semantics of the synthesised code with the given HOL function.

The whole approach is centred around a systematic way of proving theorems relating HOL functions (i.e. HOL terms) with CakeML expressions. In order for us to state relations between HOL terms and CakeML expressions, we need a way to state relations between HOL terms and CakeML values. For this we use relations (int, list _, _ $\longrightarrow$ _, etc.) which we call refinement invariants. The definition of the simple int refinement invariant is shown below: int i v is true if CakeML value v of type v represents the HOL integer i of type int.

$$\text{int } i = (\lambda\, v.\; v = \text{Litv (IntLit } i))$$

Most refinement invariants are more complicated, e.g. list (list int) xs v states that CakeML value v represents lists of int lists xs of HOL type int list list.

We now turn to CakeML expressions: we define a predicate called Eval in order to conveniently state relationships between HOL terms and CakeML expressions. The intuition is that Eval env exp P is true if exp evaluates (in environment env) to some result res (of HOL type v) such that P holds for res, i.e. P res. The formal definition below is cluttered by details regarding the clock and references: there must be a large enough clock and exp may allocate new references, $refs'$, but must not modify any existing references, $refs$. We express this restriction on the references using list append ++. Note that any list index that can be looked up in $refs$ has the same look up in $refs$ ++ $refs'$.

Eval env exp P $\iff$
$\forall$ $refs$.
   $\exists$ res $refs'$ ck.
      (evaluate (empty $with$ <| refs := $refs$; clock := ck |>) env [exp] =
      (empty $with$ refs := $refs$ ++ $refs'$, Rval [res])) $\wedge$ P res

The use of Eval and the main idea behind the synthesis algorithm is most conveniently described using an example. The example we consider here is the following HOL function:

$$\text{add1} = \lambda\, x.\; x + 1$$

The main part of the synthesis algorithm proceeds as a syntactic bottom-up pass over the given HOL term. In this case, the bottom-up pass traverses HOL term $\lambda\, x.\; x + 1$. The result of each stage of the pass is a theorem stated in terms of Eval in the format shown below. Such theorems state a connection between a HOL term t and some generated $code$ w.r.t. a refinement invariant ref_inv that is appropriate for the type of t.

general format: $assumptions$ $\Rightarrow$ Eval env $code$ (ref_inv t)

For our little example, the algorithm derives the following theorems for the subterms x and 1, which are the leaves of the HOL term. Here and elsewhere in this paper, we display CakeML abstract syntax as concrete syntax inside $\lfloor \cdots \rfloor$, i.e. $\lfloor 1 \rfloor$ is actually the CakeML expression Lit (IntLit 1) in the theorem prover HOL4; similarly $\lfloor x \rfloor$ is actually displayed as Var (Short "x") in HOL4. Note that both theorems below are of the required form.

$$\vdash \mathsf{T} \Rightarrow \mathsf{Eval}\ env\ \lfloor 1 \rfloor\ (\mathsf{int}\ 1)$$
$$\vdash \mathsf{Eval}\ env\ \lfloor x \rfloor\ (\mathsf{int}\ x) \Rightarrow \mathsf{Eval}\ env\ \lfloor x \rfloor\ (\mathsf{int}\ x) \tag{1}$$

The algorithm uses theorems (1) when proving a theorem for the compound expression $x + 1$. The process is aided by an auxiliary lemma for integer addition, shown below. The synthesis algorithm is supported by several such pre-proved lemmas for various common operations.

$$\vdash \mathsf{Eval}\ env\ x_1\ (\mathsf{int}\ n_1) \Rightarrow$$
$$\mathsf{Eval}\ env\ x_2\ (\mathsf{int}\ n_2) \Rightarrow$$
$$\mathsf{Eval}\ env\ \lfloor x_1 + x_2 \rfloor\ (\mathsf{int}\ (n_1 + n_2))$$

By choosing the right specialisations for the variables, x_1, x_2, n_1, n_2, the algorithm derives the following theorem for the body of the running example. Here the assumption on evaluation of $\lfloor x \rfloor$ was inherited from (1).

$$\vdash \mathsf{Eval}\ env\ \lfloor x \rfloor\ (\mathsf{int}\ x) \Rightarrow \mathsf{Eval}\ env\ \lfloor x + 1 \rfloor\ (\mathsf{int}\ (x + 1)) \tag{2}$$

Next, the algorithm needs to introduce the λ-binder in $\lambda x.\ x + 1$. This can be done by instantiation of the following pre-proved lemma. Note that the lemma below introduces a refinement invariant for function types, $\longrightarrow$, which combines refinement invariants for the input and output types of the function [13].

$$\vdash (\forall v\ x.\ a\ x\ v \Rightarrow \mathsf{Eval}\ (env\ [n \mapsto v])\ body\ (b\ (f\ x))) \Rightarrow$$
$$\mathsf{Eval}\ env\ \lfloor \mathtt{fn\ n\ =>}\ body \rfloor\ ((a \longrightarrow b)\ f)$$

An appropriate instantiation and combination with (2) produces the following:

$$\vdash \mathsf{T} \Rightarrow \mathsf{Eval}\ env\ \lfloor \mathtt{fn\ x\ =>\ x\ +\ 1} \rfloor\ ((\mathsf{int} \longrightarrow \mathsf{int})\ (\lambda x.\ x + 1))$$

which, after only minor reformulation, becomes a certificate theorem for the given HOL function add1:

$$\vdash \mathsf{Eval}\ env\ \lfloor \mathtt{fn\ x\ =>\ x\ +\ 1} \rfloor\ ((\mathsf{int} \longrightarrow \mathsf{int})\ \mathtt{add1})$$

Additional Notes. The main part of the synthesis algorithm is always a bottom-up traversal as described above. However, synthesis of recursive functions requires an additional post-processing phase which involves an automatic induction proof. We omit a description of such induction proofs since the solution described previously in [13] is not important for understanding this paper, and works in essentially the same way for synthesis of recursive stateful functions.

3.3 Synthesis of Stateful ML Code

Our algorithm for synthesis of stateful ML is very similar to the algorithm described above for synthesis of pure CakeML code. The main differences are:

– the input HOL terms must be written in a state-and-exception monad, and
– instead of Eval and $\longrightarrow$, the derived theorems use EvalM and $\longrightarrow^M$,

where EvalM and $\longrightarrow^M$ relate the monad's state to the references and foreign function interface of the underlying CakeML state (fields refs and ffi). These concepts will be described below.

Generic State-and-Exception Monad. The new generalised synthesis work-flow uses the following state-and-exception monad $(\alpha,\ \beta,\ \gamma)$ M, where α is the state type, β is the return type, and γ is the exception type.

$$(\alpha,\ \beta,\ \gamma)\ \mathsf{M}\ =\ \alpha\ \rightarrow\ (\beta,\ \gamma)\ \mathsf{exc}\ \times\ \alpha$$
$$\text{where}\ (\beta,\ \gamma)\ \mathsf{exc}\ =\ \mathsf{Success}\ \beta\ |\ \mathsf{Failure}\ \gamma$$

We define the following interface for this monad type. Note that syntactic sugar is often used: in our case, we write do $n\ \leftarrow\ foo$; return $(bar\ n)$ od (as was done in Sect. 1) when we mean bind $foo\ (\lambda\,n.\ \text{return}\ (bar\ n))$.

return $x = (\lambda\,s.\ (\mathsf{Success}\ x,s))$

bind $x\ f =$
$(\lambda\,s.\ \text{case}\ x\ s\ \text{of}\ (\mathsf{Success}\ y,s)\ \Rightarrow\ f\ y\ s\ |\ (\mathsf{Failure}\ x,s)\ \Rightarrow\ (\mathsf{Failure}\ x,s))$

x otherwise $y =$
$(\lambda\,s.\ \text{case}\ x\ s\ \text{of}\ (\mathsf{Success}\ v,s)\ \Rightarrow\ (\mathsf{Success}\ v,s)\ |\ (\mathsf{Failure}\ e,s)\ \Rightarrow\ y\ s)$

Functions that update the content of state can only be defined once the state type is instantiated. A function for changing a monad M to have a different state type is introduced in Sect. 3.5.

Definitions and Lemmas for Synthesis. We define EvalM as follows. A CakeML source expression *exp* is considered to satisfy an execution relation P if for any CakeML state s, which is related by state_rel to the state monad state st and state assertion H, the CakeML expression *exp* evaluates to a result *res* such that the relation P accepts the transition and state_rel_frame holds for state assertion H. The auxiliary functions state_rel and state_rel_frame will be described below. The first argument *ro* can be used to restrict effects to *references only*, as described a few paragraphs further down.

EvalM $ro\ env\ st\ exp\ P\ H\ \Longleftrightarrow$
$\forall\,s.$
  state_rel $H\ st\ s\ \Rightarrow$
  $\exists\,s_2\ res\ st_2\ ck.$
    (evaluate $(s\ with\ \mathsf{clock}\ :=\ ck)\ env\ [exp] = (s_2,res))\ \wedge$
    $P\ st\ (st_2,res)\ \wedge$ state_rel_frame $ro\ H\ (st,s)\ (st_2,s_2)$

In the definition above, state_rel and state_rel_frame are used to check that the user-specified state assertion H relates the CakeML states and the monad states. Furthermore, state_rel_frame ensures that the separation logic frame rule is true. Both use the separation logic set-up from our previous work on characteristic formulae for CakeML [5], where we define a function st2heap which, given a projection p and CakeML state s, turns the CakeML state into a set representation of the reference store and foreign-function interface (used for I/O).

The H in the definition above is a pair (h,p) containing a heap assertion h and the projection p. We define state_rel (h,p) st s to state that the heap assertion produced by applying h to the current monad state st must be true for some subset produced by st2heap when applied to the CakeML state s. Here $(*)$ is the separating conjunction and T is true for any heap.

$$\text{state\_rel } (h,p) \; st \; s \iff (h \; st \; * \; \mathsf{T}) \; (\text{st2heap } p \; s)$$

The relation state_rel_frame states: any frame F that is true separately from $h \; st_1$ for the initial state is also true for the final state; and if the references-only ro configuration is set, then the only difference in the states must be in the references and clock, i.e. no I/O operations are permitted. The ro flag is instantiated to true when a pure specification (Eval) is proved for local state Sect. 4.

$$\text{state\_rel\_frame } ro \; (h,p) \; (st_1,s_1) \; (st_2,s_2) \iff$$
$$(ro \Rightarrow \exists \, refs. \; s_2 = s_1 \; with \; \text{refs} := \; refs) \land$$
$$\forall \, F. \; (h \; st_1 \; * \; F) \; (\text{st2heap } p \; s_1) \Rightarrow (h \; st_2 \; * \; F \; * \; \mathsf{T}) \; (\text{st2heap } p \; s_2)$$

We prove lemmas to aid the synthesis algorithm in construction of proofs. The lemmas shown in this paper use the following definition of monad.

$$\text{monad } a \; b \; x \; st_1 \; (st_2,res) \iff$$
$$\text{case } (x \; st_1,res) \text{ of}$$
$$((\mathsf{Success } \; y,st),\mathsf{Rval } \; [v]) \Rightarrow (st = st_2) \land a \; y \; v$$
$$| \; ((\mathsf{Failure } \; e,st),\mathsf{Rerr } \; (\mathsf{Rraise } \; v')) \Rightarrow (st = st_2) \land b \; e \; v'$$
$$| \; \_ \Rightarrow \mathsf{F}$$

Synthesis makes use of the following two lemmas in proofs involving monadic return and bind. For return x, synthesis proves an Eval-theorem for x. For bind, it proves a theorem that fits the shape of the first four lines of the lemma and returns a theorem consisting of the last two lines, appropriately instantiated.

$$\vdash \mathsf{Eval } \; env \; exp \; (a \; x) \Rightarrow \mathsf{EvalM } \; ro \; env \; st \; exp \; (\text{monad } a \; b \; (\text{return } x)) \; H$$

$$\vdash ((assums_1 \Rightarrow \mathsf{EvalM } \; ro \; env \; st \; e_1 \; (\text{monad } b \; c \; x) \; H) \land$$
$$\forall z \; v.$$
$$\quad b \; z \; v \land assums_2 \; z \Rightarrow$$
$$\quad \mathsf{EvalM } \; ro \; (env \; [n \mapsto v]) \; (\mathsf{snd } \; (x \; st)) \; e_2 \; (\text{monad } a \; c \; (f \; z)) \; H) \Rightarrow$$
$$assums_1 \land (\forall z. \; (\mathsf{fst } \; (x \; st) = \mathsf{Success } \; z) \Rightarrow assums_2 \; z) \Rightarrow$$
$$\mathsf{EvalM } \; ro \; env \; st \; \lfloor \texttt{let } \texttt{n} \texttt{ = } e_1 \texttt{ in } e_2 \rfloor \; (\text{monad } a \; c \; (\text{bind } x \; f)) \; H$$

3.4 References, Arrays and I/O

The synthesis algorithm uses specialised lemmas when the generic state-and-exception monad has been instantiated. Consider the following instantiation of the monad's state type to a record type. The programmer's intention is that the lists are to be synthesised to arrays in CakeML and the I/O component IO_fs is a model of a file system (taken from a library).

```
example_state =
  <| ref1 : int; farray1 : int list; rarray1 : int list; stdio : IO_fs |>
```

With the help of getter- and setter-functions and library functions for file I/O, users can conveniently write monadic functions that operate over this state type.

When it comes to synthesis, the automation instantiates H with an appropriate heap assertion, in this instance: ASSERT. The user has informed the synthesis tool that farray1 is to be a fixed-size array and rarray1 is to be a resizable-size array. A resizable-array is implemented as a reference that contains an array, since CakeML (like SML) does not directly support resizing arrays. Below, REF_REL int ref1_loc st.ref1 asserts that int relates the value held in a reference at a fixed store location ref1_loc to the integer in st.ref1. Similarly, ARRAY_REL and RARRAY_REL specify a connection for the array fields. Lastly, STDIO is a heap assertion for the file I/O taken from a library.

```
ASSERT st =
REF_REL int ref1_loc st.ref1 * RARRAY_REL int rarray1_loc st.rarray1 *
ARRAY_REL int farray1_loc st.farray1 * STDIO st.stdio
```

Automation specialises pre-proved EvalM lemmas for each term that might be encountered in the monadic functions. As an example, a monadic function might contain an automatically defined function update_farray1 for updating array farray1. Anticipating this, synthesis automation can, at set-up time, automatically derive the following lemma which it can use when it encounters update_farray1.

$$\vdash \text{Eval } env \ e_1 \ (\text{num } n) \wedge \text{Eval } env \ e_2 \ (\text{int } x) \wedge$$
$$(\text{lookup\_var } \lfloor \texttt{farray1} \rfloor \ env = \text{Some farray1\_loc}) \Rightarrow$$
$$\text{EvalM } ro \ env \ st \ \lfloor \texttt{Array.update } (\texttt{farray1}, e_1, e_2) \rfloor$$
$$(\text{monad unit exc } (\text{update\_farray1 } n \ x)) \ (\text{ASSERT}, p)$$

3.5 Changing Monad Types

The possibility to change the types of the monad is useful when previously developed monadic functions (e.g. from an existing library) are to be used as part of a larger context. Consider the case of the file I/O in the example from above. The following EvalM theorem has been proved in the CakeML basis library.

$$\vdash \text{Eval } env \ e \ (\text{string } x) \wedge$$
$$(\text{lookup\_var } \lfloor \texttt{print} \rfloor \ env = \text{Some print\_v}) \Rightarrow$$
$$\text{EvalM F } env \ st \ \lfloor \texttt{print } e \rfloor \ (\text{monad unit } b \ (\text{print } x)) \ (\text{STDIO}, p)$$

This can be used directly if the state type of the monad is the IO_fs type. However, our example above uses example_state as the state type.

To overcome such type mismatches, we define a function liftM which can bring a monadic operation defined in libraries into the required context. The type of liftM r w is $(\alpha,\ \beta,\ \gamma)$ M $\rightarrow$ $(\epsilon,\ \beta,\ \gamma)$ M, for appropriate r and w.

liftM *read write op* = $(\lambda\ s.$ (let (ret, new) = $op\ (read\ s)$ in $(ret, write\ new\ s)))$

Our liftM function changes the state type. A simpler lifting operation can be used to change the exception type.

For our example, we define stdio f as a function that performs f on the IO_fs-part of a example_state. (The fib example Sect. 1 used a similar stdio.)

stdio = liftM $(\lambda\ s.\ s.$stdio$)$ $(\lambda\ n\ s.\ s\ with$ stdio := $n)$

For synthesis, we prove a lemma that can transfer any EvalM result for the file I/O model to a similar EvalM result wrapped in the stdio function. Such lemmas are possible because of the separation logic frame rule that is part of EvalM. The generic lemma is the following:

$\vdash (\forall\ st.$ EvalM $ro\ env\ st\ exp$ (monad $a\ b\ op$) (STDIO,p)) $\Rightarrow$
$\quad \forall\ st.$ EvalM $ro\ env\ st\ exp$ (monad $a\ b$ (stdio op)) (ASSERT,p)

And the following is the transferred lemma, which enables synthesis of HOL terms of the form stdio (print x) for Eval-synthesisable x.

$\vdash$ Eval $env\ e$ (string x) $\wedge$
$\quad$ (lookup_var $\lfloor$print$\rfloor$ env = Some print_v) $\Rightarrow$
$\quad$ EvalM F $env\ st$ $\lfloor$print $e\rfloor$ (monad unit exc (stdio (print x))) (ASSERT,p)

4   Local State and the Abstract Synthesis Mode

This section explains how we have adapted the method described above to also support generation of code that uses local state and local exceptions. These features enable use of stateful code (EvalM) in a pure context (Eval). We used these features to significantly speed up parts of the CakeML compiler (see Sect. 5).

In the monadic functions, users indicate that they want local state to be generated by using the following run function. In the logic, the run function essentially just applies a monadic function m to an explicitly provided state st.

run : $(\alpha,\ \beta,\ \gamma)$ M $\rightarrow$ α $\rightarrow$ $(\beta,\ \gamma)$ exc
run $m\ st$ = fst $(m\ st)$

In the generated code, an application of run to a concrete monadic function, say bar, results in code of the following form:

```
fun run_bar k n =
  let
    val r = ref ... (* allocate, initialise, let-bind all local state *)
    fun foo n = ... (* all auxiliary funs that depend on local state *)
    fun bar n = ... (* define the main monadic function *)
  in Success (bar n) end (* wrap normal result in Success constructor *)
  handle e => Failure e; (* wrap any exception in Failure constructor *)
```

Synthesis of locally effectful code is made complicated in our setting for two reasons: (1) there are no fixed locations where the references and arrays are stored, e.g. we cannot define ref1_loc as used in the definition of ASSERT in Sect. 3.4; and (2) the local names of state components must be in scope for all of the function definitions that depend on local state.

Our solution to challenge (1) is to leave the location values as variables (loc_1, loc_2, loc_3) in the heap assertion when synthesising local state. To illustrate, we will adapt the example_state from Sect. 3.4: we omit IO_fs in the state because I/O cannot be made local. The local-state enabled heap assertion is:

LOCAL_ASSERT loc_1 loc_2 loc_3 st =
REF_REL int loc_1 st.ref1 * RARRAY_REL int loc_2 st.rarray1 *
ARRAY_REL int loc_3 st.farray1

The lemmas referring to local state now assume they can find the right variable locations with variable look-ups.

$\vdash$ Eval env e_1 (num n) $\wedge$ Eval env e_2 (int x) $\wedge$
(lookup_var $\lfloor$farray1$\rfloor$) env = Some loc_3) $\Rightarrow$
EvalM ro env st $\lfloor$Array.update (farray1,e_1,e_2)$\rfloor$
(monad unit exc (update_farray1 n x)) (LOCAL_ASSERT loc_1 loc_2 loc_3,p)

Challenge (2) was caused by technical details of our previous synthesis methods. The previous version was set up to only produce top-level declarations, which is incompatible with the requirement to have local (not globally fixed) state declarations shared between several functions. The requirement to only have top-level declarations arose from our desire to keep things simple: each synthesised function is attached to the end of a concrete linear program that is being built. It is beneficial to be concrete because then each assumption on the lexical environment where the function is defined can be proved immediately on definition. We will call this old approach the *concrete mode* of synthesis, since it eagerly builds a concrete program.

In order to support having functions access local state, we implement a new *abstract mode* of synthesis. In the abstract mode, each assumption on the lexical environment is left as an unproved side condition as long as possible. This allows us to define functions in a dynamic environment.

To prove a pure specification (Eval) from the EvalM theorems, the automation first proves that the generated state-allocation and -initialisation code establishes the relevant heap assertion (e.g. LOCAL_ASSERT); it then composes the abstractly synthesised code while proving the environment-related side conditions (e.g. presence of loc_3). The final proof of an Eval theorem requires instantiating the references-only *ro* flag to true, in order to know that no I/O occurs (Sect. 3.3).

5 Case Studies and Experiments

In this section we present the runtime and proof size results of applying our method to some case studies. Performance experiments were carried out on an Intel i7-2600 running at 3.4 GHz with 16 GB of RAM. Full data is available at https://cakeml.org/ijcar18.zip.

Type Inference and Register Allocation. Both of these phases of the CakeML compiler are written with a state (and exception) monad, but were previously synthesised into pure CakeML code. We updated them to use the new synthesis tool, resulting in performant, stateful CakeML code. The allocator underwent more significant changes, because we could now use CakeML arrays via the synthesis tool. It was previously confined to using tree-like functional arrays for its internal state, leading to logarithmic access overheads. This is not a specific issue for the CakeML compiler; a verified register allocator for CompCert [3] also reported log-factor overheads due to (functional) array accesses.

Tests were carried out using versions of the bootstrapped CakeML compiler. We ran each test 50 times on the same input program, recording time elapsed in each compiler phase. For each test in the register allocation benchmark, we also compared the resulting executables 10 times, to confirm that both compilers generated code of comparable quality (i.e. runtime performance).

In the largest program (`knuth-bendix`), the new register allocator ran 15 times faster (with a wide 95% CI of 11.76–20.93 due in turn to a high standard deviation on the runtimes for the old code). In the smaller `pidigits` benchmark, the new register allocator ran 9.01 times faster (95% CI of 9.01–9.02). Across 6 example input programs, we saw ratios of runtimes between 7.58 and 15.06. Register allocation was previously such a significant part of the compiler runtime that this improvement results in runtime improvements for the whole compiler (on these benchmark programs) of factors between 2 and 9 times.

In contrast, the type inferencer became slower. We compared the performance of commit `28aba93` (incorporating the monadic inference code) against the same baseline. The slowdowns ranged between factors of approximately 3 and 1.17. However, the case with the most dramatic slowdown as a ratio still only represents a tiny proportion of the total time spent compiling. In this case (`pidigits`), the new code takes 10 ms out of a total elapsed time of 2.05 s (roughly 0.5% of the total). The best (least bad) case was in an artificial program exemplifying the worst-case for Hindley-Milner where types grow exponentially. There, the old code took 251 ms and the new took 295 ms. The extra indirection through references in the new code seems to cost performance. We intend to keep using the purely synthesised version until the compiler optimises the references better.

OpenTheory Article Checker. The type changing feature from Sect. 3.5 enabled us to produce an OpenTheory [7] article checker with our new synthesis approach, and reduce the amount of manual proof required in a previous version. The checker reads articles from the file system, and performs each logical inference in the OpenTheory framework using the verified Candle kernel [8]. Previously, the I/O code for the checker was implemented in stateful CakeML, and verified manually using characteristic formulae. By replacing the manually verified I/O wrapper by monadic code we removed 400 lines of tedious manual proof.

6   Related Work

Effectful Code Using Monads. Our work on encapsulating stateful computations (Sect. 4) in pure programs is similar in purpose to that of the ST monad [11]. The main difference is how this encapsulation is performed: the ST monad relies on parametric polymorphism to prevent references from escaping their scope, whereas we utilise lexical scoping in synthesised code to achieve a similar effect.

Imperative HOL by Bulwahn et al. [4] is a framework for implementing and reasoning about effectful programs in Isabelle/HOL. Monadic functions are used to describe stateful computations which act on the heap, in a similar way as Sect. 3 but with some important differences. Instead of using a state monad, the authors introduce a polymorphic *heap monad* – similar in spirit to the ST monad of Launchbury and Jones [11], but without encapsulation – where polymorphism is achieved by mapping HOL types to the natural numbers. Contrary to our approach, this allows for heap elements (e.g. references) to be declared on-the-fly and used as first-class values. The drawback, however, is that only countable types can be stored on the heap; in particular, the heap monad does not admit function-typed values, which our work supports.

More recently, Lammich [10] has built a framework for the refinement of pure data structures into imperative counterparts, in Imperative HOL. The refinement process is automated, and refinements are verified using a program logic based on separation logic, which comes with proof-tools to aid the user in verification.

Both developments [4,10] differ from ours in that they lack a verified mechanism for extracting executable code from shallow embeddings. Although stateful computations are implemented and verified within the confines of higher-order logic, Imperative HOL relies on the unverified code-generation mechanisms of Isabelle/HOL. Moreover, neither work presents a way to deal with I/O effects.

Verified Compilation. Mechanisms for synthesising programs from shallow embeddings defined in the logics of interactive theorem provers exist as components of several verified compiler projects [1,6,12,13]. Although the main contribution of our work is proof-producing synthesis, comparisons are relevant as our synthesis tool plays an important part in the CakeML compiler [9]. To the best of our knowledge, ours is the first work combining effectful computations with proof-producing synthesis and fully verified compilation.

CertiCoq by Anand et al. [1] strives to be a fully verified optimising compiler for functional programs implemented in Coq. The compiler front-end supports the full syntax of the dependently typed logic Gallina, which is reified into a deep embedding and compiled to Cminor through a series of verified compilation steps [1]. Contrary to the approach we have taken [13] (see Sect. 3.2), this reification is neither verified nor proof-producing, and the resulting embedding has no formal semantics (although there are attempts to resolve this issue [2]). Moreover, as of yet, no support exists for expressing effectful computations (such as in Sect. 3.4) in the logic. Instead, effects are deferred to wrapper code from which the compiled functions can be called, and this wrapper code must be manually verified.

The Œuf compiler by Mullen et al. [12] is similar in spirit to CertiCoq in that it compiles pure Coq functions to Cminor through a verified process. Similarly, compiled functions are pure, and effects must be performed by wrapper code. Unlike CertiCoq, Œuf supports only a limited subset of Gallina, from which it synthesises deeply embedded functions in the Œuf-language. The Œuf language has both denotational and operational semantics, and the resulting syntax is automatically proven equivalent with the corresponding logical functions through a process of computational denotation [12].

Hupel and Nipkow [6] have developed a compiler from Isabelle/HOL to CakeML AST. The compiler satisfies a partial correctness guarantee: if the generated CakeML code terminates, then the result of execution is guaranteed to relate to an equality in HOL. Our approach proves termination of the code.

7   Summary

This paper describes a technique that makes it possible to synthesise whole programs from monadic functions in HOL, with automatic proofs relating the generated effectful code to the original functions. Using the separation logic from characteristic formulae for CakeML, the synthesis mechanism supports references, exceptions, I/O, reusable library developments, and encapsulation of locally stateful computations inside pure functions. To our knowledge, this is the first proof-producing synthesis technique with the aforementioned features.

Acknowledgements. The second and fourth authors were partly supported by the Swedish Foundation for Strategic Research. The fifth author was supported by an A*STAR National Science Scholarship (PhD), Singapore.

References

1. Anand, A., Appel, A., Morrisett, G., Paraskevopoulou, Z., Pollack, R., Belanger, O.S., Sozeau, M., Weaver, M.: CertiCoq: a verified compiler for Coq. In: CoqPL (2017)
2. Anand, A., Boulier, S., Tabareau, N., Sozeau, M.: Typed template Coq - certified meta-programming in Coq. In: CoqPL (2018)
3. Blazy, S., Robillard, B., Appel, A.W.: Formal verification of coalescing graph-coloring register allocation. In: Gordon, A.D. (ed.) ESOP 2010. LNCS, vol. 6012, pp. 145–164. Springer, Heidelberg (2010). https://doi.org/10.1007/978-3-642-11957-6_9

4. Bulwahn, L., Krauss, A., Haftmann, F., Erkök, L., Matthews, J.: Imperative functional programming with Isabelle/HOL. In: Mohamed, O.A., Muñoz, C., Tahar, S. (eds.) TPHOLs 2008. LNCS, vol. 5170, pp. 134–149. Springer, Heidelberg (2008). https://doi.org/10.1007/978-3-540-71067-7_14

5. Guéneau, A., Myreen, M.O., Kumar, R., Norrish, M.: Verified characteristic formulae for CakeML. In: Yang, H. (ed.) ESOP 2017. LNCS, vol. 10201, pp. 584–610. Springer, Heidelberg (2017). https://doi.org/10.1007/978-3-662-54434-1_22

6. Hupel, L., Nipkow, T.: A verified compiler from Isabelle/HOL to CakeML. In: Ahmed, A. (ed.) ESOP 2018. LNCS, vol. 10801, pp. 999–1026. Springer, Cham (2018). https://doi.org/10.1007/978-3-319-89884-1_35

7. Hurd, J.: The OpenTheory standard theory library. In: Bobaru, M., Havelund, K., Holzmann, G.J., Joshi, R. (eds.) NFM 2011. LNCS, vol. 6617, pp. 177–191. Springer, Heidelberg (2011). https://doi.org/10.1007/978-3-642-20398-5_14

8. Kumar, R., Arthan, R., Myreen, M.O., Owens, S.: Self-formalisation of higher-order logic - semantics, soundness, and a verified implementation. J. Autom. Reason. **56**(3), 221–259 (2016)

9. Kumar, R., Myreen, M.O., Norrish, M., Owens, S.: CakeML: a verified implementation of ML. In: Jagannathan, S., Sewell, P. (eds.) POPL, pp. 179–192 (2014)

10. Lammich, P.: Refinement to imperative/HOL. In: Urban, C., Zhang, X. (eds.) ITP 2015. LNCS, vol. 9236, pp. 253–269. Springer, Cham (2015). https://doi.org/10.1007/978-3-319-22102-1_17

11. Launchbury, J., Jones, S.L.P.: Lazy functional state threads. In: Sarkar, V., Ryder, B.G., Soffa, M.L. (eds.) PLDI, pp. 24–35 (1994)

12. Mullen, E., Pernsteiner, S., Wilcox, J.R., Tatlock, Z., Grossman, D.: Œuf: minimizing the Coq extraction TCB. In: CPP (2018)

13. Myreen, M.O., Owens, S.: Proof-producing translation of higher-order logic into pure and stateful ML. J. Funct. Program. **24**(2–3), 284–315 (2014)

14. Owens, S., Myreen, M.O., Kumar, R., Tan, Y.K.: Functional big-step semantics. In: Thiemann, P. (ed.) ESOP 2016. LNCS, vol. 9632, pp. 589–615. Springer, Heidelberg (2016). https://doi.org/10.1007/978-3-662-49498-1_23

15. Reynolds, J.C.: Separation logic: a logic for shared mutable data structures. In: LICS, pp. 55–74 (2002)

An Abstraction-Refinement Framework for Reasoning with Large Theories

Julio Cesar Lopez Hernandez and Konstantin Korovin[✉]

School of Computer Science, The University of Manchester, Manchester, UK
{lopezhej,korovin}@cs.man.ac.uk

Abstract. In this paper we present an approach to reasoning with large theories which is based on the abstraction-refinement framework. The proposed approach consists of the following approximations: the over-approximation, the under-approximation and their combination. We present several concrete abstractions based on subsumption, signature grouping and argument filtering. We implemented our approach in a theorem prover for first-order logic iProver and evaluated over the TPTP library.

Keywords: Automated reasoning · Large theories
Abstraction-refinement

1 Introduction

Efficient reasoning with large theories is one of the main challenges in automated theorem proving arising in many applications ranging from reasoning with ontologies to proof assistants for mathematics. Current methods for reasoning with large theories are based on different axiom selection methods. Some of them are based on the syntactic or semantic structure of the axioms and conjecture formulas [15,30]. These methods select relevant axioms based on syntactic or semantic relationship between axioms and conjectures. Other methods for axiom selection use machine learning to take advantage of previous knowledge about proved conjectures [16,32,33]. What those methods have in common are two phases of the whole process for proving a conjecture: one is the axiom selection phase, and the other one is the reasoning phase. Those phases are performed in a sequential way. First, the axiom selection takes place, then using the selected axioms the reasoning process starts.

Our proposed approach based on abstraction-refinement framework [8] has the purpose of interleaving the axioms selection and reasoning phases, having a more dynamic interaction between them. This proposed approach encompasses two ways for approximating axioms: one is called over-approximation and the other one under-approximation. Those approximations are combined to converge more rapidly to a proof if it exists or to a model otherwise. There are a number of related works which consider different specific types of under

© Springer International Publishing AG, part of Springer Nature 2018
D. Galmiche et al. (Eds.): IJCAR 2018, LNAI 10900, pp. 663–679, 2018.
https://doi.org/10.1007/978-3-319-94205-6_43

and/or over approximations in different contexts [1,5,7,9,12,19,22–24,31]. Nevertheless, abstraction-refinement is largely overlooked in state-of-the-art automated theorem provers, with an exception of SPASS which was extended with abstraction-refinement into a very specialised decidable fragment to approximate general first-order reasoning [31]. Another relevant example is the Inst-Gen calculus [19] which under-approximates first-order formulas by propositional/ground abstractions and refines these approximations by model-guided instantiations. In the SMT setting, ground approximations are used in conflict and model-based instantiation methods [11,25]. In higher-order logic, over-approximations are used for efficient encodings into first-order logic [2–4], propositional logic [6] and also in higher-order patterns [10].

In this paper we take a pragmatic approach. Instead of targeting a specific decidable fragment as an abstract domain we use abstraction-refinement to simplify problems by different over and under approximations and their combinations. We present a general abstraction-refinement framework for refutation theorem proving which allows one to compare and combine different abstractions. Our framework is general enough to represent abstractions not only within the same language but also abstractions that extend or modify the language, in particular abstractions based on signature transformations. We present a number of concrete abstractions based on subsumption, signature grouping and argument filtering and discuss their combinations. In this paper we consider many-sorted first-order logic in the context of first-order theorem proving but the approach is applicable to SMT as well.[1]

2   Abstraction Functions and Refinements

Let us consider a set of formulas $\mathcal{F}$ which we call a *concrete domain* and a set of formulas $\hat{\mathcal{F}}$ which we will call an *abstract domain*. For example $\mathcal{F}$ can be the set of all first-order formulas and $\hat{\mathcal{F}}$ can be a fragment of first-order logic. Concrete and abstract domains can coincide.

An *abstraction function* is a mapping $\alpha : \mathcal{F} \mapsto \hat{\mathcal{F}}$. When there is no ambiguity we will call an abstraction function just an abstraction of $\mathcal{F}$. The identity function is an abstraction which will be called the *identity abstraction* α_{id}.

A *concretisation function* for α is the inverse mapping $\gamma : \hat{\mathcal{F}} \mapsto 2^{\mathcal{F}}$, i.e., $\gamma(\hat{F}) = \{F \mid \alpha(F) = \hat{F}\}$ for $\hat{F} \in \hat{\mathcal{F}}$.

An abstraction α is called *over-approximating abstraction* (wrt. refutation) if for every $F \in \mathcal{F}$, $F \models \perp$ implies $\alpha(F) \models \perp$. An abstraction α is called *under-approximating abstraction* (wrt. refutation) if for every $F \in \mathcal{F}$, $\alpha(F) \models \perp$ implies $F \models \perp$.

We can compose abstractions as mappings. In particular, if $\alpha_1 : \mathcal{F} \mapsto \mathcal{F}_1$ and $\alpha_2 : \mathcal{F}_1 \mapsto \mathcal{F}_2$ then $\alpha_1 \alpha_2$ is an abstraction of $\mathcal{F}$.

Proposition 1. *Composition of over-approximating abstractions is an over-approximating abstraction. Likewise, composition of under-approximating abstractions is an under-approximating abstraction.*

---

[1] Preliminary version of this work was presented at the IWIL workshop [13].

In this paper we will define several atomic abstractions and we use this proposition to compose them to obtain a large range of combined abstractions.

We define an ordering on abstractions $\sqsubseteq$ called *abstraction refinement ordering* as follows: $\alpha \sqsubseteq \alpha'$ if for all $F \in \mathcal{F}$, $\alpha(F) \models \bot$ implies $\alpha'(F) \models \bot$. Two abstractions are *equivalent*, denoted by $\alpha \equiv \alpha'$ if $\alpha \sqsubseteq \alpha'$ and $\alpha' \sqsubseteq \alpha$. The strict part $\sqsubset$ of $\sqsubseteq$ is defined as $\alpha \sqsubset \alpha'$ if $\alpha \sqsubseteq \alpha'$ and $\alpha \not\equiv \alpha'$. An abstraction is *precise* if it is equivalent to the identity abstraction. An example of a non-trivial precise abstraction can be obtained by renaming function and predicate symbols. We have that every over-approximating abstraction α_s is above and every under-approximation abstraction α_w is below the identity abstraction wrt. the abstraction refinement ordering, i.e., $\alpha_w \sqsubseteq \alpha_{id} \sqsubseteq \alpha_s$.

Weakening abstraction refinement of an over-approximating abstraction α is an abstraction α' which is below α and above the identity abstraction in the abstraction refinement ordering, i.e., $\alpha_{id} \sqsubseteq \alpha' \sqsubseteq \alpha$. *Strengthening abstraction refinement* of an under-approximating abstraction α is an abstraction α' which is above α and below the identity abstraction in the abstraction refinement ordering, i.e., $\alpha \sqsubseteq \alpha' \sqsubseteq \alpha_{id}$.

An *over-approximation abstraction-refinement process* is a possibly infinite sequence of weakening abstraction refinements $\alpha_0, \ldots, \alpha_n, \ldots$ such that $\alpha_{id} \sqsubseteq \ldots \sqsubseteq \alpha_n \sqsubseteq \ldots \sqsubseteq \alpha_0$. Similar, an *under-approximation abstraction-refinement process* is a possibly infinite sequence of strengthening abstraction refinements $\alpha_0, \ldots, \alpha_n, \ldots$ such that $\alpha_0 \sqsubseteq \ldots \sqsubseteq \alpha_n \sqsubseteq \ldots \sqsubseteq \alpha_{id}$.

3   Over-Approximation Procedure

We use ATP_S to denote an automated theorem prover which is sound but possibly incomplete (wrt. refutation) [14]. On the other hand, we use ATP_C to make a reference to an automated theorem prover which is complete but not necessary sound [5,22]. Hence, if ATP_S returns UNSAT then the conjecture is proved and if ATP_C returns SAT then the conjecture is disproved. The purpose of these ATPs is to prove or disprove conjectures more efficiently than a sound and complete ATP but with a possible loss of precision.

We consider a theory A which is a collection of axioms which we call *concrete axioms* and a set of formulas $\hat{A}^s$ called *abstract axioms*. We will assume that the negation of the conjecture is included in A, so proving the conjecture corresponds to proving unsatisfiability of A.

The *over-approximating procedure* starts by applying an over-approximating abstraction function α_s to A, to obtain an abstract representation of axioms $\hat{A}^s$, $\hat{A}^s = \alpha_s(A)$. First, the procedure tries to prove unsatisfiability of the abstract axioms $\hat{A}^s$ using an ATP_C. If ATP_C proves unsatisfiability of $\hat{A}^s$, the procedure extracts an abstract unsat core $\hat{A}^s_{uc}$ from $\hat{A}^s$, which can be obtained by, e.g., collecting all axioms involved in the abstract proof. Next, the procedure tries to prove unsatisfiability of the concretisation of the abstract unsat core $A_{uc} = \gamma_s(\hat{A}^s_{uc})$ using ATP_S. If the ATP_S proves unsatisfiability of A_{uc}, the process stops as this proves unsatisfiability of A. Otherwise, if A_{uc} is shown to be satisfiable, the set of axioms A is abstracted using a new abstraction α'_s obtained by weakening abstraction refinement of α_s. In practice, the refinement procedure refines α_s until $\alpha'_s(A_{uc})$ becomes satisfiable, which is always possible as at this point we assume A_{uc} is satisfiable. The procedure is repeated utilising the refined set of abstract axioms. This loop finishes when the conjecture is proved or disproved or the time limit of the whole procedure is reached. The diagram of the over-approximating procedure is shown in Fig. 1.

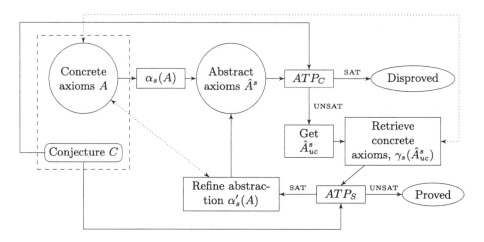

Fig. 1. The over-approximation procedure

The main parameters of this procedure are an over-approximating abstraction function and weakening abstraction refinement.

Next we define several concrete over-approximating abstractions and discuss abstraction refinement for these abstractions.

3.1   Subsumption-Based Abstraction

In this section we present abstraction-refinement based on subsumption. Informally, we partition concrete axioms based on joint literal occurrences and for each partition we define an abstract clause which subsumes all clauses in the partition.

We define the initial abstraction of A as follows. With each set of clauses A', we associate a literal ℓ_k in A' which we call a partition literal for A'. An initial partition of A is defined as $A = A^{\ell_1^+} \cup A^{\ell_1^-}$ where ℓ_1 is a partition literal for A, members of $A^{\ell_1^+}$ are all clauses containing ℓ_1 and $A^{\ell_1^-} = A \setminus A^{\ell_1^+}$. We recursively continue partitioning $A^{\ell_1^-}$ in the same way until we obtain the empty set. The result of this process is the following partition of A:

$$A = \bigcup_{k=1}^{n} A^{\ell_1^- \cdots \ell_{k-1}^- \ell_k^+},$$

where ℓ_k is the partition literal for $A^{\ell_1^- \cdots \ell_{k-1}^-}$, we assume $A^{\ell_1^- \cdots \ell_n^-}$ is empty and $A^{\ell_1^- \cdots \ell_{k-1}^- \ell_k^+} = A^{\ell_1^+}$ for $k = 1$.

For each partition $A^{\ell_1^- \cdots \ell_{k-1}^- \ell_k^+}$ literals $\ell_1, \ldots, \ell_{k-1}$ do not occur in any clause in the collection and ℓ_k occurs in all clauses in $A^{\ell_1^- \cdots \ell_{k-1}^- \ell_k^+}$. Figure 2 shows an example of such partition.

We say that ℓ_k is a *leading literal* in $A^{\ell_1^- \cdots \ell_{k-1}^- \ell_k^+}$ and each leading literal is the abstraction of their corresponding set. These abstractions form the set of abstract axioms $\hat{A}^s$. In practice, we can select the leading literal based on a heuristic criteria, e.g., the number of occurrences of a literal in the clause set.

Example 1. Consider the following set of concrete clauses A and its partition consisting of $A^{\ell_1^+}$, $A^{\ell_1^- \ell_2^+}$ and $A^{\ell_1^- \ell_2^- \ell_3^+}$. Where the leading literals are ℓ_1, ℓ_2, ℓ_3 and they form the abstract set of clauses $\hat{A}^s$, $\hat{A}^s = \{\ell_1, \ell_2, \ell_3\}$.

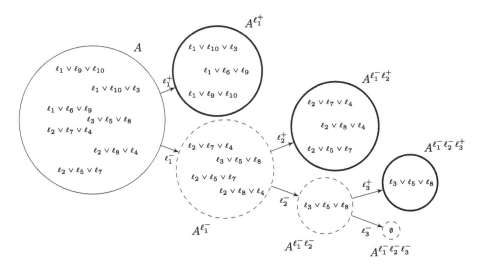

Fig. 2. Partitions of A are in bold

The mapping from sets of the form $A^{\ell_1^- \cdots \ell_{k-1}^- \ell_k^+}$ to the leading literals gives us the abstraction function α_s, which is defined as

$$\alpha_s(D) = \ell_k \text{ for } D \in A^{\ell_1^- \cdots \ell_{k-1}^- \ell_k^+}.$$

Consequently, the concretisation function γ_s is defined as

$$\gamma_s(\ell_k) = A^{\ell_1^- \cdots \ell_{k-1}^- \ell_k^+}.$$

We use the set of abstract axioms to try to prove a conjecture. If the conjecture is proved, we consider the abstract axioms from the unsat core. Those abstract axioms are refined and then replaced by their refined versions. Then, the proving process is repeated using the refined set of abstract axioms. This process continues until we get a concrete proof of the conjecture.

The refinement of abstract axioms will be defined further in this section, but first consider the following definitions. During the refinement process we will partition A into sets of the form A^σ where σ is a sequence of signed literals $\sigma = \ell_1^{s_1} \ldots \ell_n^{s_n}$, where s_j is either $+$ or $-$ for $1 \leq j \leq n$. A literal $\ell_j^{s_j}$ occurs in all clauses of A^σ if $s_j = +$ and does not occur in any of the clauses in the set if $s_j = -$.

In Fig. 2, the leaves different to the empty set are the partition of A and they have the form A^σ. The set of literals in σ with positive signs is defined as $\sigma^+ = \{\ell : \ell^s \in \sigma \text{ and } s \text{ is } +\}$. The set A^σ is abstracted with the clause $C^{\sigma^+} = \bigvee_{\ell \in \sigma^+} \ell$. Therefore, the abstraction function α_s is defined as

$$\alpha_s(D) = C^{\sigma^+},$$

where $D \in A^\sigma$. Then, concretisation function is $\gamma_s(C^{\sigma^+}) = A^\sigma$. The set A^σ is fully concretised if $A^\sigma = \{C^{\sigma^+}\}$.

For a set of clauses A', let $\mathcal{L}(A')$ denote the set of all literals occurring in clauses in A'. The refinement process is applied to an unsat core $\hat{A}_{uc}^s$ consisting of abstract clauses. The refinement process subpartitions one of $A^\sigma = \gamma_s(C^{\sigma^+})$, where $C^{\sigma^+} \in \hat{A}_{uc}^s$, such that A^σ is not fully concretised. Let $\sigma = \ell_1^{s_1} \ldots \ell_k^{s_k}$. This process starts by selecting a new partition literal ℓ_k such that

$$\ell_{k+1} \in \mathcal{L}(A^\sigma) \setminus \sigma^+.$$

Note, that since A^σ is not fully concretised, $\mathcal{L}(A^\sigma) \setminus \sigma^+$ is not empty. Using the literal ℓ_{k+1}, we obtain the starting partition $A^\sigma = A^{\sigma \ell_{k+1}^+} \cup A^{\sigma \ell_{k+1}^-}$. Then, we continue recursively partitioning $A^{\sigma \ell_{k+1}^-}$ as before until we obtain the empty set. The result of this recursive process is the partition of A^σ defined as follows:

$$A^\sigma = \bigcup_{j=1}^{m} A^{\sigma \ell_{k+1}^- \ldots \ell_{k+j-1}^- \ell_{k+j}^+},$$

where ℓ_{k+j} is the partition literal for $A^{\sigma \ell_{k+1}^- \ldots \ell_{k+j-1}^- \ell_{k+j}^+}$ and $A^{\sigma \ell_{k+1}^- \ldots \ell_{k+m-1}^- \ell_{k+m}^-}$ is the empty set. Denote $\sigma_i = \sigma \ell_{k+1}^- \ldots \ell_{k+j-1}^- \ell_{k+j}^+$ for $1 \leq i \leq m$. Then refined abstraction α_s' is defined as:

$$\alpha_s'(D) = \begin{cases} C^{\sigma_i^+} & \text{if } D \in A^{\sigma_i} \text{ for some } 1 \leq i \leq m, \\ \alpha_s(D) & \text{if } D \notin A^\sigma. \end{cases}$$

An example of this refinement is shown in Fig. 3, where the refined abstraction of A consists of $\{\ell_1, \ell_3, \ell_2 \vee \ell_4, \ell_2 \vee \ell_5\}$.

Let us note that the subsumption abstraction is an over-approximation abstraction and subsumption abstraction refinement is a weakening abstraction refinement, in particular, $\alpha_{id} \sqsubseteq \alpha_s' \sqsubseteq \alpha_s$.

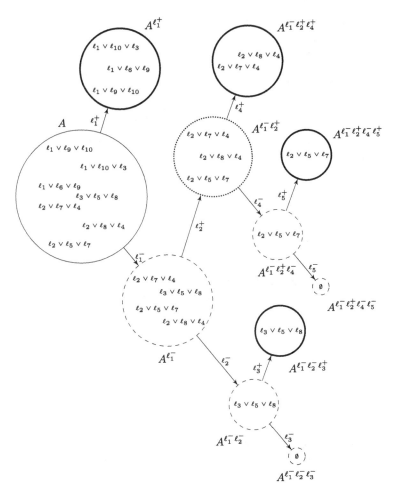

Fig. 3. Refinement of $A^{\ell_1^- \ell_2^+}$ (dotted circle); partitions of A are in bold

3.2 Generalisation Abstraction

In the generalisation abstraction we abstract clauses with their generalisations. A clause D is a generalisation of a clause C if $C = D\sigma$ for a substitution σ. Generalisation ordering on clauses can be defined as $C \sqsubseteq_g D$ if $C = D\sigma$. A generalisation abstraction α_g is a function that maps clauses to their generalisations, so we have $C \sqsubseteq_g \alpha_g(C)$. One example of the generalisation abstraction would be replacing certain non-variable terms by variables. For example, using a generalisation abstraction one can abstract the set of clauses into the Effectively PRopositional (EPR) fragment. Another abstraction strategy can be based on targeting inference positions eligible for superposition.

Example 2. Consider the following set of clauses:

$$S = \{p(g(x), g(x)) \vee q(f(g(x))); g(f(f(x))) \simeq g(f(x))\}.$$

A possible generalisation abstraction of S can be:

$$\alpha_g(S) = \{p(x, x) \vee q(f(x)); g(f(x)) \simeq g(x)\}.$$

Let us note that, e.g., superposition inference is applicable from the second into the first clause in S under any simplification ordering, which can easily lead to non-termination. On the other hand, there is no eligible superposition inferences between two abstracted clauses due to abstraction of terms headed with g in the first clause and the fact that superposition is not applied into the variable positions.

The generalisation abstraction refinement α' of α can be based on restoring abstracted terms in abstract clauses from the unsat core, i.e., $C \sqsubseteq_g \alpha'_g(C) \sqsubseteq_g \alpha_g(C)$ for $C \in \hat{A}^s_{uc}$ and $\alpha'_g(C) = \alpha_g(C)$ for $C \notin \hat{A}^s_{uc}$. We note that the generalisation abstraction is an over-approximation abstraction and generalisation abstraction refinement is a weakening abstraction refinement, in particular, $\alpha_{id} \sqsubseteq \alpha'_g \sqsubseteq \alpha_g$.

In practice, the generalisation abstraction can be naturally combined with the subsumption abstraction, by first generalising and then applying the subsumption abstraction.

3.3 Argument Filtering Abstraction

In this section we present the argument filtering abstraction. Informally, argument filtering abstraction is based on removing certain arguments in signature symbols.

Consider a signature Σ consisting of predicate and function symbols. We will represent argument selection using bit-vectors. Consider a bit-vector bv. We denote the length of bv by $|bv|$, the number of 1s in bv by $|bv|_1$ and 0s by $|bv|_0$. Let $\bar{1}_n, \bar{0}_n$ denote bit-vectors of length n, consisting of 1s and 0s, respectively. Let $\mathcal{B}_n$ denote the set of all bit-vectors of length n. We will omit index n when the bit-vector length is clear from the context or irrelevant.

With each signature (i.e., predicate or function) symbol f of arity n and a bit-vector bv of length n we associate an *abstract symbol* f^{bv} with the arity $|bv|_1$. An abstract domain for a signature symbol f, denoted $f^{\mathcal{B}}$ is the set of all abstract symbols f^{bv}, where $|bv| = arity(f)$. An abstract signature is defined as $\Sigma^{\mathcal{B}} = \cup_{f \in \Sigma} f^{\mathcal{B}}$. A *signature abstraction* is a function: $\alpha_f : \Sigma \mapsto \Sigma^{\mathcal{B}}$ such that $\alpha_f(f) \in f^{\mathcal{B}}$.

A signature abstraction can be extended to terms and atoms recursively:

$$\alpha_f(t) = \begin{cases} x & \text{if } t = x, \\ f^{bv}(\alpha_f(t_{i_1}), \ldots, \alpha_f(t_{i_k})) & \text{if } t = f(t_1, \ldots, t_n), \alpha_f(f) = f^{bv}, \text{ and} \\ & bv(i) = 1 \text{ iff } i \in \{i_1, \ldots, i_k\}. \end{cases}$$

In turn, α_f is extended to clauses and sets of clauses in an obvious way by applying α_f to atoms.

If we abstract every signature symbol f to $f^{\bar{1}}$ then we obtain a precise abstraction, i.e., equivalent to the identity abstraction. Therefore, w.l.o.g., we will identify every signature symbol f with its $f^{\bar{1}}$ abstraction.

Let us consider some special cases. If we abstract every predicate symbol p to $p^{\bar{0}}$ then we obtain a *pure propositional abstraction*, which we denote α_f^{prop}. If we abstract every function symbol f to $f^{\bar{0}}$ and every predicate symbol p to $p^{\bar{1}}$ then we obtain an *EPR abstraction*, which we denote α_f^{EPR}. If we abstract every signature symbol f to $f^{\bar{1}}$ then we obtain a precise abstraction, i.e., equivalent to the identity abstraction.

Example 3. Let us consider the following set of clauses

$$S = \{p(x, f(x, g(y))) \vee \neg p(c, x); \neg p(g(f(x, y)), g(y)); p(c, x)\}.$$

Then pure propositional abstraction will result in the following set of clauses:

$$\alpha_f^{prop}(S) = \{p^{\bar{0}} \vee \neg p^{\bar{0}}; \neg p^{\bar{0}}; p^{\bar{0}}\},$$

which is unsatisfiable. One the other hand the EPR abstraction is:

$$\alpha_f^{EPR}(S) = \{p(x, f^{\bar{0}}) \vee \neg p(c, x); \neg p(g^{\bar{0}}, g^{\bar{0}}); p(c, x)\}.$$

It is easy to see that the EPR abstraction is satisfiable and therefore the original set of clauses is also satisfiable.

In order to define abstraction-refinement we introduce a partial ordering on abstract symbols: $f^{bv_0} \sqsubseteq_{af} f^{bv_1}$ iff $bv_1(i) \leq bv_0(i)$, for all $0 \leq i < arity(f)$. Then we extend this ordering on abstractions by defining $\alpha_f^0 \sqsubseteq_{af} \alpha_f^1$ iff $\alpha_f^0(f) \sqsubseteq_{af} \alpha_f^1(f)$ for all $f \in \Sigma$. We call $\sqsubseteq_{af}$ *argument filtering ordering*. The top element in this ordering is the pure propositional abstraction.

The following proposition implies that argument filtering abstraction is an over-approximation abstraction and abstraction refinement based on the argument filtering ordering is a weakening abstraction refinement.

Proposition 2. *The argument filtering ordering is compatible with the abstraction refinement ordering, i.e., if $\alpha_f^0 \sqsubseteq_{af} \alpha_f^1$ then $\alpha_f^0 \sqsubseteq \alpha_f^1$. Moreover, every argument filtering abstraction is above the identity abstraction, i.e., $\alpha_{id} \sqsubseteq \alpha_f$.*

In the example above, the EPR abstraction is a refinement of the propositional abstraction. In practice, one can start with a propositional or EPR abstraction and define the weakening refinement process by restoring arguments of abstract symbols occurring in the unsat core, as described in the Sect. 3.

Abstracting Variable Dependencies. Let us observe how argument filtering can be used to abstract variable dependencies. As an example we consider clause splitting without backtracking [26], which can be defined as follows. Given a clause $C(\bar{x}, \bar{y}) \vee D(\bar{x}, \bar{z})$ one can split this clause into two clauses by introducing a fresh splitting predicate over joint variables $sp(\bar{x})$ and replacing this clause with two clauses $C(\bar{x}, \bar{y}) \vee sp(\bar{x})$ and $\neg sp(\bar{x}) \vee D(\bar{x}, \bar{z})$. In this way the splitting predicate represents variable dependencies between different subclauses. We can abstract such variable dependencies by restricting argument filtering abstraction-refinement to the splitting predicates. In the same way we can target formula definitions introduced during clausification and Skolem functions which encode existential variable dependencies.

3.4 Signature Grouping Abstraction

Consider a finite signature Σ and let $\mathcal{T}$ be the set of all types of symbols in Σ. In many-sorted first-order logic, a type of a symbol can be represented as a sequence of sorts in a standard way. We partition Σ into groups $\Sigma = \bigcup_{\tau \in \mathcal{T}} \Sigma_\tau$, such that symbols in Σ_τ are all symbols in Σ of type τ. With each non-empty subset of $\sigma_\tau \subseteq \Sigma_\tau$ we associate an abstract symbol f^{σ_τ} of type τ. The abstract signature Σ^S is defined as the union of all abstract symbols.

Consider partitioning Σ into groups $\Sigma = \bigcup_{i=1}^{n} \sigma_i$, such that all symbols in σ_i have the same type. We define a *signature grouping abstraction* α_{sig} as a function: $\alpha_{sig} : \Sigma \mapsto \Sigma^S$ such that $\alpha_{sig}(f) = f^{\sigma_i}$ if $f \in \sigma_i$ for some $1 \leq i \leq n$. In a similar way to Sect. 3.3, we extend α_{sig} to an abstraction over terms, atoms and clauses. We can also define an ordering on abstract symbols: $f^{\sigma_0} \sqsubseteq_{sig} f^{\sigma_1}$ iff $\sigma_0 \subseteq \sigma_1$ and extend this ordering to abstractions: $\alpha_{sig}^0 \sqsubseteq_{sig} \alpha_{sig}^1$ iff $\alpha_{sig}^0(f) \sqsubseteq_{sig} \alpha_{sig}^1(f)$ for all $f \in \Sigma$. We call $\sqsubseteq_{sig}$ the *signature grouping ordering*. Let us note that the top element in this ordering is the abstraction corresponding to the maximal partitioning $\Sigma = \bigcup_{\tau \in \mathcal{T}} \Sigma_\tau$ and the bottom element is a precise abstraction corresponding to the partitioning into singleton sets.

Example 4. Consider the following set of clauses over a signature consisting of a single non-Boolean sort:

$$\{q(f(c)) \vee p(f(c)); \neg p(f(x)) \vee s(g(z), f(a)); \neg p(g(x)) \vee r(f(z), g(a)); \neg r(x, y)\},$$

we can group symbols of the same type such as q and p which are replaced by q'. Predicates s and r are replaced by s'; functions symbols f and g are replaced by f'. The resulting abstract set is:

$$\{q'(f'(c)); \neg q'(f'(x)) \vee s'(f'(z), f'(a)); \neg s'(x, y)\}.$$

This abstraction is unsatisfiable and we can refine it by concretising certain abstract symbols occurring in the unsat core, e.g.,

$$\{q(f'(c)) \vee p(f'(c)); \neg p(f'(x)) \vee s'(f'(z), f'(a)); \neg s'(x, y)\},$$

where q' is concretised.

Proposition 3. *The signature grouping ordering is compatible with the abstraction refinement ordering, i.e., if $\alpha_{sig}^0 \sqsubseteq_{sig} \alpha_{sig}^1$ then $\alpha_{sig}^0 \sqsubseteq_{sig} \alpha_{sig}^1$. Moreover, every signature grouping abstraction is above the identity abstraction, i.e., $\alpha_{id} \sqsubseteq \alpha_{sig}$.*

Let us note that signature grouping can be naturally combined with the argument filtering abstraction. In particular, argument filtering can reduce symbol types which in turn can be used to produce larger groups of abstract symbols.

4 Abstraction by Under-Approximation

The process starts by applying the weakening abstraction function to the set of concrete axioms A, $\hat{A}^w = \alpha_w(A)$. This set $\hat{A}^w$ of weaker axioms is used to prove the conjecture, using an ATP_S. If the conjecture is proved the procedure stops and provides the proof. Otherwise, a model I of $\hat{A}^w$ and the negated conjecture is obtained. This model is used to refine the set of weaker axioms $\hat{A}^w$. During this refinement (strengthening abstraction refinement), the procedure tries to find a set of axioms $\breve{A}$ that turns the model into a countermodel but are still implied by A, i.e., $I \not\models \breve{A}$ and $A \models \breve{A}$. If the set of axioms $\breve{A}$ is empty, $\breve{A} = \emptyset$, the procedure stops and disproves the conjecture. Otherwise, the obtained set of axioms is added to the set of weaker axioms, $\hat{A}^w := \hat{A}^w \cup \breve{A}$. Using this new set of abstract axioms $\hat{A}^w$, another round for proving the conjecture starts. The process finishes when the conjecture is proved or disproved or the time limit for the quest of a proof is reached. The diagram of this procedure is shown in Fig. 4.

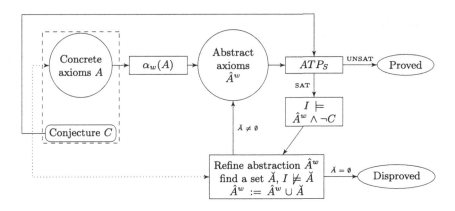

Fig. 4. Under-approximation

4.1 Weakening Abstraction Function

In the case of under-approximation, we propose two weakening abstractions: *instantiation abstraction* and *deletion abstraction*. In the case of instantiation abstraction, abstraction function generates ground instances of the concrete axioms as it is done in the Inst-Gen framework [19]. In the case of deletion abstraction we delete certain concrete axioms from the theory. This abstraction can be used to incorporate other axioms selection methods into this framework, which are based on removing irrelevant axioms. In particular, we incorporated SInE [15] which selects axioms based on syntactic relevance. In practice, different abstractions can be recombined.

4.2 Strengthening Abstraction Refinement

In the case of deletion abstraction, refinement can be done by adding concrete axioms $\breve{A}$ that turn the model I, which is obtained form ATP_S, into a counter-model, $\breve{A} \subseteq \{\breve{a} \mid \breve{a} \in A, I \not\models \breve{a}\}$. In the case of instantiation abstraction, refinement can be done by generating a set of ground instances of axioms $A\sigma$ such that $I \not\models A\sigma$, $\breve{A} := A\sigma$.

5   Combined Approximation

We can combine over- and under-approximations as follows. We use under-approximation in the outer-loop and over-approximation in the place of ATP_S (see Fig. 4). Let us note that abstractions can be shared between approximation loops. This combination allows us incorporate other axiom selection methods [15,30,32,33] as part of the under-approximation abstraction and combine them with over-approximation abstractions described in this paper.

6   Evaluation and Experimental Results

We implemented the abstraction-refinement framework described in this paper as part of the current version of iProver v2.7 [18,19][2], which is also the ATP that we utilised in our experiments.

We evaluated our implementation of the abstraction-refinement framework on the standard benchmark for first-order theorems provers: the TPTP library [29] with the set of problems from the Large Theory Batch (LTB) category in CASC-26 [17,21,28], during the competition the wall clock time limit was 90000 s per batch. All experiments described in this section were performed using a cluster of computers with the following characteristics: Linux v3.13, cpu 3.1 GHz and memory 125 GB. We used a time limit of 240 s for each attempt to solve a problem.

We experimented with different types of over-approximation abstractions: (i) subsumption abstraction, (ii) argument filtering abstraction, (iii) argument filtering restricted to Skolem functions and splitting predicates, (vi) signature grouping abstraction, and (v) signature grouping restricted to Skolem functions.

---

[2] iProver is available at: http://www.cs.man.ac.uk/~korovink/iprover/.

We implemented arbitrary combinations of these abstractions, which can be specified as a command line option to iProver, e.g.,

```
--abstr_ref "[subs;sig;arg_filter]".
```

For under-approximation abstractions we used the SInE axiom selection algorithm [15] and the Inst-Gen calculus which is the backbone of iProver. SInE is included with Vampire's [20] clausifier, which we also used for clausification.

The first set of experiments were performed over 1500 problems out of which: 716 were solved by signature grouping, 704 by signature grouping of Skolem symbols and constants, 637 by subsumption and 627 by argument filtering. Results are shown in Table 1.

Table 1. Problems solved by over-approximation abstractions with SInE.

Abstraction	Solutions
Signature grouping	716
Signature grouping Skolem/constants	704
Subsumption	637
Argument filter	627

In the next set of experiments we combined different over-approximation abstractions. In Table 2, we present the results obtained from combining different abstractions. Abstractions were applied in the same order as they are presented.

From these results, we can conclude that combination of abstractions considerably improves the performance. The best combination of abstractions is subsumption, signature grouping and argument filtering which solves around the 55% of the 1500 problems.

Table 2. Problems solved by iProver combination of abstractions and SInE

Abstraction	Solutions
subs; sig grouping; arg filter	826
subs; sig grouping	798
subs; sig grouping Skolem/constants; arg filter	733
subs; sig grouping Skolem/constants	719
subs; arg filter	630

We experimented with the top 3 strategies by restricting argument filtering and signature grouping to Skolem functions and splitting predicates and compared these to unrestricted versions. In this experiments the option --schedule was set to default. The results are shown in Table 3.

Table 3. Problems solved by iProver default schedule with abstractions and SInE

Abstraction	Solutions
subs; sig grouping; arg filter Skolem/splitting	957
subs; sig grouping; arg filter	942
subs; sig grouping Skolem/constants; arg filter	930

Table 4. Top strategies after removing overlapping solutions with **default schedule**. Where **subs** stands for subsumption, **sig** for signature, **arg-filt** for argument filtering, **SK** restriction to Skolem functions and splitting symbols in iProver.

Abstractions	Signature	Arg-filter	Until SAT	Solutions
subs, sig, arg-filt		SK	False	957
subs, sig, arg-filt	SK	Default	False	38
subs		Default	True	27
subs, sig, arg-filt		Default	True	11
subs, sig, arg-filt		Default	False	8
subs		Default	False	2
subs, sig, arg-filt	SK	Default	True	1
			Total	1044

Table 4 shows the number of solutions found by each strategy but excluding the problems solved by the previous ones. The total number of solved problems is 1044. There are several strategies from other combinations of abstractions, which solved small number of problems but turned out that those solutions are unique. If we combine these solutions with solutions shown in Table 4, the total number of solutions increases to 1070. Finally, in Table 5 we compare iProver and recent CASC-26 results. From this table we can conclude that integration of combinations of over-approximation abstractions considerably improves performance of iProver. Overall iProver considerably outperforms E-LTB [27] and gets close to the top systems Vampire [20] and MaLARea [32].

Table 5. Comparison with CACS-26 LTB results

Vampire-LTB	MaLARea	iProver-v2.7-all	iProver-v2.7	iProver-LTB-v2.6	E-LTB
1156	1131	1070	957	777	683

7 Conclusion and Further Work

In this paper, we presented a theoretical framework to abstraction-refinement for reasoning with large theories. We presented a number of concrete abstractions based on subsumption, argument filtering and signature grouping and discussed their combinations. We implemented the abstraction-refinement framework in iProver and evaluated different abstractions over the large theory problems in the TPTP library. The results are encouraging and show considerable improvements in the number of overall solved problems in the LTB category. Overall, the number of solved problems is 1070 problems out of 1500 which is considerably larger than the number of problems solved by the previous version of iProver-LTB-2.6 (777) and E-LTB-2.1 (683). Although still below the CASC winner Vampire-LTB-4.2 (1156) and MaLARea-0.6 (1144). We believe that fine-tuning abstraction parameters will help to further improve the performance.

Acknowledgements. We would like to thank anonymous reviewers for many helpful suggestions.

References

1. Alberti, F., Bruttomesso, R., Ghilardi, S., Ranise, S., Sharygina, N.: An extension of lazy abstraction with interpolation for programs with arrays. Formal Methods Syst. Des. **45**(1), 63–109 (2014)
2. Benzmüller, C., Steen, A., Wisniewski, M.: Leo-III version 1.1 (system description). In: Eiter, T., Sands, D., Sutcliffe, G., Voronkov, A. (eds.) IWIL@LPAR 2017. Kalpa Publications in Computing, vol.1, pp. 11–26. EasyChair (2017)
3. Blanchette, J.C., Böhme, S., Paulson, L.C.: Extending sledgehammer with SMT solvers. J. Autom. Reason. **51**(1), 109–128 (2013)
4. Blanchette, J.C., Böhme, S., Popescu, A., Smallbone, N.: Encoding monomorphic and polymorphic types. Log. Methods Comput. Sci. **12**(4:13), 1–52 (2016)
5. Bonacina, M.P., Lynch, C., de Moura, L.M.: On deciding satisfiability by theorem proving with speculative inferences. J. Autom. Reason. **47**(2), 161–189 (2011)
6. Brown, C.E.: Reducing higher-order theorem proving to a sequence of SAT problems. J. Autom. Reason. **51**(1), 57–77 (2013)
7. Bryant, R.E., Kroening, D., Ouaknine, J., Seshia, S.A., Strichman, O., Brady, B.: Deciding bit-vector arithmetic with abstraction. In: Grumberg, O., Huth, M. (eds.) TACAS 2007. LNCS, vol. 4424, pp. 358–372. Springer, Heidelberg (2007). https://doi.org/10.1007/978-3-540-71209-1_28
8. Clarke, E.M., Grumberg, O., Long, D.E.: Model checking and abstraction. ACM Trans. Program. Lang. Syst. **16**(5), 1512–1542 (1994)
9. Conchon, S., Goel, A., Krstic, S., Majumdar, R., Roux, M.: Far-cubicle - a new reachability algorithm for cubicle. In: Stewart, D., Weissenbacher, G. (eds.) FMCAD 2017, pp. 172–175. IEEE (2017)
10. Gauthier, T., Kaliszyk, C.: Sharing HOL4 and HOL light proof knowledge. In: Davis, M., Fehnker, A., McIver, A., Voronkov, A. (eds.) LPAR 2015. LNCS, vol. 9450, pp. 372–386. Springer, Heidelberg (2015). https://doi.org/10.1007/978-3-662-48899-7_26
11. Ge, Y., de Moura, L.: Complete instantiation for quantified formulas in satisfiabiliby modulo theories. In: Bouajjani, A., Maler, O. (eds.) CAV 2009. LNCS, vol. 5643, pp. 306–320. Springer, Heidelberg (2009). https://doi.org/10.1007/978-3-642-02658-4_25

12. Glimm, B., Kazakov, Y., Liebig, T., Tran, T., Vialard, V.: Abstraction refinement for ontology materialization. In: Bienvenu, M., Ortiz, M., Rosati, R., Simkus, M. (eds.) Informal Proceedings of the 27th International Workshop on Description Logics. CEUR Workshop Proceedings, vol. 1193, pp. 185–196. CEUR-WS.org (2014)
13. Hernandez, J.C.L., Korovin, K.: Towards an abstraction-refinement framework for reasoning with large theories. In: Eiter, T., Sands, D., Sutcliffe, G., Voronkov, A. (eds.) IWIL@LPAR 2017. Kalpa Publications in Computing, vol. 1. EasyChair (2017)
14. Hoder, K., Reger, G., Suda, M., Voronkov, A.: Selecting the selection. In: Olivetti, N., Tiwari, A. (eds.) IJCAR 2016. LNCS (LNAI), vol. 9706, pp. 313–329. Springer, Cham (2016). https://doi.org/10.1007/978-3-319-40229-1_22
15. Hoder, K., Voronkov, A.: Sine qua non for large theory reasoning. In: Bjørner, N., Sofronie-Stokkermans, V. (eds.) CADE 2011. LNCS (LNAI), vol. 6803, pp. 299–314. Springer, Heidelberg (2011). https://doi.org/10.1007/978-3-642-22438-6_23
16. Irving, G., Szegedy, C., Alemi, A.A., Eén, N., Chollet, F., Urban, J.: DeepMath - deep sequence models for premise selection. In: Lee, D.D., Sugiyama, M., von Luxburg, U., Guyon, I., Garnett, R. (eds.) NIPS 2016, pp. 2235–2243 (2016)
17. Kaliszyk, C., Urban, J.: HOL(y)Hammer: online ATP service for HOL light. Math. Comput. Sci. **9**(1), 5–22 (2015)
18. Korovin, K.: iProver - an instantiation-based theorem prover for first-order logic (system description). In: Proceedings of the IJCAR 2008, pp. 292–298 (2008)
19. Korovin, K.: Inst-Gen – a modular approach to instantiation-based automated reasoning. In: Voronkov, A., Weidenbach, C. (eds.) Programming Logics. LNCS, vol. 7797, pp. 239–270. Springer, Heidelberg (2013). https://doi.org/10.1007/978-3-642-37651-1_10
20. Kovács, L., Voronkov, A.: First-order theorem proving and VAMPIRE. In: Sharygina, N., Veith, H. (eds.) CAV 2013. LNCS, vol. 8044, pp. 1–35. Springer, Heidelberg (2013). https://doi.org/10.1007/978-3-642-39799-8_1
21. Kumar, R., Myreen, M.O., Norrish, M., Owens, S.: CakeML: a verified implementation of ML. In: POPL 2014, pp. 179–192 (2014)
22. Lynch, C.: Unsound theorem proving. In: Marcinkowski, J., Tarlecki, A. (eds.) CSL 2004. LNCS, vol. 3210, pp. 473–487. Springer, Heidelberg (2004). https://doi.org/10.1007/978-3-540-30124-0_36
23. Plaisted, D.A.: Theorem proving with abstraction. Artif. Intell. **16**(1), 47–108 (1981)
24. Reger, G., Bjørner, N., Suda, M., Voronkov, A.: AVATAR modulo theories. In: Benzmüller, C., Sutcliffe, G., Rojas, R. (eds.) GCAI 2016. EPiC Series in Computing, vol. 41, pp. 39–52. EasyChair (2016)
25. Reynolds, A., Tinelli, C., de Moura, L.M.: Finding conflicting instances of quantified formulas in SMT. In: FMCAD 2014, pp. 195–202. IEEE (2014)
26. Riazanov, A., Voronkov, A.: Splitting without backtracking. In: Nebel, B. (ed.) Proceedings of the IJCAI 2001, pp. 611–617. Morgan Kaufmann (2001)
27. Schulz, S.: System description: E 1.8. In: McMillan, K., Middeldorp, A., Voronkov, A. (eds.) LPAR 2013. LNCS, vol. 8312, pp. 735–743. Springer, Heidelberg (2013). https://doi.org/10.1007/978-3-642-45221-5_49
28. Slind, K., Norrish, M.: A brief overview of HOL4. In: Mohamed, O.A., Muñoz, C., Tahar, S. (eds.) TPHOLs 2008. LNCS, vol. 5170, pp. 28–32. Springer, Heidelberg (2008). https://doi.org/10.1007/978-3-540-71067-7_6
29. Sutcliffe, G.: The TPTP problem library and associated infrastructure. From CNF to TH0, TPTP v6.4.0. J. Autom. Reason. **59**(4), 483–502 (2017)

30. Sutcliffe, G., Puzis, Y.: SRASS - a semantic relevance axiom selection system. In: Pfenning, F. (ed.) CADE 2007. LNCS (LNAI), vol. 4603, pp. 295–310. Springer, Heidelberg (2007). https://doi.org/10.1007/978-3-540-73595-3_20

31. Teucke, A., Weidenbach, C.: First-order logic theorem proving and model building via approximation and instantiation. In: Lutz, C., Ranise, S. (eds.) FroCoS 2015. LNCS (LNAI), vol. 9322, pp. 85–100. Springer, Cham (2015). https://doi.org/10.1007/978-3-319-24246-0_6

32. Urban, J.: MaLARea: a metasystem for automated reasoning in large theories. In: CEUR Workshop Proceedings, vol. 257, pp. 45–58 (2007)

33. Urban, J., Sutcliffe, G., Pudlák, P., Vyskočil, J.: MaLARea SG1 - machine learner for automated reasoning with semantic guidance. In: Armando, A., Baumgartner, P., Dowek, G. (eds.) IJCAR 2008. LNCS (LNAI), vol. 5195, pp. 441–456. Springer, Heidelberg (2008). https://doi.org/10.1007/978-3-540-71070-7_37

Efficient Model Construction for Horn Logic with VLog
System Description

Jacopo Urbani[1], Markus Krötzsch[2], Ceriel Jacobs[1], Irina Dragoste[2], and David Carral[2(✉)]

[1] Vrije Universiteit Amsterdam, Amsterdam, The Netherlands
{jacopo,ceriel}@cs.vu.nl
[2] cfaed, TU Dresden, Dresden, Germany
{markus.kroetzsch,irina.dragoste,david.carral}@tu-dresden.de

Abstract. We extend the Datalog engine VLog to develop a column-oriented implementation of the *skolem* and the *restricted chase* – two variants of a sound and complete algorithm used for model construction over theories of existential rules. We conduct an extensive evaluation over several data-intensive theories with millions of facts and thousands of rules, and show that VLog can compete with the state of the art, regarding runtime, scalability, and memory efficiency.

1 Introduction

Rules of inference are a fundamental building block of many important algorithms in automated reasoning, and in related fields, such as artificial intelligence, data analytics, information integration, and knowledge management. They are at the core of leading tools and methods in many areas, ranging from logic programming, tableaux-based model construction, and "consequence-driven" approaches to ontological reasoning [13,14], over data integration [11] and query answering under constraints [7], to reasoning over knowledge graphs [16], and even social network analysis [18]. The optimisation of rule-based inferencing is therefore of crucial interest to automated reasoning.

In the recent past, there has been significant progress in this area, and many new rule-based systems have been presented [2,3,5,6,12,17,20]. At the core of these implementations is the most basic rule language *Datalog*, which syntactically corresponds to Horn logic without functions or existential quantifiers, while semantically it might be viewed either as a query language (reasoning = second-order model checking) or as a knowledge representation language (reasoning = first-order entailment checking) [1]. Systems nevertheless may exhibit strong differences due to the different use cases they have been designed for, which often also leads to different extensions and limitations.

This work was supported by the DFG within the cfaed Cluster of Excellence, CRC 912 (HAEC), and Emmy Noether grant KR 4381/1-1.

© Springer International Publishing AG, part of Springer Nature 2018
D. Galmiche et al. (Eds.): IJCAR 2018, LNAI 10900, pp. 680–688, 2018.
https://doi.org/10.1007/978-3-319-94205-6_44

One of the most important such extensions is support for *value invention*, manifested in the ability to handle either existential quantifiers or function terms in the consequences of rules. Equivalent formalisms are *existential rules* in ontological modelling, *tuple-generating dependencies* in database query answering, and Horn logic programs with function symbols in logic programming. The ability to create new terms during reasoning is crucial in many applications, e.g., for capturing incomplete information in databases [1], or for creating auxiliary structures in knowledge modelling [15]. But it is also much harder to implement since the resulting logic may no longer admit finite universal models, and reasoning becomes undecidable [10].

In this system description, we present our recent implementation of existential rule reasoning support in the Datalog engine VLog [20]. VLog differs from many other systems because of its column-based ("vertical") approach for storing inferred facts. This leads to high memory efficiency and competitive runtimes, but also requires specific implementation strategies and data structures. To the best of our knowledge, existential rule reasoning has never been implemented or studied in such an architecture.

Rule engines typically implement the so-called *chase* procedure – a saturation-based bottom-up model construction akin to a Horn logic tableau procedure. We implement it in two variants, the *skolem chase* and the *restricted chase*, of which the latter is more complicated but can produce smaller models in many cases. Indeed, it has recently been demonstrated that the restricted chase can compute models for many real-world ontologies where the skolem chase fails to terminate altogether [8]. This often requires Datalog rules to be preferred over existential rules, which we ensure in VLog.

We conduct an extensive evaluation to gauge the performance of our tool in comparison to the state of the art. In a recent evaluation of several chase implementations, Benedikt et al. found RDFox to be the most efficient tool in many contexts [4]. RDFox is also similar to VLog in that both conduct most of their computation in memory. We therefore compare VLog against RDFox, repeating many experiments of Benedikt et al. and adding several more using further real-world datasets. We find that, for reasoning with plain existential rules on a reasonably powerful laptop, VLog can often deliver comparable or even better performance than RDFox, while consistently needing much less memory. The former came as a surprise, since RDFox could take full advantage of its highly parallel algorithms, whereas VLog ran on a single thread on one CPU.

2   Preliminaries

We give a brief account of the relevant basic definitions and notation. Existential rules are based on a standard predicate logic vocabulary consisting of infinite, mutually disjoint sets of *predicates* $\mathbf{P}$ (each with a fixed arity), *constants* $\mathbf{C}$, and *variables* $\mathbf{V}$. A *term* is a variable $x \in \mathbf{V}$ or a constant $c \in \mathbf{C}$. An *atom* is a formula of the form $p(t_1, \ldots, t_n)$ where $t_1, \ldots, t_n$ are terms, and $p \in \mathbf{P}$ is a predicate of arity n. An *existential rule* (or simply *rule* in the context of this paper) is a formula of the form

$$\forall \boldsymbol{x} \forall \boldsymbol{y}. \big(B_1 \wedge \ldots \wedge B_k \rightarrow \exists \boldsymbol{v}. H_1 \wedge \ldots \wedge H_l \big), \tag{1}$$

where $\boldsymbol{x}$, $\boldsymbol{y}$, and $\boldsymbol{v}$ are mutually disjoint lists of variables, and $B_1, \ldots, B_k$ are atoms with variables from $\boldsymbol{x}$ and $\boldsymbol{y}$, $H_1, \ldots, H_l$ are atoms with all variables from $\boldsymbol{y}$ and $\boldsymbol{v}$, $l \geq 1$, and all variables in $\boldsymbol{y}$ occur in $B_1, \ldots, B_k$. The premise of a rule is called the *body*, while its conclusion is called the *head*. A *Datalog rule* is a rule without existential quantifiers, and a rule with $k = 0$ is called a *fact* (a conclusion that is unconditionally true). A finite set of facts is a *database*. Since all variables in rules are quantified, we often omit the explicit preceding universal quantifiers.

Example 1. The following rules capture basic part-whole relationships (meronomy), and are a typical pattern in many ontologies.

$$Bicycle(x) \rightarrow \exists v.hasPart(x, v) \wedge Wheel(v) \tag{2}$$
$$Wheel(x) \rightarrow \exists w.properPartOf(x, w) \wedge Bicycle(w) \tag{3}$$
$$properPartOf(x, y) \rightarrow partOf(x, y) \tag{4}$$
$$hasPart(x, y) \rightarrow partOf(y, x) \tag{5}$$
$$partOf(x, y) \rightarrow hasPart(y, x) \tag{6}$$

A major reasoning task of rule engines is (conjunctive) query answering. A *conjunctive query* (CQ) is a formula $\exists \boldsymbol{v}. B_1 \wedge \ldots \wedge B_k$, where B_i are atoms. Free variables (not in $\boldsymbol{v}$) are called *answer variables*. A *substitution* is a partial mapping $\sigma : \mathbf{V} \rightarrow \mathbf{V} \cup \mathbf{C}$. It is *ground* if it only maps to constants. Its application to terms and formulae is defined as usual. An *answer* to a CQ q over a set of rules $\mathcal{R}$ and database $\mathcal{D}$ is a ground substitution σ defined on the answer variables of q such that $\mathcal{R}, \mathcal{D} \models q\sigma$ under the usual semantics of first-order logic. Existential variables can be replaced by function terms. The *skolemisation* of a rule ρ as in (1) is obtained by replacing each variable $v \in \boldsymbol{v}$ by the term $f_{\rho,v}(\boldsymbol{y})$, where $f_{\rho,v}$ is a fresh skolem function symbol specific to ρ and v.

3   The Chase

The *chase* is a class of sound and complete reasoning algorithms that are widely used to implement query answering [4]. Rules are applied bottom-up until saturation, resulting in a *universal model*, which matches exactly those queries that are entailed by the original rules (and given data). For existential rules, the chase may fail to terminate (approximating an infinite universal model instead), and detecting termination is undecidable [1]. However, many decidable criteria that are sufficient for termination have been proposed and shown to be applicable in many practical cases [9]. There are many variants of the chase, depending, e.g., on which conditions are checked to determine whether the consequence of an

---

Algorithm 1. applyRule(rule $\rho = \forall \boldsymbol{x}, \boldsymbol{y}.\varphi \rightarrow \exists \boldsymbol{v}.\psi$)

Global variables : index i, index prev_ρ, previous derivations $(\Delta^k)_{k \leq i}$, bool changed

1.1 $\Delta^{i+1} = \emptyset$ $\ell = \mathsf{prev}_\rho$

1.2 **foreach** *match* σ *of* ρ *over* $\Delta^{[0,i]}$ *with* $\varphi\sigma \cap \Delta^{[\ell,i]} \neq \emptyset$ **do**

1.3   **if** $\Delta^{[0,i+1]} \not\models \exists \boldsymbol{v}.\psi\sigma$ **then**

1.4 $\sigma' = \sigma \cup \{\boldsymbol{v} \mapsto \boldsymbol{n}\}$ where $\boldsymbol{n} \subseteq \mathbf{Nulls}$ is fresh

1.5 $\Delta^{i+1} = \Delta^{i+1} \cup \{\psi\sigma'\}$

1.6 $\mathsf{prev}_\rho = i + 1$

1.7 $i = i + 1$

1.8 **if** $\Delta^{i+1} \neq \emptyset$ **then** changed = **true**

Algorithm 2. restrictedChase(rule set $\mathcal{R}$, database $\mathcal{D}$)

2.1 $i = 0$ $\Delta^0 = \mathcal{D}$ changed = **true** $\mathsf{prev}_\rho = -1$ for all rules $\rho \in \mathcal{R}$

2.2 **while** changed **do**

2.3   changed = **false**

2.4   **foreach** $\rho \in \mathcal{R}$ **do** applyRule(ρ)

2.5 **return** $\Delta^{[0,i]}$ // final result: union of all derived facts

---

applicable rule should be added. In this section, we explain the *restricted* and *skolem chase* since these are among the most studied variants.

Any chase produces a sequence of databases $\mathcal{D}^0, \mathcal{D}^1, \ldots$, beginning from the initially given database. In the cases we consider, we have $\mathcal{D}^{i+1} = \mathcal{D}^i \cup \Delta^{i+1}$, for the set Δ^{i+1} of facts derived in step $i+1$. We use abbreviations $\Delta^{[i,j]} = \bigcup_{k=i}^{j} \Delta^k$, $\Delta^0 = \mathcal{D}$ (the initial database), and $\Delta^{-1} = \emptyset$. In the chase variants we consider, only one rule is applied in each chase step, and consecutive chase steps consider different rules. We therefore store, for each rule ρ, the index prev_ρ of the chase step when it was last applied.

Algorithm 1 shows how one rule ρ is applied during the chase to compute Δ^{i+1}. Line 1.2 iterates over all *matches* of ρ: a *match* of a rule $\forall \boldsymbol{x}, \boldsymbol{y}.\varphi \rightarrow \exists \boldsymbol{v}.\psi$ over a database $\mathcal{D}$ is a ground substitution σ defined on $\boldsymbol{x} \cup \boldsymbol{y}$ such that $\mathcal{D} \models \varphi\sigma$. The additional requirement $\varphi\sigma \cap \Delta^{[\ell,i]} \neq \emptyset$ ensures that we only consider matches that were not found up to the previous application of ρ. This corresponds to a *semi-naive* materialisation strategy; we omit the details of how the matches σ can be found in practice [20]. Line 1.3 verifies that the entailments under a given match are logically relevant. Line 1.4 selects fresh labelled nulls for instantiating the newly derived fact(s), which then get(s) added. After finishing, we update ρ's step counter (Line 1.6) and global chase step (Line 1.7). Global variable changed records if any fact was derived (Line 1.8).

Algorithm 2 now shows the overall *restricted chase* procedure. It is named after the check in Line 1.3, which restricts the application of rules – when omitting this check, one obtains the *oblivious chase* instead. The restricted chase can

---

Algorithm 3. restrictedOrderedChase(rule set $\mathcal{R}$, database $\mathcal{D}$)

3.1 $i = 0$ $\Delta^0 = \mathcal{D}$ changed = **true**   prev$_\rho = -1$ for all rules $\rho \in \mathcal{R}$

3.2 **while** changed **do**

3.3 | changed = **false**

3.4 | **foreach** *Datalog rule* $\rho \in \mathcal{R}$ **do** applyRule(ρ)

3.5 | **if** ¬changed **then**

3.6 | └ **foreach** *Non-Datalog rule* $\rho \in \mathcal{R}$ **do** applyRule(ρ)

3.7 **return** $\Delta^{[0,i]}$ // final result: union of all derived facts

---

reduce the number of derived facts, which may allow it to terminate in more cases than the oblivious chase.

Example 2. Consider the restricted chase over the rules from Example 1 with database $\mathcal{D} = \{Bicycle(c)\}$. Applying rules in the given order, the first iteration of Line 2.4 yields $\Delta^0 = \mathcal{D}$, $\Delta^1 = \{hasPart(c, n_1), Wheel(n_1)\}$, $\Delta^2 = \{properPartOf(n_1, n_2), Bicycle(n_2)\}$, $\Delta^3 = \{partOf(n_1, n_2)\}$, $\Delta^4 = \{partOf(n_1, c)\}$, and $\Delta^5 = \{hasPart(n_2, n_1)\}$. Note that, when computing Δ^2, the check in Line 1.3 finds that $\Delta^{[0,2]} \not\models \exists w.partOf(n_1, w) \wedge Bicycle(w)$. No further derivations are produced thereafter; specifically the previous inferences already entail $\exists v.hasPart(n_2, v), Wheel(v)$. In contrast, the oblivious chase in this case would not terminate, since it would continue to apply rule (2) to new nulls.

Example 3. In contrast to the oblivious chase, the restricted chase is sensitive to the order of rules. For Example 2, if we apply rules in order (2), (3), (5), (6), (4), then we obtain $\Delta^0 = \mathcal{D}$, $\Delta^1 = \{hasPart(c, n_1), Wheel(n_1)\}$, $\Delta^2 = \{properPartOf(n_1, n_2), Bicycle(n_2)\}$, $\Delta^3 = \{partOf(n_1, c)\}$, $\Delta^4 = \emptyset$, and $\Delta^5 = \{partOf(n_1, n_2)\}$. Rule (3) can then be applied to match $\{x \mapsto n_2\}$ before $hasPart(n_2, n_1)$ gets inferred. The chase does not terminate.

Finally, the *skolem chase* is obtained by initially applying skolemisation to the rules in $\mathcal{R}$. This eliminates all existential variables, so that we have $\sigma = \sigma'$ in Line 1.4. Moreover, Line 1.3 in this case is merely a syntactic check for duplicates: since $\psi\sigma$ is ground, $\Delta^{[0,i+1]} \models \psi\sigma$ holds only if $\psi\sigma \subseteq \Delta^{[0,i+1]}$. The skolem chase terminates in significantly more cases than the oblivious chase, but it is still inferior to the restricted chase in this respect.

4   Chasing in VLog

VLog adopts the distinctive approach of computing each set Δ^i in bulk using an efficient "set-at-a-time" processing, storing the set of derivations column-by-column rather than row-by-row. Recent literature on columnar databases has shown that columnar data structures are very memory efficient and enable fast data access, but cannot be updated easily [20]. To avoid this problem, VLog works in an append-only mode and stores each set Δ^i into a dedicated data

Table 1. Rules and databases used in benchmarks (*MA* is maximal predicate arity)

Dataset	Number of rules	Number of facts	MA
Uniprot-005 / 010	531	4,713,207 / 9,252,708	2
Reactome-040 / 060 / 080	601	3,144,962 / 4,400,913 / 5,604,133	2
UOBM-10 / 20 / 40	426	1,926,879 / 3,980,967 / 7,843,543	2
STB-128	198	1,109,037	10
Ontology-256	529	2,146,490	11
doctors-10K / 1M	16	10,837 / 951,500	6
LUBM-010 / 100 / 1K	136	1,272,575 / 13,405,381 / 133,573,854	2
deep-100 / 200 / 300	1,100 / 1,200 / 1,300	1,000	4

structure. This strategy avoids the problem of updates altogether, and in practice has resulted in significantly shorter runtimes and lower memory consumption than the state-of-the-art – sometimes up to an order of magnitude.

The rest of this section sums up some of our main insights on implementing the restricted and the skolem chase efficiently in VLog. For the restricted chase, we make two further adjustments. First, we do not consider facts that were derived in the current (ongoing) chase step for checking if a rule application is restricted. Line 1.3 therefore checks if $\Delta^{[0,i]} \not\models \exists v.\psi\sigma$. This leads to what is called the *1-parallel restricted chase* [4].

Second, we ensure that Datalog rules are applied exhaustively before considering existential rules, as shown in Algorithm 3. This is motivated by recent studies of Carral et al., who proposed a criterion that uses this order to detect chase termination in more cases than previous works [8]. In fact, Example 1 shows a case for which VLog's restricted chase terminates, while other restricted chase implementations (e.g., of RDFox) do not.

From an implementation perspective, the execution of a rule can be split into the computation of all matches of the rule and the consequent computation of instantiations of the head. The first operation is the same regardless whether the rule contains existential quantifiers or not. Thus, we can reuse the same efficient algorithms developed for non-existential rule execution. The second operation, in contrast, requires ad-hoc operations due to the existence of unbound variables.

The exact operations differ depending on whether a restricted or a skolem chase is being computed. In the first case, we perform a series of merge joins between the set of matches and the columnar data structures that store the existing facts to remove partly instantiated matches. A merge join is very efficient here because the columnar data structures are already sorted [20]. Notice that if the head of the rule is a conjunction of multiple atoms, this procedure must be repeated for each head atom. Whenever the merge join finds a substitution to remove, it adds an entry into a positional index and use this index to skip to-be-removed matches. This strategy is adopted to avoid costly in-place removals. In the second case, we do not need to remove matches but we must retrieve the correct skolem terms. To support this operation, the system maintains a series of hash maps in main memory (one per rule/variable) with the arguments of the function and use it to return fresh IDs with average constant time.

5   Evaluation

We conducted an evaluation to gauge the performance and correctness of VLog.[1]
We compared to RDFox, which emerged as a leading tool in [4]. Experiments
were conducted in a laptop system (2.2 GHz Intel Core i7 (4 CPUs), 16 GB
1600 MHz DDR3, 512 GB SSD, MacOS High Sierra v10.13.3). The benchmark
inputs we use are shown in Table 1. UOBM, Reactome, and Uniprot are based on
data-intensive OWL ontologies,[2] which we converted to rules after removing nondeterministic axioms that do not correspond to Horn logic rules. The remaining
benchmarks are as given by Benedikt et al., where we omitted the rules with
equality, which are not supported by VLog [4].

For all tests, we measured the time and peak memory used for computing (a)
the restricted chase and (b) the skolem chase. We also verified that the size of
the skolem chase was the same for VLog and RDFox in all cases (the restricted
chase shows minor fluctuations, as expected for the different implementations).
The results are shown in Fig. 1. VLog could finish deep-300 on our laptop, but
using some OS swap space. Since we cannot measure this reliably, we only report
a lower bound.

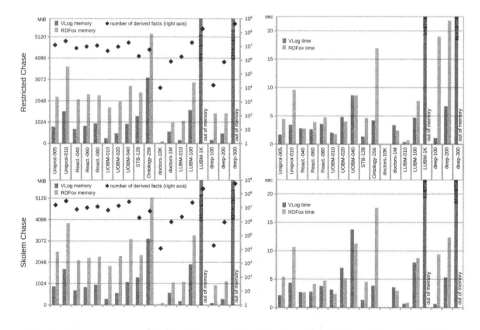

Fig. 1. Memory usage (left) and materialisation time (right) for VLog and RDFox

---

[1] All files used in this section are available at https://github.com/karmaresearch/
Chasing-VLog.
[2] Source http://www.cs.ox.ac.uk/isg/tools/PAGOdA/2015/jair/, accessed 2 Feb
2018.

VLog generally used much less memory, on average 40% of what was used by RDFox in either chase. This is expected since VLog uses highly optimised compressed data structures. In cases where only VLog finished (LUBM-1K and deep-300), RDFox ran out of memory. The times taken by VLog ranged from 5.8% (deep-100, rest.) to 137.5% (doctors-1M, rest.) of what was needed by RDFox. This is surprising, since VLog used only a single thread, whereas RDFox used maximal parallelism and often achieved above 700% CPU utilisation. Comparing the chase variants, VLog used significantly less time and memory for the restricted chase, except on deep-100, deep-200, and Ontology-256. RDFox shows similar behaviour, though the additional cost on deep is more pronounced. Nevertheless, the restricted chase seems to be the more efficient algorithm in general.

6   Conclusions

VLog is a fast and memory-efficient system for constructing models for Horn Logic. We extended its set-at-a-time and columnar approach to handle existential rules and discussed our implementation of the chase, which exhibits excellent performance.

The system is free and open source,[3] with only few dependencies for optional database connectors. Pre-compiled Docker images enable quick installation on major platforms (Docker repository *karmaresearch/vlog*). Users can control VLog through a command-line tool, a web interface (useful for demonstrating the system), and through the Java bindings of the companion project *VLog4j*.[4] The latter is available as a Maven package that includes the necessary binaries for major operating systems. In the future, we plan to add further expressive features, such as equality, negation, or aggregation. This can make VLog useful in even more scenarios, and thereby further advance our understanding of the potential of this architecture for automated reasoning in general.

References

1. Abiteboul, S., Hull, R., Vianu, V.: Foundations of Databases. Addison Wesley, Boston (1994)
2. Aref, M., ten Cate, B., Green, T.J., Kimelfeld, B., Olteanu, D., Pasalic, E., Veldhuizen, T.L., Washburn, G.: Design and implementation of the LogicBlox system. In: Proceedings of the 2015 ACM SIGMOD International Conference on Management of Data, pp. 1371–1382. ACM (2015)
3. Baget, J.-F., Leclère, M., Mugnier, M.-L., Rocher, S., Sipieter, C.: Graal: a toolkit for query answering with existential rules. In: Bassiliades, N., Gottlob, G., Sadri, F., Paschke, A., Roman, D. (eds.) RuleML 2015. LNCS, vol. 9202, pp. 328–344. Springer, Cham (2015). https://doi.org/10.1007/978-3-319-21542-6_21
4. Benedikt, M., Konstantinidis, G., Mecca, G., Motik, B., Papotti, P., Santoro, D., Tsamoura, E.: Benchmarking the chase. In: Proceedings 36th Symposium on Principles of Database Systems (PODS 2017), pp. 37–52. ACM (2017)

---

[3] C++ source code and documentation: https://github.com/karmaresearch/vlog.

[4] Java source code and documentation: https://github.com/mkroetzsch/vlog4j.

5. Benedikt, M., Leblay, J., Tsamoura, E.: PDQ: proof-driven query answering over web-based data. PVLDB **7**(13), 1553–1556 (2014)
6. Bonifati, A., Ileana, I., Linardi, M.: Functional dependencies unleashed for scalable data exchange. In: Proceedings of the 28th International Conference on Scientific and Statistical Database Management (SSDBM 2016), pp. 2:1–2:12. ACM (2016)
7. Calì, A., Gottlob, G., Kifer, M.: Taming the infinite chase: query answering under expressive relational constraints. In: Proceedings 11th International Conference on Principles of Knowledge Representation and Reasoning (KR 2008), pp. 70–80. AAAI Press (2008)
8. Carral, D., Dragoste, I., Krötzsch, M.: Restricted chase (non)termination for existential rules with disjunctions. In: Sierra [19], pp. 922–928 (2017)
9. Cuenca Grau, B., Horrocks, I., Krötzsch, M., Kupke, C., Magka, D., Motik, B., Wang, Z.: Acyclicity notions for existential rules and their application to query answering in ontologies. J. Artif. Intell. Res. **47**, 741–808 (2013)
10. Dantsin, E., Eiter, T., Gottlob, G., Voronkov, A.: Complexity and expressive power of logic programming. ACM Comput. Surv. **33**(3), 374–425 (2001)
11. Fagin, R., Kolaitis, P.G., Miller, R.J., Popa, L.: Data exchange: semantics and query answering. Theor. Comput. Sci. **336**(1), 89–124 (2005)
12. Geerts, F., Mecca, G., Papotti, P., Santoro, D.: That's all folks! LLUNATIC goes open source. PVLDB **7**(13), 1565–1568 (2014)
13. Kazakov, Y.: Consequence-driven reasoning for Horn $\mathcal{SHIQ}$ ontologies. In: Proceedings of the 21st International Joint Conference on Artificial Intelligence (IJCAI 2009), pp. 2040–2045. IJCAI (2009)
14. Kazakov, Y., Krötzsch, M., Simančík, F.: The incredible ELK: from polynomial procedures to efficient reasoning with $\mathcal{EL}$ ontologies. J. Autom. Reason. **53**, 1–61 (2013)
15. Krötzsch, M., Thost, V.: Ontologies for knowledge graphs: breaking the rules. In: Groth, P., Simperl, E., Gray, A., Sabou, M., Krötzsch, M., Lecue, F., Flöck, F., Gil, Y. (eds.) ISWC 2016. LNCS, vol. 9981, pp. 376–392. Springer, Cham (2016). https://doi.org/10.1007/978-3-319-46523-4_23
16. Marx, M., Krötzsch, M., Thost, V.: Logic on MARS: ontologies for generalised property graphs. In: Sierra [19], pp. 1188–1194 (2017)
17. Nenov, Y., Piro, R., Motik, B., Horrocks, I., Wu, Z., Banerjee, J.: RDFox: a highly-scalable RDF store. In: Arenas, M., et al. (eds.) ISWC 2015. LNCS, vol. 9367, pp. 3–20. Springer, Cham (2015). https://doi.org/10.1007/978-3-319-25010-6_1
18. Seo, J., Guo, S., Lam, M.S.: SociaLite: an efficient graph query language based on Datalog. IEEE Trans. Knowl. Data Eng. **27**(7), 1824–1837 (2015)
19. Sierra, C. (ed.) Proceedings of the 26th International Joint Conference on Artificial Intelligence (IJCAI 2017). IJCAI (2017)
20. Urbani, J., Jacobs, C., Krötzsch, M.: Column-oriented Datalog materialization for large knowledge graphs. In: Proceedings of the 30th AAAI Conference on Artificial Intelligence (AAAI 2016), pp. 258–264. AAAI Press (2016)

Focussing, **MALL** and the Polynomial Hierarchy

Anupam Das$^{(\boxtimes)}$

University of Copenhagen, Copenhagen, Denmark
anupam.das@di.ku.dk

Abstract. We investigate how to extract alternating time bounds from 'focussed' proofs, treating non-invertible rule phases as nondeterministic computation and invertible rule phases as co-nondeterministic computation. We refine the usual presentation of focussing to account for deterministic computations in proof search, which correspond to invertible rules that do not branch, more faithfully associating phases of focussed proof search to their alternating time complexity.

As our main result, we give a focussed system for MALLw (MALL with weakening) with encodings to and from true quantified Boolean formulas (QBFs): in one direction we encode QBF satisfiability and in the other we encode focussed proof search. Moreover we show that the composition of the two encodings preserves quantifier alternation, yielding natural fragments of MALLw complete for each level of the polynomial hierarchy. This refines the well-known result that MALLw is **PSPACE**-complete.

1 Introduction and Motivation

Proof systems are often a source of optimal decision algorithms for logics, theoretically speaking. We now know how to extract bounds for proof search in terms of various properties of the proof system at hand. E.g. we may compute:

- nondeterministic time bounds via *proof complexity*, e.g. [6,7,12];
- (non)deterministic space bounds via the *depth* of proofs or search spaces, and *loop-checking*, e.g. [3,11,22];
- deterministic or co-nondeterministic time bounds via systems of *invertible rules*, see e.g. [20,26].

However, despite considerable progress in the field, there still remains a gap between the obtention of (co-)nondeterministic time bounds, such as **NP** or **coNP**, and space bounds such as **PSPACE** (equivalently, *alternating polynomial time*, cf. [4]). Phrased differently, while we have many logics we know to be **PSPACE**-complete (intuitionistic propositional logic, various modal logics, etc.), we have very little understanding of their fragments corresponding to subclasses of **PSPACE**. In particular, in this work we are interested in the levels of

While conducting this research, the author was supported by a Marie Skłodowska-Curie fellowship, ERC project 753431.

the *polynomial hierarchy* (**PH**) [25], which correspond to alternating polynomial-time Turing machines with *boundedly* many alternations.

One relevant development in structural proof theory in the last 20–30 years has been the notion of *focussing*, e.g. [1,13,15]. Focussed systems elegantly delineate the phases of invertible and non-invertible inferences in proofs, allowing the natural obtention of alternating time bounds for a logic. Furthermore, they significantly constrain the number of local choices available, resulting in reduced nondeterminism during proof search, while remaining complete (the 'focussing theorem'). Such systems thus serve as a natural starting point for identifying fragments of **PSPACE**-complete logics complete for levels of **PH**.

In this work we will consider the case of *multiplicative additive linear logic* (MALL) [10], often seen as the prototypical system for **PSPACE** since its proof rules constitute the abstract templates of terminating proof search. (Indeed, MALL is well-known to be **PSPACE**-complete [16,17].) By considering a focussed presentation of the *affine* variant, MALLw, which admits weakening, we analyse proof search to identify classes of theorems belonging to each level of **PH**.[1] To demonstrate the accuracy of this method, we also show that these classes are, in fact, *complete* for their respective levels, via encodings from true quantified Boolean formulas (QBFs) of appropriate quantifier complexity, cf. [4].

One shortfall of focussed systems is that, in their usual form, they unfortunately do not make adequate consideration for *deterministic* computations, which correspond to invertible rules that do not branch, and so the natural measure of complexity there ('decide depth') can considerably overestimate the alternating time complexity of a theorem. In the worst case this can lead to rather degenerate bounds, exemplified in [8] where an encoding of SAT in intuitionistic logic requires a linear decide depth, despite being **NP**-complete.[2] In this work we keep the same abstract notion of focussing, but split the usual invertible, or 'asynchronous', phase into a 'deterministic' phase, with non-branching invertible rules, and a 'co-nondeterministic' phase, with branching invertible rules. In this way, when expressing proof search as an alternating predicate, a ∀ quantifier needs only be introduced in a co-nondeterministic phase. It turns out that this adaptation suffices to obtain the tight bounds we are after.

This paper is structured as follows. In Sect. 2 we present preliminaries on QBFs and alternating time complexity, and in Sect. 3 we present preliminaries on MALL and focussing. In Sect. 4 we present an encoding of true QBFs into MALLw, tracking the association between quantifier complexity and 'decide depth' in focussed proof search. In Sect. 5 we briefly explain how provability predicates for focussed systems may be obtained as QBFs, with quantifier complexity calibrated appropriately with decide depth (the 'focussing hierarchy'). In Sect. 6 we show how this depth measure can be feasibly approximated to yield a bona fide encoding of MALLw back into true QBFs. Furthermore, we show that the composition of the two encodings preserves quantifier complexity, and yields fragments of MALLw complete for each level of the polynomial hierarchy.

---

[1] MALLw is also **PSPACE**-complete, a folklore result subsumed by this work.

[2] In fact the same phenomenon presents in this work, cf. Fig. 3.

Finally, in Sect. 7 we give some concluding remarks regarding the case of (non-affine) MALL, and further perspectives on our presentation of focussing.

2   Preliminaries on Logic and Computational Complexity

We will recall some basic theory of Boolean logic, and its connections to alternating time complexity. Throughout this paper we omit constants (or 'units'), both for classical and linear logic, to simplify exposition and avoid clashing notations.

2.1   Second-Order Boolean Logic

Quantified Boolean formulas (QBFs) are obtained from the language of classical propositional logic by adding (second-order) quantifiers varying over propositions. Formally, let us fix some set Var of propositional *variables*, written x, y etc. QBFs, written φ, ψ etc., are generated as follows:

$$\varphi \quad ::= \quad x \mid \overline{x} \mid \varphi \vee \varphi \mid \varphi \wedge \varphi \mid \exists x.\varphi \mid \forall x.\varphi$$

The formula $\overline{x}$ stands for the *negation* of x, and all formulas we deal with will be in *De Morgan normal form*, i.e. with negation restricted to variables as in the grammar above. Nonetheless, we may sometimes write $\overline{\varphi}$ to denote the De Morgan *dual* of φ, generated by the following identities:

$$\overline{\overline{x}} := x \qquad \overline{(\varphi \vee \psi)} := \overline{\varphi} \wedge \overline{\psi} \qquad \overline{\exists x.\varphi} := \forall x.\overline{\varphi}$$
$$\overline{(\varphi \wedge \psi)} := \overline{\varphi} \vee \overline{\psi} \qquad \overline{\forall x.\varphi} := \exists x.\overline{\varphi}$$

A formula is *closed* if all its variables are bound by a quantifier ($\exists$ or $\forall$). We write $|\varphi|$ for the number of occurrences of literals (i.e. x or $\overline{x}$) in φ.

An *assignment* is a subset $\alpha \subseteq$ Var. We define the *satisfaction* relation between an assignment α and a formula φ, written $\alpha \vDash \varphi$, in the usual way:

- $\alpha \vDash x$ if $x \in \alpha$.   - $\alpha \vDash \exists x.\varphi$ if $\alpha \setminus \{x\} \vDash \varphi$ or $\alpha \cup \{x\} \vDash \varphi$.
- $\alpha \vDash \overline{x}$ if $x \notin \alpha$.   - $\alpha \vDash \forall x.\varphi$ if $\alpha \setminus \{x\} \vDash \varphi$ and $\alpha \cup \{x\} \vDash \varphi$.
- $\alpha \vDash \varphi \vee \psi$ if $\alpha \vDash \varphi$ or $\alpha \vDash \psi$.
- $\alpha \vDash \varphi \wedge \psi$ if $\alpha \vDash \varphi$ and $\alpha \vDash \psi$.

Definition 1 (Second-order Boolean logic). *A QBF φ is satisfiable if there is some assignment $\alpha \subseteq$ Var such that $\alpha \vDash \varphi$. It is valid if $\alpha \vDash \varphi$ for every assignment $\alpha \subseteq$ Var. If φ is closed, then we may simply say that it is true, written $\vDash \varphi$, when it is satisfiable and/or valid.*[3]

Second-order Boolean logic *(CPL2) is the set of true QBFs.*

---

[3] Notice that, by definition of satisfaction these two notions coincide for closed QBFs.

In practice, when dealing with a given formula φ, we will only need to consider assignments α that contain variables occurring in φ. We will assume this later when we discuss predicates (or 'languages') computed by open QBFs.

We point out that, from the logical point of view, it suffices to work with only closed QBFs, with satisfiability recovered by prenexing $\exists$ quantifiers and validity recovered by prenexing $\forall$ quantifiers; in the presence of units/constants, the definition of 'truth' above could be adapted with no reference to α. However we will also make use of open formulas in this work to describe languages/predicates, so it will be useful to have the notion of satisfaction available.

Definition 2 (QBF hierarchy). *For $k \geq 0$ we define the following classes:*

- $\Sigma_0^q = \Pi_0^q$ *is the set of quantifier-free QBFs.*
- $\Sigma_{k+1}^q \supseteq \Pi_k^q$ *and, if $\varphi \in \Sigma_{k+1}^q$, then so is $\exists x.\varphi$.*
- $\Pi_{k+1}^q \supseteq \Sigma_n^q$ *and, if $\varphi \in \Pi_{k+1}^q$, then so is $\forall x.\varphi$.*

Notice that $\varphi \in \Sigma_k^q$ if and only if $\overline{\varphi} \in \Pi_k^q$, by the definition of De Morgan duality.

We have only defined the classes above for 'prenexed' QBFs, i.e. with all quantifiers at the front. It is well known that any QBF is equivalent to such a formula. For this reason we will systematically assume that any QBF we deal with is in prenex form. In this case we call its quantifier-free part, i.e. its largest quantifier-free subformula, the *matrix*.

In this work we will not need to formally deal with any deduction system for CPL2, although we point out that there is a simple system, *semantic trees*, whose proof search dynamics closely match quantifier complexity [14].

2.2 Alternating Time Complexity

In computation we are used to the distinction between *deterministic* and *non-deterministic* computation. Intuitively, *co-nondeterminism* is just the 'dual' of nondeterminism: at the machine level it is captured by 'nondeterministic' Turing machines where *every* run is accepting, not just *some* run as in the case of usual nondeterminism. From here *alternating* Turing machines generalise both the nondeterministic and co-nondeterministic models by allowing both universally branching states and existentially branching states.

Intuitions aside, we will introduce the concepts we need here assuming only a familiarity with deterministic and nondeterministic Turing machines and their complexity measures, to limit the formal prerequisites. Our exposition is informal, but the reader may find comprehensive details in, e.g., [23].

For a language L, we write $\mathbf{NP}(L)$ to mean the class of languages accepted in polynomial time by some nondeterministic Turing machine which may, at any point, query in constant time whether some word is in L or not. We extend this to classes of languages $\mathcal{C}$, writing $\mathbf{NP}(\mathcal{C})$ for $\bigcup_{L \in \mathcal{C}} \mathbf{NP}(L)$. We also write $\mathbf{co}\mathcal{C}$ for the class of languages whose complements are in $\mathcal{C}$.

Definition 3 (Polynomial hierarchy, [25]). *We define the following classes:*

- $\Sigma_0^p = \Pi_0^p := \mathbf{P}$.
- $\Sigma_{k+1}^p := \mathbf{NP}(\Sigma_k^p)$.
- $\Pi_{k+1}^p := \mathbf{co}\Sigma_{k+1}^p$.

The polynomial hierarchy *(**PH**) is* $\bigcup_{k \geq 0} \Sigma_k^p = \bigcup_{k \geq 0} \Pi_k^p$.

We may more naturally view the polynomial hierarchy as the bounded-quantifier-alternation fragments of QBFs we introduced earlier. For this we construe Σ_k^q and Π_k^q as classes of finite *languages*, by associating with a QBF $\varphi(\vec{x})$ the class of (finite) assignments $\alpha \subseteq \vec{x}$ satisfying it. (Assignments themselves may be seen as binary strings of length $|\vec{x}|$ which encode their characteristic functions.) Σ_k^q-*evaluation* (Π_k^q-*evaluation*) is the problem of deciding, given a Σ_k^q (resp. Π_k^q) formula $\varphi(\vec{x})$ and an assignment $\alpha \subseteq \vec{x}$, whether $\alpha \vDash \varphi(\vec{x})$.

Theorem 4 ([cf. [4]). *For $k \geq 1$, the Σ_k^q-evaluation (Π_k^q-evaluation) problem is Σ_k^p-complete (resp. Π_k^p-complete).*

Corollary 5. *For $k \geq 1$, the class of true closed Σ_k^q QBFs is Σ_k^p-complete, and the class of true closed Π_k^q QBFs is Π_k^p-complete.*

Proof (Idea). Membership is immediate, evaluating under the assignment $\varnothing$. For hardness, we may simplify a QBF under an assignment to a closed formula. □

3 Linear Logic and Proof Search

In this section we introduce *multiplicative additive linear logic* (MALL) and its proof theory [10], in particular a certain *focussed* proof system for it, cf. [1,5,9].

3.1 Multiplicative Additive Linear Logic

For convenience, we work with the same set Var of variables that we used for QBFs and, as for classical logic, we omit constants/units for simplicity (though their inclusion would not affect our results). To distinguish them from QBFs, we use the metavariables A, B, etc. for MALL formulas, generated as follows:

$$A \quad ::= \quad x \mid \overline{x} \mid A \,\invamp\, B \mid A \oplus B \mid A \otimes B \mid A \mathbin{\&} B$$

$\invamp, \otimes$ are called *multiplicative* connectives, and $\oplus, \&$ are called *additive* connectives. Like for QBFs, we have restricted negation to the variables, thanks to De Morgan duality in MALL. Again we may write $\overline{A}$ for the De Morgan dual of A, which is generated similarly to the case of QBFs:

$$\overline{\overline{A}} := A \qquad \overline{(A \invamp B)} := \overline{A} \otimes \overline{B} \qquad \overline{(A \oplus B)} := \overline{A} \mathbin{\&} \overline{B}$$
$$\overline{(A \otimes B)} := \overline{A} \invamp \overline{B} \qquad \overline{(A \mathbin{\&} B)} := \overline{A} \oplus \overline{B}$$

Due to De Morgan duality, we will work only with 'one-sided' calculi for MALL, where all formulas occur to the right of the sequent arrow. This means we will have fewer cases to consider for formal proofs, although later we will also informally adopt a two-sided notation when it is convenient, cf. Notation 11.

Definition 6 (MALL(w)). *A cedent, written Γ, Δ etc., is a multiset of formulas, delimited by commas ',', and a* sequent *is an expression $\vdash \Gamma$. The system (cut-free) MALL is given in Fig. 1. MALLw, a.k.a.* affine MALL, *is defined in the same way, only with the (id) rule replaced by:*

$$wid \; \frac{}{\vdash \Gamma, x, \overline{x}} \tag{1}$$

Notice that, following the tradition in linear logic, we write '$\vdash$' for the sequent arrow, though we point out that the deduction theorem does not actually hold w.r.t. linear implication. For the affine variant, we have simply built weakening into the identity step, since it may always be permuted upwards in a proof:

$$id \; \frac{}{\vdash x, \overline{x}} \qquad \mathord{\mathscr{Y}} \; \frac{\vdash \Gamma, A, B}{\vdash \Gamma, A \mathbin{\mathscr{Y}} B} \qquad \otimes \; \frac{\vdash \Gamma, A \quad \vdash \Delta, B}{\vdash \Gamma, \Delta, A \otimes B} \qquad \oplus \; \frac{\vdash \Gamma, A_i}{\vdash \Gamma, A_0 \oplus A_1} \, i \leq 1 \qquad \& \; \frac{\vdash \Gamma, A \quad \vdash \Gamma, B}{\vdash \Gamma, A \mathbin{\&} B}$$

Fig. 1. The system (cut-free) MALL.

Proposition 7 (Weakening admissibility). *The following rule, called* weakening, *is (height-preserving) admissible in MALLw:*

$$wk \; \frac{\vdash \Gamma}{\vdash \Gamma, A}$$

Notice also that we have not included the 'cut' rule, thanks to cut-elimination for linear logic [10]. It will play no role in this paper.

3.2   (Multi-)focussed Systems for Proof Search

Focussed systems for MALL (and linear logic in general) have been widely studied [1,5,9,13]. The idea is to associate polarities to the connectives based on whether their introduction rule is invertible (negative) or their dual's introduction rule is invertible (positive). Now bottom-up proof search can be organised in a manner where, once we have chosen a positive principal formula to decompose (the 'focus'), we may continue to decompose its auxiliary formulas until the focus becomes negative. The main result herein is the *completeness* of such proof search strategies, known as the *focussing theorem* (a.k.a. the 'focalisation theorem').

It is known that 'multi-focussed' variants, where one may have many foci in parallel, lead to 'canonical' representations of proofs for MALL [5]. Furthermore, the alternation behaviour of focussed proof search can be understood via a game theoretic approach [9]. However, such frameworks unfortunately fall short of characterising the alternating complexity of proof search in a faithful way. The issue is that the usual focussing methodology does not make any account for *deterministic* computations, which correspond to invertible rules that do not branch. Such rules are usually treated just like the other invertible rules, and so 'morally' introduce extraneous quantifiers when encoding proof search as an alternating time predicate.

For these reasons we introduce a bespoke presentation of (multi-)focussing for MALL, with a designated *deterministic* phase allowing invertible non-branching rules, in this case the $\mathbin{⅋}$ rule. To avoid conflicts with more traditional presentations, we call the other two phases as *nondeterministic* and *co-nondeterministic* rather than 'synchronous' and 'asynchronous' respectively; at the same time this reinforces the intended connections to computational complexity.

In what follows, we use a, b, etc. to vary over literals. We also use the following metavariables to vary over formulas with the corresponding top-level connectives:

$$
\begin{aligned}
M &: \text{'negative and not deterministic'} & & \& \\
N &: \text{'negative'} & & \&, \mathbin{⅋} \\
O &: \text{'deterministic'} & & \otimes, \mathbin{⅋}, a \\
P &: \text{'positive'} & & \otimes, \oplus \\
Q &: \text{'positive and not deterministic'} & & \oplus
\end{aligned}
$$

'Vectors' are used to vary over multisets of associated formulas, e.g. $\vec{P}$ varies over multisets of P-formulas. Sequents may now contain the delimiters $\Downarrow$ or $\Uparrow$.

Definition 8 (Multi-focussed proof system). *We define the (multi-focussed) system* FMALL *in Fig. 2. The system* FMALLw *is the same as* FMALL *but with the* (id) *rule replaced by the rule* (wid) *from* (1).

Deterministic phase:

$$
id \, \frac{}{\vdash a, \overline{a}} \qquad
\mathbin{⅋} \, \frac{\vdash \Gamma, A, B}{\vdash \Gamma, A \mathbin{⅋} B} \qquad
D \, \frac{\vdash \vec{a}, \vec{P} \Downarrow \vec{P'}}{\vdash \vec{a}, \vec{P}, \vec{P'}} \, \vec{P'} \neq \varnothing \qquad
\bar{D} \, \frac{\vdash \vec{a}, \vec{P} \Uparrow \vec{M}}{\vdash \vec{a}, \vec{P}, \vec{M}} \, \vec{M} \neq \varnothing
$$

Nondeterministic phase:

$$
\oplus \, \frac{\vdash \Gamma \Downarrow \Delta, A_i}{\vdash \Gamma \Downarrow \Delta, A_0 \oplus A_1} \, i \leq 1 \qquad
\otimes \, \frac{\vdash \Gamma \Downarrow \Sigma, A \quad \vdash \Delta \Downarrow \Pi, B}{\vdash \Gamma, \Delta \Downarrow \Sigma, \Pi, A \otimes B} \qquad
R \, \frac{\vdash \Gamma, \vec{a}, \vec{N}}{\vdash \Gamma \Downarrow \vec{a}, \vec{N}}
$$

Co-nondeterministic phase:

$$
\& \, \frac{\vdash \Gamma \Uparrow \Delta, A \quad \Gamma \Uparrow \Delta, B}{\vdash \Gamma \Uparrow \Delta, A \& B} \qquad
\bar{R} \, \frac{\vdash \Gamma, \vec{P}, \vec{O}}{\vdash \Gamma \Uparrow \vec{P}, \vec{O}}
$$

Fig. 2. The system FMALL.

Note that the determinism of $\otimes$ plays no role in this one-sided calculus, but in a two-sided calculus we would have a full symmetry of rules. A proof of a formula A is simply a proof of the sequent $\vdash A$, i.e. there is no need to pre-decorate with arrows, as opposed to usual presentations, thanks to the deterministic phase.

The rules D and $\bar{D}$ are called *decide* and *co-decide* respectively, while R and $\bar{R}$ are called *release* and *co-release* respectively. We have not included a 'store' rule, for simplicity, but if we did we would also recover a dual 'co-store' rule.

As usual for multi-focussed systems, the analogous focussed system can be recovered by restricting to only one focussed formula in a nondeterministic phase. Moreover, in our presentation, we may also impose the *dual* restriction, that there is only one formula in 'co-focus' during a co-nondeterministic phase:

Definition 9 (Simply (co-)focussed subsystems). *A* FMALL *proof is* focussed *if $\vec{P}$ in D is always a singleton. It is* co-focussed *if $\vec{M}$ in $\bar{D}$ is always a singleton. If a proof is both focussed and co-focussed then we say it is* bi-focussed.

The notion of 'co-focussing' is not usually possible for (multi-)focussed systems since the invariant of being a singleton is not usually maintained in an asynchronous phase, due to the $\otimes$ rule. However we treat $\otimes$ as deterministic rather than co-nondeterministic, and we can see that the $\&$-rule indeed maintains the invariant of having just one formula on the right of $\Uparrow$.

Theorem 10 (Focussing theorem). *The class of bi-focussed* FMALL-*proofs* (FMALLw-*proofs) is complete for* MALL *(resp.* MALLw*).*

Evidently, this immediately means that FMALL (FMALLw), as well as its focussed and co-focussed subsystems, are also complete for MALL (resp. MALLw). The proof of Theorem 10 follows routinely from any other completeness proof for focussed MALL, e.g. [1,13]; our only change is at the level of notation.

To aid our exposition, we will sometimes use a 'two-sided' notation and extra connectives so that the intended semantics of sequents are clearer. Strictly speaking, this is just a shorthand for one-sided sequents: the calculi defined in Figs. 1 and 2 are the formal systems we are studying.

Notation 11. *We write $\Gamma \vdash \Delta$ as shorthand for the sequent $\vdash \overline{\Gamma}, \Delta$, where $\overline{\Gamma}$ is $\{\overline{A} : A \in \Gamma\}$. We extend this notation to sequents with $\Uparrow$ or $\Downarrow$ symbols in the natural way, writing $\Gamma \Uparrow \Delta \vdash \Sigma \Uparrow \Pi$ for $\vdash \overline{\Gamma}, \Sigma \Uparrow \overline{\Delta}, \Pi$ and $\Gamma \Downarrow \Delta \vdash \Sigma \Downarrow \Pi$ for $\vdash \overline{\Gamma}, \Sigma \Downarrow \overline{\Delta}, \Pi$. In all cases, (co-)foci are always written to the right of $\Downarrow$ or $\Uparrow$.*

We write $A \multimap B$ as shorthand for the formula $\overline{A} \otimes B$, and $A \multimap^+ B$ as shorthand for the formula $\overline{A} \oplus B$. Sometimes we will write, e.g., a step,

$$\multimap_l \frac{\Gamma \vdash \Delta \Downarrow A \quad \Gamma' \Downarrow B \vdash \Delta'}{\Gamma, \Gamma' \Downarrow A \multimap B \vdash \Delta, \Delta'}$$

which, by definition, corresponds to a correct application of $\otimes$ in FMALL(w).

4   An Encoding from CPL2 to MALLw

From now on we will work only with MALLw, i.e. affine MALL. In this section we present an encoding of true QBFs into MALLw. The former were also used for the original proof that MALL is **PSPACE**-complete [16,17], though our encoding differs considerably from theirs and leads to a more refined result, cf. Sect. 6.

4.1 Positive and Negative Encodings of Quantifier-Free Evaluation

The base cases of our translation from QBFs to MALLw will be quantifier-free Boolean evaluation. This is naturally a deterministic computation, being polynomial-time computable.[4] However one issue is that this determinism cannot be seen from the point of view of MALLw, since the only deterministic connective ($\mathbin{⅋}$, on the right) is not expressive enough to encode evaluation.

Nonetheless we are able to circumvent this problem since MALLw is at least able to 'see' quantifier-free evaluation as a problem in $\mathbf{NP} \cap \mathbf{coNP}$, via a pair of corresponding encodings. For non-base levels of $\mathbf{PH}$ this is morally the same as being deterministic. Indeed, the availability of both types of encodings is the main reason why we consider MALLw rather than MALL in this work.

Definition 12 (Positive and negative encodings). *For a quantifier-free Boolean formula φ_0, we define φ_0^- (φ_0^+) as the result of replacing every $\wedge$ in φ_0 for & (resp. $\otimes$) and every $\vee$ in φ_0 by $\mathbin{⅋}$ (resp. $\oplus$).*

For an assignment α and list of variables $\vec{x} = (x_1, \dots, x_k)$, we write $\alpha(\vec{x})$ for the cedent $\{x_i : x_i \in \alpha, i \leq k\} \cup \{\overline{x}_i : x_i \notin \alpha, i \leq k\}$. We write $\alpha^n(\vec{x})$ for the cedent consisting of n copies of each literal in $\alpha(\vec{x})$.

Proposition 13. *Let φ_0 be a quantifier-free Boolean formula with free variables $\vec{x}$ and let α be an assignment. For $n \geq |\varphi_0|$, the following are equivalent:*

1. *$\alpha \vDash \varphi_0$.*
2. *MALLw proves $\alpha(\vec{x}) \vdash \varphi_0^-$.*
3. *MALLw proves $\alpha^n(\vec{x}) \vdash \varphi_0^+$.*

Proof. $2 \implies 1$ and $3 \implies 1$ are immediate from the 'soundness' of MALLw with respect to classical logic, by interpreting $\otimes$ or & as $\wedge$ and $\oplus$ or $\mathbin{⅋}$ as $\vee$.

$1 \implies 2$ and $1 \implies 3$ are both proved by induction on $|\varphi_0|$. In the former case, this follows directly from the invertibility of rules, while in the latter case we appeal to the properties of satisfaction: for $\oplus$-formulas we choose an appropriate disjunct satisfied by α, and for $\otimes$-formulas we split $\alpha^n(\vec{x})$ into $\alpha^k(\vec{x})$ and $\alpha^l(\vec{x})$ s.t. k and l bound the size of their respective conjuncts, reducing to the inductive hypothesis. For both arguments we must appeal to affinity for the base case. $\square$

4.2 Encoding Quantifiers in MALLw

As we said before, we do not follow the 'locks-and-keys' approach of [16,17]. Instead we follow a similar approach to Statman's proof that intuitionistic propositional logic is $\mathbf{PSPACE}$-hard [24], modulo some improvements that are discussed, for the intuitionistic setting, in [8]. One of the main differences is that we use 'Tseitin extension variables', necessary to avoid an exponential blowup during translation, only in positive positions, not under negation, and this allows our encodings to admit similar proofs to the 'semantic trees' of the QBF we started with.

---

[4] In fact, quantifier-free Boolean formula evaluation is known to be $\mathbf{NC}^1$-complete [2].

Definition 14 (CPL2 *to* MALLw)**.** *Given a QBF* $\varphi = Q_k x_k. \cdots . Q_1 x_1. \varphi_0$ *with* $|\varphi_0| = n$, *we define* $[\varphi]$ *by induction on* $k \geq 1$ *as follows,*

$$[\varphi_0] := \begin{cases} \varphi_0^+ & \text{if } Q_1 \text{ is } \exists \\ \varphi_0^- & \text{if } Q_1 \text{ is } \forall \end{cases}$$

$$[Q_k x_k. \varphi'] := \begin{cases} ([\varphi'] \multimap y_k) \multimap ((x_k^n \multimap y_k) \oplus (\overline{x}_k^n \multimap y_k)) & \text{if } Q_k \text{ is } \exists \\ ([\varphi'] \multimap^+ y_k) \multimap ((x_k^n \multimap y_k) \,\&\, (\overline{x}_k^n \multimap y_k)) & \text{if } Q_k \text{ is } \forall \end{cases}$$

where y_k *is always fresh.*

Lemma 15. *Let* $\varphi(\vec{x})$ *be a QBF with all free variables displayed and matrix* φ_0. *Then* $\alpha \vDash \varphi$ *if and only if* MALLw *proves* $\alpha^n(\vec{x}) \vdash \vec{y}, [\varphi]$ *for any* $n \geq |\varphi_0|$, *any assignment* α *and any* $\vec{y}$ *disjoint from* $\vec{x}$.

$$
\begin{array}{c}
\overbrace{}^{IH} \\
\cfrac{\alpha^n(\vec{x}), \pm x^n \vdash \vec{y}, [\varphi]}{\cfrac{\alpha^n(\vec{x}), \pm x^n \vdash \vec{y} \Downarrow [\varphi]}{}\, R_r} \quad \cfrac{\cfrac{y \vdash y}{\Downarrow y \vdash y}\,R_l \; \overset{id}{\rule{1.5em}{0.4pt}}}{} \\
\cfrac{\phantom{\alpha^n(\vec{x}), \pm x^n \Downarrow [\varphi] \multimap y \vdash \vec{y}, y}}{\cfrac{\alpha^n(\vec{x}), \pm x^n \Downarrow [\varphi] \multimap y \vdash \vec{y}, y}{\cfrac{\alpha^n(\vec{x}), \pm x^n, [\varphi] \multimap y \vdash \vec{y}, y}{\cfrac{\alpha^n(\vec{x}), [\varphi] \multimap y \vdash \vec{y}, \pm x^n \multimap y}{\cfrac{\alpha^n(\vec{x}), [\varphi] \multimap y \vdash \vec{y} \Downarrow \pm x^n \multimap y}{\cfrac{\alpha^n(\vec{x}), [\varphi] \multimap y \vdash \vec{y} \Downarrow (x^n \multimap y) \oplus (\overline{x}^n \multimap y)}{\cfrac{\alpha^n(\vec{x}), [\varphi] \multimap y \vdash \vec{y}, (x^n \multimap y) \oplus (\overline{x}^n \multimap y)}{\cfrac{\alpha^n(\vec{x}) \vdash \vec{y}, ([\varphi] \multimap y) \multimap ((x^n \multimap y) \oplus (\overline{x}^n \multimap y))}{\alpha^n(\vec{x}) \vdash \vec{y}, [\exists x. \varphi]}}}}}}}}
\end{array}
$$

Fig. 3. Proof of $\exists$ case for left-right direction of Lemma 15.

Proof (sketch). We proceed by induction on the quantifier complexity of φ. For the base case, when φ is quantifier-free, we appeal to Proposition 13. The left-right direction follows directly by weakening (cf. Proposition 7), while the right-left direction follows after observing that $\vec{y}$ does not occur in $[\varphi]$ or $\alpha^n(\vec{x})$; thus $\vec{y}$ may be deleted from a proof (along with its descendants) while preserving correctness.

For the inductive step, in the left-right direction we give appropriate bi-focussed proofs in Figs. 3 and 4, where: $\pm x$ in Fig. 3 is chosen to be x if $x \in \alpha$ and $\overline{x}$ otherwise; the derivations marked *IH* are obtained by the inductive hypothesis; and the derivation marked ... in Fig. 4 is analogous to the one on the left of it.[5]

---

[5] Note that, for the derivations for the innermost quantifier ($\exists$ or $\forall$), the topmost R or $\overline{R}$ step of Figs. 3 or 4 (resp.) does not occur.

$$
\cfrac{
\cfrac{
\cfrac{
\cfrac{
\cfrac{
\cfrac{
\cfrac{
\bar{R}_r \cfrac{\overset{\displaystyle \triangledown_{IH}}{\alpha^n(\vec{x}), x^n \vdash \vec{y}, y, [\varphi]}}{\alpha^n(\vec{x}), x^n \vdash \vec{y}, y \Uparrow [\varphi]} \qquad \bar{R}_l \cfrac{id \; \overline{\alpha^n(\vec{x}), x^n, y \vdash \vec{y}, y}}{\alpha^n(\vec{x}), x^n \Uparrow y \vdash \vec{y}, y}
}{\alpha^n(\vec{x}), x^n \Uparrow [\varphi] \multimap^+ y \vdash \vec{y}, y} \; \multimap_l^+
}{\alpha^n(\vec{x}), x^n, [\varphi] \multimap^+ y \vdash \vec{y}, y} \; \bar{D}_l
}{\alpha^n(\vec{x}), [\varphi] \multimap^+ y \vdash \vec{y}, x^n \multimap y} \; \multimap_r
}{\alpha^n(\vec{x}), [\varphi] \multimap^+ y \vdash \vec{y} \Uparrow x^n \multimap y} \; \bar{R}_r \qquad \bar{R}_r \cfrac{\raise6pt\hbox{$\vdots$}}{\alpha^n(\vec{x}), [\varphi] \multimap^+ y \vdash \vec{y} \Uparrow \overline{x}^n \multimap y}
}{\alpha^n(\vec{x}), [\varphi] \multimap^+ y \vdash \vec{y} \Uparrow (x^n \multimap y) \& (\overline{x}^n \multimap y)} \; \&_r
}{\alpha^n(\vec{x}), [\varphi] \multimap^+ y \vdash \vec{y}, (x^n \multimap y) \& (\overline{x}^n \multimap y)} \; \bar{D}_r
}{\alpha^n(\vec{x}) \vdash \vec{y}, ([\varphi] \multimap^+ y) \multimap ((x^n \multimap y) \& (\overline{x}^n \multimap y))} \; \multimap_r
}{\alpha^n(\vec{x}) \vdash \vec{y}, [\forall x.\varphi]} \; =
$$

Fig. 4. Proof of $\forall$ case for left-right direction of Lemma 15.

For the right-left direction, we need only consider the other possibilities that could occur during bi-focussed proof search, by the focussing theorem, Theorem 10. For the $\exists$ case, bottom-up, one could have chosen to first decide on $[\varphi] \multimap y$ in the antecedent. The associated $\multimap_l$ step would have to send the formula $(x^n \multimap y) \oplus (\overline{x}^n \multimap y)$ to the right premiss (for y), since otherwise every variable occurrence in that premiss would be distinct and there would be no way to correctly finish proof search. Thus, possibly after weakening, we may apply the inductive hypothesis to the left premiss (for $[\varphi]$). A similar analysis of the upper $\multimap_l$ step in Fig. 3 means that any other split will allow us to appeal to the inductive hypothesis after weakening. For the $\forall$ case the argument is much simpler, since no matter which order we 'co-decide', we will end up with the same leaves.[6]   □

Theorem 16. *A closed QBF φ is true if and only if* MALLw *proves* $[\varphi]$.

Proof. Follows immediately from Lemma 15, setting $\vec{y} = \varnothing$.   □

5   Focussed Proof Search as Alternating Time Predicates

In this section we show how to express focussed proof search as an alternating polynomial-time predicate that will later allow us to calibrate the complexity of proof search with levels of the QBF and polynomial hierarchies. The notions we develop apply equally to either MALL or MALLw.

The following definition generalises the notions of 'decide depth' and 'release depth' found in other works, e.g. [21];

---

[6] This is actually exemplary of the more general phenomenon that invertible phases of rules are 'confluent'.

Definition 17 ((Co-)nondeterministic complexity). *The* nondeterministic *complexity of a* FMALL(w) *proof* $\mathcal{P}$, *written* $\sigma(\mathcal{P})$, *is the maximum number of alternations between* D *and* $\bar{D}$ *steps in a branch through* $\mathcal{P}$, *starting with* D. *Its* co-nondeterministic *complexity,* $\pi(\mathcal{P})$, *is defined similarly, only starting with* $\bar{D}$.

For a cedent Γ, *we write* $\sigma(\Gamma)$ $(\pi(\Gamma))$ *for the least* $k \in \mathbb{N}$ *such that there is a* FMALL(w) *proof* $\mathcal{P}$ *of* $\vdash \Gamma$ *with* $\sigma(\mathcal{P}) \leq k$ *(resp.* $\pi(\mathcal{P}) \leq k$).

Notice that the above notions of complexity are robust under the choice of multi-focussed, (co-)focussed or bi-focussed proof systems: while the number of D or $\bar{D}$ steps may increase, the number of alternations remains constant. This robustness will also apply to the other concepts we introduce in this section.

We will now introduce 'provability predicates' that delineate the complexity of proof search in a similar way to the QBF and polynomial hierarchies we presented earlier. Recall the notions of deterministic, nondeterministic and co-nondeterministic rules from Definition 8, cf. Fig. 2.

Definition 18 (Focussing hierarchy). *A cedent* Γ *of* MALL(w) *is:*

- Σ_0^f-*provable, equivalently* Π_0^f-*provable, if* $\vdash \Gamma$ *is provable using only deterministic rules.*
- Σ_{k+1}^f-*provable if there is a derivation of* $\vdash \Gamma$, *using only deterministic and nondeterministic rules, from sequents* $\vdash \Gamma_i$ *which are* Π_k^f-*provable.*
- Π_{k+1}^f-*provable if every maximal path from* $\vdash \Gamma$, *bottom-up, through deterministic and co-nondeterministic rules ends at a* Σ_k^f-*provable sequent.*

As expected, we may directly link the (co-)nondeterministic complexity of a cedent with its position in the 'focussing hierarchy':

Proposition 19. *A cedent* Γ *of* MALL(w) *is* Σ_k^f-*provable* $(\Pi_k^f$-*provable) if and only if* $\sigma(\Gamma) \leq k$ *(resp.* $\pi(\Gamma) \leq k$).

Moreover, we have a natural correspondence between the focussing hierarchy and the other hierarchies we have discussed:

Theorem 20. Σ_k^f-*provability* $(\Pi_k^f$-*provability) for* MALL(w) *is computable in* Σ_k^p *(resp.* Π_k^p). *Moreover, for* $k \geq 1$, *there are* Σ_k^q *(resp.* Π_k^q) *formulas* Σ_k^f-Prov$_n$ *(resp.* Π_k^f-Prov$_n$), *constructible in time polynomial in* $n \in \mathbb{N}$, *that compute* Σ_k^f-*provability (resp.* Π_k^f-*provability) on all formulas* A *such that* $|A| = n$.

We omit a proof of this, which is routine albeit technical, due to space constraints, but direct the reader to the analogous construction in previous work, [8]. We point out that, for the $\otimes$ rule, even though there are two premises, the rule is context-splitting, and so a nondeterministic machine may simply split into two parallel threads with no blowup in complexity.

6 An 'inverse' Encoding from **MALLw** into **CPL2**

From Theorem 20, let us henceforth fix appropriate QBFs $\Sigma_k^f\text{-Prov}_n$ and $\Pi_k^f\text{-Prov}_n$, for $k \geq 1$, computing Σ_k^f-provability and Π_k^f-provability, resp., for MALLw formulas of size n. Given these formulas, we will in this section give an explicit encoding from MALLw to CPL2, i.e. a polynomial-time mapping from MALLw-formulas to QBFs whose restriction to theorems has image in CPL2. Moreover, we will show that this encoding acts as an 'inverse' to the one we gave in Sect. 4, and finally identify natural fragments of MALLw complete for each level of **PH**.

To this end, the issue with the complexity functions σ, π introduced earlier is that they are hard to compute. Instead we give an 'over-estimate' here that will suffice for the encodings we are after.

So that the notions we define below are well defined, we will assume some arbitrary total order on formulae. The precise choice is unimportant, as long as it is polynomial-time computable; this way our ultimate encoding remains computable in polynomial time.

Definition 21 (Approximating the complexity of a sequent). *We define the functions $\lceil \sigma \rceil$ and $\lceil \pi \rceil$ on sequents in Fig. 5.*

$$
\begin{aligned}
\lceil \sigma \rceil(\vec{a}) &:= 0 \\
\lceil \sigma \rceil(\Gamma, A \,\invamp\, B) &:= \lceil \sigma \rceil(\Gamma, A, B) & A \text{ is least in } \Gamma, A \\
\lceil \sigma \rceil(\vec{a}, \vec{P}, P) &:= \lceil \sigma \rceil(\vec{a}, \vec{P} \Downarrow P) & P \text{ is least in } \vec{P}, P \\
\lceil \sigma \rceil(\vec{a}, \vec{P}, \vec{M}, M) &:= 1 + \lceil \pi \rceil(\vec{a}, \vec{P}, \vec{M}, \Uparrow M) & M \text{ is least in } \vec{M}, M
\end{aligned}
$$

$$
\begin{aligned}
\lceil \pi \rceil(\vec{a}) &:= 0 \\
\lceil \pi \rceil(\Gamma, A \,\invamp\, B) &:= \lceil \pi \rceil(\Gamma, A, B) & A \text{ is least in } \Gamma, A \\
\lceil \pi \rceil(\vec{a}, \vec{P}, P) &:= 1 + \lceil \sigma \rceil(\vec{a}, \vec{P} \Downarrow P) & P \text{ is least in } \vec{P}, P \\
\lceil \pi \rceil(\vec{a}, \vec{P}, \vec{M}, M) &:= \lceil \pi \rceil(\vec{a}, \vec{P}, \vec{M}, \Uparrow M) & M \text{ is least in } \vec{M}, M
\end{aligned}
$$

$$
\lceil \sigma \rceil(\Gamma \Downarrow A \oplus B) := \begin{cases} \lceil \sigma \rceil(\Gamma, A) & \lceil \sigma \rceil(A) \geq \lceil \sigma \rceil(B) \\ \lceil \sigma \rceil(\Gamma, B) & \text{otherwise} \end{cases}
$$

$$
\lceil \sigma \rceil(\Gamma \Downarrow A \otimes B) := \begin{cases} \lceil \sigma \rceil(\Gamma, A) & \lceil \sigma \rceil(A) \geq \lceil \sigma \rceil(B) \\ \lceil \sigma \rceil(\Gamma, B) & \text{otherwise} \end{cases}
$$

$$
\lceil \sigma \rceil(\Gamma \Downarrow X) := \lceil \sigma \rceil(\Gamma, X) \qquad\qquad X \text{ is } a \text{ or } N
$$

$$
\lceil \pi \rceil(\Gamma \Uparrow A \,\&\, B) := \begin{cases} \lceil \pi \rceil(\Gamma, A) & \lceil \pi \rceil(A) \geq \lceil \pi \rceil(B) \\ \lceil \pi \rceil(\Gamma, B) & \text{otherwise} \end{cases}
$$

$$
\lceil \pi \rceil(\Gamma \Uparrow X) := \lceil \pi \rceil(\Gamma, X) \qquad\qquad X \text{ is } O \text{ or } P
$$

Fig. 5. Approximating (co-)nondeterminstic complexities.

It is not hard to see that we have:

Proposition 22 (Over-estimation). $\sigma \leq \lceil \sigma \rceil$ and $\pi \leq \lceil \pi \rceil$.

Notice that the over-estimation for the $\otimes$ case is particularly extreme: in the worst case we have that the entire context is copied to one branch. This, along with the fact that the base case applies to only atomic cedents, means that it does not actually matter which order we compute an approximation.

Moreover, we have the following:

Proposition 23. $\lceil \sigma \rceil$ and $\lceil \pi \rceil$ are polynomial-time computable.

Finally, we have that these approximations are in fact tight for the translation $[\cdot]$ from MALLw-formulas to QBFs (cf. Sect. 4) by an inspection of its definition:

Proposition 24. $\lceil \sigma \rceil([\varphi]) = \sigma([\varphi])$ and $\lceil \pi \rceil([\varphi]) = \pi([\varphi])$.

We are now ready to define our 'inverse' encoding to $[\cdot]$:

Definition 25 (MALLw to CPL2). For a MALLw formula A, we define:

$$\langle A \rangle \; := \; \begin{cases} \Sigma_k^f\text{-Prov}_{|A|}(A) & \text{if } k = \lceil \sigma \rceil(A) \leq \lceil \pi \rceil(A) \\ \Pi_k^f\text{-Prov}_{|A|}(A) & \text{if } k = \lceil \pi \rceil(A) < \lceil \sigma \rceil(A) \end{cases}$$

Finally, we are able to present our main result:

Theorem 26. We have the following:

1. $[\cdot]$ is an encoding from CPL2 to MALLw.
2. $\langle \cdot \rangle$ is an encoding from MALLw to CPL2.
3. The composition $\langle \cdot \rangle \circ [\cdot] : \text{CPL2} \to \text{CPL2}$ preserves quantifier complexity, i.e. it maps Σ_k^q (Π_k^q) theorems to Σ_k^q (resp. Π_k^q) theorems, for $k \geq 1$.

Proof. We have already proved 1 in Theorem 16. 2 follows from the definitions of Σ_k^f-Prov and Π_k^f-Prov (cf. Theorem 20), under Propositions 19 and 22. Finally 3 then follows by tightness of the approximations $\lceil \sigma \rceil$, $\lceil \pi \rceil$ in the image of $[\cdot]$, Proposition 24.

Consequently, we may identify polynomial-time recognisable subsets of MALLw-formulas whose theorems are complete for levels of the polynomial hierarchy:

Corollary 27. We have the following, for $k \geq 1$:

1. $\{A \in \text{MALLw} : \lceil \sigma \rceil(A) \leq k\}$ is Σ_k^p-complete.
2. $\{A \in \text{MALLw} : \lceil \pi \rceil(A) \leq k\}$ is Π_k^p-complete.

7   Conclusions and Further Remarks

We gave a refined presentation of (multi-)focussed systems for multiplicative-additive linear logic, and its affine variant, that accounts for deterministic computations in proof search, cf. Sect. 3. We showed that it admits rather controlled normal forms in the form of *bi-focussed* proofs, and highlighted a duality between focussing and 'co-focussing' that emerges thanks to this presentation.

The main reason for using focussed systems such as ours was to better reflect the alternating time complexity of bottom-up proof search, cf. Sect. 5. We justified the accuracy of these bounds by showing that natural measures of proof search complexity for FMALLw tightly delineate the theorems of MALLw according to associated levels of the polynomial hierarchy, cf. Sects. 4 and 6. These results exemplify how the capacity of proof search to provide optimal decision procedures for logics extends to important subclasses of **PSPACE**. As far as we know, this is the first time such an investigation has been carried out.

It is natural to wonder whether a similar result to Theorem 26 could be obtained for (non-affine) MALL. The reason we chose MALLw is that it allows for a robust and uniform approach that highlights the ability of focussed systems to realise tight alternating time bounds for logics, without too many extraneous technicalities. Nonetheless, we briefly discuss how a similar result could be obtained for MALL, although stop short of giving formal results due to space constraints.

The main issue for MALL is the fact that there does not seem to be any 'negative' encoding of quantifier-free satisfaction, apparently only allowing characterisations of the levels Σ^p_{2k+1} and Π^p_{2k} in the same way. Apart from this, the rest of the argument can be recovered for MALL with some local adaptations. One may redefine $[\forall x.\varphi]$ as $(x \oplus \overline{x}) \multimap [\varphi]$, in order to avoid the need for weakening, and the associated coding of assignments also needs to be more structured, combining $\otimes$ and $\&$ to reflect the precise choices made in proof search. The proof of the corresponding form of Lemma 15 requires a more global analysis, for the right-left direction, due to the absence of weakening. For the inverse encoding, the definition of $\langle \cdot \rangle$ remains the same, and our main inversion result, Theorem 26, goes through as before.

In fact, by enriching the proof system with a deterministic 'evaluation' rule for positive encodings of quantifier-free satisfaction, we may recover fragments of MALL complete for *each* level of **PH**. A similar approach was followed for fragments of intuitionistic logic in [8], although this leads to further technicalities when approximating (co-)nondeterministic complexity of a sequent.

Our presentation of FMALL should extend to logics with units/constants, quantifiers and exponentials, following traditional approaches to focussed linear logic, cf. [1, 13]. It would be interesting to see what could be said about the complexity of proof search for such logics. For instance, the usual $\forall$ rule becomes deterministic in our analysis, since it does not branch:

$$\forall \frac{\Gamma, A(y)}{\Gamma, \forall x.A(x)} \; y \text{ is fresh}$$

As a result, the alternation complexity of proof search is not affected by the ∀-rule, but rather interactions between positive connectives, including ∃, and negative connectives such as &. Interpreting this over a classical setting might give us new ways to delineate true QBFs according to the polynomial hierarchy, determined by the alternation of ∃ and propositional connectives rather than ∀.

Much of the literature on *logical frameworks* via focussed systems is based around the idea that an inference rule may be simulated by a 'bi-pole', i.e. a single alternation between an invertible and non-invertible phase of inference steps. However, accounting for determinism might yield more refined simulations, where non-invertible rules are simulated by phases of deterministic and nondeterministic rules, but not co-nondeterministic ones, cf. Definition 18. In particular we envisage this to be possible for certain translations between modal logic and first-order logic, cf. [18,19].

Acknowledgements. I would like to thank Taus Brock-Nannestad, Kaustuv Chaudhuri, Sonia Marin and Dale Miller for many fruitful discussions about focussing, in particular on the presentation of it herein.

References

1. Andreoli, J.: Logic programming with focusing proofs in linear logic. J. Log. Comput. **2**(3), 297–347 (1992)
2. Buss, S.R.: The Boolean formula value problem is in ALOGTIME. In: Proceedings of the 19th Annual ACM Symposium on Theory of Computing (STOC) (1987)
3. Buss, S.R., Iemhoff, R.: The depth of intuitionistic cut free proofs (2003). http://www.phil.uu.nl/~iemhoff/Mijn/Papers/dpthIPC.pdf
4. Chandra, A.K., Kozen, D.C., Stockmeyer, L.J.: Alternation. J. ACM **28**(1), 114–133 (1981)
5. Chaudhuri, K., Miller, D., Saurin, A.: Canonical sequent proofs via multi-focusing. In: Ausiello, G., Karhumäki, J., Mauri, G., Ong, L. (eds.) TCS 2008. IIFIP, vol. 273, pp. 383–396. Springer, Boston, MA (2008). https://doi.org/10.1007/978-0-387-09680-3_26
6. Cook, S., Nguyen, P.: Logical Foundations of Proof Complexity, 1st edn. Cambridge University Press, New York (2010)
7. Cook, S.A., Reckhow, R.A.: The relative efficiency of propositional proof systems. J. Symb. Log. **44**(1), 36–50 (1979)
8. Das, A.: Alternating time bounds from variants of focussed proof systems (2017, submitted). http://anupamdas.com/alt-time-bnds-var-foc-sys.pdf
9. Delande, O., Miller, D., Saurin, A.: Proof and refutation in MALL as a game. Ann. Pure Appl. Logic **161**(5), 654–672 (2010)
10. Girard, J.: Linear logic. Theor. Comput. Sci. **50**, 1–102 (1987)
11. Heuerding, A., Seyfried, M., Zimmermann, H.: Efficient loop-check for backward proof search in some non-classical propositional logics. In: Miglioli, P., Moscato, U., Mundici, D., Ornaghi, M. (eds.) TABLEAUX 1996. LNCS, vol. 1071, pp. 210–225. Springer, Heidelberg (1996). https://doi.org/10.1007/3-540-61208-4_14
12. Krajíček, J.: Bounded Arithmetic, Propositional Logic, and Complexity Theory. Cambridge University Press, New York (1995)

13. Laurent, O.: A study of polarization in logic. Theses, Université de la Méditerranée - Aix-Marseille II (2002)
14. Letz, R.: Lemma and model caching in decision procedures for quantified Boolean formulas. In: Automated Reasoning with Analytic Tableaux and Related Methods, International Conference (TABLEAUX) (2002)
15. Liang, C., Miller, D.: Focusing and polarization in linear, intuitionistic, and classical logics. Theor. Comput. Sci. **410**(46), 4747–4768 (2009)
16. Lincoln, P., Mitchell, J.C., Scedrov, A., Shankar, N.: Decision problems for propositional linear logic. In: 31st Annual Symposium on Foundations of Computer Science (FOCS) (1990)
17. Lincoln, P., Mitchell, J.C., Scedrov, A., Shankar, N.: Decision problems for propositional linear logic. Ann. Pure Appl. Log. **56**(1–3), 239–311 (1992)
18. Marin, S., Miller, D., Volpe, M.: A focused framework for emulating modal proof systems. In: 11th Conference on Advances in Modal Logic, pp. 469–488 (2016)
19. Miller, D., Volpe, M.: Focused labeled proof systems for modal logic. In: Davis, M., Fehnker, A., McIver, A., Voronkov, A. (eds.) LPAR 2015. LNCS, vol. 9450, pp. 266–280. Springer, Heidelberg (2015). https://doi.org/10.1007/978-3-662-48899-7_19
20. Negri, S., Von Plato, J., Ranta, A.: Structural Proof Theory. Cambridge University Press, Cambridge (2008)
21. Nigam, V.: Investigating the use of lemmas (2007, preprint)
22. Ono, H.: Proof-Theoretic Methods in Nonclassical Logic – An Introduction. MSJ Memoirs, vol. 2, pp. 207–254. The Mathematical Society of Japan, Tokyo (1998)
23. Papadimitriou, C.H.: Computational Complexity. Addison-Wesley, Boston (1994)
24. Statman, R.: Intuitionistic propositional logic is polynomial-space complete. Theor. Comput. Sci. **9**, 67–72 (1979)
25. Stockmeyer, L.J.: The polynomial-time hierarchy. Theor. Comput. Sci. **3**(1), 1–22 (1976)
26. Troelstra, A., Schwichtenberg, H.: Basic Proof Theory. Cambridge Tracts in Theoretical Computer Science, vol. 43. Cambridge University Press, Cambridge (1996)

Checking Array Bounds by Abstract Interpretation and Symbolic Expressions

Étienne Payet[1] and Fausto Spoto[2]([⊠])

[1] Laboratoire d'Informatique et de Mathématiques, Université de la Réunion,
Saint-Denis, France
[2] Dipartimento di Informatica, Università di Verona, Verona, Italy
fausto.spoto@univr.it

Abstract. Array access out of bounds is a typical programming error. From the '70s, static analysis has been used to identify where such errors actually occur at runtime, through abstract interpretation into linear constraints. However, feasibility and scalability to modern object-oriented code has not been established yet. This article builds on previous work on linear constraints and shows that the result does not scale, when polyhedra implement the linear constraints, while the more abstract zones scale to the analysis of medium-size applications. Moreover, this article formalises the inclusion of symbolic expressions in the constraints and shows that this improves its precision. Expressions are automatically selected on-demand. The resulting analysis applies to code with dynamic memory allocation and arrays held in expressions. It is sound, also in the presence of arbitrary side-effects. It is fully defined in the abstract interpretation framework and does not use any code instrumentation. Its proof of correctness, its implementation inside the commercial Julia analyzer and experiments on third-party code complete the work.

1 Introduction

Arrays are extensively used in computer programs since they are a compact and efficient way of storing and accessing vectors of values. Array elements are indexed by their integer offset, which leads to a runtime error if the index is negative or beyond the end of the array. In C, this error is silent, with unpredictable results. The Java runtime, instead, mitigates the problem since it immediately recognizes the error and throws an exception. In both cases, a definite guarantee, at compilation time, that array accesses will never go wrong, for all possible executions, is desirable and cannot be achieved with testing, that covers only some execution paths. Since the values of array indices are not computable, compilers cannot help, in general. However, static analyses that find such errors, and report some false alarms, exist and are an invaluable support for programmers.

Abstract interpretation has been applied to array bounds inference, from its early days [5,8], by abstracting states into linear constraints on the possible values of local variables, typically polyhedra [3,4]. Such inferred constraints let then one check if indices are inside their bounds. For instance, in the code:

© Springer International Publishing AG, part of Springer Nature 2018
D. Galmiche et al. (Eds.): IJCAR 2018, LNAI 10900, pp. 706–722, 2018.
https://doi.org/10.1007/978-3-319-94205-6_46

```
1  public DiagonalMatrix inverse(double[] diagonal) {
2     double[] newDiagonal = new double[diagonal.length]; // local var.
3     for (int i = 0; i < diagonal.length; i++)
4        newDiagonal[i] = 1 / diagonal[i]; ... }
```

the analysis in [5,8], at line 4, infers $0 \leqslant i < diagonal = newDiagonal$; *i.e.*, index
i is non-negative and is smaller than the length of the array diagonal, which
is equal to that of newDiagonal. This is enough to prove that both accesses
newDiagonal[i] and diagonal[i] occur inside their bounds, always.

Programming languages have largely evolved since the '70s and two problems affect the application of this technique to modern software. First, code is
very large nowadays, also because object-oriented software uses large libraries
that must be included in the analysis. The actual scalability of the technique,
hence, remains unproved. Second, the limitation to constraints on *local variables*
(such as i, diagonal and newDiagonal above) is too strict. Current programming
languages allow arrays to be stored in expressions built from dynamically heap-
allocated object fields and other arrays, which are not local variables. For instance,
the previous example is actually a simplification of the following real code from
class util.linalg.DiagonalMatrix of a program called Abagail (Sect. 7):

```
1  private double[] diagonal; // object field, not local variable
2  public DiagonalMatrix inverse() {
3     double[] newDiagonal = new double[this.diagonal.length];
4     for (int i = 0; i < this.diagonal.length; i++)
5        newDiagonal[i] = 1 / this.diagonal[i]; ... }
```

The analysis in [5,8] infers $0 \leqslant i, 0 \leqslant newDiagonal$ at line 5 above, since
this.diagonal is not a local variable and consequently cannot be used in
the constraint. The latter does not entail that the two array accesses are
safe now, resulting in two false alarms. Clearly, one should allow expressions
such as this.diagonal in the constraint and infer $0 \leqslant i < this.diagonal = newDiagonal$. But this is challenging since there are (*infinitely*) many expressions (potentially affecting scalability) and since expressions might change their
value by *side-effect* (potentially affecting soundness). In comparison, at a given
program point, only *finitely* many local variables are in scope, whose value can
only be changed by syntactically explicit assignment to the affected variable.
Hence, this challenge is both technical (the implementation must scale) and theoretical (the formal proof of correctness must consider all possible side-effects).

One should not think that it is enough to include object fields in the constraints, to improve the expressivity of the analysis. Namely, fields are just examples of expressions. In real code, it is useful to consider also other expressions. For
instance, the following code, from class util.linalg.LowerTriangularMatrix
of Abagail, performs a typical nested loop over a bidimensional array:

```
1     UpperTriangularMatrix result = new UpperTriangularMatrix(...);
2     for (int i = 0; i < this.data.length; i++)
3        for (int j = 0; j < this.data[i].length; j++) {
4           // any extra code could occur here
5           result.set(j, i, this.data[i][j]); }
```

To prove array accesses safe at line 5 above, one should infer that $0 \leqslant i < this.data, 0 \leqslant j < this.data[i]$. The analysis in [5,8] cannot do it, since it considers local variables and abstracts away fields (`this.data`) and array elements (`this.data[i]`). Moreover, safeness of these accesses can be jeopardised by extra code at line 4 modifying `this.data` or `this.data[i]`: side-effects can affect soundness. That does not happen for arrays held in local variables, as in [5,8].

For an example of approximation of even more complex expressions, consider the anonymous inner class of `java.net.sf.colossus.tools.ShowBuilderHex-Map` from program Colossus (Sect. 7). It iterates over a bidimensional array:

```
1  clearStartListAction = new AbstractAction(...) {
2    public void actionPerformed(ActionEvent e) {
3      for (int i = 0; i < h.length; i++)
4        for (int j = 0; j < h[i].length; j++)
5          if ((h[i][j] != null) && (h[i][j].isSelected())) {
6            h[i][j].unselect();
7            h[i][j].repaint();  }}};
```

Here, `h` is a field of the outer object *i.e.*, in Java, shorthand for `this.this$0.h` (a field of field `this$0`, the synthetic reference to the outer object); `h[i]` stands for `this.this$0.h[i]` (an element of an array in a field of a field). In order to prove the array accesses at lines 4, 5, 6 and 7 safe, the analyser should prove that the constraint $0 \leqslant i < this.this\$0.h, 0 \leqslant j < this.this\$0.h[i]$ holds at those lines. The analysis in [5,8] cannot do it, since it abstracts away the expressions `this.this$0.h` and `this.this$0.h[i]`. This results in false alarms. Note that, to prove the access at line 7 safe, an analyser must prove that `isSelected()` and `unselect()` do not affect `h` nor `h[i]`. That is, it must consider side-effects.

The contribution of this article is the extension of [5,8] with specific program expressions, in order to improve its precision. It starts from the formalisation of [5,8] for object-oriented languages with dynamic heap allocation, known as path-length analysis [19]. It shows that its extension with expressions, using zones [10] rather than polyhedra [3,4], scales to real code and is more precise than [5,8]. This work is bundled with a formal proof of correctness inside the abstract interpretation framework. This analysis, as that in [19], is interprocedural, context and flow-sensitive and deals with arbitrary recursion.

This article is organised as follows. Section 2 reports related work. Sections 3 and 4 recall semantics and path-length from [19] and extend it to arrays. Section 5 introduces the approximation of expressions. Section 6 defines the new static analysis, with side-effect information for being sound. Section 7 describes its implementation inside the Julia analyser and experiments of analysis of open-source Java applications. The latter and the results of analysis can be found in [14]. Analyses can be rerun online at https://portal.juliasoft.com (instructions in [14]).

2   Related Work and Other Static Analysers

Astrée [6] is a state of the art analyser that infers linear constraints. For scalability, it uses octagons [12] rather than polyhedra [7]. It targets embedded C software, with clear limitations [6]: no dynamic memory allocation, no unbound recursion, no conflicting side-effects and no use of libraries. Fields and arrays are dealt under the assumption that there is no dynamic memory allocation, which limits the analysis to embedded C software [11]. These assumptions simplify the analysis since, in particular, no dynamic memory allocation means that there is a finite number of fields or array elements, hence they can be statically grouped and mapped into linear variables; and the absence of conflicting side-effects simplifies the generation of the constraints between such variables. However, such assumptions conflict with the reality of Java code. Even a minimal Java program uses dynamic memory allocation and a very large portion of the standard Java library, which is much more pervasive that the small standard C library: for instance, a simple print statement reaches library code for formatting and localization and the collection library contains hundreds of intensively used classes. The recent 2018 competition on software verification showed that a few tools can already perform bound verification for arrays in C, with good results[1].

A type system has been recently defined for array bounds checking in Java [15]. It infers size constraints but uses code annotation. Thus, it is not fully automatic.

Facebook has built the buffer overflow analyser Inferbo[2] on top of Infer[3]. Inferbo uses symbolic intervals for approximating indices of arrays held in local variables. We ran Inferbo on Java code but the results were non-understandable. After personal communication with the Infer team, we have been confirmed that Infer does work also on Java code, but Inferbo is currently limited to C only.

We also ran FindBugs[4] at its *maximal analysis effort*. It spots only very simple array access bounds violations. For instance it warns at the second statement of `int t[] = new int[5]; t[9] = 3`. However, in the programs analysed in Sect. 7, it does not issue any single warning about array bounds violations, missing all the true alarms that our analysis finds.

Previous work [19] defined a *path-length* analysis for Java, by using linear constraints over local variables only. It extends [5,8] to deal with dynamically heap-allocated data structures and arbitrary side-effects. It uses a bottom-up, denotational fixpoint engine, with no limits on recursion. It was meant to support a termination prover. This article leverages its implementation inside the Julia analyser for Java bytecode [17], by adding constraints on expressions.

This work has been inspired by [1,2], which, however, has the completely different goal of termination analysis. It identifies fields that are locally constant inside a loop and relevant for the analysis, by using heuristics, then performs a polyvariant code instrumentation to translate such fields into *ghost variables*. That analysis is limited to fields and there is no formalisation of path-length with arrays. In this article, instead, path-length with arrays is formalised, applied to

---

[1] https://sv-comp.sosy-lab.org/2018/results/results-verified/META_MemSafety.
    table. See, in particular, the results for tools Map2Check, Symbiotic and Ultimate.
[2] https://research.fb.com/inferbo-infer-based-buffer-overrun-analyzer.
[3] http://fbinfer.com.
[4] http://findbugs.sourceforge.net.

array bounds checking with an evaluation of its scalability. Our analysis identifies, on-demand, expressions that need explicit approximation and can be much
more complex than fields *e.g.*, array elements of variables or fields, or fields
of array elements; moreover, it does not need any code instrumentation, since
expressions only exist in the abstract domain, where linear variables symbolically stand for expressions. As a consequence, it has a formal correctness proof,
completely inside the abstract interpretation framework. As far as we can infer
from the papers, the analysis in [1,2] has no formal correctness proof.

3   Concrete Domain

We extend [19] with arrays. Namely, an array *arr* of type t and length $n \in \mathbb{N}$ is
a mapping from $[0, \dots, n-1]$ to values of type t; it has type *arr.t* and length
arr.length. A *memory* maps locations to *reference values i.e.*, objects or arrays.
The set of locations is $\mathbb{L}$, the set of arrays is $\mathbb{A}$. A multidimensional array is just
an array of locations, bound to arrays, exactly as in Java. The set of classes of
our language is $\mathbb{K}$ and the set of types is $\mathbb{T} = \mathbb{K} \cup \{\texttt{int}\} \cup \mathbb{T}[\,]$. The domain of a
function f is $dom(f)$, its codomain or range is $rng(f)$. By $f(x)\downarrow$ we mean that f
is defined at x; by $f(x)\uparrow$, that it is undefined at x. The composition of functions
f and g is $f \circ g = \lambda x.g(f(x))$, undefined when $f(x)\uparrow$ or $g(f(x))\uparrow$.

A *state* is a triple $\langle l \,\|\, s \,\|\, \mu \rangle$ that contains the values of the local variables l,
those of the operand stack elements s, and a memory μ. Local variables and
stack elements are bound to values compatible with their static type. Dangling
pointers are not allowed. The size of l is denoted as $\#l$, that of s as $\#s$. The
elements of l and s are indexed as l^k and s^k, where s^0 is the base of the stack
and $s^{\#s-1}$ is its top. The set of all states is Σ, while $\Sigma_{i,j}$ is the set of states
with exactly i local variables and j stack elements. The concrete domain is
$\langle \wp(\Sigma), \subseteq \rangle$ *i.e.*, the powerset of states ordered by set inclusion. Denotations are
the functional semantics of a single bytecode instruction or of a block of code.

Definition 1. *A denotation is a partial function $\Sigma \to \Sigma$ from a pre-state to a
post-state. The set of denotations is Δ, while $\Delta_{l_i,s_i \to l_o,s_o}$, stands for $\Sigma_{l_i,s_i} \to
\Sigma_{l_o,s_o}$. Each instruction* ins, *occurring at a program point p, has semantics* $ins_p \in
\Delta_{l_i,s_i \to l_o,s_o}$ *where l_i, s_i, l_o, s_o are the number of local variables and stack elements
in scope at p and at the subsequent program point, respectively. Figure 1 shows
those dealing with arrays and fields. Others can be found in* [19].

Figure 1 assumes runtime types correct. For instance, $arrayload_p$ t finds on the
stack a value ℓ which is either $\texttt{null}$ or a location bound to an array of type
$t' \leqslant t$. Such static constraints must hold in legal Java bytecode [9], hence are not
checked in Fig. 1. Others are dynamic and checked in Fig. 1: for instance, index
v must be inside the array bounds $(0 \leqslant v < \mu(\ell).length)$. Figure 1 shows explicit
types for instructions, when relevant in this article, such as t in $arrayload_p$ t. They
are implicit in real bytecode, for compactness, but can be statically inferred [9].
Figure 1 assumes that runtime violations of bytecode preconditions stop the Java

$$store_p \ k = \lambda\langle l \parallel v :: s \parallel \mu\rangle.\langle l[k \mapsto v] \parallel s \parallel \mu\rangle$$

$$inc_p \ k \ c = \lambda\langle l \parallel s \parallel \mu\rangle.\langle l[k \mapsto l^k + c] \parallel s \parallel \mu\rangle$$

$$newarray_p \ t \ n = \lambda\langle l \parallel v_n :: \cdots :: v_1 :: s \parallel \mu\rangle.\langle l \parallel \ell :: s \parallel \mu[\ell \mapsto arr, \ldots]\rangle \quad \text{if } 0 \leqslant v_1, \ldots, v_n$$

where $\ell \in \mathbb{L}$ is fresh and arr is an n-dimensional array of type t

and lengths $v_1, \ldots, v_n$, initialised to its default value,

with subarrays, if any, bound in the ... part to fresh locations

$$arrayload_p \ t = \lambda\langle l \parallel v :: \ell :: s \parallel \mu\rangle.\langle l \parallel \mu(\ell)(v) :: s \parallel \mu\rangle \quad \text{if } \ell \neq \text{null}, \ 0 \leqslant v < \mu(\ell).length$$

$$arraystore_p \ t = \lambda\langle l \parallel v_1 :: v_2 :: \ell :: s \parallel \mu\rangle.\langle l \parallel s \parallel \mu[\ell \mapsto \mu(\ell)[v_2 \mapsto v_1]]\rangle$$

$$\text{if } \ell \neq \text{null and } 0 \leqslant v_2 < \mu(\ell).length$$

$$arraylength_p \ t = \lambda\langle l \parallel \ell :: s \parallel \mu\rangle.\langle l \parallel \mu(\ell).length :: s \parallel \mu\rangle \quad \text{if } \ell \neq \text{null}$$

$$getfield_p \ f = \lambda\langle l \parallel \ell :: s \parallel \mu\rangle.\langle l \parallel \mu(\ell)(f) :: s \parallel \mu\rangle \quad \text{if } \ell \neq \text{null}$$

$$putfield_p \ f = \lambda\langle l \parallel v :: \ell :: s \parallel \mu\rangle.\langle l \parallel s \parallel \mu[\ell \mapsto \mu(\ell)[f \mapsto v]]\rangle \quad \text{if } \ell \neq \text{null}$$

$$(call_p \ \kappa.m)(\delta) = \text{see [19]}$$

Fig. 1. The concrete semantics of a fragment of Java bytecode, with arrays and fields. The semantics of an instruction is implicitly undefined if its preconditions do not hold.

Virtual Machine, instead of throwing an exception. Exceptions can be accomodated in this fragment (and are included in our implementation), at the price of extra complexity. This is explained and formalised in [13]. Instruction newarray allocates an array of n dimensions and leaves its pointer ℓ on top of the stack. When $n > 1$, a multidimensional array is allocated. In that case, the array at ℓ is the spine of the array, while its elements are arrays themselves, held at further newly allocated locations. Instructions arrayload, arraystore and arraylength operate on the array $\mu(\ell)$, where ℓ is provided on the stack. The first two check if the index is inside its bounds. getfield and putfield are similar, but $\mu(\ell)$ is an object. Objects are represented as functions from field names to field values. call plugs the denotation δ of the callee(s) at the calling point.

4   Path-Length Abstraction with Arrays

Path-length [19] is a property of local and stack variables, namely, the length of the maximal chain of pointers from the variable. It leads to an abstract interpretation of Java bytecode, reported below, after extending path-length to arrays: their path-length is their length. This is consistent with the fact that the length of the arrays is relevant for checking array bounds and for proving the termination of loops over arrays. Hence the elements of an array are irrelevant *w.r.t.* path-length and only the first dimension of a multidimensional array matters.

Definition 2. *Let μ be a memory. For every $j \geq 0$, let (1) $len^j(\text{null}, \mu) = 0$, (2) $len^j(i, \mu) = i$ if $i \in \mathbb{N}$, (3) $len^j(\ell, \mu) = \mu(\ell).length$ if $\ell \in dom(\mu)$ and $\mu(\ell) \in \mathbb{A}$, (4) $len^0(\ell, \mu) = 0$ if $\ell \in dom(\mu)$ and $\mu(\ell) \notin \mathbb{A}$, (5) $len^{j+1}(\ell, \mu) =$*

$1 + \max \left\{ len^j(\ell', \mu) \mid \ell' \in rng(\mu(\ell)) \cap \mathbb{L} \right\}$ *if* $\ell \in dom(\mu)$ *and* $\mu(\ell) \notin \mathbb{A}$, *with the assumption that the maximum of an empty set is* 0. *The* path-length *of a value* v *in* μ *is* $len(v, \mu) = \lim_{j \to \infty} len^j(v, \mu)$.

In the last case of the definition of len^j, the intersection with $\mathbb{L}$ selects only the non-primitive fields of object $\mu(\ell)$. If $i \in \mathbb{Z}$ then $len(i, \mu) = len^j(i, \mu) = i$ for every $j \geq 0$ and memory μ. Similarly, $len(\texttt{null}, \mu) = len^j(\texttt{null}, \mu) = 0$. Moreover, if location ℓ is bound to a cyclical data-structure, then $len(\ell, \mu) = \infty$.

A state can be mapped into a *path-length assignment* i.e., into a function specifying the path-length of its variables. This comes in two versions: in the *pre-state version* $\widetilde{len}$, the state is considered as the pre-state of a denotation. In the *post-state version* $\widehat{len}$, it is considered as the post-state of a denotation.

Definition 3. *Let* $\langle l \, \| \, s \, \| \, \mu \rangle \in \Sigma_{\#l, \#s}$. *Its* pre-state path-length assignment *is* $\widetilde{len}(\langle l \, \| \, s \, \| \, \mu \rangle) = [\breve{l}^k \mapsto len(l^k, \mu) \mid 0 \leq k < \#l] \cup [\breve{s}^k \mapsto len(s^k, \mu) \mid 0 \leq k < \#s]$. *Its* post-state path-length assignment *is* $\widehat{len}(\langle l \, \| \, s \, \| \, \mu \rangle) = [\hat{l}^k \mapsto len(l^k, \mu) \mid 0 \leq k < \#l] \cup [\hat{s}^k \mapsto len(s^k, \mu) \mid 0 \leq k < \#s]$.

Definition 4. *Let* $l_i, s_i, l_o, s_o \in \mathbb{N}$. *The* path-length constraints $\mathbb{PL}_{l_i, s_i \to l_o, s_o}$ *are all finite sets of integer linear constraints over variables* $\{\breve{l}^k \mid 0 \leqslant k < l_i\} \cup \{\breve{s}^k \mid 0 \leqslant k < s_i\} \cup \{\hat{l}^k \mid 0 \leqslant k < l_o\} \cup \{\hat{s}^k \mid 0 \leqslant k < s_o\}$ *with the* $\leqslant$ *operator.*

One can also use constraints such as $x = y$, standing for both $x \leqslant y$ and $y \leqslant x$.

A path-length assignment fixes the values of the variables. When those values satisfy a path-length constraint, they are a *model* of the constraint.

Definition 5. *Let* $pl \in \mathbb{PL}_{l_i, s_i \to l_o, s_o}$ *and* ρ *be an assignment from a superset of the variables of* pl *into* $\mathbb{Z} \cup \{\infty\}$. *Then* ρ *is a* model *of* pl, *written as* $\rho \models pl$, *when* $pl\rho$ *holds i.e., when, by substituting, in* pl, *the variables with their values provided in* ρ, *one gets a tautological set of ground constraints.*

The *concretisation* of a path-length constraint is the set of denotations that induce pre- and post-state assignments that form a model of the constraint.

Definition 6. *Let* $pl \in \mathbb{PL}_{l_i, s_i \to l_o, s_o}$. *Its* concretisation $\gamma(pl)$ *is* $\{\delta \in \Delta_{l_i, s_i \to l_o, s_o} \mid$ *for all* $\sigma \in \Sigma_{l_i, s_i}$ *such that* $\delta(\sigma) \downarrow$ *we have* $\left(\widetilde{len}(\sigma) \cup \widehat{len}(\delta(\sigma))\right) \models pl\}$.

In [19] it is proved that γ is the concretisation map of an abstract interpretation [5] and sound approximations are provided for some instructions such as const, dup, new, load, store, add, getfield, putfield, ifeq and ifne, as well as for sequential and disjunctive composition. For instance, there is an abstraction $getfield_p^{\mathbb{PL}} \, f$, sound *w.r.t.* the concrete semantics of getfield f at p (Fig. 1). They remain sound after introducing arrays to the language. Figure 2 reports the abstraction of array instructions. This defines a denotational fixpoint static analysis of Java bytecode, that approximates the path-length of local variables. We cannot copy from [19] the complete definition of the abstract semantics. We only observe that the analysis uses possible sharing [16] and reachability [13] analyses for approximating the side-effects of field updates and method calls.

$$newarray_p^{\text{PL}}\ t\ n = Id(\#l, \#s - n + 1) \cup \{\breve{s}^{\#s-1} \geqslant 0, \ldots, \breve{s}^{\#s-n} \geqslant 0\}$$

$$arraylength_p^{\text{PL}}\ t = Id(\#l, \#s)$$

$$arrayload_p^{\text{PL}}\ t = \begin{cases} Id(\#l, \#s - 2) \cup \{0 \leqslant \breve{s}^{\#s-1} < \breve{s}^{\#s-2}\} & \text{if } t = \text{int} \\ Id(\#l, \#s - 2) \cup \{0 \leqslant \breve{s}^{\#s-1} < \breve{s}^{\#s-2}, \widehat{s}^{\#s-2} \geqslant 0\} & \text{otherwise} \end{cases}$$

$$arraystore_p^{\text{PL}}\ t = Id(\#l, \#s - 3) \cup \{0 \leqslant \breve{s}^{\#s-2} < \breve{s}^{\#s-3}\}$$

Fig. 2. Path-length abstraction of the bytecodes from Fig. 1 that deal with arrays. $\#l, \#s$ are the number of local variables and stack elements at program point p. $Id(x, y) = \{\breve{l}^i = \widehat{l}^i \mid 0 \leqslant i < x\} \cup \{\breve{s}^i = \widehat{s}^i \mid 0 \leqslant i < y\}$.

Proposition 1. *The maps in Fig. 2 are sound w.r.t. those in Fig. 1.*   □

We do not copy the abstract method call from [19], since it is complex but irrelevant here. Given the approximation pl of the body of a method m of class κ, it is a constraint $(call_p^{\text{PL}}\kappa.m)(pl)$, sound *w.r.t.* call $\kappa.m$ at program point p. Method calls in object-oriented languages can have more dynamic target methods, hence pl is actually the disjunction of the analysis of all targets. A restricted subset of targets can be inferred for extra precision [18].

Sequential composition of path-length constraints pl_1 and pl_2 matches the post-states of pl_1 with the pre-states of pl_2, through temporary, overlined variables. Disjuctive composition is used to join more execution paths.

Definition 7. *Let* $pl_1 \in \mathbb{PL}_{l_i, s_i \to l_t, s_t}$, $pl_2 \in \mathbb{PL}_{l_t, s_t \to l_o, s_o}$ *and* $T = \{\overline{l}^0, \ldots, \overline{l}^{l_t-1}, \overline{s}^0, \ldots, \overline{s}^{s_t-1}\}$. *The sequential composition* $pl_1;^{\text{PL}} pl_2 \in \mathbb{PL}_{l_i, s_i \to l_o, s_o}$ *is the constraint* $\exists_T (pl_1[\widehat{v} \mapsto \overline{v} \mid \overline{v} \in T] \cup pl_2[\breve{v} \mapsto \overline{v} \mid \overline{v} \in T])$. *Let* $pl_1, pl_2 \in \mathbb{PL}_{l_i, s_i \to l_o, s_o}$. *Their disjunctive composition* $pl_1 \cup^{\text{PL}} pl_2$ *is the polyhedral hull of* pl_1 *and* pl_2.

5   Path-Length with Expressions

Definition 4 defines path-length as a domain of numerical constraints over local or stack elements, which are the only program expressions that one can use in the constraints. That limitation can be overcome by adding variables that stand for more complex expressions, that allow the selection of fields or array elements.

Definition 8. *Given* $l \geq 0$, *the set of* expressions *over* l *local variables is* $\mathbb{E}_l = \{l^k \mid 0 \leqslant k < l\} \cup \{e.f \mid e \in \mathbb{E}_l \text{ and } f \text{ is a field name}\} \cup \{e_1[e_2] \mid e_1, e_2 \in \mathbb{E}_l\}$. *Given also* $s \geq 0$, *the* expressions *or stack elements are* $\mathbb{ES}_{l,s} = \mathbb{E}_l \cup \{s^i \mid 0 \leqslant i < s\}$. *When we want to fix a maximal depth* $k > 0$ *for the expressions, we use the set* $\mathbb{E}_l^k = \{e \in \mathbb{E}_l \mid e \text{ has depth at most } k\}$.

Definition 9. *Given* $\sigma = \langle l \parallel s \parallel \mu \rangle \in \Sigma_{\#l, \#s}$ *and* $e \in \mathbb{ES}_{\#l, \#s}$, *the evaluation* $[\![e]\!]\sigma$ *of* e *in* σ *is defined as* $[\![l^k]\!]\sigma = l^k$ *and* $[\![s^k]\!]\sigma = s^k$ (l^k *and* s^k *is an expression on the left and the value of the* kth *local variable or stack element on the right); moreover,* $[\![e.f]\!]\sigma = \mu([\![e]\!]\sigma)(f)$ *if* $[\![e]\!]\sigma \in \mathbb{L}$ *(undefined, otherwise);* $[\![e_1[e_2]]\!]\sigma = \mu([\![e_1]\!]\sigma)([\![e_2]\!]\sigma)$ *if* $[\![e_1]\!]\sigma \in \mathbb{L}$ *and* $[\![e_2]\!]\sigma \in \mathbb{Z}$ *(undefined, otherwise).*

Section 4 can now be generalised. Path-length assignments refer to all possible expressions, not just to local variables and stack elements (compare with Definition 3).

Definition 10. *Let* $\sigma = \langle l \| s \| \mu \rangle \in \Sigma_{\#l,\#s}$. *Its pre-state path-length assignment is* $\widetilde{len}(\sigma) = [\tilde{e} \mapsto len(\llbracket e \rrbracket \sigma, \mu) \mid e \in \mathbb{ES}_{\#l,\#s}]$. *Its post-state path-length assignment is* $\widehat{len}(\sigma) = [\hat{e} \mapsto len(\llbracket e \rrbracket \sigma, \mu) \mid e \in \mathbb{ES}_{\#l,\#s}]$.

Path-length can now express constraints over the value of expressions (compare with Definition 4); such expressions are actually numerical variables of the constraints.

Definition 11. *Let* $l_i, s_i, l_o, s_o \in \mathbb{N}$. *The set* $\mathbb{PL}_{l_i,s_i \to l_o,s_o}$ *of the* path-length constraints *contains all finite sets of integer linear constraints over the variables* $\{\tilde{e} \mid e \in \mathbb{ES}_{l_i,s_i}\} \cup \{\hat{e} \mid e \in \mathbb{ES}_{l_o,s_o}\}$, *using only the* $\leqslant$ *comparison operator.*

Definitions 5, 6 and 7 remain unchanged; the abstractions in Fig. 2 and from [19] work over this generalised path-length domain, but do not exploit the possibility of building constraints over expressions. Such expressions must be selected, since $\mathbb{E}_{l,s}$ is infinite, in general. The analysis adds expressions on-demand, as soon as the analysed code uses them. Namely, consider the abstractions of the instructions that operate on the heap. They are refined by introducing expressions, as follows, by using definite aliasing, a minimum requirement for a realistic static analyser: $e_1 \sim_p e_2$ means that e_1 and e_2 are definitely alias at program point p.

Definition 12. *Let* $k > 0$ *be a maximal depth for the expressions considered below. From now on, the approximations on* $\mathbb{PL}$ *of* $getfield_p^{\mathbb{PL}} f$ *and* $putfield_p^{\mathbb{PL}} f$ *from* [19] *and* $arrayload_p^{\mathbb{PL}}$ *and* $arraystore_p^{\mathbb{PL}}$ *from Fig. 2 will be taken as refined by adding the following constraints:*

$$to\ getfield_p^{\mathbb{PL}}\ f : \{\widehat{s^{\#s-1}} = \widetilde{e.f} = \widehat{e.f} \mid e.f \in \mathbb{E}_{\#l}^k \text{ and } e \sim_p s^{\#s-1}\}$$

$$to\ putfield_p^{\mathbb{PL}}\ f : \left\{ \widetilde{s^{\#s-1}} = \widehat{e.f} \,\middle|\, \begin{matrix} e.f \in \mathbb{E}_{\#l}^k,\ f \text{ does not occur in } e, \\ e \sim_p s^{\#s-2} \text{ and } f \text{ has type } \textbf{int} \text{ or array} \end{matrix} \right\}$$

$$to\ arrayload_p^{\mathbb{PL}}\ t : \left\{ \widehat{s^{\#s-2}} = \widetilde{e_1[e_2]} = \widehat{e_1[e_2]} \,\middle|\, \begin{matrix} e_1[e_2] \in \mathbb{E}_{\#l}^k, \\ e_1 \sim_p s^{\#s-2} \text{ and } e_2 \sim_p s^{\#s-1} \end{matrix} \right\}$$

$$to\ arraystore_p^{\mathbb{PL}}\ t : \left\{ \widetilde{s^{\#s-1}} = \widehat{e_1[e_2]} \,\middle|\, \begin{matrix} e_1[e_2] \in \mathbb{E}_{\#l}^k,\ e_1 \sim_p s^{\#s-3},\ e_2 \sim_p s^{\#s-2}, \\ e_1, e_2 \text{ do not contain array subexpressions} \end{matrix} \right\}$$

Definition 12 states that getfield f pushes on the stack the value of $e.f$, where e is a definite alias of its receiver. Bytecode putfield f stores the top of the stack in $e.f$, where e is, again, a definite alias of its receiver. Similarly for arrayload and arraystore. Bytecodes putfield and arraystore avoid the introduction of expressions whose value might be modified by their same execution.

Proposition 2. *The maps in Definition 12 are sound w.r.t. those in Fig. 1.* □

Example 1. In the snippet of code from `util.linarg.DiagonalMatrix` at page 2, the compiler translates the expression `this.diagonal.length` at line 4 into

```
1   load 0              // load local variable this
2   getfield diagonal   // load field diagonal of this
3   arraylength double  // compute the length of this.diagonal
```

At the beginning $\#s = 1$, local 0 is `this`, local 1 is `newDiagonal` and local 2 is `i`, hence the latter is a definite alias of stack element 0, going to be compared against the value of `this.diagonal.length`. The next table reports the number $\#s$ of stack elements ($\#l = 3$ always), definite aliasing just before the execution of each instruction (self-aliasing is not reported) and its resulting abstraction:

instruction	$\#s$	definite aliasing	abstraction
load 0	1	$\{l^2 \sim s^0\}$	$\{\breve{l}^0 = \widehat{l}^0 = \widehat{s}^1, \breve{l}^1 = \widehat{l}^1, \breve{l}^2 = \widehat{l}^2, \breve{s}^0 = \widehat{s}^0\}$
getfield diagonal	2	$\{l^2 \sim s^0, l^0 \sim s^1\}$	$\left\{ \begin{array}{l} \breve{l}^0 = \widehat{l}^0, \breve{l}^1 = \widehat{l}^1, \breve{l}^2 = \widehat{l}^2, \breve{s}^0 = \widehat{s}^0 \\ \widehat{s}^1 = l^0.\widetilde{diagonal} = l^0.\widehat{diagonal} \end{array} \right\} (pl_1)$
arraylength double	2	$\{l^2 \sim s^0\}$	$\left\{ \begin{array}{l} \breve{l}^0 = \widehat{l}^0, \breve{l}^1 = \widehat{l}^1, \breve{l}^2 = \widehat{l}^2 \\ \breve{s}^0 = \widehat{s}^0, \breve{s}^1 = \widehat{s}^1 \end{array} \right\} (pl_2)$

The abstraction of `getfield diagonal` uses the definite aliasing information $l^0 \sim s^1$ to introduce the constraint $\widehat{s}^1 = l^0.\widetilde{diagonal} = l^0.\widehat{diagonal}$ on expression `this.diagonal` (Definition 12). The sequential composition of the three constraints approximates the execution of the three bytecode instructions: $\breve{l}^0 = \widehat{l}^0, \breve{l}^1 = \widehat{l}^1, \breve{l}^2 = \widehat{l}^2, \breve{s}^0 = \widehat{s}^0$. Unfortunately, it loses information about `this.diagonal`.

The approximation in Example 1 is imprecise since pl_1 (see Example 1) refers to $l^0.\widetilde{diagonal}$, but pl_2 does not refer to $l^0.\widetilde{diagonal}$ at all: hence their sequential composition does not propagate any constraint about it. To overcome this imprecision, one can include frame constraints in the abstraction of each instruction ins, stating, for *each* expression e whose value is not *affected* by ins, that its path-length does not change: $\breve{e} = \widehat{e}$. But this is impractical since, in general, there are infinitely many such expressions. Next section provides an alternative, finite solution.

6   Expressions and Side-Effects Information

Let us reconsider the sequential composition of pl_1 and pl_2 from Example 1. Since pl_1 refers to the expression $l^0.\widetilde{diagonal}$, not mentioned in pl_2, we could define $pl'_2 = pl_2 \cup \{l^0.\widetilde{diagonal} = l^0.\widehat{diagonal}\}$ and compute $pl_1;^{\mathbb{PL}} pl'_2$ instead of $pl_1;^{\mathbb{PL}} pl_2$. The composition will then propagate the constraints on $l_0.\widetilde{diagonal}$. This redefinition of $;^{\mathbb{PL}}$ is appealing since it adds the frame condition $l^0.\widetilde{diagonal} = l^0.\widehat{diagonal}$ only for $l^0.\widetilde{diagonal}$ i.e., for the expressions that are introduced on-demand during the analysis. However, it is unsound when pl_2 is the abstraction of a piece of code that affects the value of $l^0.diagonal$ (for instance, it modifies l^0 or *diagonal*): the constraint $l^0.\widetilde{diagonal} = l^0.\widehat{diagonal}$ would not hold for all its concretisations. This leads to the addition of side-effect information to $\mathbb{PL}$.

Side-effects are modifications of leftvalues, that is, local variables, object fields or array elements. A local variable is modified when its value changes. A field f is modified when at least an object in memory changes its value for f. An array of type t is modified when at least an array of type t in memory changes.

Definition 13. *Let $\delta \in \Delta_{l_i,s_i \to l_o,s_o}$ and $\sigma = \langle l \| s \| \mu \rangle \in \Sigma_{l_i,s_i}$. Then δ modifies local k in σ iff* [5] $\sigma' = \langle l' \| s' \| \mu' \rangle = \delta(\sigma) \downarrow$ *and either l'^k does not exist or $l^k \neq l'^k$. It modifies f in σ iff $\sigma' = \langle l' \| s' \| \mu' \rangle = \delta(\sigma) \downarrow$ and there exists $\ell \in dom(\mu)$ where $\mu(\ell)$ is an object having a field f and either $\mu'(\ell) \uparrow$, or $\mu'(\ell)$ is not an object having a field f, or $\mu(\ell)(f) \neq \mu'(\ell)(f)$. It modifies an array of type t in σ iff $\sigma' = \langle l' \| s' \| \mu' \rangle = \delta(\sigma) \downarrow$ and there exists $\ell \in dom(\mu)$ where $\mu(\ell)$ is an array of type t and either $\mu'(\ell) \uparrow$, or $\mu'(\ell)$ is not an array of type t, or $\mu(\ell).length \neq \mu'(\ell).length$, or $\mu(\ell)(i) \neq \mu'(\ell)(i)$ for some index $0 \leqslant i < \mu(\ell).length$.*

It is now possible to define a more concrete abstract domain than in Definition 11, by adding information on local variables, fields and arrays that might be modified.

Definition 14. *Let $l_i, s_i, l_o, s_o \in \mathbb{N}$. The abstract domain $\mathbb{PLSE}_{l_i,s_i \to l_o,s_o}$ for path-length and side-effects contains tuples $\langle pl \| L \| F \| A \rangle$ where $pl \in \mathbb{PL}_{l_i,s_i \to l_o,s_o}$ (Definition 11), L is a set of local variables, F is a set of fields and A is a minimal set of types i.e., for all $t, t' \in A$ it is never the case that $t < t'$.*

A tuple $\langle pl \| L \| F \| A \rangle$ represents denotations that are allowed to modify locals in L, fields in F and arrays whose elements are compatible with some type in A:

Definition 15. *Let $\langle pl \| L \| F \| A \rangle \in \mathbb{PLSE}_{l_i,s_i \to l_o,s_o}$. Its concretisation function is $\gamma(\langle pl \| L \| F \| A \rangle) = \{\delta \in \Delta_{l_i,s_i \to l_o,s_o} \mid (1)\ \delta \in \gamma(pl)\ [Definition\ 6], (2)$ if δ modifies local k in σ then $l^k \in L$, (3) if δ modifies f in σ then $f \in F$, (4) if δ modifies an array of type t in σ then $t \leqslant t'$ for some $t' \in A\}$.*

Proposition 3. *$\mathbb{PLSE}_{l_i,s_i \to l_o,s_o}$ is a lattice and the map γ of Definition 15 is the concretisation of a Galois connection from $\Delta_{l_i,s_i \to l_o,s_o}$ to $\mathbb{PLSE}_{l_i,s_i \to l_o,s_o}$.* $\square$

The abstract semantics uses now Fig. 2, [19] and Definition 12 and adds side-effects. For method calls, callees in Java cannot modify the local variables of the caller.

Definition 16. *The approximations $ins^{\mathbb{PLSE}}$ are defined as $ins_p^{\mathbb{PLSE}} k = \langle ins_p^{\mathbb{PL}} \| \{l^k\} \| \varnothing \| \varnothing \rangle$ if ins is store k or inc k c; $putfield_p^{\mathbb{PLSE}} f = \langle putfield_p^{\mathbb{PL}} f \| \varnothing \| \{f\} \| \varnothing \rangle$; $arraystore_p^{\mathbb{PLSE}} t = \langle arraystore_p^{\mathbb{PL}} t \| \varnothing \| \varnothing \| \{t\} \rangle$;*

---

[5] In Java bytecode, local variables are identified by number and their amount varies across program points. Source code variable names are not part of the bytecode.

$(call_p^{\mathbb{PLSE}} \ \kappa.m)(\langle pl \,\|\, L \,\|\, F \,\|\, A \rangle) \ = \ \langle (call_p^{\mathbb{PL}} \ \kappa.m)(pl) \,\|\, \varnothing \,\|\, F \,\|\, A \rangle; \ ins_p^{\mathbb{PLSE}} \ = \ \langle ins_p^{\mathbb{PL}} \,\|\, \varnothing \,\|\, \varnothing \,\|\, \varnothing \rangle$ *for all other* ins.

Proposition 4. *The maps in Definition 16 are sound w.r.t. those in Fig. 1.* $\square$

Definition 17. *Let* $a = \langle pl \,\|\, L \,\|\, F \,\|\, A \rangle \in \mathbb{PLSE}_{l_i,s_i \to l_o,s_o}$. *Then* $e \in \mathbb{E}_{l_i}$ *is affected by* a *iff (1)* $e = l^k$ *and* $l^k \in L$, *or (2)* $e = e'.f$ *and* $f \in F$ *or* e' *is affected by* a, *or (3)* $e = e_1[e_2]$, *the type of* $e_1 \leqslant t \in A$ *or* e_1 *or* e_2 *is affected by* a.

Abstract compositions over $\mathbb{PLSE}$ use side-effect information to build frame conditions for expressions used in one argument and not affected by the other.

Definition 18. *Let abstract elements* $a_1 = \langle pl_1 \,\|\, L_1 \,\|\, F_1 \,\|\, A_1 \rangle \in \mathbb{PLSE}_{l_i,s_i \to l_t,s_t}$, $a_2 = \langle pl_2 \,\|\, L_2 \,\|\, F_2 \,\|\, A_2 \rangle \in \mathbb{PLSE}_{l_t,s_t \to l_o,s_o}$, $U_1 = \{\breve{e} = \hat{e} \mid e \in \mathbb{E}_{l_t}$ *is used in* pl_2 *and not affected by* $a_1\}$ *and* $U_2 = \{\breve{e} = \hat{e} \mid e \in \mathbb{E}_{l_t}$ *is used in* pl_1 *and not affected by* $a_2\}$. *The* sequential composition $a_1;^{\mathbb{PLSE}} a_2 \in \mathbb{PLSE}_{l_i,s_i \to l_o,s_o}$ *is* $\langle (pl_1 \cup U_1);^{\mathbb{PL}} (pl_2 \cup U_2) \,\|\, L_1 \cup L_2 \,\|\, F_1 \cup F_2 \,\|\, \mathsf{maximize}(A_1 \cup A_2) \rangle$, *where* $\mathsf{maximize}(A) = \{t \in A \mid \neg \exists t' \in A$ *such that* $t < t'\}$. *Let* $a_1 = \langle pl_1 \,\|\, L_1 \,\|\, F_1 \,\|\, A_1 \rangle$, $a_2 = \langle pl_2 \,\|\, L_2 \,\|\, F_2 \,\|\, A_2 \rangle$ *be in* $\mathbb{PLSE}_{l_i,s_i \to l_o,s_o}$ *and* U_1, U_2 *be as above. The* disjunctive composition $a_1 \cup^{\mathbb{PLSE}} a_2$ *is* $\langle (pl_1 \cup U_1) \cup^{\mathbb{PL}} (pl_2 \cup U_2) \,\|\, L_1 \cup L_2 \,\|\, F_1 \cup F_2 \,\|\, \mathsf{maximize}(A_1 \cup A_2) \rangle$.

Proposition 5. *The compositions in Definition 18 are sound w.r.t. the corresponding concrete compositions on denotations* [19]. $\square$

Example 2. In Example 1, no instruction has side-effects, hence the last case of Definition 16 applies. The abstraction of getfield diagonal is now $a_1 = \langle pl_1 \,\|\, \varnothing \,\|\, \varnothing \,\|\, \varnothing \rangle$. That of the subsequent arraylength double is now $a_2 = \langle pl_2 \,\|\, \varnothing \,\|\, \varnothing \,\|\, \varnothing \rangle$ (pl_1 and pl_2 are given in Example 1). Expression $l^0.diagonal$ is used in pl_1 and is not affected by a_2. Hence (Definition 18) $U_1 = \varnothing$, $U_2 = \{l^0.\widetilde{diagonal} = l^0.\widetilde{diagonal}\}$ and $a_1;^{\mathbb{PLSE}} a_2 = \langle pl_1 \,\|\, \varnothing \,\|\, \varnothing \,\|\, \varnothing \rangle;^{\mathbb{PLSE}} \langle pl_2 \cup \{l^0.\widetilde{diagonal} = l^0.\widetilde{diagonal}\} \,\|\, \varnothing \,\|\, \varnothing \,\|\, \varnothing \rangle = \{\breve{l^0} = \hat{l^0}, \breve{l^1} = \hat{l^1}, \breve{l^2} = \hat{l^2}, \breve{s^0} = \hat{s^0}, \hat{s^1} = l^0.\widetilde{diagonal} = l^0.\widetilde{diagonal}\}$. The result refers to $l^0.diagonal$ now: the imprecision in Example 1 is overcome.

7   Experiments

Implementation. $\mathbb{PLSE}$ (Definition 14) needs to implement its elements $\langle pl \,\|\, L \,\|\, F \,\|\, A \rangle$, its abstract operations (Definition 16) and a fixpoint engine for denotational, bottom-up analysis. We have used the Julia analyser [17] and its fixpoint engine. Elements of $\mathbb{PLSE}$ use bitsets for L, F and A, since they are compact and with fast union (Definition 18). The pl component has been implemented twice: as bounded differences of variable pairs, by using zones (Chap. 3 of [10]) and as a hybrid implementation of zones and polyhedra, by using the Parma Polyhedra Library [3] for polyhedra. We use zones rather than the potentially more precise octagons, only for engineering reasons: zones are already

Program	Category	LoC	LoC w. Libs	Watchpoints	True Alarms
Snake&Ladder	game	794	17818	15	1
MediaPlayer	entertainment	2634	87368	28	2
EmergencySNRest	web service	3663	42540	36	2
FarmTycoon	game	4005	69659	1998	2
Abagail	mach. learn.	12270	49243	2986	126
JCloisterZone	game	19340	116858	590	49
JExcelAPI	scientific	34712	67944	2031	162
Colossus	game	77527	194994	1988	173

Fig. 3. The programs analyzed. **LoC** are the non-blank non-commented lines of source code; **LoC w. Libs** includes the lines of the libraries reachable and analyzed; **Watchpoints** is the number of arrayload or arraystore, whose bounds must be checked; **True Alarms** are index bound violations found by the analysis (*i.e.*, actual bugs).

available, tested and optimised in Julia. They cannot accomodate constraints such as those for add or sub, that refer to three variables (two operands and the result) and are dropped with zones. This keeps the analysis sound but reduces its precision. In the hybrid representation, instead, polyhedra represent them. A fixpoint is run for each strongly-connected code component. Polyhedra and zones have infinite ascending chains, hence widening [3,5,10] is used after 8 iterations. The cost of operations on polyhedra and zones depends on their dimensions *i.e.*, variables (locals, stack elements and expressions, in pre-state ($\check{v}$), post-state ($\hat{v}$) and overlined $\overline{v}$, see Definition 7). We have limited zones to 200 dimensions and polyhedra to 110; variables beyond that limit are projected away. This does not mean that the analysed programs have only up to 200 (or 110) variables: the limit applies at each given program point, not to the program as a whole. Since there are infinitely many expressions (Definition 12), we fixed a limit of 9. This does not mean that the analysis of a program considers 9 expressions only: it applies to each given program point. We fixed $k = 3$ in Definition 12. When, nevertheless, abstraction (Definition 12) or composition (Definition 18) generate more than 9 expressions, the implementation prefers those from a_2 in $a_1;^{\text{PLSE}} a_2$ and drops those beyond the 9th.

Results. We used an Intel 8-core i7-6700HQ at 2.60 Ghz, OpenJDK Java 1.8.0_151 and 15 GB of RAM. Small to medium-size open-source third-party programs have been analysed, up to 195000 lines of code, cloneable from [14]. Figure 3 reports their size, characteristics and number of index bound violations found by the analysis. The reachable libraries have been included and analysed, together with the application code. This is needed for the approximation of method calls to the library. However, warnings have been generated only on the application code. Figure 4 reports the results. Programs have been first analysed as in [5,8], with zones only (column **Zones Only**). Each alarm has been manually classified as true (*i.e.*, an array index bug) or false. True alarms range from 1 to 173 per program. If classification was impossible, since we do not fully understand the logic of the code or its invariants, alarms have been conservatively

Program	Zones Only			Zones + Poly			Zones + Exps		
	Alarms	Time	Mem	Alarms	Time	Mem	Alarms	Time	Mem
Snake&Ladder	1	13	1.7	1	14	1.7	1	14	1.7
MediaPlayer	5	52	4.8	5	64	4.8	5	52	4.9
EmergencySNRest	2	26	3.0	2	30	3.0	2	26	3.1
FarmTycoon	17	51	4.3	17	54	4.7	9	53	4.4
Abagail	664	69	4.3	662	202	11.2	339	81	7.2
JCloisterZone	116	108	7.1	116	127	8.3	90	108	7.4
JExcelAPI	1061	95	4.8	1045	317	13.8	786	141	9.7
Colossus	597	312	9.7	*out of memory*			477	328	11.5

Fig. 4. Analysis results for the programs in Fig. 3. **Zones Only** uses zones only; **Zones+Poly** zones and polyhedra; **Zones+Exps** zones with expressions (Sects. 5 and 6). Julia issues index bound **Alarms**, bounded by **Watchpoints** in Fig. 3. **Time** is full analysis time, in seconds. **Mem** is the peak memory usage, in gigabytes.

classified as false. Thus, column **True Alarms** in Fig. 3 is a lower bound on actual bugs. Comparing **Alarms** of **Zones Only** with **True Alarms** shows a major precision gap. To close it, we tried to exploit the extra precision of polyhedra through the hybrid use of zones *and* polyhedra. Column **Alarms** of **Zones+Poly** in Fig. 4 deceives our hopes: polyhedra hardly improve the precision, at the price of higher analysis time and memory footprint, up to an out of memory. Instead, columns **Zones+Exps** show that the technique of Sect. 6, with zones only, scales to all programs, with fewer alarms: precision benefits more from expressions than from polyhedra and expressions are cheaper than polyhedra *w.r.t.* memory usage. Figure 4 reports full analysis times and peak memory usage during parsing of the code, construction of the control-flow graph and of the strongly-connected components, heap, aliasing and path-length analysis. The alarms are in [14], annotated as TA when they classify as true alarms. Note that the analysis has false positives but no false negatives (true bugs that the analysis does not find), since it is provably sound.

False Alarms that Disappear by Using Expressions. Zones Only issues false alarms for all examples in Sect. 1. They disappear with Zones + Exps. In the first example, the analyser uses a variable for the expression this.diagonal; in the second, for this.data and this.data[i]; in the third, for this.this$0.h and this.this$0.h[i]. Expressions are chosen automatically and on-demand.

True Alarms. In jxl.biff.BaseCompoundFile of JExcelAPI, Julia issues a true alarm at line 3 below[6], since the constructor is public and its argument d is arbitrary, hence might have less than SIZE+1 elements[7]:

---

[6] Line numbers, conveniently starting at 1, do not correspond to the actual line numbering of the examples, which are simplified and shortened *w.r.t.* their original code.

[7] We assume that public entries can be called with any values, as also done in [15].

```
1 | public PropertyStorage(byte[] d) {
2 |   this.data = d;
3 |   int s = IntegerHelper.getInt(this.data[SIZE], this.data[SIZE+1]); }
```

Julia issues a true alarm at line 2 of class domain.Farm of FarmTycoon, for a public method whose argument options is hence arbitrary. Very likely, the programmer should have written options[pos] here, instead of options[1]:

```
1 | public static void objPrinter(String[] options) {
2 |   Storm[] objStorm = new Storm[options.length];
3 |   for (int pos = 0; pos < options.length; pos++)
4 |     objStorm[pos] = new Storm(Long.parseLong(options[1])); }
```

Julia issues a true alarm at line 5 of edu.cmu.sv.ws.ssnoc.common.logging. Log in EmergencySNRest, since the stack trace might be shorter than 4 elements (the documentation even allows getStackTrace() to be empty):

```
1 | private static Logger getLogger() {
2 |   ... = Thread.currentThread().getStackTrace()[3].getClassName(); }
```

Julia issues true alarms from line 5 of java.net.sf.colossus.webclient.Web-ClientSocketThread in Colossus, where fromServer comes from a remote server and might contain too few tokens: it should be sanitised first:

```
1 | String fromServer = getLine();
2 | String[] tokens = fromServer.split(sep, -1);
3 | String command = tokens[0]; // ok: split() returns at least one token
4 | if (command.equals(IWebClient.userInfo)) {
5 |   int loggedin = Integer.parseInt(tokens[1]);
6 |   int enrolled = Integer.parseInt(tokens[2]);
7 |   ... String text = tokens[6]; ... }
```

False Alarms: Limitations of the Analysis. In func.svm.SingleClass-SequentialMinimalOptimization of Abagail, Julia issues false alarms at line 8:

```
1 | public SingleClass...Optimization(DataSet examples, ..., double v) {
2 |   v = Math.min(v, 1); ...
3 |   this.a = new double[examples.size()];
4 |   this.vl = v * examples.size(); int ivl = (int) this.vl;
5 |   int[] indices = ABAGAILArrays.indices(examples.size());
6 |   ABAGAILArrays.permute(indices);
7 |   for (int i = 0; i < ivl; i++)
8 |     this.a[indices[i]] = 1 / vl; }
```

It is $0 \leqslant i < ivl = \lfloor v * examples.size() \rfloor \leqslant examples.size()$ and ABAGAIL-Arrays.indices(x) yields an array of size x. Thus indices[i] is safe. Also this.a[indices[i]] is safe, since the elements of ABAGAILArrays.indices(x) range from 0 to x (excluded) and permute() shuffles them. Such reasonings are beyond the capabilities of our analysis.

Julia issues false alarms at lines 3 and 4 of net.sf.colossus.util.Static-ResourceLoader in Colossus:

```
1  while (r > 0) {
2    byte[] temp = new byte[all.length + r];
3    for (int i = 0; i < all.length; i++) temp[i] = ...;
4    for (int i = 0; i < r; i++) temp[i + all.length] = ...; }
```

Here, Julia builds a constraint $temp = all + r$. Since $r > 0$, then i in the first loop is inside $temp$; since $0 \leqslant i < r$, the same holds in the second loop. Zones cannot express a constraint among three variables. Polyhedra can do it, but do not scale to the analysis of Colossus (Fig. 4).

8   Conclusion

The extension of path-length to arrays (Sect. 4) scales to array index bounds checking of real Java programs, but only with weaker abstractions than polyhedra, such as zones. Precision improves with explicit information about some expressions (Sects. 5 and 6). Experiments (Sect. 7) are promising. The analysis has limitations: it is unsound with unconstrained reflection or side-effects due to concurrent threads, as it is typical of the current state of the art of static analysers for full Java; also remaining false alarms (Sect. 7) show space for improvement.

References

1. Albert, E., Arenas, P., Genaim, S., Puebla, G.: Field-sensitive value analysis by field-insensitive analysis. In: Cavalcanti, A., Dams, D.R. (eds.) FM 2009. LNCS, vol. 5850, pp. 370–386. Springer, Heidelberg (2009). https://doi.org/10.1007/978-3-642-05089-3_24
2. Albert, E., Arenas, P., Genaim, S., Puebla, G., Ramírez Deantes, D.V.: From object fields to local variables: a practical approach to field-sensitive analysis. In: Cousot, R., Martel, M. (eds.) SAS 2010. LNCS, vol. 6337, pp. 100–116. Springer, Heidelberg (2010). https://doi.org/10.1007/978-3-642-15769-1_7
3. Bagnara, R., Hill, P.M., Zaffanella, E.: The Parma Polyhedra Library: toward a complete set of numerical abstractions for the analysis and verification of hardware and software systems. Sci. Comput. Program. **72**(1–2), 3–21 (2008)
4. Bagnara, R., Hill, P.M., Zaffanella, E.: Applications of polyhedral computations to the analysis and verification of hardware and software systems. Theoret. Comput. Sci. **410**(46), 4672–4691 (2009)
5. Cousot, P., Cousot, R.: Abstract interpretation: a unified lattice model for static analysis of programs by construction or approximation of fixpoints. In: Principles of Programming Languages (POPL), pp. 238–252 (1977)
6. Cousot, P., Cousot, R., Feret, J., Mauborgne, L., Miné, A., Monniaux, D., Rival, X.: The Astrée analyzer. In: European Symposium on Programming (ESOP), pp. 21–30 (2005)
7. Cousot, P., Cousot, R., Feret, J., Mauborgne, L., Miné, A., Rival, X.: Why does Astrée scale up? Formal Methods Syst. Des. **35**(3), 229–264 (2009)
8. Cousot, P., Halbwachs, N.: Automatic discovery of linear restraints among variables of a program. In: Principles of Programming Languages (POPL), Tucson, Arizona, USA, pp. 84–96, January 1978

9. Lindholm, T., Yellin, F., Bracha, G., Buckley, A.: The Java^{TM} Virtual Machine Specification. Financial Times/Prentice Hall, Upper Saddle River (2013)
10. Miné, A.: Weakly relational numerical abstract domains. Ph.D. thesis, École Polytechnique, Paris, France (2004)
11. Miné, A.: Field-sensitive value analysis of embedded C programs with union types and pointer arithmetics. In: Languages, Compilers, and Tools for Embedded Systems (LCTES), Ottawa, Ontario, Canada, pp. 54–63 (2006)
12. Miné, A.: The octagon abstract domain. High.-Order Symbolic Comput. **19**(1), 31–100 (2006)
13. Nikolic, D., Spoto, F.: Reachability analysis of program variables. Trans. Program. Lang. Syst. **35**(4), 14:1–14:68 (2013)
14. Payet, É., Spoto, F.: Index Checking Experiments (2017). https://github.com/spoto/Index-Checker-Experiments.git
15. Santino, J.: Enforcing correct array indexes with a type system. In: Foundations of Software Engineering (FSE), pp. 1142–1144. ACM, Seattle (2016)
16. Secci, S., Spoto, F.: Pair-sharing analysis of object-oriented programs. In: Hankin, C., Siveroni, I. (eds.) SAS 2005. LNCS, vol. 3672, pp. 320–335. Springer, Heidelberg (2005). https://doi.org/10.1007/11547662_22
17. Spoto, F.: The Julia static analyzer for Java. In: Rival, X. (ed.) SAS 2016. LNCS, vol. 9837, pp. 39–57. Springer, Heidelberg (2016). https://doi.org/10.1007/978-3-662-53413-7_3
18. Spoto, F., Jensen, T.P.: Class analyses as abstract interpretations of trace semantics. Trans. Program. Lang. Syst. **25**(5), 578–630 (2003)
19. Spoto, F., Mesnard, F., Payet, É.: A termination analyzer for Java bytecode based on path-length. Trans. Program. Lang. Syst. **32**(3), 8:1–8:70 (2010)

Author Index

Abrahamsson, Oskar 646
Acclavio, Matteo 481

Bacchus, Fahiem 134
Backeman, Peter 246
Barbosa, Haniel 591
Barrett, Clark 591
Bentkamp, Alexander 28
Benzmüller, Christoph 108
Biere, Armin 134
Blanchette, Jasmin Christian 28, 89, 370
Bodirsky, Manuel 263
Bromberger, Martin 329

Carral, David 680
Ciobâcă, Ştefan 295
Claessen, Koen 388
Conchon, Sylvain 152
Cruanes, Simon 28

Das, Anupam 689
Dawson, Jeremy 117
de Lima, Tiago 1
Declerck, David 152
Dershowitz, Nachum 117
Dragoste, Irina 680

Echenim, Mnacho 279
Egly, Uwe 161

Fazekas, Katalin 134
Finger, Marcelo 194

Ge, Cunjing 312, 354
Goré, Rajeev 117
Greiner, Johannes 263

Hannula, Miika 47
Haslbeck, Maximilian P. L. 532
Heule, Marijn J. H. 516

Hirokawa, Nao 346
Ho, Son 646
Hoenicke, Jochen 549
Huang, Pei 354

Ignatiev, Alexey 627

Jacobs, Ceriel 680
Jeannerod, Nicolas 439
Jovanović, Dejan 455

Kanovich, Max 228
Katelaan, Jens 455
Kazakov, Yevgeny 609
Kiesl, Benjamin 516
Kohlhase, Michael 575
Korovin, Konstantin 663
Kotelnikov, Evgenii 405
Kovács, Laura 405
Krötzsch, Markus 680
Kumar, Ramana 646
Kuznetsov, Stepan 228

Lagniez, Jean-Marie 1
Larchey-Wendling, Dominique 422
Le Berre, Daniel 1
Lettmann, Michael Peter 64
Link, Sebastian 47
Liu, Tian 312
Lonsing, Florian 161
Lopez Hernandez, Julio Cesar 663
Lucanu, Dorel 295

Ma, Feifei 312, 354
Ma, Xutong 312
Marques-Silva, Joao 627
Melquiond, Guillaume 178
Middeldorp, Aart 81, 346
Montmirail, Valentin 1
Moser, Georg 472

Müller, Dennis   575
Myreen, Magnus O.   646

Nagele, Julian   346
Nalon, Cláudia   498
Narodytska, Nina   627
Nigam, Vivek   228
Norrish, Michael   646

Pattinson, Dirk   498
Payet, Étienne   706
Peltier, Nicolas   64, 279, 370
Pereira, Filipe   627
Piotrowski, Bartosz   566
Platzer, André   211
Preto, Sandro   194

Rabe, Florian   575
Rapp, Franziska   81
Rebola-Pardo, Adrián   516
Reynolds, Andrew   591
Rieu-Helft, Raphaël   178
Robillard, Simon   370
Rümmer, Philipp   246

Scedrov, Andre   228
Schindler, Tanja   549
Schlichtkrull, Anders   89
Schmidt, Renate A.   19

Sellami, Yanis   279
Skočovský, Peter   609
Smallbone, Nicholas   388
Spoto, Fausto   706
Steen, Alexander   108
Straßburger, Lutz   481

Tan, Yong Kiam   646
Tinelli, Cesare   591
Traytel, Dmitriy   89
Treinen, Ralf   439

Urban, Josef   566
Urbani, Jacopo   680

Viswanathan, Arjun   591
Voronkov, Andrei   405

Waldmann, Uwe   28, 89
Weissenbacher, Georg   455
Winkler, Sarah   472
Wintersteiger, Christoph M.   246

Zaïdi, Fatiha   152
Zeljić, Aleksandar   246
Zhan, Bohua   532
Zhang, Hantao   354
Zhang, Jian   312, 354
Zhao, Yizheng   19

Printed in the United States
By Bookmasters